Biology
Living Systems

GLENCOE

Macmillan/McGraw-Hill

New York, New York   Columbus, Ohio   Mission Hills, California   Peoria, Illinois

A GLENCOE PROGRAM

Biology
Living Systems

Student Edition
Teacher Wraparound Edition
Study Guide, SE and TE
Investigating Living Systems, SE and TAE
Probing Levels of Life, SE and TAE
Assessment
Videodisc Correlations
Science and Technology Videodisc Series Teacher Guide
Transparency Package
Concept Mapping
Exploring Environmental Issues
Critical Thinking/Problem Solving
Spanish Resources
Lesson Plans
Computer Test Bank

Send all inquiries to:

GLENCOE DIVISION
Macmillan/McGraw-Hill
936 Eastwind Drive
Westerville, OH 43081

ISBN 0-02-800672-0

Printed in the United States of America

4 5 6 7 8 9 R R W / M C 00 99 98 97 96 95 94

Raymond F. Oram
is a teacher/administrator at the Peddie School, Hightstown, New Jersey, with more than 30 years of teaching experience. Mr. Oram chairs the Science Department, as well as teaching classes in introductory biology, honors biology, and advanced placement biology. Mr. Oram received his A.B. degree from Princeton University and his M.S.T. from Union College, Schenectady, New York. He is a member of several professional organizations, including NABT, NSTA, and AAAS. In 1976, Mr. Oram was recognized with the Princeton Prize for Distinguished Secondary School Teaching in the State of New Jersey. He also has been named Master Teacher, has received the Outstanding Classroom Teacher Award, and is holder of the Casperson Chair in Science, all at the Peddie School.

CONSULTANT

Paul J. Hummer, Jr.
taught seventh grade life science and high school biology from 1958 to 1986 in Frederick County, Maryland, public schools. He is currently Assistant Professor of Education at Hood College in Frederick, Maryland. Mr. Hummer received a B.S. in Education from Lock Haven State University, Pennsylvania, and an M.S. degree from Union College. He was the 1972 Maryland recipient of the Outstanding Biology Teacher Award, NSTA STAR Award recipient in 1972, and the 1984 Maryland recipient of the Presidential Award for Excellence in Teaching Science and Mathematics.

CONTRIBUTING AUTHORS

Albert Kaskel
Evanston Township High School
Evanston, IL

Linda Lundgren
Bear Creek High School
Lakewood, CO

CONTENT CONSULTANTS

Jerry Downhower, Ph.D.
Department of Zoology
The Ohio State University
Columbus, OH

Donald G. Kaufman, Ph.D.
Department of Zoology
Miami University
Oxford, OH

Melissa S. Stanley, Ph.D.
Department of Biology
George Mason University
Fairfax, VA

David G. Futch, Ph.D.
Department of Biology
San Diego State University
San Diego, CA

Peter N. Petersen, Ph.D.
Science Department
Wilcox High School
Santa Clara, CA

Richard Storey, Ph.D.
Department of Biology
Colorado College
Colorado Springs, CO

James A. Gavan, Ph.D.
Department of Anthropology
University of Missouri
Columbia, MO

Fred David Sack, Ph.D.
Department of Plant Biology
The Ohio State University
Columbus, OH

E. Peter Volpe, Ph.D.
School of Medicine
Mercer University
Macon, GA

READING CONSULTANT

Barbara S. Pettegrew, Ph.D.
Department of Education
Otterbein College
Westerville, OH

REVIEWERS

David G. Cowles
Victor Central High School
Victor, NY

Kathy Hoppin
Wheatland High School
Wheatland, CA

Mary E. Schwartz
Hamilton Heights High School
Arcadia, IN

John E. DeMary
Loudoun Valley High School
Purcellville, VA

Susan Hutchinson
Benicia High School
Benicia, CA

Demby Sejd
McLean High School
McLean, VA

John W. Fedors
Greenwich High School
Greenwich, CT

Priscilla Jane Lee
Venice High School
Los Angeles, CA

Thomas Phil Talbot
Skyline High School
Salt Lake City, UT

Lois Mayo
Pius X Central High School
Lincoln, NE

The Nature of Biology   2

UNIT TWO

Energy and the Cell   52

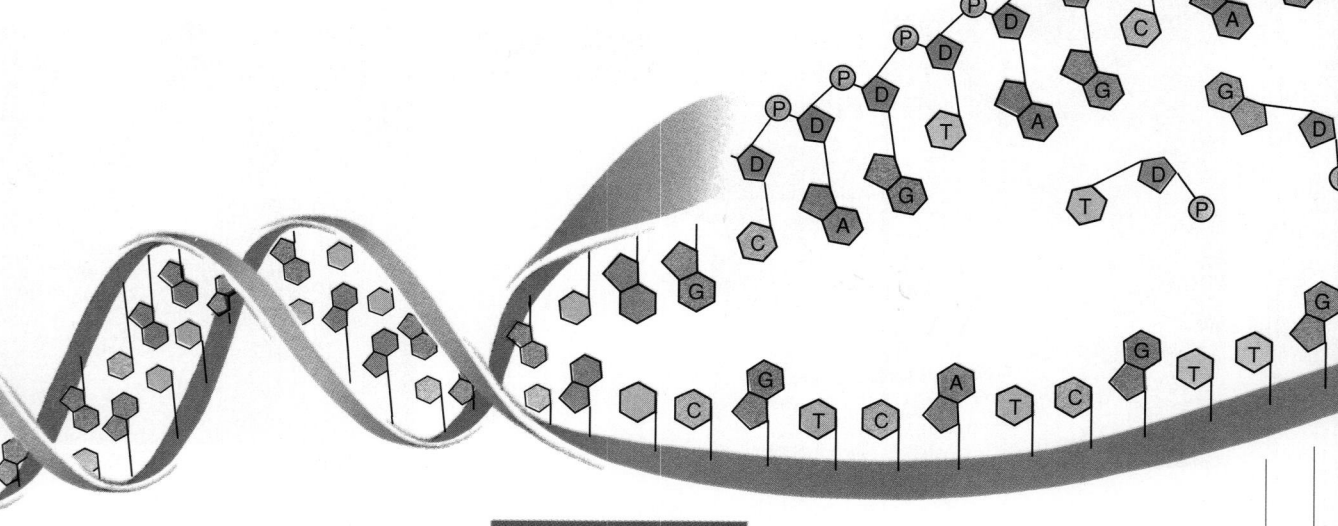

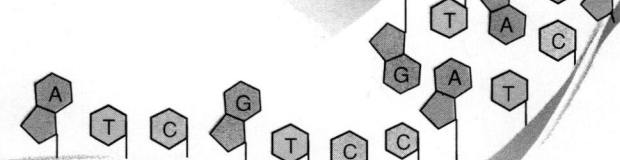
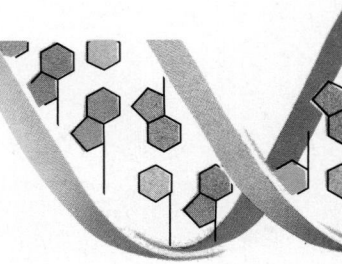

UNIT FOUR

Evolutionary Relationships 302

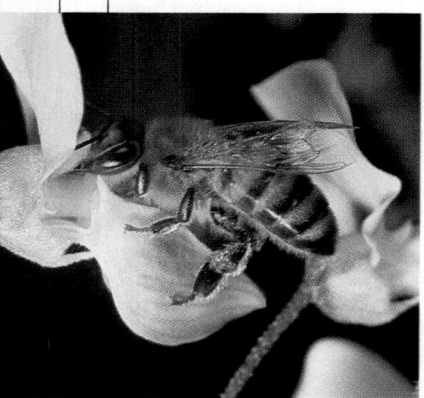

UNIT
FIVE

Life Functions of Organisms   480

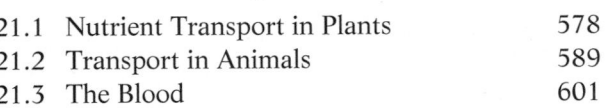

UNIT FIVE

UNIT SEVEN

Interactions in the Environment 756

INVESTIGATIONS

Here's your chance to explore your creativity. INVESTIGATIONS give you an opportunity to act like a biologist and develop your own plans for studying a question or problem. You'll find that there's more than one way to explore biology!

MINILABS

Want to see it for yourself? Then try it! Discover for yourself how something works. Just as a picture is worth a thousand words, a MINILAB may be worth hours of reading.

MINILABS

THINKING LABS

Sharpen up your pencil and your wit because you'll need them. THINKING LABS offer a unique opportunity to evaluate someone else's experiments and data without the mess.

———

ISSUES

What are some aspects of biology that have an impact on society? Develop your own viewpoint on the ISSUES that might be controversial. Learn how to live with biology and use it to enrich your life.

CONSIDERING THE ENVIRONMENT

Do you take for granted that when you flip the switch, the light will come on? Developing alternate ways to provide energy without damaging the environment is one problem that challenges scientists and people like you. Find out how people can make a difference.

GLOBAL CONNECTIONS

How are people in Mexico adapted to a high-altitude lifestyle? How do Native Americans view animals? These and other intriguing questions are answered as you explore these GLOBAL CONNECTIONS.

INTERDISCIPLINARY CONNECTIONS

It may not have occurred to you that biology is an integral part of all your courses. Learn in these features how biology is connected to art, literature, and other subjects that interest you.

Health

Science and Math

INTERDISCIPLINARY CONNECTIONS

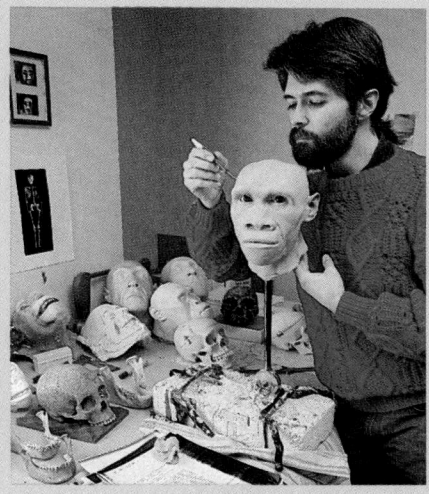

BIOTECHNOLOGY

What new advances have been developed in the last decade from applying biological principles? Explore these new developments and find out how they improve the quality of life.

CAREER CONNECTIONS

How can you use biology to develop a career? Find an idea from the CAREER features presented in your textbook. Then plan for an exciting future of discovery.

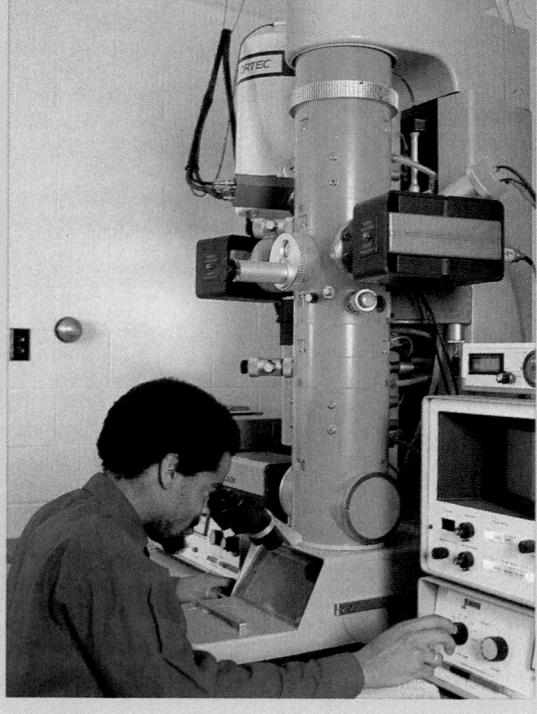

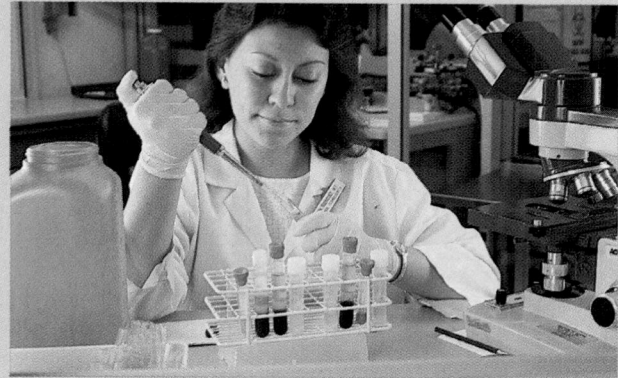

CREDITS

Page 548 From *Healing Heart*, by Gloria T. Hull,
1989, Kitchen Table: Women of Color
Press, PO Box 908, Latham, NY 12110.
Used by permission.

Page 776 From *When I Am an Old Woman I Shall
Wear Purple*, an anthology of short stories
and poetry, 1987, Papier-Maché Press,
Manhattan Beach, CA. Reprinted by permission of the author.

Page 822 "Winter Poem" from MY HOUSE by
Nikki Giovanni. Copyright © 1972 by
Nikki Giovanni. Reprinted by permission
of William Morrow & Co. Inc.

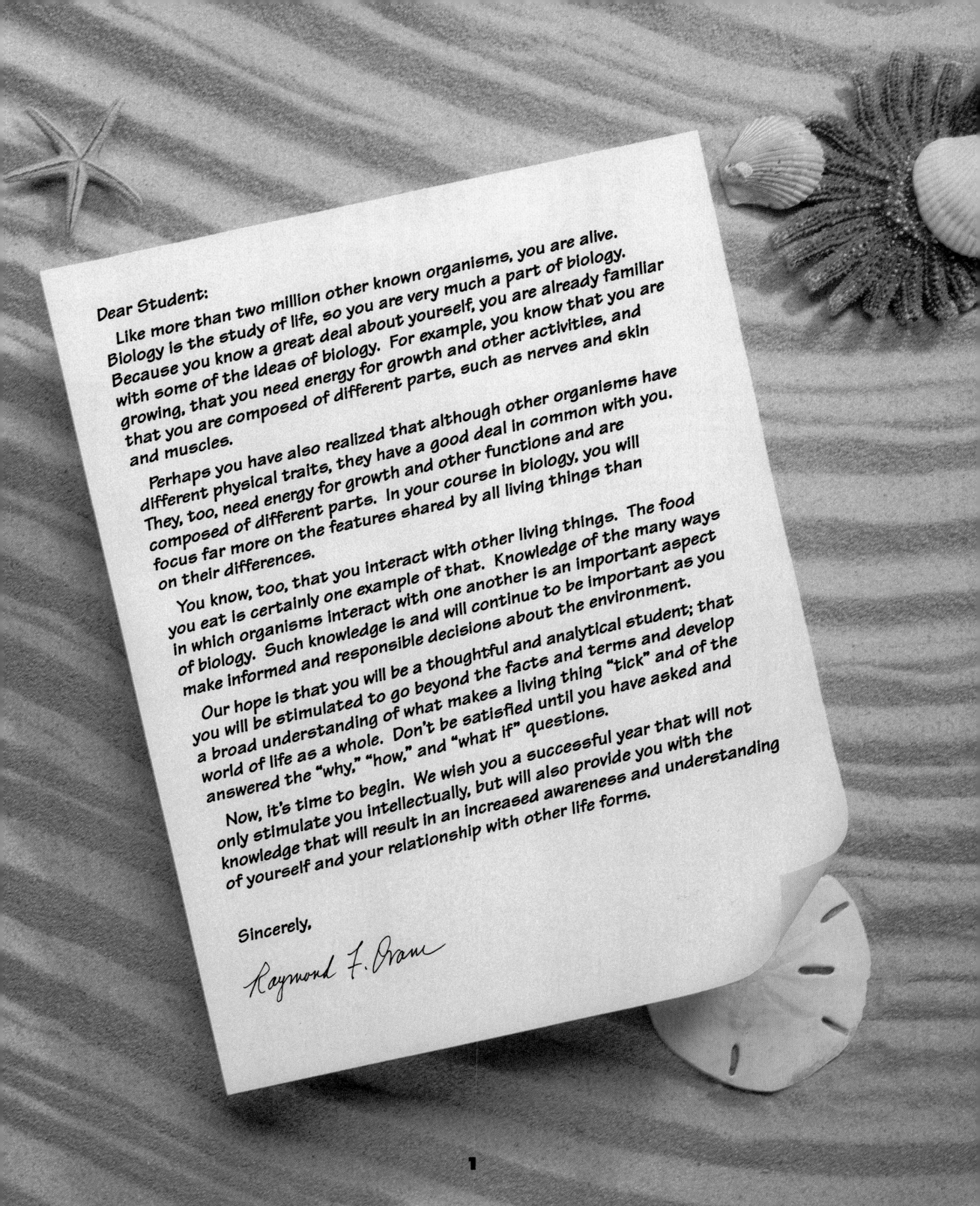

Dear Student:

Like more than two million other known organisms, you are alive. Biology is the study of life, so you are very much a part of biology. Because you know a great deal about yourself, you are already familiar with some of the ideas of biology. For example, you know that you are growing, that you need energy for growth and other activities, and that you are composed of different parts, such as nerves and skin and muscles.

Perhaps you have also realized that although other organisms have different physical traits, they have a good deal in common with you. They, too, need energy for growth and other functions and are composed of different parts. In your course in biology, you will focus far more on the features shared by all living things than on their differences.

You know, too, that you interact with other living things. The food you eat is certainly one example of that. Knowledge of the many ways in which organisms interact with one another is an important aspect of biology. Such knowledge is and will continue to be important as you make informed and responsible decisions about the environment.

Our hope is that you will be a thoughtful and analytical student; that you will be stimulated to go beyond the facts and terms and develop a broad understanding of what makes a living thing "tick" and of the world of life as a whole. Don't be satisfied until you have asked and answered the "why," "how," and "what if" questions.

Now, it's time to begin. We wish you a successful year that will not only stimulate you intellectually, but will also provide you with the knowledge that will result in an increased awareness and understanding of yourself and your relationship with other life forms.

Sincerely,

Raymond F. Oram

The Nature of Biology

Optical telescopes like this one "see" images from space by collecting light energy and focusing it onto sensors. Unlike the facets of an insect's eye, the mirrors of some optical telescopes can be moved as Earth moves. This enables optical telescopes to have a much wider field of view.

Insects with compound eyes, such as this dragonfly, see individual pictures through many facets. Each of the dragonfly's 28 000 facets perceives a slightly different picture than any other, but when merged, the pictures give the insect a much larger field of view. Compound eyes are just one type of vision found in insects.

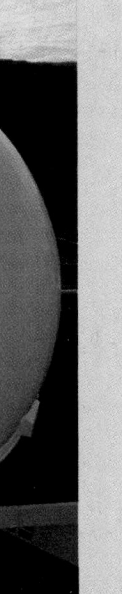

If you could look through a dragonfly's eyes, what would you see? Dragonflies, like many insects, have compound eyes made of many facets. Light rays from a small area fall on a single facet and are concentrated on light-sensitive cells below. Information from each facet is collected and interpreted as a mosaic of the insect's field of view.

In what ways do people collect information about their world? Scientists collect information about the planets and other objects in space with optical tele-scopes such as the Keck Telescope on Mauna Kea, Hawaii. The Keck Telescope has 36 hexagonal mirrors, each 1.8 meters wide, that reflect light energy from space. The light energy is focused onto sensors and collected. Gathering and interpreting information about the world around us is one important part of science inquiry. In this unit, you will learn how inquiry and other scientific methods help you explore the nature of biology.

BIOLOGY—
THE SCIENCE
OF LIFE

A FISH TANK may be a small world but it is by no means a simple one. It teems with life and life's diversity. Start with the obvious—the fish. They're all swimming around in the same water, but do you see any two that are exactly alike? How many are the same size, the same color, or have identical tails and fins? Those filmy green plants you see floating near the top of fish tanks are alive too. And what's going on under the rocks at the bottom? There may be life there too, unseen, but just as much a part of the scene as the fish and plants. How much attention do these living things seem to pay to one another? It may seem as if they pay no attention to their fellow occupants, but don't be fooled. Each one does matter to the others.

There would be no mistaking the little snail in the photo for a fish. It may live in a fish tank, but it doesn't look or behave anything like a fish. Who says life has to be all the same anyway? Did you ever think about what might happen if no one had any other forms of life to relate to? Would we even be here? In the exciting journey you are about to undertake, you will find that living things do indeed differ in many ways. That much you can tell just by looking around you. But, living things also are alike in some very fundamental ways that may not be readily apparent. Add to that an understanding of the interdependence that they share and you have the basis for deep appreciation for the science of living things—biology.

The fish, the snail, and the plants in a fish tank depend on one another—for food, oxygen, protection, support. Can you think how? Now imagine yourself sitting under the green boughs of a tall tree. The tree gives you shade on a hot, sunny day. Now, what might you be doing to return that favor?

Chapter Preview

Relate various characteristics of zebra mussels to the problems they cause.

Characterize living things based on four broad life functions.

1.1 Rowdy Newcomers in the Great Lakes

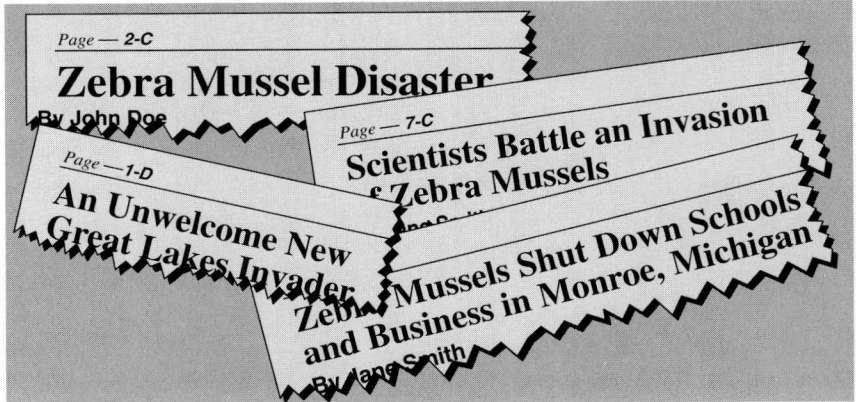

Page — 2-C
Zebra Mussel Disaster
By John Doe

Page — 7-C
Scientists Battle an Invasion of Zebra Mussels

Page — 1-D
An Unwelcome New Great Lakes Invader

Zebra Mussels Shut Down Schools and Business in Monroe, Michigan
By Jane Smith

Figure 1-1 Zebra mussels get their name from the alternating bands of brown and cream that color their shells. These small mollusks are spreading rapidly throughout freshwater lakes and streams in the United States, where they are causing a multitude of problems.

Troublemakers often make headlines. The zebra mussel crowd, which arrived in the Great Lakes in the mid-1980s, falls into the troublemaking category. It has been grabbing national headlines ever since.

Considering the zebra mussel's bad press, you might expect to confront some monstrous demon if you came face to face with one. If you live along the shore of Lake Erie or one of the other Great Lakes, you may actually have done just that. You can easily hold one of the little creatures in the palm of your hand. Even a full-grown and hefty one is petite enough to nestle in the bowl of a soup spoon with room to spare. As far as size goes, the zebra mussel seems innocent enough, even insignificant.

How does a zebra mussel stack up when it comes to looks? On the outside, which consists of a cream-colored shell striped with brown lines—much as a zebra is striped in black and white—it looks not at all unpleasant. The body inside the shell is headless, armless, legless, and boneless. It's hairless too—unless you think that tuft of 200 or so little threads sticking out near its foot looks something like hair. Yes, it has a blob of a foot—just one—but no toes. Put this way, the picture may not sound very flattering, at least to a human being. But among other zebra mussels, it's just the usual shape and order of existence.

Enter and Multiply

It's that part about "other zebra mussels" that stirs up trouble. There are so *many* of them! There may have been only a few when they first arrived, probably as stowaways. Zebra mussels have lived in the freshwater lakes and rivers of Europe for several centuries. But they are not native to the Americas. As far as anybody knows, a party or several parties of zebra mussels still in their infancy rode along in the freshwater ballasts of European cargo ships crossing the Atlantic Ocean. Ballast is water

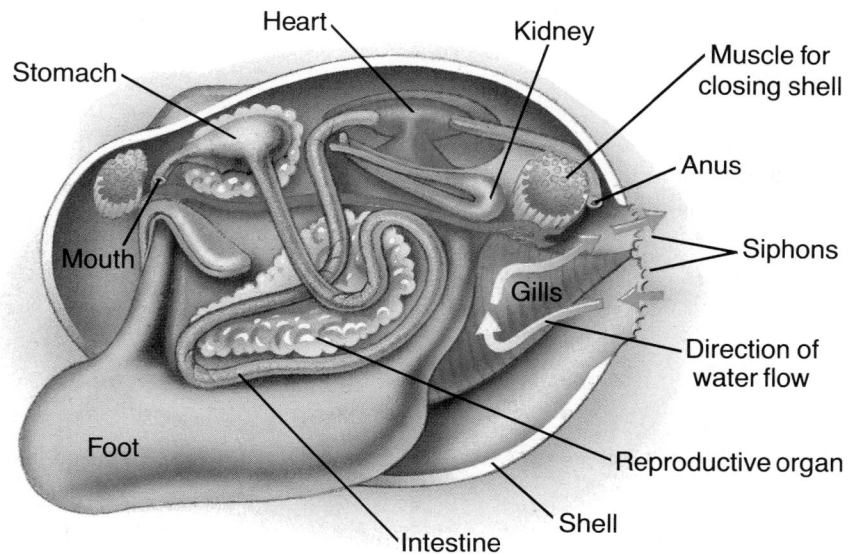

Figure 1-2 A zebra mussel has two shells for protection and a foot for burrowing into the mud or sand. Water containing food and oxygen is taken in by way of one siphon. Wastes are expelled by another siphon.

pumped into the hold of a ship to add weight and help steady the seagoing vessel. When the ships—probably in late 1985 or early 1986—emptied their holds into Lake St. Clair to take on fresh ballast, they spilled their young passengers into American waters. Nobody knew or noticed; the passengers would have been too small to attract attention even if anyone had been looking.

The transported youngsters grew up. By the time they were two years old, they were sexually mature. They became parents, grandparents, great grandparents, and so on. Keep in mind that we're talking big families here. Each female zebra mussel produces about 400 live young each year. And this is only a fraction of the eggs she produces. If all her eggs survived, she would produce an unimaginable 40 000 offspring each year. But 400 offspring a year is quite enough to swell the entire population of zebra mussels into a menace—a menace that has already spread into all the Great Lakes, most of New York's inland waterways, and the Mississippi, Ohio, and Illinois rivers.

A researcher digging for clams in Lake St. Clair was the first to identify the newcomers in 1988. By then, the first arrivals would have been full grown and recognizable. By the time another year went by, the zebra mussels were numerous enough to assume troublemaking status—in a number of ways and in a number of places.

Figure 1-3 Native to the Caspian Sea, zebra mussels are thought to have crossed the Atlantic Ocean in the ballast of cargo ships. They probably first entered the Great Lakes system from Lake St. Clair.

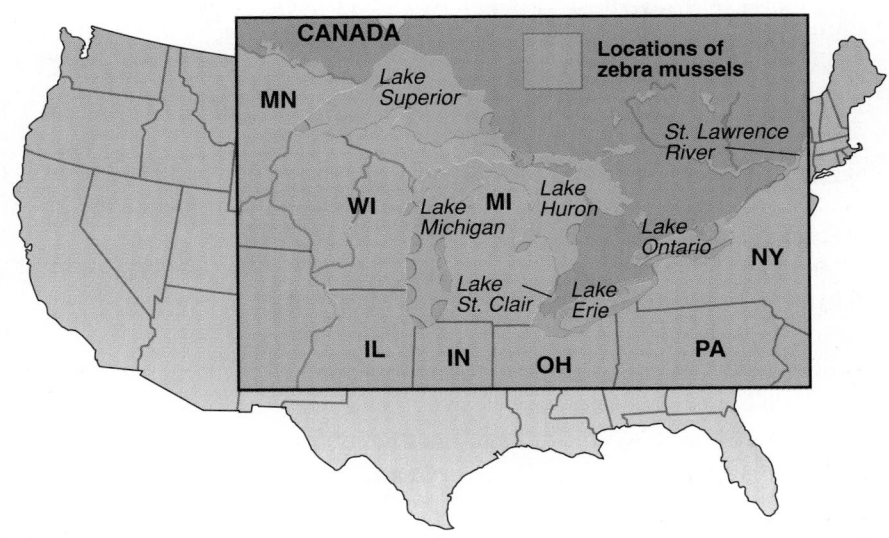

Attack

In a lake or river untouched by human settlement, nearly grown zebra mussels usually head for rocks either lying at the bottom of the water or wedged into shorelines under the surface of the water. There, those hairy little threads, which in reality are more like tiny strands of tape with incredible sticking power, grab hold on a rock. Thus attached, each young zebra mussel becomes anchored to the spot where it will spend the rest of its life.

CHARACTERIZING LIFE

What distinguishes living things?

If you expect to get a good grade in biology this year, you know you will have to perform certain actions—like studying for tests and paying

attention in class. Such functions help characterize good students in any subject. Biologists, too, refer to certain functions when they describe and define life. Broadly speaking, all organisms carry out four functions that distinguish them as living creatures.

Reproduction   The production of offspring is **reproduction,** probably the most obvious of all the characteristics of life. Living things reproduce. Nonliving things do not.

Among many kinds of organisms, including the zebra mussel, each new life begins with a fertilized egg. At appropriate times, male zebra mussels release sperm into their watery environment. Shortly thereafter, female zebra mussels expel eggs into the same waters. When a sperm and egg connect, fertilization is accomplished.

A fertilized egg contains the raw materials for life and certain chemicals. One of those chemicals holds a master code, which determines how the raw materials within the egg combine. Following the instructions of the master code, the egg will eventually turn into a new organism that will take the form of a zebra mussel—or a giraffe, whale, horse, or whatever kind of organism the parents are. Each time reproduction occurs among parents in a particular group of organisms, the same set of coded instructions is passed on from one generation to the next.

Growth and Development   Young organisms get bigger. As they do, they take on the special traits that identify them as members of a particular group of organisms. Such are the processes of growth and development. **Growth** is an increase in the amount of living material in an organism. **Development** is the series of changes an organism undergoes in reaching its final, adult form.

Fertilized eggs of the zebra mussel begin to grow at once. During their early stages of growth, the young zebra mussels, called larvae, float around freely in surface waters and currents. The zebra mussel larvae do not yet resemble adult zebra mussels. That comes later.

After about three weeks, the young zebra mussels attach to a surface. At this point, their bodies begin to develop into new, adult forms. They develop gills for obtaining oxygen, siphons for taking in food, and body parts necessary for reproduction. Two shells, hinged together so that they can open to draw in water or snap shut to protect the soft, defenseless body, also develop. At the same time, the zebra mussels continue to

Zebra mussels aren't fussy about having "all-natural" addresses. Any solid surface will do. Human-produced junk or debris that people have tossed into the water suits them just fine. So do the hard surfaces of submerged pipes and channels that humans have built for taking in or discharging water. And in the Great Lakes region, people have built many such structures along the shores. They use them to get water for drinking, running air conditioners, cooling the machinery that operates factories and electric power plants, or doing any other work that requires large quantities of water.

increase in size. The largest get to be about 4 centimeters across before they stop growing.

Homeostasis Outward growth and development end when an organism reaches adulthood, but other life functions continue. One of them is concerned with maintaining proper conditions for the internal workings of the organism regardless of what happens in the external environment. If a zebra mussel's body is to carry on life processes, it needs to be surrounded by water. What happens if the organism is removed from its watery environment? It will conserve moisture by closing its shells. Thus encased, a zebra mussel can continue to carry on normal life processes in dry surroundings for several days. It survives because it has managed to maintain **homeostasis,** the steady state of the internal operation of an organism regardless of external changes.

Your own body offers more familiar examples of homeostasis. The temperature at which your body works with maximum efficiency is 37°C. What happens when hot weather or some other source of heat raises your body temperature?

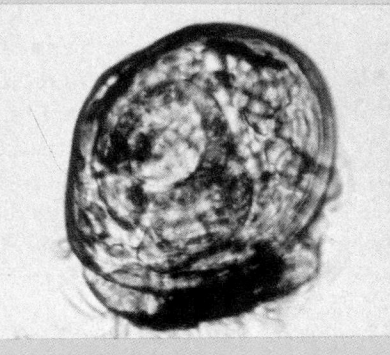

The free-floating larval stage of a zebra mussel bears little resemblance to the adult.

Your body gets rid of excess heat by perspiring. When you're outside in cold weather, your body temperature begins to fall. Your muscles may start to contract rapidly and you begin to shiver. The internal muscle action generates heat and raises your body temperature to the proper level.

Organisms face constant changes in their environments. If they are to survive these changes, their bodies must be able to respond in ways that maintain homeostasis.

Organization Each function or life process that an organism performs is vital to its existence. But none is isolated from the others. All come together and interact with

one another to create a single, orderly living system. Such precise **organization** is characteristic of all life forms. When a zebra mussel opens its shells, for example, it is coordinating its activities with its feeding needs. When it closes its shells, the organism can no longer take in food but it will have protection from danger, a more immediate need than food intake in given circumstances.

Organization shows itself in the way an organism is put together as well as in the ways the organism operates. For every body function, there is a body part—a basic body structure—to perform it. Those body structures take different forms, but they, too, are linked together in the single living system. The zebra mussel has a structure called a siphon for taking food into its body. From the siphon, the food passes to other body structures.

Cells–the basic unit of life–are also organized. Different kinds of cells grouped together make up complex structures. Altogether, these structures are organized into an even more complex system of interacting structures and functions involved in the business of life.

The Invasion Continues

Within the first year after their presence became known, zebra mussels colonized the water intake pipes of facilities all around Lake Erie on both the U.S. and Canadian sides. In the city of Monroe, Michigan, young zebra mussels about one centimeter across were attached to electric power plant intake surfaces in the summer of 1988. Only about 200 mussels clung to each square meter of surface area. A year later, an average of 700 000 clung to each square meter of surface. Zebra mussels attach to one another's shells as easily as to any other hard surface. So with this many mussels piled up on one another, the power plant's intake tubes and other surfaces were covered with a layer of zebra mussels some four centimeters thick. Water lines became clogged and flows were reduced. In some places, heated water from power-plant equipment killed and dislodged the animals, but their shells remained and accumulated. At the same time, the dead bodies emitted a foul odor.

Similar colonies built up in pipes of the drinking-water treatment facilities in Monroe. By December of that year, the pipes and other equipment were so clogged that city water intake was severely reduced. Schools and businesses had to close down for two days due to water shortages.

The zebra mussel has earned a bad name in other places, too. Swimmers and sunbathers walking along many beaches are now careful to wear sandals. They have to protect their bare feet from knicks and scrapes inflicted by sharp edges of the now-too-familiar striped shells that wash ashore. The putrid odor of decaying mussel bodies has ruined some beaches.

Figure 1-4 **Zebra mussels attach to rocks or other surfaces by means of sticky threads. The mussels are capable of sticking to anything, including each other, and can quickly clog pipes and encrust any available surface.**

Threads

Boaters find colonies growing on and weighing down hulls. Detached, dead animals wash into marine engines, where they damage parts and give off bad odors. Zebra mussel colonies stick to the underwater portions of floating channel markers placed in river or lake channels to guide boaters away from unseen dangers. The added weight of the mussel colonies sinks many channel markers and so deprives boaters of essential safeguards.

A Scientific Approach

What is to be done about the striped invaders in the Great Lakes? Probably, the first impulse is to scrape the zebra mussels away from their moorings. And that's exactly what has been done in many places under attack. In fact, scraping has been about the only workable method so far. But it's no firm solution to the problem. The zebra mussels are right back within a year. Workers who do the scraping often have to enter dangerous underwater passageways. The continual expense of scraping adds millions of dollars to the operating costs of power plants and water-treatment facilities.

Clearly, some heavy input of knowledge and investigation is needed if the zebra mussel problem is to be brought to a successful conclusion. In this case, such input can come from science. And since the zebra mussel is a living creature, that science will be biology. *Bios* is the Greek word for *life* or *way of life*. **Biology,** then, is the study of life. Its subjects are all the components of the biosphere, that thin envelope of space surrounding and projecting a short distance below Earth's surface in which all forms of life reside. The work of the biologist is to investigate and enlarge our understanding of the biosphere.

Minilab 1

How do a snail and mussel compare?

Before removing them from the aquarium, observe the movements and locations of the mussels and snails in the aquarium. Remove a snail and place it on a glass slide until it becomes attached. **CAUTION:** *Use care when handling live animals.* Invert the slide over a watch glass, and observe with a binocular microscope the structure and functioning of the snail's foot. Remove a mussel from the aquarium and place it in a culture dish with aquarium water to cover the shells. Note what happened to the shells as you made the transfer. Note any changes after the mussel has been in the culture dish for 10 minutes. Make a chart of similarities and differences between the snail and the mussel.

Analysis *What do the locations of the snail and the mussel in the aquarium tell you about where these animals might live in a lake? How do the snail and mussel move?*

Figure 1-5 **Zebra mussels are removed by the truckload from some underground water pipelines. In many areas, daily removal is necessary just to keep pipelines clear.**

Diversity

A biologist could give us some basic information about the zebra mussel without ever seeing one. First of all, the zebra mussel carries on life processes. One of these processes is reproduction. Given their growing numbers in the Great Lakes, who could doubt the zebra mussels' capacity to reproduce? In addition to reproducing, each zebra mussel will also grow and develop. Finally, a zebra mussel must maintain the internal processes that go on in its body, and be able to keep the different parts of its body working together as a unit. How can a biologist be sure of these things? He or she knows them because the core of scientific knowledge already amassed shows that these four functions are common to all organisms. *(See Characterizing Life, page 8.)* An **organism** is anything capable of carrying on life processes.

Nobody knows exactly how many different kinds of organisms inhabit our planet. So far, scientists have identified more than 2 million. How many of them can you list if you take a minute to think about it? Don't forget human beings, like yourself. Remember, too, things like oak trees and rosebushes as well as dogs, cats, and zebra mussels. How would you describe and compare the differences among the various organisms on your list? Just thinking about that question gets a little overwhelming. For now, leave the answers up to more experienced biologists. It is enough for you to remember that all organisms are alike in that they share similar life processes and requirements for life. Biologists, too, keep this in mind as they search for ways people can deal with the recently arrived organisms now troubling American waterways.

Figure 1-6 **Enormous diversity exists among the two million types of living things that inhabit Earth. Shown here are a bird of paradise flower and a pitcher plant. How do these organisms differ? How are they alike?**

Section Review

Understanding Concepts
1. Explain why zebra mussel populations in North America exploded so quickly.
2. How is the lifestyle of adult zebra mussels related to the problems they cause in water intake pipes?
3. In what ways are you like a zebra mussel?
4. *Skill Review—Observing and Inferring:* You collect organisms from a nearby pond and transfer them to an aquarium. Among the organisms is an animal that you think is a fish. Over time, though, the animal begins to form legs, and its tail seems to be disappearing. What life function(s) are you observing? For more help, see Thinking Critically in the **Skill Handbook.**
5. *Thinking Critically:* Explain the following statement about organisms: Like produces like.

1.2 Community Misfits

"Their presence could be catastrophic." That is how one Canadian scientist summed up his view of zebra mussels in the Great Lakes. His concern rests on the biological knowledge that, like any other organism, zebra mussels don't live alone.

Life and the Community

Each time zebra mussels settle down in a new American lake or river, they join a community of many other kinds of organisms. A **community** is an assortment of life forms within a given place where all the organisms interact and depend upon one another in various ways. You observe organisms interacting in a community when you see robins unearthing worms, squirrels using trees as nesting sites, or bees tracking pollen from one flower to another. Animal parents caring for their young are also interacting within a community.

The aquatic communities that zebra mussels invade teem with bacteria—the simplest and tiniest forms of life. The bacteria either lodge in bottom mud or are suspended in the water. Floating above in the surface water are tiny algae and different kinds of microscopic animal life. Of course, there are also many larger and more obvious animals—fish, waterfowl, worms, frogs, and native mollusks such as clams.

Right from the start, biologists have worried about the many ways in which zebra mussels could and probably have already begun to disrupt Great Lakes communities. For example, the shallow, rocky edges of shorelines are favored anchoring spots for zebra mussel colonies. Some fish, such as walleye and lake trout, reproduce only in these waters. Large zebra mussel colonies could interfere with the reproduction of these kinds of fish and so reduce their numbers. Walleye and trout are important to Great Lakes fishing industries. Reduction in their numbers would hurt both fishers and people who like to eat the fishers' catches.

Objectives

Discuss how energy is made available to all members of a community.

Distinguish between producers and consumers and between photosynthesis and cellular respiration.

Relate the zebra mussel's food and energy needs to the biological problems it causes.

Figure 1-7 **Zebra mussel colonies are typically found along rocky shorelines of freshwater lakes, such as the Great Lakes. Other organisms that inhabit Great Lakes communities include the walleye. What other types of communities can you name?**

Food and Energy Needs in a Community

Zebra mussels also threaten clam populations, another valuable fishing resource. Clams, though larger than the new invaders, are also bivalves. This means that they also have two shells hinged together. Zebra mussels attach to clam shells. The sheer weight of piled-on zebra mussels prevents clams from opening their shells, thus restricting their food intake.

Food as an Energy Source   The connection between food and life is obvious, as well as common to all life forms. Just what is it about food that makes it so important? In a word, the answer is energy.

Energy is the ability to do work. Work is anything that causes motion. When you read about the four common characteristics of life in the first section, you learned that all organisms carry on certain kinds of biological work—reproducing, growing, developing, and maintaining homeostasis. Organisms must have energy to keep this work going. Without energy, a zebra mussel could not expel eggs and sperm for reproduction, or close its shell to preserve moisture and maintain homeostasis. Nor could you perspire or shiver to maintain homeostasis in your own body. The zebra mussel, you, and every other organism gets energy for biological work from food. Energy fuels the work of life just as surely as gasoline fuels an automobile engine and keeps its parts functioning. Every organism uses the energy from some of the materials in food as fuel for the functions needed to keep it alive.

Producing Food   Maintaining an adequate food supply is critical in any community. Each community has an orderly system through which organisms obtain food from other organisms. Certain members of a community are food **producers.** That is, they make food. In some communities, plants—such as trees, grasses, and shrubs—and certain bacteria are the producers. In the communities invaded by zebra mussels, algae are the producers.

Among most producers, the ingredients in the food-making recipe are carbon dioxide and water. **Chlorophyll,** a green substance in the producers, traps light from the sun. Sunlight provides the energy for the work of combining carbon dioxide and water. Simple sugars are the finished product. Oxygen, given off while the manufacture is taking place, is the by-product. The whole process is called **photosynthesis.**

The light energy of the sun is converted during photosynthesis to chemical energy in the form of simple sugars. Thus, these sugars are good sources of energy and hence food for the producer itself. But producers do not use all the simple

Figure 1-8 Zebra mussels belong to a group of mollusks that have two shells. The shells offer protection and help maintain homeostasis.

Figure 1-9   During photosynthesis, carbon dioxide and water (raw materials) react to form sugars (end products) and oxygen (a by-product). Without the photosynthetic process, life on Earth would not exist.

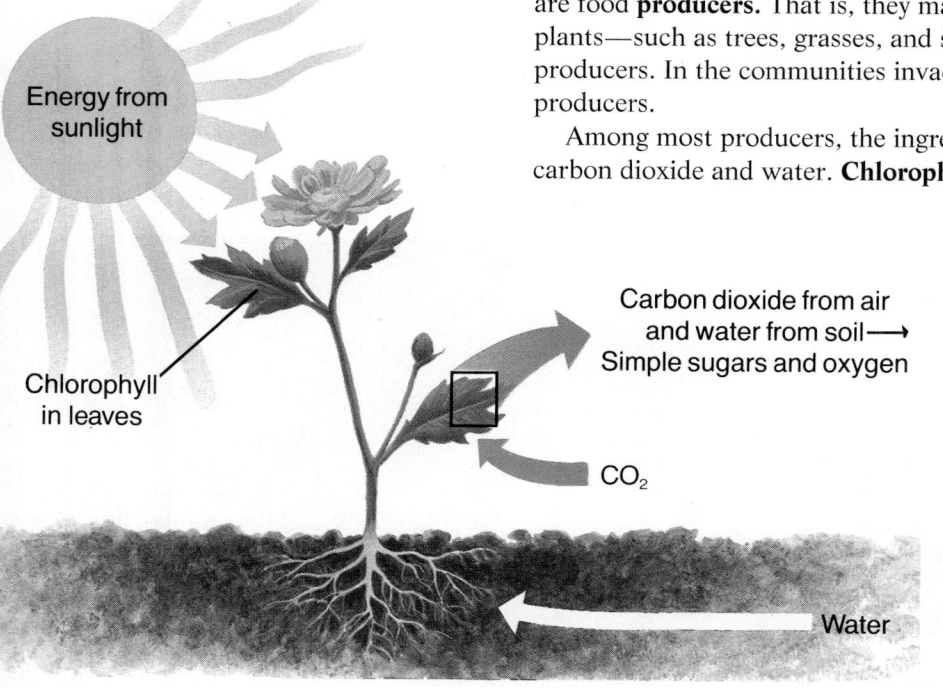

Energy from sunlight

Chlorophyll in leaves

Carbon dioxide from air and water from soil⟶ Simple sugars and oxygen

CO_2

Water

sugars they make for food. They convert some of the sugars into other, more complex substances that are rich in energy and can be stored for later use. The producer also uses some of the other complex substances to build new cells and body structures.

Consuming Food   Not only do producers make food for themselves, they also serve as food sources for other members of the community—those that are unable to produce their own. Organisms that cannot make their own food are **consumers.**

Some consumers, like leaf-grazing deer, feed directly on producers. When a consumer feeds on a producer, it converts some of the producer's complex substances into another sugar, known as glucose. Although glucose is not the *only* energy-rich food consumers use, it is the most important one. Producers and consumers convert the energy in glucose and other sugars to a usable form of energy through a process called **cellular respiration.** Energy converted during cellular respiration drives the biological work of life.

Zebra mussels are consumers that feed directly on producers—the microscopic algae in the water. As a zebra mussel draws water into its body, it filters algae out of the water through its siphon structures. Much of the algae is used as food as it passes through the digestive system of the zebra mussel. Some, however, is of no use to the zebra mussel and is expelled.

Small, shrimplike animals occupying surface waters also feed on algae. These small animals are food for small fish, which in turn are food for large fish—like walleye, trout, and bass. In this way, food is passed from producers through a series of consumers. Such a passage of food and the energy it contains is called a **food chain.** Different sets of organisms create different food chains. Each organism in the chain helps distribute energy that originates with the sun throughout the community.

One zebra mussel may not eat much, but thousands of zebra mussels together represent one hearty appetite. This many zebra mussels can eat a lot of algae. Thus, the feeding activities of zebra mussels could disrupt

Figure 1-10   **Producers convert light energy to chemical energy, which they may use or which may end up as an energy source for consumers (left). Two-shelled mollusks have two siphons, one for drawing in water, and one for expelling water (right). As water is drawn into the body, microscopic food particles contained in it are extracted by the gills and then passed along to the mouth. Using this siphoning action, zebra mussels are capable of filtering great volumes of water.**

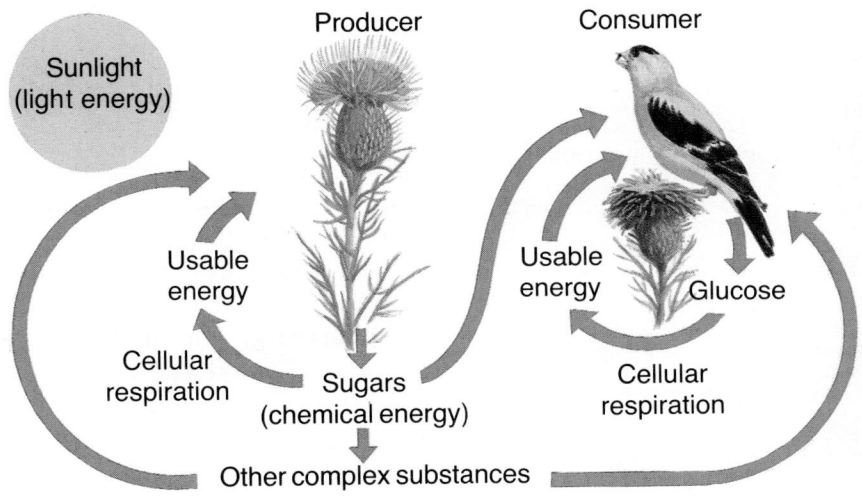

Minilab 2

How do mussels get food?

Mussels take in water through one siphon, filter out food particles, and let the water out of another siphon. Place a mussel in a beaker of dechlorinated water. **CAUTION:** *Use care when handling live animals. Do not disturb it for 10 minutes.* Carefully add an amount of carmine powder that will fit on the end of a flat toothpick. Draw a diagram of the path taken by the carmine powder through the water, into the mussel, and out of the mussel.

Analysis *How long did it take for the carmine powder to enter the mussel? What does this tell you about the filtering power of mussels?*

Figure 1-11 **Even the simplest food chain includes producers, consumers, and decomposers. Note that decomposers act on the wastes of all organisms in the chain, not just the last consumer.**

entire food chains. With fewer algae to go around, there would be fewer small, shrimplike animals, fewer small fish, and finally fewer walleye and other larger fish. In addition, the food materials that zebra mussels do not use settle to the bottom. There they add to the food supply of bottom-dwelling animals. These bottom dwellers could thus become more numerous and further disrupt the community.

Consumers categorized as **decomposers** complete every food chain. Decomposers feed on the dead remains and waste products of living organisms. Mushrooms on a rotting log, for example, are decomposers. As with all decomposers, their feeding causes decay.

Flow of Food in a Community

Planet Earth holds only a given supply of water, carbon dioxide, and other materials that living things need to carry out life processes. Even so, life goes on. There are always enough of these materials for generation after generation of producers and consumers. How can that be? Why doesn't the supply run out? It's because the same materials are used over and over again. Materials used for making food flow in a continuous cycle through the food chains of communities.

Cycling Food Materials All the links in a food chain—producers, consumers, and decomposers—constantly undergo cellular respiration. They give off the waste products of cellular respiration—water and carbon dioxide—as they use the energy in sugars and other fuels. The water and carbon dioxide are released into the environment. Water and carbon dioxide, of course, are the substances needed for photosynthesis. The food chain's producers take in water and carbon dioxide from the environment and use it to carry on photosynthesis. In the process, they make sugars

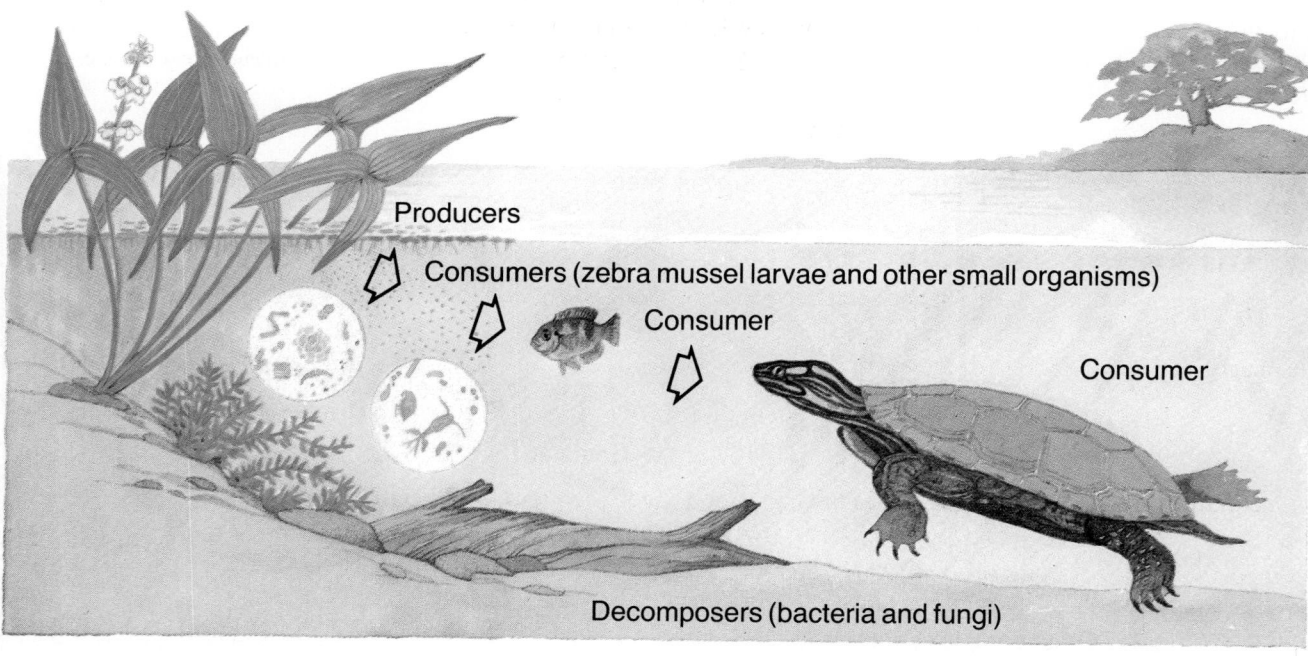

Producers

Consumers (zebra mussel larvae and other small organisms)

Consumer

Consumer

Decomposers (bacteria and fungi)

and other energy fuels, and release oxygen into the environment. Oxygen, as you have read, is needed by most organisms, including producers, for cellular respiration. Thus, the processes of photosynthesis and cellular respiration constantly interact. Producers and consumers constantly exchange the materials needed for life processes.

Unlocking Resources in Wastes   At the same time a food chain's organisms are obtaining food and using its energy for carrying on life processes, they are also producing wastes. Eventually, organisms die. Organisms' wastes and dead remains contain certain materials needed for carrying on life processes. If these materials were allowed to stay and accumulate in wastes and dead remains, there eventually would not be enough materials available to living organisms in a community. But decomposers, such as fungi and some kinds of bacteria, prevent this from happening.

Many decomposers are able to use the fuels still left in the wastes and remains of other organisms for cellular respiration. As with all organisms, the cellular respiration of decomposers releases carbon dioxide and water that can be cycled back to producers. But decomposers also carry on other processes. These processes break down and release nitrogen, sulfur, and phosphorus into the environment. All three of these substances are taken up by producers, which need them to make the more complex materials that also cycle through food chains.

In lake communities beset by zebra mussels, bacteria decompose wastes and dead organisms that settle to the bottom. If organisms deprived of food by zebra mussels begin to die in great numbers, the number of bacteria and volume of bottom decomposition will increase. That decomposition would consume great amounts of oxygen during the cellular respiration of the bacteria. Significant decreases in oxygen supplies could in turn lead to an even greater number of deaths in the community and higher volumes of decomposition on the bottom. The end result, biologists fear, would be that materials would not be cycled efficiently enough to sustain life in the community.

Flow of Energy in a Community

The path that energy takes as it passes through a community is more like a straight line than a circle. Instead of going from producer to consumer and back again, it changes form after it has been used to do work. Try this bit of work: Vigorously rub your hands together. Did you feel heat? The energy it took to rub your hands was changed to heat as you did the work

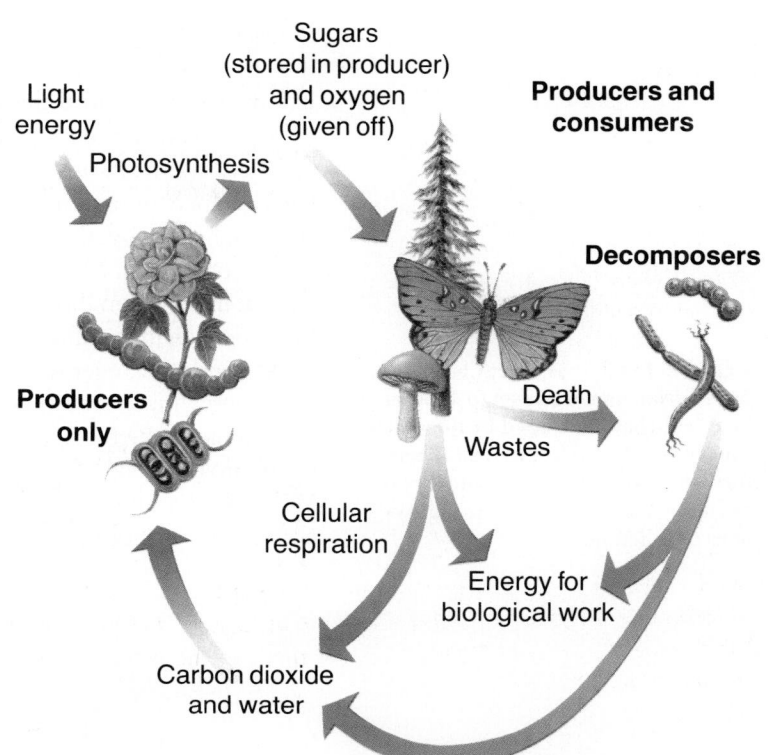

Figure 1-12   The products of photosynthesis are involved in respiration, while the products of respiration are involved in photosynthesis. Thus, materials are continuously cycled between the two processes.

of moving them. That energy still exists—in the form of heat released into the environment. You cannot reuse that energy to do additional work. You will have to take more energy from the fuel substances in food you eat.

In a food chain, energy sources are passed from one organism to another, first from producer to consumer and then from consumer to consumer. Once an organism uses a unit of energy for any kind of work, be it rubbing hands together or doing any other biological work, the energy cannot be reused. Thus—unlike such materials as water and sugars, oxygen and carbon dioxide, nitrogen and sulfur—energy is not cycled. Instead, it changes into heat, one of several forms that energy takes.

As you know, energy drives the processes, functions, and activities of life. Energy must always be available in the form of fuels made by producers. To make the fuels, producers must have a constant source of energy for photosynthesis. That energy comes from the sun in the form of light. Most communities, including those under attack by zebra mussels, could not survive without the sun and its energizing light.

Figure 1-13   As energy is trans- ferred from one organism to another in a food chain, it is used by the organ- isms for work and lost as heat. There- fore, there must be a constant supply of energy from the sun for living sys- tems to function.

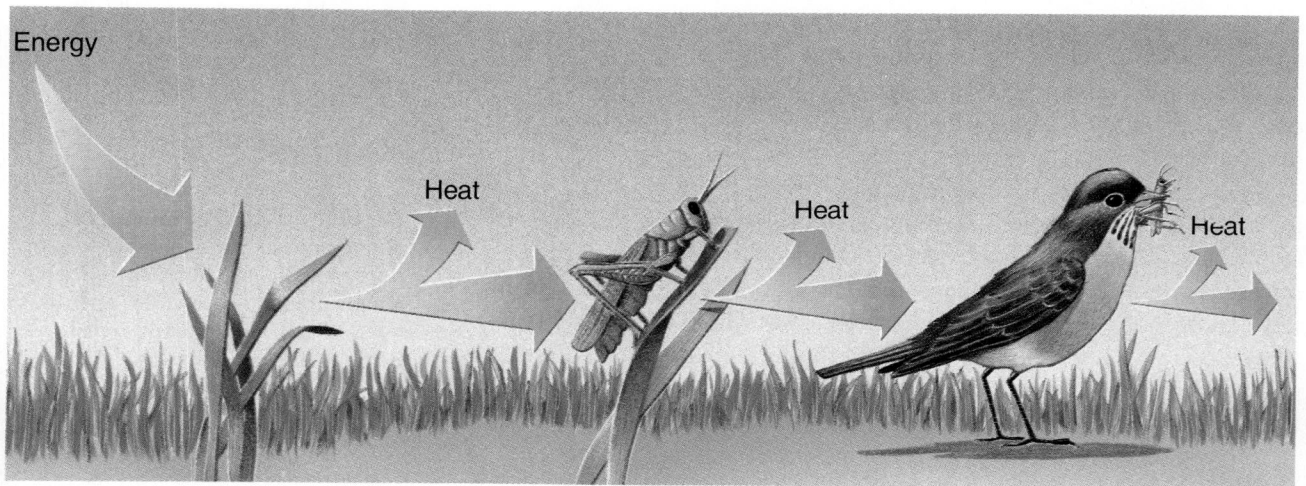

Energy

Heat   Heat   Heat

Section Review

Understanding Concepts

1. How is a candy bar you eat related to moving your muscles?
2. A catfish feeds on algae in an aquarium. Which of these organisms is a producer? A consumer? Which undergoes photosynthesis? Cellular respiration?
3. A forest community consists of trees, birds, insects, deer, bacteria, worms, ferns, mosses, squirrels, toadstools, spiders, snakes, and mice. List, in order, the organisms that might form a food chain composed of four links. Which of the organisms would act as decomposers?
4. **Skill Review—Making Tables:** Using the organisms listed in question 3, make a table that divides them into producers and consumers. For more help, see Organizing Information in the **Skill Handbook**.
5. **Thinking Critically:** If it were possible to decrease the size of algae populations, would such a decrease affect the size of zebra mussel populations? Explain.

1.3 Now That They're Here

According to one U.S. scientist, "There won't be one answer to the zebra mussel problem. We are going to need a bag of tricks." Indeed, science and industry are attacking the dilemma from a number of different standpoints. Their approaches take into account not only the biological communities invaded by the zebra mussels but also the physical environment.

Interactions in the Physical Environment

In any community, organisms respond to and interact with other living things. At the same time, they also respond to and interact with the physical environment. For example, a whole colony of zebra mussels can filter a great deal of water as each organism siphons out algae for food. A colony can filter so much water, in fact, that the water around the site becomes very clear. From the human point of view, this interaction with the physical environment could be a possible benefit.

From the zebra mussel point of view, physical factors of the North American environment help explain why the organisms have succeeded in multiplying so fast. For one thing, the zebra mussel finds North American water temperatures to its liking. The mussels can survive and reproduce at temperatures between 7°C and 32°C. Water currents in North American lakes and streams have also proved beneficial. The currents flow gently enough to allow young zebra mussels to attach easily, and evenly enough to deliver a steady food source. The waters also contain calcium, an important substance in the makeup of zebra mussel shells. In their search for ways to combat the zebra mussel, science and industry are looking at ways to change the physical environment so that it is less favorable for the zebra mussel.

Objectives

Discuss how organisms are influenced by nonliving factors of their environment.

Explain how the process of evolution has resulted in a great diversity of life forms and why there is unity within the diversity.

Apply knowledge of the biology of zebra mussels to possible methods of controlling them.

Figure 1-14 **The filtering action of zebra mussels can cause the water in which they live to become as crystal clear as this stream.**

Some Answers for Industry

Figure 1-15 Keeping pipes and waterways clear of zebra mussel buildup is an ongoing battle. One method of cleaning off the tiny invaders is to spray them with water heated to between 45°C and 55°C.

Shippers and lakeshore industries and drinking-water plants are still using scraping procedures to clear surfaces of zebra mussel clogging. They have also developed some improved screening devices that keep small larvae as well as larger forms of zebra mussels out of some water intake entrances. But these answers do not take care of all the problem areas.

Another possible control is to deprive the zebra mussels of oxygen, which you may recall they need for cellular respiration. Adding chemicals that contain sodium and chlorine removes oxygen from water that is sealed off in a given area. Lack of oxygen will kill zebra mussel larvae quickly, but adult zebra mussels can survive for several days without fresh oxygen. That often means that a plant using this method to clear clogged pipes will have to close down for several days while the affected areas are out of operation. Many factories and plants that provide drinking water to large cities cannot afford such lengthy shutdowns.

Increasing the speed of the current, if methods become available, might also prove worthwhile. A faster flow of water would prevent maturing forms of the organisms from settling and attaching. Electric currents sent

UNITY WITHIN DIVERSITY

Alike but Different

Basic life functions combined with the need for energy unite all of Earth's different kinds of organisms in a common bond that defines life. The statement you just read hints at one of the wonders of life. Life is amazingly diverse; although scientists have counted 2 million different forms, there may be another 4 million yet to be identified. All those life forms exhibit the same life functions and the same need for energy. How did such unity within such diversity come about?

Evolution   Clues to life's riddle of unity within diversity lie in the environment and how organisms interact with and respond to it. Adding change and time factors to the clues brings us right up to the answer, which in a word is *evolution*. Change in organisms over time is the meaning of **evolution.**

Environments Differ   Environments differ from place to place on Earth today, and they differ from environments of the past. In any environment, organisms have traits suited to that environment. These traits, which develop according to an organism's coded instructions, are known as **adaptations.**

Algae, for example, have the traits needed to produce chlorophyll for photosynthesis. Zebra mussels are adapted to draw in water, filtering out the algae and digesting substances from the algae they take in. Deer, which live in an entirely different environment from that of zebra mussels or algae, are adapted to browse on leaves for nourishment.

Environments Change   Changes to environments occur slowly, very slowly over the space of countless centuries. Populations of organisms living and interacting within that environment also change, slowly and over a long period of time. If enough changes occur, or if

through water have also been found to interfere with attachment. As a last resort, new pipes could be built and buried beneath the sand. Europeans, long experienced with zebra mussel harassment, put in such underground structures as long as a century ago. At that time, the Europeans were just beginning to build their first power plants. Americans expect to spend at least 4 billion dollars in the 1990s just to get rid of mussels already in intake pipes. New construction projects to install underground pipes would cost even more.

In those pipes that are exposed to zebra mussel intrusion, Europeans sometimes flush the pipes with water heated to between 45° and 55°C. At these temperatures, both adult and immature zebra mussels die; they can survive at temperatures no higher than 32°C. The flushing must be done at least three times a year. It usually requires some scraping as well, because some of the organisms remain attached. In the United States, widespread use of this method could increase water temperatures in lakes and rivers. The impact could be harmful to other communities.

Some Biological Possibilities

In the long run, probably the best solution to America's zebra mussel dilemma is to control the growth of the population rather than try to kill off and remove colonies as they arise. One of the reasons zebra mussel populations have soared so quickly in the United States is that no native consumers feed on zebra mussels.

changes are dramatic, a new type of organism may arise.

Therein lies the secret of the diversity of life forms on Earth. Earth's first life forms appeared some 3 billion years ago and have been evolving ever since. The first forms of life gradually gave rise to other, new forms, which in turn evolved into still other kinds of organisms.

The great diversity among the many kinds of organisms challenges scientists in a number of ways. First, there is the challenge of discovering and counting all the different life forms. Then there is the challenge of making sense of their differences and similarities. Take the zebra mussel. There are thousands of other animals whose bodies are boneless, like the zebra mussel's. There are also thousands of animals that have shells, some of which—like those of clams and oysters—open and close the way the zebra mussel's do. Although they do not have shells, slugs, squids, and octopuses also have many traits similar to those of the zebra mussel. Intriguing likenesses and differences exist throughout all 2 million of today's known life forms.

Kingdoms of Living Things
Biologists bring order to life's diversity by dividing and classifying organisms into a series of groups and subgroups. The first grouping is that of kingdom. A kingdom is the broadest division in the classification of organisms. There are five kingdoms, each with its own name and each encompassing organisms with certain traits. You will study the organization of and representatives belonging to these five kingdoms—Monerans, Protists, Fungi, Plants, and Animals—when you read Chapters 15, 16, and 17.

Control Through Predation   Introducing zebra mussel predators, organisms that kill and eat zebra mussels for food, might be one form of control. One such group of predators, in fact, may have already arrived—again from Europe and probably in the same way that the zebra mussels arrived. The predators are goby fish, and they seem to have an appetite for zebra mussels, shells and all. But biologists are not celebrating. The goby fish also eat the same kinds of small animals that walleye and lake trout eat. There is a danger that the goby will rob these valuable fish of their food supply and thus reduce their numbers.

Interfering with Reproduction   Another, and perhaps the most promising, biological approach focuses on tampering with the zebra mussels' reproductive pattern. If successful, both the size of zebra mussel populations and their spread to new waterways could eventually be decreased. Furthermore, no other organisms in the community would be adversely affected.

I N V E S T I G A T I O N

Mussels and Their Environment

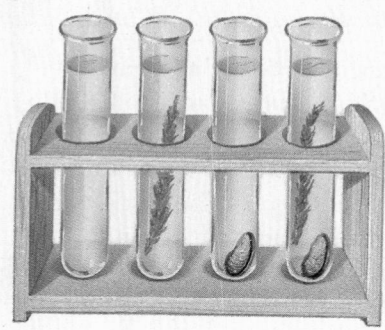

Why do you breathe? You breathe in oxygen, which is needed by your body for breaking down food and converting the energy it contains to a form your body can use. What do you breathe out? Carbon dioxide. This gas is a waste product from the breakdown of food. Does a mussel "breathe"? Like you, a mussel needs oxygen for the conversion of energy into usable forms through cellular respiration. Mussels, of course, get oxygen from the water that surrounds them. Does a mussel breathe out? If a mussel takes in oxygen and gives off carbon dioxide, how will this gas exchange affect the ecosystem in which it lives? Consider other organisms that live in the water with the mussel. They, too, exchange gases with their watery environment. What one organism gives off, another may use. In this way, organisms in a food chain make materials available to each other. In this lab, you will see how the mussel affects its environment and how other organisms might be dependent on the mussel.

Problem
How do mussels affect their environment?

Safety Precautions 🧤 ☠️ 🔥
Bromthymol blue will stain clothing and skin. Do not drink bromthymol blue. Wash your hands at the end of the lab.

Experimental Plan
1. Fill a test tube with distilled water. Add a drop or two of bromthymol blue until you have a very light blue color. Check the color by holding it against a piece of white paper.
2. Place a clean straw into the test tube and blow into the bromthymol blue solution. Do

Zebra mussels begin reproducing when they detect an explosive growth in the algae on which they feed. Reproducing at that time ensures a plentiful supply of food for the young. If the adult zebra mussels could be tricked into reproducing at a time when algae are less plentiful, far fewer mussel offspring would survive. Biologists might accomplish such trickery if they find substances that would cause males to release sperm and females to release eggs at the wrong times in the algae life cycle.

Another possibility would be to trick males and females into releasing sperm and eggs at widely spread time intervals. Because sperm and eggs live only a short time, fertilization would not occur. Yet another approach arises from the fact that adult zebra mussels stuff themselves with food just prior to reproducing. They need extra food to get the extra energy they need for producing sperm and eggs. Biologists are searching for a substance that would cause the mussels to close their shells, and thus prevent food intake. Such a substance, though, would have to be one that doesn't interfere with the feeding routines of similar animals, like clams.

Figure 1-16 The goby is a natural predator of the zebra mussel. However, it also eats the same foods that walleye and lake trout eat. What are the possible effects of an increasing goby population?

not get any of the solution in your mouth. Continue to blow into the solution until you notice a color change.

3. In your group, decide what may have caused the color change.

4. Do you think the color of bromthymol blue solution would change if a water plant were placed in the test tube overnight in bright light? Why or why not? What about a mussel? What about a mussel and a plant?

5. In your group, make a list of possible ways the mussel might affect its environment.

6. From the list you have made, select an effect you could test with bromthymol blue solution, a mussel, and a water plant.

7. Make a prediction about what you think will happen during your investigation.

8. You will be provided with several test tubes, one of which should contain only bromthymol blue solution so you will have a basis for comparison at the end of your investigation.

9. Following the style of a recipe, write a numbered list of directions that anyone could follow.

10. Make a list of the materials you will need.

11. Make a table in which to record what happens. In your table, you may want to include the type(s) of organism(s) you tested and the colors observed in the bromthymol blue.

Make sure your teacher has approved your plan before you proceed further.

Data and Observations

Carry out your investigation and record what happens in your table.

Analysis and Conclusions

1. What happened to the bromthymol blue in which you put a mussel? An aquarium plant? An aquarium plant and a mussel? Explain.

2. Compare your outcome with those of other groups. How do the outcomes compare? Explain.

3. If you were going to do this same investigation again, what would you do differently? Why?

4. Explain how mussels and plants in a pond are dependent on each other.

5. How might the balance of living things in the pond be upset if the mussels increased in huge numbers?

Going Further

Based on this lab experience, plan another investigation that helps to determine more about the relationship between mussels and their environment. Assume that you will have unlimited resources and will not be conducting this lab in class.

Change and the Environment

All of the solutions biologists are exploring involve human-made changes in the environment. Scientists want to restrict these changes to ones that conflict with the adaptations of zebra mussels only. Nature also makes changes in the environment. Natural disasters such as floods, fires, or volcanic eruptions can arise suddenly and have drastic effects on entire biological communities.

For the most part, natural disasters occur infrequently. More gentle and more predictable changes, however, go on constantly in both physical and biological environments. Responding to change in various ways is part of everyday living for organisms. As you have just read, for example, zebra mussel reproduction takes place in response to changes in algae growth. Many organisms, especially those that live on land, face seasonal changes every year. Many kinds of plants respond to cooler temperatures

Thinking Lab

INTERPRET THE DATA

How do physical factors affect zebra mussels?

Background

What do you think would happen to living things in a lake if the oxygen dissolved in the water disappeared? What would happen to the same living things if the temperature of the water were raised 10°C? Most likely, these changes in the physical factors of the lake would have an adverse effect on life in the lake. In Europe, zebra mussels building up in water pipelines have

been controlled to some degree by periodic flushing of water pipes with hot water. Oxygen can be eliminated from a sealed water system by adding specific chemicals. Zebra mussels can live for a few days without oxygen, but eventually they die.

You Try It

Interpret the data in the table below and answer the following questions.

1. In which water temperatures do zebra mussels die more quickly?
2. What is the relationship between water temperature, dissolved oxygen, and death of zebra mussels?
3. If you were the manager of a water system contaminated with zebra mussels, which method of mussel control would you use? Speculate about the advantages and disadvantages of each method.

Table 1.1 Zebra Mussel Death Rates

Water Temp (°C)	Day 0	Day 1	Day 2	Day 3	Day 4	Day 5	Day 6	Factors Studied
17-18	7.5	—	0.7	0.03	—	0	0	dissolved oxygen mg/L
	0	—	0	0	—	90	100	% dead mussels
20-21	9.6	0.08	0	—	0	—	—	dissolved oxygen mg/L
	0	0	10	—	100	—	—	% dead mussels
23-24	7.1	0.33	0	0	—	—	—	dissolved oxygen mg/L
	0	0	48	100	—	—	—	% dead mussels

of autumn by losing their leaves. For many consumers, this means a reduction in food supply. Some consumers might respond by migrating to areas where food is abundant. Others might hibernate, thus reducing their need for food during the long winter of barren plants.

Patterns of change are part of the natural world and influence the lives of all organisms. Humans are the only organisms who can make choices and decisions about how to respond to changes in the natural world. They can also make decisions about how or how not to introduce changes themselves. As you have seen, biologists apply caution as they consider changes aimed at combating zebra mussels. But one change they have persuaded government leaders to make relates to the dumping of ships' ballast. Ships' crews are now required to release ballast offshore, in saltwater ocean areas, instead of waiting until they put in at freshwater ports. In this way, they should eliminate the future risk of changing communities by introducing unwelcome misfits like zebra mussels.

Figure 1-17 **Natural disasters lead to changes in living systems. Most change, however, is regular and does not disrupt the system.**

LOOKING AHEAD

IN THIS CHAPTER, you looked over the shoulders of biologists as they predicted outcomes of the zebra mussel presence. By now, you probably appreciate how important it is that scientists find a way to prevent those outcomes from disrupting life in our lakes and rivers. In the next chapter, you will get an idea of how biologists go about finding solutions. You will see that scientists follow similar steps and procedures as they attack any scientific problem. Altogether, these steps and procedures make up a kind of formula for research and investigation—the never-ending story of science.

Section Review

Understanding Concepts
1. What physical factors promote growth of zebra mussel colonies in intake pipes?
2. Octopuses and zebra mussels are two types of mollusks. Each animal has unique characteristics. The octopus has no shell, and the zebra mussel has two shells. How did those different characteristics originate? How are the two mollusks the same?
3. Is it likely that zebra mussels could establish colonies in swiftly moving streams? Explain.
4. *Skill Review—Classifying:* You observe that spider monkeys and rhesus monkeys have many features in common. How can you explain their similarities? For more help, see Thinking Critically in the **Skill Handbook**.
5. *Thinking Critically:* You discover an organism that seems to be rooted in the soil. How could you decide whether that organism is a fungus or a plant?

BIOLOGY, TECHNOLOGY, AND SOCIETY

Biotechnology

Watching the Brain at Work

For centuries, people have puzzled over this question: What does a thought look like?

PET Medical equipment that has already been in use for several years provides clues to the answer. A PET (positron emission tomography) machine can give doctors a peek inside the brain. Here's how it works.

Energy comes in many different forms. In this chapter, you've read that sunlight is a form of energy and food contains chemical energy. Certain elements also emit energy. These radioactive elements, as they are called, have proven to be indispensable medical tools. PET is just one technique that uses radioactive elements to produce images of the inside of the body. Before the PET machine can do its job, a patient is given a simple sugar (glucose) solution tagged with a radioactive element. After this solution moves into the brain, the radioactive tracer gives off particles like X rays, thus allowing a computer to trace its location. The PET machine shows the location on a color computer monitor. The more active the brain is, the more it uses the sugar solution. And the greater the use of sugar by the brain, the brighter the spot on the monitor.

During the PET scan, the doctor may vary the patient's exposure to sound or sight in order to watch for changes in the brain's image. By comparing images of a patient's brain during different activities, researchers have identified specific areas of the brain used for seeing, hearing, speaking, and thinking.

MANSCAN A device newer than PET is the mental activity network scanner, or MANSCAN for short. This device, which fits over a patient's head like a soft helmet, takes a picture of brain activity every 4 milliseconds. (A millisecond is one-thousandth of a second.)

One application of MANSCAN is to study short-term memory. This component of thought, which involves remembering and using something that has just been learned, is often disturbed by serious head injuries or Alzheimer's

CAREER CONNECTION

Radiology technicians need specialized training to work with new technology such as PET and MANSCAN machines. People working in these careers in hospitals and medical schools must have an interest in and aptitude for science and mathematics as well as people skills.

disease. MANSCAN can record brain-wave patterns to predict whether a person is about to perform a memory task correctly. This application of MANSCAN may be useful in diagnosing brain diseases and in testing the effectiveness of drugs or methods of rehabilitation.

Another potential use of MANSCAN could be in a kind of brain monitor for people in jobs involving extremely critical decisions, such as airline pilots or brain surgeons. By putting on the MANSCAN helmet for a quick "checkup," these people could predict the onset of mental fatigue before their job performance began to decline.

Researchers expect that an even more detailed picture of the working brain may result when PET and MANSCAN, plus yet-to-be-developed technologies, are teamed up.

1. **Issue:** Exposure to radioactive elements has been linked to diseases such as cancer. How can the use of radioactive elements be justified?

2. **Connection to Social Studies:** What are some other careers or jobs in which you think a MANSCAN "checkup" would be useful? Why?

Rare Animals as Pets

How would you like to have a herd of Indian elephants in your backyard? Does a basement full of rare Rothschild's myna birds sound appealing? Or how about a snow leopard basking in the sunshine on your patio? In about 850 American homes, exotic pets like these are cared for through the Fish and Wildlife Service's Captive-Bred Wildlife Registration program.

Animal Adoption   Lovers of exotic animals have to meet some strict requirements before they can keep rare animals as pets. (Obviously, even a small herd of elephants would need more roaming room than the average suburban backyard has to offer.) However, the people who can take on such a project do it for reasons besides personal enjoyment. They are concerned about the dwindling number of animals in the wild and want to give them a safe place to grow up and reproduce. Loss of animals or plants, or any other type of organism from the wild, reduces diversity.

Problem Pet Owners   You probably know people who enjoy the attention they get from wearing unusual clothing and owning out-of-the-ordinary things. This personality trait can be fun, but when it extends to possessing rare animals, the results can be rather cruel. Take, for example, a black bear. When it is a cub, it looks like an overgrown teddy bear. People who think the bear would make a cuddly pet may be in for a surprise in a few months because a bear is and always will be a wild animal. Even if its claws and some of its sharpest teeth are removed, a bear is still capable of behaving as it would in the wild and may act aggressively toward its owners.

By the time wild animal owners realize they have a problem, they find that it is nearly impossible to give a bear, tiger, or wolf to a zoo. The animal may have developed undesirable behaviors or habits and be unfit for life in a zoo. The pet these owners once loved could end up on a kind of ranch where hunters pay a fee to shoot animals. Birds, which are easier to place in new homes, present a different kind of problem. Birds often outlive their owners. Parrots, for example, can live to be 90 years old.

Illegal Entry   With wild pets not as potentially dangerous as bears, lions, or snakes, there are other factors to consider. A wild bird is probably a safe enough animal in a home, but people planning to buy one should consider where it has come from. Up to 20 million birds are taken from the wild each year—a number that could push many species near extinction. Half a million of these birds end up legally in the United States each year, while another 100 000 are smuggled in. As many as 60 percent of captured birds are thought to die in transit to the United States. Why so many? Consider parrots. Parrots are sometimes taken across the Mexican-U.S. border in the hubcaps of cars. It isn't surprising that they can't survive the trip.

Many animal collectors justify keeping exotic pets by claiming that they are saving them from death as natural habitats are reduced or destroyed. Would these animals be better off if their habitats were preserved?

Understanding the Issue
1. Would you like to participate in the Wildlife Registration program? Why or why not?
2. Why do you think that preserving a wild animal's natural habitat is often difficult?

Readings
- Luoma, J.R. "Born to Be Wild." *Audubon,* Jan.-Feb., 1992, pp. 50-58.
- Yerton, Stewart. "Rare Species and Red Tape." *New York Times,* June 25, 1992, pp. C-1.

Summary

The more than 2 million different kinds of organisms are distinguished on the basis of physical characteristics. The focus of a modern course in biology, however, is not on differences among various forms of life, but rather the common features shared by all.

Living things do not live alone, but exist together in communities. They do so because they are interdependent. A fundamental way in which organisms depend upon one another is for food.

Through photosynthesis, most producers use the energy of sunlight to make substances that become food for other organisms. That food, and the energy it contains, is made available to all members of a community through food chains. Through cellular respiration, both producers and consumers release the energy of substances in food to perform biological work.

In any community, materials are cycled but energy is not. To exist, almost all communities must have a continual input of energy in the form of sunlight.

Energy released during cellular respiration is used for a variety of functions: reproduction, growth and development, homeostasis, and organization. These functions are common to all organisms.

In addition to their need for energy to carry out functions, all organisms share other common features. They interact with one another and with the physical aspects of their environment, and they respond to patterns of change. Although evolution has resulted in the great diversity of living things on Earth, the common features of organisms have been retained over time. Though diverse, the many forms of life can be classified into five kingdoms.

Language of Biology

Write a sentence that shows your understanding of each of the following terms.

adaptation	evolution
biology	food chain
cellular respiration	growth
chlorophyll	homeostasis
community	organism
consumer	organization
decomposer	photosynthesis
development	producer
energy	reproduction

Understanding Concepts

1. An aquarium consists of several plants, a variety of fish, and snails. How does each type of organism contribute to cycling of materials in the aquarium?
2. What other kinds of organisms must be present in the aquarium in order for cycling of materials to continue over time? Explain.
3. Could life in the aquarium continue to exist if the aquarium were placed in the dark? Explain.
4. Could zebra mussels colonize intake pipes if there were no continuous flow of water into those pipes? Explain.
5. Before it is a butterfly, an insect is a fertilized egg and then a caterpillar. What common features of life do these facts illustrate? What is needed for those features to occur?
6. Your heart beats more quickly and you breathe more rapidly when you exercise. How are such changes beneficial? Of what common feature of life are those changes an example?
7. What happens to the organization of an organism when it dies? What is the cause? Why does it not happen when the organism is alive?
8. How are the stages of development of zebra mussels an adaptation for spreading to new areas?

9. When you begin a new course in school, you usually adapt to the way your teacher wants you to do your homework. How is the meaning of the word "adapt" in everyday conversation different from the biological meaning of the word?

Applying Concepts

10. Suppose a predator of zebra mussels is found. The advantage of introducing such a predator into waters where the mussels are found is clear. What disadvantages might follow introduction of the predator?
11. One suggested method of dealing with the problem of intake pipes clogged with zebra mussels is to block the intake of oxygen. Why would this be an effective method?
12. Identify several different types of natural communities. What do they have in common? How do they differ?
13. Why can most organisms live only in a certain type of community?
14. What would be the simplest possible food chain in a community? Explain.
15. A piece of potato immersed in water absorbs some of the water and swells. Has the piece of potato grown? Explain.
16. How would each of the following physical factors be important in a forest community: availability of water; direct sunlight versus shade; type of soil?
17. Why are there no algae at the bottom of a deep lake or ocean?
18. **Biotechnology:** How is MANSCAN used to monitor brain activity?
19. **Issues:** Explain the types of problems that arise as a result of keeping wild animals as pets.

Lab Interpretation

20. **Investigation:** If you had a chemical solution that was red in the absence of oxygen and green in the presence of oxygen, what color would the following solutions be after exposure to bright light for 24 hours? Explain.

    a. Solution with mussel
    b. Solution with water plant
    c. Solution with mussel and water plant
    d. Solution alone

21. **Minilab 1:** You are walking along an ocean beach when a very unusual specimen rolls up to your feet in the waves. It looks somewhat like a rock and somewhat like a shell. It is crusted over with a hard buildup of a substance that you cannot scrape off, so you cannot see it well. What could you do to tell if it is a relative of a snail or mussel?
22. **Thinking Lab:** Various types of mussels were tested for their tolerance to heated water. Results are illustrated in the graph below. Which species might live in the tropics, which in temperate climates, and which in the far north? Explain.

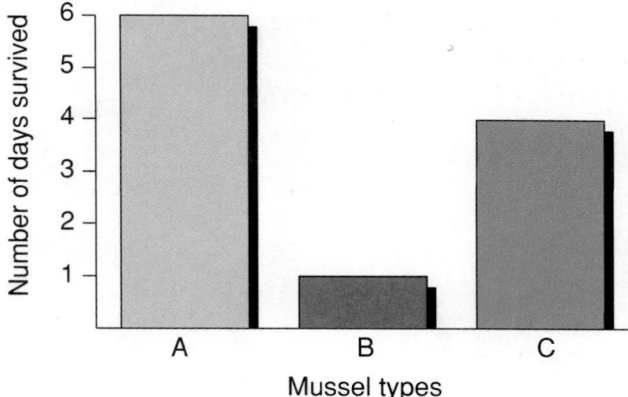

Connecting Ideas

23. How is the adaptation of filtering food related to other adaptations characteristic of adult zebra mussels?
24. What is the common requirement necessary for all the following activities? A firefly lights up. A muscle in your arm contracts. A leaf produces simple sugars. A zebra mussel closes its shells.

2

BIOLOGY AS A SCIENCE

HAVE YOU EVER watched a nature program on television and wished you could have the opportunity to travel to strange places and study exotic animals? What may have looked like a wonderful vacation and adventure to you is serious business for the scientists who spend weeks, months, and even years out in the field gathering information about the world of biology. These scientists study and collect information about organisms, such as physical characteristics, what they eat, how they reproduce, and how they interact with other living things and their environment.

The naturalist in the photograph is trying to learn more about a population of falcons that inhabits a specific area of the eastern United States. Part of the investigation includes capturing and banding individual birds. If this bird is recaptured in some other place, information recorded on the band will be collected and added to a body of knowledge that scientists throughout the world have collected on peregrine falcons. How do scientists from one country follow and understand the work of scientists who live in another country? In this chapter, you will learn about the methods most scientists follow as they conduct their investigations. You will become familiar with some of the tools that biologists use. You also will learn about the universal system of measurement that enables scientists from all over the world to communicate the results of their investigations.

Suppose you found an animal that had been banded or tagged. What information might you learn about that animal from the band?

Chapter Preview

Objectives

Relate the various methods used by scientists to learn about the natural world.

Apply your understanding of a controlled experiment to the testing of a hypothesis.

2.1 Methods of Science

Have you ever tried to figure out what an object in a closed box might be? Perhaps you shook the box or tilted it back and forth. You listened to the sound of the object as it moved within the box. You tried to determine its size, shape, weight, and perhaps the material it was made of. As you manipulated the box, you were making observations about it, and, based on your previous knowledge and experience, were ruling out certain conclusions and leaving open other possibilities. Finally, you narrowed your choices further and made a decision about what the object was. In trying to solve the puzzle of what was in the box, you were following the kinds of procedures that scientists use to learn about the natural world. Before continuing with this chapter, do the Minilab on page 33. Find out how sharp your powers of observation really are.

Development of Science

Think about the times when you tried to figure out what was inside a box. Perhaps you were curious because the object inside was a present for your birthday. Humans have always been curious about the natural world. They see effects and try to determine the causes of those effects. Why, for example, does the sun always disappear from the western sky and reappear in the east? Why do birds migrate when the seasons change, and how do they know where to go?

People long ago attempted to explain nature and make use of what they learned. Explanations based on reason were assumed correct and passed on from one generation to the next. In the 1500s people began to question the explanations of earlier generations. They developed new explanations based on the information they discovered. Much of the new information concerned earth, physical, and biological sciences. Galileo and Copernicus revolutionized the study of the planets, while Newton's laws of gravity and motion led to a new field of physics. In biology, the English scientist, Robert Hooke, used an early microscope to discover that cork is made of many boxlike structures he called cells.

Figure 2-1 **Migrating geese often fly in a V-shaped formation. Can you explain why?**

Modern biology began in the 1800s. By then scientists observed that all organisms are made up of cells, and that all life comes from preexisting life. The theory of evolution was proposed, as were the principles of heredity. Such findings were important for two reasons. They applied to all organisms, and they provided a basis for later discoveries.

What brought about these discoveries? It was a new way of learning about nature. One of the most important factors in good science is asking the right questions. Many discoveries made in the last 100 years are still valid because the scientists who made them were good detectives.

What Is Science?

What does the word *science* mean to you?

Science as a Process   You may think of science as a list of terms to be memorized or facts to be learned, but that is not what science is. **Science** is a *process* that produces a body of knowledge about nature. As you read this book, you will learn a great deal of information about what is currently known about living things. You will also learn about the research leading to some of this information. You will work like a scientist and get a feel for how science is carried out.

Applied Science   You know that science is carried out because people are curious. They want to learn more about nature. Some people enjoy gaining knowledge for its own sake. Others want to apply the knowledge for a useful purpose. Applying knowledge to real problems is **technology,** or applied science. For example, some scientists wanted to know how traits are transmitted to offspring. Their curiosity led to the field of genetics. The scientific knowledge of genetics has been used to produce plants that grow faster and bacteria that produce drugs.

Remember that scientists note effects and attempt to explain the causes of those effects. There is no fixed way to do this; scientists arrive at explanations in many ways. Each situation is different. Several factors play a role in learning about nature. Among them are careful and thorough observations, interpretation of what is observed, explaining what is observed, and testing to see if those explanations are "on the right track." These steps do not always follow a precise order, and a scientist may go back and forth many times between many steps to solve a puzzle.

Observation

Observations about the natural world are the heart of science. A scientist may notice something that sparks his or her curiosity and then decide to observe it very carefully over a period of time. Often, this may lead to reading what is already known about the subject or related subjects. Sometimes an observation may seem unimportant at first. However, it may lead to something very important. The discovery of the role of DNA came about in such a way.

Collecting Data   Observations lead to collecting bits of information about the natural world. Information can come from observations in the field, such as the average measurements of sunflowers growing in slightly different habitats. Some information can be gathered from observations made during experiments. For example, an experiment might include monitoring the development of tadpoles into frogs when they have been exposed to different chemicals. All of the information gathered from observations is called **data.** Data need not be numerical. For example, data about the shapes or colors of butterfly wings are in the form of words.

Minilab 1

How do observations help you make an educated guess?

Obtain an "object" from your teacher. Check over the "object" given to you. Make observations of the outside of this "object." Record your outside observations in a table. Diagrams may be used. Make a reasonable assumption as to what you might find inside the object, based on your observations. Open the object and carefully check its contents. Record your inside observations in your table.

Analysis   *Do your inside observations support or reject your assumption? Explain how your observations led to your assumptions about the object.*

Figure 2-2   Observation includes collecting data on living things and making observations. How do you know these organisms become frogs?

Figure 2-3 How would you record data about the objects shown in this photograph?

Making Observations Suppose you are walking around your school, a park, or the area near where you live and you notice a particular flowering plant. As you walk along admiring the flowers, you notice that some plants are taller than others. You continue your walk, forgetting the flowers as something else catches your attention.

Later, you remember something about the different sizes of the plants, and you go back for another look. Upon closer observation, you see that there are indeed marked differences in the sizes of the plants. The taller ones are the ones exposed to more direct sunlight. The plants in the shade seem, on average, to be quite a bit shorter. Returning with a tape measure, you record data—the heights of plants in sun and in shade.

Interpretation and Explanation

What do you do with data you have recorded? If you are thinking like a scientist, you would be curious about the size differences of the plants you saw on your walk and begin thinking about what might cause those differences. One possibility is that some of the plants are younger than others, but you know they were all planted a few years ago at about the same time. You also discount the idea that the plants in the sun are growing in better soil than those in the shade, because you know that the soil is the same for both groups of plants.

Using Reason What you have been doing is trying to reason things out. You have been attempting to interpret what you observed and have been doing so on the basis of previous observation and knowledge. You are being logical as you sort through possible causes of the difference in the height of these plants.

You know that plants carry out photosynthesis and that they require light in order to do so. The explanation seems clearer to you now. You assume that the plants exposed to more light over a long period of time grow taller because they produce more sugars by photosynthesis than the plants in the shade. If they produce more sugars, you reason, they can also produce more materials needed for growth. This leads you to a broader point. Light probably influences the rate of photosynthesis in all plants, not just the plants you observed.

Making a Conclusion You have reached a conclusion by analyzing your observations and by sorting through possible causes of the effect you noted. Your explanation accounts for the data originally gathered through your observations. Your explanation could be stated: "Plants exposed to brighter light over a long period of time grow taller because they undergo photosynthesis more quickly." A statement that explains and relates data is called a **hypothesis.** Scientists do not always state formal hypotheses as they carry out their work. However, they do continually develop ideas that explain their observations and guide them in their further research. Can you think of any other statement to explain why some plants are shorter than the others? Could you collect data to support your explanation?

Testing the Explanation

You have explained the difference in height of plants satisfactorily to yourself, but forming an explanation does not in itself solve the problem. Many earlier "scientists" stopped at the point of coming up with an explanation. They made the explanation the conclusion of their work. An explanation, whether a formal hypothesis or not, must be tested. Only if there is a test can a scientist be certain whether or not an explanation is realistic. A hypothesis is only a tool for further study of a problem.

A good explanation not only explains the data, but also predicts new data. Scientists ask themselves, "If sunlight causes plants to grow taller, will a lack of sunlight result in shorter plants?" Asking such questions, in other words, making predictions, is a major step in doing science. By asking questions, scientists find ways to test their explanations. Explanations are most often, but not always, tested by **experiments.**

Designing the Experiment

Let's return to your explanation. How can you test whether your idea that brighter light leads to an increase in rate of photosynthesis is correct? Think about the idea of *rate* of photosynthesis. Rate means how much photosynthesis occurs per unit of time. How can you detect rate? You know that sugars are produced during photosynthesis. Can you somehow measure how much sugar is produced in a given amount of time? That seems unlikely, because sugar is produced within the leaves of the plant. You are stumped until you remember that photosynthesis also results in the production of oxygen. Oxygen is released from a plant during photosynthesis, so, you reason, if you could somehow measure how much oxygen is given off, you would have an indirect way of measuring the rate of photosynthesis.

Figure 2-4 Some cacti in the desert grow very tall, yet other desert plants stay short. How would you explain this difference?

The Need to Reduce Doubt

Remember, you are trying to determine the cause of a particular effect. In this case, the effect is a greater rate of photosynthesis and the cause, you believe, is brighter light. In carrying out your experiment, you must eliminate all doubt about cause and effect. You have to make sure, if you can, that any differences in rate of photosynthesis you might observe are due to differences in brightness of light and no other factor. Now try the following investigation. See if you can devise and conduct an experiment that will confirm or disprove the hypothesis.

Light and the Rate of Photosynthesis

How can you measure the amount of oxygen given off by a plant? After all, you can't see oxygen coming from a plant. Or can you? You have an aquarium that contains plants, and you have noticed bubbles appearing around those plants. You never paid much attention to those bubbles before, but now you realize that those bubbles are bubbles of oxygen. The design for carrying out your experiment now becomes clear. You can estimate the rate of photosynthesis by counting the number of bubbles coming from aquatic plants in a given period of time, such as 20 minutes.

Problem

You begin by stating the problem you are trying to solve. In this case, you want to know if brighter light increases the rate of photosynthesis, so your problem statement might be: Does brighter light increase the rate of photosynthesis?

Hypothesis

Now you must prepare a hypothesis that might explain your observations. A hypothesis is usually written using an if/then format. Your hypothesis might be stated as: If the level of light influences the rate of photosynthesis, then increasing the amount (brightness) of light should increase the rate of photosynthesis.

Safety Precautions

Be careful when working with plant materials. Make sure that electrical wires do not touch the water unless they are intended for use in aquaria.

Experimental Plan

1. Now you are ready to write your experimental plan. An experimental plan details the steps you must take to collect data to test your hypothesis. Begin by listing what materials you will need. In this experiment, you will need to collect the following materials from your teacher: 2 small aquaria, 2 *Elodea* plants, an aquarium lid with a 15-watt fluorescent bulb, an aquarium lid with a 75-watt fluorescent bulb attached to a dimmer switch, water, a watch, and a thermometer.

2. Working in a group enables each group member to participate in the experiment. Scientists often work cooperatively. In this experiment, plan to work with several classmates. Each classmate may be assigned a different job, such as collecting materials, making observations, or recording data.

3. Use the materials to design an experiment. The experiment must be a controlled experiment. A **controlled experiment** is one in which all variables are the same except the one being tested (the one factor presumed to be the cause of the effect). The variable being tested is the **independent**

variable. In a controlled experiment, only this independent variable changes. All the other variables must be kept constant so that you are sure the results you observe are due to changes in the variable being tested. To have a controlled experiment, two groups must be tested. The group in which the independent variable is changed, or manipulated, is known as the **experimental group;** the group in which all the variables remain constant, including the independent variable, is known as the **control group.**

4. Make sure that your experiment tests only one variable at a time. All factors, in both control and experimental groups, must be treated exactly alike except for the independent variable. In this experiment, the independent variable will be the amount of light. The amount of water, number of plants, temperature, and location of aquaria will remain constant.

5. Allow for the collection of quantitative data (data that use numbers in reporting results). For example, you may need to count bubbles given off by a plant in a given period of time. In this experiment, the number of bubbles given off is the dependent variable. The **dependent variable** is any change that results from the manipulation of the independent varible. You could count the bubbles given off every two minutes, for example. Create a table for recording data. A data table in your experiment might

list the numbers of bubbles given off by plants in each aquaria every two minutes.

6. Discuss the following points with your group. If necessary, change your experiment to meet these criteria.
   (a) What is your control group?
   (b) Which independent variable will you manipulate?
   (c) What variables need to be controlled?
   (d) What data will you collect and how will you measure them?
   (e) How will you record your data?
   (f) How many trials will you carry out? Why?
7. Write specific numbered directions for your experiment.
8. Conduct your experiment.

Data and Observations

During the experiment, observe what happens as a result of changes you made in the independent variable. Write down in the data table what you have observed. As you complete each trial of the experiment, record the data. In this experiment, you might record the number of bubbles produced by a plant in the aquaria every two minutes up to 20 minutes.

Analysis and Conclusions

At the end of the experiment, study the data collected

in the data table for analysis and make some conclusions based on your results. The analysis may include answering questions such as these:

1. How did you determine your control group?
2. How did you determine your experimental group?
3. What variable did you test?
4. What variables did you keep the same in all groups?
5. Do your data support or reject your hypothesis? How?
6. Do your experimental results verify the hypothesis? Why?

Going Further

After you have analyzed your data and reached some conclusions, you may want to design another experiment that would answer another problem. Or you may want to revise your hypothesis or change the variable being tested to make sure that you have reached the correct conclusion.

After the Experiment

Now that you have completed the experiment, it's time to study your data. Having conducted several trials for each group, you decide to average the data for each group. You can then analyze your data and determine what they tell you.

Analyzing Your Data   In your experiment, you might find that the plants exposed constantly to a 15-watt bulb give off almost the same number of bubbles during each two-minute interval. However, the plants exposed to brighter and brighter light give off differing amounts of bubbles during each two-minute interval as the brightness is increased. The data seem to show a relationship between the brightness of light and the rate of oxygen produced. To be sure, you want to make a visual picture of the data. How can you produce this picture?

Scientists use several ways to show data, including line graphs, bar graphs, and pie charts. Line graphs plot data on an *x*-axis and a *y*-axis and show the relationship between independent and dependent variables. In this experiment, what is the independent variable? What is the dependent variable? In a line graph, the data for the independent variables are always plotted on the horizontal or *x*-axis. Data for the dependent variables are plotted on the *y*-axis. Data points show where information about the two variables meet. Bar graphs use bars instead of data points to plot the data and are a good way to show comparisons among groups when data are not dependent on one another. Pie charts or pie graphs show what percentage, out of 100 percent, a specific piece of information represents. Look at the graphs in Figure 2-5. Which one tells you what percent of Earth's water supply is found in glaciers? Which one tells you that males and females differ in the amount of calories used while engaged in the same activities. Samples of line, bar, and pie graphs are shown in Figure 2-5. For more information on graphs, turn to the **Skill Handbook.**

Figure 2-5 **Different kinds of data can be shown graphically in line graphs, bar graphs, and pie graphs.**

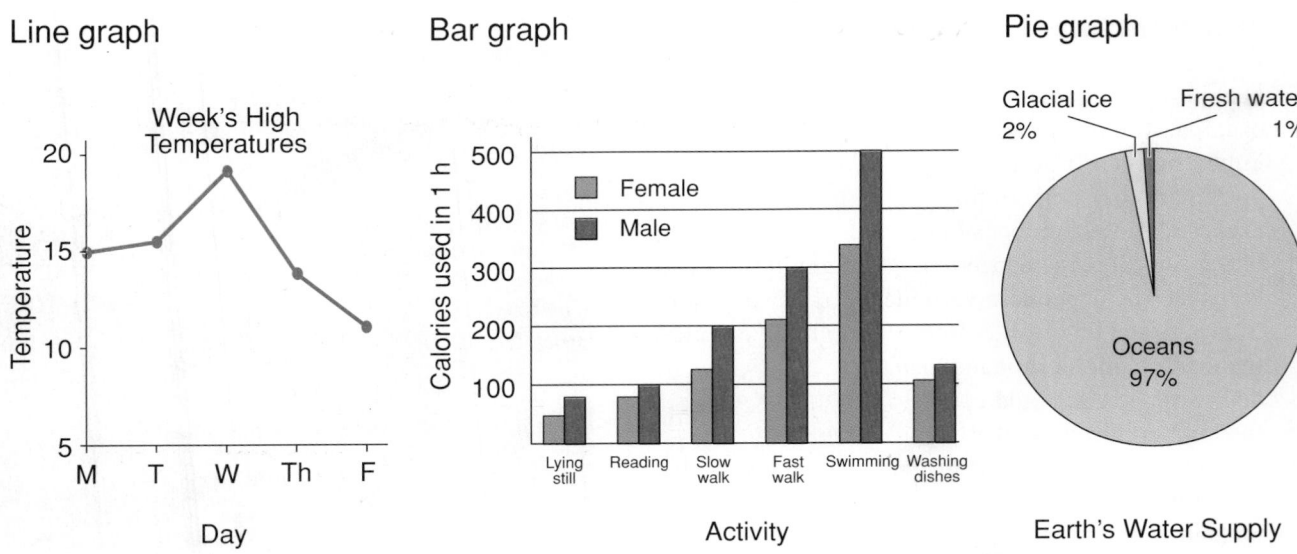

Line graph

Bar graph

Pie graph

You decide that a line graph is the best way to show your data. Your line graph shows the average total number of bubbles given off during the entire 20-minute period for each brightness setting. The data, it seems, support your explanation.

Interpreting a Graph Figure 2-6 shows a line graph representing data gathered from an experiment similar to yours. Notice what happens to the rate of oxygen production in the experimental group. As you follow the lines you should notice that the number of bubbles given off is greater at the higher brightness settings. This starts you thinking about other matters. Is there a maximum brightness of light above which no further increase in rate of photosynthesis occurs? Maybe other variables interact with light to influence rate of photosynthesis. Could temperature play a role? How about amount of carbon dioxide available to plants? These kinds of questions could lead you to search for new observations, new explanations, and new experiments to test those explanations. These are the kinds of activities that scientists do all the time. Rarely does a single experiment satisfy their curiosity.

Sometimes, even though an explanation seems logical and reasonable, an experiment yields data that show the explanation to be incorrect. In such a case, the explanation must be revised and sometimes even discarded entirely. If the explanation is changed or a new one proposed, a new experiment must, of course, be carried out to test it.

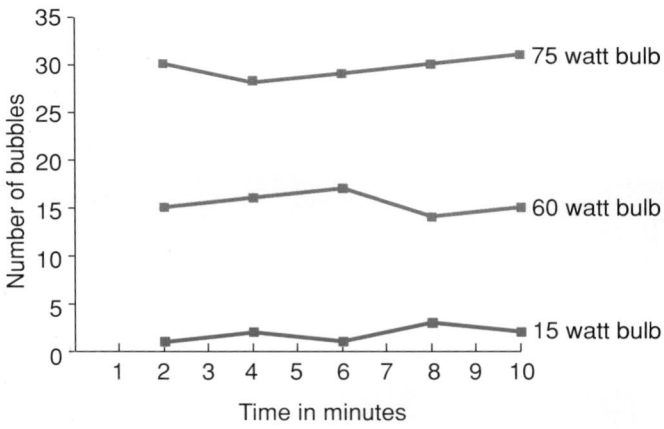

Figure 2-6 This graph shows the number of bubbles given off over a 10-minute period. Compare this graph to the graph made at the end of the Investigation. Are they similar? Can you explain why?

Thinking Lab

ANALYZE THE PROCEDURE

How can you test dog food commercial claims?

Background
You are going to test dog food commercial claims with a litter of three male and two female puppies. Five weeks after they are weaned, you start your test.

You Try It
Puppies are fed only "Puppy Chow Wow" for the next four weeks. The table shows your collected data. You conclude the following after four weeks: "Puppy Chow Wow" gives your puppies shiny coats, good health,

and good mass gain. The commercial's claims are correct. **Analyze the procedure** and describe any errors that you believe were made. How might the procedure be changed in the future if redoing this experiment?

Week	Average Mass Gain	General Appearance
1	25 g	frisky
2	38 g	healthy
3	49 g	shiny coats
4	52 g	very frisky

Research Without Experimentation

Performing an experiment to test an explanation is one type of research, but not all research lends itself to controlled experimentation. Often, scientists carry out their work by making comparisons. For example, a scientist who finds a bone of an extinct animal may make certain conclusions about that animal by comparing its bone to bones of animals that are currently alive. Many important conclusions about evolution are made on the basis of this type of research rather than controlled experiments in a laboratory. Naturalists are biologists who conduct research by making observations in nature. Naturalists often collect data "in the field"—that is, they collect organisms, observe behavior, and identify organisms as part of their research. Field work, such as collecting and identifying insects, is another type of research that does not involve experimentation. Such research, though, does involve other important elements of the process of science, including careful observation and sound reasoning.

Figure 2-7   Scientists often use collections such as this to help them identify organisms found in the field.

GLOBAL CONNECTION

The Iceman Unfreezes History

Do you think scientists and anthropologists have learned everything there is to know about the past? It's easy to feel that way because there is so much information. But every so often, something happens to remind us that the past is still a mystery.

Discovery of the Iceman   In September 1991, a hiker in the Alps along the border between Austria and Italy found the body of a man who had been frozen in a glacier for 5300 years. That discovery is changing some ideas about the Stone Age. The Iceman, as he is called, is the oldest human found with his brain and organs still inside. Scientists

think that his body was frozen by wind and cold and then enclosed in ice, which preserved him for 53 centuries. Although some of his clothing and tools were lost or damaged as he was cut from the glacier, enough remained to give anthropologists some new ideas about how humans lived in the late Stone Age.

Ideas About Stone Age Changed
Stone Age people had been thought to wear crude skin clothing, but the Iceman was dressed in leather shoes and a robe made of pieces of different animal skins sewn together in a mosaic-like pattern. The skill required to make such a robe has led anthropologists to think that the Iceman did not make the robe himself, but rather that he may have been part of a community in which people did different kinds of work.

Other things about the Iceman have also changed ideas about the Stone Age. Although no hair remained on his skull, hairs about 9 cm long found on his clothes indicate that humans were cutting their hair as early as 5300 years ago. The Iceman had tattoos on his lower spine, behind a knee, and on an ankle. Tattoos had been thought to originate about 2500 years after the Iceman lived. He also carried, strung on a leather cord, two mushrooms that are known to help fight infections. Anthropologists think they may be part of the oldest known first-aid kit.

Anthropologists are also fascinated by a fur quiver that held arrows, two of which had flint points and feathers attached by a resin-like glue. The feathers are set

Reporting on Research

Scientists publish the results of their work in scientific journals to be shared with other scientists around the world. Most of what is published does not include final answers to a question. Numerous articles, for example, have come out about various aspects of research centering on AIDS. To be acceptable and useful to other scientists, a description of an experiment and its results must be reproducible; that is, the same experiment should always yield the same results, wherever and whenever it is conducted. Scientists must be able to rely upon the work of other scientists in their field. They learn from others' work as they carry out their own research. Thus, integrity is of the utmost importance to the scientific community. Scientists must make every effort to carry out their work thoroughly and carefully and to report their results accurately and honestly.

Reporting on research often leads to a flurry of related research. An example was the discovery of penicillin. Penicillin is a chemical produced naturally by a mold. Other scientists, upon learning of the discovery of penicillin, began a search for other molds that produce similar drugs. Scientific knowledge builds upon the foundations of previous findings. An important discovery provides a basis for and stimulates further work.

---

at an angle that would help keep the arrow on a straight course. Anthropologists now suspect that Stone Age hunters knew quite a bit about the principles of flight. In addition to the quiver, which anthropologists had not known to exist this early, the Iceman carried a rucksack with a U-shaped wooden frame, not unlike a backpack, and a leather pouch that resembles the "fanny packs" worn today.

As scientists and anthropologists study the Iceman's internal organs and analyze his DNA, they hope to learn more about his diet and his immune system. They are also trying to piece together the story of how he died. The discovery of the 5300-year-old Iceman may be rewriting prehistory.

---

1. **Explore Further:** How were scientists able to determine the Iceman's age?
2. **Connection to Genetics:** Scientists think they may be able to identify the closest living descendants of the Iceman. How might this be possible?

Laws and Theories

It is not at all unusual for scientists in different labs to be working on the same basic research. It is important for you to realize they often interpret data and test explanations in different ways, and it is not unusual for them to come up with different explanations for those data. Explanations and hypotheses are not carved in stone and are always subject to further revision. It is always possible, and, in fact, often the case, that new data or novel ways of interpreting data will cause revision of an existing explanation. No explanation in science is final. Rather, you should think of science as an ongoing process in which knowledge of the world is constantly refined and improved. For these reasons, scientists shy away from using the word *proof,* which suggests something final or without other possible explanation. You may use the word *proof* in mathematics, but it really has no place in science.

Development of Theories　Many hypotheses, although subject to revision, have held up well over long periods of time. Such hypotheses, usually major ones, that have stood the test of time are known as **theories.** Major theories are ones that are the guiding principles of science. In biology, for example, Charles Darwin's explanation of how evolution occurs through natural selection is known as a theory because it guides scientists in their understanding of how change in organisms occurs. Common usage of the word *theory* suggests a guess or an uncertainty, such as, "This is only a theory, but..." The opposite meaning is true in science. A theory is as close to a complete explanation as science can offer. However, like a hypothesis, a theory is always subject to revision.

Figure 2-8　Charles Darwin, left, developed the theory of natural selection to explain how evolution occurred. Sir Isaac Newton, right, discovered the law of universal gravitation. What is the difference between a theory and a law?

Charles Darwin

Theory of Evolution

Sir Isaac Newton

Law of Gravity

Statements of Natural Events　Hypotheses and theories are explanations of observations. On the other hand, **laws** are statements about events that always occur in nature. You know, for example, that gravity will always cause an object that rolls off a window ledge to fall to the ground below. This observation and others like it about falling objects can be stated as the law of gravitation. The statement or law does not explain *why* the object falls to the ground below. It merely says that it does. Scientists use theories to explain the law.

What Science Cannot Do

A hallmark of the process of science is developing ideas that can be tested. The test results may show that the explanation is partly or totally incorrect.

Limits to Science   An explanation that is not testable poses a limit on science, at least until a means of testing becomes possible. Sometimes testing requires the development of new equipment or procedures. For example, relatively little was known about the structure and functions of cells until the electron microscope was invented and better means of chemically analyzing cell parts were discovered.

Science and Morality   Some ideas are not scientifically testable at all. The process of science cannot test and determine, for example, what is morally or socially correct or incorrect. Morality and human behavior are important issues, but decisions about those issues cannot be made by science. For example, science has produced a great understanding of how the genetic code works, and it is now possible to introduce new bits of genetic information into an organism. However, science cannot determine whether that knowledge should be used to create races of "superhumans." Similarly, science can gather information about and predict consequences of destroying a rain forest, but it cannot make the decision whether or not to do so. Science can provide information to people about the natural world. It is up to the people to decide how to make use of that information.

Figure 2-9   Observations by scientists show that clear-cutting tropical rain forests leads to soil erosion. What people choose to do with this information may be a political or moral issue, not a scientific one.

Section Review

Understanding Concepts

1. Why is the word *process* used in defining science?
2. You notice moths resting on dark tree trunks. There seem to be more dark-colored moths than light-colored ones. You also notice that birds sweep down and feed on the moths. Suggest an explanation for the observation that there are more dark-colored than light-colored moths.

3. Can you think of another equally possible explanation for the observations described in question 2?
4. *Skill Review—Designing an Experiment:* How might you design an experiment to test the hypothesis that young seedlings need light in order to develop normally? Assume you have whatever equipment you think you need. For more help, see Practicing Scientific Methods in the **Skill Handbook.**

5. *Thinking Critically:* You observe that bees seem to visit only bright-colored flowers in a garden, ignoring white ones. What might be the cause of the bees' behavior pattern?

Compare and contrast the compound light microscope, the transmission electron microscope, and the scanning electron microscope.

Measure objects using the International System of Measurement.

2.2 Tools of the Trade

Observation, as you have learned, is a key element in the methods of science. Many observations can be made directly as a scientist uses his or her senses of sight, smell, taste, touch, and hearing. Modern scientists often extend the limits of their own senses by using equipment such as telescopes to view the sky or tracking and recording instruments that detect sounds of animals. Because biology often calls for a study of extremely small objects, such as a tiny organism or certain parts of a cell, observations often require the use of microscopes.

Functions of a Microscope

One reason biologists use a microscope is to magnify a very small object that cannot be readily studied using the naked eye. Only by using a microscope, for example, can the thousands of cells that compose even a thin leaf be seen.

Resolving Power   Suppose you were examining a leaf using a microscope and you focused on an individual cell. Within that cell are many smaller parts. Many of those parts are very close to one another. Depending on the distance between the parts and the type of microscope you are using, you may or may not be able to distinguish among the parts. They might look like one object. Have you ever walked along abandoned railroad tracks? As you look off into the distance, you can distinguish the two rails of the tracks as being separate. Your eyes can distinguish them as being distinct objects. Looking as far as you can see, though, you may no longer be able to tell the two tracks apart. In the distance, they appear as one. Just as you have the ability to distinguish two objects as being separate, the microscope can also distinguish between two objects. The ability of a microscope to distinguish two objects as being separate is known as its **resolving power.**

Figure 2-10   **The compound light microscope allows light to pass through an object and through two or more lenses that enlarge the image.**

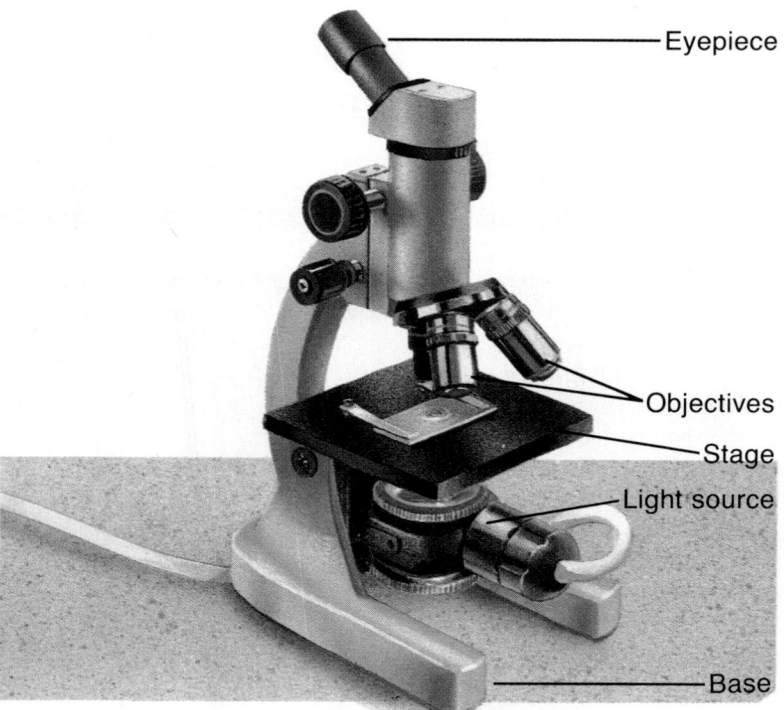

Eyepiece

Objectives

Stage

Light source

Base

The Compound Light Microscope   In your biology course, you will use a compound light microscope. Such a microscope uses glass lenses that focus light passing through an object. These microscopes can magnify up to about 2000 times and have a resolving power of about 500 nanometers (nm). A nanometer is a unit of measure that will be discussed shortly. Two objects closer together than 500 nm will be seen as one object when viewed with a light microscope.

Electron Microscopes

For greater magnification and resolving power biologists rely on electron microscopes. Such microscopes typically magnify hundreds of thousands of times and have a resolving power as great as 0.2 nm. Thus, they provide much greater detail and enable biologists to study inner and outer structures of even the smallest cells. As the name suggests, an electron microscope utilizes a beam of electrons, rather than light, to illuminate the object being viewed.

The Transmission Electron Microscope   In order to view parts of very small structures, biologists use the **transmission electron microscope (TEM).** The sample to be studied is frozen or embedded in plastic and then sliced into very thin sections. Thus, living specimens cannot be viewed. The electrons are focused by magnets instead of glass lenses. The electrons form an image that can be photographed or seen on a screen. The TEM has played an important role in the study of cell parts. It reveals details that aid in understanding how those parts function.

The Scanning Electron Microscope   Another kind of electron microscope permits a view of surfaces not possible with the TEM. This microscope, called a **scanning electron microscope (SEM),** has a lower magnification but produces a three-dimensional image. This is possible because the beam of electrons does not pass through the object. Rather, it sweeps across it and bounces off. As it does so, it causes electrons to be knocked loose from the object. These electrons produce patterns seen as an image on a television-like screen. Knowledge of the surfaces of cells, for example, is important in studying cancer. Interactions of cells with one another and with chemicals can also be studied using the SEM.

Figure 2-11   **Can you tell which photograph came from a TEM and which came from an SEM? The same organism, *Amyloodinium*, is shown as photographed by a TEM (a) and by an SEM (b). Magnifications: 3900× (a); 4200× (b)**

a
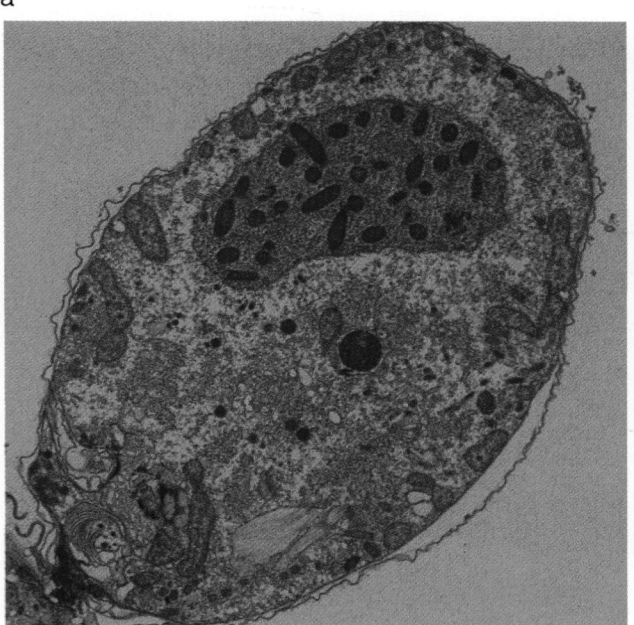

b
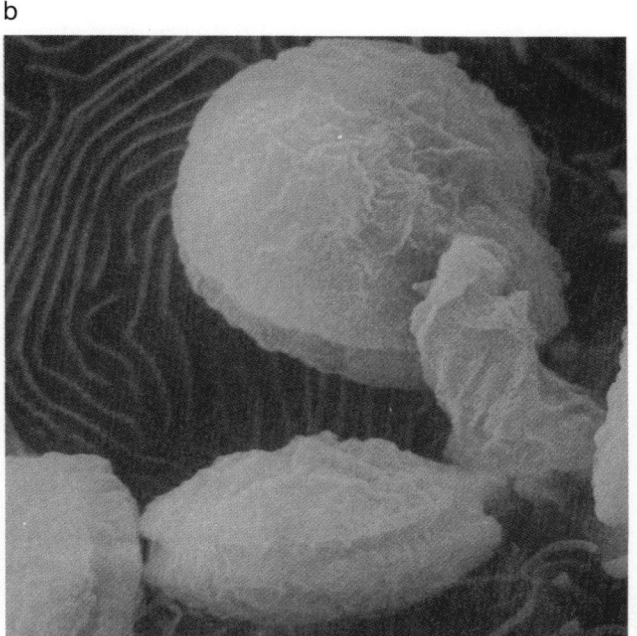

The International System of Measurement

You just learned that resolving power of microscopes is measured in terms of units called nanometers. You've probably never heard of such a unit. What does it represent and where does it come from?

Recall that scientists publish articles in journals to report on their research. Imagine the difficulty they would have interpreting the data if several different systems of measurement were used. For example, suppose a scientist in New York reported using 25 ounces of a chemical to do an experiment. A scientist in Germany would have to accurately convert the number of ounces to grams before being able to repeat the experiment. Errors could easily be made. To prevent errors and confusion, scientists worldwide use a universal language of measurements and its symbols. Those measurements and symbols are known as the **International System of Measurement,** or **SI** (from the French title, Système Internationale d'Unités). SI is a modern form of the older metric system and is now used in most countries. One major advantage of SI is that it is a decimal system. It is based on tens, multiples of tens, and fractions of tens. Because it is a decimal system, it is much easier (and more logical) to use than the English system. Although the names of the units used to measure length, mass, volume, and time differ, the decimal system applies to all of them.

Use of Prefixes

In the SI, fractions and multiples of tens are expressed by the use of prefixes. Consider, for example, measurements of length.

Measuring Length   The unit of length with which you may be familiar is the meter (m). One meter is slightly longer than a yard. In any fraction or multiple of the meter, the name of the unit—meter—remains. Consider fractions of meters. One meter is equal to ten *deci*meters, or, to put it another way, there are ten decimeters in a meter. A hundredth of a meter is a *centi*meter (cm), and a thousandth of a meter is a *milli*meter (mm). Length of cells or parts of cells are much smaller than any of these units.

Table 2.1 **Prefixes in SI**

kilo	thousand	kilometer
deka	ten	dekameter
deci	tenth	decimeter
centi	hundredth	centimeter
milli	thousandth	millimeter
micro	millionth	micrometer
nano	billionth	nanometer

Such objects are often measured in *micro*meters (μm) or *nano*meters (nm). One meter contains a million micrometers and a billion nanometers. Prefixes are also used to express multiples of meters. A *deka*meter is equal to ten meters, and a *kilo*meter equals a thousand meters. You can see these prefixes and their meanings in Table 2.1. Information about SI can be found in Appendix G.

Measuring Mass, Volume, and Time These same prefixes are used in measurements of mass, liquid volume, and time. For example, in SI, mass is measured in units called grams (g). Mass is the amount of material of which something is composed. For example, a nickel has a mass of about 5 grams. One milligram (mg) represents 0.001 grams, just as one millimeter equals 0.001 meters. The unit of measurement for liquid volume, for example the amount of cola in a clear plastic bottle, is measured in a unit called the liter (L). Many soft drinks are sold in two-liter bottles. What part of a liter is a centiliter? Time is measured in seconds. What does a nanosecond mean?

Because of the decimal basis of the metric system, the same value can be expressed in different units. One centimeter could also be written as one-hundredth of a meter (0.01 m), ten millimeters (10 mm), or even one-hundred-thousandth of a kilometer (0.000 01 km).

Minilab 2

What is the area of your classroom?

Using metersticks given to you by your teacher, measure the length and width of your classroom. Calculate the area of the classroom in square meters. Now compute the area of the classroom using a yardstick. Relate this area in square feet. Using a calculator, give the area in square centimeters and square inches.

Analysis *What is the area of the classroom in m^2? In ft^2? How do these figures compare? Is it easier to compute the area in cm^2 or in^2?*

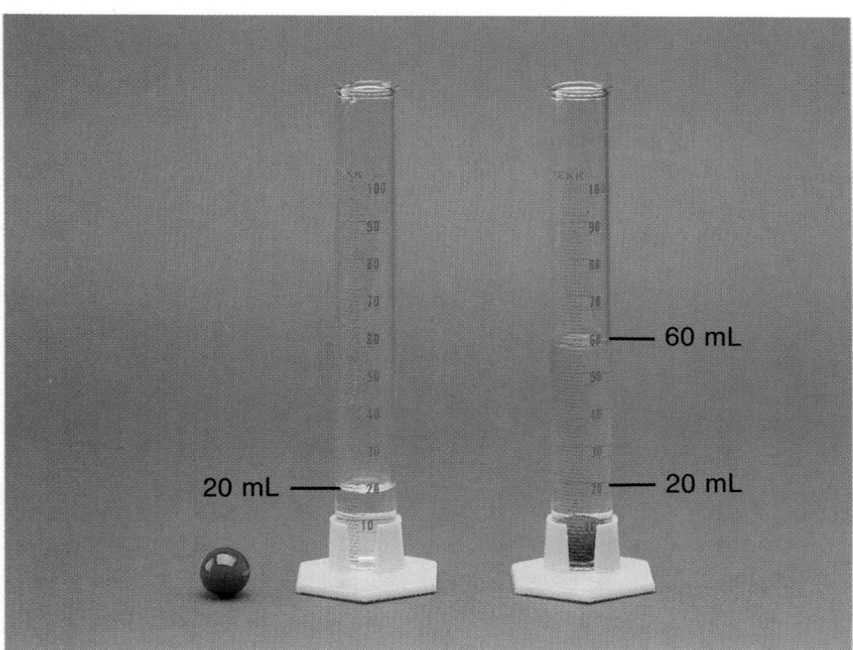

20 mL

60 mL

20 mL

Figure 2-12 **Using this photograph, figure out the volume of the object by water displacement. Use milliliters in your answer. Would it be easier, harder, or the same to figure out the volume using a cylinder marked in ounces?**

Measuring Area Areas and solid volumes can be calculated using SI measurements. To find the area of a rectangle measuring 5 cm by 3 cm, you multiply its length times its width. This rectangle has an area of 15 cm^2. To find solid volume, you multiply length times width times height. Thus, a bricklike cell measuring 100 micrometers long by 40 micrometers wide by 10 micrometers thick has a volume of 40 000 μm^3.

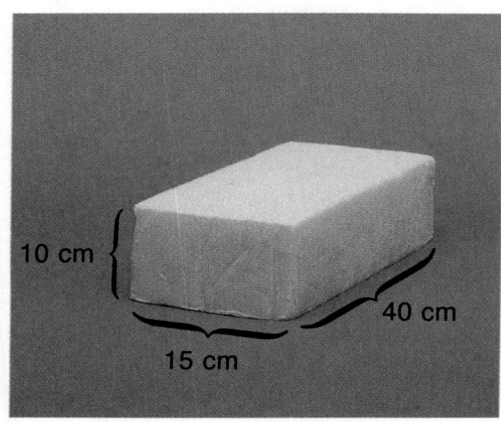

Figure 2-13 *What is the solid volume of this cheese brick in centimeters? In millimeters?*

Working with SI A little common sense will go a long way toward making you feel more confident in working with SI. Use logic to check your work. For example, suppose you were asked to convert 5.2 kilometers to meters. Do you divide or multiply? You know that a kilometer is 1000 times the length of a meter, and that a meter is only one-thousandth of a kilometer. If you are converting 5.2 kilometers to meters, it is reasonable that the answer should be a number larger than five because there are many meters in one kilometer. By simple math, you multiply 5.2 times 1000 and reach your answer: 5.2 kilometers equal 5200 meters. Now try another problem. Convert 5200 meters to millimeters. What is your answer? (Multiply 5200 by 1000 = 5 200 000 mm.) You will have opportunities to make use of and improve your skills in SI measurements as you conduct lab work throughout this course. With a little practice, you will find SI an easier system to work with than the English system.

LOOKING AHEAD

SCIENTISTS THROUGHOUT the world use the same methods, tools, and system of measurement to find answers for their questions. As you complete this course, you will use some of the same tools, follow similar methods, and use the same system of measurement to further your own understanding of biology. You will also come to the realization that biology, indeed all fields of science, cannot be studied in isolation. What is learned in one particular field of science is usually related to other areas of science as well. In the next chapter, you will study some basic principles of chemistry. Knowledge of those principles will also add to your understanding of organisms. 🙠

Section Review

Understanding Concepts

1. In a book, you see two photographs of the same kind of cell as viewed through two different microscopes. What looks like one solid mass in one photograph is seen as a cluster of dots in the other. Explain the difference.

2. As viewed with an electron microscope, a pollen grain is seen to have a bumpy surface. What kind of electron microscope is probably being used? Explain.

3. The dots seen in one of the photographs mentioned in question 1 are known to measure 2 micrometers across. Express that measurement in terms of nanometers.

4. ***Skill Review—Measuring in SI:*** Find the sum, in grams, of the following: 816 milligrams + 527 centigrams + 5.1 kilograms. For more help, see Thinking Critically in the **Skill Handbook.**

5. ***Thinking Critically:*** One mL of water has a mass of 1 g. Suppose a student has a cylinder marked off with mL lines. How can the student determine the mass of an object using such a cylinder?

BIOLOGY, TECHNOLOGY, AND SOCIETY

Physics Connection

An Unusual Idea for Testing Global Warming

A greenhouse is a glass building in which plants are grown. The glass lets in sunlight to warm the building and then keeps in the heat, so that the greenhouse stays warmer than the temperature outside. On Earth, sunlight that reaches Earth's surface is radiated back as heat. Some of this heat is absorbed by water and carbon dioxide and trapped in the upper air, where it keeps our atmosphere warm.

A Rise in Temperature Scientists think that more carbon dioxide is being released into the air than can be stored in the oceans and used by plants. As excess carbon dioxide in the air increases, more heat is trapped inside Earth's atmosphere. Some scientists speculate that the result may be a slow rise in the temperature of Earth's surface.

The idea that the temperatures of Earth's air and water are rising is a hypothesis, an idea that some scientists think to be true. In order to test their hypothesis, they have to conduct experiments. Direct measurement of surface warming of Earth is difficult, however, and it could take centuries to detect a valid trend. Recently, an international team of scientists, led by Walter Munk of the United States and Andrew Forbes of Australia, has proposed a way to measure global warming by sending sound waves through the ocean. Their unusual plan is called the Heard Island Experiment.

The Heard Island Experiment

The speed at which sound travels through water is determined mostly by the temperature of the water; the warmer the water, the faster sound travels. By transmitting sound from one point to another, you can get a measure of the average temperature of the water along the path of the sound. Global ocean warming of only 0.005 degrees Celsius a year would reduce by several tenths of a second the time it takes a sound wave to travel. Munk and Forbes think that monitoring water temperatures of ten paths for about ten years will reveal whether there is a warming trend in the oceans.

The first phase of their experiment took place in January 1991. Munk and Forbes lowered three loudspeakers into the ocean around Heard Island, 1200 kilometers north of the coast of Antarctica. They sent a series of sound transmissions to 19 receiving stations in the Atlantic, Pacific, and Indian Oceans. The purpose of the January 1991 test was to determine whether the signal could be heard by the receivers. Once they determine the exact signal needed in order to make sensitive temperature measurements, the scientists hope to begin a long-term experiment.

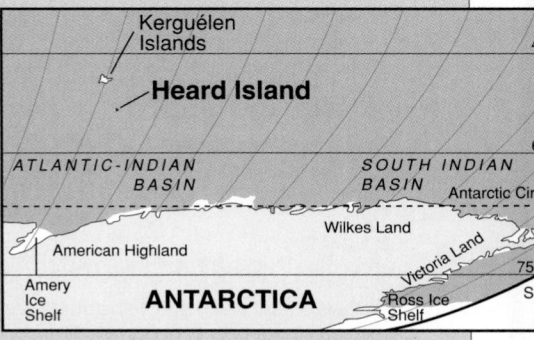

Verifying Global Warming Not all scientists think the experiment will demonstrate global warming. They say that it takes ten years for ocean water to warm to the temperature of the air. Global warming may be verified first by less costly experiments, such as measuring air temperature. They also question the accuracy of the test.

1. **Explore Further:** A buildup of carbon dioxide (CO_2) and other gases in the upper atmosphere may lead to global warming. What are the causes of increased CO_2?
2. **Writing About Biology:** Why do scientists worry about global warming?

Summary

Science is an ongoing process that produces a body of knowledge about the natural world. It is a unique field of study because of the methods by which that knowledge is gained.

The methods used by scientists vary, depending upon the questions they are attempting to answer. However, scientific research centers on careful and thorough observations, reasoning and insight, explanations of observations, and, usually, a means of testing those explanations. There is no specific order in which these activities are carried out. Rather, scientists most often go back and forth among the activities, refining their thinking and conclusions as they do so.

Where possible, scientists test their explanations, or hypotheses, by means of controlled experiments. Such experiments make use of a control group and an experimental group and should have only one variable factor. This experimental design reduces doubt about cause and effect.

If the results of an experiment do not confirm the hypothesis, a new explanation can be devised and tested. Most often, the results of one experiment lead a scientist to other, related experiments.

Results of scientists' research are published in journals where they can help other scientists and spur further work. The integrity of scientists is, therefore, of the utmost importance.

Microscopes are important to the work of many biologists, extending the limits of the observations they may make. Microscopes vary in terms of both magnification and resolving power. Use of SI to express numerical data provides a universal language of numbers and measurements understandable by scientists around the world.

Language of Biology

Write a sentence that shows your understanding of each of the following terms.

control group	law
controlled experiment	resolving power
data	scanning electron
dependent variable	microscope (SEM)
experiment	science
experimental group	technology
hypothesis	theory
independent variable	transmission electron
International System	microscope (TEM)
of Measurement (SI)	

Understanding Concepts

1. One of the things you will learn this year is how blood circulates through your body. Is such information "science"? Explain.

2. Some early "scientists" concluded that eels arise from mud at the bottom of the ocean. They came to that conclusion because, like the ocean bottom, eels are slimy. That conclusion was accepted as a true explanation of the production of eels. What methods of science were missing in reaching that conclusion? What knowledge do you possess today that would lead you to reject such a conclusion?

3. Alexander Fleming's discovery of penicillin came from an unexpected observation. He noted that bacteria growing in lab glassware died when that glassware was contaminated by a mold called *Penicillium*. He reasoned that a chemical produced by the mold spread out and destroyed the bacteria. Fleming then reasoned that the chemical alone, without the mold being present, should also kill the bacteria. He assumed he could obtain the chemical by growing the mold in a liquid medium. What formal hypothesis might Fleming have stated?

4. What prediction did Fleming offer based on his hypothesis?

5. Based on the information given in question 3 above, how would you test Fleming's hypothesis? What would your control be?

6. Suppose Fleming had explained his observations by stating that the mold sent out evil spirits that caused the death of bacteria. Why would such an explanation be useless?

7. Fleming did not set out to find substances that kill bacteria, but his experiments showed that such substances exist. How do his findings show the relationship between science and technology?

8. Was Fleming's explanation a law? Explain.

9. Suppose a scientist hurried through his research and was not careful in reporting and analyzing his data and conclusions in a paper. Why would that be a problem?

10. Why can't scientists use a TEM to study the details of how cells reproduce?

Applying Concepts

11. A student wishes to test the hypothesis that plants grow fuller if a certain mineral is added to their soil. The student uses two groups of plants, cacti and geraniums. The cacti, since they are desert plants, are given little water and are grown in a hot environment. The geraniums are given more water and are grown in a cooler environment. The mineral is added only to the soil in which the geraniums are growing. Is the student conducting this experiment properly? Explain.

12. How would you conduct the experiment in question 11 above?

13. You look through a light microscope and observe some microorganisms. The diameter of your field of vision is 1500 µm and you estimate that 50 of the organisms could fit across that diameter. How big is a single organism?

14. Express your answer to question 13 in terms of nanometers.

15. A student records the following data for the height, in centimeters, of certain kinds of plants growing in a field: 42.6, 39.4, 37.9, 44.0, 35.6, 41.7, and 40.1. What is the average height of these plants expressed in meters?

16. Convert the following: (a) 818 grams to kilograms; (b) 6.26 liters to centiliters; (c) 121 640 millimeters to meters; (d) 316 seconds to milliseconds.

17. **Global Connection:** Why do anthropologists want to know what kind of community the Iceman lived in? How do his robe, quiver, and tattoos give cultural anthropologists ideas about the social development of his community?

18. **Physics Connection:** Why do the researchers in the Heard Island Experiment need to record data on water temperature for 10 years?

Lab Interpretation

19. **Investigation:** Suppose your experiment with the rate of photosynthesis shows no difference in number of bubbles when plants are exposed to different amounts of light. Would you throw your data out and try the experiment again, or would you revise your hypothesis and test it through experimentation? Explain your answer.

20. **Thinking Lab:** Explain why it might have been better to use either male or female puppies in your study of puppy food.

21. **Minilab 1:** Suppose you were given an animal such as a turtle to observe as the "object." What observations could you make about the turtle, besides describing its outside appearance, if you didn't want to dissect it?

Connecting Ideas

22. Suppose an experiment was designed that interfered with the ability of a certain part of an organism to carry out cellular respiration. Results of the experiment show that the part can no longer carry out its normal function. Why?

23. How might you carry out an experiment to test the idea that the presence of large numbers of zebra mussels might lead to a decrease in the number of other consumers in the kinds of communities in which zebra mussels live?

Energy and the Cell

Fireflies and hatchet fish produce light by bioluminescence. In fireflies, light helps in mate recognition. Hatchet fish, which live deep in the ocean, produce light in organs under their eyes. This light may help attract prey or aid vision.

Neon lights contain gases under low pressure inside glass tubes. Each gas gives off light of a certain color when electricity is passed through the tube. For example, neon glows orange-red, and mercury vapor glows bright blue.

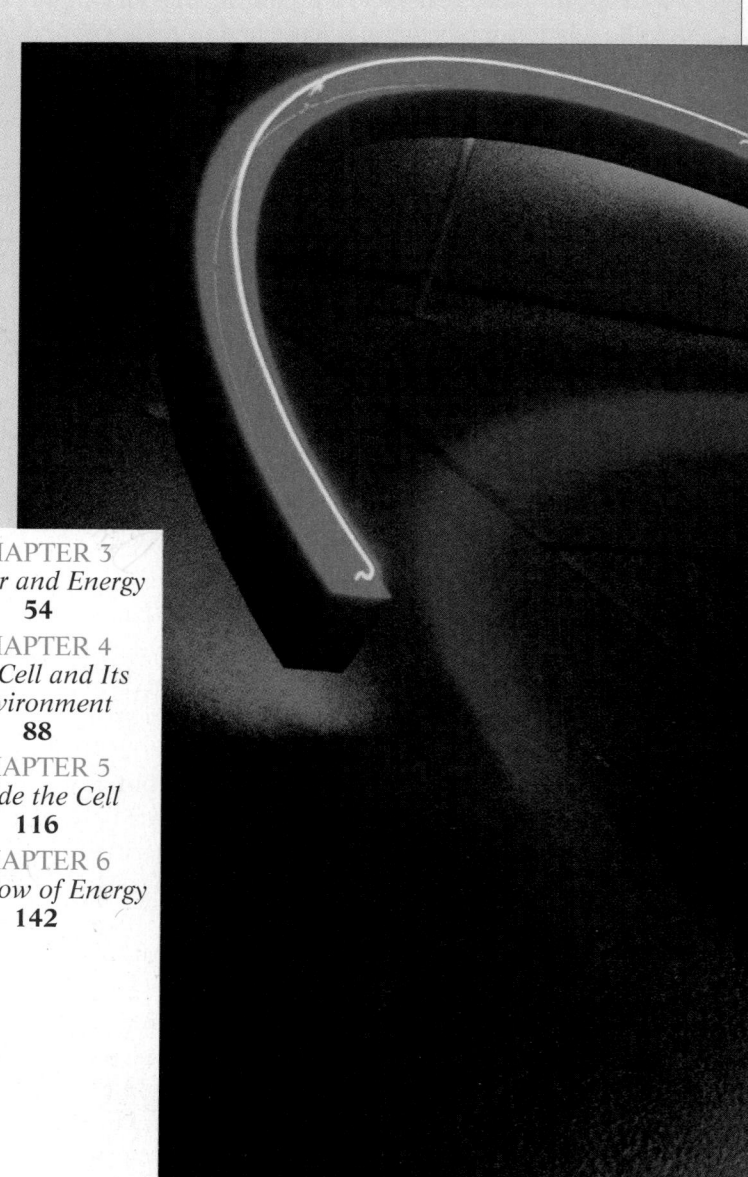

You see them everywhere, those signs on storefronts that glow in the dark. Red, yellow, green, or blue lights like these are called neon lights, but, really, they may contain any one of several types of gases including neon, argon, mercury vapor, or helium. These gases are trapped inside a transparent tube of glass under low pressure. When an electric current is passed through the tube, the gas inside is energized and begins to glow. Neon lights thus work by converting electrical energy to light energy.

Many other organisms use light to send messages. Have you ever stayed up late on a summer night to catch fireflies? By alternating flashes of light, fireflies communicate to other members of their species.

The emission of light by an organism is called bioluminescence. Bioluminescence results from a chemical reaction in which chemical energy is converted directly to light energy. How energy is used in cells, and how it is converted from one form to another, is the focus of this unit.

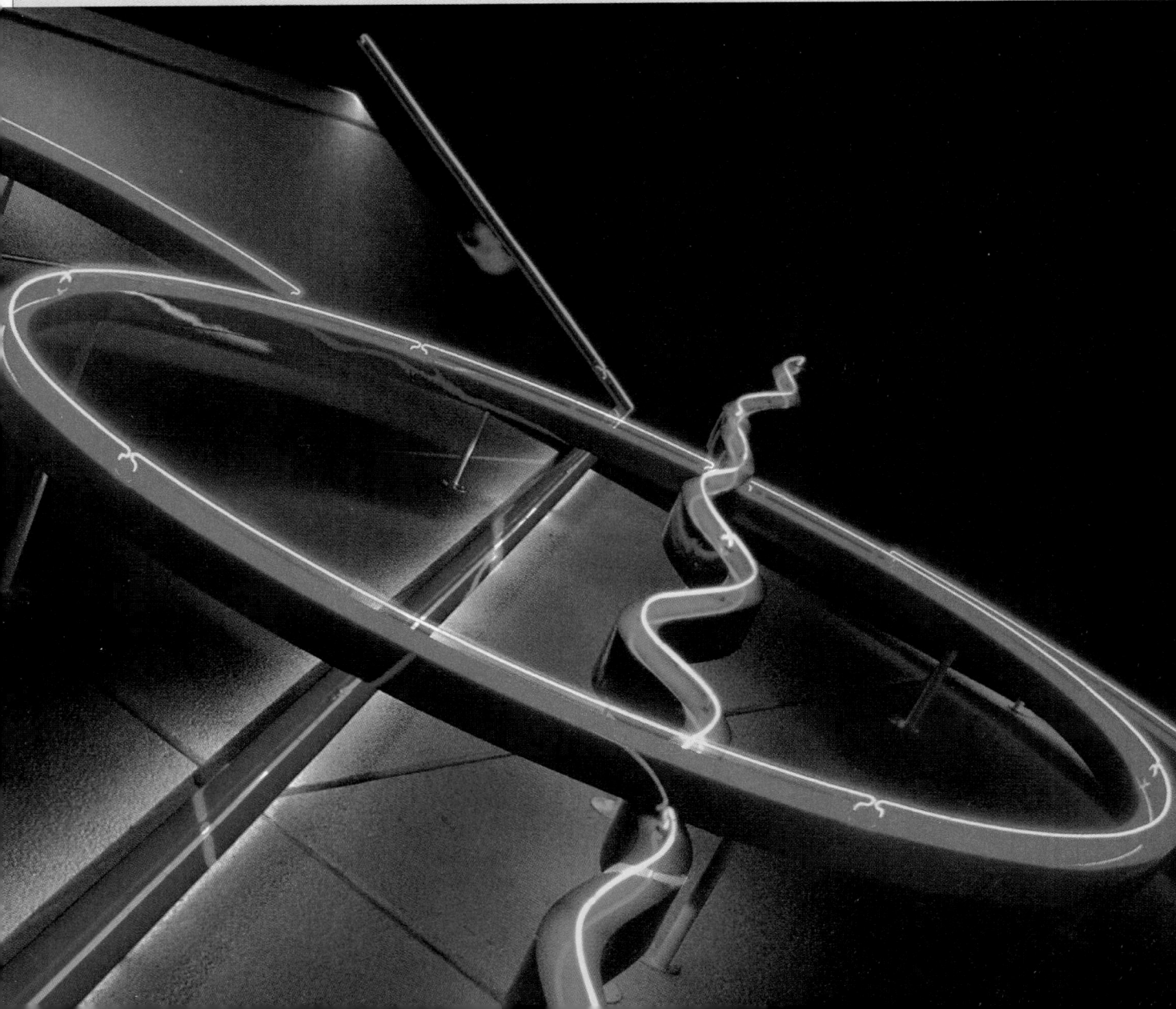

MATTER AND ENERGY

HUMMINGBIRDS—the animal world's equivalent to today's modern helicopter. If you've ever been lucky enough to spot one of these tiny creatures, you may have marvelled at the way it can dart backward and forward and then hover over a flower. Appearing more like insects than birds, hummingbirds have the most rapid wing beats of all birds—up to 100 wing beats per second! Like the huge motors in modern helicopters, hummingbirds accomplish their unique method of flight by expending large amounts of energy. In fact, hummingbirds have the highest energy output for their size of all warm-blooded animals.

You know that nonliving things such as helicopters and cars need energy to move and function. In the same way, living things need energy to carry out life processes. You need energy to think, digest food, grow, and move in gym class, just as hummingbirds need energy for their brand of aerial acrobatics. For a long time, scientists have known that the ability to carry out life processes distinguishes living things from nonliving things. However, all life processes involve changes in matter and energy. Therefore, scientists know that in order to understand life processes, they must first understand the properties of matter and how energy interacts with matter when changes take place. In this chapter, you'll explore the structure, properties, and changes of matter, including the matter involved in life processes.

Imagine being as active as a hummingbird. How much food do you suppose you'd have to eat if you used energy at the same rate that a hummingbird does?

Chapter Preview

3.1 Matter and Its Combinations

You live in a world of both living and nonliving things. In Chapter 1, you learned about the important features of organisms, and from your everyday experience you know the difference between what is alive and what is not. Although very different, both living and nonliving things are made of matter, and all matter is composed of the same building blocks—atoms.

Atoms

Helicopters and hummingbirds, stereos and spiders, textbooks and trees—these things have very different visible characteristics. All of them, though, are composed of matter. Matter is anything that has mass and occupies space. All matter, in turn, is composed of small particles called atoms. Atoms are basic building blocks that you cannot see. They are so small that even the most powerful microscopes cannot photograph their details, as Figure 3-1 shows. Most atoms measure between 0.1 and 0.5 nm in diameter. The unit nm (nanometer) is one billionth of a meter.

Atomic Structure   To understand why atoms behave as they do, let's look first at their structure. Though tiny, atoms are composed of still smaller particles, the major ones being protons, neutrons, and electrons. Protons and neutrons are packed tightly in the nucleus of an atom. Protons are positively charged and are represented by the symbol p^+. Neutrons have no charge and are represented by the symbol n^0. Moving rapidly outside the nucleus are the electrons, which are represented by the symbol e^-. Electrons are negatively charged. In an uncombined atom, the number of protons equals the number of electrons. Therefore, the atom as a whole is neutral because it has no electric charge.

Figure 3-1   **The left photo shows atoms of carbon and oxygen as seen by a scanning tunneling microscope. (Magnification: 14 000 000×) As you see, scientists are now able to arrange atoms into shapes. The right diagram is a model of an atom. Notice the small nucleus made up of protons and neutrons and the much larger cloud of electrons in three energy levels. The darkest regions in the cloud represent places where electrons are most likely to be found.**

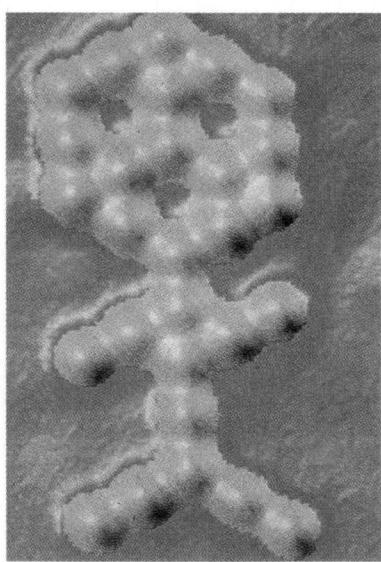

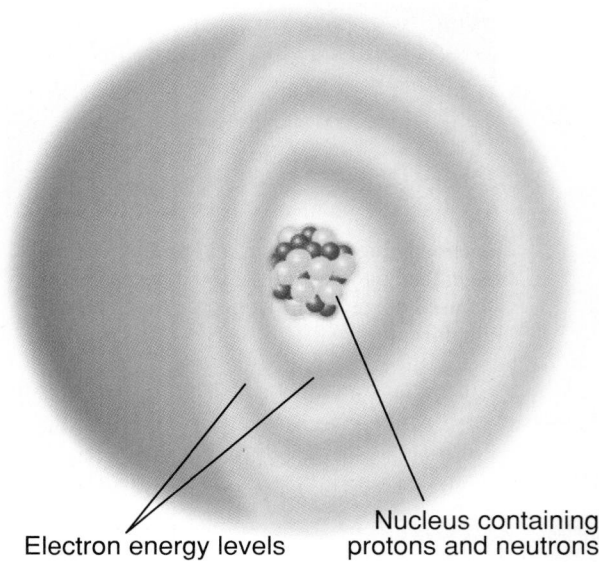

Electron energy levels

Nucleus containing protons and neutrons

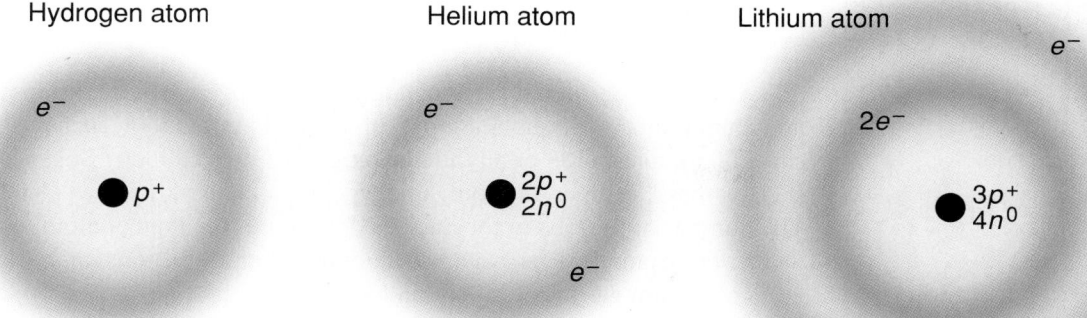

Hydrogen atom
e^-
p^+

Helium atom
e^-
$2p^+$
$2n^0$
e^-

Lithium atom
e^-
$2e^-$
$3p^+$
$4n^0$

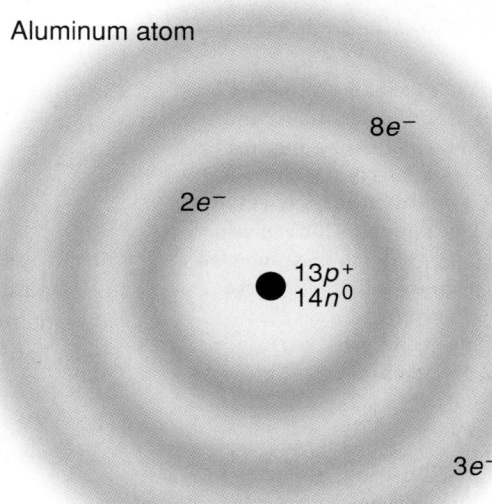

Aluminum atom
$8e^-$
$2e^-$
$13p^+$
$14n^0$
$3e^-$

A scientist's idea or picture of how something works is called a model. Models help to picture atomic structure, as seen in Figure 3-1. As you can see from this diagram, the nucleus is at the center and the electrons are moving in a region around it. Electrons do not travel in fixed paths. Rather, they move anywhere within a probable region around the nucleus. The region or space that electrons occupy while moving around the nucleus is known as the **electron cloud.**

To give you an idea of what typical atoms are like, let's look at models of some simple atoms. Hydrogen atoms are the most abundant atoms in the universe and are also the simplest because they consist of only one proton and one electron. Helium is the gas used to fill balloons that float in air. A helium atom contains two protons and two neutrons in its nucleus and two electrons in the electron cloud. Lithium is a soft, silvery metal. A typical lithium atom has three protons and four neutrons in its nucleus and three electrons in the electron cloud.

Electron Energy Levels   If you saw models of many atoms, you would note that there are several different regions in which electrons travel about the nucleus. These regions are called **energy levels**. As electrons increase in energy, they occupy energy levels farther from the nucleus. Each energy level is a region of space in which it is highly probable that an electron of a certain energy will be located. The first energy level can hold a maximum of only two electrons. Look at Figure 3-2. Note that in a helium atom, the first energy level has two electrons and, therefore, is full.

If an atom contains more than two electrons, those additional electrons must occupy higher energy levels farther from the nucleus. Look at the model of a lithium atom in Figure 3-2. Two electrons fill the first energy level, but the third is in the second energy level, a region farther away from the nucleus. The second energy level can hold eight electrons. Higher levels can hold even greater numbers of electrons. The third level holds a maximum of 18 electrons. Thus, an atom of aluminum, which has 13 electrons, has two in the first energy level, eight in the second, and the remaining three in the third level.

Figure 3-2   **Shown here are models of hydrogen, helium, lithium, and aluminum atoms. Notice that the first energy level can hold only two electrons. It is full in a helium atom. The third electron in a lithium atom must occupy the second energy level. In the aluminum atom, you can see that the second energy level is full with eight electrons. The three remaining electrons must occupy a third level. Note that in every atom the number of protons ($p+$) is the same as the number of electrons ($e-$).**

Figure 3-3 **Some radioactive isotopes are useful in diagnosing and treating illnesses because they collect in certain locations in the body. This image of a human head and neck was produced by detecting the radiation from isotopes that had been injected into the person's body.**

Figure 3-4 **A hydrogen atom is not stable by itself because its energy level has only one electron. Two electrons make the atom stable. As a result, two hydrogen atoms share electrons and bond with each other, forming a molecule written as H₂. Then the atoms are stable because, by sharing, each has two electrons—at least part of the time.**

Elements

So far, you've considered separate atoms of hydrogen, helium, lithium, and aluminum. What about a group of all one kind of atom, such as a sheet of aluminum foil or all the helium gas inside a balloon? Substances such as these, composed of only one type of atom, are known as **elements.** The number of protons in the nucleus determines the type of atom (or element). For example, every atom of aluminum has 13 protons, and every atom of helium gas inside a balloon has two protons.

Although the number of protons in a given element is always constant, atoms of the same element can differ in the number of neutrons. Such atoms are called isotopes of that element. For example, the most abundant isotope of hydrogen contains no neutrons, but other hydrogen isotopes containing one and two neutrons also exist in nature. Some isotopes are unstable and can break down, giving off radiation in the process. These radioactive isotopes are important in biological research because they can be substituted for non-radioactive atoms in molecules. However, the radiation can be dangerous to organisms.

For example, suppose a scientist wants to find out what happens to a certain weed killer when it is sprayed on plants. A sample of the weed killer can be made using atoms of the radioactive isotope, carbon-14, which has eight neutrons in the nucleus. The more abundant isotope of carbon, carbon-12, has only six neutrons in the nucleus. The weed killer is now said to be "tagged" or "labeled" because it can be detected by its radioactivity. Now, the weed killer can be traced as it enters the plant. The scientist can find out whether the chemical stays in the plant, enters the soil, travels to other plants, or is broken down into other chemicals.

Molecules

You are no doubt familiar with many other elements—gold, lead, oxygen, and copper, to name a few. In all, there are more than 100 elements. But you know from your own experience that there are many more than 100 different substances in the world. These other substances are formed when atoms combine with one another.

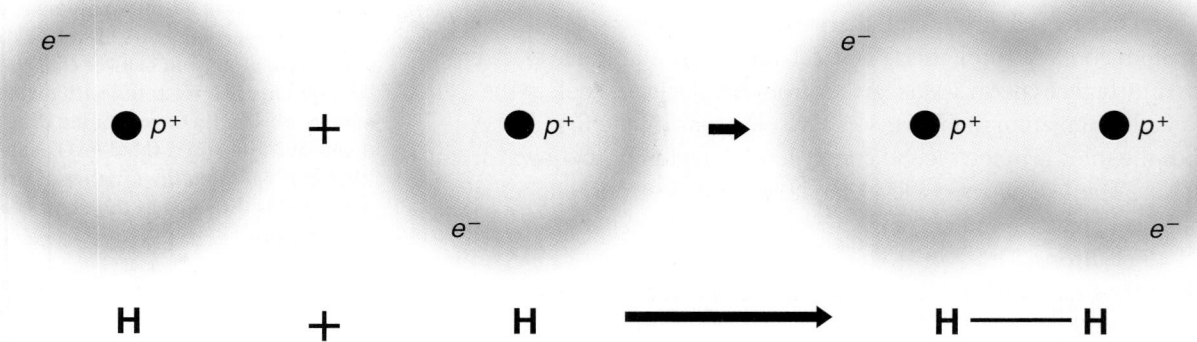

Why Atoms Form Bonds   When atoms combine, their outer energy levels become stable. They are then more resistant to change. Except for atoms of hydrogen and helium, atoms have electrons in two or more energy levels. An outer energy level becomes stable when it contains a total of eight electrons. Nitrogen, for example, has five electrons in its outer energy level. When a nitrogen atom combines with other atoms, its outer energy level must acquire three more electrons to become stable. An atom such as hydrogen, however, with only one energy level, becomes stable when that level holds its maximum of two electrons. Thus, hydrogen must acquire only one electron to become stable.

One way in which atoms combine is by sharing electrons, as in Figure 3-4. Hydrogen atoms do this when they combine with each other. The electrons travel in the space between and around the two nuclei.

Combined atoms are held together by forces called chemical bonds. When atoms combine by sharing electrons, a **covalent bond** results. A covalent bond can be represented by a line drawn between the atoms, as in Figure 3-4. A combination of two or more atoms joined by a covalent bond is called a **molecule.**

Compounds

In nature, hydrogen gas exists in the form of hydrogen molecules, composed of two covalently bonded hydrogen atoms. Oxygen gas also exists as molecules, each consisting of two atoms of oxygen covalently bonded. Figure 3-5 shows molecules of oxygen and hydrogen in motion. If these molecules collide at normal room temperature, nothing will change. However, if heat is added, the molecules will react to form molecules of water. The additional energy—heat—caused the molecules to speed up and collide with more force. As a result, the covalent bonds in the hydrogen and oxygen molecules were broken and the atoms of hydrogen and oxygen recombined in a new way to form molecules of water.

Figure 3-5   In the reaction shown here, nothing happens until enough heat is added to make a few hydrogen and oxygen molecules collide with sufficient force to break bonds. Water is the product of the reaction. After the reaction starts, the energy released causes the remaining molecules to react. The equation for this reaction is written

$$2H_2 + O_2 \rightarrow 2H_2O$$

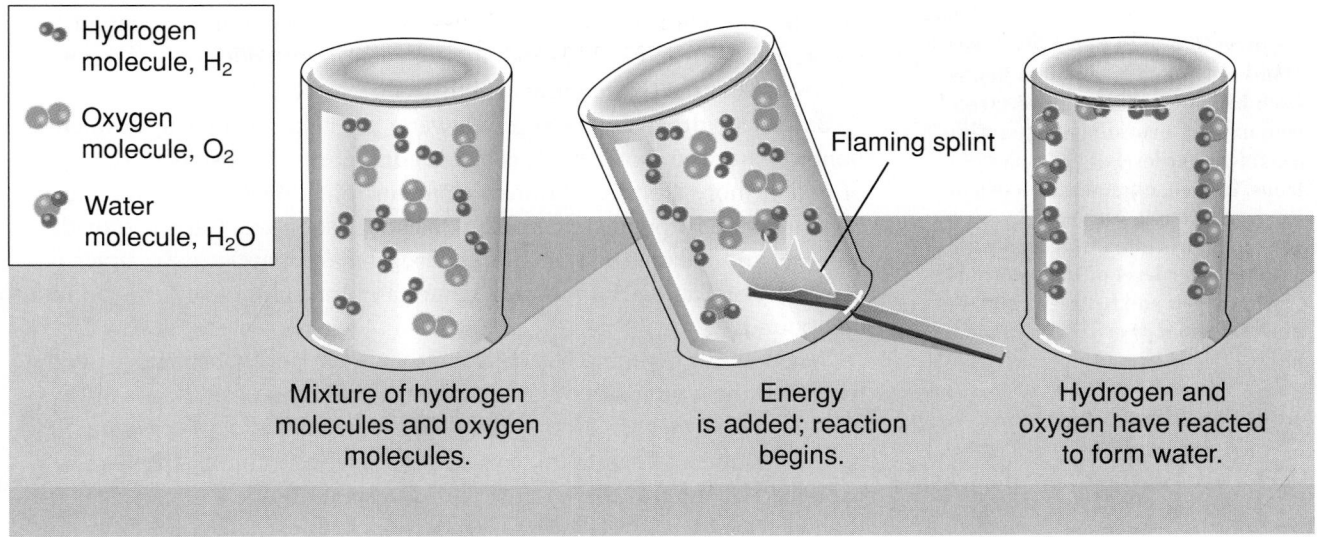

Hydrogen molecule, H_2

Oxygen molecule, O_2

Water molecule, H_2O

Flaming splint

Mixture of hydrogen molecules and oxygen molecules.

Energy is added; reaction begins.

Hydrogen and oxygen have reacted to form water.

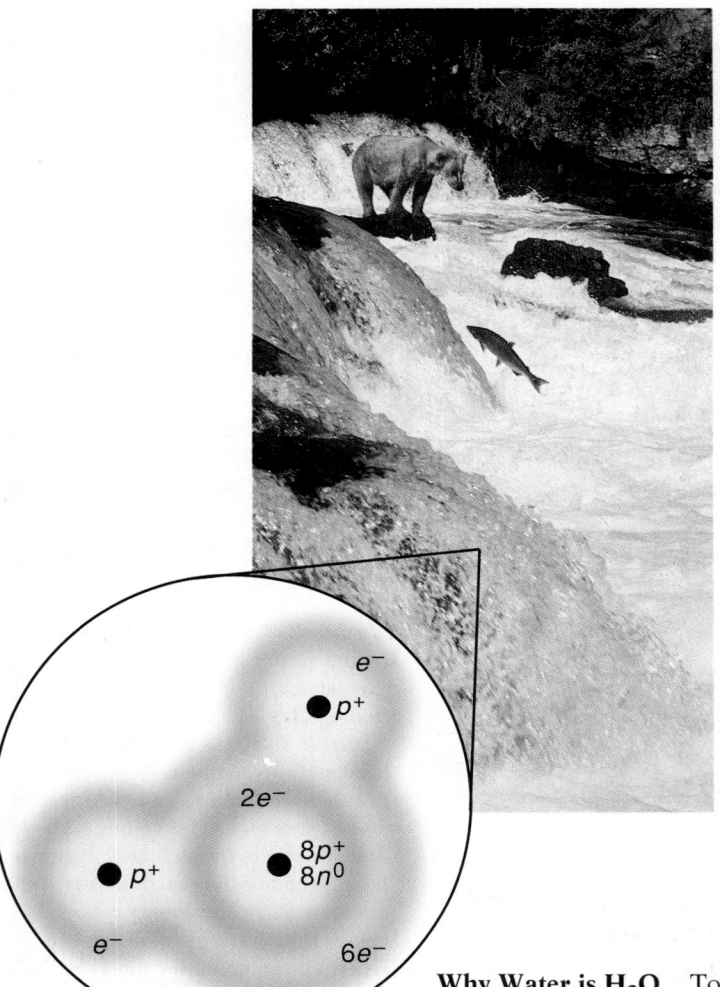

Table 3.1 **Examples of Compounds**

Name	Formula
acetic acid	CH_3COOH
ammonia	NH_3
calcium carbonate	$CaCO_3$
carbon dioxide	CO_2
chlorophyll-a	$C_{55}H_{72}O_5N_4Mg$
ethyl alcohol	C_2H_5OH
glucose	$C_6H_{12}O_6$
glycerol	$C_3H_5(OH)_3$
lactic acid	$CH_3CHOHCOOH$
octane	C_8H_{18}
sodium hydroxide	$NaOH$
sucrose	$C_{12}H_{22}O_{11}$
table salt	$NaCl$
water	H_2O

Figure 3-6 In water (H_2O), the atoms are linked by covalent bonds. Each hydrogen shares one electron with oxygen, providing oxygen with a stable outer level of eight electrons. In turn, oxygen shares two of its electrons, one with each hydrogen atom, providing hydrogen with a stable outer level of two electrons. Could oxygen and hydrogen combine to form H_3O or H_4O? Explain.

Why Water is H_2O To become stable, a hydrogen atom needs to gain one electron. An oxygen atom has six electrons in its outer energy level, and so it must obtain two more electrons. In order for hydrogen and oxygen atoms to combine so that the atoms become stable, they must combine in a ratio of two hydrogen atoms to one oxygen atom, as seen in Figure 3-6. As you can see, each hydrogen atom provides one of the two electrons that the oxygen atom needs to become stable.

Because hydrogen and oxygen in water are held together by covalent bonds, water is a molecule. It is also a **compound,** a substance composed of two or more atoms of different elements joined by a chemical bond. Not all chemical bonds are covalent, as you'll soon see. Carbon dioxide, glucose, and baking soda (sodium hydrogen carbonate) are examples of compounds. Table 3.1 shows the names and formulas of several common compounds.

When atoms combine with one another, each new substance formed has a definite composition. For example, a molecule of hydrogen always contains two hydrogen atoms, and water molecules always contain two hydrogen atoms joined to an oxygen atom. Definite composition is explained by the fact that atoms require certain numbers of electrons in their outer energy levels in order to become stable.

Ions

Not all chemical compounds result from the sharing of electrons. Another kind of bond is possible. Some atoms attract electrons much more strongly than other atoms. When these two types of atoms combine to form compounds, they do not share electrons. Instead, electrons are transferred from one atom to another as the atoms acquire stable outer energy levels. Figure 3-8 shows how sodium and chlorine atoms combine to form the compound sodium chloride, better known as table salt. The sodium atom, by losing one electron, attains a stable outer level of eight electrons. The chlorine atom obtains the electron lost from sodium, so it, too, has a stable outer level of eight electrons.

Charges on Ions   After an atom loses or gains electrons, it has an unbalanced electric charge. After losing an electron, a sodium atom has 11 positive charges (protons) but only ten negative charges (electrons). It is positive by one charge, 1+. Similarly, the chlorine atom, after gaining one electron, is negative by one charge, 1−. Such charged atoms (or sometimes groups of atoms) are called **ions.** In this case, a sodium atom has become a sodium ion, and a chlorine atom has become a chloride ion. Note that negatively charged ions formed from single elements have names ending in *-ide.* The names are slightly different from the names of the elements.

Because sodium and chloride ions are oppositely charged, they attract each other. This force of attraction between ions is called an **ionic bond.** The number and kind of charge on each ion determine the composition of a compound. In an ionic compound, the positive and negative charges must balance each other so the net charge is zero. In sodium chloride, only one of each kind of ion is needed, because the 1+ charge of a sodium ion balances the 1− charge of a chloride ion.

Comparing Covalent and Ionic Bonding   Let's compare ionic and covalent substances. You have learned that covalent substances consist of

Figure 3-7   This photo shows sodium reacting with chlorine, forming common salt. The violent reaction produces both light and heat.

Figure 3-8   When sodium and chlorine react, they form sodium chloride, an ionic compound. After the reaction, each sodium ion has a positive charge, and each chloride ion has a negative charge. As a result, there is a force of attraction between the two ions of opposite charge. This force is an ionic bond.

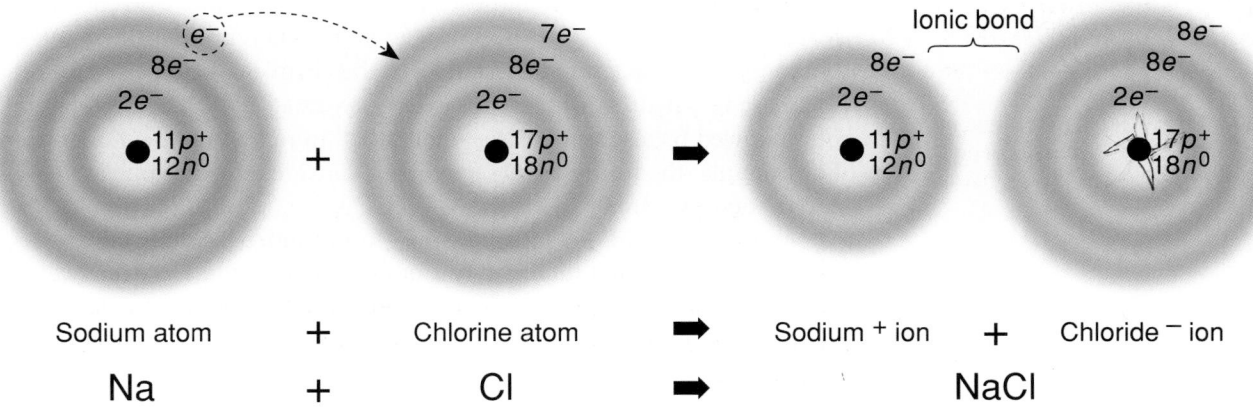

Sodium atom	+	Chlorine atom	→	Sodium + ion	+	Chloride − ion
Na	+	Cl	→	NaCl		

Figure 3-9 Crystals of common salt are cube-shaped. If you could magnify one corner of a salt crystal several million times, you would see the cubic arrangement of positive sodium ions and negative chloride ions shown below. Notice that there are no molecules of sodium chloride, only separate sodium and chloride ions.

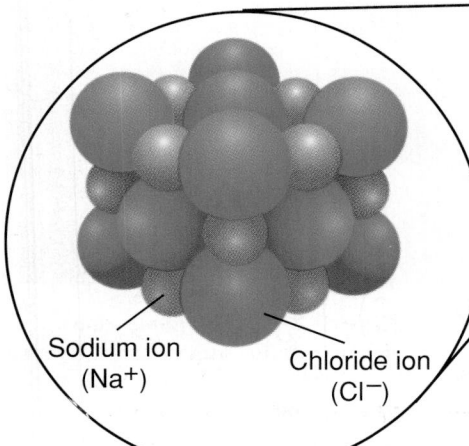

Sodium ion
(Na⁺)   Chloride ion
             (Cl⁻)

individual particles called molecules. For example, a glass of water consists of more than one septillion individual water molecules, each distinct from the others. On the other hand, ionic compounds do not consist of separate molecules. A crystal of table salt, as seen in Figure 3-9, is made up of a continuous network of alternating sodium and chloride ions held tightly by their electric attraction for one another. There are no separate and distinct units of sodium chloride, just sodium ions and chloride ions in a one-to-one ratio.

Symbols and Formulas

Though scientists around the world use different languages, they have a standard way of referring to the elements. Each element is represented by a symbol. Some symbols consist of one letter. For example, symbols of the four elements most abundant in living systems are C (carbon), H (hydrogen), O (oxygen), and N (nitrogen). Other symbols contain two letters—Ca for calcium, Zn for zinc, and Mg for magnesium. Notice that when a symbol consists of two letters, the second letter is not capitalized.

Chemical Formulas A group of symbols called a **chemical formula** shows the number and kind of each atom in a compound. The chemical formula for water is H_2O. Subscripts in a formula represent the number of atoms of an element. For example, in the chemical formula for water, there are two hydrogen atoms for every oxygen atom. When a symbol is not followed by a subscript, it is understood to represent one atom or ion. For example, the formula CO_2 for carbon dioxide indicates one atom of carbon and two atoms of oxygen.

Numbers in front of a formula apply to the entire formula. For example, $5H_2O$ means that there are five molecules of water. How many atoms are there in $5H_2O$? There are three atoms in each water molecule—two hydrogen and one oxygen. Therefore, in five molecules, there would be five times three atoms—15 atoms altogether.

Properties of Matter

Properties are the characteristics that scientists use to describe matter. For example, a piece of iron might have a certain color, hardness, and texture. It might have a mass of 1.5 kg, a volume of 190 cm³, and be molded in the shape of a gear. Such features that describe a piece of matter are called physical properties. Another characteristic of iron is that it will combine chemically with oxygen in the air to form rust. Iron will also be eaten away chemically by acids spilled on it. Characteristics such as these describe how iron behaves in contact with other elements or compounds. These characteristics are called chemical properties.

Chemical Change

Think back to the formation of water from atoms of hydrogen and oxygen. Isn't water very different from the hydrogen and oxygen gases from which it was formed? Think, too, about wood burning in a fireplace. Burning occurs as molecules in the wood combine with oxygen molecules. Invisible gases, including carbon dioxide and water vapor, are formed. Aren't these substances quite different from wood and oxygen? What has happened? In both examples, a chemical change has occurred. A chemical change involves the breaking of bonds and the formation of new ones. Chemical changes are also called chemical reactions.

Characteristics of Chemical Changes   In a chemical change, new combinations of atoms are produced, forming new substances with new properties. For example, when wood burns, large molecules that make up the solid wood break apart, combine with oxygen, and rearrange to form small molecules of gaseous carbon dioxide and water.

Chemical changes occur constantly within organisms and are essential for maintaining life. For example, digestion of a spaghetti lunch involves chemical changes. In these changes, large starch molecules that the body can't absorb are broken into small sugar molecules that the body can absorb. Then, in cellular respiration, chemical reactions break the bonds of sugar molecules, release energy, and form CO_2 and H_2O. You'll learn more about the processes of respiration and digestion later in your biology course.

Figure 3-10   In this photo, you can see several physical properties of iron. You can also see the result of one of iron's chemical properties—the fact that it combines with oxygen in the air to form rust.

Figure 3-11   Just as the rusting of iron is a chemical change, so is the digestion of starch in spaghetti. In your body, digestion chemically changes the starch to smaller sugar molecules, which can be absorbed and transported to cells. In cells, further chemical changes may break the sugar down into CO_2 and H_2O, giving off energy in the process.

63

Chemical Equations

How can you describe the chemical change that we commonly call the rusting of iron? One way is to say that a gray piece of iron became brown and flaky on the surface exposed to air. Unfortunately, this description uses a lot of words and doesn't provide much information about what reacted and what new substance formed. Instead, scientists describe chemical reactions (changes) by using a system of chemical shorthand. For example, when iron rusts, it reacts chemically with oxygen, forming iron oxide. This reaction is represented as follows.

$$Fe + O_2 \rightarrow Fe_2O_3$$

Such a representation of a chemical reaction is called a chemical equation. Iron and oxygen are called reactants, and iron oxide, Fe_2O_3, is called the product. This equation is read, "Iron plus oxygen yields iron oxide."

Why Chemical Equations Must Be Balanced Chemical equations are like algebraic equations in that both sides must represent the same quantity. In other words, the number of atoms of each element on each side of the arrow must be equal. Why is this so? Have you ever heard of the Law of Conservation of Mass? This law states that during a chemical reaction, mass is neither created nor destroyed. Therefore, the above equation must be written as follows:

$$4Fe + 3O_2 \rightarrow 2Fe_2O_3$$

Such an equation is said to be *balanced*, because the number of atoms of each element on each side is the same.

When you burn natural gas to cook food or heat your home, methane molecules (CH_4) combine with oxygen molecules (O_2) in the air. The products are CO_2 and H_2O. Study Figure 3-12, and note how the written equation represents the way the molecules change. Count the number of each type of atom in the reactants and in the products. Are the numbers equal? If so, the equation is balanced.

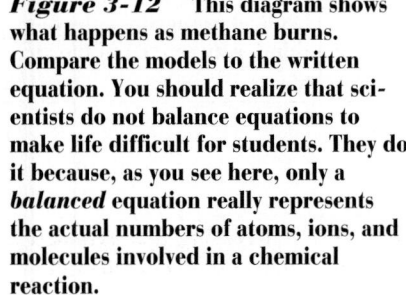

Figure 3-12 This diagram shows what happens as methane burns. Compare the models to the written equation. You should realize that scientists do not balance equations to make life difficult for students. They do it because, as you see here, only a *balanced* equation really represents the actual numbers of atoms, ions, and molecules involved in a chemical reaction.

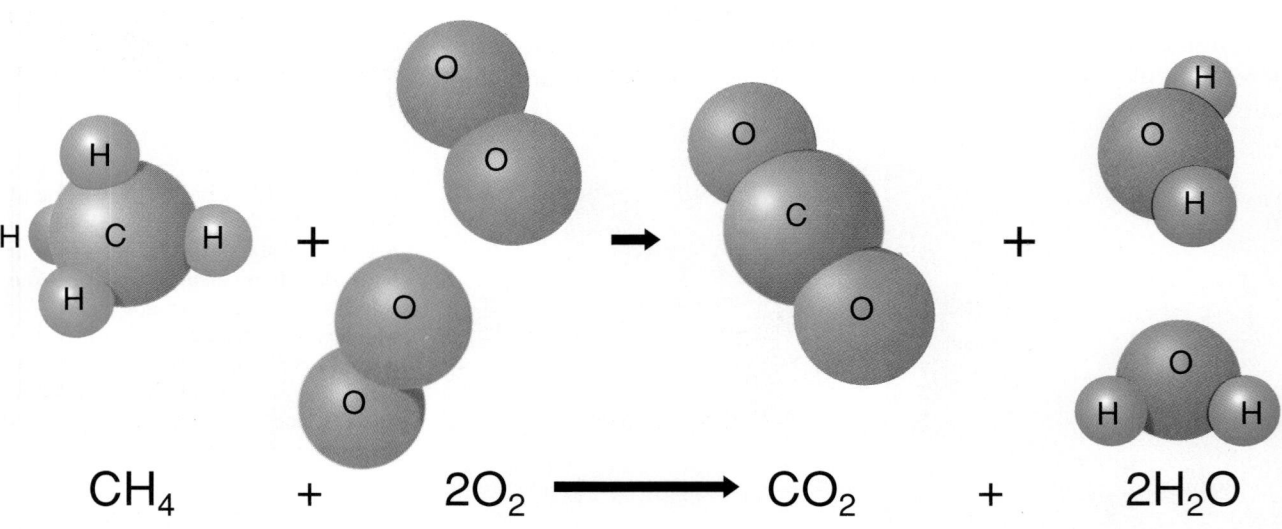

$$CH_4 \quad + \quad 2O_2 \longrightarrow CO_2 \quad + \quad 2H_2O$$

Solutions

Have you ever made a fruit-flavored drink by stirring a powdered drink mix in a glass of cold water? If so, you've mixed different substances without starting a chemical reaction. No bonds were broken and no bonds were formed. After thorough stirring, you formed a mixture that was the same throughout; that is, every part of the mixture is the same.

Now, suppose you dissolved 1 g of table salt, NaCl, a compound, in 100 mL of water. You have made salt water. If you stir 2 g of the same salt in another beaker containing 100 mL of water, you would again have salt water. Each beaker containing salt water would be the same throughout, but the two mixtures would differ in composition. The second beaker would contain a greater amount of salt in the same amount of water.

Both the fruit drink and the salt water are examples of solutions. **Solutions** are mixtures that are the same throughout but have variable compositions, depending on how much of one substance is dissolved in the other.

How Substances Dissolve   Why do some substances dissolve in water and others do not? The answer depends not only on the nature of the substance, but also on the nature of water. Look at Figure 3-13. In a water molecule, the nucleus of the oxygen atom attracts electrons more strongly than do the hydrogen nuclei. Thus, at any given time, there are more electrons near the oxygen nucleus, and electrons are pulled away from the hydrogen nuclei. As a result, the oxygen part of the water molecule has a slight negative charge, while the hydrogen part of the molecule has a slight positive charge. Molecules that have oppositely charged regions are called polar molecules.

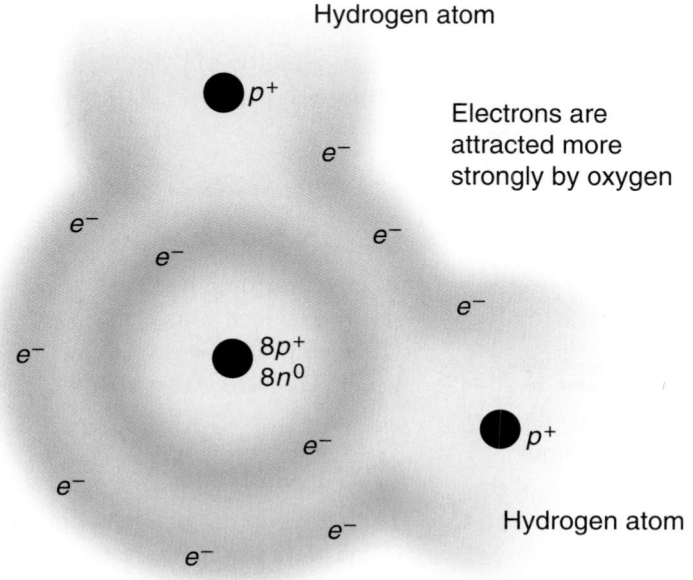

Hydrogen atom

Electrons are attracted more strongly by oxygen

Hydrogen atom

Oxygen atom

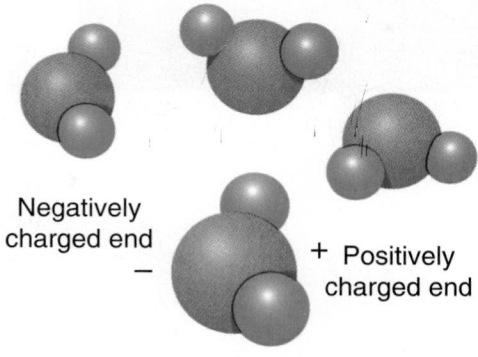

Figure 3-13   In reality, hydrogen and oxygen do not share electrons equally in covalent bonds. Oxygen "hogs" the electrons because it attracts them more strongly than hydrogen. This condition causes water molecules to be polar. Do you think that water molecules would attract each other? Why?

Negatively charged end

+ Positively charged end

Polar water molecules

Because they have charged regions, polar molecules attract other polar molecules. For example, sugar molecules, like water, are polar. The positive ends of water molecules attract the negative parts of sugar molecules and vice versa. When sugar crystals are placed in water, the attraction between oppositely charged areas of water and sugar causes the sugar molecules to separate from one another. Sugar dissolves when the sugar molecules separate and become scattered throughout the water molecules. In solution, the atoms of the sugar molecules remain together as molecules of sugar. Some covalent molecules, such as HCl, form ions when they dissolve in water.

You probably know that materials such as salad oil and wax don't dissolve in water. Fats, oils, and waxes are examples of substances whose molecules are not polar. When fats and oils are placed in water, there is very little attraction between their molecules and water molecules. As a result, fats and oils do not dissolve in water. In general, substances that are made of molecules that have very little attraction for water molecules do not dissolve in water.

Figure 3-14 **Polar water molecules do not dissolve fats and oils.**

I N V E S T I G A T I O N

The pH of Water

Is your shampoo pH balanced? Advertising would lead you to believe that it is extremely important to use pH-balanced shampoo. What is pH? The pH scale is used to measure how acidic or basic a solution is. At pH 7, a solution is neutral; decreasing from 7, a solution is more acidic; increasing from 7, it is more basic.

Pure water is neutral, with a pH of 7, but even small amounts of dissolved substances can change the pH. Rain is normally slightly acidic as a result of carbon dioxide that dissolves in the water and forms weak carbonic acid. The normal pH of rain is about 5.5. The pH of rain in some places may be very low due to pollutants reacting with water. Rain with a pH consistently below 5.5 is called acid rain.

Problem

What is the pH of water from a variety of sources?

Safety Precautions 🥽 🧤 ☣️

Be sure to wash your hands after handling water from various sources. Do not drink any of the water.

Hypothesis

With your group, make a hypothesis about which source will have water with the lowest pH and which will have water with the highest pH.

Experimental Plan

1. As a group, make a list of possible ways you might test your hypothesis using the materials your teacher has made available.

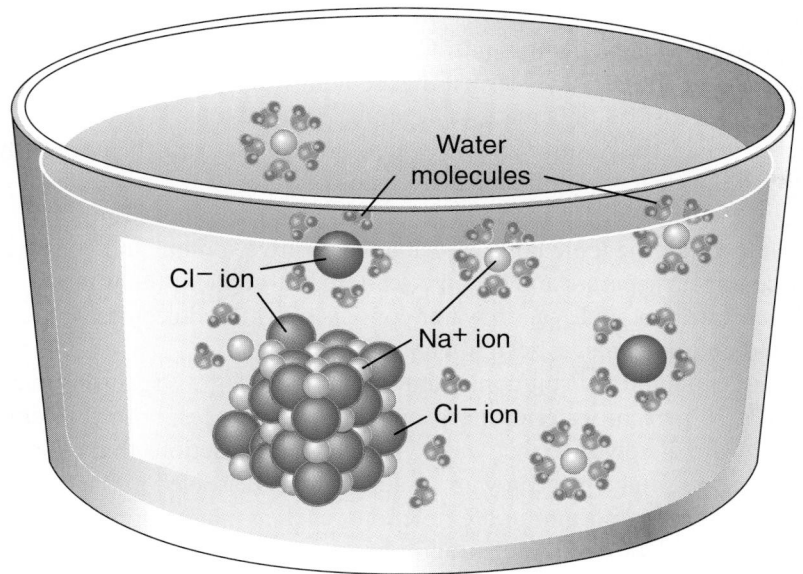

Figure 3-15 Many ionic compounds, such as common salt (NaCl), dissolve in water. When crystals of NaCl are placed in water, polar water molecules are attracted to the charged sodium and chloride ions. This force of attraction pulls the ions away from the crystal and into solution. Notice that the solution contains only separate sodium ions and chloride ions surrounded by water molecules. Compare the arrangement of water molecules around the sodium ions with that around the chloride ions. Explain the difference.

2. Agree on one idea from your group's list that could be investigated in the classroom.
3. Design an experiment that will test one variable at a time. Plan to collect quantitative data.
4. Following the style of a recipe, write a numbered list of directions that anyone could follow.
5. Make a list of materials and the quantities you will need.

Checking the Plan

Discuss the following points with other group members to decide the final procedures for your experiment.
1. What variables will need to be controlled?
2. What is your control?
3. What will you measure?
4. How much water of each type will you need?
5. Will you need more than one trial with each water sample?

6. How long will you wait before reading the results of your pH tests?
7. Have you designed and made a table for collecting your data?
8. How will you present your data in a graph?
Make sure your teacher has approved your plan before you proceed.

Data and Observations

Carry out your experiment, and complete your data table.

Analysis and Conclusions

1. What was your control?
2. Which type of water had the lowest pH, and is, therefore, the most acidic? Speculate why.
3. Which type of water had the highest pH? Speculate why.
4. Was your hypothesis supported by your data? Explain.
5. How might rainwater with a very low pH affect living organisms?

6. How might low pH in ponds, lakes, or streams affect living organisms?
7. Amphibians, such as frogs and salamanders, are declining in numbers worldwide. Some scientists think that increasing acidity of water in their habitat is the cause for this decline. Based on data you gathered, make a hypothesis about how the pH of water may be affecting amphibian numbers in your area.
8. Write a brief conclusion to your experiment.

Going Further

Based on this lab experience, design an experiment that would help answer another question that arose from your work. Assume that you will have unlimited resources and will not be conducting this lab in class.

How Ionic Compounds Dissolve   In water, ionic compounds behave differently from covalent compounds. Consider table salt (sodium chloride) as an example. Earlier, you learned that sodium chloride is not composed of molecules. Instead, it is made of separate, charged ions. When salt is added to water, the positive ends of water molecules attract the negatively charged chloride ions, and the negative ends of other water molecules attract the positively charged sodium ions. This attraction is strong enough to break the ionic bond, causing the positive sodium and negative chloride ions to separate and become scattered throughout the water. Thus, a solution of table salt in water consists of water molecules, sodium ions, and chloride ions, as seen in Figure 3-15.

An understanding of solutions is important in biology. Organisms may consist of 70 percent water or more, and many substances, both covalent and ionic, are dissolved in that water. The chemical reactions that make up life processes take place in water solution. Ions dissolved in the water of organisms are important to life functions, such as conduction of nerve impulses, contraction of muscles, and photosynthesis.

Acids and Bases

You might think that pure water would consist of only H_2O molecules. However, in about two molecules per billion, one proton of a hydrogen atom breaks away from a water molecule. This process forms a positively charged hydrogen ion (H^+) and a negatively charged hydroxide ion (OH^-). See Figure 3-16. The numbers of hydrogen and hydroxide ions that exist are equal, because one ion of each kind is produced from the breakup of every water molecule. Any solution in which the concentration of hydrogen and hydroxide ions is equal is called a neutral solution.

What happens if you dissolve in water some substance that changes the balance between the concentration of hydrogen and hydroxide ions? When hydrogen chloride gas dissolves in water, the molecules of HCl separate into hydrogen and chloride ions. The solution now has a greater concentration of hydrogen ions than hydroxide ions. Such a solution is called an **acid,** in this case, hydrochloric acid. Similarly, if the ionic substance sodium hydroxide is dissolved in water, the solution will contain a greater concentration of hydroxide ions than hydrogen ions. Such a solution is called a **base.** Solutions of acids typically have a sour taste. Lemons

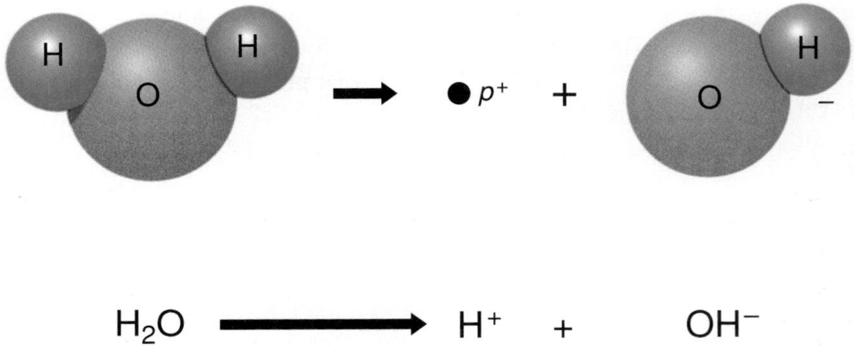

Figure 3-16   **A tiny fraction of water molecules breaks apart and forms ions. Each molecule that ionizes forms one hydrogen ion and one hydroxide ion. Therefore, in pure water, the numbers of hydrogen ions and hydroxide ions will be equal. It's important to note that a hydrogen ion is just a bare proton. In water, it is always surrounded by water molecules attracted to it because of its positive charge.**

$$H_2O \longrightarrow H^+ + OH^-$$

and grapefruit are sour because they contain citric acid. Solutions of bases, like soapy water, have a sharp, bitter taste, and they feel slippery.

The pH Scale   The concentration of hydrogen ions in solutions is indicated by a set of numbers called the pH scale, which runs from 0 to 14. See Figure 3-17. Pure water, having an equal concentration of hydrogen and hydroxide ions, has a pH of 7. A pH of 7 represents a neutral solution. Numbers below 7 represent acids. The lower the number, the greater the concentration of hydrogen ions and the more acidic the solution. Numbers above 7 represent bases. The higher the number, the lower the concentration of hydrogen ions and the more basic the solution.

The pH of solutions is important to living systems. The pH of water and soil, for example, is one of the factors that determines the kinds of organisms that can live there. Also, pH can influence the chemical reactions that can occur in an organism. For example, in order for digestion to occur properly in the human stomach, the pH there must be about 2. Human blood must be maintained at a pH slightly above 7.

Figure 3-17   Shown here are the pH values of some common materials. Which has the higher concentration of OH⁻ ions, ammonia solution or blood?

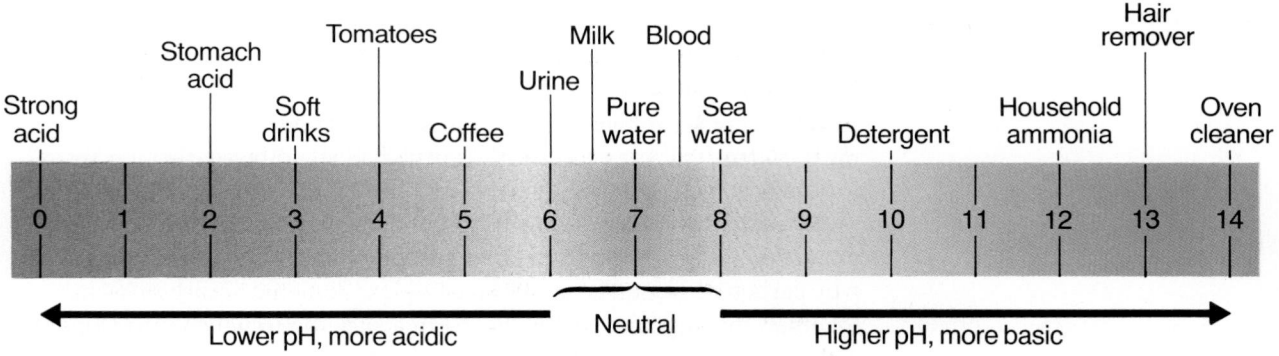

Section Review

Understanding Concepts

1. Carbon is a key element in living things. A typical carbon atom has 6 protons and 6 neutrons in the nucleus. Describe the electron arrangement of a carbon atom. How many covalent bonds would a carbon atom usually form? Explain your reasoning.
2. When an aluminum atom forms an ionic bond, will it gain or lose electrons? How many electrons does an aluminum ion have? What is its charge?
3. Calcium chloride, $CaCl_2$, is an ionic compound composed of calcium and chloride ions. If you could see the particles in a solution of calcium chloride dissolved in water, what would you see? Explain.
4. *Skill Review—Classifying:* Atoms having the same number of electrons in their outermost energy levels have similar properties. The following is a list of atoms and their total number of electrons. Divide the atoms into two groups, each having similar properties: lithium—3; magnesium—12; sodium—11; potassium—19; beryllium—4. For more help, refer to Organizing Information in the **Skill Handbook.**
5. *Thinking Critically:* Neon has a total of ten electrons. Why does neon not react with other atoms?

Objectives

Relate the structure of a carbon atom to the complexity of carbon compounds.

Compare the structures and functions of carbohydrates, lipids, and proteins.

Distinguish between condensation and hydrolysis reactions.

3.2 Biological Chemistry

What's your favorite food? Do you like pizza? Frozen yogurt? Spaghetti? A juicy steak? As you know, these foods contain parts and products of living things, which are composed of a variety of compounds. Biological compounds are much more complex than any we have looked at so far. Many of them contain thousands of atoms or more. Though they are more complex, the properties and behavior of these substances are explained by the same principles of chemistry that apply to substances in the nonliving world.

Carbon Compounds

Originally, because most carbon compounds were known to be associated only with organisms, scientists called them organic compounds. Later, chemists analyzed them and learned to make organic compounds in the laboratory. Today, organic compounds are defined as those compounds that contain carbon. All substances not classified as organic are called inorganic substances. A few carbon compounds, such as carbon dioxide, CO_2, are not classified as organic.

Carbon atoms have six electrons, two in the first level and four in the second. In order to have a stable outer level of eight electrons, carbon must obtain four more electrons. It usually does so by sharing electrons, forming covalent bonds.

Not only can carbon atoms form bonds with atoms of different elements, they can also bond to other carbon atoms. This property explains why carbon compounds can be so complex. Sometimes carbon atoms join together, making long chains that form a skeleton to which other kinds of atoms are joined. In other organic molecules, carbon atoms are bonded to form ringlike structures, and these rings may be bonded to one another.

Figure 3-18 **The simplest organic substance is methane, shown here in four different ways. Notice that carbon atoms form four covalent bonds in a pyramid shape called a tetrahedron. The diagram on the right shows how a methane molecule would look if you could see it. The two formulas on the left represent two convenient ways of writing formulas for methane.**

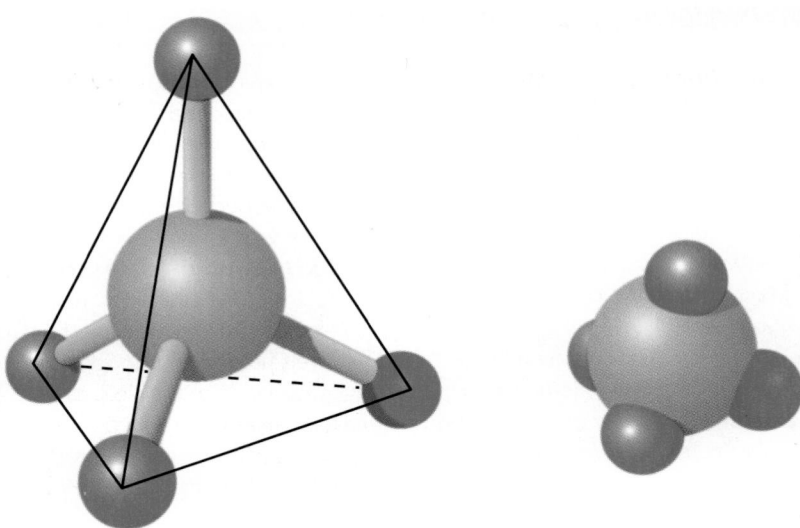

$$CH_4$$

$$H - \overset{\displaystyle H}{\underset{\displaystyle H}{\overset{|}{\underset{|}{C}}}} - H$$

Structural Formulas   Recall that chemical formulas show the kinds and numbers of atoms in a molecule. Such formulas can also be written for organic molecules. More often, though, structural formulas are used to better represent organic molecules. Structural formulas show not only the kinds and numbers of atoms in a molecule, but also their arrangement. Methane, the major component of natural gas, is the simplest organic compound. Methane has the chemical formula CH_4, as shown in Figure 3-18. Its structural formula shows that each of the four hydrogen atoms is bonded to the central carbon atom. If the formula were shown in three dimensions, it would resemble a pyramid, the true shape of the molecule. Each line drawn between atoms represents a covalent bond, the sharing of one pair of electrons.

   Some covalent bonds involve sharing of more than one pair of electrons between atoms. A structural formula with two lines between atoms, such as $C=C$ or $C=O$, indicates a double covalent bond, the sharing of two pairs of electrons.

Isomers   To see why it is important to use structural formulas to represent organic compounds, examine Figure 3-19. Count the number of each kind of atom in each structural formula and determine the chemical formula of each. As you can see from the diagram, some organic molecules may have the same chemical formula, but different structural formulas. Such molecules are called **isomers.** Isomers have different physical and chemical properties that are important to organisms. By showing structural formulas, one isomer can be easily distinguished from another.

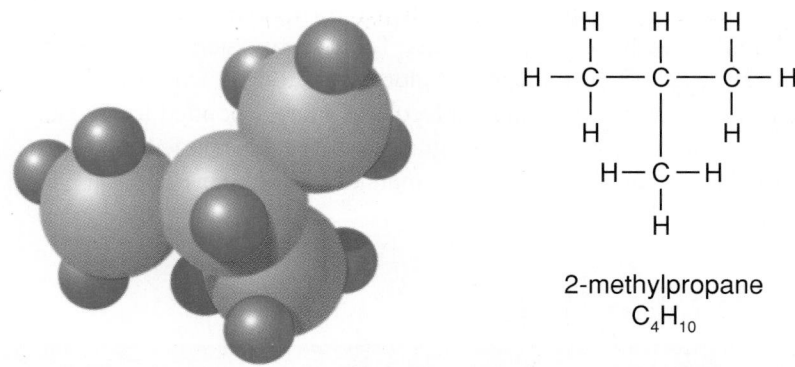

2-methylpropane
C_4H_{10}

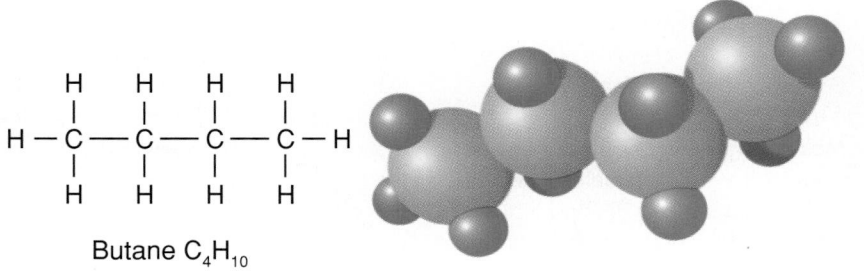

Butane C_4H_{10}

Figure 3-19   Here you can see the advantage of writing structural formulas to represent organic compounds. Both of these compounds have exactly the same chemical formula, C_4H_{10}. However, they are two different compounds because they have different structures. One is called butane, and the other is 2-methylpropane.

CH$_2$OH

Glucose

Fructose

Carbohydrates

Are you hungry? Picture a thick cheeseburger on a bun and a side order of fries. When you eat a meal like this, you're really enjoying several biologically important groups of organic compounds. One of these groups is carbohydrates. **Carbohydrates** are organic compounds composed of carbon, hydrogen, and oxygen with a ratio of two hydrogen atoms to each oxygen atom. Sugars and starches, such as those found in the hamburger bun and fries, are typical examples of carbohydrates. In living things, many carbohydrates serve as energy sources for life processes. In fact, the sugar glucose is a "fuel of life" for many different organisms.

Simple Sugars   Glucose, $C_6H_{12}O_6$, is an example of the simplest type of carbohydrate, a monosaccharide, or simple sugar. Fructose, the sugar contained in many fruits, is also a monosaccharide. It has the same chemical formula as glucose, $C_6H_{12}O_6$, but differs in structure. Another isomer of glucose is galactose, a monosaccharide in milk.

Two simple sugars can bond together to form organic compounds called double sugars, or disaccharides. Table sugar, sucrose, is a double sugar composed of one molecule of glucose and one of fructose. Lactose, milk sugar, is composed of one molecule of glucose bonded to one of galactose. Maltose consists of two glucose molecules bonded together. Sucrose, lactose, and maltose are isomers of one another. They all have the formula $C_{12}H_{22}O_{11}$.

Figure 3-20 **Look at the structural formulas for glucose and fructose. If you count the atoms of carbon, hydrogen, and oxygen, you'll discover that both sugars have the same chemical formula. Glucose is abundant in corn syrup; honey contains fructose. Simple sugars such as these are examples of carbohydrates.**

Thinking Lab

MAKE A HYPOTHESIS

Why do monarch butterflies contain two different chemicals for defense?

Background
The monarch butterfly has two chemical defense systems. As a caterpillar, the monarch eats milkweed leaves. These leaves contain a bitter-tasting toxic substance, which the butterfly stores in its body. As a butterfly, the monarch migrates to southern regions where

there are few milkweeds. There, it takes in different toxic chemicals from the nectar of certain flowers.

You Try It
Make a hypothesis to explain how toxic chemicals work as a defense system for monarchs. Keep in mind that some major predators for monarchs are birds

and reptiles. However, they are usually only made ill by the toxic substances. Describe an experiment that would test your hypothesis. What additional information would you need? Why would the presence of these chemicals benefit the species as a whole but not necessarily the individual?

Starch—a polysaccharide

Polysaccharides   Long chains of monosaccharides joined together form the most complex carbohydrates, **polysaccharides.** Hamburger buns and fries contain a polysaccharide known as starch. Starch is provided by plants. During the growing season, many plants produce more simple sugars than they use for energy. This excess sugar is converted to starch for storage. During critical periods, plants can change starch back to simple sugars, which can be used by the plant as its energy source in cellular respiration. When you digest the starch contained in the bun and fries, you convert it to glucose, which you then use as an energy source. Humans store excess glucose in the liver and muscles in the form of a starchlike compound called glycogen. Like starch in plants, glycogen can also be converted to glucose and used as an energy source when the need arises. However, not all carbohydrates are used as energy sources. For example, another plant polysaccharide, cellulose, is used for structural purposes in the cell walls of plants. Cellulose, glycogen, and starch are all made of chains of glucose molecules. They differ in the patterns by which the molecules are linked.

Lipids

Besides carbohydrates, a tasty cheeseburger meal contains another group of organic compounds, fats. The cheese in cheeseburgers and the burger itself contain fats. French fries are usually cooked in oil. Fats and oils, along with waxes, belong to a class of organic compounds called lipids. Like carbohydrates, **lipids** are also composed of carbon, hydrogen, and oxygen, but unlike carbohydrates, the number of hydrogen atoms per molecule is much greater than the number of oxygen atoms. For example, one common lipid has the formula $C_{57}H_{110}O_6$. Lipids produced by animals are usually solid fats, but those produced by plants are usually liquid oils. Waxes are produced by both plants and animals.

Figure 3-22   This photo shows some sources of lipids in our diets. In general, butter and the fat in meat contain more saturated fatty acids than do oils from plants, which contain more unsaturated fatty acids. Which examples have more double bonds?

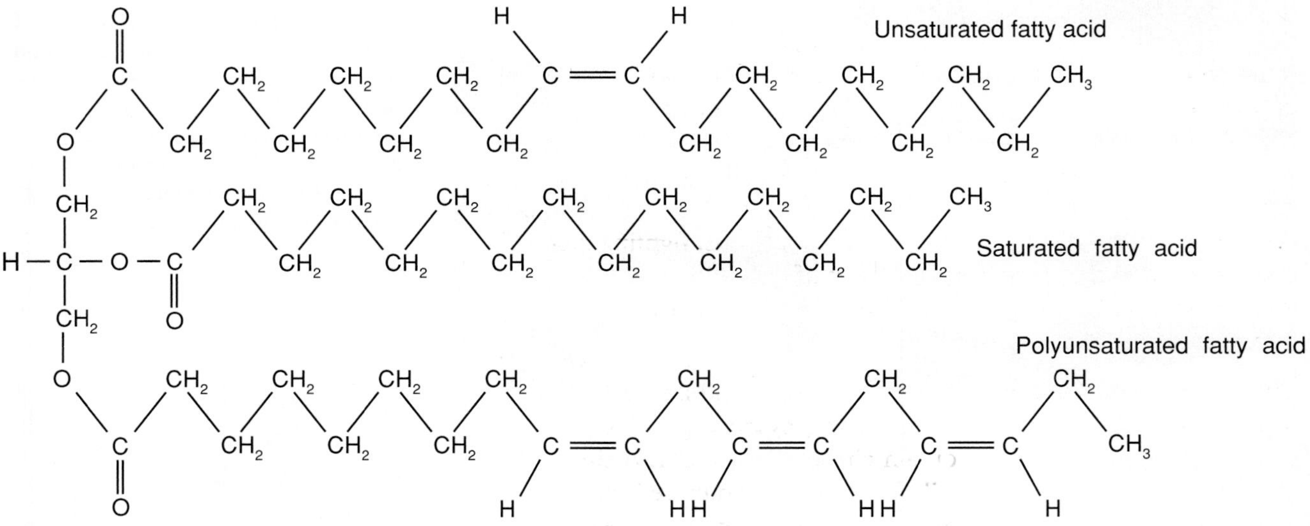

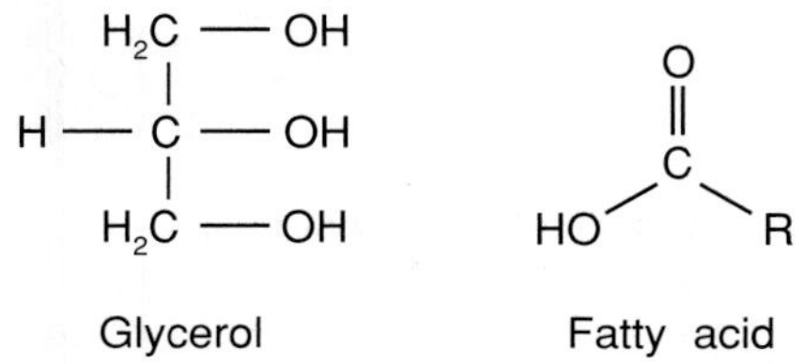

Figure 3-23 Typical fats contain three fatty acids bonded to a glycerol molecule. The structure of a fat (top) shows how the fatty acids are linked to glycerol. It also demonstrates the structural differences between saturated and unsaturated fatty acids.

In animals, excess fats in the diet, along with excess sugars that have been converted to fats, are stored and used as energy reserves. The storage of fats is an adaptation that guards against critical times when food is scarce. Lipids also form part of the structure of cell membranes (Chapter 4). In some animals, lipids protect parts of nerve cells, speed the transmission of nerve impulses, and provide insulation from the cold.

The Structure of Fats   Fat molecules are built from two types of simpler molecules—fatty acids and glycerol. There are many different types of fatty acid molecules, but all have one part in common, a group of atoms called a carboxyl group (—COOH). In the structural formula of a fatty acid, shown in Figure 3-23, the letter *R* represents a long chain of carbon and hydrogen atoms to which the carboxyl group is attached. The exact number of these atoms varies among fatty acids; thus, for simplification, the chain is written as *R*.

The kinds of bonds between carbon atoms in fat molecules also vary. Fatty acids in which carbon atoms are linked by single covalent bonds are said to be saturated, while those having one or more double covalent bonds are said to be unsaturated. As Figure 3-23 shows, saturated fatty acids are filled, or saturated, with more hydrogen atoms than unsaturated molecules. This difference is important to your health. Scientists generally agree that substituting vegetable oils that have a high proportion of unsaturated fatty acids for animal fats in the human diet lowers the risk of heart and circulatory diseases.

Cholesterol is another lipid you have probably heard of. The molecular structure of cholesterol is quite different from that of other lipids. However, cholesterol is classified as a lipid because, like other lipids, it is insoluble in water. In many organisms, especially animals, cholesterol is produced in the body. It is used in the formation of cell membranes and the manufacture of many hormones. Later, you'll learn how an excess of cholesterol in the human body can cause health problems.

Proteins

In hamburger as well as other meats and seafood, the largest group of organic molecules belongs to another class of compounds, the **proteins.** Unlike carbohydrates and lipids, proteins are not normally used as energy sources by organisms. Some are used in the building of living material, such as the muscle from which burgers are made. Others have functional roles, such as carrying out chemical reactions, fighting disease, or transporting particles into or out of cells. These functions are very specific, and each requires a protein with different structural characteristics.

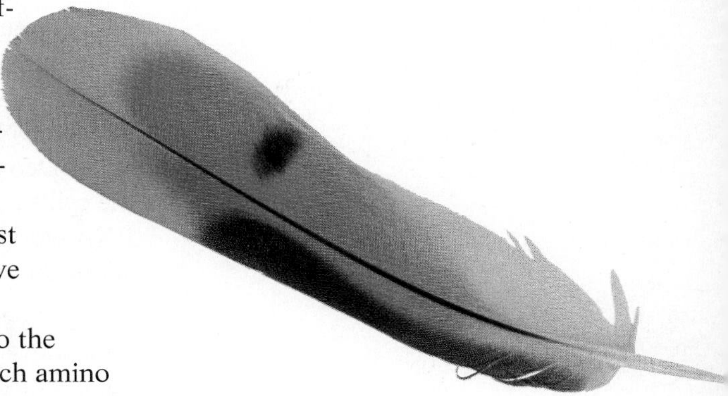

The Structure of Proteins   Like carbohydrates and lipids, protein molecules are composed of carbon, hydrogen, and oxygen atoms. In addition, they contain nitrogen and sulfur. A single protein molecule may be built from chains of hundreds or even thousands of simpler compounds called **amino acids**. Most of the 20 different amino acids that make up most proteins have a central carbon atom to which is bonded a carboxyl group, a hydrogen atom, and an amino group ($-NH_2$). Also attached to the carbon atom is the rest of the molecule, the part that makes each amino acid different. It is represented by the letter *R*, as seen in Figure 3-25.

When amino acids bond to one another, they are known as peptides. Two amino acids linked together form a dipeptide. Many bonded together are a **polypeptide.** Polypeptides differ from one another by the kind, number, and sequence of amino acids. To see why there are so many different proteins possible, try this exercise. How many other words can you make from the letters in the word STOP? In the same way that the 26 letters of the alphabet form many words with special meanings, the 20 amino acids in proteins can form an almost endless variety of polypeptides having different properties.

Figure 3-24   **Feathers, spider webs, wool, and silk are made up of proteins. In fact, feathers consist mostly of the same protein, keratin, that makes up human nails and hair.**

Figure 3-25   **Proteins are made of long chains of amino acid units. The left photo is a computer image of a molecule of immunoglobin. In the right diagram, the shaded area shows the structure that is present in almost every amino acid. Structures of two of the simpler amino acids are shown also. The R group in more complex amino acids can include longer chains, sometimes with ring structures and atoms of sulfur.**

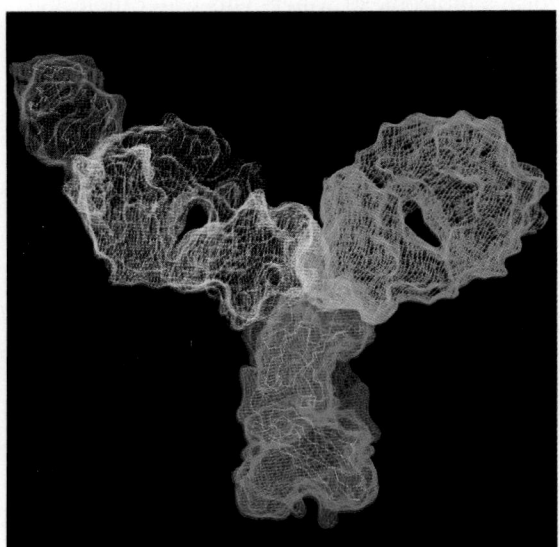

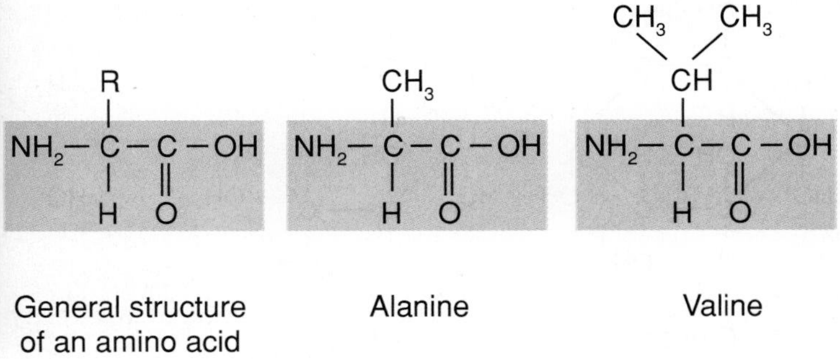

General structure of an amino acid

Alanine

Valine

Protein structure is very complex. A protein molecule may consist of just one polypeptide or two or more polypeptides bonded together. In addition, each protein has a distinct, three-dimensional shape that is crucial to the protein's function. For example, the proteins that form fingernails and hair are long and fiberlike. Some other proteins, such as hemoglobin in your blood, have a globular shape. That is, they are like a sphere with a lumpy surface.

Nucleic Acids

Every organism contains molecules that form a code that controls the organism's basic appearance and behavior. These molecules belong to a fourth class of biological compounds called **nucleic acids**. Nucleic acids include DNA and RNA and are the most complex of all biological molecules. Deoxyribonucleic acid (DNA) and ribonucleic acid (RNA) control the cell's activities. DNA usually forms the genetic code of organisms, and RNA works to carry out the instructions of that code. You'll learn more about the structure and function of nucleic acids in Chapters 9 and 10.

Reactions of Biological Compounds

As you recall, the disaccharides sucrose, maltose, and lactose all have the formula $C_{12}H_{22}O_{11}$. Does this formula surprise you? Do you think it should be $C_{12}H_{24}O_{12}$ because these molecules consist of two monosaccharides each having the formula $C_6H_{12}O_6$? If you think that two hydrogen atoms and an oxygen atom are missing, you're thinking clearly.

Condensation Reactions   Remember that in order for a chemical reaction to occur, bonds must be broken and new bonds must be formed. Figure 3-26 shows the formation of maltose from two molecules of glucose. In one glucose molecule, a bond to one hydrogen atom is broken. In the other, the bond to an —OH group is broken. During the reaction, these three atoms bond with one another to form a molecule of water. This water molecule accounts for the two hydrogen atoms and one oxygen atom that seemed to be missing. At the same time, the glucose molecules form new bonds with each other. The new disaccharide molecule is made by removing water from the original molecules. This chemical process is called *condensation*

Figure 3-26   **Large molecules in living things often consist of smaller molecules bonded to each other by condensation reactions. This diagram shows how the disaccharide, maltose, is built by the condensation of two glucose molecules. The —H and —OH that are removed form a molecule of water.**

Glucose   Glucose   Maltose

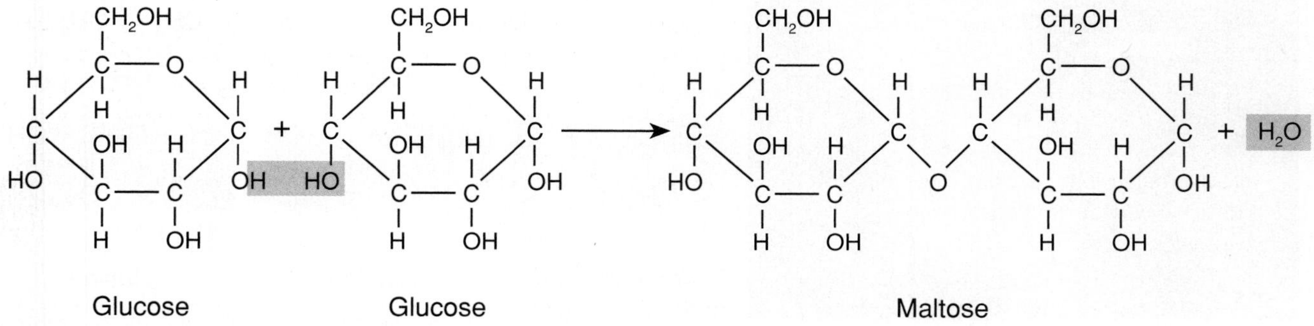

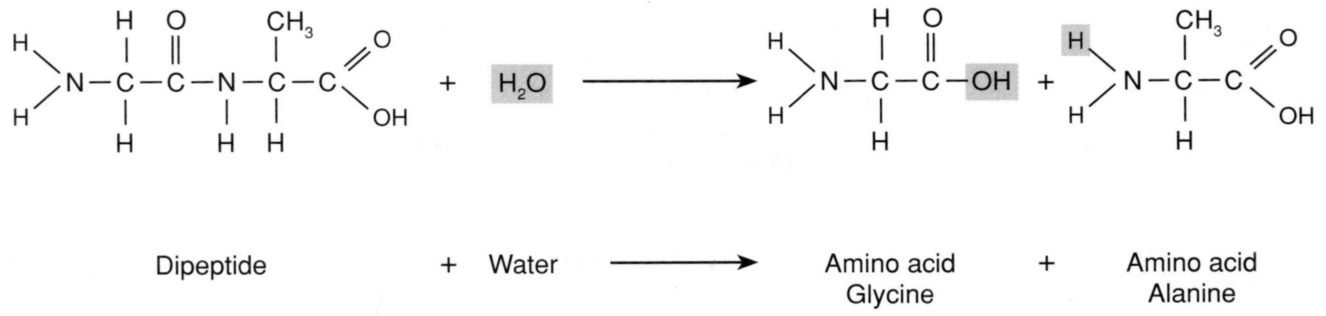

Dipeptide + Water ⟶ Amino acid + Amino acid
Glycine Alanine

and is a common reaction in organisms whenever larger molecules are made from smaller units. For example, amino acids from the foods you eat are linked by condensation reactions to form the larger proteins needed by your cells for growth and maintenance.

Hydrolysis Large molecules are converted to smaller ones by a process that is the reverse of condensation. Because this process involves the addition of water, it is called *hydrolysis*. Hydrolysis reactions are important to living things because complex molecules must be broken down into simpler ones before they can be used by cells. Digestion, for example, involves the breakdown of proteins in food into amino acids. Figure 3-27 shows such a reaction in which a dipeptide is converted to two amino acids. Notice that the reaction breaks the bond that links the amino acids together. During the hydrolysis, a nitrogen atom from one amino acid bonds to a hydrogen atom, and a carbon atom from the other amino acid bonds to an —OH group. Both the —H and —OH groups come from a water molecule. These processes of hydrolysis and condensation illustrate the importance of water to living things.

Figure 3-27 **Some biological processes, such as digestion, require that large molecules be broken down into smaller units, usually by hydrolysis reactions. Shown here is the hydrolysis of a dipeptide into two amino acids. Compare this diagram with Figure 3-26. Can you see that the two processes are the reverse of each other?**

Section Review

Understanding Concepts
1. Why are organic compounds usually more complex than inorganic compounds?
2. Why do you think the variety of proteins is so much greater than that of carbohydrates or lipids?
3. Suppose 100 amino acids are linked together in a cell, one at a time, to form a polypeptide chain. How many molecules of water will be removed in this synthesis?

4. *Skill Review—Making Tables:* Make a table comparing carbohydrates, lipids, and proteins. Compare them in terms of the elements they contain, the building blocks of which they are composed, and their functions in living organisms. For more help, see Organizing Information in the **Skill Handbook**.

5. *Thinking Critically:* One gram of fat releases 2.5 times as much energy as an equal amount of glucose. Why does the body normally use glucose for energy and store fat instead of using fat and storing glucose?

Objectives

Distinguish between potential energy and kinetic energy by using an example.

Discuss the relationship between a reaction and activation energy.

Describe the means by which an enzyme carries out a cellular reaction.

3.3 Energy and Reactions

You have probably heard that you need lots of energy for both running and thinking. What is energy? The best way to define energy is to describe what it does. Moving a car requires energy. So does melting ice. Both of these events involve motion. In a car, it's the motion of pistons and gears. With the melting ice, it's the motion of water molecules. Whenever anything moves, work is done. Therefore, energy is the ability to do work—in other words, to cause motion.

Energy is most evident to you when it is changing from one form to another. Electrical energy flowing through a light bulb changes to light, enabling you to read. Electrical energy is changed to energy of motion in the motor of a fan. The motion of the fan blades moves air, creating a breeze that cools you. In a gas oven, chemical energy changes to heat, which cooks your food. Energy changes aren't limited to the nonliving world, however. All life processes involve energy, too. Therefore, energy changes are necessary for organisms to carry out these processes.

Transformation of Energy

Does a ball have energy? Imagine carrying a ball to the top of a flight of stairs. How would you get the ball back down to the bottom of the stairs? You could push the ball over the edge of the top step so that it bounced step by step down the stairs. Or, you could drop it directly over the railing to fall quickly to the floor below. In both cases, when the ball moves downward it does work, perhaps by knocking over a lamp.

Now think about a log in a fireplace. As the wood burns, energy in the form of heat and light is released. This energy can then warm a room or pop popcorn.

Figure 3-28 You can easily see energy doing work in this antique locomotive. The energy released by the burning of wood or coal is converted to the energy of motion of the locomotive's pistons and wheels.

a

b

Types of Energy   Both the ball at the top of the stairs and the wood in the fireplace have a type of energy called potential energy. **Potential energy** is the energy of position, and is sometimes called stored energy. What is the source of the potential energy in the ball and the wood? Energy was used to carry the ball to the top of the stairs where it could fall and do work. The potential energy of molecules in wood is chemical energy—energy that is stored in the bonds of the molecules in wood cells during photosynthesis. As a ball bounces down stairs, potential energy is converted to **kinetic energy,** the energy of motion. When the ball comes to rest at the bottom of the stairs, it has lower potential energy. Similarly, the products formed from the burning of wood have a lower potential energy than the original wood molecules had.

Survival of organisms depends upon the chemical energy stored in energy-rich molecules like those in wood. When a reaction occurs, the bonds break, energy is released, and the potential energy is converted to kinetic energy that can be used by cells for biological work. For example, through a series of chemical reactions, chemical energy is transformed to electric energy in the bodies of electric eels or to light energy in fireflies.

Conservation of Energy   When energy is transformed in living organisms, is there any loss of energy? Suppose you earn $10 performing an after-school job. That $10 was paid to you by your employer, who got it from another source. You use the $10 to buy a snack that costs $1.25, receiving $8.75 in change. Of this, $5 goes into your savings account, and you gradually spend the rest. What has happened to that original $10? Has it disappeared? No. It still exists, but in different forms and belonging to different people.

Figure 3-29   This skier (a) increased his potential energy by riding to the top of the hill on a lift. Now, his increased potential energy is changing to kinetic energy as he skis downhill. In a similar way, the potential energy gained by this cheetah (b) from its last meal is stored in molecules in its cells. Now, the cheetah is converting some of that potential energy to kinetic energy by chasing another meal.

The same idea is true of energy transformations. When energy is transformed, it does not disappear. It is conserved, and the total amount of energy in a system remains constant. When wood burns, for example, energy is released as light and heat. In addition, some of the energy that was stored in the wood still exists as chemical energy in the bonds of the products formed, carbon dioxide and water. But all of the original energy is accounted for. The total amount of energy released plus the energy remaining in the bonds of the products is equal to the amount of energy originally in the wood and the oxygen molecules with which it reacted. This idea can be stated as the Law of Conservation of Energy. This law states that in chemical reactions, energy is neither created nor destroyed.

Activation Energy

Although a ball will fall down a flight of stairs, it needs a push to get started. The same is true of chemical reactions. You learned earlier that hydrogen gas and oxygen gas will not react to form water without an input of energy, usually as heat. The added energy causes some of the molecules to move faster and collide with more force. When the hydrogen and oxygen molecules hit each other hard enough, bonds break, and the reaction starts. The minimum amount of energy required to start a chemical reaction is called **activation energy.** Some reactions do not require very high temperatures and will begin at room temperature or below. In self-sustaining chemical reactions, such as burning, the activation energy needed to keep the reaction going comes from the energy given off by the molecules that have already reacted.

HEALTH CONNECTION

Proteins in Vegetarian Diets

• • • • • • • • • • • • • • • • • • • •

Have you ever seen a house with walls built of bricks that were salvaged from an old building that had been torn down? The new house may look nothing like the old building, yet it is built of the same units. This process is similar to what takes place when you eat protein in your diet.

As you know, proteins are composed of long chains of basic build-ing blocks, amino acids. In order to build the kinds of proteins that humans need, you eat proteins from a variety of sources.

Digestion breaks up the chains of amino acids in the dietary protein. The free amino acids are then used in your own cells to build new proteins for body structures or to act as enzymes. Without proteins, humans cannot survive for long. In fact, lack of protein causes a disease called kwashiorkor, which you may have seen on TV or in photographs from areas where there is famine.

Meat and fish are major sources of protein in the diet. In fact, many people in the United States consider these foods to be essential to a healthy life. Although meat and fish are good sources of protein, they're not the only ones. Plants also contain proteins. The big difference between animal proteins and plant proteins is that they contain different combinations of amino acids. Because humans are animals, animal protein has an advantage in that it contains the amino acids we need in the proportions we need them. Of the 20 common amino

Enzymes

Think again about wood in a fireplace. Although oxygen molecules are in constant motion and bouncing off the wood molecules, the wood does not burst into flame. A very high temperature, such as that of a burning match, is needed to provide enough activation energy to start the reaction. The reactions of living organisms cannot depend on heat as a source of activation energy. First, the high temperature needed would destroy cells. Second, heat would be far too random a source of activation energy. Like a flame in a fireworks factory, it could trigger many different reactions at once. To function properly, cells must carry out specific reactions at definite times.

How, then, do reactions get started in cells? Cells contain **enzymes,** proteins that lower the activation energy required and allow reactions to occur at the normal temperatures of cells. Each enzyme is specific; it guides only one type of cell reaction. In addition, enzymes are not permanently changed or used up during a reaction, so a given enzyme can be used over and over again. Most enzymes are globular proteins with particular three-dimensional shapes. The shape of an enzyme is crucial to the reaction it controls.

Reactants in cellular reactions are called substrates. Most enzymes are named by adding the suffix -*ase* to part of the name of the substrate in the

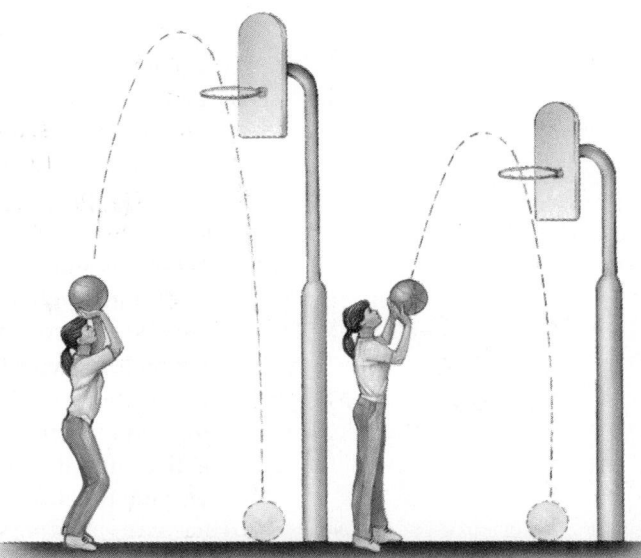

Figure 3-30 **The basketball player has to shoot with more energy when the basket is higher. When the basket is lower, the player can accomplish the same task using less energy. You can compare this situation to the action of an enzyme. An enzyme lowers the amount of energy needed to begin a chemical change.**

acids, humans can manufacture 12 from smaller molecules. The other eight must be obtained from proteins in the diet. The problem with plant protein is that no one plant protein provides all the needed amino acids.

However, a carefully planned vegetarian diet can supply the proper protein balance. By combining a cereal grain (such as rice, wheat, or corn) with a legume (such as peas, beans, or peanuts), the necessary combination of amino acids is supplied. The old favorite of rice and beans from the southern

United States is a good example of such a dish, as is a corn tortilla with frijoles (beans). This method of combining foods is called protein complementarity and is the manner by which millions of people maintain a diet balanced in amino acids. For this effort they get a bonus— plant proteins don't come bundled with animal fat that might cause additional health problems.

1. **Explore Further:** Plan a meatless meal that is balanced in amino acids. Use a vegetarian cookbook to find recipes that your family would enjoy. Which parts of the recipe supply the needed amino acids?

2. **Solve the Problem:** Imagine that you are in charge of feeding a poor country that is largely desert and has no access to the sea. What could you do to help your people eat a healthy diet?

reaction the enzyme controls. For example, the enzyme that guides the conversion of maltose to glucose is called maltase. Some enzymes, such as pepsin, which works in your stomach, were assigned names before the present naming system was adopted.

A Model of Enzyme Function

For many years, scientists worked to determine how enzymes and substrates interacted to carry out reactions. As a result of this research, scientists developed a model of how enzymes work. Enzymes were pictured as fitting together with their substrate molecules like a key fits a lock. You know that every lock has a specific key that opens it. The surfaces of the lock and key fit together. Similarly, part of an enzyme's surface, called its active site, fits together with the substrate molecule(s). This description of enzyme function is called the lock-and-key model.

Induced Fit Model More recently, scientists have modified the lock-and-key model. Research shows that the shape of the enzyme's active site and the shape of the substrate need not match exactly. When the enzyme and substrate join, the shape of the enzyme changes slightly, making the fit between enzyme and substrate more exact. In other words, the enzyme is a flexible key. This modern model is called the *induced fit model.* Modification of the lock-and-key model shows how explanations in science change when scientists make new discoveries.

Let's look at a common biological reaction—the formation of a larger protein molecule from single amino acid molecules. The reaction is shown in Figure 3-31. Suppose a cell has two amino acids that are to be joined to form a dipeptide. There is an enzyme for this type of reaction. How does it react with just the right amino acids among the many other molecules moving around in the cell? As you know, enzymes function on the basis of matching surfaces. In this case, the active site of the enzyme fits together only with the two amino acid substrates. The joining of substrates and enzymes lowers activation energy by placing the reactants in a position favorable for bonding. As the reaction continues, bonds in each amino acid are broken, a molecule of water is removed, and the two amino acids bond to one another, forming a dipeptide.

Figure 3-31 **This diagram shows the induced-fit model of an enzyme reaction. Notice how the shape of the enzyme changes and conforms to the shape of the substrates. Even so, a particular enzyme fits with only certain substrates. Thus, the induced-fit model is just a slight modification of the older lock-and-key model.**

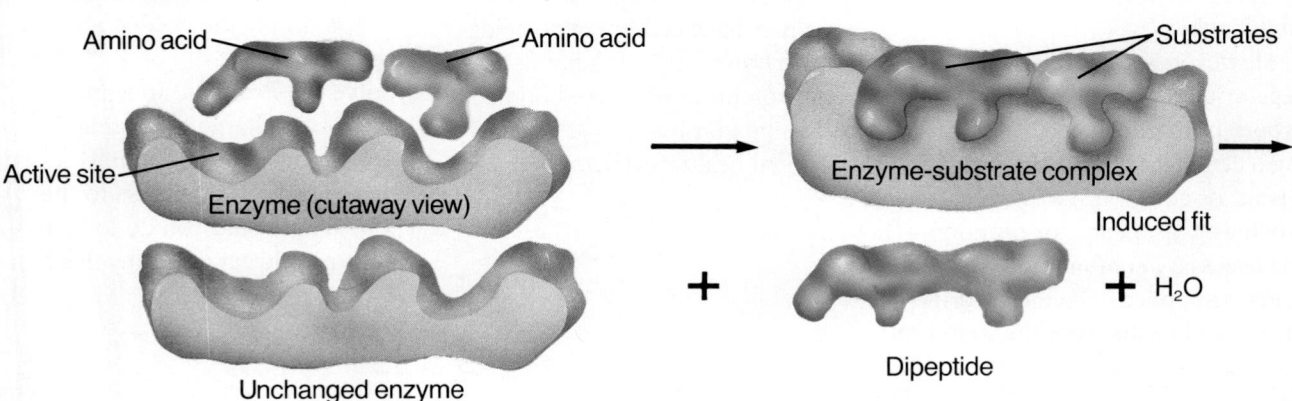

Using computers, scientists are identifying the structure of the active sites of some enzymes, as well as the part of the substrates that match those active sites. Such computer analyses not only verify the induced fit model, but also provide detailed information about how a particular reaction occurs.

Coenzymes

Just as people sometimes require helpers to carry out a job, enzymes sometimes require helpers in order to carry out reactions. These non-protein helper molecules are called **coenzymes.** Many coenzymes are made from vitamin molecules or fragments of vitamin molecules. Like enzymes, they are reusable and are needed in only small amounts.

Some coenzymes join temporarily with enzymes, changing the enzyme's active site and enabling it to fit with the substrate. Other coenzymes act as transfer agents, shuttling atoms, parts of atoms, or entire groups of atoms from one substrate to another.

LOOKING AHEAD

LIFE PROCESSES carried out by organisms involve chemical reactions and energy. Enzymes control these reactions, allowing them to occur at temperatures suitable to living systems. Most of the chemical reactions of organisms occur within cells. In the next two chapters, you will apply your knowledge of matter and energy as you explore the structure of cells and how the various parts of cells interact to carry out life processes upon which the cell and the entire organism depend.

Section Review

Understanding Concepts

1. You turn on a tape deck and the reels in the cassette begin to rotate. How does this example illustrate energy transformation?

2. Suppose you start a fire by using a magnifying glass to focus the sun's image on some paper. What is the source of activation energy for this reaction? What is the source of activation energy after the fire has started and the magnifier is taken away?

3. Explain how the enzyme maltase can control a reaction involving maltose, but not one involving sucrose.

4. *Skill Review—Sequencing:* Using Figure 3-31 as a reference, outline the sequence of events that occurs in an enzyme-controlled reaction. For more information, see Organizing Information in the **Skill Handbook.**

5. *Thinking Critically:* To function properly, cells must carry out thousands of reactions and do so at the appropriate times. How might the organization of cells be different if enzymes were not required for each of these many reactions?

Considering the Environment

Recycling Batteries

Batteries are wonderful! They provide us with a source of electrical energy that does not depend on a cord connected to a wall outlet. They make possible flashlights, cordless phones, boom boxes, digital watches, portable computers, hearing aids, smoke detectors, heart pacemakers, remote controls, and electric cars.

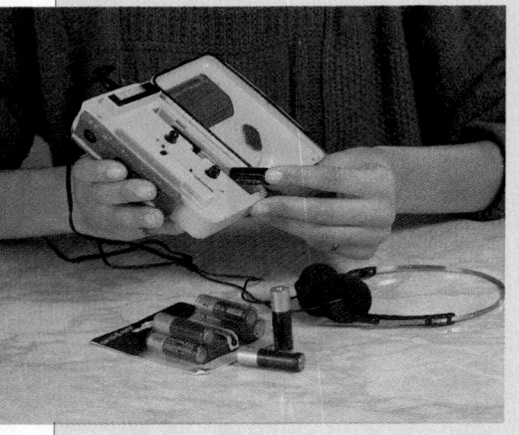

However, there is a dark side to batteries—some contain very toxic substances. Lead, mercury, and acids are just three of the most common. If these substances are ingested by humans, they can cause injury or even death.

Mercury, for example, can cause birth defects and brain and kidney damage. It doesn't take much mercury to be dangerous. The mercury in just one button battery from a wristwatch in six tons of garbage is enough to exceed government safety standards. What can be done?

One answer to this problem is recycling. When batteries are recycled, the environment is protected and, at the same time, scarce resources are conserved. Suppose you discard a car battery, which contains several lead plates and terminals. To build a new battery to replace the old one, new lead must be mined and then refined using great amounts of energy and causing considerable pollution at the mine and refinery. Even after all this, not all of the lead in the ore is recovered. However, reclaiming the lead from a recycled battery takes much less energy and causes less pollution. Moreover, the lead that would have been mined is still there for future generations.

Most stores that sell car batteries will offer a few dollars trade-in value for a battery that can be recycled. Many jewelry, camera, and department stores will recycle button batteries when they sell you new ones. At present, fewer than half of the states require stores to accept batteries for recycling. Most merchants are responsible and are proud to tell you of their efforts. Those who don't recycle may be encouraged to do so if their customers think it is important.

Another solution in some cases is to use rechargeable batteries. These batteries are usually made from nickel and cadmium. These elements are also toxic, and the battery costs more than a regular battery. However, a rechargeable battery will last up to 40 times longer. Thus, even when it is worn out, there is only about 1/40 of the waste.

1. **Connection to Literature:** The Mad Hatter in Lewis Carroll's *Alice in Wonderland* was not an entirely fictional character. People who made hats sometimes suffered from a disease that causes "madness." Research how hats were made at the time of the story and find out what caused the Hatter's madness.

2. **Explore Further:** The cause of much of the air pollution today in the United States is the internal combustion engine in automobiles. One proposed solution to this problem is electric cars powered by large, rechargeable batteries. What are some of the problems with battery-powered vehicles? Electric cars are sometimes referred to as nonpolluting. Do you think this is true?

Chemistry Connection

There's More Than Food in Your Food

Pick up any box of crackers and look at the ingredients label. Sometimes it looks as if the entire chemical industry was involved in making that cracker. Actually, there is a plan behind that list of ingredients, but it's not evil. One of the reasons for the list is that there is a problem with food—it spoils. This is not a worry if you live where you can grow food all year. You'll always have fresh food. However, if you live where there are seasons, you can eat fresh food in summer and fall, but you'll have to depend on preserved food the rest of the year. Food preserved in September will probably have to last until the following June or July.

The earliest food additives were preservatives. Natural materials, such as salt, were added to meat and fish to keep them from spoiling. Salt is still used today along with hundreds of other additives. As a result, a cracker made in October is still fresh in April. Additives combined with packaging methods, such as canning, allow us to enjoy foods like pineapple from Hawaii at reasonable prices in and out of season.

Some food additives are vitamins and minerals added to foods that do not naturally contain them. Before food additives, diseases such as rickets and scurvy were common, especially among children. Rickets

is a disease in which bones do not develop naturally and become soft. It is caused by a lack of vitamin D. Enriching milk, bread, and other foods with vitamin D has nearly wiped out rickets.

Iodine is needed for proper thyroid function. Lack of iodine causes a disease called goiter. Seafood contains iodine, so coast dwellers don't usually get the disease. People who don't have regular access to seafood can avoid goiter by eating foods that contain an iodine additive. The most common iodized food is salt.

But not all additives are safe for everyone. For example, monosodium glutamate, called MSG, is sometimes added to make foods taste better. However, a small percentage of the population is allergic to MSG. Another example is sulfites, additives used to keep green vegetables green and crisp. Sulfite-treated food can cause a serious allergic reaction in some people.

Food processing companies are regulated by a federal agency called the Food and Drug Administration (FDA). The FDA requires extensive testing of additives before they are allowed to be used. Even so, an additive may be approved, but later be found to be unsafe. In that case, the FDA can order that substance banned and all foods containing it destroyed.

1. **Explore Further:** The FDA has a special responsibility to ban food additives that might cause cancer. Find out about an additive called red dye number 2. What was it used for? What research led to its being banned?

2. **Connection to History:** In the days of sailing ships and the far-flung European colonial empires, naval vessels were at sea for many months at a time. During these voyages, many sailors used to come down with a disease called scurvy. Research the symptoms, cause, and cure of this disease.

CAREER CONNECTION

Food technologists research the chemical, physical, and biological properties of food. They do this to learn how to properly process, preserve, package, distribute, and store food. Some food technologists develop new products, while others concentrate on ensuring quality and safety standards.

Summary

Atoms are the basic building blocks of all matter, both living and nonliving. Atoms combine with one another, producing more complex arrangements of matter, such as molecules and ionic compounds, which have definite compositions. As a chemical reaction occurs, bonds are broken, new bonds are formed, and new substances having different properties are produced.

Substances dissolve in water to form solutions, which are mixtures that are the same throughout. Solutions may be acidic, basic, or neutral depending on the relative concentrations of hydrogen and hydroxide ions.

Biologically important organic compounds are covalently bonded and include carbohydrates, lipids, proteins, and nucleic acids. These molecules are formed from and can be converted to simpler molecules by condensation and hydrolysis reactions.

There are two types of energy—potential and kinetic. Energy is never lost during energy transformations.

Every chemical reaction requires a certain amount of activation energy to begin. In cells, enzymes lower the activation energy to a point at which reactions can occur at lower temperatures. Coenzymes often work along with enzymes in biological reactions.

Language of Biology

Write a sentence that shows your understanding of each of the following terms.

acid	ion
activation energy	ionic bond
amino acid	isomer
base	kinetic energy
carbohydrate	lipid
chemical formula	molecule
coenzyme	nucleic acid
compound	polypeptide
covalent bond	polysaccharide
electron cloud	potential energy
element	protein
energy level	solution
enzyme	

Understanding Concepts

1. Explain why molecules and ionic compounds have definite composition.
2. How many different kinds of atoms are there in $NaHCO_3$? How many atoms are in $4PbI_2$?
3. Why must a chemical equation be balanced? Is the following a balanced equation?

$$N_2 + 3H_2 \rightarrow 2NH_3$$

4. Potassium iodide, KI, is an ionic compound. Suppose you add some KI crystals to a beaker of water. Describe and explain what happens as the crystals dissolve. Will sugar crystals dissolve the same way? Explain.
5. A chemist tests two solutions. One has a pH of 6, the other a pH of 10. In which solution is the concentration of hydrogen ions less than the concentration of hydroxide ions?
6. Which of the following are not examples of chemical changes? Explain your reasoning.
   (a) A piece of paper burns. (b) Ice melts. (c) A twig is snapped in two. (d) A cell converts sucrose to glucose and fructose.
7. Classify each reaction below as either hydrolysis or condensation.
   (a) conversion of starch to glucose
   (b) synthesis of a polypeptide from amino acids
   (c) production of a fat from fatty acids and glycerol
8. Ethane, C_2H_6, is a gas composed of molecules. What kind of bonds are present in ethane?
9. Using diagrams, show how an enzyme would carry out the hydrolysis of maltose.
10. How does an enzyme lower the amount of activation energy needed to begin a reaction?

REVIEW

Applying Concepts

11. Ion X has a charge of 3+. Ion Y has a charge of 1−. In what ratio would the two ions combine to form a compound? Write its formula.
12. What atomic particles are most important in determining chemical reactions? Explain.
13. Pentane, C_5H_{12}, is an organic compound containing single covalent bonds. Draw structural formulas for two isomers of pentane.
14. What is incorrect in this structural formula?

$$H - \overset{\overset{\displaystyle H}{|}}{C} = \overset{\overset{\displaystyle H}{|}}{\underset{\underset{\displaystyle H}{|}}{C}} - \overset{\overset{\displaystyle H}{|}}{\underset{\underset{\displaystyle H}{|}}{C}} - H$$

15. Each enzyme operates most efficiently at a particular pH. For example, pepsin breaks down proteins in the stomach, where the pH is about 2. However, pepsin could not break down protein in the intestine, where the pH is about 8. Can you suggest an explanation of how pH might affect enzymes?
16. **Chemistry Connection:** Look up information about the sweetener aspartame. Is it chemically related to proteins, carbohydrates, or fats? Aspartame is not used in foods to be cooked because it breaks down when the food is heated. What sort of reaction do you think would cause it to break down?
17. **Health Connection:** It is generally advisable for vegetarians to eat combinations of foods that provide a complete set of amino acids at a single meal instead of eating one food in the morning and another in the evening. Why do you think this advice is valid?

Lab Interpretation

18. **Investigation:** You have tested the pH of the water of six ponds in your locality. The results you obtained are shown in the following graph. You know that a farmer in your area has been applying lime (a base) to the soil. In another location, an abandoned coal mine is producing some acid runoff. According to your results, which ponds are most likely to have been affected by the lime and the mine runoff? Explain.

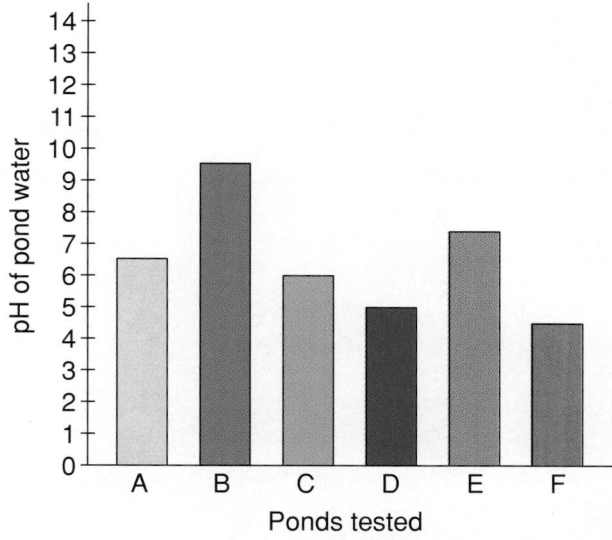

19. **Minilab 1:** You are trying to figure out whether unlabeled cans of soda pop are diet pop or regular pop. What could you do to figure it out without opening the cans?
20. **Minilab 2:** A contractor has promised to install marble tile in your kitchen. Upon examining the tile, you believe that the tile is really plastic. Describe a chemical test you could carry out to determine whether the tile might be marble. Would a positive test prove that the tile is really marble? Explain.
21. **Thinking Lab:** A certain fly makes one toxic chemical during its larval stage and another during its adult stage. What would you predict about the predators of each stage of the life cycle of this fly? Why?

Connecting Ideas

22. In Chapter 1, you encountered the word *photosynthesis*. What sort of energy transformation takes place during photosynthesis?
23. You will learn later how DNA forms a code for the manufacture of proteins. Suppose a change in an organism's DNA causes a change in the amino acid sequence of a certain enzyme. The resulting enzyme has unusual properties. How might such a change be important to a group of organisms?

THE CELL AND ITS ENVIRONMENT

SOME HOMES are surrounded by fences—tall or short, wooden or metal or brick, plain or fancy. Fences can have several functions. They can be decorative. They can define property boundaries. They can enclose and protect the home and family within. Although fences may differ in form and function, they have one thing in common—gates. A gate is a passage through which people and materials enter and leave the home. Without a gate, people within the home could not survive, because food and other supplies could not be brought in and waste materials could not leave.

Cells, too, are enclosed by "fences." Every cell is surrounded by a membrane, such as the one shown in the electron micrograph opposite. A cell's membrane func-

tions to protect and enclose the cell structures within. It has gates—some permanent, some temporary—that allow materials into and out of the cell. But not all kinds of particles can enter the cell, and not all kinds can exit. The membrane is selective. It controls the traffic of substances passing in and out of the cell. Water and dissolved substances necessary for maintaining the life of the cell enter through the membrane, and poisonous wastes exit. If needed molecules cannot pass through the membrane on their own, a cell can help get those particles through. How? In this chapter, you will read about the structure and function of the cell membrane and how this structure is vital to the survival of all organisms.

When you sweat, your body loses water and dissolved substances. Sport drinks can replenish the body's supply. Where do these substances go when you drink them? How do they get through the barriers surrounding your cells?

Magnification: 8000×

Chapter Preview

Objectives

Describe the discovery of cells and the development of the cell theory.

Relate the properties of the cell membrane to its structure.

4.1 Membrane Structure

In earlier chapters, you learned about the great diversity of organisms on Earth today. You had a glimpse of the variety of organisms in each of the five kingdoms. But remember that all the different life forms have much in common. The obvious external differences among such organisms as alligators, toadstools, and rosebushes might not suggest that different kinds of organisms share many features, but internally they share a very basic characteristic. All are composed of the same fundamental units of life—cells. As brick buildings can look very different, so, too, can organisms. The buildings are all built of bricks; the organisms are all built of cells. Bricks are lifeless clay, but cells are alive with activity. The eyes reading this page are made up of living cells. What do you suppose is going on in those cells right now?

The Cell Theory

You have probably known about cells for many years and take for granted the idea that they are fundamental parts of organisms. However, this idea took nearly 200 years to develop. The first discovery of cells was recorded in 1665. Robert Hooke, an English scientist, used a primitive microscope to examine razor-thin slices of cork, part of the bark of an oak tree. He saw, as in Figure 4-1, that cork is composed of many similar compartments, empty cubicles that he named cells. They were so tiny that he estimated there to be more than a billion cells per cubic inch of cork.

The cork cells that Hooke observed were dead. He actually saw only their stiff outer borders, which he named walls. Hooke's discovery spurred further research by others. Limited by poor microscopes, however, these researchers made slow progress. Gradually, discoveries by a number of scientists led to a more accurate picture of living cells. It was

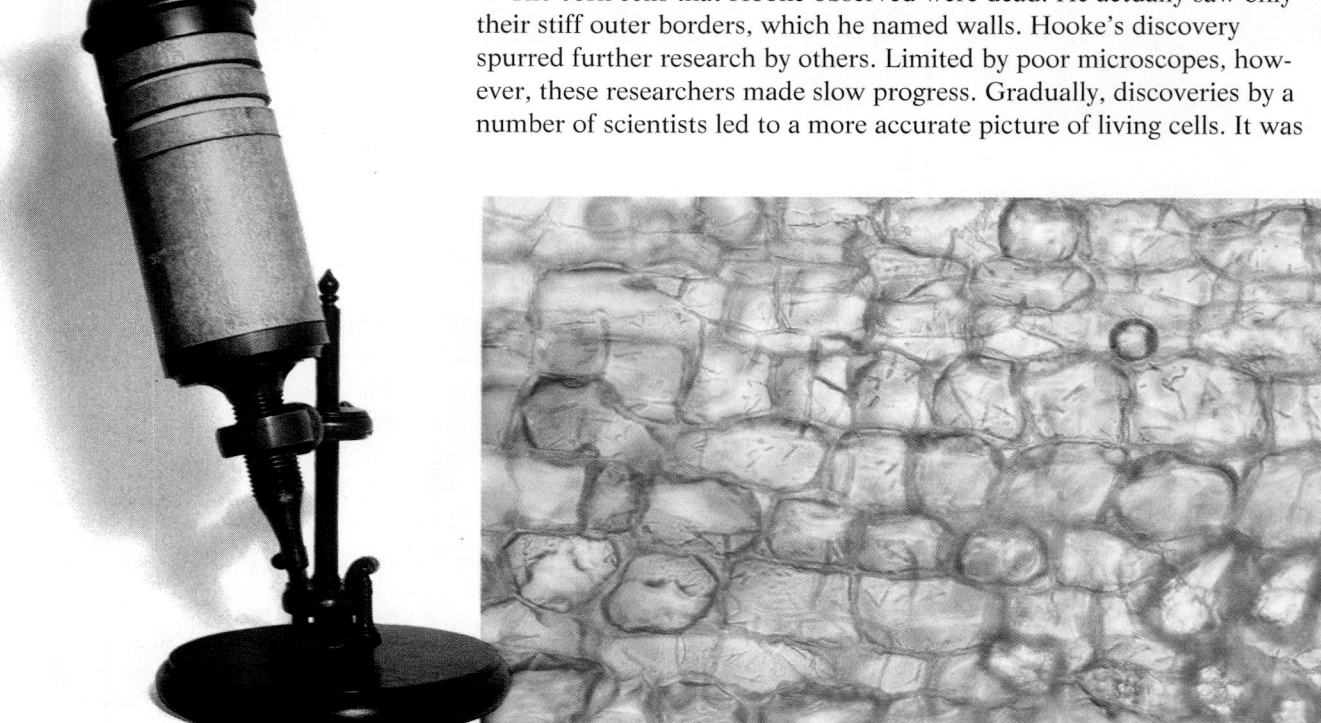

Figure 4-1 **Robert Hooke designed this microscope (left) to observe cork cells (right). The microscope has three lenses and is beautifully finished in gold-embossed leather. Magnification (right): 250×**

learned, for example, that cells are filled with a jellylike fluid and that most cells house a round, central structure that stands out from the rest of the cell. Although these individual discoveries were important, the idea of cells being the basic structural units of all organisms was a long time coming. In 1838, Matthias Schleiden, after completing a lengthy study of plant structure, concluded that all plants are composed of cells and that cells are the basis of a plant's functions. A year later, Theodor Schwann concluded the same things about animals. Then, in 1858, another important generalization was made. Rudolf Virchow stated that all cells come from other living cells.

The discoveries of other scientists and the generalizations made by Schleiden, Schwann, and Virchow came to form the basis of one of biology's most important concepts, the **cell theory.** The cell theory states that:

1. All organisms are composed of one or more cells or cell fragments.
2. The cell is the basic unit of structure and function in organisms.
3. All cells are produced from other cells.

You learned that a theory is a major hypothesis that has stood the test of time. Because it applies to all organisms, the cell theory is a cornerstone of modern biology. It is important in another way, too. The cell theory fits the theory of evolution: All organisms are descendants of an original organism, which was a single cell. As organisms evolved, they retained and shared the cell as the basic unit of structure and function.

Properties of the Plasma Membrane

Each cell, whether an entire organism or a small part of an organism, lives in a fluid environment. Amoebas are surrounded by the water of a lake or pond. Most cells of land plants are bathed by water that enters the plant from the soil. Your cells are in contact with fluids as well. Blood and tissue fluids bathe each of the cells in your body. Cells must obtain materials from and release substances into their moist environment.

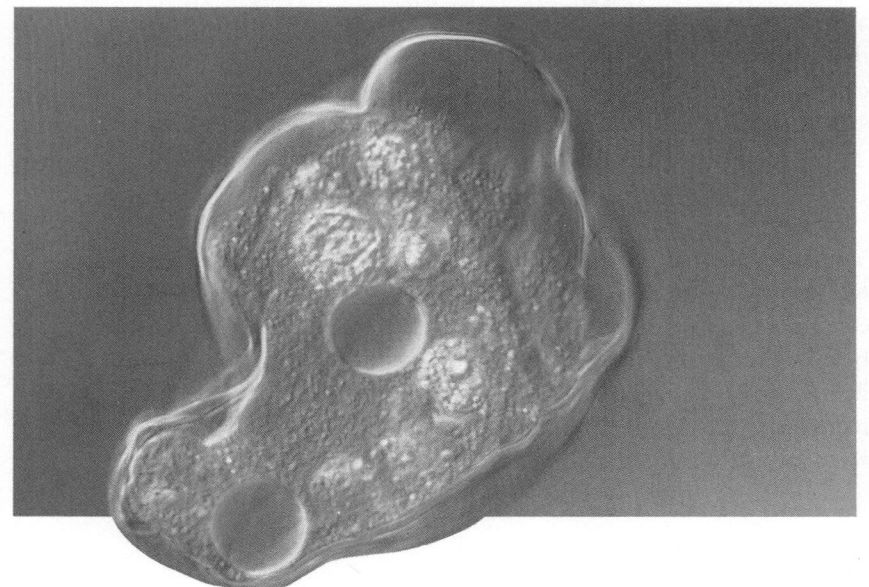

Figure 4-2 **All cells are surrounded by a membrane, which separates the cell contents from the outside environment. This amoeba is a single-celled organism. Magnification: 160×**

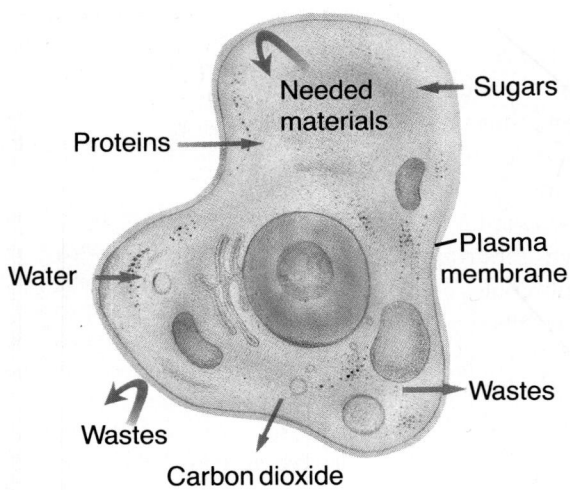

Proteins

Needed materials

Sugars

Water

Plasma membrane

Wastes

Wastes

Carbon dioxide

Figure 4-3 **A cell must remove wastes and keep other wastes from entering. Likewise, it must take in and keep needed materials while maintaining a balance with its surroundings.**

The Plasma Membrane   If each cell is a separate unit of life, it must have a boundary separating it from other cells and from its environment. The outer boundary of a cell, the **plasma membrane,** encloses the cell's contents. The membrane is also the cell's gatekeeper. It regulates which particles can enter and leave the cell. Proper cell functioning depends on passage of materials across the plasma membrane. For example, amino acid and glucose molecules move from an animal cell's environment, across the membrane, and into the cell's interior. Once inside, the amino acids can be linked to form polypeptides, and the glucose can be used in cellular respiration. Carbon dioxide produced during respiration passes from inside the cell, across the membrane, and out into the environment. These functions of the plasma membrane are directly related to its structure.

Selective Permeability   Long before much was known about the structure of the plasma membrane, biologists had learned a great deal about its properties. They knew that only certain molecules were able to pass through, or permeate, the membrane. Such a structure is called a **selectively permeable membrane.** A tea bag also is a selectively permeable membrane. It is permeable to water molecules but not to tea leaves. This property is important, because only those particles that are used by the cell or that must be expelled should cross the membrane. If all particles could pass in or out freely, the cell could not maintain its organization. The chaos would be like a parking lot without the white lines and painted arrows, with cars driving in and out anywhere and parking any which way.

Cells differ in terms of the particles they use and those that must be expelled. Thus, different types of cells are permeable to different particles. In addition, the permeability of a membrane to a certain particle may change from time to time.

Certain particles can cross the plasma membrane more freely than others. Small particles can pass through more easily than large ones. Lipids and particles that are soluble in lipids can cross the membrane more readily than particles that are insoluble in lipids. In general, small ions do not cross the membrane as easily as uncharged particles of about the same size, and some ions can permeate the membrane more readily than others.

A Model of the Plasma Membrane

You learned from your study of enzymes that scientists often propose models to explain known properties. Biologists have offered several models of plasma membrane structure over the last 50 years. The models have been revised as new ways of investigating the plasma membrane have been developed. Since the 1950s, electron microscopes and improved techniques for biochemical analysis have been especially important in studying the plasma membrane.

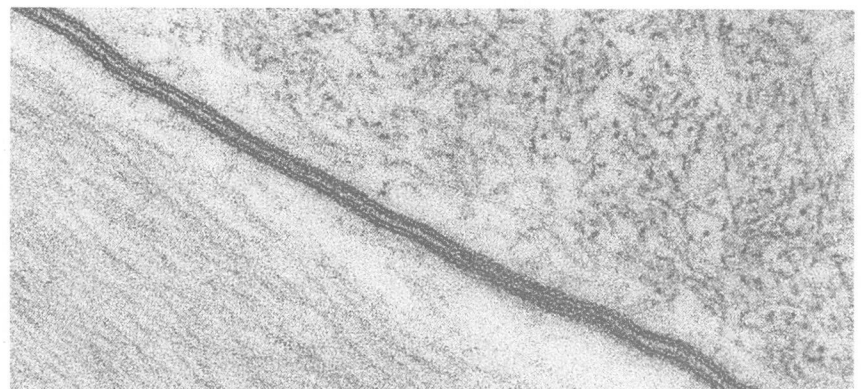

The Lipid Bilayer When viewed through an electron microscope, the plasma membrane appears as two thin lines, as in Figure 4-4. Chemical analysis has shown that the membrane consists of a lipid bilayer, made up of two layers of molecules called phospholipids. The phospholipids composing each of the two layers are different. Look back at Figure 3-23. The diagram represents a molecule of a lipid, made up of one molecule of glycerol joined to three fatty acid molecules.

Now look at Figure 4-5b. One of the fatty acid molecules has been replaced by a group of atoms called a phosphate group. This group is attached to the part of the glycerol molecule away from the fatty acids. Fatty acids are nonpolar molecules and do not dissolve in water. The phosphate group, however, is polar and easily dissolves in water. These properties determine the way the two layers of phospholipids are arranged. Because they are soluble in water, the polar phosphate heads of the phospholipids point toward the outside and inside of the cell, where water is plentiful. The nonpolar fatty acid tails of the phospholipids point away from water, toward the middle of the membrane. As a result, the two layers of phospholipids face one another, as Figure 4-5a shows. This arrangement of phospholipids into two layers is a result of their physical and chemical properties when in a watery environment. The layers form spontaneously. Once formed, the bilayer in membranes can be maintained without expenditure of energy by cells.

Figure 4-5 Two layers of phospholipids form the structure of a membrane (a). Each phospholipid has a polar head and two nonpolar tails (b). Which end of the molecule will dissolve in water?

b

Polar head

Phosphate group

Fatty acids

Phospholipid molecule

a

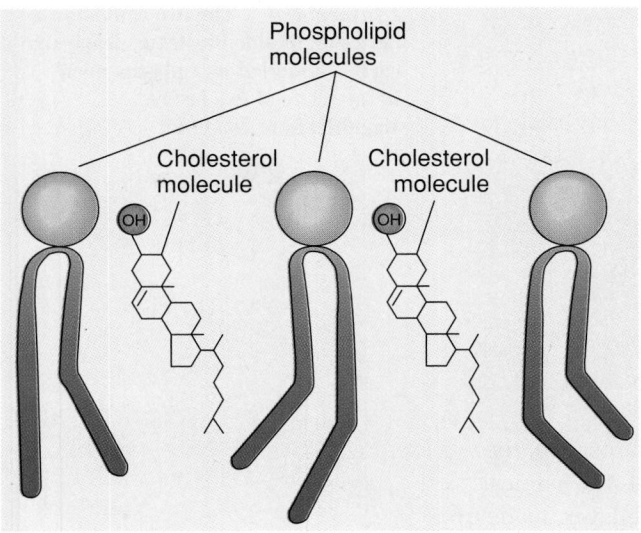

Figure 4-6 Cholesterol molecules embedded in the bilayer stabilize the plasma membranes of some eukaryotic cells and make them more rigid. Cholesterol also helps to keep the fatty acid tails apart.

This lack of energy use by the cell does not mean that the membrane is inactive. If you could see the individual molecules of the membrane, you might be surprised to see that the membrane is a dynamic structure. Within each layer, the fatty acid tails can wag back and forth, and entire phospholipids can glide sideways through their layer. These properties make the bilayer act like a very fine film of oil. Rigid molecules of cholesterol are embedded in the bilayer of many cells, where they bind to and restrict movement of the phospholipids (Figure 4-6). This binding makes the plasma membrane strong, as well as flexible.

Membrane Proteins Scientists have long known that the plasma membrane contains proteins in addition to phospholipids. Until the 1970s, it was thought that the two lines seen with an electron microscope represented protein coats on either side of the bilayer, and that the membrane was like a lipid sandwich between two "slices" of protein. A new technique showed this to be an incorrect interpretation. Plasma membranes were separated from the rest of cells and then frozen. The frozen membranes were then fractured to expose the area between the two lines. Examination of the fractured membranes revealed many bumps in the phospholipid bilayer, as seen in Figure 4-7. These bumps are proteins.

Results of the freeze-fracture technique have shown that proteins are scattered throughout the bilayer, like raisins in a slice of raisin bread. Some proteins are partly or totally embedded in the bilayer. Others lie on the surface of either the outer or inner phospholipid layer. Like phospholipids,

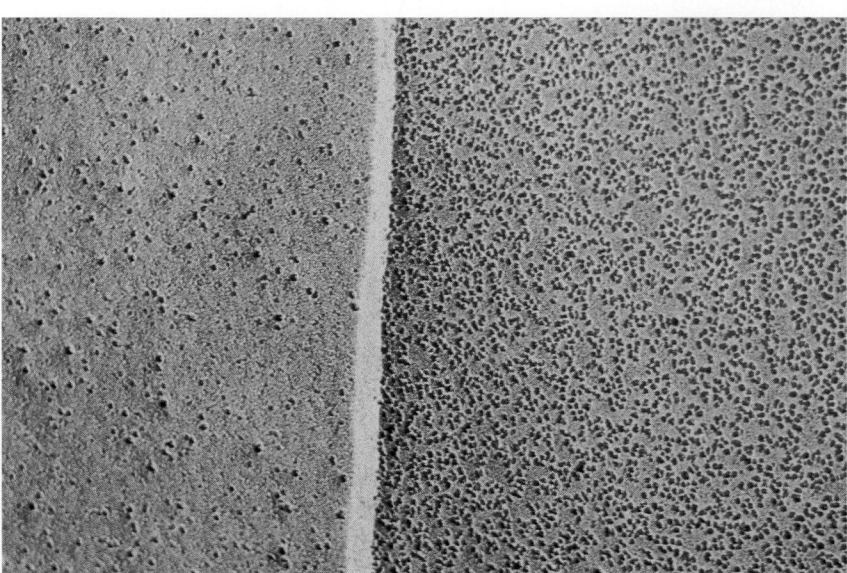

Figure 4-7 The interior surface of a red blood cell membrane can be seen using an electron microscope, after the membrane has been prepared by the freeze-fracturing technique. The bumps are proteins embedded in the membrane. Magnification: 100 000×

proteins have polar and nonpolar areas that determine their position in the bilayer. Proteins, too, can migrate sideways within the bilayer, but not as rapidly as the phospholipids. Because components of the membrane are constantly moving, like a fluid, and because the membrane has a pattern of proteins embedded in it, this current model of membrane structure is called the **fluid mosaic model.** This structure, shown in Figure 4-8, is closely related to the membrane's function.

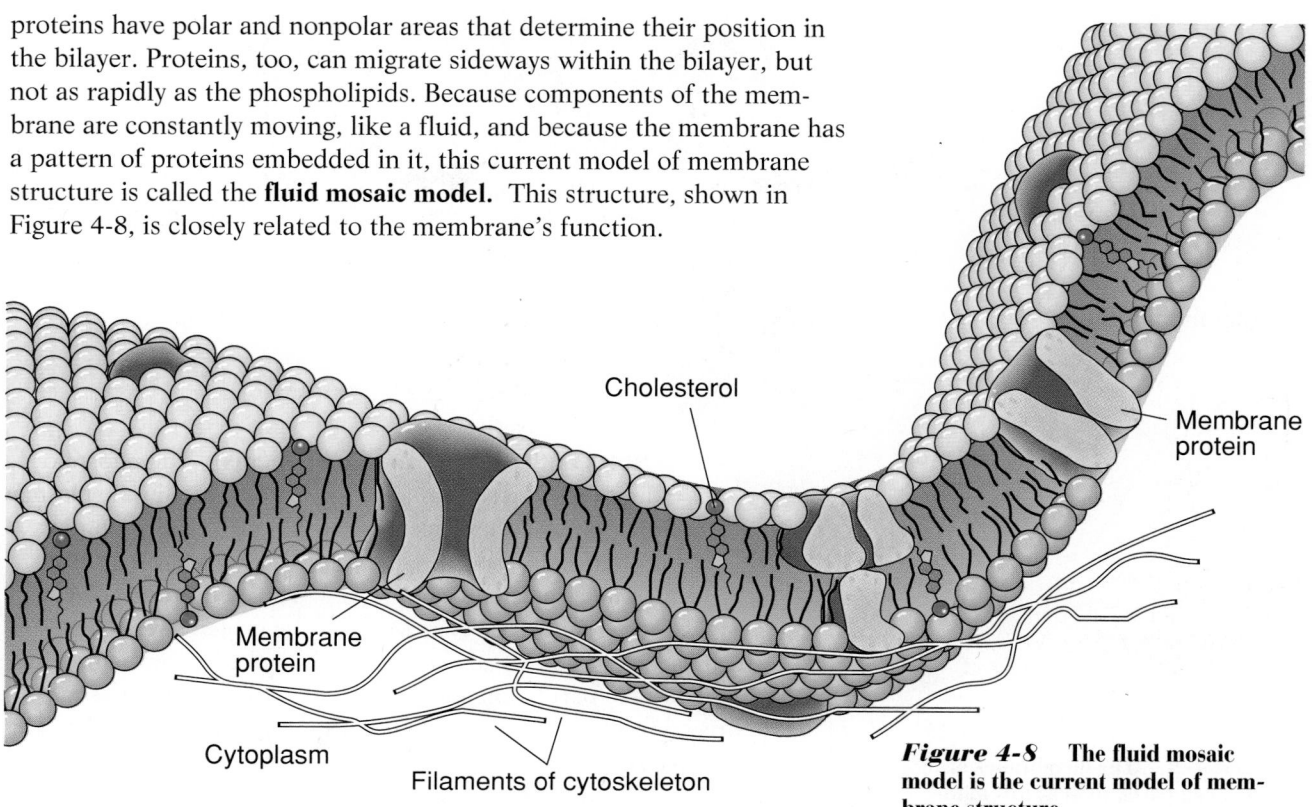

Cholesterol

Membrane protein

Membrane protein

Cytoplasm

Filaments of cytoskeleton

Figure 4-8 **The fluid mosaic model is the current model of membrane structure.**

H I S T O R Y C O N N E C T I O N

Ernest Everett Just

· · · · · · · · · · · · · · · · · · · ·

As a teenager, Ernest Just (1883-1941) left South Carolina with few academic skills other than reading. He completed high school and college in New Hampshire, graduating with honors in 1907, and joined the teaching staff of Howard University.

Just devoted his life to research in cell biology and embryology. He studied fertilization in marine organisms and the importance of the cell surface in the development of these organisms. While conducting this research, he began to realize that all parts of a cell were important to the cell's activity. Just came to believe that the plasma membrane was an active part of the cell, not just a passive covering. He also believed that the cytoplasm was as important as the nucleus in determining a cell's activities. These ideas were controversial at that time. (Remember that much of the information we have today about cells was not known at that time.)

Because of his unpopular views, Just and his work were not fully appreciated in his lifetime, even though he was later shown to be correct. He did, however, publish numerous articles and authored two books, including *The Biology of the Cell Surface.* In 1915, he was the first person to receive the Spingarn Medal, awarded annually to an African-American who has had outstanding achievement in his or her field.

———

1. **Explore Further:** Just's belief that all parts of a cell influence the cell's activities was considered controversial when it was first proposed. Find out what the popular scientific belief of the time was. Compare this belief with what scientists currently believe.
2. **Writing About Biology:** Write a short paragraph explaining why you think the plasma membrane was once considered to be a passive structure.

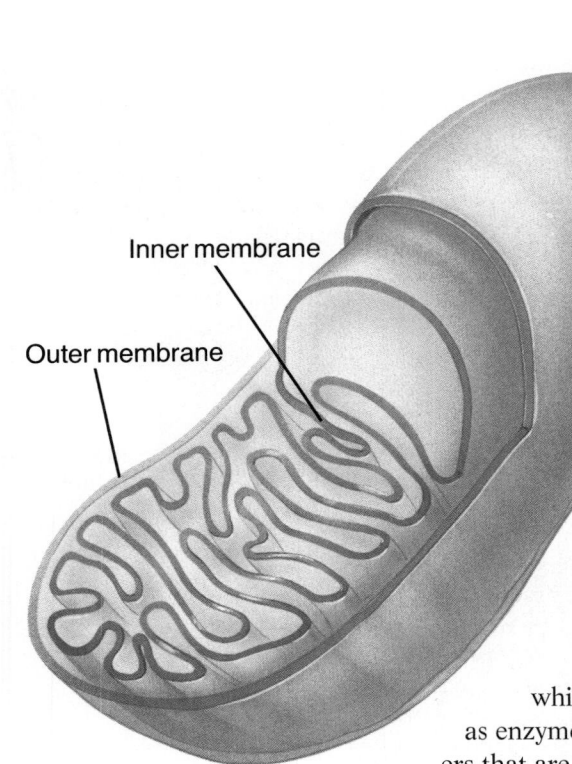

Inner membrane

Outer membrane

Figure 4-9 The membranes of internal cell parts, such as this mitochondrion, also have a fluid mosaic structure. The photomicrograph shows the internal structure of the mitochondrion, magnified about 81 000 times.

Roles of the Membrane Components The phospholipid bilayer serves mainly a structural role in the plasma membrane and forms most of the cell's outer boundary. Its chemical properties allow it to act as a barrier across which most particles cannot pass. The membrane proteins play more functional roles. Many of the proteins regulate which particles can pass across the membrane. Other proteins serve as enzymes and take part in chemical reactions. Still others act as markers that are recognized by chemicals from both inside and outside the cell. Some of these markers are involved in fighting disease. Other markers are important in recognition of other cells. Thus, in many respects, the plasma membrane can be thought of as a communication center.

Most cells have internal structures that are themselves enclosed by membranes. These internal membranes also have a fluid mosaic structure. The kinds and arrangements of the phospholipids and proteins vary from one such membrane to another. Thus, each membrane has its own distinct permeability characteristics. These differences are important because, as you will learn in Chapter 5, each cellular part carries out its own set of chemical reactions. The membrane of each cell part lets in and out those particles used or produced by that part's particular functions.

Section Review

Understanding Concepts
1. Explain how the plasma membrane is like a liquid.
2. Could different cells in your body perform different functions if their membranes were all permeable to exactly the same materials? Explain.
3. Why did it take almost 200 years after Robert Hooke's discovery for scientists to develop the cell theory?

4. **Skill Review—Making Tables:** Construct a table comparing how easily the following particles pass across membranes: lipids vs. non-lipids; small particles vs. large particles; neutral particles vs. ions; particles soluble in lipids vs. particles not soluble in lipids. For more help, refer to Organizing Information in the **Skill Handbook.**

5. **Thinking Critically:** Suppose you added some phospholipids to water. If you could see individual molecules, how would the phospholipids be arranged? Explain.

4.2 Membrane Function

Now that you have a picture of the plasma membrane and its structure, think again about its role as a gatekeeper. Keep in mind that the membrane is selectively permeable. You can make use of what you've learned about the membrane's structure to understand how particles enter and leave cells and why all membranes are not permeable to the same substances.

Diffusion

Suppose you had a collection of 500 marbles, which you kept neatly in an open box on a table. That is a very organized, or nonrandom, arrangement. Marbles do not place themselves neatly in a box. They are there because of the energy you used to put them there. Now, suppose the box is knocked over and falls to the floor. Out come the marbles, bouncing off one another, the floor, the furniture, and the walls. Eventually, the marbles stop moving. Their arrangement now is anything but organized. In fact, it is quite random. In nature, there is a tendency toward randomness. Energy is required for nonrandomness. Think about the organization of your room or locker. Is it random or nonrandom? Do your socks place themselves neatly in a drawer? Do your books line themselves up on a shelf? What must you do to achieve greater organization of your socks and books?

Random Movement The example of the marbles illustrates an important physical process. Placing some blue crystals of copper sulfate, an ionic compound, into a beaker of water shows the same process, Figure 4-10. At first, the crystals settle to the bottom, and the water is colorless. After a time, the water near the crystals appears blue, and gradually the blue color spreads throughout the water.

Figure 4-10   Copper sulfate, a blue crystalline compound, diffuses through water, distributing copper and sulfate ions evenly throughout. The solution turns bluer as diffusion proceeds.

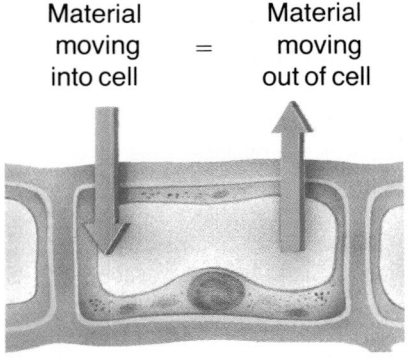

Material moving into cell = Material moving out of cell

Figure 4-11 **In dynamic equilibrium, there is motion but no net change.**

When copper sulfate is first added to water, the copper ions and sulfate ions are all concentrated in one area—in the crystals. The position of the ions in the water is nonrandom. With time, though, the ions begin to separate. They move about randomly, colliding with one another, with the water molecules, and with the sides and bottom of the beaker. At each collision, ions bounce off and move in new directions. Random movement of the ions results finally in their uniform distribution throughout the water. Unlike the marbles, which stop moving because of friction, the ions continue to move about. The water becomes and stays blue because equal numbers of copper ions are entering and leaving any given region of water at any given time. The position of the ions in the water has become more random.

Diffusion of Particles   Random movement of ions and other particles is called **diffusion.** Diffusion results in the net movement of particles from a region of greater concentration to a region of lesser concentration. With time, the particles become evenly distributed throughout the container. After this point, particles continue to move randomly and collide with one another, but the numbers of particles entering and leaving an area are equal. This condition, in which there is continuous movement but no overall change, is called **dynamic equilibrium,** Figure 4-11.

Particles may diffuse through the air, through water and other liquids, and across membranes. For example, the molecules that give toast its aroma

GLOBAL CONNECTION

Fish Out of Water

Freshly caught fish is a real treat to eat. The fresher it is, the better it tastes. It doesn't take long, however, before unpreserved fish starts to smell bad and taste worse. Most spoilage is caused by the growth of bacteria and other microorganisms. In the days before refrigeration, fish-eating people around the world came up with other ways to preserve seafood. The most common were drying and salting. These methods are still used today.

Drying   Water is vital to most biological processes. If you remove most of the water from in and around fish cells, microorganisms can't grow—at least not well. The seafaring people of Northern Scandinavia are old masters of the art of drying fish. So are the Eskimos. But perhaps the most dramatic example of preservation by drying comes from the Native Americans of pre-Columbian Peru.

Living high in the Andes Mountains, they would freeze fish in the frigid night air and dry it in the hot daytime sun. In the daytime heat, the low atmospheric pressure of the high mountains allowed

much of the fish meat's water, frozen during the night, to turn directly from ice to vapor. This was freeze-drying, a technique we consider very modern.

Salting   Salting is related to drying in that it works by affecting water in cells. In the salting process, however, the water under attack is in the microorganism. When salt is added to meat, the watery environment of the cells is loaded with a high concentration of ions. The water inside the microbes' cells contains a lesser concentration of ions. Water flows out of the microbes' cells into the surrounding environment, restoring osmotic

waft by diffusion from near the toaster, where they are more concentrated, through the air to your nose at the breakfast table, where they are less concentrated. The bumping and colliding of the air and molecules from the toast move the particles outward from their source. Eventually, dynamic equilibrium is reached and the smell of toast fills the entire kitchen.

Diffusion Through Membranes   Diffusion is a physical process that explains how some particles permeate a membrane. Particles that can diffuse across cell membranes include lipids and those particles that dissolve in lipids, such as oxygen and carbon dioxide. As an example, consider a plasma membrane surrounded by and permeable to small lipid molecules. Assume more lipid molecules are outside than inside the cell. Because they are in random motion, the molecules bump into one another and into the membrane. So many of the molecules strike the membrane that most pass through it. A greater concentration of the lipid molecules on the outside means a greater chance of their hitting and passing across the membrane from the outside than hitting and crossing the membrane from the inside. As the number of lipid molecules inside the cell increases, the chance of lipid molecules leaving the cell increases. Eventually, dynamic equilibrium occurs. The molecules continue to move, but the numbers entering and leaving the cell are equal. A net movement from a region of greater concentration outside the cell, to a region of lesser concentration inside the cell, has occurred.

balance. The microbes dry up and die.

Europeans and Americans first used salting in the late 1600s to preserve red meat during shipping. This method was later used for preserving fish. Unfortunately, salting wasn't as well suited to preserving fish as it was to preserving red meat. Most of the fish shipped came from salt waters. The microbes that attack saltwater fish are quite at home in a salty environment. So, while a 2 percent salt solution would easily preserve red meat, fish required a 12 percent salt solution.

Brine   If salt and vinegar are combined to make a strong brine, however, a much more effective fish preservative is produced. Most microbes that attack fish can't survive in such an acidic environment. Brine solutions have long been used in Eastern Europe and Scandinavia. The pickled herring you find in the supermarket is preserved in brine.

1. **Connection to Health:** If you had to preserve food without refrigeration, do you think that drying or salting would be healthier? Why?
2. **Explore Further:** The enzymes in fish cells remain active at normal room temperatures, unlike the enzymes in meat from mammals. How does this fact influence the keeping quality of fresh fish?

3. **Writing About Biology:** Another method of preserving food is by smoking. Learn more about this method of food preservation and write a short paper about the subject.

Minilab 2

Are plastic bags permeable to certain molecules?

Place the following into separate plastic bags: food coloring in water, canned beet juice in water, iodine in water, mouthwash, and cough syrup. Seal the bags. Place each bag in a separate small beaker of water. Observe the color of the beaker water 24 hours later.

Analysis *What conclusion can you make regarding permeability of plastic bags to those molecules tested? How is a plastic bag similar to a plasma membrane?*

Osmosis

The diffusion of water into and out of cells across a selectively permeable membrane is so common that it is given a special name, **osmosis.** For any cell to survive, it must be in osmotic balance with its watery environment. Osmotic balance occurs when the movement of water into and out of a cell is equal, that is, when dynamic equilibrium occurs.

Osmotic Balance in the Blood   Consider a red blood cell, which is in osmotic balance with its surrounding liquid, blood plasma. The cell and the plasma both contain water and dissolved substances. Although not necessarily the same, the substances are in equal concentrations in the plasma and in the cell. Thus, the concentration of water is the same in the plasma and in the cell. If the blood cell is removed from the plasma and placed in pure water, however, it is no longer in osmotic balance. The pure water contains no dissolved particles, so the concentration of water outside the cell is greater than inside. A net movement of water molecules into the cell begins. So much water enters so quickly that the blood cell swells and bursts, Figure 4-12b. This is why IVs given to patients are usually solutions of sugar and salts rather than pure water.

Osmotic Balance in Freshwater Plants   Red blood cells are not normally surrounded by pure water. They have no adaptations that enable them to adjust to the difference in concentration of water, and so they burst. Many cells, however, do normally live in a freshwater environment, and they have evolved ways of maintaining osmotic balance. For example, *Elodea* is a freshwater plant often seen in home aquariums. Like the red blood cell in pure water, cells of *Elodea* contain water with dissolved particles, and so water tends to move into the cells. However, the cells do not burst. The net movement of water inward is slight but does not continue, because the cell's tough outer wall balances the pressure of the small amount of water that enters. By pressing inward to counterbalance the outward pressure of the water, the cell wall prevents additional water from entering the cell. This adaptation—the cell wall—enables the cells to survive. Cells of land plants maintain osmotic balance in the same way.

What would happen if an *Elodea* plant were placed in salt water? The concentration of water molecules would be greater inside the cells than outside. As a result, there would be a net movement of water outward. Such loss of water causes the cell contents to shrink, Figure 4-12c. Continued water loss may lead to the death of cells and of an entire plant.

Diffusion and Osmosis as Passive Transport   Osmosis and diffusion are processes by which water, lipids, and lipid-soluble particles permeate membranes. These substances either are able to dissolve in the lipid bilayer or are small enough to squeeze between the phospholipids in the membrane. Their passage across membranes is due to their net random motion from a region of greater concentration to a region of lesser concentration. The cell plays no active role and does no work in moving the particles. This type of movement of particles across membranes is called **passive transport,** because the cell uses no energy to move the particles.

EFFECTS OF WATER CONCENTRATION ON CELLS

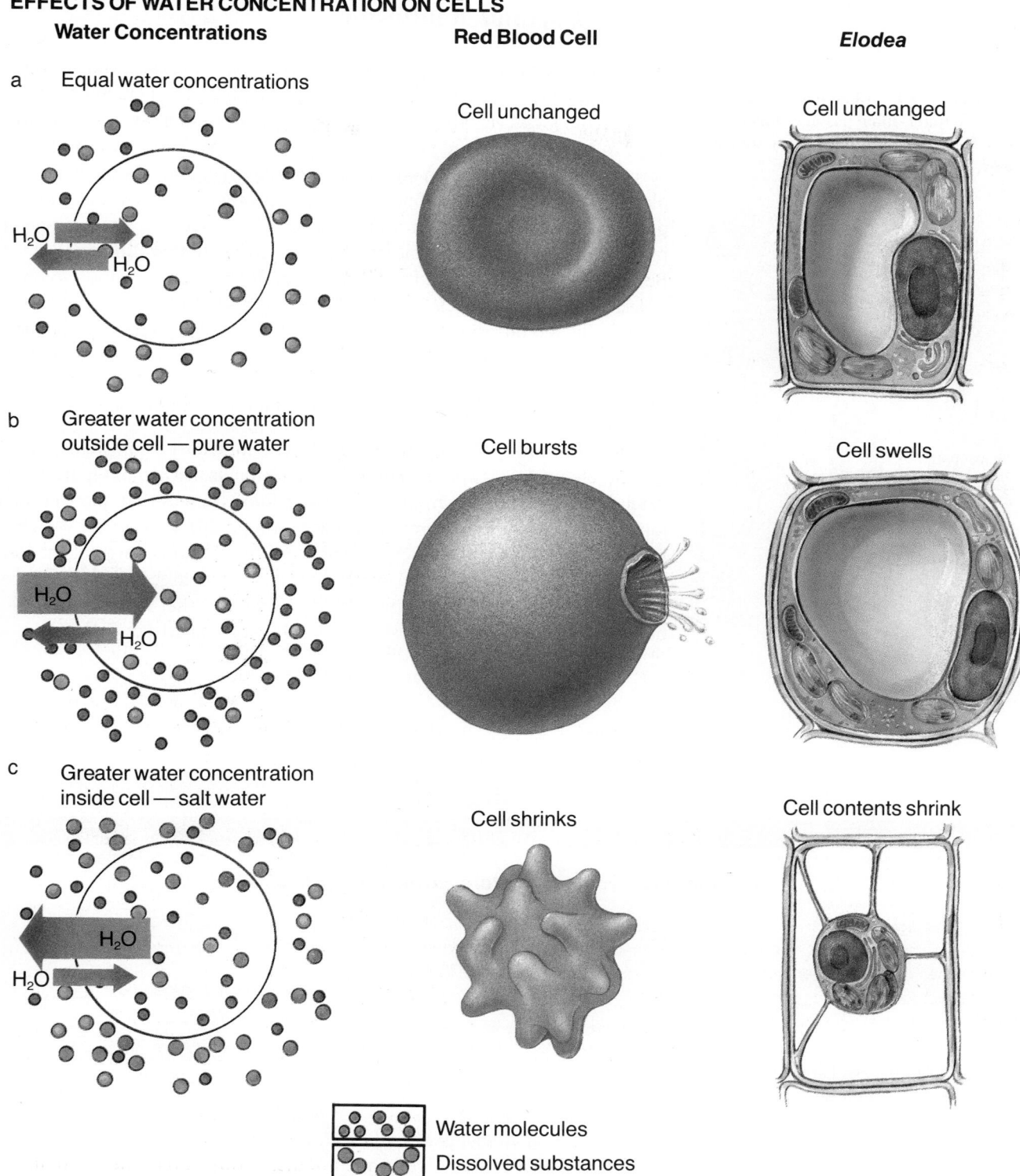

Water Concentrations **Red Blood Cell** *Elodea*

a Equal water concentrations

Cell unchanged

Cell unchanged

H₂O

H₂O

b Greater water concentration
outside cell — pure water

Cell bursts

Cell swells

H₂O

H₂O

c Greater water concentration
inside cell — salt water

Cell shrinks

Cell contents shrink

H₂O

H₂O

Water molecules

Dissolved substances

Figure 4-12   As a result of osmosis, cells respond differently when the water concentration inside the cell is equal to, greater than, or less than the water concentration in the environment.

Facilitated Diffusion

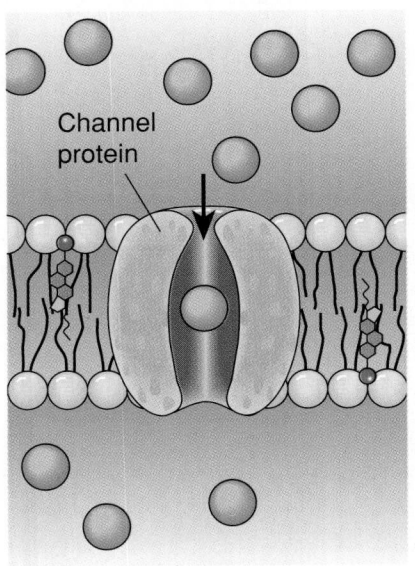

Figure 4-13 Channel proteins provide water-filled passages through which small dissolved particles, such as ions, can diffuse.

Most particles needed by cells—such as certain ions, sugars, and amino acids—may be in greater concentration outside the plasma membrane than inside. However, they cannot dissolve in the lipid bilayer and are too large to squeeze through. How can such particles, so important to cells, permeate the membrane? You read earlier that many membrane proteins play a role in the passage of particles. Such proteins are known as **transport proteins.** Use of transport proteins to aid, or facilitate, the passage of particles across the plasma membrane is called **facilitated diffusion.** Because the particles move from a region of greater concentration to one of lesser concentration, facilitated diffusion is a form of passive transport.

In facilitated diffusion, each transport protein is highly selective and aids the passage of only certain particles. Transport proteins are so selective because each one has a distinctive shape and charge that allows passage of only one kind of particle. There are several types of transport proteins.

Channel Proteins Among the simplest types of transport proteins are protein channels, also called pores. These channels, shown in Figure 4-13, extend through the bilayer and form water-filled tunnels that allow certain ions to pass through. Different ions pass through different pores. Which ion passes through a given pore depends on the size and charge of the ion and the inside diameter of the channel. The polar (charged) regions of the channel protein itself are also important. You know that like charges (++ or – –) repel each other; opposite charges (+– or –+) attract each other.

Thinking Lab

INTERPRET THE DATA

Can pore size be used to determine molecule size?

Background
Particle movement across living cell membranes may be influenced by conditions such as pore size, particle size, or particle charge. The following table shows the results of an experiment in which four molecules (W-Z) were passed through three different membranes (A-C) with varying pore sizes.

You Try It
Interpret the data to predict the size of the molecules based on their movement through the membranes. Assume that pore size is the only condition influencing molecule movement. Note: One nanometer (nm) = 1/1 000 000 mm.

Membrane	Pore Size	Can Molecule Pass Through Pore?			
		W	X	Y	Z
A	1 nm	no	yes	no	no
B	2 nm	no	yes	no	yes
C	3 nm	yes	yes	no	yes

Completing the following statements may help with interpreting the data.

Molecule W is larger than _____ nm but smaller than _____ nm.
Molecule X is smaller than _____ nm.
Molecule Y is larger than _____ nm.

Molecule Z is larger than _____ nm but smaller than _____ nm.

How could you determine the exact size of molecule Z if membranes with different known pore sizes were available?

Even if an ion is small enough to pass through a pore, the charge of the ion and of part of the channel protein may prevent it from doing so. For example, because particles of the same charge repel one another, a positively charged ion may be repelled by the positively charged part of a channel protein.

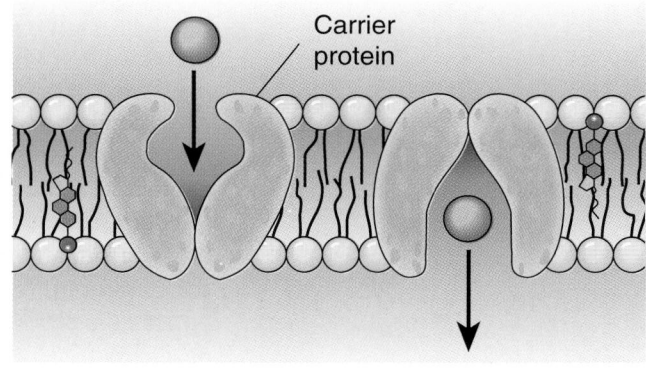

Carrier Proteins   Other transport proteins, called carrier proteins, work in a more complex way. It is thought that for carrier proteins, a particle on one side of the membrane fits into a part of the protein, like two pieces of a puzzle. Once the particle has joined the carrier protein, the shape of the protein changes so that the particle is released on the other side of the membrane. Having released the particle, the protein's original shape is restored, and it is free to combine with another of the same kind of particle. This series of events is shown in Figure 4-14.

Figure 4-14   **Carrier proteins change shape to allow certain molecules to cross the plasma membrane.**

Some carrier proteins act as gates. Figure 4-15 shows that when a certain particle combines with the carrier protein, the shape of the carrier changes so that the gate opens, allowing a different kind of particle to pass through. Many ions enter cells through gates. A gate opens when a hormone attaches to it, and the ions then pass through.

Whether passive transport occurs by osmosis, diffusion, or facilitated diffusion, it always involves a net movement of particles from a region of greater concentration to a region of lesser concentration. The particles pass across the membrane without any energy input from the cell.

Figure 4-15   **Gate proteins open to allow passage of particles, when a "signal" molecule combines with the gate. When the signal molecule is released, the gate closes.**

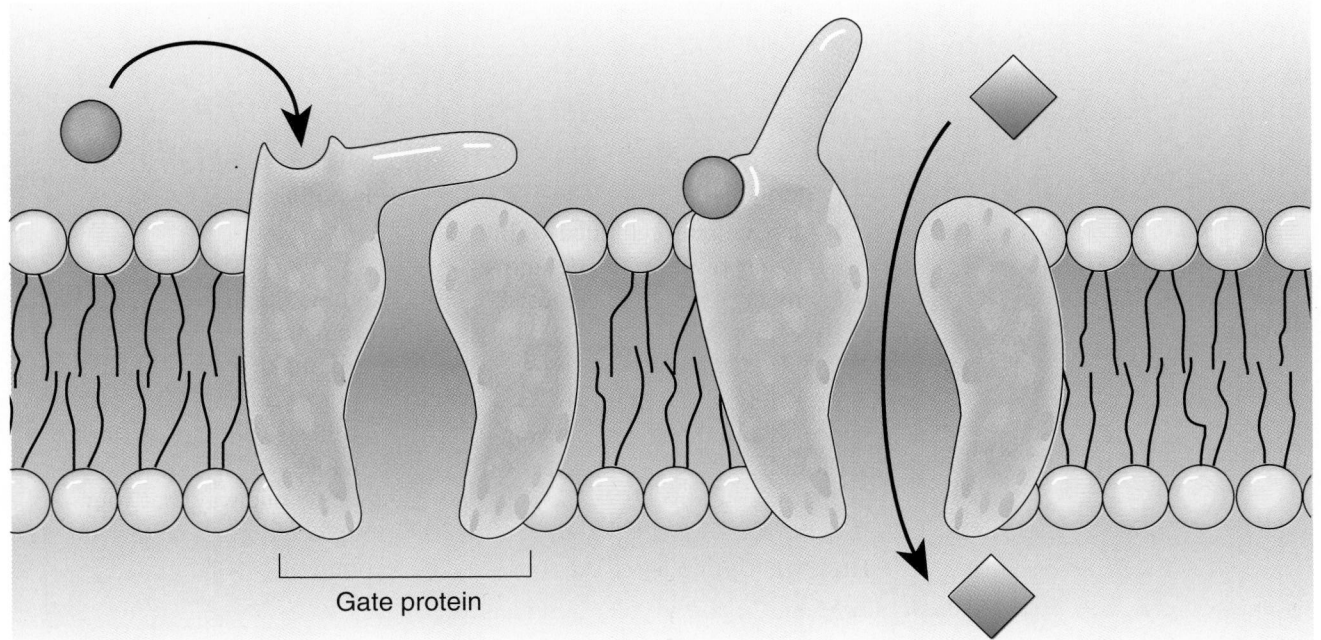

Gate protein

Active Transport

Passive transport is like the events that occur when the bell rings to end class. Students who were sitting in a nonrandom way before the bell rang now begin to move around randomly. They may even bump into one another as they move about. Eventually, they move toward and through the door, and scatter in many directions through the hallways. They have moved from an organized state (a region of greater concentration inside the classroom) to a less organized state (a region of lesser concentration outside the classroom). Like putting the marbles in a box, energy will be necessary to get the students back in the classroom—to move them from a region of lesser concentration to one of greater concentration.

I N V E S T I G A T I O N

Osmosis in a Model Cell

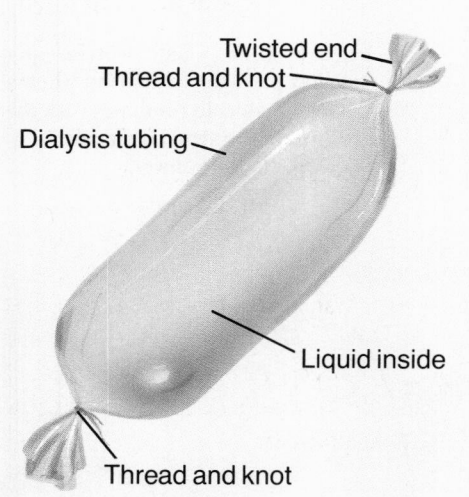

Twisted end
Thread and knot
Dialysis tubing
Liquid inside
Thread and knot

It is very difficult to measure or see osmosis actually occurring in cells because of the small size of most cells. However, if an artificial membrane that acts in some ways like a real plasma membrane could be found, then a study of osmosis using a model would be possible.

Dialysis tubing is such a membrane. It is a sausage casing-like material and is permeable to water molecules. Thus, it is possible to study osmosis using a cell model with a dialysis membrane. A dialysis bag filled with liquid appears at the left.

In this lab, you will design and then carry out an experiment to determine if certain molecules that surround a model cell result in net movement of water by osmosis. Work in a group of three or four classmates.

Problem

Will certain liquids surrounding a model cell filled with tap water result in the net movement of water by osmosis?

Safety Precautions

Wear safety goggles and an apron when mixing and pouring all liquids.

Hypothesis

Have your group agree on a hypothesis to be tested. Record your hypothesis.

Experimental Plan

1. Examine the materials provided by your teacher. Then, as a group, make a list of the possible ways you might test your hypothesis using the materials available.
2. Agree on one way that your group could investigate your hypothesis.

Pumps   Can a cell ever move particles from a region of lesser concentration to a region of greater concentration? Yes, but to do so the cell must expend energy to counteract the random motion of the particles. Energy from the cell causes the movement of the particles to become less random. Because the cell's energy is required to move the particles, this type of movement is called **active transport.** Transport proteins known as pumps are used in active transport. They are embedded in the plasma membrane. Each pump can bind to a particular particle whose shape fits it. Chemical energy is then used to alter the shape of the pump so that the particle to be moved is released on the other side of the membrane. Once the particle is released, the pump's original shape is restored.

3. The variables in this experiment might be the salt concentration outside of the bags, the length of time allowed for water movement to occur, and the amount of liquid inside and outside the bags. Design an experiment that will (a) use a control, (b) test only one variable at a time, and (c) allow for the collection of quantitative data.
4. Prepare a list of numbered directions. Include a list of materials and the quantities you will need.

Checking the Plan

Discuss the following points with other group members to decide the final procedures for your experiment. Make any needed changes in your plan.
1. What will be your control?
2. What one variable are you going to test?
3. What data are you going to collect and how are you going to measure it?
4. How long will you carry out the experiment?

5. What variables, if any, will need to be controlled?
6. How many trials will you carry out?
7. What will be the role of each classmate in your group?
8. Have you designed and made a table for collecting data?

Make sure your teacher has approved your experimental plan before you proceed further.

Data and Observations

Carry out your experiment, make any needed measurements, and complete your data table. Design and then complete a graph or visual representation of your results.

Analysis and Conclusions

1. What changes did you observe among the bags or the liquids in which the bags had been placed? How did these changes relate to the salt concentrations of the liquids in which the bags were placed?
2. What evidence do you have that net intake or loss of water due to osmosis did or did not occur?

3. What evidence do you have that dialysis tubing is permeable to water?
4. What was your control and how is it used to explain your results?
5. Was your hypothesis supported by your data? If not, suggest a new hypothesis that is supported by your data.
6. List several sources of experimental errors that may have influenced your data, and suggest ways in which each error source might be corrected.
7. What were several variables that had to be controlled? How were they controlled?
8. Write a brief conclusion to your experiment.

Going Further

Based on this lab experience, design an experiment that would help to answer a question that arose from your work. Assume that you will not be doing this lab in class. Record this information as part of your conclusion.

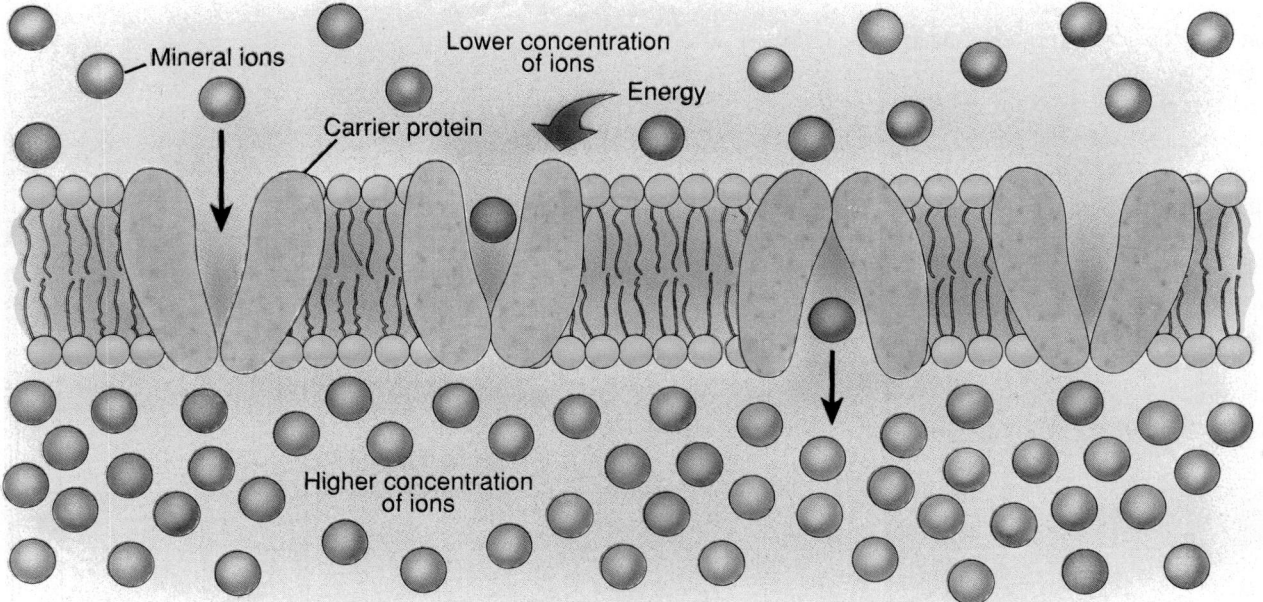

Labels in figure:
Mineral ions
Lower concentration of ions
Energy
Carrier protein
Higher concentration of ions

Figure 4-16 In active transport, energy is used to modify the shape of a carrier protein. This change allows the movement of an ion or molecule through the membrane from a region of lower concentration to one of higher concentration. Why is energy needed to move these particles through the membrane?

Active transport through a membrane may occur in either direction. Cells in the gills of a marine fish actively pump out salts even though the concentration of salts in the sea water is greater than that in the fish's fluids. Such outward movement of salts maintains the osmotic balance between the fish's cells and body fluids. In other cases, active transport is used to bring particles into a cell even when their concentration in the cell is already greater than that outside. Root cells of plants often take in large quantities of needed ions from soil by active transport.

Transport and the Fluid Mosaic Model Now that you've studied the structure of membranes and the passage of particles across them, think again about the fluid mosaic model. The model certainly explains the properties of the membrane. Lipids and lipid-soluble particles permeate easily by diffusion because they dissolve in the lipid bilayer. Smaller particles pass across readily because they can fit between the phospholipid molecules. Larger particles or those insoluble in lipids may pass across with the aid of transport proteins, either channels or carriers. Small neutral particles cross more freely than ions because they are neither attracted to nor repelled by charged parts of the molecules in the membrane. Membranes made up of different lipids and transport proteins are permeable to different particles, and permeability of a given membrane may change, depending on the position of its lipids and proteins.

Endocytosis and Exocytosis

Not all materials that enter a cell do so by passing through the membrane. Some materials can be surrounded by a portion of the plasma membrane, which then encloses the material, pinches off from the rest of the membrane, and releases the material on the other side. **Endocytosis** is the process by which the plasma membrane engulfs and then takes in substances from a cell's environment.

Phagocytosis   Endocytosis is common in unicellular organisms, such as amoebas, which depend on smaller organisms for food. Figure 4-17 shows that after food is detected, the amoeba's plasma membrane engulfs the food organism by flowing over and enclosing it. That portion of the membrane then forms the border of a small sac, called a **vesicle,** which breaks away and moves to the cell's interior. The food in the vesicle will be digested later. Endocytosis is also used by some cells of multicellular organisms. For example, some kinds of white blood cells protect you against disease by engulfing and then digesting bacteria and other foreign particles that enter your body. The form of endocytosis in which solid chunks of material are taken in is called **phagocytosis.**

Pinocytosis   Endocytosis may also involve taking in liquid droplets. The process is the same as in phagocytosis, but drops of liquid rather than solid particles are engulfed. This form of endocytosis, in which liquid droplets are taken in, is called **pinocytosis.**

Receptor-aided Endocytosis   Sometimes, specific molecules are brought into the cell by a third form of endocytosis, called receptor-aided endocytosis. Molecules of a substance outside the cell first link up with proteins called receptors in the plasma membrane. The receptors, with their bound molecules, move to one area of the membrane. An indentation, or pit, in the membrane then forms, eventually breaking away to form a vesicle. The vesicle, containing both the entering substance and the receptors, is brought into the cell. The molecules separate from the receptors and are released into the cell's interior. The vesicle, still containing the receptors, then returns to the outer part of the cell where it and its receptors become part of the plasma membrane once again. Cholesterol, necessary for the manufacture of new membranes in many cells, enters a cell this way.

Figure 4-17   **Endocytosis can involve either phagocytosis or pinocytosis, depending upon whether the material to be brought into the cell is a solid chunk or a liquid.**

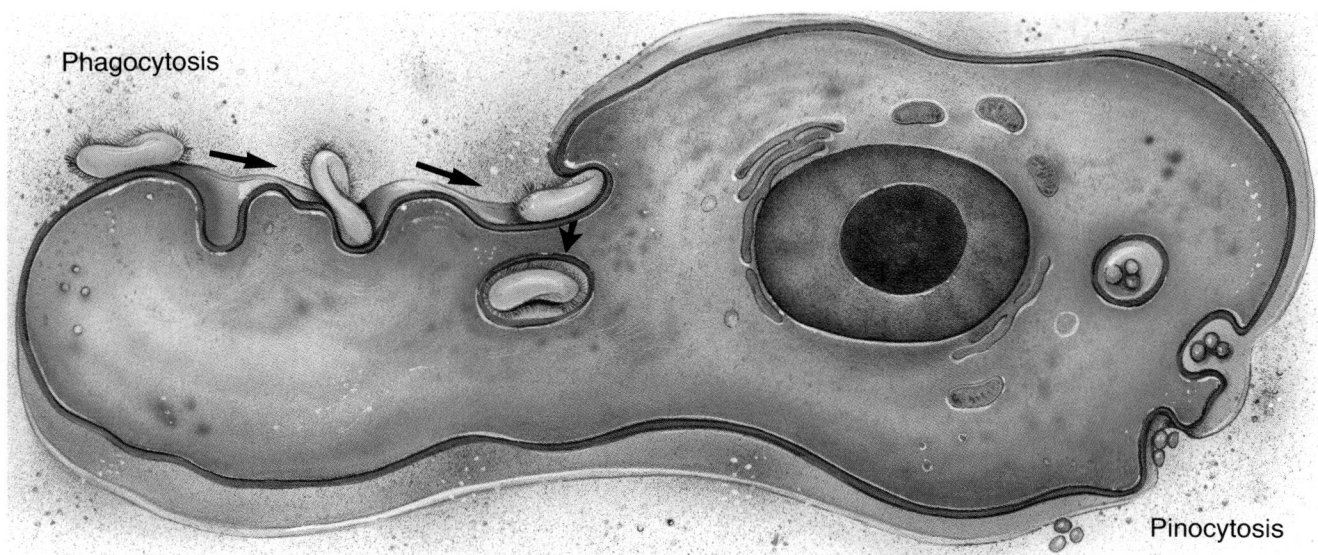

Phagocytosis

Pinocytosis

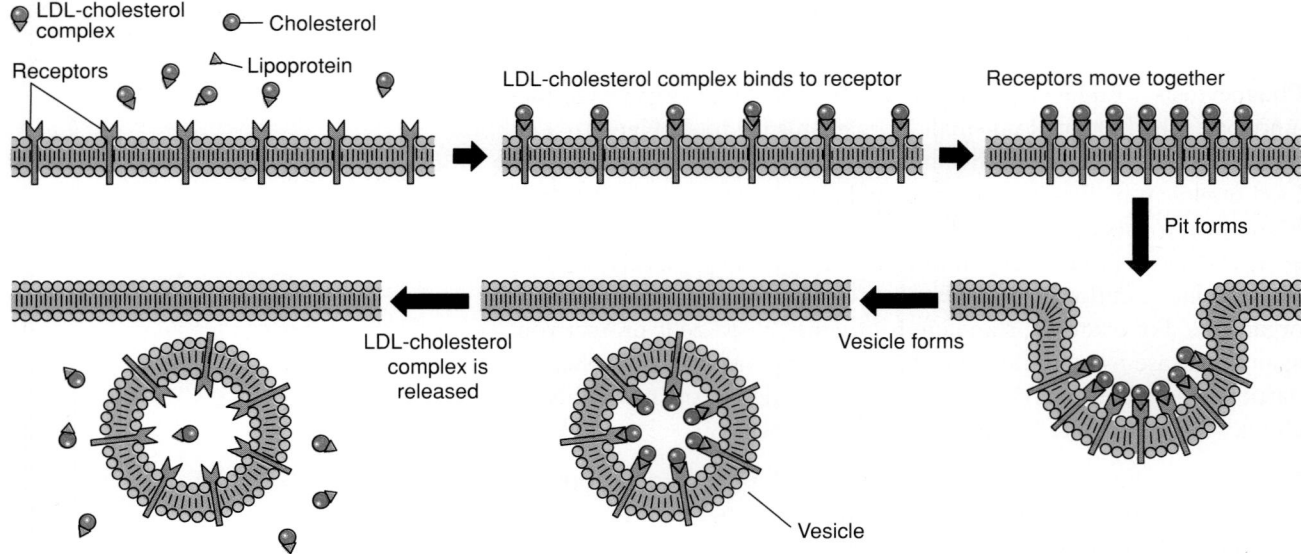

LDL-cholesterol complex
Cholesterol
Receptors
Lipoprotein

LDL-cholesterol complex binds to receptor

Receptors move together

Pit forms

Vesicle forms

Vesicle

LDL-cholesterol complex is released

Figure 4-18 **When a cell needs cholesterol, it synthesizes receptors for low-density lipoproteins and incorporates them into the plasma membrane. Each receptor binds to an LDL-cholesterol complex, which is then brought into the cell by endocytosis.**

Cholesterol molecules destined for use by body cells are found in the blood in the form of low-density lipoproteins, LDLs. Because it is insoluble in water, cholesterol cannot dissolve in the liquid part of blood. Therefore, the body wraps cholesterol molecules in water-soluble protein envelopes, forming lipoprotein molecules that can dissolve in blood. It is the LDLs that combine with receptors on cell membranes. These events are summarized in Figure 4-18.

Exocytosis A reverse process of endocytosis is used to rid some cells of wastes, such as undigestible particles, or to secrete substances, such as hormones, needed elsewhere. The substances are first enclosed in a vesicle that has a membrane with fluid mosaic structure. The vesicle moves toward the plasma membrane, breaks open, and fuses with part of the plasma membrane. As this occurs, the contents of the vesicle are expelled. This process, known as **exocytosis,** is diagrammed in Figure 4-19.

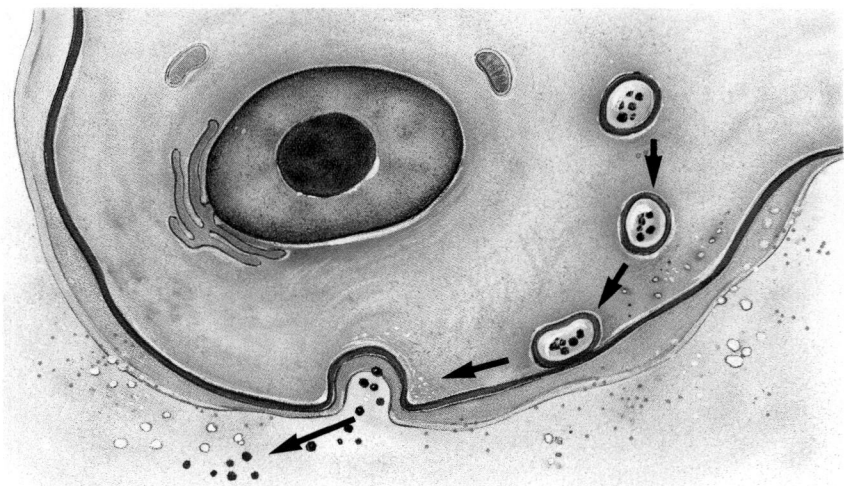

Figure 4-19 **Cell products or wastes, enclosed in vesicles, are released to the surroundings by exocytosis.**

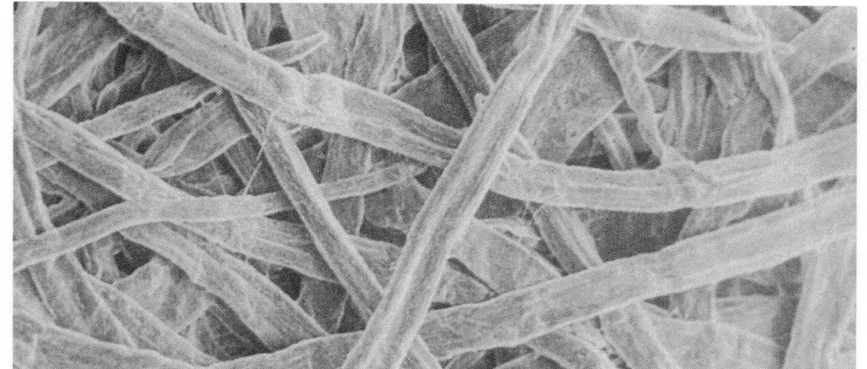

Figure 4-20 Plant cell walls are composed of cellulose. The cellulose molecules form interconnecting fibers that are interwoven in a strong network. The network protects a plant cell and gives the plant support. Magnification: 20×

Cell Walls

In many organisms, the plasma membrane is not the outermost cellular boundary. Plants, most algae, fungi, and bacteria have a **cell wall** that lies outside the plasma membrane. Remember that cell walls were the only parts of cork cells that Robert Hooke observed in 1665. Cell walls of plants and algae are composed largely of the polysaccharide cellulose. Wood products, paper, cotton, and rayon are made from the cellulose cell walls of plants. Fungal and bacterial cell walls are each composed of different materials.

Cellulose is arranged in rows of fibers stacked in layers, as Figure 4-20 shows. Spaces between the fibers allow water, ions, and other particles to pass through. Thus, the wall does not play a role in determining a cell's permeability characteristics. However, the cell wall is important in other ways. It provides protection and, in land plants, also provides support. The wood supporting a giant redwood tree consists of the cell walls of hollow, dead cells.

Figure 4-21 These giant redwood trees could not grow so large if their cells did not have tough cellulose walls.

After plant cells divide, a new cell wall forms as cellulose is deposited outside the plasma membrane. This cell wall is known as the primary cell wall. Formation of the primary cell wall continues as long as the cell continues to grow. Between the two primary cell walls of adjacent cells lies an area called the **middle lamella.** It is composed of pectin, a cellular cement that binds adjacent cells together. If the pectin breaks down, the cells become less tightly bound, a process that occurs as fruit ripens. After completing their growth, many plant cells, such as those from a redwood, deposit more cellulose, thus forming a secondary cell wall. This secondary wall, which lies inside the primary wall, makes the cell wall even stronger and provides the degree of support a large plant requires.

LOOKING AHEAD

YOU HAVE READ how the structure of the plasma membrane is related to the ways particles pass across it. The foods you eat cannot nourish your body unless the nutrients they contain enter your cells. As you study the next two chapters, you will see that whether the nutrient particles enter by passive transport or active transport, once inside a cell they are used by cellular structures in processes that maintain your body and keep it alive. Cellular wastes that could build up and kill the cell pass across the membrane and out into the surrounding fluid. In a multicellular organism, many such individual cells group together to form different tissues that work together as a team to form your body parts.

Section Review

Understanding Concepts

1. You forget to water your garden and notice that the plants have become wilted and limp. Explain what has happened to the plant cells. What will happen to the plant cells if you now water the plants?

2. Cells produce carbon dioxide during cellular respiration. Explain how carbon dioxide leaves cells.

3. A unicellular organism that normally lives in salt water is transferred to a beaker of fresh water. What will happen to the cell? Why?

4. **Skill Review—Designing an Experiment:** How could you design an experiment that would allow you to see the effect of salt water on an *Elodea* plant? For more help, refer to Practicing Scientific Methods in the **Skill Handbook.**

5. **Thinking Critically:** How are active transport and facilitated diffusion similar? How are they different?

BIOLOGY, TECHNOLOGY, AND SOCIETY

Biotechnology

Cold Storage

Now here's a chilling thought! Freeze the body of someone who has just died of a disease, then wait until medical science has progressed enough to revive and cure the patient.

When Does Death Occur? The use of this technique is attractive to many people. After all, they point out, a person was once considered dead when his or her heart stopped. Now physicians can restart a heart, and death is considered to occur when brain activity stops. But even in a person whose brain activity has ceased, the individual brain cells and body cells remain alive for a short time. So perhaps the patient may still be curable by the standards of future medicine. And, supporters continue, we've long been able to freeze and revive individual cells and even small embryos. Why not whole bodies?

The Controversy It's easy enough to freeze and preserve whole bodies, but so far, there's no safe way to defrost them. And some scientists doubt that the damage done to cells will ever allow the process to be worthwhile. Those who believe in the procedure, which is known as cryonic suspension, are betting that future medical technology *will*

be able to repair or rebuild bodies right down to the cellular level.

The Freezing Process In order for cryonic suspension to work, the body must be frozen to a temperature of $-130°C$ or lower. The freezing process puts enormous stress on the cells.

As the body temperature drops, water between cells turns to ice. All the salts, sugars, and other solids dissolved in the water become concentrated in the still-liquid water, a much higher concentration than exists inside the cell. Water flows out of the cell, keeping the concentration of salts and sugars inside and outside of the cell in equilibrium. A cell losing water is similar

to a balloon losing air—it shrinks.

Meanwhile, water between the cells continues to freeze, the concentration of solids outside the cell continues to rise, and water continues to flow from the inside to the outside of the cell. In time, the cell becomes little more than a tiny mass of membrane and extremely concentrated solids embedded in ice.

Damaged Cells The stress this process puts on the cell can damage the cell's membrane structure and distort essential cell proteins. If such a damaged cell is defrosted, it will no longer work properly.

Today's cryonic techniques use protective agents to shield the cell from damage. But some of the

CHAPTER 4 THE CELL AND ITS ENVIRONMENT **111**

protective agents themselves are toxic and damage the cells.

Even more damage occurs between cells. Because cells are organized in groups, the spacing between them is important. The space between cells is usually no more than about 5 micrometers. (A micrometer is one-millionth of a meter.) The ice crystals formed in cryonic preservation, however, are generally about 30 micrometers in diameter. So as the intercellular water freezes, it pushes the cells apart. Many vital cell-to-cell junctions are invariably torn.

Ice crystals cause similar damage to the capillaries, which provide blood flow to the cells. The average capillary is only 8 micrometers in diameter and can only stretch so much before bursting.

To date, no whole organ, let alone an entire body, has ever been fully restored after being frozen to the temperatures long-term storage requires. Yet, animal brains have survived extremely low temperatures without apparent damage.

For so cold a topic, cryonics continues to light bright fires in many imaginations.

1. **Connection to Law:** Suppose many of those people declared dead and preserved in cryonic suspension prove to be curable in the future. How might some laws have to be changed?
2. **Issue:** There is a case of a woman who was declared dead and placed in cryonic suspension. The coroner suspected that her death might have been due to foul play. He wanted her body removed from cryonic suspension for autopsy. How would you have ruled if you were the judge?

Health Connection

Fiber to the Rescue?

A few years ago, an unlikely hero began grabbing headlines around the world for supposedly saving lives and offering people the promise of a healthier future. This overnight media sensation was a long-time staple of winter breakfast tables—oatmeal, or more specifically, oat bran. According to scientific studies, oat bran seemed to possess the ability to lower levels of cholesterol in the bloodstream.

Cholesterol, a lipid component of membranes, long had been associated with clogged arteries and an increased risk of heart attack. Some nutritionists suggested that cholesterol might be brought under control with a daily dose of oats. Soon, oat bran began appearing on grocer's shelves in a variety of foods: oat-bran cookies, oat-bran muffins—even oat-bran tortilla chips!

The oat-bran frenzy began to fizzle when new studies failed to duplicate earlier promising results. Nevertheless, scientists were intrigued and began studying the nutritional role of dietary fiber more closely.

Indigestible and Indispensable
The indigestible cell walls of plant cells, found in fruits, vegetables, and whole grains, are called fiber. Fiber, an essential part of everyone's diet, comes in two forms—water-soluble and water-insoluble.

Water-soluble fiber, found in such foods as apples, carrots, and oats, appears to lower blood cholesterol. Water-insoluble fiber, such as that present in brown rice, lentils, and wheat bran cereals, provides

the roughage that stimulates the muscles of the digestive tract and prevents constipation.

Different Fibers, Different Functions   Scientists still don't understand fully how fiber affects cholesterol. Recent studies, however, provide some interesting clues.

Cholesterol is broken down in the liver to form bile acids. These bile acids help to break down fats in the stomach and small intestine.

After the bile acids have done their work, the small intestine absorbs them for reuse later.

Carrot pectin, a water-soluble fiber, acts like a sponge, absorbing bile acids and preventing the small intestine from reabsorbing them. Thus, the liver must break down additional cholesterol to make more bile acids. The result is lower cholesterol levels in the bloodstream.

Scientists still have not pinpointed all of the exact mecha-

nisms that enable dietary fiber to do its work. One thing is clear, however. Fiber is now recognized as an essential component in a balanced, nutritious diet.

———

1. **Explore Further:** Which foods provide the greatest amount of fiber? Which provide the least amount?
2. **Writing About Biology:** Explain why vegetarian diets are likely to be higher in fiber than nonvegetarian diets.

These scanning electron micrographs show a healthy, unclogged artery (left) and an artery clogged with cholesterol (right). What will happen if the artery becomes completely blocked? Magnification: left, 3×; right, 4×

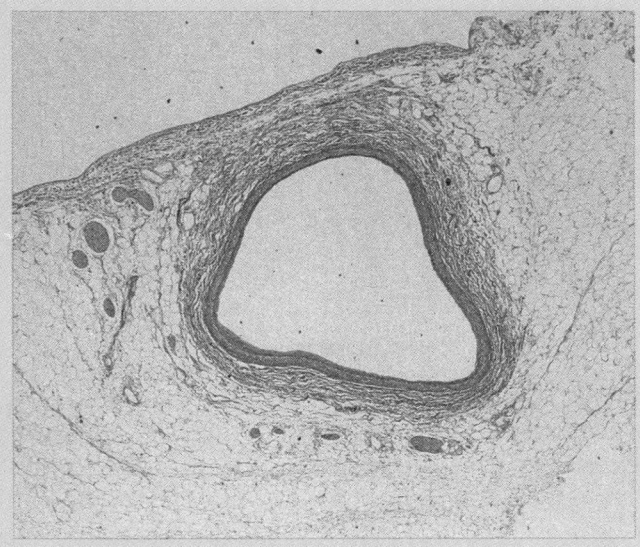
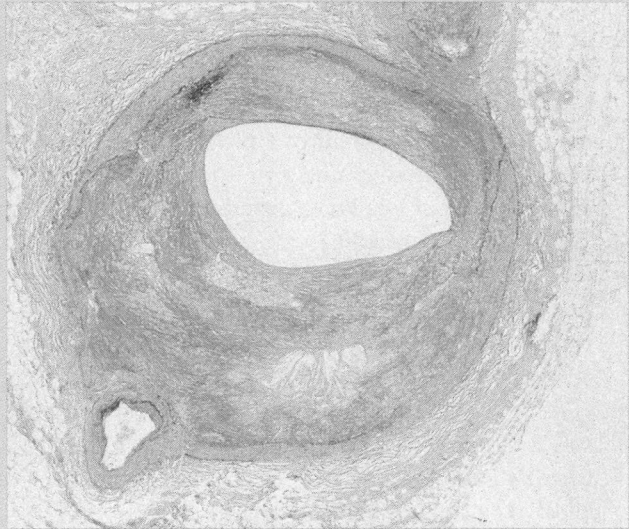

REVIEW
C H A P T E R
4

Summary

Cells were first discovered by Robert Hooke. Over a period of almost 200 years, study of a variety of cells resulted in the establishment of the cell theory.

Cells are surrounded by a liquid environment from which they obtain needed materials and in which they dump their products and wastes. Each cell is bordered by a plasma membrane, which regulates the passage of these substances.

According to the fluid mosaic model, the plasma membrane, as well as the membranes of internal parts of cells, consists of a phospholipid bilayer that provides strength and flexibility. Embedded in the bilayer are proteins. Many proteins determine to which particles a given membrane is permeable.

In passive transport, the cell uses no energy. Movement of particles across the membrane is due to the energy of the moving particles themselves and results in net movement from a region of greater concentration to one of lesser concentration. Diffusion, osmosis, and facilitated diffusion are forms of passive transport.

In active transport, the cell uses energy to move particles from a region of lesser concentration to a region of greater concentration. Some substances do not pass across membranes, but enter or leave cells by endocytosis and exocytosis.

Language of Biology

Write a sentence that shows your understanding of each of the following terms.

active transport
cell theory
cell wall
diffusion
dynamic equilibrium
endocytosis
exocytosis
facilitated diffusion
fluid mosaic model
middle lamella

osmosis
passive transport
phagocytosis
pinocytosis
plasma membrane
selectively permeable membrane
transport protein
vesicle

Understanding Concepts

1. Why is the cell theory so important to the science of biology?
2. How is the environment of your cells different from that of a plant cell? Could one of your cells exist in the environment of a plant cell?
3. Use the fluid mosaic model to explain how different cells are permeable to different particles.
4. Why is it not possible for most particles to pass across a membrane by diffusion?
5. A student on the other side of the lab accidentally spills some alcohol. Explain, in terms of molecules, why you can smell it.
6. How is energy related to random and nonrandom states of matter?
7. Identify and compare two ways that ions may enter cells by passive transport.

8. Why is passive transport often referred to as a downhill process and active transport as an uphill process?
9. Pores are often thought of as definite openings in a membrane. Does this definition fit the channel proteins of a plasma membrane? Explain.
10. How does the process of endocytosis differ from diffusion, facilitated diffusion, and active transport?

Applying Concepts

11. Do you think that the membrane of an amoeba has much, if any, cholesterol in it? Explain your reasoning. Hint: Think about how an amoeba takes in food.

12. A single-celled alga living in the ocean is found to contain ten times as many iodide ions as are present in an equal volume of sea water. The cell continues to take in even more iodide. What process must the cell be using to take in these ions? Explain the process.

13. Suppose a cell has just reproduced. One of the new cells cannot make several transport proteins that cells of this type usually have. Will the cell be able to survive?

14. A cell biologist has taken a photograph of a plasma membrane as viewed through an electron microscope. She notices that part of the membrane was closing around a bit of solid material outside the cell. What process is occurring? What would happen next if the cell were still alive?

15. You are observing a freshwater cell in pure water. You slowly add small amounts of salt to the water. Considering that the cell's membrane is permeable to water but not to salt, how will the size of the cell change as the salt is added? Why?

16. Certain cells of a snake produce venom, a large molecule. How will the venom leave the cells in which it is produced?

17. Would phospholipids move more rapidly or less rapidly if a membrane contained no cholesterol? Why?

18. **Global Connection:** Some people sprinkle salt on weeds to kill them. Why would salt have this effect?

19. **Health Connection:** What substance in cereals makes them such good sources of dietary fiber? Why do you think this substance serves as roughage in the digestive tract?

Lab Interpretation

20. **Investigation:** An experiment was done using three dialysis bags, labeled A, B, and C. Before the experiment began, all of the bags appeared like the unlabeled bag shown in the next column. Bags A, B, and C were filled with a 20 percent sugar solution and were placed into different liquids in separate beakers. Predict the nature of the liquids in which each bag was placed, compared with the liquid inside the bag. (Assume each bag is impermeable to sugar.) Explain your answers.

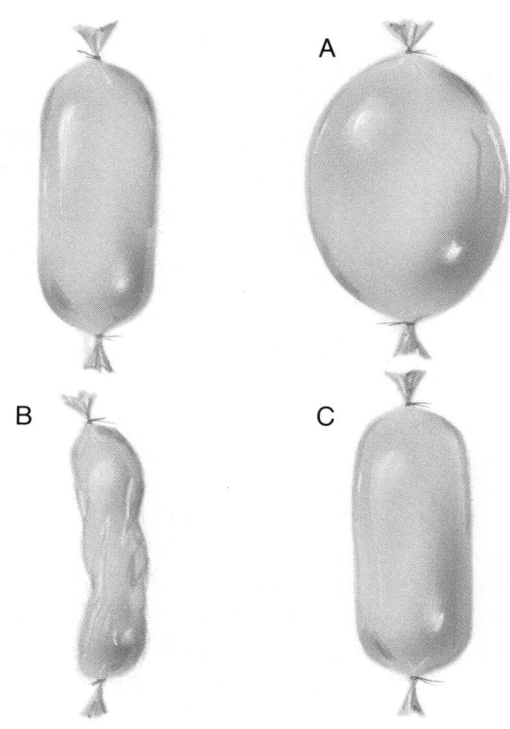

21. **Investigation:** Assume that the dialysis tubing in question 20 is impermeable to sugar but not to water. Diagram a model of the dialysis tubing showing its pore size in relation to the sizes of water and sugar molecules.

22. **Minilab 2:** If you were to repeat the experiment in Minilab 2, but place the test liquids in the beakers and the water in the bags, what would be the outcome? Explain your results.

Connecting Ideas

23. Your blood contains more oxygen than your body cells, so oxygen diffuses from the blood into the cells. However, dynamic equilibrium is not reached. Why?

24. A single-celled organism makes a transport protein that no other cell of that type has made before. How might that transport protein be important in the future of this population of organisms?

INSIDE THE CELL

A TWO-TON ELEPHANT is made of the same tiny units of life as an ant crawling on the ground at its feet. Elephants and ants, whales and shrimp, horses and humming-birds—all organisms, large and small, are made up of cells. Although cells exist in different shapes and sizes, most cells are so small they can be seen only with the aid of a microscope. Microscopic organisms, such as bacteria and *Amoeba,* consist of single cells. Larger forms of life, like ants and elephants, are composed of thousands, millions, or even trillions of cells. The size of an organism depends on the number of cells that compose it, rather than the size of those cells.

The activities that take place within cells constitute the processes of life. What are those processes? What lies inside the plasma membrane, and what happens there? In this chapter, you will take a look inside a typical cell and learn about the forms and functions of the parts within it. You will learn how the cells of single-celled organisms differ from the cells that make up many-celled organisms such as plants and animals. You also will find out how the cells of multicellular organisms are organized, and how they evolved from simpler forms.

What one thing does the organism in the photograph (magnification: 725×) have in common with green beans, mushrooms, and you? Can you think of any other trait that these organisms share?

Chapter Preview

Structure of Cells

Objectives

Recognize general differences between the cells of unicellular and multicellular organisms.

Describe how cells are arranged into increasingly complex levels of organization.

Discuss how cell organelles contribute to the efficiency of cellular functions.

You have seen photographs and drawings of cells. Perhaps you've also looked at cells that have been frozen, stained, or prepared in some other way for viewing under a microscope. These images can make cells seem dull and inactive, because they show you only a single instant in the life of a cell. Instead, imagine you have a microscope that allows you to watch what's actually happening inside a living cell—perhaps one of the cells of your own body.

An amazing number of activities are taking place inside a living cell. Molecules are passing into and out of the cell through the plasma membrane. Inside the cell, there is constant motion. Molecules are constantly being transported from one area to another. Some molecules are being joined to create new substances, while others are being broken down into smaller parts. You've entered a dynamic factory of life, where large numbers of very complex manufacturing processes are going on.

Characteristics of Cells

When you look at a photograph, you see a flat, two-dimensional representation of something you know is actually three-dimensional. Pictures of cells can make them seem like circles, squares, or rectangles. Cells, however, are actually spheres, cubes, or bricks. Many cells can change their shapes, and some also have spikes, bulges, or other strange extensions. Cells vary in size as well as shape. Bacteria are among the smallest cells; some are less than half a micrometer (written as the Greek symbol μm) across. One micrometer is equal to one thousandth of a millimeter (1μm = 1/1000 mm). Twenty thousand bacterial cells, side by side, would measure about one centimeter. A human egg cell is large by comparison, measuring nearly a millimeter in diameter. It would only take about 10 egg cells, side by side, to measure one centimeter.

Figure 5-1 **Cells can be shaped like boxes, cubes, cylinders, or spheres and have spikes, bulges, or indentations. Does the shape of a cell tell you anything about its function? Magnification: 150×**

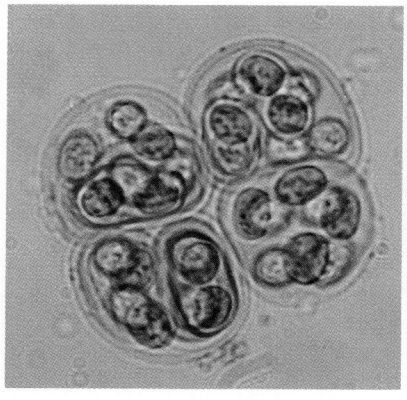

a

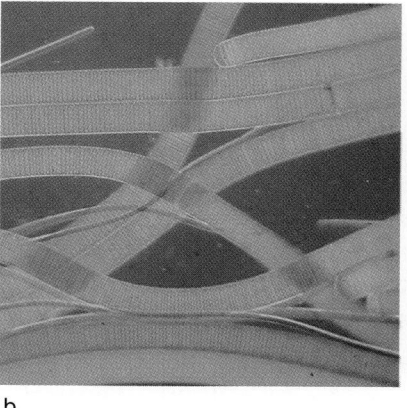

b

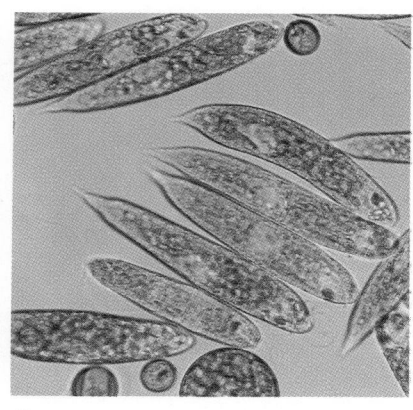

c

Unicellular Organisms   Bacteria, *Amoeba,* certain algae, and yeasts are examples of single-celled, also called unicellular, organisms. Because these organisms are made up of only a single cell, that one cell has to be able to perform all the functions necessary for life. Some of these cells, like *Amoeba,* must be able to obtain food and oxygen from the environment and, by cellular respiration, release the energy in food molecules. That energy is required for functions such as reproduction, production of enzymes, elimination of waste products, and motion. Unicellular algae must take in carbon dioxide and water and convert them to sugars in the process of photosynthesis. Then, through cellular respiration, they, too, obtain the energy needed for their functions. All unicellular organisms must be versatile, because they must do everything for themselves.

Multicellular Organisms   Many-celled, also called multicellular, organisms evolved from unicellular life forms. Almost all cells of multicellular organisms have retained the ability to carry out very basic functions, like enzyme production, that first evolved among their unicellular ancestors. However, some of the cells of multicellular organisms are adapted for performing specific tasks. For example, not all of the cells of a tree are capable of photosynthesis. Only cells that contain chlorophyll can perform this function. The food produced by these chlorophyll-containing cells is transported to other parts of the tree. Most of your cells cannot transmit messages from one part of your body to another; only your nerve cells can. Cells that are adapted for specific functions are said to be specialized.

Figure 5-2   Examples of unicellular organisms include *Gloeocapsa* (a), *Oscillatoria* (b), and *Euglena* (c). Some unicellular organisms are found in colonies or attached to one another to form long chains.
Magnifications: 400× (a), 100× (b), 180× (c).

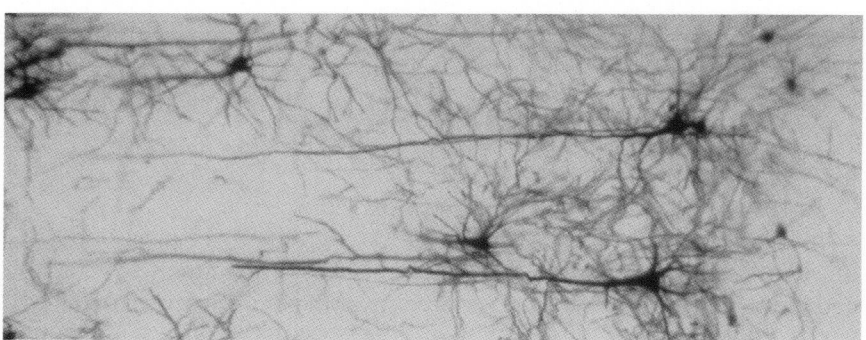

Figure 5-3   Some cells are specialized to carry out specific functions in multicellular organisms. What functions do these human nerve cells carry out? Magnification: 300×

Although multicellular organisms must be able to perform all the functions necessary for life, each cell need not be equipped to do everything. Instead, the many different kinds of specialized cells interact to keep the organism alive. Specialized cells are dependent on one another. For example, many animals depend on their blood cells to transport the oxygen all the other cells need. If the blood cells fail to function properly, the animal may not be able to survive. Specialized cells often have shapes or sizes that enable each type of cell to perform its function more efficiently. Nerve cells carry messages from one part of your body to another. What shape would you expect nerve cells to be?

Levels of Organization in the Living World

Think for a moment about what it's like to be part of a team. You may be a member of a softball, soccer, or football team. Perhaps you've joined a choir, band, or drama troupe. Your coach or director organizes the mem-

I N V E S T I G A T I O N

Sizes of Cells

How big is a cell? The first thought that comes to mind is "very small." However, nerve cells in your body can be 1.5 m long! An ostrich egg cell is about 100 mm in diameter. This egg cell is able to grow larger than other cells because its food, the yolk, is inside the cell. Other cells must take in materials for survival through their surfaces. As a cell gets larger, its need for materials may exceed the ability of the surface to let materials in. As a cell increases in size, it has proportionately less surface area to volume.

In this lab, you will determine the size of plant and animal cells. Your teacher will supply a small, transparent millimeter ruler and a variety of plant and animal cells. If you have trouble seeing the cells, you might use Lugol's solution or methylene blue stain. You can measure the diameter of the field of view under low power with your millimeter ruler. There are 1000 micrometers (μm) in 1 mm. It might be easier to make your cell measurements in micrometers because cells are very small.

Problem
How big are cells?

Safety Precautions
Stains may be permanent on clothing and last a long time on skin. Rinse immediately if spilled on skin. Wash your hands at the end of the lab.

Hypothesis
What is your group hypothesis? Explain your reasons for forming this hypothesis.

Experimental Plan
1. As a group, make a list of possible ways you might test your hypothesis using the materials your teacher has made available.

bers of your group in a way that takes advantage of each individual's talents and abilities. You do your part, and you depend on the other team members to do theirs. You work together. Similarly, the specialized cells of multicellular organisms work together to accomplish a task.

Tissue   A group of cells that have the same basic structure and function is called a **tissue.** For example, the cells that make up the muscle tissue in your legs work together to enable you to walk. Nerve cells are grouped into tissues that extend throughout your body to carry messages to and from your brain. Epithelial tissue forms the protective outer layer of your skin and the lining of your internal cavities. Plant tissues include protective tissues, such as bark, and conductive tissues that transport water, minerals, and organic molecules from one part of the plant to another.

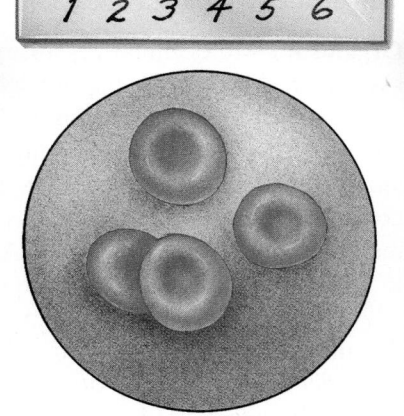

2. Agree on one idea from your group's list that could be investigated in the classroom.
3. Design an experiment in which you collect quantitative data.
4. Following the style of a recipe, write a numbered list of directions for your experiment that anyone could follow.
5. Make a list of materials and the quantities you will need.

Checking the Plan

Discuss the following points with other group members to decide the final procedures for your experiment.

1. How many cell types will you use?
2. Is it important to use both plant and animal specimens?
3. How many cells of each type will you measure?
4. Under what circumstances will you need to use stain?
5. In what way will estimation be used in your calculations?

6. If you measure the field of view with the millimeter ruler, how can this information be useful in determining the sizes of cells?
7. Have you designed and made a table for collecting data?

Make sure your teacher has approved your experimental plan before you proceed further.

Data and Observations

Carry out your experiment; make your measurements; and complete your data table. Make a graph of your results.

Analysis and Conclusions

1. What was the average size of the cells you measured?
2. Which are larger, plant or animal cells? Speculate why.
3. Which cell type was smallest? Speculate about why it is an advantage for that species of cell to be so small.
4. Which cell type was largest? Speculate about why it is an advantage for the cell of that species to be larger.
5. Why do you think cells are different sizes?

Going Further

Based on this lab experience, design an experiment that would help to answer another question that arose from your work. Assume that you will have unlimited resources and will not be conducting this lab in class.

Minilab 1

What are the differences between onion root and onion bulb cells?

Remove a small section of the inner transparent membrane of a raw onion ring cut from a bulb. **CAUTION:** *Be careful when using a razor blade.* Make a wet-mount slide of this tissue with a drop of methylene blue. With a razor, cut a paper-thin cross section of onion root from a bulb that has sprouted roots. Make a wet-mount slide of this tissue with a drop of methylene blue. Observe both of these preparations under the microscope.

Analysis *How are the two cell types alike? How are they different? Speculate about why these differences occur.*

Organ Just as the cells of a tissue work together like team members to accomplish a task, tissues also join forces to perform more complex functions. The tissues within a leaf work together to carry out the complicated process of photosynthesis. Chlorophyll-containing tissue absorbs sunlight, while other tissues obtain the carbon dioxide, water, and minerals needed to trap the energy of sunlight in the chemical bonds of simple sugars and other carbohydrates. Still other tissues move the sugars out of the leaf to other parts of the plant. All of the tissues in the leaf work together to provide food and other nutrients for the plant. A leaf is an organ of the plant. An **organ** is a structure composed of many different tissues that work together to perform a particular function. Your stomach is another organ.

System Your stomach stores food and begins the digestion of proteins. It cannot completely digest an entire meal, however. Other organs, including the liver and the large and small intestines, are needed to complete the process. A group of organs working together is known as a **system.** Your stomach is part of your digestive system. Other human organ systems include the respiratory, circulatory, and reproductive systems.

Organism Multicellular organisms are composed of a group of organ systems working together. You might think that an organism represents the highest level of organization in the living world, but it doesn't. Think about your neighborhood. It's a community made up of many people, plants, and animals. As you know, organisms live together in communities. A forest community, for example, consists of trees, bushes, flowers, birds, mammals, and other forms of life. There are even higher levels of organization in the living world. You will learn more about them later in this course.

As you can see, each level of organization—cell, tissue, organ, system, organism, and community—involves interdependence. Cells that are dependent on each other form tissues. Tissues that work together form organs. Organs depend on other organs to function as a system. Systems work together to form an organism, and organisms cooperate in forming communities.

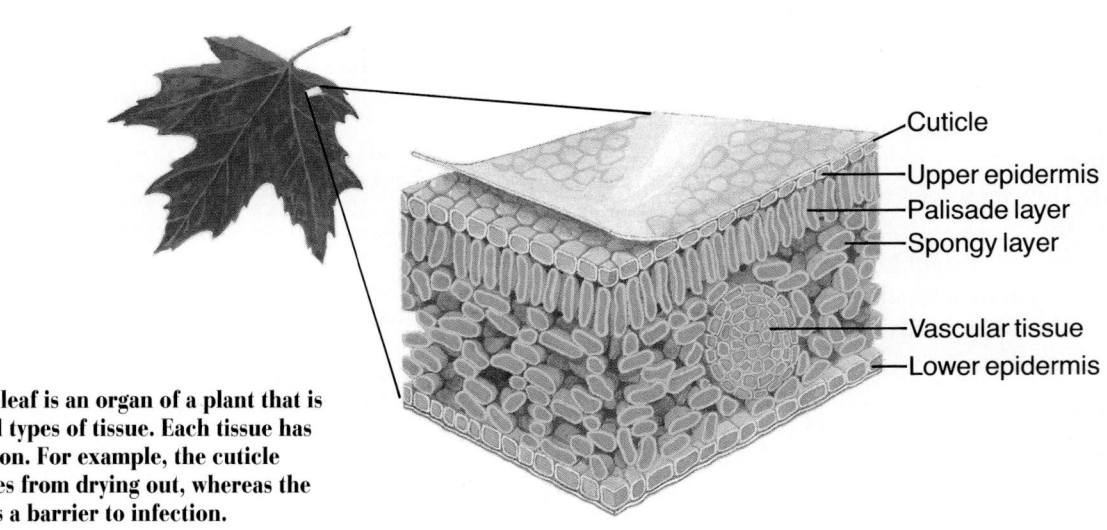

Cuticle
Upper epidermis
Palisade layer
Spongy layer
Vascular tissue
Lower epidermis

Figure 5-4 **A leaf is an organ of a plant that is made up of several types of tissue. Each tissue has a particular function. For example, the cuticle prevents leaf tissues from drying out, whereas the upper epidermis is a barrier to infection.**

Figure 5-5 A coral reef is one type of community in which organisms cooperate. The place you live is another type of community. How do organisms in your community cooperate?

Cytoplasm

When Hooke made the first observations of cells, he saw only dead, empty cells. The fact that living cells contain a substance called **cytoplasm** was not recognized until the 1830s. Analysis of cytoplasm shows that it consists of about 70 percent water and about 30 percent proteins, fats, carbohydrates, nucleic acids, and ions. Its consistency is like a gelatin dessert that has not completely solidified. The exact composition of cytoplasm varies from cell to cell and is always changing because the substances it is made of are constantly involved in chemical reactions. So much chemical change takes place within a cell that 100 000 proteins can be made in a cell each second!

Cellular reactions both build up and tear down complex molecules. The sum of all chemical changes in cells is called **metabolism.** Some chemical reactions, such as condensation or hydrolysis, occur in one or just a few steps. Often, though, a complex set of related reactions is involved, as in the processes of photosynthesis and cellular respiration.

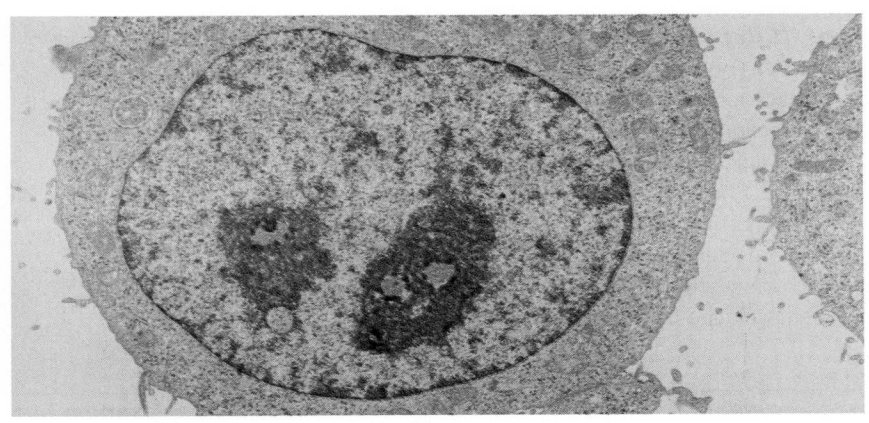

Figure 5-6 Cytoplasm contains a complex mix of enzymes, sugars, and amino acids. Cells are filled with cytoplasm, as you can see in this color-enhanced photograph. Magnification: 15 000×

Organization Within Cells

Imagine that you and some friends are planning a costume party. How would you get ready for the big night? One or two of you might gather recordings of your favorite dance music, while another group arranges for the food, and another takes charge of all the decorations. You would divide up the labor.

Division of labor is evident at all levels of biological organization. You've already seen how tissues, organs, systems, organisms, and communities illustrate this idea. Division of labor is also an important factor in the lives of individual cells. Some cellular reactions occur in the cytoplasm. Thousands and thousands of reactions take place in a cell at the same time. Would it be efficient to have all these reactions going on in the cytoplasm? Just as you would use different pans when baking a cake and cooking soup, the cell includes many structures that carry on different chemical reactions. These specialized structures found in the cytoplasm are called organelles.

The various functions performed by different organelles are interdependent and contribute to the survival of the entire cell. Recall that many internal cell parts are enclosed by membranes that have a fluid mosaic structure, and those membranes vary in terms of the particles to which they are permeable. These membranes allow organelles to be separated from one another and from the cytoplasm. The membranes maintain separate environments where particular sets of reactions can occur.

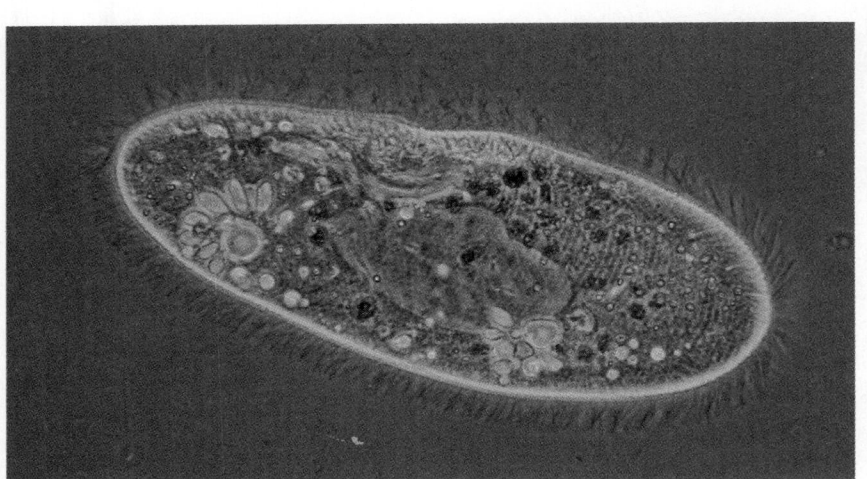

Figure 5-7 **Various functions of the cell are performed in cell organelles. Magnification: 400×**

Section Review

Understanding Concepts

1. Compare current knowledge of the basic structure of cells with what Hooke knew.
2. *Paramecium* is a unicellular consumer. What functions must it carry out that most cells of a multicellular organism would not carry out?
3. Would you expect a nerve cell and a muscle cell to have the same size and shape? Explain.

4. *Skill Review—Observing and Inferring:* You observe two photographs of cells as viewed by an electron microscope. You find many structures common to both cells, but one cell contains many additional structures. Which is probably a unicellular organism? For additional help, see Observing and Inferring in the **Skill Handbook.**

5. *Thinking Critically:* All cells of a multicellular organism carry out cellular respiration. Why don't multicellular organisms have specialized cells for this function?

5.2 Functions and Interactions of Cell Parts

You learned in Chapter 4 the various ways in which materials enter cells. Some of these materials are used to assemble new molecules that contribute to cell growth. Others are used to make molecules that regulate the cell's functions or the functions of other cells. Still others are sources of energy that fuel these activities. It is during these metabolic processes that by-products are formed. You have also learned how by-products leave the cell. Now you are ready to study how each of the organelles and other cell parts function, and how they work together to make cells the basic units of life.

Cell Types

Almost all cells found in the living world contain organelles that are surrounded by a membrane. Cells that possess a membrane-bound organelle called a nucleus are known as **eukaryotes,** meaning cells that contain a nucleus. Plant, animal, protist, and fungi cells are examples of eukaryotic cells. Cells that do not have a membrane-bound nucleus are called **prokaryotes,** meaning before a nucleus. Bacteria are prokaryotes because they lack a nucleus and other membrane-bound organelles. Prokaryotes contain one type of organelle that synthesizes proteins, but most metabolic functions take place in the cytoplasm. The vast majority of cells in the living world are eukaryotic.

Objectives

Compare the structure and function of cell organelles, including the nucleus, mitochondrion, chloroplast, lysosome, endoplasmic reticulum, Golgi body, vesicle, and vacuole.

Discuss how the functions of ribosomes, endoplasmic reticulum, and Golgi bodies are interrelated.

Describe the structure and function of other eukaryotic cell parts.

Figure 5-8 Trees, ferns, fungi, birds, flowers, insects and worms are all composed of eukaryotic cells. Eukaryotic cells are found in most multicellular organisms.

Eukaryotic Cell Organelles

Just as a newborn baby's major task is to transform food and oxygen for its own growth and development, a newly produced cell must make the proteins it needs to carry out its specialized tasks. These proteins might be used, for example, as enzymes.

Ribosomes and Endoplasmic Reticulum   Proteins are made from amino acids at organelles called **ribosomes,** tiny particles that are composed of RNA and protein. The electron microscope shows that some ribosomes float freely in the cytoplasm. Proteins synthesized at free ribosomes are used in chemical reactions that occur in the cytoplasm. Free ribosomes are found in the cytoplasm of both prokaryotic and eukaryotic cells. Ribosomes are the most common organelle in any cell. A single prokaryotic cell may have as many as 15 000 ribosomes; a eukaryotic cell contains many more. In eukaryotic cells, other ribosomes are attached to the membranes of the **endoplasmic reticulum,** or ER. The ER is a network of interconnected, flattened, or tubelike structures that end in blind alleys, as Figure 5-9 shows. The ER on which ribosomes are located has a bumpy appearance and is known as rough ER. Proteins made on ribosomes of rough ER enter channels in the ER membrane and are transported to other parts of the cell. Do you see how the tubular structure of the ER is suited to its function of transport?

Endoplasmic reticulum that lacks ribosomes is known as smooth ER. In human liver cells, enzymes located in membranes of smooth ER break down harmful substances, such as alcohol. Other proteins in the smooth ER membrane serve as enzymes to produce phospholipids. Rough and smooth ER are sometimes connected. In these cases, the smooth ER may help in the transport of proteins coming from the rough ER.

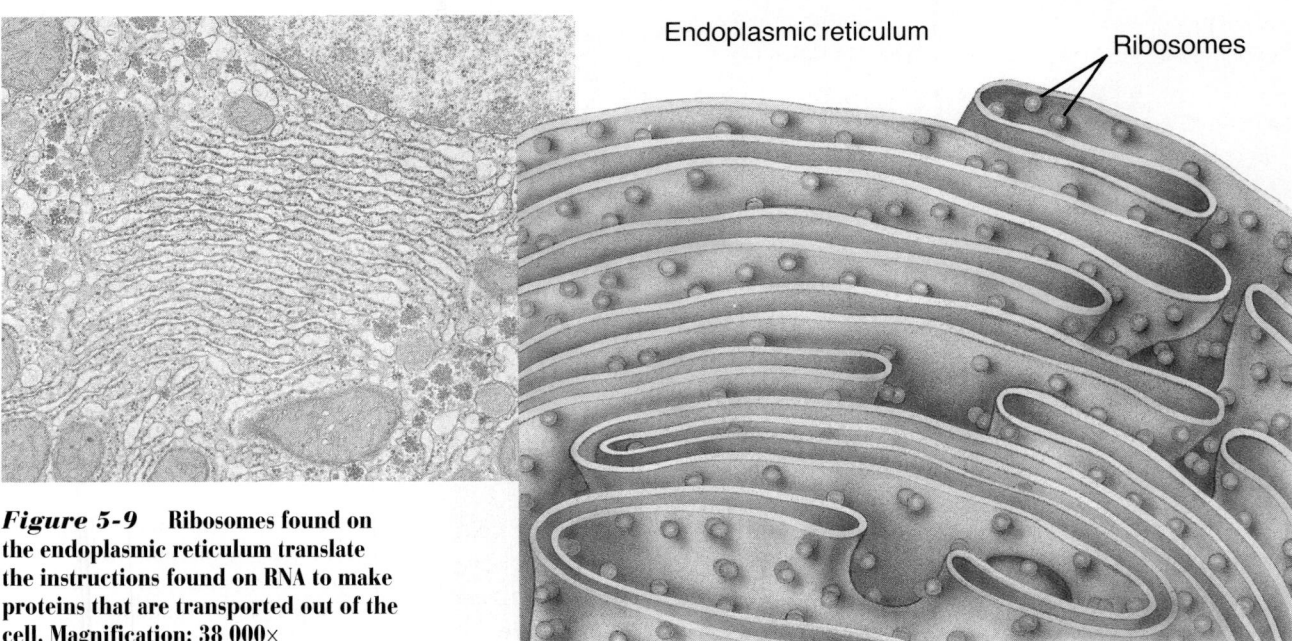

Endoplasmic reticulum

Ribosomes

Figure 5-9   **Ribosomes found on the endoplasmic reticulum translate the instructions found on RNA to make proteins that are transported out of the cell. Magnification: 38 000×**

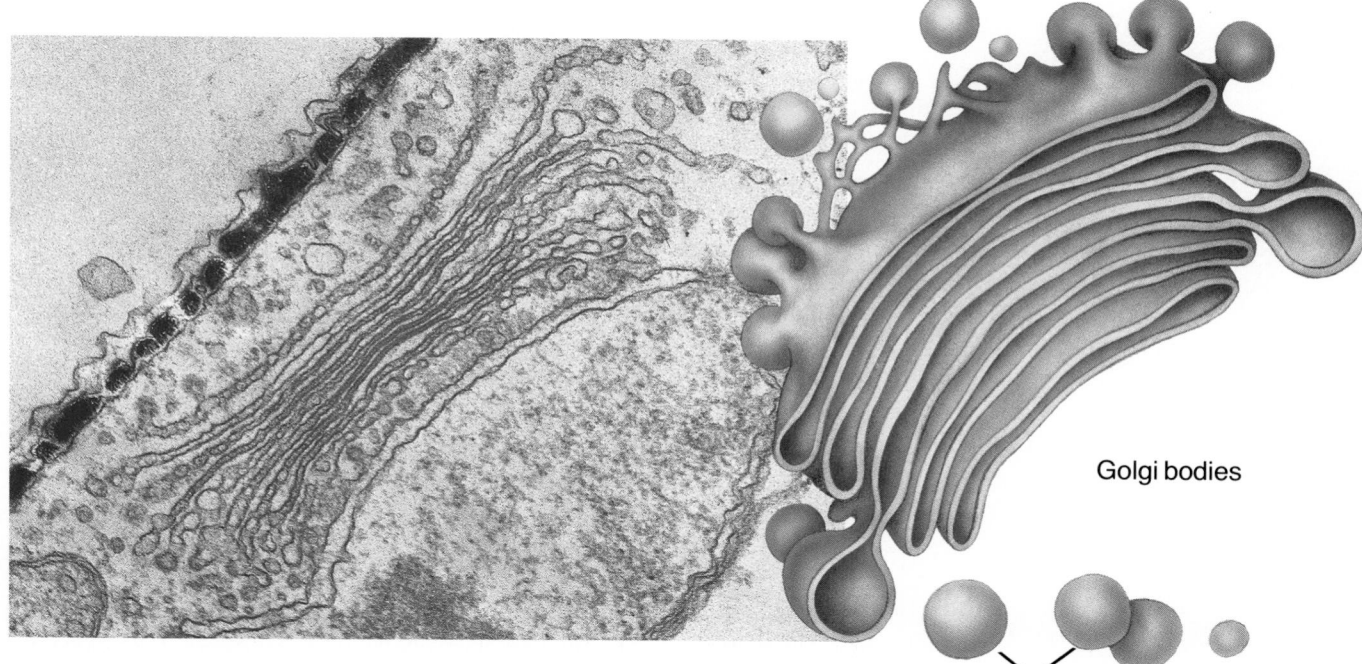

Golgi bodies

Sacs of protein moving out

Golgi Bodies   Lying close to the ER are sets of flattened, slightly curved sacs, as shown in Figure 5-10. Although they seem to resemble ER, they are more disklike than ER and much smaller. These groups of saclike structures are known as the **Golgi bodies.**

When proteins reach the end of rough ER, they are surrounded by a small sac called a vesicle formed from the ER membrane, which then breaks away from the ER. This vesicle moves toward the membrane of the nearest sac of a Golgi body, fuses with it, and empties its protein molecules into it. The proteins pass from one sac to another in vesicles formed from the Golgi membrane. These are the vesicles visible near the Golgi bodies in Figure 5-10. Like pieces of mail coming into a post office, the various proteins traveling through the Golgi body are sorted and eventually sent to their proper destination inside or outside of the cell. While the proteins are in each sac, they are modified chemically. This modification is part of the process by which proteins are properly identified and sorted. The sorted proteins are packaged in a vesicle formed from the final Golgi sac. Each vesicle contains proteins needed for a particular function. For example, some vesicles contain enzymes that will remain in the cell. Other vesicles contain hormones that will be transported to other cells in the same organism. Some may contain chemicals, such as poisons, that will eventually be used to affect other organisms. Vesicles that contain materials intended for use outside the cell fuse with the plasma membrane and release the materials through exocytosis.

Interaction among ribosomes, ER, and Golgi bodies illustrates how various parts of a cell work together. It also shows that the many membranes present in cells are interchangeable. For example, vesicles formed from ER and Golgi membranes can fuse with the cell membrane. They can be recycled from one part of a cell to another. This shuttling of membranes is possible because they all have the same basic fluid mosaic structure.

Figure 5-10   **At the edges of the Golgi bodies, vesicles containing proteins pinch off and move to other locations in the cell. Golgi bodies are the delivery system in a eukaryotic cell. Magnification: 48 000×**

Mitochondria The alarm clock rings, but you turn it off for another 40 winks. When you finally wake up, you have 15 minutes to get ready and out the door, so you skip breakfast. Two hours later, you find yourself nodding off in math class, or unable to keep up with your partner in science. What's wrong? You didn't give your body the energy it needs to perform to the best of its ability.

The cell needs energy to perform to the best of its ability, too. An organelle in eukaryotic cells provides the energy "breakfast" that cells need. **Mitochondria,** found in all eukaryotic cells, are the powerhouses of cells. Organic molecules are broken down in mitochondria to release energy. This energy is then used to produce other molecules that release energy needed for cell reactions. The more energy a particular cell needs, the more mitochondria it is likely to have. Mitochondria are often found in groups in portions of the cell that use a lot of energy. Although they are among the largest organelles in the cell, mitochondria vary in size and shape. Mitochondria may look like spheres, cylinders, or peanuts. They may be large in cells that use a lot of energy, such as heart muscle cells. A slice through a mitochondrion, as seen in Figure 5-11, reveals that it has a double membrane. The final reactions in cellular respiration take place on the inner membrane. The inner membrane is highly folded within the fluid-filled central cavity. The membrane's many folds increase the surface area for these final steps. With more surface area, more reactions can occur, and more energy can be released.

CHEMISTRY CONNECTION

Spider Silk

The U.S. Army is definitely interested. So is the Navy. Medical equipment manufacturers are conducting their own experiments, and even the world of high fashion is keeping a close watch on new developments.

What is at the center of all this interest? Some new synthetic wonder fabric, the product of advanced chemical research? No, the fabric is spider silk, and scientists are trying to unravel its chemical and genetic properties so that it can be commercially produced.

Wonder Thread The reason for all the interest in spider silk is its unique characteristics. Ounce for ounce, spider silk is five times stronger than steel. Furthermore, it's incredibly elastic; spider silk can be stretched up to 130 percent of its original length before it snaps. Spider silk is waterproof, and it's nonallergenic, too.

Commercially produced spider silk would have hundreds of uses. But before it can be produced in a laboratory, scientists first must discover how it is produced by spiders.

Internal Chemistry Spiders produce a variety of silks—dragline silk for dropping out of sight when threatened, capture silk for weaving webs to snare prey, cocoon silk for

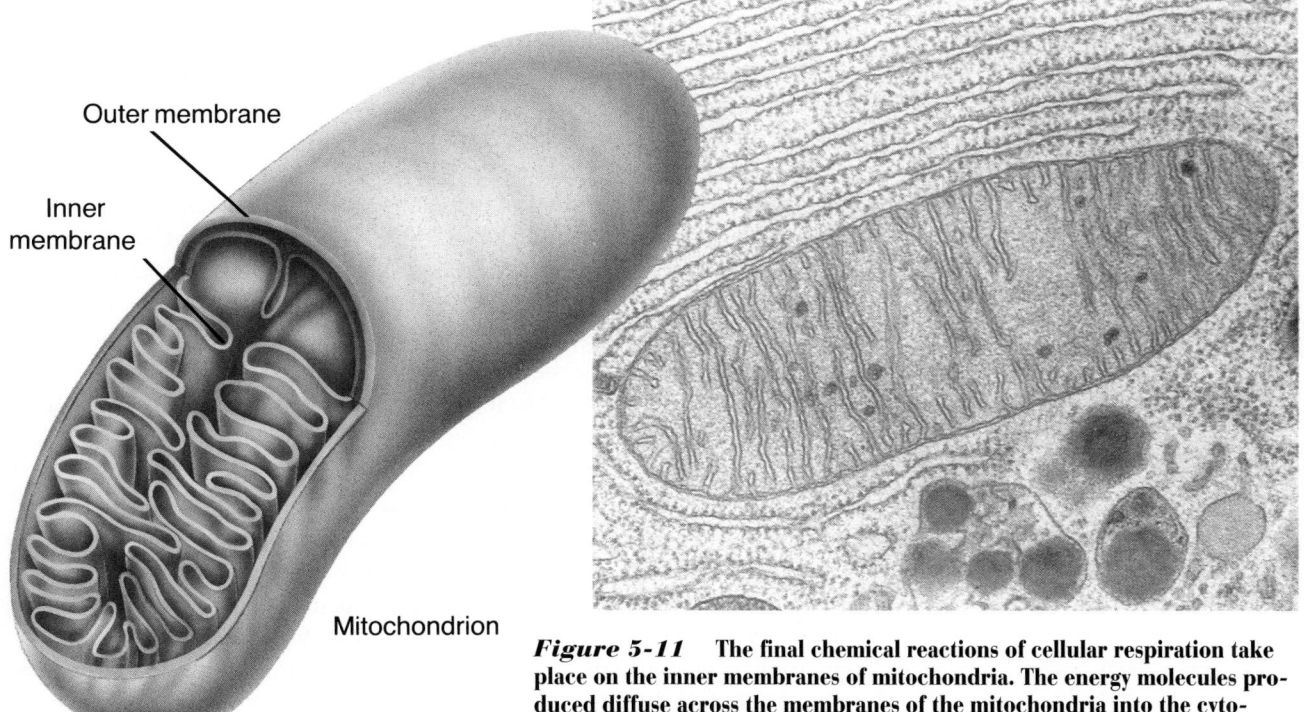

Outer membrane

Inner membrane

Mitochondrion

Figure 5-11 The final chemical reactions of cellular respiration take place on the inner membranes of mitochondria. The energy molecules produced diffuse across the membranes of the mitochondria into the cytoplasm. Magnification: 90 000×

protection and insulation, and more. Although types of silk vary, production methods are similar.

Within a spider's abdomen are several silk glands. Each gland is filled with soluble proteins made of chains of amino acids. Cells in the glands that produce these proteins use a lot of energy. High energy use in a cell means there are lots of mitochondria in the cell. These cells produce large amounts of protein, too, so they also have numerous ribosomes.

The proteins produced in one silk gland pass through a narrow tube on their way to the spinnerets, spigot-like structures on the spider from which the silk is extruded. As the soluble proteins pass through the narrow tube, the proteins' molecules are forced to align in the same direction, forming a solid thread up to 34 amino-acids long. The proteins that make up the extruded silk have stiff, rodlike regions that give the silk its strength, and curved, springy regions that give it elasticity.

In order to manufacture large quantities of spider silk, scientists will first have to engineer the genes responsible for producing silk proteins. These genes will then be implanted in cooperative organisms, such as lab bacteria or yeast. Scientists hope that these genetically altered hosts will become miniature silk factories, making commercially produced silk a reality.

———

1. **Explore Further:** How has spider silk been used by healers throughout history? What medical applications does it have for the future?
2. **Writing About Biology:** Explain why scientists can't use spiders themselves to produce commercial quantities of silk.

Minilab 2

Why are stains used when observing cells?

Prepare a slide of *Paramecium* and observe it under the microscope. Add a drop of dilute methylene blue to the edge of the coverslip. Place filter paper at the opposite edge of the coverslip to draw out the fluid. Notice that the nucleus and granules in the cytoplasm absorb the stain. Prepare another slide of *Paramecium*. Add a drop of Lugol's solution in the same way as described above.

Analysis   *What do you think the granules in the cytoplasm are? What organelles absorb the Lugol's stain?*

Nucleus   As in any organization, the cell must have an area of central control to coordinate the many activities that take place within it. In eukaryotic cells, that control area is the **nucleus,** which is separated from the rest of the cell by a double membrane. Inside the nucleus is a dense mass of material called **chromatin,** which looks something like a tangled mass of string. Chromatin is composed of individual chromosomes. **Chromosomes** are composed of proteins and DNA, the genetic code. The DNA contained in the nucleus ultimately controls all the activity that takes place in the cell. It does so by making RNA. The DNA remains in the nucleus, but the RNA can move from the nucleus into the cytoplasm. Some RNA acts as a messenger, carrying out the instructions of the genetic code by directing the synthesis of proteins at the ribosomes.

Before a cell reproduces, its chromosomes are copied. Then, when the cell begins to reproduce, the chromosomes that make up the tangled mass of chromatin become separate and distinct. During reproduction, one set of chromosomes goes to the nucleus of each new cell. Thus, each offspring cell contains the genetic information required to control its activities.

Although the nucleus is enclosed in its own membrane, it must communicate with the rest of the cell. Definite openings called pores in the nuclear membrane allow some large molecules, such as RNA, to pass between the nucleus and the cytoplasm. Other materials move within portions of ER that are joined to the nuclear membrane.

Parts of some chromosomes consist of multiple copies of DNA that make the RNA present in ribosomes. The RNA molecules produced in the nucleus stand out as prominent bodies known as **nucleoli.**

Thinking Lab

INTERPRET THE DATA

What organelle directs cell activity?

Background
Acetabularia, a type of marine algae, grow as single, huge cells 2-5 cm in height. The nuclei of these cells are in the "feet." Different species of these algae have different kinds of caps, some petal-like and others like umbrellas. If a cap is removed, it quickly regenerates. If both cap and foot are removed from the cell of one species of algae and a foot from another species is attached, a new cap will grow. This new cap will have a structure with characteristics of both species. Then, if this new cap is removed, the next cap that grows will be like the cell that donated the nucleus.

The scientist who discovered these properties was Joachim Hämmerling. He wondered why the first cap that grew had characteristics of both species, yet the second cap was clearly like that of the cell that donated the nucleus.

You Try It
Look at the diagrams below and **interpret the data** to explain the results. Why is the final cap like the one from which the nucleus was taken?

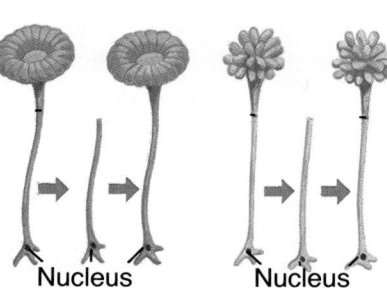

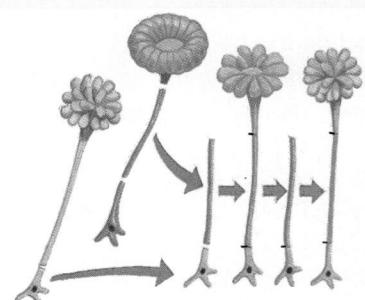

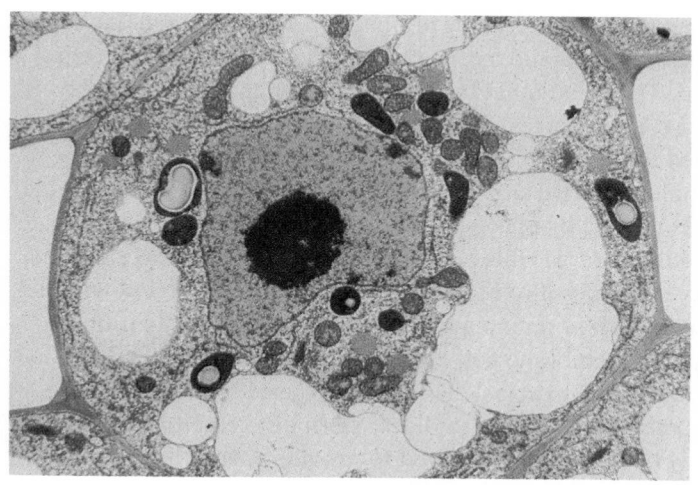

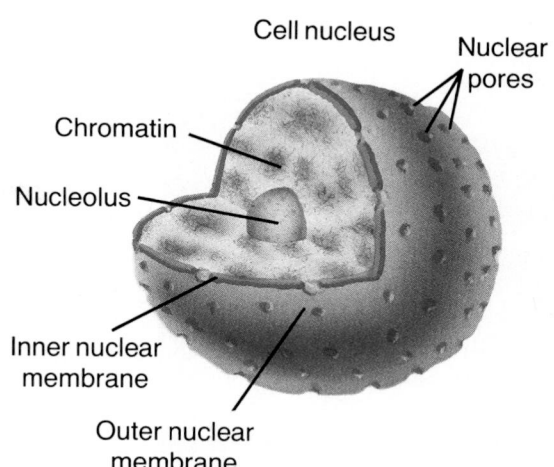

Cell nucleus

Nuclear pores

Chromatin

Nucleolus

Inner nuclear membrane

Outer nuclear membrane

Other Eukaryotic Cell Structures

In addition to the organelles discussed above, some eukaryotic cells contain additional organelles, such as plastids and lysosomes. The use of new, high-voltage electron microscopes has increased our understanding of other cell structures and led to the discovery of some new ones, including parts that help cells move.

Plastids   While all eukaryotic cells have mitochondria, a nucleus, ribosomes, ER, and Golgi bodies, only certain cells of some plants and algae contain organelles called plastids. Some plastids are used to store lipids and starches, whereas others contain pigments. The most common pigment-containing plastids are **chloroplasts.** Chloroplasts contain chlorophyll and other pigments required for photosynthesis. Remember, only those plant cells that actually carry out photosynthesis contain chloroplasts. A chloroplast consists of a double membrane surrounding an interior filled with liquid and numerous internal membranes. During photosynthesis, the internal membranes trap light energy, which is used to produce simple sugars or other organic molecules.

Figure 5-12   **The nucleus contains chromatin and a nucleolus. It communicates with the rest of the cell through pores in the outer membrane. Magnification: 1750×**

Figure 5-13   **A chloroplast is a double-walled organelle found in plants and some algae. The internal membranes of a chloroplast trap sunlight used in photosynthesis. Magnification: 40 000×**

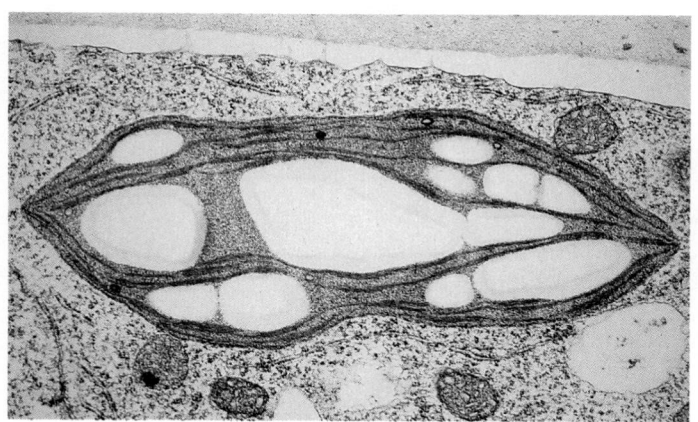

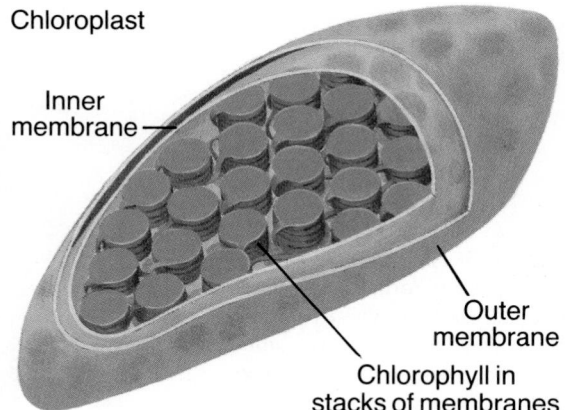

Chloroplast

Inner membrane

Outer membrane

Chlorophyll in stacks of membranes

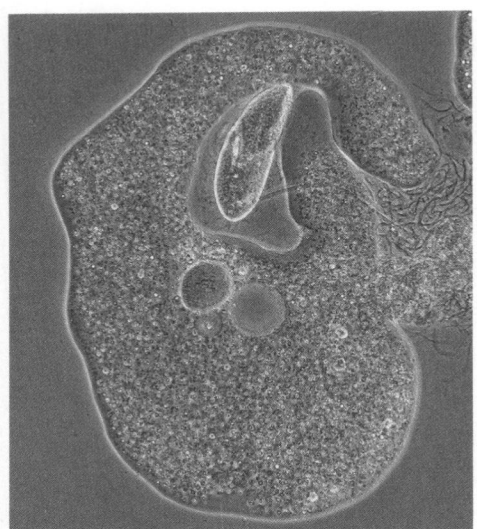

Figure 5-14 An *Amoeba* traps food and takes it in by phagocytosis. Phagocytosis continues as lysosomes fuse with vesicles containing food and release digestive enzymes to break food down. Magnification: 80×

Lysosomes   Remember the process of phagocytosis described in Chapter 4? Phagocytosis is the means by which cells like *Amoeba* trap and take in food. That food must be digested before the cell can make use of the glucose, amino acids, and other substances the food contains. Organelles called **lysosomes** are vesicles formed from the Golgi bodies. Lysosomes contain digestive enzymes that break down food. For example, as *Amoeba* takes in food by phagocytosis, the food is enclosed in a vesicle (sometimes called a food vacuole). A lysosome fuses with the vesicle and releases digestive enzymes into it. The enzymes break the food down into smaller molecules, such as amino acids and glucose, which can diffuse across the vacuole's membrane and out into the cytoplasm. What do you think happens to the amino acids? How are they used? What about the glucose? Lysosomes are also used to digest worn out cell parts. Their components may be recycled or digested to provide energy. Lysosomes are found mostly in animal or animal-like cells.

The segregation of digestive enzymes within lysosomes is an example of the importance of enclosing certain chemicals and reactions within membranes. If lysosomes were to break open, the enzymes released could destroy the entire cell. The cell is protected from the digestive enzymes by the lysosome's membrane.

Vacuoles   Cells often produce substances that are stored for later use. **Vacuoles** are fluid-filled, membrane-bound structures that store these substances. Many vacuoles act as cellular reservoirs for food, water, and minerals. In plant cells, small vacuoles merge to form a single, large vacuole, which nearly fills the mature cell. Many unicellular freshwater protists have unique contractile vacuoles, so named because as they contract they act as pumps for removing excess water. A contractile vacuole is an example of an adaptation for maintaining osmotic balance in a freshwater environment.

Figure 5-15 In plant cells, vacuoles merge to form a single, large vacuole that contains digestive enzymes, amino acids, carbohydrates, and cellular waste products. Magnification: 7800×

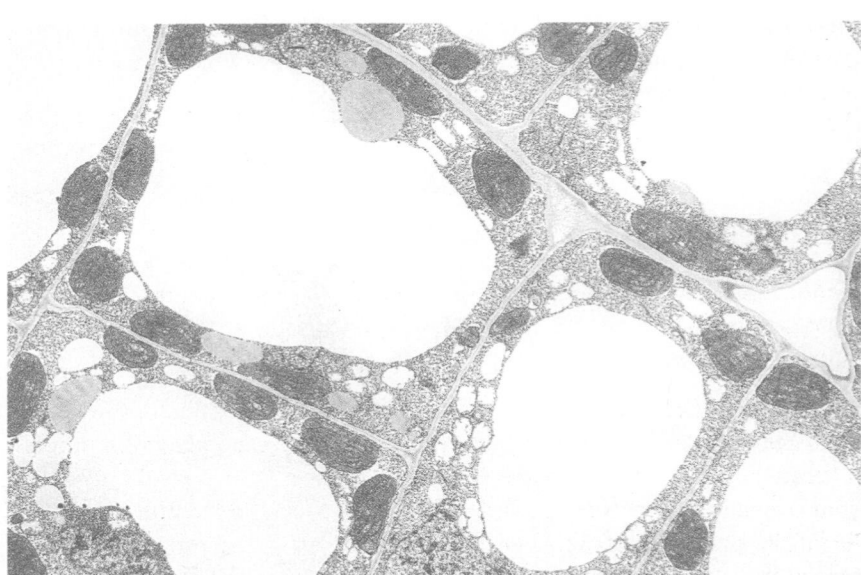

Cytoskeleton Proteins in a cell are also used in the manufacture of long, thin structures that give shape and support to the cell. These structures are referred to as the cytoskeleton. The cytoskeleton provides a framework for the cell, much as your bony skeleton provides the framework for your body.

One structure of the cytoskeleton, called a **microfilament,** is composed of the proteins actin and myosin. Microfilaments are found in many types of eukaryotic cells. In addition to providing structural support, microfilaments assist in cell movement. For example, they make it possible for you to use your muscles. When you stretch an arm or bend a knee, your muscles contract. Your muscles depend on microfilaments to make those contractions. Microfilaments are also involved in changing the shapes of cells and in the movement of some unicellular organisms, such as *Amoeba.*

Another structure of the cytoskeleton is called a **microtubule,** which is usually longer and thicker than a microfilament. Some microtubules extend all the way from the center of the cell, near the nucleus, to the area just inside the cell membrane. These microtubules help certain organelles to move from place to place within the cell. Vesicles and mitochondria, for example, travel along microtubule "tracks" rather than just drift through the cytoplasm.

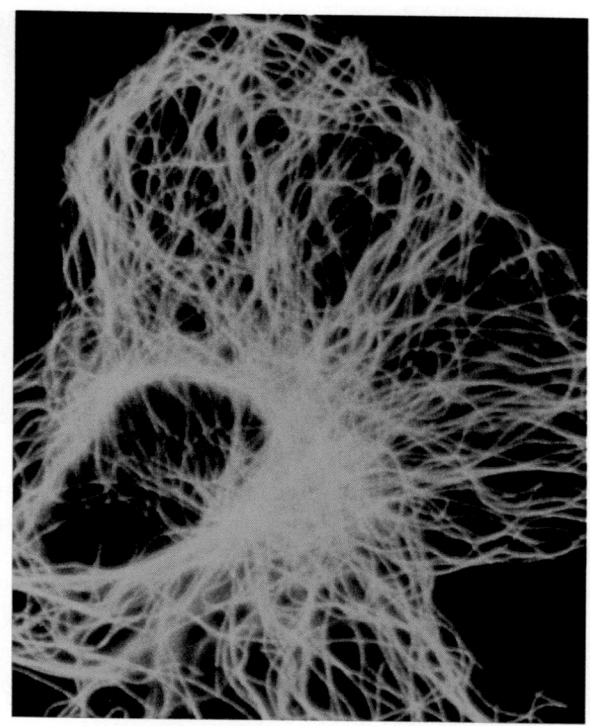

Figure 5-16 The cytoskeleton of eukaryotic cells is not stable, but is always being assembled and disasssembled. The cytoskeleton of a cell can be seen in this specially treated photograph. Magnification: 1100×

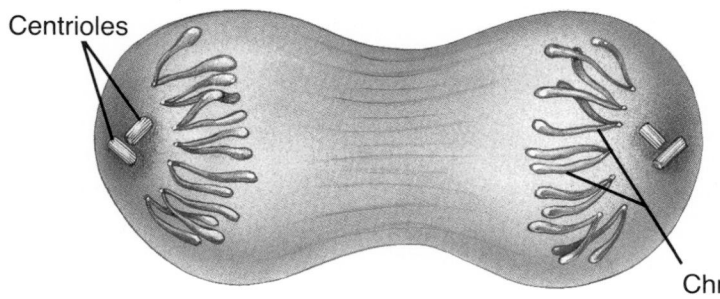

Centrioles

Chromosomes

Figure 5-17 In cells of animals and some algae and fungi, two pairs of centrioles play a role in cell division.

Centrioles Microtubules also serve as components of some organelles. For example, cells of animals and some algae and fungi contain two pairs of **centrioles,** which play an important role in cell reproduction. Each centriole is composed of sets of microtubules arranged to form a cylinder. Figure 5-17 shows how centrioles function in cell division.

Cilia and Flagella Microtubules play yet another role in the cell. They are an important component of two cell parts that are involved in motion of the cell. **Cilia** and **flagella** are flexible projections that extend outward from the cell, though they are still surrounded by the cell membrane. Flagella are longer than cilia and fewer in number. The arrangements of microtubules in cilia, flagella, and centrioles are very similar.

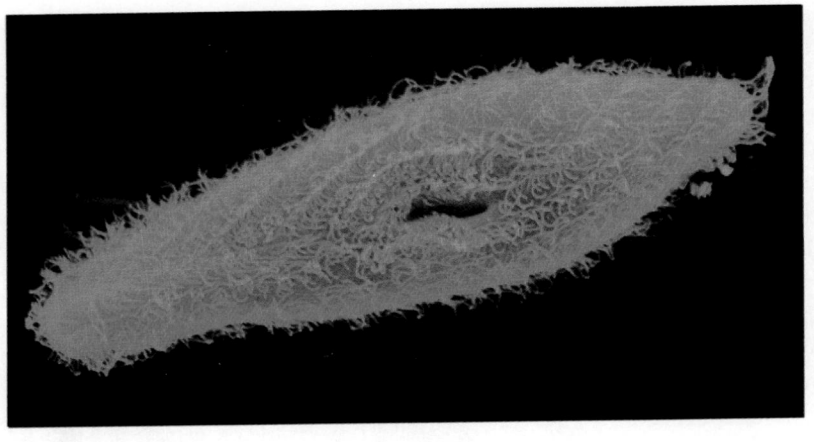

The whiplike motions of cilia or flagella enable unicellular organisms to move around in their environment. Cilia and flagella are also found in some cells of multicellular organisms. For example, portions of the cells that line your windpipe have numerous cilia. These cells do not move, but the motion of the cilia moves mucus up and out of your lungs. A sperm cell moves by means of a single flagellum.

Figure 5-19 illustrates generalized animal and plant cells. The drawings are based on observations made using an electron microscope. While such drawings are good for review of cell parts, keep in mind that they do not represent any particular cell. Cells vary in shape and size as well as the kind and number of organelles they contain. Table 5.1 lists the structures of eukaryotic cells and their functions.

Figure 5-18 This *Paramecium* moves by the whiplike motion of its cilia. What are cilia made of? Magnification: 1500×

Table 5.1 Eukaryotic Cell Structures

Structure	Found in Cells of	Functions
Cell wall	Plants; algae; fungi	Support; protection
Centrioles	Animals; some algae and fungi	Cell reproduction
Chloroplasts	Plants; algae	Photosynthesis
Cytoplasm	All eukaryotes	Metabolism
Endoplasmic reticulum	All eukaryotes	Intracellular transport of proteins (rough and smooth); smooth ER enzymes break down substances
Golgi bodies	All eukaryotes	Modification and packaging of some proteins; in plants, secretion and processing of carbohydrates
Lysosomes	Mainly animals or animal-like cells	Digestion within cells
Microfilaments	All eukaryotes	Movement; support
Microtubules	All eukaryotes	Movement; support
Mitochondria	All eukaryotes	Break down organic molecules to release energy for cell reactions
Nucleolus	All eukaryotes	Production of ribosomes
Nucleus	All eukaryotes	Control of cellular activities and reproduction
Plasma membrane	All eukaryotes	Boundary; regulates passage of materials into and out of cells
Ribosomes	All eukaryotes	Protein synthesis
Vacuoles	Most prominent in plants and algae	Storage, digestion, maintenance of osmotic balance

Typical Animal Cell

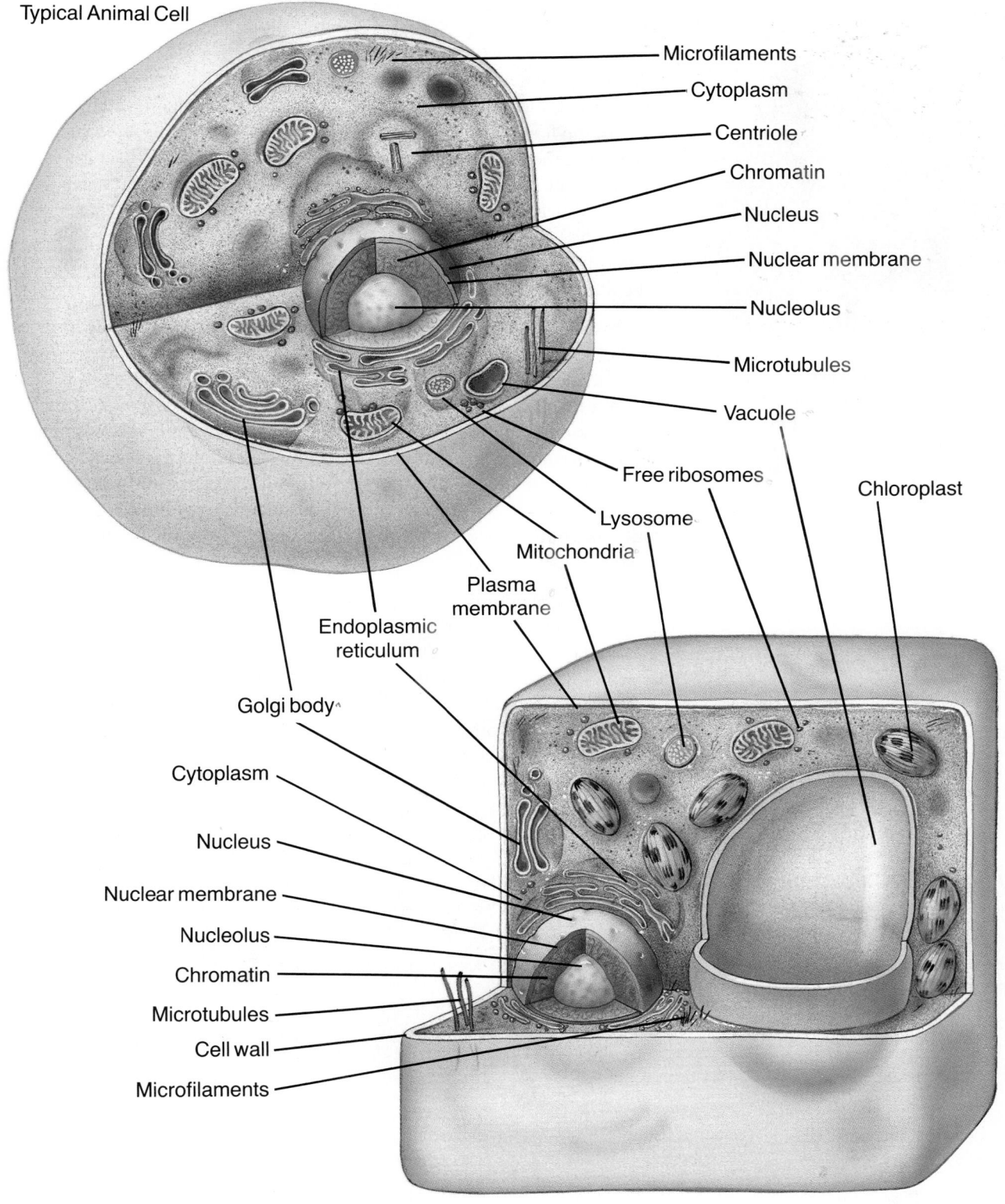

Microfilaments
Cytoplasm
Centriole
Chromatin
Nucleus
Nuclear membrane
Nucleolus
Microtubules
Vacuole
Chloroplast
Free ribosomes
Lysosome
Mitochondria
Plasma membrane
Endoplasmic reticulum
Golgi body
Cytoplasm
Nucleus
Nuclear membrane
Nucleolus
Chromatin
Microtubules
Cell wall
Microfilaments

Typical Plant Cell

Figure 5-19 **The structures of typical plant and animal cells show their similarities and differences.**

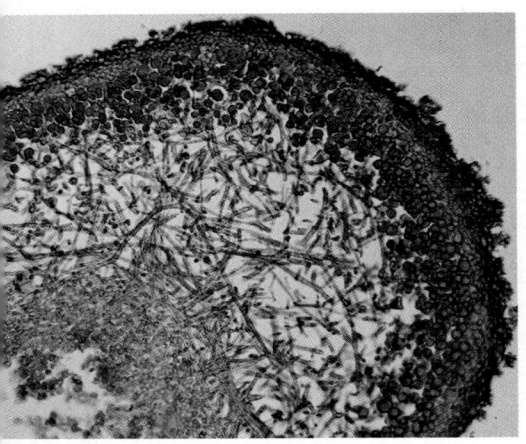

Figure 5-20 *A lichen is an example of symbiosis between fungi cells and the cells of monerans or protists. Magnification: 67×*

Evolution of Eukaryotes

Prokaryotes such as bacteria were the first organisms to evolve. As you have learned, these simple cells had no nucleus. They did have DNA, but it was in the form of a single, double-stranded, circular chromosome. Prokaryotes then and now are essentially the same. Today's bacteria lack a nucleus, but do have a single chromosome. These prokaryotes also lack organelles such as ER, Golgi bodies, and most of the other cell parts you have studied. Ribosomes, although present, are smaller than those of eukaryotic cells. Table 5.2 compares some of the features of prokaryotes and eukaryotes.

Symbiosis Scientists have combined logic and indirect evidence to piece together the puzzle of how eukaryotes might have evolved from prokaryotes. The most widely held model is called the symbiotic theory. It is based on the idea of **symbiosis,** a relationship in which two organisms live in close association. For example, certain bacteria living in a cow's intestine produce an enzyme that digests cellulose. Without the bacteria, the cow wouldn't be able to digest grass. Without the cow's intestine as a place to live, and cellulose and other foods provided by the cow, bacteria would die. Both the cow and the bacteria benefit from this symbiotic relationship.

Evidence for Symbiotic Theory The symbiotic theory states that sometime during evolution, prokaryotic cells were engulfed by other cells to become the ancestors of eukaryotes. That is, prokaryotic cells became the symbiotic partners of other cells, giving rise to the eukaryotic cells. Evidence for the symbiotic theory comes from the study of mitochondria and chloroplasts. Both these organelles contain their own DNA, RNA, and ribosomes. These structures seem to duplicate what is already present

Table 5.2 **Comparison of Prokaryotes and Eukaryotes**

Prokaryotes	*Eukaryotes*
no true nucleus or nuclear membrane	true nucleus and nuclear membrane
single circular chromosome of double-stranded DNA	several linear chromosomes of DNA and protein
no mitochondria, ER, Golgi bodies, lysosomes	mitochondria, ER, Golgi bodies, lysosomes
photosynthetic membrane, if present, not enclosed in organelle	photosynthetic membrane, if present, in chloroplasts
cell wall, if present, containing murein or other substances	cell wall, if present, containing cellulose or other substances
no microtubules in flagella	microtubules in flagella and cilia
smaller ribosomes	larger ribosomes

elsewhere in the cell. Why do mitochondria and chloroplasts have their own versions of these structures? It is thought that modern mitochondria and chloroplasts are what remains of what were once free-living prokaryotes. These prokaryotes were engulfed by and remained within other prokaryotes, forming a symbiotic relationship. If mitochondria and chloroplasts were once free-living, it makes sense that they would have equipment for synthesizing proteins and regulating their reproduction. Also, like the DNA of prokaryotes, the DNA of mitochondria and chloroplasts is circular, and their ribosomes are smaller. Furthermore, their enzymes for making nucleic acids and proteins are similar to those found in prokaryotes.

This model of the origin of eukaryotes is logical and appealing, but incomplete. How, for example, did the nuclear membrane, the various membrane-bound organelles, and all the other eukaryotic cell structures evolve? These questions have not yet been answered.

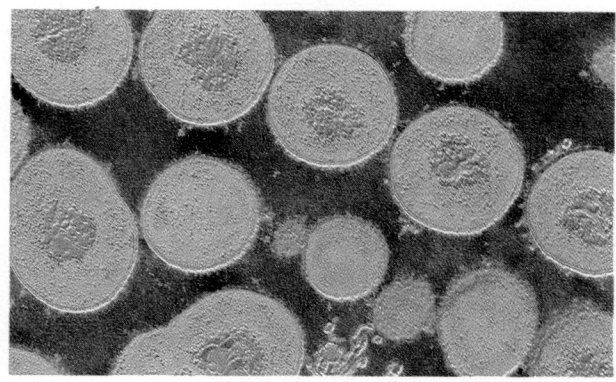

Figure 5-21 **Prokaryotic cells do not have a nucleus, but they do have double-stranded DNA. The DNA of prokaryotes is found in a circle within the cell. Magnification: 6000×**

LOOKING AHEAD

DIVISION OF LABOR is a characteristic of cells. Organelles and other cell parts carry out particular functions efficiently. Together, all the parts contribute to the survival of the cell, the basic unit of living systems. So far, the functions of the organelles have been identified, but not studied in detail. In the next chapter, you will explore two of those functions—cellular respiration and photosynthesis—in greater depth. You will learn how both processes are essential to an organism's ability to meet its energy requirements, and how the two processes are related.

Section Review

Understanding Concepts
1. A unicellular organism is discovered. It has DNA and ribosomes, but neither mitochondria nor ER. Is the cell a prokaryote or eukaryote? Explain.
2. Observing a photograph made using an electron microscope, a biologist sees what appears to be a small, saclike structure between the ER and a nearby Golgi body. What is the structure and what does it contain?
3. Why must digestion that occurs within a cell occur inside the food vacuole instead of in the cytoplasm?
4. *Skill Review—Classifying:* Divide cell parts into two categories—those found in all eukaryotic cells and those present in only some eukaryotic cells. List the function(s) of each part. For more help, see Organizing Information in the **Skill Handbook.**
5. *Thinking Critically:* Based on your answer to question 4, explain why eukaryotes have the parts in common that they do.

Biotechnology

Culturing Artificial Skin

On a warm July day in 1983, life changed suddenly for two brothers, Jamie and Glen Selby, ages five and six. The boys were using a solvent to remove paint from their bodies when the solvent ignited. By the time the flames were extinguished, Jamie and Glen had suffered third-degree burns over nearly 90 percent of their bodies.

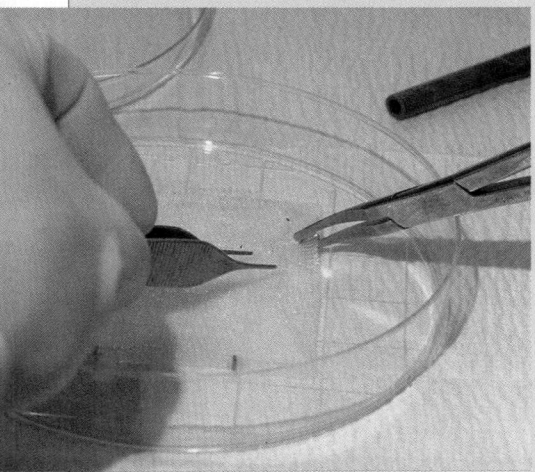

The prognosis for the two Wyoming boys was grim. Skin is the body's first defense against injury and infection. When both the outer and inner layers of the skin are burned away, as they are in third-degree burns, bacteria and other harmful organisms have access to the body's systems. Doctors expected Jamie and Glen to develop massive infections, go into shock, and die.

That's not what happened, however. The brothers were airlifted to the Shriners Burn Institute in Boston, Massachusetts, where nickel-sized pieces of their unburned skin were cultured into thin sheets of epithelial cells. This artificial skin was grafted onto the boys' burned bodies where it grew and formed new, permanent skin.

The Structure of Skin   How is artificial skin produced? Skin has two layers of cells: the dermis, which is the thick, inner layer; and the epidermis, the thin, outer layer. The dermis is made up of connective tissue. Within the dermis are oil and sweat glands, hair follicles, blood vessels, and nerves.

The epidermis consists of 20 layers of scalelike epithelial cells. These cells, called keratinocytes, are produced in the basal area of the epidermis. As new keratinocytes are produced by mitosis, the older cells begin dehydrating and dying as they are pushed upward toward the skin's surface. Finally, these cells are shed.

Growing Epidermis   The process of culturing artificial skin begins with some type of culture medium on which the skin cells can grow. Keratinocytes are placed onto the medium and washed with a liquid

mixture of proteins and nutrients similar to those found in the blood vessels of the dermis.

The keratinocytes divide and multiply. Eventually, a thin layer of epidermis is formed, and the sheet of artificial skin is ready for use.

Problems Yet to Be Overcome
There remain hurdles to be overcome before the use of artificial skin as a replacement for burned skin is perfected. For example, artificial skin is quite fragile—bacteria can attack and dissolve artificial epidermis.

Artificial skin can set off the body's immune response. If the skin has been cultured from cells from another donor, chances are the body will reject the new skin.

Nevertheless, the promise of artificial skin is quite real. Jamie and Glen Selby, and thousands of burn victims like them, are proof that tissue culturing has become much more than just an experimental procedure. It's a lifesaver.

1. **Explore Further:** What other applications exist for commercially produced artificial skin?
2. **Issue:** Should animals be used to test the effectiveness of artificial skin as a treatment for severe burns?

Issues

Don't Let the Sun Catch You Fryin'

During the Victorian era in the late 19th and early 20th centuries, no woman of manners and social standing ever would have exposed her skin to the sun's rays. Wide-brimmed hats and parasols were considered essential accessories for a woman on any sunny day.

By the middle of the 20th century, however, sunbathing had gained popularity as a favorite leisure pastime. A well-bronzed complexion was considered a sure sign of health and vitality among whites, and hats and parasols were exchanged for sunglasses and sun-tan oils.

Today, however, the "cult of the sun" finally may be coming to an end. The incidence of malignant melanoma, a virulent and often fatal skin cancer, continues to rise dramatically. In the United States today, 32 000 people are diagnosed with melanoma, and 6000 people die from it, each year. By the year 2000, some researchers estimate that one in 90 Americans will develop this disease. A major culprit, scientists agree, is ultraviolet radiation from the sun.

Two different wavelengths of light cause damage to the skin. Ultraviolet A (UVA) is long-wavelength solar radiation that penetrates the epidermis and travels down to the dermis. There, UVA attacks two proteins, collagen and elastin, which give skin a smooth, supple look. UVA destroys these proteins, and the result is wrinkled, sagging skin.

Ultraviolet B (UVB) is shorter-wavelength solar radiation that does its damage in the epidermis. UVB rays destroy certain DNA bonds in the nuclei of cells called keratinocytes. Although these cells have the ability to repair their destroyed DNA, sometimes a cell rearranges the repaired genetic material incorrectly. The result is a damaged cell that may begin multiplying uncontrollably, leading to skin cancer.

Although scientists agree that solar radiation is the cause of most skin cancers, there is some disagreement about why skin cancer, particularly malignant melanoma, has suddenly become so prevalent.

CAREER CONNECTION

If you enjoy solving mysteries, a career in cytology may be right for you. A **cytologist** discovers and analyzes previously unknown cell structures. Cytologists specialize in the study of plant cells and animal tissue to increase their understanding of the basic unit of all life—the cell.

Many scientists blame the ozone layer, which blocks much harmful solar radiation. As polluting gaseous chemicals, such as chlorofluorocarbons, rise up into the atmosphere, they destroy part of the ozone layer, allowing greater amounts of solar radiation to strike Earth.

Other scientists are skeptical of this hypothesis, claiming instead that the rise in skin cancers can be tied to lifestyle changes. More people are spending more time outdoors, often wearing less clothing than they did in the past. The result is increased exposure to UV radiation.

Despite some differences of opinion, scientists do agree on ways to prevent skin cancers. You can reduce your exposure to the sun, and use specially formulated sunscreens with sun protection factor (SPF) numbers of 30 or above. Protecting yourself *from* the sun, it seems, is the only sure way to prevent damage—potentially fatal damage—*by* the sun.

Understanding the Issue
1. Many people still consider a tan both healthy and glamorous. What are some ways these perceptions might be changed?
2. Do you think it's okay to sunbathe as long as you use sunscreens? Explain your reasoning.

Readings
- Saltus, Richard. "Genetic Damage and Skin Cancer." *Technology Review,* February-March 1992, pp. 11-12.
- Fischman, Ben. "Sun: Let the Fryer Beware." *Science World,* May 1991, pp. 4-6.

CHAPTER 5 INSIDE THE CELL **139**

Summary

Multicellular organisms evolved from versatile unicellular ancestors. As they did so, their cells became specialized and interdependent.

Cells of multicellular organisms work together in increasingly complex levels of organization. Thus, the entire organism, as well as each of its many cells, benefits from division of labor.

Division of labor in eukaryotes is made possible by the many organelles and cell parts of the cytoplasm. These parts provide for separate environments within the cell, each performing its own particular function, but working together to maintain the life of the cell.

Prokaryotes, the first cells to evolve, are very simple in structure, lacking the complexity of their eukaryotic descendants. The symbiotic model suggests how evolution of eukaryotes from prokaryotes might have occurred.

Language of Biology

Write a sentence that shows your understanding of each of the following terms.

centriole	microfilament
chloroplast	microtubule
chromatin	mitochondria
chromosome	nucleoli
cilia	nucleus
cytoplasm	organ
endoplasmic reticulum	prokaryote
eukaryote	ribosome
flagella	symbiosis
Golgi body	system
lysosome	tissue
metabolism	vacuole

Understanding Concepts

1. Nerves and muscles are two kinds of specialized cells in animals. How do muscles depend upon nerves?
2. You dissect a part of an animal and find it contains nerves, muscles, and epithelium. What is the level of organization of the part?
3. In answer to a quiz question, a student identifies the parts of a root cell: cell wall, cytoplasm, chloroplast, and nucleus. What is incorrect in that answer? Explain.
4. What organelles would be especially numerous in a cell specialized for secreting a protein hormone?
5. Based on your answer to question 4, outline the steps that would occur to produce and secrete that hormone.
6. Sperm cells must swim to an egg. What kind of internal organelle would sperm cells require in large numbers? Why?
7. When an animal cell reproduces, the parent cell membrane pinches in, and eventually the cell breaks in two. What cell structure might play a role in this pinching-in process?
8. Explain why bacteria can survive without membrane-bound organelles, but not without ribosomes.

Applying Concepts

9. A white blood cell in your body engulfs a bacterium and digests it. By what process is the bacterium brought into the cell? Where and how is it digested? Where were the enzymes needed for digestion produced and packaged? How would the undigestible parts of the bacterium be expelled from the white blood cell?
10. Can you suggest a reason why large pores are present in the nuclear membrane but absent in the plasma membrane?

11. Why are lysosomes found mainly in cells of consumers?

12. Other than protein synthesis and reproduction, what other single basic function must all prokaryotes be able to perform?

13. Can you explain how lysosomes might be related to the aging process? (Hint: What might go wrong?)

14. Given that prokaryotes have no membrane-bound organelles, where do you suppose most of their chemical reactions are carried out? Compare their efficiency to that of eukaryotes.

15. **Biotechnology:** Why would it be an advantage for burn victims to receive skin grafts of artificial skin grown from their own skin cells? Would your answer be different if it took a long time to grow enough new skin to cover the burned areas?

16. **Issues:** Your best friend jogs, drinks a lot of water, eats a balanced vegetarian diet, gets enough sleep, doesn't smoke, and has a nice tan. Is your friend healthy? Why?

Lab Interpretation

17. **Investigation:** The following are microscopic fields of view. Estimate the sizes of the objects in these diagrams. Explain how you got your answers.

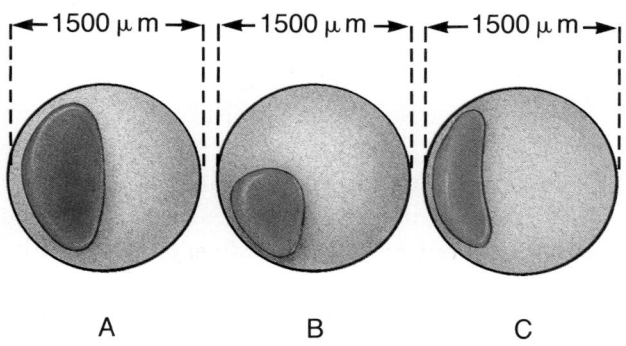

A   B   C

18. **Thinking Lab:** Examine the diagrams of the cells below. Predict what might happen if the cells are recombined as indicated. Draw diagrams of what the resulting cells might look like.

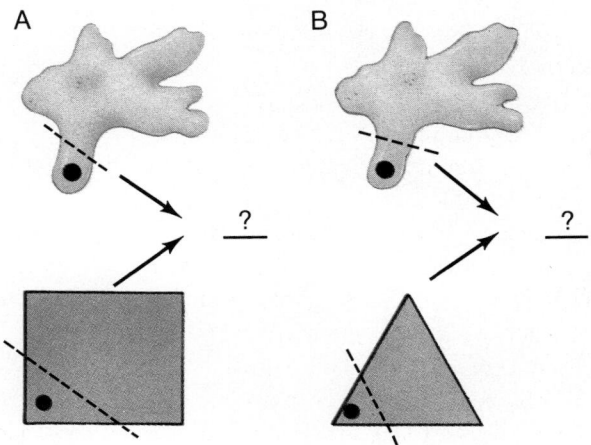

19. **Minilab 1:** You are trying to classify some protozoans and need to know whether they have cilia or flagella. They move very fast, and even slowing them with methyl cellulose does not allow you to make this determination. What could you do to determine if they have cilia?

Connecting Ideas

20. *Paramecium* is a unicellular, freshwater protist. What osmotic problem does it face? What organelle would you expect it has to solve that problem?

21. Some biologists think that the membranes inside eukaryotic cells might have evolved from parts of the cell membrane that broke off and moved inside the cell. What evidence would support that hypothesis?

THE FLOW OF ENERGY

PICTURE YOURSELF in a lush, wet, tropical rain forest like the one shown here. The hot sun is shining brightly on thick stands of tall trees. Small birds and mammals feed noisily in the canopy high above the jungle floor, and swarms of insects fill the air with an endless buzzing sound. The jungle is full of life! But tropical rain forests are not just things of beauty. And, to the billions of organisms living in rain forests, the huge, broad-leaved plants and trees provide more than a well-protected habitat. They also convert the sun's energy to energy that can be used by living things for movement, growth, and reproduction. In fact, plants in tropical rain forests convert a large portion of all the sun's energy converted by living organisms.

From the dry, hot deserts of Northern Africa to the dense pine forests of North America, all life depends on organisms that carry out photosynthesis. Through the process of photosynthesis, producers convert the energy of sunlight to chemical energy carried in organic molecules. Thus, this stored energy is passed through the food chains to all living organisms. The energy used by organisms to carry out their life processes can ultimately be traced to the biological work of producers. But how do organisms get this energy? In this chapter, you will investigate how organisms use the energy made available by photosynthesis to fuel their own processes and explore the vital process, photosynthesis, itself.

Compared to a rain forest, a desert has a very sparse supply of living things. However, just as in a rain forest, all organisms, including this desert lizard, rely on the work of photosynthesis to supply energy for life.

Chapter Preview

Objectives

Distinguish between endergonic and exergonic reactions.

Explain how ATP is used in linking exergonic and endergonic reactions in cells.

Compare the processes of aerobic respiration and fermentation.

6.1 Energy for Cells

Have you ever heard the loud booming of exploding fireworks on the Fourth of July or the crackling of a bolt of lightning in a thunderstorm? What do these events have in common? As you may have guessed, events such as these release enormous amounts of energy. What about playing a tape on your stereo, running a fan, or making toast in a toaster? These activities require energy. In the everyday world, activities that require energy get that energy from some energy-releasing event. For example, the energy that heats the coils in your toaster may have started out as the energy of water rushing through the turbine of an electric generator. Now let's see how energy-releasing and energy-absorbing events are linked in living things.

ATP

As you know, living organisms also require energy to function. You need energy to think just as a plant needs energy to grow, and as a dog needs energy to run. But where do living organisms get their energy? What are the energy-releasing events in the living world?

Free Energy   As you learned in previous chapters, in order for organisms to carry out life processes, their cells must have energy available to do biological work. This energy available to do work is called **free energy.** For example, free energy is needed for the contraction of a muscle cell, the active transport of ions and molecules across a cell membrane, and the synthesis of proteins from single amino acids. Chemical reactions, such as these, that require free energy are called **endergonic** reactions.

Figure 6-1   **An exergonic process releases free energy. This energy is transformed to light by the caterpillar on the left in a process called bioluminescence. The skunk cabbage on the right carries out the endergonic process of photosynthesis. It also generates heat in an exergonic process. This heat melts the snow surrounding the plant.**

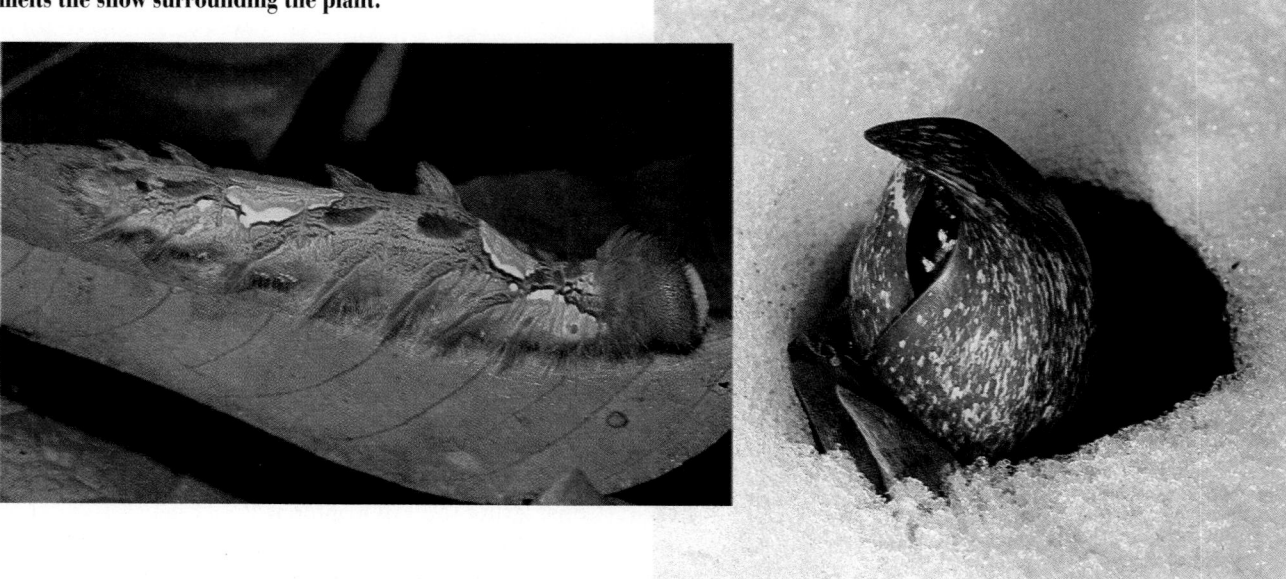

In Chapter 3, you learned that cellular reactions require enzymes to lower the activation energy needed to cause the reaction. However, endergonic reactions in organisms need more than just enzymes and activation energy. They also require an input of free energy because products of endergonic reactions have higher potential energy than the reactants.

Sources of Free Energy   The free energy required for endergonic reactions must come from energy released by other chemical reactions in the cell. These energy-releasing reactions are called **exergonic** reactions. Because free energy cannot just "wait around" until it is needed, both exergonic and endergonic reactions must be linked together for a cell to carry out biological work.

In cells, most of the free energy to drive endergonic reactions comes from an exergonic reaction involving a molecule known as adenosine triphosphate, commonly called **ATP.** The structure of ATP is shown in Figure 6-2. ATP can be represented as A–P–P–P, where the A stands for the adenosine part of the molecule and –P–P–P represents three bonded phosphate groups.

Releasing Free Energy from ATP   When the bond between two phosphate groups of ATP is broken, free energy is released. One type of exergonic (energy-releasing) reaction involving ATP is a hydrolysis reaction in which ATP reacts with water, the last phosphate group is broken off, and energy is released. The products of the reaction are adenosine diphosphate, ADP, which has only two phosphate groups, and free inorganic phosphate, P_i.

$$ATP + H_2O \rightarrow ADP + P_i + energy$$

The energy released in this reaction is used to produce heat in warm-blooded animals and to drive some processes, such as active transport. In another type of reaction, the phosphate group removed from ATP is attached to another molecule, raising the potential energy of that molecule.

Figure 6-2   **Hydrolyzing ATP to ADP and inorganic phosphate releases free energy. The reverse process stores free energy given off by other processes. This method of storing and releasing energy is called the ATP-ADP cycle. Why do you think that all successful life forms have both long-term and short-term mechanisms for storing energy?**

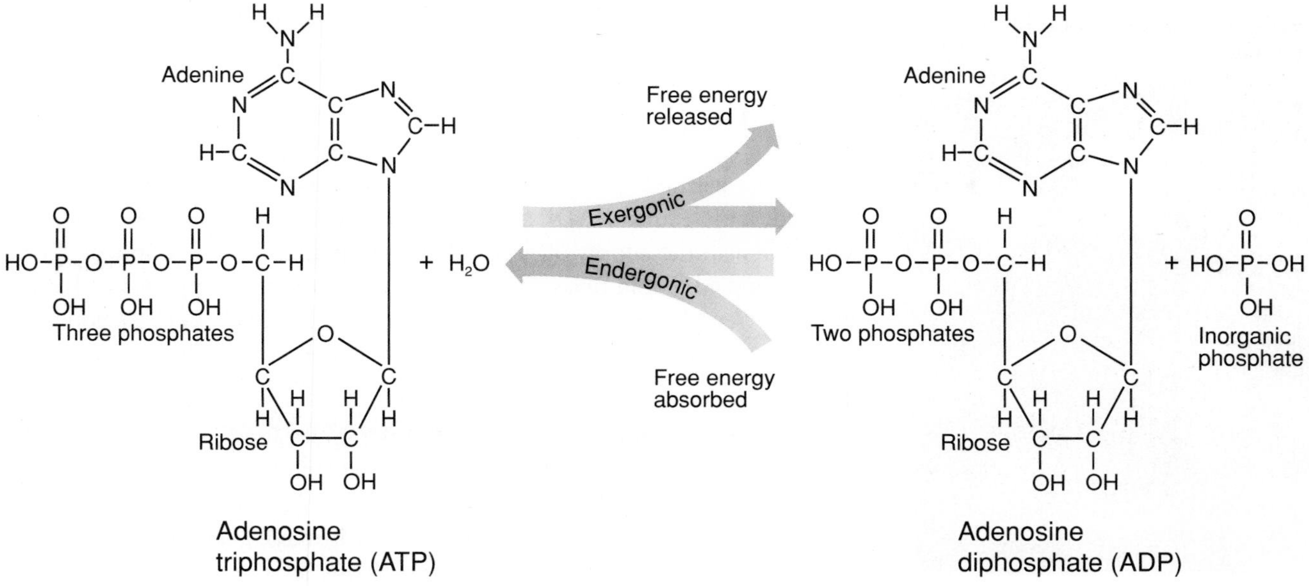

Adenosine triphosphate (ATP)

Adenosine diphosphate (ADP)

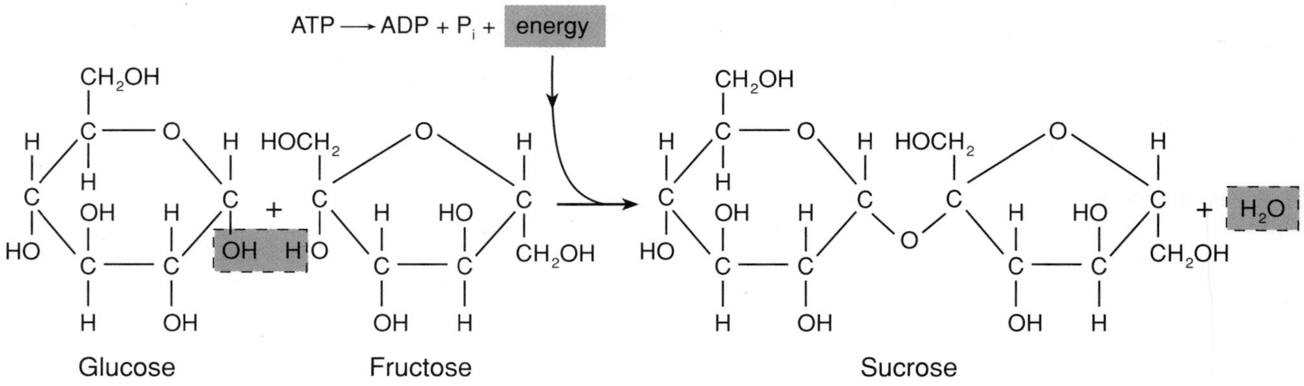

$$ATP \longrightarrow ADP + P_i + \boxed{energy}$$

Glucose + Fructose $\longrightarrow$ Sucrose + $\boxed{H_2O}$

Figure 6-3 When plants make sucrose (table sugar) from glucose and fructose, an input of energy is required. That energy comes from the breakdown of ATP to ADP and inorganic phosphate.

Linking Energy Release and Energy Use As an example of how ATP is used to link exergonic and endergonic reactions, consider a reaction in which glucose and fructose are combined to make the disaccharide sucrose. The making of sucrose is endergonic, because the product, sucrose, has more potential energy than the reactants, glucose and fructose. The reaction can be summarized in two main steps.

(1) ATP + glucose $\xrightarrow{\text{enzyme}}$ ADP + glucose – P

(2) glucose – P + fructose $\xrightarrow{\text{enzymes}}$ sucrose + P_i

Net: ATP + glucose + fructose $\xrightarrow{\text{enzymes}}$ sucrose + ADP + P_i

Do you see how one reaction is linked to the other? First, the attachment of phosphate to glucose transfers some of the potential energy of ATP to glucose. Next, the glucose phosphate combines with fructose to form sucrose and a free phosphate group. Thus, the energy from the conversion of ATP to ADP (an exergonic reaction) is used to form the chemical bond between glucose and fructose (an endergonic reaction). ATP seems to have evolved as a main energy link between exergonic and endergonic reactions, because the amount of free energy it transfers is suitable for most cellular reactions.

Respiration with Oxygen

As you just learned, after a molecule of ATP is used and energy is released, a molecule of ADP and inorganic phosphate remain. How do cells replenish the supply of ATP? Cells can maintain their supply of ATP by rejoining ADP and phosphate. But, a source of free energy is needed to bond the free phosphate to ADP and form a molecule of ATP. As you know, free energy must come from exergonic reactions. What are those reactions?

How ATP Is Replenished All organisms require food for energy. In animals, glucose is a food molecule commonly available as an energy source, although other molecules such as fatty acids can also be used. You have learned that the process by which the energy is made available for work is cellular respiration, an exergonic process. Now, let's refine our under-

Figure 6-4 All organisms require food molecules that can be broken down to release energy. In a bear, much of that energy is used to make fat molecules, which can be broken down to release heat and energy for body functions as the bear hibernates in winter.

standing of that process. During cellular respiration, some of the energy released by breaking the bonds of food molecules such as glucose is used to make ATP from ADP. These ATP molecules are then used as needed. Processes that require oxygen are called *aerobic processes.* Respiration in most cells makes use of oxygen and is called **aerobic respiration.**

Aerobic Respiration Aerobic respiration involves many reactions that follow one another in a particular sequence. The first several steps take place in the cell's cytoplasm. The products of the last of these early reactions move into the mitochondria, where the process called the citric acid cycle takes place. The process as a whole is exergonic. It can be summarized in the following equation.

$$C_6H_{12}O_6 + 6O_2 + 38ADP + 38\ P_i \xrightarrow{\text{enzymes}} 6CO_2 + 6H_2O + 38ATP$$

glucose   oxygen   carbon   water
                                              dioxide

This equation shows that aerobic respiration in some cells can release enough energy from a molecule of glucose to produce 38 ATP. Not all the steps of the process are exergonic, however. Early steps in the breakdown of glucose use the energy of two ATP. So, the maximum net ATP that can be produced is 36 molecules. ATP production from the breakdown of glucose is not always that efficient, however. Note that aerobic respiration does not produce 38 ATP in every cell. In general, about 39 percent of the energy available in glucose is transferred to molecules of ATP. The rest escapes as heat. More important than the exact number of ATP molecules produced, though, is this basic point: During cellular respiration, the energy of one glucose molecule is released bit by bit and transferred to many molecules of ATP.

Figure 6-5 summarizes the interrelationships among aerobic respiration, ATP, ADP, and the work of the cell. This series of chemical events is sometimes referred to as the ATP-ADP cycle. The diagram shows that during aerobic respiration, the energy of glucose is transferred to ADP and inorganic phosphate, forming ATP, which can then be used as a free energy source for cellular work.

Minilab 1

Can you detect energy released by a biological reaction?

Half fill a test tube with a 3 percent solution of hydrogen peroxide. **CAUTION:** *Wear goggles. Wash away any spills immediately with water.* Use a thermometer to take the temperature of the liquid in the tube. Add one or two small chunks of fresh liver to the tube. Observe what happens. After one minute, take the temperature again.

Analysis *What evidence of a reaction did you observe? Was the reaction endergonic or exergonic? What is your evidence? Considering that the reaction was carried out by living tissue, what do you think was responsible for the breakdown of hydrogen peroxide? Do you think this reaction could be used by cells to do biological work? Why or why not?*

Figure 6-5 **When sugar burns in air, it releases CO_2, H_2O, and a lot of energy at once. In respiration, the same process occurs, but step by step, releasing a little energy at a time and storing it in ATP.**

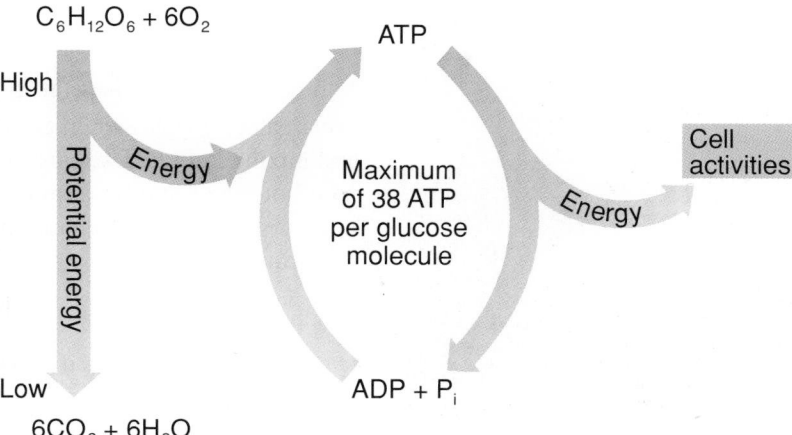

$C_6H_{12}O_6 + 6O_2$

ATP

High

Potential energy

Energy

Maximum of 38 ATP per glucose molecule

Energy

Cell activities

Low

$6CO_2 + 6H_2O$

ADP + P_i

ATP as the Cell's Energy Currency   Consider this comparison. Suppose you work for another person, and that person pays you with a $100 bill. It's not very convenient to buy lunch with a $100 bill, and you certainly wouldn't want to stuff one in the slot of a soft drink machine. Smaller bills and coins are much more convenient for daily expenses. In the same way, all the energy from the breakdown of a glucose molecule is a very inconvenient amount for a cell. Instead, respiration "makes change," storing small amounts of energy in each ATP. The energy stored in a molecule of ATP is a convenient amount for most cell reactions. That's why scientists often speak of ATP as a cell's energy currency.

I N V E S T I G A T I O N

Optimum Temperature

Humans are most comfortable with a surrounding temperature around 23°C. When it is too hot, we feel uncomfortable and start to sweat. When it is too cold, we begin to shiver and are equally uncomfortable. Thus, humans have an optimum temperature, a temperature at which they are most comfortable. Do all living things share the same optimum temperature for life activities? No. The optimum temperature for a deep-sea fish is quite different from that of a desert coyote.

Yeast cells also have an optimum temperature for life processes. In this investigation, you will design and then carry out an experiment to determine the temperature at which yeast best carries out fermentation. Carbon dioxide is given off during the process of alcoholic fermentation. Therefore, it can be used as a means for judging the rate at which yeast cells carry out this process. To help you get started, look at the diagram at the right. It shows how you could measure carbon dioxide gas given off by yeast cells during fermentation.

Problem

What is the optimum temperature for yeast cell fermentation?

Safety Precautions

Wear safety goggles when pouring and mixing liquids. Use caution with hot plates and hot liquids. Use mitts or hot pads when handling hot beakers.

Hypothesis

Have your group agree on a hypothesis to be tested. Record your hypothesis.

Experimental Plan

1. Review the discussions of aerobic respiration and fermentation in this chapter. Note particularly the reactants involved.
2. Examine the materials provided by your teacher. Then, as a group, make a list of the possible ways you might test your hypothesis using the materials available.
3. Agree on one way that your group could investigate your hypothesis.
4. Design an experiment that will (a) use a control, (b) test only

Anaerobic Processes

Have you ever lifted a heavy object and held it until you felt your muscles starting to give out? When your muscles become fatigued like this, it is because your muscle cells are not receiving enough oxygen to carry out aerobic respiration. Instead, muscle cells are undergoing anaerobic processes to release needed energy. **Anaerobic** processes take place in the absence of oxygen. Some bacteria carry out the entire process of respiration without using oxygen. More familiar, though, are two anaerobic processes called lactic acid fermentation and alcoholic fermentation.

one variable at a time, and (c) allow for the collection of quantitative data.

5. Prepare a list of numbered directions. Include a list of materials and the quantities you will need.

Checking the Plan

Discuss the following points with other group members to decide the final procedures for your experiment. Make any needed changes in your plan.

1. What materials will you use to grow your yeast?
2. What will you measure to determine the rate of fermentation?
3. What factor will you vary in your experiment?
4. What factors will you control in your experiment?
5. To what range of temperatures will you expose the yeast? Why?
6. What food will you use?
7. How will you determine appropriate amounts of yeast and food to use?
8. How long will you carry out the experiment?
9. How many tubes of yeast will you expose to each temperature?

10. Have you designed and made a table for collecting data?

Make sure your teacher has approved your experimental plan before you proceed.

Data and Observations

Carry out your experiment, make any needed measurements, and complete your data table. Design and complete a graph of your results.

Analysis and Conclusions

1. What conditions did you keep constant in the experimental group? What one condition varied?
2. What differences did you observe between experimental groups? Were these differences due to temperature differences? Explain how you know.
3. Explain how you were able to determine the optimum temperature for fermentation in yeast.
4. Was your hypothesis supported by your data? If not, suggest a new hypothesis that is supported by your data.
5. List several ways that your methods may have affected the outcome of the experiment. Suggest

ways that you might improve your methods.

6. Write a brief conclusion to your experiment.

Going Further

Based on this lab experience, design another experiment that would help to answer any questions that arose from your work. What factors might you allow to vary if you kept temperature constant? Assume that you will not be doing this lab in class.

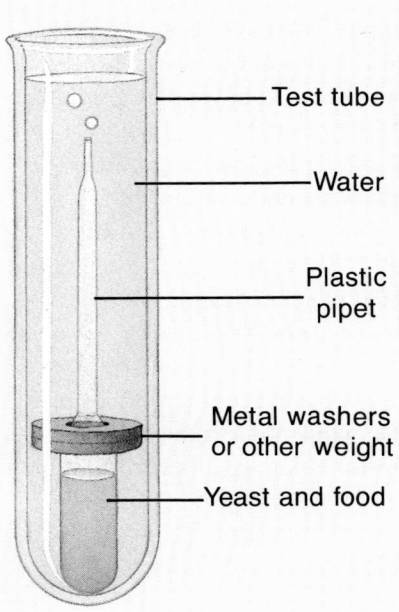

Test tube

Water

Plastic pipet

Metal washers or other weight

Yeast and food

Figure 6-6 With strenuous exercise, lactic acid fermentation begins because oxygen cannot reach the cells fast enough to meet energy demands. Why does the oxygen supply to cells become limited during exercise? Read more about this subject in the Leisure Connection below.

Lactic Acid Fermentation Ordinarily, muscles would carry out aerobic respiration, using oxygen to break down glucose and transfer energy to ATP. If not enough oxygen is available, the early steps of glucose breakdown continue to occur because they do not require oxygen. However, without oxygen, the later steps, which produce carbon dioxide and water, cannot continue. Instead, muscles carry out a process called lactic acid fermentation. **Lactic acid fermentation** is an anaerobic process in which enzymes break down a glucose molecule into two lactic acid molecules, transferring energy to ATP in the process. The following equation summarizes lactic acid fermentation.

$$C_6H_{12}O_6 + 4ADP + 4P_i \xrightarrow{\text{enzymes}} 2CH_3CHOHCOOH + 4ATP$$
glucose   lactic acid

Compare this equation with the previous equation for aerobic respiration. Lactic acid fermentation produces many fewer ATP molecules than would be produced in aerobic respiration. However, it does provide a mechanism by which muscles can keep on working when oxygen can't reach them fast enough. It is the accumulation of lactic acid in muscles that causes the pain of muscle fatigue. Besides animal cells, some microorganisms also carry out lactic acid fermentation. Examples include the bacteria used to produce sauerkraut and buttermilk.

LEISURE CONNECTION

Hitting the Wall

Perhaps you have seen marathon runners strain toward the finish line, cheeks hollow and eyes glazed. Some cross the finish line and collapse, gasping, on the ground. Others cannot even finish.

These runners have "hit the wall." The wall refers to a point in a run at which a drop in energy production causes a sudden decrease in performance. With some runners, this drop can occur within a single mile, while other runners can run for more than 40 miles without hitting the wall. Beyond the wall, the runner's legs often feel heavy and leaden. Moving the body takes so much effort that some people compare it to trying to run while carrying a bear.

You can understand what happens when runners hit the wall if you remember that ATP supplies energy to the muscles, and is produced by the cellular respiration of glucose. Most of the glucose used during long-distance running comes from glycogen stored in muscle tissue and the liver. Because the body can store only limited amounts of glycogen, long-distance runners often exhaust their stores before the end of a run.

At this point, the body begins to metabolize fat for energy. However, metabolizing fat is a more complex process than simply oxidizing sugar obtained from glycogen. Meanwhile, the runner attempts to use energy faster than the body can produce it by this slower process. At this point, the runner "hits the wall."

1. **Explore Further:** Explain how additional body fat would help or hinder a runner trying to extend the collapse point.
2. **Explore Further:** Why is hitting the wall more likely to be a problem for a long-distance runner than for a sprinter?
3. **Solve the Problem:** Suppose a runner in a marathon hits the wall at 38 km. Should that runner try to finish the race? Why or why not?

Alcoholic Fermentation   Many plant cells and microorganisms can carry out another type of fermentation called alcoholic fermentation. **Alcoholic fermentation** is an anaerobic process in which enzymes break down a glucose molecule into two molecules of ethanol and two molecules of carbon dioxide, transferring energy to ATP. The following equation summarizes alcoholic fermentation.

$$\underset{\text{glucose}}{C_6H_{12}O_6} + 4ADP + 4P_i \xrightarrow{\text{enzymes}} \underset{\text{ethanol}}{2C_2H_5OH} + \underset{\substack{\text{carbon} \\ \text{dioxide}}}{2CO_2} + 4ATP$$

In industry, yeast is used to ferment sugars from grain, fruit, or other sources in order to produce ethanol. Ethanol is the alcohol found in beer, wine, and other alcoholic beverages. It is also useful as a gasoline additive. Alcoholic fermentation is important in baking as well. Yeast added to dough carries out fermentation, breaking down carbohydrates in the flour. Bubbles of the carbon dioxide produced are trapped in the dough, causing it to rise. When the dough is baked, the alcohol produced by fermentation evaporates, raising the dough even more.

Energy Yield from Fermentation   The early steps of both types of fermentation are the same as those of aerobic respiration, requiring two ATP molecules to provide energy. So, even though fermentation produces four ATP molecules, the net ATP produced is only two molecules. This amount is only about 5 percent of the maximum ATP available from aerobic respiration. Still, by using up glucose at a rapid rate, some cells can produce ATP by fermentation almost as fast as by aerobic respiration, at least for a short time.

Some cells, such as certain bacteria, rely entirely on fermentation or other anaerobic processes for energy. Other cells, like yeast and muscle cells, switch between aerobic respiration and fermentation, depending on the amount of oxygen present.

Figure 6-7   **Fermentation by yeasts generates carbon dioxide, which makes the dough rise. The alcohol produced in the same process evaporates during baking. When yeast is used to ferment a solution containing a lot of sugar, fermentation eventually stops although sugar and yeast are still present. What do you think causes fermentation to stop?**

Section Review

Understanding Concepts

1. A friend tells you that the burning of a campfire must be an endergonic process because starting the fire required the heat from a match. Would you agree? Explain your answer.

2. Suppose a cell combines molecules A and B to produce molecule C in an endergonic reaction. Explain how ATP may be used in this type of reaction. Include the idea of free energy in your answer.

3. Suppose two cells are producing energy at the same rate by breaking down glucose—one by aerobic respiration and one by alcoholic fermentation. What conclusion can you draw about the rates at which waste would have to be removed from the two cells? Explain your reasoning.

4. *Skill Review—Designing an Experiment:* How would you design an experiment to determine whether the rate of alcoholic fermen-

tation in yeast cells differs when cells use fructose instead of glucose? For more help, see Practicing Scientific Methods in the **Skill Handbook.**

5. *Thinking Critically:* The lactic acid produced during lactic acid fermentation is transported to the liver and converted to glucose. Do you think that reaction would require ATP? Explain.

Objectives

Describe the interaction of white light with objects including chlorophyll molecules.

Relate the events of the light reactions to those of the Calvin cycle in photosynthesis.

Analyze the relationship between photosynthesis and cellular respiration.

<u>6.2</u> # Photosynthesis

The energy you need to run, think, and grow is released in the process of cellular respiration. The key ingredient in this process, of course, is the energy-rich molecule glucose. But where does glucose come from? How is it produced? What makes it energy-rich? As you know, algae, some bacteria, and plants are producers in almost every ecosystem on Earth. These producers absorb energy from sunlight and convert it to chemical energy stored in organic molecules including glucose. To understand how this process occurs, you must first learn a little about the properties of light itself.

Light

Light is a form of radiant energy. Besides light, radiant energy includes radio waves, microwaves, X rays, and gamma rays. All radiant energy travels in waves. The distance between one wave crest and the next is called the **wavelength** of the waves, as shown in Figure 6-8. The human eye is sensitive to radiant energy with wavelengths of about 400 to 700 nm. The abbreviation *nm* stands for *nanometer,* which is one billionth of a meter. This mixture of wavelengths of radiant energy that is visible to humans is called white light.

The Visible Spectrum When white light passes through a prism, the prism separates it into a range of colors. This range of colors that make up white light is known as the **visible spectrum.** The colors of the spectrum are usually given as red, orange, yellow, green, blue, and violet. Red light has the longest wavelength of the visible spectrum, and violet light has the shortest wavelength. Ultraviolet and infrared wavelengths, on either end of the visible spectrum, are invisible to humans.

Figure 6-8 **White light passing through a prism is broken into a spectrum because shorter wavelengths (violet) are bent more by the prism than longer wavelengths (red). Wavelength is the distance from one wave crest to the next. Where are ultraviolet light and infrared light located on the spectrum?**

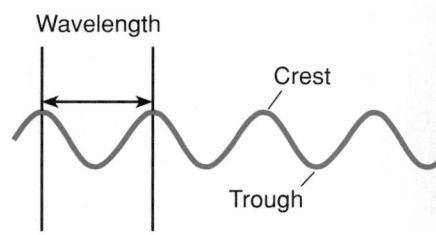

Why Objects Appear Colored   Molecules in colored objects absorb some wavelengths of light and reflect others. The color of an object depends on which wavelengths it reflects. For example, a red car looks red because it reflects more red light than other colors of light. A white shirt looks white because it reflects all colors. Likewise, a dark blue sweater looks blue because it reflects some light in the blue part of the spectrum and absorbs most other wavelengths of light.

If you've ever worn a dark-colored sweater outdoors on a sunny day, you probably remember feeling warm. The darker the material, the greater the amount of light that is absorbed. The light energy that is absorbed changes to heat. As you'll see, light energy can sometimes be changed to forms other than heat.

Figure 6-9   The color of an object depends on which colors of the spectrum it reflects. Some animals have distinct colors as camouflage or to distract and confuse predators. The bright colors of many other animals tell potential predators to beware because they are poisonous.

Chlorophyll and Other Pigments

Light energy can be changed to chemical energy. Photosynthesis is the process by which producers store energy from sunlight as chemical energy in organic molecules, mainly in molecules of carbohydrates. Organisms that carry out photosynthesis contain colored pigments that absorb light energy and convert it to chemical energy. Among these pigments, the green pigment known as chlorophyll is the most important in photosynthesis. There are several types of chlorophyll, but in plants, chlorophyll *a* is the most important. In eukaryotes, such as plants and algae, chlorophyll is found in chloroplasts. However, prokaryotes, such as photosynthetic bacteria, have no chloroplasts. Chlorophyll of photosynthetic bacteria is most often attached to membranes in the cytoplasm.

How Chlorophyll Interacts with Light   Producers usually appear green because of the presence of chlorophyll. Chlorophyll is green because it reflects light in the green and yellow parts of the visible spectrum. Chlorophyll absorbs most other wavelengths of visible light. This pattern of absorption and reflection can be demonstrated by passing white light through a solution of chlorophyll *a* and then through a prism.

Figure 6-10   In blue-green bacteria such as the *Anabaena* (left), chlorophyll is found on membranes within the cytoplasm. (Magnification: 400×) In plants and algae, such as the *Spirogyra* (right), chlorophyll is contained within chloroplasts. (Magnification: 100×)

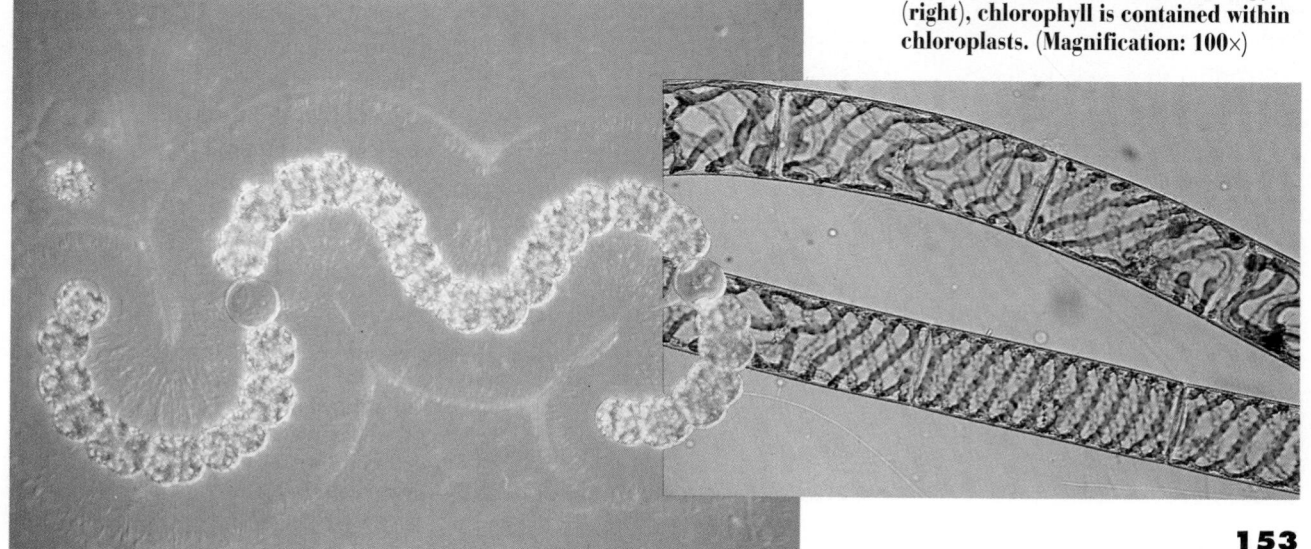

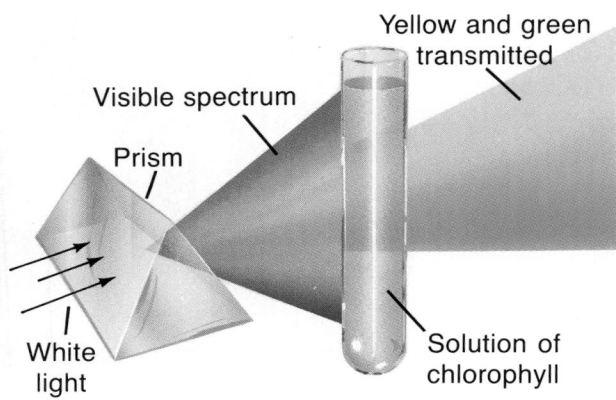

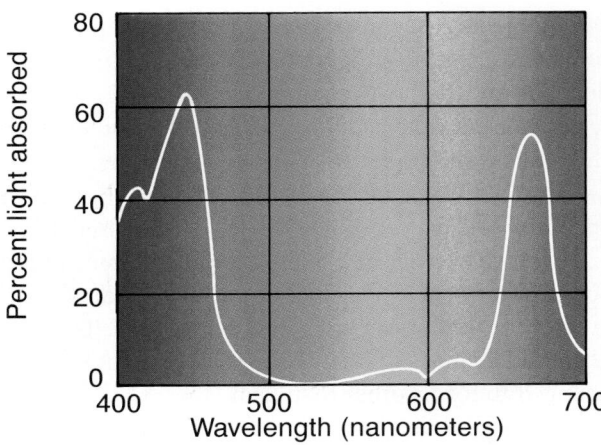

Figure 6-11 When white light
strikes chlorophyll, it absorbs colors in
the red, blue, and violet parts of the
spectrum. Thus, chlorophyll appears
yellow-green. The graph on the right
shows the absorption spectrum of
chlorophyll.

Because some wavelengths are absorbed by the chlorophyll, the spectrum produced is called an **absorption spectrum,** as shown in Figure 6-11. Violet and red wavelengths are greatly absorbed, as well as some blue wavelengths. During photosynthesis, these absorbed wavelengths are transformed from light energy to chemical energy.

Other Colored Pigments Besides chlorophyll *a* and *b,* chloroplasts also contain yellow, orange, and red pigments called **carotenoids.** Carotenoids absorb mainly blue and green wavelengths of light, and pass this energy to chlorophyll *a.* Thus, other wavelengths are also important in photosynthesis. Chlorophylls and carotenoids interact to absorb the light energy necessary for photosynthesis.

Carotenoids and other pigments are also present in other parts of a plant. These yellow, orange, and red colors are most often seen in the flowers and fruits of plants. A different assortment of carotenoids is present in leaves but is often masked by the green color of chlorophyll. In northern latitudes, the manufacture of chlorophyll usually stops in autumn. With less chlorophyll present, the carotenoids and other pigments become visible, so leaves turn from green to yellow, orange, and red.

Figure 6-12 When leaves lose
chlorophyll in the fall, the colors of
other pigments become visible. These
colors were previously masked by the
green color of chlorophyll.

Overview of Photosynthesis

Photosynthesis occurs only in producers—some bacteria, algae, and plants. In photosynthetic bacteria and most algae, photosynthesis occurs in all cells. With evolution of green plants, however, certain cells became specialized for this process. As you study photosynthesis, think about a familiar plant, such as a tree. As you will learn in Chapter 21, chloroplasts, the organelles of photosynthesis, are located in just some, not all, of the cells of a tree.

Like cellular respiration, photosynthesis is a set of many separate, related reactions. Taken as a whole, these reactions are endergonic, and the ultimate source of energy needed for the reactions to occur is external—light. The many reactions of photosynthesis can be summarized by the following equation.

$$6CO_2 + 6H_2O \xrightarrow[\text{light energy}]{\text{enzymes, chlorophyll}} C_6H_{12}O_6 + 6O_2$$

carbon dioxide   water   simple sugar   oxygen gas

This equation tells us that molecules of carbon dioxide and water are combined into simple sugars by enzymes, and oxygen is given off as a byproduct. Energy for the process of photosynthesis comes from light energy absorbed by chlorophyll. Simple sugars are represented in the equation by the formula $C_6H_{12}O_6$, which can represent fructose, glucose, or other six-carbon sugars.

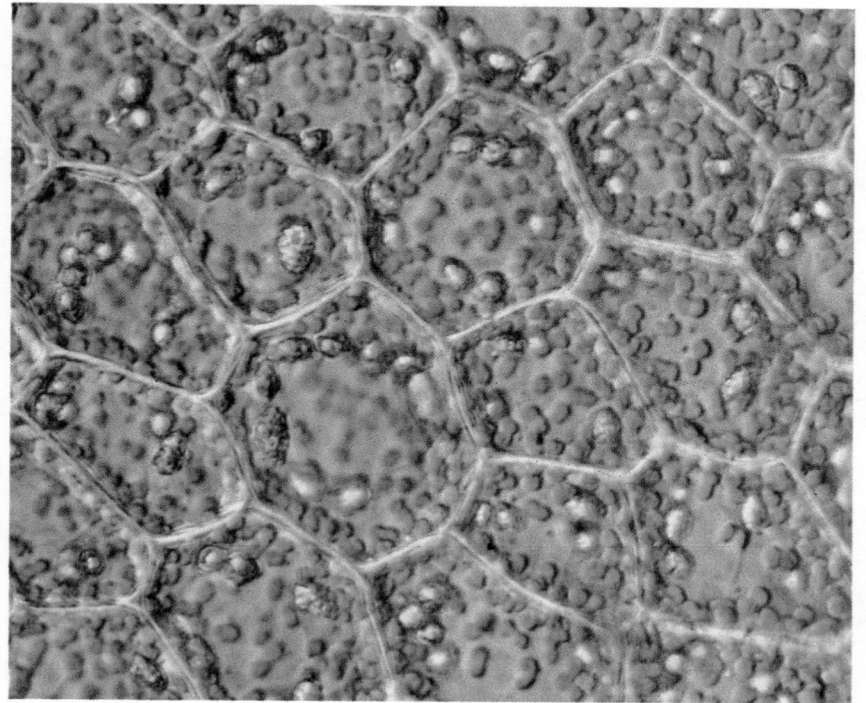

Figure 6-13 In plants and algae, photosynthesis takes place in chloroplasts. The green objects inside these leaf cells are chloroplasts. Magnification: 100×

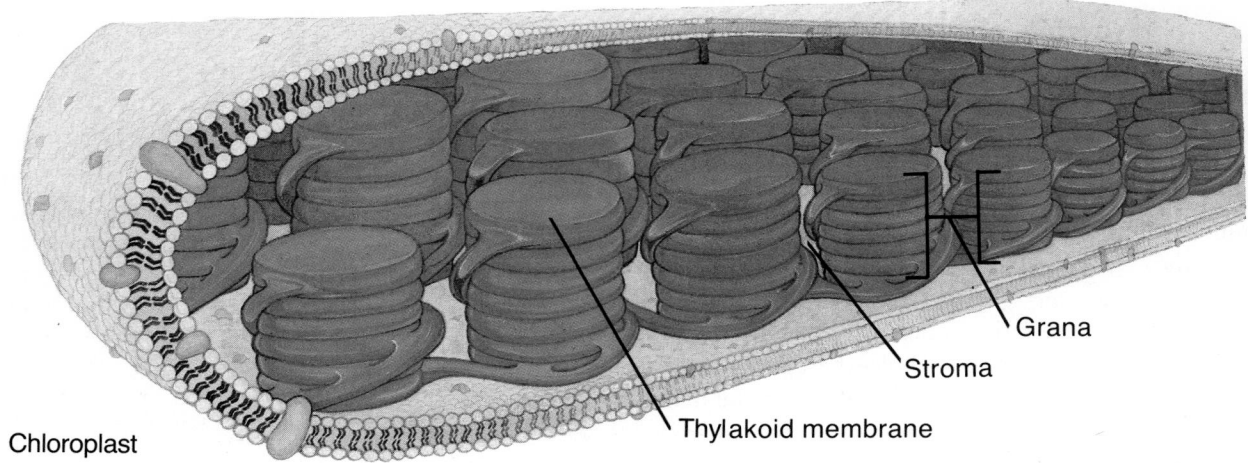

Chloroplast

Grana

Stroma

Thylakoid membrane

Figure 6-14 **Inside a chloroplast, chlorophyll molecules are located on the thylakoid membranes, which are arranged in stacks called grana. The fluid around the thylakoids is called stroma. Sugars are synthesized in the stroma. A chloroplast has a second outer membrane not shown here. What would be an adaptive advantage of having chlorophyll-containing membranes in stacks?**

The Light Reactions

The whole process of photosynthesis is divided into two main sets of reactions. The first set is known as the light reactions. The **light reactions** are so named because they involve a series of changes that convert light energy into chemical energy.

Trapping Energy from Light   When you studied cell structure in Chapter 5, you learned that a chloroplast contains internal membranes. These membranes are called thylakoid membranes. They have a fluid mosaic structure and have chlorophyll and carotenoid molecules embedded along with the proteins in their bilayers. Part of these thylakoid membranes are arranged in stacks called **grana,** seen in Figure 6-14. It is on these internal membranes that the light reactions occur.

Thinking Lab

Does light color influence the rate of photosynthesis?

Background
A student took four similar plants and exposed each plant to a different colored light for 24 hours. She then measured the amount of starch present within each plant's leaves. She chose to measure starch because starch is a major product formed from the sugars produced by photosynthesis.

You Try It
In her experiment, the student obtained the data shown in the following table.

Plant	Color of Light	Starch Present (in milligrams)
A	red	72 mg
B	yellow	10 mg
C	green	13 mg
D	blue	68 mg

Interpret the data given in the table. State a conclusion that you would draw from the data given in the table. Be sure to limit your conclusion to the conditions of the experiment. Based on the data given, what colored light bulbs would be best for growing plants indoors? What colors would be the poorest? What sort of results do you think the student would obtain if she had grown one plant under white light? List two things the student might do to improve her experiment.

During these reactions, light is absorbed by chlorophyll and carotenoids. When these molecules absorb light, their energy increases and that energy is passed along to electrons in a specific chlorophyll molecule. When electrons absorb energy, they move to higher energy levels and are said to be excited. Next, the excited electron in this chlorophyll molecule leaves the molecule. Then, by a series of steps, the energy of the excited electron is used to bond a molecule of inorganic phosphate to ADP to form an ATP molecule. Thus, energy that originally came from sunlight becomes stored as chemical energy in the bonds of ATP molecules. Energy from light is now available to do biological work.

Splitting of Water   Another part of the light reactions involves the water used in photosynthesis. As light is transformed to chemical energy and stored in the bonds of ATP, water molecules are split into hydrogen ions, electrons, and oxygen. The electrons replace those lost by chlorophyll, and the oxygen is given off as a by-product. The hydrogen ions become attached to organic carrier coenzymes called NADP and are used in later steps of photosynthesis.

Note that the light reactions of photosynthesis do not involve carbon dioxide, and no sugars are produced. In summary, the light reactions involve three main events.

1. Light energy is absorbed and converted to chemical energy in the bonds of ATP.
2. Water is split into hydrogen ions, oxygen, and electrons.
3. Hydrogen ions from water are attached to carrier coenzymes for use in later steps of photosynthesis.

Figure 6-15   **In the light reactions, there are really two events involving chlorophyll. In one event, energy from the sun boosts electrons from chlorophyll to a higher energy level. As these electrons lose energy, that energy is used to form ATP molecules. Electrons lost from chlorophyll are replaced by electrons from water. This action breaks water into hydrogen ions and oxygen gas. In the second event, the energy of other electrons is raised again by light. This time, the electrons' energy is used to attach hydrogen ions (H$^+$) to carrier coenzymes called NADP to form NADPH. The ATP and NADPH generated in the light reactions are used in the Calvin cycle, the part of photosynthesis that makes sugars.**

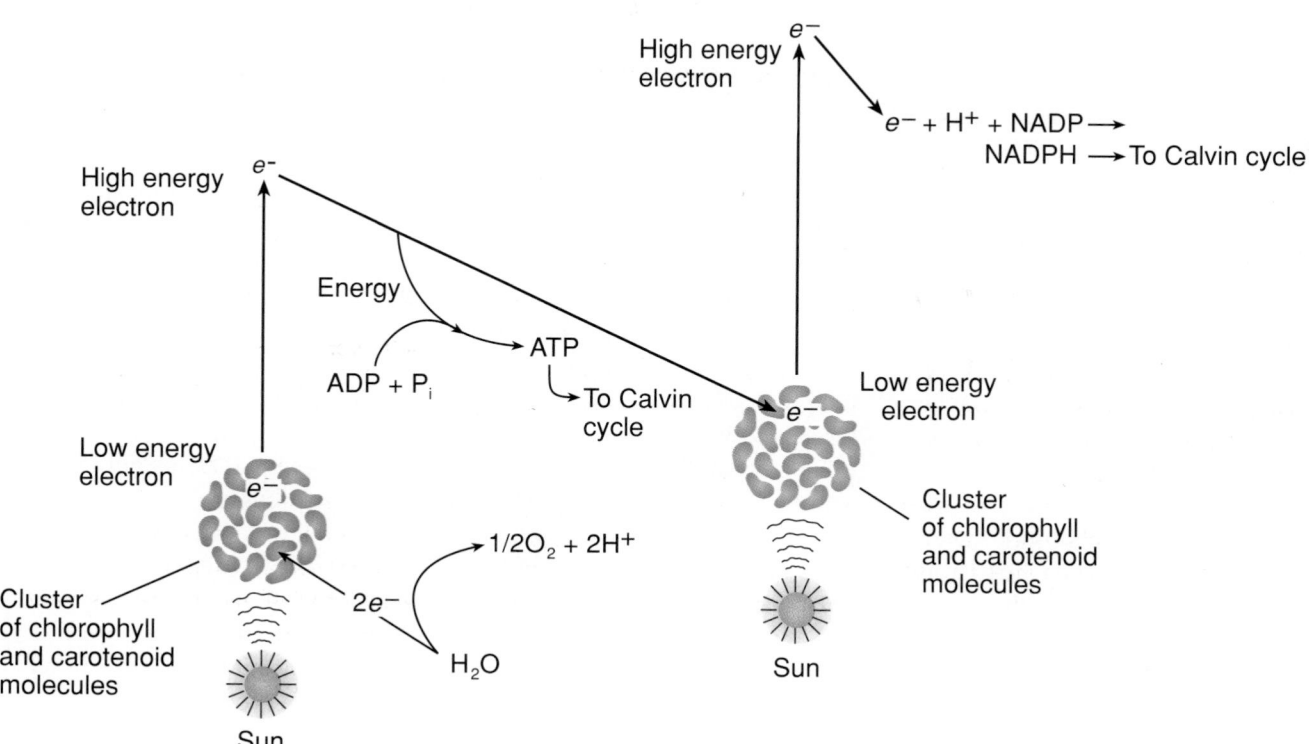

The Calvin Cycle

Thylakoid membranes, the site of the light reactions, are surrounded by a fluid that contains many enzymes and other molecules. This fluid part of a chloroplast is called the **stroma,** and the molecules within it are used in the manufacture of sugars. This process of synthesizing sugars is known as the **Calvin cycle.** In a sense, the Calvin cycle is the "synthesis" part of photosynthesis.

Events of the Calvin Cycle   Consider what must happen in the Calvin cycle. Sugars are composed of carbon, hydrogen, and oxygen atoms, so these atoms must be present in the reactants. Also, because sugars have more potential energy than the reactants from which they are made, a source of free energy is needed. If you study Figure 6-16 carefully, you can understand how the light reactions and the Calvin cycle are related. As you recall, the light reactions provide hydrogen ions as well as a source of energy, ATP. Along with these hydrogen ions and ATP, the plant uses carbon dioxide it has obtained from the environment. The carbon dioxide is a source of carbon and oxygen atoms. The reactions of the Calvin cycle convert that carbon dioxide to sugars. Appendix C provides more details about the Calvin cycle.

Products of the Calvin Cycle   The molecules produced in the Calvin cycle are not glucose, but rather molecules of three-carbon sugars. Some of these three-carbon sugars are used directly as energy sources, while others are converted to other molecules, such as lipids, amino acids, and parts of nucleic acids. Two of the three-carbon molecules can be combined to form

ECONOMICS CONNECTION

Yeasts: Raising Economic Dough

You have probably smelled bread baking at home or near a bakery. Does it seem possible that the tantalizing aroma could result from fungi in the dough? Yeasts are indeed types of fungi. Yeasts obtain their energy by fermenting sugars.

No record exists of the first use of yeast in baking. It is thought that, originally, flour and water were left standing until naturally occurring yeasts multiplied and fermented the carbohydrates in the dough. Thereafter, the baker saved a small amount of dough to use for "starter" in the subsequent batch of bread. Because varying amounts and types of yeast along with bacteria can develop, this method is unpredictable. Some bread is still fermented by using starter dough. It is called sourdough because its high acidity gives it a sour taste. The acidity results from acetic and lactic acids produced by some types of yeast and bacteria.

For most types of bread, modern bakers use pure cultures of selected strains of yeast grown in sterile, formulated liquid media. *Saccharomyces cerevisiae* are the fungi in bakers' yeast.

Yeasts leaven (raise) dough by fermenting carbohydrates such as glucose, fructose, maltose, and sucrose. One product of fermentation is carbon dioxide, which creates most of the leavening effect as it is released and forms bubbles in the elastic dough.

Alcohol is another product of fermentation. When dough is baked, the alcohol evaporates and raises the dough a little more.

The production of alcohol by fermentation is economically important,

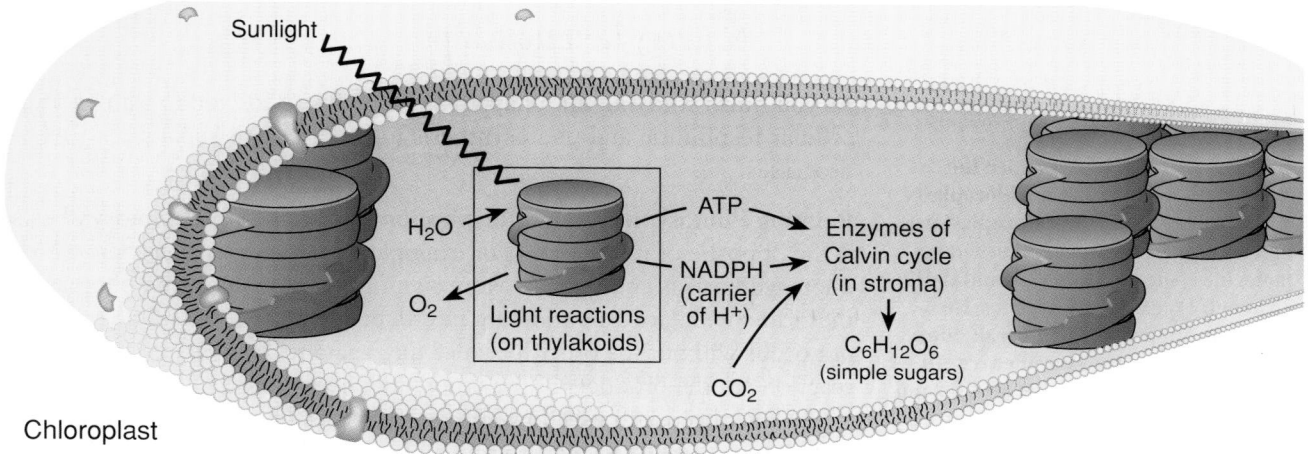

Sunlight

H_2O

O_2

Light reactions
(on thylakoids)

ATP

NADPH
(carrier
of H^+)

CO_2

Enzymes of
Calvin cycle
(in stroma)

$C_6H_{12}O_6$
(simple sugars)

Chloroplast

six-carbon sugars such as glucose and fructose. However, most plants contain very little free glucose. Why? The glucose often combines with fructose to form sucrose, which is transported to other parts of the plant. Other glucose molecules are joined to form starch, which can be stored in stems, roots, or seeds. The starch molecules can later be reconverted to glucose, which can yield energy in cellular respiration. Still other glucose molecules are used to form cellulose to build new cell walls. Thus, even though photosynthesis produces three-carbon sugars, the equation is usually written to indicate $C_6H_{12}O_6$ as one of the products. A principal reason for writing the equation this way is that sucrose, starch, and cellulose are all made of six-carbon sugar units.

Figure 6-16 The Calvin cycle takes place in the stroma. Enzymes combine CO_2 from air or water with hydrogen from NADPH and energy from ATP to form 3-carbon simple sugars. Many of these sugars are used to form 6-carbon sugars, which make up sucrose, starch, and cellulose.

too. For example, fortunes have been made by the manufacture and sale of beer, another product made possible by yeast-caused fermentation. Beer brewers use yeast to ferment a malted grain, such as barley, with hops and water. Malting involves allowing the grain to begin to germinate. In this process, an enzyme converts starch stored as a food reserve in the grain to simpler sugars, mainly maltose, a disaccharide made of two glucose units. Thus, sugar becomes available as an energy supply to the germinating seedling. In brewing, the germination process is stopped by heating

as soon as most of the starch is converted. The alcohol and carbon dioxide in beer are products of fermentation.

A by-product of beer brewing is the nutrient-rich yeast that settles out during the aging and filtering process. This yeast is such a concentrated source of nutrients that some nutritionists have proposed it as a partial solution to world hunger. Unfortunately, most people don't like the taste of nutritional yeast. Brewers' yeast has a bitter flavor. Some other types of yeast, such as torula yeast, are milder in flavor and richer in nutrients.

1. **Explore Further:** How would adding simple sugars affect the leavening action of the yeast in bread dough?

2. **Explore Further:** Which use of yeast would you judge to be more profitable: baking, brewing, or nutrition? Argue in support of your opinion.

3. **Solve the Problem:** Propose some strategies for using nutritional yeast to alleviate famine and malnutrition, in spite of people's preference for other foods.

Energy Relationships

As you finish studying this chapter, it is important for you to think about cellular respiration and photosynthesis together, not as separate processes.

Relating Photosynthesis and Respiration   In some ways, photosynthesis is the opposite of respiration. For example, respiration is exergonic while photosynthesis is endergonic. More significant, though, is that photosynthesis and cellular respiration are interdependent processes. One could not occur without the other, because the reactants (raw materials) of one reaction are the products of the other, and vice versa.

During photosynthesis, low-energy molecules of carbon dioxide and water are converted to high-energy molecules of sugar. Sunlight provides the energy to build these sugar molecules. Once formed, sugars are a rich source of energy for all organisms. Producer organisms themselves obtain energy from the breakdown of sugars. Sugars or products made from them are also passed along to consumers.

Figure 6-17   Here you see the interrelationship between chloroplasts and mitochondria. The chloroplast uses light energy to synthesize high-energy molecules (sugars) from CO_2 and H_2O, giving off O_2 gas in the process. The mitochondrion uses O_2 to break down sugars, releasing the stored energy to make ATP from ADP, giving off CO_2 and H_2O. Note that the chloroplasts and mitochondria might be in the same cell or in different cells of the same organism. In many cases, the chloroplasts are in one organism (a producer), and the mitochondria are in another (a consumer).

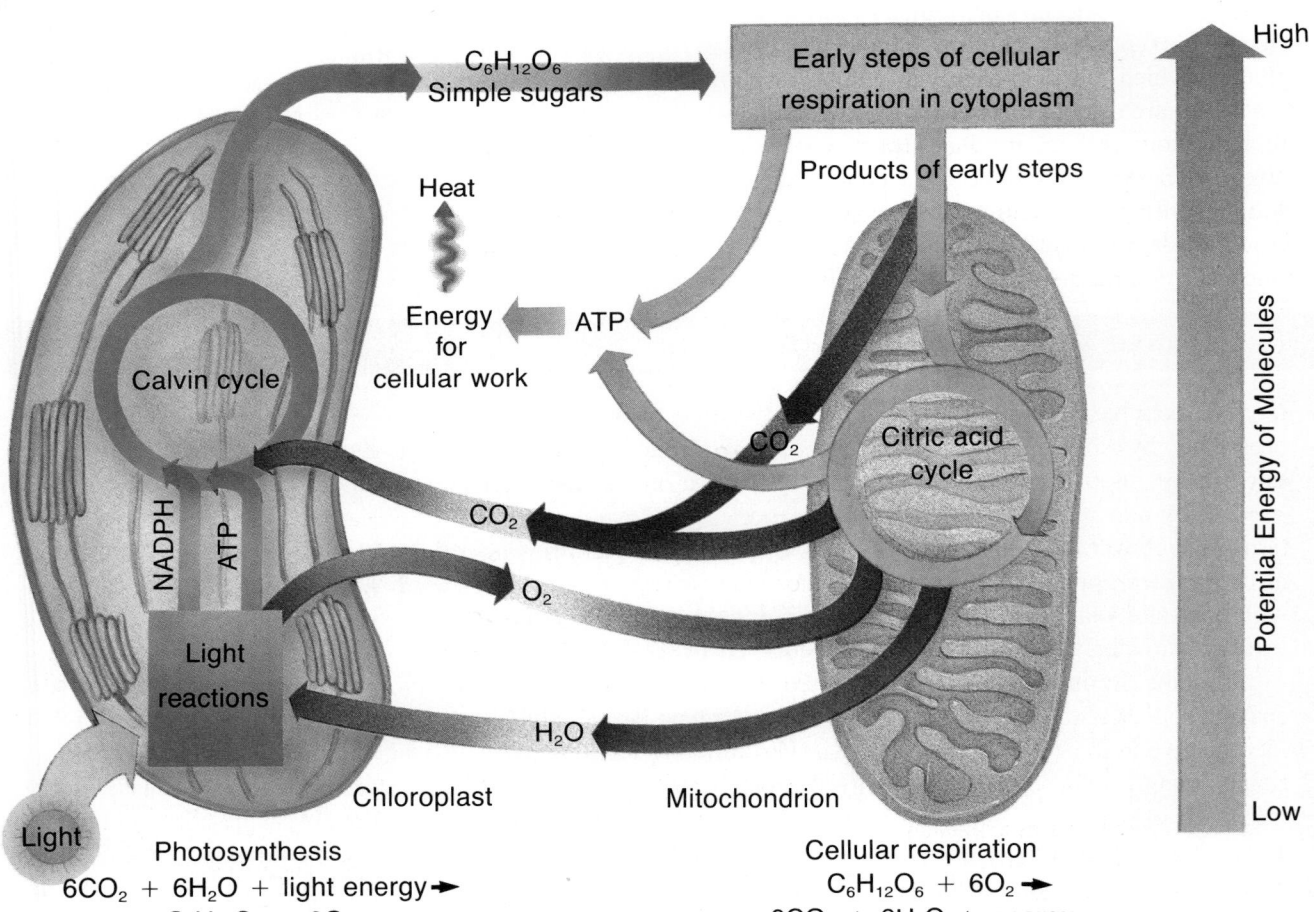

Photosynthesis
$$6CO_2 + 6H_2O + \text{light energy} \rightarrow$$
$$C_6H_{12}O_6 + 6O_2$$

Cellular respiration
$$C_6H_{12}O_6 + 6O_2 \rightarrow$$
$$6CO_2 + 6H_2O + \text{energy}$$

Why Continuous Energy Input Is Needed   Whether in the cells of producers or consumers, sugars are broken down by cellular respiration. As respiration occurs, what was originally light energy is transferred to ATP, which is then used for biological work. But energy of ATP, once used, cannot be recycled. The reason is that whenever energy is used to drive endergonic reactions in cells, some of that energy escapes as heat. This is why living systems require a continuous input of energy.

When you work or play hard, much of the energy used to contract muscle cells is transformed to heat. Also, warm-blooded animals usually require more food than cold-blooded animals of the same size because much of the energy from the food is transformed to heat to maintain body temperature.

But the low-energy carbon dioxide and water molecules produced by respiration are recycled through the environment; producers use them to make new sugars by photosynthesis. Thus, more energy-rich molecules are made available. This is why sugars are referred to as the "crossroads" of energy in living systems.

LOOKING AHEAD

Y OU HAVE NOW TRACED the flow of energy from sunlight to chemical energy in energy-rich molecules of simple sugars. Energy from the breakdown of energy-rich molecules is used to make ATP molecules, which provide energy in convenient amounts to do the biological work of cells. One very important way in which cells use energy is in the production of new cells. In the next chapter, you will find out why cells reproduce, how they do so, and how that reproduction is controlled.

Section Review

Understanding Concepts

1. You are looking at a blue mailbox outdoors in the sun. Describe what happens when the sunlight strikes the box that enables you to see it and distinguish its color.

2. Suppose you have an orange-colored substance. What colors of light would that substance absorb? Reflect? What happens to the light energy the object absorbs?

3. Describe photosynthesis in terms of energy transformations and relative potential energy of the reactants and products involved.

4. **Skill Review—Recognizing Cause and Effect:** How would photosynthesis be affected if:
   (a) a plant were placed in air having a higher concentration of CO_2?
   (b) a plant could not make enough NADP?
   (c) a chloroplast were missing carotenoid molecules?
   (d) a plant failed to absorb enough phosphate ions?

   (e) water were in short supply? Explain your reasoning for each answer. For more help, see Thinking Critically in the **Skill Handbook.**

5. **Thinking Critically:** Not all cells of a leaf carry out photosynthesis. Picture a leaf as being composed of several layers of cells. In which layer(s) do you think the photosynthetic cells would be located? Explain your reasoning.

BIOLOGY, TECHNOLOGY, AND SOCIETY

Biotechnology

Enlightened Product Testing Using Bioluminescence

No one likes to test consumer products on animals, but the public and the government demand that products be proven safe and free of harmful substances. As a result, scientists are looking for alternative methods of testing product toxicity.

One such method involves the use of bioluminescent bacteria. It is more sensitive than animal testing.

If a chemical kills or damages bioluminescent bacteria, the bacteria lose their ability to produce light. This result suggests to the researcher that the tested material could be toxic or irritating to humans.

A familiar example of bioluminescence is the emission of light by fireflies and many other organisms. Light is emitted by an organism when oxygen combines with a substance called luciferin in the presence of the enzyme luciferase and energy in the form of adenosine triphosphate (ATP). Therefore, changes in the levels of ATP can be detected by monitoring changes in the degree of bioluminescence.

Such tests, called bioluminescence assays, are used for many different purposes. For example, they were used at a medical school in Israel to detect the presence of the HIV virus, and in Sweden to detect the presence of bacteria in urine. A researcher at the University of Miami School of Medicine uses bioluminescence to monitor the effect of chemical or radiation treatment on uterine cancer cells.

Because federal laws require testing of some products on animals and humans, bioluminescence assays cannot entirely replace animal testing. However, when used as preliminary screening tests, they can avoid subjecting animal or human subjects to unnecessary risk.

1. **Issue:** Why might researchers be less concerned about harming bacteria than about harming mammals by toxicity testing?
2. **Explore Further:** Why don't researchers test substances on animals if bioluminescence assays strongly suggest that a substance is harmful?

Considering the Environment

Alternative Sources of Energy

You may have heard that air pollution regulations are getting tougher, but they are lenient compared to a law passed in the 1300s. Edward I, an English king, forbade the burning of coal because of its terrible smell. The penalty for burning it was death. Because there were few other energy alternatives in England at the time, the law was eventually repealed.

Today, emissions from burning fossil fuels—coal, petroleum, natural gas—are still a problem. These emissions can impair health, damage forests and wildlife, and even corrode marble statues. For this reason, and because fossil fuel supplies are limited, humans are seeking new energy sources.

Solar Energy As you know, the sun is the ultimate source of energy for all life forms. It is also the source of the energy in fossil fuels, which are the remains of living things. However, the use of sunlight directly, solar energy, is considered a modern idea even though it has been used directly for thousands of years.

Unfortunately, unless additional equipment is used, sunlight isn't sufficiently concentrated to produce the high temperatures needed for generating electricity and powering machinery. Also, the sun isn't a reliable energy source in places where it is cloudy or where days are short.

To achieve the high temperatures needed to generate electricity, large areas covered with flat or curved reflectors concentrate sunlight by focusing it onto a small area, usually tubes of water, which then boils to produce steam. The steam spins the blades of a turbine connected to an electrical generator.

Hydropower People don't usually think of hydropower, or energy from moving water, as derived from the sun. Yet the sun's energy evaporates water, which then falls to Earth as rain and runs from high ground to low. For centuries, humans have used the kinetic energy of moving water to turn waterwheels, which power small mills, grindstones, saws, and even looms for weaving cloth.

Today, hydroelectric power plants harness moving water to generate electricity. Usually, the plant has a dam that raises the level of the water behind it much higher than the power plant at the bottom

of the dam, which increases the gravitational potential energy of the water. As electricity is needed, water is released with tremendous force. The rushing water turns the blades of a turbine, which in turn spins generators.

Wind Power By heating Earth's surface unevenly, the sun creates wind. For example, when air over a warm region is heated, its density decreases and it rises. Cooler air from other regions rushes in, creating wind.

The energy of wind was among the first forms of energy harnessed by humans. Sailors used it to move ships; people built windmills to pump water and grind grain. Thousands of wind machines are now in use. Usually, wind farms, with many wind-powered electric generators connected together, supply electrical energy for communities. Like direct sun, the use of wind power is limited to areas where wind is relatively strong and steady. In the United States, California has the most wind farms.

Fossil fuel gradually replaced wind and water power because

fossil fuel was convenient, reliable, cheap, easily transported, and supplied energy on demand. Inventions such as the steam engine, the internal combustion engine, and the technology to generate electricity escalated demand for fossil fuels.

As oil sources became unreliable, and people began to be more concerned about environmental quality, interest in ancient energy sources such as sunlight, wind, and water revived. Engineers reached back into the distant past for "new" sources of energy to move civilization forward.

1. **Solve the Problem:** Study a map of the world. Select some areas where you think solar energy would be a useful source. Select others that you believe have potential for hydropower and still others where wind power could be developed. On what factors did you base your choice?

2. **Explore Further:** Why does burning fossil fuels release carbon dioxide?

3. **Explore Further:** What are some possible adverse environmental effects of hydropower?

Summary

Many cell processes involve endergonic reactions. Such reactions require free energy, which is released in an exergonic reaction that is linked to the endergonic reaction. Free energy needed to drive endergonic reactions in cells is transferred from the conversion of ATP to ADP molecules.

Energy required to form ATP from ADP and phosphate is transferred from the breakdown of energy-rich molecules such as glucose during aerobic respiration or fermentation. Aerobic respiration releases much more energy per glucose molecule than does fermentation. Thus, many more ATP molecules can be produced from a single glucose molecule by aerobic respiration.

In photosynthesis, light energy absorbed by chlorophyll is used to build three-carbon simple sugar molecules from the raw materials carbon dioxide and water. Oxygen is a by-product of the process. These simple sugars can be used directly as energy sources, or used to make more complex molecules, such as fats and amino acids. They can also be used to form glucose, most of which is changed to disaccharides, starch, or cellulose.

The sugars produced by the Calvin cycle of photosynthesis have higher potential energy than the raw materials. This added energy was converted from sunlight during the light reactions.

Cellular respiration and photosynthesis are interdependent processes, because the products of one process are recycled and used as the raw materials of the other. Once the energy-rich molecules that are the products of photosynthesis have been broken down to release energy for biological work, some energy is lost as heat. Thus, a continuous input of energy in the form of sunlight is necessary for life on Earth.

Language of Biology

Write a sentence that shows your understanding of each of the following terms.

absorption spectrum	exergonic
aerobic respiration	free energy
alcoholic fermentation	grana
anaerobic	lactic acid fermentation
ATP	light reactions
Calvin cycle	stroma
carotenoid	visible spectrum
endergonic	wavelength

Understanding Concepts

1. Explain why biological reactions that are exergonic require an enzyme only, but endergonic reactions require an enzyme and ATP.
2. Compare the relative potential energies of reactants and products in endergonic reactions with those in exergonic reactions.
3. An ATP molecule is used in the first step of cellular respiration, producing a molecule of glucose phosphate and a molecule of ADP. Is this process exergonic or endergonic? How do you know? Which has higher potential energy, glucose or glucose phosphate?
4. Why do you think that ATP is referred to as the "energy currency" of living organisms?
5. Tell what happens in the ATP-ADP cycle.
6. When sugar burns in air, it reacts quickly, giving off heat and light as water and carbon dioxide are formed. Aerobic respiration of sugar in cells involves the same overall reaction. How does the reaction in cells differ from the reaction in air?
7. Examine the equation summarizing photosynthesis on page 155. What is the source of each of the atoms in the molecules of sugar produced? What is the source of the oxygen given off?
8. Suppose you have a test tube containing chlorophyll and water. If you bubble carbon dioxide into the tube and shine light on it, will photosynthesis occur? Explain.

9. An experiment can be done showing that wavelengths of light, other than blue and red, the wavelengths absorbed by chlorophyll *a,* are necessary for photosynthesis. How can this observation be explained?

10. How is the oxygen produced during photosynthesis important to both producers and consumers?

11. The leaf is the organ of photosynthesis in most plants, but only some of its cells carry out photosynthesis. Other leaf parts perform functions needed for photosynthesis to take place. What are some of those functions?

Applying Concepts

12. The title of this chapter is "The Flow of Energy." Summarize the flow of energy in most living systems beginning with light energy.

13. An important scientific law states that the total amount of free energy in the universe is declining. How does the process of cellular respiration illustrate that law? Hint: Remember the definition of free energy.

14. Imagine a sealed container that light can enter. It contains producers, but the only consumers are those that carry out lactic acid fermentation. Could the life in that container continue to exist? Explain.

15. Would the survival period of the organisms in the container in the previous question be longer if consumers that undergo aerobic respiration were present? Explain.

16. **Economics Connection:** What simple clue, requiring no special equipment, would indicate that yeast in a sugar solution are undergoing fermentation rather than aerobic respiration?

17. **Considering the Environment:** Large arrays of photovoltaic cells can convert sunlight into electric energy to power homes. What provision would be necessary in such a system to be able to have power on cloudy days?

Lab Interpretation

18. **Investigation:** Suppose you want to determine the optimum pH for fermentation of glucose by yeast cells. State a hypothesis and describe an experi-

ment that you could do to test it. What factors would you keep constant?

19. **Minilab 1:** A student carries out an experiment similar to the one described in Minilab 1. The student uses two test tubes, records the temperature in each tube every 30 seconds, and prepares the following graph from the data. Use the graph to answer the following questions. (a) Which tube probably contained liver chunks? How do you know? (b) According to the graph, when were the liver chunks added? How do you know? (c) Which tube was the experimental control? How would the student's conclusion have differed if there had been no control? Why? (d) Predict the temperatures of tubes A and B if they were allowed to stand for 20 minutes more. Explain your reasoning.

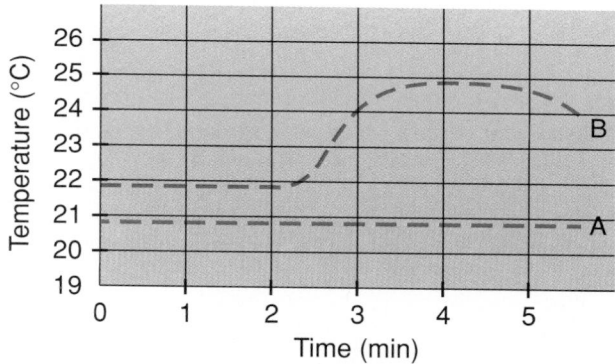

20. **Minilab 2:** Suppose you prepare two test tubes with Elodea sprigs and bromothymol blue solution. The blue color indicates that there is almost no CO_2 present. Then, instead of placing the tubes in the light, you place both in the dark for 24 hours. Predict what you would see after 24 hours. Explain your observations. Hint: Review Section 6.1 of your text.

Connecting Ideas

21. Active transport requires ATP. What role does ATP play in this process?

22. During aerobic respiration, a molecule binds to the product of one reaction, pulls it through the membrane of a mitochondrion, and transfers it to another molecule. What kind of molecule does this? How do you think the molecule passes through the membrane?

The Continuation of Life

The Rosetta Stone, shown below, provided an important clue to the deciphering of Egyptian hieroglyphic writing because it had three translations of the same text side by side. Other codes have been broken when scientists called cryptologists learned to recognize patterns of letters or proper names, such as the name and title of a king.

The model of DNA is based on the work of many people. The clues used to decipher this code were discovered over a period of 80 years, and were contributed by scientists from many disciplines. How do scientific methods enable scientists to decipher a code such as this one?

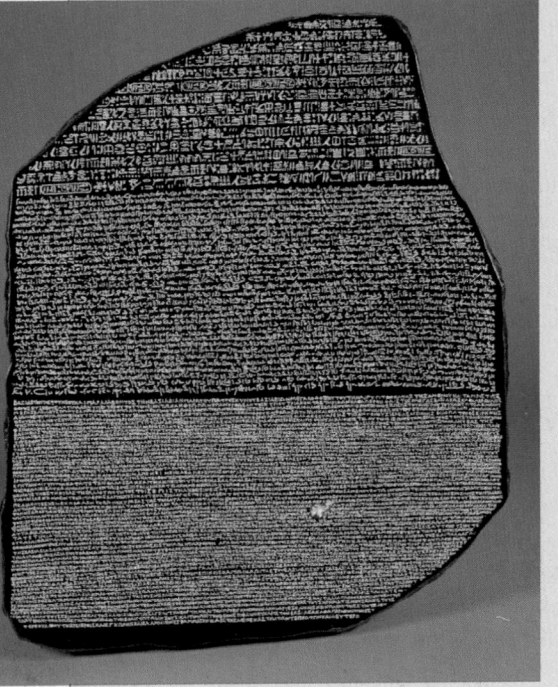

The discovery of the Rosetta Stone was a dramatic breakthrough for linguists, scientists who study and translate languages. Linguists recognized one of the languages on the stone as Greek, and the other two as forms of ancient Egyptian. At that time, linguists were not able to translate either of the Egyptian languages. The Greek text stated that the same information was written on the tablet in all three languages. Working with this valuable clue, linguists over the next 30 years were able to decipher the hieroglyphic writing that appeared on many of the ancient Egyptian monuments.

Just as translating hieroglyphic writing depended upon breaking a code, identifying and translating the code of life depended upon many different lines of investigation over a long period of time. Learning how the code of life was deciphered, and understanding how the code is translated, is the focus of Unit Three, The Continuation of Life.

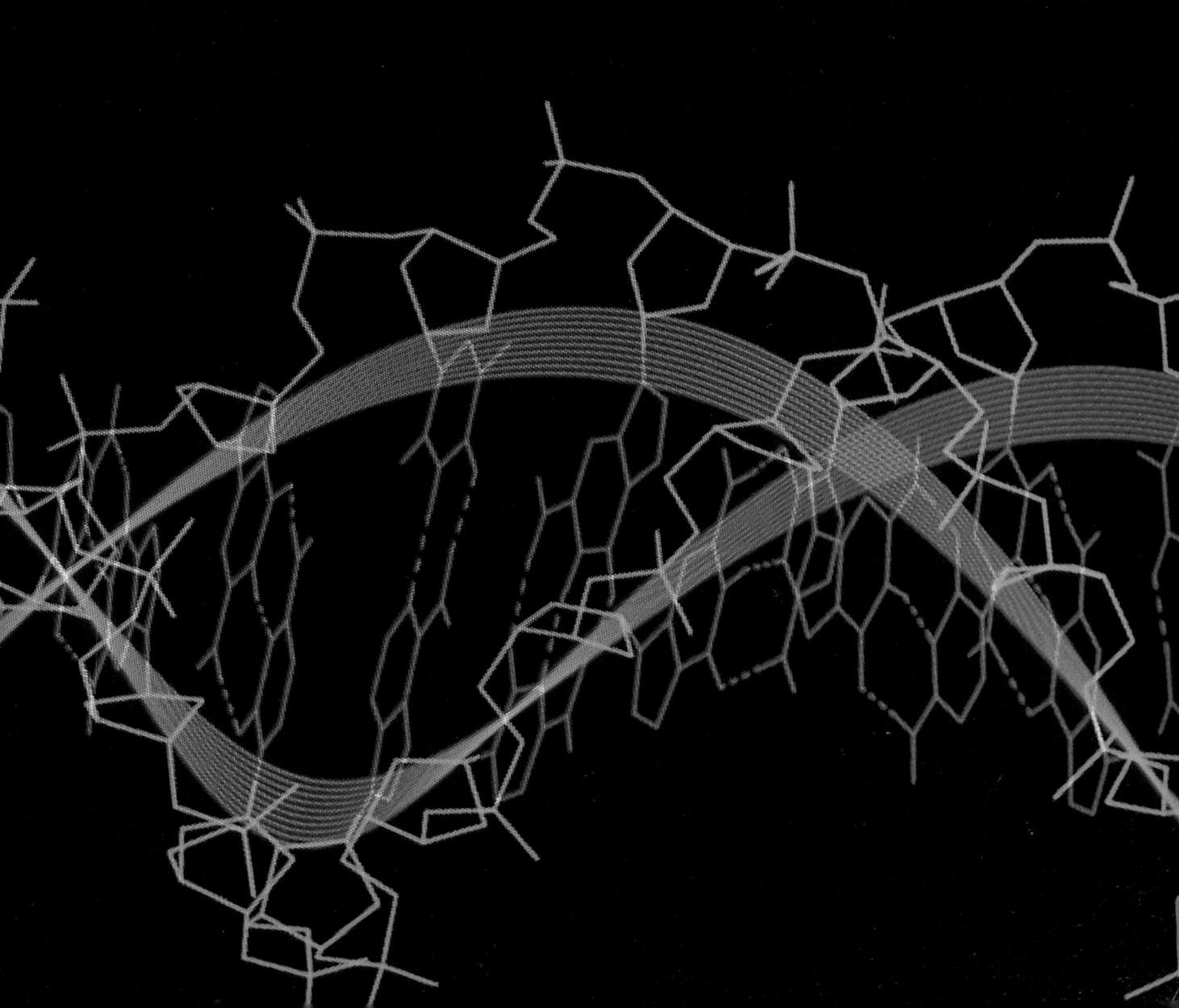

CELLULAR REPRODUCTION

ALL ORGANISMS, great and small, start with a single cell. Then they begin to grow. Young animals, such as bear cubs and human teenagers, seem to spend much of their time eating and growing. Their growth is especially apparent as they increase in body size and weight. Less obvious is the growth that occurs in adults. They may not be growing larger or heavier, but their cells are continuously being renewed and replaced. Cuts are healed, the lining of the intestine is regrown, and worn blood cells are replaced.

Growth is the result of a cell's ability to reproduce itself. New body cells are produced from other, already existing cells when one cell becomes two. Whether growth occurs in young animals or in adults, new cells are produced that contain the same genetic information as the cells from which they came. But when a new individual, such as you or a bear cub, is first produced, specialized reproductive cells come together. This joining of reproductive cells, one from each parent, produces an individual with a unique combination of genetic material. In this chapter, you will study the process that produces and replaces body cells and the process that produces reproductive cells needed for the formation of new individuals. You will see that, although the two processes differ, they both ensure *continuance*. The first process of cell reproduction results in continuance of an individual's growth and life. The second—the formation of reproductive cells—assures continuance of the species to which that individual belongs.

Is this a cell of a cub or kitten, starfish or guppy, worm or butterfly? Whatever the identity of the organism, we know that it's growing because cell reproduction is underway. Magnification: 315×

Chapter Preview

Describe the debate surrounding spontaneous generation and how Redi's and Pasteur's experiments ended that debate.

Sequence the events of the cell cycle in which new body cells are produced.

Analyze the ways in which events of the cell cycle are controlled.

7.1 Life from Life—Cells from Cells

Suppose you were to plant a partly eaten apple, or perhaps just drop it on the ground. Under the right conditions, a seedling sprouts and eventually grows into a mature apple tree. Each new growing season, pollen from blossoms on the tree reaches the eggs of other blossoms and more seeds are formed. Thus, the potential for more apple trees continues.

Continuity exists among all life forms. Butterflies mate, the female deposits her fertilized eggs on a leaf, and those eggs hatch into caterpillars that change form and eventually become mature butterflies. These examples, though different in their details, illustrate a major idea: Living things are produced from other living things of the same kind.

Debating Spontaneous Generation

In the past, people were not all so certain that like produces like. Many myths prevailed to explain the origin of some organisms. According to some myths, eels came from the slime at the bottom of rivers and oceans, and maggots arose from decaying meat.

Testing the Myths A number of scientists in the past accepted the myths because the scientists were not applying some very basic methods of science. Often, the myths themselves were the result of insufficient observations and data. In modern terms, the myths were hypotheses; and hypotheses, however logical they may be, are only explanations of what a scientist has observed. As you know, the validity of a hypothesis must be determined by controlled experiments. Some early scientists did perform experiments, but they did not use proper controls. The results of such faulty experiments convinced those scientists that some nonliving things could give rise spontaneously to living things, an idea that is known as **spontaneous generation.**

In 1668, Francesco Redi performed a simple, but well-thought-out, controlled experiment to test the commonly held belief that maggots were spontaneously generated from decaying meat. Redi had observed that maggots became flies and that flies were always found near decaying meat. He suspected that the maggots in meat actually arose from the eggs of flies that landed on the meat, rather than from the meat itself. To test his suspicion,

Figure 7-1 **Redi's experiment showed that maggots do not spontaneously generate from decaying meat.**

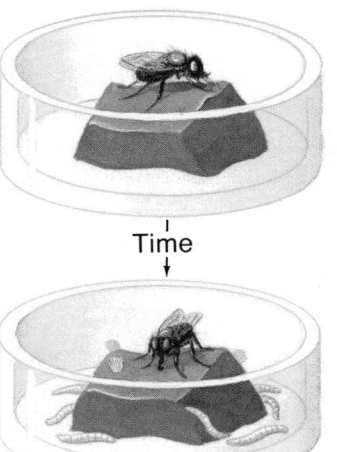

Air circulates through cloth.

Time

Eggs and maggots do not appear.

Time

Eggs and maggots appear on meat.

Redi placed a piece of meat in each of two jars and tightly covered one jar with a piece of cloth so that air, but not flies, could reach the meat. The other jar was left open. In a few days, Redi observed that maggots were crawling on the meat that was exposed to flies. In the sealed jar, maggots never developed. The experiment showed that maggots are the offspring of houseflies, that they develop into flies just like the parents that produced them, reproduce, and, in turn, produce more maggots.

Two More Centuries Over the next few years, the conflict about how maggots and other familiar organisms come into existence gradually subsided. Then, in 1675, the world of microorganisms was discovered. A major event in itself, this discovery reopened the split between scientists who believed in spontaneous generation and those who did not. Those who believed quickly assumed that these never-before-seen organisms were spontaneously generated. The debate, which was to last for almost another 200 years, raged again.

Disproving Spontaneous Generation

Finally in 1864, Louis Pasteur, a great French scientist, devised an elegant experiment that ended the debate. Pasteur's findings from earlier research had revealed the existence of reproductive cells called spores. Pasteur hypothesized that these spores were carried on dust particles in the air, where they were inactive, but that they developed into active microorganisms when nutrients were available.

At the time, air was believed to be an essential ingredient necessary for spontaneous generation; it was supposed to contain a vital force necessary to generate life. Pasteur realized that to disprove spontaneous generation, he would have to conduct an experiment in which spores were kept from entering the nutrient-rich solutions, called infusions, in which microorganisms grow, but air was allowed to enter. To conduct such an experiment, he designed a novel flask that had a long, S-shaped neck, as seen in Figure 7-3. He prepared infusions and boiled them thoroughly, allowing steam to flow throughout the flask. The heat killed any organisms that were present in the infusion, and steam killed any spores clinging to the glass. Pasteur predicted that air entering the neck of the flask would contain dust and spores, but that they would be trapped in the curve of the neck and never reach the infusion. The infusion would remain clear and free of microorganisms.

Pasteur's prediction proved to be correct. His experiment showed that microorganisms, like other, more familiar organisms, are not spontaneously generated, even in the presence of air.

Figure 7-2 Fungi, such as this Earth star, shoot millions of tiny spores into the air. If a spore lands where nutrients are available, it starts to grow. Why might this occurrence have seemed to be spontaneous generation to early scientists?

Figure 7-3 Pasteur's experiment is diagrammed here. His flasks are on display at the Pasteur Institute in Paris. They are still free from growth of microorganisms after almost 150 years!

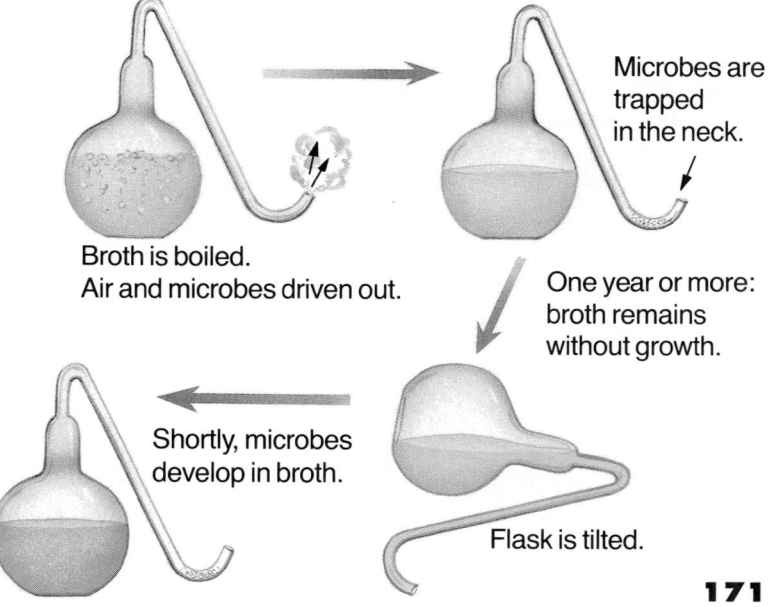

Broth is boiled.
Air and microbes driven out.

Microbes are trapped in the neck.

One year or more: broth remains without growth.

Shortly, microbes develop in broth.

Flask is tilted.

Pasteur's work led to a major biological theory: At the present time and under present conditions on Earth, all organisms are produced from other organisms. That statement, known as the theory of biogenesis, tied in nicely with the cell theory, formulated at about the same time. The cell theory states that cells are the basic units of life, and that all cells are reproduced from other cells. Thus, by the late 19th century, scientists were working with two theories indicating that all life must come from life.

Reproduction of Body Cells

Think back to the apple tree or the butterfly discussed earlier in the chapter. The lives of the two organisms, like those of every other kind of organism, involve a cycle. Offspring are produced. They grow and take on their familiar traits as they mature, and at some point they reproduce, giving rise to descendants that repeat the pattern.

The cells that make up organisms go through a similar cycle. Cell reproduction occurs when parent cells divide. Two new daughter cells arise from each parent cell. Ongoing production of new cells serves several vital functions. Early in the development of a multicellular organism, an increasing number of cells contributes to overall growth and to the production of the many different cell types that will make up the mature organism. After reaching full development, the organism needs new cells to repair damaged tissue, replace cells that are lost from outer surfaces, and resist disease.

Small Is Better   Cells come in different sizes, but almost all are microscopic, as you know. Why must cells be so small? Why doesn't each parent cell just keep growing, instead of dividing over and over again into small daughter cells?

Picture a balloon as you begin to inflate it. At first, the surface area of the balloon is large compared to the volume of air inside. But what happens as more air enters? The volume of air gets larger and larger and increases faster than the surface area of the balloon.

Figure 7-4   **A coconut is the seed of a coconut palm tree. The seed sprouts and begins to grow. Eventually, a mature multicellular tree is formed.**

The same relationship between surface area and volume exists in a cell. As a cell grows larger, the volume of its cytoplasm increases more quickly than the surface area of its plasma membrane. Now recall that needed materials must enter, and cell products and wastes must leave, through a cell's plasma membrane. As the volume of cytoplasm increases, more and more particles must pass through the membrane to keep the cell functioning. If a cell continued to grow unchecked, it would eventually reach a point where the surface area of its plasma membrane would not be great enough to meet the cell's needs. The cell either would starve because of a lack of nutrients or would become poisoned because of a buildup of wastes. Cells stop growing or they reproduce before that point is reached.

The Cell Cycle   When you view cells through a microscope, you will most often see them in a non-reproducing state. The nuclear membrane is clearly visible, and inside the nucleus, one or more nucleoli can be readily seen. If proper stains have been used, the chromatin, a mass of long, thin intertwined chromosomes, will stand out. This non-reproducing stage, in which a cell spends most of its life, is called **interphase.**

Interphase begins when cell reproduction is completed. Early in interphase, the newly produced cells are using energy to carry out the basic cell functions you have studied. In addition, new cell parts such as ribosomes and mitochondria are produced. New centrioles are made in animal cells, and chloroplasts are made in certain plant or algal cells.

In eukaryotes, the DNA of chromosomes is wound like a thread around spools of proteins, as seen in Figure 7-5. These wound spools, in turn, are highly coiled and arranged to form thick strands. It is these strands that make up the chromatin present in an interphase nucleus. Later, these strands will condense (shorten and thicken) and will become visible as distinct chromosomes.

Remember that the genetic information in DNA is used by the cell to make proteins. Some of these proteins function in cell parts, and others function as enzymes needed for cellular reactions. To survive, each daughter cell must have the same DNA instructions as its parent cell. Because DNA resides in the chromosomes, each daughter cell must have the same number of chromosomes as its parent cell. How can that happen?

About midway through interphase, each chromosome and the DNA it contains replicates. To replicate is to make an exact copy. Chromosome replication cannot be seen directly, but chemical tests show that it occurs. This replication provides a second set of chromosomes. Following chromosome replication, the final events of interphase occur. Then, two identical sets of chromosomes will be equally distributed to daughter nuclei during another part of the cell cycle called **mitosis.** Mitosis is usually, but not always, accompanied by division of the rest of the cell to produce two daughter cells. After mitosis and cell division are completed, the new daughter cells enter interphase. Together, interphase and mitosis make up the **cell cycle.**

Figure 7-5 The DNA in eukaryotic chromosomes is wrapped around spools of proteins and is highly coiled.

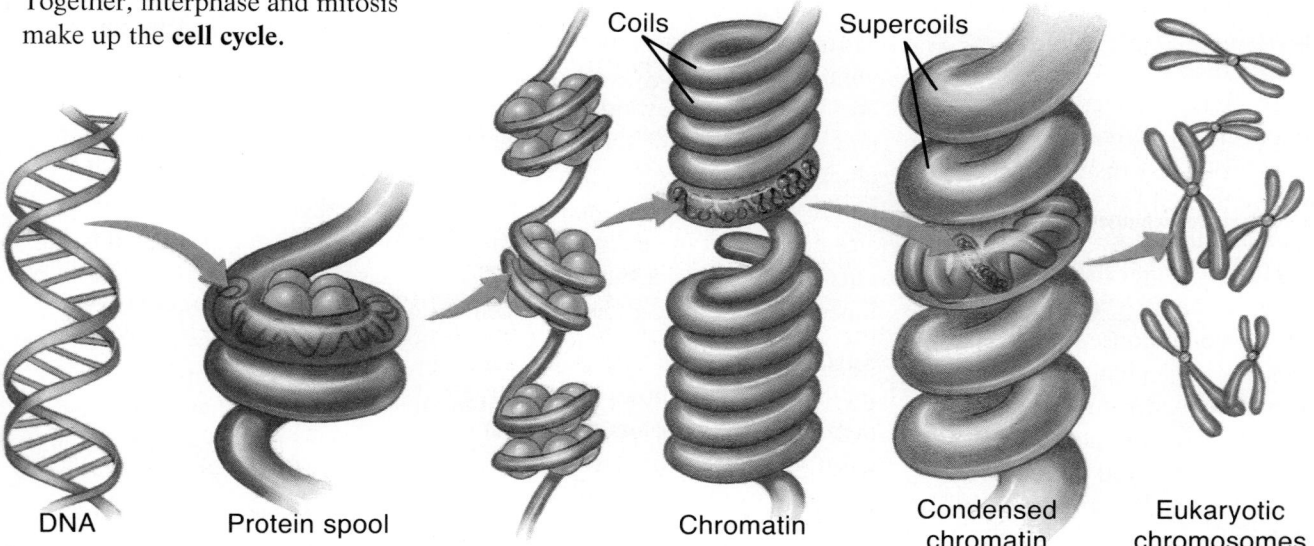

Coils   Supercoils

DNA   Protein spool   Chromatin   Condensed   Eukaryotic
                                                                  chromatin   chromosomes

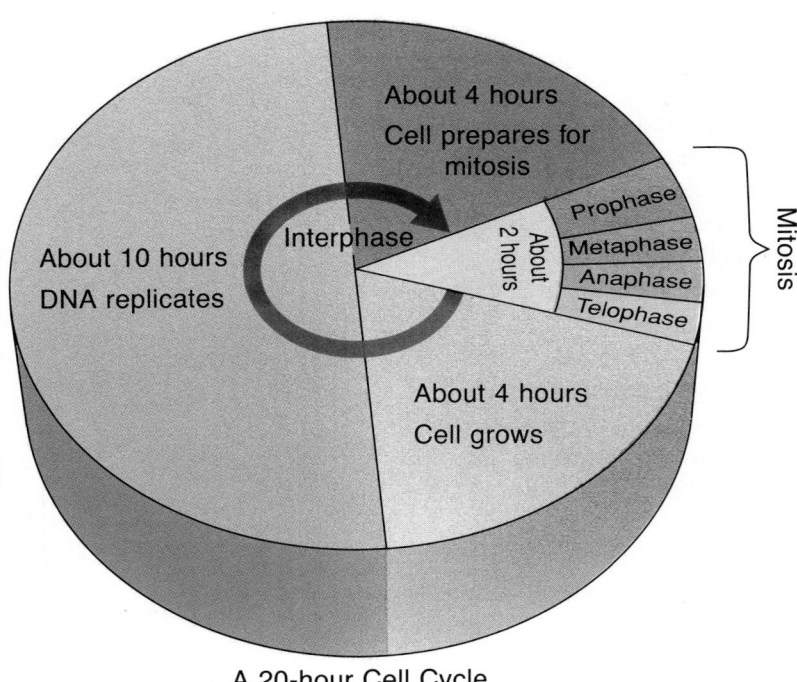

About 4 hours
Cell prepares for
mitosis

Prophase
Metaphase
Anaphase
Telophase

About 2 hours

Mitosis

Interphase

About 10 hours
DNA replicates

About 4 hours
Cell grows

A 20-hour Cell Cycle

Figure 7-6 This cell cycle is typical for some actively dividing cells of mammals. Note that the largest portion of time is spent in interphase.

Mitosis

Suppose, during an investigation with your microscope, you were to catch a living cell in the mitosis portion of its cell cycle. You might think of yourself as having a seat for a four-act drama. Each act, however, follows the previous one without a pause or curtain close. Your program tells you that the drama you are witnessing occurs in most of the cells of eukaryotes. The program also reminds you that the main characters, the chromosomes, prepared for the drama during interphase when they replicated. You are already familiar enough with the plot to know that the drama will end when the chromosomes have distributed the replicas equally among new nuclei of new cells. So turn down the lights and watch the players in action.

MATH CONNECTION

Why Cells Divide

One hundred trillion cells—that's approximately how many make up your body. As you can imagine, each cell is very small. If you could line up 1000 of those cells, they would total less than 2 centimeters in length—only about the width of your thumbnail.

How does cell size in other living things compare? You might expect the cells of a mouse to be tiny and those of an elephant to be gigantic. Surprisingly, these animals have something in common with each other and with you. The cells in mice, elephants, and humans are all about the same size.

Why Small? Whatever the size of the whole organism, the cells remain small. That's because it is advantageous for cells to have as much surface area as possible. Through this all-important surface area, cells absorb oxygen and nutrients and give off carbon dioxide and wastes. A larger surface area allows more of this interchange to go on. In addition, the smaller cell size means that the incoming nutrients have to travel a shorter distance to reach the center of the cell.

Surface-to-Volume The diagram on the next page shows how surface area is related to volume. The first cube, which is 1 centimeter on each side, has a volume of 1 cubic centimeter (height × width × depth).

Its surface area is 6 square centimeters (height × width × 6 sides of the cube). Comparing the surface area to the volume produces a ratio of 6/1, or 6.

Using the same math on the second cube, which is 4 centimeters on each side, yields a surface area of 4 × 4 × 6, or 96, and a volume of 4 × 4 × 4, or 64. Thus, this cube has a smaller surface-to-volume ratio—96/64, or 1.5. Even though it is larger, the second cube has a relatively smaller surface area than the first cube.

Now look what happens when the larger cube is divided into 64 small cubes. The large cube still has the same volume, 64, but it now has a total surface area of 384 (1 × 1 × 6 × 64 cubes), and its surface-

Act One—Prophase   The first act among the four stages of mitosis—**prophase**—opens with disintegrating nucleoli. The thick strands of chromatin are forming loops and condensing, growing thicker and shorter in the process. Separate chromosomes cannot yet be seen clearly.

A little later, the nucleoli have disappeared completely. Chromosomes become clearly visible, and the nuclear membrane is beginning to break down. In animal cells, the two pairs of centrioles begin to separate and migrate toward opposite poles of the cell. Microtubules form between the two pairs of centrioles and become assembled into tiny fibers. These fibers, located between the centrioles, form a football-shaped structure called the **spindle.** Some fibers are quite short, extending just a short distance from the centrioles. Many of the longer fibers overlap one another at the equator (center) of the spindle. Although most non-animal eukaryotic cells lack centrioles, they do have spindles.

Still later in prophase, the nuclear membrane has completely disappeared. Chromosomes appear even more distinct. The fact that replication has occurred is noticeable because the chromosomes are double-stranded structures. Each strand is a replica of the other and is called a **chromatid.** The two chromatids of a chromosome, called sister chromatids, are joined at a region of attachment, the **centromere.** The centromere region of the chromosome serves not only to attach the two chromatids to each other but also to attach the chromosome to the microtubules of the spindle.

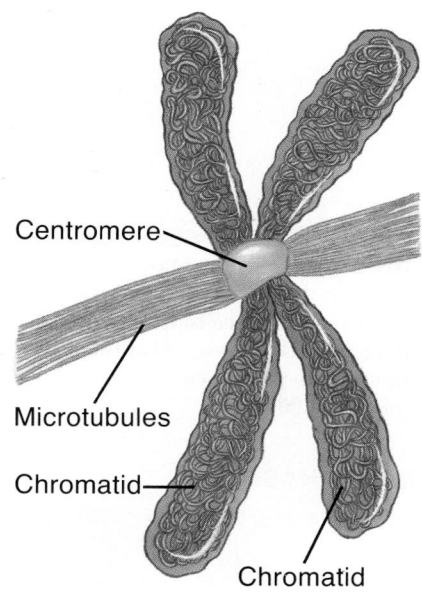

Centromere

Microtubules

Chromatid

Chromatid

Figure 7-7   **This metaphase chromosome has two sister chromatids held together at the centromere and attached to spindle microtubules.**

---

to-volume ratio increases from 1.5 to 6. The divided cube has a much greater surface area.

To get an idea of relative surface area, pretend that the large cube and the individual parts of the divided cube are gift boxes. Suppose that you are wrapping each gift box in fancy paper. You'd need a lot more paper to wrap all of the small boxes than to wrap the one large box. However, you could store either the one large box or all of the small boxes in the same space because their volumes are equal.

1. **Solve the Problem:** Figure out the surface-to-volume ratio of a cube that is 2 centimeters on each side.

2. **Connection to Health:** Keeping in mind surface-to-volume ratio, why do you think it would take less time to digest food that is well chewed?

3. **Solve the Problem:** A nucleus can exert control over only a limited volume of cell contents. What adaptation might a large cell have to overcome this problem?

Volume	1	64	64
Surface Area	6	96	384
Surface-to-Volume Ratio	6	1.5	6

Act Two—Metaphase The players proceed immediately with the short second act, **metaphase.** The chromosomes become attached to the spindle when the microtubules of the spindle "capture" the chromosomes by their centromeres. Interaction of the spindle fibers causes the chromosomes to move agitatedly, as if they were dancing. Eventually, the chromosomes move to the equator of the spindle and become aligned. The dangling chromatids may face in any direction. Once the chromosomes are thus aligned on the spindle's equator, the next act proceeds directly.

Act Three—Anaphase The centromeres split at the beginning of **anaphase,** the third stage of mitosis. The two sister chromatids of each chromosome begin to separate from one another and move apart. Each one heads toward a pole of the spindle. Although the details are not fully understood, action of microtubules facilitates the movement of the chromatids. The microtubules slide past each other, causing the spindle as a whole to become flattened and elongated. The poles move farther apart and the sister chromatids are pulled to opposite poles. Each sister chromatid now behaves as a separate chromosome. As anaphase ends, there is one set of single-stranded chromosomes at each end of the cell.

Act Four—Telophase The final stage of mitosis, **telophase,** opens as two cells begin to form from one. The plasma membrane of animal cells begins to pinch together at the cell's center, as a ring of microfilaments attached to the membrane contracts. Rigid cell walls, however, prevent plant cells from pinching in. In most plants, a **cell plate** begins to appear, growing outward from the middle of the cell. The cell plate may form from membranes of Golgi bodies or from endoplasmic reticulum (ER). From the cell plate will come the plasma membranes of the two new plant cells. Later, cell walls form between the membranes.

Figure 7-8 **Identify the characteristics of each stage in the cell cycle, and note the differences between mitosis in animal cells (diagrams) and in plant cells (photos). What two differences do you notice? Magnification (photos): 1000×**

INTERPHASE

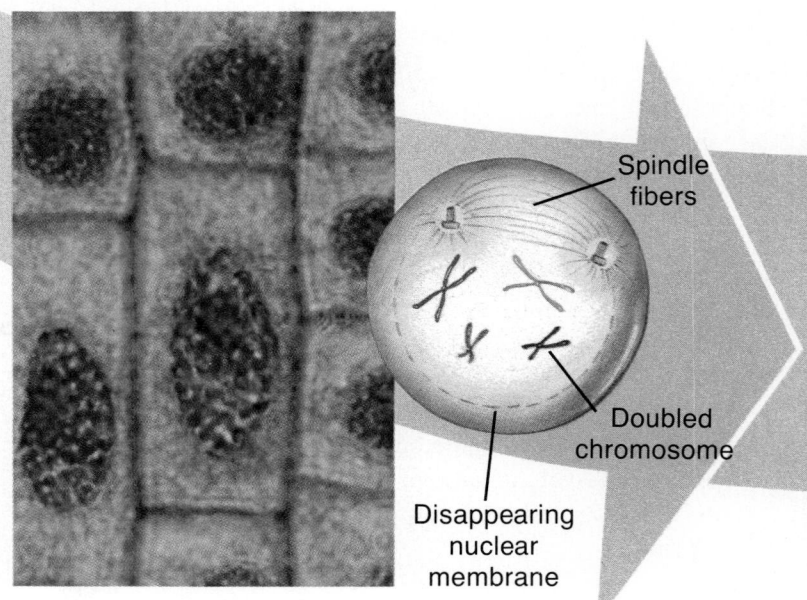

Nucleus

Centrioles

Chromatin

PROPHASE

Spindle fibers

Doubled chromosome

Disappearing nuclear membrane

Telophase ends with events in reverse order from those with which mitosis began. The nucleoli reappear, and a nuclear membrane encloses each set of chromosomes. The nuclear membrane forms from ER. Meanwhile, chromosomes lose their distinct outlines and once again appear as a mass of chromatin.

In animal cells, when cell division is completed, the cell membrane pinches together so that the single parent cell is separated into two daughter cells. In plants, the cell plate is completed to form the daughter cells. Thus, mitosis accomplishes its mission. New cells are produced, each of which contains the same set of genetic information as the parent cell.

Control of the Cell Cycle

Most of the cells in your body are in one or another stage of the cell cycle at this very moment. At least some of your skin cells, stomach-lining cells, and countless other cell types are probably producing daughter cells. It has been estimated that 25 million cell divisions occur every second in the adult human body.

Rate of Cell Division   Different types of cells divide at different rates. Some types of cells, such as those in bone marrow that form red blood cells, reproduce rapidly to keep up with the need for new blood cells. In these cells, the cycle lasts less than an hour. At the other extreme, certain nerve and muscle cells never reproduce once they mature. They remain in interphase as long as they live. Human liver cells divide only when new cells are needed to repair a wound. It is important in a multicellular organism that a balance be maintained among the many kinds of cells in the body. If one kind of cell begins to reproduce too rapidly, the normal organization of the organism will be disrupted. This kind of disruption occurs in cancer, when the rapid, uncontrolled division of one kind of cell invades and disrupts cells in other tissues.

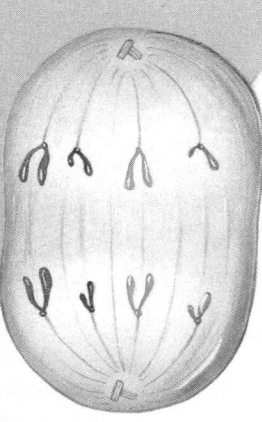

Nuclear membrane reforms

TELOPHASE

ANAPHASE

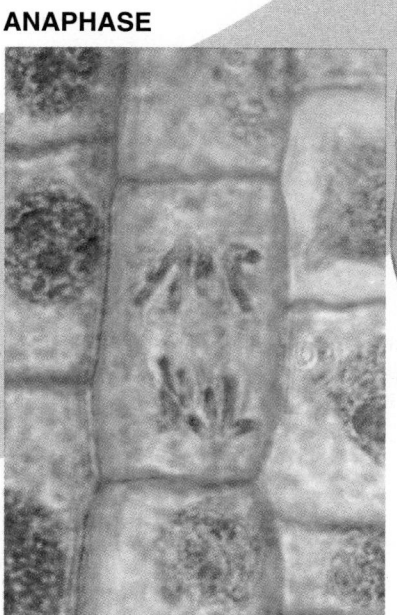

Chromatids pull apart at centromere

METAPHASE

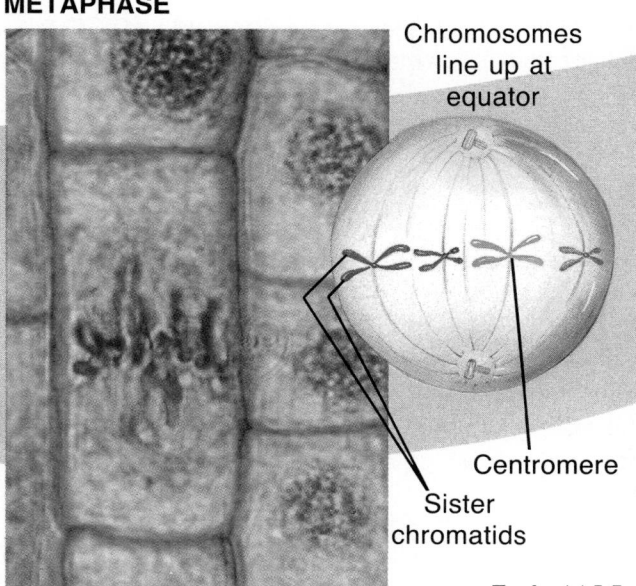

Chromosomes line up at equator

Centromere

Sister chromatids

Regulation of the Cell Cycle   How is the cell cycle regulated? Scientists are just now beginning to understand some of the many mechanisms that seem to be involved. One factor that influences cell division is cell-to-cell contact, Figure 7-9. Most normal cells will divide repeatedly until they come in contact with another cell. Then cell division stops. If cells are isolated from one another and grown in a nutrient medium, they will reproduce until the spaces between them are filled with cells. As soon as a continuous layer of cells is formed, they stop dividing.

When cells stop dividing, they always seem to stop at the same point in the cell cycle—just before DNA replication begins in interphase. Once a cell passes this point, it is committed to completing the cell cycle. Scientists are trying to identify the control mechanisms that operate at this point in interphase. Because many factors that affect the cell cycle have already been identified as being chemical in nature, scientists are concentrating their search on chemical controls. Several different proteins have been discovered and identified. In frog eggs, a protein was discovered that controls the onset of mitosis. This control protein is present during interphase but not during mitosis. Later studies showed that the

I N V E S T I G A T I O N

Growth of Onion Root Tips

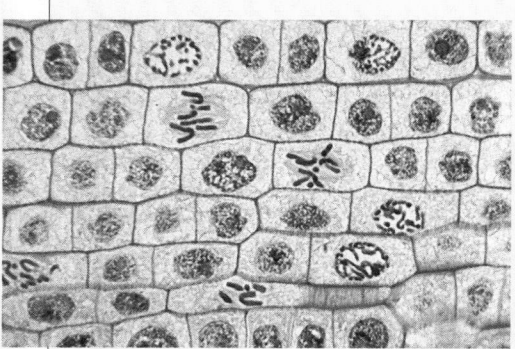

Cells in this cross section of onion root tissue are in various stages in the cell cycle. Magnification: 1000×

When you were born, you probably weighed about 7 pounds. If you gained 2 pounds in the first month, you increased your weight by 29 percent. If you weigh 125 pounds now and gain 2 pounds in a month, you are increasing your weight by 1.6 percent. Clearly, at the beginning stages of life, growth is very rapid. Certain parts of organisms also tend to grow more rapidly than others. In this lab, you will examine growing onion roots and determine what part of the root is growing most rapidly. You will observe the root with a microscope, as well as observe growth of particular sections of the root over a number of days. Your teacher will supply methylene blue for staining any microscope slides you may wish to prepare.

Problem

What part of an onion root grows most rapidly?

Safety Precautions

Use care when handling a razor blade. Methylene blue stains clothing permanently and skin temporarily. Wash your hands at the end of the lab.

Hypothesis

What is your group's hypothesis? Explain your reasons for making this hypothesis.

Experimental Plan

1. As a group, make a list of possible ways you might test your hypothesis using the materials your teacher has made available.

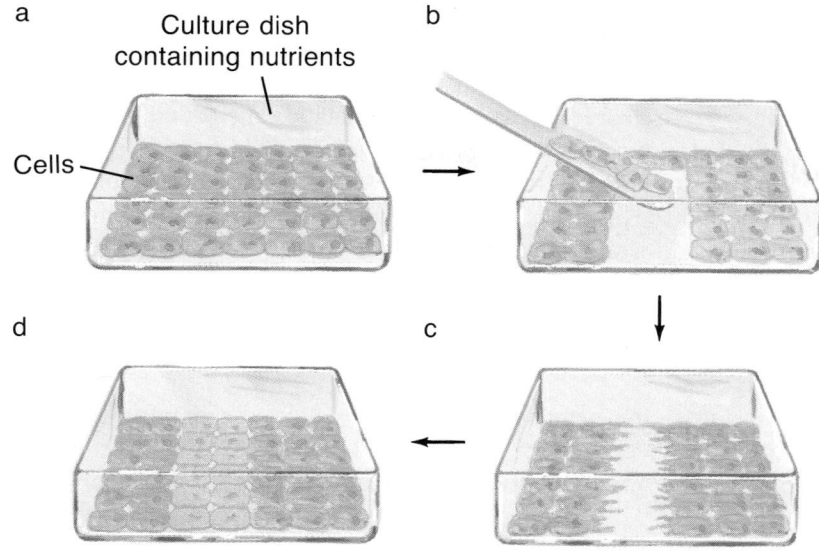

a

Culture dish
containing nutrients

Cells

b

d

c

Figure 7-9 When cells are grown in a culture dish, they will divide only until they come into contact with other cells (a). If some of the cells are removed (b), the cells at the edge of the gap begin to divide again (c). Cell division stops when the dish is covered by a single layer of cells (d).

2. Agree on one idea from your group's list that could be investigated in the classroom.
3. Design an experiment that will test one variable at a time. Plan to collect quantitative data.
4. Following the style of a recipe, write a numbered list of directions that anyone could follow.
5. Make a list of materials and the quantities you will need.

Checking the Plan

Discuss the following points with other group members to decide the final procedures for your experiment.
1. What variables will need to be controlled?
2. What is your control?
3. What will you measure or count? How will you determine which part of the root tip is growing fastest?
4. How many cells or roots will you count, measure, or examine?

5. Have you designed and made a table for collecting data?

Make sure your teacher has approved your experimental plan before you proceed further.

Data and Observations

Carry out your experiment, make your measurements, and complete your data table. Make a graph of your results.

Analysis and Conclusions

1. According to your results, which part of the onion root is growing fastest?
2. What is the experimental evidence for this conclusion?
3. Was your hypothesis supported by your data? Explain.
4. If your hypothesis was not supported, what types of experimental error could be responsible?
5. Consult with other groups that

used different procedures to examine this problem. Were their conclusions the same as yours? Why or why not?
6. Speculate about why onion roots grow rapidly in some regions and not in others.
7. Some scientists have a keen interest in finding out why certain cells undergo rapid division and others do not. Speculate about why research on rapidly growing cells is important.
8. Write a brief conclusion to your experiment.

Going Further

Based on this lab experience, design an experiment that would help to answer a question that arose from your work. Assume that you will have unlimited resources and will not be conducting this experiment in class.

control protein must be activated by still other proteins during interphase before it can do its job. Once the control protein is activated, it can begin the events of mitosis by acting as an enzyme that transfers a phosphate group from ATP to another molecule. The added phosphate group activates this molecule which, in turn, can begin a series of events that activates other molecules. It is this complex chain of activations that sets in motion the events of mitosis, such as breakdown of the nuclear membrane and formation of the spindle.

Several genes that appear to be required for cell division have been discovered and identified in both yeasts and humans. Researchers are trying to determine what role these genes play and how they exert control over the cell cycle.

In most cells, the events of the cell cycle will not begin until various checkpoints have been properly completed. For example, mitosis will not begin unless DNA has been replicated and, where necessary, repaired.

Figure 7-10 **Prokaryotes reproduce by binary fission. Each new cell receives a copy of the single chromosome.**

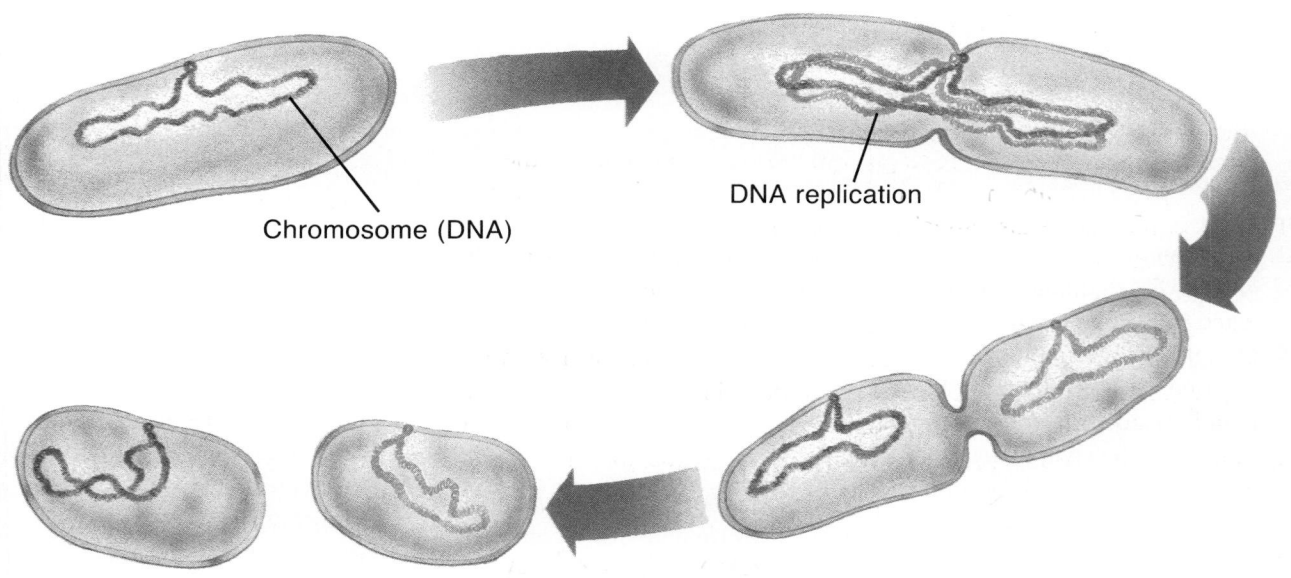

Chromosome (DNA)

DNA replication

Thinking Lab

MAKE A HYPOTHESIS

Where is the nuclear membrane during mitosis?

Background

During prophase of mitosis, the nuclear membrane disappears. It reappears at the end of telophase. Where is the nuclear membrane during mitosis? Electron microscope studies show that during prophase, small saclike vesicles appear in the cytoplasm. In late anaphase, these vesicles disappear.

You Try It

Make a hypothesis to explain what might be happening to the nuclear membrane during mitosis. Draw a diagram to show these events, making sure that your diagram fits both the electron microscope data and your hypothesis. Then, describe an experiment that would test your hypothesis. What results would you expect if your hypothesis were correct?

Similarly, separation of chromatids will not take place until all chromosomes are properly attached to spindle fibers. And, the formation of two daughter cells will not occur until the chromosomes have arrived at opposite poles of the cell.

New knowledge about control of the cell cycle comes from basic research carried out in many labs by many different scientists. As more and more is learned about this subject, that knowledge may be used to replace cells damaged by injury or disease, to allow organisms to live longer lives, or to control the spread of cancers.

Cell Reproduction in Prokaryotes

Prokaryotes have but a single chromosome. Unlike eukaryotic chromosomes, that single chromosome is circular rather than linear. Prokaryotic chromosomes have little or none of the proteins associated with the DNA of eukaryotic chromosomes, so there are no protein spools either.

Prokaryotes cannot be said to have a cell cycle. In fact, their simple structure does not include the cell parts necessary for mitosis to occur. However, like all cells, they must replicate their DNA and distribute the two copies equally to daughter cells. Prokaryotes do so by a process called **binary fission,** shown in Figure 7-10.

As binary fission begins, the single chromosome is attached to the inside of the cell membrane. There a duplicate chromosome is formed. The new chromosome also attaches to the cell membrane, at a point near the first chromosome. While this attachment occurs, growth of new cell membrane and cell wall material between the attachment sites separates the two copies of the chromosome and elongates the cell. At the midpoint in the length of the cell, which is now about twice its original size, more new membrane and wall form and begin to push inward. As this inward growth continues, the two chromosomes become separated. Two separate daughter cells are produced, each with the same genetic material.

Section Review

Understanding Concepts

1. You prepare a nutrient infusion, which you pour into two flasks of the same size. You boil both infusions for the same amount of time. You seal one flask tightly with plastic wrap and place a wad of cotton loosely in the neck of the other. What hypothesis are you testing, and what results do you expect? Why do you expect these results?

2. A parent cell has a total of 12 chromosomes. How many chromosomes will each of its daughter cells contain? Why must they contain that number?

3. You are observing a prepared slide of many cells of one type of animal tissue. You notice that the cells are in various stages of the cell cycle. In what stage will most of the cells be? Why?

4. **Skill Review—Sequencing:** Sequence the activities of chromosomes during the cell cycle. For more help, refer to Organizing Information in the **Skill Handbook.**

5. **Thinking Critically:** Review Pasteur's experiment. What kind of control might he have carried out? Explain the reasoning behind that control.

Objectives

Sequence the series of events by which reproductive cells are produced in complex plants and in animals.

Analyze the significance of meiosis with respect to adaptation and evolution.

7.2 Production of Reproductive Cells

How many chromosomes does it take to make a cell? Would you guess that it depends on the kind of organism? If that's your guess, your thoughts are headed in the right direction. The number of chromosomes per cell does vary from one kind of organism to another.

Your cells have, and always have had, 46 chromosomes. The cells of fruit flies have eight. Does this mean, then, that every cell cycle in mature humans passes along 46 chromosomes to daughter cells? Or that every cell cycle in mature fruit flies passes along eight? The answer would be yes, in every cycle except among one class of cells—namely the cells that form reproductive cells. The job of reproductive cells is to produce offspring like the parent organism. The way chromosomes are distributed to reproductive cells differs from mitosis. And it's largely a matter of numbers.

Chromosome Numbers and Characteristics

Think for a moment about what happens when a male and female butterfly mate. Their reproductive cells unite to form a single cell that is the beginning of a new butterfly. What would happen if each of those reproductive cells contained the same number of chromosomes as all other cells in the butterfly? The offspring's cells would have twice the number of chromosomes as its parents' cells. In reality, however, the number of chromosomes does not double. Instead, it remains constant from one generation to the next. Understanding how this happens begins with a close-up look at cells and chromosomes.

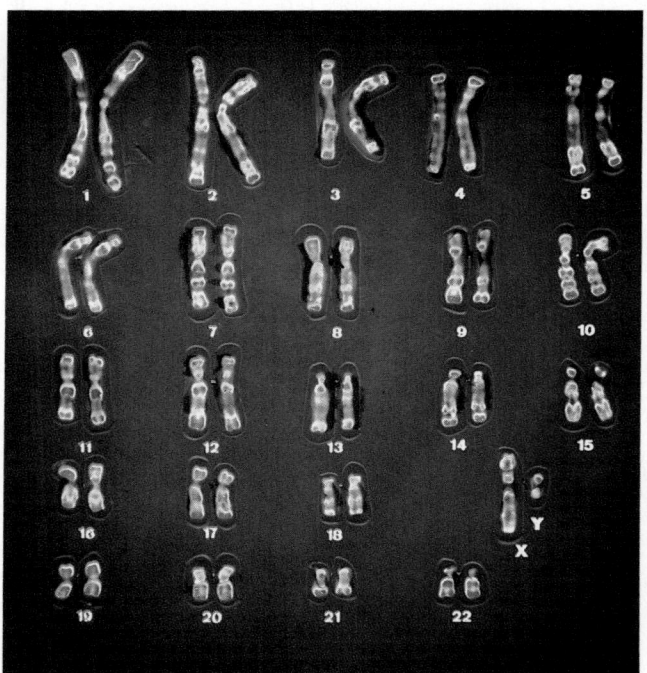

Figure 7-11 **Human cells contain 23 pairs of chromosomes, or 46 chromosomes. Here, enlarged photographs of stained chromosomes have been cut up and arranged by pairs.**

Diploid Cells   Chromosomes vary in a number of ways, including size and shape. All body cells of a given kind of organism normally have the same kinds and numbers of chromosomes. In the diagrams in Figure 7-8, you saw the stages of mitosis in a cell having four chromosomes. Did you notice that among those four, there were two large and two small chromosomes? In the cells of animals, chromosomes usually come in pairs, as shown by the two shades of the same color in Figure 7-8. Cells of many simple plants such as mosses, and cells of other simple organisms such as fungi, do not have such paired chromosomes. More complex plants, such as carrots and cabbages and oak trees, have paired chromosomes as animal cells do. Human cells have 23 pairs of chromosomes, as shown in Figure 7-11; fruit fly cells have four pairs. Cells having two of each chromosome are said to be **diploid.** The diploid number of chromosomes is represented by $2n$, where n = the number of different pairs. Humans have 23 different pairs, so $2n = 46$.

Homologues   In diploid cells, each chromosome of a pair has the same basic structure. The two members of each pair are called homologous chromosomes, or simply **homologues.** The DNA of each homologue carries information for the same traits as its partner, although the specific information may differ. For example, each homologue may carry the instructions for eye color. The instructions on one homologue may be for brown eyes; the instructions on the other homologue may be for blue eyes. The combination of the two instructions determines the eye color of the individual. That same homologous pair of chromosomes carries information for perhaps thousands of other inherited traits as well.

Meiosis Rather than Mitosis

Where do diploid plant and animal cells get their two homologues of each chromosome pair, and why does it matter? The answer to both questions goes back to the beginning—when two parent organisms mate to produce a single cell. That single cell is a fertilized egg called a **zygote.**

Haploid Sex Cells   A zygote results from the union of two different kinds of **gametes,** which are the sex cells. One gamete is a sperm from the male parent. The other is an egg from the female parent. The fusion of egg and sperm to form a zygote is called **fertilization.** All your body cells originated from a single zygote and were produced as a result of mitosis and cell division. Your body cells each contain 46 chromosomes. Cells of your parents also contain 46 chromosomes, and cells of their parents, too.

Sex cells of complex plants and animals, which must unite to form a zygote, differ from body cells in an important way—they do not have the diploid number of chromosomes. Sex cells contain only *one* chromosome from each homologous pair. Cells such as eggs and sperm that contain one chromosome of each homologous pair are said to be **haploid.** The haploid number is represented by n. In human gametes, $n = 23$.

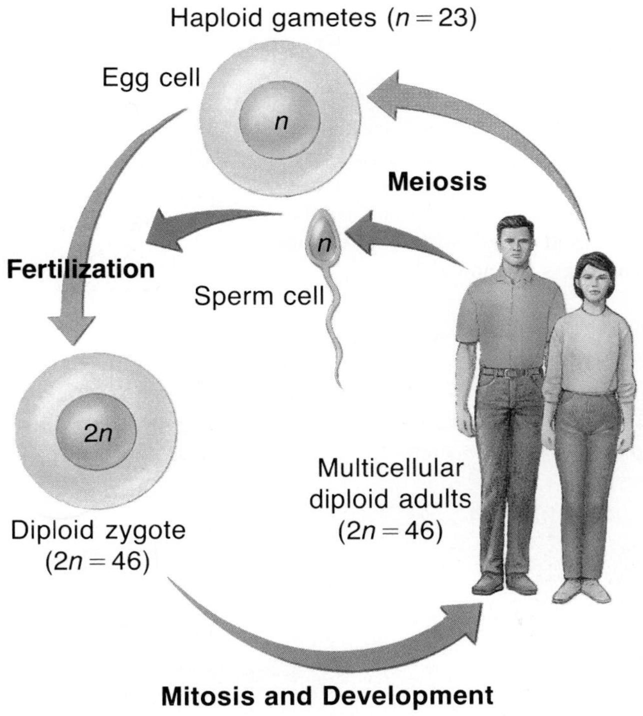

Haploid gametes ($n = 23$)
Egg cell
n
Meiosis
Fertilization
n
Sperm cell
$2n$
Diploid zygote
($2n = 46$)
Multicellular diploid adults
($2n = 46$)
Mitosis and Development

Figure 7-12   In a sexual life cycle, the doubling of the chromosome number that results from fertilization is balanced by the halving of chromosome number that results from meiosis.

Reproduction that involves the production and subsequent fusion of sex cells is called **sexual reproduction.** Fusion of haploid gametes to form a diploid zygote, which then develops to produce a mature multicellular adult, is the key feature of sexual reproduction. If you think about it, you will realize that fertilization restores the diploid number of chromosomes.

Haploid from Diploid   Haploid cells cannot be produced by mitosis. Instead, they are produced by **meiosis,** the process by which haploid

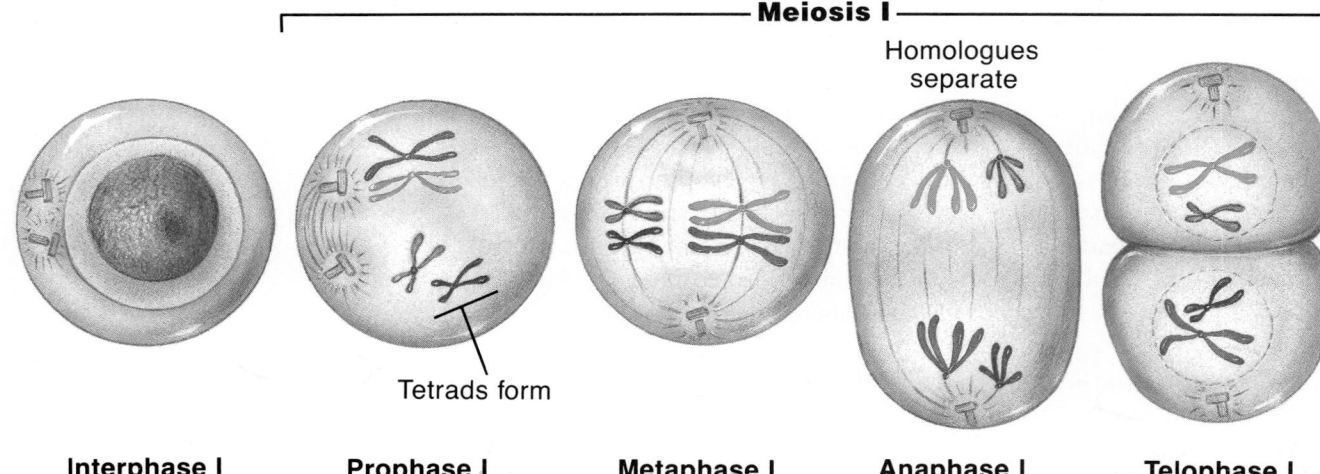

Meiosis I

Homologues separate

Tetrads form

| Interphase I | Prophase I | Metaphase I | Anaphase I | Telophase I |

Figure 7-13   These diagrams show meiosis for a cell with a diploid number of 4. Is this an animal or a plant cell? How do you know? Compare the diagrams of meiosis with those of mitosis on pages 176 and 177.

nuclei are formed from diploid nuclei. Meiosis usually is accompanied by cell division. In animals, meiosis results in production of haploid gametes, sperm and egg.

Meiosis is not limited to animal cells. It also occurs in the life cycles of plants, multicellular algae, fungi, and in some unicellular organisms. In plants and some algae and fungi, meiosis results in the production of haploid spores, not gametes. A **spore** is a small reproductive cell that can develop into a new haploid organism. Unlike a gamete, a spore may give rise to a multicellular individual without first fusing with another cell. In unicellular organisms, meiosis often produces haploid nuclei instead of spores or gametes. In any case, meiosis always involves two successive divisions—meiosis I and meiosis II. The controls operating when meiosis occurs are much the same as those operating when mitosis occurs.

Unlike mitosis, which occurs in most kinds of body tissues, meiosis occurs only in certain reproductive tissues. Correct meiosis ensures that gametes will have the right *number* and *kinds* of chromosomes. As a result, the zygote formed from the union of two gametes will also contain the proper number and kinds of chromosomes.

Meiosis in Male Animals

The main events of meiosis are diagrammed in Figure 7-13 on these two pages. As in the case of mitosis, you can think of meiosis as a series of acts in a drama that progresses without a pause from beginning to end. You will recognize the players and many of the scenes, but the plot takes some extra twists and ends differently.

Meiosis I   Early in meiosis I, chromosomes do not appear to be double stranded. However, studies show that the chromosomes have replicated during interphase, as they do before mitosis. A bit later in prophase I, a crucial event occurs—the homologous chromosomes move close together. Each pair of homologues comes together so that matching sections along the length of their chromatids lie side by side. During this time, the non-

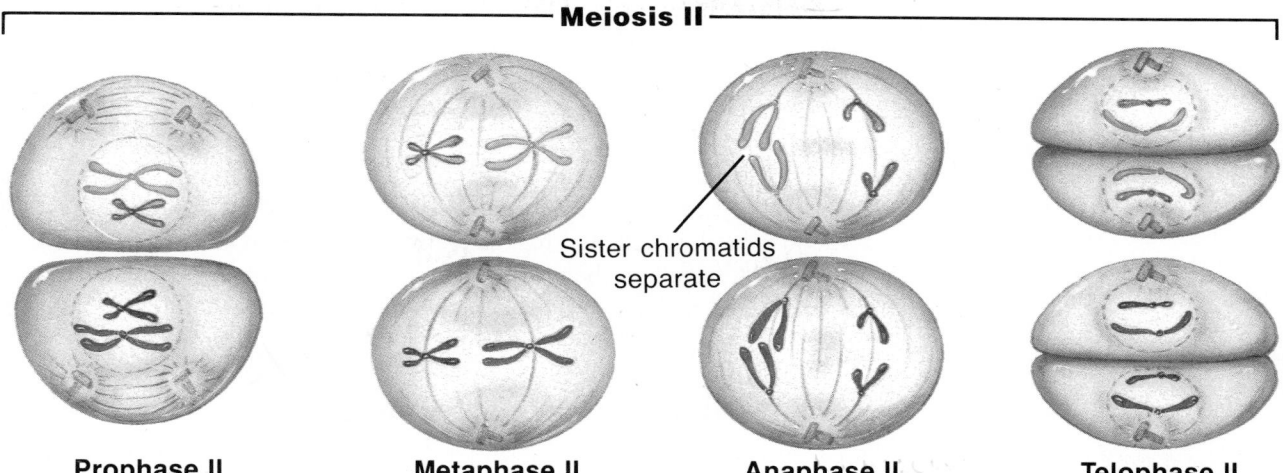

Meiosis II

Prophase II   **Metaphase II**   **Anaphase II**   **Telophase II**

Sister chromatids separate

sister chromatids of two homologues may be wound around one another, Figure 7-14. Soon, the double-stranded condition of the chromosomes, indicating that replication has occurred, becomes clear. Because two homologous chromosomes pair, and because each chromosome of a homologous pair is double stranded, a quadruple chromatid structure called a tetrad is formed.

While the chromosomes are coming together, some events similar to mitosis occur in the cell. The nucleoli and nuclear membrane disappear, the centrioles move apart, and the spindle forms. The major difference between this prophase and prophase of mitosis is that the homologous chromosomes are paired in prophase I of meiosis.

At metaphase I of meiosis, tetrads consisting of homologous pairs of chromosomes line up at the equator of the spindle. This is a second difference between mitosis and meiosis because, as you read, homologous chromosomes do not pair during metaphase in mitosis.

A third important difference between mitosis and meiosis occurs at the beginning of anaphase I and involves the centromeres. Remember that anaphase in mitosis begins by a splitting of the centromeres, so that the sister chromatids can be separated and pulled to opposite poles. At anaphase I of meiosis, the centromeres do not divide, and sister chromatids do not separate. Instead, one homologous chromosome of each pair goes to each pole, and the sister chromatids stay attached to each other. Thus, meiosis I separates the homologues of a homologous pair of chromosomes, not the sister chromatids of individual chromosomes.

With the tetrads thus separated, the cell membrane begins to pinch inward. In telophase I, the original cell divides to form two cells, each of which contains one member of every homologous pair (*n*). However, each chromosome is still double stranded and consists of two chromatids that are still joined together.

Meiosis II   After a short interphase, during which no replication of DNA or chromosomes occurs, meiosis II begins. Meiosis II is much like mitosis. Each of the two cells may skip early events, and the chromosomes move

Figure 7-14   Late in prophase I, homologous chromosomes come together to form a tetrad. Arms of non-sister chromatids may wind around each other.

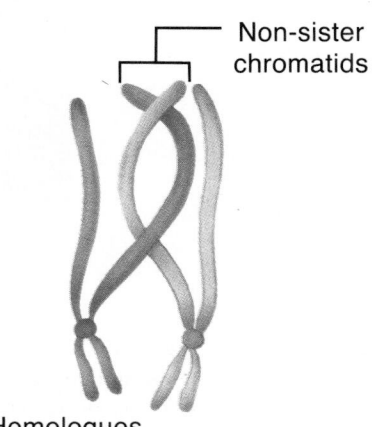

Non-sister chromatids

Homologues

directly to the equator. The centromere of each chromosome splits, and sister chromatids are pulled toward opposite poles. Thus, separation of the chromatids occurs in the second meiotic division. By the time meiosis II is completed, each new nucleus contains single-stranded chromosomes, as seen in Figure 7-13, telophase.

The overall outcome of meiosis in male animals is the formation of four haploid cells, immature sperm, from one diploid cell, Figure 7-16. Each of the four haploid cells has half as many chromosomes as the original parent cell. A flagellum then forms from one of the centrioles of each cell as it matures. The flagellum helps the mature sperm to swim toward an egg, thus aiding fertilization.

Meiosis in Female Animals

Production of eggs by meiosis in a female animal follows the same general pattern as sperm production in males. But there are some interesting differences. Meiosis I in human females begins before birth, as early as the twelfth week of fetal development. Meiosis then stops until sexual maturity is reached many years later. At that point, several cells continue meiosis I on a cyclic basis. Usually, however, only one cell survives per cycle. The surviving cell divides during meiosis I to produce two cells of unequal size. The larger cell enters meiosis II. The smaller cell may do so as well, but if it does, the cells produced from it die. During meiosis II, the larger cell divides unequally once again. The smaller cell dies, but the larger one matures into an egg. Figure 7-16 compares the major events of meiosis in male and female animals.

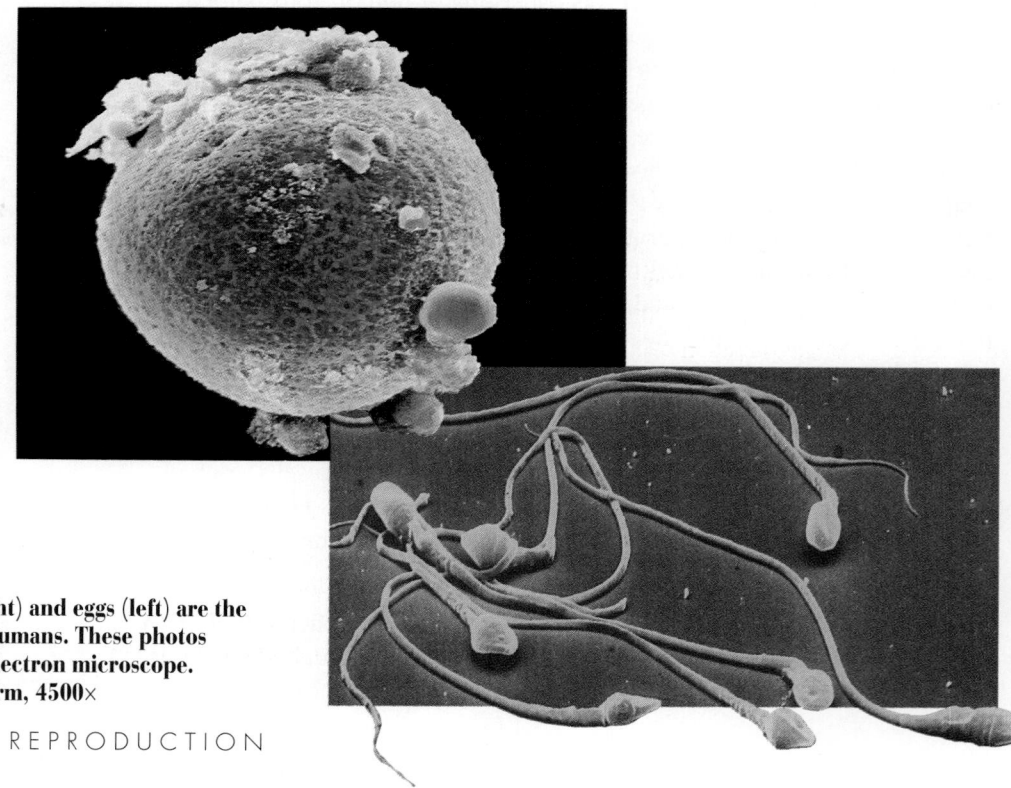

Figure 7-15 Sperm (right) and eggs (left) are the final products of meiosis in humans. These photos were taken with a scanning electron microscope. Magnification: egg, 470×; sperm, 4500×

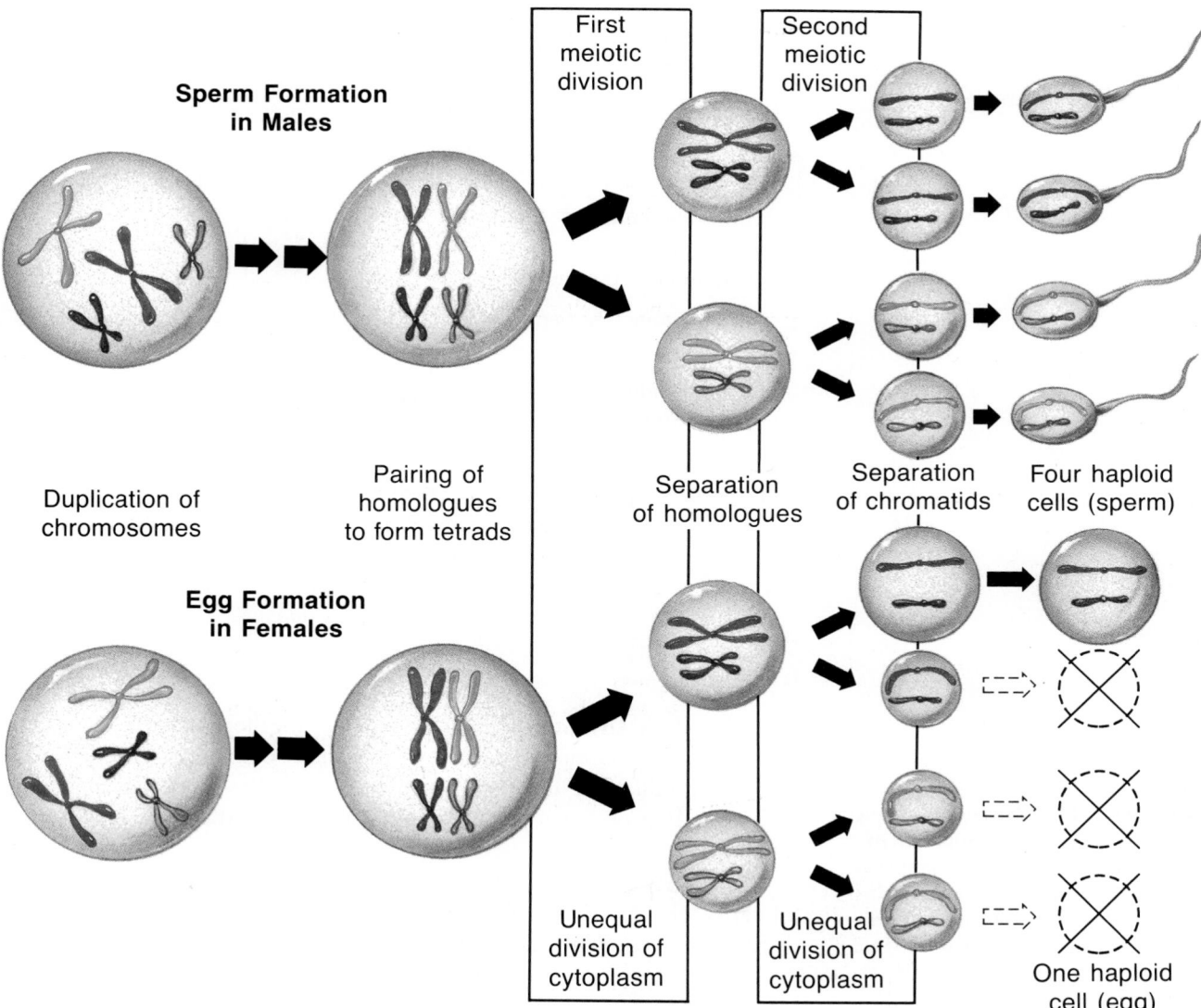

Sperm Formation in Males

Duplication of chromosomes

Pairing of homologues to form tetrads

First meiotic division

Separation of homologues

Second meiotic division

Separation of chromatids

Four haploid cells (sperm)

Egg Formation in Females

Unequal division of cytoplasm

Unequal division of cytoplasm

One haploid cell (egg)

Figure 7-16 Compare the products of meiosis in male and in female animals.

Although a sperm must be in final form before fertilization can occur, an egg may be fertilized before it is fully mature. In some animals, fertilization triggers the second meiotic division in a female. In a male, four sperm develop from each cell that undergoes meiosis, but in a female, only one egg develops from each cell that undergoes meiosis. In what way might unequal cell division resulting in production of one large egg be adaptive?

Importance of Meiosis

Meiosis may produce sperm or eggs in animals, or spores in plants, algae, and fungi. No matter which type of organism it is, each of these reproductive cells contains the haploid number of chromosomes. Thus, each carries genetic information in a form that can be passed to the next generation. If meiosis did not occur, the life cycle could not continue.

Table 7.1 **Comparison of Mitosis and Meiosis**

	Mitosis	Meiosis
Occurs in	all body cells	certain cells of reproductive organs
Number of cells produced per parent cell	two	four (three may die)
Chromosome number of parent cell	diploid (2*n*) or haploid (*n*)	diploid (2*n*)
Chromosome number of daughter cells	same as parent cell	haploid (*n*)
Kind of cell produced	various cells, such as skin and bone	animal gametes or certain spores
Function	genetic continuity from cell to cell	genetic continuity between generations; promotes variation

Figure 7-17 **Traits such as height and skin color can be reshuffled by genetic recombination. If changes in the environment occur, an individual with a different combination of traits may have a better chance of survival.**

The Variation Advantage   Meiosis provides for more than the transfer of genetic information. Combined with fertilization, which is needed for sexual reproduction, meiosis also provides the means for variation among offspring. Reproduction by mitosis alone produces little or no variety in offspring.

The variety arising from meiosis and fertilization is a little like a game of cards. The deck is shuffled before it is dealt. Shuffling rearranges the cards so that you play each hand with a different set of cards.

Meiosis is a process that reshuffles chromosomes—and the genetic information they carry. In this case, the hands played contain the genetic information that will be passed to offspring. In any population, there are variations among organisms. Such differences arise from various combinations of chromosomes.

How is this variation important to a species? Variation provides a better chance for survival of some organisms if changes in the environment should occur over time. By ensuring that individuals will have slightly different traits, nature keeps many hands in the game. Thus, the chance of producing too many individuals with undesirable combinations of genes is decreased. Adaptations of organisms result from genetic variations that enable them to survive in a particular environment.

Chromosome Combination Possibilities   How does meiosis increase variety? Consider an animal whose diploid number is four—two pairs. Such an animal will produce gametes having two chromosomes, one of each pair. How many possible combinations of chromosomes may appear among the gametes? Look at Figure 7-18. The parent cell, shown in metaphase I, has two pairs of homologous chromosomes—*A* and *a*; *B* and *b*.

When the cell at the left completes meiosis, the gametes produced from it contain two possible chromosome combinations—*A* with *B* or *a* with *b*. The cell at the right produces gametes with different combinations—*a* with *B* or *A* with *b*. (Notice that there is always one member of each homologous pair in each gamete.) The particular combination of chromosomes in a gamete depends on how the chromosomes lined up at the equator during meiosis I. In this example, four combinations of chromosomes are possible among the gametes.

Now consider the possible combinations of chromosomes in zygotes produced by the union of these gametes. Figure 7-19 on page 190 shows that 16 different combinations of chromosomes may be produced. You have read that homologous chromosomes carry information for the same traits, but that the information may be in a different form. For example, one homologue may carry information for brown eyes; the other may carry information for blue eyes. Also, a single chromosome may carry information for thousands of traits. Thus, combinations of chromosomes resulting in many different features are possible among offspring. Some combinations may be more useful to an organism's survival than others. This reshuffling of chromosomes and the genetic information they carry is one of the mechanisms for what is called **genetic recombination.** Genetic recombination is a source of important raw material in the process of evolution—inheritable variation.

Figure 7-18   **How chromosomes are combined in gametes depends on how homologues line up during metaphase I of meiosis, an event that is a matter of chance.**

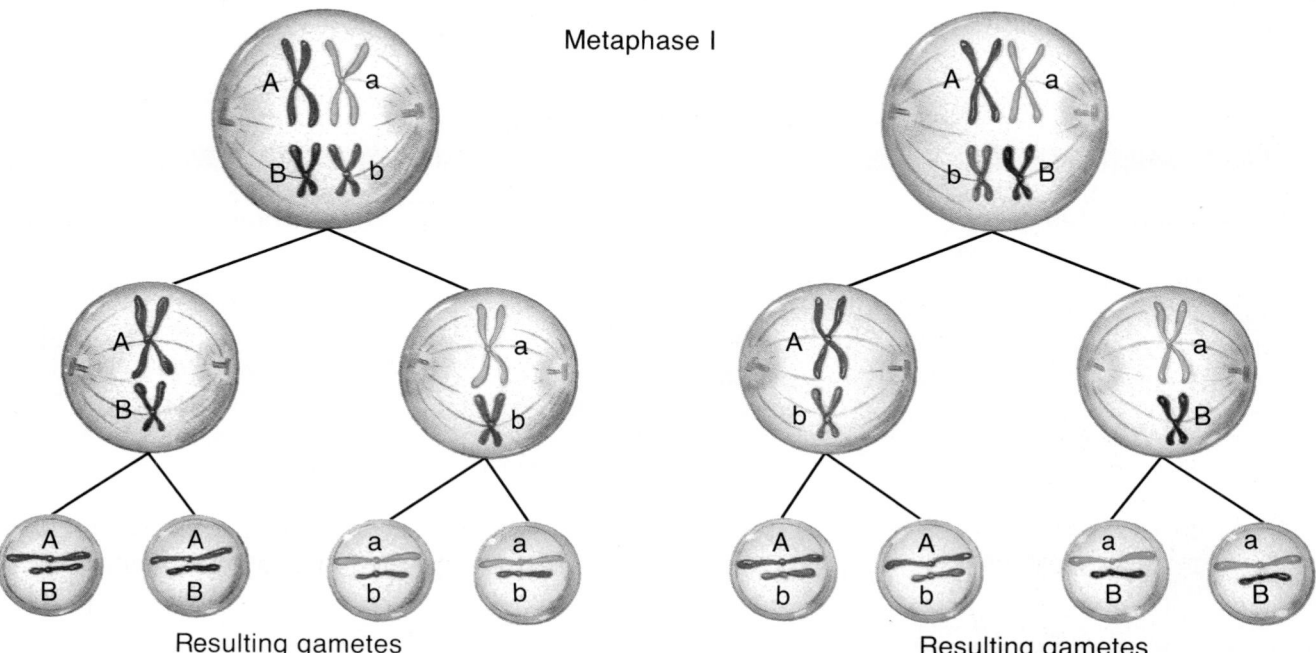

Metaphase I

Resulting gametes

Resulting gametes

Figure 7-19 When gametes contain different combinations of chromosomes, fertilization leads to genetic recombination and increased variety among offspring.

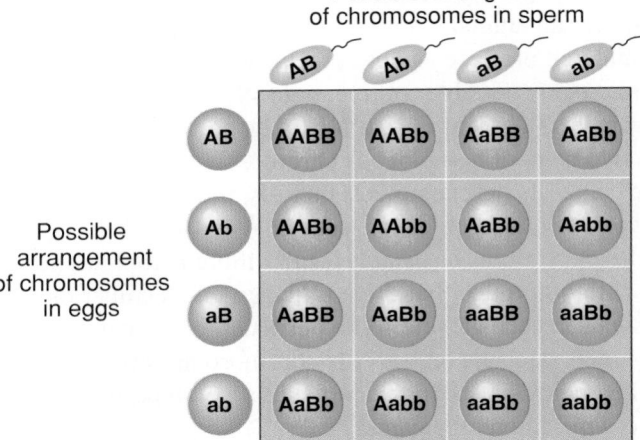

Possible arrangement of chromosomes in sperm

Possible arrangement of chromosomes in eggs

Possible combinations of chromosomes in zygotes (in boxes)

LOOKING AHEAD

IF MEIOSIS IS the *process* of ensuring that genetic information is passed from one generation to the next, then what is the *content* of that information? You will find answers to this question in the next chapter. You will discover how traits are passed on from parents to offspring, and why offspring have combinations of characteristics different from their parents. Then you will solve problems in order to predict or explain results of matings between different parents.

Section Review

Understanding Concepts

1. Most cells of a certain animal contain 20 chromosomes. Some of its cells, though, are found to contain 10 chromosomes. What kinds of cells contain 10 chromosomes and by what process were they produced? Compare the 10 chromosomes in those cells to the 20 chromosomes in the other cells.

2. A certain cell has two pairs of chromosomes—*C* and *c*; *D* and *d*. Gametes produced from this cell are found to have the following chromosome combinations: *C* and *D*, *C* and *d*; *c* and *D*, *c* and *d*. How can these different combinations be explained?

3. Most members of a population of animals have dark fur. Some offspring are produced, though, with light fur. Under what conditions might this difference be important to these animals in the future?

4. **Skill Review—Sequencing:** Sequence the activities of chromosomes during meiosis I. For more help, refer to Organizing Information in the **Skill Handbook.**

5. **Thinking Critically:** What events of meiosis I are critical if each cell produced by the end of meiosis is to have the haploid number of chromosomes? In what way are these events important in the production of genetic recombination?

Biotechnology

Aging: Can We Stop The Clock?

How old do you think "old" is? Thirty? Forty? Seventy? Your answer might be similar to that of Bernard Baruch, an advisor to several U.S. presidents. Baruch said, "I will never be an old man. To me, old age is always 15 years older than I am." Baruch died at 95, possibly with the same attitude about age.

However you define old age, the reality is that you, just like everyone else, are on your way toward it. In fact, aging has already begun to dull your senses of taste and hearing. You no longer have taste buds on the roof of your mouth, as a baby has. Nor can you hear as well as a baby, who is able to hear the "silent" noise of a high-pitched dog whistle.

Of course, the signs of aging are much easier to spot in people of your grandparents' generation. Graying hair, stronger glasses, and weaker muscles are just some of those signs.

As more people live longer, they want to understand the changes they face as they grow older. They also want to know why aging occurs and whether or not it is inevitable. Scientists are providing some answers as they study the aging process.

The Wear-and-tear Model   One of the older models of aging could be called the "wear-and-tear" model. This model compares the human body to a machine, such as an automobile. You expect a car to last a certain length of time before its parts break down or wear out. Like a machine, the human body can be overworked and its "repairs" neglected. Under the wear-and-tear model, it seems logical that too much work and stress could shorten a person's life span.

The Ticking-clock Model Other researchers claim that aging seems to be programmed into the genes of a human body. Scientist Leonard Hayflick tried to demonstrate this idea by growing fibroblasts, a type of skin cell, under accelerated conditions in his laboratory. He discovered that cells taken from fetuses divided about 50 times, then slowed their rate of division, stopped, and died. Fibroblasts from adults divided fewer times before dying. Repeating the experiment with other types of body cells yielded about the same number of divisions. Under normal conditions within the human body,

these 50 divisions would take place over 100 to 115 years, which Hayflick decided is probably the upper limit of a life span. Hayflick hypothesized that the entire human body may be tuned to a biological time clock that ticks off the years of life. He left for other scientists the question of whether or not this clock could be reset.

The Malfunctioning-body Model Another factor that probably contributes to aging is the increasing chance that an older body's immune system will malfunction. As the defenders of the body age, they become less efficient in recognizing and responding to invasions of organisms that can harm the body. These aging defenders can confuse disease-causing invaders with the body's own cells and mistakenly attack the "good guys."

Other mistakes could involve the molecules that carry the genetic instructions for making proteins in the cell. If these molecules malfunction, the wrong directions would be used as the cell builds its proteins. To understand this breakdown in communications, imagine a contractor who plans to use 16-penny nails in building a house. If the message were garbled on the way to the carpenter and became "Use 16 nails," the house certainly wouldn't be sturdy. With incorrect instructions for its building activities, a body cell wouldn't be sturdy, either.

The Waste-buildup Model
Another factor interfering with cell construction could be the cellular wastes that build up inside a cell. With wastes lying around, it becomes more difficult for a cell to carry out its functions, just as it is hard to prepare dinner in a kitchen cluttered with unwashed dishes from lunch. If there is too much of this unneeded cellular material clogging a cell, the cell could "choke" on its own waste and die.

The Free-radical Model
One of the newest models of aging involves free radicals. Not a political group, as the name might suggest, these substances are unstable particles that are formed in the cell during normal metabolism, especially

when oxygen combines with unsaturated fats. Free radicals are fast moving and have one free, unpaired electron. Because of this unpaired electron, a free radical is attracted to and can steal an electron from other molecules. The resulting union of molecule and free radical can cross-link the molecules of a cell. Cross-links may prevent DNA from passing on its information and result in cancer, or they may cause breakdowns in the normal functioning of the body. If a free radical steals an electron from the protein, collagen, in connective tissue of skin or joints, this cross-linking can show up as wrinkled skin or stiff joints. And if an electron is stolen from a phospholipid molecule in a cell membrane, a hole could be punctured in the membrane. This would kill the cell.

Because the number of free radicals increases with exposure to sunlight, air pollution, tobacco smoke, and the normal breakdown of fat within the cell, damage can be extensive. Scientists are studying the action of some compounds, such as vitamins A, C, and E. These vitamins seem to work by scavenging unpaired electrons from free radicals before they cause damage.

As scientists continue to study cells looking for additional clues to the aging process, they have encountered a contrast that might someday be put to use. While most human cells run out of dividing time on the biological clock, cancerous cells may keep on dividing indefinitely. This contrast poses a challenge for biologists: Can they find a way to exchange some properties of normal and cancerous cells? If such a trade is possible, aging cells might be given a new lease on life. They may continue to divide, while the dangerous, uncontrolled division of cancerous cells is brought to a halt.

1. **Explore Further:** What is the average life expectancy of males and females born today in the United States? In some Asian countries? In some African countries? What factors might account for the differences among countries?
2. **Connection to Math:** When cells divide, one cell becomes two, and two become four. Calculate how many cells there would be after 20 divisions. If you can, continue to 50 divisions.
3. **Connection to Social Studies:** The Hopi Indians of the American Southwest have this saying: "It is well not to count the years. That makes one old." Do you agree with this saying? Why or why not?
4. **Issue:** Should biotechnology intervene to lengthen the average life span? What positive and negative effects might result?

Hope for Spinal Cord Injury Patients

It was a hot day, and the lake looked so cool and inviting. Fourteen-year-old Dan dived into what appeared to be eight feet of water. It was only three. Dan became one of the estimated 10 000 people in the United States who suffer spinal cord injuries every year.

Spinal Cord Injury The future for people like Dan could be a lifetime of paralysis. Some of the paralysis associated with spinal cord injury seems to be caused by the body's own responses to the trauma. When the spinal cord is injured, the body responds in a series of events. First, free radicals released from injured cells attack the membranes of surrounding cells. Then, endorphins, the body's natural painkillers, cause a sharp drop in the amount of blood flowing to the damaged area. A reduced flow of blood means that nerve cells in the spinal cord are deprived of crucial nutrients and oxygen. Finally, the molecules that transmit nerve impulses pour out of injured cells and bind to surrounding cells, causing cellular swelling and damage.

Drug Treatment Some drugs are able to counteract each of these bodily responses. If the drugs are administered quickly enough, within a few hours of injury, they could mean a difference between total paralysis below the site of injury and being able to move.

These drug treatments were developed in accord with the long-held belief that nerve cells cannot replace themselves. Unlike skin and stomach-lining cells, nerve cells were thought never to divide after a person is born. Scientists believed that once the nerve pathways were injured, they would never heal. Recently, researchers discovered that this belief is not always true. Under some conditions, nerve cells *can* divide and grow.

Because of this discovery, a new drug seems to hold promise for spinal cord injury patients. This newcomer, presently called GM-1, is a chemical extracted from the

CAREER CONNECTION

Occupational therapists help people with serious injuries to regain independence in their daily lives. For example, these professionals may teach a paralyzed patient how to turn the pages of a book by using a motor that is activated by breathing into a tube.

brains of cows. In laboratory tests on animals, GM-1 has been shown to protect nerve cells after damage to the spinal cord. Even more encouraging, GM-1 showed in similar tests that it could actually stimulate regeneration of injured nerve cells and cause new nerve cells to grow. Researchers are especially excited because the drug seems to restore the function of paralyzed muscles.

Much research and testing remains to be done, but doctors say they are hopeful that soon they can begin to consider spinal cord injury as treatable rather than incurable.

1. **Connection to Health:** Why is an injury to the spinal cord so serious?
2. **Connection to Genetics:** How might new drugs like GM-1 help people born with spinal cord defects?

REVIEW
C H A P T E R
7

Summary

Pasteur's experiment leading to the downfall of spontaneous generation also led to the theory of biogenesis, a theory that ties in nicely with the idea that all cells come from other cells.

New eukaryotic cells arise during cell cycles that are regulated by interaction of many proteins and external agents. When a cell cycle includes mitosis, as in the reproduction of most body cells, the chromosomes are first replicated and then distributed equally to daughter cells. Each daughter cell receives the same set of genetic instructions originally present in the parent cell. Prokaryotic cells distribute their replicated chromosomes to daughter cells by binary fission.

A cell cycle involving meiosis also requires chromosome replication, but results in the formation of specialized reproductive cells that contain only half the number of chromosomes as their parent cells. This reduction in the number of chromosomes ensures that the chromosome number will remain constant from generation to generation. Meiosis and fertilization result in genetic recombination, important in the variation of organisms, and therefore important in evolution.

Language of Biology

Write a sentence that shows your understanding of each of the following terms.

anaphase	interphase
binary fission	meiosis
cell cycle	metaphase
cell plate	mitosis
centromere	prophase
chromatid	sexual reproduction
diploid	spindle
fertilization	spontaneous generation
gamete	spore
genetic recombination	telophase
haploid	zygote
homologue	

Understanding Concepts

1. A pot of leftover soup is not refrigerated and sits out overnight. A student takes some of the soup into his science class and examines a drop under the microscope. He discovers microorganisms in the soup and concludes that they formed from the soup itself. What could you teach him about his conclusion?
2. Make a series of labeled diagrams illustrating mitosis in an animal cell in which 2n = 6.
3. Why did Pasteur believe that in order to disprove the idea of spontaneous generation, he had to design a flask that allowed air to enter but kept spores out?
4. How would the cell cycle be affected if DNA were prevented from replicating? Why?
5. What is the relationship between a cell's surface area and its volume as the cell grows? How does this limit cell size?
6. Make a series of labeled diagrams illustrating meiosis in an animal cell in which 2n = 6.
7. Would binary fission be an efficient means of cell reproduction if prokaryotes had several chromosomes instead of just one?
8. Explain how two fruit flies having long wings can produce some offspring having short wings.
9. Which part of meiosis—meiosis I or meiosis II—is sometimes called the reduction division? Why?
10. What is accomplished by mitosis in terms of the genetic information of a cell?

Applying Concepts

11. How might you explain the production of a human cell that had only 45 chromosomes?

12. If a nucleus from a nondividing type of cell, such as a nerve cell, is transplanted into a skin cell whose nucleus is undergoing DNA replication, within minutes DNA replication begins in the chromosomes of the transplanted nerve cell nucleus. What do the results of this experiment suggest about control of the cell cycle?

13. Some organisms reproduce in such a way that their offspring arise from just a single cell of the parent. Which method of cell reproduction would such reproduction involve? Compare the variety among offspring produced in this way with the variety among offspring resulting from fertilization of an egg by a sperm.

14. Suppose a cell had four homologous chromosomes of each type instead of two. How many homologous chromosomes of each type would be in the daughter cells produced by mitosis? By meiosis?

15. In what ways does cell reproduction in prokaryotes and eukaryotes differ? Could mitosis or meiosis occur in prokaryotes? Why or why not?

16. Colchicine is a chemical that prevents the formation of microtubules. What would be the effect of adding colchicine to cells just before mitosis were to begin?

17. At one time, interphase was referred to as the "resting phase" of mitosis. Why do you think this description is no longer used?

18. Damage to nerves is often permanent. Why is it sometimes not possible to recover from such damage?

19. In some organisms, mitosis is not followed by cell division. Describe what the cellular structure of these organisms might look like.

20. **Math Connection:** Why is it important for incoming nutrients to reach the interior of a cell quickly?

21. **Medicine Connection:** Why do you think it is important to administer drugs quickly after spinal cord injury?

Lab Interpretation

22. **Investigation:** Bean roots were marked into three equal sections (A, B, and C), with A being closest to the root tip and C being farthest from the root tip. The length of each section was measured each day for five days. Total growth in each section is indicated in the graph below. In which section of the root is growth fastest? Explain.

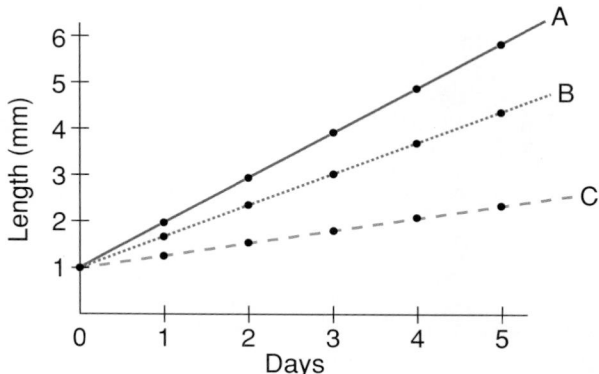

23. **Thinking Lab:** A scientist studying certain cells finds that Golgi bodies disappear at the same time that there is an increase in the numbers of vacuoles. Hypothesize what might be happening in this cell.

24. **Minilab 1:** A cell undergoes a problem during mitosis in that spindle fibers form in only one half of the cell. The spindle fibers that do form are associated with centrioles at only one pole of the cell. If this cell continues in mitosis, predict what might happen. Is this a plant cell or an animal cell? How do you know?

25. **Minilab 2:** When a large-scale model is made of something that occurs on a small scale, it is often easier to understand. Why? Explain the parts of meiosis that were easier to understand after making a model.

Connecting Ideas

26. How do certain events in the control of the cell cycle illustrate that energy is needed for the biological work of cell reproduction?

27. You learned in Chapter 5 how various parts of cells interact. What parts interact during mitosis?

HEREDITY

HAVE YOU EVER stopped to admire a butterfly flitting from flower to flower? Butterflies belong to a group of insects called Lepidoptera, from the Greek words *lepis,* which means "scale," and *pteron,* which means "wing." The scales on a butterfly's wing form the beautifully colored, sometimes iridescent pattern you see. The different kinds of butterflies can be identified by the particular patterns on their wings. Monarch butterflies, for example, have an orange and black pattern, while tiger swallowtails are yellow, edged with black, blue, and orange.

What causes a monarch butterfly's wings to be orange and black? The same thing that causes a zebra to be striped and causes you to have blue, brown, or green eyes. Heredity. The pattern of scales on a butterfly's wing is inherited; it is passed from parent to offspring. While many traits are shared by all organisms of the same kind, the *forms* of traits may vary among offspring. All traits are transmitted from one generation to the next by microscopic bits of matter on the chromosomes of gametes. The same bits of matter that control the butterfly's intricate and brilliantly colored pattern of scales also help determine the color of your eyes, the texture of your hair, and your sex. And what sex you are, male or female, may, in turn, affect which other traits you inherit. What may seem most surprising about heredity is that the occurrence of many traits is largely determined by probability—the "roll of the dice."

What would monarch butterflies look like if the pattern of scales on their wings changed with each generation? What problems might this cause for these butterflies?

Chapter Preview

Objectives

Discuss Mendel's experiments and his results.

Explain the three principles of genetics Mendel obtained through his experiments.

Solve genetics problems using Punnett squares.

8.1 Principles of Genetics

You know that like produces like. Cats produce cats, fruit flies produce fruit flies, and maple trees produce maple trees. Offspring are like their parents in that you can easily recognize them as the same kind of organism. But are offspring and parents exactly alike? Certainly the colors of kittens in a litter are not identical. Some kittens may have a color very much like one of the parents, but others may have a color not seen in either parent. Whenever offspring are produced by sexual reproduction, they vary from one another and from each of their parents.

Why is the last statement true? Recall from Chapter 7 that each parent contributes a set of DNA instructions, located on chromosomes, to its offspring. These two sets of chromosomes can combine in many ways, depending upon how they line up at the equator during meiosis I. The particular combination of the two sets of parental chromosomes in the zygote determines the traits the offspring will develop. This passing of traits from parent to offspring is called **heredity**. Can we predict what traits the offspring are likely to have if we know the traits of their parents? Are we able to determine what kinds of genetic information parents must have by analyzing the traits of their offspring? Answers to these questions can be formulated through genetics experiments. **Genetics** is the science of heredity.

Origin of Genetics

The basis for what we know about genetics came from the work of Gregor Mendel (1822-1884). Mendel was a monk in a monastery in Czechoslovakia. He went to Vienna to study science and mathematics

Figure 8-1 Kittens from the same litter can have similar or very different colors. The chromosomes each kitten inherited from its parents determine its unique color pattern.

Figure 8-2 Through experiments with common pea plants, Mendel discovered basic laws of inheritance.

198 🐝 HEREDITY

to become a teacher. After two years, Mendel returned to the monastery, where he tended the monastery garden and experimented with plants. He also taught at a local school. In 1866, Mendel published the results of eight years of experiments and analysis. His work was ignored until 1900, when it was finally recognized as a major new development in biology. As you read about Mendel, keep in mind the methods of good science. All of them are shown in Mendel's creative work.

Mendel's Experiments

Mendel worked with the common pea plant. Pea plants are grown easily and produce large numbers of offspring in a short time. Certain varieties of pea plants have sets of clearly different characteristics that remain unchanged from one generation to the next. Mendel found that these traits or characteristics are hereditary—they are passed from generation to generation. For example, pea flowers are either purple or white. The stems of pea plants are either short or tall. Mendel's work was to determine how these traits were transmitted from parent to offspring.

Mendel planned his experiments very carefully. He chose traits of pea plants that had clear-cut differences to test. He studied the offspring of many generations, not only the first, and he worked with a large number of plants, realizing that this would increase the chances that his results would be meaningful. He counted the offspring and analyzed the numbers mathematically. This quantitative data helped him support his findings. Mendel was looking for general trends to form a basic set of rules about the transmission of traits.

Mendel chose a total of seven traits to work with in his first experiments on pea plants. Those traits are shown in Figure 8-3.

Minilab 1

How does sample size affect results?

The ratio of males to females born in a population should be 50/50. Your class should show this balance, but does it? Count the number of males and females in your class. What is the ratio? The ratio may not be 50/50. One reason is that your class is a very small sample size. Now count the brothers and sisters of your classmates and total the number of males and females. How do these totals compare to the expected 50/50 ratio? How does this sample size compare to just your class? Redo the count and this time include all brothers, sisters, and first cousins of the members of your class. How do these totals compare to the 50/50 ratio? How does this sample size compare to your original sample size?

Analysis *How does sample size affect results?*

Figure 8-3 **Mendel chose seven traits of peas with clearly different forms for his experiments.**

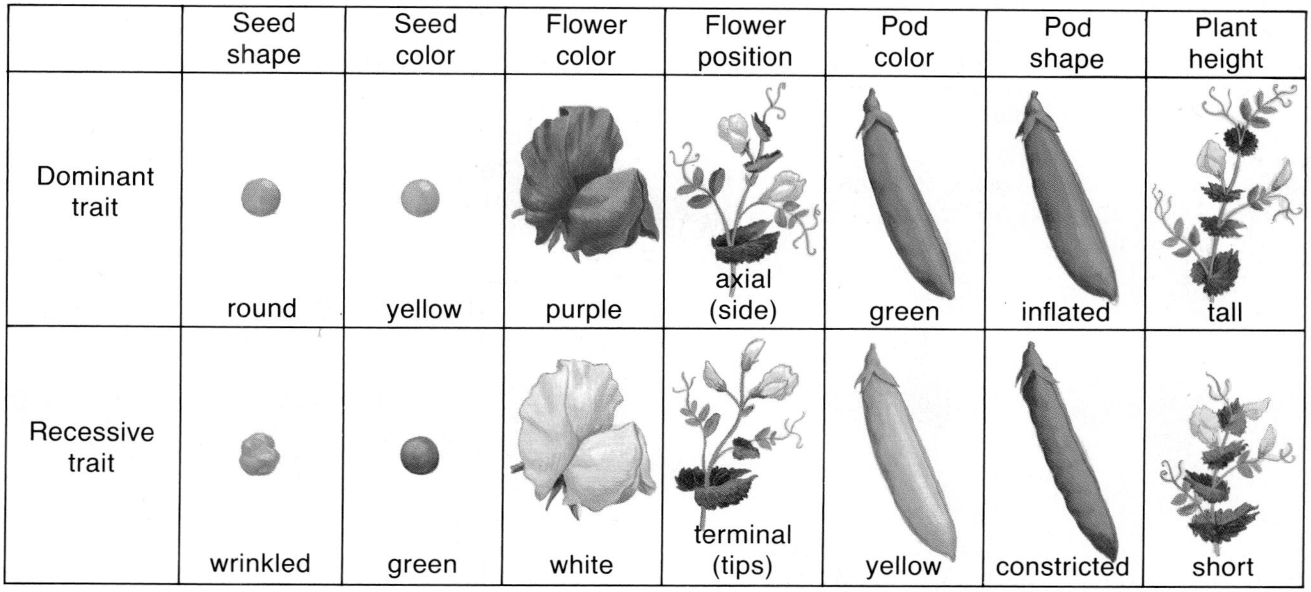

	Seed shape	Seed color	Flower color	Flower position	Pod color	Pod shape	Plant height
Dominant trait	round	yellow	purple	axial (side)	green	inflated	tall
Recessive trait	wrinkled	green	white	terminal (tips)	yellow	constricted	short

Pea Flower Structures

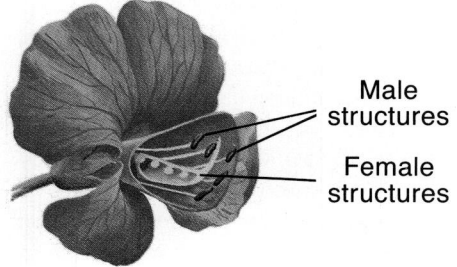

Male structures

Female structures

Pollen Transfer

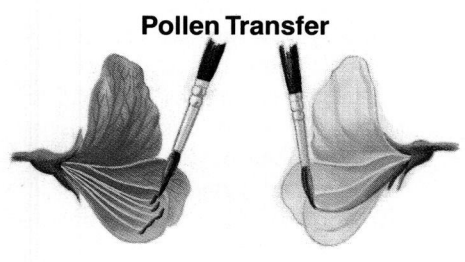

Figure 8-5 Mendel's cross of pure purple-flowered plants with pure white-flowered plants resulted in an F$_1$ generation of all purple-flowered plants.

Mendel's Results

Pea plants reproduce sexually. Both male and female sex organs are in the same flower. Normally, male gametes fertilize eggs of the same flower. After many generations, offspring still have the same features as the parents. Such plants are said to be pure. In his experiments, Mendel crossed a plant having a certain trait with a plant having the opposite trait. For example, he crossed pea plants that produced purple flowers with plants that produced white flowers. He transferred the pollen of one plant to another plant as shown in Figure 8-4.

In some cases, the pollen was obtained from purple-flowered plants and transferred to plants with white flowers. The opposite combination was also done. Mendel found that in every case, the offspring of this cross yielded offspring that all had purple flowers. The offspring of a parental cross are called the first filial, or F$_1$ generation.

Mendel was impressed that there were no plants that produced white flowers in the F$_1$ generation. He allowed members of the F$_1$ generation to reproduce in the usual fashion.

The offspring of the second cross are called the second filial, or F$_2$ generation. Mendel's results for this cross were very revealing. Of the 929 offspring produced, 705 plants produced purple flowers and 224 plants produced white flowers. These numbers give a ratio very close to three purple-flowered plants to one white-flowered plant.

Mendel noted that for each trait there is one form that dominates the other. Based on the results for the F$_1$ generation, the trait for purple flowers is the **dominant** trait. The other trait, in this case white flowers, disappears in the F$_1$ generation and is called a **recessive** trait. Mendel conducted this experiment using several other traits of pea plants. In each case, he obtained the same results. From his experiments, he concluded that one form of a hereditary trait, the dominant trait, dominates or prevents the expression of the recessive trait. This is now known as the principle of dominance.

Parent Plants

F$_1$ generation

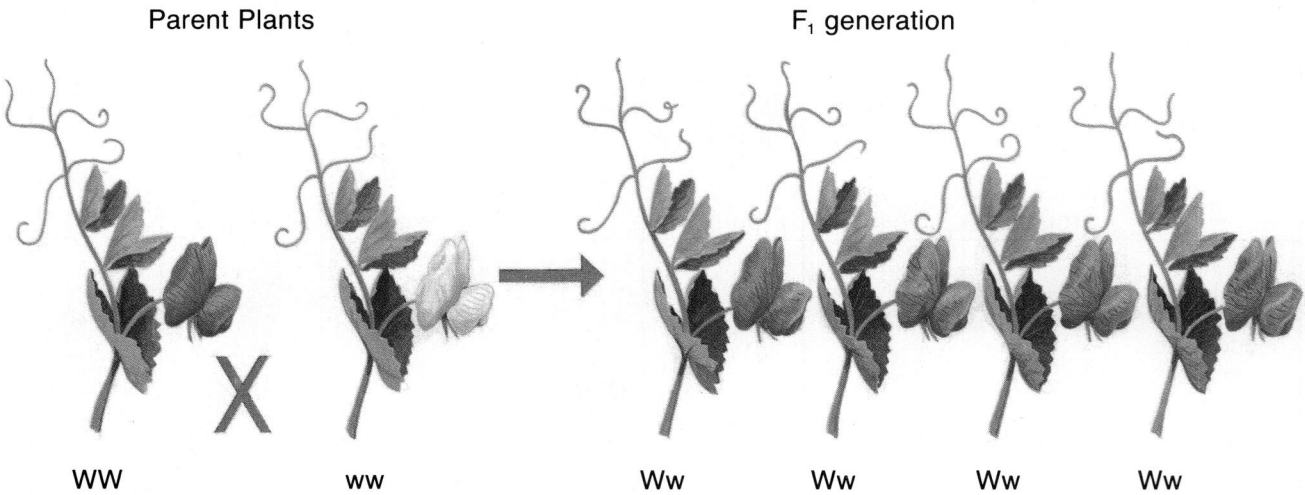

WW   ww   Ww   Ww   Ww   Ww

Mendel's Hypothesis

In Mendel's crosses, the recessive trait disappeared in the F_1 generation but reappeared in the F_2 generation. To explain this result, Mendel made certain assumptions. Each pea plant is produced as a result of the union of a sperm and an egg. He reasoned that for every trait, there must be a pair of factors. One must come from the sperm and one must come from the egg. He called the factors *characters,* but we now call them **genes.** Mendel represented the genes by symbols. A dominant gene is represented by a capital letter. A recessive gene is represented by the same letter but in lowercase. For example, if the gene for purple flowers (dominant) were *W,* then the gene for white flowers (recessive) would be *w.* The two genes combine in the fertilized egg so that each new pea plant contains two genes for each trait.

Gene Segregation   In Mendel's experiments, the pure parental plants each had two identical genes for a trait. Purple-flowered plants (dominant) were *WW.* White-flowered plants (recessive) were *ww.* Mendel reasoned that the two genes segregate (separate) during gamete formation. Sperm and eggs would have just one gene for each trait. Therefore, the gametes of purple-flowered parents should have the *W* gene, and gametes of white-flowered parents should have the *w* gene. As a result, all F_1 plants should be *Ww,* the only possible combination. Because the *W* gene is dominant to the *w* gene, all F_1 plants would have purple flowers.

If Mendel's assumptions were correct, all F_1 plants *(Ww)* should produce half their gametes with the *W* gene and half with the *w* gene. There would be three possible combinations in the F_2 generation—*WW, Ww,* and *ww.* The chance of *W* and *w* gametes combining would be twice as great as other possible combinations. These possibilities can be expressed as a ratio—1 *WW*:2 *Ww*:1 *ww.* Both *WW* and *Ww* plants will produce purple flowers, while *ww* plants will produce white flowers. This result represents a ratio of 3 purple-flowered plants:1 white-flowered plant. This is, in fact, what Mendel observed.

Figure 8-6   **The cross of individuals from the F_1 generation resulted in approximately one white-flowered pea plant for every three purple-flowered plants.**

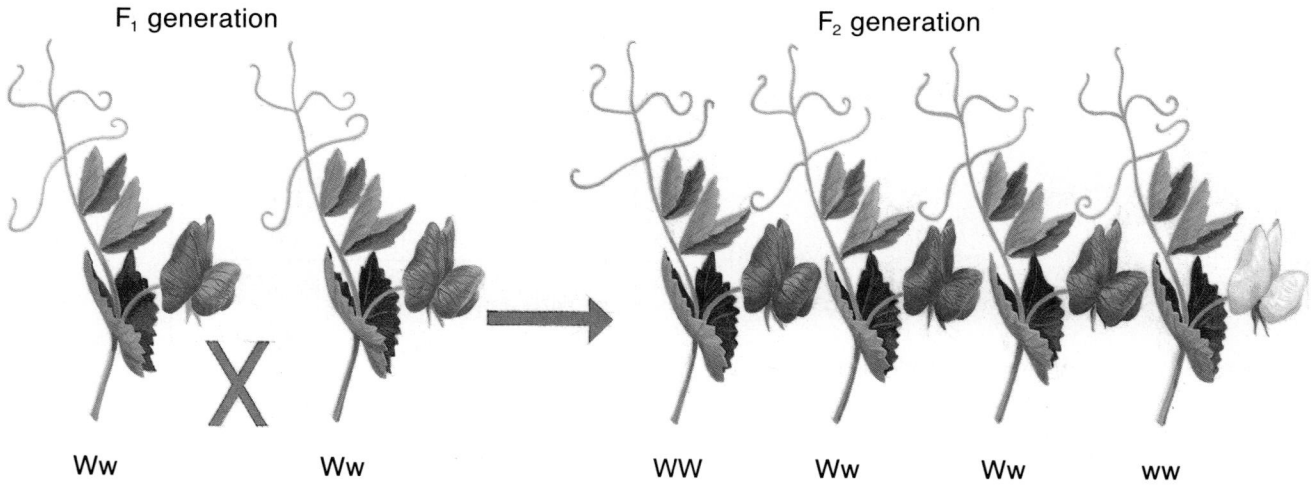

F₁ generation   F₂ generation

Ww   Ww   WW   Ww   Ww   ww

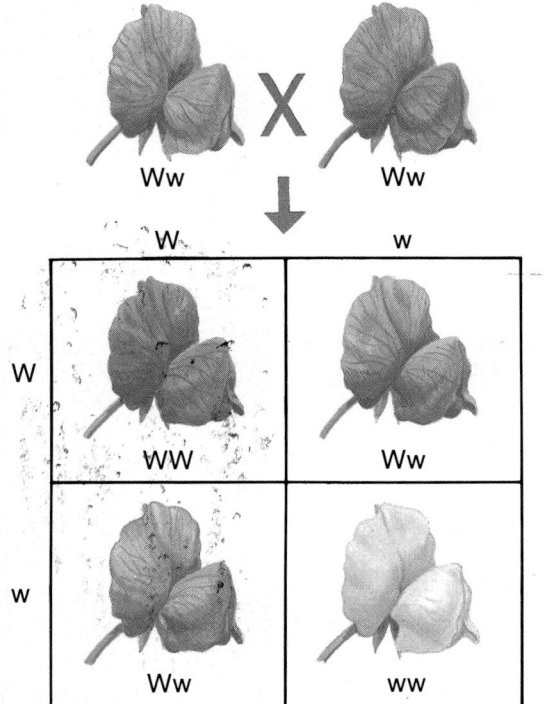

Figure 8-7 The results of Mendel's F₁ cross are shown in this Punnett square. Here there are three possible gene combinations: *WW*, *Ww*, and *ww*.

Punnett Squares Results of crosses can be predicted with Punnett squares (Figure 8-7). A Punnett square is a chart used to determine the possible combinations of genes among offspring. Across the top of the square are written the gametes produced by one parent. The gametes produced by the other parent are listed along the left side of the square. The square is then filled in by listing all the possible combinations of gametes. Each square in the chart represents a possible zygote.

The Test of Segregation

Mendel was able to explain what he observed in his experiments. His conclusions about dominant and recessive traits and his hypothesis that genes segregate when gametes are formed grew from his experiments. His first experiments alone did not mean that his explanation was correct; another test was needed to confirm his hypothesis. Mendel did this test after predicting the outcome of yet another cross.

Mendel predicted that if F₁ generation purple-flowered plants were crossed with white-flowered plants, he would get ratios different from any of his previously obtained ratios. According to his hypothesis, F₁ purple-flowered plants are *Ww* and white-flowered plants are *ww*. Purple-flowered plants produce both *W* and *w* gametes, while white-flowered plants produce only *w* gametes. Mendel predicted that the possible combinations in offspring resulting from a cross between *Ww* and *ww* parents would be *Ww* (purple) and *ww* (white). He predicted also that the numbers of each would be about the same. When Mendel performed the experiment, he obtained about half (50%) purple-flowered and half (50%) white-flowered plants. The same ratio was observed when he did the same experiment using plants having other traits.

Figure 8-8 A cross between a purple-flowered plant with the genes *Ww* and a white-flowered plant with the genes *ww* results in 50% purple-flowered plants and 50% white-flowered plants. What are the genotypes of the offspring?

As a result of his experiments, Mendel assumed his hypothesis was correct. His hypothesis and data led him to form the principle of segregation—during gamete formation the pair of genes responsible for each trait separates so that each gamete receives only one gene for each trait. The gametes unite to produce predictable ratios of traits among the offspring.

Terminology

Many inherited traits, like the color of pea plant flowers or the length of a fruit fly's wings, take different forms. For many traits, such as those of the pea plants in Mendel's experiments, there are two forms. Each alternative form of a gene for a certain trait is called an **allele.** For example, the trait expressed as seed shape is determined by the alleles *R* and *r*. The trait of flower color in the previous examples was determined by the alleles *W* and *w*.

The combination of alleles for a given trait is referred to as the **genotype.** For example, *RR*, *Rr*, and *rr* are possible genotypes controlling seed shape in pea plants; the two seed shapes possible are round or wrinkled, with round being dominant to wrinkled. The appearance of a trait as determined by a given genotype is called the **phenotype.** Round and wrinkled are possible phenotypes of seed shape. In this example, the genotype *RR* would produce a round seed. The genotype *Rr* would also produce a round seed. The genotype *rr* would produce wrinkled seeds.

Did you notice that different genotypes in the example above may result in the same phenotype? Some phenotypes, such as the shape of pea seeds, are visible. However, many, such as human blood type, are not.

When each cell of an organism contains two of the same alleles for a given trait, the organism is **homozygous** for that trait. *RR* and *rr* are homozygous genotypes. To avoid confusion, the terms *homozygous dominant* and *homozygous recessive* are used. When each cell of an organism contains two different alleles for a trait, such as *Rr,* that organism is **heterozygous** for the trait.

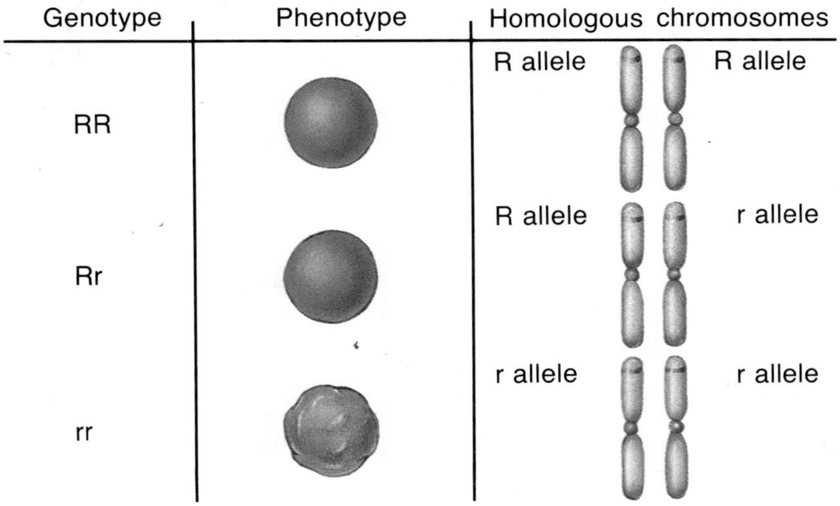

Genotype	Phenotype	Homologous chromosomes	
RR		R allele	R allele
Rr		R allele	r allele
rr		r allele	r allele

Figure 8-9 **The combination of alleles in the chromosomes of an individual, its genotype, produces its physical appearance, its phenotype.**

Two Traits

After Mendel's first set of experiments, he studied the inheritance of two traits at once. For example, what types of plants would develop from a cross between a parent plant that produced peas that were round and yellow *(RRYY)* and a parent plant that produced peas that were wrinkled and green *(rryy)*? Remember that the law of segregation states that each gamete must contain one allele for each trait. Therefore, the gametes of the yellow, round parent should have the alleles *Y* and *R* and the wrinkled, green parent would produce gametes with *y* and *r* alleles. If they do, F_1 plants would all have the genotype *RrYy* and the phenotype round and yellow. Mendel found these results.

INVESTIGATION

Predicting Plant Genotypes

It may be difficult to predict traits for tobacco plants if provided with only their seeds. But, if these seeds were planted and allowed to grow, certain traits would appear. You might even be able to predict the genotypes and phenotypes of the parents based on the appearance of traits in the offspring.

In this lab, you will design and then carry out an experiment to determine the genotypes of three tobacco seed batches. It should be pointed out in advance that the two seed batches to be used will yield plants that appear either all green in color or some green and some albino (white). Use these genotypes for help: *CC* = green, *Cc* = green, and *cc* = albino (white). Another batch will yield some green, some yellow-green, and some yellow. Use these genotypes: *GG* = green, *GY* = yellow-green, and *YY* = yellow.

Work in a group of three or four students.

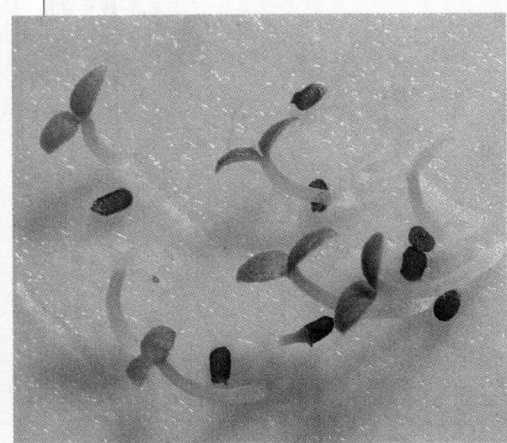

Problem

Predict the genotype for each tobacco seed batch based on the appearance of color in the mature plants.

Hypothesis

Have your group agree on a hypothesis to be tested. Record your hypothesis.

Experimental Plan

1. Examine the materials provided by your teacher. Then, as a group, make a list of the possible ways you might test your hypothesis using the materials available.

2. Agree on one way that your group could investigate your hypothesis.

3. Design an experiment that will allow for the collection of quantitative data.

4. Prepare a list of numbered directions. Include a list of materials and the quantities you will need.

The Principle of Independent Assortment   The F_1 cross presented another problem to Mendel. The F_1 plants had the genotype *RrYy*. He knew that gametes produced by these plants must contain one of each kind of allele. But, do the alleles segregate independently of each other during gamete formation? Or do they stay in the same combination, *RY* or *ry*? Mendel hypothesized that they segregate independently and that the possible gametes would be *RY, Ry, Yr,* and *ry.* Given these possible gametes, the F_2 generation would show a phenotypic ratio of 9:3:3:1, or 9 round and yellow:3 round and green:3 wrinkled and yellow:1 wrinkled and green. These results are shown in the Punnett square in Figure 8-10. Again, Mendel's experiments verified his prediction. From his results, he formulated the principle of independent assortment. It states that genes for different traits segregate independently during gamete formation.

Checking the Plan

Discuss the following points with other group members to decide the final procedure for your experiment. Make any needed changes in your plan.

1. What data are you going to collect?
2. How long will you carry out the experiment?
3. How many seeds of each batch will you use?
4. What variables, if any, will have to be controlled? (Hint: Think about the amount of light, water, and temperature.)
5. What will be the role of each classmate in your group?
6. Have you designed and made a table for collecting data?

Be sure your teacher has approved your experimental plan before you proceed.

Data and Observations

Carry out your experiment, make any needed observations, and complete your data table. Design and then complete a graph or visual representation of your results.

Analysis and Conclusions

1. Why was it necessary to grow the seeds into plants in order to "read" the seeds' phenotypes?
2. Using the information provided in the introduction regarding genotype and phenotypes, describe the difference between the inheritance pattern of the green gene C and the green gene G.
3. What were some of the variables that you had to consider in this experiment? Explain how they could have influenced your data if not taken into account.
4. For the batch of seeds that provided all green plants, are you able to predict exactly the genotypes of (a) the parents that formed these seeds? Explain. (b) Each plant observed? Explain.
5. For the batch of seeds that provided some green and some albino plants, are you able to predict exactly the genotypes of (a) the parents that formed these seeds? Explain. (b) Of each plant observed? Explain.
6. For the batch of seeds that provided some green, yellow-green, and yellow plants, are you able to predict exactly the genotypes of (a) the parents that formed these seeds? Explain. (b) Each plant observed? Explain.

Going Further

Based on this lab experience, design an experiment that would help to answer another question that arose from your work. Assume that you will be doing this lab in class. Record this information as part of your conclusion.

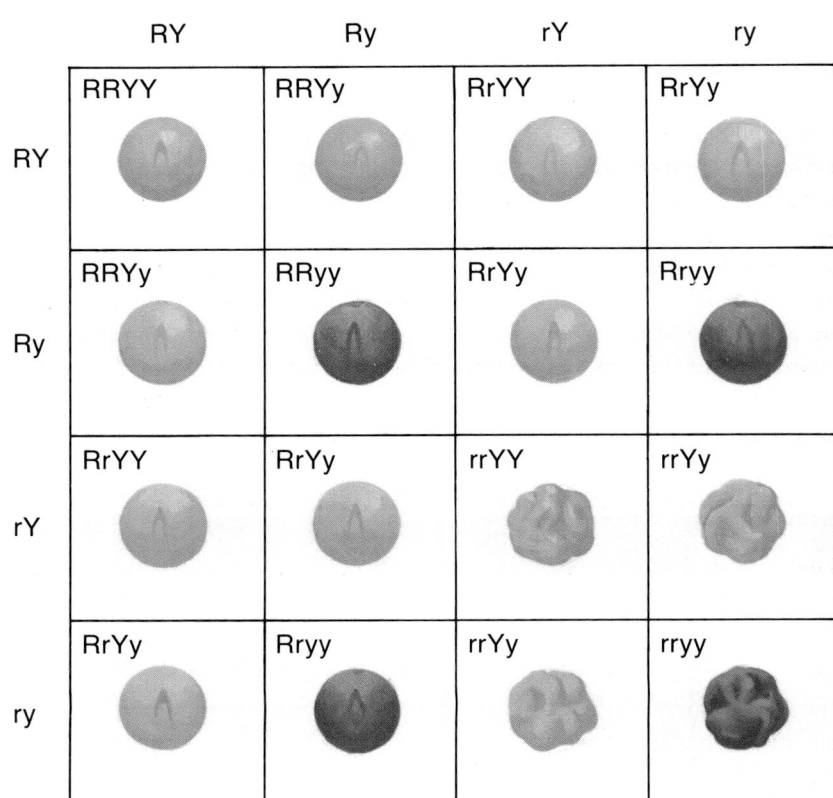

F₁ cross
RrYy × RrYy

round yellow

round green

wrinkled yellow

wrinkled green

Figure 8-10   The F₂ generation of a two-trait cross clearly shows that alleles segregate independently of one another. How many different phenotypes and genotypes result from this cross?

A Two-Trait Problem   Try another problem involving two traits. In guinea pigs, the gene for rough coat (*R*) is dominant over the gene for smooth coat (*r*). The gene for black color (*B*) is dominant over the gene for albino (*b*). A heterozygous black, smooth-coated male is bred to a heterozygous black, rough-coated female. Use a Punnett square to solve this problem. What genotypic and phenotypic ratios could you expect among their offspring?

Section Review

Understanding Concepts

1. Explain how two black mice, when mated, can produce some offspring that are white.
2. Apply the law of segregation to explain the kinds of gametes produced by an animal that has the genotype *Aa*.
3. One parent is homozygous for a certain trait, and the other is heterozygous. What fraction of their offspring would you expect to be heterozygous?
4. **Skill Review—Observing and Inferring:** The offspring of a cross between two red-flowered plants show a ratio of three red-flowered plants to one white-flowered plant. Which trait is dominant? What must the genotypes of the parents be? For more help, see Observing and Inferring in the **Skill Handbook.**
5. **Thinking Critically:** In pea plants, tall is dominant over short. Suppose you have a tall plant but don't know whether it is homozygous or heterozygous. How could you find out?

8.2 Solving Genetics Problems

When you toss a coin, the chances of getting a head and the chances of getting a tail are equal, or about 50-50. Stated another way, there is an equal probability of tossing a head and tossing a tail. This probability can also be written as a ratio—1 head:1 tail or 1/2 heads:1/2 tails. These ratios are what you expect, but the results do not always come out that way. The probability of getting a 1:1 ratio of heads to tails increases with the number of tosses. You will probably come closer to the expected ratio if you toss the coin ten times rather than twice. And you will do even better if you toss the coin 100 times. You can use these principles of probability to solve genetics problems.

Probability and Genetics

What ratio of head and tail combinations might you expect if you toss two coins together? For each toss, the chance of getting a head or the chance of getting a tail is 1/2. The tossing of the two coins together must be considered as two independent events. The probability that both coins will land heads is the product of the independent events, or $1/2 \times 1/2 = 1/4$.

Figure 8-11 shows the possible combinations of heads and tails when you toss two coins together. You can see that the chance of a heads-tails combination is twice as great as the chance of either a heads-heads or a tails-tails combination. Thus, the expected ratio is 1/4 heads-heads:1/2 heads-tails:1/4 tails-tails. Looking at this mathematically, you can see that the chances of heads and tails of one coin toss have been multiplied by the chances of the second coin toss.

Multiplying the chances of independent events occurring to determine the chance of those occurring together is a basic law of probability. It is also the basis for solving problems in genetics.

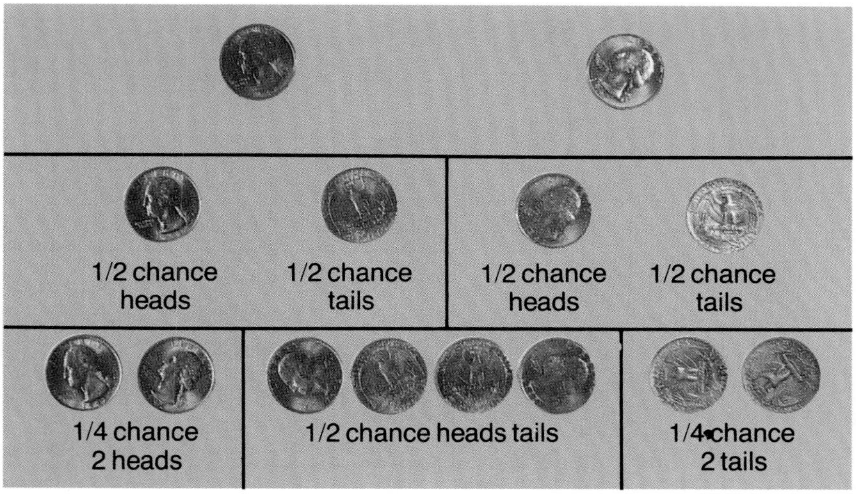

> (1/2 heads + 1/2 tails) × (1/2 heads + 1/2 tails) =
> 1/4 heads-heads + 1/4 heads-tails +
> 1/4 heads-tails + 1/4 tails-tails =
> 1/4 heads-heads + 1/2 heads-tails + 1/4 tails-tails

Objectives

Apply the rules of probability to solve genetics problems.

Demonstrate the inheritance of traits resulting from incomplete dominance, codominance, and multiple alleles.

| 1/2 chance heads | 1/2 chance tails | 1/2 chance heads | 1/2 chance tails |

| 1/4 chance 2 heads | 1/2 chance heads tails | 1/4 chance 2 tails |

Figure 8-11 Each toss of a coin has a probability of being a head or a tail. To find the probability of combinations of tosses in a row, multiply the individual probabilities, as shown above.

In the last section, you used Punnett squares to predict the results of crosses. This chart helps determine the possible combinations of genes among offspring. In effect, the Punnett square helps you multiply the chances of different alleles combining. The following example will show you how probability works in solving genetics problems.

In fruit flies, the gene for straight wings *(S)* is dominant over the gene for curled wings *(s)*. Two heterozygous straight-winged flies are mated. What ratios of genotypes can you expect from this cross? Below are the steps to follow in solving this kind of problem.

1. Write the genotype of each parent.
   male heterozygous straight = *Ss*
   female heterozygous straight = *Ss*
2. Use the law of segregation to list the different alleles that may be carried by the gametes of each parent.
   male *Ss* = *S* and *s*
   female *Ss* = *S* and *s*
3. Write the different alleles from the male parent along the top of the Punnett square. Write the alleles from the female parent down the side.
4. Fill in the boxes with the possible combinations of alleles in the offspring.
5. Determine the ratio of genotypes. The denominator of your fraction should be the total number of boxes in your square. Make sure to group similar genotypes. In this case, the genotypic ratio is 1/4 *SS*:1/2 *Ss*:1/4 *ss*.

What are the ratios of phenotypes in the offspring? To determine the phenotypic ratio, interpret the genotypic ratio:

1/4 *SS* = straight winged
1/2 *Ss* = straight winged
1/4 *ss* = curled winged

Therefore, the phenotypic ratio is 3/4 straight winged:1/4 curled winged.

Figure 8-12 Combinations of characters among offspring can be determined by using Punnett squares. Sperm and egg possibilities are listed along the top and the left side of the square. Possible offspring are shown inside the Punnett square.

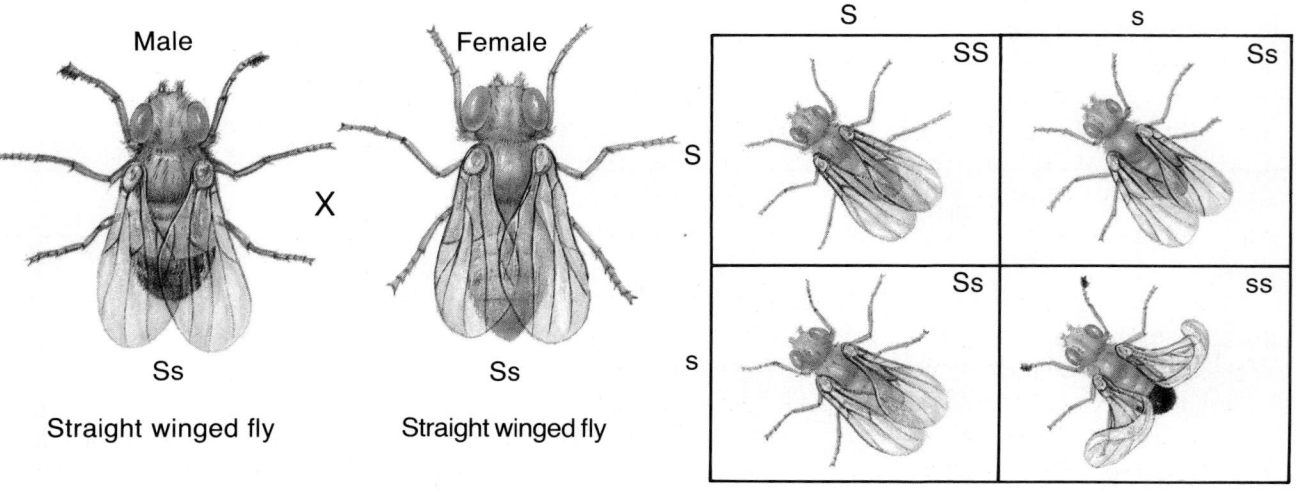

Male

Ss

Straight winged fly

X

Female

Ss

Straight winged fly

Solving genetics problems with Punnett squares applies the laws of probability. When you write the alleles for each parent, you are actually representing probabilities. For the problem above, you write S and s along the top and side because it is probable that half of all gametes that go into producing the next generation will contain the S allele and half will contain the s allele. Whether a gamete with an S allele combines with a gamete containing an S allele or one containing an s allele is also a matter of chance. When you fill in the boxes of the Punnett square, you are really multiplying independent chance events to determine the chances of their occurring together. The results are genotypic and phenotypic ratios.

Remember that these are *expected* genotypic and phenotypic ratios. For the problem above, those ratios tell you that out of every four offspring produced, three are expected to have straight wings and one is expected to have curled wings. But, as with tossing a coin, expected results and actual results do not always agree. It is possible—but less probable—that two heterozygous straight-winged flies could produce four offspring with straight wings. It is even less probable—but still possible—that the offspring all would have curled wings. The larger the number of offspring produced, the more likely it is that the actual ratios will be close to the expected results.

Incomplete Dominance

In Mendel's original experiments, he studied seven different pairs of traits in garden pea plants. In every case, one form of the trait always completely dominated the other. Dominant and recessive traits are not always as clearly defined as the seven pea plant traits Mendel studied.

For example, plants called four o'clocks have alleles for red flowers and white flowers, but neither of these colors is dominant. In these plants, heterozygous genotypes result in pink flowers. This pattern of inheritance is known as **incomplete dominance.** It appears that the parental genes blended together to form an intermediate color. Actually, incomplete dominance of two alleles results in the possibility of three different phenotypes because neither allele is dominant over the other. The third phenotype, which results when the two alleles combine in an offspring, is an intermediate form of the other two phenotypes, not a blending of the two. If heterozygous pink flowers are allowed to self-pollinate, the colors red and white return in the offspring.

Punnett squares can be used to solve problems involving incomplete dominance. For example, try to solve this problem: What genotypic and phenotypic ratios could you expect from a cross between a red four o'clock and a pink four o'clock? The two alleles for flower color are represented as R (red) and R' (white). Use a Punnett square to solve the problem.

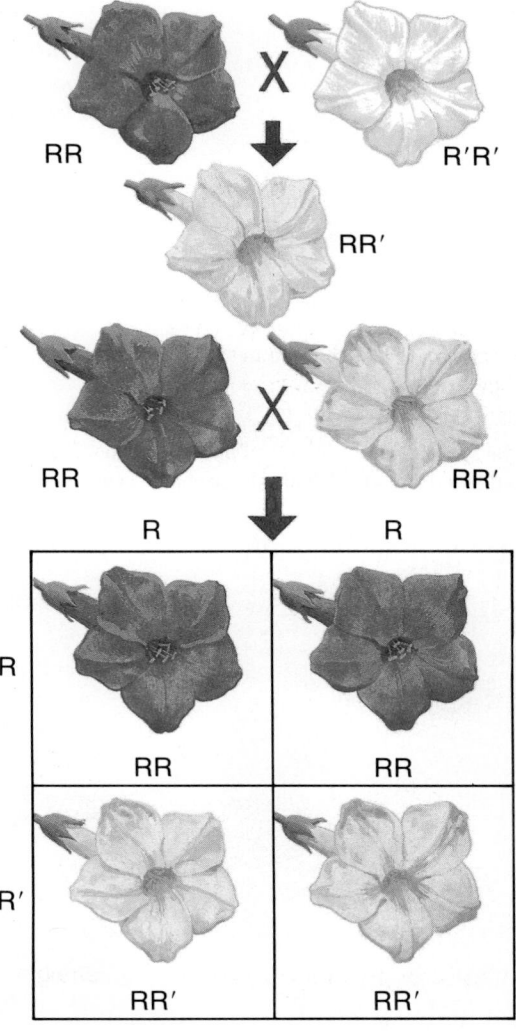

Figure 8-13 In incomplete dominance, neither allele is dominant to the other. Three completely different phenotypes are possible. The third phenotype is intermediate between the other two.

Multiple Alleles

You have learned that two alleles act together to produce a phenotype. Although each *organism* normally has just two alleles for a trait, more than two alleles may be possible for that trait in the *population*. In the human population, for example, three alleles govern blood type: I^A, I^B, and i. Human blood types are therefore determined by **multiple alleles,** a set of three or more different alleles controlling a trait. In humans, both I^A and I^B are dominant over i. But I^A and I^B are codominant. Like incomplete dominance, **codominance** results when one allele is not dominant over the other. Unlike incomplete dominance, in which the presence of both alleles results in an intermediate third phenotype, codominance results in both alleles being expressed equally.

From Table 8.1 you can see that six genotypes and four phenotypes are possible for human blood types. The four blood phenotypes are *A, B, AB,* and *O.*

Table 8.1 Blood Groups in Humans

Phenotypes (Blood Types)	Genotypes
A	I^AI^A or I^Ai
B	I^BI^B or I^Bi
AB	I^AI^B
O	ii

Figure 8-14 The three alleles in humans that govern blood type are a familiar example of multiple alleles. If all these children had blood type O, what are the probable genotypes of the parents?

Using the information above, try to solve this problem: A woman having type *O* blood marries a man having type *AB* blood. What are the expected genotypic and phenotypic ratios among their children? Use a Punnett square to determine your ratios.

You can also use Mendel's laws to determine information about parents based on knowledge of the offspring they have produced. Consider this problem, for example: A woman who has type *B* blood marries a man who has type *A* blood. They have five children, all with type *AB* blood. What are the most *probable* parental genotypes?

To solve this problem, first think about the children. Each child has type *AB* blood. Therefore, each genotype must be I^AI^B. What is the source of each allele in each child? One allele came from its mother, and the other from its father. Now consider the parental possibilities.

The mother could be either I^BI^B or I^Bi. The father could be I^AI^A or I^Ai. But if all five children are I^AI^B, both parents are most probably homozygous. Each child's I^A gene came from the father, while the I^B gene came from the mother. If either of the parents were heterozygous, some other blood types would probably occur among the children. Using this reasoning, the mother's genotype is probably I^BI^B, and the father's is probably I^AI^A.

Coat color in rabbits is a visible trait that is governed by multiple alleles. Different combinations of these four alleles produce four different coat colors as seen in Figure 8-15. In this case, the alleles involved have a dominance relationship with each other.

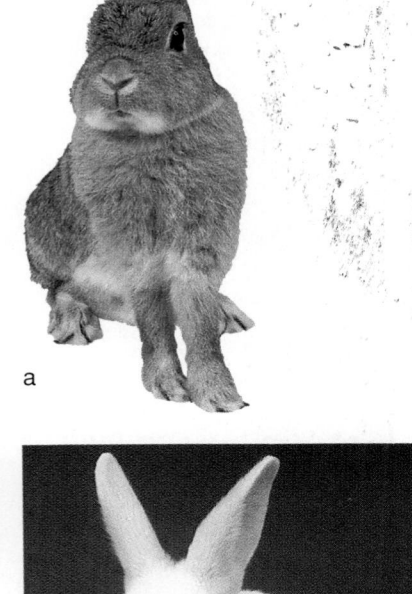

Figure 8-15 **Different combinations of four alleles produce different coat colors in rabbits: the wild type or agouti rabbit (CC, Cc^ch Cc^h, or Cc) (a); light grey rabbit (c^chc^ch, c^chc^h or c^chc) (b); an albino with black points (c^hc^h or c^hc) (c); and an albino rabbit (cc) (d).**

a

b

c

d

Section Review

Understanding Concepts

1. What are the chances of tossing four pennies at the same time and having them all come up tails?

2. What genotypic ratio could you expect from a cross between an organism having the genotype *Dd* and an organism with the genotype *DD*?

3. Watermelons with a round shape are crossed with those having a flatter shape. Some offspring have an oval shape. Explain.

4. ***Skill Review—Making and Using Graphs:*** In guinea pigs, rough coat (*R*) is dominant over smooth coat (*r*), and black (*B*) is dominant over albino (*b*). Make a pie graph showing the ratio of different genotypes expected when a guinea pig having the genotype *BbRr* is crossed with one having the genotype *bbRr*. For more help, see Organizing Information in the **Skill Handbook.**

5. ***Thinking Critically:*** A man with blood type *B* marries a woman with blood type *A*. Their first child has blood type *O*. Explain. What other blood types are possible for their future children?

Discuss the development of the chromosome theory of heredity.

Apply the laws of probability to solve genetics problems involving sex-linked traits.

Explain how the expression of a pair of genes may be influenced by other genes as well as by the environment.

8.3 The Chromosome Theory of Heredity

Gregor Mendel explained simple patterns of inheritance without knowing of the existence of chromosomes or of how they are packaged in gametes by means of meiosis. Not until the early 1900s did scientists begin to connect cellular processes and genetic principles to understand how genetic information is transmitted. New knowledge not only helped us understand how traits are physically passed from one generation to the next, but also enabled us to explain different patterns of inheritance that do not follow Mendelian principles.

Mendel had concluded that there are two characters, which we now call genes, for each trait, both of which were transmitted to offspring. He had also concluded that these characters segregate during gamete formation—each gamete has just one character per trait. Later, when new staining techniques allowed scientists to observe chromosomes in cells, it was revealed that each diploid cell has one homologous pair of chromosomes of each type and that a zygote inherits both homologues. Furthermore, homologous chromosomes segregate during meiosis—each gamete has only one homologue of a pair.

In the early 1900s, a young scientist named Walter Sutton took these new observations and the results of Gregor Mendel and began making his own observations. His work ultimately resulted in a new theory.

Mendel's Work Rediscovered

Walter Sutton, while a student at Columbia University, proposed that Mendel's hereditary factors (genes) are carried on chromosomes. Sutton made this proposal after observing meiosis in grasshoppers. Sutton saw that during meiosis the chromosomes in each grasshopper cell lined up in pairs and that each pair of chromosomes was the same size and shape.

Sutton next observed that the homologous pairs of chromosomes segregate during meiosis so that each gamete receives one chromosome from each pair. He then reasoned that when fertilization occurs, the resulting zygote has a full set of homologous chromosomes.

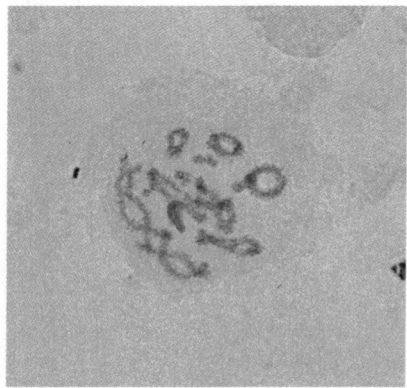

Figure 8-16 **Grasshopper chromosomes are shown during a mitotic cell division. Homologous pairs of chromosomes can be found.**

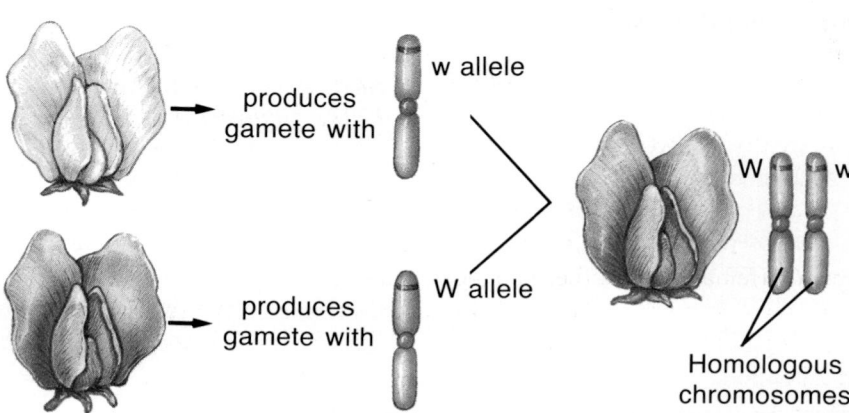

Figure 8-17 **Each new organism receives one allele for a gene from each parent. The phenotype of an organism results from the combination of alleles it receives.**

produces gamete with — w allele

produces gamete with — W allele

W   w

Homologous chromosomes

From his observations, Sutton found that chromosomes do indeed behave like Mendel's factors. During meiosis, chromosomes follow each of Mendel's principles of heredity. This led Sutton to propose the chromosome theory of heredity—Mendel's hereditary factors (genes) are carried on chromosomes.

Scientists soon realized that organisms have many more hereditary characteristics than they have chromosomes. For example, humans have 46 chromosomes, but we certainly have more than 46 inherited characteristics. Because of this, Sutton reasoned that each chromosome must carry hundreds of genes.

Although Sutton's observations give good evidence that genes are located on chromosomes, many scientists were not convinced. The work of Thomas Hunt Morgan, a biologist from the United States, supported Sutton's work and ultimately confirmed the chromosome theory of heredity.

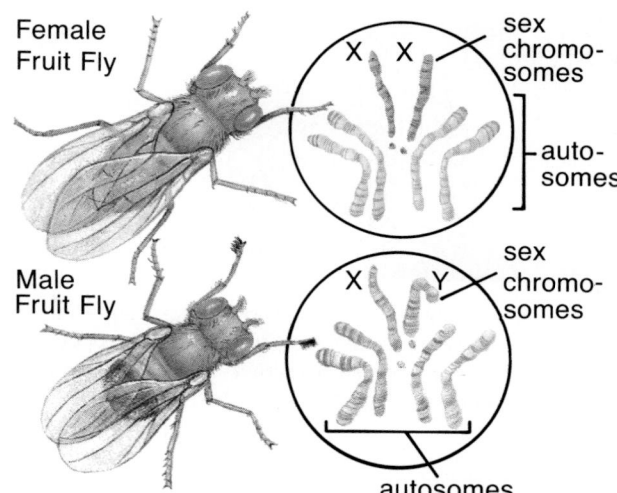

Figure 8-18 Fruit flies (**Drosophila melanogaster**) have four pairs of chromosomes. Three pairs are autosomes and the other two chromosomes are sex chromosomes.

Sex Determination

Drosophila, commonly known as fruit flies, are small (about 2 mm in length), easily handled, and produce many offspring in about two weeks. For these reasons, *Drosophila* have played an important role in the study of genetics.

Thomas Hunt Morgan (1866-1945) helped pioneer the use of fruit flies to study genetics. Early in Morgan's research, he discovered that chromosomes of male and female *Drosophila* cells are slightly different, as seen in Figure 8-18.

Because the *X* and *Y* chromosomes differ between the sexes, they are called **sex chromosomes**. Other chromosomes are called **autosomes**. Thus, both female and male *Drosophila* have three pairs of autosomes and one pair of sex chromosomes. A female normally has three pairs of autosomes and two *X* chromosomes. A male normally has three pairs of autosomes, an *X* chromosome, and a *Y* chromosome.

All the gametes produced by a female fruit fly will contain an *X* chromosome. The male fruit fly's *X* and *Y* chromosomes are not homologous, but they behave as a homologous pair during meiosis. They separate from each other during gamete formation. Thus, half the male's gametes will contain the *X* chromosome, and half will contain the *Y* chromosome. The possible combinations of *X* and *Y* chromosomes among offspring can be determined as with other traits, as seen in Figure 8-19. Using probability methods, you can determine that half the offspring would have the genotype *XX* (female). The other half would have *XY* (male).

Such a pattern of sex determination is evident in many organisms, including humans. However, the pattern is different in some organisms. In grasshoppers, for example, males have only one *X* chromosome but no *Y* chromosome, while females have two *Y* chromosomes.

Figure 8-19 This Punnett square shows that half the possible offspring in fruit flies are female and half are male.

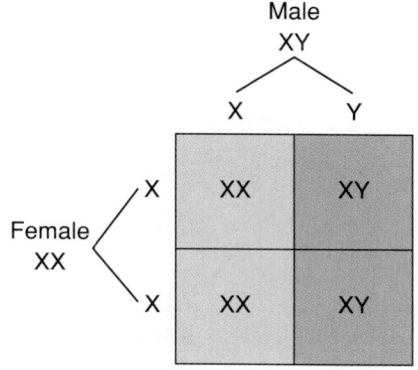

Genotypic ratio = 1/2 XX : 1/2 XY
Phenotypic ratio = 1/2 female : 1/2 male

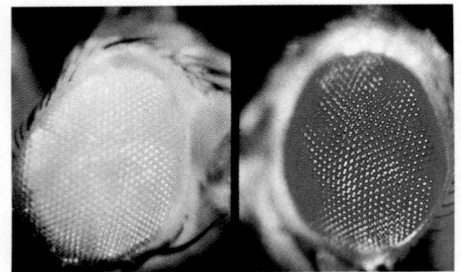

Figure 8-20 In fruit flies, red eyes (right) are dominant to white eyes (left). The allele for eye color is carried on the X chromosome.

Sex-linked Inheritance

In 1910, Morgan made an unexpected discovery about how the sex of an organism can be related to inheritance of a trait. He made this discovery studying eye color in fruit flies. He found a male fly with white eyes. This fly was produced from a pure line having only red eyes.

Morgan decided to cross this white-eyed male with red-eyed females. Assuming that red eye color (R) was dominant over white eye color (r), Morgan expected to find all red-eyed offspring in the F_1 generation. He also expected the F_1 flies, when interbred, to produce a ratio of three red-eyed flies to one white-eyed fly. He did, in fact, get these results.

However, there was something unusual about the F_2 generation. Morgan noted that the white-eyed flies in this generation were all males. There were no white-eyed females. Because white eye color seemed to be linked with sex, it was called a **sex-linked trait.**

Morgan's Explanation Morgan knew of the hypothesis that genes are located on chromosomes and used that idea to form a hypothesis of his own. Morgan's hypothesis was that the alleles for eye color are carried only on the X chromosome, and that there are no alleles for eye color on the Y chromosome. If this were true, then the parental cross had been between red-eyed females (X^RX^R) and white-eyed males (X^rY), as illustrated in Figure 8-21. The results of the cross—the F_1 generation—had been red-eyed males (X^RY) and red-eyed females (X^RX^r). And the F_2 generation had been red-eyed females (X^RX^R or X^RX^r), red-eyed males (X^RY), and white-eyed males (X^rY). Morgan's hypothesis explained his previous data by showing how all white-eyed flies in the F_2 generation had to be males.

However, a hypothesis should also predict new facts. Morgan predicted that he could produce white-eyed females by breeding an F_1 red-eyed female (X^RX^r) with a white-eyed male (X^rY). When such a cross was done, Morgan's prediction was confirmed. There were white-eyed females among the offspring. This confirming evidence also helped support the hypothesis that genes are carried on chromosomes, because Morgan's explanation was based on that concept.

Figure 8-21 Assuming eye color is sex-linked, results predicted with Punnett squares agree with Morgan's experimental results. The parental and F_1 crosses he did are shown here.

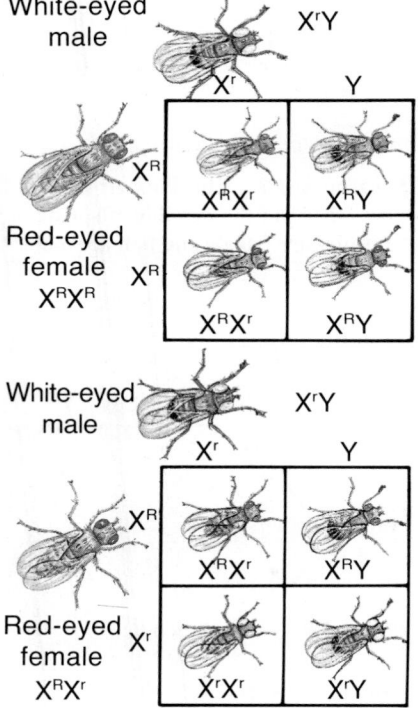

White-eyed male X^rY

Red-eyed female X^RX^R

White-eyed male X^rY

Red-eyed female X^RX^r

Sex-linked Traits and Punnett Squares Morgan's discovery of sex-linked traits and his explanation of how they are inherited show that problems involving such traits can also be solved using Punnett squares. If information given in a problem tells you that a trait is sex-linked, you assume that the trait is carried on the X chromosome because it carries more information than the smaller Y chromosome. In humans, very few genes are located on the nonhomologous portion of the Y chromosome, and therefore few traits are sex-linked this way. A few genes in humans are Y-linked. These traits are only expressed in males and are transmitted from father to son. Consider the sex chromosomes and the genes they carry together as a unit (for example, X^R) when determining the kinds of gametes produced by each parent. When you determine phenotypic ratios, do so by including the sex as well as the trait of the offspring.

Try solving the following problem: In a certain imaginary animal, the gene for black coat color (*B*) is dominant over the gene for orange coat color (*b*), and the trait is sex-linked. Determine the expected genotypic and phenotypic ratios among the offspring produced by the mating of a heterozygous female and a black male.

Many Genes—One Effect

Inheritance involving incomplete dominance, codominance, multiple alleles, and sex-linkage does not follow the patterns first discovered by Mendel. However, in each case, the phenotype is determined by a single pair of interacting alleles. Some other patterns of heredity are not controlled that way. Traits in humans, such as height, hair, eye and skin color, are all controlled by more than one pair of alleles. Many traits in other organisms are also controlled this way.

For example, consider the length of ears of corn. If ear length in corn were governed by a simple case of complete dominance, only two lengths would be found: long and short. If ear length were controlled by a single pair of genes showing incomplete dominance, three phenotypes could be expected: long, medium, and short. However, ears of corn are not just long or short, or even long, medium, or short. Instead, a whole range of phenotypes exists, from long to short. The presence of many phenotypes, from one extreme to another, is an example of **continuous variation.**

The graph in Figure 8-22 relates ears of corn and their lengths. How could you explain these data? Assume that, in this case, ear length is controlled by two pairs of genes (*A* and *a; B* and *b*) located on different pairs of homologous chromosomes. Both pairs of genes contribute to ear length in an additive fashion. Suppose that *AABB* represents the longest ears, and *aabb* produces the shortest ears. A cross between *AABB* and *aabb* parents would result in F$_1$ offspring that are *AaBb* with medium-length ears.

Figure 8-22 **When two pairs of genes determine a trait, many different phenotypes (continuous variation) are possible in the F$_2$ generation (left). A graph of the distribution of ear lengths in corn (right) is bell-shaped. Medium-length ears occur most often.**

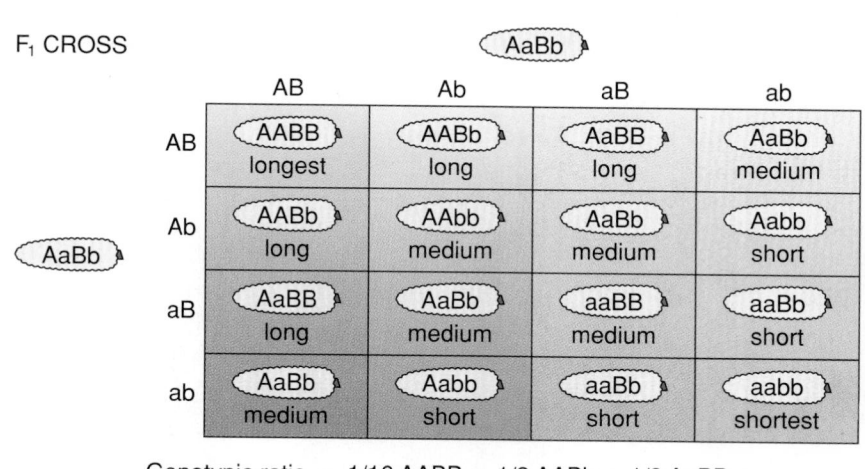

F$_1$ CROSS

Genotypic ratio = 1/16 AABB : 1/8 AABb : 1/8 AaBB :
1 1/4 AaBb : 1/16 AAbb : 1/6 aaBB :
1/8 Aabb : 1/8 aaBb : 1/16 aabb
Phenotypic ratio = 1/16 longest : 1/4 long : 3/8 medium :
1/4 short : 1/16 shortest

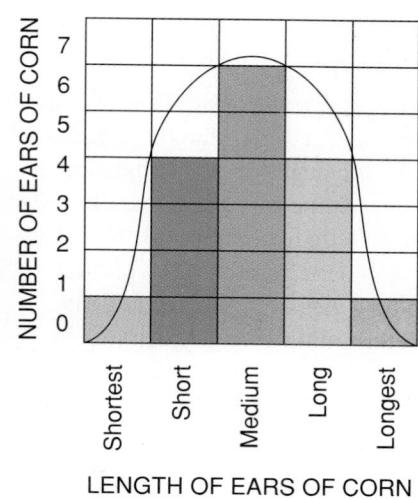

If you cross these F_1 corn plants, the F_2 offspring exhibit a broad range of genotypes and phenotypes, as Figure 8-22 shows. Relate the phenotypic ratios to the graph in the figure. The genotype *AaBb,* medium-length ears of corn, is the one most often expected. *AABB* (longest) and *aabb* (shortest) are seldom expected. The rest of the expected genotypes correspond to various intermediate lengths. Some ears would be longer than medium; some would be shorter. When these crosses are actually made, the results largely agree with those predicted by the Punnett square.

Geneticists explain continuous variation in terms of **multiple genes,** a pattern of inheritance in which many genes affect a single trait. Human skin color shows continuous variation and is also thought to be determined by multiple genes.

MATH CONNECTION

Crossing Over as a Tool to Determine Gene Location on Chromosomes

• •

You are asked to find the location for cities B, C, and D on a map. Try it. The solution is not difficult but requires a little logic. You are told the following:

Distance Between Cities
A and C is 50 km
D and C is 25 km
D and B is 100 km
B and A is 175 km

This same type of application can be used to determine the location of specific genes on a chromosome if you assume that genes are like cities and the map corresponds to a chromosome. The problem with mapping genes on a chromosome is determining how you can measure distance between genes. The size of chromosomes is very small and the distance between genes is, there-fore, almost impossible to measure by direct observation. Luckily for scientists, an event that occurs between chromosomes called crossing over allows them to make these types of measurements indirectly. Thus, they are able to prepare chromosome maps showing positions and distances between genes.

What is crossing over, and how does it allow for mapping of genes on chromosomes? Remember that during early events of meiosis, the four chromatids from each homologous pair of chromosomes line up

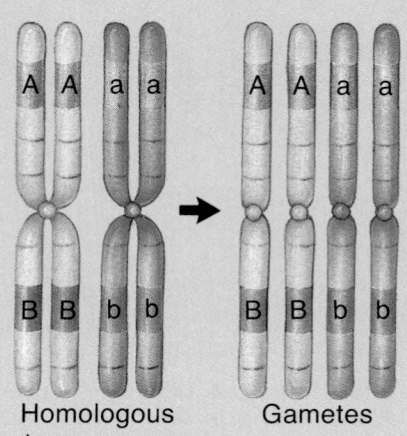

Homologous chromosomes   Gametes

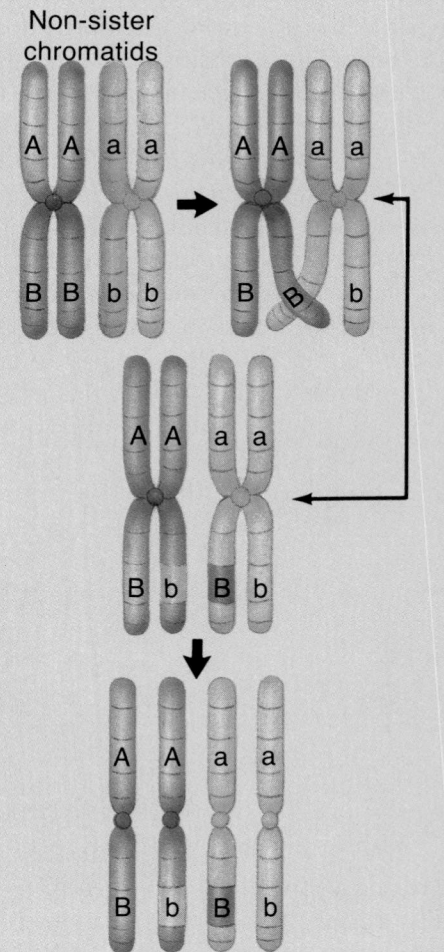

Non-sister chromatids

Gametes

Expression of Genes

Genes can interact with one another to control various other patterns of inheritance. Actually, most characteristics that make up an individual's phenotype are not inherited in Mendel's simple patterns. For example, consider the color of your eyes. Brown color in human eyes (B) is due to the presence of a pigment called melanin. People with blue eyes (b) have no melanin in their eyes. You know, however, that many other eye colors—such as gray, hazel, and green—exist. Those other colors are due to the presence of other genes inherited along with the B and b alleles. These other **modifier genes** affect eye color by influencing the amount, intensity, and distribution of melanin in cells of the eye.

with corresponding genes appearing side by side. To help simplify this idea, only one set of homologous chromosomes is shown. As meiosis continues, the homologous pairs are pulled apart and the resulting four gamete cells with their chromosome and gene combinations are shown. However, the arms of non-sister chromatids may overlap as shown on page 216. Overlapping may lead to an actual breaking off and exchange of chromosome ends during this early phase of meiosis. This overlapping and exchange of chromosome parts, along with their genes, is called crossing over. Note that the gametes have a different arrangement of genes on their chromosomes as a result of crossing over. The percent of these crossed-over traits appearing in offspring is then used as a means for determining gene location on a chromosome. This is true because the frequency or amount of cross-over that occurs corresponds to the location of genes on a chromosome. Genes that cross over more often with one another are far-

ther apart on the chromosome than are genes that cross over less often.

On a chromosome map, the distance between genes is stated in map units. Thus, the distance between map units is an indication of the amount of or percentage of crossing over occurring between genes.

1. **Solve the Problem:** If one map unit = 1% crossing over, determine the location of the following genes on the chromosome map:

Percent of Cross Over
Cut wing and bar eye = 37
Cut wing and ruby eye = 12.5
Mini wing and cut wing = 16
Mini wing and ruby eye = 28.5
Ruby eye and bar eye = 49.5

Draw a diagram of this fruit fly chromosome and locate its genes with their correct distances in map units on your diagram. Note that the location of the ruby eye gene is already positioned for you in map units.

2. **Explore Further:** Build a model using tape and wool strands to

show homologous chromosomes and their four chromatids. Add labels to indicate which are non-sister chromatids. (a) Show how these chromatids appear during prophase 1 of meiosis if no crossing over were to take place. (b) Show how these chromatids appear during early phases of meiosis if crossing over does occur.

3. **Writing About Biology:** Offer a possible explanation as to why genes located on the same chromosome will have higher percentages of cross-over when farther apart than if they are closer together. Use your model in question 2 to help illustrate and support your explanation.

CAREER CONNECTION

Genetic counselors are individuals who are trained in the field of genetics. Their training allows them to test, advise, and counsel people on the possibility of genetic diseases appearing in their families.

Minilab 2

What are some human variations?

Read this list of human traits and their alleles:

- Oriental eye shape = O
- Non-oriental eye shape = o
- Widow's peak = W
- No widow's peak = w
- Tongue roller = T
- No tongue rolling = t

Record a list of all the possible combinations of these three traits in one person. Then, tally the number of students in your class fitting each category. Predict the possible genotypes for each category.

Analysis *Explain why it is almost impossible for two people, except for identical twins, to look exactly alike.*

It is the rule, not the exception, that many genes interact to control a phenotype. Because most traits are controlled by several genes working together, an individual may have alleles for a particular trait but not show that trait. For example, a person may have the genotype *Bb* but have blue eyes if a modifier gene prevents melanin from being produced.

Another factor that determines the expression of genes is the environment in which an organism develops. In fruit flies, for example, the gene for vestigial wings is expressed differently at different temperatures. Flies with small, vestigial wings bred at 29°C will have offspring with larger wings. But vestigial-winged flies bred at temperatures below 29°C will have offspring with small wings. The difference probably results from how enzymes operate at different temperatures. A higher temperature may deactivate an enzyme, preventing a reaction from occurring and thereby changing a phenotype.

In general, the development of any organism is affected by the environment. At low altitudes, trees may grow to be several hundred feet tall. At high altitudes, with fewer nutrients and harsher conditions, the same trees will only reach a height of several feet. Less dramatic conditions can also affect an organism's development. A corn plant may have the genotype for long ears of corn. But the soil in which the corn is planted may be poor, or a drought may occur during the growing season. Each of these events can affect the phenotype, with the result that the corn produces short ears or even no ears.

Thinking Lab

INTERPRET THE DATA

What pattern of inheritance allows for the prediction of a genotype when phenotypes are known?

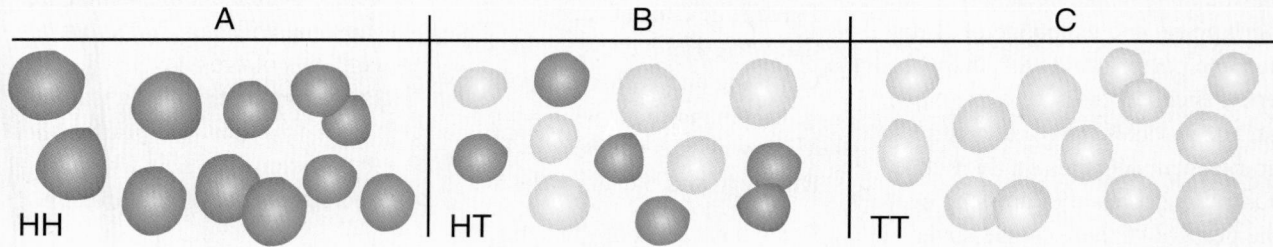

A   HH

B   HT

C   TT

Background

An inherited blood disease called thalassemia affects red blood cells. A person with thalassemia minor has normal cells and abnormal cells. The abnormal cells lack hemoglobin. Thalassemia major results in all abnormal cells.

You Try It

The following diagrams show the blood cells and genotypes of three people. **Interpret the following data:** If HH = normal red cells, HT = thalassemia minor, and TT = thalassemia major, predict the (a) phenotypes of persons A, B, and C; (b) possible genotypes of

A, B, and C's parents; and (c) phenotypes of children born to parents when both have thalassemia minor, and when one is normal and one has thalassemia major. Which pattern of inheritance tells you genotypes when phenotypes are known? Why?

Figure 8-23 Eye color is determined by the interaction of many genes.

In summary, phenotypes—the expression of traits—are more complicated than early biologists thought. Whether a certain trait is expressed depends upon the particular alleles for that trait, all other alleles carried by the organism, and the environment in which the organism develops.

LOOKING AHEAD

THE LAWS of probability can be used to predict the likely results among all the offspring of a genetic cross. The genes themselves carry the code that determines which traits an individual offspring inherits. Just what is the structure of a gene? And what is the nature of the genetic code it helps carry? You'll discover the answers to these questions in the next chapter.

Section Review

Understanding Concepts

1. Colors of wheat kernels range from white to dark red, with various shades of pink in between. By what pattern of inheritance is this trait controlled? Sketch a graph that would represent this distribution of colors.
2. In fruit flies, the gene for red eyes (R) is dominant over the gene for white eyes (r) and is sex-linked. If an offspring is a white-eyed female, and her female parent has red eyes, what must the genotype of her male parent be?
3. Explain how the environment can affect the expression of various traits in an organism.
4. **Skill Review—Observing and Inferring**: Sutton proposed his chromosome theory of heredity after observing meiosis in grasshopper cells. Explain what Sutton observed and what inferences he made on which he based his theory. For more help, see Practicing Scientific Methods in the **Skill Handbook**.
5. **Thinking Critically:** Certain rabbits are white except for black fur on their feet, tail, ears, and nose. If a patch of fur is removed from a rabbit's back, and that area is covered by an ice pack, the fur that grows back will be black. How can you explain this fact?

History Connection

What if Mendel had picked different traits?

Mendel provided us with our basic understanding of genetics and established several laws, which you have studied in this chapter. Mendel also was very lucky or very smart with the choice of traits that he used in his studies. Why was he lucky?

Assume that yellow pea seed color = *Y* and green color = *y*, round pea seed shape = *R* and wrinkled = *r*. What Mendel may have observed in offspring after making a cross of *YyRr* × *YyRr* parents is shown in Table A.

Table A

Offspring phenotypes	Number observed	Ratio
Yellow, round	1841	9
Yellow, wrinkled	611	3
Green, round	618	3
Green, wrinkled	202	1

YR, Yr, yR, and *yr.* The observed appearance of these two traits in the offspring and the ratio of phenotypes supports the law established by Mendel. A phenotypic ratio of 9:3:3:1 when both parents are heterozygous for both traits being studied is what Mendel predicted and observed. Mendel was lucky because he picked genes located on different pairs of chromosomes.

But how lucky was Mendel? What if the two gene traits that Mendel worked with, pea seed color and seed shape, were both located on the same chromosome? Would he have observed the same ratios in the offspring? Would he have been able to demonstrate his law of independent assortment?

Genes for different traits located on the same chromosome are called

linked genes. This refers to the fact that the genes are found on the same chromosome. How would this have altered Mendel's observations, data, and formation of the law itself? Let's see. The diagram shows a pair of homologous chromosomes with the two traits used before. This time, however, they are linked. The parent genes are still heterozygous for both traits. The difference will be in the distribution of genes into gamete cells and the resulting combinations in the offspring. This is where gametes containing linked genes differ from gametes in which genes are not linked. The diagram shows the gametes formed when the genes are linked. Table B shows the data that one might observe for these two linked traits.

His data were the foundation for the formation of his law of independent assortment. This law states that genes for different traits segregate independently during gamete formation. This law is the basis for predicting the gene combinations in gametes. For example, when both parents are heterozygous for two traits, the following gene combinations are possible in gamete cells:

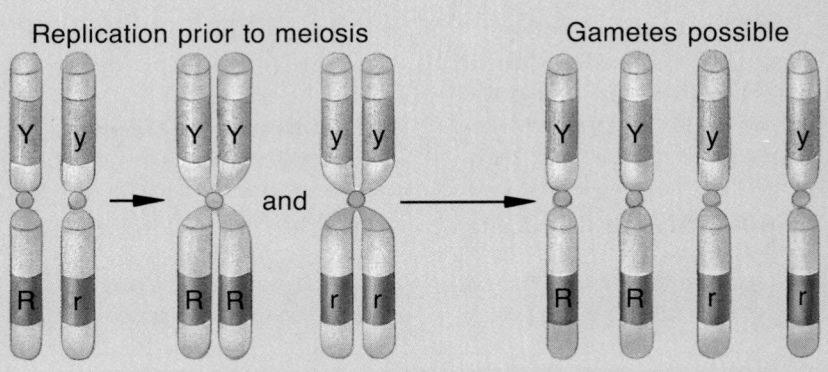

Replication prior to meiosis and Gametes possible

Table B

Offspring phenotypes	Number observed	Ratio
Yellow, round	1361	3
Green, wrinkled	452	1

Compare these data to those in Table A. Note that the ratios and phenotypic results are different with genes that are linked versus those that are not linked. Mendel was lucky because the traits he used were not linked. How might he have explained his data if the genes were linked?

1. **Solve the Problem:** Could Mendel have used data from linked genes to help formulate the law of independent assortment? Explain.
2. **Explore Further:** Describe the experimental procedure that you would use to determine whether or not the following genes are linked: *B* = black fur in guinea pigs, *b* = white fur, *S* = short fur length in guinea pigs, *s* = long fur. Describe the phenotypic ratios that would result in each situation, and explain how you would interpret your data.
3. **Writing About Biology:** Why is it impossible to tell, looking at a parent plant, if the genes for two heterozygous traits are linked or not?

Issues

Inbreeding: Helpful or Harmful?

There are laws in many states that forbid the marriage of first cousins. These laws are intended to prevent inbreeding. Inbreeding is the mating of individuals who are closely related. Why should such a law exist? What are the effects of inbreeding?

Inbreeding has led to offspring that show many harmful traits. This can be explained genetically. Certain recessive traits are prevented from showing because an organism may be heterozygous for these traits. It can be assumed that close relatives will also be heterozygous for these same traits. Mating two individuals who are both heterozygous for the same traits will increase the chance of both recessive genes appearing in the offspring. Thus, a higher number of recessive harmful traits may show up in the offspring.

Inbreeding is frequently used by animal and plant breeders. It is not uncommon for a male dog to be mated with his female offspring during breeding programs. The chance for harmful recessive traits in the offspring is increased.

Examples of inbreeding are plentiful in domesticated animals and especially dogs. Miniature poodles are good examples of problems associated with inbreeding. Have you ever noticed the dark brown stains appearing in the corners of a white poodle's eyes? These stains are the result of plugged tear ducts. Tears usually flow naturally from eye to tear duct to inside the nose. It's the same in humans. That's why you have to blow your nose after crying. Tear stains are certainly not harmful to the life of a poodle, but other inbred problems such as dislocated shoulders, epilepsy, diabetes, and collapse of the windpipe often are.

The healthiest dogs are usually those with the least inbreeding. Mutts rarely show the problems seen in bred varieties. Humans breed dogs and other animals for the traits they wish to have in their pets or in their farm animals or crops. However, these traits bring with them the many problems associated with inbreeding.

1. **Solve the Problem:** Several dogs in a new litter show genetic problems at birth. What steps should the breeder take to prevent these same problems from appearing in future litters?
2. **Writing About Biology:** Wolves, foxes, and coyotes rarely show any genetic problems that are typical of inbreeding. Write a paragraph explaining why this is true. How is this fact a benefit to a population?

Summary

Mendel's laws of dominance, segregation, and independent assortment can be applied to analyze many patterns of inheritance. Punnett squares, based on the laws of probability, can be used to solve a variety of genetics problems dealing with both Mendelian and other patterns of inheritance.

Observations that tied together Mendel's laws and the activities of chromosomes during meiosis led to the hypothesis that genes are on chromosomes. Discovery of how sex is determined and of sex-linked traits strengthened that hypothesis.

Expression of genes as a particular phenotype depends not only upon the alleles governing that phenotype, but also upon all the other genes of the organism and the environment in which the organism develops.

Language of Biology

Write a sentence that shows your understanding of each of the following terms.

allele	homozygous
autosome	incomplete dominance
codominance	modifier gene
continuous variation	multiple alleles
dominant	multiple genes
gene	phenotype
genetics	recessive
genotype	sex chromosome
heredity	sex-linked trait
heterozygous	

Understanding Concepts

1. Suppose that in outer space there exist creatures whose traits are inherited by Mendel's laws. Purple eyes (*P*) are dominant over yellow eyes (*p*). Two purple-eyed creatures mate and produce six offspring. Four of them have purple eyes, and two have yellow eyes. What are the genotypes of the parents and the offspring?

2. Two long-winged fruit flies produced 49 short-winged and 148 long-winged offspring. Which trait do you assume is dominant? Why? What were the probable genotypes of the parents? About how many of the long-winged offspring would you expect to be heterozygous?

3. You inspect an ear of corn and find the following number of kernels: 461 purple and starchy, 142 purple and sweet, 156 yellow and starchy, 53 yellow and sweet. What genotypes did the parent corn plants probably have? Support your answer by working out the cross.

4. In humans, brown eyes (*B*) are dominant over blue eyes (*b*). Two heterozygous brown-eyed parents produce eight children, all brown eyed. Do these results surprise you? What idea about genetics do the results illustrate?

5. The probability of any one child being a boy is 1/2. What is the probability that six children in a family will all be boys?

6. In fruit flies, the gene for tan body color (*T*) is dominant over the gene for ebony color (*t*). The gene for red eyes (*R*) is dominant over the gene for white eyes (*r*). The latter trait is sex-linked. Calculate the expected genotypic and phenotypic ratios resulting from a cross between a homozygous ebony female with red eyes and a heterozygous tan male with white eyes.

7. Explain why Morgan reasoned that the gene for eye color in fruit flies is carried on the *X* chromosome but not the *Y* chromosome.

8. Why are few sex-linked traits in humans located on the *Y* chromosome? How are *Y*-linked traits transmitted? Draw a Punnett square showing the transmission of a *Y*-linked trait.

Applying Concepts

9. A cross between a homozygous red horse and a homozygous white horse produces an offspring with a coat coloring called roan. Examination of the roan's coat reveals the presence of both red and white hairs. How is this pattern of inheritance explained? What results would you expect if two roan horses were crossed? How do you know that the genes controlling coat color are not completely dominant over each other?

10. Doing experiments with fruit flies, you unexpectedly find a male fly having narrow eyes. Normal flies have round eyes. How could you determine whether the gene for narrow eyes is autosomal or sex-linked?

11. In chickens, the gene for a rose comb (R) is dominant over the gene for a single comb (r). The gene for a pea comb (P) is dominant over the gene p, which also produces a single comb. When R and P are both present, they combine to produce what is called a walnut comb. Calculate the phenotypic ratio expected from a cross between two chickens with the genotypes RrPp.

12. **History Connection:** Describe the experimental procedure you would use to determine whether or not the following genes are linked: B = black fur in guinea pigs; b = white fur; S = short fur; s = long fur. Describe the phenotypic ratios that would result in each situation and explain how you would interpret your data.

13. **Issues:** Suggest a way a breeder of a particular show dog could obtain the traits he wanted in the dogs without inbreeding.

Lab Interpretation

14. **Minilab 1:** Your high school is co-ed with 82 females and 67 males in the sophomore class. The entire school has 322 females and 317 males. Explain why the numbers for sophomores and the entire school may or may not be close to 50:50 for males and females.

15. **Minilab 2:** List your phenotypes for the three traits described in the Minilab. Predict the possible genotypes that you might have for these three traits. Explain why you cannot be certain about having a homozygous dominant condition for any of the three traits.

16. **Thinking Lab:** The roots of radish plants are either long, round, or oval. (a) What type of gene inheritance pattern might this suggest? (b) Assign genotypes to the traits if an oval radish parent crossed with an oval yields 121 long, 246 oval, and 117 round. (c) How could you grow only oval radish plants?

17. **Investigation:** How many generations of tobacco plants would you have to observe to determine the actual genotype of a parent plant?

Connecting Ideas

18. Relate the law of segregation to the events of meiosis.

19. Explain the scientific methodology Mendel used in his experiments. Why is it important to have carefully thought out procedures when conducting experiments?

Single combs

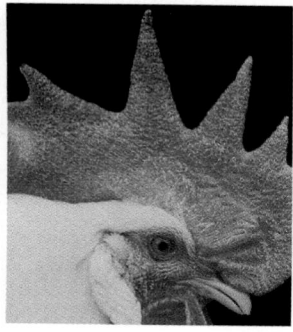

Rose comb

Pea comb

Walnut comb

CHEMISTRY OF THE GENE

TAP A KEY on a computer keyboard, and a symbol appears on the screen. Tap another key, and a different symbol appears. With more than 100 keys on an expanded computer keyboard, almost anything the human mind can think of can be created—in words or in pictures. The trick, however, is to tap the keys in the correct sequence.

All organisms are created using the same chemicals. Almost any trait you can think of— color, size, temperament—can be coded for by these chemicals. The trick is to get these chemicals arranged in a particular order.

All the people in this photograph are made up of the same parts—arms, legs, eyes, hearts, and so on. Each of those parts is made up of tissues. Each of those tissues is made up of cells. Each of those cells is made up of chemicals. All people are made up of the same chemicals. It's the way these chemicals are put together that makes us different. Just as combinations of the 26 letters of our alphabet can make up many thousands of words, combinations of chemicals are used to produce the many thousands of characteristics that make us human. We are alike in many ways because we are made of the same chemicals. We look different because we are made up of different combinations of those chemicals.

As letters can be combined to form sentences, so chemicals can be combined to form molecules of DNA. What is DNA? It's what makes you, you.

This model of DNA was generated by a computer. How does your body generate actual DNA?

Chapter Preview

Structure of DNA

Objectives

Sequence the experimental evidence that led to the conclusion that DNA is the chemical of which genes are composed.

Diagram the double helix model of DNA.

Demonstrate the replication of DNA.

A new house is about to be built. The house will consist of many rooms. Those rooms will be similar in many ways, but will vary according to the activities that will take place within them—cooking, sleeping, or relaxing. Before building can begin, a set of plans called blueprints must be created. The blueprints provide the instructions for building the entire house and each of its parts. Building a house requires the coordinated efforts of many workers. Each worker needs a set of plans to build a particular part of the house, so the blueprints must be duplicated and given to each worker before work can begin.

Like the blueprints needed to build a house, the plans for making a new organism must be copied from the parent organisms and passed on to newly produced cells. Each cell of an organism then has instructions for its development in the form of a genetic blueprint. The cells of an organism are, like the rooms of a house, the same in many basic ways. However, just as a kitchen is specialized for cooking, cells are also specialized to perform certain functions.

What are the genetic blueprints made of? Although you know that genes direct the expression of traits that an organism will develop, what are genes, exactly? What chemicals are they made of? How can a chemical be copied? We'll explore the answers to these questions, and how the answers were discovered, in this chapter.

Heredity and Chemistry

In 1928, researcher Fred Griffith was studying the disease pneumonia, caused by a bacterium called *Streptococcus pneumoniae*. One strain of these bacteria consists of cells enclosed within a jellylike outer capsule and is referred to as smooth. Another strain has cells that are not enclosed by a capsule. Cells without a capsule are called rough. Smooth cells cause pneumonia, but rough cells do not.

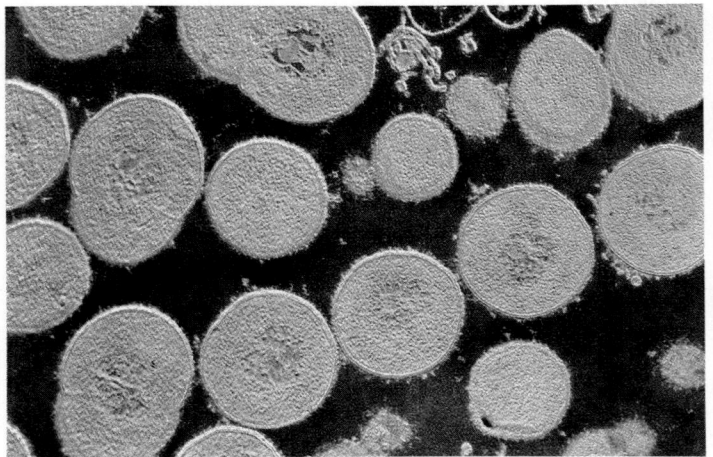

Figure 9-1 **The disease pneumonia is caused by encapsulated *Streptococcus pneumoniae* bacteria. Magnification: 5880×**

Griffith injected a mixture of dead smooth cells, killed by heat, and living rough cells into mice. He expected the mice to remain healthy, because rough cells do not cause pneumonia and because the smooth cells were dead. However, some of the mice died of pneumonia. On examining the blood of the dead mice, Griffith made an unexpected observation—living smooth *Streptococcus pneumoniae* cells were present. He concluded that somehow the dead smooth cells influenced the rough cells to become smooth. An inherited trait of rough cells must have changed in the presence of dead smooth cells. This kind of change in a bacterial trait is called a bacterial transformation.

Changes in Traits   How could a trait change as it had done in these experiments? Scientists suspected that a chemical in the dead smooth cells caused the change. To test this hypothesis, colonies of smooth cells were ground up to release their chemicals. The chemicals were removed from the cells in a solution called an extract. It was thought that a chemical in the extract would cause transformation.

The extract from the smooth bacteria was added to a new set of culture dishes containing only rough cells. After many colonies had formed, examination of the dishes showed that many of these colonies now contained both rough and smooth cells. The new smooth cells caused pneumonia when they were injected into healthy mice. The extract had changed the inherited traits of the rough cells, just as the hypothesis had predicted. A chemical in the extract changed the form of one strain of bacteria, and this change in traits was passed on to the offspring. This chemical was thus called the transforming principle.

Identifying the Transforming Principle   The work of many scientists helped to identify the transforming principle. Earlier chemical analyses of cells showed that the chromosomes of eukaryotic cells contain about equal amounts of proteins and a chemical, deoxyribonucleic acid (DNA). Most scientists believed hereditary information in a cell was contained in the proteins, because they are more complex than the chemical components of DNA. In 1943, O.T. Avery demonstrated that DNA was the transforming principle in bacterial cells, but this information did not convince scientists that DNA is the primary hereditary mechanism in *all* cells.

Figure 9-2   A mixture of heat–killed smooth cells and living rough cells caused mice to die of pneumonia (a). Rough cells growing in a culture with an extract of living smooth cells were changed to smooth cells, an example of bacterial transformation (b).

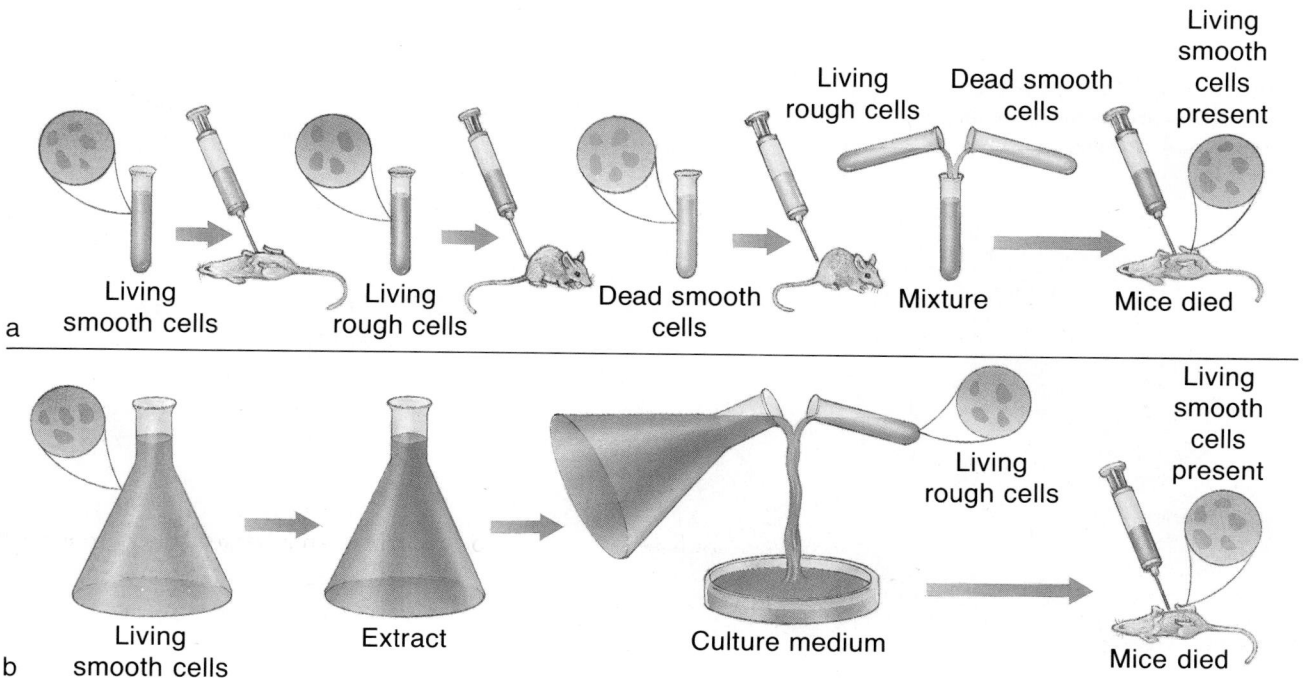

Figure 9-3 **Bacteriophages are viruses that attack bacteria. They consist of a hexagonal head and a complex tail. Magnification: 60 000×**

Evidence for DNA   Scientists studying heredity began to focus on DNA as the genetic material as a result of many experiments. One of the most important experiments was performed by Alfred Hershey and Martha Chase in 1952. Hershey and Chase experimented with viruses. Viruses can reproduce only by invading specific cells. Viruses that attack bacteria, known as **bacteriophages,** or phages, consist of an outer coat of protein and an inner core of DNA. That simple structure, seen in Figure 9-3, seemed ideal for resolving the debate about whether protein, DNA, or a combination of the two controlled heredity.

Bacteriophages attach to cells, and within an hour the cells contain hundreds of new viruses. The cells soon burst, releasing the viruses, which can then attack other cells. Hershey and Chase knew that the protein coat of a bacteriophage contains sulfur but not phosphorus, and that a phage's DNA contains phosphorus but not sulfur. Using this information, these researchers infected bacterial cells with bacteriophages. After infection had begun, they shook the infected cells in a blender, then spun them in a centrifuge to separate the cells from materials remaining outside the cells. What did they discover? Sulfur from the bacteriophages remained outside the bacterial cells, but phosphorus was found *inside* the bacterial cells. In other words, the protein coats of the bacteriophages remained outside the cells, but the phage's DNA had moved into the bacterial cells. It was the phage's DNA that entered cells and directed them to create new bacteriophages. This experiment clearly demonstrated that DNA, not protein, carries the information needed for reproduction and that DNA is the chemical genes are made of.

A Model of DNA

Figure 9-4   **The four nitrogen bases of DNA are adenine, guanine, thymine, and cytosine.**

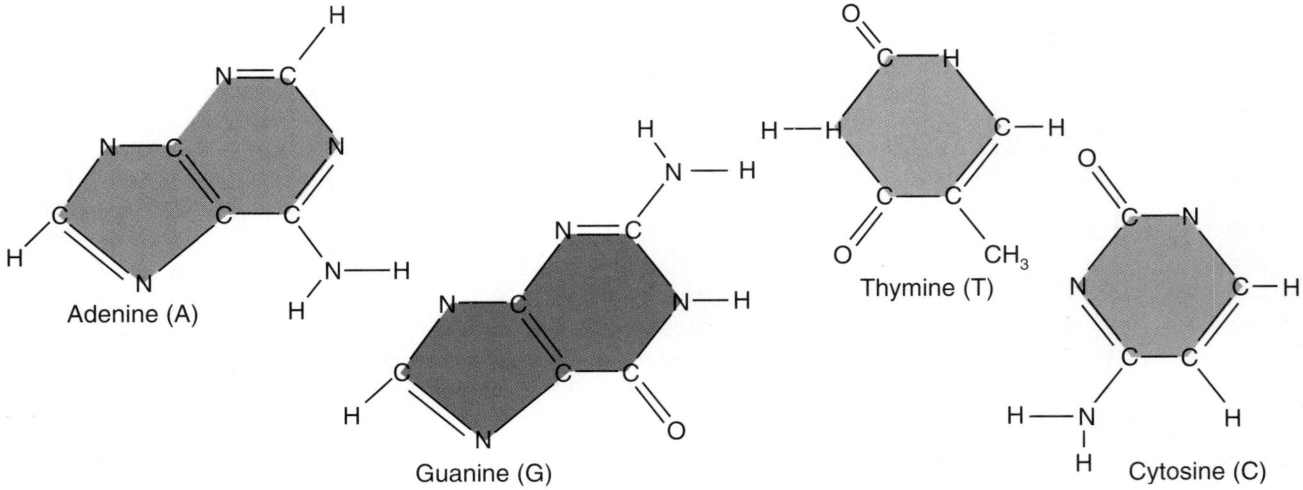

Although DNA was first discovered in a laboratory in 1869, its chemical components were not known until the 1920s. P.A. Levine, a biochemist, showed that DNA can be broken down into a sugar (deoxyribose), a phosphate group, and four nitrogen-containing bases. These bases are **adenine** (A), **guanine** (G), **thymine** (T), and **cytosine** (C). In Figure 9-4,

Adenine (A)

Guanine (G)

Thymine (T)

Cytosine (C)

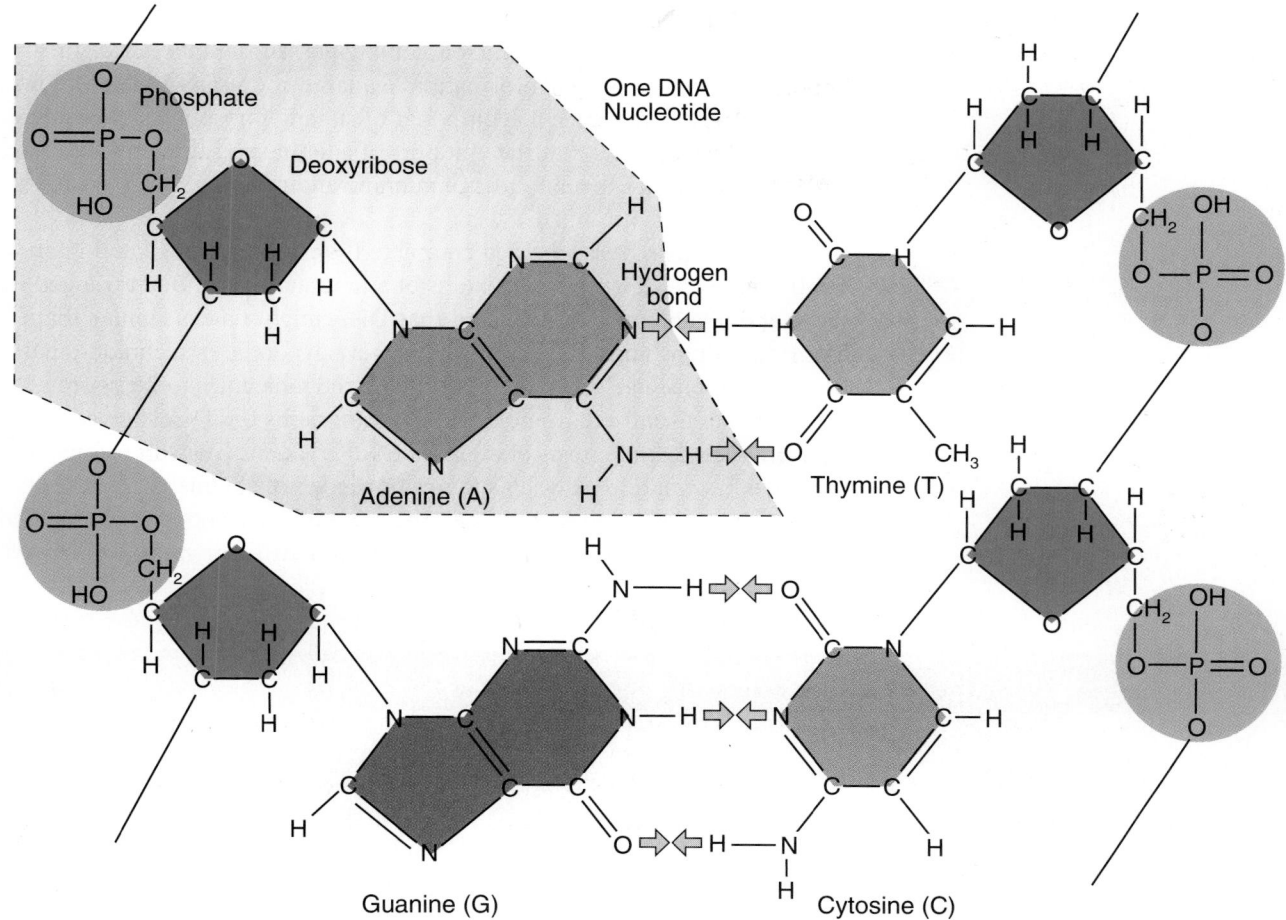

the chemical structure of each base is shown. Notice the colors of the shapes. Each color refers to that specific base in the figures that follow. It will be easy to tell what chemicals are shown in the drawings if you keep in mind what the colors represent as you examine the other figures in this chapter.

Levine deduced that each base was attached to a molecule of sugar, and each sugar was then attached to a phosphate group to form a single molecule called a **nucleotide.** Each nucleotide is named for the base it contains. For example, an adenine nucleotide contains adenine. Figure 9-5 shows the arrangement of the three parts of a nucleotide. Many nucleotides are linked together to form a long chain of DNA.

Bases Appear in Matching Quantities   Other scientists studying the chemical structure of DNA made some important discoveries. Erwin Chargaff analyzed the content of many different kinds of DNA. He found that the nitrogen bases do not occur in equal proportions. In the DNA of any given species, the amounts of adenine and thymine are always the same, and the amounts of guanine and cytosine are always the same. This piece of information helped solve the problem of how DNA could be complex enough to be the chemical controlling heredity.

Figure 9-5   A nucleotide consists of a phosphate group, a sugar molecule, and one of four nitrogen bases. The shaded area shows an adenine molecule.

Describing DNA How is genetic information carried in DNA? To answer this question, James Watson, a biologist, and Francis Crick, a physicist, compiled all the information available about DNA. They knew that DNA is a large molecule composed of sugars, phosphate groups, and four bases. They knew that the amounts of adenine and thymine are always equal, as are the amounts of guanine and cytosine. But how are these components arranged?

Two other discoveries helped complete the picture. Linus Pauling, in 1950, showed that a protein's chains of amino acids are often arranged in the shape of a helix. He speculated that DNA might have a similar shape. Maurice Wilkins and Rosalind Franklin were experts in a technique that bombards molecules with X rays, allowing the molecules to be photographed and visualized. Some of their photographs of DNA showed patterns that looked like turns of a giant helix. Crick and Watson, using all available information as well as their knowledge of chemistry, set about building a model of DNA that fit all the known facts. They first concluded that the DNA molecule must be in the form of a helix, or spiral, composed

I N V E S T I G A T I O N

A Classroom Model of DNA

Have you ever been in a restaurant with your friends and suddenly had a thought you didn't want to forget? To keep from forgetting, you probably wrote your thought on a paper napkin. In the early 1950s, Rosalind Franklin provided, in her X-ray pattern studies of DNA, the clues James Watson and Francis Crick needed to construct a three-dimensional model of DNA. Where and how did they think about the possibilities for the structure of DNA? They brainstormed with pencils on paper napkins in restaurants. They tinkered in their lab with bits of wire, plastic, and string. In this lab, you will use colored construction paper and your fellow students in the class to make a working model of DNA replication.

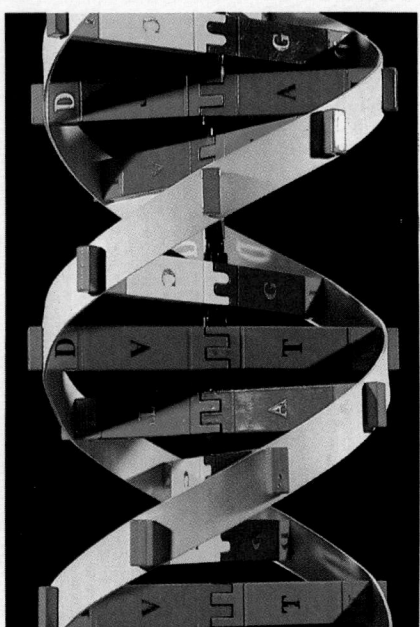

Problem
How can a model of DNA replication be made?

Safety Precautions
Use care with scissors.

Experimental Plan
1. As a group, you will receive ten pieces of each of six colors of construction paper.
2. The six colors will correspond to the six different molecules that make up the DNA molecule.
3. Decide what shapes you will use to represent each of the six molecules.
4. Using paper and pencil, brainstorm, in your group, a list of possible ways various shaped placards could be held by students in the class as they move through a representation of the process of DNA replication.
5. Select the best idea and, following the style of a recipe, write a numbered list of directions that anyone could follow.

of a single chain of nucleotides. Each nucleotide must be joined to the next by the sugar-phosphate backbone.

One Chain or Two   However, Watson and Crick calculated that a single chain of DNA would have only half the density known for DNA. They realized that a molecule of DNA must be composed of two chains of nucleotides. But how are the two chains joined? The fact that, in any DNA molecule, the numbers of adenine and thymine nucleotides are always equal, and the numbers of guanine and cytosine nucleotides are always equal, gave them the answer.

   Watson and Crick built a model of DNA in which the two chains of nucleotides are joined. Using wire and tin, they constructed a ladder-like model in which the uprights of the ladder are composed of the sugar and phosphate parts of the nucleotides. The rungs are composed of the nitrogen bases of the nucleotides. Each rung contains two nitrogen bases held together by what are called hydrogen bonds. Only two bonding arrangements are possible: adenine with thymine, and guanine with cytosine.

6. Cut out the number of hand-held, large placards you will need to model DNA replication by students in your class. Label each placard with the name of the molecule it represents.
7. Decide how you will identify the bonds that hold your molecule together.
8. Decide how you will demonstrate your model. Decide what each group member will contribute during the demonstration.

Checking the Plan
Discuss the following points with other group members to decide the final procedures for your model.
1. Have you incorporated all the parts of the DNA molecule into your model and most, if not all, of the students in class?
2. Does your model come apart, just as DNA "unzips" during replication?

3. Does your model include the idea of new molecules being added on to each half of the DNA molecule once it is unzipped?
4. Do you have two molecules identical to the original when you have completed your modeling of DNA replication?

Make sure your teacher has approved your model before you proceed further.

Observations
Use the time given to you by your teacher to demonstrate your model. Be effective and efficient in your distribution of shapes and your directions for movement of students.

Analysis and Conclusions
1. What did you learn about DNA by making a model of its replication?
2. Did you use the "ladder" comparison to make your model? If so, what kind of molecules made

up the sides of the ladder? What molecules made the rungs?
3. Explain how, during replication, DNA is able to make an exact copy of itself.
4. What was the easiest part of making your model?
5. What was the most difficult part of making your model?
6. Write a brief summary of how DNA replicates.

Going Further
Based on this lab experience and Table 9.1 on page 241, prepare a model of a DNA strand that codes for the following molecules: phenylalanine, tryptophan, and argenine in that order. You may use the same materials for this model or expand on an idea that you think might work better. Assume that you will have unlimited resources and will not be conducting this lab in class.

Adenine and guanine belong to a group of nitrogen bases called purines. Thymine and cytosine belong to a different group of nitrogen bases known as pyrimidines. Bonding in the DNA molecule only occurs between a purine and a pyrimidine because of their chemical structures. In fact, the chemical structure of adenine enables it to bond *only* with thymine.

ART CONNECTION

Craig Schaffer— Sculptor

Take a double helix, add two parallel lines, and mix. What do you get? If you are sculptor Craig Schaffer, you get six-sided stars that double as candlesticks. By looking at the double helix of DNA in a different way, Schaffer interpreted the structure of life itself into art with a religious context for a Jewish house of worship. Interpreting the structure of DNA as art is only one of the ways in which Schaffer has been inspired by natural forms. He

uses stone, wood, metal, and even plastic to explore the artistic possibilities of the natural world.

"People think that art and science are at opposite ends of the poles," says Schaffer, "but really they are very similar." The sculptor believes that artists and scientists share several traits—both are very curious about the world around them, and both perform experiments to test their ideas. Like a scientist, an artist often begins with a question, or hypothesis. How does shape influence function? Are there repeating patterns in seemingly chaotic structures? These are some of the questions artists may ask as they begin to explore new possibilities. Artists follow the scientific method, Schaffer explains, because they begin each new work with a different question. They try something new, then learn from the results. As in scientific inquiry, the results of an experiment are not always what you expect. Often an artist may move along a different path as a result of one experiment, just as a scientist's experiments often lead to a new hypothesis.

As a sculptor, Craig Schaffer tries to explore the artistic possibilities of both science and mathematics. He has created sculpture for clients such as Princeton University and the Mathematical Association of America, and has had various

other commissions around the country. Schaffer is especially interested in combining various disciplines in his work. One of his double helix designs graces the Congregation Beth Tikvah in Columbus, Ohio. The double helix structure of DNA offered the sculptor an opportunity to combine a natural form with religious interests to create a six-sided star shape, which is a Jewish symbol.

Schaffer feels that you can learn how an artist evolves by viewing work from various times in the artist's life. At a retrospective show, you can see how an artist's perspective has changed over time. Why, for example, would an artist paint the same mountain over and over again throughout his or her life? Schaffer explains that, each time, the artist wanted to know what would happen if one variable was altered. Scientists operate in exactly the same way.

1. **Writing About Biology:** Schaffer believes that art and science share common traits. Do you agree with him? Explain why.
2. **Explore Further:** Many artists are interested in natural forms. Find out what made the work of Frank Lloyd Wright so original, and how natural forms influenced his work.

The same is true for guanine and cytosine. Do you see why the amounts of adenine and thymine in a DNA molecule are always the same?

Either base of a pair may be on either nucleotide chain, and the sequence of base pairs may vary along the entire length of the DNA molecule. The two chains of nucleotides run in opposite directions and are twisted around one another to form a double helix. The twisting results in indentations along the molecule, known as major grooves and minor grooves. These grooves play a role in the activation of genes. Figure 9-6 shows the double helix model constructed by Watson and Crick.

Study the model of DNA shown in Figure 9-6. Notice the base pairing in the structure of DNA. The order of these bases makes up the genetic code. Differences in the order of bases are what make one species different from another. Make certain you understand DNA structure. As you look at the drawing of a DNA molecule, think again about the role of DNA. How does a chicken egg become a chicken? Remember that DNA is the chemical that genes are made of. A gene is one sequence of nucleotides along a DNA molecule. One DNA molecule, therefore, contains many genes. The simplest known virus has 5000 paired bases in its DNA. A single human cell contains about 6.6 billion base pairs in its nuclear DNA. The amount of DNA in a single human cell carries as much information as 600 000 printed pages of 500 words each. If all this DNA were stretched out in a line, it would measure almost two meters. Not all of the DNA in a cell carries genetic information. The portion that does contains about 50 000 to 100 000 genes.

Figure 9-6 **The double helix structure of DNA results from pairing of the four bases and the angles of their chemical bonds.**

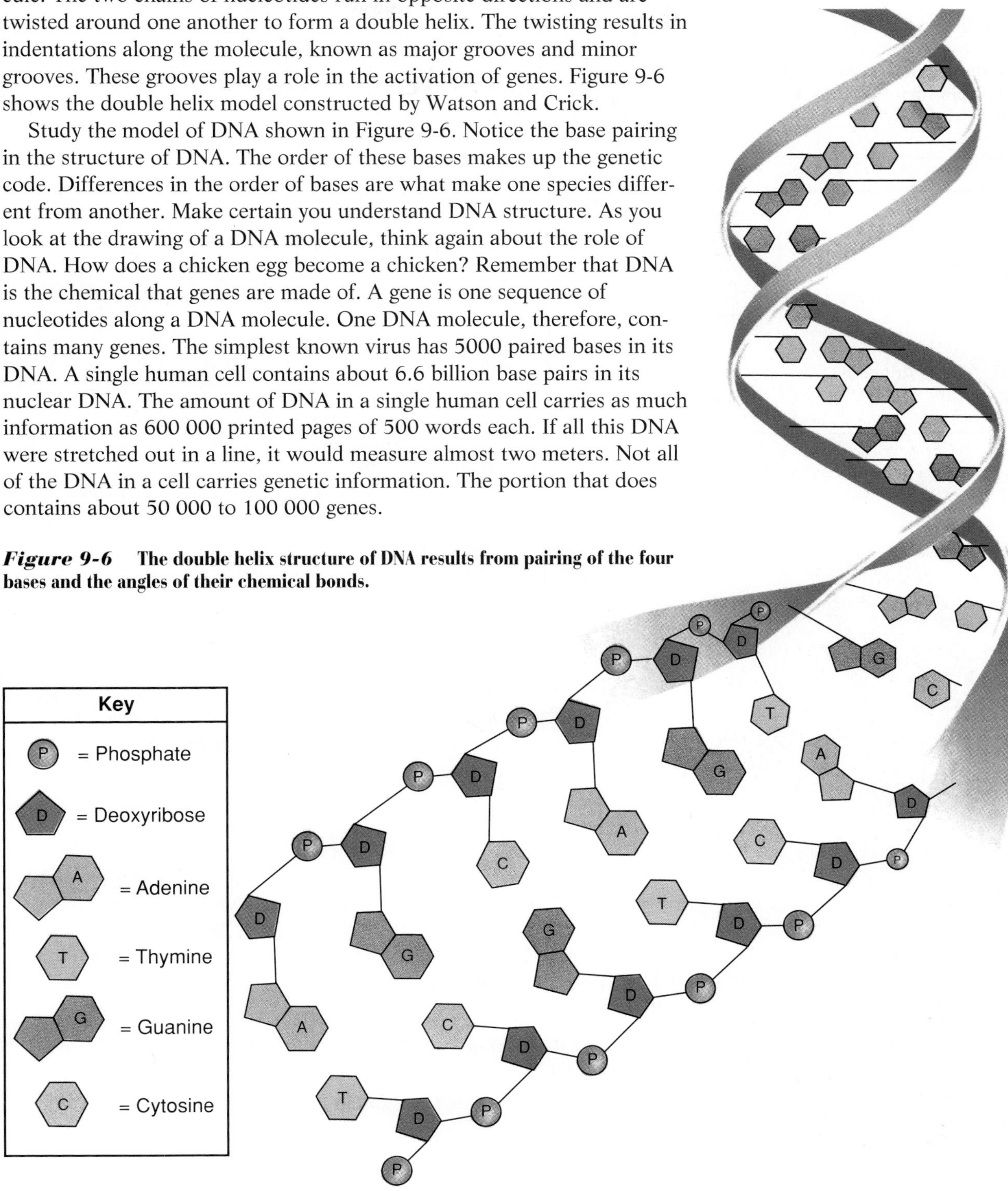

Key

P = Phosphate

D = Deoxyribose

A = Adenine

T = Thymine

G = Guanine

C = Cytosine

Replication of DNA

How are genes passed on to offspring when a cell reproduces? Recall from Chapter 7 that in a eukaryote, many chromosomes are duplicated during mitosis. A prokaryote does not undergo mitosis. Instead, its single chromosome is duplicated in the cytoplasm as binary fission occurs. In spite of these differences, replication of the DNA of both eukaryotic and prokaryotic chromosomes is basically the same. It is also the same in mitochondria and chloroplasts.

Figure 9-7 DNA replication results in exact copies of genetic information that can be passed on to offspring. Replication is possible because of base pairing.

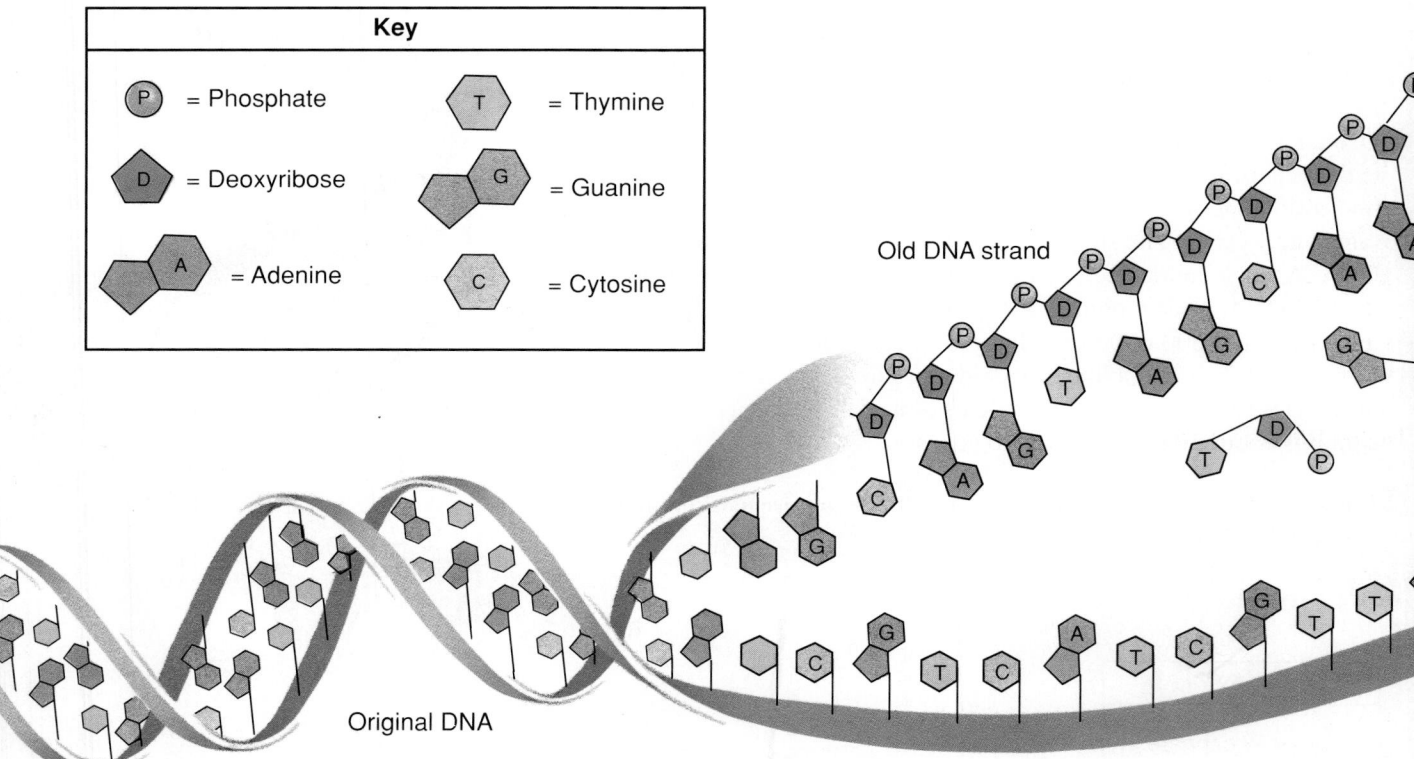

Watson and Crick's model suggested how DNA replication might occur. They hypothesized that each original chain of nucleotides in DNA acts as a template, or mold, for making a new chain. Because of base-pairing rules, A with T and G with C, new nucleotides would be put in place according to the sequence of bases in each parent chain. Watson and Crick's hypothesis was tested experimentally by other researchers, and found to be correct.

How DNA Is Replicated One requirement for genetic material is the ability to provide exact copies of itself. How can DNA accomplish this? Replication of DNA begins as a protein binds to a section of DNA called the origin. Once that binding occurs, an enzyme begins breaking the hydrogen bonds between nitrogen bases that hold the two chains of nucleotides together, and the double helix begins to unzip. These actions

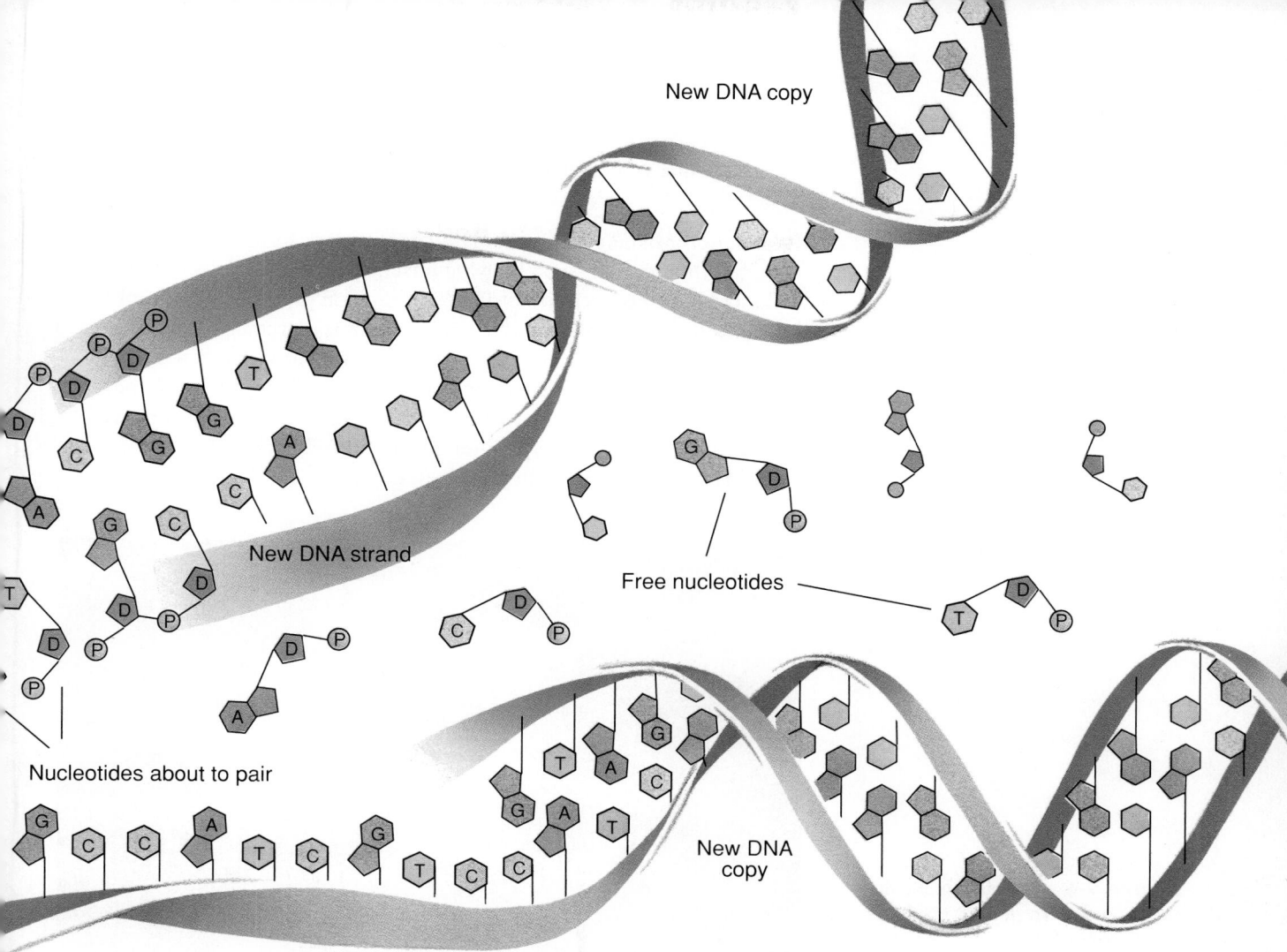

New DNA copy

New DNA strand

Free nucleotides

Nucleotides about to pair

New DNA copy

begin at several places along the DNA molecule; as sections unzip in both directions, the unzipped sections eventually merge and the result is an entirely unzipped molecule. As the chains of nucleotides separate, a complex enzyme (DNA polymerase) binds to each of the separated parent chains. The enzyme sweeps along, bonding new nucleotides to the nucleotides of the parent chain. Addition of nucleotides occurs in opposite directions on the two chains—from top to bottom on one chain, from bottom to top on the other. The nucleotides are matched by base pairing in the same order as one of the parent chains. That is, if one nucleotide on the parent chain has adenine as a base, the new nucleotide added will have the base thymine. As a result, each new chain is a complement of one of the two parent chains.

The replication of the DNA in a human cell takes about four hours. An enzyme then bonds the nucleotides together in the new replicated chain. ATP energy is needed for many of these steps, such as breaking the bonds between nucleotides and adding new nucleotides to the growing chains. Each new DNA molecule consists of one parent chain and one new chain and is a replica of the parent molecule. Recently, a protein that inhibits replication has been discovered. It attaches to the origin and prevents the breaking of bonds between nucleotide chains. This protein, along with the one that starts replication, plays an important role in regulating the replication part of the cell cycle.

Replication Mistakes   During DNA replication, some mistakes may occur. Although rare, such copying mistakes may be serious because each cell must have exact copies of DNA to function properly. The number of mistakes that are actually passed on, though, is extremely small because a cell can repair its DNA. Special repair enzymes "proofread" the newly made DNA for errors. They remove any incorrect nucleotides. Other enzymes then insert the correct ones. DNA repair reduces the chance of errors to about one in every billion nucleotides. Mistakes that are not corrected result in variations in genetic material.

Figure 9-8   **A variation in chickens called the frizzle trait results in weak, stringy feathers. Caused by a single mistake in DNA, this trait also results in chickens with enlarged hearts and spleens, changes in kidneys, and decreased egg–laying abilities.**

Section Review

Understanding Concepts
1. How did the results of Hershey and Chase's experiment finally change the notion that genes were made of proteins?
2. Would Watson and Crick's first idea that DNA is a single chain of nucleotides offer an explanation of how DNA replicates? Explain.

3. One chain of a DNA molecule has the nucleotide sequence C, C, G, C, T. What is the sequence of nucleotides on its partner chain?
4. *Skill Review—Sequencing:* Sequence the steps that occur in the replication of a DNA molecule. For more help, see Organizing Information in the **Skill Handbook**.

5. *Thinking Critically:* Why is it important that a signal for stopping DNA replication be part of the cell cycle?

9.2 The Role of DNA

Blueprints are drawings that show the dimensions of each room of a house, how one room is joined to another, and the locations of plumbing and electrical wires. DNA, like a blueprint, contains information for cells to function and for the organism made up of those cells to develop as a whole. That information is in the form of a code. Before you can decipher the code, you need to know exactly what kind of information the code carries. What instructions does DNA give to a cell? The answer to that question involves proteins.

Genes and Proteins

Proteins are molecules made up of 20 kinds of amino acids. In cells, hundreds of amino acids may be linked to form long chains called polypeptide molecules. Two or more polypeptides may be joined to make a particular protein. The kind and sequence of amino acids make the shape of one polypeptide different from another. Shapes of proteins are related to their particular functions. Recall from Chapter 4, for example, the importance of the shape of enzymes, which govern every cellular reaction.

All organisms make proteins. Consider two simple organisms, *Amoeba* and *Paramecium.* Both of these unicellular forms take in basically the same raw materials. What makes one cell different from the other? The answer is the kinds of enzymes and other proteins each contains. Because cells have different sets of enzymes, they use their raw materials in different ways. Their chemical reactions vary; as a result, they have unique features.

Multicellular organisms differ as well, and for the same reason. Individual organisms of the same species have many of the same proteins. Your hemoglobin is the same as the hemoglobin of any other human being; your pet dog's hemoglobin, however, is different. Similarity of proteins is one way scientists identify the species of newly discovered organisms. However, even among species there can be variations in proteins. Some human beings, for example, have sickle cell anemia. Their hemoglobin differs from normal hemoglobin, but it is still human hemoglobin and not like that of any other species.

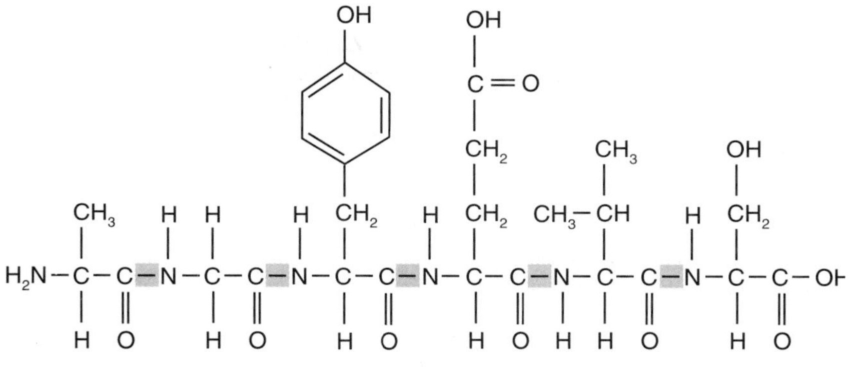

Objectives

Discuss the evidence for the fact that DNA codes for proteins.

Demonstrate an understanding of the DNA code.

Figure 9-9 This polypeptide was formed by six amino acids. The shaded areas show where amino acids are bonded chemically to form the polypeptide.

Relationship Between Proteins and Genes Proteins are unique. Genes are unique. Could there be a relationship between genes and proteins? Perhaps the role of genes is to carry information for the synthesis of proteins. That idea was first proposed early in this century, when a physician noted that several inherited diseases result from the absence of a necessary enzyme or the presence of a defective enzyme (remember, enzymes are proteins). Normal individuals who have at least one dominant gene make normal enzymes and do not have the disease. Individuals who have two recessive genes do not produce the normal enzyme and have the disease. Thus, there was evidence, based on analysis of family histories, that genes are associated with synthesis of proteins.

Genes Direct Protein Synthesis In 1941, researchers learned that each gene contains information for the synthesis of a particular enzyme. Since that time, it has been learned that genes contain information for making not only enzymes, but also all other kinds of proteins. Because many proteins are made of two or more different polypeptides, each gene must control the synthesis of a particular polypeptide. Hemoglobin, the protein that carries oxygen in your blood, is a protein composed of four polypeptide chains—two of one kind and two of another. Two genes are involved, each coding for the synthesis of one kind of polypeptide. After the polypeptides are synthesized, they join to form hemoglobin.

Thinking Lab

DESIGN AN EXPERIMENT

How can crop plants be genetically engineered to produce higher yields?

Background
Scientists can remove DNA from bacteria that kill insect pests and inject it into plant cells. They coat microscopic metal particles with bacterial DNA and shoot them into plant cells with a tiny gun. As these cells reproduce, this bacterial DNA is incorporated into the new plant cells. The bacterial DNA codes for a substance that kills only the insects that are plant pests, such as moth larvae. When moth larvae feed on plant tissue that includes cells with bacterial DNA, they die. Other insects, however, are not affected. By incorporating bacterial DNA into plant cells, scientists can give plants traits that they do not ordinarily have in order to protect them from pest organisms.

You Try It
Weeds can reduce crop yields by 70 percent. **Design an experiment** similar to the one described to produce crop plants resistant to herbicides, chemicals that kill plants. Farmers would benefit from these plants because they could selectively kill weeds without harming crop plants. Fewer chemicals would result in less harm to the environment. Think about how cells take in materials, how cells reproduce, and how DNA replicates to guide your design. What would be the advantages of crops that are resistant to herbicides? What would be the disadvantages of such plants? Think about how they would affect other organisms in the environment.

The DNA Code

Now you know that genes code for polypeptides. But what exactly are genes? At this point, let's define a gene as a segment of a DNA molecule that carries the code for the synthesis of a particular polypeptide. Polypeptides are composed of amino acids. The sequence and kinds of amino acids make one polypeptide different from another. Therefore, a gene must carry a code for putting particular amino acids together in a certain order. What part of DNA might carry the code? Only the sequence of base pairs along a DNA molecule varies. Experiments have shown that it is this sequence of bases along one of the two chains of a DNA molecule that carries the code.

Coding for Proteins   There are 20 amino acids, but DNA contains only four kinds of bases. Thus, a single base cannot represent an amino acid. Likewise, a sequence of two bases in a row such as AA, CG, TA, or GT cannot represent an amino acid. The four bases taken two at a time yield only 16 possible sequences. Considering three bases at a time, however, gives more than the 20 combinations needed. In fact, it provides 64 possible combinations of bases. Think of the 26 letters in the English alphabet. They can be combined in various ways to make thousands of words. Experiments have shown that three bases in a row do act as the code for amino acids. Each set of three bases representing an amino acid is known as a **codon.** Because there are 64 possible codons and only 20 amino acids, some amino acids have more than one codon. The cell "reads" these codons, and links together certain amino acids to form a particular polypeptide. DNA language in the nucleus is converted to protein language in the cytoplasm.

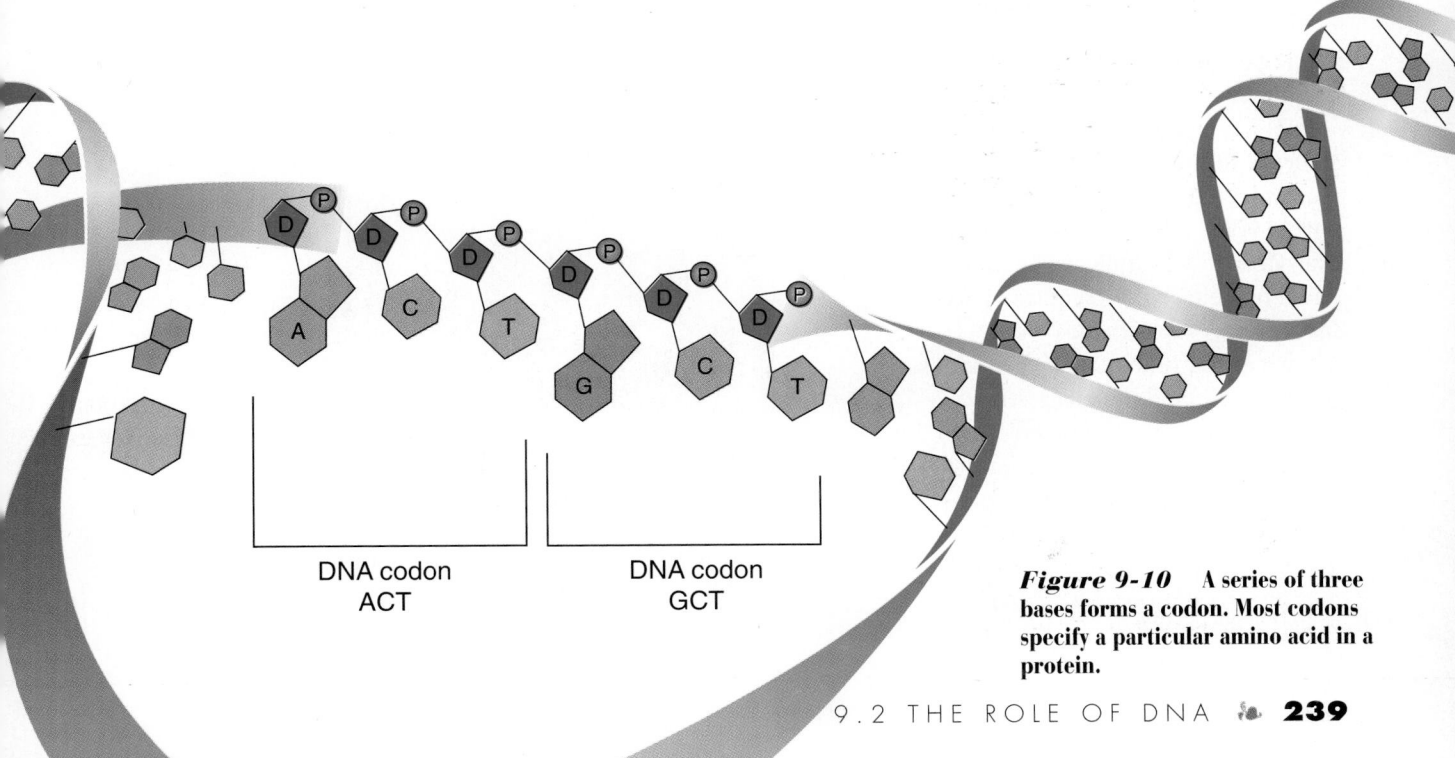

DNA codon
ACT

DNA codon
GCT

Figure 9-10   **A series of three bases forms a codon. Most codons specify a particular amino acid in a protein.**

The Genetic Code   Table 9.1 shows the various codons and the amino acids they represent. All but two amino acids are represented by two or more codons. In codons that are synonyms, the first two nucleotides are usually the same. Valine, for example, is coded for by four different codons, but all begin with the two nucleotides cytosine and adenine. Notice that three of the 64 codons do not represent amino acids. These are the codons at the end of a gene. They act as stop signals, like periods at the end of sentences, that signify the end of the code for a particular polypeptide. It is important to understand that the codons are read in only one direction. Reading the codon in reverse does not result in the same amino acid. In Table 9.1, the codon for serine is AGC, whereas the codon for alanine is CGA. The nucleotides are arranged in a specific order and read in one direction only.

The DNA code is a universal one. That is, with rare exceptions, a codon representing a given amino acid in one organism represents the same amino acid in another organism. This tells us that the DNA code must have arisen very early in the evolution of life and then been passed down unaltered to all later life forms. How was DNA passed down to later life forms? Replication of DNA ensured that exact copies of DNA were made. These copies were passed on to offspring unaltered.

HISTORY CONNECTION

Har Gobind Khorana

Cracking any code can be a monumental task or a simple one, depending on how complex the initial code happens to be or what equipment is available for decoding. For scientists who were trying to crack the genetic code, the task was indeed monumental. In the early 1960s, a biochemist named Har Gobind Khorana synthesized small nucleic acid molecules in the laboratory. Nucleic acids were known at the time to carry genetic information, and the exact structure of some nucleic acid molecules was already determined. Khorana was able to produce an artificial molecule in which two or three nucleotides were repeated over and over in a known sequence. For example, one sequence was

AGAGAGAGAGAG. When Khorana combined his synthetic nucleic acids with the proper materials, these acids caused proteins to be synthesized just as proteins are

made in the cell. By comparing the proteins made with the known structure of these synthetic acids, Khorana was able to determine which portions of the nucleic acid were the codes for each protein. For example, the molecule in which A and G were repeated over and over resulted in the production of two amino acids made by the codons AGA and GAG. For this work on the genetic code, Har Gobind Khorana shared a Nobel Prize for Medicine in 1968 with two other researchers.

1. **Writing About Biology:** Explain why many scientific discoveries are made by several people working together.
2. **Explore Further:** Find out who shared the Nobel Prize with Khorana, and what each person contributed to the final result.

Table 9.1 **The DNA Code**

First Base in Codon	Second Base in Codon				Third Base in Codon
	A	**G**	**T**	**C**	
A	phenylalanine	serine	tyrosine	cysteine	A
	phenylalanine	serine	tyrosine	cysteine	G
	leucine	serine	stop	stop	T
	leucine	serine	stop	tryptophan	C
G	leucine	proline	histidine	arginine	A
	leucine	proline	histidine	arginine	G
	leucine	proline	glutamine	arginine	T
	leucine	proline	glutamine	arginine	C
T	isoleucine	threonine	asparagine	serine	A
	isoleucine	threonine	asparagine	serine	G
	isoleucine	threonine	lysine	arginine	T
	methionine	threonine	lysine	arginine	C
C	valine	alanine	aspartate	glycine	A
	valine	alanine	aspartate	glycine	G
	valine	alanine	glutamate	glycine	T
	valine	alanine	glutamate	glycine	C

Figure 9-11 **What do you and this leaf insect have in common? Both the insect and you have DNA that codes for the same amino acids.**

Figure 9-12 These fossil organisms passed on their DNA code to their offspring. Why is this ability important to the preservation of species?

LOOKING AHEAD

DISCOVERING THE STRUCTURE of DNA and the nature of genetic information is only one step in understanding how cells produce proteins. Although you know that genes code for the assembly of polypeptides, and that polypeptides make up proteins, how does this information get from the DNA in a cell's nucleus to the cytoplasm or the organelles where proteins are synthesized? In the next chapter, you will study the means by which cells decipher the code to make polypeptides and where this synthesis takes place. You will also see how subtle changes in DNA can lead to variation among organisms.

Section Review

Understanding Concepts

1. Identical twins have exactly the same sets of DNA. How does that fact support the idea that genes code for synthesis of proteins?
2. Several human diseases are associated with production of hemoglobin. How would you expect the hemoglobin of diseased people to compare with that of normal people?

3. What would be the minimal number of different genes between normal people and those having a disease associated with hemoglobin? Explain.
4. *Skill Review—Sequencing:* Use Table 9.1 to sequence the amino acids coded for by the following sequence of codons: GAG CTC AAC. For more help, see

Organizing Information in the **Skill Handbook.**
5. *Thinking Critically:* How can you explain that two genes can vary in terms of one nucleotide pair yet code for exactly the same polypeptide?

Biotechnology

Transgenic Organisms—What Are They and Who Owns Them?

Suppose you were given an assignment to redesign any type of living thing. What would you redesign? Would you design a human who is resistant to all disease, or a tomato the size of a basketball? What other possibilities can you think of? Is this something that will never be possible? What do you think? The possibility of redesigning living things may sound like science fiction, but the opportunity for accomplishing this is here today, thanks to genetic engineering. Scientists already have the ability to redesign living things. They can remove genes from one organism and insert them into another. This provides them with an opportunity to place desirable gene traits from one living thing into another living thing that is missing those traits. Organisms that have genes derived from other organisms are called transgenic organisms.

What are some examples of this type of technology? Scientists have perfected transgenic tobacco and tomato plants that are resistant to virus attack. A virus infection could reduce crop yield or destroy an entire crop. Resistance to a disease-

causing virus, therefore, increases the harvest of a particular crop. Potato, cotton, and corn plants are now designed with built-in insect resistance. Moth and butterfly larvae and adult beetles that once fed upon these plants no longer do so. Tomatoes also are now available with genes inserted that slow down the process of spoilage. This makes storage, shipping, and stocking of tomatoes in supermarkets less of a problem. The chances of seeing a rotten tomato may be reduced in just a year or two. Are scientists using this new technology with animals, too? Animals such as goats,

sheep, and cows are being used to produce milk that contains high levels of chemicals that can be used against hemophilia, cancer, and emphysema. These transgenic animals may even be used to form a protein that helps to break up blood clots.

An interesting question arises from the formation of all transgenic organisms. Who owns them? Do scientists or the companies that have done the research on producing them own the organisms that they have created? The Supreme Court of the United States was asked this very question. According to the court's decision, the ownership or patent for these organisms does indeed belong to the company or scientist who designed them through research. The court stated that a "patent can be granted on anything under the sun which can be made by man." This decision is one reason that companies today are investing millions of dollars in research to produce transgenic organisms. It is hoped that the tools of biotechnology will help scientists produce new drugs or improve food production and quality for all of us.

1. **Writing About Biology:** Why is the term "transgenic" very appropriate for genetically engineered organisms?
2. **Issue:** Should scientists use biotechnology to develop new organisms?
3. **Explore Further:** Scientists have been selectively breeding new plants and animals for centuries. Is the formation of transgenic organisms different or similar? Explain.
4. **Solve the Problem:** Design product names for the following genetically engineered organisms: tomatoes that do not rot, oranges with more juice, corn cobs with more kernels.

Issues

Genetic Profiles and Denying Health Care and Employment

You are ready to try out for the football team. You have been working out over the summer and have gained enough weight not to make a fool of yourself on the football field. You go to the doctor for a physical exam, and you are ready to start practice on the first day of school. But wait. The doctor ran a test from a sample of your blood and found out that you carry a gene for bone cancer. The school, after receiving this report, is not willing to insure you against injury. No insurance means that you cannot play football.

The situation just described is fictional, but it may become a possibility in the future. How? Scientists will soon be able to read the genes on all 46 human chromosomes. This will enable them to analyze each person's genome (all their genes) for genetic diseases.

Is this new knowledge about your genetic makeup good news or bad news? Let's look at the good news first. If you find out at an early age that you are at high risk genetically for a particular disease, you may be inclined to have checkups more often. You might even wish to have a specific organ removed before it becomes a problem. This will work, of course, only if the organ is not vital for life. You may wish to change or alter your diet or begin certain therapy programs that may help reduce the effects of a certain disease. The possibility of passing this genetic disease on to your offspring is also an important piece of information to help you decide whether to have children. Doesn't it seem that knowing in advance about the possibility of having a certain genetic disease is a good thing?

Unfortunately, such information may not be all to your benefit. Once your genetic makeup becomes known, it could be used against you. For example, say you apply for life or health insurance. The insurance company may decide not to insure you because you have a gene that results in a disease called Huntington's chorea. This disease usually does not show up until a person is about 40 years old. But in the meantime, you may break your arm in a fall, or come down with a viral disease, and have no insurance to pay for your medical bills. What if you are denied insurance coverage because you may come down with a genetic disease in your forties, but meanwhile you contract cancer in your twenties? What if a future employer obtains your genetic profile and finds out about genetic diseases in your genome? Will this employer hire you, or hire someone who is less at risk of contracting a genetic disease? Often, the cost of health insurance is the responsibility of the employer. It eventually may cost employers more if they hire employees with higher risks of becoming ill than those who have lower risks. Employers could discriminate against people because of their genetic makeup.

Should a person be informed about his or her own genome? How will a person react to knowing that he or she carries the gene for some lethal genetic disease? Is your genome your own private business or are others entitled to know all about your genome, like they know your social security number?

Some researchers believe that information about genomes should only be available to individuals who are at risk of passing a genetic

disease on to a future child. These people may have a parent or relative with a certain disease, and wish to know what their chances are of carrying a defective gene. Genetic counselors work with families to help them make decisions about whether to have children or not. However, many people may carry a defective gene without being aware of it. Researchers don't believe a person should be forced to find out information about his or her genome if such information is not wanted.

Understanding the Issue

1. Should a person's genome be protected by law as something private? Explain your answer.
2. Huntington's chorea, a genetic disease, causes early death. By the time a person finds out that he or she has the disease, the gene that causes the disease may have already been passed on to his or her children. (a) Would

you want to know if you have the gene for Huntington's chorea? Explain your answer. (b) Would you want to know if your children have received this lethal gene from you? Explain your answer. (c) As a child from a parent with this disease, would you want to know if you are or are not carrying the gene? Explain your answer.
3. If you are told that you have the gene for Huntington's chorea, would you want your health insurance company or your employer to know this? Explain why.

Readings
- Brownlee, Shannon. "The Assurances of Genes: Is Disease Prediction a Boon or a Nightmare?" *U.S. News and World Report,* July 23, 1990, p. 57.
- Zoler, Mitchel L. "Genetic Tests: Can We Afford the Answers?" *Medical World News,* Jan. 1991, p. 32.

CAREER CONNECTION

Research often requires laboratory experimentation. The people "behind the scenes" in all laboratories are **lab technicians.** Under a scientist's guidance, lab technicians perform experiments, record data, and compile the results of experimental trials. They report experimental findings to the scientist in charge. Lab technicians may be high school graduates with an interest in biology, chemistry, earth science, or physics, or they may have some college training in these subject areas.

Summary

Griffith's discovery of bacterial transformation led to later experiments showing that DNA seemed to be the chemical of which genes are composed. Hershey and Chase's experiments made use of the reproduction of phages to show finally that genes are indeed made of DNA, not protein.

Watson and Crick's model shows DNA to be a double helix composed of two chains of nucleotides joined together by hydrogen bonds between nitrogen bases. Adenine can bond only to thymine, and cytosine can bond only to guanine. During replication of DNA, the two chains separate, and each chain acts as a template, guiding the formation of a new partner chain. Because of base-pairing requirements, two replica DNA molecules are produced from the parent molecule.

Evidence based first on logic and later on experimentation revealed that genes, segments along a DNA molecule, carry a code for the synthesis of a polypeptide. Each codon represents one amino acid in the polypeptide chain.

Language of Biology

Write a sentence that shows your understanding of each of the following terms.

adenine	guanine
bacteriophage	nucleotide
codon	thymine
cytosine	

Understanding Concepts

1. In terms of DNA, explain what happened to living rough cells when they were mixed with heat-killed smooth cells. (Hint: DNA is stable under high temperatures, whereas other organic molecules are not.)
2. Explain why the phages that were dislodged from bacterial cells in a blender consisted of just their outer coats.
3. How is the idea of DNA as a ladder related to the double helix model?
4. Use a series of drawings to illustrate the events of replication of a DNA molecule, assuming one chain of the parent molecule has the nucleotide sequence TAG GAG ACT.
5. Which two people, brother and sister or two cousins, would be more similar in terms of proteins? Why?
6. Explain why it is more precise to say that a gene carries a code for the synthesis of a polypeptide than for a protein.

7. Why is it logical to assume that the sequence of base pairs of DNA, rather than the sugars and/or phosphates, carries the code for synthesizing polypeptides?
8. Explain the relationship among codons, amino acids, and a polypeptide.
9. Why must stop codons be present along a DNA molecule?

Applying Concepts

10. Suppose an error occurred in replication of a gene so that one nucleotide was added. How would that error affect the polypeptide coded for by that gene?
11. A gene carries the code for a polypeptide consisting of 100 amino acids. How many codons are in the gene? How many nucleotides?
12. PKU is a genetic disease resulting in a form of mental retardation. It occurs in people having two recessive genes. Normal people are either homozygous dominant or heterozygous. What is the likely cause of the disease?

13. Most bacteria of a certain type can synthesize an amino acid they need from raw materials that enter their cells. Some forms of these bacteria cannot. Assuming you had the equipment necessary to extract DNA from bacteria, how could you demonstrate the idea of bacterial transformation?

14. Bacteria are growing and reproducing on culture dishes in a lab. Colonies of the bacteria are clearly visible on the medium in the dishes. A biologist discovers one morning that the bacteria in many dishes have disappeared and that no other organisms are visible on the medium. What is a likely reason for the disappearance of the bacteria?

15. Suppose a molecule of DNA has replicated but one error occurred and was not corrected or repaired. What will the DNA molecules produced after the next replication and in future replications be like? What might the result of this error be?

16. **History Connection:** Why is an understanding of chemistry often important in biology? Use the information in the History Connection to explain your answer.

17. **Art Connection:** Sculpture is one art form in which the shapes of natural forms such as chemicals can be expressed. Name an art form in which the following could be expressed: colors, light, language, relationships.

Lab Interpretation

18. **Investigation:** Based on your paper model of DNA, design on paper another model of DNA. Use toothpicks for chemical bonds and various kinds of candy or dry cereal for the molecules. Give a brief set of directions for putting together your DNA molecule. Trade your directions with another group. See if they can visualize and draw the DNA molecule you designed.

19. **Thinking Lab:** Corn plants are killed by a specific bacterial disease. A scientist discovers a weed that is resistant to this bacterial disease. Using genetics, what could the scientist do to make corn plants have the same resistance to the disease as the weeds?

20. **Minilab 1:** You are starting your own business, a greenhouse nursery, in which you will grow and sell houseplants. You have very little money to invest in this business. How could the principle of cloning help you save money?

21. **Minilab 2:** DNA can be more easily extracted from rapidly dividing cells. If you were going to extract DNA from cells other than the ones used in this Minilab, what other cells might you try?

Connecting Ideas

22. Transformation is an accident in which a bacterium receives a new gene. Because bacteria are haploid, all offspring of the transformed cell will have the trait coded for by the new gene. How might transformation be important to a population of bacteria?

23. Is replication of DNA an endergonic reaction? How do you know?

10

FROM GENES TO PROTEINS

ALBINO ORGANISMS, such as the rabbits shown in the large photo, have no color. Their tissues do not contain the pigments needed for normal coloration. Complete albinos have white or very pale fur, feathers, or hair. Their skin and eyes look pink, because there is no pigment to mask the red color of the blood vessels in the iris of the eye and the top layers of skin. You may be more familiar with partial albinos, such as domesticated ducks and geese, which have white feathers and pale skin but show some color in their eyes, beaks, or legs. Although albinism is unusual, it does occur in many organisms, including flowering plants, insects, rabbits, koalas, and humans.

The genes of albino organisms do not code for a normal enzyme that makes pigment. Since albinism is a recessive trait, an organism that shows normal coloration could have a gene for albinism, and that gene could be passed to the organism's offspring. Offspring that inherit a gene for albinism from each parent will express an albino phenotype.

How did genes for unusual traits like albinism develop? How do the cells of an organism follow the instructions contained in its DNA to manufacture proteins such as the enzyme that makes a pigment? In this chapter, you will find out how DNA is involved in protein synthesis. You will learn how genetic material can be altered, and what can happen as a result. You will also find out how geneticists have learned to alter DNA to develop organisms that have specific traits.

If a normal brown rabbit, like the one above, and an albino rabbit were side by side in a field, which would be more likely to escape the eye of a hungry hawk? Why do you think albinism is extremely rare in the wild?

Chapter Preview

Objectives

Analyze the process of transcription in prokaryotes and eukaryotes.

Sequence the events that occur during translation.

Relate genes and polypeptide synthesis to the expression of phenotypes.

10.1 Polypeptide Synthesis

As you grow, the cells that make up your body increase in number. When each cell reproduces, the genes it contains are replicated and passed to its daughter cells. You have learned that genes carry the instructions needed by each new cell to make its polypeptides. Each gene is a segment of a DNA molecule and consists of a chain of nucleotides. The information for making a polypeptide lies in the sequence of those nucleotides. You also know that the DNA needed for making most of a cell's proteins is located in the nucleus, but that proteins are synthesized outside the nucleus, at the ribosomes. Once scientists understood this much about the role of DNA in heredity, they began looking for the answers to two basic questions. First, how does the information in DNA get from the nucleus to the ribosomes? And once the information arrives, how is it used by the ribosomes to make the polypeptide it represents?

Transcription

Geneticists began to understand how the information in DNA moves into the cytoplasm when they discovered that another kind of nucleic acid, ribonucleic acid (RNA), is found not only in the nucleus, but also in the cytoplasm. They also learned that RNA forms part of the ribosomes and that cells actively making polypeptides contain more RNA than cells less active in polypeptide synthesis. These facts suggested that RNA, as well as DNA, is involved in the manufacture of proteins.

Structure of RNA   Like DNA, RNA is a nucleic acid made up of a sugar, a phosphate, and four bases. Although there are several forms of RNA, each form shares three basic characteristics that distinguish it from DNA. First, RNA is usually composed of a single chain of nucleotides, rather than the double chain characteristic of DNA. Second, RNA contains the sugar ribose, rather than deoxyribose. Third, instead of the base thymine, RNA contains **uracil.**

Formation of RNA   Cells synthesize proteins as they are needed. When it is time to build a protein, the gene for each polypeptide in that protein is activated. Near each gene is a region of DNA that acts as a start signal for the synthesis of the needed polypeptide. A complex enzyme, like the one that unzips DNA during replication, binds to this region of DNA. Once the enzyme is attached, a series of events much like those of DNA replication begins. The two chains of the DNA begin to "unzip" at the start signal. Notice in Figure 10-2 that only one chain of the DNA molecule acts as a template for RNA. RNA nucleotides match up with the DNA template until a stop signal is reached. The RNA nucleotides bond together to form a single-stranded molecule of RNA, which breaks away from the DNA chain. The two DNA chains then rejoin.

Figure 10-1   **The difference between thymine and uracil and between ribose and deoxyribose involves only a few atoms. Uracil and ribose in RNA take the places of thymine and deoxyribose in DNA.**

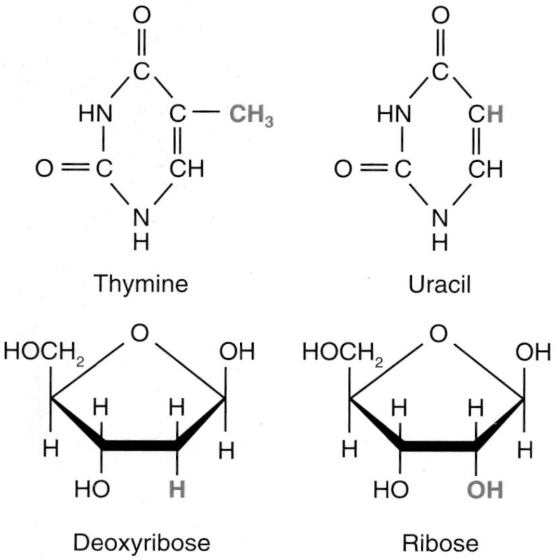

Thymine

Uracil

Deoxyribose

Ribose

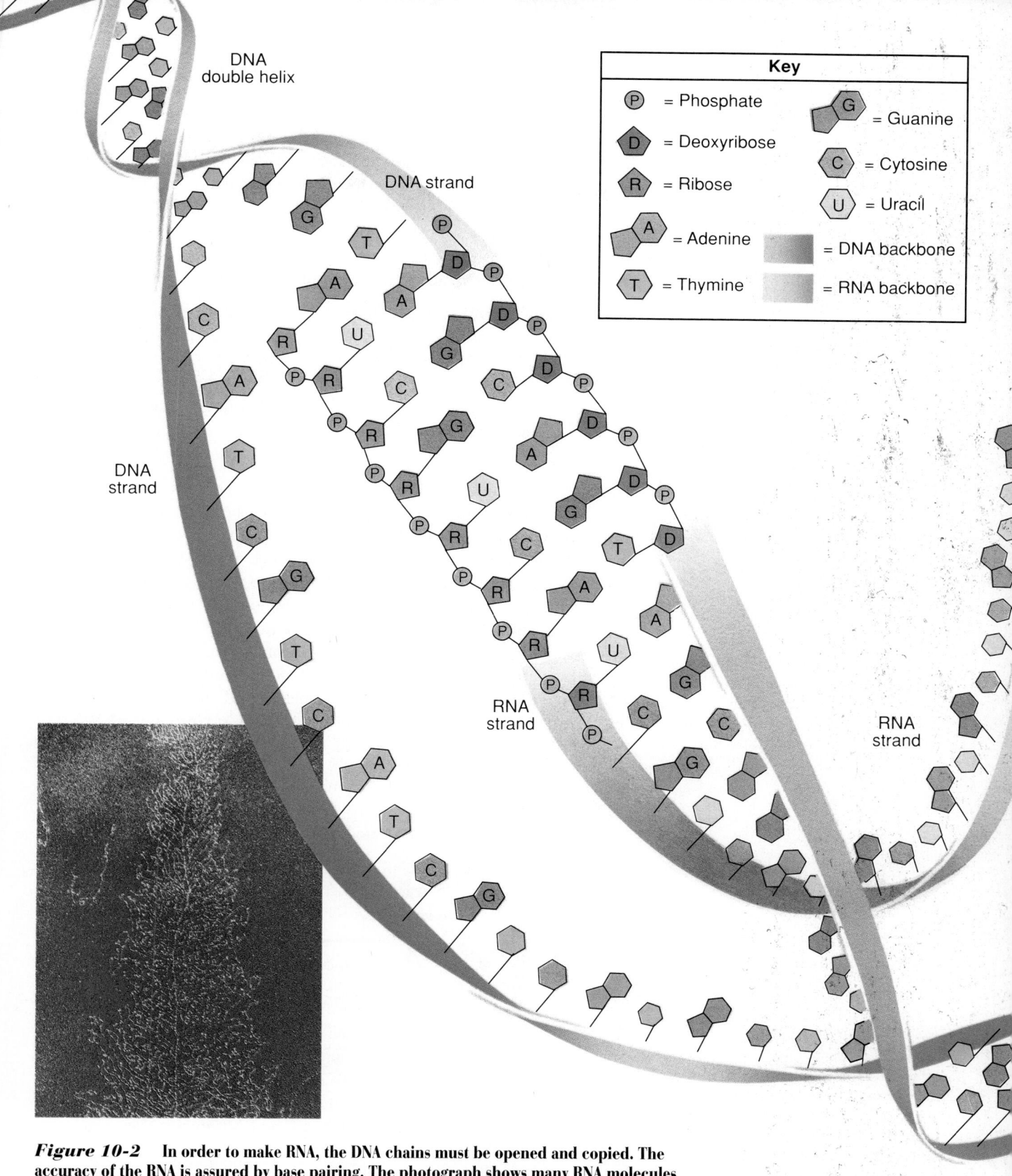

Figure 10-2 In order to make RNA, the DNA chains must be opened and copied. The accuracy of the RNA is assured by base pairing. The photograph shows many RNA molecules being produced from the same section of DNA. Magnification: 6700×

The sequence of bases composing the new RNA molecule is complementary to the base sequence of the DNA template on which it was made. For example, a DNA chain with the sequence AGC TTA TCC AGG will code for RNA with the sequence UCG AAU AGG UCC. (Notice that in RNA, U takes the place of T.) The process of making RNA from DNA is called **transcription.** RNA transcribed from DNA that codes for a polypeptide is called **messenger RNA (mRNA).** Not all DNA, however, codes for polypeptides. For example, some DNA codes for the RNA that makes up the ribosomes. This type of RNA is called **ribosomal RNA (rRNA).** Both mRNA and rRNA are transcribed from DNA.

Transcription in Prokaryotes   In prokaryotes, transcription is a relatively simple process. Prokaryotes have no nucleus, and the ribosomes lie close to the DNA in the cytoplasm. As a result, the first part of the mRNA molecule attaches directly to the ribosomes even as the rest of the molecule is still being transcribed (Figure 10-3). Assembly of amino acids into proteins begins right away.

Transcription in Eukaryotes   In eukaryotes, the transcription process is more complex. You learned in Chapter 7 that the DNA of eukaryotic chromosomes is wound around protein spools. The DNA must be unwound from the spools before transcription can begin. Otherwise, the enzyme that begins the process could not attach to the start signal region.

It took the careful work of many scientists to figure out the details of polypeptide synthesis. As this work was going on, one researcher made an unexpected discovery. In eukaryotes, after a new mRNA molecule is transcribed from its DNA template, the mRNA breaks up into pieces. When these pieces reassemble to form the finished mRNA molecule, portions of the original are left out, making the finished strand shorter than the original. Why does the original mRNA molecule break up into pieces? What is discarded when the pieces are spliced back together?

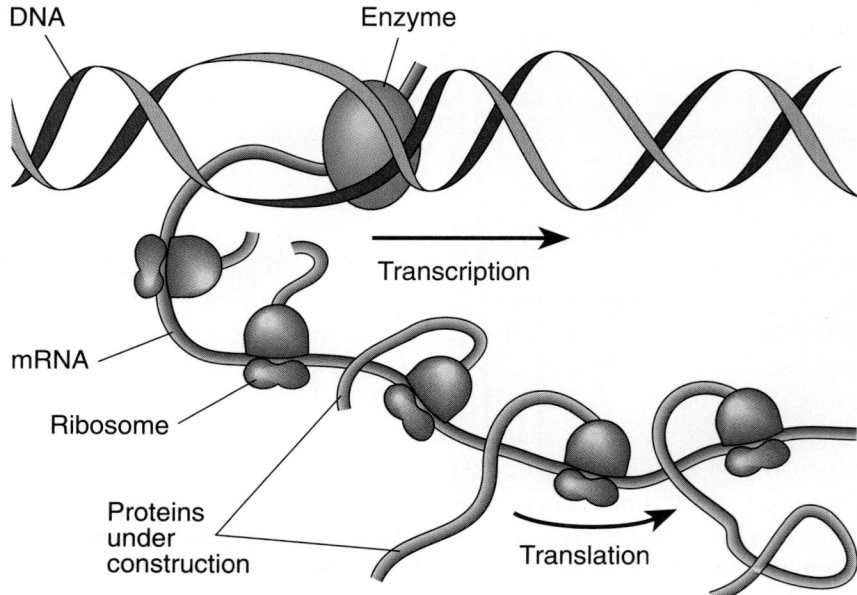

Figure 10-3   **In prokaryotes, as the genetic message is transcribed from DNA to mRNA, ribosomes attach to the forming mRNA, and translation begins simultaneously. What characteristic of prokaryotes makes simultaneous transcription and translation possible?**

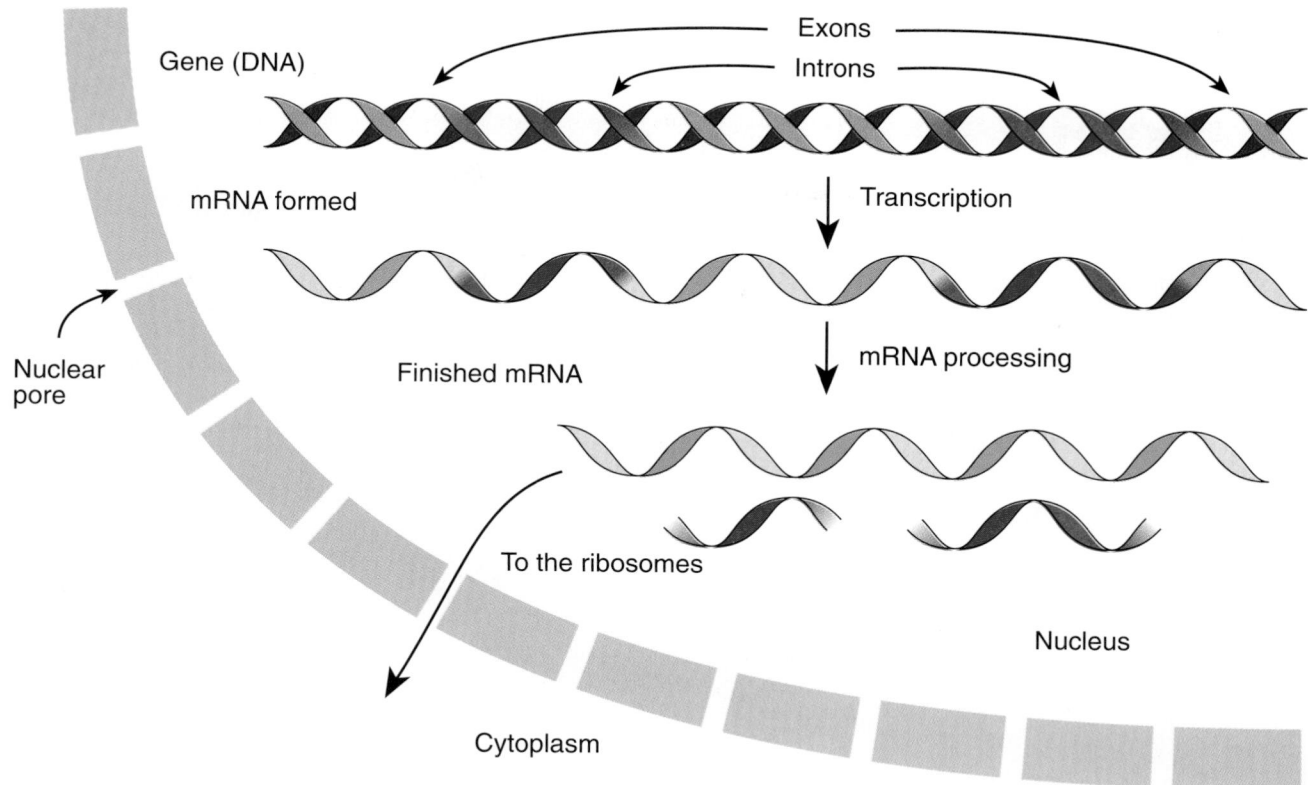

Gene (DNA)

Exons

Introns

mRNA formed

Transcription

Nuclear pore

Finished mRNA

mRNA processing

To the ribosomes

Nucleus

Cytoplasm

Further research revealed that the original mRNA molecule produced by transcription contains many sets of nucleotides that do not code for amino acids. These *in*tervening sets of nucleotides are called **introns.** The sets of nucleotides that do code for amino acids are called **exons,** because they are *ex*pressed. After a new mRNA molecule is transcribed, and before it can function in protein synthesis, the introns must be removed and the exons spliced together. The removing of introns from mRNA is known as mRNA processing and is diagrammed in Figure 10-4. When processing is complete, the finished mRNA molecule moves through pores in the nuclear membrane and travels to the ribosomes.

Figure 10-4 **During mRNA processing, introns are cut out and exons are spliced together. These events occur in the nucleus of the cell. What happens to the mRNA molecule after processing is completed?**

Translation

In eukaryotes, processed mRNA attaches to a free ribosome in the cytoplasm and polypeptide synthesis begins. In many cases, synthesis is completed on free ribosomes. In other cases, the free ribosome and its attached strand of mRNA bind to rough ER, where protein synthesis is completed.

Messenger RNA carries the genetic information—a sequence of RNA codons—that represents a particular order of amino acids. Making the proper protein requires deciphering the information and assembling the amino acids it calls for. Synthesis of a polypeptide from the information carried by an mRNA molecule is called **translation.** How is mRNA "language" translated into protein "language"?

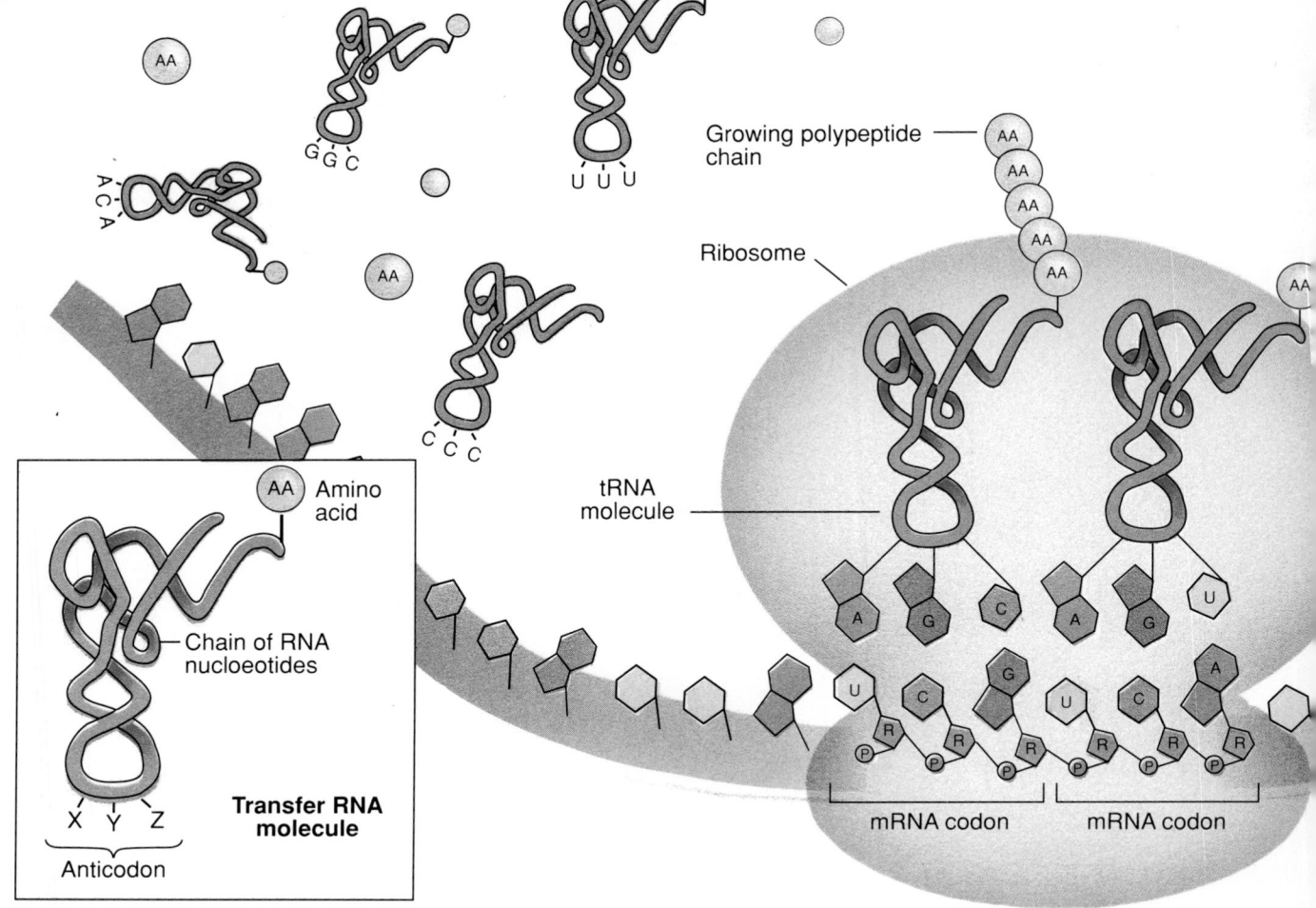

Figure 10-5 A polypeptide is made when the mRNA code is translated. Each mRNA codon pairs with the anticodon of tRNA carrying an amino acid. When the first two amino acids bond together, the tRNA attached to the first amino acid is released. The ribosome then moves along the mRNA strand to the next codon, continuing the process of translation.

The Role of Transfer RNA Picture an mRNA molecule attached to a ribosome. In the surrounding cytoplasm are molecules of all 20 amino acids. If a polypeptide is to be made, the amino acids must be carried to the ribosomes. This job of bringing amino acids to the ribosomes is performed by yet another type of RNA, **transfer RNA (tRNA).** Like other types of RNA, transfer RNA molecules are transcribed from certain regions of DNA in the nucleus. A tRNA molecule is a single chain of nucleotides folded into a cloverleaf shape and then bent over on itself. Each tRNA combines with and carries only one type of amino acid. Enzymes and ATP are used to join an amino acid to one end of the tRNA molecule. At the other side of the tRNA molecule is a set of three bases that is specific to the type of amino acid the tRNA carries. This set of bases is called an **anticodon.** Each type of tRNA has a different anticodon, which can bind only to the complementary mRNA codon for that amino acid.

The Joining of Codons and Anticodons Correct translation depends upon proper joining of mRNA codons and tRNA anticodons. Follow Figure 10-5 as you read how translation occurs.

As translation begins, a ribosome attaches to the first codon of the mRNA strand. At the same time, a tRNA molecule approaches, carrying its amino acid. If the tRNA anticodon recognizes the mRNA codon, the

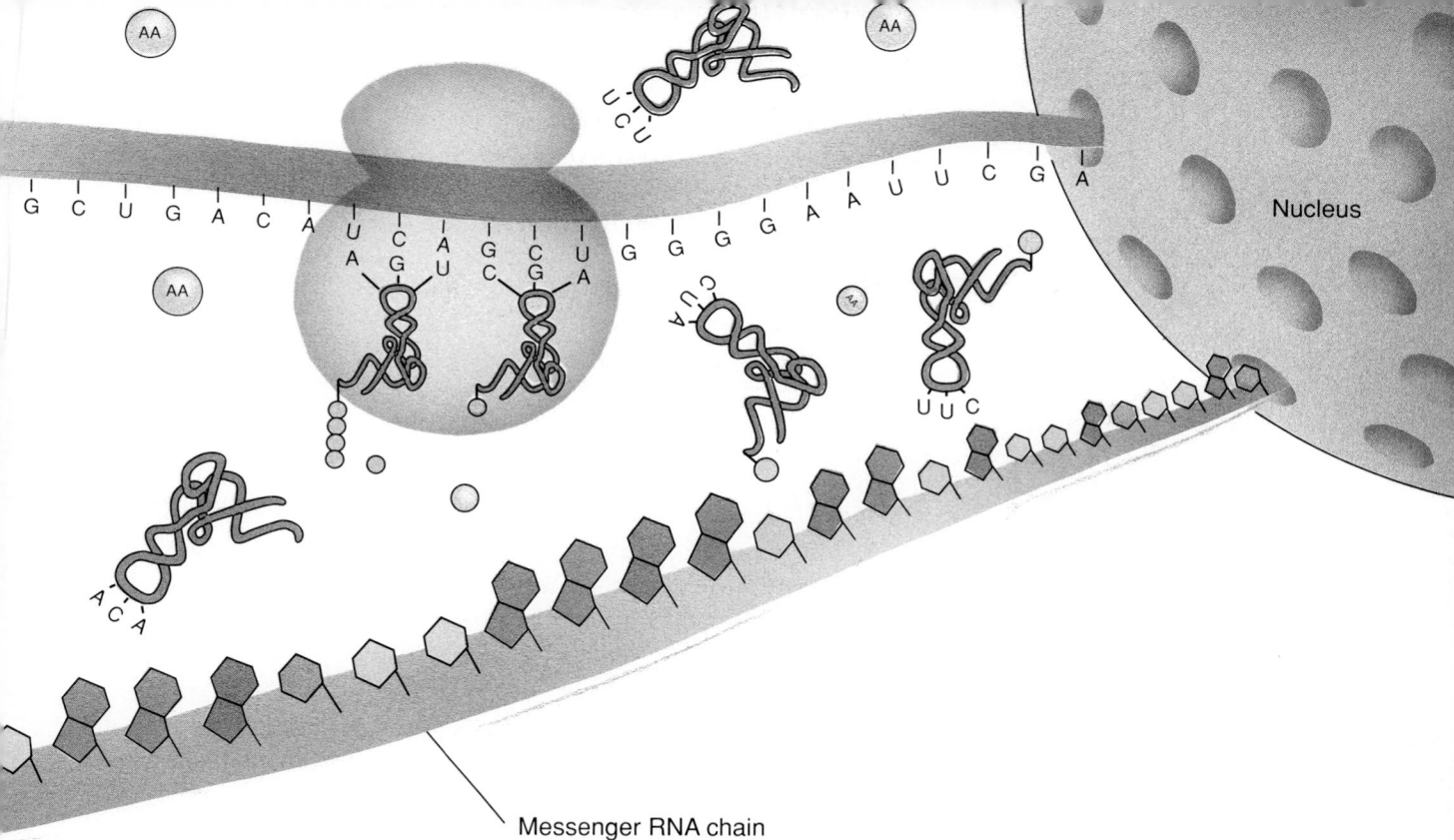

G C U G A C A U C A G C U A G G G G A A U U C G A

Nucleus

U C U

AA

AA

AA

A G
C
G
U A C
G A U
U
C
G A

U C
A C
A

C
A U

U U C

A
C
A

Messenger RNA chain

two molecules join. For example, the first codon, which acts as a start signal, is usually AUG. AUG codes for the amino acid methionine, so a tRNA molecule that carries this amino acid will have the anticodon UAC. The UAC anticodon and the AUG codon match up and join together. The tRNA with its attached amino acid remains joined to the first mRNA codon as the ribosome moves along the mRNA strand, exposing the next mRNA codon. Again, the proper tRNA with its amino acid joins the mRNA strand.

Formation of a Polypeptide   Once the first and second amino acids are in place, they bond together. The first tRNA is then released, making it available to bring another molecule of its amino acid to a ribosome. The process continues as the ribosome moves along the mRNA strand; thus, a polypeptide chain grows. Amino acids are added to a chain at the rate of about 15 per second. At the end of the mRNA strand is one of three codons—UAG, UAA, or UGA—that stops the translation process. There are no tRNA molecules with anticodons to match these codons. When the ribosome reaches one of these codons, translation stops and the completed polypeptide chain breaks away from its assembly line. The new polypeptide molecule may be ready to function in cellular processes, or it may join with other polypeptides to form a more complex protein molecule.

The translation process is basically the same for both prokaryotes and eukaryotes. Recall, however, that in prokaryotes translation may begin even before transcription is completed. Also, prokaryotes have no ER. All translation takes place on free ribosomes.

You may remember that the DNA and ribosomes found in mitochondria and chloroplasts are very similar to those of prokaryotes. Mitochondria and chloroplasts synthesize some of the polypeptides needed for their own functioning by carrying out the processes of transcription and translation in the same way prokaryotes do. However, most of the proteins needed by these organelles are coded for by nuclear genes. The proteins are synthesized on ribosomes in the cytoplasm and then move into the organelles.

Summary   The path from genes to proteins can be summarized briefly:

1. DNA stores instructions for polypeptide synthesis.
2. The DNA instructions are transcribed into mRNA, which carries the information from the nucleus into the cytoplasm.
3. Ribosomes and tRNA interact with mRNA to translate the information and build polypeptides.

Each cell of an organism has the same set of genes as every other cell in that organism. A cell doesn't make every protein for which it has instructions, however. Proteins are made to order—they are synthesized only when they are needed and in the quantity needed. This means that only certain genes are expressed at any one time. The kinds of proteins synthesized by a cell differentiate it from other kinds of cells in an organism. For example, many of the proteins made by a muscle cell are different from those made by a nerve cell.

Thinking Lab

INTERPRET THE DATA

Which parent provides the genes in organelle DNA?

Background
DNA has been found in mitochondria and chloroplasts. The DNA in chloroplasts controls traits such as the appearance of white patches in a leaf caused by a lack of chlorophyll. This trait is known as a variegated leaf.

You Try It
The following genetic crosses were performed many times, with the same results each time.

1. A variegated female plant crossed with a variegated male plant produces all variegated offspring.
2. A variegated female plant crossed with a non-variegated male plant produces all variegated offspring.
3. A non-variegated female plant crossed with a variegated male plant produces all non-variegated offspring.
4. A non-variegated female plant crossed with a non-variegated male plant produces all non-variegated offspring.

Interpret the data to determine which parent is responsible for giving the variegated trait to its offspring. Explain how the data support your answer. Draw a diagram tracing the inheritance of this trait through cell division.

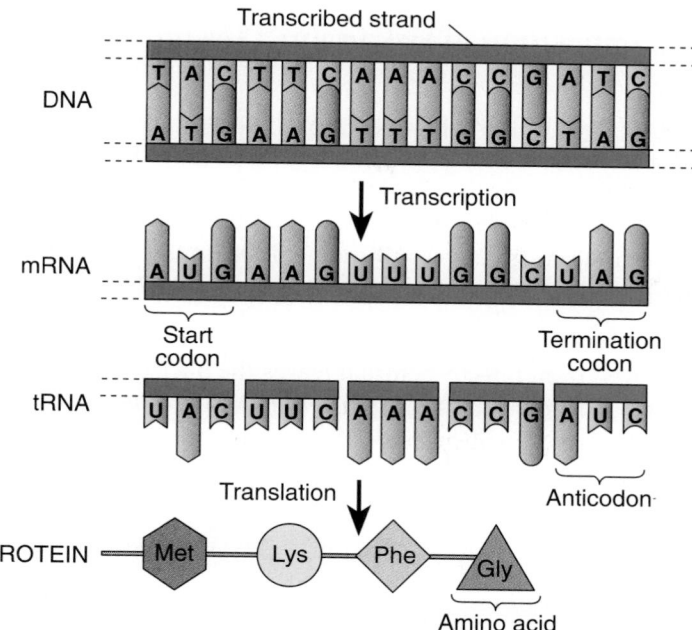

Transcribed strand

DNA

T A C T T C A A A C C G A T C
A T G A A G T T T T G G C T A G

↓ Transcription

mRNA

A U G A A G U U U G G C U A G

Start codon

Termination codon

tRNA

U A C U U C A A A C C G A U C

Translation ↓

Anticodon

PROTEIN — Met — Lys — Phe — Gly

Amino acid

Figure 10-6 When the genetic message in one of the two DNA strands is transcribed into mRNA, the resulting mRNA has codons that are complementary to those in the DNA strand. The tRNA anticodon, then, has the same base sequence as the original DNA codon, except for having uracil instead of thymine.

DNA and Phenotypes

What you have learned about DNA and protein synthesis can explain some concepts of genetics. Consider coat color in guinea pigs. Black fur (*B*) is dominant to albino (*b*). Guinea pigs having the genotype *BB* or *Bb* will be black. Guinea pigs with the genotype *bb* will be albino. Can these patterns of inheritance be explained at a chemical level?

Genes consist of DNA that codes for specific proteins. Suppose the *B* allele carries a code that directs the synthesis of a particular enzyme, an enzyme being a protein. That enzyme aids in the conversion of a substance in the cells of the guinea pig to melanin, a black pigment. Thus, a guinea pig that is *BB* or *Bb* will be able to make melanin and will be black. The *b* allele is a mutant form of the *B* allele. Its nucleotide sequence, and so its code, is slightly different. Because of the difference, the enzyme needed to make melanin is defective and cannot function properly. Therefore, a *bb* guinea pig cannot make melanin. Its phenotype will be albino.

You have seen how genetic information is expressed by transcription and translation into specific proteins, which in turn bring about an organism's phenotype. The phenotype of an individual is the result of the particular proteins that are produced as the individual's genes are expressed. Thus, all phenotypes have a chemical basis.

Figure 10-7 Phenotypes result from the combination of proteins produced by a given set of alleles. The black guinea pig has a black coat because its DNA codes for the enzymes that make the pigment melanin. The white guinea pig cannot make melanin because its DNA contains a defective gene. What are the possible genotypes of these two guinea pigs?

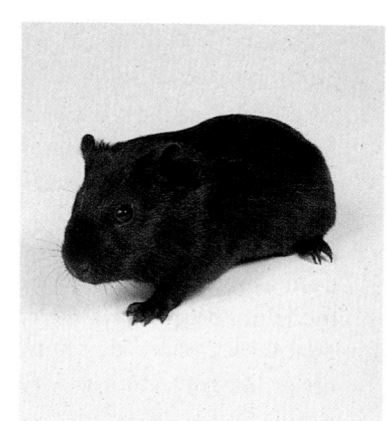

Composition of Eukaryotic Chromosomes

Using a variety of new techniques to analyze chromosomes, geneticists have continued to discover surprising facts about these structures. One unexpected finding is that the vast majority of the DNA in a eukaryotic chromosome does not code for polypeptides. Some of this DNA codes for other kinds of molecules, such as ribosomal RNAs and transfer RNAs. But much of the DNA doesn't code for any functional RNA.

Nonfunctional DNA   As you have seen, some DNA codes for introns that are removed from mRNA before it leaves the nucleus. Other DNA serves as start and stop signals for transcription. Still other types of noncoding DNA have base sequences similar to, but not exactly like, genes

INVESTIGATION

Modeling Cell Processes

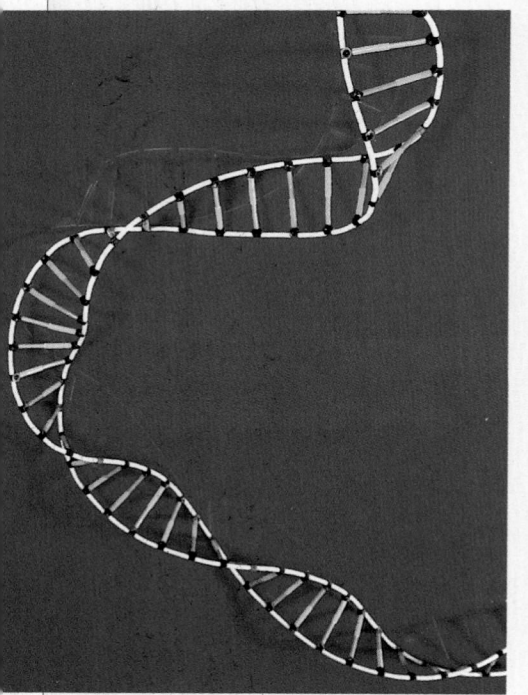

Scientists often use models to explain complex processes or to aid in better understanding of an event. The events of transcription, translation, and protein formation within a cell are difficult to see. The understanding of how and when they occur is critical to the understanding of normal and even abnormal cell functioning. Models will be designed to better illustrate these complex cellular processes.

Problem

How can models using your classmates be designed to illustrate the processes of transcription, translation, and protein formation as they occur within a eukaryotic cell?

Experimental Plan

1. As a group, make a list of possible ways you might design your models using the materials your teacher has made available. Remember, classmates are to be used as an actual part of the modeling process. HINT: A student with a paper sign hung over his or her neck can easily represent a DNA or RNA nitrogen base. Holding hands can represent molecular bonds.
2. Agree on one idea from your group's list that would be a correct model showing each process listed in the Problem section.
3. Write a numbered list of directions that could be followed to build and execute your models. You may also want to diagram the models to be used with each directional step.
4. Make a list of materials and the quantities you will need.

that code for polypeptides. These DNA regions are called pseudogenes (false genes) and do not code for functional polypeptides. And some DNA has no known function.

Repeating DNA Sequences   The picture becomes even more puzzling when we learn that each cell has multiple copies of many DNA sequences. Some of these repetitious regions consist of pairs of bases that are repeated over and over again, thousands and even millions of times. Other repetitious regions appear to be multiple copies of genes.

In some cases, the presence of repetitive DNA is an advantage to a cell. For example, there are many copies of the genes that code for tRNAs and rRNAs. Having multiple sets of these genes allows the cell to make these RNAs quickly and in great quantities, an advantage when a busy cell must synthesize proteins rapidly.

Checking the Plan

Discuss the following points with other group members to decide the final procedures for your model building.

1. How many students will actually be needed in your model?
2. Can the same student assume a second or third role in the model-building process at a later time to conserve on the number of students needed?
3. Your model may have to use all the students in your class. Make sure that the model or plan designed by your group will provide specific enough directions so that other students not in your group will be able to execute your plan with little confusion.
4. Who will be in charge of actually directing the student models?
5. Who will be in charge of explaining the events that are occurring to either the students participating or to any students observing the modeling?
6. Who will be in charge of designing and drawing any needed signs?

Make sure your teacher has approved your experimental plan before you proceed further.

Data and Observations

Carry out the model building one group at a time. For example, if there are five class groups, each group will be allowed to build and demonstrate the workings of its model. Again, realize that each group may wish to use the entire class as part of its model.

Analysis and Conclusions

1. How did your group represent DNA and how many base pairs were used? How did your group model differ from other models at this step?
2. How did your group represent mRNA and transcription? How did your group model differ from other group models at this step?
3. How did your group represent tRNA, translation, and protein or polypeptide formation? How did your group model differ from other group models at this step?
4. How did your group determine the specific amino acid to use with a particular mRNA codon? How did your group model differ from other group models at this step?
5. Based on all group presentations, explain any errors that were made and how these errors should be corrected. List suggestions that could improve your model or other groups' models.
6. Explain the role of models in science.

Going Further

Based on this activity, design other models that would illustrate (a) transcription and translation in prokaryotic cells, (b) how a mutation might alter the formation of a polypeptide, and (c) the possible sites where mutations might occur.

Figure 10-8 Only a very small percentage of the DNA in a eukaryotic chromosome actually codes for polypeptides. The rest of the DNA has other functions, or no known function at all. In contrast, about 90 percent of a prokaryotic cell's DNA codes for polypeptides.

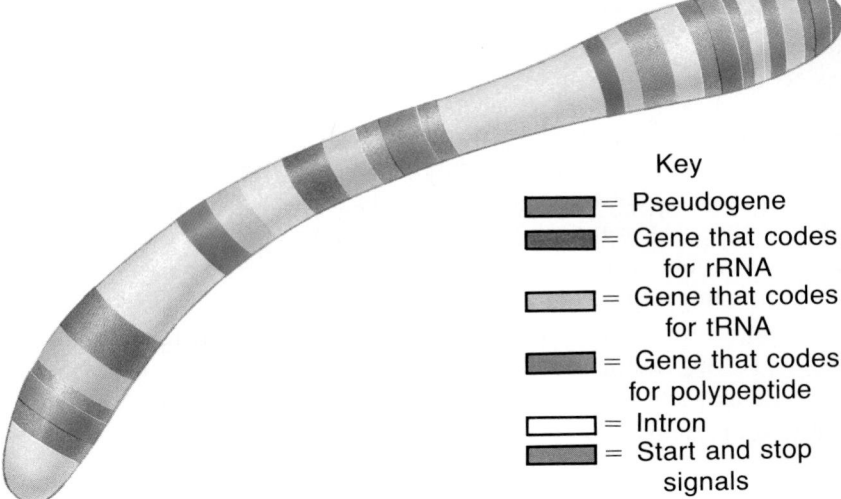

Key

= Pseudogene

= Gene that codes for rRNA

= Gene that codes for tRNA

= Gene that codes for polypeptide

= Intron

= Start and stop signals

Repetitive DNA segments represent about 30 percent of a chromosome's total DNA, while the other 70 percent consists of single copies of genes. However, most of these single copies are pseudogenes that are never transcribed. It has been estimated that only about one percent of the total DNA in a eukaryotic cell actually codes for polypeptides.

Clearly, these recent findings indicate that some of our initial ideas about heredity were far too simple. For example, we now know that there may be hundreds or thousands of alleles of some genes in a single cell, not just the two proposed by Mendel. These findings also show that much remains to be explained about the DNA of eukaryotes. Why is so much of the DNA nonfunctional? How did it arise? Why does it exist if it is not needed? Possible answers to some of these puzzling questions will be proposed in the next section of this chapter.

Section Review

Understanding Concepts

1. Why are transfer RNAs necessary for translation? What critical role do they play?
2. Suppose a chain of DNA has the base sequence GCC TAT ACA CCT. Describe transcription of that chain and list the sequence of nucleotides in the mRNA chain produced.
3. A fertilized frog egg produces an mRNA molecule consisting of 2000 nucleotides. That same mRNA molecule, when it reaches the ribosomes, has only 951 nucleotides. Explain the difference in the size of the two mRNA molecules.

4. **Skill Review—Observing and Inferring:** You are examining two different types of cells. One type of cell is part of a gland specialized for producing an enzyme needed for digestion. The other is part of a tissue specialized for transmitting nerve impulses, a process that does not involve proteins. In which type of cell would you expect to find more RNA? Why? For more help, see Thinking Critically in the **Skill Handbook.**

5. **Thinking Critically:** Suppose that, in a certain plant, the gene for green seeds, G, is dominant to the gene for colorless seeds, g. In terms of DNA function, explain why the genotype GG or Gg results in green seeds, while the genotype gg results in colorless seeds.

10.2 Sources of Variation

You probably copy homework assignments off the board each day in school. Have you ever copied one incorrectly? Perhaps instead of copying an assignment that called for answering questions 1-15, you misread the 5 and wrote a 3 instead. When you did your homework, you answered questions 1 through 13, leaving out two of the assigned questions. DNA is usually replicated correctly, but mistakes do occur. Some mistakes arise during the copying process, while others result from external factors. Such mistakes can have important consequences for organisms and their offspring.

Objectives

Infer several ways in which genetic variation may arise.

Sequence the events in a cell that lead to cancer.

Relate techniques used in recombinant DNA technology and the uses of that technology.

Mutation

The mechanisms that control transfer of hereditary traits from a cell to its daughter cells usually produce accurate results. However, when mistakes in the replication of genetic material occur, a daughter cell will contain genetic material different from that of its parent cell. Such mistakes are called **mutations,** and can arise from changes to individual genes or changes in the arrangement of genes on a chromosome.

Gene Mutations   Many gene mutations arise spontaneously as DNA is replicated. Although most replication errors are corrected, some escape detection. In one type of gene mutation, called a **deletion,** a nucleotide is left out. As Figure 10-10 shows, leaving out a nucleotide shifts the reading of the DNA message over by one nucleotide, resulting in synthesis of a polypeptide with an altered amino acid sequence. The same result can occur with an **insertion,** a type of mutation in which an extra nucleotide is added to a chain during replication. Deletion and insertion mutations

Figure 10-9   Mutations occur in all organisms. Curly wings in *Drosophila*, the short legs of a bassett hound, seedless fruit such as oranges and grapes, and a protruding "Hapsburg" lip in humans each arose as the result of a mutation.

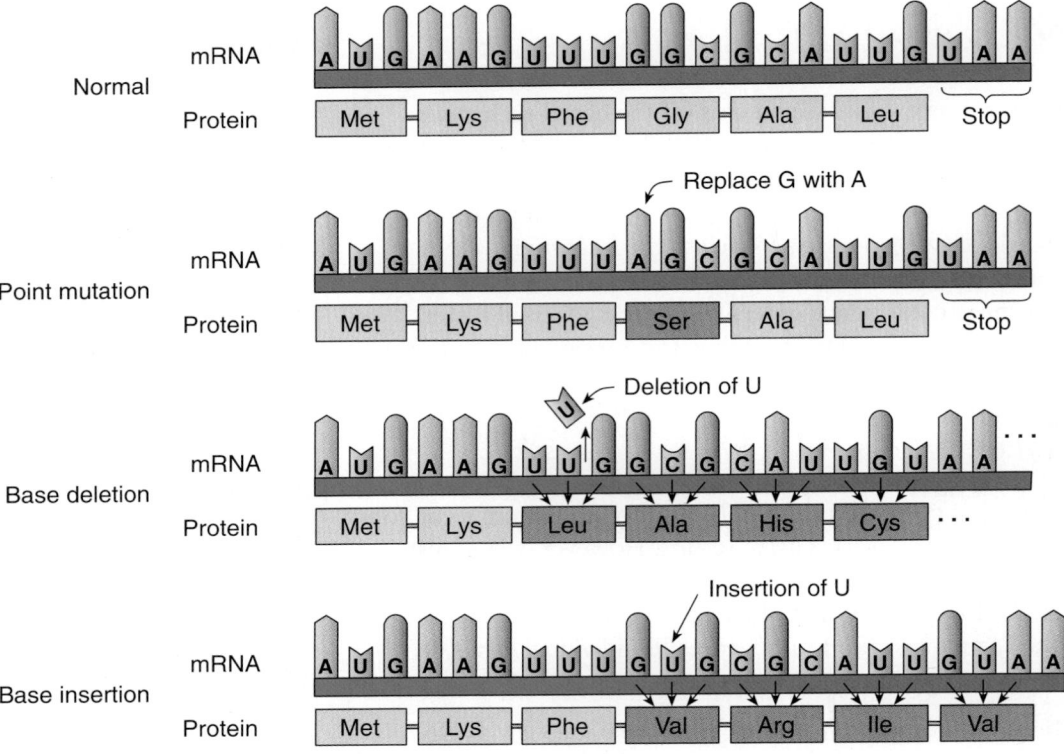

Figure 10-10 The protein formed as a result of a point mutation may be able to function because only one amino acid is affected. Deletions and insertions, however, affect all amino acids that are coded beyond the point of the mutation. The effect on the protein is usually disastrous. If three nucleotides are inserted or deleted, the protein will probably be functional. Why?

result in amino acid sequences that are very different from those coded for by normal genes because all the amino acids in the polypeptide beyond the point at which the mutation occurred will be affected. The polypeptide produced by the mutated gene, if one can be produced at all, is usually not functional.

In a **point mutation,** a base substitution occurs, and one nucleotide is substituted for another. This type of gene mutation affects only one amino acid in the polypeptide. With such a small change in its structure, the polypeptide may be able to function normally or nearly normally.

Deletions, insertions, and point mutations all result spontaneously from errors in copying. Gene mutations may also result from exposure to environmental factors, such as radiation, high temperature, and a variety of chemicals. External agents that cause mutations are called **mutagens.** Many mutagens exert their effect by breaking the chemical bonds of DNA molecules or by causing unusual bonds to form. Some of these mutagens are known to play a role in the development of many forms of cancer.

Once a gene mutation has taken place, regardless of its cause, the altered gene is copied as if it were correct. After all, DNA replication is a predetermined series of chemical events. A cell has no way of knowing that it is copying a mistake. Mutations that occur in a body cell of an organism may damage the organism or perhaps even cause its death, but they are not important to the organism's descendants. However, mutations that occur in the sex cells may affect an entire population of organisms, because the information can be passed from one generation to the next.

New Alleles It may have occurred to you that gene mutations result in new forms of genes, or, in other words, new alleles. Polypeptides produced from the transcription and translation of new alleles are often so different that they are useless or even harmful to the cell. Rarely is a new allele immediately beneficial to an organism, and most new alleles have little effect at the time they occur. They are also usually, but not always, recessive to the original allele, especially if they produce a useless product or none at all. The normal allele can produce enough useful product to mask the presence of the useless product. Because most organisms are diploid, even harmful or lethal recessive alleles may be passed to offspring and gradually spread through a population. This can happen because the effect of the recessive allele is masked by the dominant allele. Over time, many new alleles can spread through a population, adding to a storehouse of variety that may eventually become important to the species.

Chromosome Mutations Changes may occur in chromosomes as well as in genes. Chromosome mutations occur in several different ways. For example, parts of chromosomes can be broken off or lost during mitosis or meiosis. Sometimes, these broken parts rejoin the chromosome from which they came, but they may attach backwards or at the wrong end of the chromosome. Also, they may reattach to a different chromosome. Any of these changes can result in abnormal information in the cell's genetic material. Because each chromosome consists of many genes, chromosome mutations can have serious effects on the distribution of genes to gametes during meiosis. Gametes that should have complete haploid sets of genes may end up with double doses of some genes and a complete lack of others. Most chromosome mutations are not passed on, either because the zygote fails to develop, or, if it does, the mature organism is incapable of producing offspring.

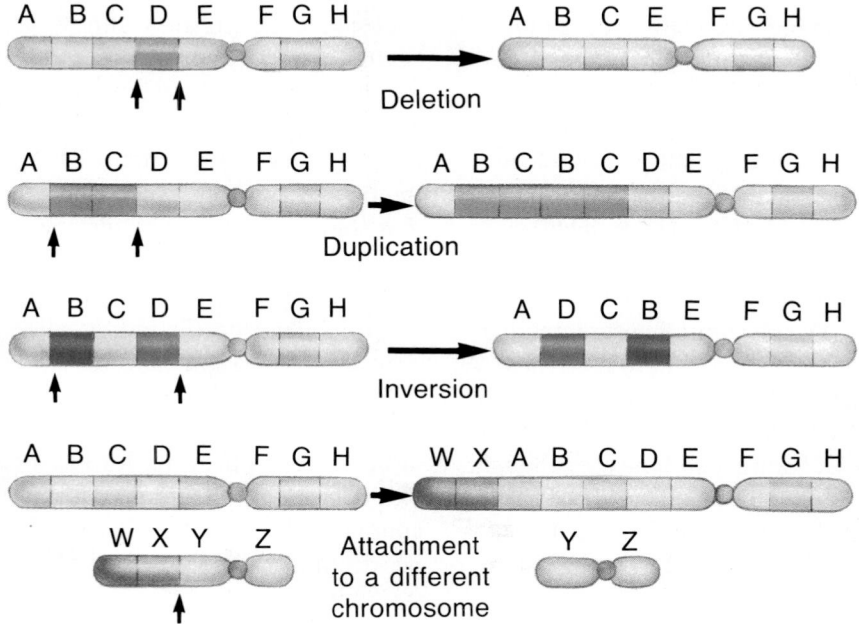

Figure 10-11 **Chromosomal mutations occur in many different ways. A deletion removes a segment of a chromosome, a duplication repeats a segment, and an inversion is caused by breakage followed by attachment backwards. Breakage and attachment of the fragment to another chromosome can also occur.**

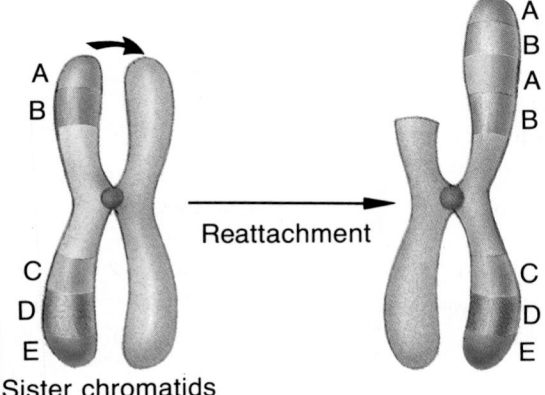

A
B

→ Reattachment →

A
B
A
B

C
D
E

C
D
E

Sister chromatids

Figure 10-12 **When a piece of a chromatid breaks off and attaches to its sister chromatid, duplication of genes occurs.**

Other Causes of Genetic Variation

You have learned that a cell has multiple copies of many DNA sequences and that some of these sequences are duplicate genes. New genes can arise when these duplicate genes undergo slight changes in their nucleotide sequence. Because the new gene is quite similar to the original, it is likely to code for a polypeptide that may be able to function in a new way.

Types of Gene Duplication   Gene duplication may occur in different ways. During meiosis, homologous chromosomes may fail to segregate properly, causing both homologues to move to the same pole. This failure to separate can lead to production of an offspring having an entire extra chromosome. That offspring would have duplicates of all the genes on that chromosome. Another type of duplication occurs when a piece of a chromatid breaks off and attaches to its sister chromatid, as Figure 10-12 shows. One chromatid would have several duplicate genes. If this duplication occurred during meiosis, one of the offspring would inherit those extra copies.

You have doubtless pictured genes as fixed sections of DNA, located at particular points along a chromosome. That is usually true, but as Barbara McClintock discovered four decades ago, some genes can actually move from one area of a chromosome to another or from one chromosome to another. Such genes, called jumping genes, often leave a copy of them-

HISTORY CONNECTION

Jumping Genes

It was not until almost 40 years after Barbara McClintock first discovered what she called the jumping gene that the scientific community recognized its importance to genetic theory. McClintock began her research in the 1940s by studying the effects of crossing corn plants that had different kernel colors and textures. She observed kernel color patterns that could only be explained if genes moved around on chromosomes. This was a star-

tling interpretation. At the time, genes were thought to be fixed at a specific place on a specific chromosome. The results of McClintock's experiments led her to conclude that some genes, instead of being fixed, move around on a chromosome, affect each other's functionings, and cause massive mutations.

Dr. McClintock presented her experimental data to her fellow scientists. They initially did not accept her conclusion because it was contrary to how they thought genes should behave. Then, in the late 1960s, jumping genes were discovered in fruit flies, yeast, and

bacteria. Scientists are just now beginning to realize how important her work was and how jumping genes bring diversity to a species. Jumping genes have actually led to the further study and suspicion that some movable genes may be the descendants of viruses. Jumping genes may even be responsible for producing specific tumor cells that result in certain cancers.

Because of the importance of her work, Barbara McClintock was awarded the Nobel Prize for Physiology and Medicine in 1983. She continued her work with corn until her death in 1992 at the age of 90.

selves before jumping to another position. Thus, the cell would have two or more copies of the same gene. Some of those copies could later undergo slight change, resulting in new functional genes.

Importance of Gene Duplication   The importance of gene duplication as a source of new genetic variation has yet to be determined. There is evidence, though, that it does occur. For example, the form of hemoglobin produced by a human fetus is different from that produced by the adult. The genes that code for the two forms of hemoglobin are located close to one another on the same chromosome, and are only slightly different from one another. One may have arisen by duplication from the other, and changed slightly over time so that it now codes for a slightly different polypeptide chain.

Creation of new genes as a result of duplication would explain some of the puzzling features of the DNA of eukaryotic chromosomes. It would certainly explain the many copies of certain genes, such as those that code for rRNA. It may also explain the presence of pseudogenes. It is likely that pseudogenes are duplicate genes that have been slightly modified so that they no longer code for functional polypeptides.

A slightly different result of duplication may be another source of new genes. It is the exons of a gene that code for amino acids. Each exon can be thought of as a set of nucleotides that codes for a series of amino acids. It is possible that several exons from different genes along a chromosome could be duplicated, then move to and become inserted into another chromosome. Putting exons together in new combinations would form new sequences of nucleotides resulting in codes for new polypeptides.

1. **Explore Further:** Find out what a virus is and how its genetic material differs from that of living things.
2. **Writing About Biology:** Find out how tumor cells that are not cancerous differ from those that are. Write a short report.
3. **Explore Further:** What evidence led Dr. McClintock to conclude that genes could jump from one chromosome to another?

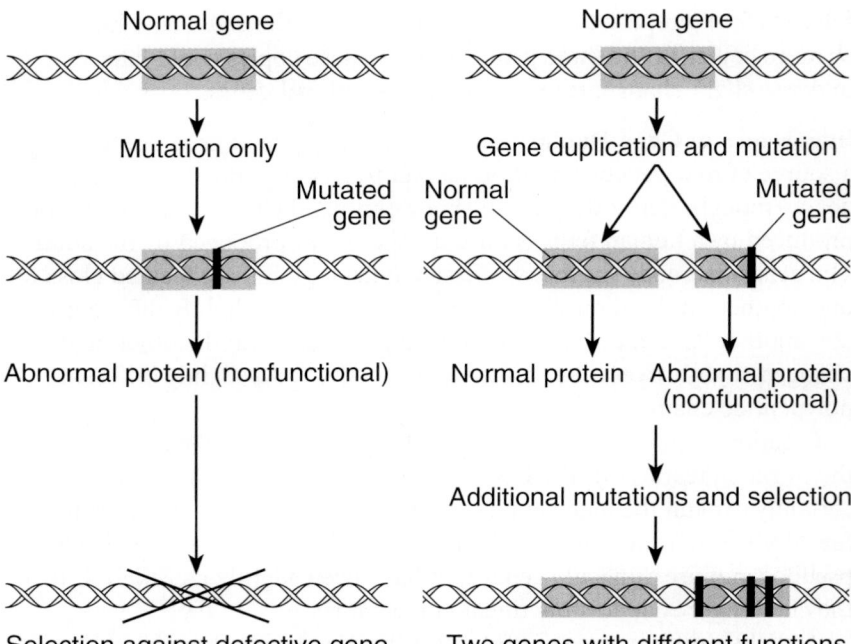

Figure 10-13 Duplication of a gene followed by random mutations and natural selection can lead to two genes with different functions. In the absence of gene duplication, the mutation on the left would be lethal if it produces a nonfunctional protein. With a duplicated gene, the same mutation that would otherwise be lethal is not disadvantageous, because the original gene continues to produce the normal protein. With time, several mutations can accumulate in the duplicated gene.

Normal gene

Mutation only

Mutated gene

Abnormal protein (nonfunctional)

Selection against defective gene

Normal gene

Gene duplication and mutation

Normal gene   Mutated gene

Normal protein   Abnormal protein (nonfunctional)

Additional mutations and selection

Two genes with different functions

You read in Chapter 8 that meiosis and fertilization are processes that reshuffle alleles, leading to genetic variation. You now know that new alleles may be produced by mutation and new genes by duplication of previously existing genes and parts of genes. Over time, the new genes and alleles may spread through a population. The continual production of new alleles and their reshuffling during meiosis and fertilization provide for genetic variation among the individual organisms in a population. The production of new functional genes provides for the evolution of greater complexity.

Cancer

Earlier you studied the cell cycle and learned of the events that control cell growth and reproduction. Although the cell cycle usually results in production of normal cells having the proper genetic information, sometimes the cycle is altered and abnormal cells are produced. These abnormal cells differ in shape from normal cells and have unique proteins in their membranes. Some of these cells have unusual numbers of chromosomes, and many have abnormal chromosomes that have been produced through chromosome mutations. The abnormal cells reproduce, forming more cells like themselves. Unlike other mature cells, many can migrate around the body and take hold in other organs. Such cells are said to be malignant, or cancerous. They reproduce in an uncontrolled way, eventually crowding out normal cells and destroying the tissues and organs to which they have spread. Because almost any body cell can become malignant, cancer is not a single disease. Rather, there are many forms of cancer, depending on which type of body cell has become malignant.

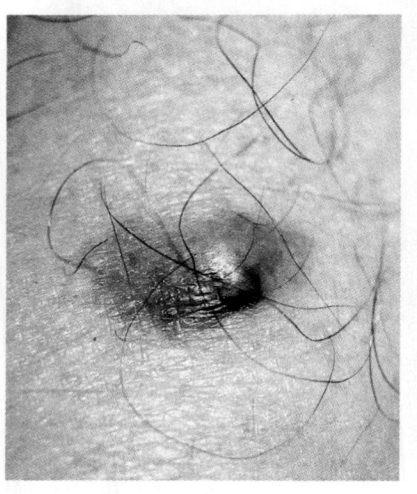

Figure 10-14 Melanoma is a deadly type of skin cancer that is caused by excessive exposure to sunlight.

Causes of Cancer: Mutation   What causes normal cells to be transformed to malignant ones? It is now known that genetic changes are ultimately to blame. Genes that control the normal events of the cell cycle mutate. When such mutations occur, normal cell cycle controls no longer operate. The mutations may be inherited, or they may be triggered by a **carcinogen,** a cancer-causing agent in the environment. Tobacco smoke, X rays and other forms of radiation, and a variety of chemicals are known carcinogens. No matter how the mutation occurs, the genes that mutate are ones that code for proteins involved in controlling normal cell growth and reproduction. When such a gene mutates, it becomes a cancer-causing gene, called an **oncogene.** Because mutations are passed on, descendant cells inherit the oncogene, and a tumor, a large mass of cancer cells, forms. Several genetic changes must occur to transform a healthy cell to a cancerous one, and some of the steps involve oncogenes. Each change probably affects the cell cycle in a different way.

Causes of Cancer: Viruses   Some viruses are known to cause cancer in animals. These viruses may cause cancer simply because their presence in the animal cell disrupts the function of normal genes. Viruses can also cause cancer if they carry an oncogene into an animal cell when they infect the cell. The oncogene becomes part of the animal cell's chromosome, where it disrupts the normal events of the cell cycle, causing cancer.

HEALTH CONNECTION

Dr. Jane C. Wright

You could say that practicing medicine was in her genes, but after reading this chapter and the next, you'll know that isn't true. But it was a very distinguished and powerful heritage that pulled Jane Wright (1919-) into medicine. Her father, an eminent surgeon, was one of the first African-American graduates of Harvard Medical School, and her grandfather was an early graduate of Meharry Medical College, a school for former slaves.

After practicing for several years as a family doctor, Dr. Wright was offered the opportunity to do cancer research and began the work that has brought her national and international fame. Her research into chemotherapy, the use of chemicals to treat cancer, has made her an international authority on the subject.

Chemotherapy is often the last resort in the treatment of cancer, after surgery and radiation therapy fail. In chemotherapy, various drugs are used, either alone or in combination, to try to kill or stop the spread of cancer cells. Every year, new chemotherapy drugs are developed. Dr. Wright tests these drugs on cancer cells in test tubes and in animals to see if the drugs are effective in killing the cancer cells. If a drug shows promise, it is tested in humans and eventually made available to doctors for use in treating cancer.

Dr. Wright has published more than 100 articles on cancer chemotherapy. She believes chemotherapy treatments are becoming more effective with the development of new and more powerful drugs. But she thinks new drugs still need to be discovered, and perhaps she will be the researcher to do it.

1. **Explore Further:** Find out how surgery and radiation therapy are used in the treatment of cancer.
2. **Solve the Problem:** Why are increasingly more powerful drugs dangerous to use in chemotherapy?

Cancer researchers have begun to identify different oncogenes, and they are attempting to determine the proteins for which the oncogenes code. When these proteins are discovered, their effects on the cell cycle will be analyzed. Knowing the effects of these abnormal proteins will also further understanding of proteins that control the cell cycle in a normal way. All this knowledge will perhaps lead to more effective treatment and cure of cancers.

Variation Through Technology—Recombinant DNA

In Chapter 9, you learned that discovery of DNA as the genetic material of all organisms came from a series of experiments in which genetic transformation took place. Transformation is one of the ways genetic variation arises in bacteria. It occurs when a bacterial cell absorbs a gene from another cell. The cell that gains the gene exhibits the phenotype of the cell that donated the DNA.

Scientists have made practical use of their knowledge of transformation by learning how to transfer a gene artificially from a donor organism to a bacterium. The DNA of the bacterium that has been altered by insertion of a foreign gene is called **recombinant DNA.** The bacterium transcribes and translates the foreign gene to produce the protein for which the foreign gene codes.

Plasmids   How is this genetic engineering accomplished? The cytoplasm of a bacterium contains, in addition to its main chromosome, small circular pieces of DNA called **plasmids.** Plasmids are the structures with which the foreign DNA is recombined. The technique involves making many recombined plasmids and putting them together with bacteria that will

Figure 10-15   Using recombinant DNA techniques, *E. coli* bacteria can be engineered to produce a desired protein such as human insulin. How are recombinant plasmids produced?

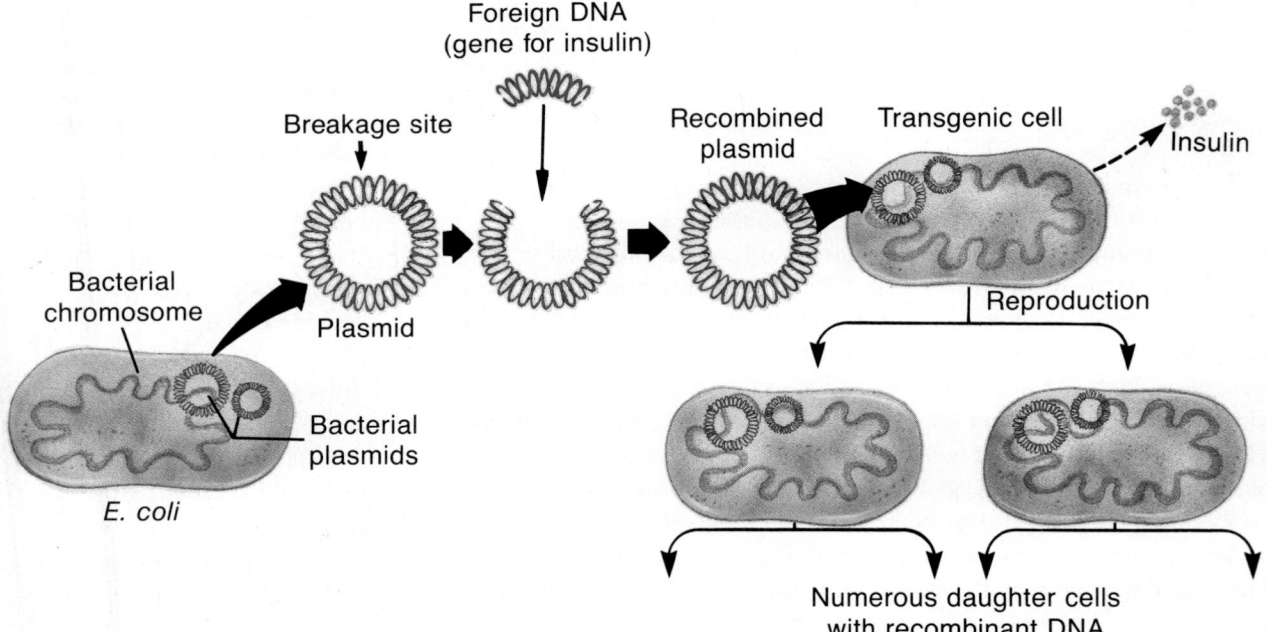

absorb them. The bacteria that take up the recombined plasmids will begin to produce large quantities of the protein for which the foreign gene codes.

For example, insulin, a hormone important in the treatment of a form of diabetes, is produced commercially this way. Recombinant DNA techniques have enabled researchers to produce this protein in purer form and in much larger quantities than was formerly possible.

Obtaining the Foreign Gene How is the foreign gene obtained? There are several approaches, but one of the most reliable is to synthesize it. First, a cell that specializes in producing the desired protein must be found, because such a cell will contain many mRNA molecules that carry the genetic code for the protein. These mRNA molecules are identified and isolated. With the help of an enzyme obtained from a virus, a single-stranded DNA molecule is produced from the mRNA molecule by a process that is the reverse of transcription. The single-stranded DNA is converted to double-stranded form, using an enzyme normally used by cells for DNA replication. This double-stranded DNA is the desired gene. Next, thousands of copies of the gene must be made. Making many copies of a gene by these methods is known as gene cloning. **Clones** are genetically identical copies. You will learn more about clones when you study reproduction.

Inserting the Foreign DNA Before the foreign DNA can be inserted into the plasmid, the plasmid must first be broken open at a specific point and changed from circular to linear form. The circle must be broken at that point where the nucleotides of the plasmid's DNA will match the nucleotides of the foreign DNA. This is accomplished using enzymes that are known to break DNA only at specific points. These enzymes are used on both the plasmid and the foreign DNA. The two segments of DNA with matching ends then join, and the plasmid regains its circular shape. The recombinant plasmids are then placed into a liquid medium containing billions of bacterial cells and chemicals that favor the cells' absorption of the plasmids. A few bacteria take up the plasmids and, if the recombinant DNA has located properly, begin to produce the desired protein, which they secrete into the medium. The protein is extracted from the medium, purified, and put to use.

Applying Recombinant DNA Techniques Although early work in recombinant DNA was limited to bacterial cells, recent advances have enabled geneticists to develop modified techniques that can be applied to eukaryotic cells as well. In some cases, foreign genes are inserted into whole organisms. In other cases, the genes are inserted into isolated cells grown in a culture medium. For example, scientists recently inserted into fertilized goat eggs a human gene for a protein that dissolves blood clots. The recombined fertilized eggs were then implanted into female goats for development. Two of the eggs picked up the foreign gene and developed into baby goats, one male and one female. These animals are called transgenic organisms because they contain genetic material from two species. The transgenic female goat secretes the clot-dissolving protein in her milk.

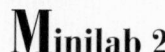

Minilab 2

Do biotechnology products help us?

The enzyme alpha-amylase can be recovered efficiently through means of biotechnology from a fungus. This enzyme is used to speed up the chemical change of starch to glucose. Test a starch solution for the presence of glucose using Benedict's solution. Combine starch with alpha-amylase, wait 15 minutes, and test again for glucose. Test a sample of the enzyme alone for glucose. Explain your results. Offer a possible commercial use for this enzyme, knowing that starch is not sweet while glucose is.

Analysis *How might the recovery of an enzyme from a microorganism illustrate biotechnology?*

Figure 10-16 **This bacterial DNA plasmid contains genes that make bacteria resistant to certain antibiotics. Magnification: 50 000 ×**

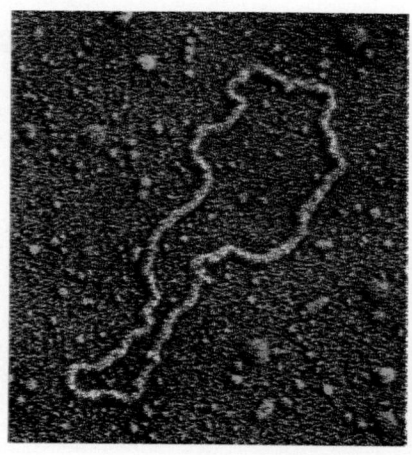

Scientists have produced transgenic plants that resist disease and spoilage. Special recombinant DNA techniques have been needed for use with plants because the tough cell walls of plant cells make it difficult for the foreign genes to be inserted. Foreign genes can be inserted into these cells by one of several methods. In one method, the foreign genes are inserted into a bacterium or a virus, which infects the plant cells and transfers the gene to the plant's genetic material. In a second method, "bullets" of metal particles coated with foreign DNA are shot into the plant cells using a device known as a "gene gun." Other methods punch temporary holes in the cell wall or dissolve the cell wall altogether. In any of these ways, the cells that successfully pick up the foreign gene may reproduce and develop into mature plants. Newly developed techniques such as these hold promise for the production of transgenic crops that are both hardier and more nutritious than conventional crops. The techniques will also enable people to produce more food at less expense.

Transgenic organisms are used not only to produce a wide variety of desired proteins, but also as a basic research tool to study genes and their control mechanisms. Recombinant DNA techniques will doubtless continue to yield benefits that will improve health, alter food crops, and benefit the economy.

LOOKING AHEAD

YOU HAVE now explored how the genetic information in DNA is used to synthesize polypeptides and how DNA is related to expression of phenotypes. In the next chapter, you will relate this information to human genetics and the causes and symptoms of hereditary disorders. Geneticists are using knowledge of DNA and protein synthesis in their efforts to diagnose, treat, and cure those disorders.

Section Review

Understanding Concepts

1. Explain how a gene mutation can be lethal, yet be passed on from one generation to another. How can such mutations prove lethal to a later generation?

2. Which would likely be more of a problem—an insertion mutation that occurs near the beginning of a gene or toward the end of a gene? Why?

3. Many fertilized eggs begin, but never complete, development. These eggs usually contain abnormal genetic material. Which is more likely to be responsible, a gene mutation or a chromosome mutation? Explain.

4. **Skill Review—Sequencing:** Sequence the steps used in recombinant DNA work with bacteria. For more help, see Organizing Information in the **Skill Handbook**.

5. **Thinking Critically:** Why would a new polypeptide produced as a result of the rearrangement of exons be more likely to be functional than a polypeptide produced by a random gene mutation?

Biotechnology

DNA Fingerprinting—A New Weapon Against Crime

In violent crimes such as murder and rape, physical evidence is often left at the scene of the crime. Blood, skin, hair, or sperm cells may be the only evidence that the police have. A suspect can be arrested and found guilty long after a crime has been committed. How do police use the cells they have found to prove that a suspect is guilty? They are able to identify the DNA in the cells as having come from the suspect, using a technology known as DNA fingerprinting. How does DNA fingerprinting work and how is it used?

Why DNA Fingerprinting Works
Genetically, we are what we are because of our DNA. We are all human because we all have 46 chromosomes made up of sequences of DNA. Half of these sequences come from our mother and half from our father. Every one of us is distinct because each new combination of egg and sperm is different. Thus, except for identical twins, every human has DNA that is different and unique. But because the original sequences come from our parents, our DNA will show similarities to our parents' DNA.

DNA fingerprinting takes advantage of the differences in base sequences in individuals. There are about 3 billion base pairs that make up the DNA of a human. Most of the DNA, such as the genes that control cellular respiration and other life processes, are common to all humans. However, some areas of DNA, the repetitive noncoding sequences found between genes, are highly variable. It is this variability that makes the DNA of each person unique.

The Technique DNA, both from the cell sample found at the crime scene and from a suspect's blood cells, is first isolated and purified. The DNA is then treated with enzymes that chop it up at specific places in the molecule. This produces about one million DNA fragments of differing lengths. The fragments are separated according to size by electrophoresis, a technique that uses an electric field to separate molecules or charged particles embedded in a jellylike material.

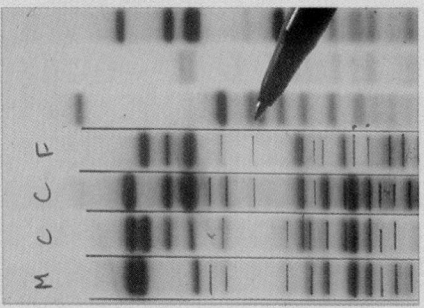

After separation, the fragments can be identified by treating them with various radioactive probes. These probes are pieces of radioactively labeled, single-stranded DNA with base sequences that may be complementary to those in the repeating sequences of the samples. If a probe matches the repeating sequences in the DNA being tested, the two pieces of DNA bind together. The location of the DNA fragments that have bound to the probe can be detected because they are radioactive. A pattern of bands is produced that resembles the bar codes you see every day on groceries and other products. It is this bar code pattern that is the DNA fingerprint.

Analysis The bar code produced by the DNA found at the scene of the crime is compared with that of the DNA from the suspect's blood cells. If the patterns match, the two samples of DNA most likely came from the same individual. The probability of two individuals having matching DNA fingerprints has been calculated to be about one chance in 9.34 billion—about double the present population of the world. Therefore, it is almost a certain conclusion that the suspect is guilty.

How Is DNA Fingerprinting Used?

You have just read how DNA fingerprinting can be used to prove that a particular suspect is guilty of a crime. The technique can also be used to prove that a suspect is innocent. If the suspect's DNA fingerprint does *not* match that of the evidence, he or she is almost assuredly innocent.

DNA fingerprinting has several other uses. It has been used to determine paternity by comparing the DNA of a child, its mother, and two or more possible biological fathers. It also can be used to establish family relationships between individuals living in different countries. These family relationships are important in solving immigration disputes. And finally, DNA fingerprinting has been used in studying animal populations to determine the amount of genetic diversity in the population. These data should be useful in studying ecology and evolution of animal populations.

1. **Issue:** Discuss what you think the proper role for DNA fingerprinting should be in court cases. Should it be the deciding evidence in cases that involve the death penalty? What about cases of assault or rape? What about cases that establish paternity?
2. **Connection to Health:** To treat certain diseases, some patients require transplants of organs or bone marrow. One problem with transplants is that the body will reject those whose cell surface proteins do not closely match the recipient's. Hypothesize about how DNA fingerprinting might be used to treat these patients.
3. **Connection to Government:** The Fifth Amendment to the Constitution of the United States prohibits forcing people to testify against themselves. Discuss whether using a person's blood or other body part that was obtained involuntarily as evidence constitutes a person testifying against him or herself.

Issues

Dying with Dignity: Who Decides?

The 85-year-old woman was rushed to the hospital on New Year's Day. She could not breathe, so doctors placed her on a respirator. After five months of trying to wean her from the machine that was helping her breathe, her doctors transferred her to a chronic care facility. In only a few weeks, however, the woman was taken back to the hospital when her heart stopped. Doctors managed to restart her heart, but the patient had lapsed into a coma that doctors believed was permanent.

The woman was connected to many machines that breathed for her, fed her, and kept her organs functioning. She lay in her hospital bed with tubes and wires attached to her body. After several weeks, doctors became convinced that her treatment was doing no good and that she would never recover. Yet, her family refused to allow doctors to remove the machines.

After many months, the hospital went to court, asking for an independent opinion on whether treatment should be continued. A hearing was held, and the court decided that the husband should make medical decisions for his incapacitated wife. The machines remained, and the woman finally died 18 months after her illness began.

Medical Futility This case illustrates the concept of medical futility. Are doctors morally obligated to administer treatment that they believe will not benefit the patient? For example, if a patient suffering from untreatable lung cancer needs a kidney transplant because of kidney failure, must a doctor perform one? Or would a transplant be futile? Some people might argue that a transplant would prolong life. Others might say that, in the end, the patient would still die of cancer. In our society, doctors are not obliged to do what will be futile, but legal decisions have shown that futility is difficult to define.

Who Decides? If the woman in the above case had been conscious, she could have communicated her wishes about whether life-sustaining treatment should be used to prolong her life. It is when a patient is no longer mentally competent to make such decisions that problems arise. In December, 1991, the first United States law to address this issue went into effect. This law, the Patient Self-Determination Act, requires all hospitals and other medical care facilities to inform patients of the laws of that state regarding the making of advance health care decisions before patients are no longer capable of making the decisions themselves.

There are two ways a person can plan in advance in case such medical decision-making becomes necessary. First, a person can write a living will, in which he or she makes a statement outlining preferences for treatment or nontreatment under different conditions. Second, a person can sign a power of attorney for health care, in which the decisions are entrusted to someone who knows the person well and knows the person's feelings.

Assisted Death Some terminally ill patients feel that they should have control over the conditions of their death. They feel that prolonging their life by technological means is inhumane and robs them of their right to die with dignity. Many of these patients have asked their doctors to assist them in dying.

Assisted death, called euthanasia, can be either passive or active. Passive euthanasia is the withholding of life-prolonging treatment. If a doctor does not attempt to resuscitate a terminally ill patient whose heart has stopped beating, the doctor is practicing passive euthanasia. On the other hand, if the doctor administers a lethal dose of a drug to end a patient's painful life at the patient's request, that is active euthanasia. Many people concerned with the right to die feel that, if a doctor knows the patient's feelings well and knows that the patient is competent to make this decision, active euthanasia should be allowed. Other people believe that active euthanasia is "killing," whereas passive euthanasia is "letting die." At the present time, active euthanasia is illegal in the United States.

Active euthanasia has been performed in The Netherlands since the 1980s. Although it is illegal, the practice has been tolerated as long as doctors follow strict guidelines. A British medical institute recently published a report in which it appeared to endorse active, as well as passive, euthanasia. In the United States and elsewhere in the

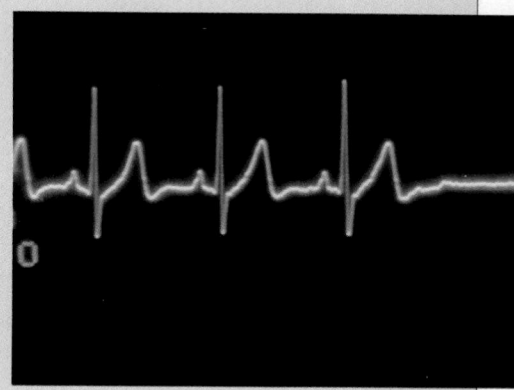

Western world, passive euthanasia is widely accepted. However, active euthanasia is a morally, politically, and legally debated issue. As the right-to-die movement becomes more widespread, laws no doubt will change.

1. **Understanding the Issue:** Should a person have the right to decide when to die? Explain. Some state laws dictate what must be done for terminally ill patients who cannot make decisions for themselves. Discuss why such laws are or are not desirable.

2. **Writing About Biology:** Research the idea of a living will. What is its purpose? Write a living will for yourself that clearly states what you would like done without unnecessarily restricting a doctor or relative from using his or her judgment in an uncertain situation.

Readings

- Gustaitis, R. and E.W.D. Young. *A Time to be Born, a Time to Die.* Reading, MA: Addison-Wesley, 1986.
- Orentlicher, David. "The Right to Die After Cruzan." *The Journal of the American Medical Association,* Nov. 14, 1990, p. 2444.

Summary

Messenger RNA carries information from the DNA in the genes to the sites of protein synthesis in the cytoplasm. The DNA information carried by each gene on the many chromosomes of a eukaryotic cell, or on the single chromosome of a prokaryote, is first transcribed into mRNA. Transfer RNA then acts as an interpreter to translate the nucleic acid language into the protein language, and a polypeptide is synthesized at the ribosomes.

Polypeptides are synthesized by a cell only when they are needed and only in the quantity needed. The phenotype of an organism is determined by the polypeptides produced by its cells, which in turn is determined by its DNA. Only a tiny fraction of the DNA of eukaryotes carries information for polypeptides.

Mutations can occur in single genes or in chromosomes and may occur spontaneously or as a result of external agents. In addition to random mutations, which produce new alleles, entirely new genes may arise from duplication and modification of pre-existing genes or by rearrangement of gene parts. Production of new alleles and genes adds to the variations present within a population.

Cancer is a disease that is caused by mutations in the genes that control the cell cycle. Viruses play a role in causing some kinds of cancer.

A variety of recombinant DNA techniques can be used to mass-produce quantities of useful proteins. These techniques are also important tools used to study cell mechanisms.

Language of Biology

Write a sentence that shows your understanding of each of the following terms.

anticodon	mutation
carcinogen	oncogene
clone	plasmid
deletion	point mutation
exon	recombinant DNA
insertion	ribosomal RNA (rRNA)
intron	transcription
messenger RNA (mRNA)	transfer RNA (tRNA)
	translation
mutagen	uracil

Understanding Concepts

1. A short chain of processed mRNA has the nucleotide sequence AUA CCG GAC UCA. The last codon is a stop signal. Explain the series of events by which a polypeptide is synthesized from this mRNA sequence.
2. Based on your knowledge of DNA and polypeptide synthesis, how might you explain incomplete dominance?
3. A tRNA molecule has the anticodon GGA. What DNA base sequence was used to produce the codon that will bind to this anticodon?
4. A new cell is produced that has just a few copies of each of the genes needed for making rRNA. What cell function would be affected? Could the cell survive?

5. Sometimes the exons of a newly synthesized mRNA molecule can be spliced together in several different ways. Each processed mRNA molecule then codes for a different polypeptide. How does this fact fit in with the one gene-one polypeptide hypothesis?
6. Why would a mutation in a haploid organism have a much more immediate effect on the entire population than a mutation in a diploid organism?
7. How might a pseudogene be important in the future of a population?
8. How could you determine whether a single chain of nucleotides is RNA or DNA?

REVIEW

Applying Concepts

9. The difference between a normal hemoglobin molecule, coded for by a dominant allele, and the abnormal hemoglobin molecule that causes the disorder sickle-cell anemia, coded by a recessive allele, is only one amino acid in one of the hemoglobin polypeptide chains. In terms of DNA, how do the two alleles differ?

10. Assume the abnormal allele for sickle-cell anemia was produced by a random mutation. What kind of mutation was it? How do you know?

11. Ultraviolet light causes two successive thymine nucleotides along a chain of DNA to bond with one another instead of to the adenine nucleotides of the complementary chain. What effects would result from this mistake?

12. An allele in chickens, when present with the normal allele, results in the phenotype called creeper, in which the chicken has short legs and walks improperly. Two heterozygotes are crossed, but 1/4 of the eggs fail to develop. What does that tell you about the creeper allele?

13. Sometimes foreign genes are inserted into plants along with another gene that produces light (as in a firefly). How might the gene for light production be useful in recombinant DNA work?

14. Many digestive enzymes have been found to have very similar nucleotide sequences. How might the genes that code for these enzymes have originated?

15. You might think that a piece of a chromosome that breaks off but attaches backwards to its original chromosome would not cause a problem because all the genes are still present. Suggest reasons why such a chromosome mutation could, in fact, cause a problem.

16. What is cancer? How do cancer cells differ from normal cells?

17. **Issues:** The deaths of many terminal cancer patients occur when vital organs stop functioning. These organs may be located far from the original site of the cancer. How can cancer spread to distant organs?

18. **Biotechnology:** Forensic lab technicians compare the blood of a rape suspect with sperm recovered at the crime scene. What concept forms the basis of this method of identifying crime suspects?

Lab Interpretation

19. **Investigate:** The following diagrams are "birds-eye" views of student models. Describe the errors that you are able to detect.
   (a) Model of DNA

   Tape
   A-T-T-C-G-G-A-C-U = row of students facing below
   ‖ ‖ ‖ ‖ ‖ ‖ ‖ ‖ ‖ } Hands
   T-A-A-G-G-C-T-G-T = row of students facing above

   (b) Model of mRNA undergoing transcription, using the bottom row from above as a template

   A-T-T-C-C-G-U-C-U

   (c) Model of tRNA for amino acid X

   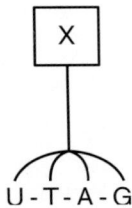

20. **Minilab 1:** Suggest two ways in which a gene mutation might slow or prevent the process of germination from occurring in a seed.

21. **Minilab 2:** Explain why, in this experiment, it was necessary to test the enzyme itself for the presence of glucose.

22. **Thinking Lab:** Explain why traits determined by organelle DNA are inherited in the pattern observed in this experiment.

Connecting Ideas

23. Suppose a cell synthesizes a protein that will leave the cell and be used by other cells. How does the protein travel from the ribosomes where it is made, through the cell, and out?

24. Different blood types are due to the presence of different proteins on the surfaces of red blood cells. Red cells of people with type O blood have no proteins of this type. In terms of DNA function, explain (a) why the genotype $I^B i$ results in type B blood and (b) why the genotype $I^A I^B$ results in type AB blood.

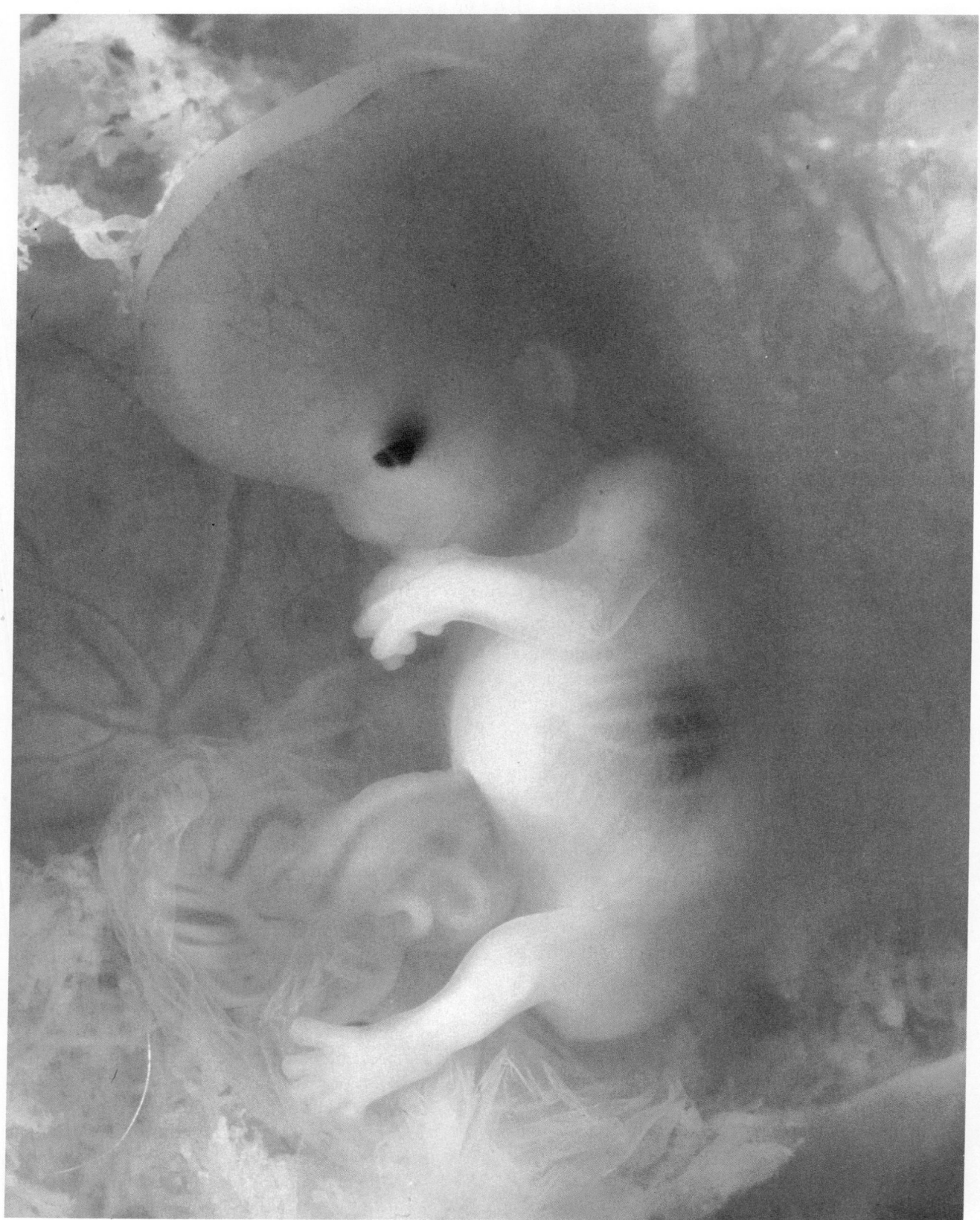

11

HUMAN GENETICS

GROWING WITHIN its mother's uterus, this developing human appears to have the potential for living a long and healthy life. Even now, eye and hair color have already been determined by its genes. This individual's genes may also have determined that it will develop a disorder that makes breathing difficult, weakens the lungs, and shortens a person's normal life expectancy. How can a parent, who has a history of this disorder in the family, be assured that this new individual is healthy? A test can be done using the white blood cells of the developing human. The cells are studied for signs of the abnormal gene that causes the disease. Results showing that this individual has normal genes bring welcome relief to the parents.

You and each one of your friends are the expression of many thousands of genes. If two children received their genes from the same parents, why does one child inherit a fatal genetic disorder while the other does not? What is the probability that a new baby sister will suffer from the same disorder? How soon after conception can the parents find out about the health of the child and what, if anything, can be done to correct an abnormality? In this chapter, you will read about the symptoms of some inherited disorders and how some genetic disorders are passed on. You will learn why some disorders affect mostly males. You will also learn how doctors discover if an individual has a genetic disorder even before its birth.

Color blindness is a genetic disorder that affects a person's ability to distinguish the colors red and green. How might these apples look to someone who is colorblind? What problems might a person with this disorder encounter every day?

Chapter Preview

Objectives

Explain the symptoms of cystic fibrosis.

Demonstrate how modern knowledge and techniques have led to understanding the cause of genetic disorders.

Discuss examples of other inherited disorders, their symptoms, and their causes.

11.1 Human Genetic Disorders

Like other organisms, you inherited genes that have determined your individual traits—your blood type, your skin color, whether your earlobes are attached or free, and the enzymes your body uses for cellular respiration. Inheritance of human traits can be explained by the same principles used to explain hereditary patterns in other organisms. So, too, can the inheritance of many human disorders. Scientists have known for many years that disorders, like other human phenotypes, are associated with production of abnormal proteins. Now, knowledge of DNA and the use of modern research techniques are permitting scientists to begin analyzing disorder-causing genes and pinpoint how the abnormal proteins actually cause the disorders.

Cystic Fibrosis

How often have you read a book and found a misspelled word among the hundreds or thousands of words used? Even with careful proofreading, a mistake will sometimes occur. A similar situation may occur with human genes. Unfortunately, those mistakes may result in a person who suffers from an inherited disorder. Cystic fibrosis (CF) is the most commonly inherited genetic disorder among white people. It is estimated that one white person in 20 carries the recessive allele that causes this disorder. As a result, one in 2000 babies is born with CF. This statistic translates into four or five babies being born with this fatal disorder in the United States every day.

Figure 11-1 Cystic fibrosis results from inheriting a recessive allele for the disorder from each parent. The inheritance pattern for two heterozygous parents is shown here.

Symptoms of CF Several body systems are affected by CF. Normally, a thin, free-flowing fluid is secreted that lubricates the passageways of the lungs and intestinal tract. CF represents a malfunctioning of the mucus-secreting glands of the body. Thick mucous secretions build up in the air passages and lungs. The mucus blocks the passage of air into and out of the lungs, causing lung damage. Almost all affected individuals have a history of wheezing and coughing. Recurrent infections weaken the lungs even more. In most CF patients, thick mucous secretions damage the pancreas and prevent or reduce the flow of digestive enzymes to the intestines. These enzymes are essential for the breakdown and absorption of food.

Scientists know that CF affects the transport of ions such as sodium and chloride across the cell membrane. This affects the movement of salt and water in and out of cells that line the lungs and digestive system. As a result, the usually thin mucus that coats these surfaces dehydrates, becomes thick and sticky, and accumulates.

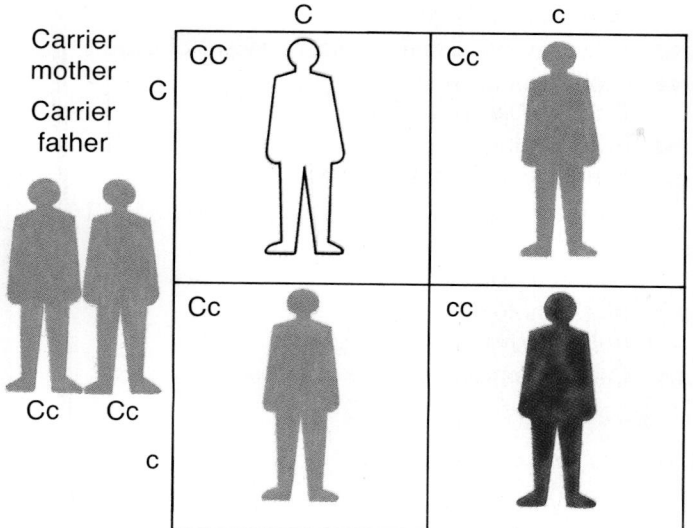

Carrier mother
Carrier father

The Cause of CF   The discovery that CF patients have defective mechanisms for transporting chloride ions across the cell membrane led to the idea that the recessive CF allele might code for an abnormal cell membrane protein. This abnormal protein might be part of a membrane protein, or another protein that is needed for passage of chloride ions out of the cell. Eventually, the existence of such a membrane protein was established, and it was found that it doesn't operate properly in CF patients. In 1989, the gene that codes for this protein was found.

A person without CF has an allele that codes for a normal cell membrane protein. In a person with CF, the recessive allele codes for an abnormal protein. Actually, there are several different CF alleles. Depending on which two recessive alleles they may have, CF patients suffer from different symptoms, some more severe than others.

Recent Developments in CF   In recent years, increased knowledge of CF and advances in medicine have improved the quality of life for people with the disorder. Physical therapy, breathing exercises, modern antibiotics, and special diets allow many patients to live longer and lead more normal lives. However, even with these recent advances, the disorder continues to be fatal. Identification of the CF alleles, along with analysis of the defects in the proteins for which they code, offers possibilities for better means of prevention and treatment. Perhaps even a cure may be found in the future. We will return to these exciting possibilities later in the chapter.

Figure 11-2   **Normal cell membranes have membrane proteins that allow cells to maintain normal levels of chloride. CF-affected cell membranes have membrane proteins that do not function properly and prevent the movement of chloride ions across the membrane. People affected with CF need daily therapy to loosen mucus buildup in the lungs.**

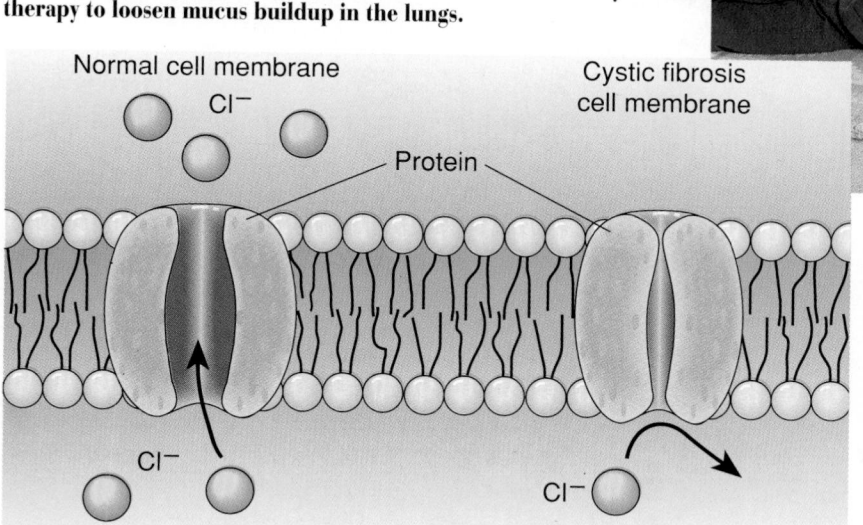

Normal cell membrane

Cystic fibrosis cell membrane

Cl⁻

Protein

Cl⁻

Cl⁻

Huntington's Disease

Cystic fibrosis is caused by a recessive gene. In fact, most human genetic disorders result from a homozygous recessive genotype. Huntington's disease, however, is caused by a dominant gene. Thus, a person heterozygous for Huntington's disease will have this fatal disease. Symptoms of Huntington's disease usually appear by the time a person is about 40 years old. Deteriorating brain cells may cause the person to become clumsy, irritable, depressed, and suffer memory loss. As the disease progresses, these symptoms become more severe, resulting in the uncontrollable jerking of arms and legs. Further loss of muscle coordination, memory, and the ability to speak are symptoms of advanced stages of the disease. Death usually occurs within 20 years after the onset of the first symptoms.

Finding the Huntington's Gene   The year 1983 provided an important milestone in the search for the cause of Huntington's disease. Until then, scientists could only reason that this disease was caused by a dominant gene by using the laws of probability when applied to genetics. Remember that offspring of a parent having a trait expressed by a dominant gene have a 50 percent chance of also having that trait. Experience showed that children of a person with Huntington's disease had a 50 percent chance of getting the disease. In 1983, a geneticist was finally able to show where the Huntington's gene is located. Scientists hoped that discovery of the exact location of the gene on the chromosome would quickly follow. A special foundation consisting of workers in six different laboratories was formed specifically to achieve this goal. In early 1993, the exact location of the gene that causes Huntington's disease was found.

Figure 11-3   **A pedigree of five generations of a family with Huntington's disease shows family relationships, the sex of each individual, and whether or not an individual is affected with Huntington's.**

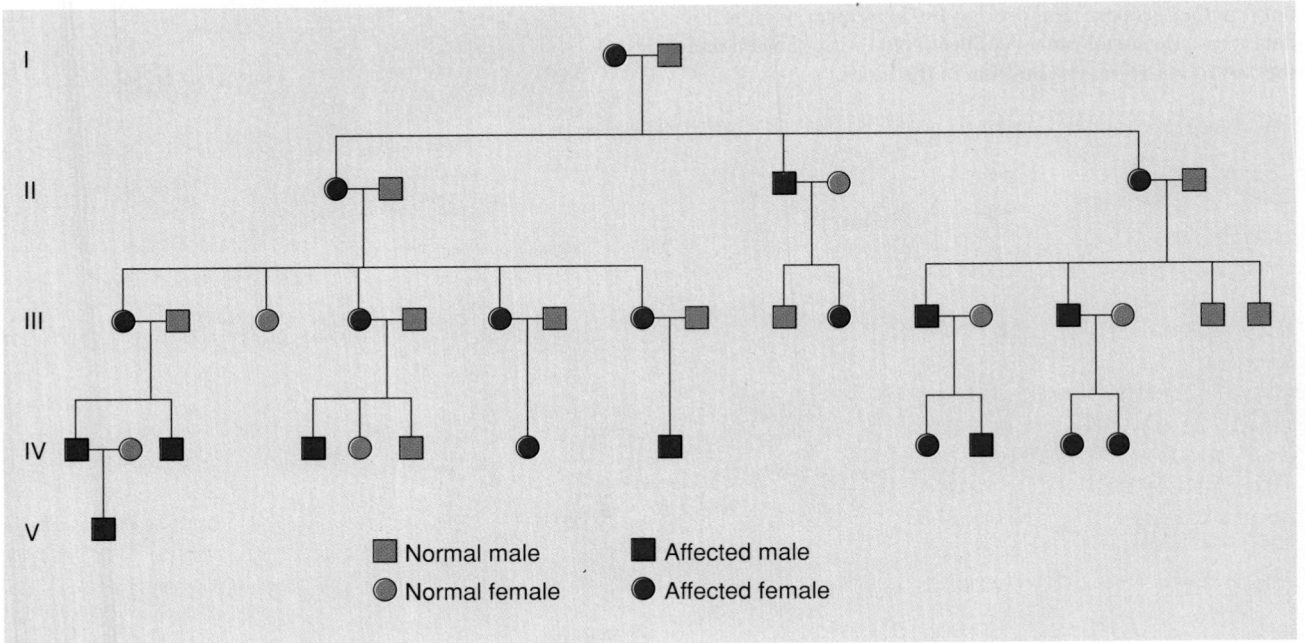

Normal male   Affected male
Normal female   Affected female

The discovery is credited to the foundation of scientists in the United States, England, and Wales who were working toward this goal—the Huntington's Disease Collaborative Research Group. For the estimated 25 000 Americans who currently have the disease, the discovery of this gene's exact location is an important one.

Now that the location of the Huntington's gene is firmly established, scientists can begin analyzing its nucleotide sequence. They can also compare the dominant allele to the recessive allele, and then, perhaps, the proteins coded by each allele can be analyzed and compared. The crucial variations in the proteins that lead to destruction of the brain cells can be identified.

Knowing the Future   Have you ever waited for the results of a major exam knowing that your participation in a special activity depended on the results? Notification of a passing grade brings happiness and relief. Failure brings disappointment. Imagine how you might feel if the exam you needed to pass was whether or not you carried the allele for Huntington's disease. Today, persons with a family history of the disease can be tested to see if they have the allele. Finding out that they don't have the allele provides relief from the fear of developing the disease and passing it along to children. On the other hand, learning that they have the allele and will face a slow death can be devastating. Many people prefer to live with uncertainty rather than experience the fear created by knowing the disease will eventually strike. These people decide against being tested. Woody Guthrie, one of America's most beloved folk singers and songwriters, died of Huntington's disease. His son , Arlo Guthrie, and his daughters have elected not to be tested for the disease that claimed their father.

Figure 11-4   **Woody Guthrie (1912-1967) died of Huntington's disease.**

Thinking Lab

DESIGN AN EXPERIMENT

Is snoring inherited?

Background
Have you ever been kept awake at night because someone near you was snoring? For many years, evidence has pointed to the fact that snoring was associated with weight gain, alcohol consumption, a disorder in which people pause in breathing during sleep, and other unknown causes. Extra fat in the walls of air passages makes them "rattle" when people assume certain positions during sleep. Alcohol's effect on the muscles of the tongue, other respiratory structures,

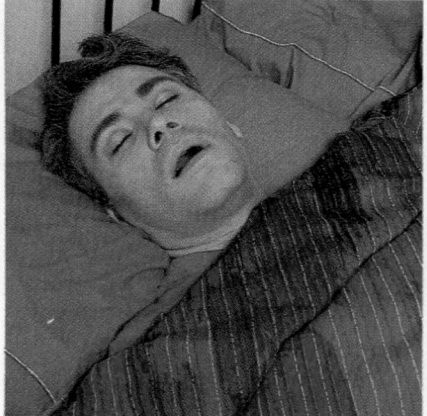

and the brain can also produce snoring. More recent evidence is also pointing to genetic causes of snoring. Researchers report that relatives of people who snore are twice as likely to snore as relatives of people who do not snore.

You Try It
Make a hypothesis and design an experiment that would investigate the idea that snoring has some genetic basis. Think about parts of the mouth and throat that you could investigate that might be involved in snoring.

Minilab 1

How do normal red blood cells compare to red blood cells of patients with sickle-cell anemia?

Sickle-cell anemia is a genetic disorder in which the gene for one chain of hemoglobin, the molecule in red blood cells responsible for carrying oxygen, is mutated. Normal red blood cells have a spherical disc shape. Examine, under low and high power, prepared slides of normal red blood cells and sickled red blood cells. Draw diagrams of both types of cells.

Analysis *Explain what problems are caused by mutated hemoglobin. Make two simple diagrams—one showing how normal red blood cells would pass through capillaries and another showing what happens when sickled cells pass through capillaries.*

Sickle-Cell Anemia

Just as the recessive CF allele causes a fatal disorder most often in white people, another recessive allele causes the disorder sickle-cell anemia most often in blacks. About one of every 500 Americans of African descent has the disorder. Sickle-cell anemia results when the structure of hemoglobin is genetically changed. Hemoglobin, a protein found in red blood cells, is made of four polypeptide chains—two alpha chains and two beta chains. Hemoglobin combines with oxygen in the lungs and transports the oxygen to body cells for use in cellular respiration. While most of us take for granted that our cells will receive the amounts of oxygen they need, people suffering from sickle-cell anemia are not so fortunate.

Sickle-cell anemia is the result of a mutation in the gene that codes for one of the polypeptide chains of hemoglobin. The mutant hemoglobin allele differs from the normal allele in the sixth codon. Just one nucleotide of that codon is different, but that difference results in one amino acid difference in the beta polypeptide chains coded for by the two alleles. The slight difference in the structure of abnormal hemoglobin molecules makes them less soluble and causes them to form crystals within the red blood cells. These crystals distort the red cells, changing their shape from the normal disc shape to sickled, as seen in Figure 11-6.

The formation of crystals of abnormal hemoglobin prevents hemoglobin from carrying enough oxygen to cells. In addition, the sickled red cells are rigid and cannot pass through capillaries as easily as normal red cells. As a result, those capillaries can be blocked. Severe pain, fever, and weakness are some of the symptoms caused by the blocked capillaries. Strokes that result in paralysis may also occur in young children.

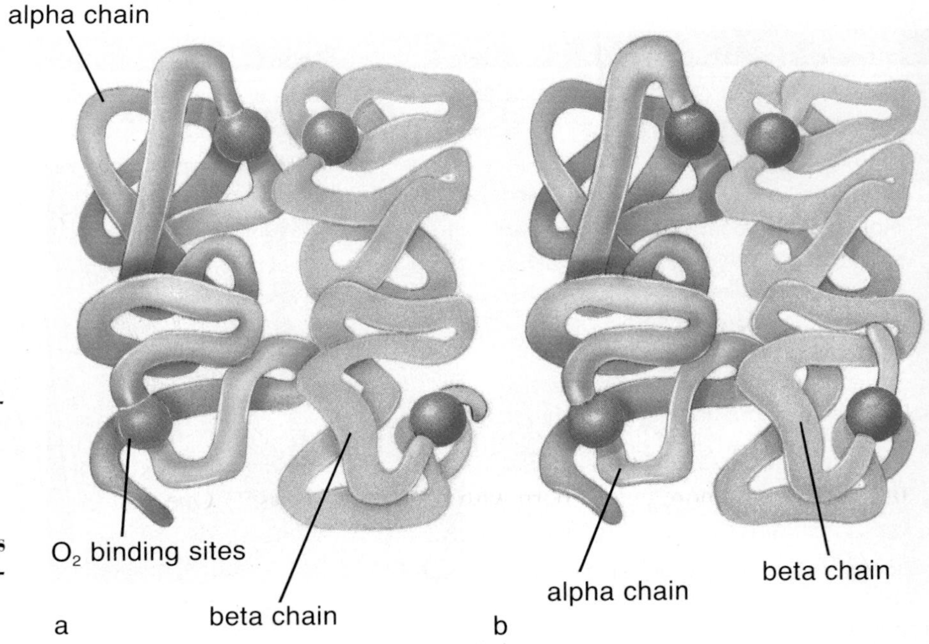

Figure 11-5 **Normal hemoglobin molecules (a) are composed of four protein chains and four iron-containing chemical groups, each of which can hold one oxygen molecule. Sickle-cell hemoglobin (b) is slightly different in structure from normal hemoglobin. This difference prevents these cells from carrying as much oxygen and results in their sickled shape.**

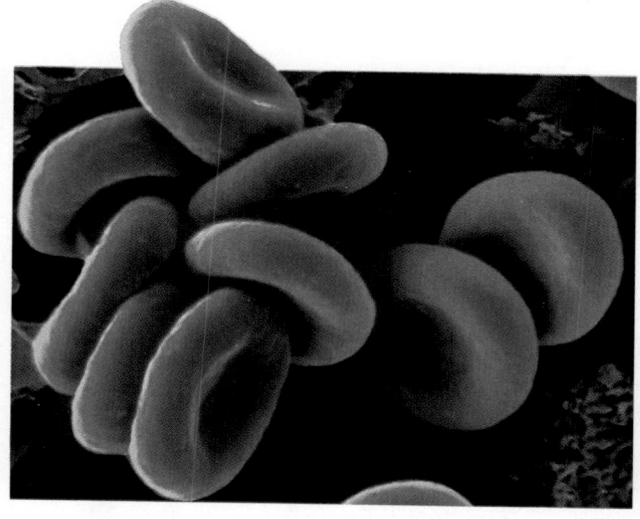
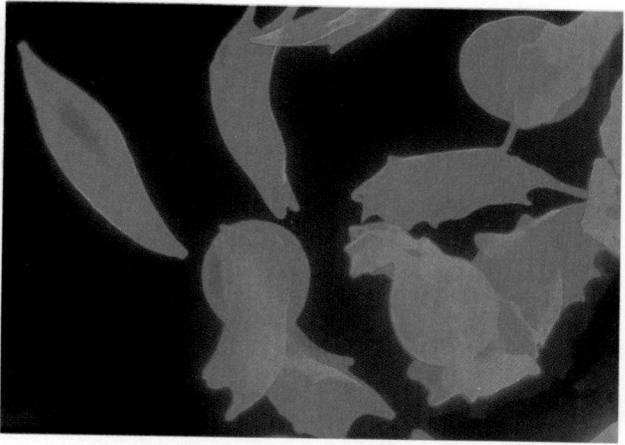

While many affected with sickle-cell anemia die while still young, some sickle-cell patients now live past age 40. Treatment for sickle-cell anemia involves a variety of approaches, including transfusions, antibiotics to treat infections, and drugs that increase the oxygen-carrying capacity of the red cells. Recently, scientists have been using mice to help study sickle-cell anemia. Fertilized mouse eggs are injected with DNA containing the normal allele for one of the hemoglobin chains together with the allele that codes for the other abnormal chain. The mice produce human sickle hemoglobin. Scientists hope that studies using these mice will provide useful information about sickle-cell anemia. Future experiments should provide new information that produces better means of dealing with the disorder.

Figure 11-6   Scanning electron micrographs of normal red blood cells (left) and sickled cells (right) shows the difference in their appearances. Magnification: 11 550×

Analysis of Genetic Disorders

Finding the cause of genetic disorders might be compared to successfully putting together the pieces of a puzzle. Disorders such as CF, Huntington's, and sickle-cell anemia are the result of mutant alleles. Locating the defective allele for each disorder might be compared to completing a very simple jigsaw puzzle. However, not all genetic disorders are explained in terms of a single pair of alleles. Many, such as diabetes mellitus, heart disease, and certain personality disorders, are believed to be the result of multiple genes. Complicating the problem of multiple genes is the fact that expression of genes is influenced by all the other genes an organism possesses. Successfully analyzing these disorders might be compared to completing a three-dimensional jigsaw puzzle with 100 interlocking pieces.

There are more harmful or lethal alleles in the human population than you may think. Just as every book may have a few misspelled words, it is estimated that every person carries five to eight harmful genes. Given that fact, why aren't more people born with hereditary diseases? One answer is that most of the harmful alleles are recessive; in other words, their effects are hidden by the presence of the normal allele. Another possibility is that although many zygotes carry harmful alleles, most of these zygotes never complete development.

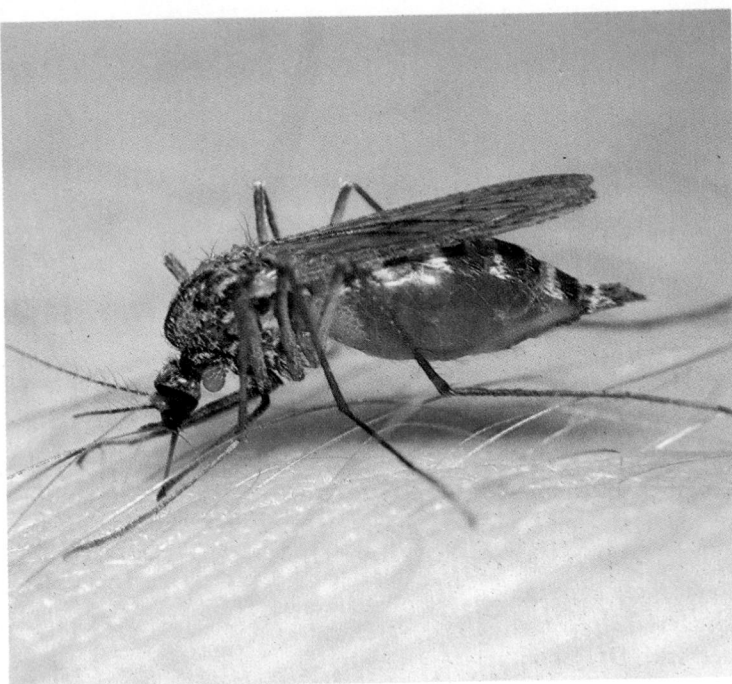

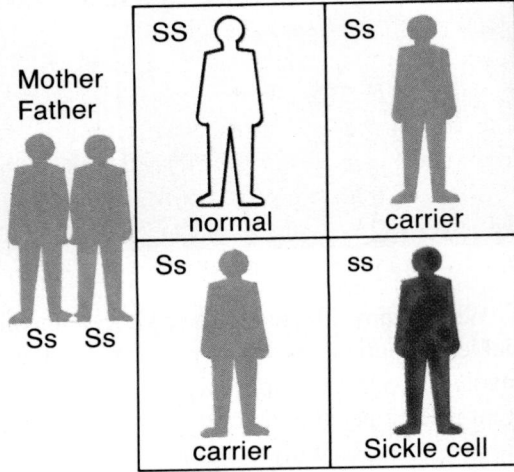

	Mother
	SS normal / **Ss** carrier
Father	**Ss** carrier / **ss** Sickle cell

Ss *Ss*

Figure 11-7 The mosquito in the photograph, the female *Anopheles*, is responsible for the spread of malaria. The Punnett square shows the inheritance of the sickle-cell trait. The offspring that are homozygous recessive or heterozygous for the sickle-cell trait are more resistant to malaria.

Heterozygote Superiority In some cases, although a certain allele is harmful in the homozygous state, that same allele actually presents an advantage to a heterozygous individual. Let's take another look at sickle-cell anemia. In the United States, one in ten African Americans carries the abnormal allele. However, depending on the area of Africa in which they live, as many as four out of ten Africans carry the allele for sickle-cell anemia. How can the difference be explained?

Malaria is a debilitating disease prevalent in parts of Africa, but uncommon in the United States today. Malaria is caused by a unicellular organism that invades red blood cells. Here the parasites multiply and soon rupture the blood cells. The newly released parasites invade other red blood cells and repeat the cycle. Periodic destruction of blood cells produces chills and fever and chronic illness. Some types of malaria can be fatal if untreated. People homozygous for normal hemoglobin are susceptible to malaria. Those who are heterozygous or homozygous for the sickle-cell hemoglobin resist malaria because the parasite does not thrive in cells with sickle hemoglobin. Therefore, in parts of Africa where malaria is prevalent, heterozygotes tend to live longer than people having the other genotypes. Unfortunately, those homozygous for sickle-cell usually die quite young. The likelihood of two heterozygotes producing children is great, and genetics tells us that every two out of four of those children will be heterozygotes, like the parents. Thus, the incidence of the recessive allele in African black populations remains high even though such a high incidence also results in many offspring having sickle-cell anemia. Because malaria is uncommon in the United States, heterozygotes have no such advantage, and over the years, the frequency of the recessive allele has declined due to natural selection.

The condition in which heterozygotes have an advantage over both homozygous genotypes, as in the case of sickle-cell anemia in Africa, is known as heterozygote superiority. This phenomenon may apply to other human genetic disorders as well.

The Results of a Higher Survival Rate   Another reason that the frequency of some harmful alleles remains high has developed in recent years. In those countries with modern health care, people having some genetic disorders are living long enough to reproduce. Their genes, therefore, are passed on. For example, earlier in this century there was no treatment for people with a severe form of diabetes mellitus. Then, in 1922, insulin injections to control diabetes became available. Instead of dying young, diabetics began to lead more normal lives, to live longer, to marry, and to pass their genes to their descendants.

Sex-linked Disorders

You know that genes for many sex-linked traits are carried on the X chromosome, and fewer are carried on the Y. Some of those genes are the cause of disorders or abnormalities in humans.

Hemophilia   How often have you gotten a small cut and noticed that very little blood was lost because a clot formed quickly? Clotting of the blood that leaks from a scratch or cut is familiar to almost everyone. In persons with hemophilia, however, a protein necessary for clotting is absent. Consequently, the smallest injury can lead to a life-threatening loss of blood. Hemophilia is due to a recessive allele on the X chromosome. If a boy's mother is heterozygous, he has a 50 percent chance of inheriting the X chromosome with the hemophilia allele from his mother. If a male inherits an X chromosome with the hemophilia allele, he will be a "bleeder." The Y chromosome does not carry the normal allele to provide the clotting factor.

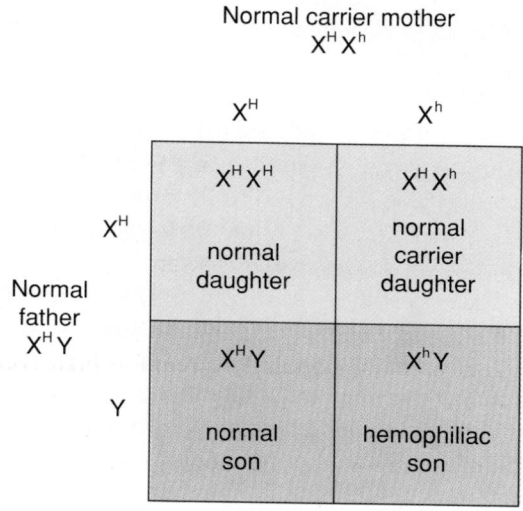

Normal carrier mother
$X^H X^h$

Figure 11-8   A cross of parents with normal blood clotting shows the probability of hemophiliac offspring if the mother carries the hemophilia gene. A female hemophiliac is rare because she must receive a gene for hemophilia from each parent. What are the phenotypic and genotypic ratios of this cross?

As with other diseases, modern technology has helped lengthen the life span and improve the quality of life for hemophiliacs. One treatment consists of blood transfusions to restore lost blood. Another involves giving the patient the missing clotting factor. Research based on recombinant DNA techniques is being done to try to make the clotting factor available in greater quantities and at a practical price.

Color Blindness   Imagine looking at a bowl of shiny red and green apples and not being able to distinguish which was which. Some people don't have to imagine such a situation because they aren't able to see the colors red or green. For them, anything that is red or green appears brown. These people have red-green color blindness, another sex-linked

HISTORY CONNECTION

Heredity and the Amish

• •

The Amish of Lancaster County, Pennsylvania are part of a remarkable group of people. The Amish were founded in the 1690s as an offshoot of the Swiss Mennonite Church. The group now living in Lancaster County is descended

from a handful of Swiss Amish who came to America in the 1720s. Like most other Amish communities in the United States, they have maintained a way of life that has changed

little in 300 years. Unlike other Amish communities, they also have the misfortune to maintain an unusually high rate of Ellis-van Creveld syndrome in their population.

Ellis-van Creveld syndrome is an autosomal recessive genetic disorder. People born with this disorder have short arms and legs. They may also have extra fingers or toes. What is it about the Lancaster Amish population that makes them more susceptible to this syndrome? The answer may lie in Amish culture and history, and in two genetic principles known as the founder effect and genetic drift.

The Amish form a relatively closed community. They live a very simple life. They wear the same style of plain, homemade clothing that their ancestors wore three centuries ago. Such things as buttons are rejected as too fancy. Amish clothes are fastened with hooks and eyes or straight pins.

The Amish live a life close to the earth and the basic needs of daily life. They don't use electricity in their homes. You won't find any computer programmers or electrical engineers in Amish communities. Even though they are famous as fine farmers and craftsmen, tractors and power tools are usually forbidden. Hand tools are used for woodworking and horses for farming. While some Amish men will drive cars or trucks as part of their work for non-Amish employers, they do their own traveling in horse-drawn carriages.

This way of life keeps the Amish outside the mainstream of American society. They take care of each other so well that poverty is almost unknown. In most cases, they don't socialize or marry outside of the

trait. Red-green color blindness is caused by a defect in either of two alleles located on the X chromosome. Dominant forms of these alleles code for red and green receptor molecules that are present in light-capturing cells of the eye. These molecules absorb red or green wavelengths of light and transform their energy into nerve impulses to the brain. In a male, if either recessive allele is present on the X chromosome, abnormal receptors are made. Red and green will not be properly distinguished. In females, both X chromosomes must have at least one of the two recessive alleles for her to be color-blind. In some cases, either the red or green receptor allele is actually missing entirely.

Amish community. This type of closed community does not allow for much genetic variability. The same genes are continually passed along in the population.

The vast majority of the Lancaster County Amish are descended from about 200 common ancestors who came to the U.S. in the late 1720s. Two hundred people is not a lot to start a community. At least one of those 200 was almost certainly a carrier of Ellis-van Creveld syndrome.

Imagine how such a gene would act in a small, closed community. If only one of the original 200 carried the gene, he or she made up just 1/2 of 1 percent of the population. But if that person passed the gene along to half of his or her children, and those children passed the gene along as well, it wouldn't be long before a large number carried the gene. After several generations, carriers would inevitably marry each other, and 1/4 of their children would be born with the syndrome. This phenomenon is called the founder effect.

Imagine now what would happen if the gene-carrying founder had many children. The percentage of carriers would probably rise, increasing the number who would eventually be born with the disorder. On the other hand, what if the gene-carrying founder had few children? The percentage of carriers would probably drop. If it dropped to zero, the gene would be lost from the population. Such change in the gene pool, resulting from chance, is known as genetic drift. Amish families are usually large, so chance favors having genes—such as the one causing Ellis-van Creveld syndrome—passed along from generation to generation in fairly high numbers.

Fortunately for geneticists, family history is very important to the Amish, so they keep good records of their ancestry. They also seek good medical care for themselves and each other. By allowing geneticists to study their medical records and family trees, the Amish have contributed a great deal to the understanding of Ellis-van Creveld syndrome and how genetic disorders affect populations through the founder effect and genetic drift.

1. **Writing About Biology:** Suppose you were studying a genetic disorder and found two excellent examples of populations with a history of that disorder. One was in an Amish community and the other in your town. Both had good family records. Would one be easier to study than the other? Why?

2. **Explore Further:** Find another example of the founder effect and genetic drift in the Amish or any other population.

Chromosome Abnormalities

Some inherited defects are due to abnormalities in the structure or number of chromosomes rather than particular alleles. For example, the most common cause of Down syndrome is nondisjunction. Nondisjunction is the failure of paired chromosomes to separate during meiosis. This results in one gamete having too many chromosomes and the other too few. The presence of three of one kind of chromosome is known as **trisomy.** This cause of Down syndrome is known as trisomy 21.

Karyotyping   Certain chromosomal abnormalities can be detected by looking at a picture of human chromosomes called a **karyotype.** Karyotypes are made by taking a blood sample and separating out the white cells. A chemical is added that stops the white cells in metaphase of mitosis. Water is added to the cells to cause them to swell and burst, releasing chromosomes. The chromosomes are then stained and photographed. The photograph is enlarged and each individual chromosome is cut out and matched with its homologue. The chromosomes are arranged by size and numbered. Visible chromosomal abnormalities can be easily identified by looking at the chromosomes. Geneticists can see large chromosome abnormalities such as extra chromosomes or added or deleted pieces of chromosomes. Additional chromosomes often cause multiple abnormalities. In a deletion, a portion of a chromosome is missing. Sometimes, a deleted portion of a chromosome attaches to another chromosome. This is called a translocation. The karyotype in Figure 11-9 shows the karyotype of an individual with Down syndrome.

Figure 11-9   **The photograph shows human chromosomes in late prophase of mitosis. The karyotype shows an extra 21st chromosome. This indicates an individual with Down syndrome.**

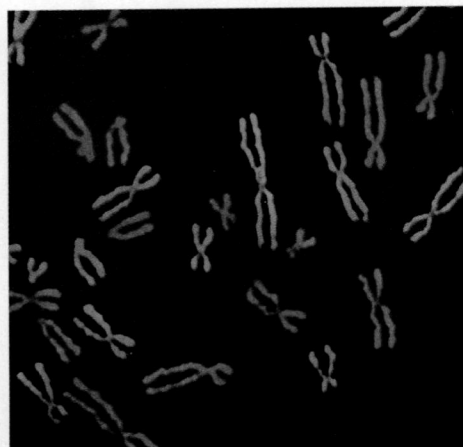

1 2 3 4 5 6

7 8 9 10 11 12

13 14 15 16 17 18

19 20 21 22 23

Characteristics of Down Syndrome

People with Down syndrome have varying degrees of mental retardation, from mild to severe. They are usually shorter than normal. Many are born with heart, intestinal, or vision problems. They are susceptible to infection and have an increased risk of developing a form of blood cell cancer known as leukemia. Sixty years ago, most people with Down syndrome died before the age of ten. Now, with better care, the average life span has increased to 30 years. About 25 percent of people with Down syndrome now live to be 50.

Some cases of Down syndrome are due not to nondisjunction, but to a different chromosome abnormality. This abnormality is caused when most of the 21st chromosome breaks off and fuses to another chromosome, usually chromosome number 14. This is a translocation. When a gamete containing this part of chromosome 21 and another normal gamete combine, the zygote has three copies of chromosome 21. This event is the cause of about four percent of Down syndrome cases. This cause of Down syndrome is most likely to occur in infants born to mothers over the age of 40. The exact reason for this is not yet known.

Geneticists are currently studying the DNA of chromosome 21 in an effort to answer many questions about Down syndrome. They need to identify which of the estimated 1500 genes on that chromosome are the cause of Down syndrome. They want to know the proteins coded for by those genes in an effort to understand why the various symptoms occur. They also want to determine why the presence of three of each gene, rather than the usual two, is so harmful.

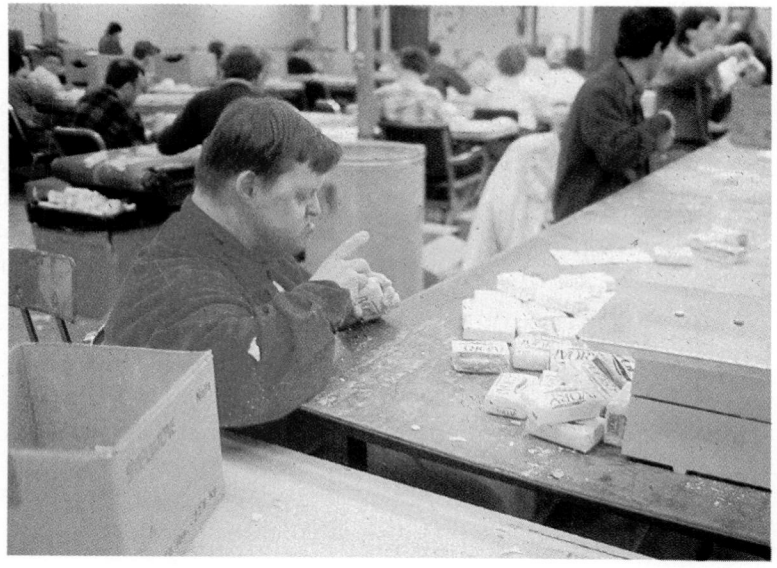

Figure 11-10 Many individuals with Down syndrome live very productive lives.

Section Review

Understanding Concepts

1. Explain how the defect in the membrane protein responsible for the transport of chloride ions is related to symptoms of CF.
2. How are the cause and onset of symptoms in Huntington's disease different from those of CF and sickle-cell anemia?
3. Relate the structure of abnormal hemoglobin to the symptoms of sickle-cell anemia.
4. **Skill Review—Making Tables:** Make a table that compares the following genetic disorders in terms of their causes and symptoms: CF, Huntington's disease, sickle-cell anemia, hemophilia, red-green color blindness, and Down syndrome. For more help, see Organizing Information in the **Skill Handbook.**
5. **Thinking Critically:** The organism that causes malaria requires oxygen carried by red blood cells. Why might heterozygotes for sickle-cell anemia be resistant to malaria?

Objectives

Demonstrate how genetic counseling can be used to provide parents with information about the possibility of having a child with a genetic disorder.

Describe techniques that permit diagnosis of a genetic disorder in the unborn.

Discuss gene therapy and how it might be used to cure genetic disorders.

11.2 Prevention and a Possible Means of Cure

Research is providing knowledge of the causes of many genetic disorders. Researchers are identifying and analyzing the genes that cause genetic disorders and defects in the proteins they encode. But being able to relate genes and proteins to the symptoms of a genetic disorder is just a beginning. In the United States alone, more than 200 000 babies each year are born with genetic disorders or congenital disabilities—disabilities that are acquired during development and not through heredity. Although many of these disorders are minor, others cause severe health problems, and some are eventually fatal. The knowledge gained through basic research must be applied to improve and extend the lives of people afflicted with these disorders and, perhaps in the not-too-distant future, to prevent or cure them.

Genetic Counseling

When a couple is expecting a baby, a question they hear often is, "What do you want, a boy or a girl?" Most prospective parents answer, "I don't care as long as it's healthy." It used to be that the parents had to wait until the baby was born to find out the answers to both of these questions. Today there are techniques available to give parents not only information about the sex of their baby but also information about the health of their unborn child.

A person trained in genetics can give parents information about the chances of their child having a genetic disorder. This process is called **genetic counseling.** This type of counseling is very valuable to couples with a history of hereditary disorders in their families. The counselor studies the family histories of the couple, and uses knowledge of basic patterns of heredity and newly developed diagnostic techniques to inform the parents of possible disorders that may be passed on.

Suppose, for example, that a couple is concerned about the possibility of having a child with Tay-Sachs. Tay-Sachs is an autosomal recessive disorder that is not apparent at birth. Babies homozygous for Tay-Sachs appear normal for the first few months. By eight to nine months of age, however, the first symptoms appear. This disorder results in the degeneration of the nervous system. Neither parent knows if he or she is heterozygous, or a carrier of the recessive allele for the disorder. A blood test has been developed that will tell parents whether they carry the recessive allele. Individuals who are heterozygous have half the normal level of the enzyme missing in children with Tay-Sachs.

Figure 11-11 **With today's diagnostic techniques and genetic counseling, parents can have information about the health of their child long before it is born.**

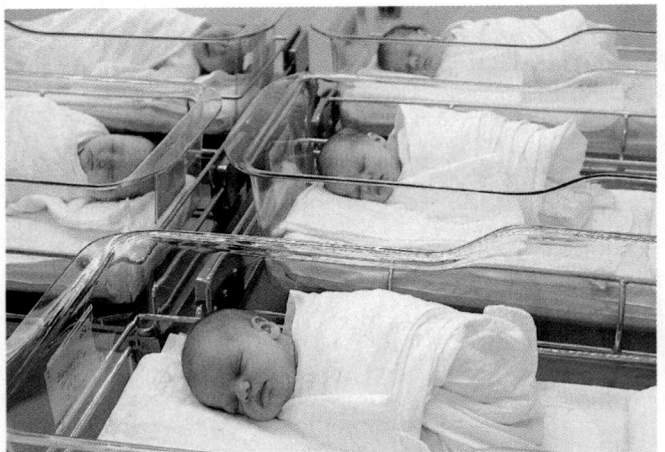

The counselor would provide the couple with information based on this test. If neither parent is a carrier, or if only one is, they would be told that there is no chance for them to have a child with Tay-Sachs. However, if both parents carry that allele, the counselor would explain to them that there is a one in four chance of their having a child with the disorder. With genetic counseling, parents would know the possibility of having a child with a genetic disorder.

While genetic counseling can provide important information about the probability of a child being born with a genetic disorder, it can't tell if a specific individual will be afflicted. Parents must turn to other methods for that answer.

Diagnosis in the Uterus

Down syndrome is a genetic disorder that is known to occur significantly more often in children born to mothers over the age of 35. Is there any way for parents to know whether their child will have Down syndrome?

Amniocentesis   When Down syndrome is a result of trisomy 21, examining the chromosomes of a developing fetus could determine whether the fetus is normal or not. Techniques are available for doing just that. The most frequently used technique is called **amniocentesis.** This is a process in which a sample of the fluid surrounding the fetus is withdrawn through a long, thin needle. The fluid contains fetal cells that are then grown in the correct conditions in a lab. A karyotype of the chromosomes is then made to determine whether the child has three, or the normal two, number 21 chromosomes. While amniocentesis can confirm the presence of Down syndrome, it can't determine the severity of the disease. However, the parents at least know if the fetus is affected.

Figure 11-12   In amniocentesis, amniotic fluid is withdrawn, and the fetal cells contained in it are studied. Ultrasonography (shown in the photograph) determines the position of the fetus before amniocentesis. This is a 25-week fetus.

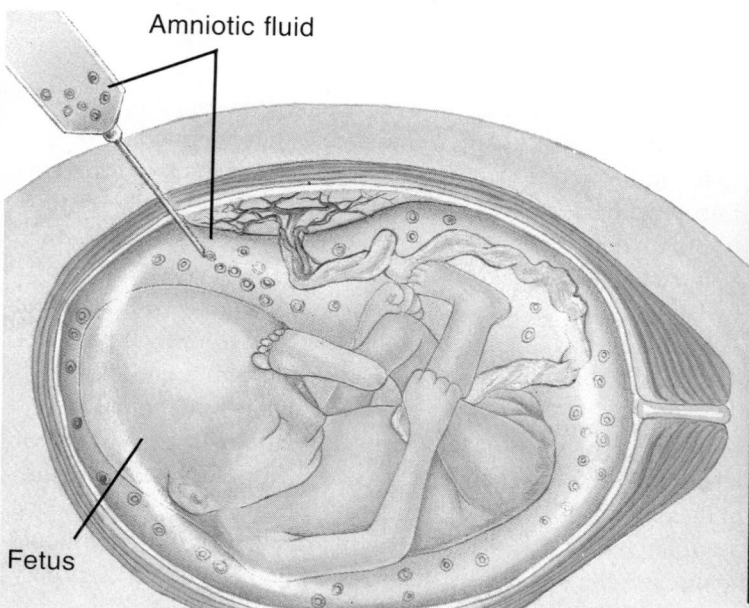

Amniotic fluid

Fetus

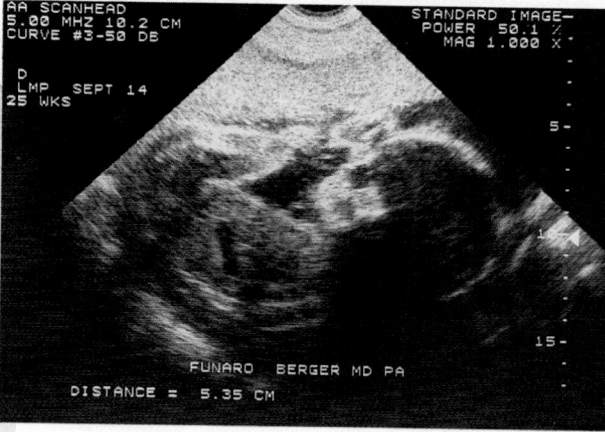

AA SCANHEAD
5.00 MHZ 10.2 CM
CURVE #3-50 DB

STANDARD IMAGE-
POWER 50.1 %
MAG 1.000 X

D
LMP   SEPT 14
25 WKS

5 -

15 -

FUNARO   BERGER MD PA

DISTANCE = 5.35 CM

Amniocentesis can also be used to diagnose several other disorders, such as sickle cell anemia and hemophilia, caused by defective alleles rather than chromosomes. For these diseases, a karyotype is not useful. Instead, the fluid is analyzed for the presence or absence of certain molecules that indirectly indicate presence of the disease.

Results of tests done after amniocentesis may not be available for ten days in cases where a karyotype is required. That amount of time is needed to grow enough cells for microscopic examination. A new variation of the test is being tried that will yield results in as little as two days. Artificial chromosomes known as probes are inserted into the nuclei of the cells extracted from the fluid. The probes, which contain a substance that glows under fluorescent light, attach themselves to the fetal chromo-

GLOBAL CONNECTION

Tracking Deadly Medicine

A bottle of medicine can be a priceless cure—or a deadly poison. Which it will be is a question of dosage. Digitalis, for example, is a medicine that has given additional years of life to people with weak hearts. Yet an overdose will end a life quickly.

Sometimes a population will develop a natural defense against a disease. And sometimes individuals can overdose on that natural defense. In this chapter, you've already seen one classic example in sickle-cell trait and sickle-cell anemia. Being heterozygous for the trait protects against malaria while being homozygous recessive for the anemia—an "overdose" of the protection—can kill. Less well understood is Tay-Sachs, which wasn't recognized as probably being the deadly form of a natural defense until the early 1970s.

Just as sickle-cell anemia is most common among people descended from Central Africans, Tay-Sachs is most likely to haunt those of Eastern European Jewish ancestry, known as Ashkenazic Jews. It was by studying this population that medical detectives tracked down clues to much of Tay-Sach's nature.

Babies born with Tay-Sachs have a genetic defect that prevents their bodies from making a vital enzyme called hexosaminidase A. In some forms of the disease, the enzyme is made but isn't active. Hexosaminidase A breaks down a fatty substance, called G_{M2} ganglioside, produced in your nerve cells.

Without the hexosaminidase A, G_{M2} ganglioside builds up in the nerve cells until it reaches toxic levels. Tay-Sachs babies seem normal at birth, but over a period of several months, they become listless, suffer convulsions, and lose whatever control of their bodies they may have learned. They go blind and lose touch with the world around them. It's rare for a Tay-Sachs baby to survive to its fourth birthday.

You have two copies of the gene that creates hexosaminidase A, one inherited from each of your parents. You only need one to code for the enzyme to prevent Tay-Sachs. A Tay-Sachs baby must inherit two recessive hexosaminidase A-producing genes—one from each parent.

Worldwide, about one baby in every 400 000 is born with Tay-Sachs. But among Eastern European Jews and their descendants, the disorder historically affected about one birth out of every 3600. Why should this population be 100 times more

somes. Fetal cells are then examined under a microscope using fluorescent light. The number of visible probes and the chromosomes to which they attach can be used to determine whether a fetus has a chromosomal defect. As more and more individual disease-causing genes are located on chromosomes, probes may be used to spot them as well.

Chorionic Villus Biopsy   Because of possible injury to the fetus, the earliest amniocentesis can be conducted is the 14th week of pregnancy. Another method, called **chorionic villus biopsy,** can be performed as early as the ninth week. Cells from a tissue called the chorion, which is part of the structure by which the fetus is linked to its mother in the uterus, can be removed. These cells have the same chromosomes as the fetus and can be karyotyped and analyzed as in amniocentesis.

prone to the disorder than the rest of the world?

Medical science got its first peek at a possible answer to that question in a 1972 study of how grandparents of Ashkenazic Tay-Sachs babies died. Since the babies had Tay-Sachs, they must have inherited one defective hexosaminidase A gene from each parent. All the parents had to be heterozygous for Tay-Sachs. That, in turn, meant that each parent had at least one parent who was also heterozygous. At least 50 percent of the babies' grandparents had to have a single defective gene.

The study showed that only one of the 306 grandparents in the study had died of one of the most common killers of their day: tuberculosis (TB). This was at a time when as much as 20 percent of the deaths in Eastern European cities were caused by TB. Was it possible that being heterozygous for Tay-Sachs gave some protection against tuberculosis?

More evidence appeared in 1983 when studies showed that the Ashkenazic population with the highest concentration of Tay-Sachs heterozygotes was also the population with the highest rate of TB. At first glance, that might make it seem as though being heterozygous for Tay-Sachs offers no protection against TB at all. But remember— it's the population with the greatest frequency of TB that would be most likely to evolve a defense. After all, sickle-cell trait evolved in Central Africa, where malaria caused the greatest number of deaths. It didn't in Norway, where malaria isn't a problem.

Still, that evidence doesn't answer the question of why a defense against TB would arise among Eastern European Jews but not among the general population. To solve that mystery, we have to look at history.

For nearly 2000 years, the Jews of Europe made their living in various businesses in an urban setting. Large numbers of them were crowded into ghettos, separated from the rest of the city.

The Jews, confined to their ghettos, remained genetically distinct. When a defense arose, it was passed along within this stable population, spreading throughout the generations.

So far, nobody knows how carrying one defective hexosaminidase A gene helps defend against TB. In fact, the suspicion that the defect is such a defense has yet to be proven. There is, however, still more evidence that being heterozygous for Tay-Sachs provides some advantage.

Despite the fact that Tay-Sachs is incurable, it no longer strikes as many Ashkenazic families as it once did. That's because it's now possible to screen for the defective gene. As a result, carriers have been able to avoid having babies. Today, thanks to the routine testing of high-risk couples, Tay-Sachs disease is no longer more likely to strike an American Ashkenazic family than it is to appear in any other.

1. **Find Out:** What populations other than those of Ashkenazic descent are prone to Tay-Sachs?
2. **Connection to Medicine:** PKU is another disorder caused by an autosomal recessive allele. What treatment is available for this disorder?

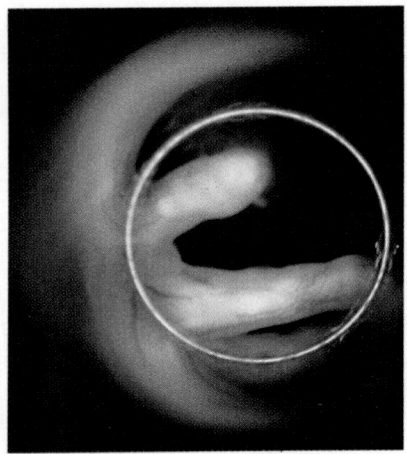

Figure 11-13 An 18-week fetus viewed through an endoscope.

Another kind of test is being developed that can be performed even earlier. During gestation, a few fetal blood cells enter the mother's bloodstream. The fetal and maternal blood cells can be distinguished from each other and the fetal blood cells examined for chromosome abnormalities.

Ultrasonography So far, all of the techniques discussed depend on analysis of cells or fluid to determine fetal abnormalities. Two other techniques can be used to observe the fetus for abnormalities. One technique called **ultrasonography** is often used to determine the position and anatomy of a fetus. During the process, an ultrasound probe is passed back and forth over the mother's abdomen. The probe emits high-frequency sound waves that "echo" from the tissues of the fetus. Different tissues have different densities. These differences in density cause the sound waves to echo back at different wavelengths that, when put together by a computer, form an image on a screen. These images, called echograms, allow physicians to spot abnormalities, such as certain forms of heart disease. Ultrasound is also used prior to amniocentesis to determine fetal position. This reduces the risk of injuring the fetus when the needle is inserted into the mother's abdomen.

I N V E S T I G A T I O N

Variations in Human Hands

Do you have a "hitchhiker's thumb"? Can you bend just the first knuckle of your fingers without bending others? Do you have hair on the middle section of your fingers? How large is your hand span? Can you easily push your thumb down to touch the side of your wrist? These traits as well as all other features of your body are determined in part by genes. Almost all human traits are determined by multiple gene pairs rather than by one pair of genes. In addition, some traits may be influenced by the environment. For these reasons, it is difficult to determine which traits are dominant and which are recessive. In this investigation, you will compare features of classmates' hands and hypothesize which features of the hand are inherited and which may be produced by the environment.

Problem
Which features of the hand are probably inherited and which may be produced by environment?

Safety Precautions
Do not force bending your hands and fingers beyond what is comfortable.

Hypothesis
What is your group's hypothesis? Explain your reasons for making this hypothesis.

Experimental Plan
1. As a group, make a list of possible ways you might test your hypothesis using the materials your teacher has made available.

Fetoscopy Direct observation of the fetus and surrounding tissues is also possible. After determining the position of the fetus by means of ultrasonography, a technique called **fetoscopy** can be used. The technique makes use of a device called an endoscope. The endoscope can be inserted through a small incision in the mother's abdomen. The fetus is viewed directly through the endoscope tube. At the same time, an image of the fetus is seen on a screen. Using instruments inserted into the endoscope, small samples of skin or blood may be withdrawn for study as in amniocentesis. Fetoscopy has been used to give a transfusion to a fetus. It also has been used to remove excess fluid around the brain of a fetus.

Gene Therapy

The amount of information geneticists have discovered in the last 20 years is staggering. It was only in 1973 that the first recombinant DNA methods were successfully used. Since then, recombinant DNA technology has led to the development of techniques that allow geneticists to locate the positions of some genes that cause genetic disorders. Nucleotide sequences of normal alleles and the harmful counterparts can be compared, as can

2. Agree on one idea from your group's list that could be investigated in the classroom.
3. Design an experiment that will test one variable at a time. Plan to collect quantitative data as well as make other observations.
4. Following the style of a recipe, write a numbered list of directions that anyone could follow.
5. Make a list of materials and the quantities you will need.

Checking the Plan

Discuss the following points with other group members to decide the final procedures for your experiment.

1. What are the variables in your experiment?
2. How have you set up a control?
3. How many traits will you count?
4. What traits will you measure or count?

5. How many students' hands will you measure, observe, or examine?
6. Have you designed and made a table for collecting data?

Make sure your teacher has approved your experimental plan before you proceed further.

Data and Observations

Carry out your experiment; make your measurements and observations; and complete your data table. Make a graph of your results.

Analysis and Conclusions

1. Was your hypothesis supported by your data? Explain.
2. According to your results, which traits are most likely inherited and which are most likely produced by environment?
3. What is your experimental evidence for your conclusion in question 2?

4. Consult with another group that used a different procedure to examine this problem. Were their conclusions the same as yours?
5. Do you know of other human traits scientists are studying to determine which is more important, heredity or environment, in influencing their development?
6. Write a brief conclusion to your experiment.

Going Further

Based on this lab experience, design an experiment that would help to answer another question that arose from your work. Assume that you will have unlimited resources and will not be conducting this lab in class.

the amino acid sequences of the proteins for which they code. Such analysis enables scientists to understand the basic causes of the symptoms of a genetic disease, as in the case of CF.

Suppose a person suffering from CF could receive the normal allele for the protein that is not functioning correctly. The patient's cells could then begin making the proper protein, chloride ions could be pumped normally, and the thick mucus associated with the disease would no longer be present.

Although all of this sounds like a plot from a science fiction novel, it isn't. The treatment of genetic diseases by introducing normal genes into body cells is called **gene therapy.** Gene therapy has already been successfully done in a few cases. More tests have been approved and are planned for the near future.

LOOKING AHEAD

YOU NOW KNOW how abnormalities of genes and chromosomes can cause genetic disorders in humans. Such abnormalities and the disorders they cause can be considered variations within the human population that tragically affect the health of specific individuals. While you may focus on the situation of a specific individual, variations are also important to the long-term health of an entire population. The information being discovered by scientists may someday lead to methods that eliminate inherited disorders. Such discoveries may also lead to a human population that is generally more healthy. In the next chapter, you will explore how changes in a population's genes can be important to the evolution of that population over long periods of time.

Section Review

Understanding Concepts

1. PKU is a disease caused by the failure of an enzyme to break down the amino acid phenylalanine. This can result in mental retardation. It is caused by a recessive allele on an autosome. What information would a genetic counselor provide to the two parents, each of whom was found to carry that allele?

2. Explain why ultrasonography is used before amniocentesis is done.
3. Could chorionic villus biospy be used to determine the existence of a disease caused by defective alleles? Explain.
4. **Skill Review—Forming a Hypothesis:** Scientists have developed several methods for introducing genes into body cells. White blood cells and viruses are two

methods already being used. Form your own hypothesis about how scientists use one of these two methods. For more help, see Practicing Scientific Methods in the **Skill Handbook.**
5. **Thinking Critically:** Could the technique of gene therapy described in this section be useful in the case of Huntington's disease? Explain.

Issues

Gender on Demand

People like to have some choice about the important things in their lives. What kind of work they do, who they marry, and whether or not they'll have children are all good examples. But when it comes time to have a child, the decision of whether to have a son or a daughter has always been out of our control.

Over the centuries, people have come up with all sorts of methods for influencing the sex of their offspring. Techniques have involved such things as getting pregnant when a north wind is blowing, eating a raw egg, and trying to conceive while wearing boots. You've probably already figured out the trouble with these methods—they don't work.

In most cases, there's no reason for couples to choose a sex for their baby except that they'd prefer a son or a daughter. But the ability to select a child's gender could be a great blessing to couples at risk of passing along a sex-linked disorder, such as hemophilia. Usually only males suffer from the disorder, but females carry it. A woman carrying the hemophilia gene has a very good reason to want daughters instead of sons.

There's still no way to guarantee either a daughter or a son, but modern medicine has found ways

of improving the odds. The most successful methods involve ways of separating sperm carrying the X chromosome from those carrying the Y chromosome. The Y-carrying sperm tend to swim faster than X carriers, so one method involves separating the faster sperm from the slower ones. Another method uses a centrifuge and differences in the sperms' masses to separate them.

Both these techniques boast a success rate of 70 to 85 percent, but both are far from being a sure thing. They also have the disadvantage of requiring the sperm to be artificially placed in a woman's reproductive tract.

Another recently popular method doesn't require so much

outside help, but is far more controversial. It's based on the theory that chemical changes inside a woman's reproductive system favor the success of either the X- or Y-carrying sperm at different times. Supposedly, the X sperm have an advantage from two to 24 hours before ovulation and Y-carrying sperm are more likely to fertilize the egg 36 hours after ovulation.

This theory led to a home kit intended to help prospective parents with their timing. It was sold in many drug stores, with the manufacturer professing a 77 to 98 percent success rate. The Food and Drug Administration disagreed, however, saying that there was no evidence to back up such a claim.

The manufacturer withdrew the product after being investigated for misleading consumers.

Something more definite may someday come from a discovery announced in mid-1990. A team of genetic researchers found a tiny string of DNA on the Y chromosome that they believe to be the gene that determines whether an embryo will become a male or a female. The gene, known as SRY, functions much like a switch. If it's present, it "turns on" a long sequence of reactions that determines the development of sex

organs. The hormones produced by the sex organs trigger the rest of the process: hormones from ovaries result in a female body while those from testes cause a male body to develop.

The researchers who discovered SRY think that they might be able to use the gene to influence the sex of mice and cattle. They are opposed, however, to using it on humans. Besides a host of technical problems, they believe the procedure would be unethical.

They aren't alone. Many people are uncomfortable with the idea of being able to select a baby's gender as easily as choosing his or her first pair of shoes. One great concern deals with the problem of sexual equality.

Many Eastern cultures value sons more than daughters. Some people worry that, once gender selection becomes simple, people from those cultures will choose to have far more boys than girls. That could cause social problems when those many boys and few girls reach marrying age. They also note that women tend to have fewer rights in many of these countries. Any movement towards women's rights could be seriously endangered if men make up the majority of the population.

Others argue that the problems could be easily avoided with regulation. Simply allow every couple to select a son once and a daughter once. After that, they must take the luck of the draw. That would keep the sexes from ever becoming far out of balance.

But that, according to some, might only lead to more subtle problems. The threat of men numerically overrunning women isn't a real concern in the United States. Here families tend to want balance—both sons and daughters. Still, some studies have indicated that, if they had their choice, more American families would choose to have a son first and daughter second than the other way around. Other research has suggested that the order of birth has some effect on personality traits, with firstborns

tending to be more assertive and achievement-oriented than later children. Therefore, the argument goes, there's a risk that the overall male population would grow increasingly more aggressive than the overall female population.

Proponents of gender choice consider that risk small compared to the benefit of being able to help control the birth rate in many poor and overpopulated countries. A family with three daughters wouldn't have to keep trying to have a son. A family with four sons wouldn't have to keep trying to get a daughter. Everyone could have both a son and a daughter in just two childbirths.

Yet, while the controversy has already begun, so far it's all largely theoretical. The reality of guaranteed gender selection is some time away. For now, everyone is still assured of getting either their first or second choice.

Understanding the Issue

1. Treatment to increase the chance of having either a boy or girl is expensive. Do you think this cost should be covered by medical insurance? Why or why not? What if the purpose is to avoid having a son with a sex-linked genetic disorder?

2. What problems do you see developing in a country that values sons over daughters 20-25 years after gender selection becomes available?

3. Suppose a country that now values sons over daughters developed a population with three times as many men as women. How do you think that would affect the way parents selected their children's genders?

Biotechnology

In Utero Treatments for Genetic Disorders

The defeat of many diseases has long depended upon early diagnosis and treatment. For families prone to severe genetic disorders, however, early diagnosis just meant early heartbreak and little chance for hope. Take the recessive disorder known as Hurler's syndrome, for example. Symptoms appear within two years of birth, but nothing can be done about them. Death before age 10 is almost certain.

When medical science devised ways to test embryos for Hurler's syndrome, parents carrying the genetic defect that causes the disease gained the option of aborting a pregnancy. But there was still no treatment in sight.

Today, it's possible that Hurler's syndrome and many similar genetic disorders can actually be cured. The key, as in other types of disorders, lies in early diagnosis and treatment. But "early" means earlier than ever before—not just before symptoms appear, but about 22 weeks before the child is even born.

Hurler's syndrome is caused by a defective gene unable to make an enzyme called alpha-iduronidase. Body cells need this enzyme to break down a group of complex carbohydrates known as mucopolysaccharides. Without the enzyme, these carbohydrates accumulate to toxic levels, creating the lethal symptoms of Hurler's syndrome.

Alpha-iduronidase is one of several vital enzymes produced in the body's stem cells. These are the cells that manufacture blood cells. In adults, they're located in the bone marrow. But they first develop in a fetus's liver. They start migrating to the bone marrow when the fetus is about 14-16 weeks old.

There's no way known to repair defective alpha-iduronidase genes—or any other genes—at any age. If a fetus has Hurler's syndrome, its stem cells will never produce the needed enzyme. But a technique called stem-cell transplantation may make up for the defect. The idea is to give the fetus cells that will make substances its flawed genes keep it from making for itself.

When healthy stem cells are transplanted into the fetus's liver in time, the healthy cells migrate to the bone marrow along with the rest of the cells. If they take, they may produce enough healthy blood cells to prevent the disorder from appearing.

Stem-cell transplants hold great hope for curing a wide variety of genetic disorders. Enzyme deficiencies, such as Hurler's Syndrome, are prime candidates for the treatment. Almost any disease caused by defective stem cells could be targeted, as long as the disease can be diagnosed within 14 weeks of conception.

However, stem-cell transplants are still both experimental and controversial. The greatest controversy is based on the problem of finding healthy cells to transplant. The best source of these cells is tissues from miscarried fetuses. Many people are opposed to using the tissues of these fetuses for experimental procedures.

1. **Issues:** Why might some people be opposed to using tissues from miscarried fetuses for research? How might society benefit from research using these tissues?
2. **Find Out:** Find other diseases or disorders that are candidates for treatment with stem-cell transplants.

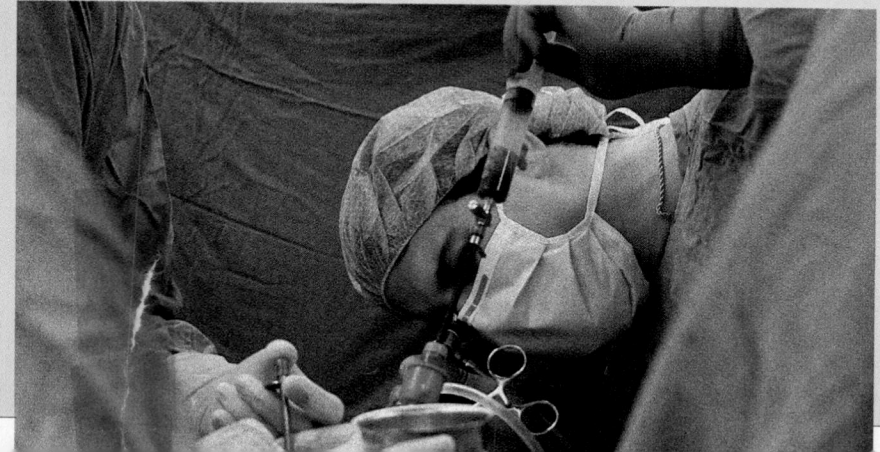

CAREER CONNECTION

Neonatologists are physicians specializing in the care of ill and premature newborns. Among the most common difficulties neonatologists treat—aside from prematurity—are lung problems, failure of newborns to stabilize their temperatures, nutritional problems, and severe skin ailments. Neonatologists generally don't perform surgery on newborns, however. That's left to a pediatric surgeon.

Summary

Using modern techniques, the locations of many genetic disorder-causing genes have been determined. In some cases, the nucleotide sequences of the genes have been worked out and compared to those of normal alleles, and the proteins coded for by the two alleles have been compared. Such work provides information that explains the symptoms of the disorder.

Many human genetic disorders are caused by defective alleles, both autosomal and on sex chromosomes. Although these alleles cause disorders in certain combinations, they may also benefit a population when in other combinations.

Genetic counseling can help prospective parents determine the chances of having children with an inherited disorder. A variety of techniques can determine whether a fetus will have a disorder or not. Gene therapy offers the promise of replacing defective genes with normal ones.

Language of Biology

Write a sentence that shows your understanding of each of the following terms.

amniocentesis
chorionic villus biopsy
fetoscopy
gene therapy

genetic counseling
karyotype
trisomy
ultrasonography

Understanding Concepts

1. Why is it important to know the nucleotide sequences of harmful genes and the abnormalities in the proteins for which they code?
2. Muscular dystrophy is an inherited disorder in which the muscle tissue gradually breaks down. It almost always occurs in males. What does that suggest about its genetic basis? Explain.
3. How might you explain the genetic basis of a disorder that occurs with equal frequency among both sexes and that does not follow basic patterns of Mendelian inheritance?
4. Explain the idea of heterozygote superiority with respect to human genetic disorders.
5. In a rare form of color blindness, a person cannot distinguish the color blue. This abnormality, when it does occur, affects both males and females and in equal numbers. Explain this pattern of inheritance and compare it to that involved in red-green color blindness.
6. Males having Klinefelter syndrome are found to have two X chromosomes and one Y. What is the genetic basis of that chromosome abnormality?
7. Besides the presence of chromosome abnormalities, what else could amniocentesis reveal about chromosomes of a fetus that parents might be eager to know?
8. What advantage does chorionic villus biopsy have over amniocentesis?
9. How does the defective cell membrane protein in CF patients affect transport of materials across the cell membrane? What is the result?
10. How can a pedigree help a geneticist get information about a certain genetic disorder?

Applying Concepts

11. A very common use of ultrasonography is to determine the size and position of a developing fetus. What other organs do you think could be examined using ultrasonography? What could it show?

12. Can you suggest a reason why certain human genetic diseases, such as CF and sickle-cell anemia, occur more frequently among one group of people than another?

13. Compare amniocentesis with chorionic villus biopsy.

14. Why are transfusions useful in treating people with sickle-cell anemia?

15. Human eggs fertilized outside the body will develop normally for a time. If separated at the eight-cell stage, each cell can still develop properly when placed in the mother's womb. Do these facts suggest a means for early detection of genetic abnormalities? Explain what might be done.

16. What is the likelihood that a female would inherit hemophilia? Draw a Punnett square to support your answer.

17. **History Connection:** Explain how the founder effect works in the transmission of a trait in a small, closed population.

18. **Global Connection:** What evidence is there that being heterozygous for Tay-Sachs is a protection against TB?

Lab Interpretation

19. **Investigation:** You hypothesize that having the second toe longer than the big toe is an inherited trait. What experimental procedure would you follow to test your hypothesis? What else would you need to do before you could say your hypothesis was confirmed?

20. **Minilab 1:** Examine diagrams of microscopic views of human red blood cells. Tell which has sickle-cell anemia, which has sickle-cell trait, and which one is normal. Explain how you know.

21. **Minilab 2:** Examine the microscopic views of diagrams of human cheek cells. Which one is male and which one is female? How do you know?

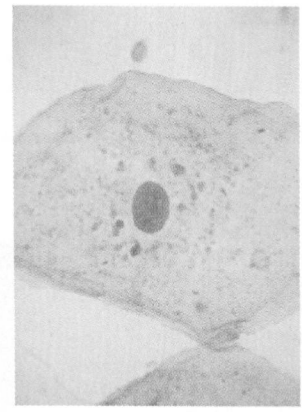 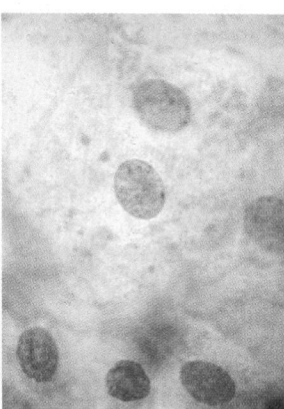

A   B

22. **Thinking Lab:** A scientist would like to test the hypothesis that grinding teeth during sleep has a genetic basis. Current research suggests that it is due to an improper bite. How might the scientist begin the investigation? What might be involved?

Connecting Ideas

23. Healthy people have cell membrane proteins that do not function properly in CF patients. ATP is necessary for these membrane proteins to function. By what process are the ions moved across the membrane?

24. Explain the relationships between DNA, genes, alleles, and chromosomes. How does specific information about chromosomes and their parts help in perhaps one day finding a cure for a specific genetic disorder?

Evolutionary Relationships

Chickens have been artificially selected for different traits for many years. This Rhode Island Red is bred for its good taste and good egg-laying ability. Other breeds of chicken may produce more eggs or more meat. Why isn't the Rhode Island Red a different species than the White Leghorn?

What kind of food do you think this finch eats? This is only one of the 14 species of finches that Darwin identified on the Galapagos Islands. Each species has a different type of beak that is related to the type of food it eats. Think of a woodpecker and a sparrow. Can you tell from their beaks that they eat different kinds of foods?

The Galapagos Islands are home to many unusual organisms, such as giant tortoises, flightless cormorants, and marine iguanas. But to Charles Darwin, the most interesting organisms were the many species of finches. How could so many species of finches exist in such a small area? Darwin's thoughts about these finches and other organisms he saw on his voyage in the 1830s led to his ideas on natural selection.

Variations among organisms have enabled humans to produce desirable organisms that would not occur in nature. Artificial selection for desirable traits has resulted in cows that produce more milk, disease-resistant corn, and seedless grapes.

Selection, either artificial or natural, only works because of the great genetic diversity that exists among species. In this unit, you will learn how variations among organisms enable evolution to occur, and how relationships among organisms of the present can be determined by exploring the evolutionary relationships of the past.

EVOLUTION

WHAT YOU SEE here is like a photograph of a photo-graph. Someone took a picture of the fossil embedded in the rock, developed the film, and made a print on paper—all within a matter of days or possibly even hours. The image in the rock is actually the preserved form of the bones from an ancient organism's body, one that lived on Earth perhaps millions of years ago. The process of fossilization may have taken thousands of years for the surrounding sediments to become rock-hard. Does it remind you of any living organisms you have seen?

The canyon walls shown in the small photo could be called a big-screen image of the past. Like ribbons of time, layers of different rock formations run across the cliff faces. Each layer records an episode in Earth's geo-logic history. A few organisms that lived out their lives during various geologic episodes may have become encased in the layers. The fossilized organisms are entries in the record of life. As you will see in this chapter, such traces of past organisms give us clues about how and when life evolved. They help us under-stand how organisms are related to one another and how the present is tied to the past.

What are the records of your past life? How will your great-great-grandchildren know what you looked like and what your life was like?

Chapter Preview

Objectives

Recognize how various lines of evidence suggest that evolution occurs.

Apply your knowledge of evolution to explain resistance of bacteria to penicillin.

12.1 Evidence for Evolution

You didn't have to sign up for a biology course to discover that a great many different life forms inhabit Earth. You had already experienced them all around you—in trees and smaller plants, in birds and insects, in animals and humans. You may even have puzzled over the existence of such great diversity. How did elephants and elm trees, or mushrooms and mice end up on the same planet? On the other hand, you may have wondered how horses and zebras could be so different from each other and yet so much alike. Perhaps you wondered about chimps and humans in the same way.

Fossil Evidence

So far, you have learned that there are millions of different kinds of life forms on Earth. They all evolved. Remember from Chapter 1 that evolution means change over time. During the course of evolution, some life forms retained closer relationships with one another than others. But how do we know that evolution continues to occur? How does evolution account for the relationships that we observe among existing organisms?

A Fossil Is Formed   Most organisms depart this world without a trace. They die; their remains decay; nothing is left to show that they ever lived. But there are exceptions. Occasionally, one or more actual parts, such as bones, of an organism or even a whole organism is preserved or petrified. Sometimes, a mold or cast of an organism is left in a rock. Any such trace of an organism that lived long ago is a **fossil.**

Some organisms were fossilized when they became trapped and quickly frozen in ice or enclosed in amber, which is resin from ancient coniferous trees. Such fossils can be very clear and revealing. They also are very rare. Most commonly, sand or clay surrounds a tooth, a bone, or other hard part of an organism soon after the organism dies. The sand or clay sediments prevent the body part from decomposing. In time, the sediments around the dead organism turn to rock. Sometimes a shell or skeleton may be preserved with very little change to its chemical makeup. Sometimes, minerals filter into the body of the preserved organism's part or parts and the organism also turns to rock. The organism has become petrified.

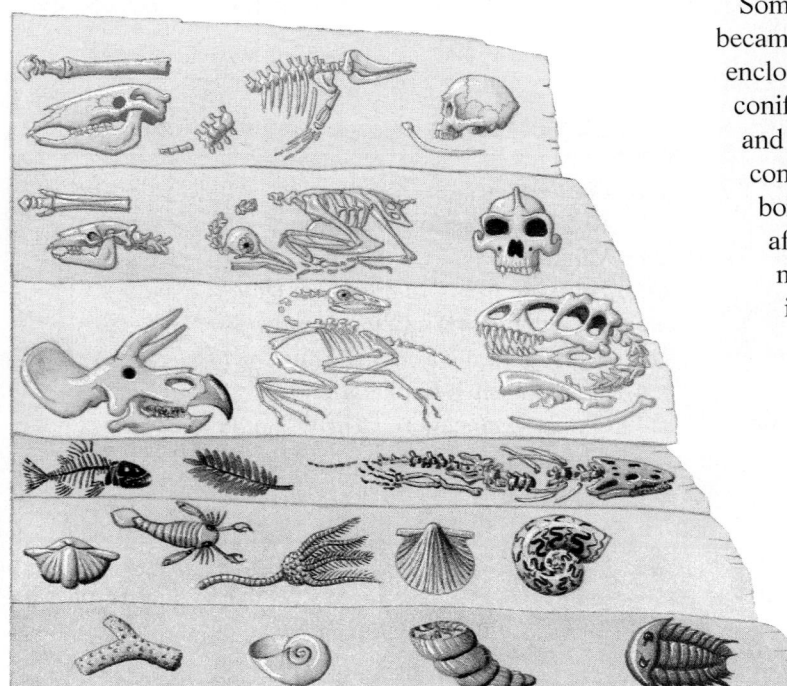

Figure 12-1   **Fossils in the lower layers of sedimentary rock are older than those found in upper layers. Often, the layers of rock can be dated by types of fossils they contain.**

The Fossil Record   Fossils give us information about kinds of organisms that have existed on Earth. Thanks to the layering in sedimentary rock, fossils also give us clues as to when the organisms may have lived. Think of it this way: You stack up newspapers by adding each day's edition to the top. As long as the stack remains undisturbed, the oldest paper is on the bottom and the newest is on the top. In much the same way, particles of weathered rocks settle in layers like the layers in a cake and form sedimentary rocks. Fossils found in the undisturbed lower layers of sedimentary rocks are usually older than those found in undisturbed upper layers. Scientists have determined the relative times of appearance and disappearance of many kinds of organisms from the locations of their fossils in sedimentary rock layers.

Often many fossils are found together at the same location, probably because all the organisms died together in a flood, volcanic eruption, or some other disaster. One site discovered in China consists of sediments from an ancient seafloor dating back 570 million years. More than 70 previously unknown fossil animals were discovered at that site.

Fossil studies often show that the majority of organisms from older layers are generally simpler in body structures than ones from younger layers. The types of fossil organisms change from one layer to the next. Thus, we can conclude that life forms became more complex over time. However, many of the simple life forms found as fossils still exist. Many others have become extinct.

Important as it is, the fossil record remains incomplete. Because fossil formation has occurred so rarely, the record may never be complete enough to show every step in the evolution of Earth's life forms. Because soft tissues are rarely preserved, it is certain that not every kind of organism that ever lived has left a record. Nevertheless, many fossils that show strong similarities to existing organisms have been found. For example,

Minilab 1

What are microfossils?

Certain fossils are so small that they can only be observed with a microscope. They are called microfossils. Prepare a wet mount of fossil "powder" taken from sedimentary rock. Examine the sample under low and high power. Make diagrams of what resembles shells of organisms. Using Appendix A, attempt to identify these organisms. You have three clues for help: size, where it lived, and the structure being viewed.

Analysis   *How do you know that fossils once lived in water? Explain why these fossils may not have been destroyed over time.*

Thinking Lab

INTERPRET THE DATA

Do rock layers provide clues for judging their age?

Background
The law of superposition states that most sedimentary rocks are laid down in horizontal layers. The fossils in these layers increase in age from top to bottom as long as the layers have not been disturbed. The two formations shown here are formed from sedimentary rocks and have not been disturbed in position. Rock layers of like colors are the same age.

You Try It
Use the law of superposition to answer the following questions. **Compare** rock formations 1 and 2.
1. Which layer in each formation is the oldest? Explain how the fossil records in these two layers compare.
2. Which layer is the youngest? Explain how their fossil records will compare.
3. Which layers are the same age? Compare their fossil records.

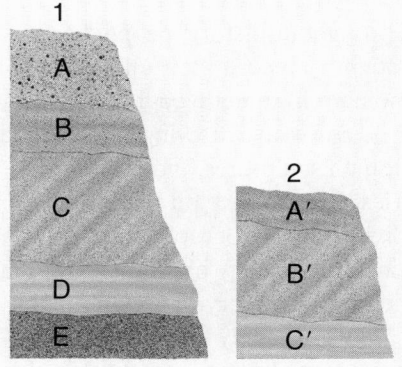

Mammoth

Mastodon

modern-day elephants are considered to be related to mastodons, fossils dating back 35 million years, and to mammoths, extinct elephants dating back to 2 million years ago. The fossil record as a whole indicates that organisms have changed over time; they have evolved.

Evidence from Anatomy

Have you ever been goofing around with friends and had somebody challenge you to walk like a chimp or flap your wings like a chicken? Your imitations may have been convincing—as well as entertaining—but they could never duplicate the real thing. Why? Your body structure is similar enough to that of a chimp or a chicken to get across the idea, but too different to give an exact rendition.

Elephant

Figure 12-2   The fossil record indicates that the ancestors of elephants were wooly mammoths that lived 2 million years ago. Mammoths are probably the descendants of mastodons that existed 35 million years ago.

ART CONNECTION

Sculpting the Past

Throughout history, people have sought to record their unique moment in time through the visual arts. Early cave drawings, Greek and Roman sculptures, medieval tapestries, Renaissance oil paintings, Civil War photographs, and, now, video technology have left us with a rich visual history of life on Earth.

Unfortunately for anthropologists, some crucial gaps exist in this

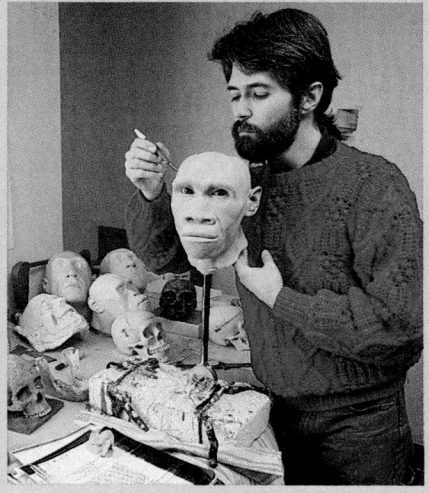

visual record. Specifically, no one knows for certain what the precursors of *Homo sapiens* looked like. Today, however, anthropologists and forensic artists have formed a partnership to attempt the reconstruction of extinct humans as they might have looked more than a million years ago.

This partnership has joined sculpture and science to create a technique that can turn a skull into a fully formed human head with scientific accuracy and artistic realism. Here's how the process works.

Look now at the forelimb structures of three familiar animals, as shown in Figure 12-3. Notice first the overall similarity, then the differences in details. What you are doing is what scientists do in **comparative anatomy**—the study of the structures of different organisms. Comparative anatomy provides further evidence of evolution. Different kinds of organisms—such as a bird, horse, or human—share similar structures.

Homologous Parts As you have just observed, each animal in Figure 12-3 has a forelimb structure that is a variation of a common pattern. The commonality suggests that these and other vertebrate animals are all related. They probably evolved from a common ancestor that had the same basic limb structure. Over many eras, these animals moved into different environments, and their limb structures gradually changed in form as each group became adapted to its new surroundings and ways of life. The differences in vertebrate forelimb structures today are suited to the way each kind of animal moves. Hence, a human couldn't walk exactly like a chimpanzee any more than you can flap your arms like a chicken.

All organisms inherit structures from their ancestors. As organisms evolve, any of their inherited structures may become modified. Some descendants may inherit one modified form of a structure; other descendants may inherit other modified forms. Yet, the overall pattern of the structure still resembles that of the common ancestor, as in the leg structure of the bird, horse, and human. Such modified structures among different groups of descendants are said to be **homologous**. The greater the

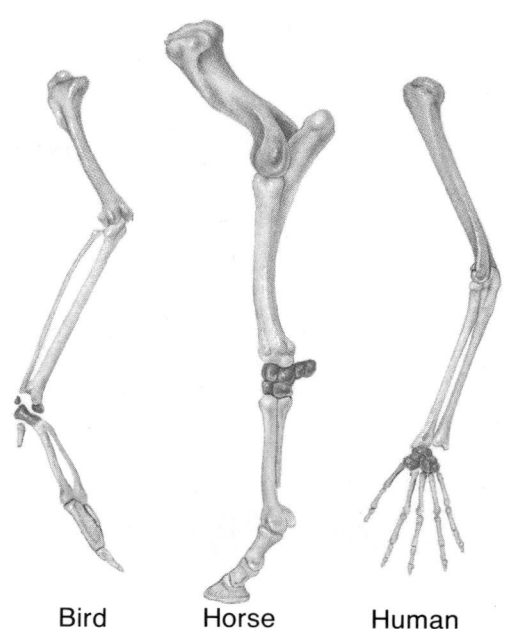

Bird   Horse   Human

Figure 12-3 **Compare the forelimbs of a bird, a horse, and a human. The bones are colored to show which ones are homologous.**

Beginning with a skull, the artist looks for telltale impressions left on the skull indicating where muscles were attached to bone. The size and depth of these impressions tells the artist how large or small the muscles were. By studying these impressions, as well as the musculature of present-day humans and the closely related great apes, the artist can sculpt in clay the underlying muscle layer of the face.

Next, the smaller muscles of the face are added. Again, comparative anatomy plays a role. If the chewing muscles of one skull are smaller than those of another, the artist can conclude that the other facial muscles were probably smaller as well. Once the muscles are in place, the artist adds the nasal cartilage, deep fat layers, and facial glands. Finally, the skin is added, and the sculpted head is complete. All that remains is the addition of some texture, color, and hair to the skin.

Of course, no one can say for certain just how close forensic artistry has come to reproducing the actual look of extinct humans.

However, evidence from forensic artists' work on contemporary murder victims for whom photos do exist suggests that their techniques can be quite accurate in reconstructing faces from the past.

———

1. **Explore Further:** Research to discover how forensic artists have assisted law enforcement in identifying murder victims.
2. **Connection to Paleontology:** How might the techniques of forensic artists aid in the study of dinosaurs?

number of homologous structures that different kinds of organisms share, the closer the relationship among them. Which do you think is a closer evolutionary relative of a cat, a dog or a lizard? Why?

Homologous parts among descendants of a common ancestor are often similar in structure, in function, or in both structure and function. However, similarities in function alone do not indicate common ancestry. For example, you know that the wings of both birds and insects have the same function in flight, but it is also clear from comparative anatomy that they are not homologous. The wings of birds and insects developed from entirely different structures in two completely unrelated groups of ancestors, as supported by fossil evidence.

Vestigial Organs   Examine the drawing of the vestigial structures in Figure 12-4. Notice the difference in the sizes of the caecum in a horse

I N V E S T I G A T I O N

Forming Your Own Fossil

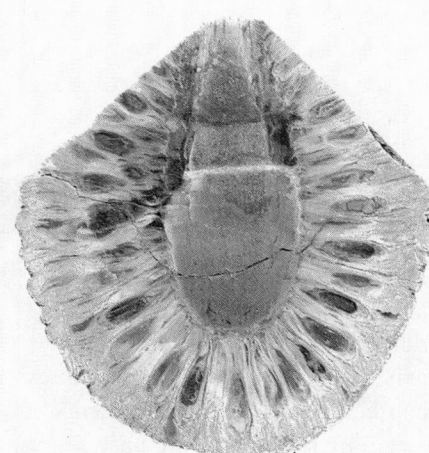

Trace Fossil

Fossils form when an organism part or an entire organism is preserved in sediments before it has a chance to decay. One common type of fossil is formed from petrified bones or shells. Body parts such as bones or shells often become harder with chemical changes to the structure's minerals. Trace fossils are simply tracks or impressions of plants or animals left in soft sediment that eventually hardens. A soft body part of an organism may become buried in sediment that hardens with time. If the organism then dissolves and decays, a mold remains in the hard sediments. A mold can sometimes become filled with minerals that harden into a cast or copy of the original organism. Fossils, regardless of the type and how formed, provide a chance to look back in time. They serve as a source of evidence for evolution.

Problem
How is a fossil formed?

Safety Precautions
Handle all materials safely. Wear an apron when measuring, mixing, and pouring all powders.

Experimental Plan
1. As a group, make a list of possible ways you might design your procedure using the materials your teacher has made available.

Your teacher will provide your group with any needed information concerning correct handling of materials.

2. Agree on one method from your group's list that could be tried in the classroom.
3. Following the style of a recipe, write a numbered list of directions that anyone could follow.
4. Make a list of the materials and the quantities you will need.

and in a human. Structures that have no function in the living organism, but may have been used in the ancestors, are said to be **vestigial.** Vestigial structures may be homologous with still-used structures in other related organisms. For example, the human appendix, a small wormlike tube connected to the large intestine, is vestigial in humans. However, in other animals that graze on plants, a similar but much larger structure, the caecum, is used for digesting this tough, fibrous plant material.

How do vestigial organs come about? At one time in the past, a vestigial structure was useful for the organism's way of life. As time passed, the organism changed in form and behavior as it became adapted to a new lifestyle. The structure, although reduced in size and function, continues to be inherited as part of the organism's body plan.

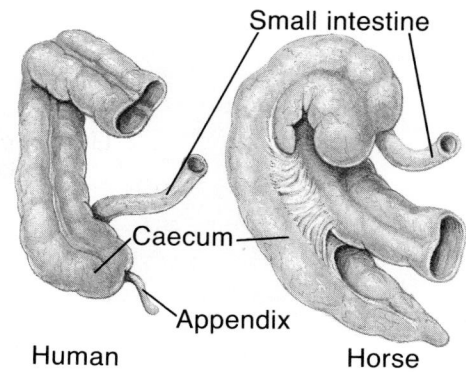

Figure 12-4 **When compared with the caecum of a horse, the caecum and appendix of humans is thought to be vestigial.**

Checking the Plan

Discuss the following points with other group members to decide the final procedures for your simulation.

1. What will be used as the original object to be fossilized?
2. Will the original object be destroyed in the procedure or will it be recovered? This is an important consideration if the object you choose to fossilize is a personal possession of one member of your group.
3. Will your fossil be a petrified original, a trace, a mold, or a cast?
4. How much time will be needed to prepare your fossil?
5. Assign specific roles and jobs for each member of your group.
6. If unsuccessful in forming a fossil the first time, be prepared to record and correct any errors that were made. Make modifications in your original procedure, and repeat the experiment if time permits.

Cast Fossil

Make sure your teacher has approved your experimental plan before you proceed further.

Data and Observations

Carry out your experiment. Diagram and label the original object used for the fossil as well as the final fossil produced.

Analysis and Conclusions

1. Based on this activity, record your own definition of a fossil.

2. Determine why the fossil you formed fits the category of petrified fossil, trace, mold, or cast.
3. Describe how the fossil that you formed is both alike and different from the original object used to form your fossil.
4. Suggest how the procedure that you used to form your fossil may have differed from the procedure that takes place in nature.
5. Assume that your fossil is real and had formed millions of years ago. What data could be taken from your fossil that might help scientists in their studies of evolution?
6. If your fossil-forming technique was not successful, explain why and describe the procedure changes you would make in a second trial.

Going Further

Based on this lab experience, design another procedure that would allow you to form another fossil type. Record this information as part of your conclusion.

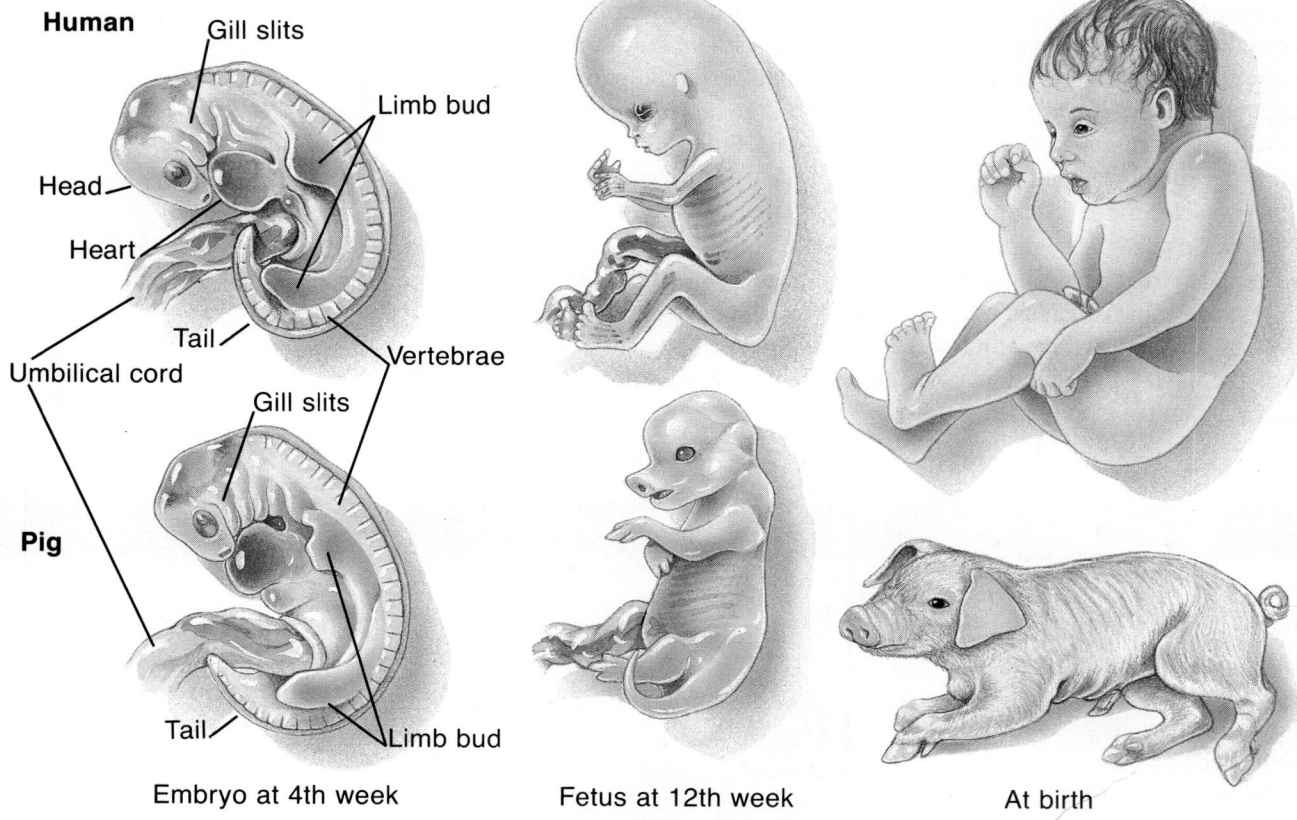

Human
- Gill slits
- Limb bud
- Head
- Heart
- Tail
- Umbilical cord
- Vertebrae

Pig
- Gill slits
- Tail
- Limb bud

Embryo at 4th week   Fetus at 12th week   At birth

Figure 12-5 Although at the fourth week of development it is difficult to tell which embryo is a pig or a human, by the 12th week the two fetuses are easily distinguished.

Evidence from Embryology and Biochemistry

Anatomy of mature organisms isn't the only area where comparisons point to evidence of evolution. Scientists also compare and contrast **embryos,** which are the early stages of developing plants or animals. The study of developing organisms, known as **comparative embryology,** shows a number of relationships not obvious in the fully grown organism.

If you viewed an early human embryo and an early pig embryo side by side under a microscope, you'd have a hard time telling them apart. The more alike the development of two organisms is, the more closely related they are thought to be. Thus, the pig and the human must be closely related. They probably inherited the same basic body plan from a common ancestor. As the pig and human embryos progress in their development, their patterns become more and more different. In their later stages before birth, they attain their distinct and more familiar forms.

Studies of organisms on a biochemical level, which constitute **comparative biochemistry,** also provide evidence of evolution. Biochemical comparisons show, for instance, that the structure of hemoglobin in a chimpanzee strongly resembles the structure of human hemoglobin. That is, the sequence of amino acids in chimpanzee and human hemoglobin is almost the same. The amino acid sequence of human hemoglobin and dog hemoglobin shows less similarity. It appears then that humans and chimpanzees have a more recent common ancestor than do humans and dogs. Much research today focuses on comparing sequences of amino acids in organisms to reveal relationships. Some scientists say that these studies provide the most fundamental evidence for evolution.

Genetic Evidence

Our understanding of evolution deepens with the expanding fossil record, with new studies in amino acid sequencing, and with all the other lines of evidence you have just read about. The fact that so many different lines of evidence exist strengthens the conclusion that evolution occurs. We have seen that change in life forms occurs over time, but how are these changes brought about?

A Basis for Change   Each generation of a given kind of organism is linked to the next by genes (DNA). But genes and the chromosomes that carry them can both mutate. Mutations, as explained in Chapter 10, are mistakes that occur in the genetic material. The mutation and duplication of existing alleles can give rise to new alleles or genes, and, thus, to new proteins. In addition, meiosis and fertilization reshuffle alleles. In other words, within any breeding group, or **population,** of organisms, there is constant change over time. Production of new alleles and genetic recombination could be called the raw materials of the process of evolution.

Humans unknowingly used the potential for change in organisms thousands of years ago when they first domesticated plants and animals. Working with wild plants in their environment, humans developed crops such as barley, wheat, and corn. They also learned to tame wild animals such as cattle and dogs for food and protection. In each case, people selected only individuals with the most desirable traits and bred them so the traits would be passed on to the offspring. They repeated the steps generation after generation. Without realizing the effects of their breeding practices, humans were directing the evolution of many kinds of plants and animals. Today we call such practices **selective breeding.** Farmers, horticulturists, pet breeders, and scientists still use selective breeding to improve domestic plant and animal varieties.

Figure 12-6   Beef cattle are a product of selective breeding. There are six major breeds in the United States. They are the Aberdeen Angus (shown here), Brahman, Charolais, Hereford, Polled Hereford, and Sinmental.

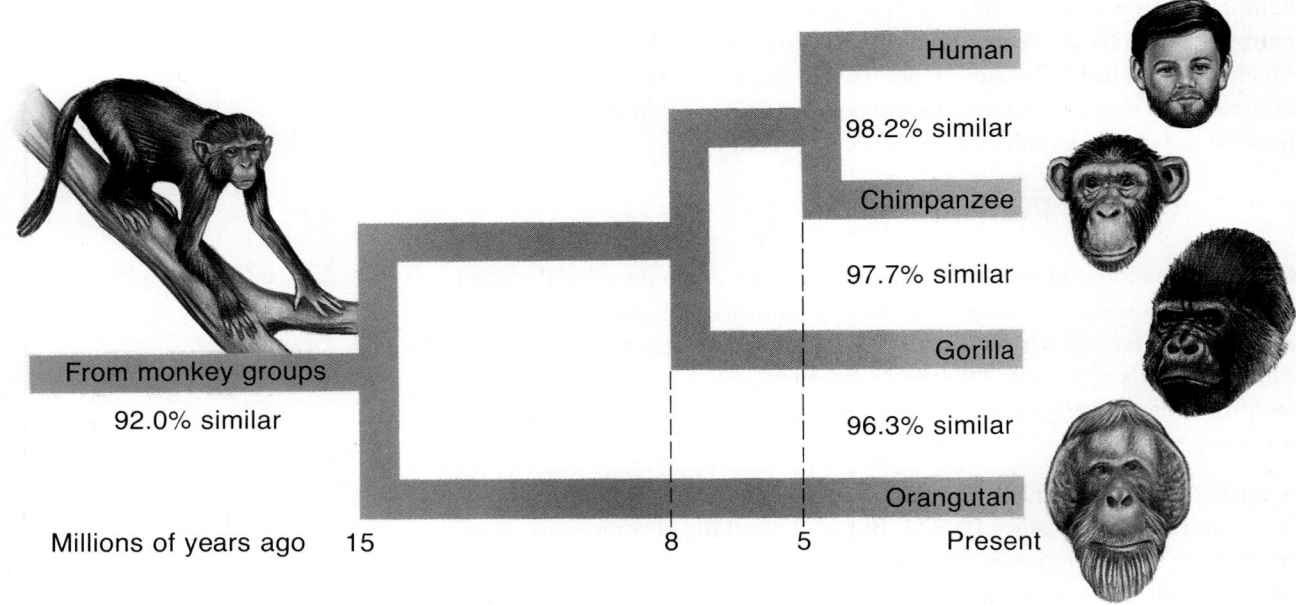

Human

98.2% similar

Chimpanzee

97.7% similar

Gorilla

From monkey groups

92.0% similar

96.3% similar

Orangutan

Millions of years ago   15   8   5   Present

Figure 12-7 Comparisons of DNA nitrogen based sequences show that chimpanzees are more closely related to humans than to gorillas or other apes.

A Basis for Relationships   Knowledge of genetics can help us understand more than how changes in genotypes and phenotypes come about. Genetics can also tell us how different groups of organisms are related back through time. Biologists first analyze the base sequences of DNA in the genes of one kind of organism. Then they examine genes controlling the same or similar traits among different kinds of organisms. As in other comparative studies, the greater the similarity, the closer the relationship. The DNA of humans and chimpanzees, for example, is 99 percent identical. Between humans and other mammals, 80 percent of the DNA is identical. Again, the evidence points to a close relationship between humans and chimpanzees.

Biologists have developed a timetable that they believe reflects the rate at which mutations occur over given time spans. Using these mutation rates, scientists have estimated the time when certain descendants began to go their separate genetic ways from a common ancestor. Mutation rates, like other lines of evidence, show that chimpanzees and humans branched off from a common ancestor at a comparatively recent point in evolution.

Direct Observation

Do you now find it easier to picture evolution? You won't be alone if you answer that the evidence that organisms have changed over time described so far seems somewhat indirect. If so, it's because the changes you were reading about took place so slowly over such a long time that no one could possibly have observed them directly. (Regardless of the fact that humans were not yet around for most of the changes!) There is, however, other evidence for evolution that can be observed directly. The evidence comes from evolutionary changes that occur rapidly and on a regular basis.

Rapid Evolution   The reaction of bacteria to penicillin often results in rapid evolutionary change. Penicillin, the chemical from the *Penicillium* mold, kills many kinds of bacteria. When penicillin is added to bacteria in a culture dish, a clear zone forms indicating death of the bacteria in that zone. Sometimes, however, a few bacteria may survive and reproduce to form visible, living colonies. Each colony will consist of millions of cells that are resistant to penicillin.

When penicillin is added to a culture dish, it changes the environment of the bacteria. Resistance to penicillin becomes vital for the survival of a bacterium. All the bacteria that have no resistance will die. As in any other group of organisms, there are variations in the genetic material among bacteria. Those few individuals with a gene for penicillin resistance will survive and go on to reproduce. Thus, this group of bacteria has evolved, or changed over time. The organisms that continue to live and reproduce after the change in the environment are adapted to the new environment.

Evolution Causes Health Problems   As you may have already guessed, penicillin-resistant bacteria create a serious health problem. Bacteria cause many diseases in humans and animals. You probably have taken penicillin or other antibacterial drugs, commonly called antibiotics, to combat an illness. On average, each bacterium in a colony will divide every 20 minutes. It is easy to see why evolution in bacteria can be rapid. When resistant forms of bacteria evolve, existing drugs will no longer kill them. For this reason, scientists need to search continually for new antibiotics to treat bacterial diseases.

Rapid evolution has also occurred among pest organisms such as weeds and insects exposed to pesticides. As these organisms have evolved a resistance to these chemicals, stronger chemicals have been used and are often applied in greater amounts. But, pesticide use endangers non pest organisms too. For example, DDT interferes with reproduction in large birds, such as falcons and pelicans. Some of these powerful chemicals have also caused health problems among the workers who apply them.

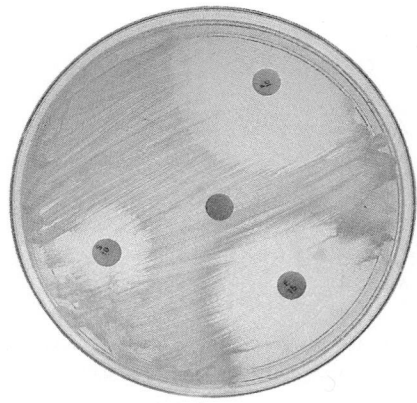

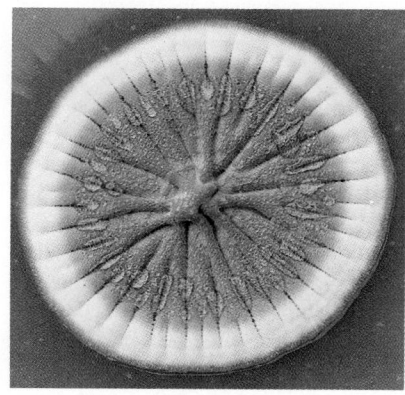

Figure 12-8   **The discovery that bacteria could be killed (the clear areas in the culture in the top photo) by a chemical released by the penicillin fungus (bottom) was made by Sir Alexander Fleming in 1928. The bacteria not killed by this or other antibiotics are resistant strains.**

Section Review

Understanding Concepts
1. Which organism would you expect to have evolved first, a spider or a lizard? Why?
2. Why should you complete a prescribed course of antibiotics to treat a bronchial infection?
3. Would vestigial structures make any sense if organisms did not evolve? Explain.

4. **Skill Review—Predicting:** You are part of a team collecting fossils at a new site. The fossils are embedded in sedimentary rock. You discover a fossil of an animal in a low layer of rock that looks something like a modern insect. Knowing about the fossil record, what would you expect to see in similar fossils found in upper layers of rock? For more help, see Representing and Applying Data in the **Skill Handbook**.

5. **Thinking Critically:** Why is it logical to say that if two animals have similar proteins, they are probably closely related genetically?

Objectives

Analyze how Darwin's observations led him to conclude that life forms change over time.

Explain the process of natural selection.

Apply modern knowledge of genetics to the theory of natural selection.

Figure 12-9 **Darwin's voyage in *HMS Beagle* (left) took him south from England, around Cape Horn, and north to the Galapagos Islands; a journey (right) that led to the conclusion that evolution occurs.**

12.2 Darwin's Explanation

What will you be doing six or seven years from now? Perhaps you will be working at a new job, or maybe still going to school. A little less than 200 years ago, a young Englishman named Charles Darwin (1809-1882) was still trying to decide his future. By the time he was in his 20s, he had tried both medical school and law school, and liked neither. Long nature walks interested him more.

Darwin: Gathering Data

While attending Cambridge University, Darwin became friends with a professor of natural history. The professor encouraged the young student to apply for a job aboard *HMS Beagle,* a British survey ship used for making voyages of scientific discovery. Darwin's knowledge of biology and geology impressed the captain of the *Beagle.* Darwin got the job of ship's naturalist. At the age of 22, he began the first and only sea voyage of his life. No one, including Darwin, guessed what that voyage would one day mean to science.

Observing Biodiversity   The *Beagle* left England for a five-year, around-the-world voyage in December of 1831. As ship's naturalist, Darwin began immediately to collect, study, and store specimens of the ocean's marine life. After crossing the Atlantic Ocean, the *Beagle* made many stops along the east and west coasts of South America. Darwin went ashore to study the land environment and to collect fossils.

Based on studies of fossils and observations in comparative anatomy, some people of Darwin's time had already guessed at the idea of evolution. Few scientists of the day, however, accepted evolution as fact. They did not have the body of evidence that scientists have now. In fact, Darwin avoided using the term *evolution* because it was used at that time

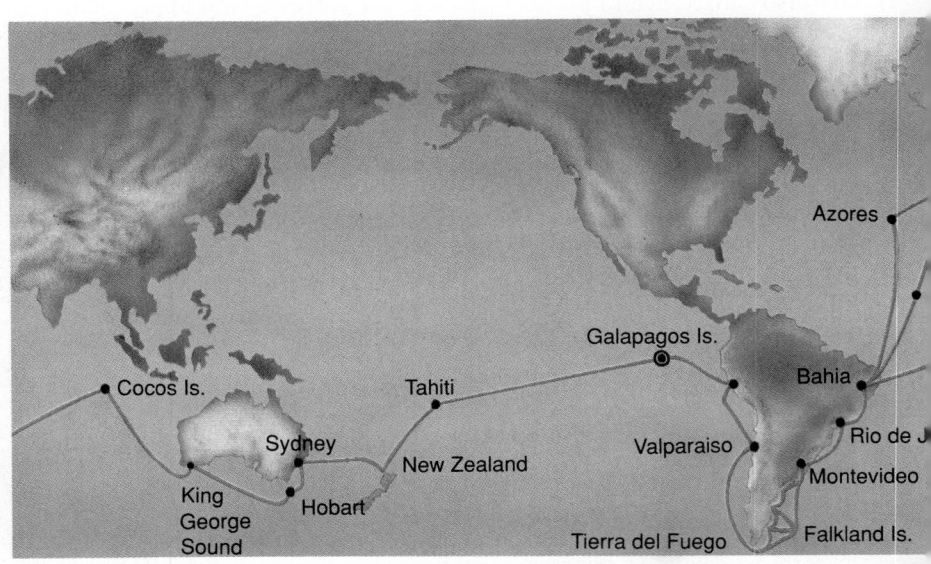

to describe the theory that embryos contained fully formed but miniature versions of the adult. As you can guess, the light microscope was not sufficiently high-powered to test this hypothesis.

During his voyage, Darwin read a book by a contemporary scientist, Charles Lyell. Lyell discussed how changes occur over time in geology. Lyell's ideas, together with Darwin's own observations of both geology and living organisms studied during his voyage, caused Darwin to think and ask questions. With the knowledge that environmental features of Earth change, it was clear to Darwin that organisms adapted to their environments must also change to remain adapted. If populations of organisms do change, he considered, how does the change occur?

In the Galapagos For Darwin, the answers to his questions pointed to evolution, or, as he described it, descent with modification. In other words, structural or functional changes occur from one group of descendants to the next, and so on. Studies into comparative anatomy that Darwin made during his voyage seemed to confirm his guess. Organisms he studied on the Galapagos Islands, about 600 miles west of South America, offered especially convincing evidence. Forms of life, such as giant tortoises and marine swimming lizards (iguanas), not found anywhere else on Earth, lived on the Galapagos. These animals were at once unique and yet similar to other forms of the same animals common in other parts of the world. The similarities between Galapagos plants, birds, amphibians, and reptiles and those of distant lands persuaded Darwin that despite their uniqueness, the Galapagos organisms were all related to the more common forms. By the end of the trip, Darwin was convinced that evolution occurs—that life forms can and do change. But as yet, he still needed to test these ideas before he could or would attempt to explain how or why such changes happen.

Figure 12-10 The Galapagos Islands are renowned for their unusual animal and plant life. The giant tortoises (left) are the largest on Earth. The marine iguanas (right) have become adapted to feeding on algae underwater.

Darwin's Explanation: Natural Selection

Charles Darwin spent the next 20 years after his voyage studying his collections as well as living organisms to help him explain to the world how evolution occurs. His methods of inquiry followed sound principles of scientific investigation. He collected still more organic specimens, he studied breeds of different animals, and he made countless additional observations in his greenhouse. After many years of evaluating his data, two facts led Darwin to hypothesize that change occurred resulting in descent with modification.

A Struggle for Existence   The idea for the first fact Darwin investigated came originally from an essay about population growth by Thomas Malthus. Malthus proposed that the human population was growing faster than the food supply needed to feed it. Darwin applied Malthus's idea not to humans, but to other forms of life. He knew that in nature there is commonly an overproduction of offspring. In other words, many of the offspring produced by organisms die before reaching maturity. For example, a mouse will give birth to six young. Imagine how many mice there would be in the world if every litter of six young mice survived, and every female mouse gave birth to 150 young and none died! Earth would be a large, mouse-infested planet within a short time. Darwin realized that there must be a struggle for existence among individuals. Darwin envisioned many kinds of struggles—competition for food and space, escape from predators, and ability to find shelter. Only some of the individuals in any population survive the struggle long enough to produce offspring of their own. But which ones?

Figure 12-11 **Two factors lead to gradual changes in new generations of species. (a) Traits in individuals that can compete successfully and reproduce will be passed on. (b) Traits in individuals that have been artificially selected will be passed on to new generations.**

a

b

Artificial Selection   The second fact that Darwin examined came from his inquiries into selective breeding. Darwin knew that plant and animal breeders generated new varieties of desirable plants and animals by selecting parents with desired traits. He observed that in any population, individuals have different variations of traits that can be inherited. For example, plants vary in color, height, mass, and number of seeds produced. Among animals, cows vary in amount of milk produced, sheep vary in the thickness of their wool, and chickens vary in the numbers of eggs they produce. Plant and animal breeders select parents having desired characteristics and breed them to produce preferred offspring. Darwin observed that breeders do not use individuals as parents if they have less desirable traits. He wondered if there were some force in nature similar to this *artificial selection* practiced by breeders.

Evolution Explained   The two facts, a struggle for existence and artificial selection, gave Darwin the information he needed to explain descent with modification. He observed that variations exist among all groups of organisms. Those individuals with the traits that aid them in coping with their environment have a better chance of surviving and, thus, leave more offspring. The offspring, in turn, tend to have the same traits and so are adapted to the environment.

As an example of this evolution, imagine that long ago among a population of spadefoot toads, a few had a louder mating croak and longer toes on the back feet. The population lived in a desert where rain comes only a few times each year. As you know, the young of toads can survive only in water. Two things are vital to a present-day desert spadefoot toad. First, it must be able to bury itself in the ground when there is no available water for breeding. The toad uses its back feet to dig itself down into the soil. The toads can survive buried in the caked mud for many months. Second, when rain falls, the toads must be fast to emerge and mate before the water is evaporated by the hot daytime sun of the desert. You may now begin to understand why the toads with the louder mating croak and longer back toes were the ones best suited to the environment in which they lived. Any toads with less-adapted traits would have left fewer offspring. The water would be gone before the tadpoles were fully mature and able to bury themselves. Darwin noted that individuals with traits not suited to their environment leave fewer, if any, offspring. In time, those individuals lose in the struggle for existence.

Figure 12-12   **The spadefoot toad is just one animal that has successfully evolved to survive in desert conditions.**

Figure 12-13 Many plants, such as these in the squash family, overproduce seeds. There is a greater chance that some seeds will survive to produce their own offspring. What are some animals that produce offspring in these enormous numbers?

Since the struggle for existence goes on unaided in nature, Darwin realized that any selection that occurs in a population is natural. The organisms that survive the day-to-day struggle are best fitted or adapted. Darwin related changes in organisms to a natural force or selection pressure that his lifetime work had set out to discover. In the course of **natural selection,** the best adapted individuals in a population survive and produce offspring that are likewise well adapted. The least adapted individuals produce fewer offspring, which, along with their parents, may die young and leave still fewer offspring.

According to Darwin's model, natural selection is an interaction of organisms and environment. Evolution, then, does not occur according to a plan or goal, as in artificial selection. It is rather a matter of variations and chance. Variations that arise either do or do not fit the environment. Suitable variations are passed on from one generation to the next; other variations are eventually lost to the population. The following sequence summarizes Darwin's explanation of how organisms evolve. They make up what we now call Darwin's **theory of natural selection.**

(1) In nature there is a tendency toward overproduction.
     For example, mice, watermelons, ants, fish, dandelions, and mosquitoes are well-known overproducers.

GLOBAL CONNECTION

The Question of Color

Take a walk down a busy boulevard in any major city in the United States and look closely at the people around you. You'll notice a rainbow of skin colors, from the palest white to the darkest black-brown. Why do humans, who all belong to the species *Homo sapiens,* have such a broad variation in skin color? The reason can be traced to a combination of events in history and human cellular biology. However, to fully understand this interaction, you will need to know the importance of skin color to human health.

When human skin cells are exposed to ultraviolet (UV) radiation, lipids in the cells are converted to vitamin D. Vitamin D is essential to human health because it aids the body in absorbing calcium from foods. Too little vitamin D can lead to rickets, a bone disease from calcium deficiency that results in brittle, misshapen, and easily broken bones. Too much vitamin D can result in excess calcium in the blood plasma, causing kidney stones and extra bone formation in the tendons.

How does your body regulate the production of vitamin D from sunlight? The key is melanin, the pigments in skin cells that help turn the skin dark as it is exposed to rays of the sun. Melanin absorbs UV radiation. The more melanin pigments in the skin, the less UV radiation is absorbed by lipids for the production of vitamin D.

How, then, can we account for differences in human skin colors? We know that people who live within equatorial latitudes have dark skin due to the large amounts of melanin pigments in their skin cells. Thus, despite constant exposure to intense UV radiation, these people do not manufacture too much vitamin D because their melanin pigments absorb much of the sunlight.

People who live in the northern latitudes have fewer melanin pigments and are exposed to much less sunlight. The lower levels of UV are absorbed by lipids, and, therefore, sufficient vitamin D is produced.

All the available anthropological evidence suggests that the species *Homo sapiens* originated as dark-skinned people in the tropical regions of Africa. So, how did blond-haired, blue-eyed Scandinavians evolve? If humans were all originally dark-skinned, why didn't everyone remain dark?

(2) Not all offspring that are produced survive.

For example, many bacteria treated with antibiotics will die. Not all seeds from a watermelon will grow and thrive.

(3) Variations exist in any population.

For example, some spadefoot toads mature faster than others.

(4) Variations are inherited.

For example, the long fur of a cat will be passed on to the cat's offspring. The color patterns of many insects are the same from generation to generation.

(5) Those individuals with variations that are suitable for their environment will live longer and leave more offspring on average than will individuals not having the variations. Thus, suitable variations will tend to be passed on and unsuitable ones lost.

For example, penicillin-resistant bacteria will survive and reproduce after treatment with this antibiotic.

(6) The resulting population as a whole will change as it becomes better adapted to its environment.

For example, an entire population of mosquitoes evolves a resistance to an insecticide.

When people with dark skin from the lower latitudes first began migrating north, they probably didn't fare well because of limited levels of sunlight. Their melanin pigments would have absorbed too much sunlight, and their bodies may have produced too little vitamin D. Diseases like rickets have commonly been found in northern human fossils.

Lighter-skinned people, however, would have been at an advantage. Natural selection clearly would favor these paler humans. With fewer melanin pigments to absorb UV radiation, their bodies were able to continue manufacturing enough vitamin D to prevent disease. In short, lighter-skinned people were adapted to this new environment; darker-skinned people were not. As natural selection continued to operate over tens of thousands of years, the number of people with lighter skin grew in the northern latitudes. Today, the availability of improved medical care has allowed for people of all skin colors to reside safely together in any region of Earth.

1. **Explore Further:** Find out how vitamin D deficiency in humans is avoided in areas of low sunlight. How do light-skinned people protect themselves from high levels of UV radiation?

2. **Issue:** Throughout history, skin color has been used as a justification for hatred, prejudice, and discrimination. Write a brief essay explaining how your own attitudes about skin color have been affected by your knowledge of evolution.

Wallace's Same Conclusion   Darwin was not alone in his investigation into evolution. A fellow scientist, Alfred Wallace, had made his own observations on journeys to other remote parts of the world. While resting from malarial fever in the West Indies, Wallace happened to read Malthus's essay on population. Within two days, he arrived at the same basic explanation as Darwin.

Wallace did not know that Darwin had also been developing a theory of the process behind evolution. He wrote to share his views with Darwin and to seek Darwin's opinions. Wallace's letter arrived shortly before Darwin planned to publish *On the Origin of Species, by Means of Natural Selection,* a book describing his years of observations and the supporting evidence that led to his explanation of evolution.

Lyell and other scientists suggested that Darwin and Wallace both present short summaries of their ideas at a joint meeting of scientists, which they did. After that, Darwin went on to publish his book in 1859. Darwin is the scientist most often remembered as the pioneer in evolution. But the idea of natural selection should properly be credited to both Wallace and Darwin.

Darwin himself could not explain the cause of variations or how they are passed on. He knew nothing about genes, much less about mutations and duplications, and nothing about meiosis and recombination. This lack of knowledge, however, did not detract from his lifelong work of explaining descent with modification. What we know about genetics today, in fact, supports Darwin's theory of natural selection.

Population Genetics

Have you ever wondered if individuals could evolve? By now you have probably recognized, as Darwin did, that it is a group of related individuals—a population, not an individual organism—that evolves. The genotype or genetic material of an individual remains constant. But, over time, the genes within a population may change. New alleles arise and are recombined, thus producing novel phenotypes.

Geneticists picture the entire collection of genes among a population as its **gene pool.** Each new offspring in the population draws its genes from this collection. It is the gene pool that evolves. The study of gene pools and the changes they undergo is called population genetics.

GENETIC ANALYSIS LAB

YOUR FAMILY GENE POOL OBVIOUSLY HAD A BRAIN DRAIN.

11-18
THAVES
© 1992 by NEA. Inc

Observing Gene Pool Changes   Imagine a population of insects in which alleles *A* and *a* exist. The number of times you can expect to see any allele in a population is often described as its frequency. In our imaginary population, the frequency of allele *A* is 80 percent and the frequency of allele *a* is 20 percent. When the population's mature individuals reproduce, they will produce countless numbers of gametes—80 percent carrying the *A* allele and 20 percent the *a* allele. Those gametes represent the gene pool—for *A* or *a* alleles—from which the genotypes of the next generation will be drawn. Because the individuals reproduce sexually by meiosis, some genotypes will be *AA,* some *Aa,* and others *aa.*

Suppose that after ten generations, the frequency of the two alleles changes; now only 60 percent of the alleles are *A* while 40 percent are *a.* The change in percentages indicates that the gene pool has evolved.

Natural Selection in Gene Pools   How do gene pools evolve? Gene pools evolve in a number of ways. One important way is by the process of natural selection. The genotypes—*AA, Aa,* or *aa*—dictate certain phenotypes. Phenotypes are the expression of the alleles that exist in populations. Suppose, for example, that the alleles *A* and *a* control the color of an insect. Insects with the *A* allele are greenish brown; insects with the *a* allele are dark brown. Because the *A* allele is dominant, most of the insects will be greenish brown. Some phenotypes are better suited to the environment than others. Suppose the phenotypes resulting from the genotypes *AA* and *Aa* are not as adapted to the environment as the ones resulting from *aa.* If, for example, the greenish-brown insects are more easily seen than the dark brown insects, they will be reduced in numbers much faster by hungry birds. Individuals having the *AA* or *Aa* phenotypes will not live as long or leave as many offspring as organisms that are *aa.* In this case, the number of *A* alleles or greenish-brown insects in the population will decline, and the number of dark brown insects with *a* alleles will increase. By analyzing gene frequencies, biologists can determine if a species is near extinction.

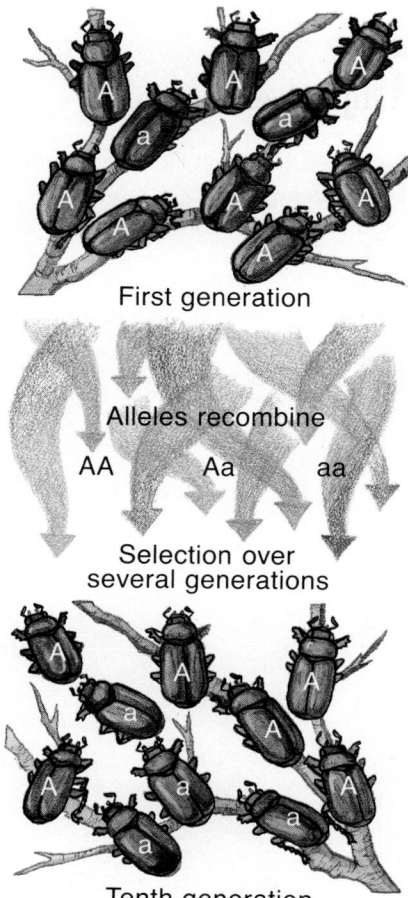

First generation

Alleles recombine

AA   Aa   aa

Selection over
several generations

Tenth generation

Figure 12-14   The evolution of a population can be determined by comparing the gene pools of different generations. The gene frequencies in the first generation are 80 percent *A* and 20 percent *a.* What are the gene frequencies of the tenth generation?

Section Review

Understanding Concepts

1. Darwin discovered flightless cormorants on the Galapagos Islands. How might the discovery of this unusual bird have influenced his thinking about evolution?
2. Suppose snakes evolved from reptile ancestors that had legs. Explain how that evolution might have occurred.

3. How does evolution of resistance to a pesticide illustrate natural selection?
4. **Skill Review—Designing an Experiment:** How would you conduct an experiment to test the hypothesis that insects resistant to a low dose of insecticide may also become resistant to higher doses? For more help, see Practicing Scientific Methods in the **Skill Handbook**.

5. **Thinking Critically:** Cauliflower, brussels sprouts, and cabbage are closely related garden vegetables. They are all descendants of one type of wild plant. How might these plants have evolved?

Sequence the events that might have led to the evolution of complex organic molecules and the first prokaryotes.

Discuss the events leading to evolution of autotrophs and the consequences of that evolution.

12.3 **O**rigin of Life

Whose imagination hasn't been kicked into overdrive by video Earthlings making space visits to imaginary planets, where they mingle with incredible creatures in strange environments? Rev up your fantasies now and climb into an imaginary time machine. You are going to visit a planet every bit as strange—and hostile—as any you have seen on a video screen. The planet is Earth. You won't recognize it, though, because it is still at the beginning of its time.

Formation of Organic Compounds

Your time of arrival is about 4.6 billion years ago, which is when scientists believe Earth first came into being. Advisedly, you put on your protective clothing and respirating equipment. You could not survive a second without them. The overall scene is one of violence. Prolonged lightning and erupting volcanoes surround you. Meteorites and ultraviolet radiation (UV) make surface heat fearsome. There are no signs of life anywhere. You decide to leave and come back later—a long time later, perhaps after life has come to the planet. You wonder how that life will evolve, but you don't care to stick around in these unfriendly conditions long enough to find out.

Simple Molecules from the Atmosphere   All the many forms of life on Earth today are descended from a common ancestor, found in a population of primitive unicellular organisms. What were those first cells like? How do we know? What events led up to their formation? No traces of those events remain, and scientists can't travel backward in time to witness what happened. Instead, they turn to scientific methods of observing data, forming hypotheses, making predictions, and constructing experiments to test their predictions.

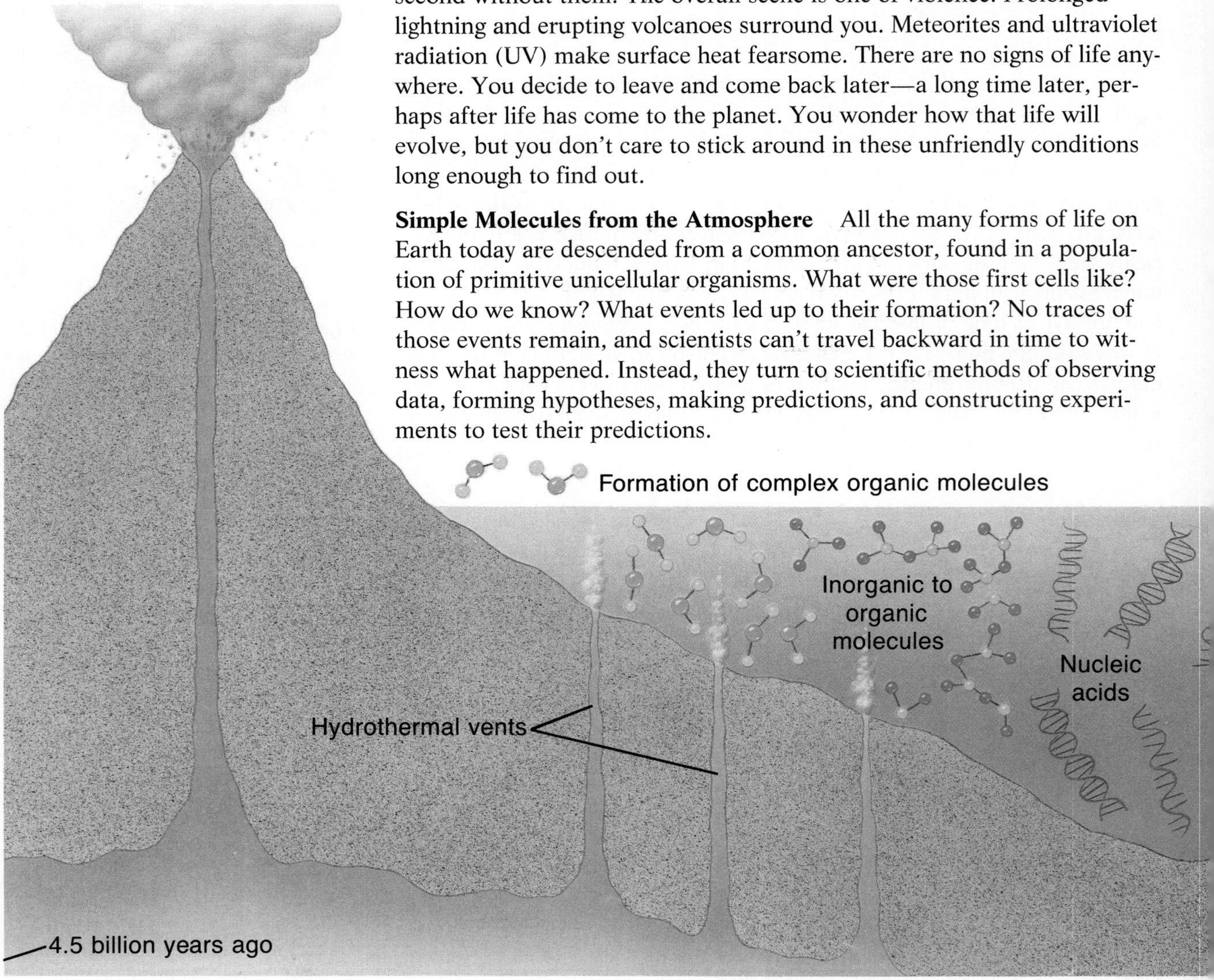

Formation of complex organic molecules

Inorganic to organic molecules

Nucleic acids

Hydrothermal vents

4.5 billion years ago

Earth's primitive atmosphere was far different from today's. In 1936, Alexander Oparin, a Russian biochemist, proposed that the primitive atmosphere consisted of the gases methane (CH_4), ammonia (NH_3), hydrogen (H_2), and water vapor (H_2O). Other scientists have since suggested that the atmosphere consisted mainly of water vapor and small amounts of nitrogen gas (N_2). Did you miss oxygen gas, so important to life as we know it today? Most of Earth's oxygen 4.6 billion years ago was probably in the form of metal oxides or water vapor. Oxygen was, therefore, not available as a gas.

Whatever the exact makeup of the primitive atmosphere might have been, the four elements most common in today's organic compounds—carbon, hydrogen, oxygen, and nitrogen—are thought to have been part of it. With time, Earth's crust gradually cooled. Runoff from torrential rains gathered to form oceans. Meanwhile, ultraviolet radiation and lightning continued to bombard the atmosphere. Scientists assume that these energy sources broke the bonds in the gas molecules, thereby releasing atoms to form simple organic compounds.

If scientists are correct in their assumption, then it should be possible to predict that organic compounds will form if primitive gases are exposed to high energy sources. Stanley Miller, a graduate student at the University of Chicago, tested that hypothesis in 1953 when he simulated conditions of the early atmosphere. In a setup like the one shown in Figure 12-15, Miller circulated steam through ammonia, methane, and hydrogen and subjected the mixture to electricity. At the end of one week, he found amino acids and a variety of other simple organic molecules in the

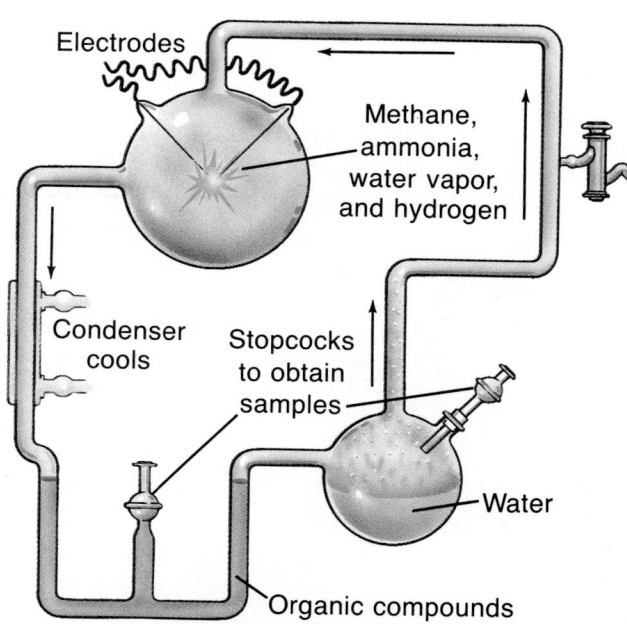

Figure 12-15 At about the same time that Miller discovered that the red goo at the bottom of his flask was rich in amino acids, it was also learned that DNA carries the code for amino acids.

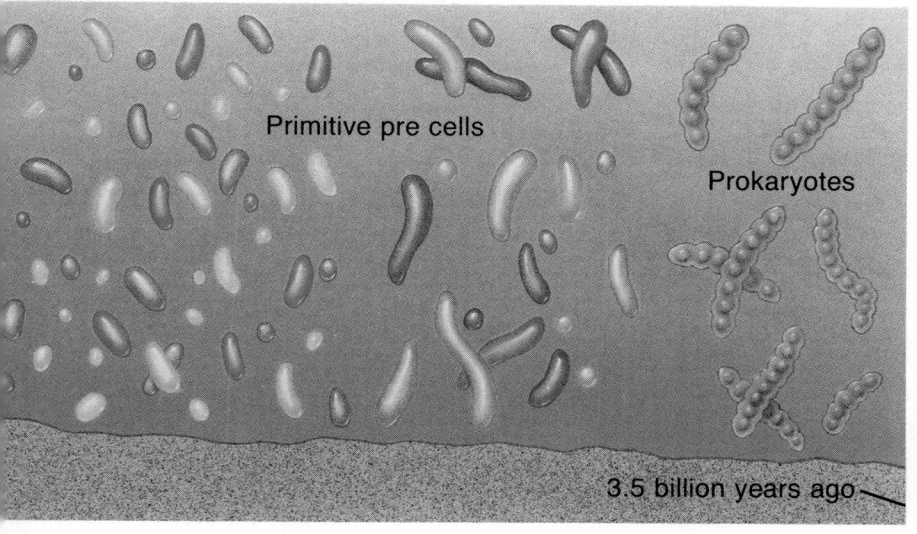

Figure 12-16 In one billion years from the beginning of time on Earth, prokaryotes such as present-day photosynthetic bacteria had emerged.

condensed water vapor at the bottom of the chamber. The assumption that organic compounds were produced in the early atmosphere, then, seems to be a valid one. It is the one most widely held among researchers in the field today.

Complex Molecules from Simple Molecules?   Simple organic molecules are necessary for life, but so are more complex molecules. How did complex molecules form? One assumption holds that the heavy rains washed simple organic molecules into the early oceans. In some places, say in puddles of water along a shoreline, the simple molecules became concentrated and were exposed to surface warmth. Under these conditions, the simple molecules may have reacted to form larger molecules. Amino acids, for example, might have joined to form short polypeptide chains. Later, the molecules could have been washed farther out into deeper ocean waters.

The First Organisms: Heterotrophs

The ocean was probably the cradle of life. That is, Earth's first organisms are believed to have evolved in the ocean. Modern scientists use knowledge of biochemistry to help them see through the window of this watery nursery.

Primitive Pre-cells   Picture a situation, as modeled in Figure 12-16, in which complex organic molecules, polypeptides, nucleotides, and simple and complex carbohydrates are floating in the water. Various inorganic ions complete the mixture and form a chemical soup. If some sort of membrane were to form around groups of these molecules and ions, a kind of pre-cell condition would result. Models explaining how this could happen have been designed and tested.

The primitive cell could not truly be alive, however, without a means of obtaining energy for its functions. And once life processes started, the cell would need a means of reproduction to keep the processes going. Scientists have hypothesized about early energy sources and reproduction.

Anaerobic respiration   Oparin and later scientists have suggested that simple organic molecules in the ocean, such as sugars, may have crossed the membranes and entered the pre-cells. At first, some inorganic molecules may have acted as enzymes that break the bonds in the sugars and release energy. A compound similar in function to ATP could have trapped that energy to fuel activities within the pre-cell. Later, polypeptide chains that formed in the soup may have taken over the role of enzymes. Because oxygen gas was not yet available, respiration in these early stages of cellular evolution must have been anaerobic.

DNA and Reproduction   All cells, as scientists observe them today, depend upon DNA to code for their enzymes and other proteins. How and when DNA first evolved and became linked to protein synthesis is not known. But some of the very earliest cells must have been the scene of DNA activity, because all of their descendants inherited a universal

Figure 12-17 **The species of photosynthetic bacteria that cover these mounds on the west coast of Australia have existed for over three and a half billion years. The mounds are made up of the fossils (inset) of these ancient prokaryotes.**

genetic code. The first cells able to obtain energy and reproduce under the guidance of DNA were prokaryotes. According to Oparin, these early prokaryotes were **heterotrophs,** which means that they required a supply of organic material from the environment. These early heterotrophs were like some present-day bacteria. They obtained their energy from simple chemical reactions.

Later Organisms: Autotrophs

As the supply of simple molecules for heterotrophs was used up, life forms that used the available supplies more efficiently were at an advantage. This competition for food may have led to the evolution of cells that had the ability to make their own food. Organisms that make their own food are **autotrophs.** Autotrophs produce their own food—through photosynthesis or chemosynthesis. You learned in Chapter 6 that photosynthesis is a process that uses carbon dioxide, water, and energy from the sun to make food. **Chemosynthesis** is a process in which an organism uses the energy from chemical reactions to produce food. For example, chemosynthesis occurs in present-day bacteria that live in places where no sunlight can reach but also where there is no oxygen. It is, therefore, possible that life forms in the early anaerobic conditions on Earth used chemosynthesis to obtain energy. The oldest fossils found, dated at 3.5 billion years, are similar to today's bacteria that live in extreme habitats such as deep ocean vents, highly salty coasts, or hot sulfur pools.

Early prokaryotes that became adapted to living in light would have had an advantage as the supplies of organic molecules were used up. They must have eventually evolved a process of making their own complex molecules. Any genetic variations enabling certain cells to use light energy to make food would have been naturally selected. The first photosynthetic autotrophs evolved.

Minilab 2

Can a calendar represent geologic time?

Draw an outline of a calendar for the entire year. Calculate how one day corresponds to the total time scale being used in Figure 12-19. For example, if the first event occurred 550 000 000 years ago, one day on the calendar equals about 1 500 000 years (365 divided into 550 000 000).

Analysis *How many years are represented by one day? How does your calendar help you to judge the rate of evolutionary change in different groups of organisms?*

Photosynthesis and the Oxygen Revolution   You will recall that the process of photosynthesis results in the production and release of oxygen. After photosynthesis evolved to the state in which we know it today, oxygen (O_2) would have become available.

It's too bad no one was around to applaud. The increase of oxygen in the drama of evolution changed the course of life forever. The change was so drastic that scientists call it the oxygen revolution. Once oxygen became abundant, aerobic respiration could evolve. As a result,

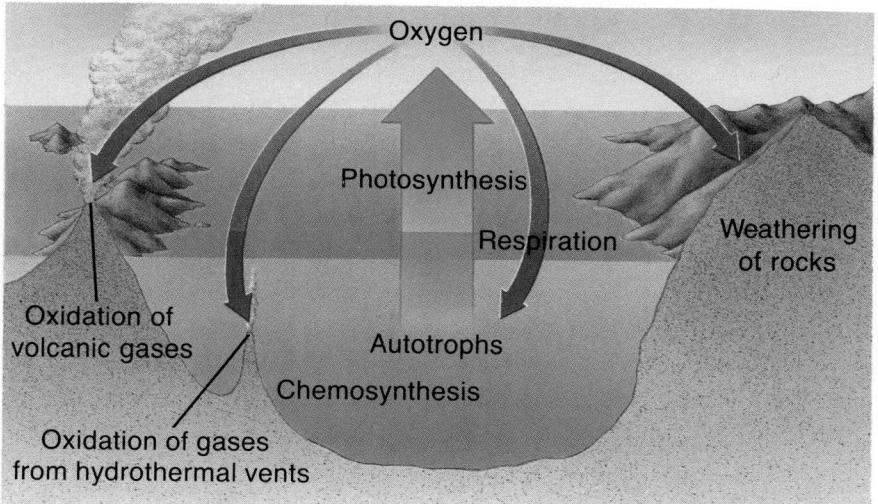

Figure 12-18   Early oxygen buildup resulted from photosynthesis by early autotrophs. The oxygen was then available for respiration, the oxidation of volcanic gases, and the weathering of rocks.

Figure 12-19 This representation of the Geologic Time Scale makes clear that the enormous diversity of ancient and present-day life evolved over a comparatively short duration. In which era did humans first appear?

more energy could be extracted from food molecules. With increased energy, greater organization and complexity among primitive cells could evolve. Still later, eukaryotes, with their complex organelle system and DNA enclosed in a nucleus, made an entrance that led to the enormous diversity of multicellular organisms today.

4.6 billion BC

ARCHEOZOIC ERA
4.6 billion to
570 million BC

PALEOZOIC ERA
570 to 250 million BC

MESOZOIC ERA
250 to 65 million BC

CENOZOIC ERA
65 million BC
to present

2.5 million BC

An Explosion of Life   Over time, oxygen would have been present in large amounts in the atmosphere. Lightning converted much of the atmospheric oxygen to ozone molecules (O_3). The molecules collected in an **ozone layer** ten to 15 miles above the surface of Earth. The ozone layer blocked most UV radiation, and so prevented any further synthesis of organic molecules in the atmosphere. At the same time, it shielded emerging life forms from UV's harmful effects and thus allowed an explosive evolution of new organisms.

More than 500 million years ago, the early eukaryotes in the sea already included red, green, and brown algae, and a host of invertebrates from sponges to crabs. Later, by 300 million years ago, fishes, amphibians, and reptiles, and massive treelike ferns and large forests of coniferous trees had evolved.

Examine Figure 12-19 to compare the fossil record of life forms found from different periods in geologic history. Note that organisms during the Archeozoic Era, which represents nearly 90 percent of Earth's history, were very simple in form. The explosion of diverse life forms that occurred in the Paleozoic Era probably reflects the oxygen revolution. Notice how the fossil record provides evidence for evolution.

LOOKING AHEAD

T HE THEORY of natural selection is central to any understanding of evolution. As you dig deeper into your studies of evolution, you will see how natural selection can result in new adaptations within a given population of organisms. New evidence for relationships among organisms is continually being presented in the areas of molecular biology, population studies, and genetics. You will trace the origin of completely new populations and species with the help of these modern studies. The most fascinating story of the history of a species is probably that proposed for humans to be described in Chapter 13.

Section Review

Understanding Concepts

1. Why are techniques such as examination of the fossil record and comparative studies of structures or genes not useful in learning about the origin of life?
2. Why could organisms today not exist in conditions like those thought to be present on early Earth?

3. Why would chemosynthesis have been an advantage for early prokaryotes?
4. **Skill Review—Sequencing:** Sequence the major events leading to the evolution of the first cells. For more help, see Organizing Information in the **Skill Handbook.**

5. **Thinking Critically:** How could the evolution of aerobic respiration have permitted evolution of more complex organisms?

Issues

Biodiversity: Which Species Should Be Saved?

They stare out at you from emotion-charged public service advertisements. They star in wildlife documentaries chronicling the disappearance of many forms of life on Earth. "They" are the endangered plant and animal species of Earth.

For decades, biologists have endeavored to identify plant and animal species headed toward extinction and to devise ways of saving them. Lately, however, dissenting voices have asked, where will conservationists find the time, the money, and the technology to save all the threatened species on Earth? And, even if this rescue effort were possible, the voices add, is it necessary?

Consider the Facts: The total number of species on Earth is still unknown. Millions of species have yet even to be identified. Furthermore, many species share such a similar evolutionary history that their genetic differences are slight. Perhaps it is time, some scientists are saying, to adopt a new goal: the preservation of biodiversity, which describes the broad variety of living things on Earth.

Earth is home to many different species. Preserving this biodiversity doesn't mean preserving all species; rather, it means protecting species that and represent a broad spectrum of variation.

For example, suppose scientists have identified ten species of falcons. Five of these species are endangered and face extinction. Three of the five endangered falcon species are close genetic cousins of a falcon that is not endangered. Which of the endangered falcon species should we protect?

Not too long ago, the answer might have been "all of them." Today, however, some scientists would suggest that the two genetically distinct falcons should be saved. If the close cousins become extinct, much of their genetic heritage will be preserved in the group of nonendangered species.

Investing in the Future: Some conservationists reject this approach. They argue that even the subtle differences between close genetic cousins are important and worth

preserving. The loss of any species through human neglect is, they say, a potential biological tragedy. A plant species allowed to disappear may be the one that contained the chemistry to cure disease, and an animal species may have secreted a substance for some other technological breakthrough.

Although the issue is likely to be debated for decades to come, the process of evolution may have already shown us the way. Throughout Earth's history, species have continuously appeared and disappeared. Nature, it seems, has been preserving biodiversity for as long as species have existed.

Understanding the Issue

1. How would you respond to someone who suggested that saving plant species wasn't as important as saving animal species?

2. Today, vast areas of tropical rain forests are being destroyed. How will this affect biodiversity?

Readings

- "So Much to Save." *The Economist,* June 13, 1992, p. 94.
- Mlot, Christine. "The Science of Saving Endangered Species." *Bioscience,* Feb. 1989, pp. 68-70.

Biotechnology

Dating Fossils

If you think you have problems finding dates, imagine the problems of the paleontologist who's just unearthed a fossil skull. The skull appears to be quite old, but the question is, "exactly how old?" The answer can come from a variety of dating techniques that scientists use to date fossils precisely.

Radioisotope Dating: In the past, paleontologists have dated bones and fossils by two standard methods—radiocarbon dating and potassium-argon dating. Radiocarbon is an unstable form of carbon, carbon-14, that changes or decays into nitrogen-14 atoms at a known rate. When an organism dies, the levels of radiocarbon continue to decrease within the bones. The age of a fossil can, therefore, be estimated from its present-day levels of carbon-14. Radiocarbon dating is useful for fossils less than 40 000 years old.

Other isotopes are necessary to date older fossils. Potassium-argon dating depends on the decay of potassium-40 to argon-40 and calcium-40. Half of the original atoms of potassium in a rock sample will decay to argon-40 in 1.3 billion years. The age of the rocks can, therefore, be estimated from present-day levels of potassium-40. The decay of potassium-40 is so slow that this dating method is useful only for fossils in surrounding rocks more than 500 000 years old.

Comparing Amino Acids: There remains a 460 000-year gap for which other dating techniques are needed. At one time, amino acid racemization (AAR) gained some popularity. Amino acids manufactured during an organism's life do not disappear when the organism dies. Instead, they slowly change from one isomeric form to another (a process called racemization). By measuring the ratio of the levels of the two isomeric forms, scientists can estimate the age of the organism.

Measuring Trapped Electrons:

Two other techniques that developed from research in chemistry and physics are becoming more popular in the dating of fossils up to two million years old. Luminescence dating and electron spin resonance (ESR) are both based on the measurement of buildup of the trapped electrons caused by radiation. The difference in the two techniques is that luminescence dating can only be used to date flint tools or pottery—items that have been subjected to high temperatures at the time of their formation; ESR can be used to date bones, coral, shells, or tooth enamel. The development of these two methods could probably lead to the unravelling of the mystery of the evolutionary history of humans.

1. **Issue:** One group of anthropologists says that humans evolved 200 000 years ago in Africa. Another group says that humans evolved later in Asia. Why is it important for scientists to use more than one method of dating fossils?

2. **Explore Further:** Use the library to find out how mitochondrial DNA has been used in determining the place of human origins.

Summary

Study of the fossil record and comparisons of anatomy, embryology, chemistry, and genes of organisms provide a wealth of evidence that evolution occurs. Results of selective breeding and direct observations of changes in populations strengthen that assumption.

Darwin and Wallace recognized the importance of variations among organisms and that variations can be inherited. The two scientists independently formed the theory of natural selection to explain how evolution occurs. Modern understanding of genetics confirms and expands that theory.

Although direct evidence to explain the origin of life is not available, scientists have proposed models and tested hypotheses about how the first cells and organisms evolved. One hypothesis proposes that the first cells were heterotrophic prokaryotes and that autotrophs evolved later.

Language of Biology

Write a sentence that shows your understanding of each of the following terms.

autotroph	gene pool
chemosynthesis	heterotroph
comparative anatomy	homologous
comparative biochemistry	natural selection
	ozone layer
comparative embryology	population
	selective breeding
embryo	theory of natural selection
fossil	vestigial

Understanding Concepts

1. Blind cave fish have eyes but do not use them to find their prey. Explain why they have these apparently useless structures.
2. How might an understanding of genetics be useful to an animal breeder?
3. Two populations of wildflowers are found in nearby meadows. There are slight differences among members of each population, such as flower color and leaf shape. How might Darwin have interpreted these facts?
4. Relate the situation faced by the first heterotrophs to the idea proposed by Malthus.
5. Why is it more correct to speak of the evolution of populations rather than of individual organisms?
6. Compare the terms *artificial selection* and *natural selection*.

7. Using Figure 12-19 as a reference, determine when plants began to evolve on land? Animals?
8. Explain the cause and consequences of the oxygen revolution.

REVIEW

Applying Concepts

9. Suppose a population of tan-colored insects inhabits a sandy area. Over time, the area becomes grown over by a population of green plants. Use Darwin's ideas to explain how and why the insect population might evolve to be green in color.

10. Explain the same evolution of insects asked about in the question above, but this time in terms of population genetics.

11. Is the evolution of the first living thing an example of spontaneous generation as described in Chapter 7? Defend your answer.

12. The theory of biogenesis, as you read in Chapter 7, states that all organisms are produced from other organisms. Why does it also state that biogenesis only refers to organisms at the present time and under present conditions on Earth?

13. Some biologists have proposed that organic molecules or even simple life forms reached Earth by hitching rides on meteorites. Even if that were true, why is that suggestion not very useful in the study of the origin of life?

14. How might mutations and immigration of the same kind of organism from another population affect a population's gene pool?

15. Darwin recognized the existence of variations and the fact that they are inherited. However, he knew nothing about genetics. How do (a) events of mitosis, meiosis, and fertilization, (b) mutation and crossing over, and (c) the role of DNA support his theory?

16. Explain why further synthesis of simple organic compounds in the atmosphere is not possible.

17. The DNA code is basically the same in all organisms. For example, a bacterium is able to use the information from human genes to make proteins when the human genes are inserted into it using genetic engineering techniques. What does this fact suggest about evolution?

18. In a population, flower color is represented by two alleles. The population consists of 10 *BB* flowers and 10 *bb* flowers. What are the allele frequencies?

19. If a population of palm trees were planted on a mountain in the Rockies, would you expect them to evolve into a new species? Explain.

20. **Global Connection:** How do humans avoid the effects of natural selection in adverse environmental conditions?

21. **Issues:** You know that diversity means a condition of being different or having various forms or qualities. In your own words, explain what biodiversity is and why maintaining it is important.

Lab Interpretation

22. **Minilab 1:** A layer of microfossils is found in nature. Directly above it is a layer of sedimentary rock with no microfossils. How might you explain this difference?

23. **Thinking Lab:** A young rock layer is found exposed at Earth's surface with a much older layer exposed nearby. What might explain why younger layers are missing from the top of the older layer?

24. **Investigation:** Amber is a hardened, resinlike material that sometimes may hold trapped insects. Describe a simple way of making an amber-trapped fossil without using amber.

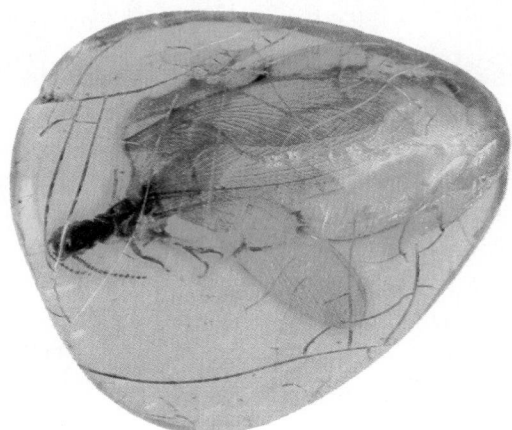

Connecting Ideas

25. Explain the frequencies of the alleles for sickle-cell anemia in populations of African Americans and native black Africans in terms of population genetics.

26. Relate the model of the origin of life presented in this chapter to the model of evolution of eukaryotes by symbiosis as described in Chapter 5.

13

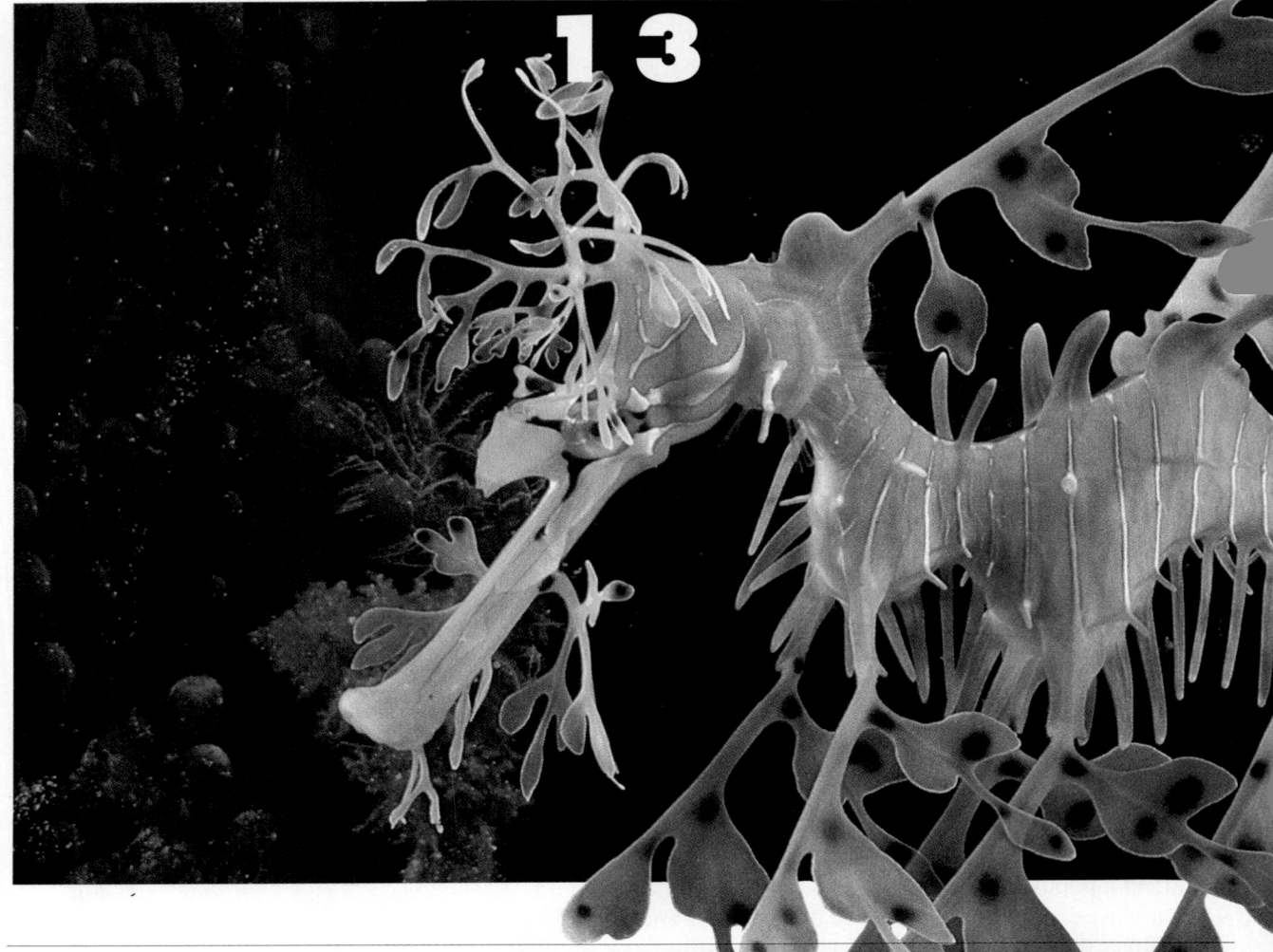

ADAPTATION AND SPECIATION

IS THIS a fish or a seaweed in this photograph? Perhaps you weren't fooled, but a predator, such as a larger fish, might have missed this tasty morsel. The fish you see hiding among the plants is called a leafy sea dragon. The sea dragon possesses large, fanlike fins that resemble the leaves of kelp seaweed, in which it generally hides. In addition, the sea dragon can hold itself upright to further camouflage itself from predators. These fish use the underwater kelp as a home base for feeding and reproduction. The sea dragon's close relatives, sea horses and pipefishes, also have the ability to swim vertically among the seaweed in order to hide from predators. Sea horses have a grasping tail, which they use to hold firmly onto underwater plants, and the long, thin body of the pipefish very much resembles the grasses it calls home. Can you see how these features enable sea dragons and their relatives to survive in this watery environment?

All organisms possess features that enable them to survive and eventually reproduce in a particular environment. Trees possess leaves that capture the sun's energy so they can produce food and survive. Mammals have fur or hair that insulates them from the cool temperatures of the environment. In Chapter 12, you learned how the process of natural selection causes the enormous variety of life forms on Earth. In this chapter, you'll consider how adaptations evolve and how they eventually can lead to the evolution of species.

When you're in a crowd, do you feel hidden? When you wear brightly colored clothes, do you expect to frighten people away? Do you have predators to run from?

Chapter Preview

13.1 Adaptation

Have you ever wondered why some trees have pointed needles while others have broad, flat leaves? Why do dogs and cats have sharp, needlelike teeth while humans have rather flat teeth? As you learned in the last chapter, the answers to these questions can be explained by Darwin's theory of natural selection. In this section, you will apply Darwin's theory to explain the differences among different kinds of organisms.

Origin of Adaptations

The pointed needles of trees such as pines, firs, and spruces and the flat leaves of trees such as maples and magnolias are features that enable them to survive in different environmental conditions. A broad-leaved tree would not survive very well in a hot, dry desert or in a cold, dry region of the far north. Too much water would be lost faster than it could be replaced. The sharp teeth of dogs and cats are also traits that are important to survival. You have read that any trait that aids the chances of survival and reproduction of organisms is called an adaptation. Because sharp teeth in dogs and cats and needle-shaped leaves in pine trees enable these organisms to survive, adaptations can also be thought of as traits that increase an organism's fitness to the environment or that increase the probability of an organism passing its genes on to the next generation. How do adaptations arise?

Figure 13-1 **The beautiful snow leopard has deadly teeth and claws. What other adaptations for survival does it have?**

Distinguishing Variations and Adaptations

Adaptations may evolve as certain variations increase in frequency within a population of organisms. But note that the words *variation* and *adaptation* do not mean the same thing. A variation may improve fitness, but it may also reduce it. For example, the sharpness and length of claws in a population of leopards is an adaptation that allows them to capture the foods they need to survive. Occasionally, some of these cats may have very long, needlelike claws, or some cats may have shorter and less sharp claws. If a leopard develops claws that are too short, this cat may not be able to catch its prey. The fitness of this animal has been reduced. Only variations that aid survival up to and including the reproductive stage will be preserved by natural selection. Over time, all members of a population will have inherited that variation, at which point the original variation becomes an adaptation. That adaptation will be a general characteristic of the entire population, like needles on pine trees or eyespots on fish. Adaptations, then, are products of evolution by natural selection. Variations are the raw materials upon which natural selection acts.

The Evolution of Complex Adaptations One of the more common misconceptions about evolution by natural selection is that adaptations arise all at once, as a result of some giant mutation or massive mistake in the genetic message. Some of you may have considered it difficult to comprehend, for example, how a complex structure such as the eye could possibly be the product of a mistake. Your confusion would be understandable. However, complex adaptations do not evolve instantly by giant mutations.

Adaptations do not arise overnight. Keep in mind that evolutionary changes occur over many, many generations of a species. Adaptations evolve slowly as a result of a series of small adaptive changes or modifications, each change representing a slight variation from the traits of the previous generation. Remember from Chapter 12 that Darwin referred to this idea as "descent with modification." But, it is important to recognize that natural selection does not invent. It only acts to modify in each successive descendant what already exists. Adaptations in organisms you see today are the result of chance variations that arose at a particular time in the evolutionary history of that type of organism. For example, if the variations for piercing mouthparts in mosquitoes had not arisen, mosquitoes probably would not be as successful as they are today.

When we marvel at a structure as complex as an eye, we are looking at a product of the ongoing process of evolution. You can better understand how an eye might have evolved if you picture a series of changes during the evolution of the eye. Each step along the way provided organisms with slightly better vision (Figure 13-2).

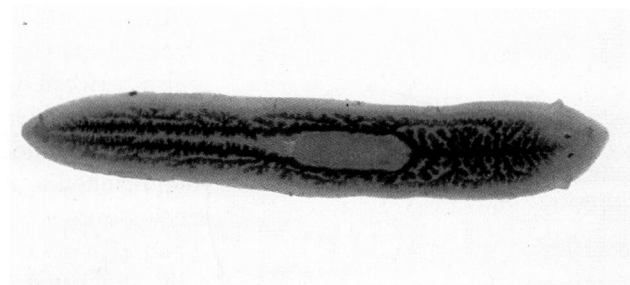
a

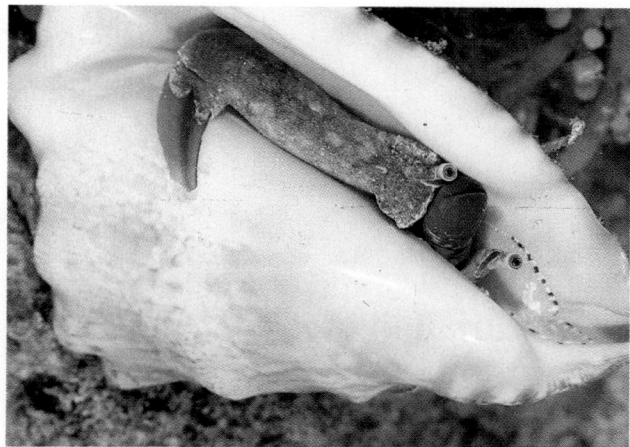
b

c

d

Figure 13-2 Simple animals such as flatworms (a) have groups of light-sensitive cells to detect direction of light. More complex animals have visual systems that form images. In mollusks (b) the images are blurry. The images are focused for a turtle (c) and are enhanced by color in the eagle (d) and many other vertebrates.

By studying simple through complex organisms, biologists know that the eye began only as a group of light-sensitive cells. The ability to distinguish between light and dark was an advantage because it enabled organisms to avoid danger and exploit new food resources more efficiently than organisms without this adaptation. Later, over time, new variations arose that resulted in the formation of a simple lens providing a blurry image. Seeing even a blurred image is an advantage over seeing no image at all. Later changes led to a sharpening of focus, and, probably at the same time, still other changes permitted color vision. Can you see how each step along the way was due purely to chance appearances of variations? The structure becomes more adaptive and improves an animal's chances of survival and, therefore, the passing of its genes to offspring. The same process was involved in the evolution of each and every complex structure of plants and animals.

I N V E S T I G A T I O N

Variation in Numbers of Stomata

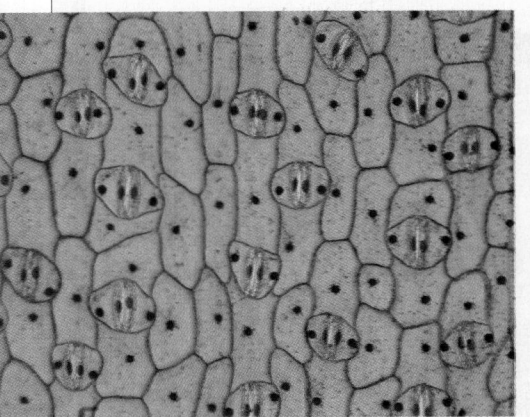

Have you ever been outside early in the morning and found the grass wet with dew? Where does this water come from? You know it didn't rain the night before, yet the grass is wet as if it did. Some of this water condensed from the air as it touched cool surfaces. Some of the water came from inside the plant. Plants give off water through stomata, which are tiny openings in their leaves. In many plants, the majority of stomata are on the lower surface of the leaves. Each stoma is surrounded by two guard cells. When the plant has access to plenty of water, the guard cells are plump and swollen with water. In this condition, the stomata are open and water can escape from the plant. When the plant does not have enough water, the guard cells become limp and the stomata close. In this lab, you will try to determine how numbers of stomata vary within the same species of plant.

Problem

How do the numbers of stomata on leaves of plants of the same species vary?

Safety Precautions

Make sure you can identify and avoid poisonous plants in your area such as poison ivy and poison oak.

Hypothesis

After observing the prepared slide of stomata provided by your teacher, hypothesize how much variation you might expect in numbers of stomata on leaves of one kind of local tree or shrub. What might produce this variation?

Experimental Plan

1. As a group, make a list of possible ways you might test your hypothesis using the materials your teacher has made available.
2. Decide on a common local species of tree or shrub from

Types of Adaptations

If you think about it, any organism, as it presently exists, can be thought of as a bundle of adaptations. For example, the shape of a bear's teeth and face, the length of its fur, and the way it behaves are adaptations that interact to promote its fitness in the environment in which it lives. Although the many adaptations of an organism may work together, we can examine them individually and classify them by type.

Structural Adaptations   The most obvious adaptations are those involving structure or anatomy. Traits such as the structure of a bird's beak, the hoof of a horse, or the shape of a tooth are called **structural adaptations.** Many obvious structural adaptations in animals are for obtaining food. A woodpecker's tongue, for example, is narrow and very long. It can probe

which to collect leaves. It should have leaves with smooth rather than fuzzy surfaces.

3. Make sure your collection procedures will provide uniform samples. To assure this, remember that a tree might have different numbers of stomata at different ages, in different habitats, at different heights, on different sides of the tree, and on different positions on the twig.
4. Agree on one idea from your group's list that could be investigated in the classroom.
5. Following the style of a recipe, write a numbered list of directions that anyone could follow.
6. Make a list of materials and the quantities you will need.

Checking the Plan

Discuss the following points with other group members to decide the final procedures for your experiment.
1. What variables will need to be controlled?
2. What is your control?

3. What will you count or measure?
4. How many leaves from how many trees will you examine?
5. Have you designed and made a table for collecting data?

Make sure your teacher has approved your experimental plan before you proceed further.

Data and Observations

Carry out your experiment, make your counts, and complete your data table. Make a graph of your results.

Analysis and Conclusions

1. Was your hypothesis supported by your data? Explain.
2. Write a brief conclusion for your experiment.
3. What is your experimental evidence for your conclusion in question 2?
4. What types of experimental error could have affected your results?
5. Compare your results with those of another group. Write a brief summary of this comparison.

6. What factors might affect the numbers of stomata on leaves of the plant species you studied?
7. If the climate in your area gradually became drier over the next ten years, how might the number of stomata on leaves be affected? Explain in terms of adaptation.

Going Further

Based on this lab experience, design an experiment that would help determine variation in numbers of stomata in houseplants. Assume that you will have unlimited resources and will not be conducting this lab in class.

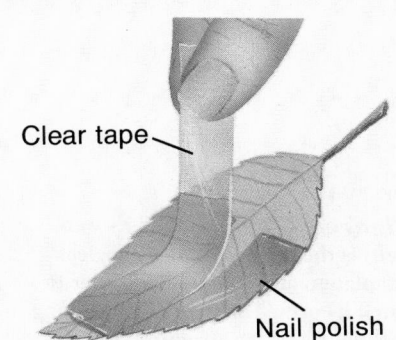

Clear tape

Nail polish

Minilab 1

Is blubber an adaptation?

Spoon 40 g of shortening into a plastic sandwich bag. Place a test tube filled with room-temperature water and a thermometer into this bag. Then place another test tube with room-temperature water and a thermometer into another plastic bag with no shortening. Fasten the bags close to the top of the test tubes with rubber bands. Working on the outside of the bag, gently spread the shortening evenly over the surface of the test tube. Place these two bagged test tubes and another unbagged test tube with the same contents into a beaker of ice water. Record the temperature of each tube as you begin, then every minute for 30 minutes.

Analysis *Compare the time needed for each tube of water to cool to the temperature of the ice water. What is the purpose of the tube with no insulation? The tube with only the plastic bag?*

small openings in trees that have been pecked away by the bird's sharp beak. Insect larvae living beneath the bark are picked out by the tongue and eaten by the woodpecker. An anteater's sticky tongue inserted into an anthill attracts ants as it moves. If the ants touch the tongue, they stick to it. When the anteater pulls its tongue in, the ants are easily trapped.

Structural adaptations for obtaining food do not always involve parts of the mouth. The adaptations of the angler fish provide a good example of this. Protruding from the top of the angler fish's skull is a long filament that is broad and flat at its tip. It hangs in front of the fish's mouth and attracts prey, much like a fishing lure. When the prey bites at the lure, the angler fish sucks the prey into its mouth. Some deepwater angler fish of the North Atlantic also have fins that are modified to glow in the dark waters. Prey are attracted by the glowing fins and so are easily trapped.

Of course, not all structural adaptations involve food-getting, and many are internal rather than external. For example, the strong muscular walls of your heart are an adaptation that enables it to pump blood throughout the body. Some single, microscopically thin nerve cells extending from your spinal cord to your leg are adapted to carry rapid nerve impulses.

Physiological and Behavioral Adaptations Other adaptations are associated with functions in organisms. The enzymes needed for digestion, clotting of blood, or muscular contractions in animals fall into this category, known as **physiological adaptations.** Such adaptations have a chemical basis. Other examples include the poison venom of a snake, the proteins used in a spider's web, and the ink of an octopus.

Organisms are also adapted in their responses to the environment. Birds migrate in search of food, squirrels hunt and store nuts, wolves track their prey, and plant stems grow toward the light. Such responses are examples of **behavioral adaptations.**

Figure 13-3 The spider's web (left) is the result of a physiological adaptation and enables the spider to catch its prey. The woodpecker's long tongue (right) is a structural adaptation for obtaining food.

Figure 13-4 Many kinds of animals migrate. Some animals, including fish and birds such as these red-winged blackbirds, migrate to find food and a less severe climate. Some animals migrate to breed. The longest journey by any migratory animal is traveled by the Arctic tern, which flies more than 35 000 miles each year.

Interactions of Adaptations   Isolating and classifying adaptations is really an artificial process, because in nature, adaptations often depend upon each other. For example, the migration of birds can be classified as a behavioral adaptation. These regular, seasonal, and mass movements of birds could not occur without a set of structural adaptations such as feathers, lightweight bones, and strong flight muscles. The migration also requires the interplay of many physiological adaptations—nerves that signal the muscles, the ability of the muscles to contract, and cellular respiration to provide the ATP energy needed. Before the actual migration even begins, other physiological adaptations, such as the release of chemicals from sex glands before a spring migration, must signal the birds that the time for migration is approaching.

$\mathbf{S}$ection Review

Understanding Concepts

1. The thorns on a rosebush protect the plant from animals that might eat it. What type of adaptation are thorns?
2. Name two human adaptations not described in this section. Describe how each is adaptive for humans.
3. In a population of brown-colored insects, an occasional variation of body color is yellow. List the stages in the evolution of a population of yellow-colored insects following a change in the environment to more desertlike conditions.
4. *Skill Review—Classifying:* Classify each of the following adaptations as either behavioral, structural, or physiological: (a) bees build a hive; (b) a young duckling follows its mother; (c) a wood-pecker's beak is pointed and sharp; (d) use of ATP; (e) photosynthesis; (f) the flat shape of a leaf. For more help, see Organizing Information in the **Skill Handbook.**
5. *Thinking Critically:* What might happen to plants that are pollinated by flies if insecticides are constantly used in their community?

Origins of Species

Objectives

Contrast various ways in which speciation may occur.

Compare gradualism and punctuated equilibrium.

Determine the survival value of mimicry and camouflage.

If the flowers in a population of horsemints become more brightly colored over time, are the plants still the same kind of plant? If the necks in a population of turtles get longer over time, are the animals still the same species? You have seen that new adaptations arise as a result of the process of natural selection. A species can include a broad range of variations. But how do plants, animals, or any other species evolve in the first place? How do new forms of life evolve? Can the evolution of new species be explained by natural selection?

Defining a Species

Horsemints are one kind of organism. Turtles are another. It's easy to distinguish between these two organisms because of some very obvious physical differences. However, if you compared a wood turtle, a spotted turtle, a bog turtle, and a western pond turtle, the differences are not so obvious. Each of these kinds of turtle is a separate species. It is easy to say that a plant and an animal are not the same species. It is easy to distinguish a cat from a turtle. But, when you study the four turtles shown in Figure 13-5, the differences are not as striking as the similarities. Now consider a golden retriever and an Irish setter. You will recognize these dogs as different, but would you call them separate species? Actually, both dogs are in the same species. But why are they considered the same species when they have such obvious differences?

Biological Species Early biologists working on the classification of organisms faced a similar dilemma. Organisms could be assigned to different species on the basis of physical features alone. Organisms are today more often classified according to a biological species concept. According to this concept, a **species** is a group of organisms that can interbreed and produce fertile offspring in nature. A species is the expression of all the genes in every population of that species. There is a gene pool for each species. Because mating between different species is rare, genes seldom pass from one gene pool to another.

Figure 13-5 **The spotted horsemint is just one species of plant in the mint family. It is, clearly, a distinct organism from a turtle. But, can you be sure that the four turtles shown here are separate species? They are all classified in the same genus.**

Spotted horsemint

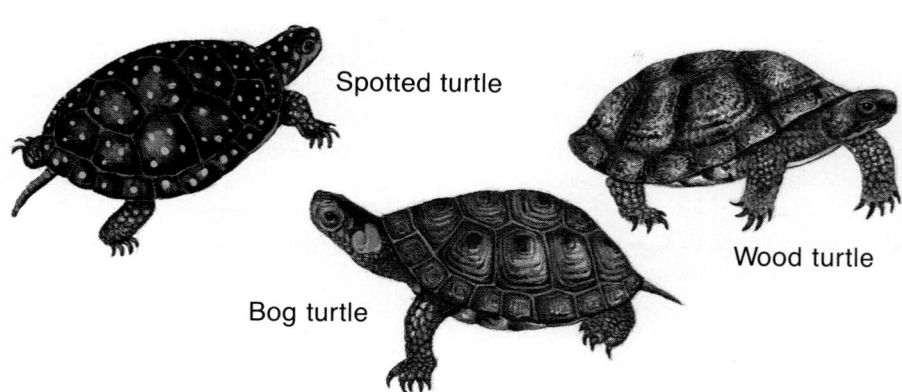

Spotted turtle

Wood turtle

Bog turtle

Western pond turtle

Some species have been seen to interbreed, but they rarely produce fertile offspring. A well-known example is in the breeding of a female horse and a male donkey to produce a mule. Horses and donkeys are separate species, but mules are sterile, and therefore unable to mate with other mules to produce more mules. Therefore, according to our biological species definition, horses and donkeys are distinct species. Another exception to the biological species concept is when a lion and a tiger are brought together in captivity to produce a liger (or tiglon). From the many similarities shared by lions and tigers, it is clear that they are related. However, in the wild, lions and tigers do not live together and are separated by thousands of miles. As you know, tiglons are not found in the wild, and lions and tigers are recognized as separate species.

When Darwin visited the Galapagos islands, he discovered a group of birds that have since come to be known as Darwin's finches. Darwin recognized 14 species divided among four genera of finches on the islands. The Galapagos finches differ mainly in their feeding behaviors, beak structures, and body sizes. Biologists have shown that the beaks are structural adaptations for the different foods they eat. For example, the three species of ground-living, seed-eating finches all have rather short and strong beaks useful in breaking the hard shells of seeds. On the other hand, the insect-eating finches have pointed beaks used for probing into holes in trees or flowers. There is also a very unusual finch with a stout beak that is used to grasp twigs or cactus thorns for use as a tool to probe into the insect's burrow until it emerges.

Because the different species of Galapagos finches share so many features, Darwin concluded that they must have had a common ancestor. Since the birds had many similarities with a species of finch commonly found on the South American mainland, he argued that the ancestor of the island finches came from the mainland group.

Figure 13-6 The beaks of Darwin's finches can be compared to different types of tools. Each tool, or beak, is adapted to dealing with a different job.

Woodpecker finch

Small tree finch

Cactus ground finch

Large ground finch

Evolution of Species

What starts the evolution of an ancestral species into new species? Often, the evolution begins with a geological change that isolates one part of a population from another.

Geographic Isolation   When a river, canyon, or mountain is formed, the populations of plants and animals in the region can often be split apart. This type of separation results in what is known as **geographic isolation.** Because the gene pools of the newly separated groups are isolated, individual organisms of the two groups cannot interbreed. If the new environments have differing conditions, different variations may become naturally selected, and the gene pools of the two new populations gradually become distinct. Mutations arise randomly in each population, genes are recombined in different ways, and as time passes, each isolated group will evolve different traits in response to its own environment. The evolution of the lion and the tiger may have occurred in this way. If the groups remain isolated, they might eventually reach a point where they can no longer interbreed, even when the geographic barrier is removed. They are now separate species only because their gene pools are isolated. Evolution of a new species is called **speciation.**

This sequence of events has been applied to the explanation of the evolution of the Galapagos finches. Somehow, possibly by being blown off course during a tropical storm, some ancestral finches got to the islands. Unable to return because of great isolation from the mainland, the finches became permanently separated from others of their own species. Because the two environments were different, the island birds evolved differently from their mainland relatives. Geographic isolation also exists between each island and between birds on different parts of each island. As a result, many gene pools evolved independently of one another, producing the 14 species of finches existing today in addition to the more common mainland finch. When one species evolves into two or more species with different characteristics, this process is called **divergent evolution.**

Reproductive Isolation   Perhaps the greatest factor in the evolution of the Galapagos finches was that the gene pools of the 14 groups were geographically isolated, preventing interbreeding. As the gene pools became

Figure 13-7   The Abert squirrel (top), of the southern rim of the Grand Canyon, and the Kaibab squirrel (bottom), of the northern rim, are two distinct species that probably evolved from a common ancestor as a result of geographic isolation.

Figure 13-8   Speciation of the Galapagos island finches occurred by geographic isolation and natural selection.

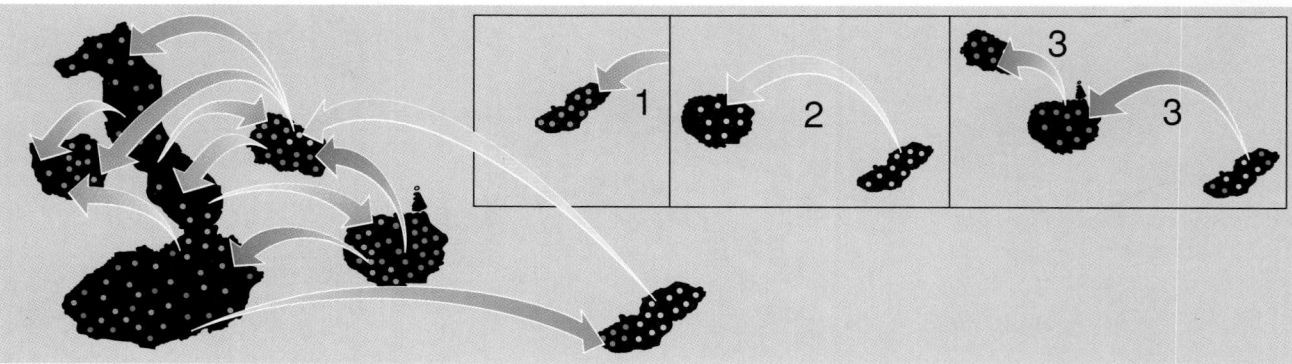

more and more different by chance mutations, a number of other factors prevented interbreeding among the finches. The present 14 species of finches have different mating habits, and mating occurs at different times of the year.

When there is no interbreeding, and gene exchange among species is prevented, a condition known as **reproductive isolation** exists. In some cases, sperm of one species cannot fertilize eggs of another species. Even if eggs are fertilized, gene and chromosome differences between the species are so great that they almost always prevent normal development of an embryo. Those offspring that do develop, however, are often sterile.

Genetic Drift   Another driving force in speciation can arise following geographic isolation. Imagine a large population of leafhoppers that lives throughout a forest on a remote island somewhere in the Pacific Ocean. One day, a local volcano erupts and hot lava covers all but small and remote portions of forest where some leafhoppers survive. The genetic variation in the individual leafhoppers' gene pools, and therefore the range of variations, will have been drastically reduced. In small populations, the chances are high that some alleles will be lost. For example, there is a greater chance that predators will completely remove one kind of phenotype from the population if it has only a few individuals. It is a matter of chance which alleles are lost to the population. They may not have any adaptive value. However, if the alleles control variations that are adaptive, the species may become endangered, Figure 13-9.

Rapid changes in the numbers and kinds of genes in a small, isolated population are known as **genetic drift.** Such random changes in gene pools may lead to the evolution of new species—populations of distinct organisms that are reproductively, isolated.

Figure 13-9   The range of *Ursus americanus,* the American black bear, once spanned across the whole of the United States. It is now listed as threatened or endangered in the few states where it can still be found.

Thinking Lab

MAKE A HYPOTHESIS

Why are finches that survive drought larger with bigger beaks?

Background
One of the Galapagos islands is inhabited by a ground finch that eats seeds. Normally, ground finches feed on a variety of grass and herb seeds that are small. Sometimes, as the supply of grass seeds is used up, the finches will feed on larger seeds of larger plants. Occasionally, drought hits the island and grasses and herbs disappear. The ground finches that survive are larger and have bigger beaks than the ones that do not survive.

You Try It
Make a hypothesis about why surviving ground finches are larger and have bigger beaks than the ones that die during years of drought. Make a complete title and caption for the graph shown here. Explain, based on the graph, what is happening to beak size of ground finches. Why?

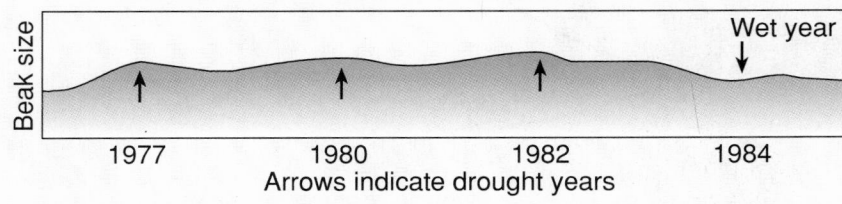

Arrows indicate drought years

Although geographic isolation plays an important role in most cases of speciation, it is not the only factor that leads to the origin of a new species. A new species may evolve in the same geographic area as the ancestral species as a result of changes in chromosome number. This kind of speciation is far more common in plants than in animals.

Speciation by Polyploidy   How can changes in chromosome number lead to the evolution of new species? As you learned in Chapter 8, a set sequence of events must occur during meiosis for organisms to reproduce successfully. What would happen in a diploid ($2n$) plant if, during meiosis, the pairs of homologous chromosomes did not separate at the first division? The gametes produced would each have two sets of chromosomes. The gametes would not be haploid (n), but diploid ($2n$).

As you may have guessed, if two diploid ($2n$) gametes were to fuse in fertilization, the offspring would then have four of each kind of chromosome. In other words, they would not be the usual diploid ($2n$) plants. They would be tetraploid ($4n$). These offspring, if they survived, would be able to undergo a normal meiosis because each pair of homologous chromosomes would line up as usual on a spindle fiber. One chromosome of each pair would be pulled to each pole in the first division, as shown in Figure 13-10. The second division would proceed normally, and gametes produced would be diploid ($2n$). At fertilization, the diploid gametes would unite to produce tetraploid ($4n$) offspring.

When organisms contain some multiple of the normal number of chromosomes, they are called polyploid organisms. Tetraploid ($4n$) plants are often bigger and produce larger flowers and fruits than their diploid ($2n$) ancestors. Suppose that a tetraploid plant was fertilized by one of its diploid parent plants. Not all chromosomes would have a partner when gametes are formed by meiosis in this hybrid offspring plant. There is, therefore, reproductive isolation between the two kinds of plants. They behave as separate species. Many important crops, such as wheat, cotton, and apples, and beautiful garden flowers, such as chrysanthemums and daylilies, have been developed by breeders who can engineer the doubling up of chromosomes and hybridize the resulting polyploids.

Figure 13-10   **Follow the stages of speciation by polyploidy. Many cultivated garden flowers, such as the chrysanthemum (inset), have been developed by selecting and hybridizing polyploid individuals.**

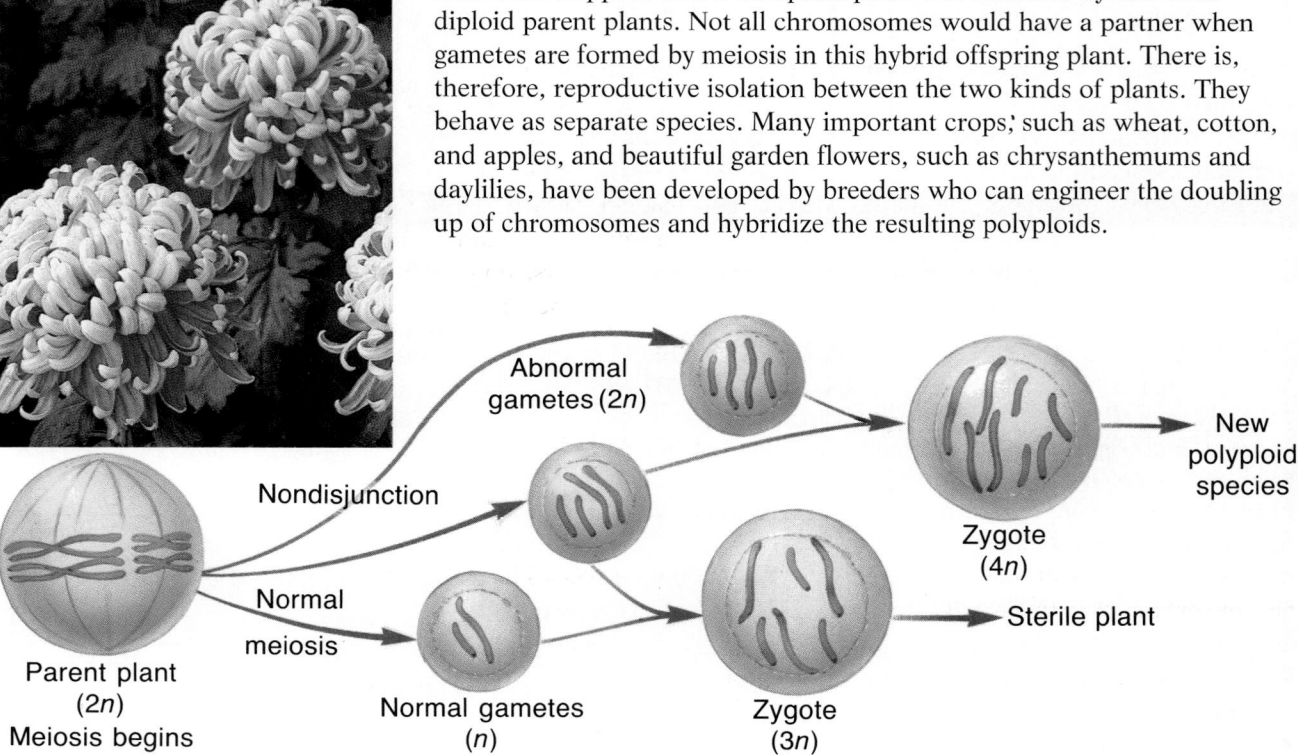

Abnormal gametes ($2n$)

Nondisjunction

Normal meiosis

Parent plant ($2n$)
Meiosis begins

Normal gametes (n)

Zygote ($3n$)

Zygote ($4n$)

New polyploid species

Sterile plant

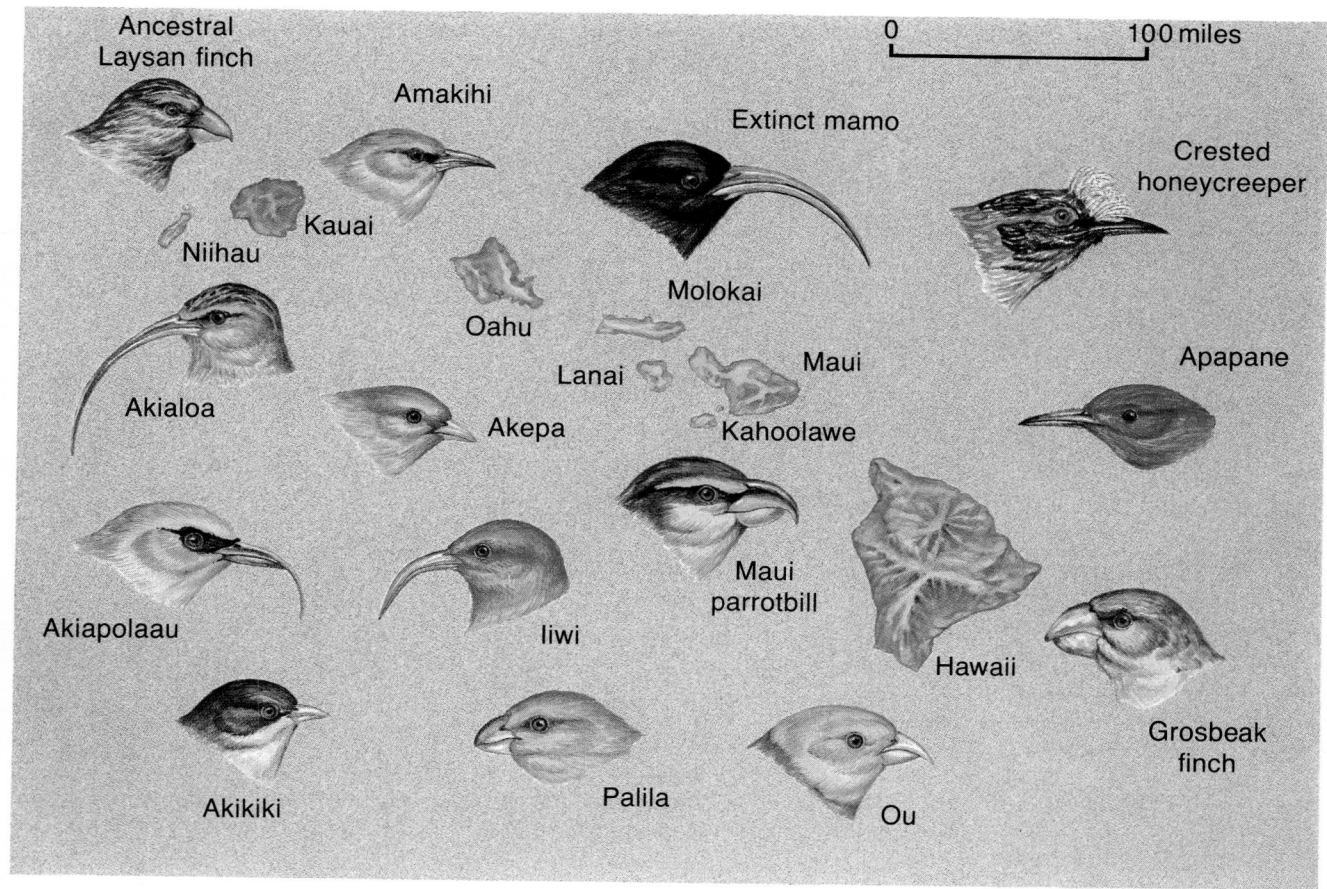

Ancestral Laysan finch

Amakihi

Niihau

Kauai

Extinct mamo

Crested honeycreeper

Oahu

Molokai

Akialoa

Lanai

Maui

Apapane

Akepa

Kahoolawe

Akiapolaau

Iiwi

Maui parrotbill

Hawaii

Grosbeak finch

Akikiki

Palila

Ou

0   100 miles

Adaptive Radiation As you learned earlier, Darwin's finches evolved into many new species within different environments of the Galapagos islands from one kind of ancestor. This pattern of divergent evolution is common among populations on isolated islands, such as the Galapagos and Hawaiian islands. How did the ancestors of all island species get there in the first place? Biologists have suggested that a small population of birds, seeds, or other organisms may have been blown off course onto isolated islands or other remote places during heavy storms. If the new environment were not too different from that of the original habitat, and the competition for food and space from other island populations were not too severe, the organisms would have had a high chance of survival.

The Hawaiian islands, like the Galapagos, have an extraordinary wealth of rare species of plants and animals not found anywhere else on Earth. Similar to the broad array of species of Darwin's finches, there have existed in Hawaii 22 species of Hawaiian honeycreepers—beautiful birds that feed on nectar, seeds, and insects. Nine of the 22 species have become extinct over the last 200 years. The plumage of the Hawaiian honeycreepers displays colors of plain greens and greys to bright reds, yellows, and black, Figure 13-11. The ancestor of all the honeycreepers in Hawaii probably crossed the 2000 miles of Pacific ocean from the American mainland some 5 million years ago.

Figure 13-11 If Darwin had landed on the Hawaiian islands, he would have observed a similar, though more dramatic, example of adaptive radiation in the honeycreepers as the example he found in his finches. For what kinds of foods do you think each beak is adapted?

Minilab 2

How do adaptations aid survival?

Some insects are better adapted than others for survival. Imagine you are a bird predator of insects. Your insects are colored either white or black. Use a paper hole punch to punch 100 dots from a sheet of white paper. Repeat with a sheet of black paper. These dots will represent white and black insects. Place both white and black insects on a black sheet of paper. Work in pairs and see how many insects of each color your partner can collect in two minutes.

Analysis *How does color affect the survival rate of an insect? What might happen over time to a similar insect population in the wild?*

The islands of Hawaii, the Galapagos, and many other island groups around the world provide large numbers of distinct and separate habitats. When a population of immigrants increased, split up, and radiated out into separate habitats, new species evolved. Each species became adapted to a particular role in the new habitat. Similar to Darwin's finches, the Hawaiian honeycreepers have many roles. They display a broad variation in bill shape, feeding behavior, and body size. For example, one species has a straight bill like that of a woodpecker. There are no woodpeckers in the Hawaiian islands. The role of a woodpecker has been filled by this species of honeycreeper.

The divergent evolution and adaptation of species to different roles in new habitats is called **adaptive radiation.** For adaptive radiation to occur, three conditions must be met.

   (1) The ancestral form must reach the new environment.
   (2) The ancestral form must have some basic adaptations suited to the new habitat.
   (3) The new environment must be free from competition with similar, better, or equally adapted species.

Darwin's finches, Hawaiian honeycreepers, and a vast array of other island plants and animals have met these conditions and have successfully evolved by adaptive radiation.

Other Patterns of Evolution

Most often, evolution results in new species that are different from one another as described in divergent evolution. Sometimes, though, convergent evolution occurs. **Convergent evolution** is a process in which species that are not closely related evolve similar traits. Convergent evolution usually occurs in unrelated species that have similar roles and live in similar environments.

Figure 13-12   **Although dolphins (right) have the same streamlined body shape as many kinds of fishes (left), their body structures are homologous with those of other mammals, not with fishes. Which of these two animals needs air to breathe?**

Similar But Not Related   Examples of convergent evolution are easy to spot. Think about fish and dolphins. The ancestor of all fish was a simple marine organism. Dolphins, though, are descendants of land mammals. Fish and dolphins, then, are not at all closely related. How are fish and dolphins similar? Both have a streamlined shape for swimming. Fish have fins, but the finlike structures of dolphins are not homologous; they are really modified limbs. The similarities between fish and dolphins can be explained in terms of their environment. Both organisms evolved in water. In the two groups, natural selection favored variations suited for an aquatic environment. But the two groups did not undergo convergent evolution in all their traits. For example, fish have gills, but dolphins have lungs. Each type of animal can live in water with these organs, but dolphins must come to the surface for oxygen. Can you think of other differences between these sea animals?

Figure 13-13   **The honeybee is a valuable insect pollinator. Many popular fruits, such as apples and oranges, would not develop without them. What kinds of flowers might be pollinated by butterflies?**

Hiding from Predators   As you know, in the world of nature, the rule is survival. If one animal is a tasty meal for another larger and stronger animal, then the only way it will survive is if it can hide. The ability to hide from predators is a well-known adaptation observed among animals. One seemingly clever strategy in animals is to deceive.

The process of natural selection has provided many animals with structural or behavioral adaptations that copy or mimic their surroundings to avoid detection. Look back at the chapter opener. The fish disguised by nature rests among the seaweed, unnoticed by hungry predators that pass by. The insect that looks like a leaf will have a good chance of surviving when a passing bird stops to rest for a midday meal. The animal that looks like its surroundings or like something a predator would avoid is preserved by natural selection.

Evolving Together   Populations of organisms do not evolve isolated from one another. Rather, as you know, different organisms live together within communities and interact in many ways. Often, these interactions involve complex relationships and affect the evolutionary history of the organisms. Are you familiar with flowering plants that are pollinated by insects, bats, or birds? Plants benefit by having pollen transferred from one flower to another, thus increasing the chances for successful reproduction. The pollinators benefit by obtaining nectar or pollen for food. Over time, this interaction has led to some interesting, mutual adaptations in many plant and animal species.

The Tempo of Speciation

Up to this point, you have seen that sometimes speciation can occur rapidly over a few generations. But how long did it take for the enormous variety of organisms, from *amoeba* to human, or from alga to oak tree, to evolve? Biologists have proposed that the evolution of all existing species occurred over very long periods of time—millions of years—during which time small adaptive changes slowly built up in populations. That idea of slow and steady change in species is known as **gradualism.**

Other biologists have described evidence suggesting that some speciation occurs much more suddenly. They explain that species probably remain constant for long periods of time. However, a sudden change in environmental conditions—such as warmer temperatures or drought—will interrupt, or punctuate, the steady rate of evolution. New species become established and remain constant for millions of years or until new changes in the environment occur. This idea of slow evolution punctuated by short events of rapid evolution was given the name **punctuated equilibrium.** It is important to note that the term *rapid* must be thought of in terms of geological time—maybe as long as 100 000 years.

LITERATURE CONNECTION

How the Crow Came to Be Black

According to Good White Buffalo of the Brule Sioux nation, crows were once white. They were also friends of the buffalo, and warned them when Native American hunters approached the herd, much to the resentment of plains Indians who depended on buffalo for food.

A young brave captured the leader of the crows and hurled him into the fire. The crow escaped, promising never to warn the buffalo again. He was singed black, and ever since, all crows have been black.

Folklore includes a variety of stories passed orally from person to person and generation to generation within a given culture. Many prescientific explanations for the origins and conditions of the world appear in folklore as myths.

The myths of a given culture center on what that culture considers important. "How the Crow Came to Be Black" illustrates the importance of buffalo to the plains tribes.

Biologists today know color has evolved through natural selection and species adaptation. In many species, color serves important survival functions. For example, Arctic foxes turn white in winter and can hide from predators against the snow.

Although evolution may provide us with many a tale of origins of weird and wonderful life forms, probably none would be as simple, or simplistic, as the myths and legends passed on from generation to generation in folk tales.

1. **Explore Further:** Give examples of species in which color aids an animal's survival.
2. **Connection to English:** Use your dictionary to distinguish among the meanings for myth, legend, fable, parable, allegory, and tale. Use your library to discover the Greek fables by Aesop that explain how other animals came to be.

Evidence of gradualism has been pointed out in the fossil record. For example, as you have read in Chapter 12, modern-day elephants can be traced back through mammoths to mastodons that lived more than 30 million years ago. This evidence clearly shows how some species have gradually changed over a long period of time. Other fossil evidence indicates that some species have remained the same over long periods of time and then, quite suddenly in geological terms, they became extinct and new species appeared. Punctuated equilibrium would explain these patterns in the fossil record. For example, the fossil record indicates that about 570 million years ago, there was a sudden appearance of enormous numbers of multicellular invertebrates. Cambrian fossils seem to indicate that, before that time, only monerans and protists existed.

Figure 13-14 **There are many more kinds of organisms now than there were 300 million years ago. Many of those ancient forms, such as the extinct arthropods called trilobites (left), show strong similarities to present-day forms, which, in this case, are the horseshoe crabs (right).**

Discussions over the rate at which evolution occurs are ongoing among biologists. The major problem with interpreting the fossil record is that only hard parts of organisms are preserved. There may have been changes in the soft tissues of evolving species that obviously cannot be studied. The most likely explanation may be that some species have evolved gradually, while others evolved rapidly. It should not surprise you that scientists offer alternate hypotheses to explain observations. As you know, hypotheses are subject to change in light of new evidence and as biological research continues.

Section Review

Understanding Concepts

1. Why is geographic isolation necessary for most cases of speciation?
2. Does the idea of punctuated equilibrium mean that the theory of natural selection is incorrect? Explain.
3. Explain why it was not possible for Darwin to find animals such as wolves or cats on the Galapagos islands.

4. **Skill Review—Observing:** You discover two groups of butterflies living in fields not too far from one another. The butterflies are similar in appearance, but their markings are not identical. What clue would you need to determine whether they are members of the same or different species? For more help, see Practicing Scientific Methods in the **Skill Handbook**.

5. **Thinking Critically:** A population of spiders is divided by a river, and the two gene pools begin evolving differently over time so that each group of spiders has some unique adaptations. A drought causes the river to dry up before the two groups have become distinct species. Explain what may happen to the two groups.

13.3 Human Evolution

Everyone's familiar with the behavior of humans. Humans plant crops and gather food. Humans breed animals to use for work, food, and protection, and humans play video games, work on computers, and build cities. Like all other forms of life, humans are products of evolution by natural selection. What evidence is there for the evolution of humans, and what adaptations have led us to the unique place we have on Earth today?

Primate Adaptations

To begin to answer the question of what makes humans unique, we must step back in time. Humans, like any other organism, can be grouped to show relationships. Humans are members of a group of mammals known as primates—animals that include lemurs, monkeys, chimpanzees, and gorillas. Humans and other primates share several traits.

One such trait is a relatively high level of problem-solving ability. Relative to body weight, primates have the largest brains of any mammals. This adaptation for large brain size has allowed primates to survive and compete in a variety of environments, such as tropical rain forests, dry savannahs, and cool, northern forests. All primates have flexible shoulders and forelimbs that can be rotated. These adaptations allow them to swing from branch to branch among the trees.

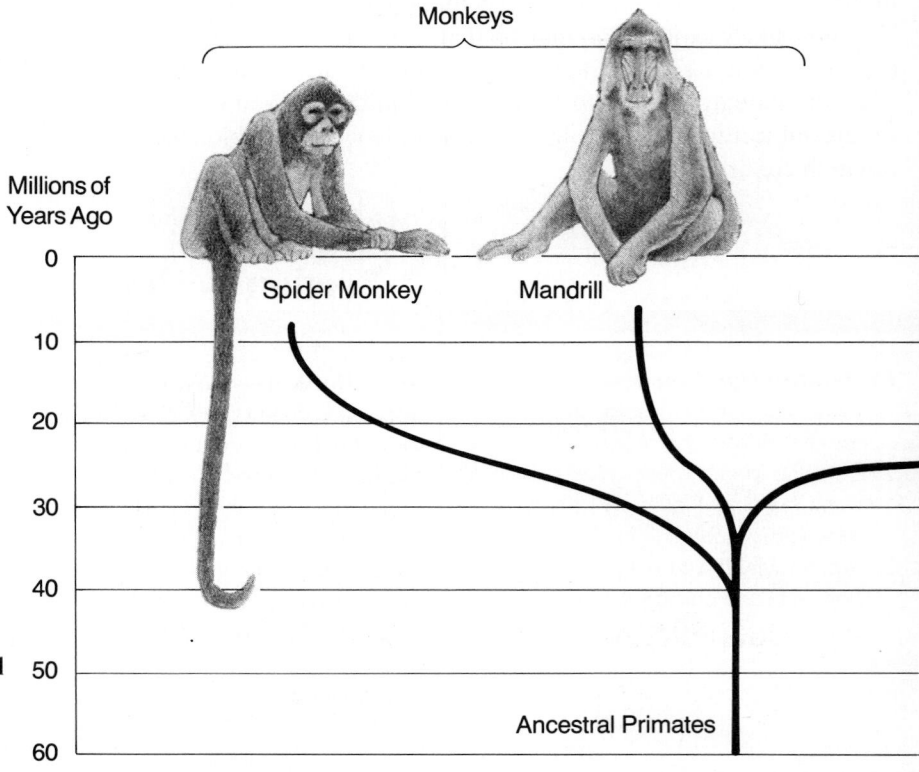

Figure 13-15 The ancestors of present-day primates probably evolved from primates such as the lemur (left photo). Primates share traits such as fingernails, four pairs of incisors (biting teeth), and opposable thumbs.

Another unique adaptation among primates is the structure of the hand. Unlike non-primate mammals, which have paws with claws, primates have extremely flexible hands consisting of five independent digits with fingernails and toenails. Primate hands have an opposable thumb—one that touches the fingers of the same hand and can grasp objects. The overall flexibility of the primate hand allows it to be used as a structure of locomotion and to manipulate objects.

Large brains, opposable thumbs, and rotating forelimbs are just some of the adaptations common to all primates. Because these traits are shared by all primates, biologists agree that the traits were probably present in the earliest primate ancestors. As monkeys and then apes evolved, there were further modifications that added to the success of these primates in a variety of environments. Apes include gibbons, orangutans, gorillas, and chimpanzees. There are more characteristics shared among humans, chimpanzees, and gorillas than among any other primates (Figure 13-15).

African Origins

The story of human evolution begins approximately 8 million years ago in Africa. During this time, scientists propose, a population of ancestral apes diverged into two groups. One group would eventually evolve into the African apes—chimpanzees and gorillas. The other group, a population of small-bodied apes, would lead to modern humans. This group gained the ability to walk upright on two legs.

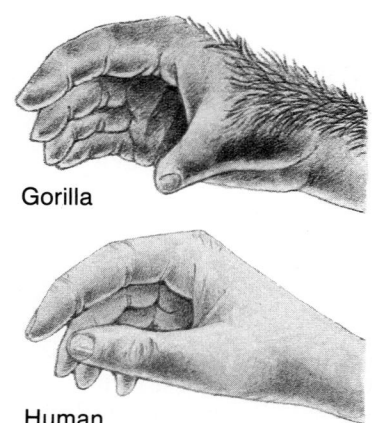

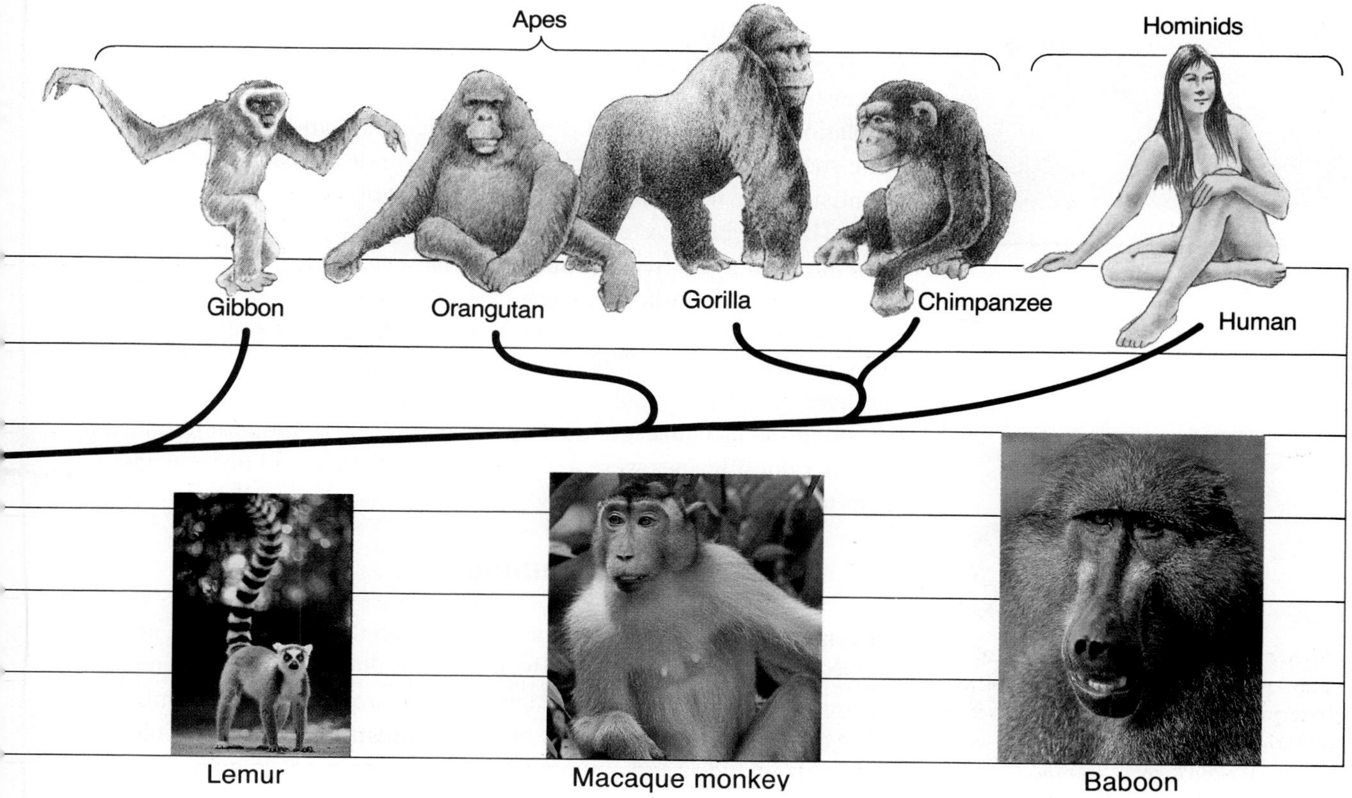

Gorilla

Human

Figure 13-16 The ability to grasp objects between a thumb and the other digits is a characteristic common to all primates, and not unique to humans.

Apes

Hominids

Gibbon   Orangutan   Gorilla   Chimpanzee   Human

Lemur   Macaque monkey   Baboon

Human Ancestors   Thus far, there is no complete fossil record for the time period when humans first began to diverge from apes. However, there is now a large collection of fossils for the humanlike apes that lived in Africa from about 5 million years ago. Most often, only small fragments of bone and teeth are discovered. However, anthropologists can tell a great deal from these meager findings. In attempting to trace the course of human evolution, the fossils are analyzed and a variety of structural adaptations are compared. Among them are skull traits including brain capacity, relative sizes of bones, and joint structure. Based on evidence such as this, anthropologists classify each and every fossil of human bone. As you might imagine, classifying extinct humans on the basis of a few bones or teeth is difficult at best, and interpretations often differ among groups of anthropologists.

Southern Ape   The oldest fossils of human ancestors were discovered in several areas of Africa. These fossils were named australopithecines because *austral* means southern and *pithekos* is Greek for ape. Several species of southern ape have been discovered. *Australopithecus afarensis* was first discovered in East Africa and has been dated at 3.5 million years old. This ancient primate was called Lucy by her discoverer, who was a fan of the pop group The Beatles. He was playing their hit song that included this name at the time of his find. Studies of Lucy and other fossils of this species indicate they had an average brain capacity approximately the size of chimpanzee brains. Lucy was three feet tall, and it was determined that she must have had a mass of 66 pounds, about the same as you weighed when you were in fourth grade. Since Lucy, other specimens of the southern ape have been found, and they date back to 5 million years ago. These early human ancestors, although they walked on two legs, were probably clumsy by modern standards. Their arms were longer in relation to body size than those of modern humans, and their legs were quite short. From the study of these limb proportions and joint structures, scientists agree that this species may have spent a great deal of time in the trees. They probably returned there for safety and feeding.

In addition to Lucy, anthropologists have recognized three more species of australopithecines. The evidence suggests that there was adaptive radiation among these southern apes in Africa. They all evolved and lived in Africa from 5 million to 1.5 million years ago. The skulls of southern apes had a smaller brain capacity than skulls of modern humans. Although the face and teeth were similar to those of modern humans, if the australopithecines were alive today, scientists would not consider them to be human.

The Origin of Humans

Modern humans, meaning you and all the people on this planet, are classified as *Homo sapiens.* During the last half million years of the time of the australopithecines, the first human ancestor arose. Fossils recognized as belonging to the same group as modern humans are called hominids. Hominids include handy man, upright man, and modern man.

Figure 13-17   The oldest fossils of Australopithecus make up the skeleton of a young female. Using this 3 million-year-old skeleton, an artist provides us with an impression of these human-like apes.

Rocking the anthropological world, a second "Lucy" is discovered in southern Uganda.

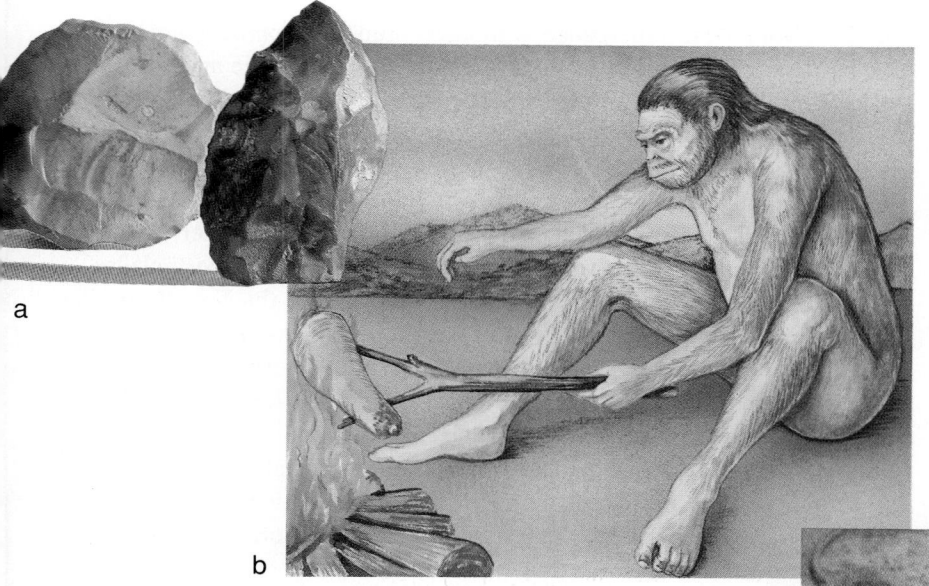

a

b

c

Figure 13-18 Early hominids first learned how to make and use tools from rocks (a). Upright man, *Homo erectus*, learned how to use fire (b). Neanderthals, an early form of modern humans, used ceremonies to bury their dead, as evidenced by these remains (c).

Handy Man After southern apes, the next oldest known hominid is dated at more than 1.8 million years old. The brain capacity was larger than that of any earlier southern ape. Stone tools were found with these fossils, indicating that these hominids were handy at making and using tools for their everyday lives. These hominids were named *Homo habilis,* meaning handy man.

There is evidence that handy man collected stones from many miles away from where they lived. They used other stones to reshape them into tools for chopping, scraping, cutting, and hammering. They probably used the tools to make weapons for hunting purposes and for butchering the meat once caught and killed. Toolmaking is an important human behavioral trait rarely seen in other animals. The ability to make and use tools is a strong indicator of the handy man's learning abilities.

Upright Man Anthropologists have suggested that *Homo habilis* gave rise to a new species that appeared more than 1 million years ago. More recent than handy man, this species is called *Homo erectus,* meaning upright man. This hominid had a rather humanlike face, flat, but with very large brow ridges and a lower jaw that sloped back with no chin. The skulls have a larger brain capacity than those of handy man.

What makes upright man different from handy man? They had learned to make and use fire! The first evidence that people had discovered the importance of fire is found with remains of the *Homo erectus* people. There is ample fossil evidence that the upright man was an excellent hunter, and discovery of hearths in caves indicates that these people used fire and lived together in groups. There is fossil evidence that these ancestors of modern humans began to migrate about 1 million years ago, and they spread throughout Africa, Asia, and possibly into Europe. Upright man became extinct about 300 000 years ago.

Modern Man   The emergence of modern humans, *Homo sapiens,* remains a tricky puzzle today due to scanty fossil evidence and controversial dating. Yet there is ample evidence to sketch out a basic picture for the origin of our species. It is known that by about 100 000 years ago, populations of *Homo sapiens* appeared in Europe, Africa, and Asia. These first people of our species lived between 100 000 and 40 000 years ago and possessed features typical of both *H. erectus* and of ourselves.

Best known among the first *Homo sapiens* are the Neanderthals, so named because the original fossils were discovered in the Neander Valley in Germany. Neanderthal fossils have been found in Europe and Southwest Asia. Neanderthals were a short, powerfully built group. Their thick skulls were somewhat flat with heavy brow ridges, and like *H. erectus,* they lacked a chin. Neanderthals and modern man show similar brain capacities, and they both developed ceremonies for the burial of the dead. The view that Neanderthals were simply brutish cavemen is incorrect. Rather, there is excellent evidence which suggests that Neanderthals were efficient hunters and creative artists.

What happened to the Neanderthals, and how did modern humans emerge? Until recently, many scientists believed that populations of Neanderthals in Europe and Asia evolved directly into the different groups of modern *Homo sapiens.* However, recent archaeological and

GLOBAL CONNECTION

Cultural Evolution

Whether the ability to communicate with language first evolved among the Cro-Magnons or earlier, there is certainly no doubt about its impact on the lives of humans. Speech was made possible by evolution of a large and complex brain.

The development of language coincided with accumulation of knowledge. Older people could teach younger ones about all they had discovered, thus avoiding the need for people to learn everything on their own.

Using Tools   At first, improved patterns of speech may have led to ideas for making better tools. With new tools came an increase in the ability to hunt larger animals and to make clothing, shelters, and new inventions such as boats. With each of these advances, humans had a better chance of surviving. Notice, however, that the changes in humans were not changes that

affected their gene pools, but were because of changes in human cultures. Instead of being molded by changes in their environment, humans began to alter or avoid the effects of that environment. Unlike all other organisms, learning and intuition began to surpass mutations and recombination as causes of change in the human species.

Learning to Farm   Early hominids were hunters and gatherers. They hunted large mammals and gathered grains, nuts, berries, and other fruits. From about

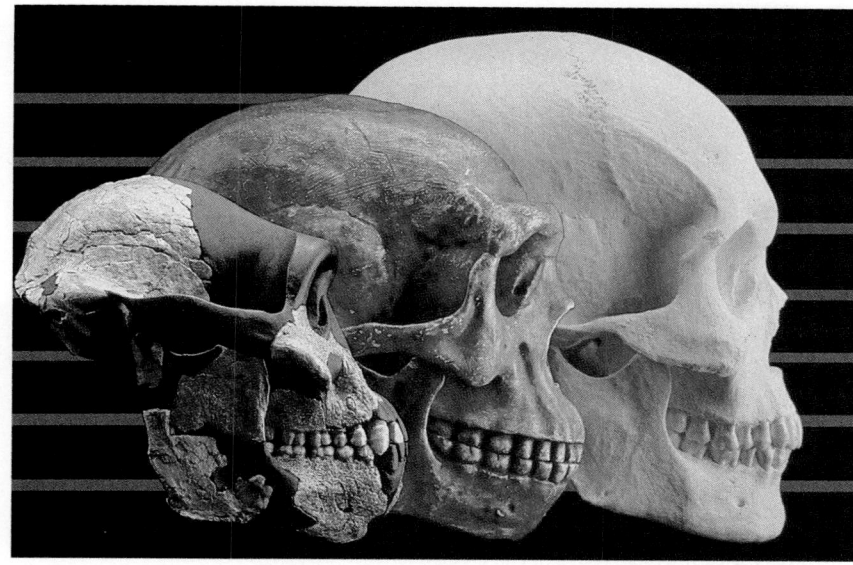

Figure 13-19 The skull of *Australopithecus afarensis* (left) has a cranial capacity of 400 cm³. In *Homo erectus* (center) it is 850 cm³, and in modern *Homo sapiens* (right) the average volume is 1360 cm³. The evolution of the human brain spans 3 million years. By how much has brain size increased during this time?

10 000 years ago, our ancestors began to sow seeds, tend crops, and protect livestock. This change to an agricultural lifestyle led to permanent settlements and different cultural practices. Growing of crops reduced competition for food; thus, population size increased and permanent homes could be established.

Intellectual Explosion

Cultural evolution also led to a much more recent development in the history of humans, the Industrial Revolution. In the early 19th century, many societies became more technological than agricultural. Factories produced new products, new methods of transporta-

tion were invented, sanitation was improved, and electricity changed the lives of humans in countless ways. The products of modern technology have become important parts of our daily lives. They include not only CD players and air conditioners, but more importantly, drugs and medical equipment that save and prolong lives. But cultural evolution and an increased human population has its downside, too. As you are aware, the ways humans interact with one another and with the environment are often destructive.

Will our future cultural advances keep our planet and its species safe from extinctions?

1. **Connection to Genetics:** What new species of plant foods have humans produced within the last few centuries?
2. **Writing About Biology:** Describe social organizations in a group of animals other than humans.

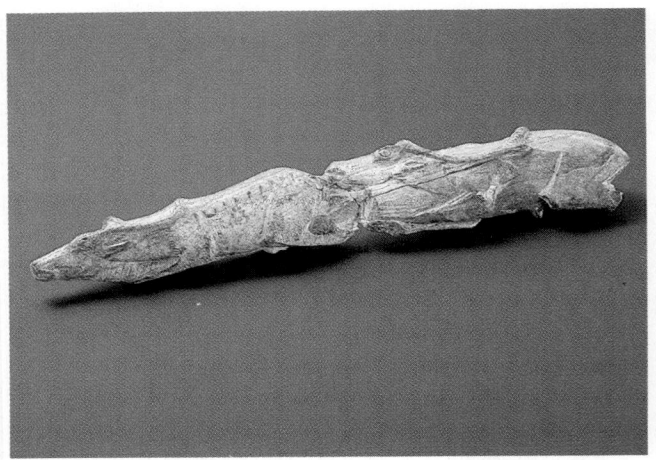

Figure 13-20 Artistic creation is an anthropologist's clue to ancient hominid levels of abstract thought. This beautiful carving of mammoth ivory depicting two reindeer is about 15 000 years old. What can you deduce from such an artifact? Magnification: 0.5×

genetic evidence suggests that this may not be true. The new evidence supports an origin of modern humans, *Homo sapiens*, in Africa perhaps as early as 200 000 years ago. Scientists now believe that this population of modern humans quickly spread throughout Europe, Africa, and Asia. By 30 000 years ago, Neanderthals were extinct and modern man had spread throughout most parts of the world. The most famous among the early modern peoples are the Cro-Magnons. Discovered in France, they were fully modern in height, skull structure, skeleton, teeth, and brain size. Not only were they advanced toolmakers, they were also talented artists. It is fairly certain that these people also used language.

The evolution of complex language and the use of symbols must have led to an increased ability to communicate ideas and to make plans with others. These capabilities might have made the difference between success and extinction as the groups competed with one another for survival.

In 30 000 years, modern humans have adaptively diverged into a wide range of groups differing physically and culturally. Probably the most obvious differences to have evolved among peoples of the world include more than a thousand languages and hundreds of cultural and ethnic social practices. These differences, however, reflect only minute cultural differences. With widespread global travel among all groups of humans, speciation by geographic isolation has not occurred. The species *Homo sapiens* has remained as one.

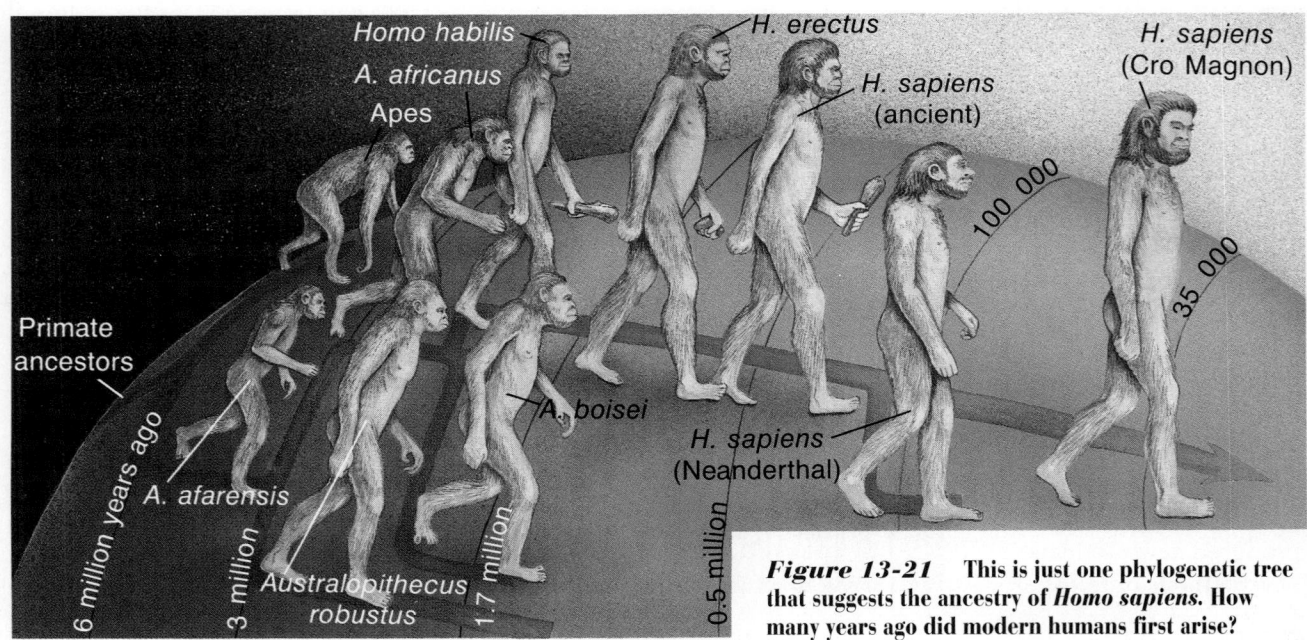

Figure 13-21 This is just one phylogenetic tree that suggests the ancestry of *Homo sapiens*. How many years ago did modern humans first arise?

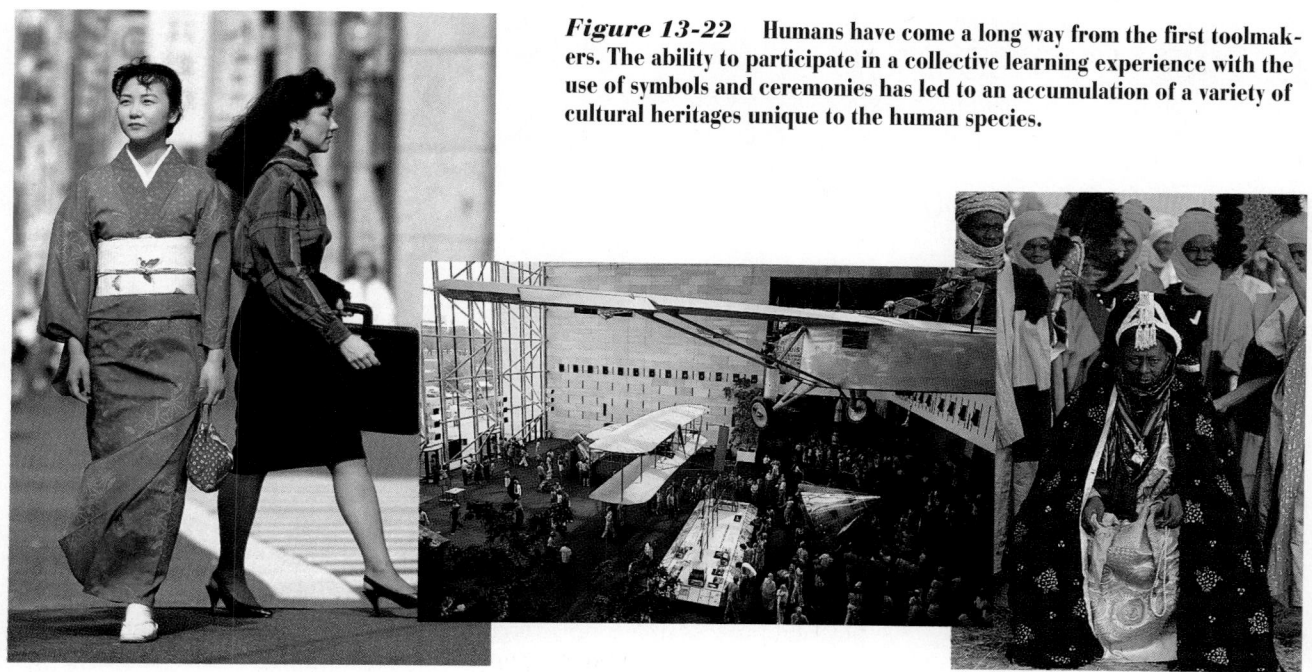

Figure 13-22 Humans have come a long way from the first toolmakers. The ability to participate in a collective learning experience with the use of symbols and ceremonies has led to an accumulation of a variety of cultural heritages unique to the human species.

LOOKING AHEAD

SPECIES BECOME adapted to their environments. As the environment changes, variations better suited to these new conditions are naturally selected and the population as a whole evolves. Information from evolution also helps to show relationships among species, which can then be used to classify organisms. You will find out that classification of species is a way of presenting their evolutionary history. 🐾

Section Review

Understanding Concepts

1. Why is it not biologically accurate to say that fossil humans are the missing link between apes and modern humans?
2. What evidence would scientists use to determine the diet of fossil humans? Explain your answer.
3. How can we determine relationships of modern humans to australopithecines, *Homo habilis,* and Neanderthals?
4. **Skill Review—Making Tables:** Make a table that compares australopithecines, upright man, Neanderthals, and Cro-Magnons in terms of where and when they lived, their brain size, and their general appearance. For more help, see Organizing Information in the **Skill Handbook.**
5. **Thinking Critically:** The average height of humans in the United States has increased by several inches during this century. Explain how that change in height might have evolved.

B I O L O G Y,
T E C H N O L O G Y, A N D
S O C I E T Y

Physics Connection

Honeybee Heating and Cooling

Would it surprise you to learn that honeybees have cooling systems similar to air conditioning and refrigeration? Humans have used water for cooling for millennia, and so have honeybees.

Water helps cool for several reasons. It draws thermal energy from its surroundings to change its physical state from solid to liquid or from liquid to vapor. Therefore, evaporating water or melting ice draws heat from the air around it. Water is a better conductor of heat than air, so water will conduct heat away from a body or hot object faster than any surrounding air.

Humans use these physical principles in refrigeration and air conditioning systems, which draw heat from the surroundings to evaporate liquid. Honeybees use the same physical process in a simpler form. In hot weather, honeybees spread water over their honeycomb, then use their wings to fan it.

Honeybees also use strategies for staying warm in cold weather. Unlike other social insects, they remain active throughout winter in the hive, actively metabolizing. To maintain body temperature, they huddle together. As external temperatures drop, the bees conserve

thermal energy by clustering more tightly. The cluster of bees moves and feeds as a unit over the honeycomb.

1. **Issue:** Why is air conditioning harmful to the environment, and how can we reduce the problems?
2. **Explore Further:** How are honeybees adapted to pollinating flowers?

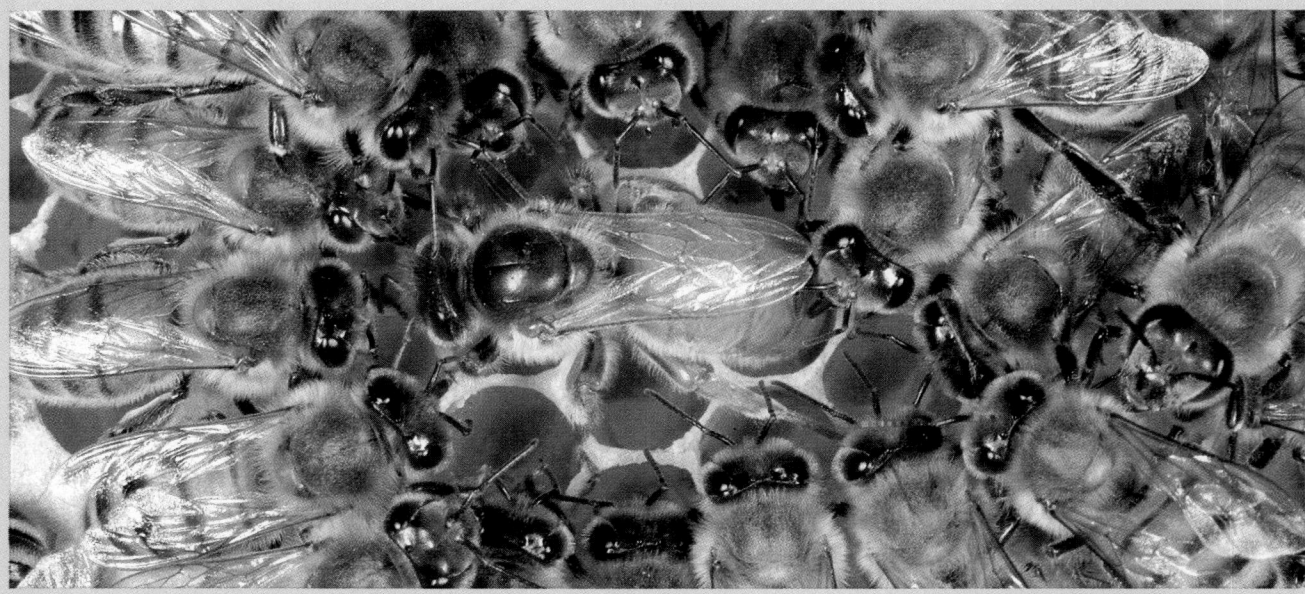

Interbreeding to Save a Species

The Florida panther and the gray wolf are registered as endangered species. Once a species has been recognized as endangered, its members receive protection under the law. They may not be killed, and in many cases their habitats are protected from development. However, the protection of the Florida panther may be questioned based on claims that it is no longer a pure species. Although only 50 panthers remain in the Everglades habitat, genetic analysis has indicated that some panthers have interbred with South American cougars that were released into the wild.

The gray wolf's endangered status was challenged by state farm bureaus in Indiana, Montana, and Wyoming because some had interbred with coyotes. Therefore, the petitions charged, the gray wolf species was not pure, but contained many hybrids.

What Is a Species? If you have heard *species* defined as a group of organisms that will not interbreed with another group, you might wonder how a species can be impure. If breeding between species is possible, what, then, is a species? The answer depends upon the definition of species.

Most taxonomists use outward appearances for their classifications. Field biologists who want to classify an organism cannot immediately identify the other organisms with which it can breed. It is much easier to classify by structure. Only with extensive breeding programs can a research scientist discover the

limits of the breeding group. When offspring are neither healthy nor able to reproduce, the species limits may have been identified.

Too Pure to Save? Without complete knowledge of what constitutes a species, you can probably appreciate the challenge facing those who are responsible for registering a species as endangered.

Another problem facing those in charge of protecting endangered species is that sometimes the efforts to keep these species genetically pure may work against the goal of preserving the species. For example, in the case of a group of organisms

with a very small population, breeding among individuals only within that population can result in genetic weaknesses that threaten the group's survival. Most Florida panthers have few offspring and many display abnormalities. In such a case, interbreeding with the cougars may produce offspring that are healthier and more resistant to disease.

Understanding the Issue

1. If you were in charge of protecting the small, inbred population of an endangered subspecies, would you try to keep the population pure, or interbreed if necessary to ensure the survival of its members' descendants? Explain your reasons.
2. How might protecting a species be a threat to local economics?

Readings

- Mlot, Christine. "The Science of Saving Endangered Species." *Bioscience* 2, 1989, p. 68.
- Radetsky, Peter. "Cat Fight." *Discover*, July 1992, pp. 56-63.

CAREER CONNECTION

As a **naturalist**, you might monitor species in the wild, study an ecosystem, or work to save an endangered species. You might also monitor pollution levels and their effects on organisms in the wild. You probably would work for a government agency such as a park or forestry service, a university, or a museum of natural history.

Summary

Adaptations, which promote fitness, evolve slowly as previously existing adaptations are modified through changes in the environment and chance events.

Most instances of speciation result after geographic barriers split an original species into two or more isolated groups. Some cases of speciation occur without isolation as a result of chromosome mistakes.

Adaptive radiation is movement of individuals into new environments where they become separate species. In convergence, organisms that are not related evolve similar traits due to living in similar environments.

Fossil evidence of hominids reveals much about the evolution of humans in Africa and their migration to all parts of the world. As time passed, evolution of language enabled humans to become adapted culturally.

Language of Biology

Write a sentence that shows your understanding of each of the following terms.

adaptive radiation
behavioral adaptation
convergent evolution
divergent evolution
genetic drift
geographic isolation
gradualism
physiological adaptation
punctuated equilibrium
reproductive isolation
speciation
species
structural adaptation

Understanding Concepts

1. Is most human behavior an adaptation in the same sense as migration of birds? Explain.
2. Do structures having similar functions necessarily indicate close evolutionary relationships? Explain.
3. Many species of millipedes are found to inhabit the same forest. Could they all have evolved there? Explain.
4. The thousands of land vertebrates (amphibians, reptiles, birds, and mammals) are descended from a type of fish. What pattern of evolution does this example illustrate? What conditions had to be met in order for the land vertebrates to evolve?
5. Based on their classification, is it possible that Neanderthals and Cro-Magnons could have interbred? Explain. Could *Homo erectus* and *H. habilis* have interbred? Explain.

6. Explain how the evolution of various hominids illustrates the way in which speciation most often occurs.
7. Suppose a small part of an original population is isolated after a disaster and is in an environment quite different from that before the disaster. Why might it very rapidly evolve into a new species?
8. Trace the evolutionary steps from the beginnings of increased brain size in southern apes to the development of complex language in Cro-Magnons.
9. The digestive tracts of cows contain bacteria that can digest the cellulose in the grass the cows eat. Cows lack an enzyme that would help digest cellulose. Can you explain how the cow/bacteria relationship might have originated?

Applying Concepts

10. Consider the various geographic groups of humans such as Pygmies, Inuits, Polynesians, and Hottentots. What factors may have led to their evolution? Will these geographic groups someday evolve into separate species? Discuss.

11. Colchicine is a chemical that can be used to induce polyploidy. How might such a chemical be used to benefit society?

12. Suppose another ice age develops in the distant future. How would the response to such a change differ between humans and other organisms if both are to survive?

13. Organisms that have variations that can be used in new ways in new environments are said to be preadapted. How was preadaptation important to Hawaiian honeycreepers?

14. Many animals have adaptations known as courtship behavior. This kind of behavior enables males and females of the same species to recognize one another and stimulates them to mate. Besides the fact that courtship behavior promotes reproduction of a species, how else is it important?

15. **Global Connection:** Why does cultural evolution result in more rapid changes in humans than does biological evolution?

16. **Issues:** If a nearly extinct species is released into a new habitat with an unlimited food supply and no competition, what might happen to the species?

Lab Interpretation

17. **Thinking Lab:** Flies are sprayed with a new chemical pesticide on Monday for five weeks. The flies that remain alive are counted each Monday. Based on the graph below, predict how many flies will be left in week five. Explain these results.

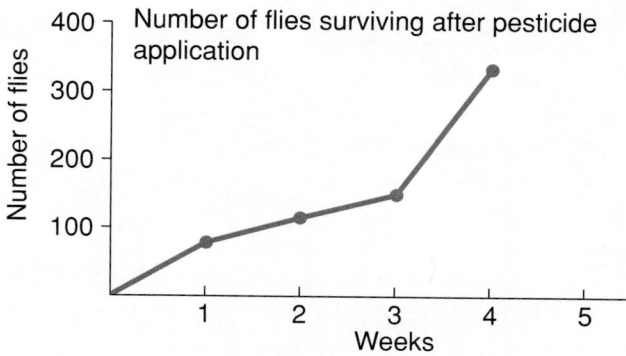

18. **Minilab 2:** A marshland in New Jersey has plants with bright green vegetation, dark brown stems, and bright red berries. Describe how a frog that lives in this marsh and is difficult to spot might look.

19. **Investigation:** Analyze the following data of numbers of stomata on leaves of one tree species. What might the data tell you about the rainfall in the areas where the data were collected?

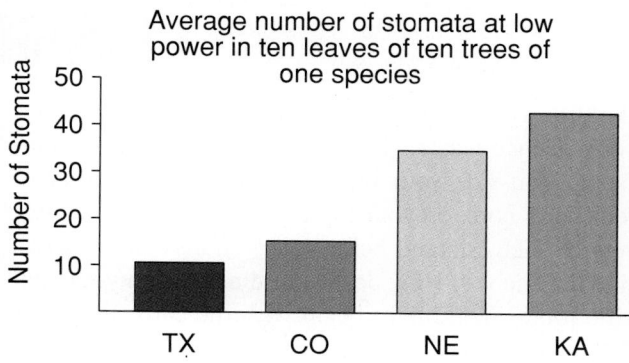

Connecting Ideas

20. Give reasons why a mule would be sterile.

21. How might viruses and the host cells they invade have evolved?

CLASSIFICATION

IMAGINE IT'S the beginning of a new semester, and your science teacher has given you a research assignment. You ride your bicycle to the public library to look for a book on your topic. In the library, row after row of shelves stand before you. Thousands of books line the shelves. How do you find a book on your specific topic? You might begin with the card catalog or the computerized catalog system. You look for your research subject and find a book that looks interesting. The Dewey decimal number tells where a book is located. When you go to the shelf to find your specific book, you will also find other books on the same topic. That's because similar books are classified together.

The same kind of situation faces biologists. Biologists deal with millions—not just thousands—of organisms. How do they locate information about a certain organism? How do they communicate with one another about their findings? Scientists have devised a classification system for all organisms, just as a classification system was created for all books. Instead of a number, the biological classification system assigns each organism a two-part name. Similar names are assigned to related organisms. Of course, the organisms aren't neatly arranged on a shelf for the biologist to study, but information about the organisms can be. That is the true benefit of classification—it enables us to organize information about the world.

How could you find a book about biology among the thousands of books in the library if there were no classification system?

Chapter Preview

Objectives

Explain the concept of binomial nomenclature.

Discuss how various lines of evidence are used to classify organisms.

Develop the concept of a phylogenetic diagram.

14.1 The Concept of Classification

Have you ever played in a school band or listened to one perform? The various musical instruments in a band are placed into several groups—brass, percussion, or woodwind. Or think about the last motor vehicle you rode in. Was it a car, a truck, a bus, or a van? Each of these broad groups can be subdivided into several more specific groups. For example, woodwind instruments include clarinets, saxophones, and oboes. Cars may be sedans, coupes, or convertibles. As a new musical instrument is developed, it can be placed in the proper category in the classification system. The same is true of a new motor vehicle. A classification system brings order and logic to instruments, vehicles, or any group of related objects.

The Need for Classifying Organisms

Just as musical instruments or motor vehicles can be classified, so, too, can organisms. Biologists classify organisms in categories called **taxa,** (*sing.,* taxon). The science of classification is, therefore, called **taxonomy.** Just as a classification system for instruments brings order to a band, a classification system for organisms brings order to the great diversity of life forms. It also serves as a basis for identifying unfamiliar organisms. New organisms are being discovered continually. Such organisms may be placed in existing taxa, or, if they are unique, a new taxon may be formed. For example, recent exploration of marine caves has led to the discovery of more than 20 new kinds of crustaceans, the taxon of organisms that includes shrimp, lobsters, and crabs.

Figure 14-1 New taxa may be created in order to classify newly-discovered organisms, such as these marine tube worms.

Classifying organisms not only results in order, but also provides a logical means of naming organisms. Common names are inadequate for biologists' purposes. The word *frog,* for example, probably conjures up a certain mental image for you, but it is inaccurate as a scientific label. A biologist would need to know what *kind* of frog it is. Is it a grass frog, a tree frog, or a bullfrog? Consider also the common word *worm.* You probably think of a long, slimy animal that lives in soil, but biologists know of various kinds of worms, including roundworms, flatworms, and segmented worms. Also, organisms such as ringworms, mealworms, and acorn worms are not worms at all. A ringworm is a fungus, a mealworm is an insect in the larval stage, and an acorn worm is a simple relative of vertebrates.

In addition, common names vary from country to country and from language to language. They may even vary among people who speak the same language. For example, a puma, a cougar, and a mountain lion are all the same organism. The need for one scientifically accepted name for each kind of organism is clear. Scientific names can be used to identify a particular organism even when the scientists studying that organism do not speak a common language. All scientific names given to organisms are in Latin, or in a Latin style, as Latin is no longer used in everyday conversation.

Binomial Nomenclature

Many early biologists devised classification schemes. Aristotle divided organisms into two groups—plants and animals. In Aristotle's system, plants were classified on the basis of structure and size—herbs, shrubs, and trees. Animals were subdivided on the basis of where they lived—air, land, or sea. Yet a classification system should have the same basis for all groupings. In the 18th century, Carolus Linnaeus developed a classification system based on just one characteristic—structural features.

Figure 14-2   **The word *frog* can be applied to a tree frog (a), a grass frog (b), or a bullfrog (c), even though they are quite different organisms.**

a

b

c

According to Linnaeus, each type of organism was a distinct species. If two organisms had the same set of features, they belonged to the same species. Different species having similar features were classified together in broader groups. Although Linnaeus thought that species were unchangeable—that they did not evolve—his decision to group organisms on the basis of structure was significant. In fact, many of his groupings are still used today.

Linnaeus made a second contribution to the study of life. He introduced a two-part system called **binomial nomenclature** for naming and classifying organisms. In this system, each organism is given a two-word Latin name. The first word, a noun, is the **genus** (*pl.,* genera) to which an organism belongs. The second word, an adjective, is the specific name. Genus and specific name together constitute the species name of an organism.

Genus and species are different taxa. A genus is a broader category than a species. Thus, a single genus may contain many different species. For example, cats belong to the genus *Felis,* but there are many species of

ART CONNECTION

John James Audubon

John James Audubon (1785-1851) could hardly have been happier. The Pennsylvania woods teemed with birds. Nearly every day in 1803, Audubon went into the woods with his gun. However, unlike other hunters, he didn't put the contents of his game bag into the stew pot. Instead, he arranged his birds in front of his drawing board and carefully sketched their features.

Audubon's lifestyle was not acceptable to his French father, who had sent his son to America with instructions to manage a mine on family property. The older man could see little future in drawing birds. Next, he tried to make a businessman of Audubon, setting him up as a partner in a general store in Kentucky. Predictably, the venture failed because Audubon found new birds to study closer to the frontier territory.

Audubon was dissatisfied with the way other artists depicted birds. They usually stuffed a dead bird and used it as a model. Audubon thought that the resulting paintings looked stiff and unnatural, so he tried something different. He built a wooden frame and stretched wire screen across it. Arranging a dead bird in a lifelike position, he wired it to the screen. He then made his initial drawing on paper lined with squares the same size as those in the screen. This method allowed him to draw the bird life-size, with each part in exact scale. He found a way to make his paintings look even more lifelike by attaching string to the bird's wings and tail so he could

change their positions. Audubon also included pictorial information about the birds by painting them in their natural settings, often showing their prey and nests.

cats. A wildcat is *Felis sylvestris,* an ocelot is *Felis pardalis,* a cougar is *Felis concolor,* and a house cat is *Felis catus.* The same is true of plant genera. Oak trees belong to the genus *Quercus.* A red oak is *Quercus rubra;* a white oak is *Quercus alba.* Note that in writing a scientific name, the first letter of the genus is capitalized, but the first letter of the species is not. Genus and species names are always underlined or italicized. If you are discussing a particular genus, the genus name may be abbreviated by using only the first letter. When discussing cats, for example, *Felis sylvestris* may be written as *F. sylvestris.* Botanists discussing oak trees would discuss *Q. rubra* and *Q. alba.*

Figure 14-3 The wildcat *Felis sylvestris* (left) shares a genus with the cougar *Felis concolor* (right).

In 1820, Audubon began an 1800-mile journey by flatboat down the Ohio and Mississippi rivers, along what was then the western boundary of the United States. His plan was to study and paint the birds he found between Cincinnati and New Orleans, and eventually to document all the different kinds of birds found in the country.

Audubon wrote a book, *The Birds of America.* Unable to find a publisher in the United States, Audubon went to England to have the book published. It immediately became popular in the United States and Audubon was a celebrity, even being entertained by President Andrew Jackson.

To accompany his book of bird paintings, Audubon wrote an ornithological biography. He included his frank opinion of the bald eagle, the species chosen as the national bird. Audubon wrote that he agreed with Benjamin Franklin, who accused the bird of "bad moral character" because of its habit of stealing fish caught by the hawk. (Franklin suggested that the wild turkey would make a better symbol for the nation.)

In another part of the biography, Audubon described watching migrating passenger pigeons in 1813. He said, "The air was literally filled with pigeons; the light of noon-day was obscured as by an eclipse." Enthusiastic hunters, in search of both food and sport, lowered the pigeon population drastically, although Audubon was confident that the numbers would remain high. A century later, the world's last surviving passenger pigeon died in a Cincinnati zoo.

The passenger pigeon, the nearly extinct whooping crane, and more fortunate species that continue to survive are all meticulously pictured in Audubon's works. Thanks to Audubon, the world has a record of the vanished American wilderness. His drawings have also been the basis for countless guidebooks that allow today's birdwatchers to identify what they see nesting in trees and on the wing.

1. **Issues:** In Audubon's time, there were about 700 species of birds on the North American continent. Today, 33 are extinct and 150 others are threatened. What do you think are some reasons for the decline?

2. **Mathematics Connection:** Audubon made a rough calculation of how many passenger pigeons there were in a flock. He figured that the birds flew 60 miles per hour in a column about one mile wide. If a flock took three hours to pass overhead, and two birds occupied one square yard of space, how many birds would that flock contain?

Determining Relationships

The science of taxonomy became more exact when evolution and the theory of natural selection were explained. Since that time, classification has been based upon ancestry of organisms. Organisms that have a common ancestor are considered related and are grouped together. For example, the various species of cats are all classified in the same genus because they have evolved from a common ancestor. Dogs are in a different genus, *Canis,* because they are descended from a different ancestor. The closer the evolutionary relationships of organisms, the more similar their classification.

Classifying Based on Structure   Once evolution was recognized, taxonomists began to use two main lines of evidence for classification: the fossil record and homologous structures. Fossils often provide clues that help scientists determine relationships among organisms. The fossil record shows clearly the evolution of the horse, for example. Scientists also have ample fossil evidence to show that horses, tapirs, and rhinoceroses evolved from a common ancestor. However, the fossil record is rarely this complete, so other lines of evidence are needed for classifying most organisms. Figure 14-5 shows how homologous structures were used to determine that walruses, seals, and sea lions belong to the same group. The reason so much of Linnaeus's classification system still remains valid is that he used the idea of homologous structures without realizing it. By using structure as his basis, Linnaeus was grouping organisms on the basis of evolutionary relationships.

However, fossils and homologous structures can sometimes provide information that is misleading or inconclusive. Also, the presence of structures similar in function that have resulted from convergence can and do mislead taxonomists. As time went on, other lines of evidence came into use.

Classifying Using Biochemistry and Development   As new scientific tools such as microscopes became more precise, taxonomists began to use

Figure 14-4   **Relationships among organisms are easy to see when the fossil record is complete, as in the evolution of the horse.**

Eohippus
55 million years ago

Mesohippus 40 million years ago

Merychippus
25 million years ago

Pliohippus
5 million years ago

Equus
2 million years ago

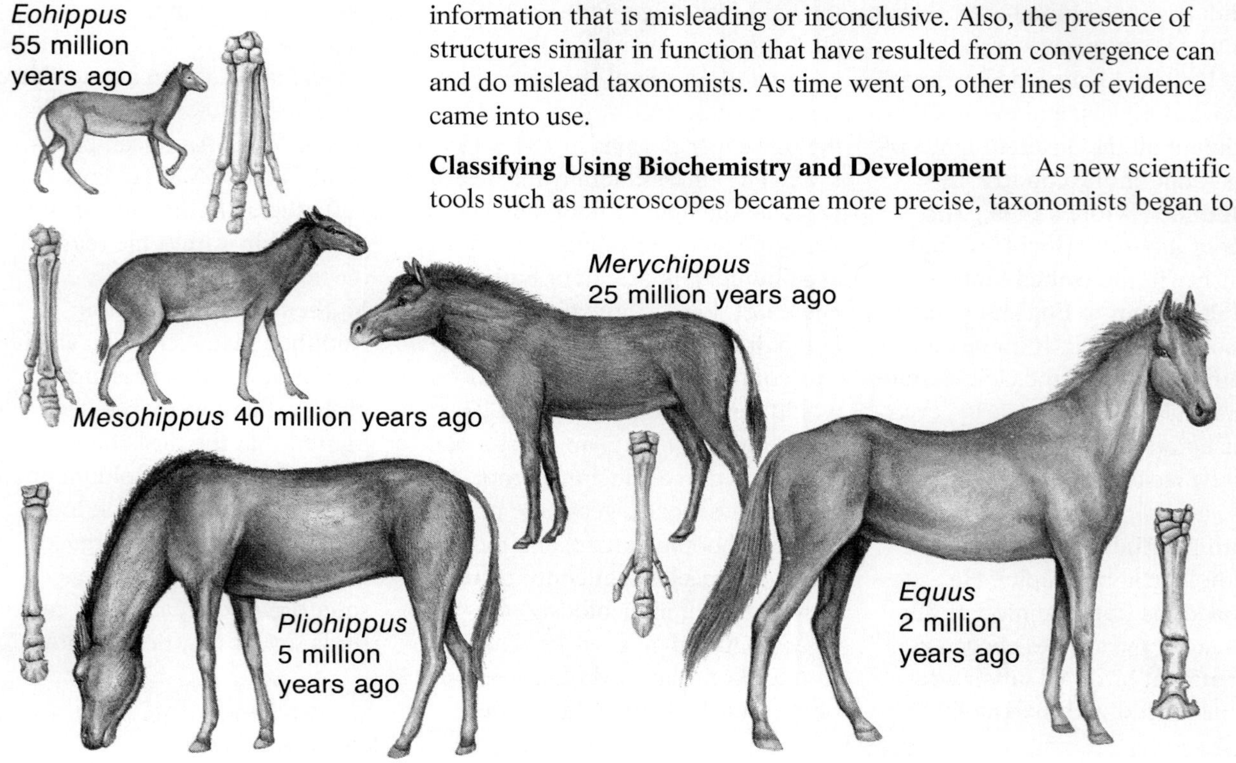

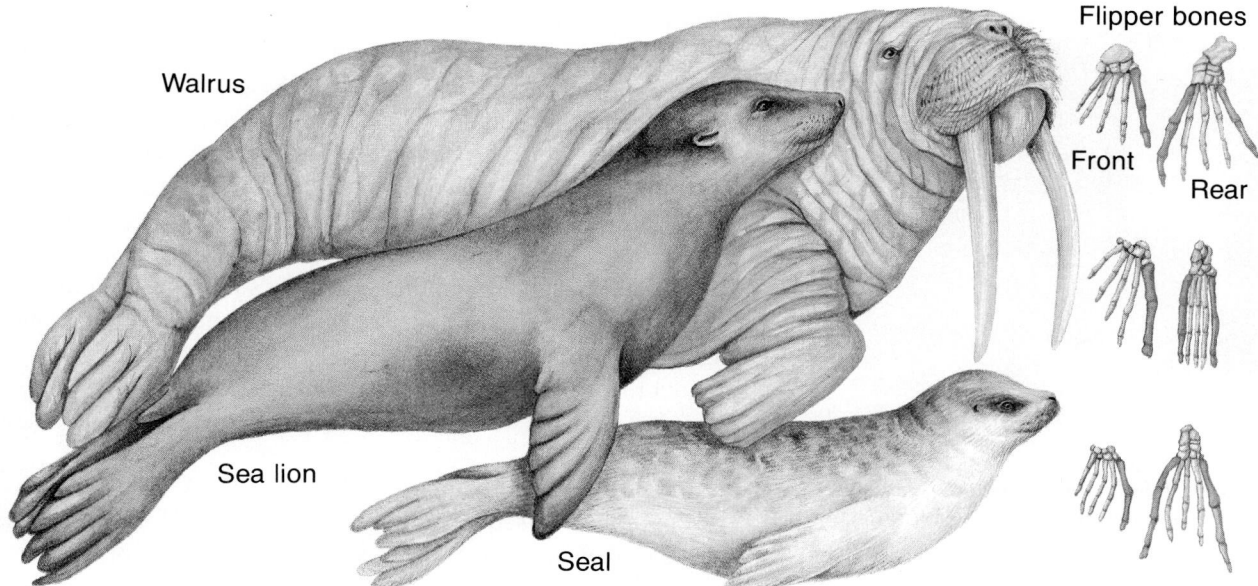

Walrus

Flipper bones

Front

Rear

Sea lion

Seal

new information to identify relationships among organisms. For example, comparative embryology has revealed that the major group to which vertebrates—including humans—belong is most closely related to the group that includes the starfish and its relatives. Such a conclusion would be difficult to reach on the basis of adult structure alone. Comparative biochemistry finally established the ancestry of the horseshoe crab, seen in Figure 14-6. As the name suggests, this animal was thought to be a true crab. Yet blood studies showed differently. It is now classified as a relative of spiders.

Figure 14-5 The similarity of bone structure in the flippers of the walrus, sea lion, and seal provides evidence of common ancestry.

Figure 14-6 Biochemical studies show that the horseshoe crab is related to spiders, not crabs.

Figure 14-7 **Guinea pigs may resemble rodents, but they have been reclassified based on DNA studies.**

Classifying on a Molecular Basis Today, studies at the molecular level play a major role in determining evolutionary relationships. Analyses of proteins and/or DNA can often reveal relationships that could not be determined in any other way. The idea is quite simple: Organisms having many proteins or genes in common are closely related.

For example, guinea pigs, shown in Figure 14-7, were classified for many years with mice and squirrels as rodents. However, when scientists compared the guinea pig's proteins with those of other rodents and those of other types of mammals, they found significant differences in the sequences of amino acids. Taxonomists concluded that the guinea pig does not share a common ancestry with rodents, and it is now classified in a new group of its own.

In one method of comparing DNA, the DNA from each of two organisms is separated into single strands. The DNA of one of the organisms is made radioactive, and the DNA of both organisms is then mixed. Base pairing causes the single strands to bond with each other, forming double-stranded molecules. Bonding will occur only where the genes, or sequences of bases, are complementary. The bonding is detected because one set of DNA is radioactive. The greater the amount of bonding, the more closely related the two organisms. Using this method, scientists have been able to match the DNA of humans with other primates. For example, 93 percent of the base pairs of human and macaque DNA match up, showing a close evolutionary relationship between these species.

Not only can DNA reveal close evolutionary relationships, it can also be used to determine *when* two organisms began to diverge from their

HISTORY CONNECTION

William Montague Cobb

Six hundred human skeletons may sound like something from a horror movie, but the collection was instead an important teaching tool for Dr. William Montague Cobb (1904-1990). He used the skeletons to help train more than 6000 physicians while he worked for more than 40 years as an anatomist at the Howard University Medical School in Washington, D.C. Each of the skeletons, which came from cadavers used in the school's dissecting

lab, is documented by race, ethnicity, and place of birth. This careful documentation makes the collection invaluable in the study of comparative anatomy. Dr. Cobb became especially well-known for his research on the physical growth and development of African Americans.

Dr. Cobb's expertise in comparative anatomy put his name in the headlines on sports pages in the 1930s. At the time, Jesse Owens, an African American track star, was breaking world records at college athletic meets. He excelled in the 100- and 200-meter dash as well

as the broad jump and other events. Two of the records he set stood for more than 20 years.

Some jealous people contended that Owens had an unfair advantage because of what they called the natural athletic ability of blacks. A specific contention was that blacks had a longer heel bone, which gave them greater leverage. After making extensive measurements, Dr. Cobb was able to conclude that hard work and courage, not physical characteristics, were at the heart of Owens's remarkable accomplishments.

common ancestor. Mitochondrial DNA is used in this type of study. Because scientists theorize that mitochondrial DNA mutates at predictable rates, it may provide a molecular clock for dating evolutionary events.

Mitochondrial DNA from different species can be analyzed and compared. The number of differences in base sequences is a function of how long the organisms have been evolving. By comparing the differences in the base sequences of the DNA and using predicted rates of mutation, scientists can establish how long the two organisms have been diverging. Using this approach, along with evidence from the fossil record and other biochemical evidence, scientists have determined that chimpanzees and humans began to evolve from a common ancestor only 5 million years ago.

It may surprise you to learn that DNA of ancient organisms can also be analyzed, if scientists are lucky enough to find it. Usually, DNA is destroyed as a dead organism decomposes, but some bits of it may survive under certain conditions. Many researchers hope to be able to analyze DNA from dinosaurs. They expect to get samples of dinosaur blood from biting insects preserved in amber. Some unusually well-preserved leaves of a magnolia tree that lived 20 million years ago provided an opportunity for DNA analysis. Scientists used a new technique that produces many copies of DNA in a short period of time, providing enough DNA to be studied. One particular gene coding for an enzyme in photosynthesis was analyzed and compared with the same gene in a modern magnolia. What was discovered? Out of a total of 820 base pairs, 803 remained the same. Such studies help determine evolutionary relationships.

Figure 14-8 Humans and chimpanzees evolved from a common ancestor. Various lines of evidence show that gibbons, orangutans, and gorillas also evolved from the same common ancestor.

In the 1960s, Dr. Cobb became troubled that African Americans did not have equal access to the hospitals in Washington, D.C., and that the city's medical society was also segregated. He organized and led the negotiating team that corrected both of these injustices. As president of the National Association for the Advancement of Colored People from 1976 to 1982, Dr. Cobb continued to champion civil rights in the medical profession and beyond.

1. **Issue:** Medical schools need human bodies to teach medical students about anatomy. What do you think are the pros and cons of donating one's organs or body after death to a medical school?

2. **Explore Further:** Why do you think it was so personally embarrassing to Adolf Hitler, dictator of Germany from 1933 to 1945, to see Jesse Owens excel in the Olympics held in Berlin?

Phylogeny

When taxonomists classify an organism, they are really providing the evolutionary history, or **phylogeny,** of the organism. Each taxon can be thought of as representing a different step in an organism's phylogeny.

For more than 100 years, the phylogeny of the giant panda has been in doubt. First studied in 1869 in western Asia, the giant panda was classified as a bear on the basis of its physical appearance or structure. Later, it was thought to behave more like the red panda and was classified with raccoons. As the debate continued, some scientists argued for the bear classification; others for the raccoon. Structure and behavior suggested to some taxonomists that the panda be placed in a group of its own. To confuse the issue further, the giant panda has a thumb, a structure found only in primates. In pandas, however, that thumb is not a true opposable thumb, but a sixth digit that evolved from a wrist bone. The panda's thumb and primates' thumbs are the result of convergence, so the presence of a thumb cannot be used to establish common ancestry.

Recent work shows just how important the use of several lines of evidence can be in classification. Using the technique of studying radioactive DNA strands described previously, scientists learned that the giant panda

Figure 14-9 Red pandas have long been classified with raccoons. When the giant panda was classified with red pandas, it became a "relative" of raccoons, too. The actual relationships among the raccoon (a), red panda (b), giant panda (c), and black bear (d) were determined using several lines of evidence.

a

b

c

d

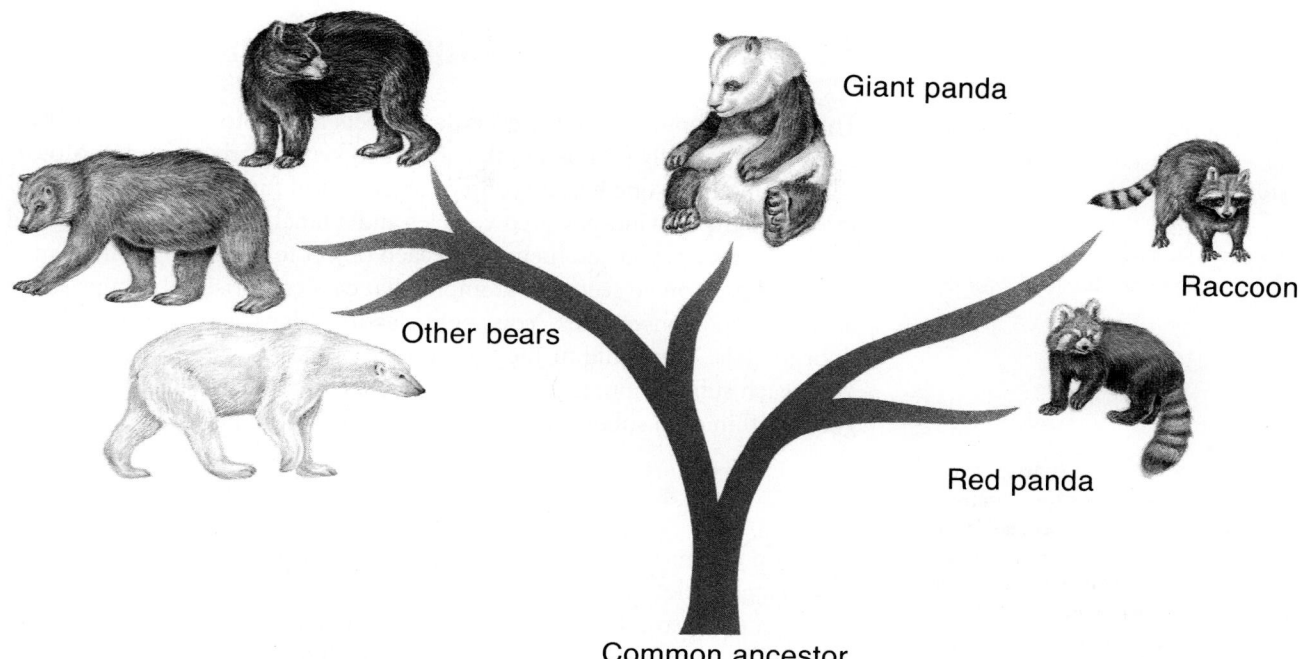

Giant panda

Other bears

Raccoon

Red panda

Common ancestor

is more closely related to bears than to raccoons. Biochemical comparisons of enzymes and other proteins led to the same conclusion. With these techniques, along with fossil evidence, the evolutionary histories of the giant panda, the bear, and the raccoon have now been more clearly established.

Figure 14-10 represents those histories in the form of what is called an evolutionary or a phylogenetic diagram. You can see that both bears and raccoons descended from a common ancestor. You can also see that the giant panda is a bear and the red panda is a raccoon. Each fork in the branches of an evolutionary diagram represents a point at which offspring of a common ancestor gave rise to different descendant groups. You can also see from this story how misleading common names can be.

Figure 14-10 The phylogenetic diagram of giant pandas shows that pandas are more closely related to bears than raccoons.

Section Review

Understanding Concepts
1. Consider a horse, a cow, and a wolf. Which two of them would have the more similar classifications? Why?
2. Give several reasons to explain why binomial nomenclature is useful.
3. Cite one example of how different lines of evidence can be used to establish evolutionary relationships.

4. ***Skill Review—Classifying:*** Based on your own knowledge, set up a scheme to place the following organisms in groups and subgroups: apple tree, baboon, butterfly, eagle, earthworm, fern, frog, human, jellyfish, moss, oak, toad, tulip, wasp. For more help, see Organizing Information in the **Skill Handbook.**

5. ***Thinking Critically:*** Suppose the molecular clock technique indicates that an organism began evolving from its ancestor a certain number of years ago. What other line of evidence should yield the same basic data?

Objectives

Demonstrate how each successive taxon is more specific than the previous taxon.

Discuss why the number of kingdoms of organisms has changed over time.

14.2 A System of Classification

Think again about classifying everyday objects like motor vehicles. In any system, you usually first assign the objects to very broad categories. Motor vehicles might be one broad group of objects used for travel. Other such groups might be wind-powered vehicles and animal-drawn devices. Beyond the broad group, you can then assign each object to increasingly more specific groups. You've seen, for example, that cars comprise subgroups such as sedans and convertibles. You could classify sedans as two-door or four-door models. You could further group each of these by type of engine. With each subdivision, the object is classified into a more specific category. Within each subdivision, the members are more like each other.

Taxa

When an organism is assigned to a species, it has been classified as specifically as possible. The genus to which that species belongs is a slightly more general taxon because several species may be members of that genus. The complete classification system includes five more taxa, for a total of seven.

The seven taxa form a series. The broadest taxon is the **kingdom.** Kingdoms are divided into increasingly more specific taxa. These are, in order, **phylum, class, order,** and **family.** Genus and species are last. Each taxon narrows the number of organisms of the previous group. Therefore, the number of organisms included in each taxon becomes smaller and smaller. Each species is one certain type of organism.

That sequence is classification in its simplest form. But classification is often more complex because each of these major groups may be further subdivided. For example, each phylum may be divided into several subphyla. The phylum that vertebrates belong to contains three subphyla, of which one is Vertebrata. A single species may be made up of several subspecies. Subspecies are sometimes called breeds, varieties, or races. The various breeds of domestic dogs, for example, are all subspecies.

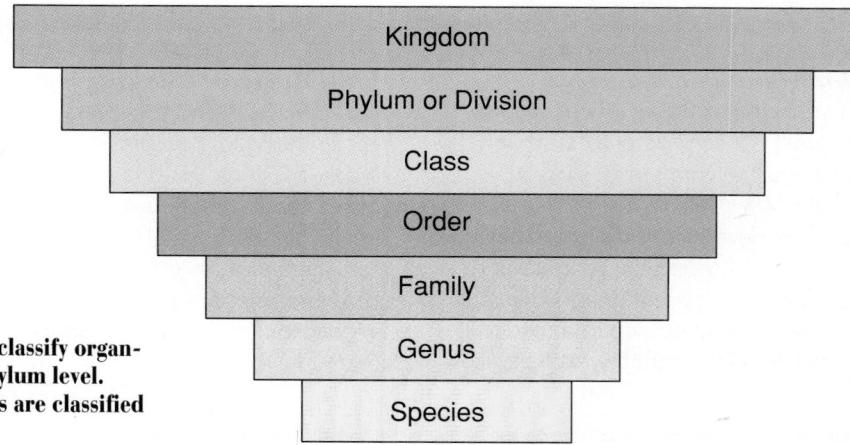

Figure 14-11 The seven taxa used to classify organisms include the division, a taxon at the phylum level. Organisms in the Plant and Fungi Kingdoms are classified in divisions rather than phyla.

Some Examples of Classification

What would the complete classification of a common house cat (*Felis catus*) be? First, a cat is classified as an animal based on the very broad characteristics of all animals. Like all other animals, for example, a cat doesn't make its own food, but eats other organisms. Cats are also multi-cellular organisms. Using these and other characteristics, the cat is placed in Kingdom Animalia.

A cat is then put into the Phylum Chordata and, because it has a backbone, into the subphylum Vertebrata. This subphylum includes the seven classes of fish, amphibians, reptiles, birds, and mammals. You can see the rest of the classification of a common house cat in Table 14.1. Each group represents more specific characteristics than the previous group.

Now compare the classification of a cat with that of a dog. Because cats and dogs share the same broad characteristics, they are in the same kingdom, phylum, subphylum, class, and order. But at the family level, the classification diverges. The different families represent different branches of animal evolution. Each of the remaining groups is, of course, also different from the others.

Table 14.1 Classification of Some Animals

Taxon	House Cat	Dog	Human	Grasshopper
Kingdom	Animalia	Animalia	Animalia	Animalia
Phylum	Chordata	Chordata	Chordata	Arthropoda
Subphylum	Vertebrata	Vertebrata	Vertebrata	
Class	Mammalia	Mammalia	Mammalia	Insecta
Order	Carnivora	Carnivora	Primates	Orthoptera
Family	Felidae	Canidae	Hominidae	Locustidae
Genus	*Felis*	*Canis*	*Homo*	*Schistocerca*
Species	*Felis catus*	*Canis familiaris*	*Homo sapiens*	*Schistocerca americana*

Now compare the cat and dog with humans. How closely do humans seem to be related to dogs and cats? How many levels of classification do humans have in common with dogs and cats? At what level do the classifications differ?

Finally, compare these three animals with the grasshopper. The only common taxon among the grasshopper, cat, dog, and human is kingdom—all these organisms are classified as animals. Because the grasshopper's characteristics are different from those of the other animals, so, too, is its classification.

The grasshopper is classified in the Phylum Arthropoda. What major difference among organisms would cause a taxonomist to place a newly-discovered animal into the Phylum Chordata rather than the Phylum Arthropoda? How are you different from a grasshopper?

I N V E S T I G A T I O N

Making a Key

You wish to identify a friend in a crowd. You do it by looking for certain key traits that you associate with your friend. If your friend has red hair or is very tall, you might first use hair color or height to screen the crowd and narrow down your search. It's usually an easy task if the crowd is not too large. However, what if you were asked to pick out a person from a crowd but had never seen this person before? That complicates things a bit. How would you do it? You would need a list of that person's traits in order to help identify him or her. How do scientists identify an organism that they are not familiar with? They use almost the same technique as you would use to pick out a person you have never met before. The scientist uses a tool called a biological key. These types of keys list traits of living things, which allows you to identify the organism in question down to its scientific name. Below is an example of what a key might look like if you were attempting to identify jaybirds.

Key to Jay Birds

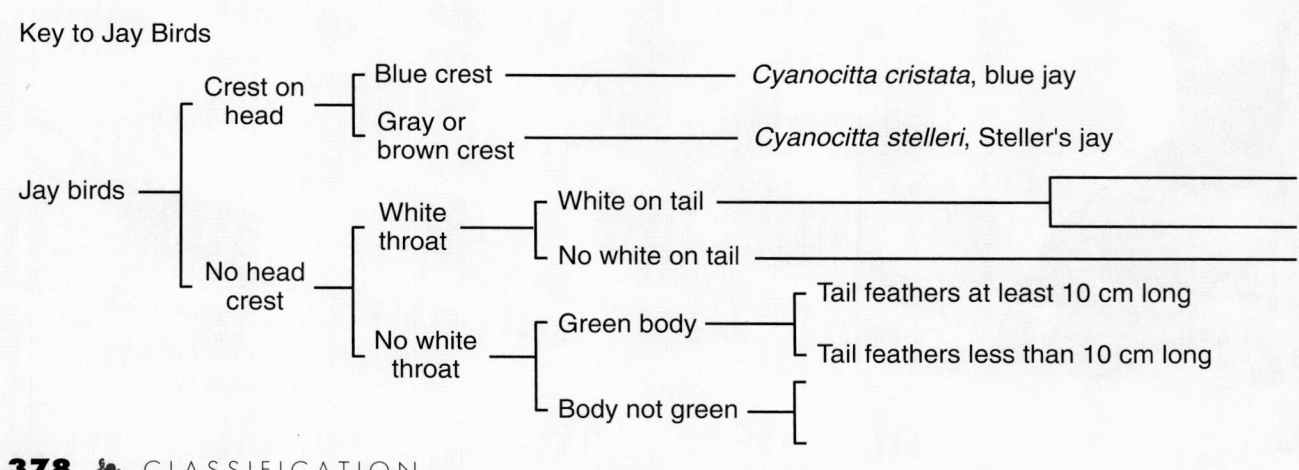

Jay birds — Crest on head — Blue crest ———— *Cyanocitta cristata*, blue jay

Gray or brown crest ———— *Cyanocitta stelleri*, Steller's jay

No head crest — White throat — White on tail

No white on tail

No white throat — Green body — Tail feathers at least 10 cm long

Tail feathers less than 10 cm long

Body not green

The Kingdom Problem

You know from Chapter 2 that most biologists today use a five-kingdom system of classification. These kingdoms include the monerans, protists, fungi, plants, and animals. Examples of organisms in each kingdom are shown in Figure 14-12 on page 381. All organisms once were classified as either plants or animals. This two-kingdom scheme was reasonable for most familiar, multicellular organisms. After all, it is not difficult to distinguish a plant like a cactus from an animal like a coyote.

However, the discovery of microscopic organisms posed a problem. Some unicellular forms could easily be classified into one kingdom or the other, but many did not fit neatly into either group. For example, an organism called *Euglena* has characteristics common to both kingdoms.

Problem

How can you design a key?

Hypothesis

What is your group's hypothesis? Explain your reasons for forming this hypothesis.

Experimental Plan

1. As a group, make a list of possible ways you might construct a key to identify by name the objects given to you by your teacher.
2. Agree on one idea from your group's list.
3. Assign responsibilities within your group for the various tasks to be done.

Checking the Plan

1. Your key must identify all objects given to your group.
2. All objects must be identified by name (or at least by the owner's name).
3. The key ends for a specific item when its name is given.

4. Note that each set of traits at each fork in the key tend to be opposites of one another. For example, the first fork in the example key compares a crest on the head to its opposite (no crest).
5. A person not familiar with the objects used by your group should be able to use your key and correctly key out any object from your group.

Data and Observations

Carry out your design by writing a key that identifies the objects given to your group.

Analysis and Conclusions

1. What is actually used as the basis for designing each different subcategory or fork on your key?
2. Try your key out with other groups. If it works, what does this tell you about your key design? If it does not work, what does this tell you about your key design?

3. Assume that all groups in the class began with the exact same objects. Might all keys be exactly alike? Explain. Is it proper to say that only one key is correct? Explain.
4. Explain the value of a biological key.
5. List several examples of when keys may be useful tools.

Going Further

Based on this activity, design another key that would also work for the identification of the objects provided you. Record this key as part of your experimental report.

Euglena is mobile, like an animal, yet autotrophic, like a plant. To confuse the matter further, at certain times *Euglena* may lose its chlorophyll and become heterotrophic.

Other problems arise with a two-kingdom system. For example, how should fungi such as mushrooms and molds be classified? They are like plants in that they do not move, but they are heterotrophic like animals. What about slime molds? They are heterotropic and move like *Amoeba*, but they produce spores like fungi. Sponges, sea anemones, and coral all are animals, but they resemble plants.

Based on current knowledge of evolutionary relationships among organisms, the five-kingdom system has solved some of these problems. Identifying cells as prokaryotes or eukaryotes helped classify many microscopic organisms. Remember from Chapter 5 that prokaryotic cells have no nucleus. All monerans are prokaryotes. However, other classification systems have been proposed. That's not surprising because biologists often interpret data in different ways. One proposed system is based on six kingdoms. The sixth kingdom consists of a group of extremely ancient bacteria that are different in certain respects from other bacteria. Some taxonomists, therefore, argue that these bacteria represent a different major branch of evolution. You'll learn about these bacteria in the next chapter. However, this book will use the five-kingdom system, as presented in Appendix A. Use it as a reference when needed.

Thinking Lab

INTERPRET THE DATA

What traits are used to classify organisms into kingdoms?

Background
Today's classification scheme uses a five-kingdom system. The separation of all life forms into these five main groups is based on shared characteristics.

You Try It
Copy the outline provided. Complete it by writing the listed terms in the correct bracket spaces. When complete, the outline should provide a logical and

correct guide to the traits of the five kingdoms. Terms: usually unicellular, autotrophic, prokaryotic, eukaryotic, multicellular, heterotrophic, ingest food, absorb food.

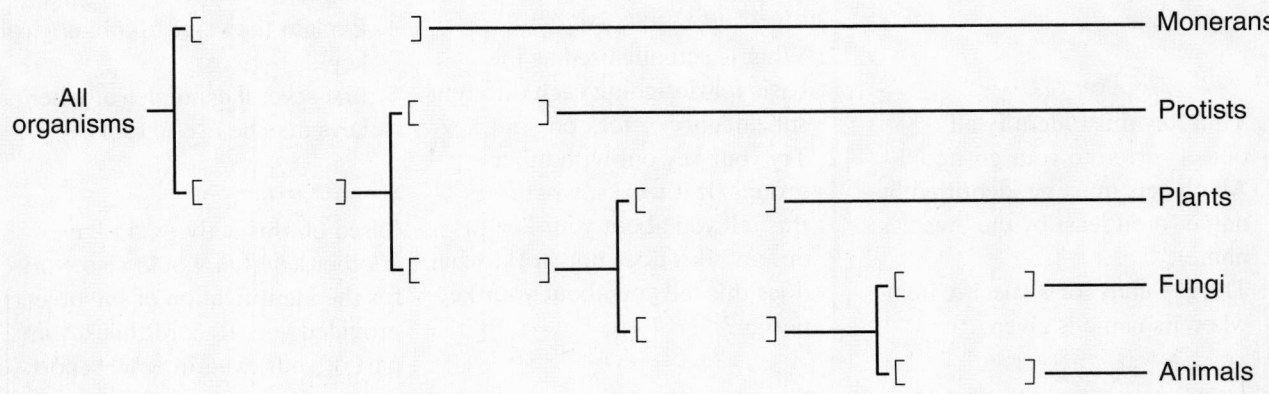

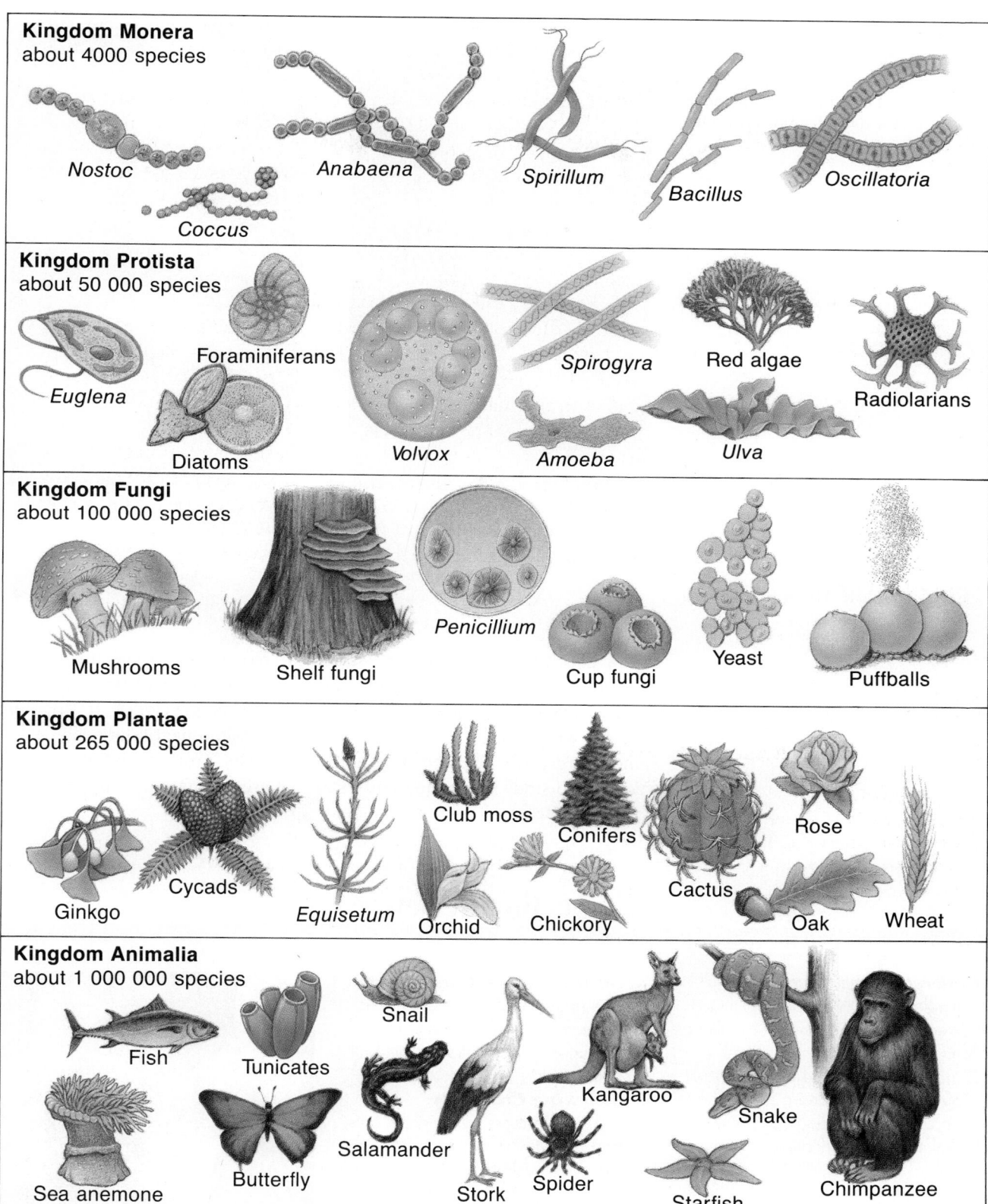

Kingdom Monera
about 4000 species

Nostoc

Anabaena

Spirillum

Bacillus

Oscillatoria

Coccus

Kingdom Protista
about 50 000 species

Foraminiferans

Euglena

Diatoms

Volvox

Spirogyra

Amoeba

Red algae

Ulva

Radiolarians

Kingdom Fungi
about 100 000 species

Mushrooms

Shelf fungi

Penicillium

Cup fungi

Yeast

Puffballs

Kingdom Plantae
about 265 000 species

Club moss

Conifers

Rose

Ginkgo

Cycads

Equisetum

Orchid

Chickory

Cactus

Oak

Wheat

Kingdom Animalia
about 1 000 000 species

Snail

Fish

Tunicates

Salamander

Kangaroo

Snake

Sea anemone

Butterfly

Stork

Spider

Starfish

Chimpanzee

Figure 14-12   The five kingdoms of life include the Monera, Protista, Fungi, Plant and Animal Kingdoms.

Figure 14-13 Which of these is a plant and which is an animal? Structure alone is not enough to classify an animal like the sponge (a) and a loofah (b), the dried fruit of a plant in the gourd genus *Luffa*.

a

b

LOOKING AHEAD

Various lines of evidence—structural, biochemical, molecular—are used to establish evolutionary relationships. From this evidence, a five-kingdom system for classifying organisms has been developed. In Chapter 15, you will learn about the diversity of and evolutionary relationships among organisms of three kingdoms—monerans, protists, and fungi. You'll also study the unique viruses, which are not classified in any kingdom.

Section Review

Understanding Concepts

1. Explain why two organisms classified in the same family must also be in the same order.
2. Based on Table 14.1, which taxa would a gorilla and a human have in common? Defend your answer.
3. Relate the classification of an organism to its phylogeny.

4. **Skill Review—Classifying:** Use Appendix A to classify a bee down to order. For more help, see Organizing Information in the **Skill Handbook.**
5. **Thinking Critically:** What structural features put humans, cats, and dogs in the same taxa through the class level? What characteristics place humans in a different order from cats and dogs?

Biotechnology

Modern Tools for Classification

For years, biologists working in the rain forests gazed longingly upward at the rain forest canopy looming 30 m above their heads. High up in what one naturalist called "an undiscovered continent" live up to 40 percent of the world's species of plants and animals. Previously, except for a few daredevils using ropes to climb to the top, scientists could only crane their necks to peer through binoculars at a world that was tantalizingly out of touch.

Parts of that world can now be explored at arm's length from a 42-meter tower on Barro Colorado Island, near the Panama Canal. Allowing even greater access is a construction crane set up near Panama City. A gondola on the end of the crane's 36-meter arm can be maneuvered to any height up to 30 meters and to any position on a circular area more than 60 meters in diameter.

A team of French biologists took a more lofty approach to reaching the rain forest canopy in French Guiana. Deciding to reverse previous methods and work from top to bottom, scientists rigged up an inflatable raft to sit atop the towering trees. They used a hot-air dirigible to carry aloft a set of rubber pontoons interlaced with fiber netting and balance it on the canopy.

The resulting 600 m² of work space is shared by teams of scientists from eight countries. It takes some daring to make the 15- to 30-minute ascension via ropes and harnesses, but once they have arrived at the "treehouse," scientists can move about in the canopy with relative safety.

Back on the ground with their samples from the canopy, biologists can use a new computer database to identify and classify the species they have collected. Appropriately called "Linnaeus," the system uses distinguishing features of taxonomic groups to guide identification. It directs a user to make a series of binary choices, such as:

1a. Animal with a carapace

1b. Animal without a carapace

Each succeeding choice narrows the possibilities until the species is finally identified.

Another useful tool in classification is the scanning electron microscope, which allows study of a living thing's DNA. This genetic material is unique for each species.

units. Looking at a DNA sequence, a biochemist who is an amateur musician decided to change Ts to Es and tried playing the resulting "notes" on a music synthesizer. Try playing or singing a random series of the notes A, E, G, and C to see how DNA "sounds."

2. **Writing About Biology:** Some scientists call taxonomy "a window backwards into evolution." What do you think this description means?

3. **Issue:** Most scientific studies carry some risk of harming the subject of the study. What risks do you think could be associated with studying the rain forest canopy?

1. **Connection to Music:** In writing out DNA sequences, scientists use the letters A, T, G, and C to refer to molecules' chemical sub-

In Competition: Fish or Dam?

In the mid-1970s, the construction of a $120 million dam on the Tennessee River was halted due to a fish only 8 cm long. This fish, called the snail darter, was not much to look at. In fact, it had been completely overlooked for centuries. Yet this unremarkable little fish kicked off a campaign that threatened to cancel a dam construction project that would provide hydroelectric power for a large area of Tennessee, create a lake, and increase recreational opportunities for state residents and visitors.

"Fish or Dam?" shouted headlines across the United States. Environmentalists and developers took their corners in one of the first tests of the Endangered Species Act passed by Congress in 1973. The act seemed to cover all bases, making it illegal for an endangered species to be "killed, hunted, collected, harassed, harmed, pursued, shot, trapped, wounded, or captured." In addition, it provided for the protection of such a species'

critical habitat, the living area a species needs in order to survive.

Soon after the Endangered Species Act went into effect, scientists discovered that the snail darter only existed in a portion of the Tennessee River scheduled to be flooded by dam construction. Efforts by conservationists succeeded in getting the snail darter declared an endangered species. In spite of protests, the United States Supreme Court halted construction of the dam—but only for a while.

The Dam is Built   Proponents of the dam convinced their congressional representatives to pass a special bill to exempt this particular dam from compliance with the act. The dam was built, and some snail darters were transported to another river. Humans and their needs for electrical power, a larger water supply, recreation, and jobs had prevailed over the needs of a tiny fish.

Owls Versus Loggers   In 1987, another small animal gained prominence in the news. The northern spotted owl was threatened by the loss of its habitat in the old-growth forests of the Pacific northwest. Timber companies wanted to continue logging these forests. Loggers feared they would lose their jobs if the northern spotted owl was given legal preference over their needs for jobs and income. Environmentalists tried to generate support for the owl by selling stuffed toy replicas of the bird, emphasizing its big, dark eyes. Loggers reacted by donning T-shirts that claimed, "I love spotted owls—on a sandwich" and "Save a logger. Eat an owl." Jobs

were at stake, and loggers were infuriated that "outsiders" would threaten their way of life and their livelihoods.

Seeking to capitalize on the feelings of loggers and their families, President Bush campaigned hard in the area in 1992. He threatened to refuse to sign a renewal of the Endangered Species Act. Like many people in the northwestern United States, the President believed that it was time to make people more important than animals, even endangered ones. Some people felt that it was time to rewrite the act to allow economic considerations to override the preservation of species.

The Earth Summit   Similar sentiments came up at the first Earth Summit, held in 1992 and attended by representatives from all over the world. Efforts by developed countries such as the United States to dictate conservation practices met with disdain from developing countries. After all, they reasoned, the United States had prospered by cutting down its own forests and exploiting its other natural resources. To them, it seemed hypocritical of the U.S. to deny other countries the same route to industrialization and prosperity.

Protection and Economics   Environmental scientists have been trying to show people that preserving the environment also benefits people in monetary terms. A researcher from the Massachusetts Institute of Technology recently finished a study that weakens the jobs-versus-environment argument. Professor Stephen Meyer looked at the relationship between jobs and environmental regulation in each state between 1982 and 1989. Although he expected to find a negative relationship, the results were quite the opposite: Environmental protection and economic development were mutually supportive.

Some American companies are also discovering on their own that they can save money by protecting the environment. For example, one huge corporation started pollution-prevention measures in 1970. Those measures translated into big savings—$670 million over 20 years.

Protecting Tropical Forests

Today, the public's attention is focused on the rain forests of South America and Asia. These rain forests cover only seven percent of Earth's surface but contain at least half of all plant and animal species. Unfortunately, rain forests are disappearing at the rate of about 70 000 square kilometers each year. They are cleared for subsistence farming and pasture for cattle. Ironically, the land these farmers clear remains useful for grazing or farming for only a few short years because most of the nutrients in rain forests are in the abundant plants, not in the soil. When the plants are cleared or burned, the soil that is left is deficient in essential nutrients such as minerals. After two or three years of planting crops, the few nutrients in the soil are used up. The soil will no longer support crops. So farmers are forced to clear more land, and the cycle of destruction continues.

The best hope for the conservation of rain forests is to find ways

for people to use them productively without destroying them. Some efforts, such as collecting certain nuts from rain forest trees, are under way right now. The race is also on to collect and test rain forest plants for medicinal uses. At least 1400 tropical plants may contain substances active against cancer. In addition to medicines, the rain forests may also contain new types of foods that could help to feed a hungry world. Scientists believe that rain forests contain more than 4000 species of edible fruits and vegetables. Only a tiny fraction of these foods are cultivated today.

Many scientists have read in the fossil record a tragic future for Earth's plant and animal species if the loss of endangered organisms accelerates. They believe that mass extinction is possible.

CAREER CONNECTION

Taxonomists study and classify organisms. They work in museums, research laboratories, universities, and scientific societies. Taxonomists often travel to all parts of the world to study endangered organisms. Taxonomists have a broad-based education, including biology, chemistry, anatomy, physiology, evolution, biochemistry, ecology, and genetics.

---

1. **Understanding the Issue:** What response could you make to Brazilians who insist that their country has a right to use its resources in any way it wants?

2. **Solve the Problem:** Imagine you were a congressional representative from a northwestern state in which loggers would lose their jobs if the spotted owl were protected to the full extent of the law. How could you save the owl and jobs for the loggers?

Readings

- Korn, Peter. "The Case for Preservation." *The Nation*, March 30, 1992, pp. 414-417.
- Adams, Robert. "Can We Find a Way to Balance the Survival of Endangered Species with the Livelihoods of People?" *The Smithsonian*, March 1992, p. 8.

Summary

Classification of organisms brings order to the great diversity of life. Each organism is assigned a two-part name—genus and species—recognized by scientists everywhere.

Taxonomy today classifies organisms on the basis of their evolutionary relationships. Using evidence based on fossils, homologous structures, and comparative studies of chemistry, embryos, proteins, and DNA, scientists can determine an organism's phylogeny.

Organisms are classified in a series of taxa, each of which represents a set of more specific characteristics. The broadest taxon is the kingdom; the most specific, the species. Most classification systems today are based on five kingdoms.

Language of Biology

Write a sentence that shows your understanding of each of the following terms.

binomial nomenclature	order
class	phylogeny
family	phylum
genus	taxa
kingdom	taxonomy

Understanding Concepts

1. Explain why common names of organisms are not scientifically acceptable.
2. How are the genus and species names of organisms related?
3. How did knowledge of evolution and natural selection improve the science of taxonomy?
4. How does knowledge of DNA make for more accurate classification of organisms?
5. Two organisms were found to have begun evolving from a common ancestor 15 million years ago. Other than fossil evidence, explain how that fact could be determined.
6. How were various lines of evidence combined to establish the giant panda's phylogeny?
7. Do you think various subspecies exist within *Felis catus*? Explain.
8. Why do you think a grasshopper is classified in a different phylum from the other animals classified in Table 14.1?

Applying Concepts

9. Certain lemurs and raccoons have similar patterns of coat coloration. Why would that similarity *not* be useful in determining the closeness of their evolutionary relationship?
10. All the following organisms are animals. Based on your everyday knowledge of these animals, place them into three different phyla. No technical terms are expected: ant, clam, crow, lobster, octopus, pig, shark, snail, spider, turtle.
11. *Amoeba* has many characteristics in common with the animals listed in question 10 above. Why is it not classified in the same kingdom? Refer to Chapter 2 or Appendix A for help.
12. Guess the common name of each of the following organisms: *Rattus norvegicus; Equus zebra; Elephus maximus; Pinus ponderosa; Camelus bactrianus.*
13. Several species of plants are known to have evolved from a common ancestor beginning 26 million years ago. Suppose the groups of plants are known by colors. One group includes species

known as yellow, golden, and red. Fossil evidence and molecular clock techniques indicate that yellow evolved 8 million years ago, golden evolved 12 million years ago, and red evolved 17 million years ago. Another group includes green, which evolved 14 million years ago, and blue, which evolved 22 million years ago. Construct an evolutionary diagram for these species.

14. Sharks, skates, and rays have internal skeletons made of cartilage. How does this fact help taxonomists place these organisms into the correct taxa? Without any other information, to what taxa can you identify these organisms in the classification scheme?

15. **Biotechnology:** Study Table 14.2, then answer the following questions: What can you infer about the relationships of these organisms? From their family names? From their order names? If a mastodon were alive today, could it mate with an African elephant to produce fertile offspring? Explain your answers.

16. **Issues:** The California condor is an endangered bird species, once limited to only 39 individuals. At enormous expense, the last few surviving condors were captured and have been bred successfully in captivity. Should money be spent to save species on the verge of extinction? Why?

Lab Interpretation

17. **Investigation:** Complete the following incomplete key for classifying writing instruments. Key out the following: ballpoint pen, pencil, felt-tip marker.

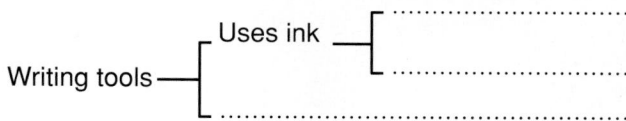

18. **Thinking Lab:** What traits are shared by the Animal and Fungi Kingdoms? How are you able to separate organisms into their correct kingdoms?

19. **Minilab 2:** An organism is found to be a consumer and can move about. Will these traits be enough to classify it into its proper kingdom? Why?

Connecting Ideas

20. Could members of different genera mate and produce fertile offspring? Could members of different subspecies? Explain.

21. How can the similar characteristics of two different species, such as tiger and cheetah, be explained?

Table 14.2 **Elephants and Their Relatives**

	African elephant	*Asian elephant*	*Woolly mammoth*	*American mastodon*
Kingdom	Animalia	Animalia	Animalia	Animalia
Phylum	Chordata	Chordata	Chordata	Chordata
Subphylum	Vertebrata	Vertebrata	Vertebrata	Vertebrata
Class	Mammalia	Mammalia	Mammalia	Mammalia
Order	Proboscidea	Proboscidea	Proboscidea	Proboscidea
Family	Elephantidae	Elephantidae	Elephantidae	Mastodontidae
Genus	Loxodonta	Elephas	Mammuthus	Mastodon
Species	Loxodonta africana	Elephas maximus	Mammuthus primigenius	Mastodon americanus

CHAPTER

1 5

VIRUSES AND MICROORGANISMS

THESE MUSHROOMS are members of the Kingdom Fungi. Like all fungi, they have no chlorophyll, so they cannot perform photosynthesis. Instead, they obtain energy by digesting organic material from the forest floor. The cap of a mushroom is only a small part of the organism. Most of a fungus consists of masses of tiny hairlike filaments that live within soil and decaying matter. The small photo shows some of these filaments.

In the early 1990s, an enormous fungus was discovered in the state of Washington. It covers an area of 2.5 square miles, the equivalent of more than a thousand football fields! Most fungi, of course, are not nearly as large.

Mushrooms exist in many different shapes, sizes, and colors. The fungus kingdom also contains organisms other than mushrooms. The mold that grows on stale bread is a fungus. So is the yeast that makes bread dough rise. In this chapter, you will learn the characteristics all fungi have in common, and find out what kinds of organisms belong to the Kingdoms Monera and Protista. You will discover that organisms from these two kingdoms and from Kingdom Fungi have similarities as well as differences. In addition, this chapter will describe viruses and give you a chance to think about whether or not viruses should be considered members of the living world.

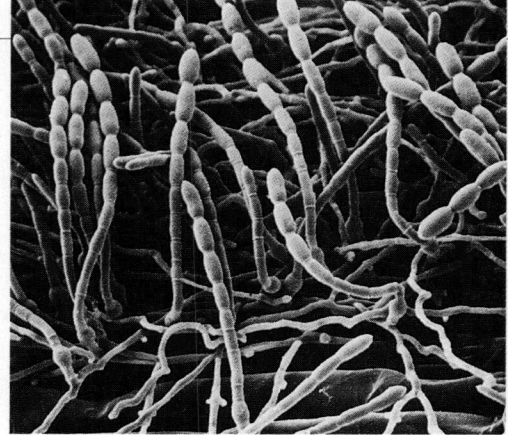

The phrase, "There's a fungus among us" is certainly true when you realize that soil contains billions of these fungal filaments.
Magnification: 1500×

Chapter Preview

Infer how the structure of a virus is related to its function.

Compare the reproductive cycles of viruses.

15.1 **V**iruses

When's the last time you had a cold or the flu? Do you remember being vaccinated against polio or measles when you were younger? Those diseases—as well as AIDS, mumps, and chicken pox—are caused by viruses. Viruses are extremely tiny particles that reproduce themselves but are not cells. Therefore, viruses cannot truly be called organisms and are not classified in any kingdom. Yet, viruses play a significant role in the lives of humans and all other organisms.

Structure of Viruses

More than a century ago, scientists discovered that many bacteria can cause disease. These scientists developed ways to isolate and identify disease-causing bacteria. They learned how to filter bacteria from liquids, grow them in the laboratory, and observe them under the microscope. The causes of some diseases, though, could not be isolated by these procedures. These mysterious disease-causing agents passed through the filters that would trap bacteria. They could not be grown in the laboratory, and they were invisible even with the aid of a microscope. These undiscovered causes of disease were named *viruses,* from the Latin word for poison. By 1935, laboratory methods and equipment had become more sophisticated, and the first virus was isolated from diseased tobacco plants. Since then, hundreds of viruses have been isolated and identified.

Chemical analysis shows that all viruses are composed of an outer protein coat that encloses a nucleic acid. Some viruses contain DNA; others contain RNA. Viruses are so small they cannot be seen with a light microscope. However, the first electron microscopes were developed about the same time viruses were discovered. These new instruments enabled biologists to see viruses for the first time. The shapes of viruses vary widely and include spherical, rodlike, cubical, and many-sided forms, Figure 15-1.

Unlike all true organisms, even the most simple bacteria, viruses have no cellular structure. They have no cytoplasm, organelles, or membrane to isolate them from their environment. Viruses do not carry out respiration or other common life processes, except that they do reproduce themselves. For this function, they require the chemicals and ribosomes of the healthy cells they invade.

Figure 15-1 **Virus particles have a variety of shapes.**

Bacteriophage (polyhedral head with tail fibers)

Tobacco mosaic virus (rod shaped)

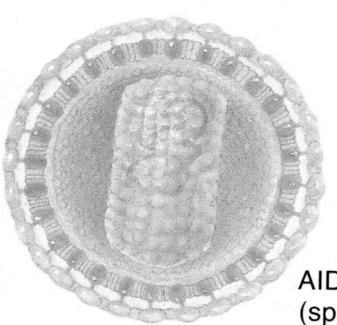

AIDS virus (spherical)

Polio virus (poly-hedral)

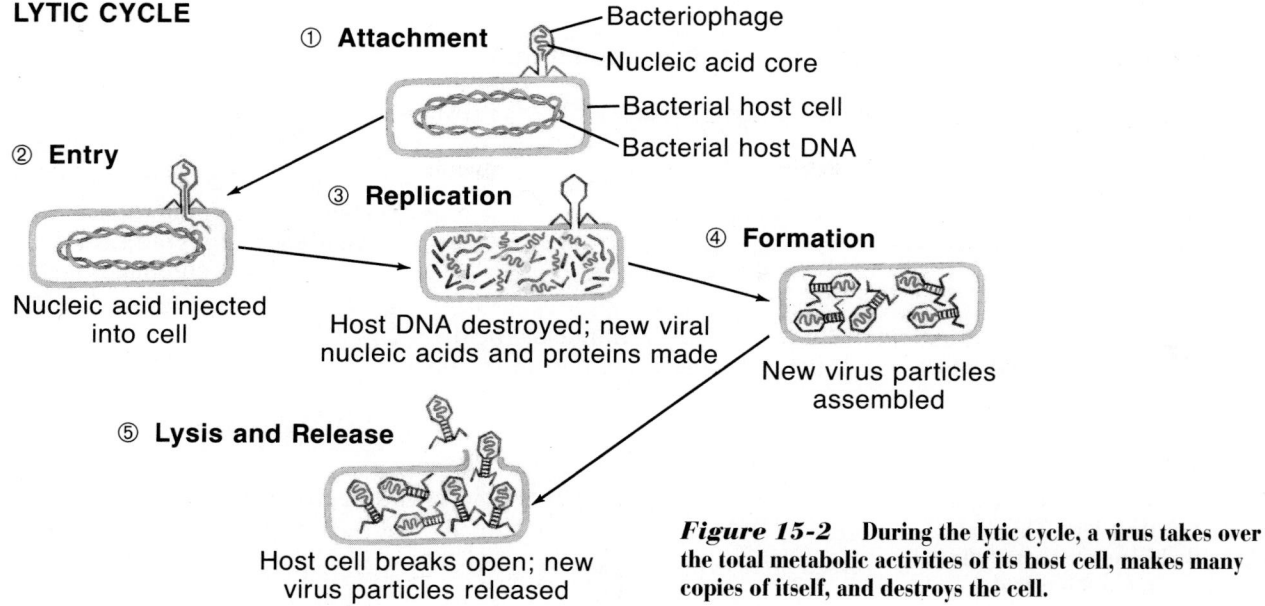

LYTIC CYCLE

① **Attachment**
Bacteriophage
Nucleic acid core
Bacterial host cell
Bacterial host DNA

② **Entry**
Nucleic acid injected into cell

③ **Replication**
Host DNA destroyed; new viral nucleic acids and proteins made

④ **Formation**
New virus particles assembled

⑤ **Lysis and Release**
Host cell breaks open; new virus particles released

Figure 15-2 During the lytic cycle, a virus takes over the total metabolic activities of its host cell, makes many copies of itself, and destroys the cell.

Viral Cycles

When a virus invades a cell, new viruses are produced. The invaded cell, called a host cell because it provides materials to the virus, usually is destroyed. Let's examine this reproductive process in greater detail, using a virus containing DNA as an example.

Lytic Cycle A virus can identify and infect a potential host cell only if part of the viral coat can match a specific receptor site on the membrane surface of the host cell, much like jigsaw puzzle pieces must match in order to fit together. For example, the virus that causes AIDS can attack only certain white blood cells, tobacco mosaic virus can invade only the leaf cells of the tobacco plant, and bacteriophages can infect only certain bacterial cells. Once the virus recognizes its host, it is able to attach itself to the host cell membrane. After attachment, the viral DNA is usually injected into the host cell. In some cases, rather than injecting its DNA into the host, the entire virus enters the cell. The DNA is then released from the protein coat. In either case, once the viral DNA is inside, it takes over the workings of the cell using the host cell's ribosomes, amino acids, enzymes, and ATP to make more viruses. Even the host cell's DNA is broken down to provide raw material for making new viruses. The viral DNA is replicated to provide genetic material for new viruses. Viral DNA is also transcribed to make mRNA, which codes for viral coat proteins and several enzymes. Some of these enzymes are used to assemble new viruses from the new coats and DNA copies. Others are used to lyse, or break open, the host cell after the new viruses have been assembled. After lysis, which can occur as quickly as 20 minutes after attachment, the newly made viruses are released to attack other host cells. This process is called the **lytic cycle** and is diagrammed in Figure 15-2.

LYSOGENIC CYCLE

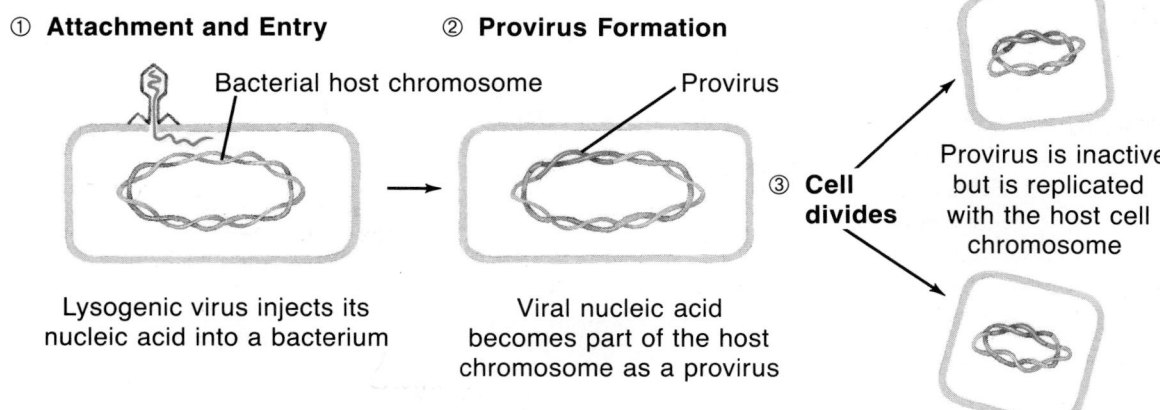

① **Attachment and Entry**   ② **Provirus Formation**

Bacterial host chromosome

Provirus

③ **Cell divides**

Provirus is inactive but is replicated with the host cell chromosome

Lysogenic virus injects its nucleic acid into a bacterium

Viral nucleic acid becomes part of the host chromosome as a provirus

Figure 15-3 **Lysogenic viruses don't destroy their host. The viral nucleic acid becomes a part of the host cell chromosome and is called a provirus.**

Lysogenic Cycle   A different kind of viral reproductive cycle was discovered in 1953, when scientists noticed that certain bacteria unexpectedly lysed after being exposed to external factors such as UV light or X rays. This event led researchers to discover that viral DNA may sometimes become integrated into the host cell's chromosome. When this occurs, the viral DNA is known as a **provirus.**

Cells containing proviruses, rather than immediately being taken over and destroyed, continue to carry on their normal functions for a time. As a host cell reproduces, the provirus it contains is replicated along with the host's DNA. A provirus may cause some phenotypic changes in its host because the viral DNA is transcribed and translated to make proteins along with the host's genes. The provirus also prevents other viruses from entering and destroying the host. When a cell that contains a provirus is

G L O B A L CONNECTION

Who were the real conquerors?

• •

For centuries, history books have described how the Americas were invaded by European explorers. Columbus arrived in 1492, and others soon followed. The powerful Aztec empire, whose largest city had about 100 000 inhabitants, was conquered by only 1000 soldiers under Cortes in 1519-21. Twelve years later, Pizarro and an even

smaller army conquered the Incan empire in Peru. The native people they met had never seen horses or guns before, but they were good warriors. And they far outnumbered the Europeans. How were the Spanish able to conquer them?

The Real Conquerors   Some historians believe that disease played a more important role than horses and guns. In the 50 years after Columbus landed, smallpox reduced the native population by half—some say by as much as 90 percent. Between a third and half of

the Aztecs died of smallpox when the Spanish first invaded their empire. Only about 1000 Native Americans survived in Cuba and the Antilles after an epidemic in 1518. And smallpox wiped out thousands of Incas in the 1530s.

Selective Killers   Why did smallpox kill only the Native Americans? Smallpox had been common in Europe, killing many children. Those who lived to be adults had developed an immunity to the disease. But, they carried the virus that causes smallpox in their bodies.

exposed to environmental factors like X rays, UV light, or certain chemicals, the provirus is triggered to become active. The viral DNA takes over the machinery of the cell. New viruses are manufactured, the host cell lyses, and the new viruses are released to invade other cells. It was this series of events that scientists observed in 1953 when they exposed bacteria to radiation. Cells containing proviruses are said to be lysogenic, in other words, subject to lysing. This pattern of viral reproduction is called the **lysogenic cycle,** Figure 15-3.

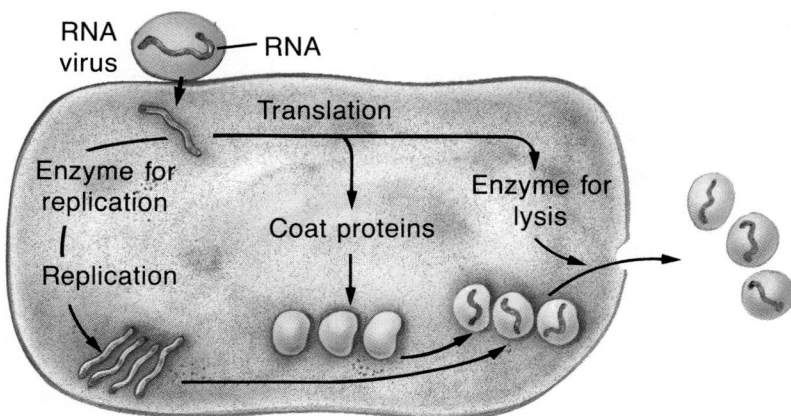

RNA LYTIC CYCLE

RNA Viruses

Viruses containing RNA go through the same lytic and lysogenic cycles as DNA viruses. In most RNA viruses, the viral RNA both replicates to make new viral RNA and serves directly as mRNA to form new viral proteins by translation. When new viruses have been assembled, the host cell lyses.

RNA viruses that enter the lysogenic cycle are very unusual. They contain RNA, but their host cells' genetic material is DNA. How, then, can the viral RNA attach to the host's DNA to become a provirus? These viruses contain a unique enzyme, reverse transcriptase, which reverses the normal transcription process. Instead of transcribing DNA into RNA, reverse transcriptase transcribes RNA into DNA. Once the DNA has been transcribed, it can become double stranded by replication and join the

Figure 15-4   **The RNA in RNA viruses replicates itself to make more viral RNA and serves as mRNA to make viral proteins by translation. Note that production of this type of RNA virus does not involve the host nucleus. What types of viral proteins are made by translation?**

There had been no epidemics of smallpox in the Americas before 1492. Native Americans are thought to have come to North America by migrating across the Bering Strait from Siberia to Alaska. They lived for long periods in the freezing Arctic region before moving south. Any viruses and bacteria they brought with them would have been killed by the cold while being transmitted from host to host. Thus, the Native Americans had no immunity to smallpox and the other diseases that existed in Europe.

When the Europeans landed, the smallpox viruses they carried swept through the native cities and villages like fire through a dry forest. They killed children and adults alike. Sick and dying, the people could not defend themselves.

There were other epidemics as well. Cholera, plague, diphtheria, and other diseases caused by bacteria also killed many Native Americans. Like the smallpox virus, these bacteria were new to the Americas. Some historians think that after the Pilgrims landed in

Plymouth in 1620, 90 percent of the Native Americans died from diseases brought by the English colonists.

1. **Connection to Medicine:** Why are there no smallpox epidemics in the world today?
2. **Issue:** Germ warfare is the use of harmful microorganisms as weapons in war. Why is germ warfare so dangerous? What do you think should be done to prevent its use?

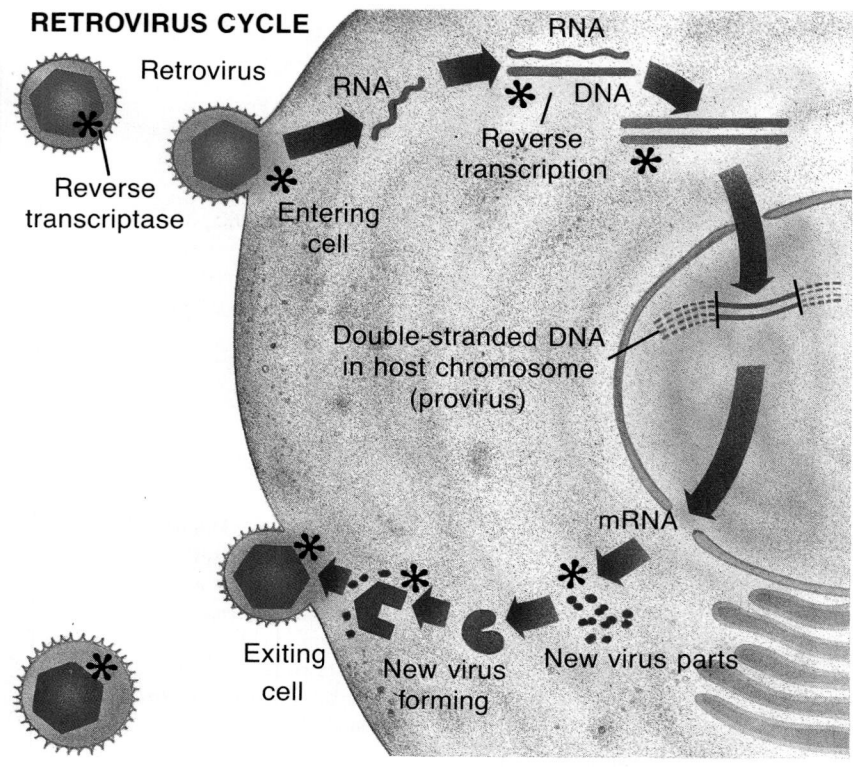

RETROVIRUS CYCLE

Retrovirus

Reverse
transcriptase

RNA

Entering
cell

RNA

DNA

Reverse
transcription

Double-stranded DNA
in host chromosome
(provirus)

mRNA

Exiting
cell

New virus
forming

New virus parts

Figure 15-5 **A retrovirus enters
a host cell and its RNA undergoes
reverse transcription. Eventually, a
double-stranded molecule of DNA is
produced and becomes a provirus,
remaining in the host cell's nucleus.**

host's chromosome. The viral DNA
may remain dormant as a provirus
for a long period of time, or it may
start directing the production of
new viruses very soon. Because they
perform transcription backwards,
these viruses are called **retroviruses**
(*retro* means backward). The virus
that causes AIDS is an example of a
retrovirus.

Some Questions

Is a virus alive? The answer is
debatable. Because viruses don't
have most of the basic features
associated with living systems, you
could conclude that a virus does
not qualify as a living organism.
However, viruses can perform the
most universal function of life.
They can reproduce.

A question that may be more
important is, how did viruses
evolve? It might seem reasonable to think that viruses, because of their
simple structure, resemble the very first forms of life that appeared on
Earth. But that explanation doesn't seem likely when you consider that
viruses require true cells in order to reproduce. According to one hypothesis, viruses may be descendants of a more complex living system that lost
certain features as it became highly specialized for one function only—
reproduction. Another hypothesis states that viruses may have arisen after
DNA escaped from a cellular organism. In either case, viruses now consist
mainly of the bare chemical necessities for reproduction, nucleic acids.

Section Review

Understanding Concepts
1. Why must viruses take over host
cells in order to reproduce?
2. Explain how viral nucleic acid can
enter a host cell and not destroy it.
3. Why can't a bacteriophage infect
human cells?
4. ***Skill Review—Observing
and Inferring:*** A microbiologist

added some viruses to a bacterial
culture. At that time and every hour
for 8 hours, he removed a sample
of liquid from the culture and determined the number of viruses present
in the sample. The number of viruses
in each sample was 15, 17, 49,
128, 385, 386, 386, and 387.
Explain what was happening in the

bacterial culture. For more help,
refer to Thinking Critically in the
Skill Handbook.
5. ***Thinking Critically:*** What
kinds of human cells must be hosts
for the viruses that cause colds?

15.2 Kingdom Monera

We know that living forms have dwelt on Earth for at least 3.5 billion years, because that is the age of the oldest fossils discovered. Those fossils are monerans. All monerans are bacteria. Although bacteria are simple in structure, they carry out the familiar functions of life and so are included in the classification of living organisms. Monerans are classified in a kingdom of their own because they are prokaryotes. All other organisms are eukaryotes. Two main groups of monerans, the ancient bacteria and the true bacteria, exist on Earth today. Some monerans have probably evolved very little over billions of years. The ancient bacteria give us some ideas about what the very first living organisms may have been like and what conditions were like on primitive Earth.

Characteristics of Bacteria

There are bacteria everywhere you look—in air, soil, water, on the surface of your desk, on your skin, even inside your body. A single bacterium is too small to be seen without the aid of a microscope. Bacteria often grow in colonies, which can become visible if they grow large enough, Figure 15-6. Let's find out what a typical bacterium looks like.

Structure   You have already learned that, as prokaryotes, monerans have no true nucleus and no membrane-bound organelles. Monerans have cell walls that protect the bacterium and help to maintain osmotic balance between the bacterium and its environment. Many bacterial cell walls are made of a substance found only in monerans. The antibiotic penicillin destroys these bacteria because it hinders the production of this substance. The bacterium cannot reproduce because it can no longer make new cell walls.

Often, the cell wall is surrounded by another structure, an outer capsule that offers the bacterium additional protection. This capsule may make it possible for bacteria to infect humans, that is, to escape destruction by the body's defenses. For example, cells of *Pneumococcus*, a pneumonia bacterium, can cause the disease only if they have capsules. *Pneumococcus* cells lacking capsules are not infectious.

Figure 15-6   **Each bacterial colony on this streak plate arose from a single cell, which underwent many generations of cell division. There are millions of cells in a bacterial colony.**

Objectives

Describe the basic characteristics of bacteria.

Compare the various means of nutrition found among bacteria.

Demonstrate knowledge of how bacteria are beneficial in nature and useful to humans.

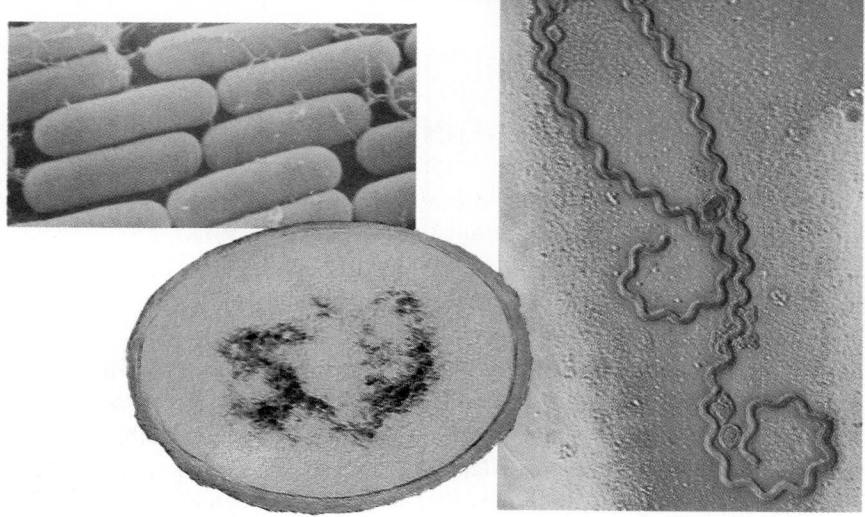

Figure 15-7 The rod-shaped bacillus, round coccus, and spiral-shaped spirillum are common shapes of bacteria. Magnification: 3500×, 26 000×, 19 000×, respectively

Some bacteria possess flagella and are able to move around. Bacterial flagella are simpler in structure than are eukaryotic flagella and are made of different materials.

Most bacterial cells have one of three shapes. They may be spherical, rod-shaped, or spiral-shaped. Many bacteria, especially the rod-shaped ones, are adapted for withstanding harsh conditions. When living conditions become unfavorable for one of these bacteria, a tough, protective wall forms around its DNA and a small bit of its cytoplasm, producing a highly resistant, dormant structure called an **endospore.** The rest of the cell may then die. However, the endospore can resist long periods of boiling, or years of freezing or drought. When conditions once again become favorable, the endospore develops into an active cell.

Reproduction   A bacterium reproduces by simply dividing into two cells. This method of reproduction, called binary fission, is a form of **asexual reproduction.** Asexual reproduction is the production of one or more genetically identical offspring from a single parent. In many types of bacteria, the new cells separate from one another. However, in some species, the new cells remain attached to form pairs, chains, or clusters. Even though these cells are grouped together, each one acts as an independent organism.

Figure 15-8 Bacteria are prokaryotes with a cell wall and a single chromosome. Cells of some species are surrounded by a protective capsule. Some may have one or more flagella.

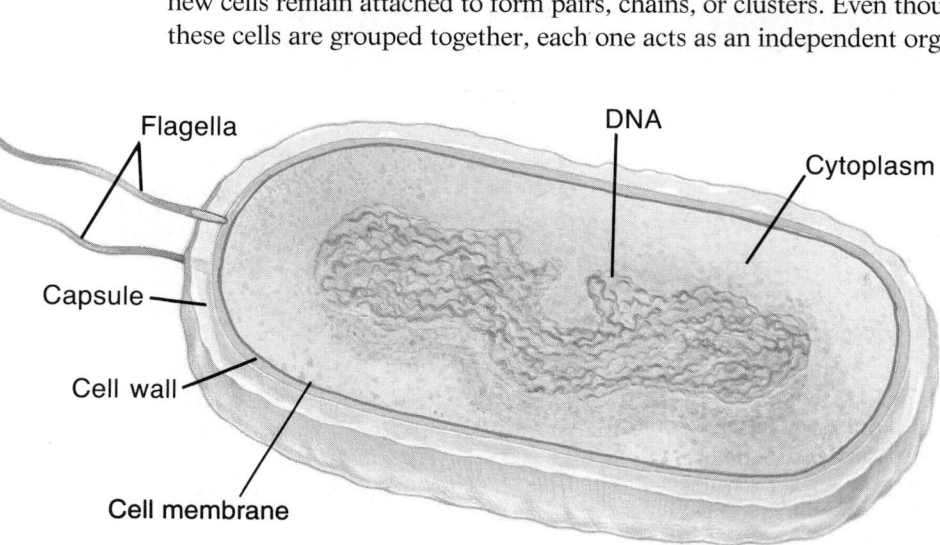

Flagella

DNA

Cytoplasm

Capsule

Cell wall

Cell membrane

Nutrition in True Bacteria

Bacteria are successful not only because of their ability to withstand harsh conditions, but also because they have evolved a wonderful variety of ways to obtain energy. It seems that every potential food source can be used by some kind of bacteria. Though most true bacteria are heterotrophs and obtain energy from the organic molecules they take in, many are autotrophs. True bacteria have also developed different methods of respiration to break down stored food and release energy. Some respire aerobically and can grow only in the presence of air. Others release energy anaerobically and cannot grow when air is present. Still others can release energy either way. If oxygen is available, they respire aerobically. If not, they become anaerobes, organisms that release energy in the absence of air.

Heterotrophs Heterotrophic bacteria are found everywhere. Because they require complex organic molecules as their energy source and are not adapted for trapping the food that contains these molecules, these bacteria must absorb nourishment from their surroundings. Many heterotrophs are **parasites,** organisms that live in or on other organisms. The organisms that are invaded, and on which parasites live, are called hosts. Parasites always cause harm to their hosts in some way. Other heterotrophic bacteria are **saprophytes,** organisms that feed on dead organisms or other organic wastes. These bacteria are extremely important in nature because they help recycle the nutrients contained in decomposing organisms and waste materials. Were it not for saprophytes, dead matter would soon envelop Earth's surface. Eventually, life would stop as carbon and other elements became unavailable for new growth.

Photosynthetic Autotrophs Several types of true bacteria are autotrophic; they use inorganic molecules or light to obtain energy. One type, the blue-green bacteria, traps the energy in sunlight by photosynthesis. During photosynthesis, they convert carbon dioxide and water to organic material and release oxygen. This group of monerans was probably very important in generating oxygen early in life's history. The blue-green bacteria do not always look blue-green in color. In addition to chlorophyll *a*, they usually contain a blue pigment, though some may contain red or yellow pigments as well. So, depending on the combination of pigments present in a particular species, it may be blue-green, black, yellow, green, or red. The Red Sea was named for the large masses of red-pigmented bacteria that grow on its surface. Because they are prokaryotes, the blue-green bacteria have no chloroplasts. Their chlorophyll and other pigments are located on simple, flattened membranes within the cytoplasm.

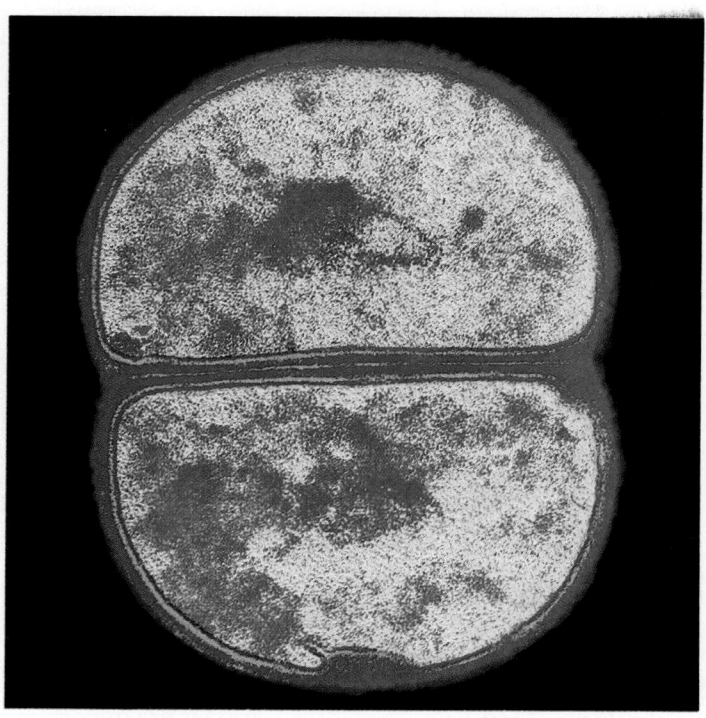

Figure 15-9 **Bacterial cells reproduce by binary fission, producing two cells of equal size. Why can't bacteria undergo mitosis? Magnification: 26 000×**

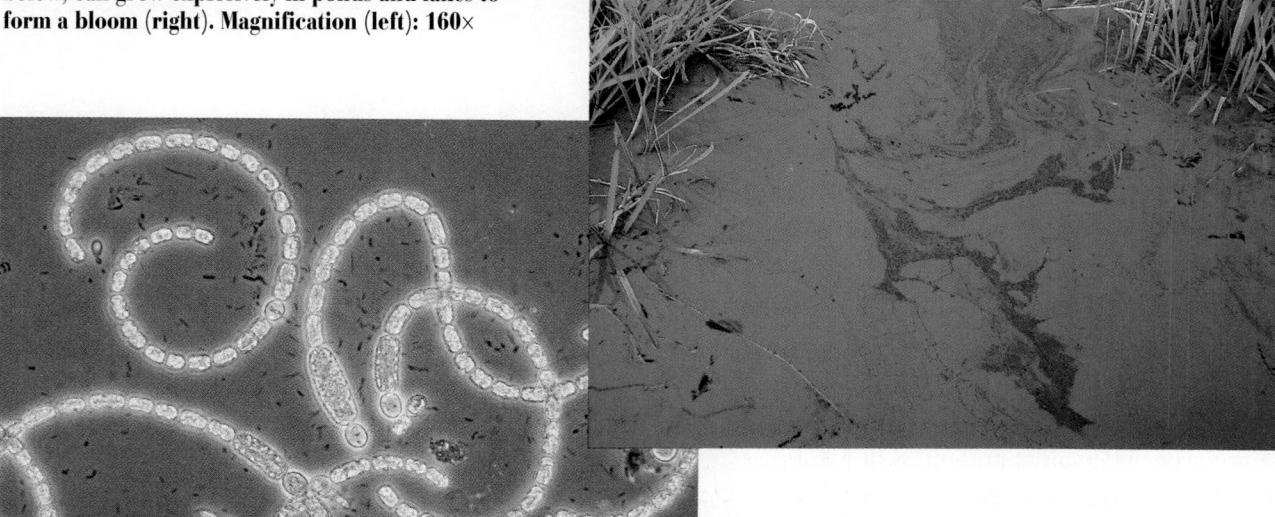

Figure 15-10 Large populations of blue-green bacteria, such as *Nostoc* shown in the photograph below, can grow explosively in ponds and lakes to form a bloom (right). Magnification (left): 160×

Blue-green bacteria are common in ponds, lakes, puddles, streams, and even moist places on land. They are important producers in aquatic communities, and many species are important in converting atmospheric nitrogen into a form that plants can use.

Chemosynthetic Autotrophs   Some true bacteria are autotrophic but not photosynthetic. They do not derive energy from sunlight. Instead, they use a process called chemosynthesis to trap energy released from the breakdown of inorganic materials such as sulfur and nitrogen compounds. Some chemosynthetic bacteria are important in converting nitrogen and sulfur compounds in the environment to forms that can be used readily by plants.

Nutrition in Ancient Bacteria

The ancient bacteria live under conditions that may be very similar to those found when life was first evolving on Earth. All of the known ancient bacteria live without oxygen and are autotrophs. Some live in places where no other organisms can exist today.

Methane-Producing Bacteria   Members of one group of ancient bacteria inhabit swamps, marshes, and other environments where there is a great deal of decaying plant matter. Bacteria from this group also inhabit the digestive tracts of humans and other mammals, particularly grazing mammals such as cattle, horses, and deer. Some of these bacteria are found in sewage treatment plants. These bacteria obtain energy by chemosynthesis. They convert carbon dioxide and hydrogen to methane, which is sometimes called marsh gas. For this reason, they are known as methane-producing bacteria. Since methane can be used as a fuel, it is possible that these bacteria could be put to practical use. They would not only aid in the decomposition of dead organisms and organic waste, but also convert those substances into a usable product.

Salt-Loving Bacteria   Other groups of ancient bacteria inhabit even harsher habitats. Members of one group, the salt-loving bacteria, live in salt lakes. They are the only prokaryotic organisms that inhabit the Great Salt Lake and the Dead Sea.

Heat- and Acid-Loving Bacteria   Another group of ancient bacteria occupies very hot and acidic areas. They are known as heat- and acid-loving bacteria. Some members of this group are producers that live deep in the ocean near underwater volcanoes and thermal vents. Others live in hot sulfur springs, such as those in Yellowstone National Park, which have temperatures near 80°C and a pH as low as 2, Figure 15-11.

Comparing Ancient Bacteria and True Bacteria   The ancient bacteria differ from the true bacteria in several ways. Their cell walls and the lipids in their plasma membranes have a different structure. The sequences of bases in the tRNAs and rRNAs of ancient bacteria are quite different from those of true bacteria. Ancient bacteria also react differently to antibiotics. You can see why ancient bacteria are classified separately from the true bacteria. Some biologists feel the ancient bacteria should be classified in a separate kingdom. Perhaps the ancient bacteria, the true bacteria, and the eukaryotes share a common ancestor that was a very early form of life on Earth.

The Importance of Bacteria

What would life on Earth be like if there were no decomposers to break down dead plant material, the bodies of dead animals, and the waste products generated by all living things? Eventually, all the nutrients needed for life would be bound up in this dead matter and waste. The raw materials needed to support new life would no longer be available.

Figure 15-11   Ancient bacteria are the only organisms that can inhabit hot sulfur springs, such as this one in Yellowstone National Park.

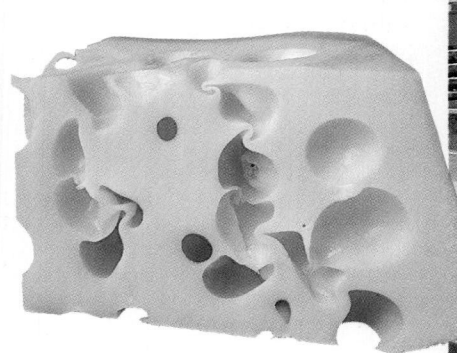

Figure 15-12 The holes in Swiss cheese are caused by bubbles of carbon dioxide produced by the bacteria that give the cheese its flavor. Cheese is made in large vats such as this one.

Bacteria play an important role in nature. Most of Earth's decomposers are bacteria that break down organic matter into basic components that can be reused by other organisms. Without bacteria, this recycling of life-sustaining materials could not take place.

Bacteria are also very important to human economy because they are needed for a variety of industrial processes. For example, the dairy industry makes yogurt and buttermilk by adding certain bacteria to milk. As the bacteria multiply, they secrete substances that flavor the milk. Production of cheese depends on bacteria that cause milk products to turn to solids. Bacteria are indirectly involved in milk production. Certain bacteria that live in the stomachs of cows aid in the digestion of plant materials. Without this help, cows would not be able to obtain the nutrients they need to produce milk. Bacteria break down the material that holds together flax and hemp fibers, freeing the fibers for use in making linen or rope. Animal skins are tanned into leather with the help of bacteria. Many of today's antibiotics are produced by bacteria.

While many bacteria are very helpful, others can be harmful and dangerous. The role of bacteria in disease is discussed in Chapter 23.

Section Review

Understanding Concepts

1. You are examining a unicellular organism with an electron microscope. What clues would enable you to determine whether the organism is a bacterium?

2. Some bacteria can break down cellulose in plants. How might this be beneficial to humans?

3. Is endospore formation a form of reproduction among bacteria? Explain.

4. **Skill Review—Classifying:** Set up a classification of bacteria based on their structural features and means of obtaining energy. For more help, refer to Organizing Information in the **Skill Handbook.**

5. **Thinking Critically:** A bacterium is discovered that does not obtain energy from organic molecules and does not undergo photosynthesis. How does it obtain its needed energy?

15.3 Kingdom Protista

By now, you understand that classification of Earth's living organisms requires more than two kingdoms. When biologists realized that some organisms were neither plants nor animals, the first new kingdom they added to their classification system was the Kingdom Protista. Protists are a diverse group of organisms. They may appear to have little in common, and they are difficult to classify. However, protists do share several features.

Protists are all eukaryotes, and they probably arose from prokaryotes. Most are unicellular, but some have a simple multicellular structure. Some are heterotrophic, others autotrophic, and some may switch from one form of nutrition to the other. The Kingdom Protista is made up of three groups of organisms: plantlike protists, which are called algae; animal-like protists, known as protozoans; and funguslike protists.

Plantlike Protists: The Algae

You may have seen seaweeds washed up on a beach, or the greenish film that often forms in still ponds or on the sides of a home aquarium. These are just a few examples of algae. Algae range in size from microscopic, single-celled forms to multicellular seaweeds that may reach 100 meters in length. Algae first appeared on Earth about 550 million years ago. The vast majority are autotrophs and are important sources of food for both freshwater and marine animals. Algae are classified on the basis of their color and structure.

Euglenoids Euglenoids are unicellular algae that have characteristics associated with both plants and animals. Like animals, they lack cell walls and move by means of a flagellum. Like plants, they are photosynthetic. One species, *Euglena,* provides a good example of why protists are difficult to classify. This organism, pictured in Figure 15-13, is usually autotrophic. However, if exposed to a period of darkness, it will lose its chloroplasts

Compare the features of the various phyla of algae.

Distinguish among the protozoan phyla.

Relate the characteristics of slime molds to their life cycle.

Figure 15-13 **Euglenoids contain many specialized structures, including contractile vacuoles that pump excess water out of the cell and an orange-red eyespot that detects light. Magnification (photo): 450×**

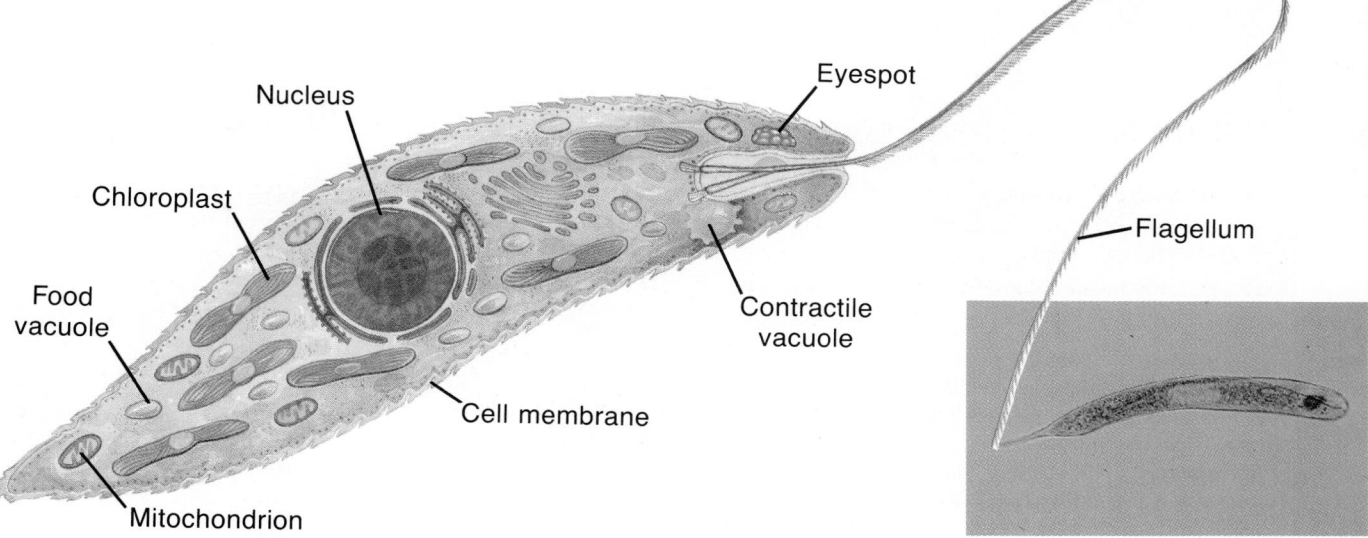

Nucleus

Eyespot

Chloroplast

Food vacuole

Flagellum

Contractile vacuole

Cell membrane

Mitochondrion

Figure 15-14 The glassy cell walls of diatoms are covered with many beautiful markings. Magnification: 300×

and become heterotrophic. You have learned that unicellular organisms must be very versatile, because they must carry out all life functions in a single cell. Note the many specialized structures in the diagram of *Euglena*.

Golden Algae   Most members of this phylum are unicellular. Their colors range from yellow-green to golden-brown, depending on the pigments they contain. Most common and best known among them are the diatoms, pictured in Figure 15-14. Diatoms inhabit both fresh and salt water and are the most numerous of the many microscopic producers that float near the ocean surface. Their many shapes and colors make diatoms some of nature's most beautiful organisms. Each diatom is composed of a two-part outer shell, one part of which overlaps the other like the lid of a box. The shell consists of a cell wall impregnated with a glasslike material. When diatoms die, their glass shells do not decay. They drift to the ocean bottom, where large masses of them collect into deposits called diatomaceous earth. Diatomaceous earth can be mined and has many uses. It serves as an abrasive ingredient in scouring powders, cosmetics, and toothpaste. It can also be used to filter small particles from juices and other liquids.

Green Algae   The green algae belong to a very diverse phylum that includes both unicellular and multicellular species. Most are found in freshwater habitats, though many live in moist areas on land. Some live in salt water, and some even inhabit other organisms. The green algae are thought to be the ancestors of the plant kingdom, partly because the cells of green algae and plants contain the same kinds of chlorophyll and other pigments. Note the great diversity of forms within this phylum, as shown in Figure 15-15. Some, like *Spirogyra*, are filaments. Others, such as *Volvox*, are globe-shaped groups of cells called colonies. Species like *Ulva*

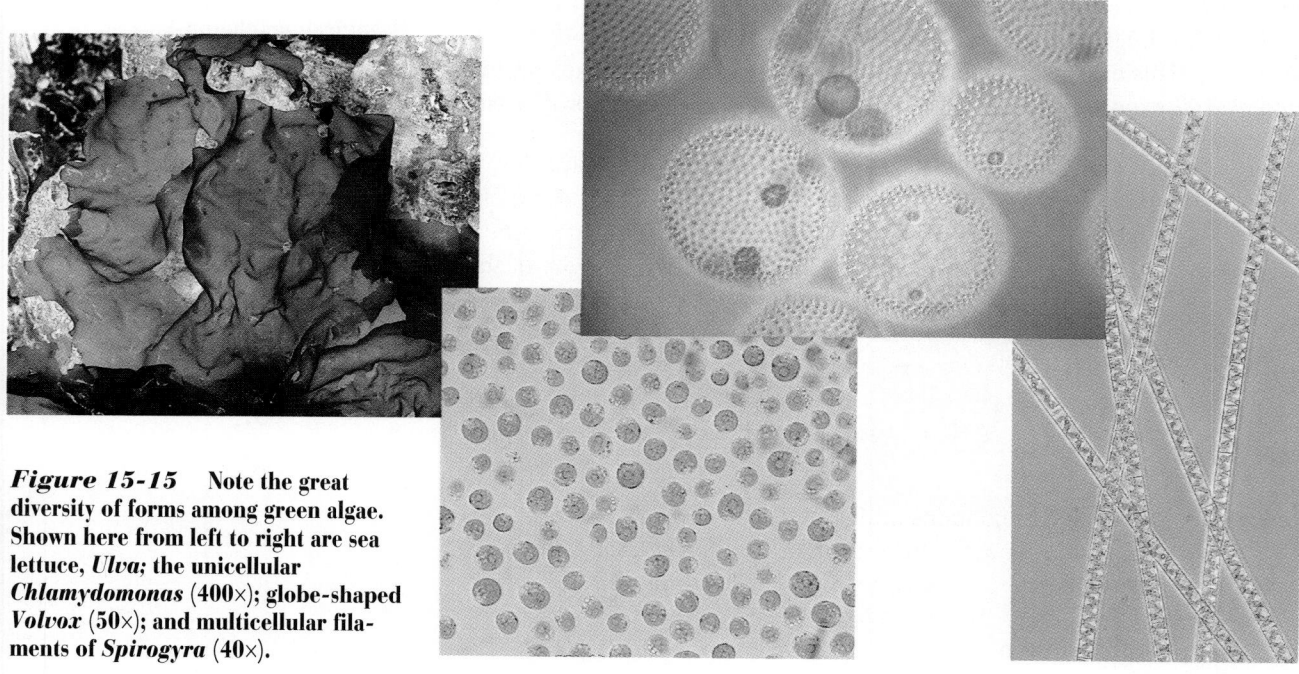

Figure 15-15 Note the great diversity of forms among green algae. Shown here from left to right are sea lettuce, *Ulva;* the unicellular *Chlamydomonas* (400×); globe-shaped *Volvox* (50×); and multicellular filaments of *Spirogyra* (40×).

Figure 15-16 These photos of kelp, a brown alga, show the structures that resemble roots, stems, and leaves. The round structures are filled with air and help the alga float on the water's surface, where it can photosynthesize.

resemble a lettuce leaf. The bodies of some larger species have parts that resemble the stems, leaves, or roots of true plants. However, algae do not have tissues or organs as plants do, though some division of labor among cells may occur. For example, most or all of the cells in a multicellular alga may contain photosynthetic pigments and perform photosynthesis. Some of those same cells may also participate in reproduction, while others do not. As the green algae evolved, they became increasingly complex, as division of labor among groups of cells increased. While asexual reproduction still occurs, most species are also capable of reproducing sexually, and many have evolved life cycles similar to those of plants.

Brown Algae Also known as seaweeds, brown algae are complex multicellular organisms that live mainly in salt water. In addition to chlorophyll, they contain a pigment that gives them their brown color. Some species are among the largest of all algae. Many brown algae have a thick body with specialized parts. *Laminaria,* for example, has a rootlike structure, a stemlike portion, and leaflike parts, Figure 15-16. Many species grow in cool water along rocky coasts. They are anchored to the rocks by their rootlike parts. Some, like *Sargassum,* float unattached at the surface of warmer waters. *Sargassum* grows so densely that it covers much of the surface of the Atlantic Ocean near Bermuda, forming an area of millions of square kilometers known as the Sargasso Sea.

   Brown algae are an important commercial source of iodine. Some brown algae, called kelp, can be used as fertilizer or as food for humans. Other species are a source of algin, which is used to thicken ice cream and other foods.

Red Algae This phylum contains multicellular forms that are more complex in structure than the other phyla of algae. Most live in salt water, but a few occupy freshwater or land environments. Their bright red color, which is due to a red pigment, and their feathery shapes make many species beautiful.

Figure 15-17 **The branched filaments of this red alga are hooked, enabling it to cling to other algae and to rocks and mud.**

The pigments present in red algae enable them to live and photosynthesize at greater ocean depths than any other group of algae. Most wavelengths of light, such as the red and violet wavelengths that are absorbed by chlorophyll, do not penetrate ocean water very deeply. This is why organisms that have chlorophyll cannot live deep in the ocean. However, some of the pigments in red algae are able to trap blue light, the wavelengths that penetrate the water most deeply. That light energy can then be used in photosynthesis. A species of red algae discovered in the Bahamas grows at a depth of nearly 270 meters. Not all red algae live at such great depths. Many live anchored in mud along the shore.

Red algae are used in a variety of ways. Some species are used as food by humans. Agar is a substance made from red algae and used to make culture media for growing bacteria in the laboratory. Other substances obtained from red algae are used in making ice cream, puddings, and cake icing.

Animal-like Protists: The Protozoans

The second major group of protists is made up of unicellular, animal-like organisms known as protozoans. Most protozoans can move, and their classification is based on their method of locomotion. Protozoans are heterotrophs and have evolved a variety of adaptations for obtaining food. Some actively trap food, while others are parasites and live at the expense of a host.

Rhizopods One of the simplest and most often studied of the protozoans is *Amoeba,* a member of the phylum known as rhizopods. Members of this phylum move by means of extensions of the plasma membrane called **pseudopodia,** which means "false feet." Formation of pseudopodia is made possible by a flexible plasma membrane and the constant movement of cytoplasm. Pseudopodia are used not only for locomotion, but also for engulfing food. By means of phagocytosis, food is surrounded by pseudopodia and brought into the cell for digestion.

Figure 15-18 **The shape of an amoeba changes constantly as pseudopodia form. Note the small protist about to be captured, and the protist in the food vacuole that is ready to be digested. Magnification (photo): 100×**

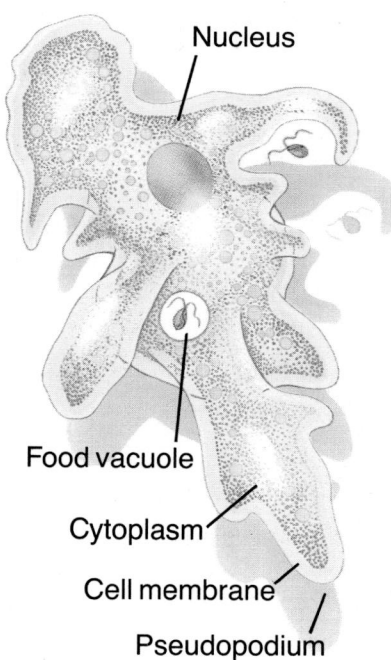

Nucleus

Food vacuole

Cytoplasm

Cell membrane

Pseudopodium

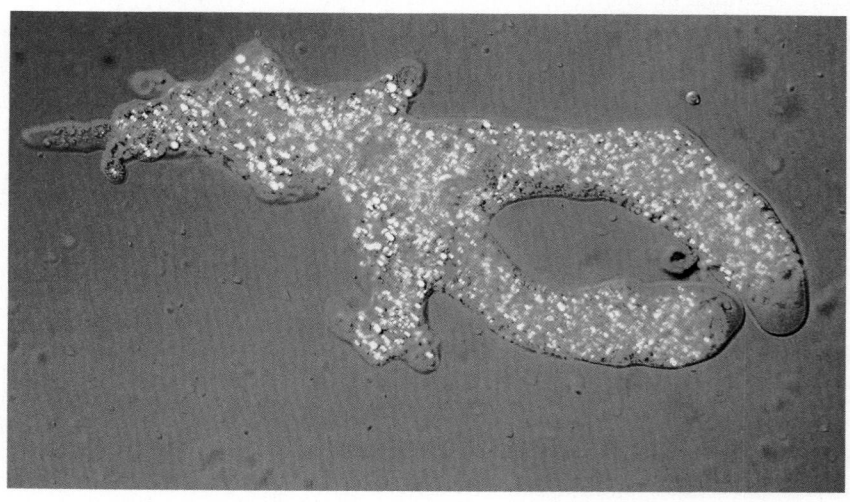

Two groups of rhizopods that inhabit the ocean are interesting because they have mineral skeletons. Members of one group, the forams, have chalklike outer skeletons. Those in the other group, the radiolarians, have internal, glasslike skeletons. Skeletons of forams make up much of the limestone and chalk on Earth, including the white cliffs of Dover, England.

Ciliates Members of another phylum of protozoans, the ciliates, move by the beating of the many cilia that project from their surfaces. The cilia also play a role in food-getting by sweeping food into the cell. This phylum includes the familiar *Paramecium.* Examine Figure 15-19. You can see that a paramecium has many specialized parts. It is, in fact, one of the most complex of all unicellular organisms, containing specialized organelles for taking in and digesting food, eliminating undigestible materials, and pumping out excess water to maintain osmotic balance. Ciliates are so complex that they have two types of nuclei. A large macronucleus controls the cell's metabolism. A smaller micronucleus is involved in reproduction. Some ciliates have many micronuclei.

Flagellates Members of another phylum of protozoans, the flagellates, move by means of flagella. Some flagellates are free-living and inhabit either fresh or salt water. Many species, though, live within other organisms. One such flagellate, a member of genus *Trypanosoma,* is transmitted to humans by the tsetse fly. This protist lives in the blood of humans and releases a poisonous substance that attacks the nervous system, causing weakness and then death. This disease is called African sleeping sickness. Another flagellate inhabits the digestive tract of termites, where it secretes enzymes that digest the wood the termites eat. This relationship, which benefits both organisms, is known as **mutualism.** The termite benefits by having its food digested, and the flagellate benefits by receiving shelter and a constant supply of wood.

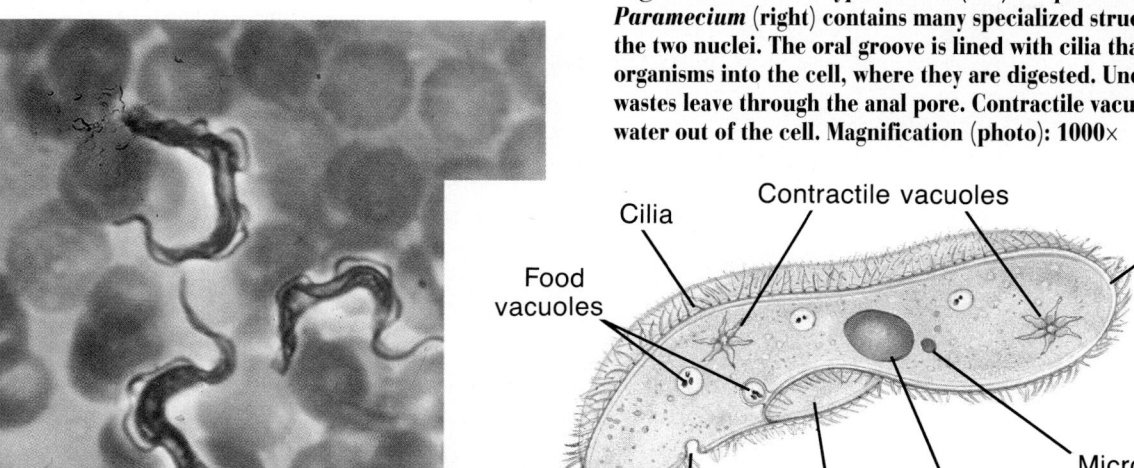

Figure 15-19 *Trypanosoma* (left) is a parasitic flagellate. *Paramecium* (right) contains many specialized structures. Note the two nuclei. The oral groove is lined with cilia that sweep tiny organisms into the cell, where they are digested. Undigested wastes leave through the anal pore. Contractile vacuoles pump water out of the cell. Magnification (photo): 1000×

Cilia
Contractile vacuoles
Cell membrane
Food vacuoles
Micronucleus
Macronucleus
Oral groove
Anal pore

Sporozoans Some protozoans, the sporozoan phylum, reproduce by tiny, sporelike structures. Remember that spores are small reproductive cells that can develop into new organisms.

Sporozoans have no cilia or flagella. Because they have no means of movement, they cannot actively trap food. Therefore, they live as parasites and obtain their food from the host's tissues. The sporelike reproductive structures are specialized for penetrating host cells and tissues. In this way, sporozoans obtain food.

Some sporozoans cause diseases in humans and animals. Most dreaded of all the sporozoans are members of the genus *Plasmodium*. These sporozoans are transmitted by certain mosquitoes and are responsible for causing malaria, a disease that has caused millions of human deaths. Every year, more that 200 million people are infected with malaria, mostly in tropical climates. In Africa alone, at least a million people die of malaria each year.

I N V E S T I G A T I O N

The Effect of Chemicals on Protozoans

How many different kinds of cologne, eyeshadow, and mascara are on the market? All cosmetics, fragrances, sunscreens, hair sprays, and similar products must be thoroughly tested to be sure they are safe for human use. A variety of animal experiments are conducted as a part of the testing process. Rabbits were commonly used to test new eye products. Because of questions raised about the use of animals in these tests, laboratories are developing different methods of testing cosmetics. Some researchers believe that all cells behave in similar ways. If a new cosmetic causes problems for protozoan cells, it may cause problems for cells in the human body also. In this lab, you will test the effects of chemicals on the life span of protozoan cultures.

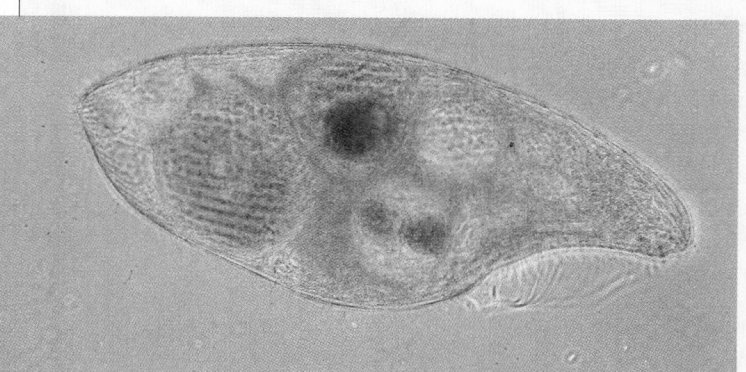

Magnification: 300×

Problem
How do chemicals affect the life span of protozoan cultures?

Safety Precautions 🔬 🧤
When handling chemicals, use guidelines given by your teacher. Wash your hands at the end of each lab period.

Hypothesis
Make a hypothesis about how a specific chemical, either a specific cosmetic or a pure compound or element, would affect the life span of a protozoan culture. Explain your reasons for making this hypothesis.

Funguslike Protists

While algae clearly resemble plants, and protozoans are very much like animals, the third group of protists is similar in many respects to fungi. Funguslike protists are all heterotrophs.

Slime Molds   One phylum, the slime molds, has a fascinating and unique life cycle with features not only of fungi, but also of protozoans. At some stages of life, a slime mold is unicellular, while at other times it is multicellular.

Some species of slime molds begin their life cycle as a colorful, slimy mass containing many nuclei. This mass, called a **plasmodium,** slowly oozes along the forest floor over rotting logs and other decaying matter. Like a giant amoeba, the plasmodium engulfs bacteria and small particles of organic matter by phagocytosis. Thus, slime molds are decomposers.

Experimental Plan

1. As a group, make a list of possible ways you might test your hypothesis using the materials your teacher has made available.
2. Agree on one idea from your group's list that could be investigated in the classroom.
3. Design an experiment that will test one variable at a time. Plan to collect quantitative data.
4. Following the style of a recipe, write a numbered list of directions that anyone could follow.
5. Make a list of materials and the quantities you will need.

Checking the Plan

Discuss the following points with other group members to decide the final procedures for your experiment.

1. What variables will need to be controlled?
2. What is your control?
3. What dilutions of your chemical will you use?
4. What will you measure or count?
5. What portion of a field of view will you count?
6. How many fields of view will you count each time?
7. For how many days will you conduct your experiment?
8. Have you designed and made a table for collecting data?

Make sure your teacher has approved your experimental plan before you proceed further.

Data and Observations

Carry out your experiment, make your measurements, and complete your data table. Make a graph of your results.

Analysis and Conclusions

1. According to your results, was your protozoan sensitive to the chemical you tested?
2. What is your experimental evidence for your conclusion in question 1?
3. Was your hypothesis supported by your data?
4. If your hypothesis was not supported, what types of experimental error could be responsible?
5. Consult with another group that used a different chemical. Were their conclusions the same as yours? Why or why not?
6. Do you think a larger or smaller amount of your chemical might produce different results? Why or why not?
7. How did your chemical affect the life span of your protozoan culture?
8. Write a brief conclusion to your experiment.

Going Further

Based on this lab experience, design an experiment that would help to answer a question that arose from your work. Assume that you will have unlimited resources and will not be conducting this lab in class.

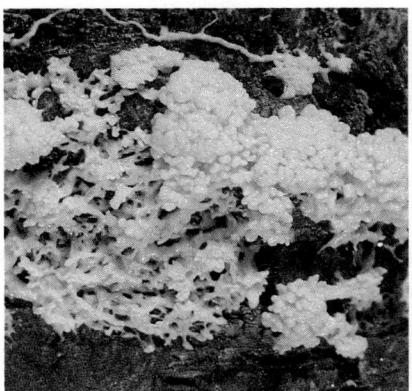

Figure 15-20 Slime molds produce an amoeba-like feeding stage called a plasmodium (left). When food becomes scarce, reproductive structures, called fruiting bodies, form (right). Magnification: (right) 10×

When food or moisture becomes scarce, the plasmodium may slow down and undergo an amazing change. It develops many reproductive structures that produce spores by meiosis. The spores are released and spread by wind and rain. When food and moisture become available, the spores germinate and develop into flagellated swarm cells. These cells are attracted to each other by a chemical they secrete. Two flagellated swarm cells fuse to form a zygote that loses its flagella and begins to move like an amoeba. The production of a new plasmodium involves division of the nucleus by mitosis, and growth without cell division. Thus, many nuclei are formed within the plasmodium. Sometimes several plasmodia fuse to form a larger plasmodium.

HISTORY CONNECTION

Protists Change a Country

In the 1800s, most people in Ireland were farmers. Potatoes were the most important crop and were the main part of the Irish diet. In the fall of 1845, a large part of the potato crop molded. The following year, the entire potato crop was wiped out in one week. The potatoes had been infected with potato blight.

Potato blight is caused by a water mold, a funguslike protist. The spores of this protist multiply most rapidly in cool, damp weather.

It rained constantly in Ireland from May 1845 until March 1846. The rainy weather continued during the summer and fall of 1846. Conditions were perfect for the protist to grow rapidly.

The potato blight changed the course of Ireland's history. There were more outbreaks of potato blight in Ireland until 1860, and a quarter of a million people starved to death or died of typhoid fever in the epidemics that followed. Ireland's population dropped by one-third. With fewer people to farm, Ireland's economy suffered.

The disease changed America's history, as well. A million and a half Irish farmers fled to the United States to begin new lives. These people and their descendants have enriched American life in many fields, including industry, politics, and entertainment.

1. **Writing About Biology:** Another water mold damaged and almost destroyed the French wine industry in the 1870s. Find out more about this disease and write a short report. What famous scientist played a part in discovering the cause of this disease?
2. **Explore Further:** Find out how the water molds are similar to true fungi and how they differ.

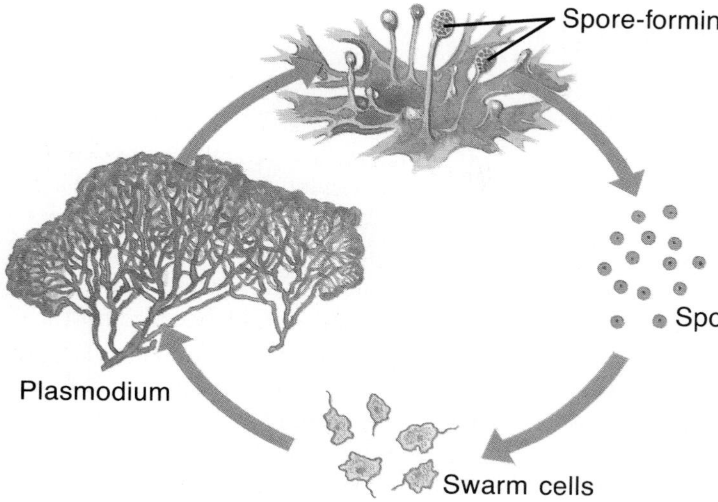

Spore-forming structures

Spores

Swarm cells

Plasmodium

Figure 15-21 The life cycle of a slime mold has several stages.

The Interest in Slime Molds   The slime molds are interesting to biologists because their life cycles involve many changes in form. These different forms resemble other forms of life. Spore production resembles reproduction in fungi. The individual, flagellated swarm cells and the plasmodium are like protozoans. Biologists study slime mold development to try to better understand the development of more complex organisms.

Evolution of Other Kingdoms   Based on structural similarities, many taxonomists argue that certain organisms in each of the three groups of protists gave rise to the other kingdoms. According to this hypothesis, algae evolved into plants, protozoans into animals, and funguslike protists into fungi. Chemical analysis, however, suggests other possibilities. Perhaps a common ancestor gave rise to protists as well as to the fungi, plant, and animal kingdoms. Because the fossil evidence is so sparse, it is difficult to know the exact origins and relationships of the various kingdoms.

Section Review

Understanding Concepts
1. Euglenoids and flagellates have much in common. Explain why both are considered protists but also why they are classified in different groups.
2. In what ways are brown algae more complex than most other phyla of algae?

3. What would be the first clue you might look for to determine whether a newly discovered unicellular protist is an alga or a protozoan?
4. *Skill Review—Recognizing Cause and Effect:* If a photosynthetic protist lost its chloroplasts, under what conditions would it be able to survive? For more help, refer

to Thinking Critically in the **Skill Handbook.**
5. *Thinking Critically:* Based on information presented in this section, what biological method might be devised to rid a house of termites?

Compare the characteristics of the different phyla of fungi.

Explain how the organisms of which a lichen is composed benefit one another.

15.4 Kingdom Fungi

You are probably familiar with many types of fungi. Some you eat, such as mushrooms; others, like molds on bread or oranges, eat the food you want to eat. You may also have seen fungi growing from the trunk of a tree or a fallen log. In these examples, the fungi are obtaining food from other organisms or parts of organisms. In nature, many fungi play important roles in decomposing dead organisms and wastes. By breaking down dead material and waste products, fungi release nutrients needed for the growth of plants. However, fungi also can cause severe economic damage when they feed upon valuable crops. Although fossils of fungi are scarce, evidence suggests that they probably first appeared about 500 million years ago and that they evolved later than the algae. Fungi are plantlike in that they are stationary, and animal-like in that they are heterotrophic. Because most fungi cannot move to trap their food, they are either parasites or saprophytes; that is, they absorb their food from the living or dead tissues of other organisms.

Structure of Fungi

Most true fungi are made up of very thin filaments called **hyphae**. In some fungi, each hypha is a tube-shaped mass of cytoplasm containing many nuclei. No crosswise cell walls divide the hypha into separate cells. In other fungi, cross walls divide the hyphae into distinct cells.

Hyphae grow and branch out to form a mat of filaments called a **mycelium.** The visible part of a fungus is only a small part of the mycelium. Most of the mycelium is embedded in the soil or in the tissues of the host organism and cannot be seen.

Most fungi reproduce by forming thick-walled spores both by mitosis and by meiosis. Thick cell walls of both hyphae and spores are adaptations that prevent drying. This adaptation permits fungi to live and reproduce on land. Most fungi produce asexual spores by mitosis in structures located at the tips of specialized hyphae. These hyphae extend outward from the food source and give a fungus its fuzzy appearance.

Fungi are classified according to the way they produce spores during sexual reproduction. Each group is named for the sexual structures it forms.

Zygote Fungi

The common bread mold, *Rhizopus,* is a member of the phylum known as zygote fungi, as shown in Figure 15-23. Zygote fungi are so named because specialized hyphae from two different organisms

Figure 15-22   **Hyphae form the basic structure of fungi. The reproductive part of the organism, known as a mushroom, is composed of densely packed hyphae. Spores are produced in the mushroom.**

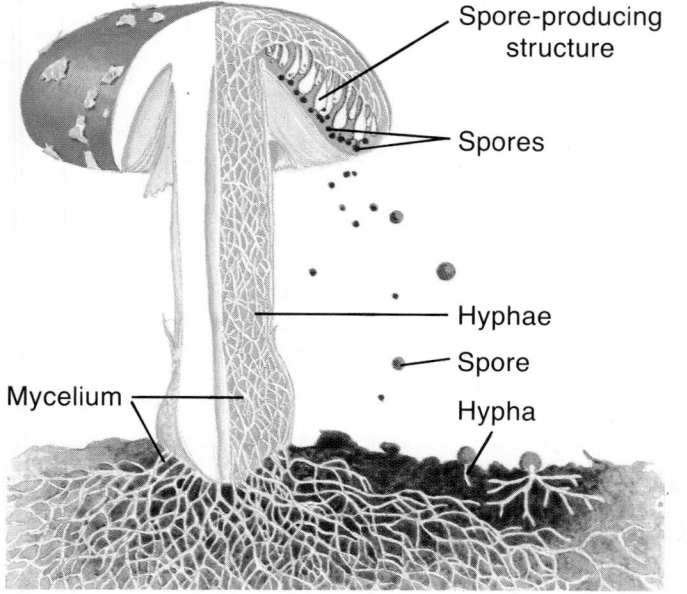

Spore-producing structure

Spores

Hyphae

Spore

Hypha

Mycelium

can meet and fuse to form a zygote. This zygote becomes a thick-walled spore, called a zygospore, that is resistant to harsh conditions. Other hyphae, called stolons, spread along the surface of the organism's food supply or substrate. Still other hyphae, **rhizoids,** anchor the fungus to the food source. Thus, rhizoids resemble roots in function. Enzymes secreted by the rhizoids break food molecules into simpler molecules, which are absorbed by the rhizoids and diffused throughout the mold.

Most relatives of bread mold are saprophytes. As a result of their activities, valuable materials are returned to the soil and atmosphere. However, some species are parasites and cause diseases by feeding on plants such as potatoes and cereal grains. These species produce rhizoids that penetrate the tissues of the host plant, thus robbing it of nutrients.

Club Fungi

This phylum includes mushrooms, shelf fungi, rusts, smuts, and puffballs. Members of this phylum are called club fungi because they possess club-shaped structures, called **basidia,** in which the sexual spores are formed.

When you eat a mushroom, you are actually consuming only the reproductive part of the organism. In the soil beneath the mushroom is a branching network of hyphae, the mycelium, that obtains nourishment in much the same way as bread mold. The cap portion of the mushroom begins as a small outgrowth of the mycelium in the soil. As the hyphae push through the soil, they develop a stalklike section and an umbrella-shaped cap that are composed of tightly packed masses of hyphae, as seen in Figure 15-22. On the underside of the cap, arranged like spokes of a wheel, are gills on which basidia are located. Each basidium produces four spores. When the spores mature, they are released from the basidium and dispersed by the wind.

Figure 15-23 *Rhizopus* **forms asexual spores by mitosis in black spore cases called sporangia, located at the tips of upright hyphae. These hyphae and sporangia give the fungus a fuzzy appearance.** *Rhizopus* **also produces sexual spores. By what process are sexual spores produced? Magnification: 5×**

Thinking Lab

DESIGN AN EXPERIMENT

Why would farmers put noodles on their crops?

Background
Why does William Connick make pasta with fungi? Connick is a researcher who was working on a project to develop a natural weed killer that would kill a weed called hemp sesbania. This weed can take over fields of cotton, soybeans, and rice. Connick knew that a particular fungus kills this weed. However, it would be difficult to spread the spores of the fungus on fields because spores are very

light in weight and would blow away. The fungus would have to be packaged, without using killing chemicals or high temperatures, into a stable material with a long shelf life. An effective number of fungal spores would have to be used, and they would need a food source to get them to start growing.

You Try It
Speculate about the characteristics of pasta that inspired Connick to make it

into what he called "pesta." **Design an experiment** that Connick might have done to determine how many fungal spores should be put into pesta to make it an effective weed killer. Think about at what stage in the growth of the crop the pesta should be applied for maximum effectiveness. What characteristics of pasta make it a suitable substance for packaging fungi for use as a weed killer?

Figure 15-24 The poisonous mushroom on the left and the edible mushroom on the right look similar. Wild mushrooms should never be eaten unless they have been identified by an expert.

While many mushrooms are edible, many more are not. Poisonous mushrooms are commonly called toadstools. Toadstools can cause illness, and some can cause death, if eaten. Many poisonous mushrooms closely resemble edible species, as shown in Figure 15-24. It takes a great deal of study and training to learn how to tell the difference between edible and poisonous species. An untrained person should never eat unidentified mushrooms found growing wild.

Whereas mushrooms are valuable to humans, many other club fungi are harmful. Mycelia of rusts, for example, invade the internal parts of wheat, oats, rye, and other plants, and can destroy entire crops. Geneticists have developed some strains of plants that are resistant to these parasites.

ART CONNECTION

Three Mushrooms by Roberto Juarez

Fungi come in many shapes and colors. Some are round or circular, resembling cups or plates. Bird's nest fungi look just like a nest with small white eggs inside. Fungi come in many colors, too: white, pink, yellow, orange, green, blue, even black. It is no wonder, then, that the shapes and colors of fungi have inspired artists in their work.

The Artist One such artist is Roberto Juarez. His father is Mexican American, his mother Puerto Rican. From the time he was small, he drew what he saw around him. It wasn't until college, however, that he realized he could become an artist. Since 1979, Juarez has exhibited his drawings and paintings in galleries across the United States and in Europe.

The Painting In several of his paintings, Juarez uses mushrooms as his main subject. In one, called *Three Mushrooms* (1986), the mushrooms float or fly above a sea of fungi. The mushrooms are dark, perhaps menacing. Their gills can be seen through the cap. Light seems to shine on the shapes behind, but not on the mushrooms. The background looks a little like a slide under a microscope. The colors are the colors of fungi. Juarez painted the picture on a piece of burlap, which gives the shapes a rough surface.

Many of Juarez's paintings have an ironic quality. There is often a background like the one in this painting that forms a pattern of shapes and colors. Hanging above this pattern is something that goes with it, and also does not. In *Three*

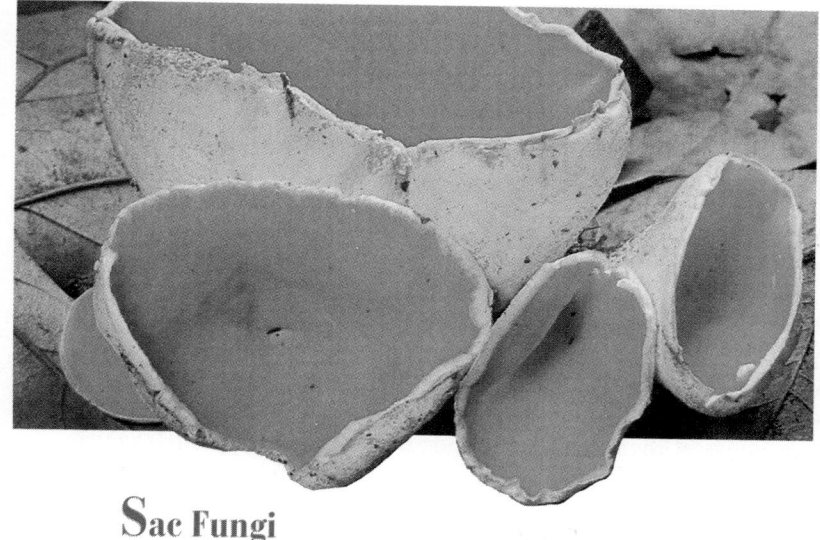

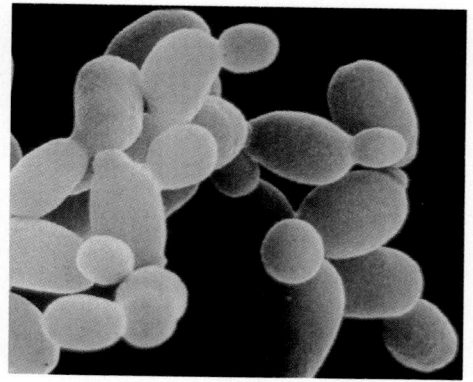

Figure 15-25 Cup fungi (left) and yeasts (right) are both sac fungi. Yeasts are unusual fungi in that they only have one cell. Magnification (right): 18 000×

Sac Fungi

Yeasts, cup fungi, and powdery mildews are all members of the phylum known as the sac fungi. Each of these organisms possesses a saclike structure called an **ascus**. Sexual spores are produced by meiosis in the ascus.

Yeasts are unusual fungi in that they are unicellular and lack hyphae. They are important in both the brewing and baking industries. They are also very important in scientific research, because they are simple eukaryotes and can be studied easily. Yeasts have proven to be of value in research on recombinant DNA and in the study of diseases such as cancer and AIDS. Cellular processes such as respiration also have been studied by using yeasts as research organisms.

Mushrooms, the mushrooms go with the background shapes because they are all circular. Yet, the mushrooms are separated from the background by their dark color. They seem to hang in the air with no connection to anything around them.

1. **Writing About Biology:** What do the shapes and colors in Juarez's picture mean to you?
2. **Connection to Agriculture:** Many kinds of mushrooms are grown for food. What are some of the different kinds of mushrooms you can eat?

Minilab 2

What do fungi need to grow?

Sprinkle a slice of bread lightly with water. Place it in a self-sealing plastic sandwich bag. Place a dry slice of bread in a second plastic bag. Seal both bags and put them in a warm, dark place for a week.

Analysis *What do you observe? What do fungi need for growth?*

Other types of fungi are also important in industry and medicine. One species of *Penicillium* mold is used in the production of penicillin. Other species of *Penicillium* are used in making cheeses such as Roquefort and Camembert. Enzymes produced by these molds give the cheeses their distinctive flavors.

Lichens

Most noticeable in desolate regions, but found throughout the world, are organisms called lichens. A lichen is a "dual organism" consisting of a fungus and a green organism, either a blue-green bacterium or a green alga. The mycelium of the fungus surrounds the cells of the green organism. Lichens grow on soil, rocks, and trees. Some grow flat and close to the surface, Figure 15-26, while others grow upward and may appear shrublike.

The Living Arrangement of Lichens   Lichens can live in barren places such as on bare rock or on Arctic ice. Neither the fungus nor the green organism could survive alone in such harsh environments. What kind of living arrangement do the two organisms have? The green organism is autotrophic; it provides food for itself and the fungus by means of photosynthesis. Exactly how the fungus aids the green organism is not clear.

Table 15.1   **Summary of Major Characteristics of Monerans, Protists, and Fungi**

Kingdom	*Common Name*	*Structure*	*Nutrition*	*Importance*
Monera	true bacteria	prokaryotic mainly unicellular	heterotrophic; autotrophic (photosynthetic, chemosynthetic)	producers; disease; industry; decomposers
	ancient bacteria		chemosynthetic	producers
Protista	euglenoids golden algae	unicellular	autotrophic (photosynthetic)	producers in aquatic habitats; industry
	green algae brown algae red algae	mostly multicellular		
	rhizopods ciliates flagellates sporozoans	unicellular	heterotrophic	consumers; parasites; disease
	slime molds	unicellular and multicellular stages	heterotrophic	decomposers
Fungi	zygote fungi club fungi sac fungi	multicellular; hyphae	heterotrophic	decomposers; disease; medicine

Perhaps the fungus provides it with protection and moisture, because the lichen can survive drying, whereas the green organism by itself cannot. The fungus also may be a source of inorganic materials. A lichen is an example of a living arrangement based on mutualism.

The Role of Lichens   Lichens sometimes serve as food for animals. An example is reindeer moss, an Arctic species that is eaten by reindeer and caribou. Lichens serve yet another important ecological function. They can actually help create soil in rocky or barren regions. The fungus portion of the lichen secretes acids that begin the breakdown of rock into particles of soil. When lichens die, they decompose, thus enriching the soil they helped produce. Eventually, the soil becomes rich enough to support the growth of more complex, true plants.

Summary   Many of the phyla contained in the Kingdoms Monera, Protista, and Fungi have been described in this chapter, but there are many more. It would take several chapters to describe them all. Table 15.1 summarizes the major characteristics of the phyla that have been included here. The common names of the phyla have been used rather than the technical names. You can see how diverse these three kingdoms are. For more information on these groups, see Appendix A.

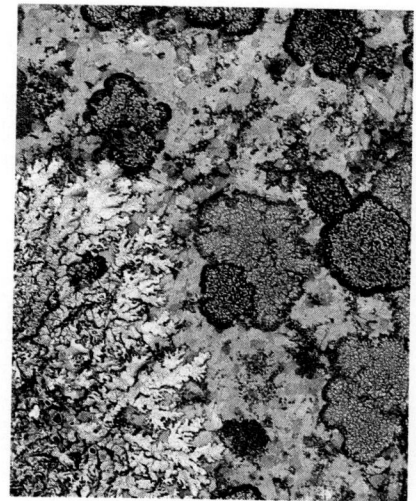

Figure 15-26   Some lichens are leaflike, others are shrublike, and still others, such as the lichens shown here, form crustlike growths on rocks.

LOOKING AHEAD

Y OU HAVE explored important features of the viruses and of the organisms that make up the Kingdoms Monera, Protista, and Fungi. Certain of these simple organisms are thought to be the ancestors of more complex and familiar living things. In the next chapter, you will study the major groups of another kingdom, the plants. You will learn about the evolution of adaptations that have made different groups of plants suited to life in particular habitats.

Section Review

Understanding Concepts
1. Explain why fungi must be either parasites or saprophytes.
2. Why do you think fungi were previously classified as plants?
3. How can you explain the fact that mushrooms just seem to pop up out of the ground and do so in about the same place each year?
4. **Skill Review—Forming a Hypothesis:** Why can't the separate organisms of which a lichen is composed live in places where the lichen can live? For more help, refer to Practicing Scientific Methods in the **Skill Handbook.**
5. **Thinking Critically:** How could you determine whether a newly discovered fungus is a relative of *Rhizopus* or of a mushroom?

BIOLOGY,
TECHNOLOGY, AND
SOCIETY

Considering the Environment

Fighting Pollution with Bacteria

You may have thought all bacteria were harmful. Think again. Some bacteria are working to clean up the damage we have done to our environment.

Oil Spills In 1989, the ship *Exxon Valdez* was carrying crude oil south from the Alaska pipeline. When the ship hit ground, a hole was ripped

in its hull and millions of gallons of oil spread along the Alaskan coast. In some places, the oil soaked two feet deep into the beaches. There seemed no way to clean up the spill. Then scientists decided to enlist the help of bacteria that live on the beaches.

Scientists knew that some bacteria naturally break down hydrocarbons, the main type of molecule found in oil, and that these bacteria lived on the Alaskan beaches. The problem was that there were not enough of these bacteria to handle the huge load of oil. To make the bacteria multiply faster, the scientists sprayed a fertilizer along 70 miles of coastline. Within 15 days, the number of bacteria had tripled. The beaches that had been treated were much cleaner than those that had not. The bacteria continued to work for five months. Without this bacterial "scouring," Alaska's beaches might still be covered with oil.

What makes bacteria such good cleaning agents? They break down toxic chemicals, such as the hydrocarbons in oil, into simple harmless substances like carbon dioxide and water. The bacteria that do this dirty cleanup work are found in soil and water—anywhere we have left our pollution. With the right fertilizer, bacteria grow quickly. And when

they have done their work, they die.

Bioremediation This process of using living things to get rid of toxic materials is called bioremediation. It is being used to clean up gasoline that leaks into the soil under gas stations. At plants that process wood pulp, scientists are using microorganisms that break down phenols, a poisonous by-product of the process. The bacteria reproduce quickly and convert the phenols into harmless salts. Recently, scientists have found bacteria that break down polychlorinated biphenyls (PCBs), the deadly toxins that can get into our water supplies as waste products of certain industries. Other scientists are experimenting with bacteria that break down explosives, such as TNT, and with bacteria that can clean up the acid drainage that seeps out of abandoned coal mines.

Given that Americans make more than 600 million tons of toxic waste a year, bioremediation may soon become a big business. Microbiologists are studying ways to speed up the work of bacteria. The most common method of bioremediation is to increase the growth rate of bacteria found naturally at the pollution site, as the scientists did in Alaska. If the right bacteria are not present, scientists can grow them in a laboratory and then add them to the soil.

Other Uses of Bioremediation The soil is not the only place bioremediation is being tried. Bacteria are used in sewage treatment plants to clean water and are even going to work against acid rain. When coal containing sulfur burns, the sulfur is released into the air, where it forms acid rain. Bacteria that "eat" sulfur are being used to treat the coal before it is burned and change it into a cleaner-burning fuel.

If scientists can identify micro-organisms that attack all the kinds of waste we produce, expensive treatment plants and dangerous toxic dumps might be put out of business.

1. **Solve the Problem:** What new problems might result from using bacteria to fight pollution?
2. **Explore Further:** Scientists are also experimenting with common plants, such as ragweed, to take lead and other metals out of soil. This is called green remediation. How does green remediation work and where is it being tried?

Medicine Connection

A Lucky Accident

In 1928, Alexander Fleming, an English bacteriologist, was studying the bacteria that cause infections on the skin and in the throat. Fleming put some bacterial cells in a petri dish and put the dish aside. When he examined the bacterial growth, he saw that a fungus had appeared as a contaminant in the culture, and the bacteria surrounding the fungus had died. Where there was no fungus, the bacteria were growing. Fleming studied the fungus and discovered that it produced a substance that was toxic to bacteria. Because the fungus belonged to the genus *Penicillium*, Fleming named the substance penicillin. Growth of the fungus was an accident, but Fleming's curiosity and careful investigation took advantage of that accident.

Penicillin as a Medicine Ten years passed before anyone thought of using penicillin for medicine. Howard Florey, another English scientist, knew that penicillin killed bacteria without harming humans or animals. At the time, England was at war with Germany and many soldiers were dying from gangrene

infections in wounds. Florey thought that penicillin might kill the bacteria that cause gangrene. Because England's factories were busy producing weapons and supplies for the war, they were not able to mass-produce penicillin. In 1941, Florey came to the United States, where he convinced American drug companies to make penicillin for use on the battlefield.

Production Begins Production of penicillin was one of the fastest successes in medical history. Most of the drug companies in the United States worked together to produce and test the new antibiotic. By 1943, they were able to produce penicillin in large enough quantities to help the English and American soldiers.

The First Uses Florey went to North Africa to oversee the first uses of penicillin on the battlefield. Penicillin was combined with sulfa drugs and put directly in the wound. Then the wound was closed. The penicillin and sulfa drugs worked together to kill the bacteria, and the wound healed. Because of Fleming's accidental discovery, thousands of soldiers' lives were saved during World War II.

1. **Explore Further:** What diseases are treated with penicillin today?
2. **Solve the Problem:** How does penicillin kill bacteria? Why isn't it harmful to human cells?

Summary

Viruses are particles of protein and nucleic acid that resemble living things only in their ability to reproduce. To do so, however, they must invade and take over the materials and energy sources of specific host cells. As the viruses reproduce, the host cells are sometimes destroyed.

Monerans are mostly unicellular, prokaryotic organisms known as bacteria. Included in this kingdom are the older, ancient bacteria and the true bacteria. A major reason for the success of bacteria is the diversity of ways they have evolved for obtaining energy.

The most diverse kingdom of organisms is the protists. Included in this eukaryotic kingdom are algae, protozoans, and funguslike species. These groups may have given rise to plants, animals, and fungi, respectively.

Kingdom Fungi includes multicellular heterotrophs that absorb nutrients from living organisms, from the wastes produced by other organisms, or from dead organisms. While many parasitic fungi cause harm, other fungi are very important in the recycling of nutrients. Many fungi are useful to humans.

Language of Biology

Write a sentence that shows your understanding of each of the following terms.

ascus	parasite
asexual reproduction	plasmodium
basidium	provirus
endospore	pseudopodium
hypha	retrovirus
lysogenic cycle	rhizoid
lytic cycle	saprophyte
mutualism	
mycelium	

Understanding Concepts

1. Compare how a retrovirus and the other type of RNA virus invade and replicate within their host cells.
2. How can the symptoms of viral diseases such as the flu be related to the reproductive cycles of the viruses that cause the diseases?
3. Which do you think might have evolved first, chemosynthesis or photosynthesis? Explain your answer.
4. In some classification systems used today, green, brown, and red algae are classified as plants rather than algae. Can you suggest a reason for that classification?
5. How are bacterial endospores different from fungus spores?
6. How do characteristics of euglenoids illustrate some of the problems of classifying organisms?
7. Some green algae are colonial forms that are made up of many individual cells connected together. The cells have developed specialized functions, but still remain independent. Explain how colonial algae may represent a step in the evolution of multicellular algae.
8. In what ways do *Euglena* and *Paramecium* show that division of labor is important in unicellular organisms?
9. How do slime molds resemble other protists as well as fungi?
10. Why do you think fungi are now classified in a kingdom of their own?

Applying Concepts

11. Mycoplasmas are bacteria that lack cell walls. How might that feature be related to the fact that all mycoplasmas are parasites within other organisms?

12. How might you develop drugs that prevent the reproduction of a retrovirus, using knowledge of its reproductive cycle?

13. Some species of euglenoids are always heterotrophic. Do you think such species have a light-sensing eyespot? Explain.

14. How do you think protozoans were classified in a two-kingdom system? Explain.

15. How could you easily distinguish between a unicellular alga and a blue-green bacterium?

16. Fungi may be harmful to crops. In what way are they helpful?

17. Is a virus a parasite? Explain.

18. Can penicillin be used to treat a viral infection? Why or why not?

19. Sometimes, during a lytic cycle, a small part of a host bacterium's DNA may be enclosed in a new phage. If this phage then attacks another bacterium, it will carry a portion of the first host's DNA into the second host. How is this process a possible source of variation among bacteria?

20. **Considering the Environment:** Why might it be better to clean up a site by using bacteria that naturally grow there rather than by bringing new bacteria to the site?

21. **Medicine Connection:** Today, the United States government must approve the use of all drugs. Sometimes it takes years before a drug is proved to be effective and safe for use. Is this time lag justified, or should new drugs be used, as penicillin was, as soon as they are developed?

Lab Interpretation

22. **Investigation:** What would the following graph indicate about a new sunscreen being tested for safety?

Protozoan numbers in 0.01% sunscreen vs. control

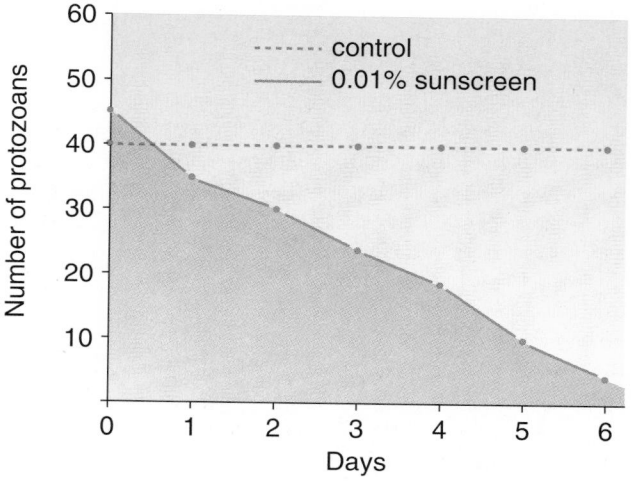

23. **Thinking Lab:** How can a fungus be used as an effective weed killer?

24. **Thinking Lab:** Why might using a fungus to control weeds be more desirable than using chemical controls?

25. **Minilab 2:** Why did you place the plastic bags containing slices of bread in a warm, dark place? Would you have gotten similar results if you had placed the bags in the refrigerator? Why or why not?

Connecting Ideas

26. Thinking back to conditions that were believed to have existed on primitive Earth, why is it reasonable to assume that ancient bacteria may have been the first organisms to evolve?

27. Lichens are sometimes called pioneer organisms because they are among the first organisms to move into a barren area. Why are lichens well suited for this role?

This British soldier lichen forms shrublike growths. Each "soldier" is 1 to 2 cm tall.

16

PLANT ADAPTATIONS

MOSSES LIVE IN damp, shady places. Individual moss plants are very small and easy to overlook, but they often grow together in dense mats covering soil, bark, or gaps between stones. Mosses are probably very similar to some of the first plants that appeared on Earth, and may have evolved from green algae that lived in ancient swamps, lakes, and oceans. Mosses, unlike their ancestors, can live on land. But they cannot survive or reproduce unless they are surrounded by at least a thin film of moisture.

Many other plants, including trees, are also thought to have descended over millions of years from ancient green algae. However, trees have evolved a number of adaptations that enable them to survive under conditions unsuitable for mosses. For example, some trees can grow and reproduce in hot, dry climates. Trees also have adaptations that enable them to grow much larger than any moss. What are these adaptations, and how did they arise? In this chapter, you will learn about the differences between mosses, ferns, coniferous trees, and flowering plants. You will find out that the plant kingdom contains a huge variety of organisms adapted for life in all kinds of environments, from sandy deserts and coastlines to snowy mountains, fertile valleys, vast flatlands, and even lakes.

Imagine you are planting a garden. What kinds of plants would grow best in the shade? Where would you plant a magnolia tree? What survival needs do the tree and shade plants have in common? How do their needs differ?

Chapter Preview

Objectives

Determine the adaptations to conditions on land found in plants.

Analyze how the traits of mosses and liverworts limit their ability to live in land habitats.

16.1 From Water to Land

Where do plants live? Have you seen lily pads floating on a pond, vines clinging to the side of a building, trees towering hundreds of feet from the forest floor, or ferns growing along a stream bank? These are just a few examples of the many environments plants can occupy. Plants are an important part of any community of living organisms. They are the producers that use the energy of sunlight to turn water and soil nutrients into food needed by themselves and their consumers. They provide shelter for caterpillars, owls, pythons, koalas, and a multitude of other organisms. Plants are also indispensable to human economy. We build homes from wood, weave cloth from cotton or other plant fibers, and make medicines from a variety of roots, leaves, and flowers. Plants were used to make the paper this book is printed on, and perhaps the desk or chair you're sitting on, and probably some of the clothes you are wearing.

Origin of Plants

All the plants on Earth today are probably descendants of green algae that dwelt in ancient oceans. This conclusion is based on three pieces of evidence. First, both green algae and plants have cell walls containing cellulose. Second, their cells contain the same kinds of chlorophyll and carotenoid pigments. Third, the product of photosynthesis in both green algae and plants is starch, which is stored in plastids in the cytoplasm.

Figure 16-1 The storage of sugars in plants and green algae takes the form of polysaccharides that make up starch, whereas storage of sugars in bacteria, fungi, and animals takes the form of glycogen. The scanning electron micrograph (left) shows starch grains in a potato cell. (Magnification: 100×) The transmission electron micrograph on the right shows the starch grain in *Chlamydamonas*, a unicellular green alga. (Magnification: 6000×)

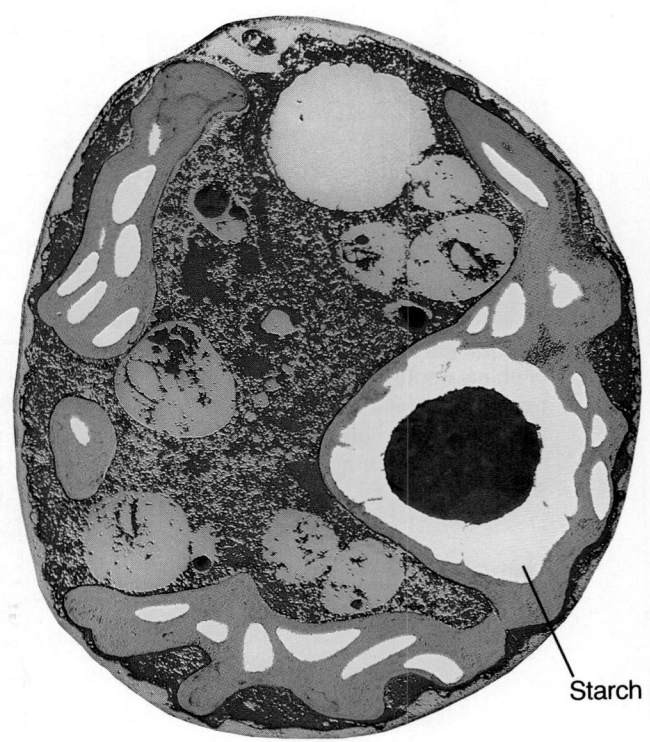

Starch

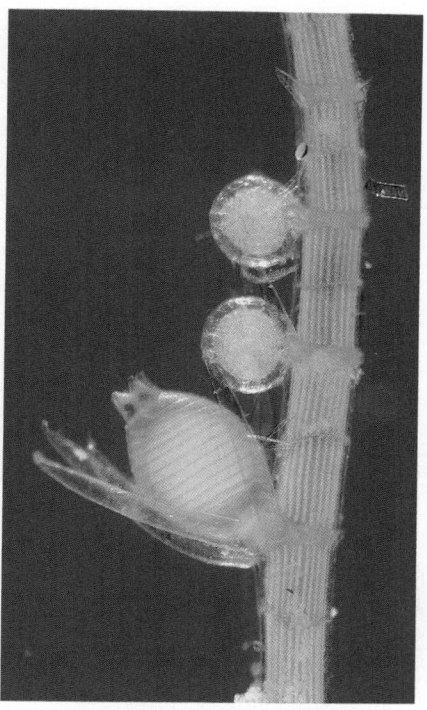

Life in the Water   As you read in Chapter 15, green algae are protists that live in water or very moist places. Water generally provides a very stable environment; temperatures remain fairly even, dissolved nutrients are plentiful and easily absorbed into cells, and, of course, there is never a shortage of water. As a result of this stability, green algae, which first appeared on Earth about a billion years ago, have survived as fairly simple organisms. They don't need to transport water or nutrients from one part of the organism to another because algal cells can obtain these materials directly from the surrounding water or from other cells that are in direct contact with the water. Similarly, each algal cell is either capable of photosynthesis or located near a cell that can carry out photosynthesis. Gametes or developing embryos don't need to be protected from drying out. They are simply released into the water, where fertilization and development occur. Some green algae have soft tissues that allow these protists to float, but none that provide structural support. Their support is the watery environment. They are perfectly adapted to their environment.

Life on Land   Life on land is more demanding than life in the water. On land, there are wide variations in temperature and in the amount of water and nutrients available. What are some of the problems that plants overcame as they became adapted to life on land?

(1) They needed protection from drying out.
(2) They needed to exchange gases with the surrounding air.
(3) They needed a transport system to provide all the cells in a
    multicellular body with nutrients.
(4) They needed support to grow upright on land.

Figure 16-2   The cells of a green alga, such as the filamentous *Chara*, can obtain nutrients directly from their watery environment (left). The gametes of *Chara* need no protection from drying out (right). Magnification: 20×

Through natural selection, plants have evolved a variety of adaptations for coping with the harshness of land conditions. These adaptations involve the evolution of different tissues that perform tasks for survival on land. A variety of plant adaptations have evolved that protect the plant body, gametes, and embryos from drying out during fertilization and development. Some tissues absorb water and dissolved nutrients from the soil, and exchange gases with the air. Some are adapted for photosynthesis, and others distribute water, nutrients, and the food produced by photosynthesis to all parts of the plant. Cells with rigid cell walls provide structural support, which enables the plant to grow upright.

I N V E S T I G A T I O N

Leaf-Cuticle Models

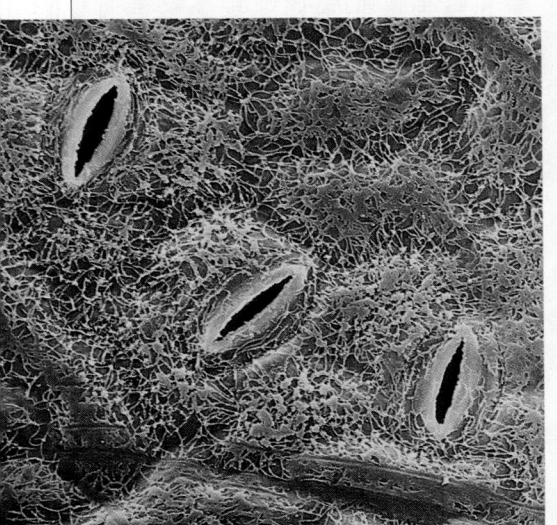

A waxy cuticle, magnification: 286×

For water-dwelling plants to evolve into land dwellers, many problems had to be overcome. For example, how might land plants prevent their drying out when once out of a water environment? A waterproof covering over their leaf surfaces might have been an adaptation that would certainly help. Land plants do have such an adaptation, called a cuticle. A cuticle forms a thin surface coating on most land-dwelling plants. Its waterproof quality is provided by a waxy secretion of cutin from the outer cells of a plant. The cuticle acts as a barrier to water evaporation, much as a covering of plastic wrap keeps your food from drying out when placed in the refrigerator.

Problem

How can a model be built to show that a plant's cuticle reduces water evaporation?

Experimental Plan

1. As a group, make a list of possible ways that you might design a model to test the idea that a cuticle does reduce evaporation. Keep in mind the materials that your teacher has made available.
2. Agree on one model from your group's list that could be built in the classroom.
3. Refine your model to:
   (a) Construct a control model without a cuticle and an experimental model with a cuticle.
   (b) Make sure that your models and the actual conducting of the experiment will test only one variable.
   (c) Plan to collect quantitative data from your models for comparison.
   (d) Design a table that will compare the data of the control model with the experimental model.
4. Following the style of a recipe, write a numbered list of directions that anyone could follow in building and using your models. Include diagrams of the two models that you plan to build.
5. Make a list of materials and the quantities you will need.

Preventing Water Loss   Water that moves into a plant cell by osmosis will also move out of the cell. It is important that land plants have mechanisms for retaining water in their cells. Most plants have a waxy layer, called the **cuticle,** that covers the outer surface of the plant body. The waxy substance is secreted by the epidermal cells. The cuticle helps prevent water loss by forming a watertight seal to the outer cell walls.

Exchanging Gases with the Air   All plants take in carbon dioxide and release oxygen during photosynthesis. While the waxy cuticle covers most of a plant's surface, there are small openings, or pores, through which these gases can pass. Before carbon dioxide, or any other gas, can diffuse

---

Checking the Plan

Discuss the following points with the other group members to decide the final procedures for building and using your models.

1. Which will be your control model? How will it differ from the experimental model?
2. What steps in the design and procedure must be taken to make sure that only one variable is being used?
3. How will you measure the evaporation of your model?
4. How many trials will you carry out?
5. Have you designed and made a table for collecting data?
6. What specific role will each member of your group perform?
7. Make any needed changes in your list of directions or model diagrams.

Make sure your teacher has approved your models and experimental plan before you proceed further.

Data and Observations

1. Build your models.
2. Use your models to collect data and complete your data table.
3. Make a graph of your data or present your data using some other visual format.

Analysis and Conclusions

1. What model parts correspond to actual plant parts?
2. Explain how the function of model parts did or did not differ from the functions of their corresponding actual plant parts.
3. (a) What were you measuring in the models?
   (b) How does this compare to what actually occurs in a plant?
4. (a) Describe and explain how data from your control model and experimental model differed.
   (b) What were some of the variables that had to be held constant during the experiment? Explain.
5. Does your model data help support the conclusion that a cuticle reduces water evaporation from a plant? Explain how.
6. Use your group's data and models while consulting with other groups to answer the following:
   (a) How did other group models differ from yours?
   (b) How did other group data compare with yours?

   (c) What conclusions might be made if other group models and data show similar or different trends from yours?
7. Write a brief conclusion about the value of models in science.

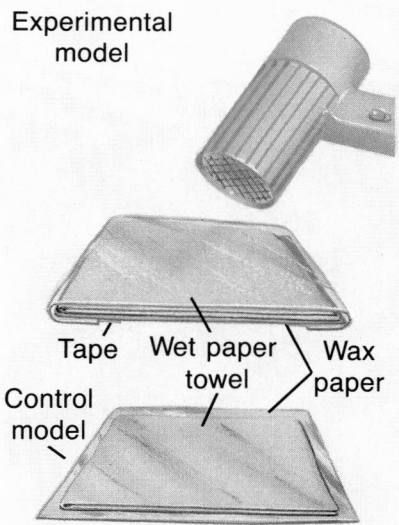

Experimental model

Tape   Wet paper towel   Wax paper

Control model

Going Further

Based on this lab experience, design another experiment or model that would help to compare a different plant adaptation to life on land. Assume that you will have unlimited resources and will not be building the model in class.

through a cell membrane and into a cell, it must be dissolved in water. This necessity poses no problem for algae, since they are not coated with a cuticle and they take in carbon dioxide that is already dissolved in the water surrounding them. But plants must obtain carbon dioxide from the air. All plants have evolved in such a way that their internal cells are surrounded by a film of water. As a gas passes through this film, it is dissolved in the water and can then be absorbed by the cell.

Transport of Nutrients   Think for a moment about the difference between a cell belonging to a thin filament of a green alga floating in a pond and a cell located deep inside the stem of a palm tree (Figure 16-3). The algal cell absorbs water, nutrients, and carbon dioxide directly from the water. It contains chloroplasts and produces its own food. But what about the tree cell? It contains no chloroplasts, nor is it exposed to sunlight, so it cannot produce its own food. It is nowhere near the soil, so it can't absorb water and nutrients. To survive, this cell needs a way to get food, water, and other nutrients from the leaves and roots of the tree.

Tissues that transport materials from one part of a plant to another are called **vascular tissues.** Plants that possess vascular tissues are called vascular plants, and those that do not are called nonvascular plants. These are the two major groups that make up the plant kingdom. Vascular plants are far more numerous than nonvascular plants. The trees, shrubs, flowers, and ferns you are familiar with are vascular plants. Nonvascular plants are much less well known.

Nonvascular Plants

In the plant kingdom, the term *division* is usually used in the classification of major groups. You will remember that in the animal kingdom, the major groups are phyla. There are several divisions of vascular plants, but only one division of nonvascular plants. Most nonvascular plants can be grouped into one of two classes—the mosses or the liverworts.

Figure 16-3   **Compare the cellular organization in a palm tree and a filamentous green alga. Why are vascular tissues essential for one organism but not the other?**

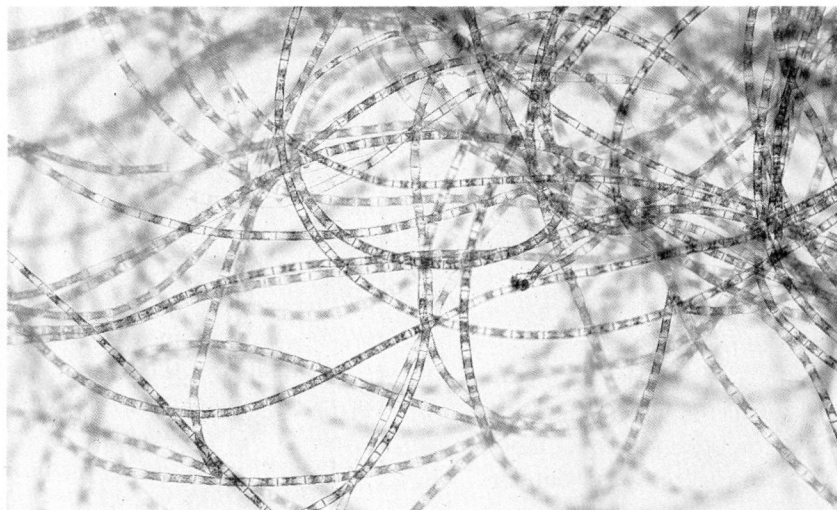

PLANT ADAPTATIONS

Identifying Nonvascular Plants Mosses and liverworts are quite similar. They contain chlorophyll and, therefore, are green in color. Individual plants are almost always very small, often no more than two or three inches high, and are usually found growing in clumps or dense carpets. They grow on soil, rocks, or rotting wood, and a few are adapted to life in water. Mosses are feathery or velvety in texture, and have thin, leaflike structures attached to a tiny upright or creeping stalk, as shown in Figure 16-4. Some liverworts are creeping plants that look like mosses, but most liverworts are flat and leathery in appearance, also shown in Figure 16-4. The flat, liverlike body of a liverwort is called a **thallus.**

Most species of mosses and liverworts inhabit shady, damp areas. Moisture retained by a covering of mosses or liverworts helps promote the decay of rotting wood or other organic matter. In this way, mosses and liverworts help build soil. They also help stabilize soil and prevent erosion by wind and water.

Figure 16-4 Mosses (top) and liverworts (bottom) are land plants that have a low resistance to drying out. They are most often found in the damp regions of the world such as the bogs of temperate regions and rain forests of the tropics.

Thinking Lab

INTERPRET THE DATA

What adaptations were necessary for plants to move onto land?

Background
Most biologists would agree with the theory that the green algae are the ancestors of plants. Most present-day green algae spend all of their life cycle in water. The following traits show how they are well adapted to this environment.

(a) Water helps to support the algal plant body.
(b) All algal cells have access to carbon dioxide and water.
(c) Algal cells cannot dry out.

(d) Most photosynthesis occurs along the top surface of a multicellular alga because only these cells receive direct light.

You Try It
Describe the traits that evolved in land plants which helped overcome the problem of moving to life on land. Use the features of algae described above for comparison. **Explain** how each trait may have led to adaptations in land plants.

Transporting Water and Nutrients   Mosses and liverworts do not have tissues that make up the roots, stems, or leaves like those of vascular plants. The leafy structures of mosses and leafy liverworts are generally made up of several layers of tissues only one or two cells thick. The thallus of a liverwort is, likewise, just a few cells thick. Instead of roots, mosses and liverworts have rootlike structures called rhizoids that anchor them to the soil, wood, or water where they live. Like roots, the rhizoids absorb water by osmosis and also take in dissolved nutrients.

Once water and nutrients have been absorbed into the cells of the rhizoids, these molecules move by osmosis and diffusion directly into the cells of the moss stalks or cells of the liverwort body. Sugars produced during photosynthesis in the leafy structures of the moss or thallus of the liverwort also move through the plant by diffusion. This type of transport does not move molecules very far or very quickly. However, transport by diffusion is adequate for nonvascular plants because they are small and the distances molecules must be transported are not great. For the same reason, nonvascular plants can grow only very close to or directly upon a source of nutrients, such as soil on the ground or on the branches of a tree in a wet, tropical forest.

Like all plants, mosses and liverworts have pores through which air enters the plant body. The cells lining these pores are covered with a thin film of water. Carbon dioxide from the air dissolves in the water, then diffuses through the cell membranes. Oxygen is released from the plant by the same process operating in reverse.

Figure 16-5   Sections through the leaflike structure of a moss (left) and the thallus of a liverwort (right) reveal the simple organization of cells and lack of vascular tissues for transport of water and nutrients.

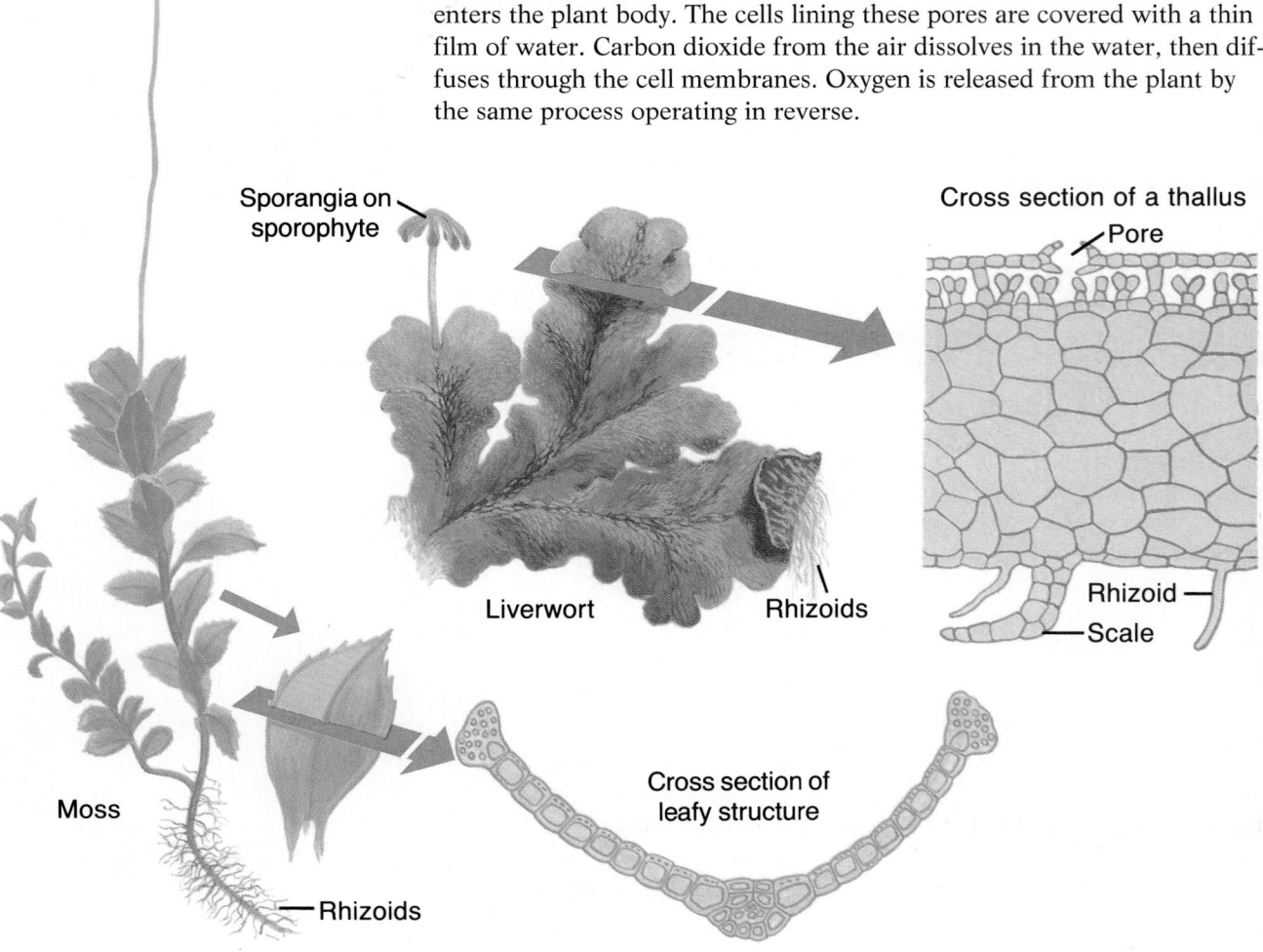

Spore capsule

Sporangia on sporophyte

Liverwort

Rhizoids

Cross section of a thallus

Pore

Rhizoid

Scale

Moss

Rhizoids

Cross section of leafy structure

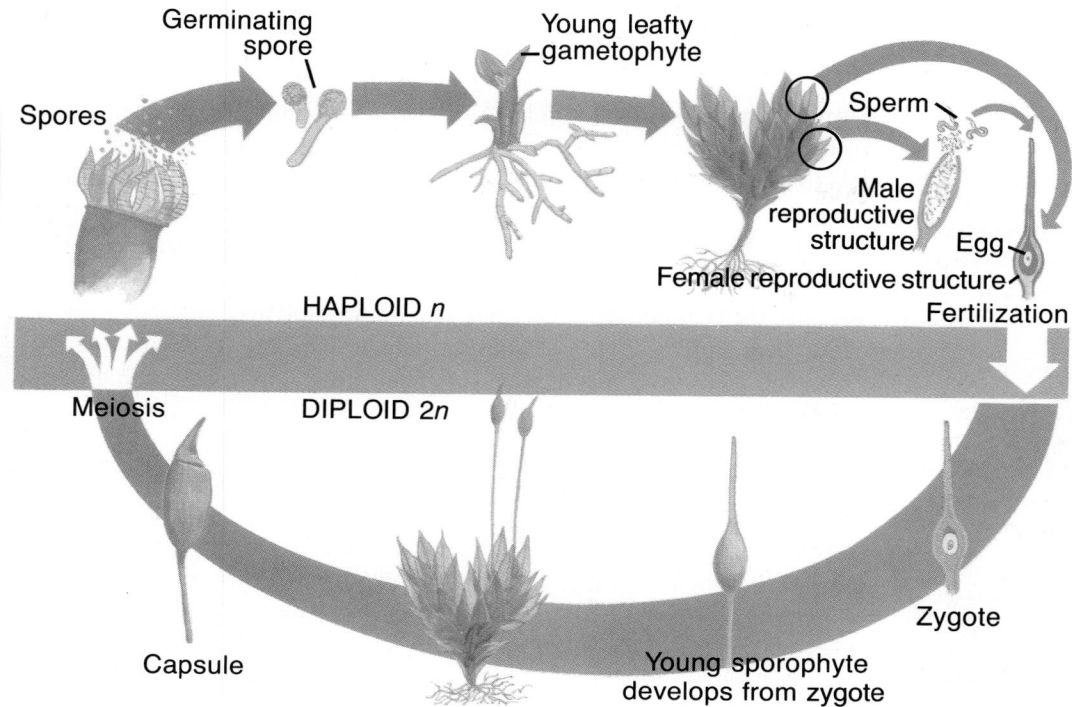

Spores

Germinating spore

Young leafty gametophyte

Sperm

Male reproductive structure

Egg

Female reproductive structure

Fertilization

HAPLOID *n*

Meiosis

DIPLOID 2*n*

Capsule

Young sporophyte develops from zygote

Zygote

The Plant Life Cycle

The lives of all plants including liverworts and mosses consist of two alternating stages, or generations. The **gametophyte generation** in plants is the stage responsible for the development of gametes. All the cells of the gametophyte are haploid (*n*). The second stage is called the **sporophyte generation,** which is responsible for the production of spores. The sporophyte generation is made up of diploid (2*n*) cells. A typical plant life cycle that is made up of this **alternation of generations** is shown in Figure 16-6.

In a moss, the gametophyte generation makes the green, leafy carpets with which you are probably most familiar. The sporophyte generation of a moss makes up the stalks and the capsules where spores are produced by meiosis. Plants of the moss gametophyte generation are haploid and have male and female reproductive organs. Sperm produced in the male organ of a moss swim, by means of flagella, through surface rain or dew to a female organ, which contains the egg. The moisture required by mosses and liverworts for sexual reproduction explains why these nonvascular plants still live in wet habitats. The female reproductive structure provides a moist, protected environment in which fertilization can take place and the embryo can develop.

The embryo develops into a diploid stalk, which is the beginning of the sporophyte generation. A capsule forms at the tip of the stalk and cells inside undergo meiosis to form spores. Spores are adapted to life on land. They have walls that are tough and resist evaporation. When moss spores settle on moist land, they may develop into new haploid plants, thus continuing the plant life cycle.

Figure 16-6 **The typical alternation of gametophyte and sporophyte generations in a plant is illustrated here by a moss. What are the products of meiosis in a plant?**

Origin of Nonvascular Plants

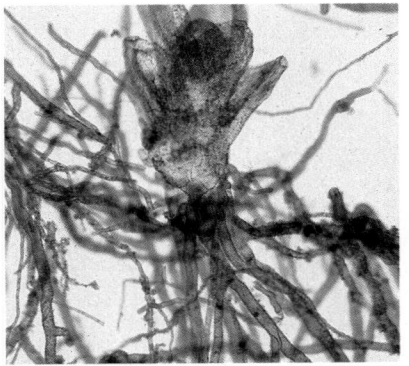

Figure 16-7 **The young gameto-phyte of a moss (top) shows close simi-larity to a filamentous green alga. A few mosses, such as *Funaria* (left), show traces of vascular tissues in their leaves and stems (right). How are these char-acteristics useful as indicators of relationships?**

Although adapted for life on land, nonvascular plants are not as common or widespread as vascular plants. They need a more moist environment in which to carry on reproduction, photosynthesis, and other life functions. Because adequate water is not available everywhere, nonvascular plants are limited to moist habitats. Their size is also limited because they lack vascular tissues or thick-walled cells for transport or structural support. Because they cannot grow very tall, neighboring plants can easily over-grow them, cutting them off from sunlight and gases.

As you can see, nonvascular plants have more similarities with algae than with vascular plants. For this reason, many biologists consider non-vascular plants the most ancient plants alive today. The close resemblance of the newly developing gametophyte generation to filamentous green algae helps support the theory that nonvascular plants evolved originally from ancient green algae. But there is also another theory that could explain the origin of nonvascular plants. Biologists have observed that some nonvascular plants have traces of vascular tissue. The presence of this tissue could indicate that nonvascular plants are just another relative of vascular plants.

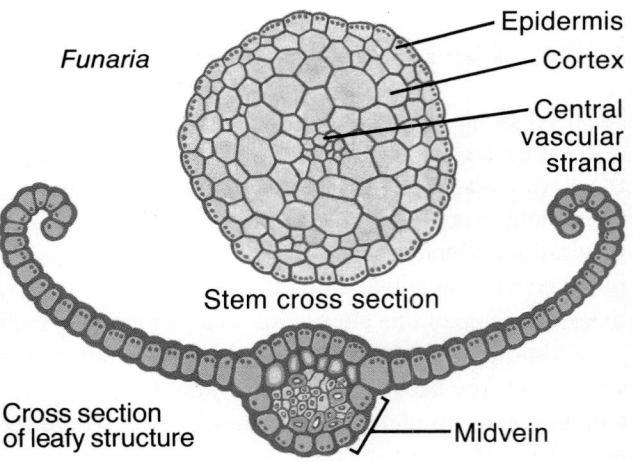

Funaria

Epidermis
Cortex
Central vascular strand

Stem cross section

Cross section of leafy structure

Midvein

Section Review

Understanding Concepts

1. List the reasons why land is a more demanding environment than water.
2. You collect several plants growing near the edge of a pond. How can you tell which are mosses?

3. How are rhizoids similar to and dif-ferent from roots?
4. **Skill Review—Making Tables:** Construct a table listing how mosses are adapted for life on land. For more help, refer to

Organizing Information in the **Skill Handbook**.

5. **Thinking Critically:** How are tropical or temperate rain forests per-fect habitats for mosses?

16.2 Vascular Plants

According to the fossil record, the earliest vascular plants appeared on Earth about 400 million years ago. Since then, land conditions have changed, new habitats have appeared, and plants adapted for survival in those new habitats have gradually evolved. As a result, the number and diversity of plant species has increased vastly. One of the adaptations that enabled plants to survive in new environments was the evolution of vascular tissues. Vascular plants have roots, stems, and leaves that are supplied by vascular tissues.

Possession of vascular tissue is the major difference between vascular and nonvascular plants. But, there is also another important difference—in vascular plants, the diploid, sporophyte generation is larger and more conspicuous than the haploid, gametophyte generation. As vascular plants have evolved, the gametophyte generation has become so much smaller than the sporophyte that most vascular plants no longer have an independent gametophyte generation. The gametophyte remains only as a few cells dependent on the body of the sporophyte. The sporophyte is the stage that contains vascular tissues.

Simple Vascular Plants

Imagine traveling back almost 400 million years in time to observe Earth's ancient vascular plants. As you look around the damp, swampy forest, you see huge flying insects, but no birds, no mammals—not even any dinosaurs. These animals don't appear until much later in Earth's history. You also notice that there are no flowers—just leafy plants and many extremely tall but unfamiliar-looking trees (Figure 16-9). These plants, after they die, will gradually be transformed into the vast deposits of coal that will provide humans with fuel hundreds of millions of years later.

Club Mosses There are plants today that strongly resemble fossils of vascular plants that formed ancient forests. Among these plants are the club mosses, which first appeared about 390 million years ago. Some grew as tall as 100 feet and had rootlike stems at the base of the trunk. Most species died out about 280 million years ago, when Earth's climate became drier and cooler.

Describe the characteristics of vascular plants.

Contrast the adaptations of ferns and mosses for life on land.

Evaluate why seed plants have become the dominant plants on Earth.

Figure 16-8   The gametophyte and sporophyte plant structures of a simple vascular plant such as a club moss are distinct and separated by the processes of fertilization and meiosis.

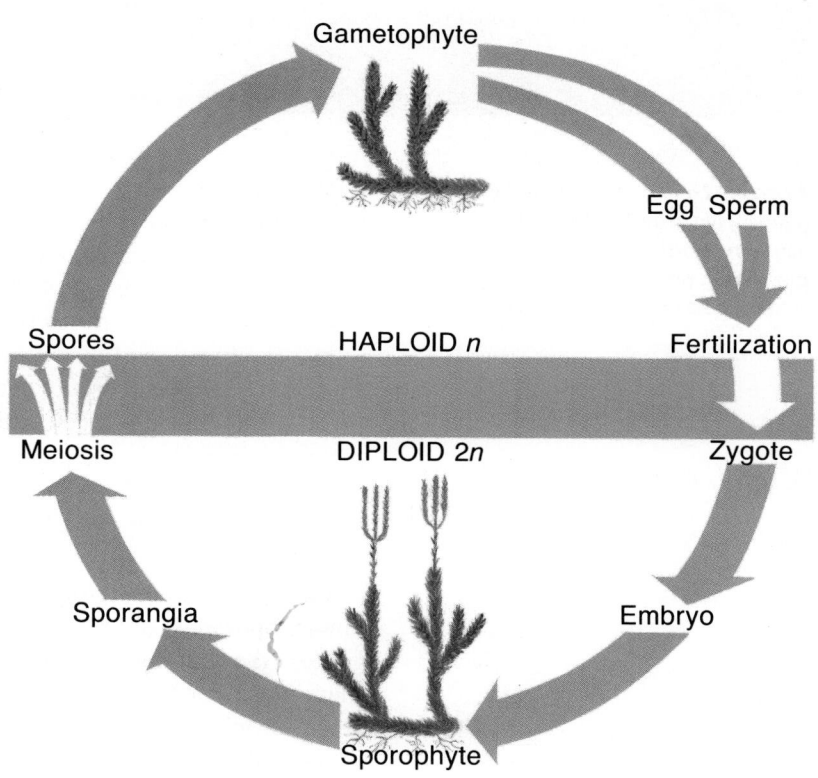

Gametophyte

Egg Sperm

Spores   HAPLOID *n*   Fertilization

Meiosis   DIPLOID 2*n*   Zygote

Sporangia   Embryo

Sporophyte

Figure 16-9 Ancient and extinct plants from around 320 million years ago were mostly relatives of present-day simple vascular plants and were the source of Earth's nonrenewable supply of fossil fuels.

Modern club mosses are much smaller than their ancestors. They grow close to the ground and are found mostly in damp forests, though some live in desert or mountain climates. One species found in forests in the northern United States is called the ground pine because of the plant's evergreen and pine-treelike growth habit. The name *club moss* comes from a combination of their resemblance to true mosses and the clublike appearance of their reproductive structures. The plant shown in Figure 16-10 is the sporophyte generation of the club moss. The gametophyte generation is smaller and much less conspicuous, and sometimes remains buried in the soil. Like other vascular plants, the sporophyte of a club moss has roots, stems, and leaves.

A major advance in this group of vascular plants was the evolution of leaves that are adapted for reproduction and hold structures called sporangia. **Sporangia** are reproductive structures that produce spores. In club mosses, the sporangia-bearing leaves cluster together to form a clublike shape. As in mosses, the spores grow into new gametophyte plants. For new sporophytes to be produced, sperm must swim through a film of water to the eggs.

Horsetails The horsetails represent another group of simple vascular plants. Like the early club mosses, early horsetails were tree-sized members of the ancient forest community. Today's horsetails are much smaller and are usually found growing in damp soil. As in club mosses, the sporophyte generation is the most conspicuous, with small, triangular-shaped and paper-thin leaves that circle each joint of the slender, hollow stem. The name *horsetail* refers to the bushy appearance of the plant. Sporangia-bearing leaves form a conelike structure at the tips of some stems. Some horsetails are called scouring rushes because they are impregnated with silica, an abrasive substance, and have been used to scour cooking utensils.

Ferns

Appearing in the fossil record at about the same time as club mosses and horsetails, the ferns evolved into more species than any of these other groups of vascular plants. The ferns were especially prominent members of Earth's ancient plant population, their tall treelike forms creating vast fern forests. Some tree-sized ferns still exist today, primarily in the tropics. Although ferns are still abundant, most modern species are much smaller than their ancestors. You may have seen shrub-sized ferns on the damp forest floor or along stream banks. Some species float in water or are rooted in mud, whereas others live in dry areas. Some even cling to the sides of rocky cliffs. Ferns that inhabit dry regions become dormant when moisture is scarce, but resume growth and reproduction when water is available.

Figure 16-10 The spores in club mosses (left) and horsetails (right) are formed in sporangia on the upper surfaces of small leaves, which are clustered together in conelike structures at the tips of sporophyte stems.

Like the club mosses and horsetails, it is the sporophyte generation of the fern that has roots, stems, and leaves with vascular tissues and is the plant commonly recognized as a fern. In most ferns, the main stem forms an underground clump under the soil surface. This thick, underground stem is called a **rhizome.** Rhizomes are underground food storage organs in which many of the cells are filled with starch. The apparently upright stems of tree ferns that extend aboveground are in reality formed by the masses of leaf bases that grow from the rhizome underground. Roots descend from the rhizome into the soil. The fern's leaves, usually called fronds, are often highly subdivided and sometimes have a lacy appearance, as shown in Figure 16-11.

Life Cycle   The sporophyte generation of a fern is the stage that is most familiar. Spores are produced inside sporangia, which are usually located in clusters on the undersides of the fronds, but sometimes are formed on separate stalks (Figure 16-11). As in the club mosses, haploid fern spores are produced by meiosis and released into the air. If a spore comes to rest in favorable conditions, it develops into a tiny, nonvascular gametophyte in which the gametes are formed. As in club mosses and nonvascular plants, the sperm must swim through water for fertilization to take place. The zygote develops into the new, diploid sporophyte fern plant.

Adaptations for Life on Land   The sporophyte generation of a fern is not limited in size because it has vascular tissues that provide both transport and structural support. As with nonvascular and vascular plants alike, water loss is reduced in ferns by the presence of a waxy cuticle covering the stems and fronds. Ferns have an additional adaptation that helps prevent water loss. They have cells that can control the size of the pores on the plant's surface. When water is scarce, the pores decrease in size.

Figure 16-11   Dryopteris (top) is a fern common throughout the temperate regions of the world. It has the typical fern structures of fronds with sporangia, rhizomes, and fiddlehead shoots. The ostrich fern (left) is unusual in having sporangia only on distinct fertile fronds. The trunk of a tree fern (right) is formed by upright growth of the rhizome.

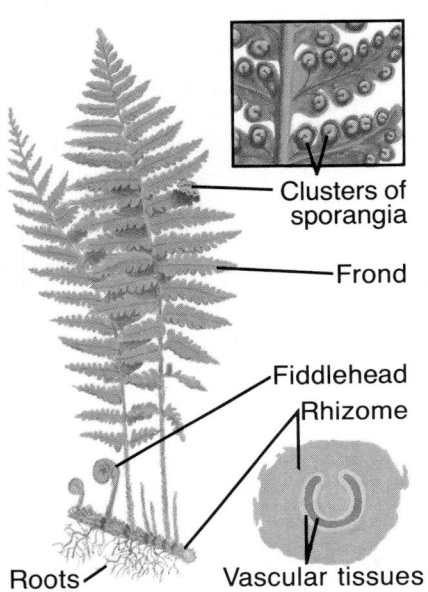

Clusters of sporangia

Frond

Fiddlehead

Rhizome

Roots

Vascular tissues

Ferns, club mosses, and horsetails are like nonvascular plants in that they must live in a moist environment where there is enough water to allow sperm to swim to the egg during sexual reproduction. Because the gametophyte generation of these plants is nonvascular, water must be readily available for absorption and transport by osmosis. These spore-bearing plants are not completely adapted to a life on dry land.

The Seed-bearing Plants

Although spore-producing vascular plants dominated the ancient forests 300 million years ago, they were not the only plants present. Early seed-bearing plants, though fewer in number, also had begun to emerge. About 280 million years ago, when club mosses, ferns, and other spore producers had reached their greatest numbers and diversity, Earth's climate changed. There began long periods of drought and freezing weather, and glaciers began to form. Many spore-producing plants became extinct during these times of unsettled conditions. But, a few seed-bearing plants had adaptations that enabled them to survive and reproduce even in areas where there was little water. As the numbers of spore producers decreased, the seed-bearing plants began to take over. As Earth's climate continued to change, these ancient plants underwent adaptive radiation and evolved into many new and different species.

What is a seed-bearing plant? It is one that bears a structure called a seed, which protects the young plant embryo. A **seed** consists of an embryo with a food supply enclosed in a tough, protective coat. It is one of the adaptations that has enabled seed-bearing plants to be so successful in the many environmental conditions now found on Earth. The embryo is protected during harsh conditions, and when those conditions improve, the embryo can continue its development. The seed-bearing plants are far more diverse and numerous than any other plants living today. They can be classified into two major groups: those that produce seeds enclosed in a fruit, and those with seeds that are not enclosed.

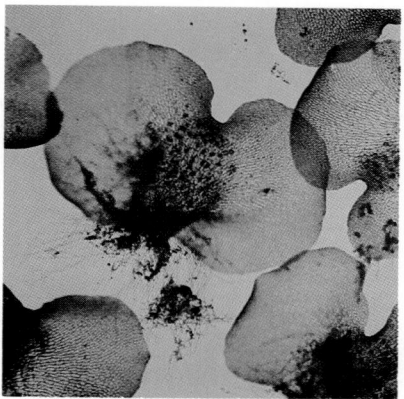

Figure 16-12 The gametophyte of a fern is heart-shaped. The gametes are formed in organs within or near the notch of the heart. In what kind of habitat would you expect to find a fern gametophyte? Magnification: 6×

Figure 16-13 The major distinction between an angiosperm (left) and a gymnosperm (right) can be made by examining their seeds. In the mountain ash (left), the seeds are enclosed within berries. In the cycad (right), the seeds are completely exposed. Why are both of these plant species so successful?

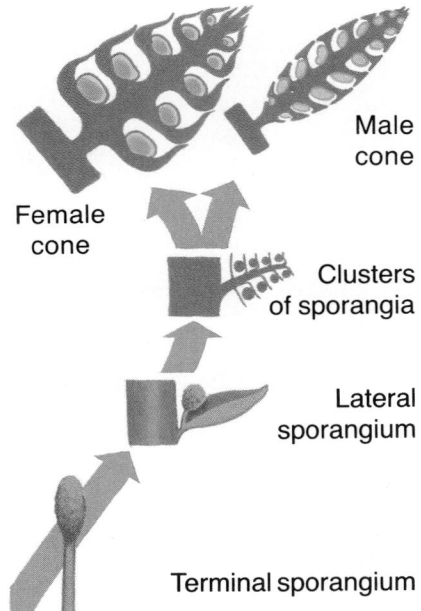

Male
cone

Female
cone

Clusters
of sporangia

Lateral
sporangium

Terminal sporangium

Figure 16-14 **The evolution of the male and female cones of gymnosperms probably began with a terminal sporangium. The sporangia then developed in the angles of leaves. The stems became condensed, and the sporangia and leaves formed a woody cone.**

Gymnosperms According to the fossil record, the first seed-bearing plants to appear were those that produce seeds with no enclosing tissue. This group is commonly called the **gymnosperms,** a name that means "naked seeds." One group of gymnosperms are the cycads—true relics of the past. Two hundred million years ago, cycads were widespread and existed with the dinosaurs. Now, only nine genera survive in remote places around the world, and they are often grown as curiosities in botanic gardens.

Conifers The most familiar and numerous of the gymnosperms are the **conifers,** so named because their seeds develop on the woody scales of reproductive structures that form cones. Modern conifers include pine, fir, juniper, spruce, and giant redwood trees. It is hypothesized that cones evolved from spore-bearing leaves or branches of ancestral gymnosperms. Each scale on the cone is actually a small leaf or tiny, reduced branch, and along with their attached reproductive structures, they have become woody and clustered together to form a male or female cone.

Conifers are distributed over most of Earth. They were widespread 150 million years ago. Huge petrified trunks of conifers from that time can be seen today at the Petrified Forest National Park in Arizona. Huge coniferous forests were once widespread in all temperate regions. The greatest coniferous forest once ranged from Alaska along the Pacific coast, and south to northern California. Ninety percent of this forest in the United States has been felled. What remains is the Pacific Northwest coniferous forest, which is the habitat of the largest and oldest trees on Earth. Today, coniferous forests are found mainly in northern latitudes of North America, Scandinavia, and Russia, where they provide much of the lumber used in human commerce.

Figure 16-15 **The petrified trunks of trees that are scattered throughout the Petrified Forest National Park in Arizona are the fossils of ancient and now extinct gymnosperms that lived 200 million years ago. When the trees were buried by volcanic ash and mud, the original wood tissues were slowly replaced with silica deposits.**

The Protection of Sperm   In all gymnosperms, the sporophyte generation is dominant, and cells of the gametophyte develop within the tissues of the sporophyte. This characteristic has proved to be a vital adaptation to the development of plants capable of surviving in harsh, dry climates. You know that in the plants so far studied, moisture is essential for sperm to reach an egg. Conifers and all other seed-bearing plants have a mechanism that accomplishes fertilization without the need for a watery environment. The sperm of all seed plants is enclosed in protective structures called **pollen.**

In conifers, the many grains of pollen develop from the male gametophytes inside the male sporangium of a male cone. A pollen grain is composed of the male gametophyte with four haploid cells housed inside a tough outer wall, which resists evaporation and provides protection for the sperm that will later develop within. Pollen grains released from the male cone are scattered by wind or air currents. Two of the haploid nuclei disintegrate. The pollen grains that land inside a female cone continue their development, first by producing two sperm nuclei and then by fertilizing the egg inside each female sporangium. After fertilization, the embryo remains attached to the female cone, where it develops into a seed. When mature, most gymnosperm seeds are released into the wind. Both the pollen grains and the seeds of a pine have wings that help them stay aloft when they are carried by the wind away from the parent plant (Figure 16-16). The pollen grain and the seed are two more adaptations of plants for life on land.

Other Gymnosperm Adaptations   How are gymnosperms able to grow so tall? Vascular tissue that transports water to all parts of a plant is made of cells with very hard cell walls. In trees, these cells are surrounded by thick-walled fibers that help in providing support to woody plants. This woody tissue of trees, especially of coniferous trees, makes up the wood that we burn as fuel and make into houses, furniture, and paper. The support

Figure 16-16   The Scotch pine, *Pinus sylvestris* (right), originated in Scotland but has been planted throughout Europe, Asia, and the United States. The production of pollen and seeds in the Scotch pine (left) is typical of all conifers.

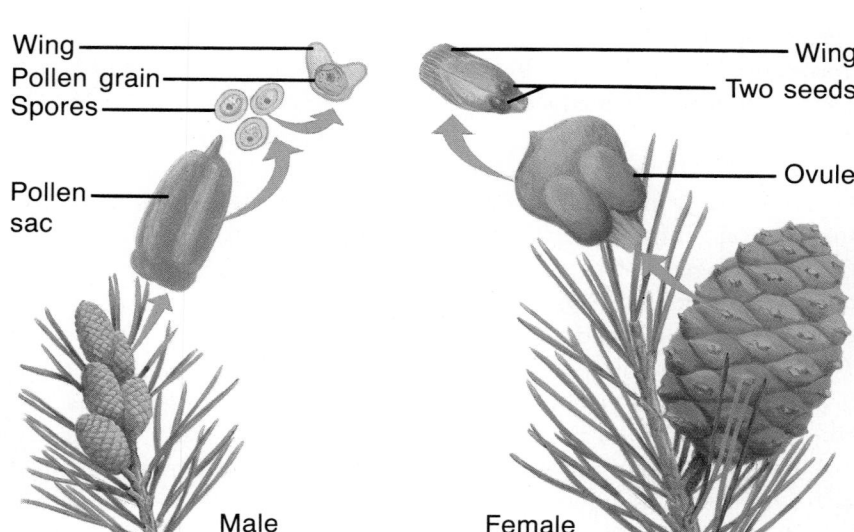

Wing
Pollen grain
Spores

Pollen sac

Male cones

Wing
Two seeds

Ovule

Female cone

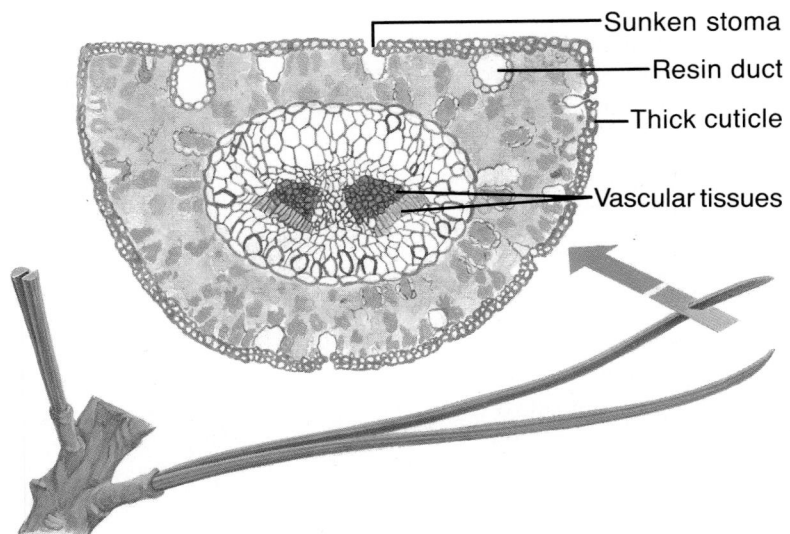

Sunken stoma
Resin duct
Thick cuticle
Vascular tissues

Figure 16-17 The leaves of pines are always in groups of two, three, or five needles. Because of the needlelike leaf's small surface area, water loss is reduced.

provided by woody tissue enables giant redwoods, for example, to grow to heights of more than 100 meters. Tall, upright growth offers several advantages to a plant. For example, a plant that grows taller than its neighbors does not have to compete with other plants for access to sunlight. Tall, upright growth can also help in the broad dispersal of pollen and mature seeds.

Adaptations for reducing water loss include the development of bark, a feature that is shared by all seed-bearing trees. Bark serves as a protective covering that helps prevent water loss through the stem. The needle-shaped leaves characteristic of conifers also have adaptations for preventing water loss. They are shaped somewhat like rolled-in leaves, and pore openings are sunken beneath a thick layer of cuticle, as shown in Figure 16-17. Conifer needles are covered with a thick, waterproof cuticle, and their long, thin shape reduces the surface area from which water can evaporate.

GLOBAL CONNECTION

Corn: The Native American Grain

Before humans learned to save seeds from year to year to grow annual crops, they lived as hunter-gatherers. Living in this way meant constant traveling in search of animal herds and wild fruits, vegetables, and grains. Often, when food was scarce, people starved. In time, people discovered that seeds could be saved and planted each year to provide a regular supply of food. Corn was one of the earliest crops to be domesticated.

The development of corn as a major human food source began with the Toltecs and Aztecs in Mexico. About 7000 years ago, a wild corn that produced edible kernels was discovered. Over many generations, the Aztecs and others selected and traded corn. Corn is wind-pollinated, so the crop required little care other than weeding and watering. *Tripsacum,* a wild grass, also produced large edible grains, and it was planted along with corn.

When the wind stirred the corn tassels, cross-pollination between the two grain crops occurred. The result was the establishment of a new plant called *Teosinte,* a cross between a wild corn and *Tripsacum.* When *Teosinte* was planted with true corn, another new and stronger plant was produced.

Corn was so important to these early farmers that their culture was built around it, and they worshiped a god of corn. Largely as a result of their development of corn and other agricultural practices, the Aztecs developed a great empire.

By the time Columbus came to America in 1492, Native Americans of both North and South America were growing many kinds of corn. Flint corn, which has a very hard, smooth kernel, was the kind most widely grown. Flour corn was grown by the Aztecs and Incas. Its soft kernels made it easy to grind into flour, the main ingredient in tortillas, a major part of the regional diet today.

Because corn was an easy crop to grow and it was plentiful, it had a symbolic place in the culture of many Native American tribes. In the Tewa Tribe of New Mexico, for instance, summer is the time of the

Angiosperms

About 200 million years ago, yet another group of plants began to emerge. Within the next 80 million years, the flowering plants became the most prominent plant group in the fossil record. The flowering plants are commonly called **angiosperms,** a term that means "seeds in a vessel." The vessel that contains the seeds of angiosperms is what you know to be a **fruit.** There are more than 250 000 species of angiosperms.

Most of the plants you are familiar with are probably angiosperms, partly because this plant group provides most of the plant food that humans eat. Angiosperms include wild and cultivated flowers such as dandelions, poppies, and chrysanthemums; they produce grains such as oats, wheat, rye, and rice; they form fruits such as strawberries, tomatoes, peaches, and cucumbers; and they provide us with vegetables such as spinach, broccoli, onions, and carrots. Angiosperms are also important to the survival of many other animals besides humans. Without flowering plants, biodiversity on Earth would disintegrate.

Figure 16-18 Flower form is used for classifying plants into families. Can you point out the flowers here that belong to the daisy family?

Blue Corn Chief, while winter is the time of the White Corn Chief. When a child is named, its mother offers two perfect ears of corn, one white and one blue, to the six sacred directions, and prayers are said for good fortune.

Many Native American tribes consider corn a sacred plant. Much rain is needed for corn to grow well. Before the Hopi and Zuni tribes developed the practice of irrigation, corn dances were part of a ritual to bring rain, and other dances were performed to thank the gods for a bountiful harvest.

The early English colonists were taught to plant corn by Native Americans who lived in northeastern North America. The colonists and the Spanish explorers not only used it as a food source, they exported it to Europe and beyond. From these beginnings sprang a plant that is now grown all over the world and is a major food source for both humans and their livestock.

1. **Connection to History:** Another plant from South America that influenced world history has the scientific name *Solanum tuberosum.* Research and report on how this plant influenced the history of Europe and North America in the mid-1800s.
2. **Connection to Genetics:** One of the most influential scientists working with corn in the 20th century was Barbara McClintock. Find out the variety of corn she worked with and what she discovered.

Flower Structure   Angiosperms have many traits in common with the gymnosperms, including vascular tissues and pollen. However, the reproductive structures of angiosperms and gymnosperms differ. Angiosperms have reproductive organs in flowers rather than in cones.

Like the scales of a cone, a flower consists of several rows of structures on a short stem that are homologous with leaves and which function as reproductive organs. Use Figure 16-19 to follow the description of a typical flower. The outer row of leaves that make up a flower are called **sepals;** they enclose and protect a flower in its bud stage. When the flower bud opens, the sepals may fall off. Sometimes, though, you can still see them surrounding the base of the open flower. You are probably familiar with the colorful petals of garden flowers. **Petals** make up the next inner row or rows of leaflike structures of the flower. Petals have many functions important to the reproductive success of a flowering plant. Often, they are brightly colored, highly perfumed, provide a convenient landing platform for insects, and dispense nectar, a sugary solution that provides many insects and birds with a high-energy food supply. Some flowers do not have showy petals. For example, grass plants, maple trees, and corn have small, inconspicuous flowers.

ART CONNECTION

Affection by Patricia Gonzalez

● ●

This oil-on-canvas painting hangs in a private collection in New York. To understand this picture, imagine what the scene would be without the flowering vine. The stark, forbidding view is transformed into a lush garden covered with brilliant blooms. The title, *Affection,* seems to suggest a metaphor of a clinging vine and the closeness of friendship. Perhaps the artist used the vine to express close family ties that are a major part of the Hispanic culture. Love and affection can make lives very beautiful, just as the flowering vine has enriched the painted landscape. Utilizing the colors and forms of Earth, the artist has produced a striking portrayal of the strength that exists within the family and within the Hispanic culture.

Educated in England and Colombia, Gonzalez now lives and paints in Houston, Texas. Most of the images she creates come from fantasy rather than living models.

1. **Explore Further:** There are many different flowering vines. Research to find out how these plants hold onto their supports.

2. **Chemistry Connection:** What are some sources of pigments used in paintings?

3. **History Connection:** Find out which plants Vincent Van Gogh preferred to paint.

At the center of the rows of petals are male and female reproductive organs that also are homologous with leaves. Many plants have both male and female structures. Sometimes only **stamens,** the male reproductive organs of a flower, are present. Many trees, such as oaks and willows, have male flowers consisting only of sepals and stamens (Figure 16-19). A stamen produces the male gametes and pollen inside the **anther,** which sits at the top of a short stalk called the filament. Sometimes only a **pistil,** the female reproductive organ, sits in the center of a flower. Within the swollen base of a pistil is the **ovary,** where the female gametes are produced. The neck of the ovary is called the **style** and at the tip of the style is a rough or sticky surface called the **stigma.** Usually, however, all four kinds of flower structures—sepals, petals, stamens, and pistils—are present. You will learn how all these structures are important to the reproduction of flowering plants in Chapter 18.

Minilab 2

What do flower parts really look like?

Use a flower to observe and record the following in a table: number of sepals, number of petals, number of stamens, pistil. Use a scalpel to make a cross-section cut through the ovary. **CAUTION:** *Be careful when using a scalpel.* Record numbers and arrangement of young seeds. Prepare a wet mount of the anther and record the shapes of pollen grains as seen under low power. **Analysis** *What are the functions for all the parts in your table? How did the numbers of sepals, petals, and stamens compare? How many compartments could you see in your cross section of the ovary? What was the shape of the pollen grains?*

Figure 16-19 **Flowers are the organs of reproduction in angiosperms. The diversity of flower structures indicates their enormous success. From the male and female catkins of a walnut (left) to the spiny cluster of small tubular flowers of the thistle (center) to the showy floating heads of a water lily (right), flowering plants have evolved many ways to ensure the continuation of the species.**

Pistil — [Stigma
Style
Ovary

Petal
Sepal
Young seeds

Stamen
Filament] Anther

Figure 16-20 Butterflies and moths visit flowers for their nectar, which is often produced at the bottom of long tubular flowers.

Other Angiosperm Adaptations The angiosperms are the most successful plants on Earth. Most of that success is due to adaptations that have made reproduction possible under a variety of conditions. For example, the dispersal of pollen by wind is a fairly random method. There is little control over whether a pollen grain lands on any female reproductive structure, much less on one belonging to the same species. Therefore, a flower that depends on wind produces a great deal of pollen. Many flowering plants, as you read in Chapter 13, depend on insects to transport pollen from one flower to another flower of the same species. Less pollen is wasted, and less is needed.

Another adaptation in flowering plants that provides success in a variety of conditions is the protection and dispersal of seeds inside a fruit. You will read in Chapter 18 how fruits are formed following fertilization and how important fruits are to the survival of the individual species.

As you finish this chapter, think about the major features of each group of plants you've studied. What kinds of environments does each group inhabit? What is their typical size, now and in the past? How have their reproductive structures changed over time? Can you recognize members of each group growing near your home or school? Use Appendix A to help you distinguish the divisions and classes of the plant kingdom.

LOOKING AHEAD

YOU HAVE learned how plants descended from aquatic algae and evolved adaptations that enabled them to occupy a broad range of habitats on land. Did these same trends occur in animals? Features that enable animals to exploit both land and water environments will be described in the next chapter. You will also see that animals had to overcome some of the same problems as plants as they began to fill more habitats on land. ‚à

Section Review

Understanding Concepts
1. How are the cones of club mosses and horsetails similar to those of conifers?
2. In what ways are ferns better adapted for life on land than are nonvascular plants?
3. How has the evolution of pollen been important to the survival of land plants? What other traits of plants are important to their survival on land?
4. **Skill Review—Classifying:** Set up a system to classify the divisions of vascular plants on the basis of their adaptations for survival on land. For more help, see Organizing Information in the **Skill Handbook**.
5. **Thinking Critically:** Why does it seem probable that angiosperms evolved from primitive gymnosperms?

Biotechnology

Plants: Earth's Major Energy Trappers

Almost all of the energy used here on Earth comes from the sun. Very small amounts come from nuclear reactions and geothermal sources such as volcanoes, hot springs, and natural steam vents. However, approximately 90 percent of our energy needs today are met with solar energy that was captured by ancient and now extinct plants and stored within their cells millions of years ago. These ancient plant remains are called fossil fuels and are found buried deep in Earth. One fossil fuel that affects almost everyone's life today is petroleum. But, there's a problem—we're using petroleum faster than it can be formed.

A Finite Supply   The total world petroleum consumption since we first began to use it has been about 550 billion barrels. Known reserves of oil and gas have dropped rapidly since 1970. Today's reserves are estimated at about 990 billion barrels. It is also estimated that undiscovered, recoverable quantities of oil may total about 500 billion barrels. In other words, at the present rate of usage, the supply of fossil fuels may cease to be a dependable source of energy not long after the year 2000. This fuel shortage is critical because fossil fuels are used for

more than heating buildings and moving vehicles down the road.

Our Energy-dependent Society
Fossil fuels are the raw materials from which plastics, fabrics, some medicines, and other products are made. When fossil fuels are burned,

their stored energy is no longer available for use in other products. The modern world is dependent on easy access to energy. Computers, for example, are totally dependent on energy, and most industrial processes are very energy-intensive. Without abundant energy, today's society could not function. What can be done?

Solutions   One answer to the fuel shortage could be to use an energy resource that can be constantly

renewed. For example, you may have seen service stations advertising gasohol as a fuel for automobiles. This is a mixture of 10 percent ethanol and 90 percent gasoline. Ethanol is a resource that can be replenished yearly. It is a by-product from the processing of foods such as potatoes, sugar beets, sugar cane, grain, or corn. Presently, the United States grows a surplus of these crops. By blending alcohol into a fuel, gasoline can be conserved. For example, 10 gallons of a 10 percent gasohol blend contains one gallon of ethanol and nine gallons of gasoline. Thus, the gasoline supply could be stretched by approximately 10 percent if everyone used gasohol.

New Energy Supplies   Another future option might be provided by a desert plant called jojoba (ho HO bah). The jojoba produces peanut-sized seeds that are almost 50 percent oil. This oil is very versatile and could possibly be used as a substitute for fuel and other products that are made from petroleum today. Currently, it is used to make products as diverse as lubricating oil, face creams, and low-cholesterol salad dressing.

At present, jojoba oil is very expensive because it is extracted from beans that usually are hand gathered from scattered wild plants in the desert. It has only recently been grown efficiently as a crop.

1. **Issue:** In about 100 years, humans have consumed approximately 25 percent of the recoverable petroleum that took millions of years to form. Propose and defend some guidelines that could govern the use of nonrenewable resources such as petroleum, iron, gold, copper, zinc, and lead.

2. **Connection to Physics:** Plants aren't the only things that can convert and store solar energy. Find out and describe how a silicon solar cell works.

Issues

How shall our forests be harvested?

When Europeans first arrived in the Americas, millions of acres of forest greeted them. Most were full-grown stands containing millions of mature trees. While Native Americans used few trees in their lives, the Europeans viewed them as a commercial resource to be exploited. In addition, the colonists used large amounts of wood for housing and fuel. At first, this wasn't a problem because there were so few colonists and so many trees. But the population grew, cities made of wood sprang up , and mechanical methods of harvesting trees were developed.

Clear-cutting   Today, our ability to harvest trees exceeds the ability of the forests to regenerate. Additionally, the economics of mechanical harvesting makes it cheaper and more efficient to cut down every tree in an area and then replant, rather than select out a few fully mature trees at a time. The replantings are usually of one species and spaced to facilitate later mechanical harvesting. Thirty to 40 years later, the area can easily be harvested again. This method of forestry is called clear-cutting.

Sustained Yield   At one time, forests were managed for sustained yield, where only the mature trees were cut on a regular schedule. This allowed the forest to yield smaller amounts of wood at each cutting, but it did so steadily every few years. In addition, sustained-yield harvesting allowed for a diverse forest in which many species could flourish, including animals and birds that require forests to survive.

Which Method to Choose?
Recently, there has been much con-

troversy over what philosophy should govern how our forests are managed. This is important because much of the land being logged is public land, and some of the organisms that live there are rare or endangered. Some large companies claim that they cannot compete in the lumber business if forced to selectively cut trees. They claim that it is too expensive and inefficient.

Opposing groups take the position that forests are a resource we hold in trust for future generations to use and enjoy and that they should be preserved. They claim that a variety of species is important to the survival of a community.

Understanding the Issue
1. Write down the main points for and against clear-cutting and sustained-yield forestry. Prepare a balanced presentation.

2. Many large groups of poor people are homesteading in the forests of South America. They clear-cut and then burn a region of forest land on which they grow crops until the soil is depleted of nutrients. They then move on to another forest region. What is the long-term effect of this practice? What are some alternatives you can suggest?

Readings
- Sanoff, A.P. "The Greening of America's Past." *U.S. News & World Report,* October 19, 1992.
- Wilson, E.O. "Rain Forest Canopy: The High Frontier." *National Geographic,* December, 1991.

Considering the Environment

Controlling Pests

All life competes for survival. Food, water, space, sunlight—they're all objects of competition by organisms struggling to live and reproduce. Some of the best at this competition are insect pests.

Problems with Control   Some of the earliest methods of insect pest control were mechanical. Farmers simply picked harmful bugs off their plants. However, this method only works over small areas. One of the early and most effective methods, still in use today, was to kill pests with chemical poisons. This method had several drawbacks.

First, some of the chemicals used, such as arsenic for killing aphids, were as dangerous to the farmers as they were to the pests. Second, some of the chemicals did not break down in the soil and were transferred to other untargeted species, such as bees and birds, that are helpful to humans. Third, as the pests were affected by the chemicals, the less resistant ones died. The survivors reproduced and passed their resistance on to their offspring. Thus, the pests adapted and the chemicals had less and less effect.

Biological Control   Recently, biologists have been studying biological control of insect pests. Biological control is the use of one organism to control the population of another. A common example of this method of biologically controlling pests is the use of ladybird beetles to eat aphids that destroy crops and flowers.

The Neem Tree   Another example of biological control is using the leaves and seeds of the neem tree from India and Burma. The leaves are placed in grain bins to repel insects. The seeds, when ground up and dissolved in water, are sprayed on plants as an insect repellent. Japanese beetles, a major plant pest, will starve to death rather than eat a neem-protected plant.

The insect-repellant chemical compounds produced by the neem tree are called limonoids. There are about 20 limonoids . One closely resembles a hormone that causes caterpillars to shed their skin during metamorphosis. When this limonoid is eaten, the caterpillar does not make enough of its own hormone, fails to mature, and dies.

The chances of any pest adapting to all 20 limonoids is very remote. Scientists hope to produce an insecticide that targets the pests but does not harm useful organisms, and that breaks down into harmless compounds in the soil.

1. **Explore Further:** The bacterium *Bacillus thuringiensis* has been used for years to control garden pests. Find out how this organism controls pests and what the pros and cons of its use are.
2. **Issue:** One source of new chemicals is the tropical rain forests. Right now, these forests are rapidly being destroyed. Discuss the long-term effects of this behavior.
3. **Literature Connection:** Read Rachel Carson's *Silent Spring.* Find out what problems she identified and what pesticides were the cause of the problems.

Summary

Plants are thought to have evolved from green algae. As they did so, they became adapted to life on land, a much harsher environment than water.

Nonvascular plants include mosses and liverworts. Though not complex, these plants have adaptations that provide the ability to survive on land. In many ways, though, they are like algae in that they are restricted to a moist habitat and are limited in size.

Evolution of vascular tissue enabled later plants to transport materials more efficiently and grow larger in size. Club mosses, horsetails, and ferns have adaptations more suited to life on land than do the nonvascular plants, but must live in moist areas because of their mode of sexual reproduction.

The evolution of pollen, seeds, and woody tissues has enabled the seed plants to live in drier areas. Because of their efficient means of pollen and seed dispersal, flowering plants have become the most successful group of plants today.

Language of Biology

Write a sentence that shows your understanding of each of the following terms.

alternation of generations	pollen
angiosperm	rhizome
anther	seed
conifer	sepal
cuticle	sporangium
fruit	sporophyte generation
gametophyte generation	stamen
gymnosperm	stigma
ovary	style
petal	thallus
pistil	vascular tissue

Understanding Concepts

1. Explain why algae are thought to be the ancestors of mosses and liverworts.
2. Identify and explain the importance of new structures that evolved among club mosses, horsetails, and ferns.
3. Compare the dominant stages of the life cycle in nonvascular and vascular plants.
4. Based on external structure, how could you distinguish between a fern and a gymnosperm?
5. How are the scales of a cone and the petals of a flower homologous structures?
6. How was the modification of leaves for reproduction important in the evolution of plants?
7. Compare the adaptations for support and reproduction in conifers and ferns.

8. Why was the evolution of seeds so important?
9. Give two reasons why angiosperms are more successful than gymnosperms.

Applying Concepts

10. Why is it beneficial to have seeds dispersed away from parent plants?

11. Construct a phylogenetic tree that shows the evolution of plants from their common ancestor. Hint: The tree should have two major branches. What are they?

12. Coniferous forests are the major kinds of communities in Canada. Why do you think conifers are so abundant in that country?

13. How might flowers with pollen dispersed by wind differ from flowers with pollen transported by insects?

14. What might happen to nonvascular plants and plants like club mosses and ferns if exposed to long periods of drought?

15. How might the distribution of plants on Earth be affected by a significant increase in worldwide temperature?

16. **Biotechnology:** What might be the effect on agriculture of using large amounts of corn to produce fuel? How might it affect price and availability of corn as a food?

17. **Issue:** Forests all over the world take in carbon dioxide and give off oxygen. What could be some possible effects on other life forms if the forests are destroyed by forestry practices?

Lab Interpretation

18. **Minilab 1:** What evolutionary similarities do mosses and ferns share?

19. **Minilab 2:** Guard cells and stomata should have been seen when observing the wet mounts of the plant's food-making structures. Which plant showed these structures?

20. **Investigation:** Some plants, such as water lilies, have a cuticle only on their top leaf surface. A student applied petroleum jelly to the underside of such a plant and used a second plant as a control. Use the following graph to predict answers to these questions.

(a) Which plant may have had petroleum jelly applied to its underside? How do you know?

(b) Which plant was the control? Explain.

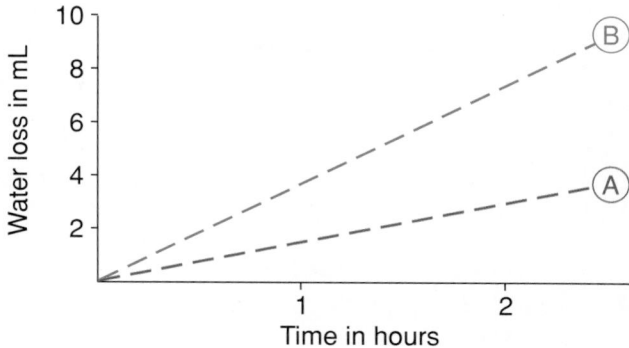

Connecting Ideas

21. How would you explain the origin of a plant with no chlorophyll, such as the Indian pipe found in northern forests?

22. Two new plants are discovered in a tropical rain forest. They are similar in many respects. How could you determine if they are descendants of a common ancestor?

ANIMAL ADAPTATIONS

EARTH IS HOME to an unbelievable number of different types of animals. Some are as familiar as your pet dog or a robin. Others are unfamiliar. Who would think that this coral reef is a marine mountain made of living animals? Unlike mountains that rise from the ground when forces within Earth uplift layers of rock, this reef is built by individual coral animals that live on the skeletal remains of earlier coral generations. The reef itself is shelter for many other amazing animals. Sea anemones, clown fish, sponges, crabs, and moray eels are at home on the reef. As you study each of these animal species, you will find a great diversity in body forms, body functions, and behavior. The crab shown on the next page has a very different form than any of the other reef animals. Its body functions and behaviors are also different. Yet, in spite of the differences, this crab and the other animals have many similarities. They all reproduce, respond to stimuli, grow, and depend on other organisms for their source of energy.

Looking at all of the different animals may cause you to wonder how this great diversity came about. In this chapter, you will learn why certain animals are considered to be more advanced than others. You will trace the evolution of different forms, functions, and behaviors from the simplest animals to the most complex. You will discover how the adaptations allow each kind of animal to be successful within its own environment.

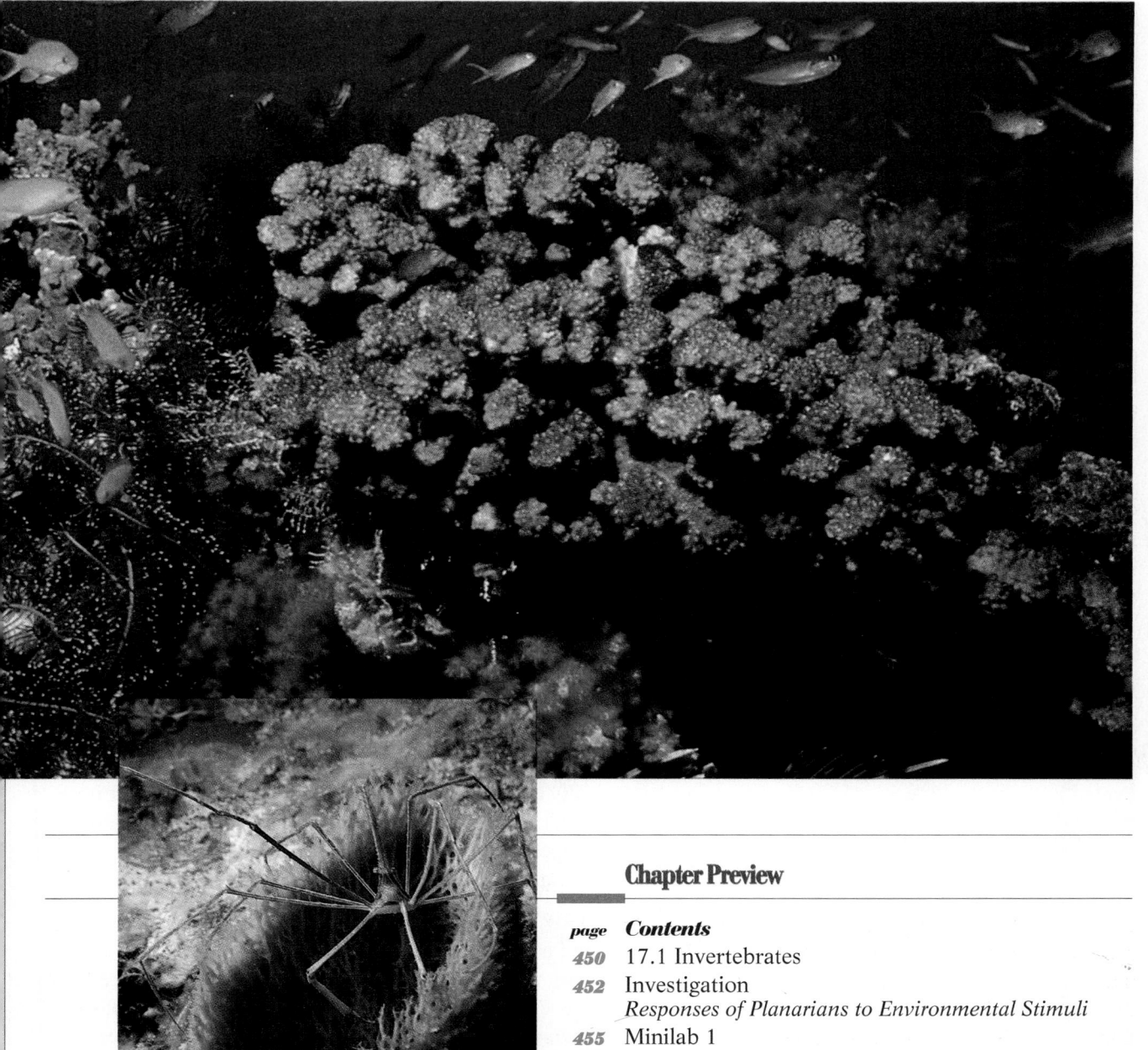

Suppose you awoke one morning and found yourself transformed into an organism such as the crab pictured here. What changes would you have to make to adjust to this organism's lifestyle?

Chapter Preview

Objectives

Compare the features of sponges and cnidarians.

Explain how flatworms are more complex than sponges and cnidarians.

Diagram the symmetry of an organism.

17.1 Invertebrates

Suppose a visitor from another solar system happens to land in your neighborhood. This visitor wants to study the various life forms on Earth. You suggest that the local zoo and aquarium are great places to find out about animals. So the visitor spends a day with gorillas, snow leopards, macaws, cobras, killer whales, dolphins, sharks, and sea lions. At the end of the day, the visitor journeys back to its home planet thinking it has learned all about Earth's animals. Actually, the visitor has learned very little, because most of the animals in zoos and aquariums belong to a group of animals that represent only five percent of the entire animal kingdom.

You know that organisms are classified according to their body structures. An informal way of classifying animals is based on whether or not they have a backbone. Most of the animals with which you are familiar have a backbone, yet about 95 percent of all animals have no backbone. Animals with backbones are called **vertebrates** and are grouped into one subphylum within the classification system. Animals without backbones are called **invertebrates** and are grouped into about 30 different phyla.

Figure 17-1 **The natural sponge used by a potter is the "skeleton" of a once-living animal.**

Sponges

Have you ever watched a vase being shaped on a potter's wheel? As the wheel spins and the vase rises out of the clay, the potter uses a natural sponge to keep the clay moist. It's difficult to believe that the natural sponge was once a living animal. Sponges are the least complex invertebrate. Most of the 8000 species live in saltwater environments, although a few do inhabit fresh water. The various species exhibit a variety of sizes, shapes, and bright colors such as yellow, red, and orange.

Body Structure A sponge is made of two layers of cells and a central cavity. While a sponge doesn't have tissues, organs, or systems, it does have specialized cells that carry out specific functions. The outer layer of the sponge is composed of flat cells that provide protection for the animal. The inner layer has cells with flagella that constantly move back and forth, creating water currents in the sponge. The water currents help to draw food and oxygen into the sponge and carry wastes out of the sponge. Water enters through the hundreds of pores that penetrate the sponge's body. It exits from one or more larger pores, usually located near the top of the sponge. This method of obtaining nutrients in sponges is adaptive because sponges are sessile animals. Sessile animals are attached to some object, such as a rock or shell, on the ocean bottom. Adult sponges are not able

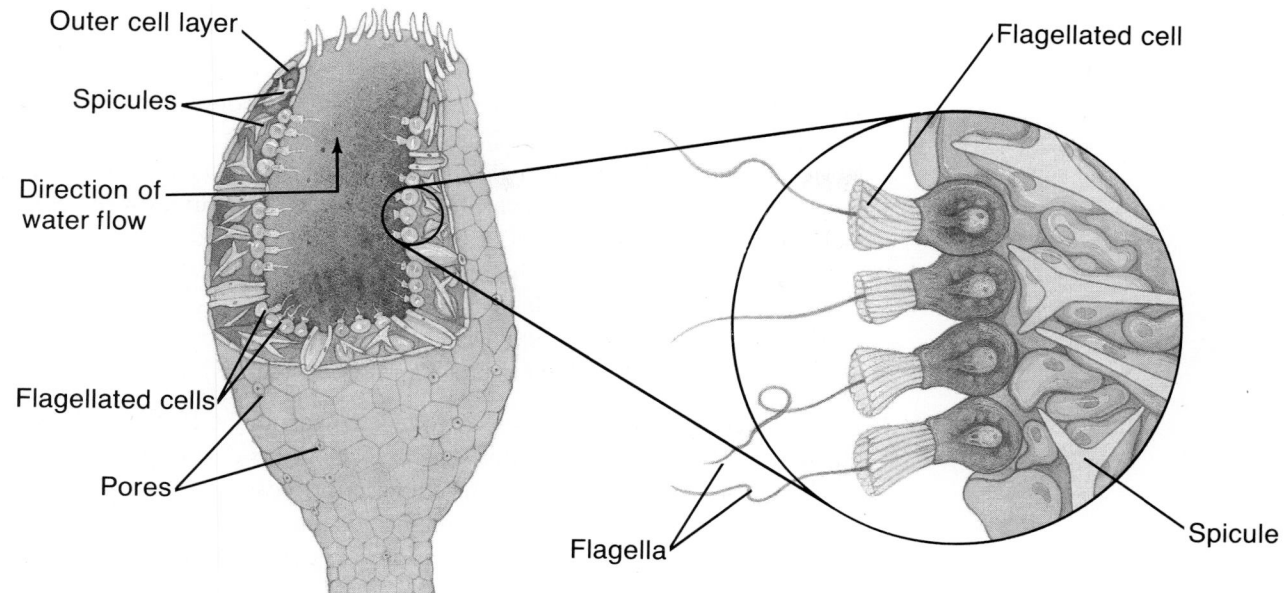

Outer cell layer

Spicules

Direction of water flow

Flagellated cells

Pores

Flagella

Flagellated cell

Spicule

to move about looking for food or shelter. Would a sessile animal need a means of support? Think again about the potter's sponge. What the potter uses to shape the clay is actually the remains of the sponge's support framework. This elastic substance appears as a meshlike network when viewed very closely. Not all sponges, however, have this meshlike elastic network. In fact, sponges are classified according to the mineral content and shape of their support structures called spicules. Some sponges have glasslike structures that form a lattice. Other sponges have massive support structures made of calcium, the same mineral found in bone.

Reproduction   Because sponges are sessile, they need a way of reproducing that doesn't require movement of individuals. To solve this problem, sponges reproduce both sexually and asexually. Asexual reproduction occurs when small fragments of the parent sponge break off, float away, settle to the ocean floor, and grow into a new sponge. While this type of reproduction results in a new sponge, it doesn't allow for any genetic variability. Because the parent sponge and the new sponge have exactly the same genes, the new sponge is an identical copy of the parent. In asexually reproducing sponges, only mutations provide for variability.

Many sponges are hermaphrodites. A hermaphroditic animal produces both sperm and eggs. However, hermaphroditic sponges do not fertilize their own eggs. The sperm are released into the water and fertilize eggs that are retained between the two cell layers of other sponges. After a period of development within the sponge, embryos are released and swim about until they attach themselves to a rock or shell. The advantage of sexual reproduction lies in the recombination of genetic material, which results in genetic variability. An additional advantage to being hermaphroditic is that an individual animal can mate with any other individual of its own species because it can produce both types of gametes.

Figure 17-2   Although sponges do not have true tissues, they are composed of several different types of cells. The flagella of the flagellated cells help water move through the sponge. Food particles are pulled from the water by these flagella for use by the sponge. Sponges also have an inner support system of spicules. Why do you think this support system is beneficial?

Cnidarians

The rhythm of a jellyfish's undulating body has a hypnotic effect as you watch it move through the water of an aquarium exhibit. The delicate tentacles and transparent body invite the curious observer to reach out and touch the strange body. However, such an act can be painful, for the tentacles are long, armlike structures containing stinging cells that shoot out poisonous filaments. These stinging cells are called cnidocytes, hence the phylum name *cnidaria*. While the sting of small jellyfish is painful and irritating, the sting of large cnidarians, such as the Portuguese man-of-war, is strong enough to kill large fish and seriously harm humans.

Body Structure   The cnidarians include jellyfish, corals, *Hydra*, and sea anemones. Most live in salt water. The two cell layers of the sponges have evolved into two tissue layers in the cnidarians. Recall that tissues are organized groups of cells carrying out specific jobs. This organization into tissues is a major evolutionary trend leading to the development of more

INVESTIGATION

Responses of Planarians to Environmental Stimuli

When the mail carrier comes into your yard, your dog barks. When you use your can opener, your cat comes running. When you see an advertisement for a pizza bubbling with cheese, your mouth waters. All animals react to stimuli in their environments. Planarians are freshwater flatworms that live on organic debris. They glide over rocks or gravel in the water on a slippery trail of mucus that they make. In this lab, you will determine how planarians respond to a variety of stimuli.

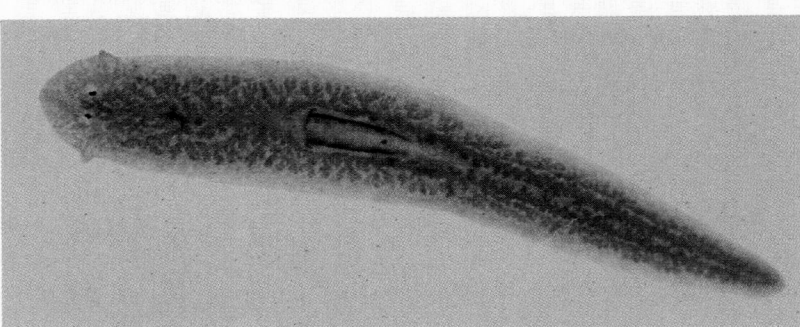

Problem
How do planarians respond to environmental stimuli?

Safety Precautions
Make sure you do not let electrical cords get wet. Wear goggles when working with acids or bases. Immediately rinse spills on skin.

Wash your hands at the end of the lab.

Hypothesis
Make a hypothesis about how planarians would react to a specific environmental stimulus. Explain your reasons for making this hypothesis.

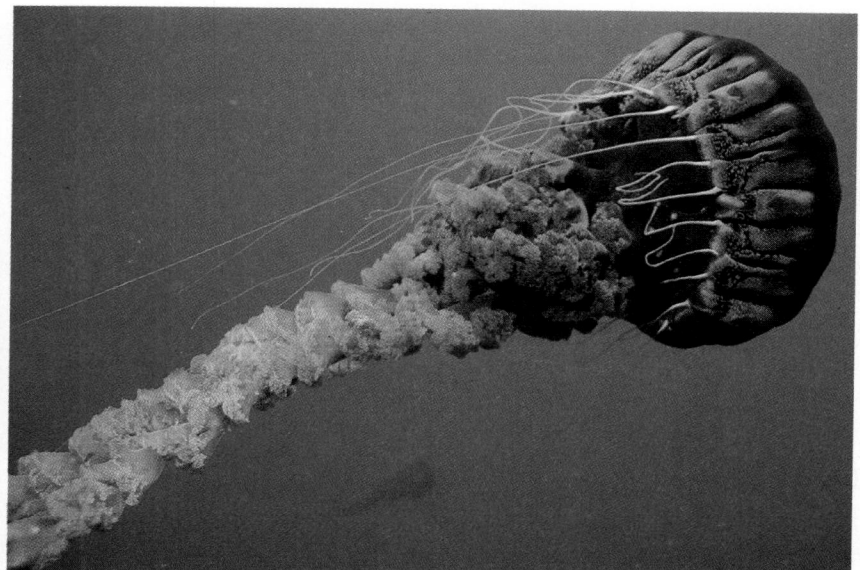

Figure 17-3 Some jellyfish, such as this giant pelagic jellyfish, can be extremely large. This jellyfish can have tentacles up to 15 feet long. Some tiny fish and crabs, impervious to the stinging cells of the jellyfish, hide among its tentacles.

Experimental Plan

1. As a group, make a list of possible ways you might test your hypothesis using the materials your teacher has made available.
2. Agree on one idea from your group's list that could be investigated in the classroom.
3. Design an experiment that will test one variable at a time. Plan to collect quantitative data.
4. Following the style of a recipe, write a numbered list of directions that anyone could follow.
5. Make a list of materials and the quantities you will need.

Checking the Plan

Discuss the following points with other group members to decide the final procedures for your experiment.

1. What variables will need to be controlled?
2. What is your control?

3. What will you measure or count?
4. How many worms will you test?
5. How many trials will you run?
6. Have you designed and made a table for collecting data?

Make sure your teacher has approved your experimental plan before you proceed further.

Data and Observations

Carry out your experiment; make your measurements; and complete your data table. Make a graph of your results.

Analysis and Conclusions

1. According to your results, how do planarians respond to the stimulus you selected?
2. What is your experimental evidence for your conclusion in question 1?
3. Was your hypothesis supported by your data? Explain.

4. If your hypothesis was not supported, what types of variation in your method could be responsible?
5. Consult with other groups that tested different environmental stimuli. What were their conclusions?
6. In what way is the response of the planarians you tested an adaptation to their lifestyle?
7. Explain the planarian responses tested by other groups in terms of adaptive value to the animal.

Going Further

Based on this lab experience, design an experiment that would help to answer another question that arose from your work. Assume that you will have unlimited resources and will not be conducting this lab in class.

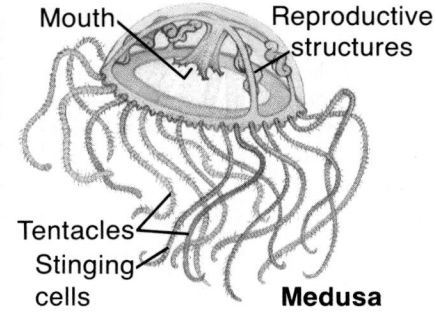

Mouth

Reproductive structures

Tentacles

Stinging cells

Medusa

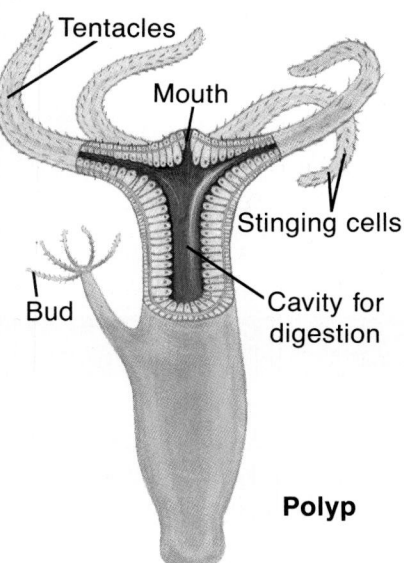

Tentacles

Mouth

Stinging cells

Bud

Cavity for digestion

Polyp

Figure 17-4 **Cnidarians come in one of two body shapes. The motile medusa (top) is umbrella- or bell-shaped. The sessile polyp (bottom) is vase-shaped. Each form has the phylum-distinguishing stinging cells located in the tentacles.**

Figure 17-5 **A radially symmetrical animal can be cut lengthwise through the center in any direction and result in two equal halves. A bilaterally symmetrical organism shows mirror symmetry; the right and left sides are nearly identical. Organisms such as amoebas and sponges show no symmetry.**

complex animals. The outer layer of cells is called the **ectoderm,** and the inner cell layer is called the **endoderm.** In cnidarians, a jellylike layer separates the ectoderm and endoderm. True nerve cells are arranged within the bodies of cnidarians and serve to coordinate the activities of the animal.

Cnidarians have a hollow body that is either vase-shaped as in a polyp, or umbrella-shaped as in a medusa. A single body opening called the mouth passes food into the hollow body and ejects wastes. Surrounding the mouth are numerous tentacles, the cnidarian's means of capturing prey and defending itself. Each tentacle has a number of stinging cells, which are used to poison or entangle prey. When prey brush against the tentacles, the stinging cells discharge and immobilize the prey. The tentacles then surround the food and transfer it to the mouth. Digestion occurs within the hollow body.

Reproduction Cnidarians reproduce both sexually and asexually. Many polyps reproduce asexually by producing buds, which are small outgrowths that eventually grow into new polyps. Sexual reproduction is accomplished by the production of gametes. All medusa-shaped cnidarians reproduce sexually. Like sponges, cnidarians are hermaphroditic. Each individual is capable of producing both eggs and sperm. Eggs and sperm are released into the water, where they fuse and develop into a new individual.

Body Symmetry

Animals exhibit one of two types of body symmetry. Some animals, such as sponges, have no definite body shape. Animals without a specific body shape are called asymmetrical. Cnidarians have a body plan called radial symmetry. Picture a bicycle wheel with the spokes arranged around a central point. No matter where the wheel is divided through the central point, a mirror image is produced. In **radial symmetry,** body structures of the animal are arranged equally around its center like a bicycle wheel.

Most animals have bilateral symmetry. In **bilateral symmetry,** the body is divided through the midline, from head to tail, to produce roughly identical left and right halves. Other cuts would not produce identical halves, as they would in a radially symmetrical animal. Bilateral symmetry is an adaptation for movement. Thus, animals with bilateral symmetry have definite areas named in relation to the direction of movement. Generally the front end, or **anterior,** has a head while the back end, or **posterior,** has a tail. There is also a definite top or back called the **dorsal** side, and a bottom or belly called the **ventral** side.

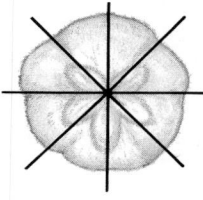

Radial symmetry

Bilateral symmetry

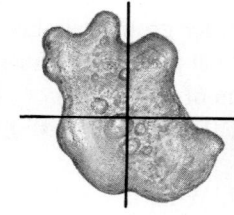

Asymmetry

Flatworms

Have you ever taken your cat or dog to the veterinarian because it had a tapeworm? Those parasitic worms belong to a phylum of flat, elongated animals known as flatworms. Flatworms are the simplest animals with bilateral symmetry. They have an advantage over sponges and cnidarians in being able to move forward in a certain direction. Flatworms are also the simplest animals with organs and organ systems. These systems are described below.

Body Structure   In addition to the inner and outer tissue layers that cnidarians have, flatworms have a middle, or third, tissue layer called the **mesoderm.** The evolution of mesoderm allowed for increasing complexity in organisms. The anterior end of a planarian has a primitive brain and sensory organs that allow it to sense danger and food more easily. Two nerve cords run the length of the body, in addition to a primitive digestive system. Like cnidarians, flatworms have a single opening, the mouth, through which food enters the body and undigested materials are removed. The planarian uses muscles and cilia located on its ventral surface to move toward food or away from danger. Muscles allow for controlled movement, even though movement is slow. Balance of internal fluids is maintained by a network of cells that draw excess fluids out of the tissues. However, systems for gas exchange, waste removal, and transport have not evolved in this phylum.

In addition to free-living forms such as planarians, this phylum also includes many parasitic animals. Flukes and tapeworms, for example, have complex life cycles during which they invade several different hosts, including humans.

Reproduction   Flatworms reproduce both sexually and asexually. Some members can reproduce asexually by breaking apart, with each part growing missing body parts to become a complete animal. Each part will grow, or regenerate, into an identical animal. Flatworms are also hermaphroditic. Two individuals will join together and exchange sperm to cross-fertilize the eggs. Genetic variability is insured by this exchange of genetic material.

Minilab 1

What are the characteristics of animals?

Observe carefully the group of animals given to you by your teacher. What are some of the characteristics of these organisms? Describe characteristics of their bodies, senses, movement, and behaviors. Think of these same characteristics of your own body.

Analysis   List the ways that you and these organisms are the same and ways that you differ. What characteristics do you and these organisms share? Does each of the organisms have distinct head and tail regions? Does each have a method of locomotion? Do all the organisms have the same number of appendages?

Section Review

Understanding Concepts
1. Compare the food-getting adaptations of sponges and cnidarians.
2. In what ways are cnidarians more complex than sponges?
3. What kind of symmetry do you have? Explain.

4. **Skill Review—Classifying:** Set up a classification key that would enable you to identify animals as sponges, cnidarians, or flatworms. For more help, see Organizing Information in the **Skill Handbook.**

5. **Thinking Critically:** Some species of sponges, as adults, become attached to the backs of crabs. How might that attachment benefit the sponge?

Describe the features of roundworms, segmented worms, mollusks, and arthropods.

Discuss how evolution of many features has contributed to the success of arthropods.

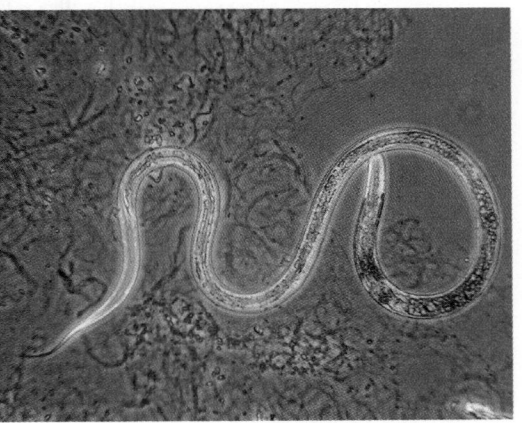

Figure 17-6 **Roundworms were among the first organisms to have two body openings.**

Figure 17-7 **The basic body plans of the animal phyla are (1) no body cavity, (2) a false body cavity, and (3) a true body cavity. In organisms with a true coelom, the digestive tube is suspended in the body cavity.**

17.2 Divergence of Animal Phyla

So far, you have followed the evolution of increasingly complex body plans beginning with sponges and moving to the cnidarians and the flatworms. The phyla discussed in this section show continued advances in the complexity of body structure.

Roundworms

Roundworms are one of the most numerous types of animal. They live in both salt and fresh water, in the soil, and even as parasites within other animals. These cylindrical animals are tapered at each end and have two distinct body openings—a mouth and an anus. The presence of a mouth, gut, and anus represent a much more efficient adaptation than the one-opening, mouth-only body plan of the cnidarians and flatworms. The passage of food in one direction through the worms allows for regional specialization. Specific areas of food intake, digestion, absorption, and excretion are developed in the digestive system. In roundworms, there are separate male and female sexes, and reproduction is always sexual. What can you say about the genetic variability possible in these animals?

Patterns of Development

Cnidarians, as you will recall, have only two tissue layers—endoderm and ectoderm. There is no body cavity between these layers. Flatworms also lack a body cavity, but do have a third layer, the mesoderm, in which primitive organ systems are located. Roundworms show further advancement in the body plan. The roundworm body plan has sometimes been referred to as a tube within a tube. These animals are the first to have a body cavity, which develops between the endoderm and mesoderm. This fluid-filled cavity gives the animal a firmness similar to a long, water-filled balloon. The body cavity is only partially lined with mesoderm, and organs are suspended within the cavity. The fluid provides some protection and support for the organs. It also provides resistance for the muscles, which run only along the length of the body, to work against.

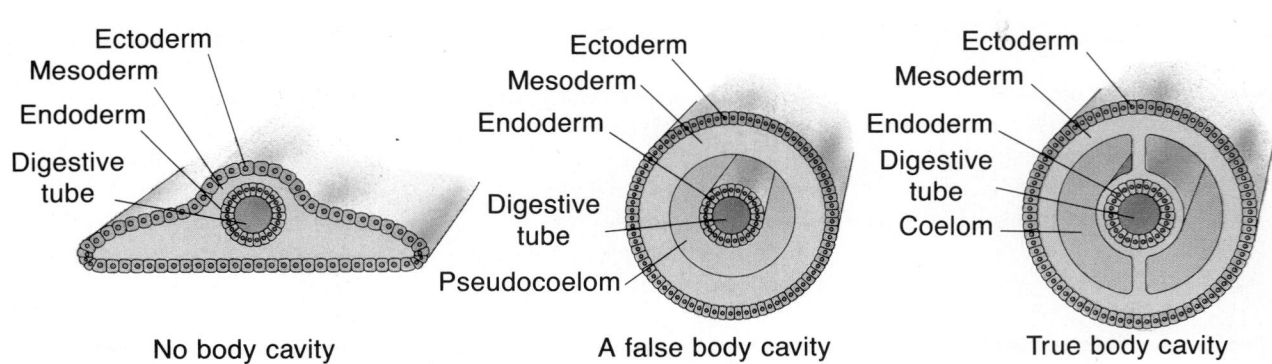

Ectoderm Mesoderm Endoderm Digestive tube	Ectoderm Mesoderm Endoderm Digestive tube Pseudocoelom	Ectoderm Mesoderm Endoderm Digestive tube Coelom
No body cavity	A false body cavity	True body cavity

The rest of the phyla in the animal kingdom have a true body cavity called a coelom. The **coelom** is a body cavity that forms within the mesoderm. As it develops, the coelom becomes completely lined with mesoderm. Notice the differences between the cross sections shown in Figure 17-7. In animals with coeloms, some organs develop from the mesoderm, becoming surrounded by a thin layer of this tissue in the process. This arrangement supports and protects the organs. More complex organ systems are therefore possible. The fluid within the coelom keeps the organs moist and may also provide some support for the body.

Segmented Worms

You've seen that major differences exist among animal groups, particularly in body plan. Other differences in developmental patterns also can be observed among groups. Animals with coeloms—segmented worms, mollusks, arthropods, echinoderms, and chordates—can be divided again into two major evolutionary groups based on differences in early development. Segmented worms, mollusks, and arthropods represent one group. Echinoderms and chordates represent the other.

Did you ever see an earthworm crawling across a parking lot or sidewalk after a rainstorm? If you look carefully at the worm, you will see individual segments on its body. Most of these segments contain the same internal structures. A common earthworm can have about 100 segments and be several centimeters long.

Earthworms and other segmented worms have several advantages over the flatworms and roundworms. Their body cavity is a true coelom. They also have several organ systems for controlling body functions. Their complex digestive system includes an esophagus, an intestine, and an anus. The circulatory system has two main blood vessels that transport materials to and away from body cells. Most segments contain a pair of excretory organs that remove body wastes. While earthworms exchange gases through their skin, aquatic members of this group have gills in paddlelike structures that stick out from each segment. The nervous system includes a primitive brain and nerve cord. The muscular system is composed of outer circular muscles and inner longitudinal muscles. Some species of segmented worms have separate sexes although the earthworms are hermaphroditic.

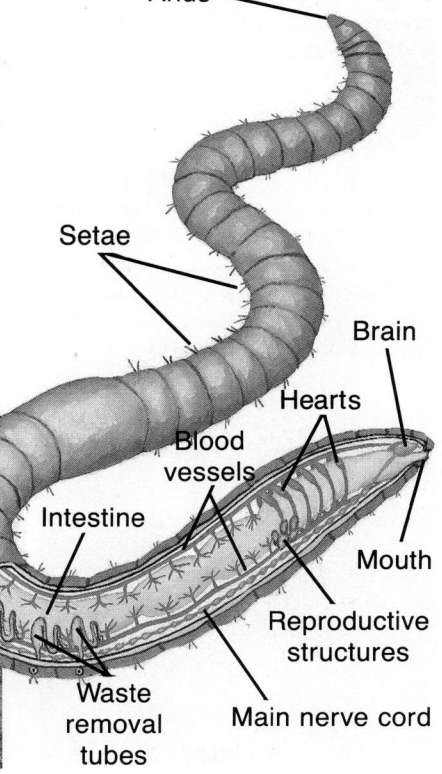

Figure 17-8 **The segmentation of an earthworm is internal as well as external. Many structures repeat in each segment. The external bristles, setae, help the earthworm move.**

Anus

Setae

Brain

Hearts

Blood vessels

Intestine

Mouth

Reproductive structures

Waste removal tubes

Main nerve cord

Mollusks

Have you ever walked along the seashore collecting shells that have washed up on the beach? Or, have you read a restaurant's menu that offered steamed clams, fried squid, and escargots (snails) in garlic and butter? If so, then you are familiar with some of the diverse group of animals called mollusks—organisms with a soft body enclosed by a shell or shells. Although many different kinds of mollusks exist, they all have certain features in common. They are bilateral, have three tissue layers and a coelom, and also have both a mouth and an anus.

Most mollusks, except for the group that includes clams and oysters, have a well-developed head. Sensory organs such as eyes and tentacles are located on the head, along with a tongue-like organ—the radula—that has hard teeth used for scraping algae off rocks. A muscular organ called the foot moves mollusks across the bottom of their ocean or freshwater habitats or even over dry land. The single foot has been modified into tentacles or arms in the squid and octopus and is used for food-getting and movement. A mass of organs for digestion, reproduction, excretion, and circulation is located above the foot. This is known as the visceral mass. A unique folded tissue called the mantle hangs down over the visceral mass and creates a cavity where the gills are located. Calcium carbonate, which is used to form the shell, is secreted from the mantle. On the basis of both internal and external features, it would seem that mollusks and segmented worms are not at all closely related. However, comparison of embryos indicates that they probably evolved from a common ancestor.

Scientists have found that the shell-less mollusks, the squid and octopus, are both exceptionally intelligent animals. They are capable of learning behavior and adapting their actions to locate and capture prey. Their behavior is more like the behavior of vertebrates than invertebrates.

Figure 17-9 Mollusks are a group of animals that are very diverse in size and shape. However, all mollusks have the same basic body plan. The body plan of a mollusk is shown below. A bay scallop (top left), day octopus (bottom left), banded snail (right).

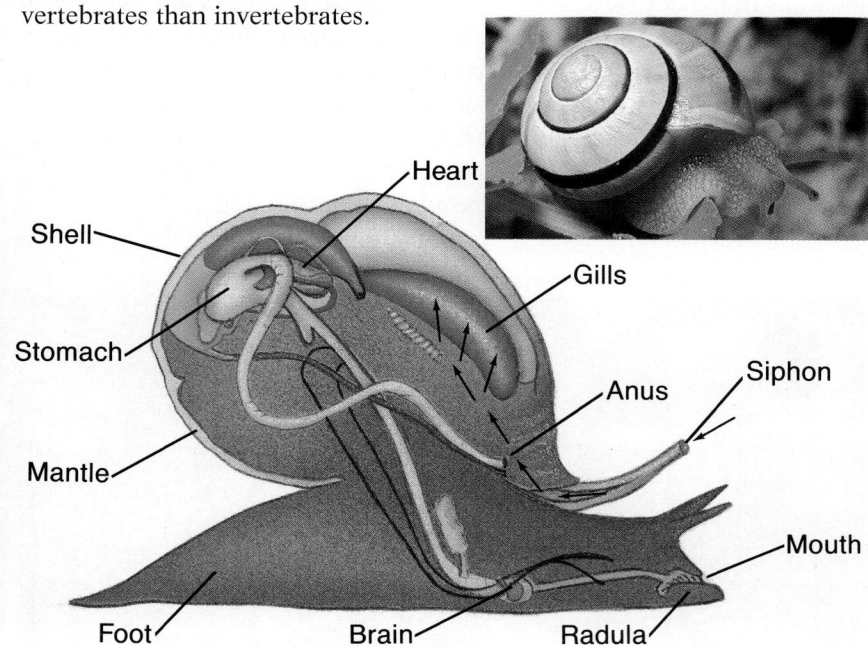

Arthropods

You watch as the heroine walks slowly down a dark staircase in an old castle. The only light comes from a smoky torch she holds in one hand. A sudden noise causes her to swing around. As she turns, something brushes across the side of her face. A sigh of relief escapes her when she realizes that the large spider hanging from its silky thread is only interested in finding a juicy insect and not in harming her. This scene is *not* typical of how Hollywood has presented spiders, bees, shrimp, lobsters, and other arthropods as major villains in science fiction and horror movies. Hollywood's arthropods usually lose the battle, while in real life, they are one of nature's most successful groups. About a million species of these animals have been found. They live in more habitats than any other group of animals. Indeed, if numbers of species or diversity indicate success, then the arthropods win hands down.

Figure 17-10 *Peripatus,* the most likely link between segmented worms and arthropods, can be found in the rain forests of South America.

A Link Between Two Groups Arthropods have several features in common with the segmented worms, the most prominent being body segments. Scientists have also found an animal that seems to be a link between the segmented worms and arthropods. Figure 17-10 shows *Peripatus.* This animal is segmented and has a soft outer skin like the segmented worms, but it also has stumpy legs with claws and a transport system like that of some arthropods. The phylum to which *Peripatus* belongs may have branched off from the evolutionary line leading to arthropods.

Novel Adaptations Several important adaptations have led to the amazing success of arthropods. A hard outer covering, or **exoskeleton,** protects the animal from predators, prevents loss of water from the body through evaporation, and provides support for tissues and organs. The exoskeleton also provides a point of attachment for muscles. The ends of the muscles are attached to different segments, thus enabling movement.

Figure 17-11 All arthropods have pairs of jointed appendages. This moth is an arthropod. What pairs of appendages do you observe?

Arthropods have pairs of jointed appendages. An **appendage** is a structure—such as an arm, leg, or antenna—that grows out of the body. In arthropods, the jointed appendages have evolved into a variety of sizes and shapes based on their functions. Some are used for walking, swimming, feeding, spinning webs, or gathering sensory information. Because of the exoskeleton, muscles, and jointed appendages, arthropods have evolved a variety of movements much more complex than those of any animal discussed so far. Consider the speed, height, and suddenness of a grasshopper's jump, or the graceful flutter of a butterfly's wings during flight.

Arthropods differ from segmented worms in that the segments of arthropods are fused together into distinct body regions or compartments. Evolution of such regions is associated with greater specialization, with particular functions being carried out in specific sections. This is an advance over duplicating a function in each individual segment, as in the excretory organs of annelids.

A variety of well-developed sensory organs confers another evolutionary advantage to the arthropods. The complex nervous system receives information from these sensory organs. Antennae provide information about sounds, odors, tastes, and touch sensations.

GLOBAL CONNECTION

Fellow Beings on the Land—The Native American View of Animals

There is no one people called "Native Americans" or "American Indians." There are hundreds of different Native American peoples, all with differing beliefs and cultures. So there is no single Native American way of considering animals. There are, however, some broad, underlying ideas that various Native Americans hold in common, though they may express them differently.

Native American beliefs generally include a strong tie between people and the land. They stress the unity of the entire natural world, including both animate and inanimate objects and such phenomena as the wind and the rain.

In such a world view, it's only natural that animals have an important place. Rather than seeing animals as lower creatures fit only for work or food, Native American traditions often portray animals as beings as important as or even more important than humans. Many cultures stress the importance of giving great respect even—or especially—to animals hunted as food.

Some Northwestern Native American peoples, for example, considered salmon, a major food source, to be higher beings who took the form of fish for the purpose of feeding people. If such beings weren't treated respectfully, they might become angry and never return as salmon again.

Often, animals were depicted as the descendants of great animal spirits responsible for the creation of the world or for the origination of social customs. The roles they play in Native American myths and legends not only illustrate how certain natural things came to be, but also communicate cultural values.

An Iroquois legend, for example, tells about how the owl made the Creator angry by greedily demanding too much and arrogantly sticking his beak in where it didn't belong. Rather than giving the owl what he asked for, the Creator gave him a short neck so he wouldn't be able to crane his neck to look where he shouldn't and big ears so that he could better hear what he was told.

A Papago myth tells of the Creator being saddened by the thought that the beautiful children,

Visual images are gathered by sets of simple eyes and a unique pair of compound eyes. The multiple lenses or facets allow an arthropod to view images from all around it at the same time. This is one reason why it is so difficult to successfully swat at and hit a fly.

Reproduction in arthropods is mainly sexual, many species with internal fertilization. Males form sperm that are deposited into the female's body. This process provides a greater chance of success because the sperm are deposited in the same area where the eggs are released. Animals that have internal fertilization are no longer dependent on a water environment to carry the sperm to the eggs.

---

the lovely leaves, and the colorful flowers would all grow old and die. To preserve some of the beauty, he took the beautiful sights and sounds around him and created butterflies, only to be reproached by the birds. The birds were upset that the Creator had given butterflies the melodious songs that belonged to them. Realizing that it wasn't right to take from the birds what was meant to be unique to them, the Creator let the butterflies keep their colors but took back their songs.

In the preceding legends, there is a creator distinct from the animals. But often it's an animal, or an animal spirit, who is responsible for some or all of creation.

Much Northwestern Native American lore deals with Coyote— an anthropomorphic hero and ancestor of the animal that shares his name. One story from the Nez Percé tells how Coyote tricked and killed a monster who was eating all his fellow animals. To celebrate his victory, Coyote created a new animal from the monster's remains. He cut up the monster's body, throwing the pieces in all directions.

Wherever a piece landed, a tribe of Native Americans came into being.

Note that this myth doesn't show humans being created as a dominant species, but simply as a new animal. They are to share the world, not control it. And while other Native American tribes had different myths, this one depicts an attitude shared by virtually all.

1. **Writing About Biology:** Compare the story of Coyote creating people to the creation story of some other culture. What similarities and differences are there in the way the two tales portray the relationship of humans and animals to nature and each other?

2. **Issues:** Even today, animals play a large part in Native American religious practices. The American Indian Religious Freedom Act, signed into federal law in 1978, was supposed to assure Native Americans' rights to exercise their religions. Even

so, some Native Americans have been prosecuted for using eagle feathers in their ceremonies. Eagles are also protected by federal law as an endangered species. What do you think should take precedence: Native Americans' rights to religious freedom or maximum protection of an endangered species?

Table 17.1 Summary of Major Characteristics of Invertebrates

Phylum	Sponges	Cnidarians	Flatworms	Roundworms	Segmented Worms	Mollusks	Arthropods
Locomotion	none	sessile; free-swimming	muscles; cilia	muscles	muscles	muscles	muscles, skeleton
Symmetry	none	radial	bilateral	bilateral	bilateral	bilateral	bilateral
Number of body openings	pores	one	one	two	two	two	two
Number of tissue layers	none	two	three	three	three	three	three
Nervous system	none	nerve net	present	present	present	present	present
Digestive system	none	none	present	present	present	present	present
Excretory system	none	none	none	present	present	present	present
Circulatory system	none	none	none	none	present	present	present
Respiratory system	none	none	none	none	present	present	present
Skeletal system	none	none	none	none	none	none	exoskeleton

Sponge Sea anemone Leech Mussel Crayfish

Section Review

Understanding Concepts

1. What features seen in roundworms are not present in less complex invertebrates?
2. Why is a true body cavity an important evolutionary development?
3. What evidence links the segmented worms, mollusks, and arthropods as closely related phyla?

4. **Skill Review—Sequencing:** Develop a sequence for the new features found among the roundworms, segmented worms, and arthropods. For more help, see Organizing Information in the **Skill Handbook.**

5. **Thinking Critically:** You have only microscopic slides of cross sections of two animals to study. What clues would help you decide which is a flatworm and which is a roundworm?

17.3 Echinoderms and Chordates

Segmented worms, mollusks, and arthropods represent one major path of evolutionary development. The next organisms you will learn about are those animals scientists think have followed another path of evolutionary development.

Echinoderms

Its seems unlikely, from just looking at them, that the feather star, sea cucumber, and starfish all belong to the same animal phylum. Yet they are all members of the group of marine animals called echinoderms, or spiny-skinned animals. It also seems unlikely that echinoderms are considered the most advanced group of invertebrates. After all, these animals exhibit radial symmetry, a characteristic not seen since the very simple invertebrates. Scientists know, however, that echinoderms have bilateral symmetry in their larval stage. Echinoderms also have several well-developed organ systems. One exception to these well-developed systems is their nervous system, which is not centralized. A nerve ring is usually located around the mouth with nerves radiating to other parts of the animal's body. This decentralization works well in animals with radial symmetry because it allows them to respond to stimuli coming from any direction.

As mentioned, the name of this phylum means "spiny skinned." The name is appropriate because spines project through the skin of these animals. The spines are made of calcium carbonate plates that link together to form an **endoskeleton,** which is an internal framework or support system. Reproduction in echinoderms is usually sexual. Gametes are released into the water, and fertilization occurs externally. Echinoderms have an amazing ability to regenerate parts. Some reproduce asexually this way by breaking apart, with each part then regenerating the missing body parts.

Objectives

Compare the features that evolved in echinoderms and chordates.

Describe the characteristics of vertebrates.

Analyze why certain adaptations were necessary before vertebrates could live on land.

Figure 17-12 Starfish (left) and sea cucumbers (right) are two members of the phylum Echinodermata.

Echinoderms have a unique water-pumping system that includes a series of internal canals connected to external body parts called tube feet. Echinoderms utilize this system to catch food, move, and for gas exchange. While the movements caused by this system of locomotion are not very fast, animals such as the starfish prey on clams and oysters. Speed isn't necessary to locate these stationary prey. However, this system does allow the starfish to exert a constant, pulling force to open the shells of its prey. You will learn more about the water-pumping system of echinoderms in Chapter 26.

Hemichordates

Once classified in the chordate phylum, these animals are now classified in a phylum of their own. In fact, hemichordates are so named because of the characteristics they share with chordates. These characteristics include a nerve cord running along the dorsal side of the animal and gill slits (discussed later). However, the nerve cords of hemichordates and chordates are not homologous structures, a fact which suggests that they do belong in separate phyla.

Hemichordates also share certain characteristics with echinoderms. Most notably, the larval stage shows bilateral symmetry. The similarities among the three groups lead biologists to believe that echinoderms, hemichordates, and chordates evolved from a common ancestor.

Chordates

Remember when the alien visitor went to the zoo and aquarium to learn about animals? Most of the animals the alien was able to study in these places belong to a phylum known as the chordates. In fact, it's likely that most of the animals with which you are familiar belong to this group. You, too, are a member of the chordate phylum.

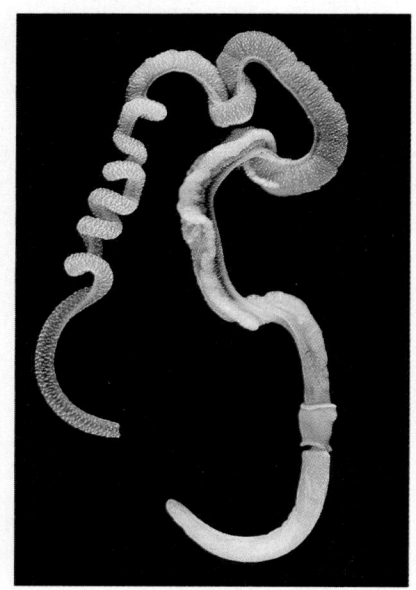

Figure 17-13   **This acorn worm (*Saccoglossus kowalewskii*) belongs to the phylum Hemichordata. This species is common to the Atlantic coast of North America.**

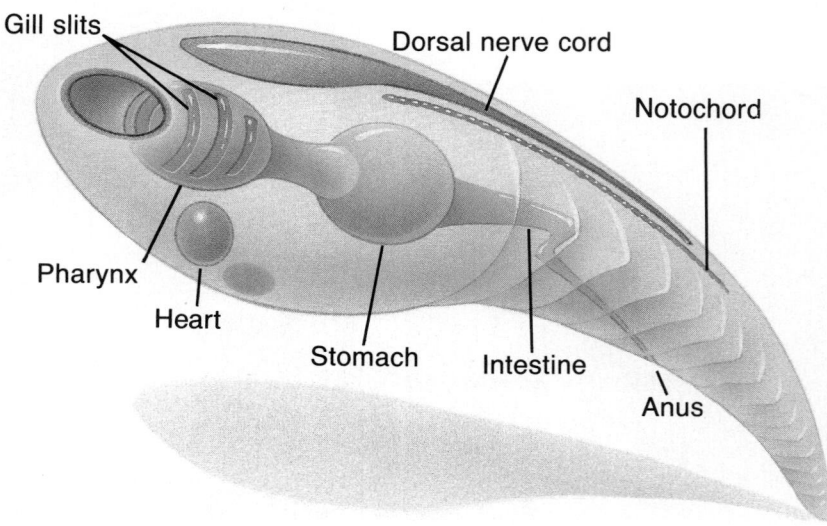

Figure 17-14   **This generalized chordate shows all the chordate characteristics: a dorsal nerve cord, notochord, and gill slits.**

All chordates have three common characteristics: a notochord, a dorsal nerve cord, and gill slits or gill pouches. Each characteristic is present at some time during the life cycle of a chordate. However, in some animals the characteristic might appear only during the embryonic stage.

The first characteristic, the notochord, is a dorsal rod made of cartilage. This rod is flexible, yet firm enough to provide support for the animal's body. The dorsal nerve cord is a group of nerves that form a hollow tube. It is located above the notochord. The gill slits or pouches are perforations in the pharynx or throat. In some animals, these slits develop into gills that are used for gas exchange. Other chordates have gill slits only in the larval stage. Which of these three characteristics do you still have?

Chordate Subphyla   The chordate phylum is divided into three subphyla. Two subphyla include animals that do not have backbones. One of these two subphyla includes marine organisms known as tunicates. Tunicates are sessile animals that grow singly or in colonies attached to rocks. They have a tough outer covering, called a tunic, and two body openings. When disturbed, contractions of the body force water out of the openings, thus earning these animals the common name of sea squirt.

The second subphylum of chordates without backbones includes the lancelets. These marine animals have streamlined bodies that are pointed at each end. In fact, they bear some resemblance to a small fish. Although lancelets can swim, they usually live partly buried in sand. There, they feed on microscopic organisms that are drawn into the mouth by the beating of cilia that line the oral cavity.

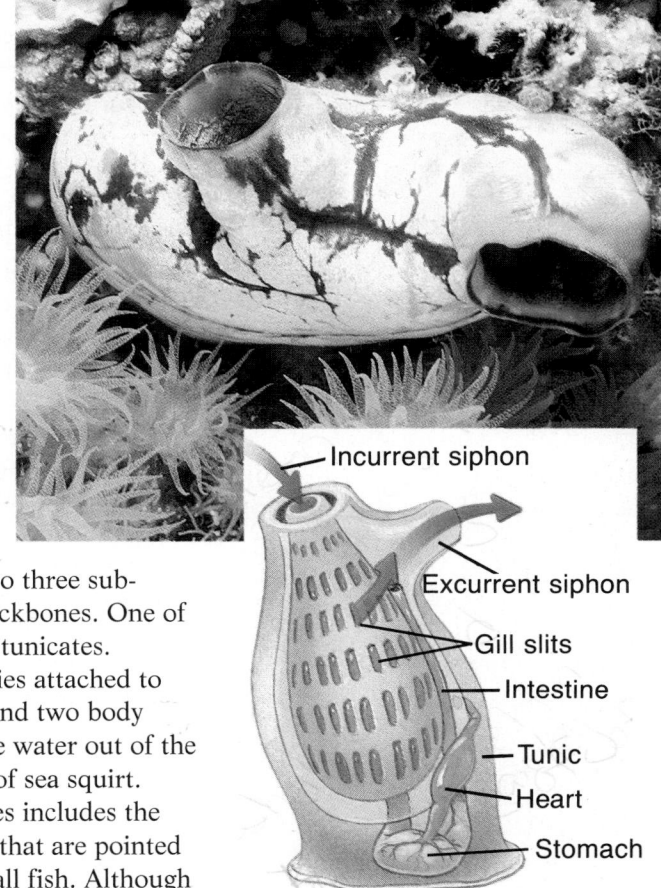

Figure 17-15   Tunicates, also called sea squirts, exhibit all chordate characteristics only in their larval stage. What chordate characteristic remains in the adult?

Figure 17-16   Unlike tunicates, lancelets retain all chordate characteristics throughout their lives.

Vertebrate Characteristics The third and largest of the chordate subphyla are the vertebrates. Vertebrates have a backbone, made of bony segments or cartilage, that develops around the notochord or replaces it. The bony or cartilaginous segments making up the backbone are called **vertebrae.** Vertebrae provide protection for the animal's spinal cord. They also fit together in a way that allows great flexibility of movement. It is from these structures that the group receives its name. In addition, the backbone is the major support structure of the endoskeleton. Unlike the endoskeleton of echinoderms, the endoskeleton of vertebrates is made of living cells and tissue. This adaptation allows the support structure to grow with the animal—which means that vertebrates can grow to enormous sizes. Whales, elephants, and the extinct dinosaurs are proof of the success of this adaptation.

The nervous system of vertebrates has evolved into a highly efficient system with a complex brain that is separated into different regions, each with different functions. The regionalization allows for more specialization. Greater sensitivity to sounds, odors, images, and vibrations, as well as increased intelligence, resulted in vertebrates having much more complex behaviors than the invertebrates. As in arthropods, the nervous system also interacts more with other systems of the body. For example, it interacts with the muscles that are attached to the endoskeleton. The interaction of the muscular, skeletal, and nervous systems means vertebrates have greater speed and agility than most other animals.

Thinking Lab

INTERPRET THE DATA

What is the function of ringed seal whiskers?

Background
You have seen cats carefully washing their faces and whiskers. Many mammals have whiskers. Ringed seals that live in the cloudy water of Lake Saimaa in Finland may use their whiskers for finding the six pounds of fish they need to eat each day. With only six feet of visibility in the water, seals must use some other means besides sight to find fish. Biologists think that the seals may be using echolocation, emitting sounds and detecting the echoes, with their whiskers. The seal makes vocal clicking sounds that may be picked up by whiskers as the sound bounces back off a nearby object or fish. Biologists know that the whiskers of most mammals have about 150 nerve fibers attached to them, while ringed seals' whiskers have 1200 nerve fibers attached. Biologists hypothesize that the extra nerve fibers may be involved in echolocation.

To test their hypothesis, a team of biologists placed several ringed seals in an indoor pool. The pool had reduced visibility similar to that of their natural habitat. One hundred fish were added to the pool. The biologists noted that each seal caught six fish. Next, the biologists blocked the seals' whiskers with a cover and conducted the same experiment. Here, each seal caught an average of two fish. The biologists concluded that their hypothesis was correct.

You Try It
Interpret the biologists' results. Were they correct in their conclusions? What was missing? Suggest other methods biologists might use to determine the purpose of the extra nerve fibers.

Another important adaptation is the evolution of jaws. As jaws evolved, new and more efficient ways of eating were developed. Rather than depending on sucking, rasping, or filter feeding as most invertebrate groups do, animals with jaws and teeth can grasp and hold on to prey. They have a distinct advantage in the competition for food.

Evolution of Vertebrate Classes

Vertebrates are subdivided into two major groups, the fish and the tetrapods—vertebrates with legs. Fish are streamlined, aquatic organisms that breathe by means of gills and move about by means of fins. There are three living fish classes, one of which lacks jaws.

Jawless Fish   Jawless fish include the lampreys, a familiar animal to people who live along the Great Lakes. These parasites have been responsible for great losses to the sport and commercial fishing industries. Lampreys have a sucker-shaped mouth that is lined with sharp teeth. They use their mouth to pierce the flesh of trout or other fish, and then feed on the blood and tissue of the host.

Cartilagenous Fish   The two remaining fish classes have jaws. One, in fact, was immortalized for this distinctive characteristic in a movie by the same name. We are talking, of course, about sharks, skates, and rays—the cartilagenous fish. These fish have skeletons composed entirely of cartilage and skin that is covered with toothlike scales. Although they breathe by means of gills, the gill openings are not covered. The evolution of hinged jaws in this group is associated with a more predatory way of life. Cartilagenous fish probably evolved from a group of armored fish that lived more than 400 million years ago.

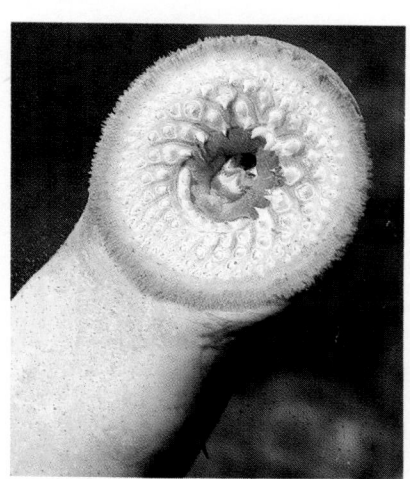

Figure 17-17   The round, jawless mouth of a lamprey is lined with sharp, toothlike structures.

Figure 17-18   The many sharp teeth, great size, and quickness of sharks make them fierce predators.

Bony Fish   Bony fish probably arose from the same armored ancestor that gave rise to the cartilagenous fish. As their name implies, these animals have skeletons of bone. The flattened, bony scales along with the slimy covering of mucus on the body protect and streamline the animal. Instead of the exposed gills that cartilagenous fish have, these fish have gill coverings composed of hard plates.

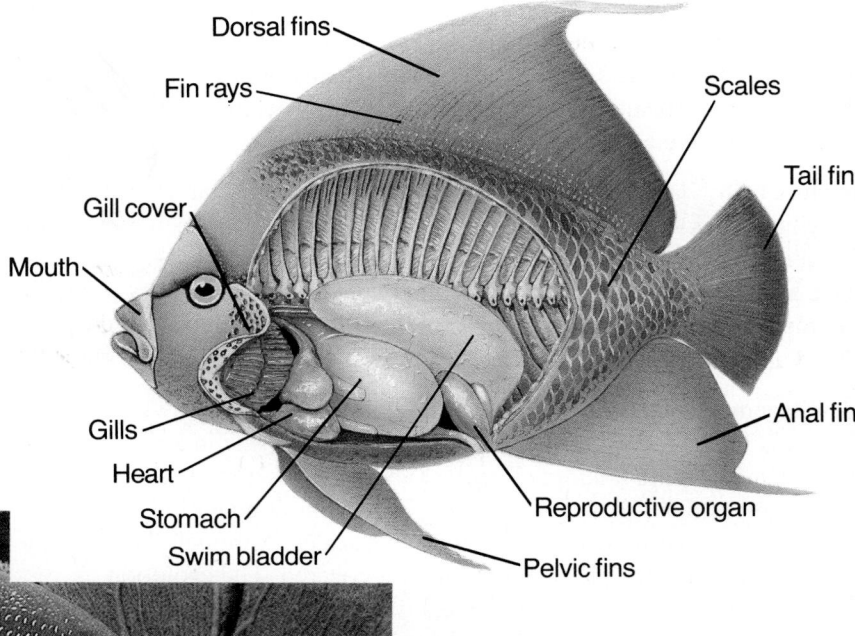

Figure 17-19   Although bony fish come in a wide variety of sizes and shapes, they all have the same basic structures, as shown in this French Angel fish.

Amphibians   Four other classes of vertebrates are grouped together based on the presence of legs, as opposed to fins. Most primitive among these classes are the amphibians—frogs, salamanders, toads, and newts.

The ancestor of today's amphibians was the lobe-finned fish that lived about 350 million years ago. The lobe-finned fish had two characteristics that favored their ability to survive on land, at least for short periods of time. The lobed fins enabled movement on dry surfaces. Primitive lungs supplemented the gills in delivering oxygen to the body. Scientists speculate that millions of years ago, there was a drought that caused shallow lakes and stream beds to dry up. Lobed-finned fish would have been able to survive the drought and reproduce by seeking out new, deeper sources of water. Natural selection would favor the survival of these animals, and their traits for surviving on land would be passed to their offspring.

Over the course of millions of years, genetic mutations and natural selection would lead to further modifications in traits, and animals with primitive legs would appear. The evolution of primitive legs set the stage for the adaptive divergence of land vertebrates. That divergence led not only to the evolution of amphibians, but also to reptiles, birds, and mammals.

Most modern amphibians, like their early ancestors, are still chained to the water for reproduction. Water provides a protective, moist environment for developing embryos. And, because fertilization is external in most water-dwelling vertebrates, the water provides a medium for sperm to reach the eggs. Most amphibians have gills in the larval stage and lungs as adults. Thus, they spend part of their life cycle in the water and part of it on land.

Figure 17-20 This Tiger salamander (*Ambystoma maculatum*) can be found across most of North America from the plains to high-elevation forests. Although it has a wide range, it must have water for reproduction.

Reptiles If vertebrates were to move completely away from water, adaptations for different modes of reproduction and development, as well as for preventing dehydration, were necessary. In vertebrates, such adaptations first appeared among the reptiles. Reptiles evolved from an amphibian ancestor nearly 300 million years ago. These animals, which include lizards, snakes, turtles, and alligators, don't require water for reproduction or development. Major advancements that allowed for the separation from water were internal fertilization of the eggs and protection of the eggs by an outer covering—the shell. A reptilian egg can be compared to a hen's egg. Each has a yolk—the food supply for the developing embryo—and a water supply. As adults, reptiles have well-developed lungs for gas exchange. Their dry, scaly skin prevents evaporation of body fluids. Most reptiles have legs that are positioned under the body, an adaptation that allows quicker movement on land.

Figure 17-21 The English viper of Costa Rica (right) and an American alligator (left) are representatives of the reptile class.

Birds   Birds are believed to have evolved from a small, two-legged dinosaur more than 150 million years ago. Today's birds have several features in common with reptiles. Protective scales cover their legs, and they have a tough beak that is similar to that of some reptiles. Like reptiles, birds also have shelled eggs.

Birds have one feature that no other group of animals possesses—feathers. Feathers, which are modified scales, are associated with the ability of birds to fly. Flight is a major innovation among the vertebrates. Other vertebrates, such as flying fish or squirrels, can glide for short distances. However, only birds and bats (mammals) are capable of sustained flight. Birds are by far the strongest fliers, some species migrating for thousands of miles from breeding grounds to wintering grounds. Large, strong muscles in the breast and lightweight, hollow bones also allow for flight.

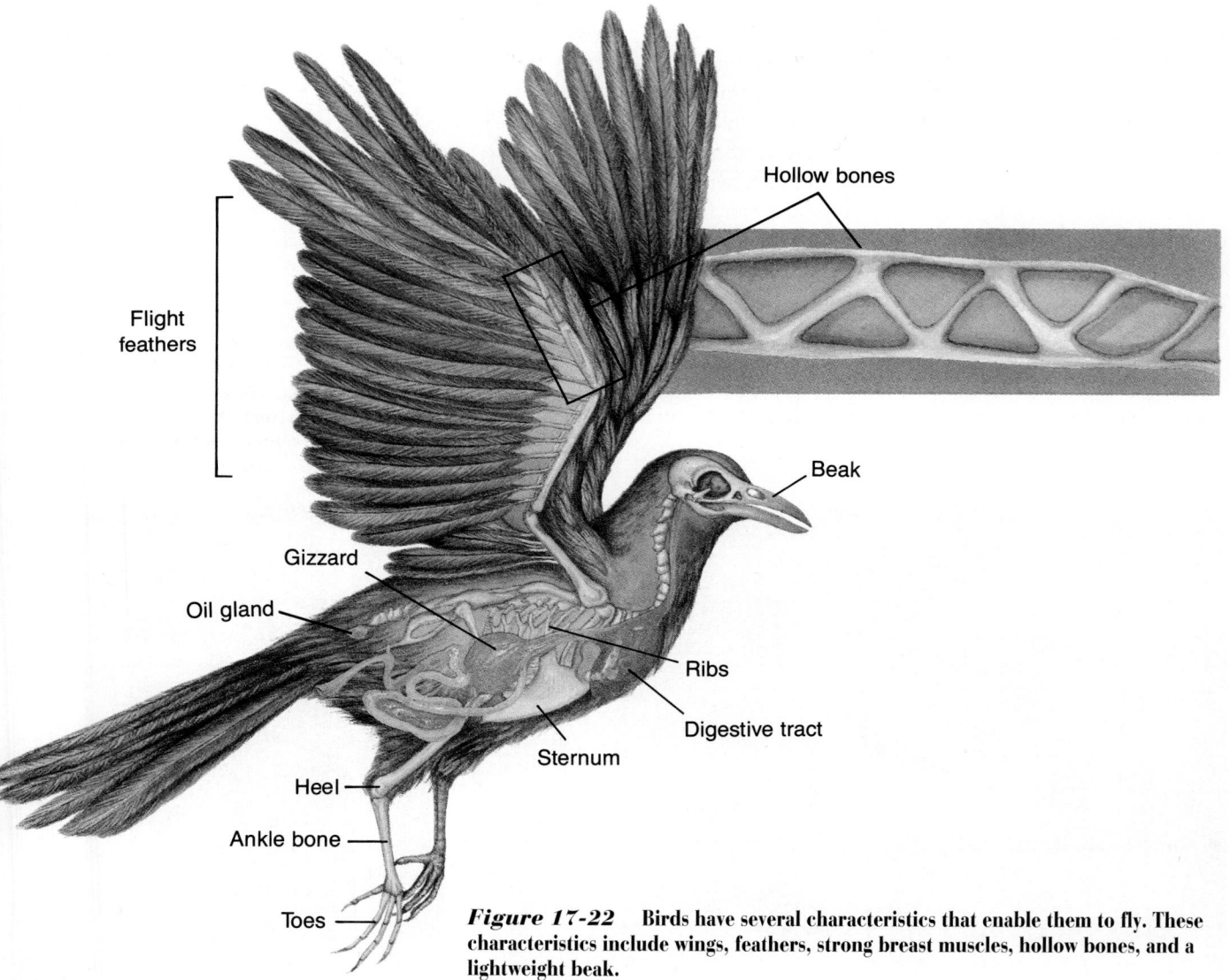

Flight feathers

Hollow bones

Beak

Gizzard

Oil gland

Ribs

Digestive tract

Sternum

Heel

Ankle bone

Toes

Figure 17-22   **Birds have several characteristics that enable them to fly. These characteristics include wings, feathers, strong breast muscles, hollow bones, and a lightweight beak.**

Quite possibly one of the most important adaptations that enabled birds to conquer the air is one that can't be observed directly. This adaptation involves the method by which birds regulate their internal body temperature. Birds have high energy requirements due to their aerial lifestyle. Maintenance of a constant body temperature is necessary for the internal chemical processes that are required to sustain high levels of activity. Animals that maintain a constant body temperature through internal metabolic actions are called **endotherms.** Regardless of the external temperature, an endothermic animal is able to remain active. For example, an owl is able to hunt even in much cooler nighttime temperatures, but a snake must seek shelter once the sun sets.

Endothermy is in direct contrast to temperature regulation in fish, amphibians, and reptiles, which are ectotherms. **Ectotherms** are animals whose body temperature can change with the temperature of the environment. Unlike endotherms, ectotherms generally have low metabolisms and are less active. To help regulate internal body temperature, ectotherms rely on a variety of behaviors. You may have seen a turtle sunning itself on a log in the cool morning, or retreating to the shady undergrowth of the shore during the hot afternoon. Both actions
to regulate its body temperature. Sunning itself increases its
perature, while retreating to the shade lowers its temperatur

Why is temperature regulation so important in the first p
enzymes necessary for metabolism function best at tempera
40°C. Vertebrates must keep their internal temperature clos
optimum enzyme functioning.

Minilab 2

What types of food do different birds prefer?

Observe birds in your backyard, schoolyard, or nearby park for several days. Obtain a field guide and identify several of the most common species you see. Obtain several different sizes and types of birdseed. Place the seed out for the birds and record the type of food each species of bird prefers.

Analysis   *Which types of seed did the birds prefer? Relate the bird's size, beak shape, and any other physical characteristics to the food*

Mammals   Mammals are thought to have evolved from a group of endothermic, mammal-like reptiles. Exactly when the first true mammals appeared is difficult to tell, but about 70 million years ago there was a sudden, dramatic increase in the number of mammal species. This increase in mammal diversity followed the extinction of the dinosaurs, leading scientists to believe that the two events were directly related. Since then, mammals have come to occupy every conceivable habitat—on land, in water, and even in the air.

LITERATURE CONNECTION

Gorillas in the Mist

In the 18 years she lived among them, Dian Fossey came to know mountain gorillas better than any other human being did. Her book, *Gorillas in the Mist,* tells how she transformed herself from an occu-pational therapist in America to a field zoologist in the wilds of Rwanda, as well as one of the fore-most authorities on mountain gorilla behavior. In the process, she survived violent weather, life on dangerous moun-tain slopes, deal-ings with witch doctors (friendly and unfriendly), and political upheaval.

The book is not only Fossey's story. It's also the tale of four gorilla families. The reader gets to know the gorillas as Fossey did, each as an individ-ual with its own personality.

Blended in with Fossey's exploits as a zoological researcher, the adventures of daily gorilla life, and scientific informa-tion on gorilla behavior is the story of the fight for the gorillas' very existence. Fossey tells of her strug-gle to protect the mountain gorillas from illegal hunting and the destruction of their habitat. Her efforts included hiring private anti-poacher patrols to remove traps and make poaching as difficult as possible in and around her research area.

Dian Fossey was found mur-dered in her Rwanda camp in 1985—two years after the publica-tion of her book. Many believe that she was killed by poachers angered by her forceful attempts to protect the gorillas.

Gorillas in the Mist is a stirring legacy left by a dedicated researcher whose work led her down many unexpected paths. Fossey's life and work stand as an example and inspiration to biologists, naturalists, and lovers of animals.

1. **Explore Further:** Obtain from your local library copies of the book and movie versions of *Gorillas in the Mist.* First read the book, then view the movie. How do the two differ? Which did you prefer?
2. **Writing About Biology:** Many people view conservation versus development to be an all-or-none issue. Write a paragraph describ-ing a middle ground on this type of issue.

Figure 17-24 Mammals have many adaptations for survival. Several of these adaptations are represented here. The fur of the Arctic fox (top left) allows these organisms to survive Arctic temperatures. Beavers demonstrate complex behaviors in dam building (top right); and a horse's greatest protection against predators is its speed.

There are several reasons for their success. First, mammals feed their young with milk from mammary glands. This adaptation helps to ensure survival of the offspring by providing them with a readily available, nutrient-rich food source. A second adaptation that aids offspring survival is internal development of the young. This protection during development gives mammals the best chance of having a fertilized egg mature into an adult. Third, mammals have hair for insulation. This adaptation, in addition to endothermy, helps them to maintain a constant body temperature. Fourth, mammals are extremely agile. The legs are positioned under the body, allowing for greater speed. Finally, mammals have complex nervous systems, thus allowing for complex behaviors and increased intelligence.

Table 17.2 summarizes the characteristics of vertebrates. As you review this table, make sure that you can explain how the evolution of new features allowed each class to be successful.

Table 17.2 Summary of Characteristics of Vertebrates

Class	Examples	Outer Covering	Body Temperature	Limb Structure	Gas Exchange	Fertilization
Jawless fish	lamprey, hagfish	slimy skin	ectotherm	unpaired fins	gills	external
Cartilagenous fish	shark, skate, ray	scales	ectotherm	2 pairs of fins	gills	internal
Bony fish	perch, bass, trout	scales and slimy skin	ectotherm	2 pairs of fins	gills	external
Amphibians	frog, toad, salamander	slimy skin in most forms	ectotherm	2 pairs of legs; no claws	gills; lungs	external
Reptiles	turtle, lizard, snake, alligator	dry, scaly	ectotherm	2 pairs of legs; claws	lungs	internal
Birds	robin, eagle, pelican	feathers; scales on legs	endotherm	1 pair of wings; 1 pair of legs with claws	lungs	internal
Mammals	bear, whale, kangaroo	hair/fur	endotherm	2 pairs of legs; claws in most forms	lungs	internal

LOOKING AHEAD

T HE NEARLY 2 million different species making up all of the five kingdoms illustrate the great diversity of life on Earth. Remember, though, that there is unity within that diversity. The many forms of life carry out the same basic life processes. One of the most fundamental of these processes is reproduction. In Chapter 18, you will learn about the basic patterns of reproduction that occur and study specific examples of how each pattern is carried out.

Section Review

Understanding Concepts

1. In which way are echinoderms and chordates similar?
2. What characteristics of vertebrates allow them to be such a successful group?
3. What advantages do endotherms have over ectotherms?
4. **Skill Review—Classifying:** Construct a classification key that could be used to identify the various classes of vertebrates. For more help, see Organizing Information in the **Skill Handbook.**
5. **Thinking Critically:** Based on features discussed in this section, explain why reptiles do not have to eat as much or as often as birds.

Considering the Environment

White Death in the Coral Reefs

Around the world, oceanographers and other environmental scientists are becoming alarmed as a record number of coral reefs turn white and sometimes die. The delicate color we think of as coral doesn't belong to the coral itself. Individual coral animals, known as polyps, have a clear body over a limestone skeleton. Polyps live in tight colonies and, as each dies, it leaves its skeleton behind for another polyp to grow on. Where, then, does the coral color come from?

The Color of Coral   The algae, zooxanthella and their chloroplasts give coral polyps their green, yellow and brown colors. This algae lives symbiotically with the coral. During the day, the algae produce oxygen and simple sugars through photosynthesis. These products become available to the coral polyps for food. In return, the algae get carbon dioxide and nitrogen produced by the polyps as waste. The cooperation works beautifully for both organisms, creating—over millions of years—the coral reefs of today. But the system is delicate and easily disrupted.

If the polyps' environment becomes poor and they become stressed, they will eject their algae. The coral then turns white because the limestone skeletons become

visible through the polyps' clear body tissue. Because it looks as if the coral reef has been bleached, the effect is known as coral bleaching.

Causes of Bleaching   A lot of things can cause the stresses that lead to coral bleaching; pollutants and poisons in the water, damage from hurricanes, changes in water salinity, and natural diseases are among them. But a major cause is overly warm water.

Coral can survive only in a relatively narrow range of temperatures. A difference of only one degree above normal annual high temperatures for only a few weeks can cause coral reefs to bleach. A jump of only four degrees above normal highs can cause bleaching in a few days.

> **CAREER** CONNECTION
>
> **Oceanographer** is the general name for scientists who study the sea. They may study the composition or movements of the water itself, the plant or animal life that lives there, or underwater geology. Many have recently developed specialties in such topics as ocean conservation, public policy, and archaeology. Virtually any study having to do with the sea is fair game for an oceanographer.

Worse, if a reef is already stressed by something, a small additional stress from another source can cause coral bleaching. For example, a slight warming that would otherwise pose no danger can start bleaching if the coral is already stressed from a recent hurricane.

Many oceanographers and environmental scientists consider the increased incidence of coral bleaching a possible early indicator of global warming. But, because there aren't yet any long-term data on coral, it will be some time before definite conclusions are possible.

There have been many other factors in recent years that may have greatly contributed to the frequent bleachings, including epidemics of coral diseases, large numbers of strong hurricanes, and surges in the populations of animals that prey on coral.

Fortunately, coral reefs can survive short periods of bleach-ing. If the stress on a reef ceases, polyps will regain their algae and color in a matter of weeks. Frequent or long-lasting bleach-ings, however, have already killed several reefs.

1. **Find Out:** Name three other marine plants or animals that are endangered when a coral reef dies.

2. **Writing About Biology:** Some scientists think it is possible that loss of the coral reefs may result in a noticeable drop in the amount of carbon dioxide being made in the world. What effects on global ecology might such a drop have?

Issues

Drawing the Line on Animal Pain

There are few issues in the world of science as emotionally charged as the issue of experimentation on animals. Each side has horror stories to back up its point of view.

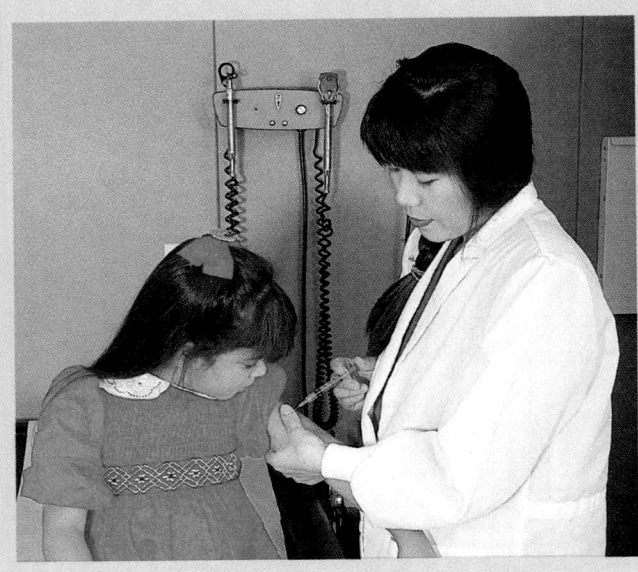

The literature of those who fight for animal rights is filled with tales of cruelty to animals. There are accounts of animals blowtorched in burn studies, deafened or addicted to drugs in studies of social behavior, systematically and painfully poisoned to determine levels of tox-icity, and much more—all without anesthetics.

Scientists who work with animals have their own tales to tell. They often hear stories of children dying from diseases that can't be cured today, but which may someday be eliminated through animal research. They also tell stories of people who are alive today because of treatments developed through animal research.

Cystic fibrosis (CF) is a case in point. Cystic fibrosis is a hereditary disease caused by a defective gene. Until recently, most people born with CF suffered painful death in childhood. By the mid-1980s, treatment had progressed enough to allow more than half of the victims to survive into their twenties. The treatment, say experimenters, could not have been developed without animal research. As this book is being written, there's finally hope that the disease can be cured. The cure is being tested on monkeys.

Most advanced surgical techniques have been developed with animal tests. Coronary bypass operations and organ transplants are examples. Animal experiments led to the eradication of smallpox, a dreaded killer that became the first disease to be entirely wiped out by modern medicine. It's impossible to know how many lives have been saved by penicillin and the polio vaccine we've come to take for granted—both of which were tested and proven on animals. Thousands of people with diabetes are alive today because of the animal research that led to our understanding of insulin.

Most animal rights proponents are willing to allow research necessary to save lives, provided that the animals are treated as humanely as possible. But many question whether much of the research is truly necessary. They claim that a great deal of the work could be done with computer simulations, and they suspect that animals are only used because it's less expensive than developing alternate methods.

Experimenters counter that computer simulations are only useful in studying a system that's fully understood. Medical research, they say, deals with so many unknown factors that no accurate simulation is possible.

But there's more to animal testing than the medical research that might one day save your life. There's cosmetic research as well. One of the most notorious tests involves putting cosmetic ingredients directly onto the eyeballs of live rabbits to see just how irritating they are. Rabbits are used because their eyes don't make tears to wash the irritating substances away. Public opposition to such practices has lead several cosmetic companies to abandon animal tests. There are now stores devoted entirely to selling cosmetics and other products developed without research on animals.

A highly controversial area is military research that tests weapons on animals. Many animal rights activists are appalled by the idea of killing and mutilating animals for the sole purpose of testing the ability to destroy. On the other side is the view that military actions are an unavoidable evil in our world. Should war develop, having adequate knowledge of the destructive power of our weapons could save the lives of countless soldiers and civilians. Should our country be in danger, the knowledge might be critical in making decisions that could mean the difference between victory and defeat.

The questions don't stop at the treatment of animals in the laboratory. Even when animals are treated humanely, as most are, there's the problem of what to do with test subjects once research is finished. Many are routinely killed. Some can theoretically be released into the wild, but preparing them to survive on their own can be an extremely

expensive process lasting years. Others will never be able to live alone in the wild and are unfit to be pets. Caring for them can be costly. Furthermore, many animal rights activists maintain that dooming animals to a life in captivity is in itself cruel.

There are two extreme positions—one that any activity that causes animals to suffer is wrong, the other that any animal suffering leading to improved conditions for humans is justified. In between there are no easy answers.

Understanding the Issues

1. Under what conditions, if any, do you think experimentation on animals is justified? Defend your view.

2. Suppose you discovered that your favorite shampoo was tested on animals in painful experiments to determine how irritating its ingredients might be. Would you change brands? Think of whatever it is you're most looking forward to buying right now. Suppose it was developed with animal tests. Would you change your mind about buying it?

3. Should the criteria for animal testing be the same for all types of animals? Is experimentation on a rat more justifiable than experimentation on a chimpanzee?

Readings

- Ritvo, Harriet. "Toward a More Peaceable Kingdom." *Technology Preview,* February/March, 1992, p. 54.

- Karpati, Ron. "A Scientist: 'I Am the Enemy.'" *Newsweek,* December 18, 1989, p. 12.

Summary

About 95 percent of all animals are invertebrates, animals with no backbone. Sponges are the least complex of the invertebrates. They are made of two layers of cells and lack tissues, organs, and systems. Sponges are sessile heterotrophs.

The cnidarians have two tissue layers with a jellylike substance between them. They have a primitive nerve network, a single body opening, and some type of locomotion. Cnidarians are radially symmetrical—all their body parts are arranged equally around a center point.

The flatworms have bilateral symmetry and distinct head and tail regions. They have three tissue layers and distinct organs. Their nerves are arranged into a ladderlike system and they have a primitive digestive system.

Based on their pattern of early development, one group of animals, the segmented worms, mollusks, and arthropods, are thought to have diverged from a common ancestor. New adaptations that evolved in these organisms include a coelom, separate sexes, two body openings, true systems, segmentation, body sections, and features that permitted life on land.

The other main group of animals includes echinoderms, hemichordates, and chordates. The chordates include the subphylum of vertebrates, which have evolved a variety of adaptations for life in water and on land.

Language of Biology

Write a sentence that shows your understanding of each of the following terms.

anterior	endotherm
appendage	exoskeleton
bilateral symmetry	invertebrate
coelom	mesoderm
dorsal	posterior
ectoderm	radial symmetry
ectotherm	ventral
endoderm	vertebra
endoskeleton	vertebrate

Understanding Concepts

1. How is being a hermaphrodite both a benefit and a disadvantage to an organism?
2. How did evolution of tissues allow for more complexity in an organism?
3. Other than segmentation, what other features would allow you to distinguish between a roundworm and a segmented worm?
4. How is being radially symmetrical a benefit to a sessile organism?
5. What adaptations would an endotherm from a cold environment have to make if it were relocated to a tropical climate?
6. Broadly speaking, what adaptations evolved in both insects and reptiles that allowed them to be successful away from water?
7. Based on information in Table 17.2, explain how you could distinguish between reptiles and amphibians. How could you distinguish between the jawless fish and bony fish?
8. Identify the features that evolved among vertebrates and explain the importance of each.
9. Explain the adaptations that led to the evolution of land vertebrates.
10. Compare and contrast the body plans of segmented worms and arthropods.

Applying Concepts

11. In what ways are mollusks with two shells, such as clams, similar to sponges?
12. Why are arthropods such a successful group?
13. How are a starfish and a hydra similar? How are they different?
14. Arthropods and mollusks are very successful groups of animals. In what ways are they similar?
15. Why do many ectotherms migrate or hibernate? Do endotherms do so for the same reason?
16. What hypothesis could you make to explain the advantages of sessile animals and parasites having both male and female sex organs even though their free-living relatives have separate sexes?
17. **Considering the Environment:** Why do many oceanographers and environmental scientists consider an increase in coral bleaching to be a possible early indicator of ocean pollution?
18. **Global Connection:** Why do you think animals became so important in Native American culture?

Lab Interpretation

19. **Investigation:** You would like to know if a particular species of hydra is temperature sensitive. The graph below shows numbers of these organisms active in samples observed under a microscope at various temperatures. From the data on the graph, determine the optimum temperature range for this hydra. What experimental procedure would you follow to check if the data shown on this graph are accurate?

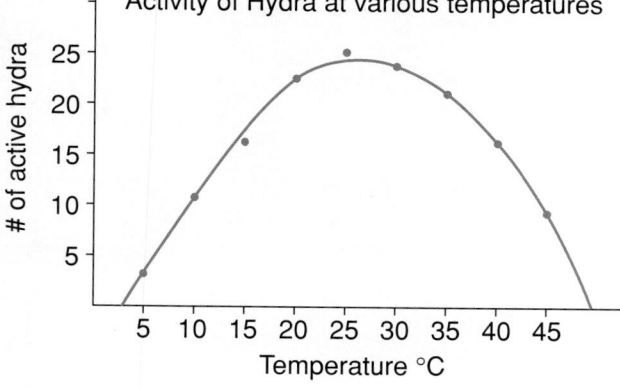

Activity of Hydra at various temperatures

20. **Thinking Lab:** Assume that an animal's ability to use its whiskers for echolocation would require that more nerve fibers be attached to the whiskers. A scientist finds that animal A has 200 nerve fibers attached to each whisker; animal B, 150 nerve fibers; animal C, 900 nerve fibers; and animal D, 1000 nerve fibers. Which animals, would you hypothesize, might use their whiskers for echolocation? Why do you think that?
21. **Minilab 2:** Observe the teeth of the animals pictured below. From your observations, what types of food do you think each eats? Explain your answer.

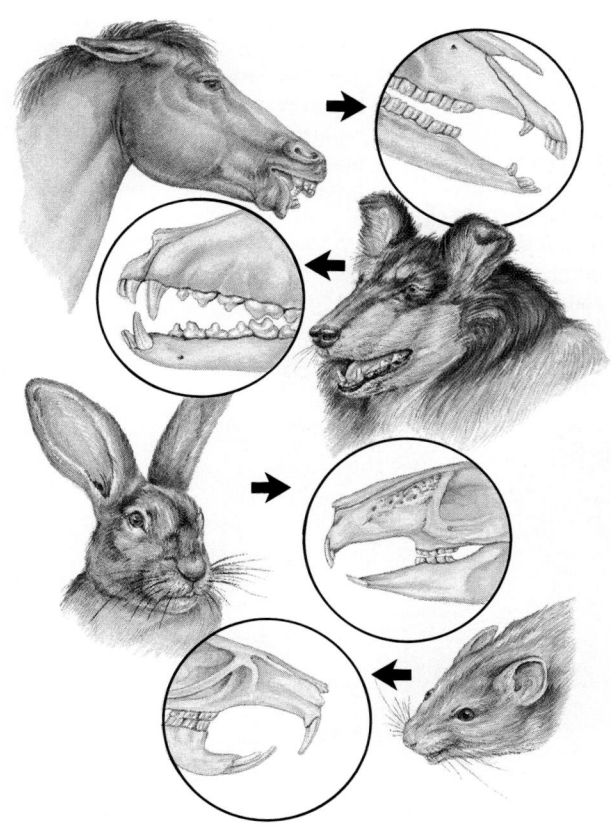

Connecting Ideas

22. In what ways are animals similar to protozoa? How are they different?
23. In what ways was the evolution of insects and reptiles similar to that of plants?

Life Functions of Organisms

This close-up view of the hooks and eyes of de Mestral's locking tape shows how the two strips connect. The Swiss mountaineer hoped that his invention would replace metal zippers as a clothing fastener. Why do you think this has not happened?

Cocklebur seeds have hooks that enable them to cling to animal fur and other types of materials. Other plants have seeds that float in the wind or on water.

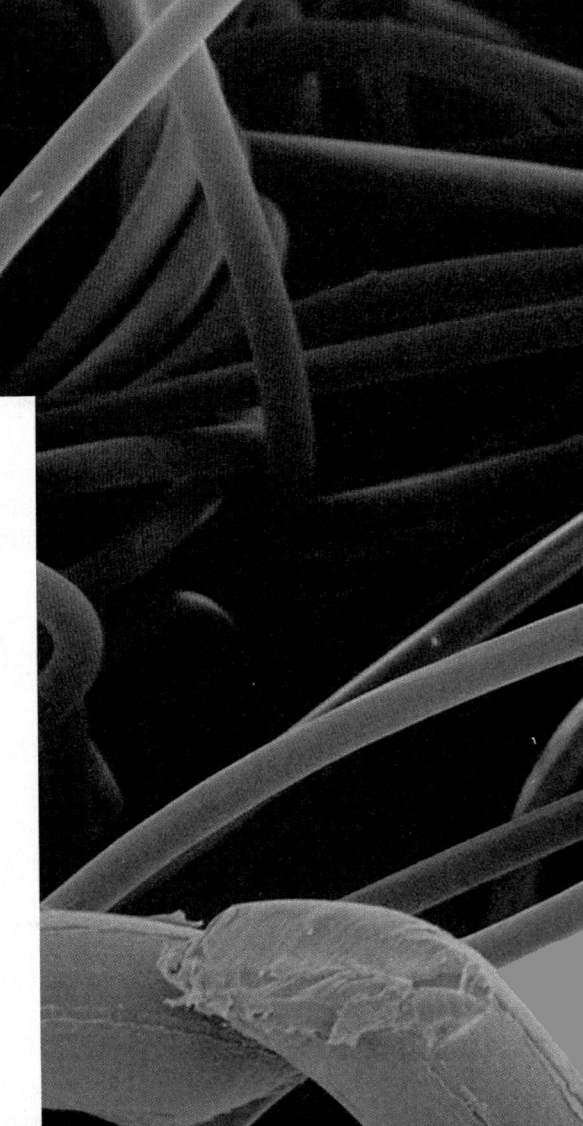

Why do seeds like the cocklebur have hooks, spines, barbs, or bristles? The seeds of flowering plants are dispersed by wind, water, or as hitchhikers on organisms that move from place to place. The dispersal of seeds far away from the parent plant reduces competition with the parent for sunlight and water.

In 1948, a Swiss mountaineer named George de Mestral discovered that cockleburs attached to fibers by way of tiny hooks and wondered if this same technique could be used to fasten clothing. Working with a weaver, de Mestral developed two strips of cloth, one with tiny hooks and the other with smaller eyes. Today, de Mestral's locking tape is used to secure items in space, seal the chambers of artificial hearts, and, of course, to fasten clothing from diapers to tennis shoes. The reproductive strategies of a plant led to an invention that has changed how people do things. Reproduction is just one of the life functions that you will examine in this unit.

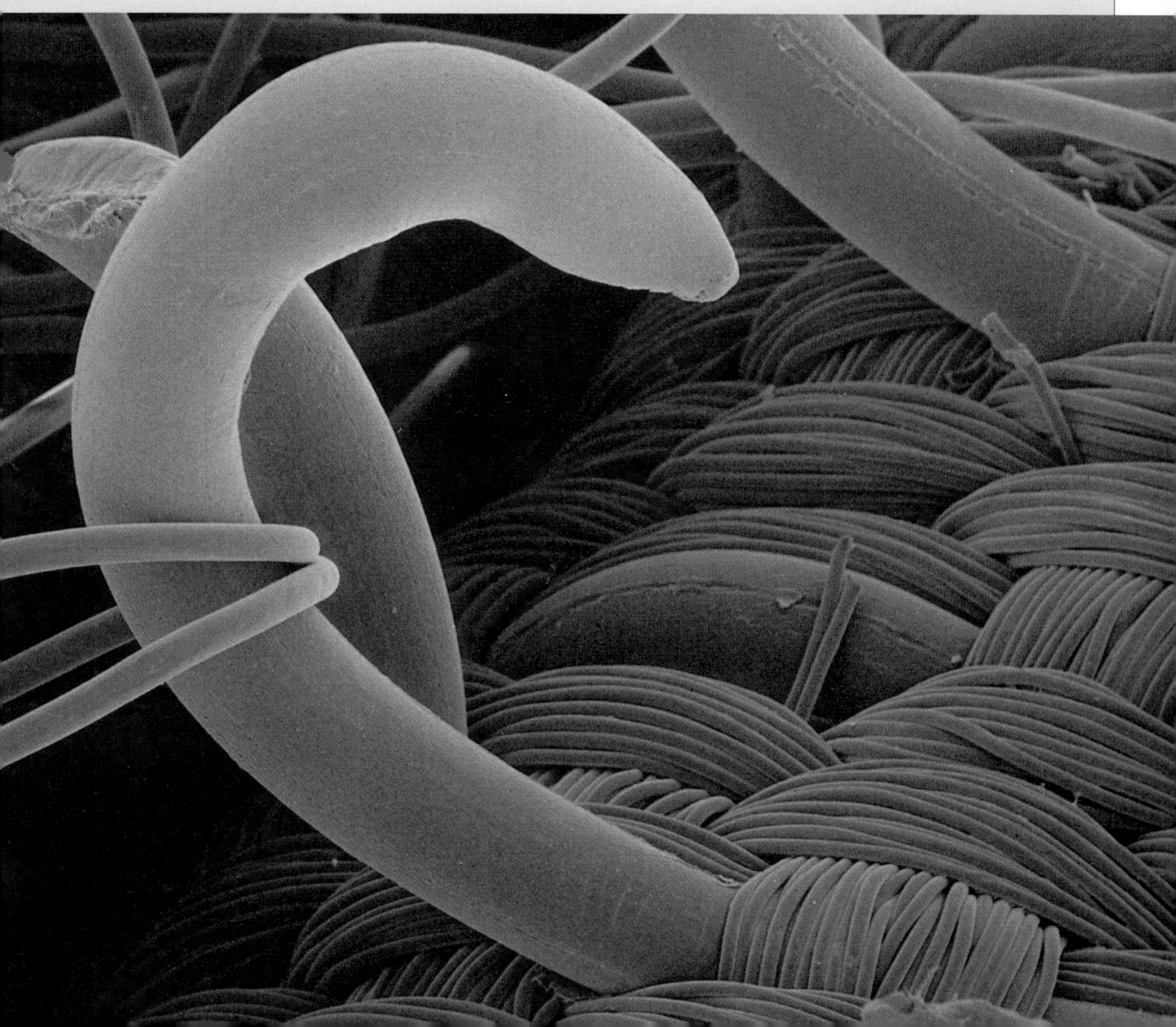

REPRODUCTION

ONE MOMENT, a large, milky white seed pod hangs from the stem of a milkweed plant. The next moment—poof! The pod suddenly bursts open, and feathery white seeds are cast to the breeze. Hundreds of such seeds may be scattered by the wind—some perhaps to favorable spots on the ground where they germinate and grow into new milkweed plants. In fact, the enormous number of seeds that a single milkweed plant produces increases the chances that the plant will successfully reproduce.

Many plants overproduce seeds; this helps ensure continuation of the species. After all, the chances that a single milkweed seed or tomato seed or watermelon seed will take root may be limited by many environmental factors. The chances that at least one out of hundreds will survive and grow is much better. Likewise, many animals overproduce offspring. A toad, for example, may produce as many as 10 000 tadpoles in one clutch. Some large spiders produce more than 2000 offspring at a time. In contrast, some organisms—including humans—normally produce just one offspring at a time.

Regardless of the numbers, the result of reproduction is new members of the species. Reproduction involves the transfer of genetic information from parent to offspring—information that tells how the new species member will be structured and how it will function. In some organisms, the genetic information in the offspring is exactly the same as in the parent; in others, the information is a combination of information from two parents. In this chapter, you'll explore the different ways in which organisms from each of the five kingdoms reproduce.

If your school grounds had ten milkweed plants, and every plant produced 500 seeds each year, and every seed grew to be an adult plant, how long would it be before you would have to move to another school?

Chapter Preview

Objectives

Compare and **give examples** of different forms of asexual reproduction.

Hypothesize how asexual reproduction is an adaptation in plants and animals.

18.1 Asexual Reproduction

Did you ever see a tulip begin to emerge from beneath the soil in early spring? Have you ever noticed new grass growing in what had been almost bare ground? Did you ever see a green shoot start to develop from the eye of an old potato? If you answered yes to any of these questions, you've seen evidence of asexual reproduction. In each case, a new organism was produced from a single parent, with no new combination of genetic material. The tulip, the grass plants, and the sprouting potato are all genetically identical to the individual parents that produced them.

You'll find organisms that reproduce asexually in each of the five kingdoms. In fact, in monerans, asexual reproduction is the only method by which offspring are produced. Many other organisms can undergo either asexual or sexual reproduction, which involves the combining of two sets of genetic information. As you have seen in Chapter 16, plants have a life cycle in which there is an alternation of haploid and diploid generations. One generation reproduces by mitosis; the next generation reproduces by meiosis.

Reproduction by Splitting

There are several kinds of asexual reproduction. In one kind, the parent organism divides into two new offspring by splitting into two. Monerans normally reproduce by binary fission. In binary fission, the DNA in the parent cell duplicates, then the cell divides into two identical cells. Under ideal conditions, some monerans can divide into two new organisms every ten to 20 minutes. In unicellular protists such as *Amoeba* and *Euglena,* the process of asexual reproduction occurs by mitosis and cell division and takes several hours. First, the DNA in the nucleus duplicates, then the nuclear membrane breaks down, and eventually the cell divides into two.

Figure 18-1 **Asexual reproduction by binary fission in prokaryotes (left, magnification: 30 000×) and by mitosis and cell division in unicellular eukaryotes such as *Paramecium* (right) results in offspring and parents with identical genetic information.**

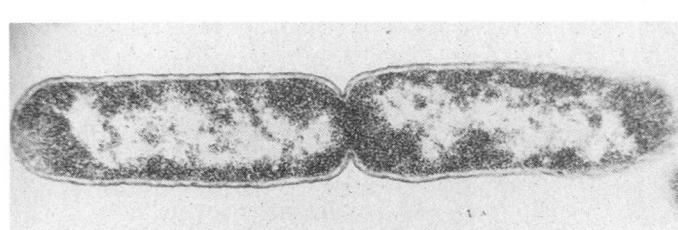

macronucleus
micronucleus
Oral groove
Contractile vacuole
Mitosis of micronucleus
Division of macronucleus
Paramecium

Cell division in the protist *Paramecium* is more complex because, as described in Chapter 15, it has two kinds of nuclei. The smaller micronucleus controls reproduction, while the large macronucleus controls other cellular processes.

Reproduction by Budding

A second method of asexual reproduction involves the growth and separation of a small part of a parent organism to form a new organism. In some organisms, a miniature and complete version of the parent grows out from the parent body, breaks off, and develops into an adult. In animals, such as *Hydra* and sea anemones, and in fungi, such as yeast, this process is called **budding.** In plants, this kind of asexual reproduction is called **vegetative reproduction.** For example, in strawberry plants, spider plants, and many kinds of grasses the new plants form at the end of thin, creeping stems that grow out from the parent plant. When the young plants become rooted to the ground, the stem that joins them to their parent eventually rots away. The new plants will then grow to be independent adult plants.

There are other forms of vegetative reproduction. Have you ever planted tulip bulbs, gladiolus corms, or potato tubers? Bulbs, corms, and tubers are food storage organs that allow a plant to survive over winter or during a drought. When the temperature and water supplies are good for growth, the organs begin to produce leaves. The food that is stored in the organs is used as an energy source for development of new parts such as stems and leaves. Food made during the growing season is stored in new underground storage organs at the end of the season. Many new bulbs, corms, or tubers are produced each year. If you have ever dug up bulbs, corms, or tubers, you may have noticed that you had more at the end of the season than what you planted in the spring. Your plants have reproduced by vegetative reproduction.

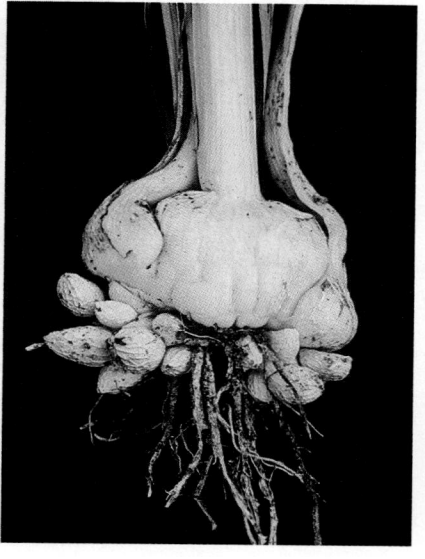

Figure 18-2 Miniature *Hydra* buds grow out from one parent (left). In a similar way, tiny *Gladiolus* corms form from a single parent during its growing season (right).

Figure 18-3 **Repair of damaged tissue is common to most organisms. In starfish (left), whole new arms can be regenerated. Regeneration of tissue helps a gardener to graft beautiful rose plants to the stems of disease-resistant rose plants (right).**

Reproduction by Fragmenting

A third kind of asexual reproduction is actually a form of repair to damaged or broken parts of organisms. Sometimes a small part of an organism breaks off, and the piece grows by mitosis to replace all the missing, major body tissues. In this process, called **fragmentation,** each small, broken piece of organism develops into a whole, new adult, exactly like its parent. For example, a sponge, planarian, or starfish can grow from a small fragment of a parent. The replacement or regrowth of missing parts is called **regeneration.** Planarians and starfishes show remarkable powers of regeneration as you can see in Figure 18-3. If a starfish is damaged by predators or harsh weather conditions, many of the broken pieces will often regenerate all the missing parts and eventually form a small population of starfishes, all exactly alike. When fragmentation occurs in a filamentous prokaryote, such as the cyanobacterium *Nostoc,* regeneration occurs by binary fission.

Gardeners often use the natural process of regeneration to form new plants artificially from small pieces of selected plants. One way to do this involves taking cuttings from a plant stem of the kind of plant you want to reproduce. If you place a stem cutting in moist sand, the piece of stem develops roots, which anchor the cutting and begin absorbing water and minerals. Other plant parts then develop by regeneration. In the same way, you can use roots, such as those of chrysanthemums, and leaves, such as those of begonias, for cuttings.

Many garden shrubs and fruit trees are artificially produced. For example, a cutting of a good fruit-producing tree is bound into a cut on the branch of a young branch or trunk of another tree that may be more resistant to disease. If this grafting process, in which the tissues of two stems of different but related plants regenerate and grow together, is successful, a shoot of the cutting will eventually produce healthy fruits. Apples, cherries, roses, and peaches are often reproduced in this artificial way.

Spores and Parthenogenesis

The fourth method of asexual reproduction involves the production of spores. Spores are produced by a variety of organisms—algae, some protozoans, many fungi, and plants. Some spores are produced by mitosis; others by meiosis. Some are diploid (2n), whereas others are haploid (n). All spores have features in common. Each spore contains DNA and a small mass of cytoplasm surrounded by a tough outer wall. This wall protects the spore until it encounters conditions that are favorable for growth. It also prevents the contents within from drying out, an important adaptation for organisms that live on land. Some organisms produce and release large numbers of spores, which are carried far and wide by water currents or wind. Production of many spores and their dispersal further increases the chances that some spores will find suitable places for growth.

You've examined asexual reproduction in some simpler animals, plants, and monerans. More complex animals can undergo asexual reproduction too. For example, some insects, such as aphids and honeybees, may sometimes reproduce by means of parthenogenesis. During this process an unfertilized egg develops into an adult. For example, the queen honeybee lays both fertilized and unfertilized eggs. The fertilized eggs develop into females, mostly workers. The unfertilized haploid eggs become males called drones. The only function of drones appears to be that they mate with the queen.

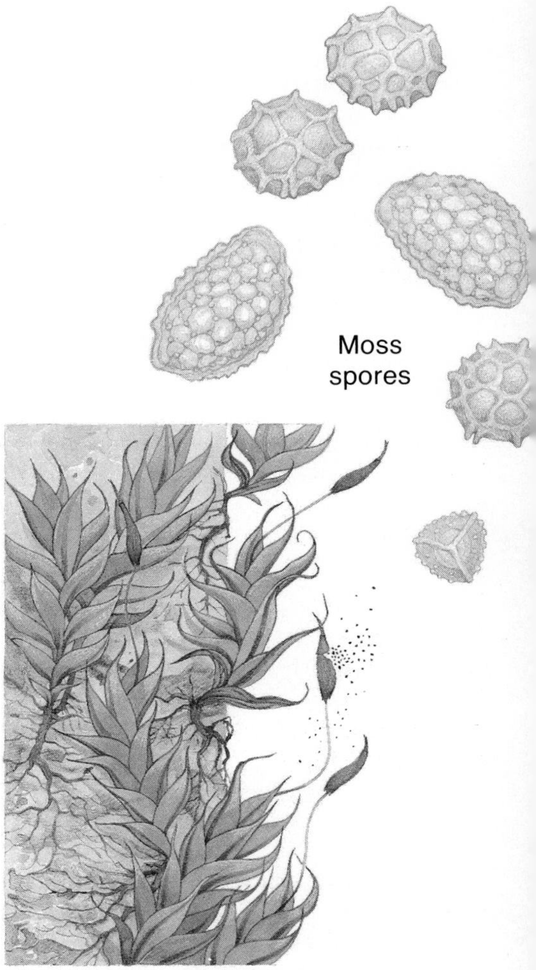

Moss
spores

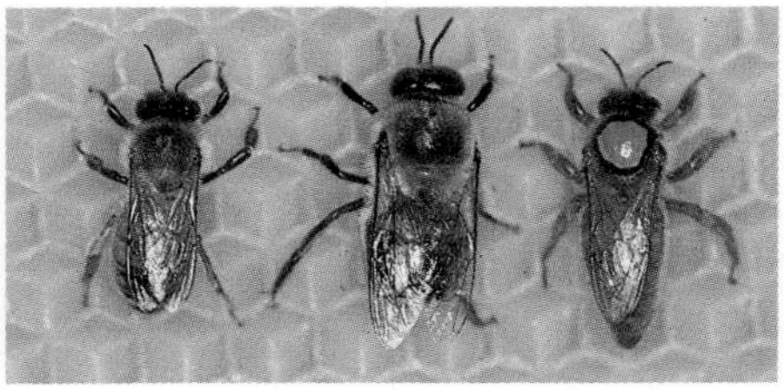

Figure 18-4 A queen bee (left) lays fertilized eggs that develop into workers (right), and unfertilized eggs that develop into drones (center) by parthenogenesis. Mosses are one kind of plant that reproduces by spores (above).

Section Review

Understanding Concepts

1. Compare asexual reproduction in a bacterium and a unicellular alga.
2. How are fragmentation and budding similar? How are these two methods different?
3. Explain how the term *budding* could be applied to asexual reproduction of a flowering plant.

4. **Skill Review—Forming a Hypothesis:** You discover an enormous population of clover in your lawn and observe that the plants are all identical. You know that just two weeks ago there were only one or two clover plants. How could you explain the rapid increase in the production of clover? For

more help, see Practicing Scientific Methods in the **Skill Handbook**.
5. **Thinking Critically:** What trait of spores makes them more efficient than any other means of asexual reproduction?

Objectives

Compare conjugation in different organisms.

Explain the generalized life cycle of plants.

Relate the idea of alternation of generations to the life cycle of a typical flowering plant.

18.2 Sexual Reproduction

In the previous section, you studied different means of asexual reproduction. You saw how to get new potatoes from an old potato and how two starfishes develop from one. You know that new plants can develop from shoots of mature plants, and that spores help many organisms reproduce asexually. While all these organisms may undergo asexual reproduction, they are also capable of reproducing sexually.

Conjugation

Whereas asexual reproduction is a splitting process that always involves one parent, sexual reproduction is the fusion of two different sets of DNA. Over evolutionary time, sexual reproduction has become the major form of reproduction among organisms. Although there are differing views on how it first evolved, the reason that sexual reproduction has become more common is generally agreed upon. When two different sets of DNA fuse, genetic recombination occurs, resulting in variety among offspring. As you learned in Chapter 12, variations are the raw materials for adaptation and evolution. With the exception of monerans, almost all forms of life are capable of some means of sexual reproduction.

Conjugation in Algae Most kinds of organisms reproduce by fusion of two kinds of gametes, eggs and sperm. During the process of fertilization, a zygote is formed. Eggs are often much larger in size than sperm. In most algae, sexual reproduction is by the fusion of two gametes that are similar in size. However, in a few algae, there are different mating strains that come together and form connecting tubes through which genetic material is transferred. The sexual process by which genetic material is transferred from one cell to another by cell-to-cell contact is called **conjugation.** During conjugation, the DNA of the two strains of an alga is combined.

Figure 18-5   Chlamydamonas is a unicellular green alga with one cup-shaped chloroplast (left, magnification: 500×). Sexual reproduction is by conjugation (right).

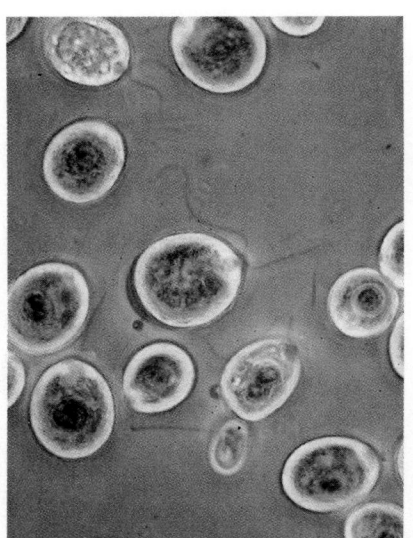

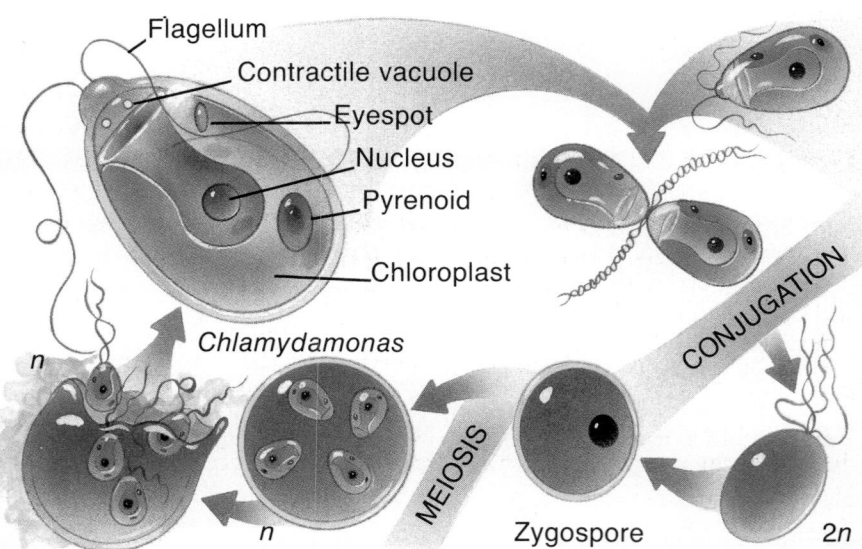

488 REPRODUCTION

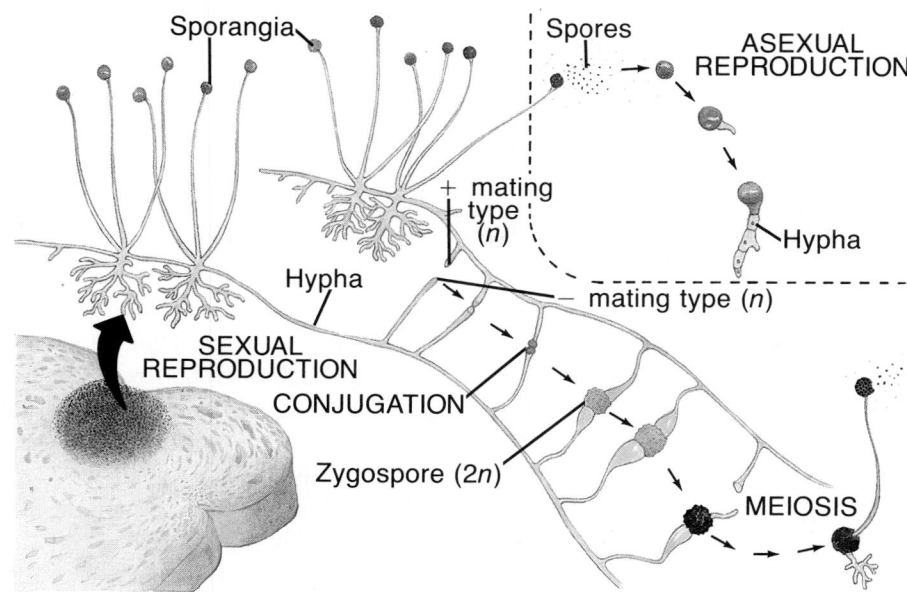

Sporangia

Spores

ASEXUAL REPRODUCTION

+ mating type (n)

Hypha

Hypha

– mating type (n)

SEXUAL REPRODUCTION

CONJUGATION

Zygospore (2n)

MEIOSIS

Figure 18-6   The bread mold *Rhizopus* is a mass of fine hyphae that can reproduce asexually by spores or sexually by conjugation.

Conjugation in Fungi   Conjugation can also be seen in fungi, such as the *Rhizopus* shown in Figure 18-6. Conjugation in this bread mold occurs between different types of hyphae, often labeled *plus* and *minus.* The many nuclei in fungal hyphae are all haploid (*n*). The plus and minus hyphae don't look different, but chemicals produced by each type of hypha cause them to be attracted to each other. Many nuclei acting as gametes move into the tip of each branch. Then, a cell wall forms behind the nuclei. The walls between the tips of the branches join and break down, allowing pairs of gametes to come together and fuse. The result is numerous diploid (2*n*) zygotes. A tough outer wall then develops around all the zygotes and forms a single zygospore.

Zygospores are adapted to extreme temperatures or dry conditions. After a period of dormancy, however, all but one of the zygotes breaks down. The remaining zygote undergoes meiosis, producing four haploid (*n*) spores. In turn, three of these spores die. The surviving spore develops a new hypha with a terminal sporangium and many asexual spores, which eventually produce new plus or minus strains of *Rhizopus.*

Conjugation in Bacteria   Although bacteria have not been observed to reproduce sexually, a few bacteria are known to transfer genetic material from one cell to another in conjugation. As in bread mold, there are two different mating types that can be called plus and minus.

Only plus and minus bacteria can conjugate. During conjugation, a bridge of cytoplasm forms between the two bacteria. A plasmid moves from the plus cell through the bridge and into the minus cell. Therefore, the minus cell has received genes it did not have before, and it now becomes a plus cell. Genetic recombination has occurred, but not reproduction because there has been no increase in the number of cells. However, bacteria receiving genes from other bacteria can pass those genes on to offspring via binary fission.

Figure 18-7   Bacteria exchange genetic material by forming connecting bridges of cytoplasm.

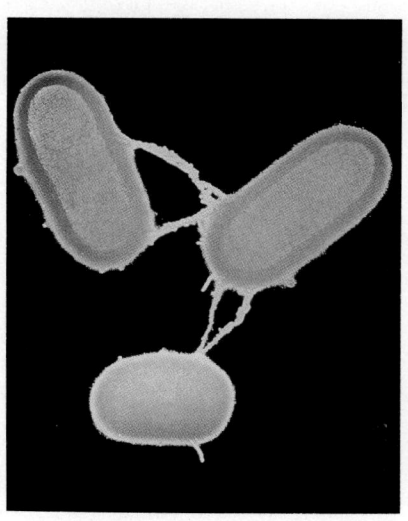

Plant Life Cycles

You've learned that plants reproduce asexually in a number of ways. Perhaps you already know that plants can also reproduce sexually with male and female gametes. You may recall from Chapter 16 that plants have a life cycle in which a gametophyte stage and a sporophyte stage alternate. The two stages make up two generations of plants. One generation is haploid (n) and produces gametes and the other generation is diploid ($2n$) and produces spores.

How exactly does the alternation of generations occur? Figure 18-8 shows the life cycle in a flowering plant. You will remember from Chapter 16 that during the gametophyte stage of the cycle, the plant is haploid. Its cells have one of each pair of chromosomes. Through mitosis, it produces eggs and sperm that also have half the normal number. Two gametes, produced from one or two parent plants, fuse to form a zygote, which develops into the next, or sporophyte, stage of the plant life cycle.

The sporophyte stage is diploid because its cells have two of each kind of chromosomes. Through meiosis, the sporophyte produces spores, each with half the number of chromosomes. The spores develop into new gametophyte plants and the plant life cycle continues.

I N V E S T I G A T I O N

Growing Bread Mold

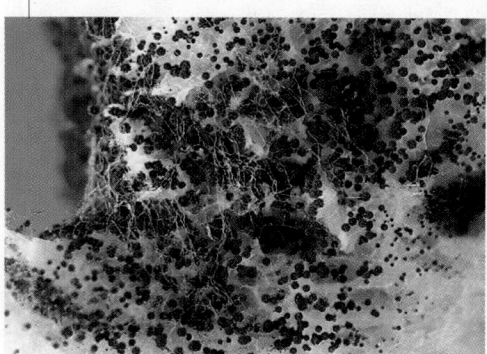

Magnification: 12×

Spores are reproductive cells that germinate under proper conditions to form new offspring. Spores are a common means of reproduction in fungi, protists, and some kinds of plants, such as mosses. The air is full of spores. Many are so small, they may seem like nothing more than dust particles. Not all spores find suitable conditions to grow, which probably accounts for the fact that the world is not covered in mold!

Mold spores are easily grown from the bread mold fungus *Rhizopus nigricans*. The spores give the mold its characteristic black color. Spores may be collected from a parent fungus with a damp cotton-tipped swab.

Problem

What conditions are best for the germination and growth of mold spores?

Safety Precautions

Wash your hands after handling fungi. Dispose of mold growths according to your teacher's directions.

Hypothesis

Working in groups, develop a hypothesis that could be tested to demonstrate the most suitable conditions for spore growth in bread mold.

Experimental Plan

1. Agree upon a way to test your two variables. Keep in mind the available materials.

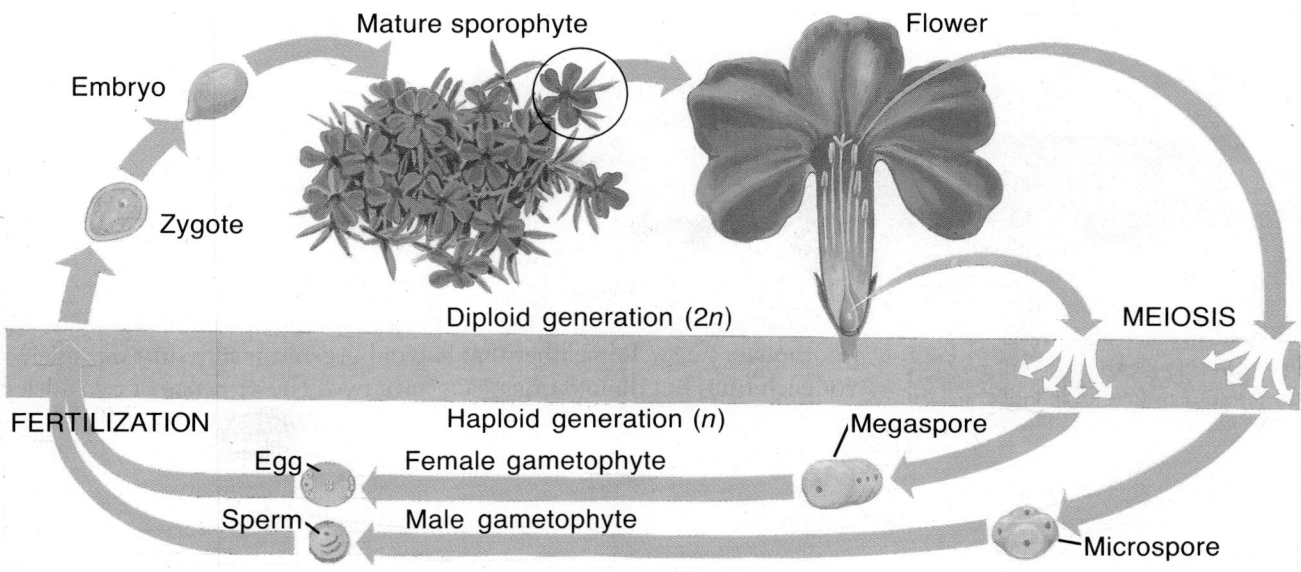

Figure 18-8 The alternation of sporophyte and gametophyte generations is typical for all plants. What two cell processes separate the two generations?

2. Determine how many spore samples you will need.
3. Remember to test only one variable at a time and use suitable controls for each condition.
4. List the conditions for possible mold growth.
5. Choose the conditions you wish to use to test mold growth.
6. Record your procedure and list materials that you will need.

Checking the Plan

1. Be aware of other possible variables that will have to be controlled.
2. What will be your control?
3. What quantitative data will you collect and how will you record these data?
4. How long will you allow the experiment to run?

5. How many trials will be used to test each variable?
6. Assign roles for your plan.

Make sure your teacher has approved your experimental plan before you proceed further.

Data and Observations

Carry out your experiment and develop a table to record your data.

Analysis and Conclusions

1. State your hypothesis. Describe the way in which your first variable was tested.
2. Which variables had to be controlled? What was your control?
3. Use quantitative data to compare the control group and the experimental group.
4. Based on your answer from

question 3, draw a conclusion for how the condition being tested influenced mold growth of *Rhizopus nigricans*.
5. Locate another group in your class that tested the same condition. How did their procedure differ from your group's? How did their data differ and how did their conclusions differ? Explain a possible reason for any differences.
6. Repeat questions 1-5 for your second variable being tested.

Going Further

Based on this lab experience, design an experiment that would determine the conditions of growth for yeast cells. Assume that you will have limited resources and will be conducting the lab in class.

Minilab 2

How do spores and pollen compare?

Obtain a male pine cone, and a fern frond with sporangia, or a moss with capsules. Prepare wet mounts of pine pollen and fern or moss spores. Observe each cell type under a microscope and make diagrams of each. Prepare a data table that compares structure and function for pollen and spores. Use your text to include the following information.
Analysis *By what process is each cell formed? Is each cell type haploid or diploid? Is each cell a part of a sporophyte or gametophyte generation?*

The relative importance of gametophyte and sporophyte generations varies from one type of plant to another. For example, mosses spend more time in the haploid gametophyte stage; following fertilization, the diploid sporophyte grows out from the body tissues of the gametophyte parent and lives for just a short time. As vascular plants evolved more complex tissues and systems, the diploid sporophyte became predominant and the haploid gametophyte became short lived, nonvascular, and quite small. In seed plants, which you'll examine shortly, the gametophyte is short lived and consists of only a few cells embedded within the sporophyte.

What advantage is there to a life cycle spent mostly in the diploid sporophyte stage? Remember that haploid organisms have just one allele for each trait, but diploid organisms have two. The presence of two alleles has several advantages. First, the effect of a harmful or lethal allele can be masked if it is recessive. Second, a greater variety of phenotypes can be expressed with two alleles for each trait as, for example, in incomplete dominance, codominance, and multiple genes. Having a greater number of phenotypes provides greater opportunities for differences upon which natural selection can act.

Life Cycle of a Flowering Plant

When you look at a flowering plant, you are observing a sporophyte. All of its many parts—roots, stem, leaves, and flowers—are composed of diploid cells. Recall from Chapter 16 that the flower is the organ of reproduction in these plants.

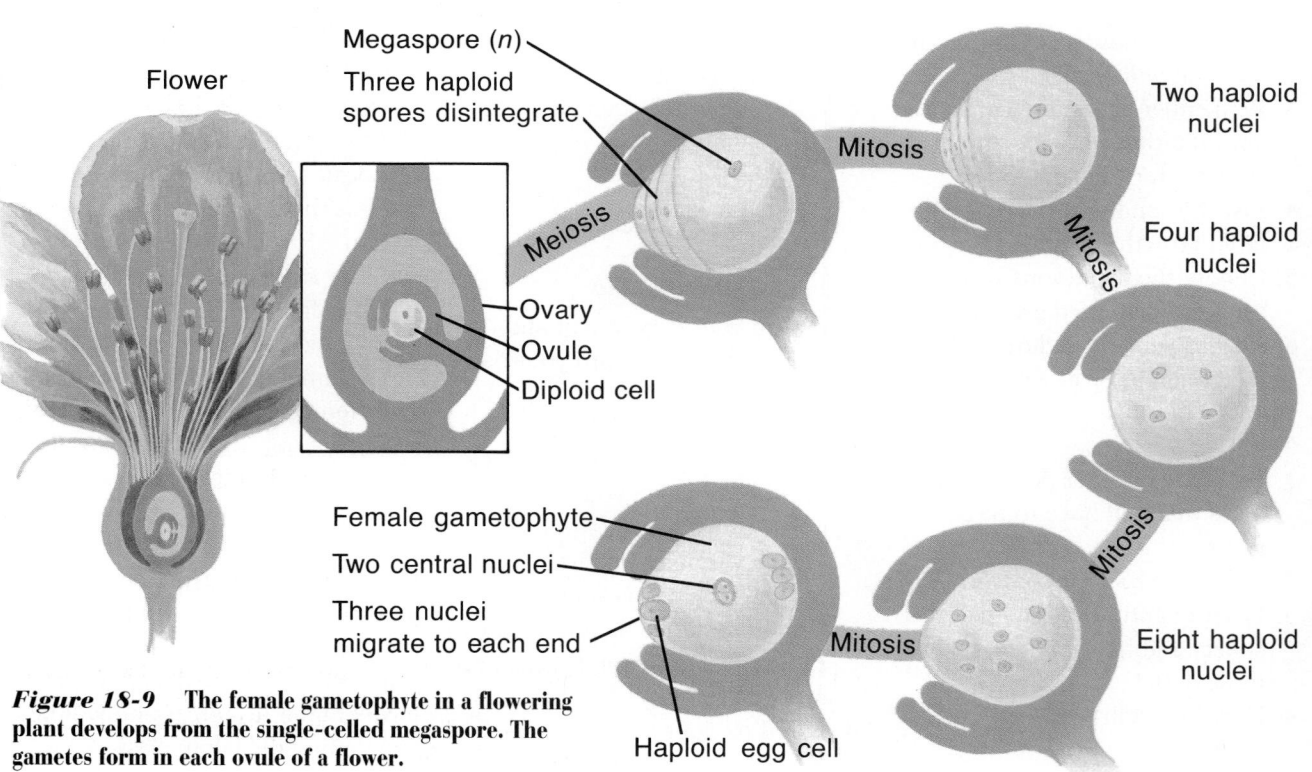

Figure 18-9 **The female gametophyte in a flowering plant develops from the single-celled megaspore. The gametes form in each ovule of a flower.**

Gametes in a Flower   Keep in mind that all the parts of a flower are composed of diploid cells. Cell divisions in the anthers and ovaries of a flower lead to the production of spores. Look at Figures 18-9 and 18-10 as you read how eggs and sperm of flowering plants are produced.

Meiotic cell divisions occur within the ovary. You have read in Chapter 16 that the ovary is the structure in the pistil where the seeds are eventually formed. Within the ovary are one or more **ovules** that will one day become seeds if fertilization occurs. A diploid cell in the ovule divides by meiosis to produce four haploid spores. However, only one large spore—a megaspore—survives. Within this megaspore, the nucleus then divides by mitosis to form two haploid nuclei. Mitosis continues until a total of eight haploid nuclei are produced. These nuclei and the cytoplasm around them form the embryo sac or female gametophyte. Eventually, one of the nuclei becomes an egg cell. Two others move to the center of the embryo sac, and a membrane forms around them.

Within each anther, diploid cells also divide by meiosis to produce four small haploid cells known as microspores. Within each microspore, the nucleus then divides by mitosis to form two haploid nuclei; one becomes the nucleus that will help the growth of a pollen tube into the stigma, and the other divides to form two sperm. The three cells of the microspore make up the male gametophyte. When the outer wall of the microspore hardens, the structure becomes a pollen grain. The walls of pollen grains have beautifully intricate patterns that can be used to identify individual species of plants.

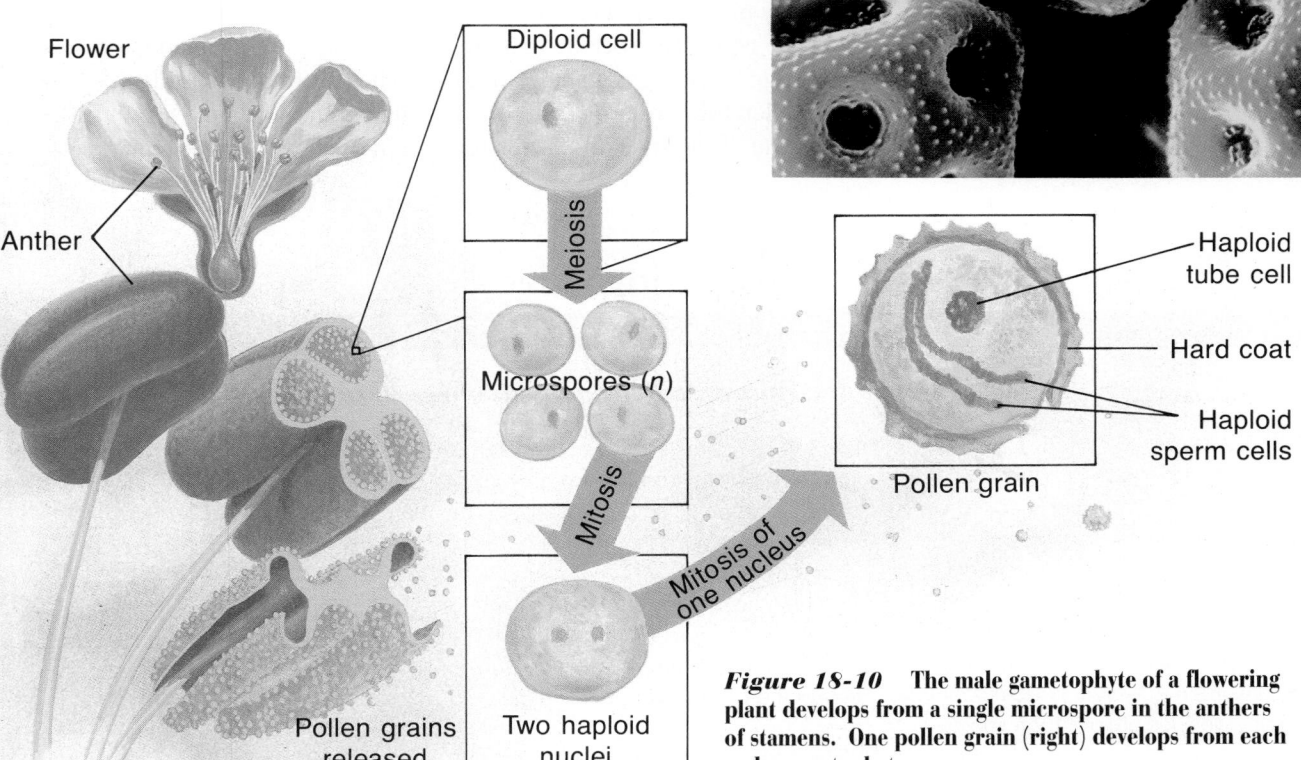

Flower

Anther

Diploid cell

Meiosis

Microspores (*n*)

Mitosis

Two haploid nuclei

Mitosis of one nucleus

Pollen grains released

Pollen grain

Haploid tube cell

Hard coat

Haploid sperm cells

Figure 18-10   **The male gametophyte of a flowering plant develops from a single microspore in the anthers of stamens. One pollen grain (right) develops from each male gametophyte.**

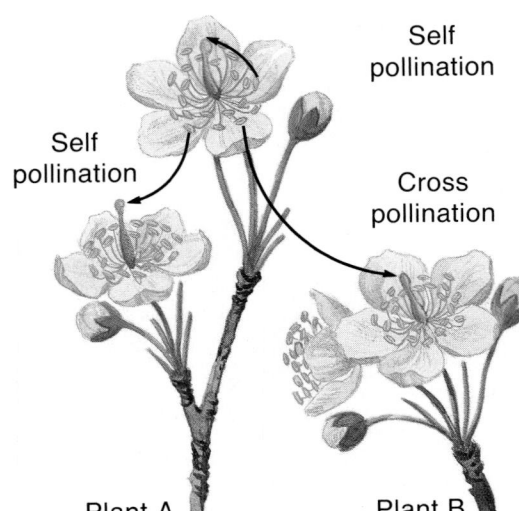

Self pollination

Self pollination

Cross pollination

Plant A

Plant B

Figure 18-11   When pollen is transferred from one plant to another in cross-pollination, genetic material is exchanged. Self-pollination involves the genetic material of one parent.

The three nuclei of the ovule and the two sperm nuclei of the pollen all contribute to the fertilization of a flowering plant, as you will read about next.

Pollination   In order for plants to reproduce sexually, the gametes in a pollen grain and an ovule must meet and fuse. The pollen grain must first be transferred from an anther of a stamen to the stigma of a pistil. This transfer is called **pollination.**

It is possible that a plant that has flowers with both male and female sex organs can be pollinated by its own pollen. This self-pollination can occur in a single flower or between two flowers on the same plant, as shown in Figure 18-11. On the other hand, cross-pollination occurs when the pollen from one plant is transferred to the stigma of another plant. Both self- and cross-pollination involve genetic recombination; however, cross-pollination results in more variety in the offspring because the two gametes come from different parents.

How is pollen transferred? In some cases, insects are the pollinators. The rugged walls of pollen grains and the hairy bodies of insects will often easily stick together. Pollen grains on the body of an insect that visits a new flower may then be transferred to the sticky or rough stigma as the insect passes over. In flowers that are pollinated by insects, the petals are often large, attractively colored, or scented. The flowers that are not showy are usually pollinated by wind or flies. In wind-pollinated flowers, there is often an abundant production of pollen. In fly-pollinated flowers, the flowers often have the smell of rotting meat. Other flowers are pollinated by birds, bats, and even mice.

Once pollination occurs and one or more pollen grains land and stick on a stigma, each grain begins to grow. The tube nucleus leads the growth of a pollen tube from one end of the pollen grain. The tube grows through the tissue of the style and heads toward an ovule within the ovary. The

Thinking Lab

INTERPRET THE DATA

How can you determine chromosome numbers in plants?

Background

In the plant kingdom, a pattern of reproduction occurs that is called alternation of generations. Each generation looks different and serves different functions. *Sporo*phyte generations form *spores* by meiosis while *gamet*ophyte generations form *gametes* by mitosis. Spores form directly into gametophytes. Gametophytes always form sporo-

phytes after gametes fuse at fertilization. Sporophyte generations are always diploid while gametophyte generations are always haploid.

You Try It

Examine the table shown here. Information is provided for four plants, A to D. **Interpret the data** and fill in any of the missing chromosome numbers.

Chromosome numbers

	Sporo-phyte	Gameto-phyte	Spore	Gamete
A	20			
B			12	
C				8
D		34		

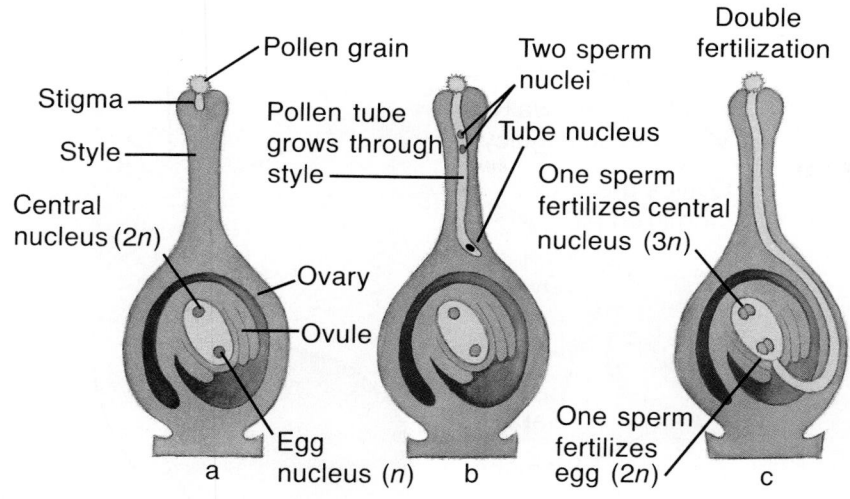

Pollen grain

Stigma

Style

Central
nucleus (2n)

Ovary

Ovule

Egg
nucleus (n)

a

Double
fertilization

Two sperm
nuclei

Tube nucleus

One sperm
fertilizes central
nucleus (3n)

Pollen tube
grows through
style

b

One sperm
fertilizes
egg (2n)

c

two sperm nuclei follow the tube nucleus down the pollen tube and through a small opening into the ovule.

Double Fertilization   Within an ovule, one of the sperm nuclei fuses with the two nuclei in the center of the ovule, forming a triploid ($3n$) nucleus that eventually develops into a food supply for the new plant. The other sperm nucleus fertilizes the egg to form a zygote. This double fertilization is typical for all flowering plants. Zygote formation marks the beginning of a new diploid sporophyte generation.

The triploid nucleus divides many times to form a mass of tissue called the **endosperm.** Endosperm is the food source that provides energy for the new plant's early development. The outer covering of the ovule hardens into a protective coat. At this point, the ovule has become a young seed. The seed contains an embryo sporophyte plant and the food source needed for its early growth.

Figure 18-12   Each seed in a flowering plant results from double fertilization within the female gametophyte (left). Without insects, the majority of flowering plants would not be pollinated (right).

Fruits: Nature's Way of Dispersal

Soon after fertilization, most of the flower parts of the parent plant begin to die. However, the ovary, in which the developing seeds are located, enlarges rapidly. The enlarged ovary becomes the fruit. You're certainly familiar with fleshy sweet fruits, such as oranges, apples, grapes, and peaches. Many of the plant foods you call vegetables are actually fruits, too—tomatoes, green beans, corn, and cucumbers. You will realize from this list that sometimes fruits are not all fleshy, and that some have only one seed. Besides corn, other grains such as wheat, rice, and oats are single-seeded, dry fruits. Flowering plants can be identified just by examining the forms of their fruits.

Figure 18-13   Many fruits, such as cereal grains, and a variety of nuts have a dry, not fleshy, wall and hold only one seed.

Figure 18-14 The fleshy tissue of a fruit usually develops from the wall of the ovary as shown here in a blueberry. The seeds develop from each fertilized ovule.

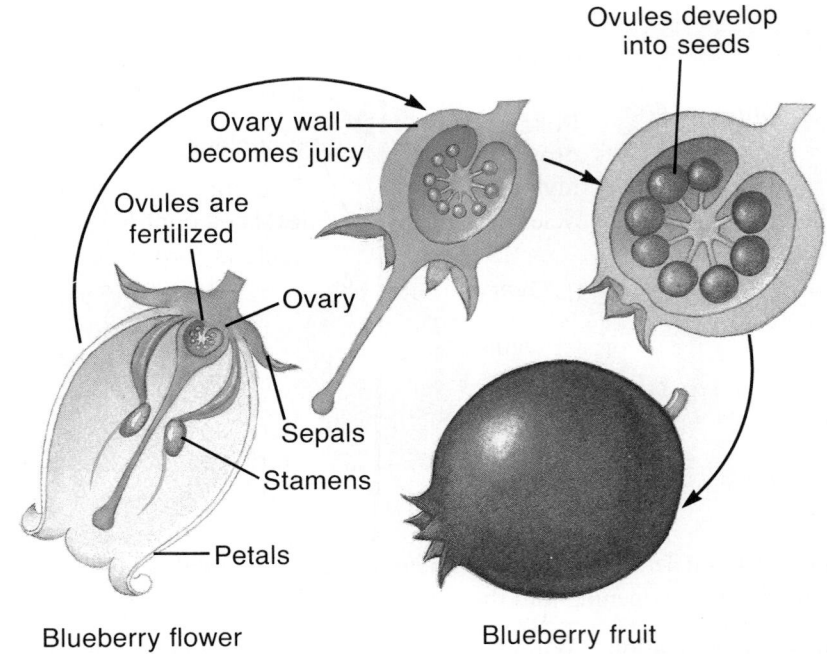

Ovules develop into seeds

Ovary wall becomes juicy

Ovules are fertilized

Ovary

Sepals

Stamens

Petals

Blueberry flower

Blueberry fruit

Fruits initially protect the seeds, but they are also important in dispersing seeds. Many fruits are eaten by animals. The hard seeds pass unharmed through the digestive systems and are deposited in the animal's wastes away from the parent plant. Animals that gather nuts may drop and lose many of these dry fruits. The nuts, such as acorns, will then grow away from the parent plant. Some fruits have their own dispersal mechanisms. For example, when seeds of the squirting cucumber are ripe, the pressure inside the cylindrical fruit builds up until the end wall explodes. The seeds are shot out over great distances. The winged fruits of maples, dandelions, and elms are carried by wind. Other fruits, such as water lily nuts, float away from the parent plant. Still others with hooks will stick to clothes of humans or the fur of passing animals. Dispersal of seeds is important. If seeds began to germinate in the shadow of the parent plants, they would need to compete with the parents for sunlight, oxygen, carbon dioxide, and water. Dispersal of fruits and seeds reduces this competition.

Section Review

Understanding Concepts

1. In what ways are conjugation in bacteria and bread mold similar? How are they different?
2. How is the life cycle of a moss different from that of a dandelion?
3. Explain why the gametophyte generation of a flowering plant cannot become an independent plant.
4. **Skill Review—Sequencing:** Sequence the series of events from pollination through seed formation.

For more help, see Organizing Information in the **Skill Handbook**.

5. **Thinking Critically:** Why are seeds good food sources for humans and other animals?

18.3 Sexual Reproduction in Animals

Objectives

Distinguish between external and internal fertilization in animals.

Relate human reproductive structures to functions in females and males.

Compare estrus in mammals with the menstrual cycle in humans.

You've learned about the intriguing life cycle of plants in which diploid and haploid stages alternate. Is the same true of the bird outside the window, your family's pet dog, and other members of the animal kingdom? The answer is no. Life cycles of animals differ in several ways from those of plants. Unlike plants, animals have a life cycle in which there is no alternation of haploid and diploid generations. For almost all animals, the organism is always diploid. Only an animal's gametes are haploid. As you've already learned, simpler animals such as sponges may reproduce either asexually or sexually. In more complex forms, though, sexual reproduction is predominant.

Fertilization

In animals, as in plants, gametes are produced in sex organs. Sperm are produced from cells of **testes,** and, just as in plants, eggs are produced from cells within ovaries. Among many simpler animals, both kinds of sex organs develop in the same individual. These animals are called **hermaphrodites.** However, as animals became more complex over time, separate sexes—males and females—became more usual. Why would separate sexes be an adaptation? For the same reason you learned cross-pollination in plants was adaptively beneficial, the fusion of DNA from two separate parents leads to more variation among the offspring. As you know from Chapter 12, variation is necessary for adaptation and evolution to occur. For the most part, in hermaphrodites, such as earthworms and clams, two different parents exchange gametes.

Figure 18-15 **This young elephant developed within the body of its mother after the fusion of male and female gametes from separate parents.**

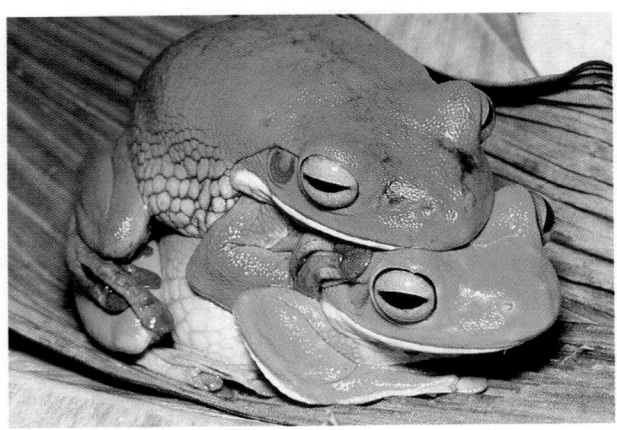

Figure 18-16 **Mating behaviors in animals can ensure the fertilization of eggs even though this occurs outside the parent's body.**

Successful sexual reproduction in animals first depends on fertilization of the egg by a sperm. In animals such as sponges, cnidarians, most worms, and many fishes and amphibians, fertilization occurs outside the female. But in insects, reptiles, birds, mammals, and some fishes and amphibians, fertilization occurs within the female. Do you remember from Chapter 17 that the evolution of internal fertilization was an adaptation that may have helped animals to survive on land?

Whether external or internal, fertilization is no sure thing. Because gametes are delicate, they must be protected. The life span of gametes is short; therefore, the timing of their release must be matched so that both sperm and eggs are present together. Also, because sperm and eggs would dehydrate in a dry environment, they must remain in a moist environment. A liquid medium is also important for another reason—for sperm to be able to swim to an egg. Animals have a variety of adaptations that ensure fertilization.

External Fertilization In general, animals with external fertilization meet the need for moisture by reproducing in water. But an ocean, a lake, or even a pond is a vast amount of water and is full of risk and danger. The chances of sperm finding eggs are decreased in so large a volume of water. In addition, exposed eggs may be eaten by other animals. Unfavorable physical factors such as lack of oxygen, extreme temperatures, and pollutants may destroy the gametes. How, then, do these animals survive? Animals with external fertilization release huge numbers of

HEALTH CONNECTION

Fetal Alcohol Syndrome

Seven-year-old Jason has lots of friends. He is outgoing, loving, and very energetic. Unfortunately, he is off to a poor start at school because he has trouble remembering his numbers and letters and he can't pay attention to his work for as long as most children his age. Jason's problems are not really his fault; they're his mother's.

Jason's mother loves him very much. She was thrilled when she

found out she was going to have a baby. Her pregnancy was a time of celebration, including parties with beer, wine, and other alcoholic drinks. Although she never drank enough to get drunk, she unknowingly did damage to Jason. She didn't realize that alcohol crossed over into her developing baby's blood.

Fortunately for Jason, his symptoms of FAE (Fetal Alcohol Effect) are not as severe as they might be. Some children whose mothers drink heavily during pregnancy are

born mentally retarded and physically malformed. They may have a whole range of physical problems, including a deformed skull, heart defects, poor eyesight and hearing, bad teeth, and stunted growth. Most of these problems, called FAS (Fetal Alcohol Syndrome) will last a lifetime.

Each year, more than 50 000 babies with alcohol-related defects are born in the United States. One-fourth of these babies have damage severe enough to be termed FAS. Some researchers think that FAS is

gametes at one time. A female salmon, for example, lays from 2000 to 10 000 eggs during spawning. There is safety in numbers in that at least some of the eggs are fertilized and have a greater chance of survival.

Other adaptations help ensure fertilization. The eggs of some animals such as frogs are enclosed in a jellylike layer that protects and insulates them. Also, many animals follow patterns of courtship behavior that lead to the release of sperm and eggs at precisely the same time. In frogs, for example, the male clasps tightly to the back of a female and stimulates the release of both their gametes.

Internal Fertilization   Internal fertilization removes the dangers of predators and harsh environmental conditions for the gametes. Inside the female parent's body, eggs are retained in reproductive organs. When sperm enter the female's body, they move in a direct path toward the eggs. The reproductive organs provide a moist and protective environment for sperm and eggs. The only time sperm are exposed is when they leave the male body and before fertilization. During these times, they are protected from drying out by a fluid in which they are able to swim to the egg. When the chances of fertilization are high, fewer eggs are produced.

Figure 18-17   **When a male insect deposits sperm into the body of a female, fertilization is internal and loss of eggs to predators is reduced.**

---

the cause of 20 percent of all cases of mental retardation in the U.S.

Michael Dorris learned this statistic in a heartbreaking way. Half Native American himself, Dorris adopted a three-year-old boy whose Sioux parents had died from alcohol abuse. The boy, now grown up, is still undersized, a slow learner, and unlikely to lead a truly independent life. Dorris, who tried everything to help his son, says sadly, "If you wouldn't give your kid a bottle of gin the day after birth, why give it one the day before?"

Perhaps the most tragic fact about FAS is that it is entirely preventable. Even an alcoholic woman can have a baby free of FAS—if she stops drinking before she becomes pregnant.

------

1. **Issue:** Some people think that mothers who refuse to stop drinking should be jailed to control their access to alcohol. What arguments for and against this proposal can you provide?
2. **Writing About Biology:** What advice would you give a friend

who says, "I'm worried about FAS, so I'll stop drinking as soon as I find out I'm pregnant"?

Although sperm may be deposited within the female's body, there is no guarantee that her eggs are ready to be fertilized. Adaptations for proper timing are still important. Earthworms and many female insects have storage organs where sperm can be kept alive for months or even years. As the female's eggs mature, they can be fertilized with immediately available sperm. In birds, proper timing is ensured by elaborate courtship behaviors.

The Estrous Cycle

In many mammals, including dogs, rats, and whales, the female is receptive to mating at only a few times each year. These receptive times are known as periods of **estrus** or heat. What controls this cycle of events in mammals? The production and release of eggs and sperm would never happen without chemicals called hormones released by the body. A **hormone** is a chemical that is produced in one part of an organism and then is transported to another part, where it causes a response. During estrus, sex hormones control the release of eggs and the receptive behavior of the female. At the peak of the estrous cycle, when the female is in estrus, changes in her behavior are perceived by the male. He senses these changes as her readiness for mating. It is at that point in the cycle that her eggs are released by the ovary and can be fertilized.

Human Reproduction

Humans, like other mammals, have internal fertilization. Figure 18-18 shows the human female reproductive system. On each side of a female's abdominal cavity is an ovary. Within the ovary are groups of cells called **follicles,** where eggs develop from cells that divide by meiosis.

Figure 18-18 **Eggs develop in the ovaries, organs on either side of the female's abdomen. Notice that the female reproductive organs are separate from the urinary and digestive tracts. What is the name of the opening to the uterus?**

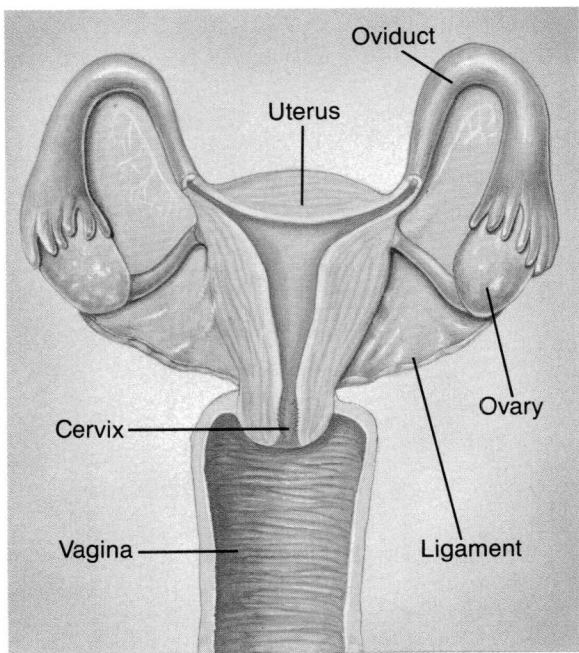

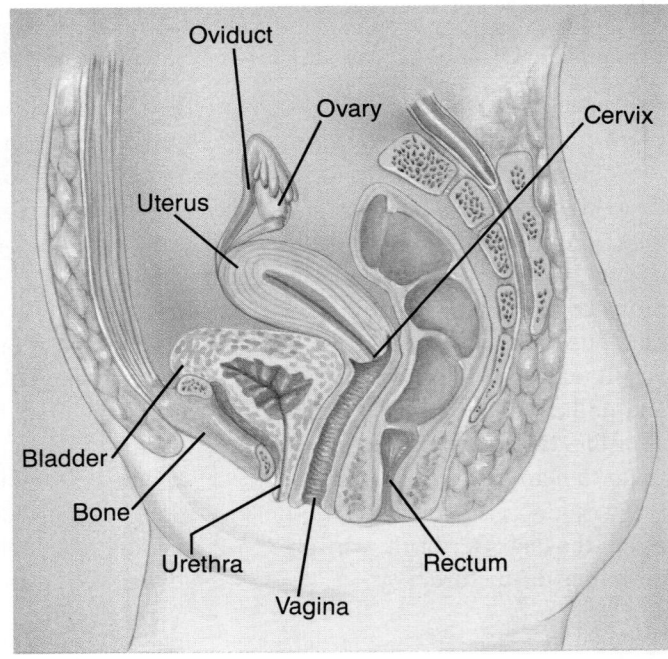

Ovulation   Very close, but not attached, to each ovary is an opening to a tube called an **oviduct.** When an immature egg is released from a follicle, a process called **ovulation,** it is usually pulled into the oviduct by tiny cilia that line the opening of this tube. The left and right oviducts join to form a thick-walled, muscular organ called the **uterus.** Within the hollow uterus, a fertilized egg will develop. The uterus narrows to a small neck called the cervix, which becomes plugged during the development of a fetus. Usually, the cervix opens to the **vagina,** which leads to the outside of the female's body and is the passageway along which sperm pass on their way to the uterus.

Sperm Formation   Figure 18-20 shows the human male reproductive system. The testes of the male are two oval structures within an external sac called the **scrotum.** In many animals, the testes are within the abdomen. But in humans and some other mammals, sperm require a temperature slightly lower than that of the body. The position of the testes away from the body in an external structure is therefore adaptive.

   Sperm are produced from cells of the testes and then transferred along a series of highly coiled tubes. As sperm leave these tubes, they pass several structures that add fluids. Each sperm has a flagellum that helps it swim in these fluids. The combination of sperm and the fluids is called **semen.** Semen leaves the male's body through the penis.

Conception   During sexual intercourse, semen is ejected by a series of muscular contractions from the male's urethra into the female's vagina. Although millions of sperm may enter the vagina, only a few thousand of them will pass through the cervix into the uterus. Many of these enter the oviduct, where fertilization usually occurs if an egg is present. However, only one sperm fertilizes the egg. As soon as one sperm penetrates the

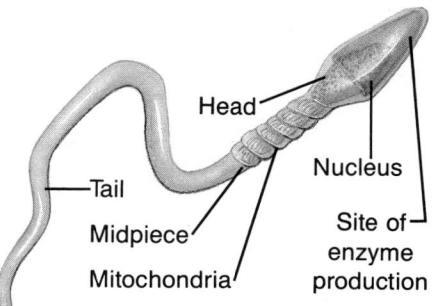

Figure 18-19   **A mature sperm has a head containing the nucleus and a tail with mitochondria, which is shed after the sperm head penetrates the egg.**

Figure 18-20   **Sperm develop in the testes, organs within sacs outside the male's abdomen. Notice that the urinary and reproductive tracts have one shared passage. Why do sperm need to be stored outside the body cavity?**

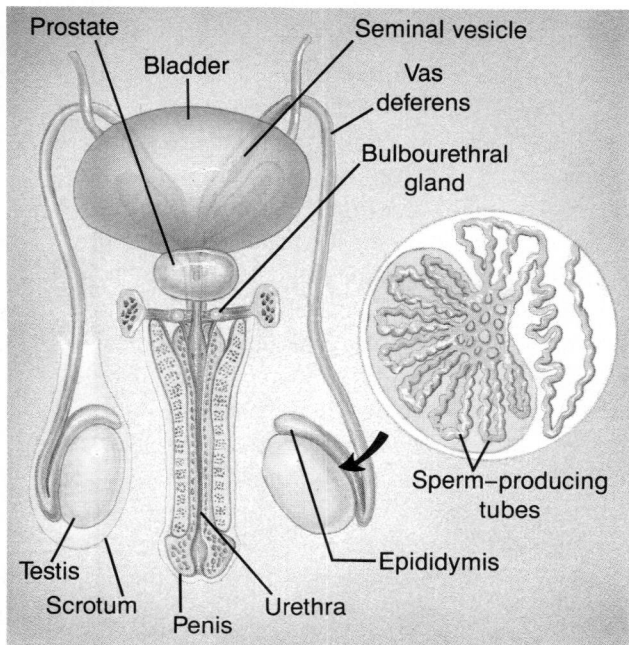

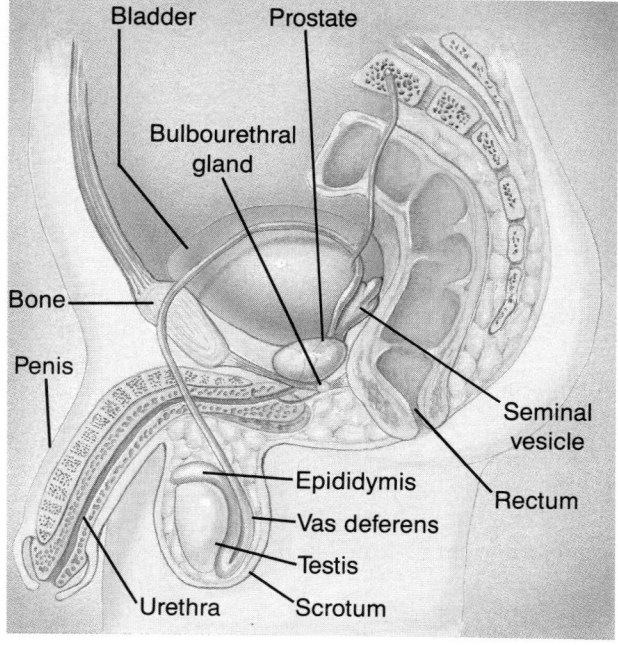

egg, the cell membrane of the egg becomes unresponsive to the attempts of other sperm to break the egg's outer barrier. Penetration of the egg by the sperm triggers the almost-mature egg to complete meiosis. Then the haploid nuclei of sperm and egg combine to form the diploid zygote.

If a zygote is formed, it moves along the oviduct toward the uterus, where development will occur. A pregnancy begins when the embryo becomes embedded in the uterine wall, which has thickened and become enriched with blood vessels since just prior to the time of ovulation. Ovulation occurs in the average, sexually mature female on a near-monthly basis as a result of the female's menstrual cycle.

The Menstrual Cycle

Beginning on average around age 12, and continuing until about age 50, human females undergo a series of changes similar to those of the estrous cycle. These hormonally controlled changes result in the **menstrual cycle.** During this cycle, an egg matures and is released

LITERATURE CONNECTION

Native American Names

Columbus started the confusion. Thinking he had discovered a new trade route to India, he called the people living in the Americas "Indians." More confusion arose when the European settlers tried to learn the personal names of the Native Americans they encountered. The Europeans discovered that the original inhabitants of the land often changed their names as they grew older. A childhood name might be exchanged for a new one at adolescence, or after a first battle or an important dream. Another

cultural mystery to Europeans was that some Native Americans considered their names so personal and sacred that only a bad-mannered person would address them directly by those names.

Naming the Baby The Native American tradition of naming babies is a practice that survives today among many tribes. A baby might be called simply "boy" or "girl" until he or she is several months old. By that time, the baby has developed some personality. A tribe elder observing the baby will then know enough about his or her personal traits to suggest a name that seems to fit. That name might be one given to honor a dead ancestor, or it might come from a dream

of the baby's mother or father. You can read an account of how the famous Shawnee chief, Tecumseh, received his name in the book *Panther in the Sky,* by James Alexander Thom. Thom writes that Tecumseh's father saw a birth sign, a yellow-green shooting star, and that sign determined the child's name, Panther in the Sky or Tecumseh.

Respect for a Name Even after a Native American child grows up, his or her name can change. In some tribes, names are considered a personal possession that can be loaned, pawned, given away, or even thrown away. Following the death of a family member, all the surviving members of the family

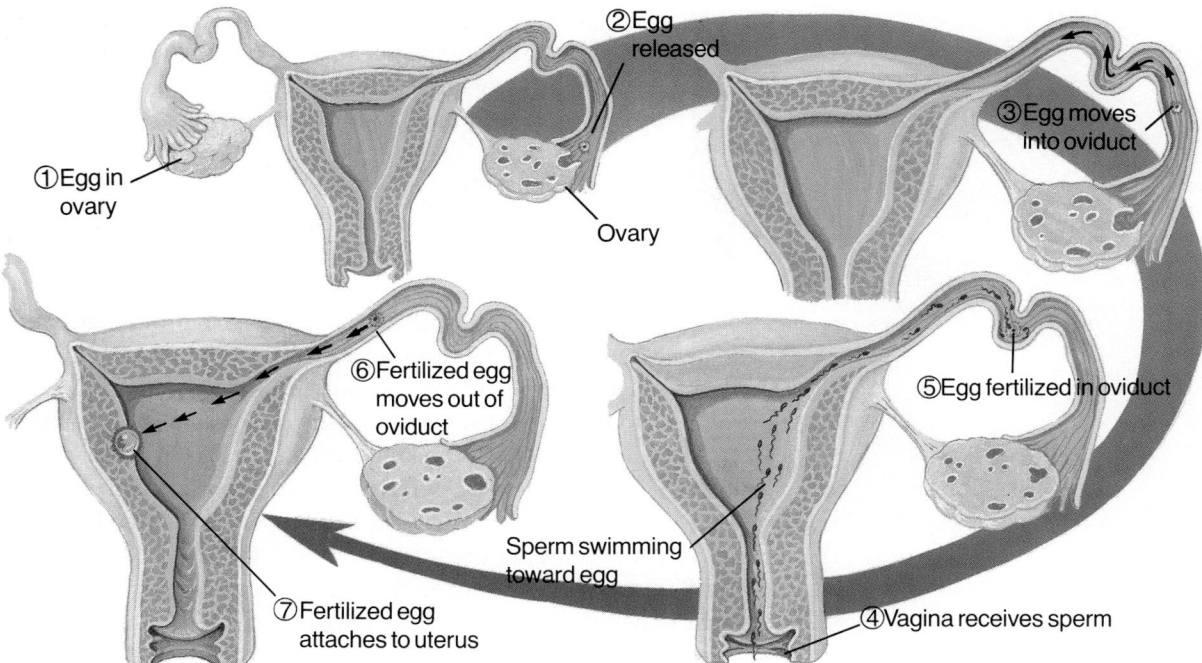

Figure 18-21   Follow the steps in the movement of an egg from the ovary to when it becomes embedded in the wall of the uterus following fertilization.

① Egg in ovary

② Egg released

Ovary

③ Egg moves into oviduct

⑥ Fertilized egg moves out of oviduct

⑤ Egg fertilized in oviduct

Sperm swimming toward egg

⑦ Fertilized egg attaches to uterus

④ Vagina receives sperm

may take new names and avoid saying any word that suggests the name of the deceased for many years to come.

Because of misunderstandings, many Native American names have been made fun of. For instance, the name "Stinking Blanket" may have caused snickers among Europeans. However, attitudes about that name changed significantly after hearing an explanation of its source: An Indian brave fought so long and hard against his enemies that he never had time to stop and change his horse's saddle blanket. Similarly, the name "Crazy Horse" sounds different when the word "crazy" is correctly translated as "recklessly brave."

Today, many Native Americans have two personal names. One name might be an ordinary one used for convenience. The other and more important name is the tribal name, which establishes membership in the tribe.

1. **Explore Further:** Do a survey in your school to learn the most common first names for boys and girls. Ask older family members about popular first names when they were young.
2. **Connection to Social Studies:** Nicknames often reveal more about a person than his or her real

name does. What are some nicknames among your friends or among sports stars? Why do you think these nicknames were given?

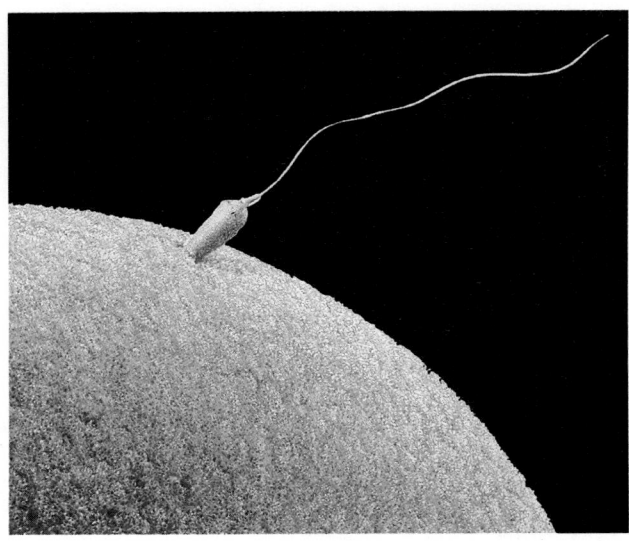

by an ovary at ovulation, and the uterus is prepared for a possible pregnancy. However, unlike estrus in other mammals, the human female's readiness for mating is not limited to this or any one particular time.

Each menstrual cycle lasts for about 28 days and consists of four different phases: thickening of the uterine tissue, ovulation, passage of the egg into the uterus, and loss of egg and uterine tissue via the vagina. This last phase, called **menstruation,** marks the onset of a new cycle and lasts about five days. Once menstruation has ended, a new egg begins to mature, and the uterine tissue once again is prepared for pregnancy. Ovulation usually occurs about midway through the cycle. It is at this time and for perhaps a day or two more that if sperm are already present in or are introduced into the vagina, fertilization can occur. If an egg is fertilized and the embryo becomes embedded in the uterine wall, pregnancy begins. The menstrual cycle ceases throughout pregnancy. You'll discover how hormones interact to control the events of the menstrual cycle and pregnancy in Chapter 24.

Figure 18-22 **This false-colored illustration of a human egg and sperm shows their relative sizes. Magnification: 1500×**

ART CONNECTION

Family Group

For all of history, the survival of our species has depended on our ability to reproduce. However, we have learned that just producing offspring is not enough. Without constant care and attention, even well-fed babies will not develop well, nor will they gain weight.

Early representations of females in art stressed fertility by portraying pregnancy because of the importance of continuation of the family line by reproduction. Some more recent artists have gone beyond just showing fertility. Henry Moore, a 20th century English sculptor, is one such artist. In 1949, he developed a bronze sculpture called

"Family Group," which shows a nurturing family spirit. The parents are providing protection and care for the child. Henry Moore was influenced early in his career by Mexican and African art. His works in stone, wood, and bronze have simple lines, and many look weathered. Several of his sculptures depict families or mothers with their children.

1. **Writing About Biology:** How does Moore's sculpture relate to what you know about today's families?
2. **Issue:** The number of single-parent families has increased greatly over the past few decades. What are some of the problems these parents encounter as their children grow up?

3. **Connection to Social Studies:** Talk to an adult about his or her view of the ideal family. Write a brief report on your impressions.

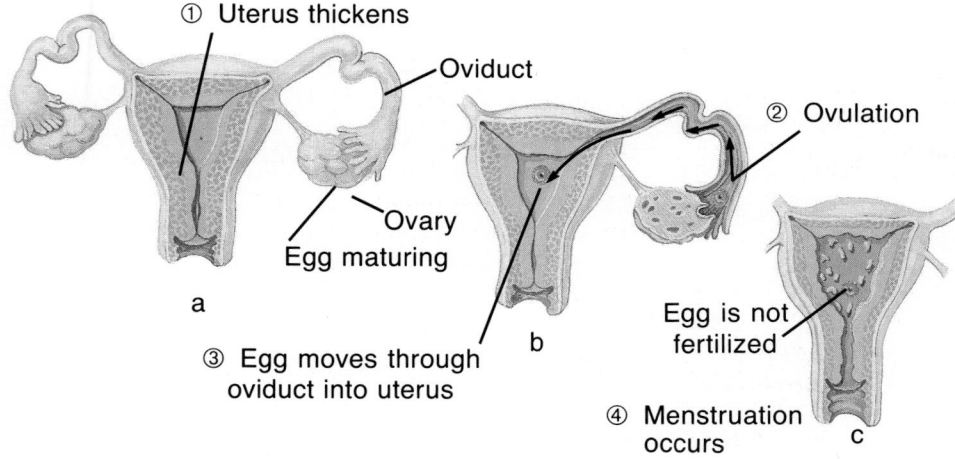

① Uterus thickens

Oviduct

② Ovulation

Ovary

Egg maturing

a

Egg is not fertilized

b

③ Egg moves through oviduct into uterus

④ Menstruation occurs

c

Figure 18-23 The wall of the uterus thickens and becomes engorged with blood in preparation for pregnancy. The wall lining falls away during menstruation if fertilization does not occur.

LOOKING AHEAD

THERE ARE various patterns of reproduction among different kinds of organisms. Obviously, there is a good deal of diversity. However, it's also important to focus on what the many different kinds of organisms have in common. How does reproduction in different kinds of organisms compare? Whether reproduction is asexual or sexual, the result is newly produced offspring. These offspring must then undergo a period of development. You'll discover that development in plants and animals follows similar patterns, and you'll also explore how genetics controls that development. 🐾

Section Review

Understanding Concepts

1. Compare the numbers of eggs produced at one time by a cat and a fish. Explain the reason for the difference.

2. How are humans adapted for solving the problems of fertilization?

3. Explain why it is important for animals with internal fertilization to have precise timing for the release of their gametes.

4. *Skill Review—Making Tables:* Make a table listing how animals with external and internal fertilization are adapted to the conditions required for fertilization. For more help, see Organizing Information in the **Skill Handbook**.

5. *Thinking Critically:* A species of frogs in Puerto Rico lives in trees, where they reproduce by external fertilization. What conditions must be present in those trees that enable these frogs to reproduce?

Issues

And Baby Makes Three?

How many people does it take to conceive a baby? The question may seem silly, but sometimes the answer isn't two. The couple must turn to a third person—a doctor—to help them conceive. Between eight and 15 percent of couples cannot conceive a child despite years of trying. The man's sperm count may be low. The woman may not ovulate, her oviducts may be blocked, or the fertilized egg may not implant in her uterus. In all of these cases, pregnancy seems impossible. Or it did, that is, before in vitro fertilization, a procedure during which sperm and egg are brought together outside the body and then returned to the uterus.

A Miracle Baby When a couple decides to try in vitro fertilization, the woman is given fertility drugs so that more than one egg will ripen during ovulation. Using modern techniques, a doctor can remove the ripe eggs from the ovaries and fertilize them in a glass laboratory dish. Placing sperm and egg in direct contact increases the chances of fertilization. One or more fertilized eggs, called zygotes, are then placed in the uterus and the rest are usually frozen. If a zygote implants in the wall of the uterus, an embryo will develop. If it does not, the pro-

cedure can be repeated using the remaining fertilized eggs.

What if the woman no longer has a uterus or is unable to produce eggs? In the first case, the zygote

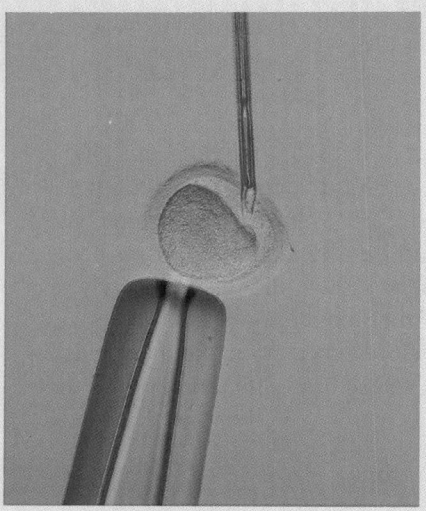

may be placed in the uterus of another woman, called a surrogate, who agrees to carry the fetus to birth and then return the fully developed baby to the couple. If the woman is unable to produce eggs, an egg from another female, called a donor, can be fertilized with the man's sperm and then placed in the woman's uterus. In both cases, two parents have now become three.

For the couple who has been unable to conceive a baby, in vitro

fertilization may seem like a miracle worth any price. They have the baby they always wanted. But is it all that simple? Some couples find there are unforeseen complications.

Risky Business To increase the chances of pregnancy, more than one fertilized egg is generally introduced into the uterus. One woman who received four eggs found herself pregnant with quadruplets.

Sometimes, couples freeze fertilized eggs for later use. One couple who did this divorced shortly afterwards. The woman asked the court to give her custody of the frozen zygotes. The husband objected, however, saying he, too, should have a say in what happens to the zygotes fertilized by his sperm.

When a couple receives eggs from a donor or pays a surrogate mother to carry the fetus, the question of who is the real mother may come up as a legal issue. Several couples have found themselves in court fighting for custody of the baby they thought was theirs.

While there are more than 200 centers in the United States and Canada that offer in vitro fertilization, and they have given thousands of couples their only hope of having a child, there are many legal and ethical questions yet to be resolved.

Understanding the Issue

1. What, if any, regulations on in vitro fertilization do you think are necessary?
2. Doctors can now test zygotes for genetic defects. Explain why you think this is good or bad.
3. Should the current ban on use of federal funds for in vitro fertilization be lifted? Why or why not?

Readings

- "Mixed Blessings." *Health,* Vol. 22, Nov. 1990, pp. 64-67.
- "And Donor Makes Three." *Newsweek,* Sept. 30, 1991, pp. 60-61.
- "The Rights of Frozen Embryos." *Time,* July 24, 1989, p. 63.

Biotechnology

Cloning Humans: Is It Possible?

It is the twenty-first century. A young women walks out of a medical clinic with a new liver, one her body will not try to reject because it came from her clone. Several years before, the woman had been cloned, or genetically duplicated. The fetus developed in a test tube womb, and once the clone grew, it was frozen in liquid nitrogen. When the woman developed liver cancer, the liver was removed from her clone and put into her body. If the cancer spreads to other organs, if her heart or lungs become diseased, or if she loses an arm or leg in an accident, these body parts can also be replaced from her clone. "This way I will always be young," she thinks.

This scene may not be as impossible as it sounds today. The concept fascinates the human mind, and many science fiction writers have explored the idea since.

Starting with Frogs Cloning is the making of an exact copy of an organism; thus the genes—and all the parts—of the clone are identical to those of the original organism. Scientists have been working on reproducing exact genetic copies of animals since the 1950s. By pulling the nucleus out of the cell of a frog embryo and injecting it into an egg cell from which the chromosomes had been removed, scientists were first able to grow a tadpole. Since then, they have successfully cloned sheep, rabbits, and cows in the same way. By cloning cows that give the best milk, for example, the quality of the herd can be kept genetically consistent.

So far, scientists have been able to clone animals using only cells from embryos. Scientists understand that all of the genes present in the cell of an embryo are still present in the cells of the different tissues, but they are not all turned on for action. When they discover how the genes are switched on and off, they may be closer to understanding how to clone humans.

Stay Forever Young Some biologists are already working on ways to make adult body cells of vertebrates revert back to their original embryonic stage, when all the genes were active. One biologist has successfully placed the nucleus of a red blood cell from a frog into a frog's egg cell that isn't quite mature. She

found that the metabolic process that occurs in the egg helped to activate the genes in the introduced adult nucleus. The biologist has so far been successful in cloning a tadpole from a blood cell until it reached the stage of forming limb buds.

Cloning began with frogs. This latest research on frogs could mean that the power to clone human adult cells is not far away.

1. **Issue:** Not everyone thinks that biologists should try to clone humans. What are some of the arguments against cloning?
2. **Connection to Genetics:** In 1990, the United States organized the Human Genome Project. A genome is all the genes in one cell. What is the goal of this project?

Summary

Asexual reproduction occurs among members of all kingdoms and, in general, is the only means available to monerans. It may occur by binary fission, budding or vegetative reproduction, fragmentation, or spore formation. Each of these processes results in offspring that are genetically identical to the parent and to one another.

Sexual reproduction involves gametes that combine DNA and produce offspring different from one another and from the parents. Conjugation is a form of sexual reproduction carried out by bacteria, some algae, and fungi. More complex organisms, though, produce male and female gametes.

Plants have a life cycle in which the sporophyte and gametophyte generations alternate with one another. The diploid sporophyte generation is the major stage in angiosperms.

Animals do not undergo alternation of diploid and haploid generations, and only their gametes are haploid. Eggs of animals are fertilized either internally or externally, and animals have many adaptations that ensure fertilization.

Language of Biology

Write a sentence that shows your understanding of each of the following terms.

budding	ovulation
conjugation	ovule
endosperm	pollination
estrus	regeneration
follicle	scrotum
fragmentation	semen
hermaphrodite	testes
hormone	uterus
menstrual cycle	vagina
menstruation	vegetative reproduction
oviduct	

Understanding Concepts

1. What is the adaptive advantage of sexual reproduction?
2. How are a seed coat, a pollen grain, and a spore wall adaptive structures?
3. How can you explain the fact that many fruits have more than one seed?
4. Fungi release spores in great numbers. What kinds of animals produce gametes in great numbers? What are the advantages of high production of spores or gametes?
5. What do most amphibians and mosses have in common in terms of reproduction? How does that similarity affect both kinds of organisms in terms of their habitats?

6. How is sexual reproduction in a flowering plant similar to that in a reptile, bird, or mammal? Why?
7. What is the adaptive advantage for hermaphrodites to exchange sperm and eggs with other individuals?

8. Why is sperm production in animals such as amphibians or reptiles located internally, whereas in mammals sperm is produced in an external structure?
9. Compare estrus in a cat with the menstrual cycle of a human.
10. A sporophyte plant has cells containing 16 chromosomes. How many chromosomes are in the spores of this plant? The gametes? The zygote?

Applying Concepts

11. Recall that many cnidarians have a life cycle in which polyp and medusa stages alternate. Is this an example of alternation of generations? Explain why or why not.
12. How may vegetative reproduction be an advantage to plants invading a new habitat?
13. How are reproductive hormones important to the survival of a species?
14. A gametophyte plant dies because it has inherited a lethal allele. The sporophyte plant that produced it had the allele but did not die. Why?
15. Grasshoppers mate in spring, and eggs are fertilized in the summer. What adaptation for proper timing of release of gametes must grasshoppers have?
16. Bacteria normally produce substances essential for their growth. Two cultures of mutant bacteria are discovered that cannot grow without the addition of these substances. The two cultures require two different substances. If the two cultures are mixed and then placed on a solid medium without the added substances, a few bacteria begin to grow. Explain how this could happen.
17. Parasites are often hermaphrodites. How is that trait an advantage?
18. How are the conditions for fertilization in animals also met by flowering plants?
19. Does fragmentation and regeneration always involve reproduction of an organism? Explain.
20. **Issues:** What are some reasons why women may not be able to conceive?
21. **Biotechnology:** As you know, your skin can regenerate after it has been damaged. How is this characteristic important to doctors who attend severely burned patients?

Lab Interpretation

22. **Minilab 1:** How is budding in yeast similar to the formation of potatoes?
23. **Minilab 2:** Describe how each type of cell, spore or pollen, would have eventually developed if you could continue to observe it.

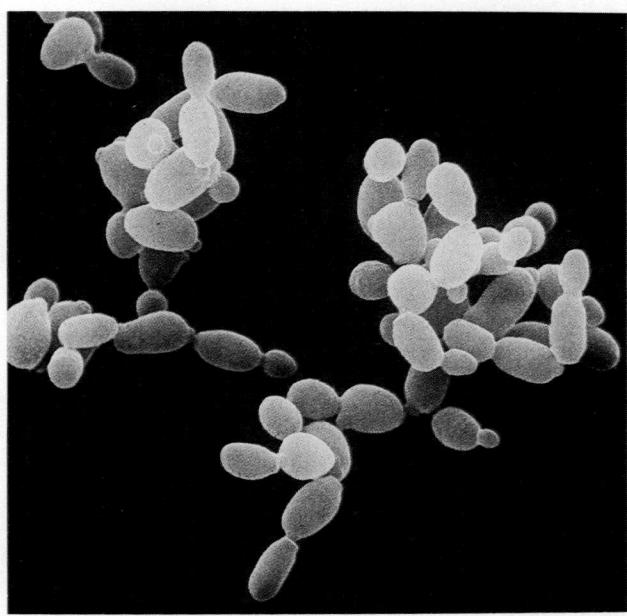

24. **Investigation:** Offer a possible reason why: (a) dry crackers do not mold when lying out of their wrapper for months, (b) moist dog food is sealed in individual meal packets.

Connecting Ideas

25. When self-pollination occurs, the two sets of DNA that fuse come from the same plant. How will this fertilization result in variety among the offspring?
26. Why was the requirement for moisture not a problem in the process of fertilization for early life forms?

CHAPTER

19

DEVELOPMENT

PERHAPS THE most unusual animals are the diverse group of mammals that live mainly in Australia—the marsupials. Most remarkable about marsupials, such as the kangaroo pictured opposite, is the presence of a skin pouch that is used to protect and nurture the new offspring. Marsupial offspring are born very tiny and physically immature. In some kangaroo species, they spend only 30 days inside the mother's uterus. Once born, kangaroo babies crawl into their mother's pouch and attach themselves to a nipple inside. There they remain—feeding, growing, and developing—for five to eleven months until fully developed, preparing for life in the outside world.

The evolutionary success of a species depends, in large part, on adaptations related to development from a fertilized egg to a mature organism. Animals and plants have a variety of adaptations to ensure the survival of offspring to maturity. In this chapter, you will learn how the offspring of some animals and plants grow and develop. You will find out how their adaptations work to bring offspring to maturity and prepare them for the outside environment.

Baby kangaroos are only about the size of honey-bees when they are born and crawl into their mother's pouch. Can you imagine what your development would have been like if your mother had a pouch?

Chapter Preview

19.1 Early Stages of Development

You have probably seen buds forming on trees, leaves or flowers opening, and fruits enlarging and ripening. These events signal the development of new plant parts. And as you've probably observed, these events occur gradually and in many separate stages. From a single cell—the zygote—a new organism is produced. At first, repeated cell divisions produce a mass of similar looking cells. Later, the embryo starts to take on a definite shape, and specialized parts begin to form—tissues, organs, and systems. Similar changes occur as the body plan of a newly produced animal unfolds. What series of events results in the production of a new organism, and how are those events guided and coordinated?

Animal Development—Cleavage

As you know, development is a series of changes leading to formation of a mature organism. Scientists first studied animal development using animals that develop in water. Animals such as frogs, starfish, and sea urchins were studied because they were easy to raise and observe. With new technological advances, it became possible to study development in birds, reptiles, and mammals as well. Perhaps the most important finding of these early studies was that, although there are differences in the early stages of development among different types of animals, there is a general scheme that applies to all of them.

Figure 19-1 **Frog eggs have a darker, pigmented animal pole and a lighter colored vegetal pole. The eggs are surrounded by a jelly coat that swells when they are deposited in water. This swollen jelly protects the eggs.**

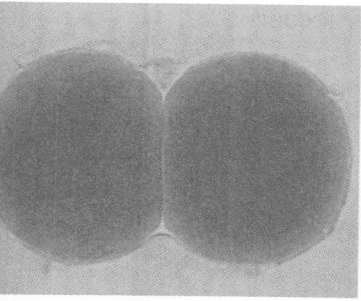 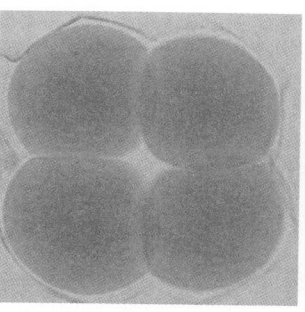 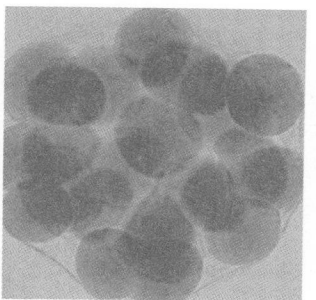 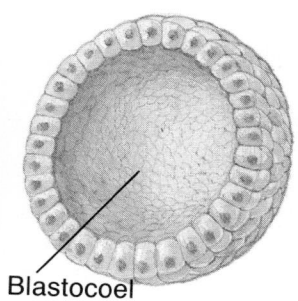

Blastocoel

Cleavage

In sexually reproducing animals, immediately after the fertilization of an egg by a sperm cell, development begins with a rapid series of cell divisions. Because the original zygote is split up, or cleaved, into many smaller cells during initial development, this process is known as **cleavage**. The group of cells produced as a result of cleavage is called a **blastula**. As you will learn, the pattern of cleavage varies among animals and is related to the amount and distribution of yolk in the zygote. Yolk is a thick protein and lipid substance that serves as a food source for the developing embryo. Cell division occurs more slowly in areas of the embryo where there is a lot of yolk because the thick yolk interferes with cleavage.

Cleavage in a Starfish Figure 19-2 shows cleavage in a starfish. The small amount of yolk in the zygote is evenly distributed. As cell division occurs without growth, the newly produced cells become smaller and smaller. The original zygote finally becomes a ball of cells, like a mass of soap bubbles, surrounding a hollow area called the **blastocoel**.

Cleavage in a Frog Now examine Figure 19-3, which shows cleavage in a frog. The frog zygote has a darker, pigmented animal pole and a lighter-colored vegetal pole. Most of the light-colored yolk is in the vegetal pole, and its presence has an effect on the pattern of cleavage. Early cleavage results in production of cells of about equal size, but as cleavage progresses, cells in the animal pole divide more rapidly than those in the vegetal pole. As a result, the cells in the animal pole are smaller and more numerous than those of the vegetal pole. Unlike the starfish, the frog blastula at this point resembles a more solid ball of cells, but it, too, contains a blastocoel filled with fluid (Figure 19-3, upper right).

Figure 19-2 Cleavage in a starfish results in a blastula (far right) that resembles a hollow ball of soap bubbles. The cells in the blastula are all about the same size. Magnification (photos): 400×

Figure 19-3 The first two cell divisions in the frog embryo are from pole to pole. The third division is perpendicular to the polar axis and results in the formation of four smaller cells at the animal pole and four larger cells at the vegetal pole. The far right photo and the diagram show the blastula stage. Magnification (photos): 20×

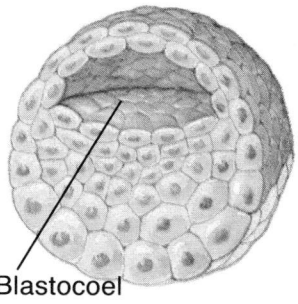

Blastocoel

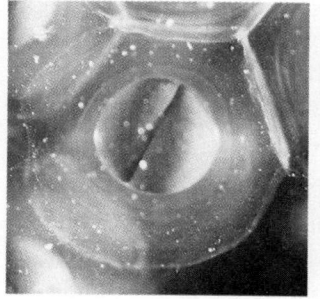  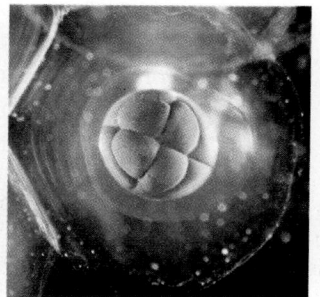 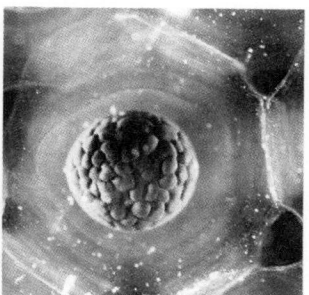

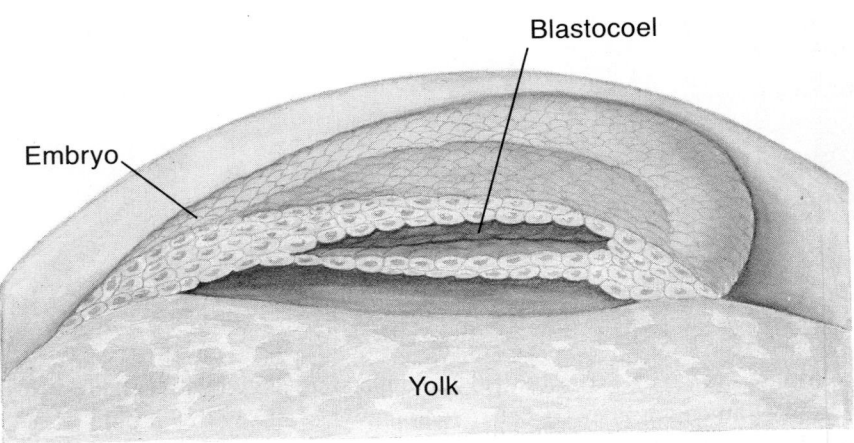

Figure 19-4 Cleavage in a chick embryo is limited to a small area at the top of the yolk. There is so much yolk that it is not cleaved through, and only a small disk of cells is produced.

Cleavage in a Chick   Figure 19-4 shows cleavage in a chick embryo. As you might expect, this cleavage pattern is quite different from those of both starfish and frogs because the chick egg contains a large amount of yolk. The embryo is a small area on top of the yolk. Because cleavage cannot cut through the yolk easily, a disk of cells, and not a ball, forms on top of the yolk like a cap, as the embryo undergoes cell division.

In all these examples, cleavage produces a mass of many cells from one original cell. Cell division occurs so quickly that there is little or no growth of the embryo at first. Why does cleavage occur first, and why is it necessary? One reason is that cleavage produces the raw materials—many cells—needed for production of a new multicellular animal. Division of the zygote into many smaller cells increases the surface-to-volume ratio of each cell in preparation for the rapid uptake of oxygen and nutrients that will be needed during the explosive growth of the embryo. Formation of the blastocoel is another reason why cleavage is important. The blastocoel is necessary to provide a space for movement of cells in the next stage of development.

Animal Development—Morphogenesis

If you think about it, you will realize that an embryo cannot just acquire more and more cells. It also begins to take on a form characteristic of its species. The term *morphogenesis* comes from two Greek words meaning "beginning structure." **Morphogenesis,** then, literally means the beginning of form or structure. That form begins to take shape during the next stage of animal development and occurs as blastula cells begin moving from one place to another.

Starfish Morphogenesis   In a starfish blastula, for example, morphogenesis begins when the outer layer of cells surrounding the blastocoel, called the ectoderm, begins to push inward at one end. This inward movement produces a small indentation. As cell division continues and more and more cells move inward, the blastocoel begins to disappear, and a new cavity forms in its place. This cavity, the archenteron, will later develop into the digestive tract. The archenteron opens to the outside through an opening called the blastopore. Later, the blastopore will become the anus, and the mouth will form from a second opening that develops at the opposite end of the archenteron. Thus, with this early movement of cells comes the first evidence of body systems seen in mature organisms. Although the shape of the embryo has changed, it is no larger than the zygote was. Real growth still has not occurred.

Notice from Figure 19-5 (center) that instead of one layer of cells, the blastula now has two layers. The ectoderm cells that migrated inward have created a second layer, called the endoderm, that lies up against the inside of the ectoderm. You can picture how this happens if you think about collapsing an uninflated basketball so that the wall on one side of the ball is pushed in against the inside of the wall on the opposite side. Part of the outer wall of the basketball will form a second layer as you push it inwards, and the original interior of the basketball will disappear as a new hollow area is created.

Later, a third layer, the mesoderm, will form between the ectoderm and endoderm, further adding complexity to the developing embryo. This rearrangement of cells changes the blastula to a cup-shaped form called a **gastrula.** The entire process of cell movement, and the formation of a triple-layered embryo, is called gastrulation.

Frog Morphogenesis   Gastrulation doesn't occur in exactly the same way in a frog, because a frog blastula is a different shape and contains a greater amount of yolk concentrated in the vegetal pole. Instead of ectoderm cells pushing in, some animal pole cells first move inward through a slitlike blastopore that forms midway between the animal and vegetal

Figure 19-5   **Gastrulation in the starfish occurs when the outer layer of cells begins to form an indentation (left). As the cells push inward, the blastocoel becomes smaller and the archenteron forms in its place (center). A third layer of cells, the mesoderm, forms between the ectoderm and the endoderm (right).**

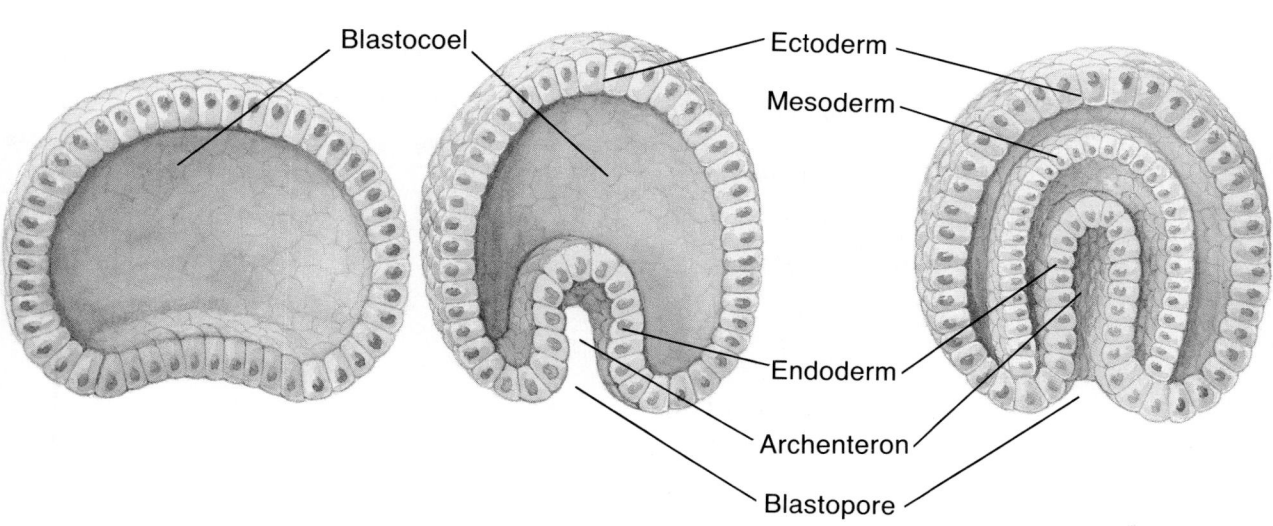

Blastocoel
Ectoderm
Mesoderm
Endoderm
Archenteron
Blastopore

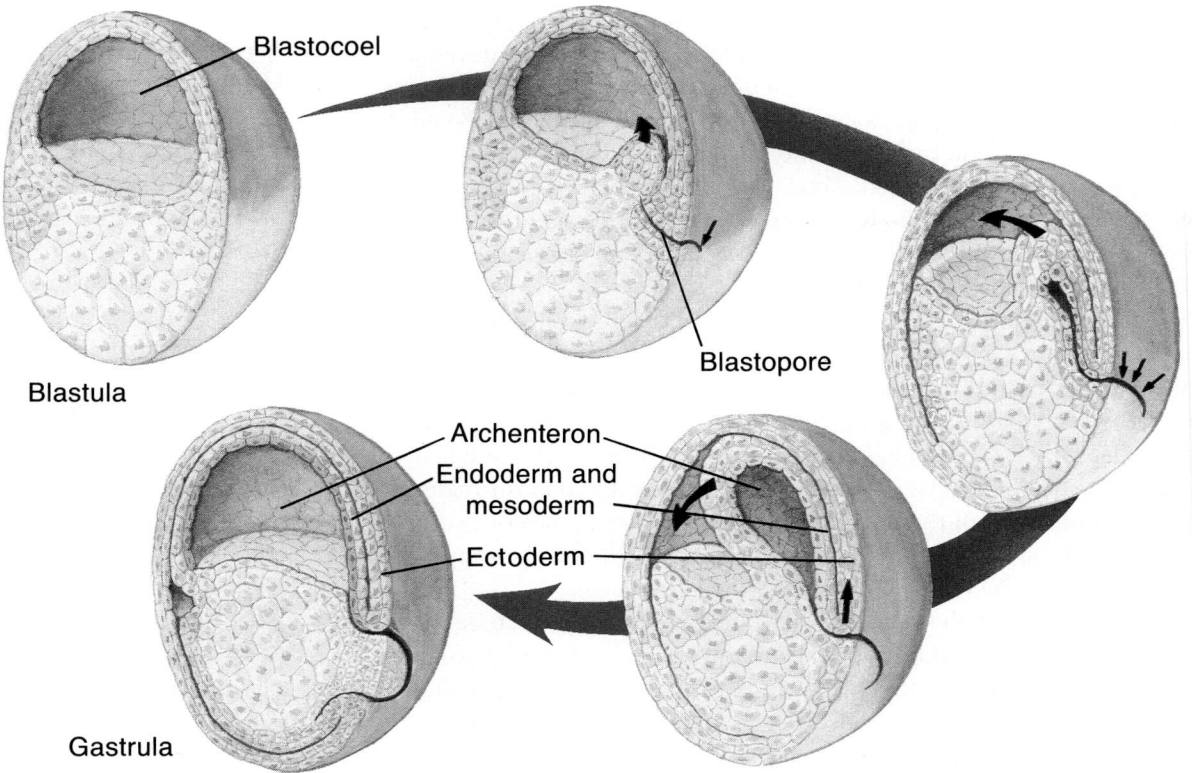

Blastocoel

Blastula

Blastopore

Archenteron
Endoderm and
mesoderm
Ectoderm

Gastrula

Figure 19-6 **Gastrulation in the frog embryo involves movement of cells through the slit-like blastopore into the blastocoel. Note the formation of the archenteron.**

poles, Figure 19-6. Later, cells on the surface of the embryo move to the blastopore until, eventually, the entire sheet of ectoderm cells that forms the surface of the embryo streams inward, as if being pulled in by an unseen pulley. As with gastrulation in the starfish, the blastocoel is replaced by the archenteron, and three tissue layers are eventually formed. If a frog gastrula is viewed from the outside, the entire embryo is dark-colored as a result of dark-pigmented ectoderm cells moving down from the animal pole and then back up and over the light-colored cells of the vegetal pole to the blastopore.

Table 19.1 **Structures Produced by Tissue Layers**

Ectoderm	*Mesoderm*	*Endoderm*
brain	skeleton	pancreas
spinal cord	muscles	liver
nerves	gonads	lungs
skin (outer layer)	excretory system	lining of digestive system
eye cup and lens	skin (inner layer)	
nose and ears		

The cell movements and changes that occur in gastrulation mark the beginning of a definite shape for the embryo. At this point of development in the frog and starfish, there is no visible evidence of any differences in the cells of the embryo, other than size and yolk content. The cells are merely arranged into three tissue layers. However, those three tissue layers are crucial for giving rise to specialized parts later in development. Table 19.1 lists some structures that develop from each of the layers in a vertebrate.

Animal Development—Neurulation

After the completion of gastrulation in a frog (and other vertebrates), further changes can be seen along the back, or dorsal, surface of the embryo. Morphogenesis continues as ectoderm cells along the dorsal side of the embryo divide rapidly and elongate to form a flat plate. Then the sides of this plate rise up, buckling into two folds along the entire length of the embryo. These folds meet and fuse together to form a hollow structure called the **neural tube,** which pinches off from the rest of the ectoderm. These events mark the beginning of the nervous system in the developing embryo. The brain and spinal cord soon form from the neural tube. Figure 19-7 shows the development of the neural tube, called neurulation, in the embryo.

Formation of the brain and spinal cord is an example of **differentiation,** changes that result in the formation of specialized parts. During further development, all the various parts of the embryo develop. Cell division, morphogenesis, and differentiation will continue until development is completed. Because the supply of food and raw materials available to the early embryo is limited, the embryo changes shape but little real growth occurs. When differentiation is completed and the young organism can feed, a period of rapid growth begins.

Figure 19-7 As the sides of the neural plate fold upward and inward, a hollow tube is formed. This tube later becomes the spinal cord and brain. Magnification (photo): 250×

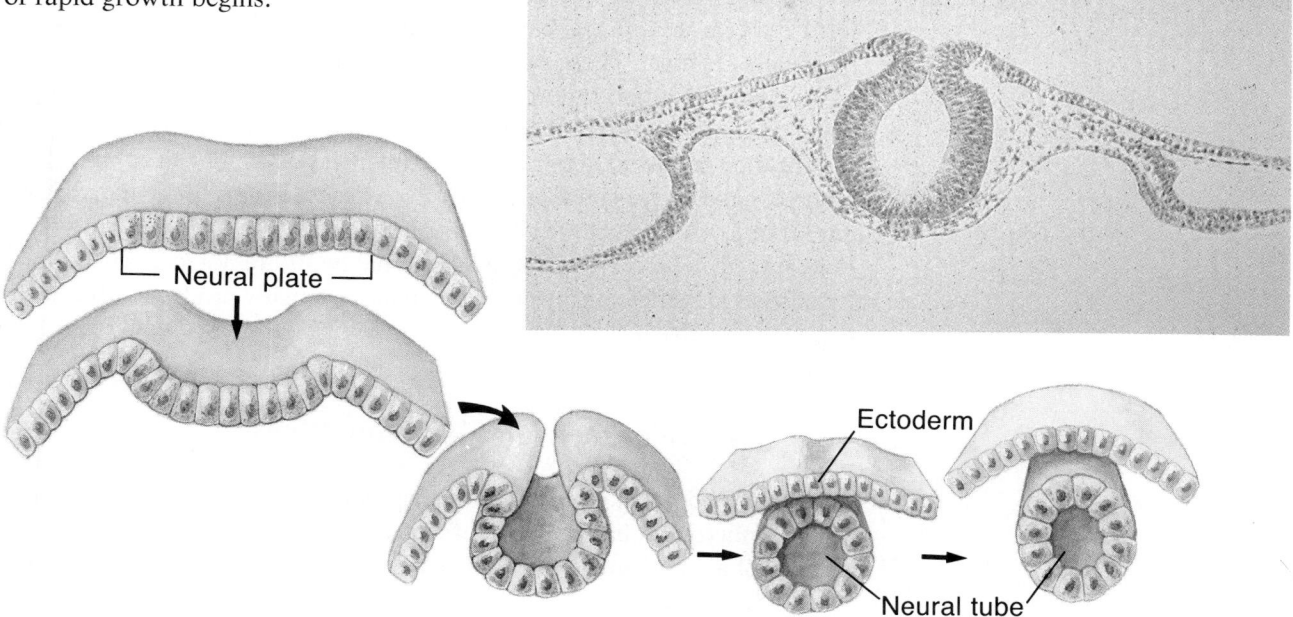

Neural plate

Ectoderm

Neural tube

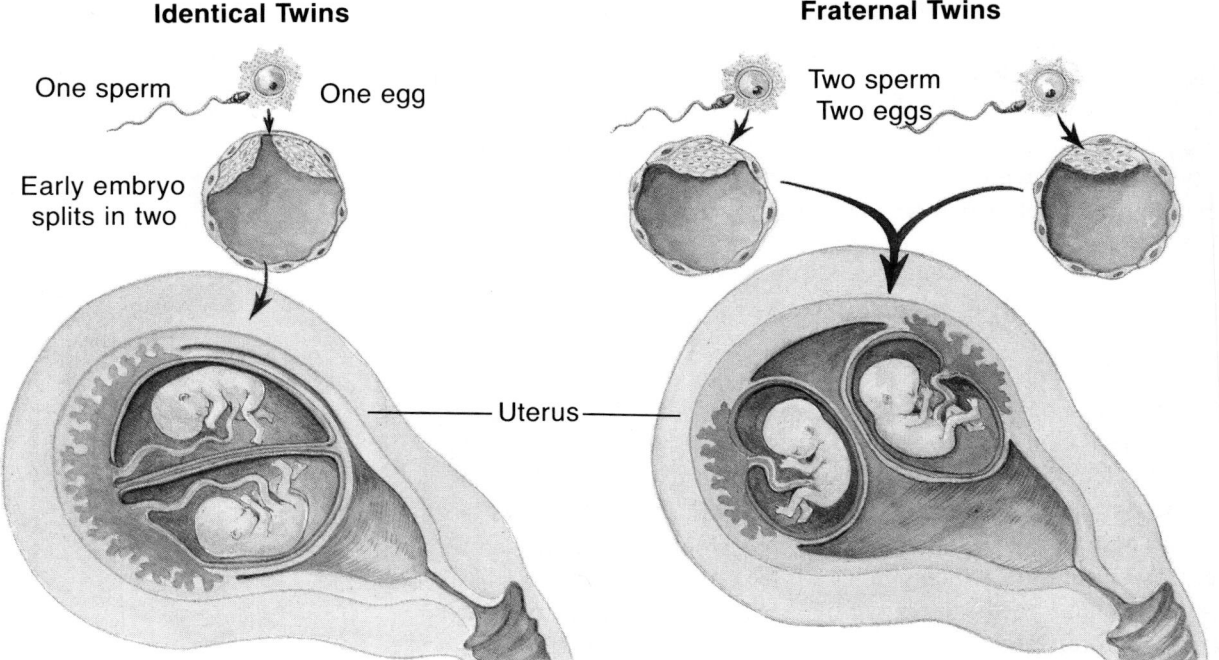

Identical Twins

One sperm   One egg

Early embryo
splits in two

Uterus

Fraternal Twins

Two sperm
Two eggs

Uterus

Figure 19-8 Identical twins form
from a single fertilized egg that divides
into two cell masses. Fraternal twins
develop from two fertilized eggs.
Why are identical twins identical in
appearance?

Twinning

Occasionally, as an early embryo begins to develop, the cells of the embryo
separate into two cell masses, rather than remaining as one embryo. Each
of these cell masses, however, still undergoes the processes of gastrulation
and differentiation to develop into a new and complete animal. Thus, two
animals are produced instead of one. Because they came from the same
fertilized egg, the animals have the same sets of DNA and, therefore, are
physically identical. These two animals are identical twins.

The more common fraternal twins are not alike, because they develop
when each of two different eggs is fertilized by a different sperm.
Fertilization occurs at about the same time, and development of the
embryos takes place simultaneously. Each zygote begins life with a different set of DNA instructions. Therefore, fraternal twins are no more alike
than any other brothers or sisters. Fraternal twins are not as common in
humans as are single births because females normally produce only one
egg each menstrual cycle. Must fraternal twins be the same sex? Must
identical twins be the same sex? Why?

Development in Flowering Plants

As you have learned, seeds from angiosperm plants are structures containing an embryo that has developed from a zygote. Events leading to production of a plant embryo are somewhat different from the early stages of
animal development.

Cleavage and Morphogenesis After the zygote is formed, it divides by mitosis, producing two cells of different sizes. One of these cells, through continued cell division, gives rise to the embryo. The second cell gives rise to an elongated structure that will function in transport of nutrients from the parent plant to the forming embryo. The embryo continues to undergo cell division and growth, forming three kinds of tissues that will later develop further. Unlike cleavage in an animal embryo, early cell division and growth in the embryo of a flowering plant occur at the same time. Nor does morphogenesis occur as a result of cell movement as it does in animals. Instead, morphogenesis comes about as a result of different rates and patterns of cell division in the different parts of the embryo. To understand development in plants, you need to think about the structures seen in mature plants. Embryos of vascular plants contain structures that will eventually develop into roots, stems, and leaves. Like animal embryos, the embryos in seeds of flowering plants also contain a food source that provides the energy needed for early growth and cell division. Remember from Chapter 18 that the food source of a flowering plant embryo is the endosperm. As the food supply in the endosperm is used up, the cotyledons become the major food storage organs of the seed. A **cotyledon** is a leaflike storage organ that develops from the upper part of the embryo and to which the embryo is attached. The embryos of some flowering plants have one cotyledon; most have two.

Embryo Structure Figure 19-9 shows the fully developed embryo of a bean seed. It consists of three regions. The **hypocotyl** is the stalklike part of the embryo attached just below the cotyledons. The hypocotyl will eventually form the stem and part of the root. At the upper end of the hypocotyl, attached above the cotyledons, is the **epicotyl.** The epicotyl will form a pair of small leaves, which open early in development. At the lower end of the hypocotyl is the **radicle,** which forms the root. The epicotyl and radicle develop from regions of cell division known as **meristems.** Meristems are found in many places of growth in mature plants, as well as in the embryo. They produce cells that develop into new parts each growing season. Thus, unlike animals, many plants continue to develop new structures throughout their lifetimes.

Seed Germination Once the embryo has been formed, a seed enters a period of inactivity or dormancy that may last for long periods of time, even years in some cases. Only when a seed is placed in a suitable environment will the dormancy be broken. **Germination,** the development of a seed into a new plant, occurs if conditions such as the amounts of oxygen and moisture and a suitable temperature are favorable for growth.

Minilab 2

Will seeds grow without stored food?

Soak six bean seeds in water overnight. The following day, carefully remove the seed coat from five of the seeds. Separate the two halves of the seeds. Using a toothpick, very gently remove the embryo from each of the five seeds. Place the embryos on a moist paper towel in a petri dish. As a control, place the sixth seed, which has not been disturbed, on the moist paper towel. Cover the petri dish. Store the petri dish in a warm place and examine it every day for several days.
Analysis What did you observe? Can a plant embryo grow without food? What is the food source for a bean embryo?

Figure 19-9 The large fleshy cotyledons of bean seeds store food that was absorbed from the endosperm when the seed developed. Note that the plant embryo inside the seed is not yet green. Why is that so?

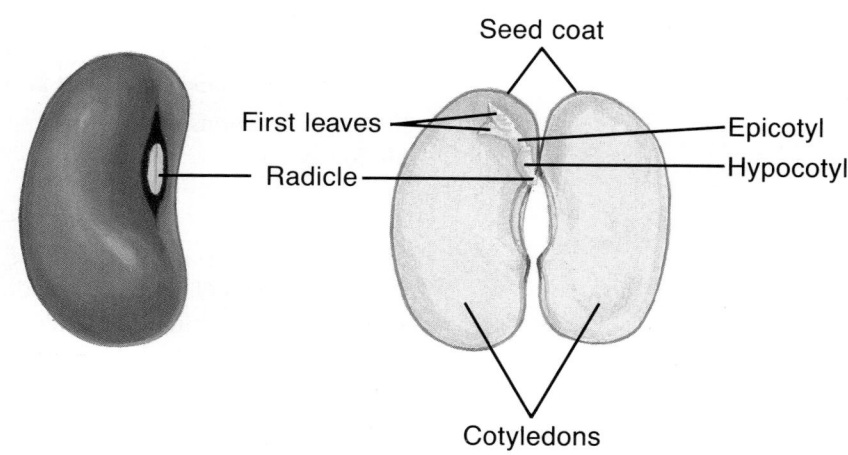

The radicle is the first part of the embryo to emerge from the seed. It becomes the primary root of the plant. The area behind the radicle, the hypocotyl, lengthens, forming an arch that pushes the bean cotyledons above the soil. The part of the hypocotyl that remains in the soil becomes the upper part of the primary root. The aboveground portion becomes the lower part of the stem. The two cotyledons separate and the epicotyl emerges, forming part of the stem and the first leaves. While these events take place, the cotyledons shrink in size as the food they contain is absorbed and used by the developing plant.

I N V E S T I G A T I O N

The Effect of Microwave Radiation on Seed Germination

Did you ever grow plants from seeds? The backs of the seed packages said that germination would take five days for some kinds of seeds, seven days for some seeds, two weeks for others, and three weeks for still others. Farmers select crops based on how long it takes for seeds to germinate and plants to grow. The length of time that growing conditions in the area are suitable for a particular crop is also important. If these conditions could be altered in specific ways, farmers would be able to alter planting schedules and types of crops to make their harvests larger and more profitable. In this lab, you will try to alter the germination time of seeds by treating them with microwave radiation.

Problem

What is the effect of microwave radiation on the germination of seeds?

Safety Precautions

Follow instructions for using the microwave oven.

Hypothesis

Make a hypothesis about the effect of varying doses of microwave radiation on seed germination. Explain your reasons for making this hypothesis.

Experimental Plan

1. As a group, make a list of possible ways you might test your hypothesis, using the materials your teacher has made available.
2. Agree on one idea from your group's list that could be investigated in the classroom.
3. Design an experiment that will test one variable at a time. Plan to collect quantitative data.
4. Following the style of a recipe, write a numbered list of directions that anyone could follow.
5. Make a list of materials and the quantities you will need.

Plant Development   You have learned that plant development doesn't involve movement of cells, as occurs during animal development. Rather, plants develop as cell division and growth occur, and more and more layers of cells build up. As an example of further plant development, consider a developing plant root. At the tip of the root is a root cap that protects the delicate root as it pushes downward through the soil. Just behind this cap is a meristem region that continually produces new cells. Cells forming near the root tip replace root cap cells as they are worn away. At the same time, cells on the opposite side of the meristem expand

Germination of a Bean Seed

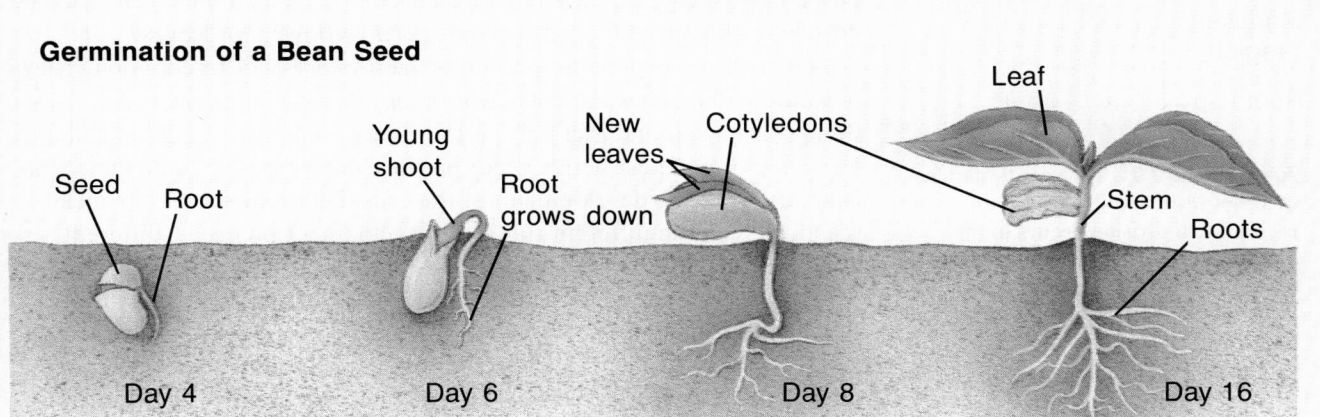

Day 4   Day 6   Day 8   Day 16

Checking the Plan

Discuss the following points with other group members to decide on the final procedures for your experiment.

1. What variables will need to be controlled?
2. What is your control?
3. What will you measure or count?
4. What power levels will you use in the microwave?
5. How many seconds will you use at each power level?
6. How many groups of seeds will you use?
7. How many seeds will you use in each group?
8. Have you designed and made a table for collecting data?

Make sure your teacher has approved your experimental plan before you proceed further.

Data and Observations

Carry out your experiment, make your measurements, calculate percentages of germination, and complete your data table. Make a graph of your results.

Analysis and Conclusions

1. Was your hypothesis supported by your data? Explain.
2. What is the effect of microwave radiation on seed germination?
3. Consult with other groups and compare your results with theirs. How do you account for the similarities? The differences?

4. Speculate about why your results turned out as they did.
5. How might the size of your seeds determine the outcome of your experiment?
6. In what way might farmers benefit from the kind of lab results you obtained in this experiment?
7. Write a brief conclusion based on the results of your class or several classes.

Going Further

Based on this lab experience, design an experiment that would help to answer a question that arose from your work. Assume that you will have unlimited resources and will not be conducting this lab in class.

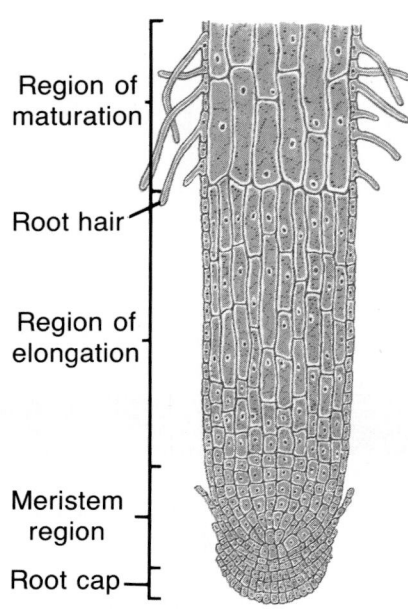

Region of maturation

Root hair

Region of elongation

Meristem region

Root cap

Figure 19-10 The growth pattern of a root results in three definite regions. Cell division occurs in the meristem region. Lengthening of the root occurs in the region of elongation. Cell specialization occurs in the region of maturation.

Figure 19-11 Shown here is part of the experiment described on this page. Nuclei from blastula and gastrula cells were also transplanted into frog eggs, with similar results.

in length, adding to the total length of the root and aiding the root in pushing more deeply into the soil. Next to this lengthening region are older, more mature cells. It is here that differentiation of plant tissues begins. Tissues for uptake of water and minerals; transport of water, minerals and food; food storage; and protection all form here. In some angiosperms, another kind of meristem forms, producing cells that will develop into more transport tissue as the plant ages. The addition of such transport tissue causes the root to grow in diameter as well as in length.

Control of Differentiation

A zygote or, in the case of asexual reproduction, a single cell, contains a set of DNA that guides its development. All the many cells of the developing organism inherit that same set of DNA. If all cells contain the same DNA, how, then, does differentiation occur? How do animal cells become muscle or skin, or plant cells become parts of flowers or leaves?

Scientists are beginning to unravel the mystery of differentiation. They now know that genes are "switched on or off" during development. Biologists have found that only certain genes are active in a given cell at any given time. These active genes produce enzymes that guide the chemistry, and thus the development, of the cells. Different sets of chemical reactions result in different traits. As development proceeds, different sets of genes become activated, and cells become more and more specialized.

Actually, biologists have suspected for many years that genes may be switched on and off. Experiments conducted with frogs supported this idea. In one experiment, nuclei from blastula cells, gastrula cells, and intestinal cells of a tadpole were transplanted into unfertilized eggs from which the nuclei were destroyed by ultraviolet radiation. Many of these eggs developed into normal frogs. These offspring were clones because they were genetically identical to their parents. This experiment showed that the nuclei of cells in various stages of development still have the full set of DNA instructions. The same DNA was present in the nuclei when

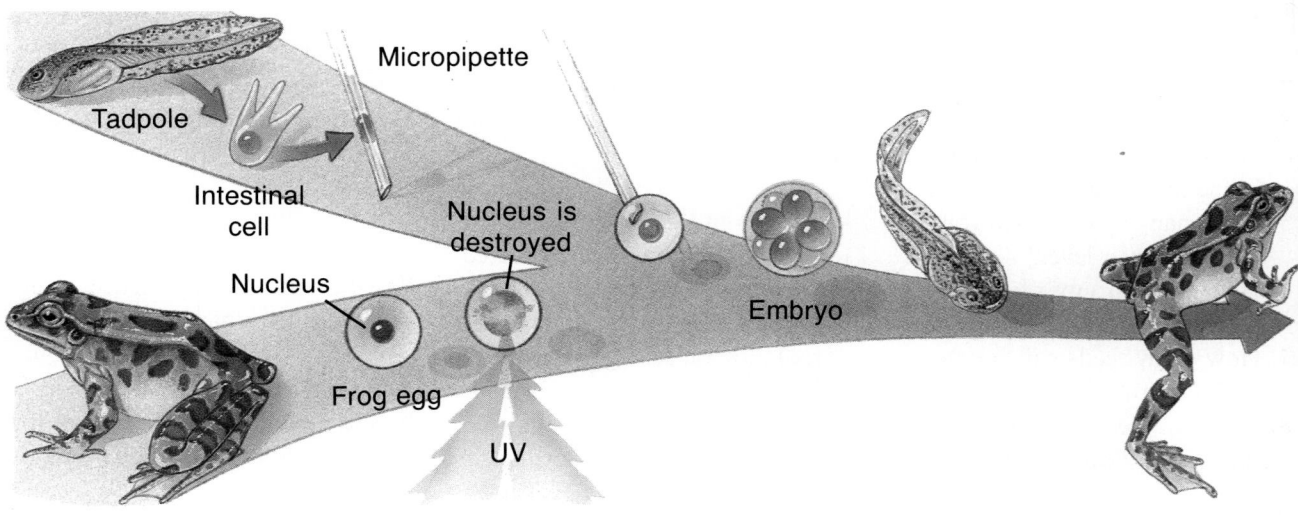

Tadpole

Micropipette

Intestinal cell

Nucleus

Nucleus is destroyed

Frog egg

UV

Embryo

they were parts of blastula, gastrula, or intestinal cells, so some of the DNA must not have been operating. But, when the nuclei were placed in the egg cells, the DNA was switched on again. Something in the cytoplasm must have interacted with the DNA to switch it on.

Today, scientists are working to understand what causes genes to be switched on and off. Such gene regulation is probably involved in many developmental processes. How is a salamander limb regenerated? How do cells become changed in cancer? These processes are thought to be the result of different genes being switched on and off. Thus, these problems are important challenges of modern biological research.

Embryonic Induction

Using evidence gathered from studies in the 1920s, scientists know that if genes are to be switched on and off during development, some kind of chemical signal is needed. Genes respond to the signal by becoming either activated or inactivated. Scientists have learned that the relative position of embryo parts is an important factor in gene regulation.

Thinking Lab

INTERPRET THE DATA

What factors are important in sea urchin development?

Background

At one time, you were a single fertilized egg cell. How did the complexity of your body with millions of cells develop from such a simple beginning? The information coded in the genes of this single cell directs descendant cells to change in form and function in countless ways.

Fertilized sea urchin eggs contain stored food in the lower half of the egg, called the vegetal pole, and specific mRNA and protein molecules in the upper, or animal, pole of the egg. The first two cell divisions occur from pole to pole. The third division occurs at right angles to the polar axis. When an eight-cell sea urchin embryo is divided into two sections, the resulting embryos are different, depending on how the division is made.

You Try It

The following diagram illustrates the results of an experiment in which eight-cell sea urchin embryos are divided in two ways. **Interpret the data,** in terms of the importance of the nucleus and cytoplasm, in development of sea urchins. What do you think the animal and vegetal poles contribute to the developing embryo?

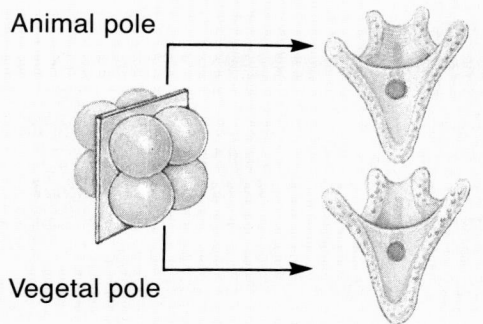

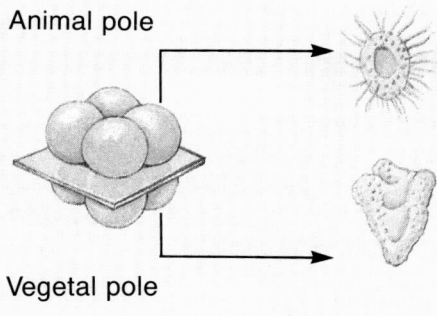

Figure 19-12 The development of many structures is induced by other tissues lying near the structures. Formation of the frog tympanic membrane, the external hearing organ, is induced by the cartilage beneath it.

Tympanic membrane

Inducers The results of these early studies led to the development of the concept of embryonic induction. **Embryonic induction** occurs when one part of an embryo somehow influences, or induces, the development of another part of the embryo. During induction, one tissue releases a chemical that passes to another tissue. That chemical, called the inducer, acts as the signal for differentiation. In the development of nervous tissue, for example, the mesoderm cells release an inducer that activates a gene (or genes) in the ectoderm. Ectoderm cells then develop into nervous tissue.

Although the specific chemical involved in nervous system development has not been identified, other inducers are known. Perhaps the best known inducers are hormones. For example, in humans, differentiation of testes or ovaries depends upon the presence of male sex hormones. The hormones induce the tissue to become testes. If the male hormones are absent, the tissue becomes ovaries.

The Role of Cytoplasm Factors other than induction also play a role in guiding differentiation of an embryo. During cleavage, division of cells can occur in such a way that the chemical composition of cytoplasm begins to vary from cell to cell. For example, some cells may contain more yolk than others. Also, different areas of the embryo become surrounded by slightly different chemical and physical environments. Such differences may act as signals that regulate genes. A domino effect can occur, where signals received by some cells result in production of chemicals that act as signals for other cells. As more and more signals are received, cells may become increasingly committed to a specialized structure and function. Most likely, these signals act by switching on genes in a specific order at specific times. Those genes, in turn, guide the overall unfolding of the animal's body plan.

Section Review

Understanding Concepts

1. How does the cleavage stage in animal development prepare for gastrulation?
2. How is neurulation an example of morphogenesis?
3. In what ways does early development in a plant differ from that in animals?
4. **Skill Review—Sequencing:** Sequence the steps in germination of a bean seed. For more help, refer to Organizing Information in the **Skill Handbook**.
5. **Thinking Critically:** Suggest a reason why morphogenesis in plants cannot occur by the movement of cells, as it does in animals.

19.2 Patterns of Animal Development

Although animal embryos at the gastrula stage have a long way to go before they develop into mature organisms, they are still living organisms. As living organisms, they must be adapted for obtaining food, exchanging gases, and excreting wastes. As delicate organisms, they also require a means of protection and must be kept moist. Though the pattern of development up through gastrulation is basically the same in all animals, the patterns of later development vary among different groups of animals.

Metamorphosis

Most animals that develop outside the female go through a series of remarkable changes as they reach adult form. Animals such as frogs, toads, insects, crabs, clams, and starfish all share this pattern of development, called **metamorphosis,** in which they go through a series of changes in body structure. The changes that occur during metamorphosis aid developing organisms in meeting their life needs.

Metamorphosis in Frogs   Shortly after the gastrula stage, frogs, many insects, and many marine animals quickly develop into a **larva,** an immature stage that can live on its own. Larvae usually bear no resemblance to the final, mature form, but will change into that final form with time.

Fertilized frog eggs develop for one or two weeks within their jellylike blanket, obtaining energy from the yolk, and exchanging gases and getting rid of wastes by diffusion. The color pattern of their eggs, dark on top and light on the bottom, provides protection by camouflaging them from predators. To predators swimming above the eggs, the dark color they see blends with the bottom of the stream or pond. Predators swimming below the egg mass see the light color that blends with the sky.

Objectives

Relate the events of metamorphosis and development within an amniotic egg to the needs of animal embryos.

Describe the formation and functions of the placenta.

Figure 19-13   After hatching from the egg, many animals develop into a larva. These caterpillars are the larvae of insects.

The larval form of the frog—the tadpole—is an animal that is probably familiar to you. Very young tadpoles are inactive, but changes soon begin to occur. A long, tapering tail and three pairs of external gills develop. Because the embryo is now too large to exchange gases and excrete wastes by diffusion, this exchange takes place in the gills. Gills pick up oxygen that is dissolved in the water and release carbon dioxide into the water. Until the larval structures are fully formed, the embryo must rely on its attached yolk, originally stored in the egg, as a source of both raw materials and energy. The fully formed larva has no yolk remaining as a food source, but it can now feed, using its small teeth to scrape algae and plant matter from rocks in the water. Tadpoles are protected from predators by their coloration, which enables them to hide, and by their ability to swim rapidly.

As tadpoles grow and mature, further changes occur. First, the hind legs, then the front legs, begin to appear, and the tail begins to disappear. Internally, a pair of lungs develop, and the circulatory system becomes coordinated with the lungs and skin. Eventually, the lungs and skin will be able to obtain oxygen directly from the air and get rid of carbon dioxide. There are also changes in the digestive system as the tadpole's diet switches from plants to insects. As development proceeds, new structures are formed to replace old structures, which are broken down. For example, the tadpole's tail is broken down and its components are taken into the body. Many of the materials from which these old structures were composed are used as an additional energy source for the developing frog. The many steps in metamorphosis take about three months in the common leopard frog, Figure 19-14, but the exact length of time depends on the temperature of the water.

Figure 19-14 **Metamorphosis in the frog involves an aquatic tadpole stage and a terrestrial adult stage. Note the formation of gills and eyes in the tadpole.**

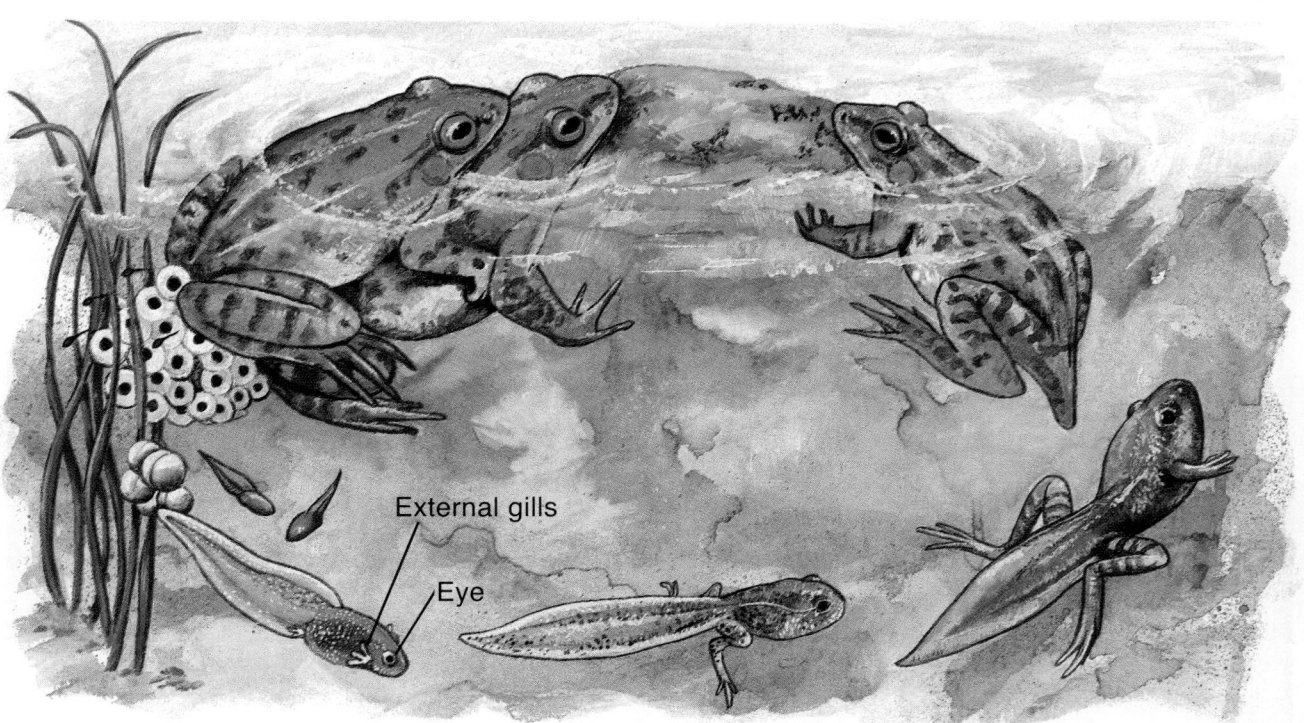

External gills

Eye

Figure 19-15 Complete metamorphosis in insects is shown in the development of a butterfly. The egg undergoes changes, becoming a larva and then a pupa. The adult emerges from the pupa.

Complete Metamorphosis in Insects   After internal fertilization, the eggs of insects are deposited by the female in places such as water, soil, plants, or other animals, where they develop and grow to maturity. If you've ever looked at the undersides of leaves, you may have seen such eggs. The young embryos enclosed in the eggs obtain food from the yolk. They exchange gases and release wastes by diffusion through a porous shell, which also protects the embryo. Eventually the embryo hatches, begins to feed, and undergoes metamorphosis.

Most insects undergo complete metamorphosis, which occurs in four stages—egg, larva, pupa, and adult. In complete metamorphosis, a fertilized egg first develops into a wormlike larva such as a fly maggot or the caterpillar of a butterfly or moth. Protected by its exoskeleton and camouflage coloration, an insect larva is free-living and secures its own food. This food is, of course, needed as an energy source for the larva, but it is also important for the changes that follow.

After a period of rapid growth, the larva becomes inactive and develops into a **pupa,** a stage in which tremendous change occurs. Because it is inactive, the pupa is usually enclosed within some type of protective case. During the pupal stage, the tissues of the animal are completely reorganized. The source of energy needed for this change is the food eaten during the larval stage. After a period of time, a fully formed adult emerges from the pupa.

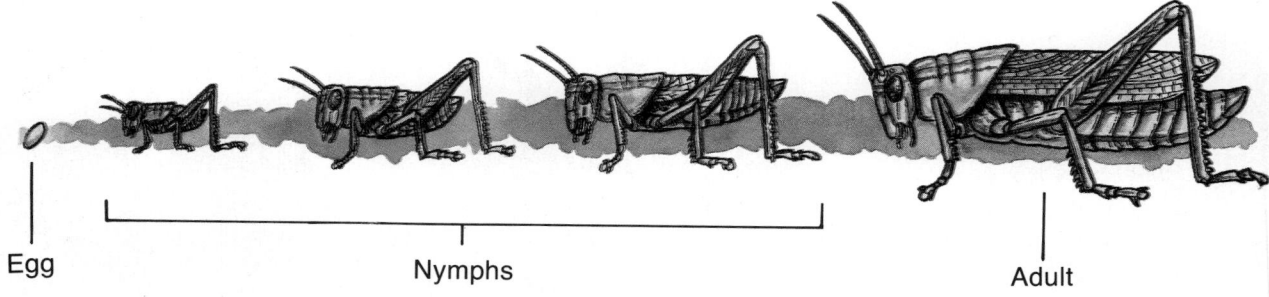

Egg   Nymphs   Adult

Figure 19-16 Grasshoppers are insects that exhibit incomplete metamorphosis. A female deposits fertilized eggs in the ground. A nymph hatches from the egg and grows in size. Eventually, the nymph will develop wings and a reproductive system, and become an adult.

Incomplete Metamorphosis Some insects, such as grasshoppers, undergo a gradual or incomplete metamorphosis, as seen in Figure 19-16. Their life cycle includes only three stages—egg, nymph, and adult. The female grasshopper digs a hole in the ground and deposits her fertilized eggs in the hole. A fertilized grasshopper egg hatches into a **nymph,** an immature stage that closely resembles the adult but is smaller and wingless. Note in Figure 19-16 that the nymph also has a large head, compared to the adult insect. As a nymph develops into an adult, it grows in size and usually forms wings. Internally, a reproductive system also develops.

The Amniotic Egg

If you've ever seen a mass of frog eggs in the water, you've probably noticed that there are hundreds, perhaps thousands, of such eggs. The laying of such large numbers of eggs is an adaptive strategy of some animals to ensure survival of offspring. Because the eggs and the developing larvae of these animals are eaten by so many predators, many eggs must be produced to ensure survival of a few. And because the eggs are fertilized externally, they must be laid in water. Water keeps the eggs moist and serves as a medium for transporting sperm.

Survival on Land An important adaptation that allowed reptiles and birds to survive on land was internal fertilization. After fertilization occurs within the female, the zygote travels down the oviduct as a shell develops around it. In reptiles, the shell is tough and leathery; in birds, it is brittle and porous.

Membranes In addition to the shell, four membranes later develop around the embryo. The innermost of these, the **amnion,** is a fluid-filled sac in which the embryo floats. The amniotic fluid keeps the embryo moist, and the waterproof shell offers protection and prevents evaporation. Because of the presence of the amnion, eggs of birds and reptiles are known as amniotic eggs. Evolution of the amnion, along with that of the shell and internal fertilization, allowed development to occur on land.

Connected to the digestive region of the embryo is a membrane sac called the **allantois.** Because the developing embryo is a living organism, it produces waste products, and these are excreted into and collected in the allantois. When a reptile or bird hatches, it leaves behind the allantois and its collected wastes. Thus, the allantois can be thought of as an embryonic garbage bag.

Also attached to the developing digestive system of the embryo is the yolk sac, a membrane that encloses the yolk. The fatty yolk, along with the protein-rich white of the egg, is a food source for the developing embryo. The yolk sac contains blood vessels through which the food passes to the embryo's cells.

Beneath the porous shell lies a very thin membrane, the **chorion,** which encloses the other three membranes. Part of the chorion lies close to the allantois. Together, these two membranes function in gas exchange. Oxygen passes across the porous shell and through the membranes of the chorion and the allantois. The oxygen diffuses into blood moving through the vessels of the allantois and is carried back to the embryo. Carbon dioxide leaves the embryo, traveling in the opposite direction.

You may have noticed a small pocket of air trapped near the blunt end of an egg. It is especially noticeable when you remove the shell from a hard-boiled egg. Near the end of its development period, a bird obtains additional oxygen by poking its beak through the membrane into this air pocket. Using this air, the bird's lungs start working. The bird then begins to break open its shell and breathes air from outside.

Figure 19-17 Some reptiles lay amniotic eggs, an adaptation for life on land. These Eastern hognose snakes are just hatching. Note the leathery shell of reptile eggs.

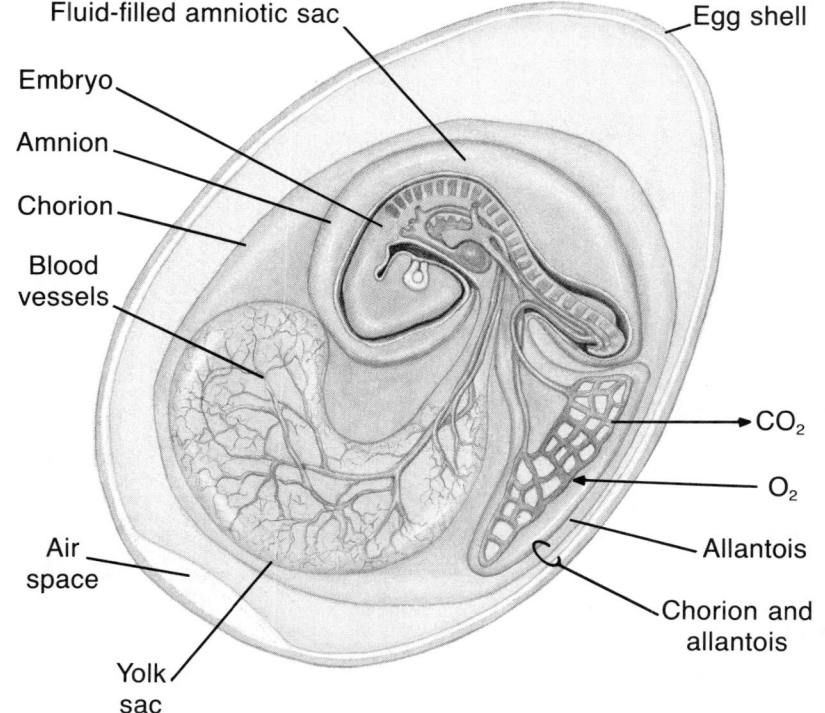

Fluid-filled amniotic sac
Egg shell
Embryo
Amnion
Chorion
Blood vessels
Air space
Yolk sac
CO₂
O₂
Allantois
Chorion and allantois

Figure 19-18 An amniotic egg provides protection, a food source, and a means of gas exchange and waste removal. What is the food source?

Internal Development

Figure 19-19 Placental mammals feed their young milk that is produced in mammary glands.

Figure 19-20 The eggs of placental mammals are fertilized in the oviduct. The zygote then begins development before implantation in the uterus.

In the chapter opener, you learned that the offspring of the pouched group of mammals, the marsupials, develop inside the female's uterus for only a short period of time. Because marsupials have no means of providing the embryo with food and oxygen, and removing carbon dioxide and wastes, for a longer period of time, the offspring develop in the uterus only briefly and are born very tiny and physically immature. In most mammal groups, however, more complete development of the embryo occurs within the mother. If complete internal development is to occur, there must be methods for a continuous food supply, gas exchange, and waste removal. These methods are found in another group of mammals—the placentals. In these mammals, tissue called the **placenta** is formed from both the mother's uterus and the embryo itself. Exchange of gases between the embryo and the mother occurs in the placenta. Also, the placenta is the site where nutrients are obtained and wastes are removed. Humans, dogs, rats, horses, and deer are examples of placental mammals.

Development of the Placenta   Prior to fertilization in placental mammals, the uterus becomes thickened, and its blood supply increases. After an egg is fertilized in the oviduct, the newly formed zygote divides to form a blastula that travels down the oviduct to the uterus. During this time, the embryo obtains nutrients from a small amount of yolk. On the sixth or seventh day after fertilization, the blastula attaches to the spongy wall of the uterus. Cells of the blastula produce enzymes that digest away some of the thick uterine lining, and the embryo becomes embedded in the uterine wall. The attaching and embedding of the embryo within the uterus is called implantation. This event marks the start of pregnancy.

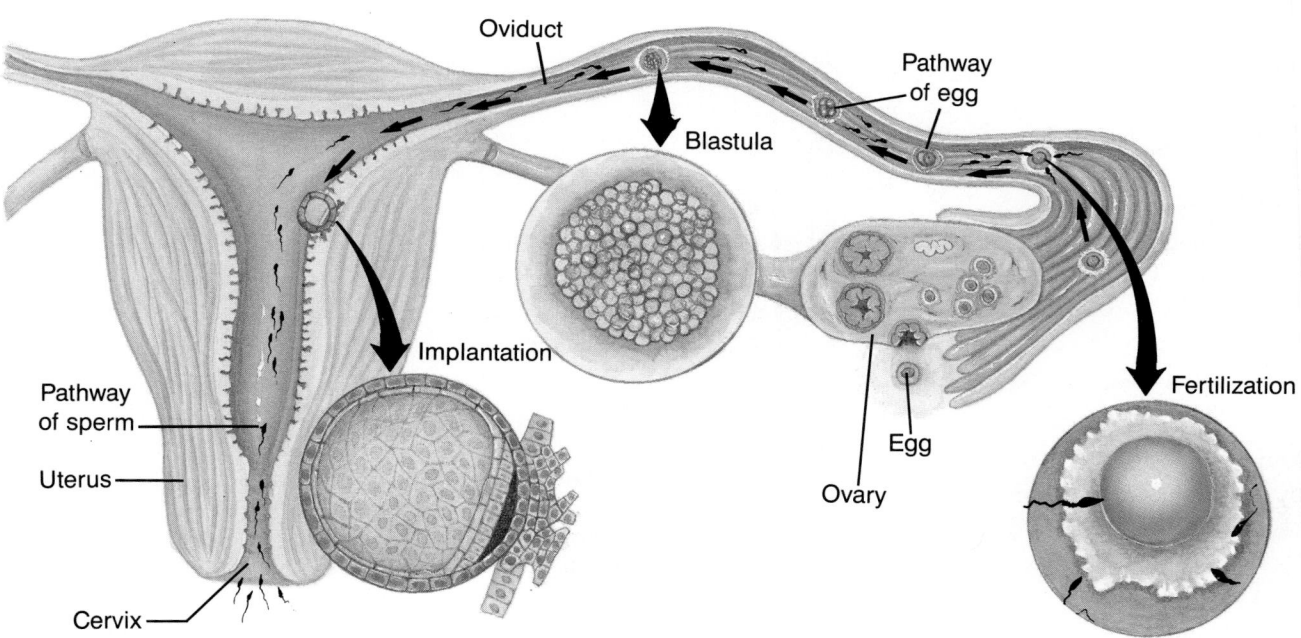

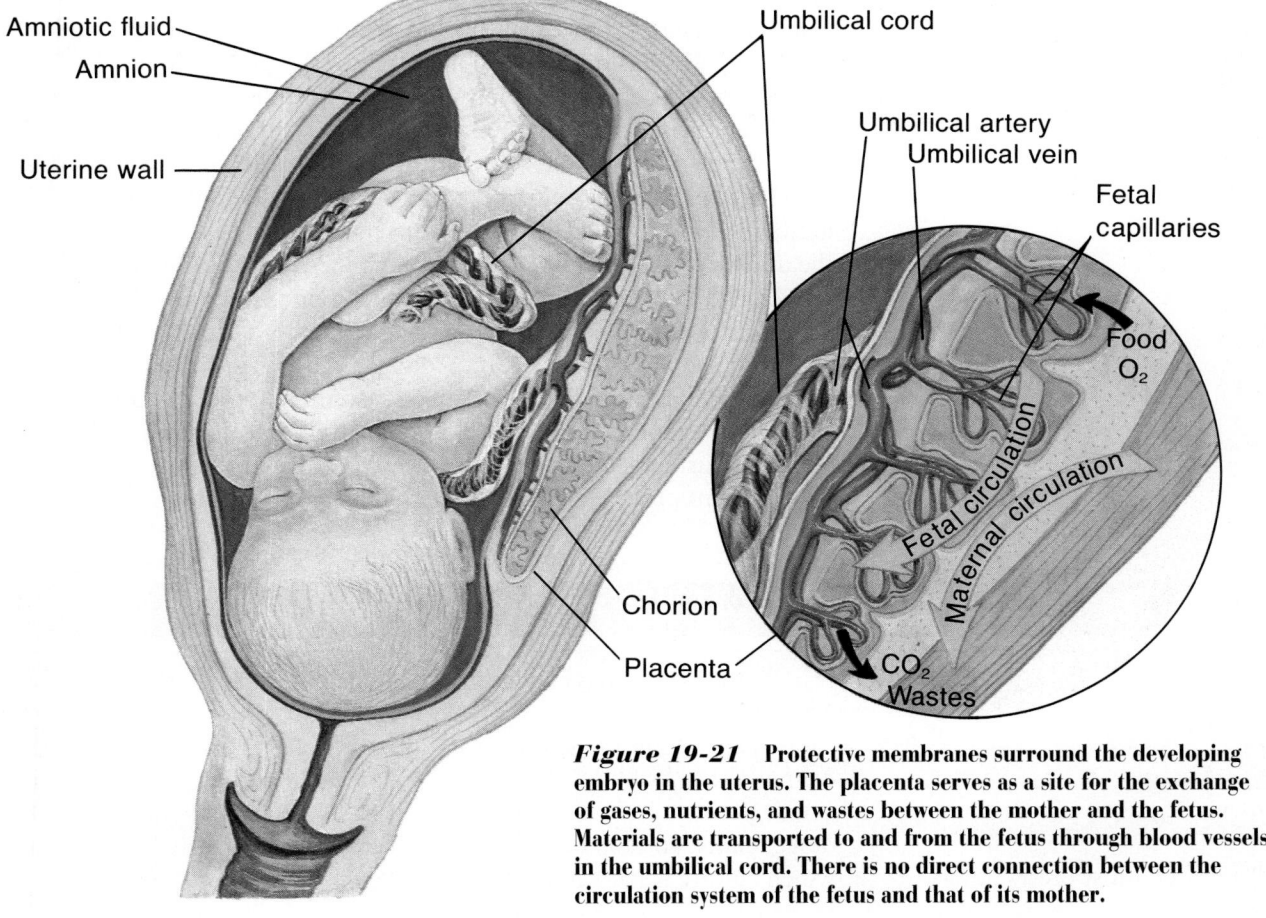

Amniotic fluid
Amnion
Uterine wall
Umbilical cord
Umbilical artery
Umbilical vein
Fetal capillaries
Food O₂
Fetal circulation
Maternal circulation
CO₂ Wastes
Chorion
Placenta

Figure 19-21 Protective membranes surround the developing embryo in the uterus. The placenta serves as a site for the exchange of gases, nutrients, and wastes between the mother and the fetus. Materials are transported to and from the fetus through blood vessels in the umbilical cord. There is no direct connection between the circulation system of the fetus and that of its mother.

As implantation occurs, an amnion, similar to the amnion of bird and reptile eggs, begins to form around the embryo. In addition to the amnion, the other membranes found in the amniotic egg develop. In placental mammals, however, these membranes are modified for different functions and some become part of the placenta. For example, as the placenta develops, fingerlike projections of the chorion begin to grow into the uterine wall. In these projections, small blood vessels—capillaries—develop from the allantois and become part of the placenta. The rest of the placenta is made of uterine tissue. The capillaries in the placenta are part of the embryo's circulatory system. As these capillaries form, the embryo's heart and circulatory system are also developing. In humans, these changes occur during the fourth week of development.

Functions of the Placenta The embryo is connected to the placenta by the **umbilical cord,** which contains large blood vessels that transport blood between the embryo and the placenta. These blood vessels are formed from the allantois. There is a separate placenta and umbilical cord for each developing embryo, except for most identical twins, who share a common placenta.

Blood pumped by the embryo's heart enters arteries of the umbilical cord and circulates through the capillaries of the placenta. The blood comes into close contact with the blood in the mother's uterine wall. The two blood supplies normally do not mix together, but they are close enough so that certain materials diffuse between them. The mother's blood carries nutrients digested by her system and oxygen taken in by her lungs. Some of these molecules diffuse from her blood to the embryo's blood in the capillaries of the placenta. Water and vitamins also pass from mother to embryo. The blood then travels to the embryo through a vein in the umbilical cord and circulates through the embryo's blood vessels. The oxygen and nutrients diffuse from the blood into the embryo's body cells. At the same time, carbon dioxide leaves the embryo's cells and enters its blood. At the placenta, the carbon dioxide diffuses into the mother's blood and is later exhaled by her lungs. Excretory products are also transferred to the mother at the placenta to be excreted later by her. The amniotic fluid cushions the embryo and keeps it moist. Thus, placental development provides an embryo with a favorable environment until it is

ART CONNECTION

Storytellers: The Birth and Development of an Art Form

"They aren't just pretty things I make for money," says Pueblo artist Helen Cordero of her storyteller figures, adding that her work comes from her heart. In making her first clay figure of a storyteller, Cordero created a new form to express two common themes in art: the interaction between humans and their descendants, and the systems by which culture is passed on. In the Pueblo pottery of New Mexico, these themes were traditionally expressed in images of mothers singing to their children. Although most Pueblo pottery has taken the form of functional pots and vessels decorated with geometric designs, clay figures have flourished in the last century, especially at Cochiti Pueblo.

Late in life, Cordero, who lives at Cochiti Pueblo, began to make clay figures. At one of her first exhibits, her figures attracted the attention of a collector, who bought all of her work. He later ordered a number of commissioned pieces, including a singing mother with her children.

Shaping the singing mother reminded Cordero of her grandfather, a famous Pueblo storyteller. She formed a clay image of him telling stories to his five grandchildren. Because Cordero herself was one of the grandchildren, the piece was a self-portrait as well as a portrait of her grandfather. The figure of her grandfather was the first of many storyteller figures by Cordero.

ready to exist independently. All the needs of the developing embryo are provided for within the mother's body. Because placental development meets an embryo's needs so well, the embryo's chances of survival are very good. Fewer eggs need to be produced to ensure continuity of the species.

Birth

As development proceeds, the embryo grows and causes the uterus to expand. After about eight weeks, the embryo is called a fetus. The developing fetus is nourished by the placenta for the entire gestation period, the period of internal development. In humans, the gestation period lasts about 266 days, about nine months. This time span is divided into three trimesters, each equal to three months. During the first trimester, all the body systems of the embryo begin to form. During the second trimester, growth and maturation of fetal tissues continue. The fetus triples its mass during the third trimester. Figure 19-22 on page 536 shows human development.

Her storytellers soon achieved widespread popularity.

During the next 30 years, hundreds of artists made clay storytellers. Storytellers are now displayed in nearly every exhibit of southwestern Native American art. They exist in a variety of forms—storyteller statues, storyteller pots, even storyteller Christmas tree ornaments—and bear the distinct styles of the different pueblos from which they originate. Storyteller figures are loved and appreciated by the people who buy them and see them in museums and exhibits. Perhaps these people feel the same sense of love that inspired the artists to create these works of art.

1. **Connection to Anthropology:** What common elements of life do singing mothers and storytellers depict?
2. **Writing About Biology:** Find out more about the art forms of the Pueblo Indians in northern New Mexico. Are there other popular art forms, in addition to storytellers? Write a short report on your findings.
3. **Connection to Economics:** How has creating what she loves made Cordero an economic success? Do you think an artist is more likely to be successful by creating what is marketable or by creating from the heart? Explain.
4. **Explore Further:** What forms of storyteller figures exist? How do they differ from pueblo to pueblo?

At the end of the gestation period, hormones secreted by the mother interact to start labor, a series of contractions of the uterine muscles. The amniotic sac bursts, and the cervix and the opening of the vagina dilate. As contractions become more frequent and intense, the fetus is pushed downward in the uterus. Usually the fetus is in a head-down, face-down position. Further contractions push the fetus headfirst out of the mother's body through the vagina. This expulsion of a fetus from the uterus is called birth, and is followed by a final series of contractions that expels

PSYCHOLOGY CONNECTION

Infant Learning

"The assumption used to be that there was practically nothing going on in the heads of these infants," says UCLA psychologist Randy Gallistel. "Now we can see that many of the foundations of adult thinking are present at a very early age." The findings of numerous studies support his statement.

In fact, research conducted at Johns Hopkins University indicates that human babies are born capable of making logical inferences and perceiving simple cause-and-effect relationships. When babies a few hours old were repeatedly given sugar from the left side after their foreheads were stroked by researchers, the babies began to turn to the left whenever their foreheads were stroked. If no sugar was provided, they cried.

Cognition

Cognition is the process by which humans gain meaning from what the senses perceive. The senses provide the brain with raw information. During infancy, humans develop the ability to learn about, draw meaning from, and act upon this sensory information. They learn to relate events, experiences, and sensory perceptions.

Sense Perception at Birth
The ability to see, hear, smell, taste, and touch all develop in the fetus before birth. Although infants are born with all five senses, perception and discrimination are not well developed. Even so, babies can recognize mothers by voice or scent within their first month of life.

Newborns seem to seek information about their environment, and they prefer certain kinds of sensory stimuli. For example, they prefer vivid colors, sweet tastes, language sounds, and certain odors. Babies examine patterns methodically, as though trying to learn about them.

Sense Perception in the First Six Months
By the time infants are three or four months old, they begin to understand that objects have constant shapes and sizes. They can tell objects apart by their size, shape, and color. They also discriminate between complex sounds.

Before six months of age, infants can tell whether or not a person's lip movements match the speech sounds they hear. They also know that objects continue to exist even when hidden from view, and that two objects cannot occupy the same space.

the placenta, a stage called afterbirth. In humans, the umbilical cord is tied and cut immediately after birth. The remainder of the cord dries up and falls off the baby after several days. The navel is the place where the umbilical cord was attached to the baby's body.

In reptiles, birds, and mammals, most differentiation is completed before the young animal is hatched or born. After that, mostly growth occurs. The period of growth varies from animal to animal. In humans, it lasts for about 18 to 21 years.

In 1992, a study at the University of Arizona suggested that infants as young as five months of age can count and perform simple addition and subtraction. For example, if an infant watched a researcher place first one, then another doll behind a screen, the infant expected to see two dolls when the screen was removed.

You might wonder how a researcher determines what infants know, believe, or expect, given that they cannot speak. When infants see something different from what they expect, they stare at it longer than when they see something they predicted. Researchers use this tendency to evaluate infants' predictions. For example, in the study just mentioned, an infant watched a researcher place two dolls behind the screen. If the researcher then removed the screen, revealing only one doll, the baby registered surprise by staring longer than when it saw the two dolls it expected to see.

Informally and in research settings, infants reveal the connections they make in various ways as they interact with their environment. You may have watched babies smile, laugh, cry, and explore objects as they made connections. Perhaps researchers who establish that infants are capable of complex cognitive processes are only proving what mothers have always known.

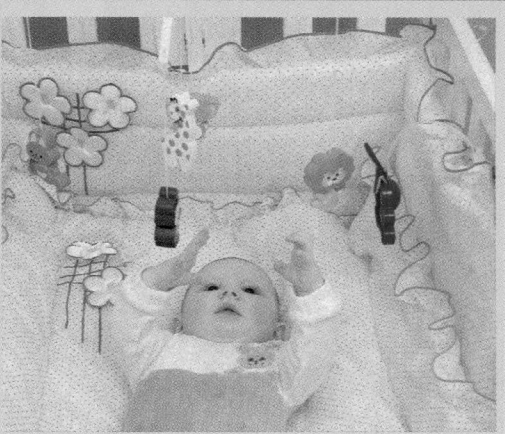

1. **Connection to Neonatology:** The experiment at Johns Hopkins University suggests that infants made incorrect cause-and-effect connections and incorrect inferences. How, then, does the research support the statement that babies have the ability to extract cause and effect and make logical inferences?

2. **Solve the Problem:** How could the methods used at the University of Arizona be used to demonstrate that babies know objects continue to exist even when hidden from view?

3. **Solve the Problem:** If an infant did not understand that objects exist even when hidden from view, how would this affect his or her performance on the test for mathematical ability at the University of Arizona?

4. **Connection to Health:** Based on the information in this feature, what could a parent do to provide sensory stimulation and information for a young infant? How can their interaction with families help babies learn about their environment?

5. **Explore Further:** What role does sensory stimulation play in infant development? What effect does sensory deprivation have on young humans and animals?

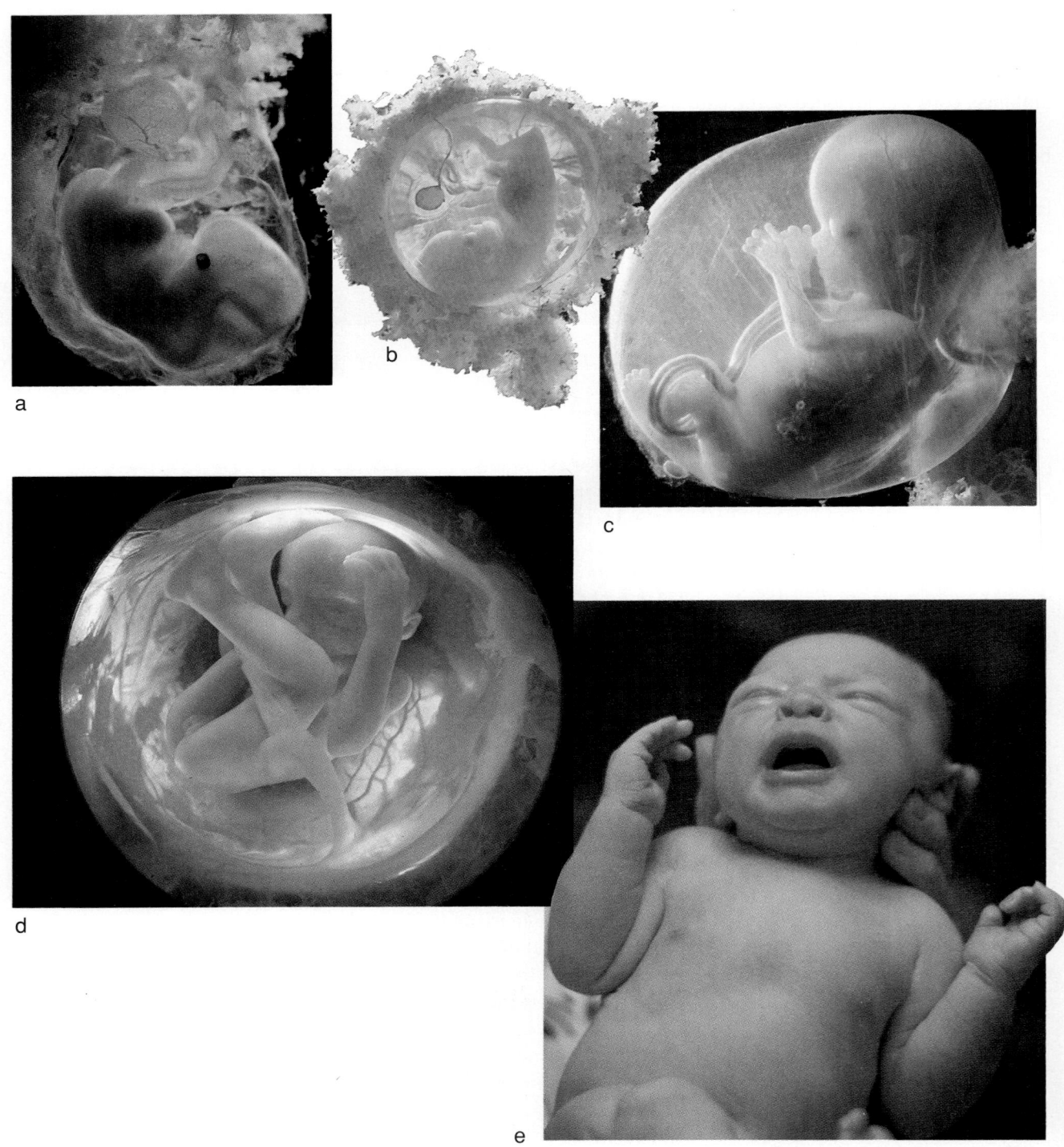

Figure 19-22 These photos show selected stages of the 9-month development that produces a human baby. A six-week human embryo (a) shows paddles for hands and feet on limb buds, and a large head with forming eyes and ears. (Magnification: 10×) At eight weeks (b), facial features become more obvious and hands and feet more developed. (Magnification: 7×) At eleven weeks (c), fingers and toes are clearly visible, and the rib cage can be seen just beneath the elbow. At 22 weeks (d), a human fetus is well-developed and grows rapidly. If born prematurely at this time, however, it can survive only with special care. After nine months (e), a baby is born.

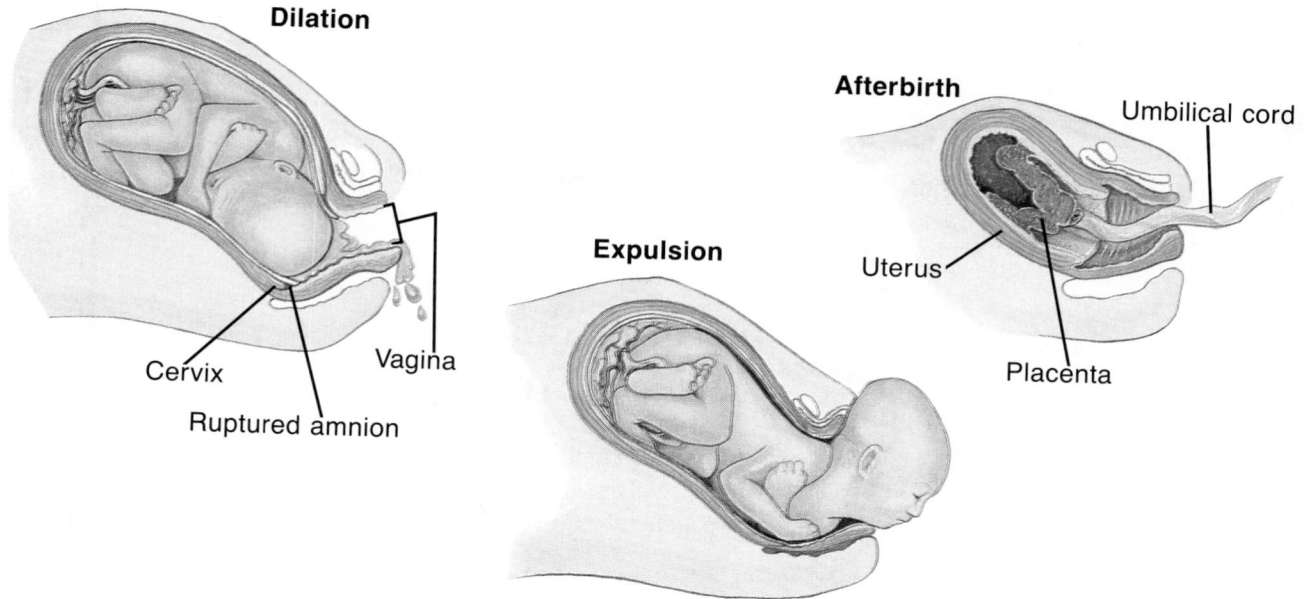

Dilation

Cervix

Vagina

Ruptured amnion

Expulsion

Afterbirth

Umbilical cord

Uterus

Placenta

Figure 19-23   The stages of birth are dilation, expulsion, and afterbirth.

<u>LOOKING **A**HEAD</u>

IN THIS CHAPTER and the last, you have studied two related processes—reproduction and development. In the next chapter, you will learn how organisms obtain the materials they need for growth and for energy, and what happens to those materials once they enter the organism. You will compare the needs of autotrophs and heterotrophs, and examine why heterotrophs must have specialized means of processing the materials they obtain from the environment. Finally, you will examine different ways in which heterotrophs are adapted to carry out that processing. ❧

Section Review

Understanding Concepts

1. Why must the embryo of a frog quickly become a larva?
2. Both insects and birds develop within shelled eggs. Do they have the same pattern of development? Explain.

3. Explain why frogs, reptiles, and placental mammals, in that order, produce decreasing numbers of eggs at one time.
4. *Skill Review—Making Tables:* Make a table listing the adaptations that butterfly, bird, and human embryos have for obtaining food,

exchanging gases, excreting wastes, protection, and keeping moist. For more help, refer to Organizing Information in the **Skill Handbook.**
5. *Thinking Critically:* Why do you think fertilized insect eggs are often deposited in soil or water?

Considering the Environment

Seed Banks

There are banks to hold your money, sperm banks, and even banks for blood and body organs. Scientists also have formed seed banks. The deposits made in these banks are exactly what you would expect from its name—seeds. Seeds from all over the world are frozen and stored in vats of liquid nitrogen. Someday, they will be removed from the vats and germinated.

Why do seeds need to be stored? Seeds are the link from one generation of plants to the next. One in ten plant species is in danger of becoming extinct, due to agricultural practices, development by humans, and changes in climate and ecosystems. As Earth changes over time, species that cannot adapt to the changes become extinct, and new species evolve. If a plant species becomes extinct, there is no way to get it back. Extinction is forever.

Groups and individuals are making efforts to save the habitats of endangered species, with varying degrees of success. Fortunately, plant species themselves are much easier to preserve than are entire ecosystems. A seed bank is a means of saving plants that are in danger of becoming extinct. To ensure the survival of existing plant species, many groups have established seed banks to store seeds from plant species, with special attention to those that are endangered.

How Seeds Are Stored More than 3 million seed samples are frozen and stored worldwide in about 100 seed banks. Seed storage has become a global enterprise, with groups collaborating on storage facilities for seeds from many nations. Should plant species become extinct in the future, stored seeds may be used to reintroduce the species in suitable habitats. Scientists of the future could also crossbreed plants that have desirable traits with plants from the past. This technique could be used to produce new species of crop plants that grow faster or contain more nutrients, in case of escalating demand for food in the future.

Unfortunately, many seed storage projects are beset by problems. For example, the temperatures at which seeds are stored in some seed banks are too high to protect them from spoilage. In some banks, plant samples are lost, and so is identifying information.

A Global Initiative To solve these problems, an international group of researchers, environmentalists, and agricultural industry representatives met in 1991. They issued a global initiative calling for establishment of a nonprofit organization that will fund and support research for preserving plant germ plasm to safeguard future generations of existing plant life. Without plants for food, most living things cannot survive. The "interest" that seed banks pay on seed deposits is that all living things benefit. Seed banks will help make the best seeds available for growing food for the future. Scientists will also be able to place extinct plant species back into the wild when a new and suitable environment is found.

1. **Connection to Ecology:** Before an extinct plant species could be introduced into a new environment, what conditions would need to be present?
2. **Explore Further:** Research a plant species that is classified as endangered. Write a paragraph or two about the reasons for its endangerment and the existing efforts to save it from extinction.

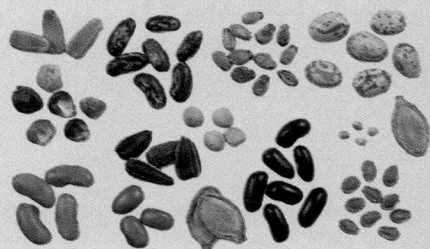

Biotechnology

Fetal Surgery—An "Out-of-Body" Experience

Imagine that you (or your spouse) are just 18 weeks pregnant when your doctor informs you that the fetus has an obstruction in its urinary tract. The obstruction is causing a backup of urine into the kidneys. If the obstruction is not cleared, it could lead to the deterioration of the fetus's kidneys, a fatal condition known as hydronephrosis. Normally, babies with this condition are operated on soon after birth to clear the blockage. But what if the blockage is so severe that normal fetal development is

being threatened and the life of the fetus is in danger? The only possible solution to the problem is to operate on the fetus before it is born. But there are risks for both mother and fetus if this is done. What would you do? In 1986, a couple named Williams had to make such an impossible choice.

The Williams's fetus was diagnosed as having hydronephrosis. Blockage was so severe that the fetus would almost certainly die before it could be born. Even though they knew that there was

some risk involved, the Williamses believed that the blockage had to be removed if there was even a small chance of saving their baby. The Williamses were good candidates for a new type of treatment — surgery on a fetus while it is still in its mother's uterus. Called fetal surgery, this new technique is helping to save fetuses with congenital defects by operating on them before they are born.

What is involved in fetal surgery? In the Williams's case, doctors made incisions in the mother's abdomen and uterus, siphoned out the amniotic fluid, then lifted the tiny fetus halfway out of the mother's uterus. Making an incision in the fetus's lower abdomen, doctors created an external opening between the bladder and the abdomen so that the urine could drain out of the fetus's body and into the amniotic fluid. Then they placed the fetus back into the uterus, replaced the amniotic fluid, and sewed the mother's uterus and abdomen closed. A few months later, the Williamses welcomed a normally developed baby girl into their family. Later, surgery permanently repaired the blockage.

There are several types of congenital defects that can be corrected through fetal surgery. Hydrocephalus, a condition in which excess fluid builds up in the brain, can lead to brain damage unless the fluid is drained as the fetus grows. Fetal surgery allows doctors to place a drainage shunt in the brain of such a fetus to prevent such damage. Congenital diaphragmatic

hernia (CDH) is a condition in which there is a hole in the diaphragm, the muscle that helps the lungs inflate. In severe cases of CDH, other organs move through the hole into the chest area, and the lungs have no room to develop properly. Babies born with CDH are routinely operated on at birth, but if the condition is severe, they invariably die for lack of well-developed lungs. Fetal surgery allows doctors to repair the hole in the diaphragm before lung development is affected.

Are there any adverse effects of fetal surgery? The surgery itself may lead to premature birth, and the newborn dies of complications unrelated to the fetal surgery. But for parents of fetuses with life-threatening congenital defects, the reward is worth the risks.

1. **Issue:** Some medical researchers are opposed to the use of fetal surgery. They feel that the mother will lose her right to make decisions about her own body if the rights of the fetus are considered to be more important than those of the mother. Should parents have the right to decide whether or not to proceed with fetal surgery if it is an option?

2. **Explore Further:** Some congenital disorders can be corrected during pregnancy by treating the mother. Find out about fetal cardiac dysrhythmia, in which there are irregularities or disturbances in the fetal heartbeat. Why can this disorder be treated with medication given to the mother?

Summary

Early development in animals begins with cleavage. Morphogenesis occurs during gastrulation, producing in most animals an embryo consisting of three tissue layers. The pattern of cleavage and gastrulation is influenced by the amount and distribution of yolk present in the zygote.

Development of plant embryos is characterized by cell division and growth that occur at the same time. Morphogenesis in plants comes about by differences in the rates and patterns of cell division, rather than by the movement of cells. Earlier events set the stage for differentiation, most of which comes later in development.

Development is guided by genes. Cells in different parts of embryos receive chemical signals that switch some genes on or off. The particular set of active genes in a cell guides its chemistry and thus its characteristics.

All animal embryos require food, a means of gas exchange and excretion, protection, and moisture. Different patterns of development—metamorphosis, the amniotic egg, and internal development—have evolved to meet these requirements.

Language of Biology

Write a sentence that shows your understanding of each of the following terms.

allantois	hypocotyl
amnion	larva
blastocoel	meristem
blastula	metamorphosis
chorion	morphogenesis
cleavage	neural tube
cotyledon	nymph
differentiation	placenta
embryonic induction	pupa
epicotyl	radicle
gastrula	umbilical cord
germination	

Understanding Concepts

1. With respect to protection and food during development, what do flowering plant and bird embryos have in common?

2. In some animals, early embryos consisting of as many as eight cells can be separated into two groups, and each will develop into a normal organism. If cells are separated later, though, normal development of two animals will not occur. Why?

3. Compare cleavage and gastrulation in a starfish and a frog. Explain the reasons for the differences.

4. A single cell from the root of a carrot is placed on a culture medium. The cell begins dividing, producing an unorganized mass of cells at first. Later, though, an entire carrot plant develops. What do these results indicate about how development is controlled?

5. Why do you think that certain drugs can result in birth defects?

6. Suppose an area of an embryo that normally develops into part of the eye is transplanted next to the area that develops into skin, before the eye area begins differentiating. What will probably happen to it? Why?

7. Suppose the area that normally develops into part of the eye is transplanted after it has begun differentiating. What will probably happen to it? Why?

8. Suppose you marked evenly spaced lines on a very young developing plant root. Would the lines still be evenly spaced after the root has grown? Explain.

9. Describe the function of each of the four membranes in an amniotic egg.

REVIEW

10. Why is development more efficient in placental mammals than in any other group?

Applying Concepts

11. Why does the primary root of a flowering plant develop first?
12. How can a young angiosperm continue to develop after its cotyledons shrivel and fall away?
13. Hormones that inhibit germination are diluted by water moving into the seeds. As the hormones are diluted, their effect decreases. Do you think desert seeds would have high or low concentrations of such hormones? Explain.
14. What kind of twins might be produced if gastrulation occurred at two places in the same embryo? Explain.

15. Explain how a baby can be born addicted to drugs if its mother were a drug abuser during pregnancy.
16. How might a double-yolked hen's egg be produced?
17. What might be an adaptive advantage to the dormancy period of seeds before germination?
18. **Considering the Environment:** Why are seeds easier to preserve than whole plants?
19. **Psychology Connection:** What assumptions was the University of Arizona researcher making in analyzing the behavior of the infants she tested for understanding of addition? What could she verify for certain by observation?

Lab Interpretation

20. **Investigation:** A scientist hypothesizes that ionizing radiation at low levels will not affect seed germination. The results of the experiment are illustrated in the graph below. Do the results support the hypothesis? Why or why not? Interpret the results of the experiment.

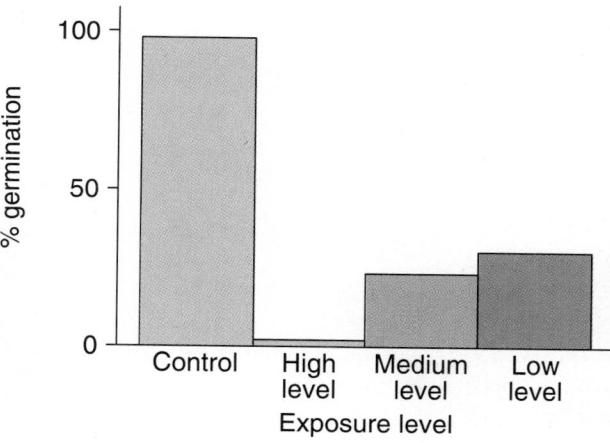

21. **Thinking Lab:** Identical twins are formed when the fertilized egg divides into two separate cells. If the early stages of human development are similar to those of sea urchin development, what must happen during this cell division to ensure that each twin is an exact copy of the other?
22. **Minilab 2:** Why was the sixth seed needed in this experiment? What was its purpose?

Connecting Ideas

23. In what way do larvae serve a function similar to that of seeds?
24. Relate the events of fertilization and development in a human, paying attention to adaptations for each.

CHAPTER 19 DEVELOPMENT **541**

20

NUTRITION AND DIGESTION

THE HUMPBACK WHALE you see above can grow to a length of 15 meters, literally as long as some houses. Its mass is thousands of kilograms. How is this enormous animal able to nourish its massive bulk? Believe it or not, humpback whales are filter feeders. As they suck in water, they use their baleen to filter small invertebrates from the water. Baleen, shown in the smaller photo of a gray whale, are comblike structures that trap small crustaceans and other plankton. The whale then licks the baleen and swallows its meal.

Every organism requires nutrients to survive. How does a tree obtain food? How are cheetahs and clams alike in getting food? Do sponges and whales have anything in common? In this chapter, you will read about nutrition in organisms that make their own food and in organisms that must take in food. You'll learn how some consumers simply absorb nutrients and how others, like the blue whale, filter food from their watery environment. You'll see how still other consumers, such as the *Hydra* and the human, obtain food in chunks and must break it apart to digest it. Finally, you'll discover that although heterotrophs get food in different ways, they all need to break down the complex molecules of food into smaller molecules that can enter cells.

Imagine having combs in your mouth and having to make a meal out of whatever became stuck in them. It's a system that works well for whales because of the environment they live in.

Chapter Preview

Objectives

Distinguish between the nutrient requirements of autotrophs and heterotrophs.

Compare intracellular and extracellular digestion.

Explain the need for digestion.

20.1 Patterns of Nutrition and Digestion

A giraffe browses on the leaves of a tree and a horse grazes on grass. A hawk captures and feeds on a snake, and a tapeworm absorbs some of the food that moves through a cow's intestines. A mushroom soaks up organic matter from the soil, an amoeba traps a smaller protozoan, and you munch a pizza. You know that you and all other heterotrophic organisms must take in food. The trees and grass that giraffes and horses eat are autotrophs, which make their own food. The need for food is a universal characteristic of life.

Nutrient Requirements

Everything that lives needs food. If the cells of living organisms are to grow and divide, carry on respiration, synthesize proteins, and perform all other life functions, they need a constant source of energy and raw materials. Food contains complex organic molecules that can supply both energy and raw materials. These complex molecules and the simpler substances that are used in life processes are called **nutrients.**

Nutrition in Autotrophs   Autotrophs need the same kinds of organic substances that heterotrophs obtain by eating food. But autotrophs don't eat. Instead, they build the needed organic substances by using water, carbon dioxide, and minerals absorbed from their surroundings. For example, in the case of an apple tree, water and minerals present in the soil easily pass through plasma membranes and into cells of the plant's roots and are transported to the leaves. Carbon dioxide enters the plant through openings in the leaves, where photosynthesis converts water and carbon

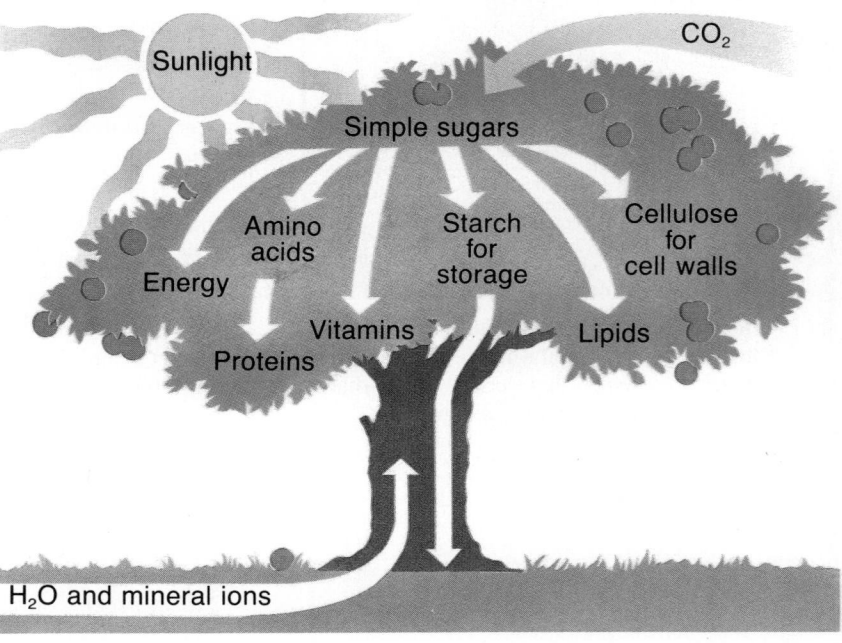

Figure 20-1   **Autotrophs produce simple sugars by photosynthesis. These sugars are the raw materials for building nearly all other substances the plant needs. The word *autotroph* literally means "self feeder." Explain how the drawing illustrates the meaning of *autotroph.***

dioxide into simple sugars. Some of these simple sugars are used as a source of energy for the plant cells. However, most of them are used to build more complex molecules, such as starch and cellulose, or combined with minerals to form lipids and amino acids which are needed for growth and other life functions.

Nutrition in Heterotrophs   Heterotrophs obtain their nutrients by eating autotrophs and/or other heterotrophs. Many heterotrophs, including humans, hawks, and sea anemones, consume fairly large chunks of food. To make use of the nutrients in the food, each chunk must be broken down into smaller and smaller particles until it is in the form of molecules small enough to pass through a plasma membrane. The process of converting complex organic molecules to simpler ones that can pass through a cell membrane is called **digestion.** Digestion is a chemical process involving enzymes that carry out a series of hydrolysis reactions. Review hydrolysis on page 77 in Chapter 3. Large, complex food molecules, such as fats, carbohydrates, and proteins, must be converted by hydrolysis to smaller molecules of fatty acids, glycerol, monosaccharides, and amino acids. These products of digestion are molecules that are small enough to pass through plasma membranes.

Not all nutrients required by heterotrophs are in the form of large, complex molecules. They also include simpler molecules, such as water, minerals, and vitamins, that readily pass across cell membranes with little or no need for digestion.

Figure 20-2   In contrast to autotrophs, heterotrophs must take in food, most of which is in the form of complex molecules. These molecules are broken down by digestion to yield simple molecules which are used as building blocks to form materials that the heterotroph needs.

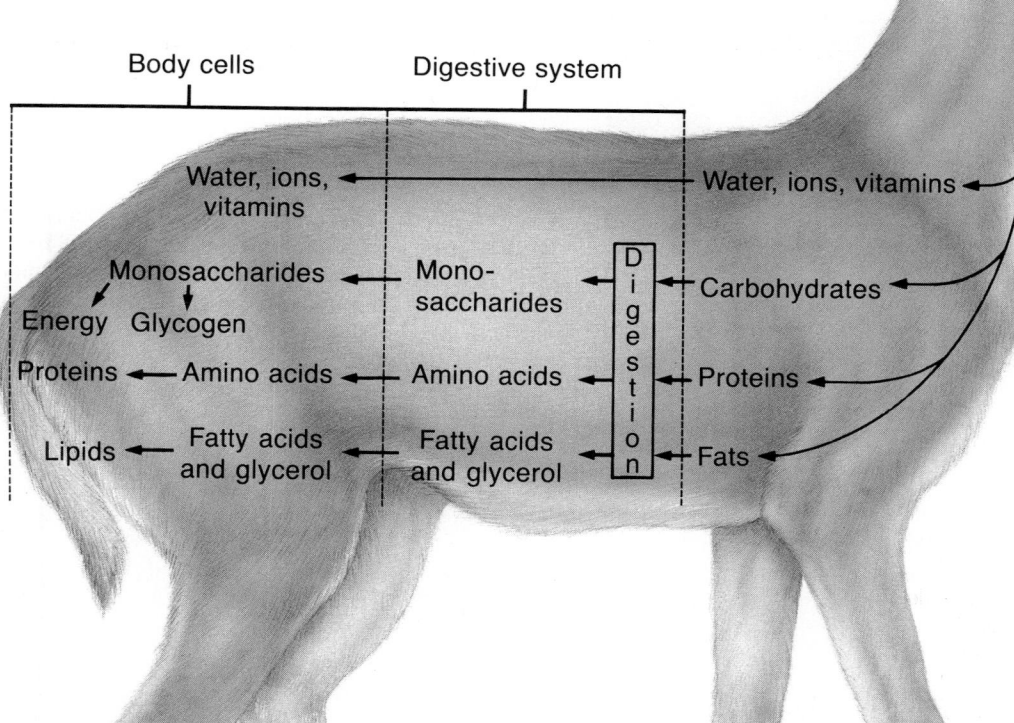

Digestion in Heterotrophs

Because heterotrophs obtain food by eating other organisms, they have a variety of adaptations that enable them to get and digest food. These adaptations include specialized structures along with physical and chemical processes.

Intracellular Digestion A variety of specialized structures for the crucial functions of food-getting and digestion have evolved in heterotrophs. Even organisms as simple as water-dwelling protozoans are well adapted for taking in and digesting food efficiently.

Amoeba and *Paramecium* both feed on microorganisms, which are digested inside a food vacuole. In *Amoeba,* the food vacuole is formed as food is being captured. This process, called phagocytosis, is shown in Figure 20-3. In *Paramecium,* the action of cilia creates a current that draws food into the organism's oral groove. The food is swept to the bottom of the groove, where a food vacuole forms around it. In both of these protozoans, digestion begins when a lysosome fuses with the food vacuole. Enzymes from the lysosome convert large, complex food molecules into smaller molecules that can pass through the membrane of the vacuole and enter the cytoplasm of the cell. These molecules are then available for use in cell metabolism. In *Paramecium,* the food vacuole slowly moves through the cytoplasm as digestion proceeds. Such movement ensures that nutrients are distributed to all parts of this unusually large and complex protozoan. Digestion that takes place inside a cell is called **intracellular digestion.**

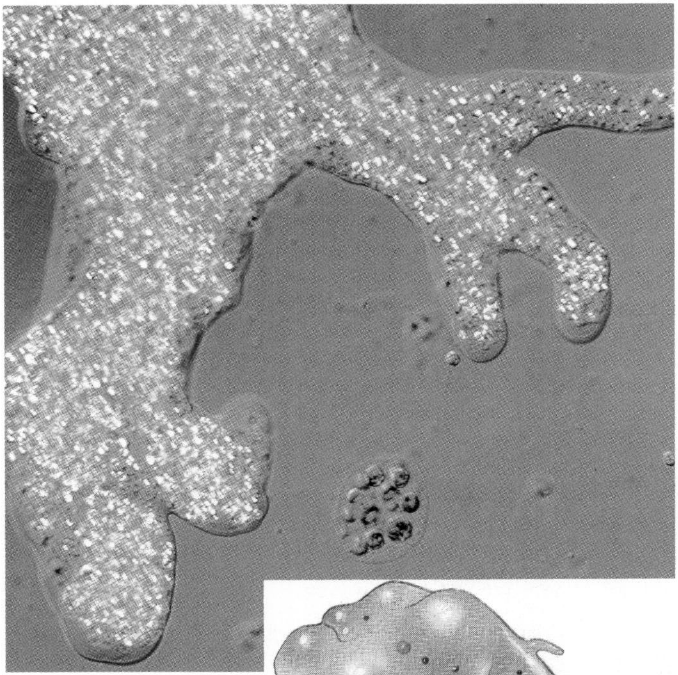

Figure 20-3 The photo on the left shows an amoeba trapping prey. (Magnification: 100×) The prey will be taken in by the process of phagocytosis, as shown below. Enzymes digest the food inside a food vacuole, and the nutrients are absorbed into the cytoplasm.

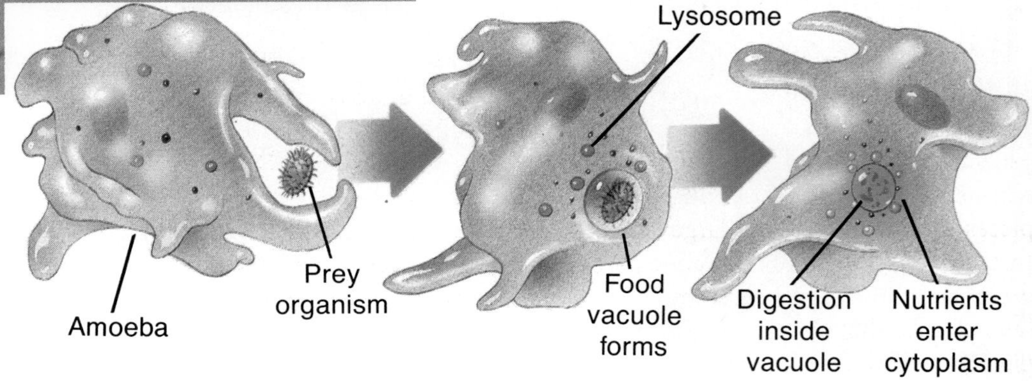

Lysosome

Amoeba

Prey organism

Food vacuole forms

Digestion inside vacuole

Nutrients enter cytoplasm

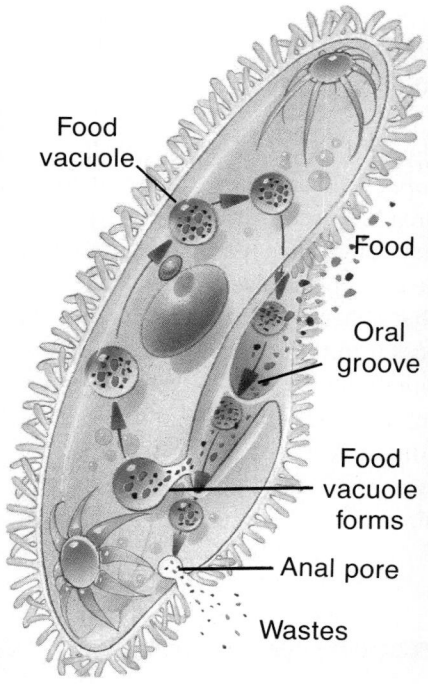

Food vacuole

Food

Oral groove

Food vacuole forms

Anal pore

Wastes

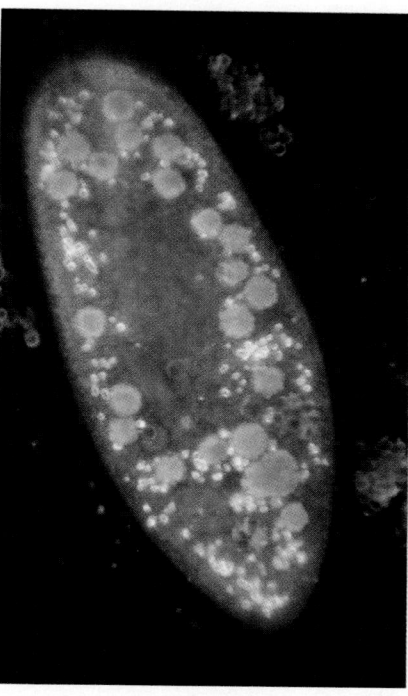

Figure 20-4 In *Paramecium,* a food vacuole forms at the bottom of the oral groove. The vacuole moves throughout the cytoplasm as its contents are digested. This mechanism helps distribute nutrients evenly throughout the cell. Solid wastes are expelled through the anal pore. The photo shows a paramecium that has been fed dyed yeast cells so that its food vacuoles can be seen more easily. Magnification: 250×

Not all material consumed by heterotrophs is digestible, so adaptations for eliminating undigestible material are important. In *Amoeba,* elimination takes place by exocytosis, a process opposite to phagocytosis. In elimination by exocytosis, the food vacuole moves to the plasma membrane and merges with it. Then, the indigestible particles that remain in the food vacuole are ejected from the cell. In *Paramecium,* a vacuole containing indigestible food fuses with an opening in the cell membrane called the anal pore. The food waste is ejected from the cell through this pore.

Although autotrophs obtain organic nutrients primarily by building them from simpler materials, they sometimes need to digest complex molecules. For example, many of the sugars produced by an apple tree during spring and summer is not needed for that year's growth. It is transported to the tree's roots as sucrose, where it is converted to starch and stored in plastids within the root cells. During the following spring, those starch molecules are digested. The resulting sucrose molecules are then transported to the upper parts of the tree, where they supply energy for growth and development of new leaves and flowers.

Extracellular Digestion   Intracellular digestion is an efficient means of breaking down starch molecules in root cells as well as the tiny microorganisms consumed by an amoeba. But the diets of humans, birds, earthworms, clams, and other multicellular heterotrophs consist of much larger pieces of food that are too large to be taken into cells for digestion. Instead, large pieces of food are completely or partly digested outside the organism's cells. This process is called **extracellular digestion** and may take place within a cavity inside an organism, or it may take place completely outside the organism.

Some heterotrophs, such as protozoans and animals, have the mobility needed to find, take in, and digest food. Fungi, although they are also heterotrophs, are generally not mobile and cannot capture food. Instead, fungi obtain food in other ways.

Have you ever seen a piece of bread covered by a blanket of black fuzz? The fuzz is a mold, usually *Rhizopus,* which obtains nutrients from the starch and other materials in the bread. The short, rootlike rhizoids of the mold penetrate the bread where they can absorb food. But starch molecules are too large to pass through the cell's plasma membrane. The rhizoids release enzymes into the bread that convert the starch into glucose. The rhizoids absorb the glucose and pass it to the rest of the mold, as shown in Figure 20-5. Other saprophytic fungi, such as mushrooms and cup fungi, digest and absorb organic matter from the soil in much the same way. Many bacteria also secrete digestive enzymes into their surroundings. These bacteria and fungi decompose dead materials and recycle the nutrients those materials contain.

LITERATURE CONNECTION

Akasha (Gloria) Hull

. . . .

Make no mistake about it:
Mother love has taste.
That's what
    seasons the beans
    and salts the roast
    and makes the cabbage taste
    good
    (even when they're just
    cooked in bacon grease).

These lines are from "The Taste of Mother Love," a poem written in the early 1970s by Akasha Hull (1944-   ). As she wrote it, she was remembering the days she spent in her mother's sunny yellow kitchen in Shreveport, Louisiana. That kitchen, with its shabby, broken-back chairs but an abundance of affection, was the center of family life in Hull's Louisiana home. Hull recalls, "Meals were never anything fancy, but my father always made sure we had enough to eat. Whatever there was, my mother always managed to make it taste special, to feed both our bodies and our spirits."

As she grew up, Hull faithfully followed her mother's recipes, but the resulting dishes never tasted quite the same. They were missing that one special ingredient: her mother's love.

Nearly a quarter century later, Hull shudders a little at the mention of roast and bacon grease. Although she is now a strict vegetarian, she says, "I still have that Black Southern good cooking in my palate. It's a challenge to make vegetarian fare satisfy me in the same way." A lot of garlic, tamari, and cayenne pepper, plus a generous portion of mother love for her own family, meets that challenge. Hull emphasizes that she is maintaining the connection to her mother's cooking while growing in her own way.

Hull is now a Professor of Women's Studies and Literature at the University of California, Santa Cruz. She is an inspiring symbol of African American womanhood to her students. She advises students who share her heritage to "honor our culture, and nurture both the body and the spirit because they make one."

──────

Explore Further
1. Ask older family members if their diets have changed over the years. If there have been changes, find out the reasons.
2. Think ahead to the time when you will be grown and living on your own. What food do you think you will most want to eat when you visit your old home?
3. **Connection to Psychology:** What elements besides flavor contribute to a person's enjoyment of a meal?

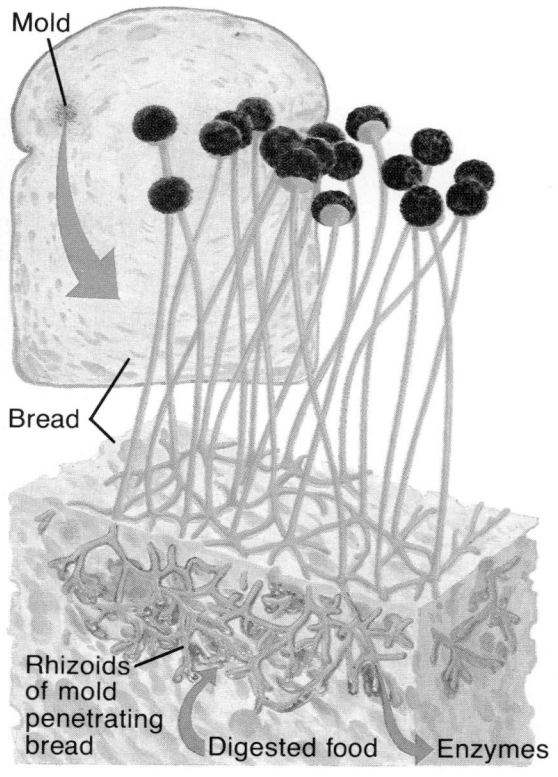

Some plants are not only capable of photosynthesis, but are also capable of digesting insects. These insectivorous plants are often found in soils that are low in nitrogen and other important nutrients. Insects provide a needed supply of these nutrients. Venus's-flytrap, pitcher plant, and sundew are plants that have unique adaptations for the trapping and extracellular digestion of insects.

In a Venus's-flytrap, each leaf is bordered by spines and is hinged along its middle. When sensory hairs on the leaf are stimulated by the touch of an insect, the leaf quickly closes. The crossed spines trap the insect inside. Digestive enzymes are then secreted into the hollow formed by the folded leaf, where digestion occurs. Small molecules such as amino acids are absorbed by the leaf cells and transported throughout the plant. In the sundew and the pitcher plant, as in Venus's-flytrap, natural selection has resulted in the modification of leaves to form insect traps.

Figure 20-5 A Venus's-flytrap secretes enzymes onto trapped insects and absorbs the resulting nutrients. Bread mold carries out extracellular digestion by releasing enzymes from rhizoids into the bread.

Section Review

Understanding Concepts

1. Why do autotrophs get most of their nutrients in the form of simple molecules rather than complex organic molecules?

2. Why do heterotrophs have specialized structures for digestion whereas autotrophs usually do not?

3. How is digestion in fungi and insectivorous plants similar?

4. **Skill Review—Making Tables:** Construct a table that compares the digestion of starch by a plant, a bread mold, and *Amoeba*. For each organism, indicate where the digestion occurs, the kind of molecule produced by digestion, and whether the digestion is intracellular or extracellular. For more help, refer to Organizing Information in the **Skill Handbook.**

5. **Thinking Critically:** A starfish can pry open the shells of a clam. It then pushes its stomach through its mouth into the clam. The stomach secretes enzymes that begin digesting the clam. What type of digestion is this and why is it necessary?

Objectives

Compare patterns of digestion in animals having one body opening with those having two body openings.

Describe the process of filter feeding in several different animals.

Relate the processes of physical digestion and chemical digestion.

20.2 Digestion in Animals

From high atop a tree, an owl spots a field mouse. With precision, it swoops down and takes the mouse in its talons. How does the owl digest its meal? Once digested, how does the food reach all of its body cells? Because animals are complex organisms, elaborate adaptations have evolved for taking in and digesting food and then distributing nutrients to cells. You'll see, though, that there are only a few general patterns of digestion.

Two-way Traffic

You learned in Chapter 17 that cnidarians and flatworms have only one body opening, the mouth. That single opening is used both for taking in food and eliminating material that isn't digested. As a result, there is two-way traffic within the animal. While cnidarians lack organs specialized for digestion, they do have a variety of cells with different functions for trapping and digesting food.

Hydra, as shown in Figure 20-6, is a typical cnidarian. When mature, *Hydra* is sessile, so it can't chase after prey. Like other cnidarians, though, *Hydra* has tentacles to trap prey organisms that pass close by. *Hydra* does not have organs of vision, so it must employ other senses to locate prey. It has sensory cells that are sensitive to touch and can detect chemicals given off by potential prey organisms moving through the water.

To catch prey, *Hydra* extends its tentacles, which are covered with stinging cells. The filaments of the stinging cells are forcefully ejected toward the prey. Some filaments wrap around the prey, trapping it. Filaments of stinging cells also have tips that look like barbed harpoons,

Figure 20-6   Like humans, *Hydra* is a chunk feeder and carries out extracellular digestion. Unlike humans, *Hydra* has only one body opening which both takes in food and eliminates wastes. Magnification: 25×

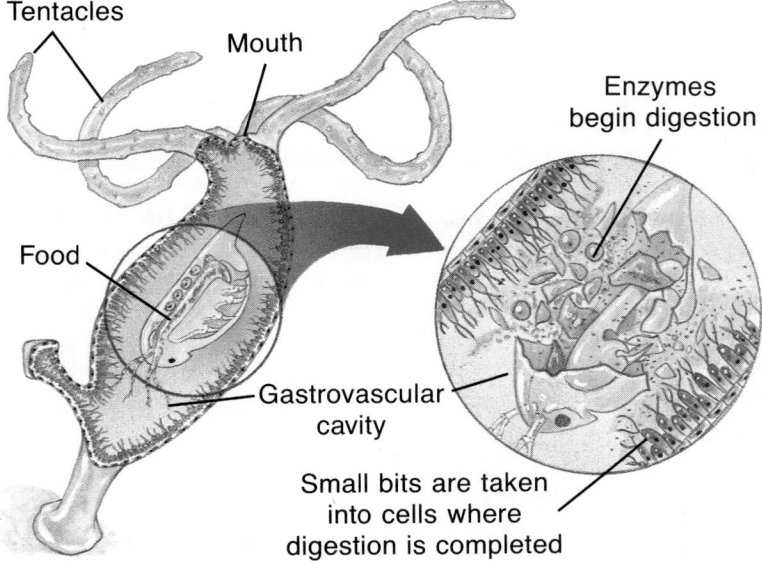
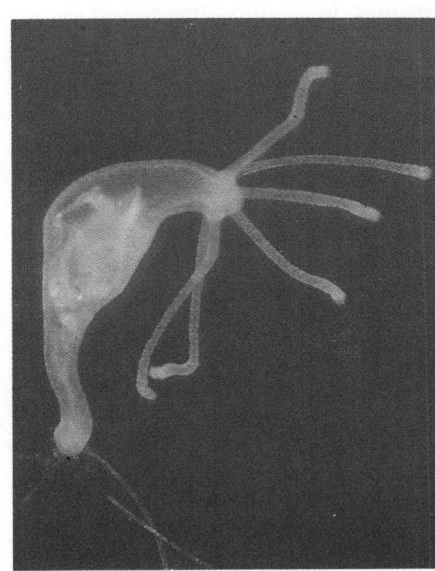

which pierce the prey and discharge a poison that paralyzes or kills it. Once the prey is captured, the *Hydra* uses its tentacles to draw the prey to its mouth and push the prey through.

Digestion in *Hydra* begins in the hollow, water-filled **gastrovascular cavity,** which extends into the tentacles. Lining the cavity are numerous gland cells that secrete digestive enzymes. In the cavity, the enzymes begin breaking food into smaller bits. This process is an example of extracellular digestion. Animals, like *Hydra,* that take in relatively large pieces of food are known as chunk feeders. Because *Hydra's* prey is generally larger than any one of its cells, you can see how extracellular digestion is important for chunk feeders.

The *Hydra's* digestive process does not end in the body cavity. Some of the cells lining the cavity take in smaller bits of food by phagocytosis. Food vacuoles are formed, intracellular digestion takes place, and nutrients diffuse into the cytoplasm. Because *Hydra* consists of only two layers of cells, digested food can diffuse readily to all of the animal's other body cells. No cell is far from a source of nutrients.

One-way Traffic

Think about an automobile assembly line. As a partially built car moves along the line, workers attach parts in a logical order. An assembly line is organized and efficient because the process moves in one direction, and specialized tasks can be performed at each stop.

The digestive systems of most animals function something like automobile assembly lines, but in reverse. In fact, they might better be called "disassembly lines." The digestive pathway through which the food moves is essentially a hollow tube known as the digestive tract. Food enters through the mouth and travels along the digestive tract, where it is broken down physically and chemically into smaller and smaller pieces. Undigested material does not travel back to the mouth, but is eliminated through a second body opening, the **anus.** The anal opening, which first evolved among wormlike animals such as roundworms, makes it possible for food to travel through the animal in one direction only. One-way traffic also allows for a long digestive tract providing thorough, efficient digestion. As a result, the digestive process could be divided into a series of stages. At each stage, food particles are broken down until they are the right size for the next stage, continuing through a series of steps until the particles are small molecules. Then, digestion is complete.

Figure 20-7 Shown here is a diagram of a digestive tract with two openings. This type of digestive system exists in animals from roundworms to humans. Note, however, that not every animal has all of these regions. Also, functions are not always as clearly separated as in this diagram. Why does a digestive system with two openings generally work more efficiently than a system with only one opening?

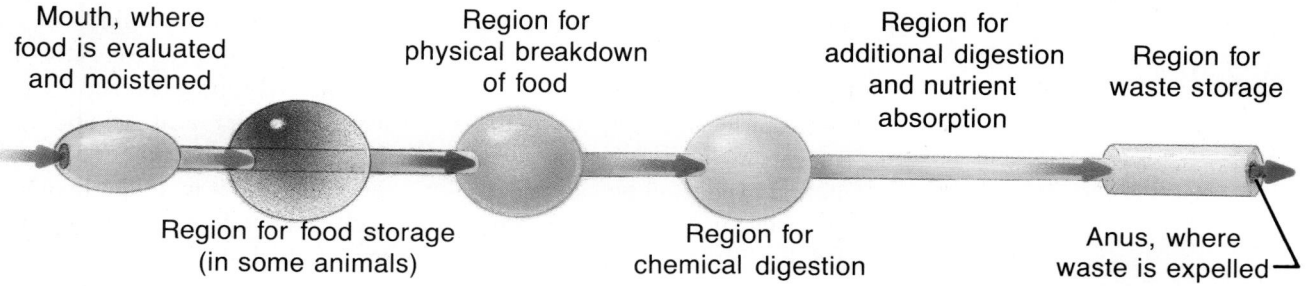

Mouth, where food is evaluated and moistened

Region for physical breakdown of food

Region for additional digestion and nutrient absorption

Region for waste storage

Region for food storage (in some animals)

Region for chemical digestion

Anus, where waste is expelled

One-way Traffic in an Earthworm An earthworm is an example of an animal that has two body openings. Use Figure 20-8 to identify the parts of the digestive system of an earthworm. The earthworm's mouth moistens the food, which is then sucked in by muscle contractions of the **pharynx.** From the pharynx, the food moves through the **esophagus,** where continued muscle contractions push the food into a stretchable organ, the **crop,** where the food can be stored. From the crop, the food passes into the thick-walled **gizzard.** With the aid of coarse soil particles taken in with the food, the muscular contractions of the gizzard grind the food into smaller particles.

I N V E S T I G A T I O N

Modeling Digestion

The lining of the intestine is a selectively permeable barrier. It allows the passage of nutrients from digested foods through its cells and into the bloodstream or lymph, but molecules of undigested foods cannot pass across this barrier. Dialysis tubing may be used as a model for the intestinal lining. You will use starch as food and amylase as a digestive enzyme. You can then determine if the products of starch digestion indeed pass through the membrane.

Starch, in the presence of amylase enzyme, changes chemically to maltose, a disaccharide. Benedict's solution reacts with maltose. To carry out this test, mix equal amounts of the solution in question with Benedict's solution in a test tube and heat the tube in a hot water bath for 5 minutes. A color change from the original blue to either green, yellow, red, or rust indicates that maltose is present. The presence of starch can be detected by the blue color produced when iodine is added.

Problem

Is dialysis tubing (a model for human intestines) permeable to the products of starch digestion?

Safety Precautions

Wear goggles and an apron. Benedict's and iodine solutions are toxic. Avoid contact with skin and eyes. Wash away spills immediately.

Hypothesis

With your group, make a hypothesis regarding what substances you will find inside and outside the dialysis tubing after starch digestion has taken place.

Experimental Plan

1. Work with a group assigned by your teacher. Form a hypothesis that could be tested using starch, dialysis tubing, Benedict's solution, iodine solution, beakers, and the enzyme amylase.

2. Decide on a way to test your group's hypothesis. Keep the available materials in mind as you plan your procedure. Note that dialysis tubing can be soaked in water until soft and then opened into a tube. To fill the tubing, tie it in a tight knot at one end, then use a dropper or small funnel to add liquid.

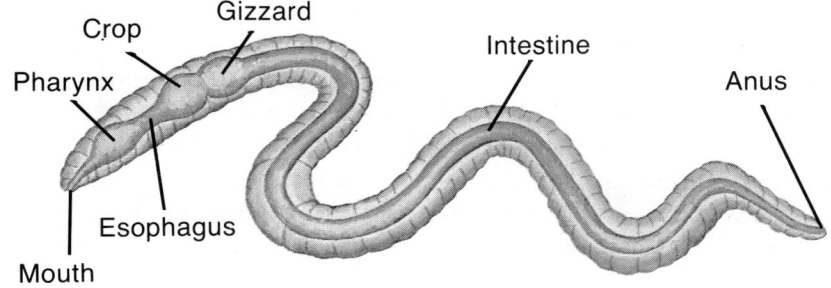

Crop

Gizzard

Pharynx

Intestine

Anus

Esophagus

Mouth

Figure 20-8 The digestive system of the earthworm provides one-way traffic during digestion. Why do you suppose the intestine is so long compared to the rest of the digestive tract?

Finally, tie the open end tightly. Work over a pan or several layers of newspaper to catch spills. The filled tubing should look like a small sausage. You'll need about 15 cm of tubing, and the filled part should be about 10 cm long. Remember that you can test for maltose with Benedict's solution and for starch with iodine solution.

3. Remember to test only one variable at a time and use suitable controls.

4. Record your procedure and list the materials that you will need.

Checking the Plan

1. What controls will be used and what factors will have to be controlled?

2. How will you be certain that no maltose is present in any of the solutions before the experiment begins? How will you determine that maltose diffused through the dialysis tubing and that the tubing did not just leak some of its contents into the surrounding water?

3. What data will you collect and how will it be recorded?

4. How long will you experiment

run and how many trials will be needed?

5. Assign roles for this investigation.

Make sure your teacher has approved your experimental plan before you proceed further.

Data and Observations

Carry out your experiment. Prepare a data table to record your findings.

Analysis and Conclusions

1. After digestion had taken place, what substances were in the water surrounding the dialysis tube? What substances remained inside the dialysis tube? What is your evidence?

2. Write a conclusion to your experiment stating what your data showed about the permeability of the dialysis tubing. Explain how your observations support your conclusions. Base your conclusions entirely on your observations.

3. Did your data support your original hypothesis? Explain. If the data did not support your hypothesis, write a revised hypothesis that takes into account what you observed.

4. Assume that the tubing is a reasonably accurate model of the human intestine. Make a statement about what must be taking place during starch digestion in the intestine.

5. Explain the role of the enzyme in this investigation.

6. How might your data have been different if protein had been used instead of starch but the same enzyme had been used? Explain. How might your data have been different if a different enzyme had been mixed with the starch? Explain.

7. Does your experiment offer evidence that all foods must be digested in order to pass across dialysis tubing? Explain.

8. Considering your answer to question 7, explain the value of using models in science.

Going Further

Based on the results of this experiment, design an experiment that would help to answer another question that arose from your work. Assume that you will have limited resources and will be conducting the lab in class.

Physical processes such as chewing or grinding are important first steps in the digestive processes of many animals including humans. These actions break chunks of food into many smaller particles. Breaking up food in this way increases the total surface area of the food many times. For instance, breaking a chunk of food into 1000 smaller chunks increases the surface area about ten times. The increased surface area allows more enzyme molecules to come into contact with the food, thus breaking it down more efficiently.

Chemical Digestion   In the earthworm, the ground food passes from the gizzard into the intestine. The intestine is the longest portion of the digestive tract, and carries out the final stages of digestion. In the intestine, enzymes break the food particles into simple nutrient molecules that can pass into cells through a plasma membrane. In addition, some water and dissolved minerals are absorbed in the intestine. Undigested material is passed out of the earthworm through the anus.

Absorption and Distribution   Once the food has been digested, it must be absorbed and transported to all of the earthworm's body cells. Initially, the nutrient molecules pass across the membranes of cells lining the intestine. It is important that the earthworm absorb as many of the nutrient molecules as possible from the intestine. Nutrients that are not absorbed before they reach the anus will be passed out of the body and lost to the worm that expended the energy to digest them.

What does the inside of an earthworm's intestine look like? If you were to slice through a tube and look at the cut end, you would see a cross section of the tube's interior, which forms a circle. Now picture the same circle but with a fold that extends downward from the top into its interior, as shown in Figure 20-9. Which circle would have the greater interior surface area for absorbing food? If you said the circle with the fold, you would be right. An earthworm has a fold of tissue extending along its entire length, increasing the inner surface area of the intestine. This increased surface greatly increases the absorption of digested food.

Figure 20-9   Compare the drawings of an intestine with and without an infolding of tissue. Can you see how this adaptation increases the surface area available for absorbing nutrients? The photo on the right shows a cross section through the intestine of an earthworm. Magnification: 10×

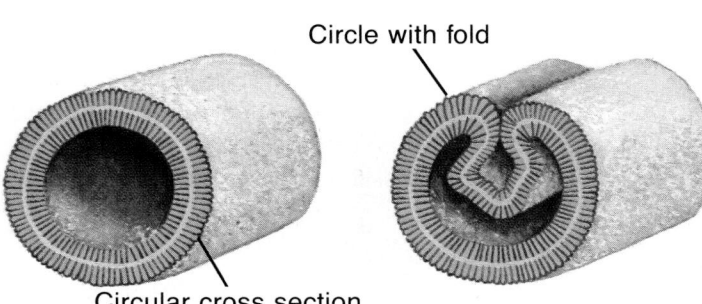

Circle with fold

Circular cross section

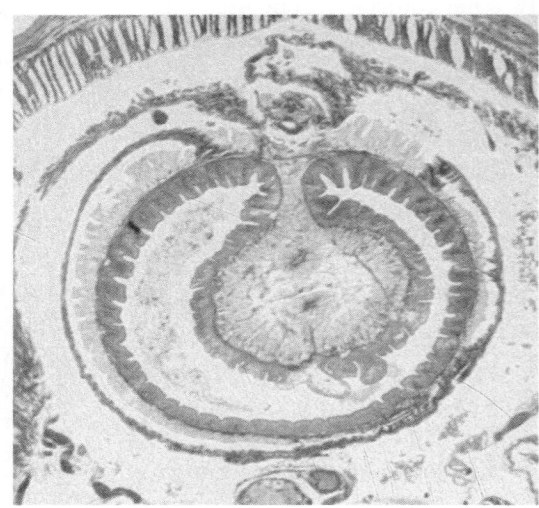

Once nutrients have been absorbed by the cells of the intestine, they cannot be distributed to other cells by simple diffusion, as in *Hydra.* The body of an earthworm is much thicker than the body of a *Hydra,* and most of an earthworm's cells are too far from the intestine to receive nutrients directly by diffusion. Instead, nutrients from the cells of the intestine pass into the bloodstream by means of small blood vessels in the tissues of the intestine. As the blood flows through the rest of the earthworm's body, nutrients pass from it and into the body cells.

Filter Feeding

The animals you are most familiar with are probably chunk feeders. Most birds, fish, mammals, reptiles, and amphibians consume fairly large pieces of food that are digested as they move through the digestive tract. But there are a number of aquatic animals that feed on tiny organisms or microscopic bits of organic matter suspended in water. These animals obtain their food by passing water through a part of their body that strains out the food. They are known as filter feeders.

What kinds of animals are adapted for filter feeding? This method of obtaining food makes sense for sessile or slow-moving species that are not adapted for chasing prey or taking in large bits of food. Sponges and certain mollusks fit this category.

Digestion in Sponges   Sponges are sessile and do not have a mouth opening or digestive tract for taking in and breaking down large food particles. They have cells with flagella that draw water into the sponge through its many pores. As the water passes through the sponge, microorganisms and other food particles are filtered out, captured, and ingested by individual cells. The food is digested inside these cells, and the nutrients pass on to other cells by diffusion.

Figure 20-10   Sponges can be very large, as shown in the photo below. Even so, all sponges obtain food in the same way. The arrows in the diagram show how water flows through the sponge. Food particles become trapped in the mucus of the collar cells, are taken in by phagocytosis, and digested in food vacuoles. Sometimes, entire food vacuoles are transferred to amoebocytes which travel throughout the sponge.

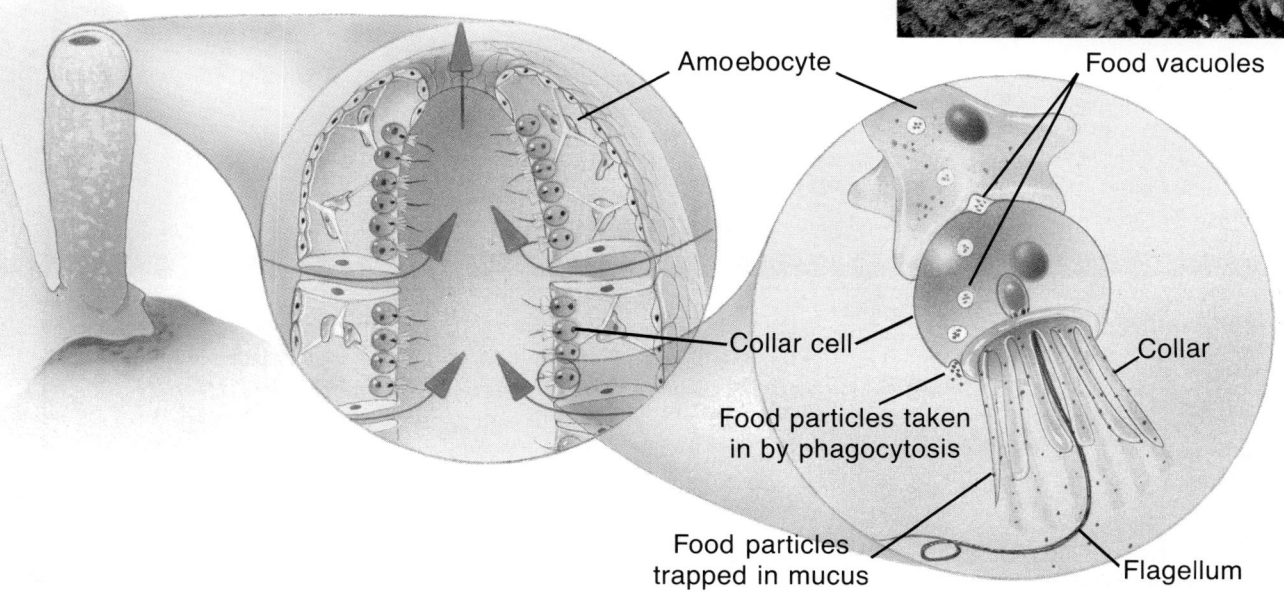

Amoebocyte

Food vacuoles

Collar cell

Collar

Food particles taken in by phagocytosis

Food particles trapped in mucus

Flagellum

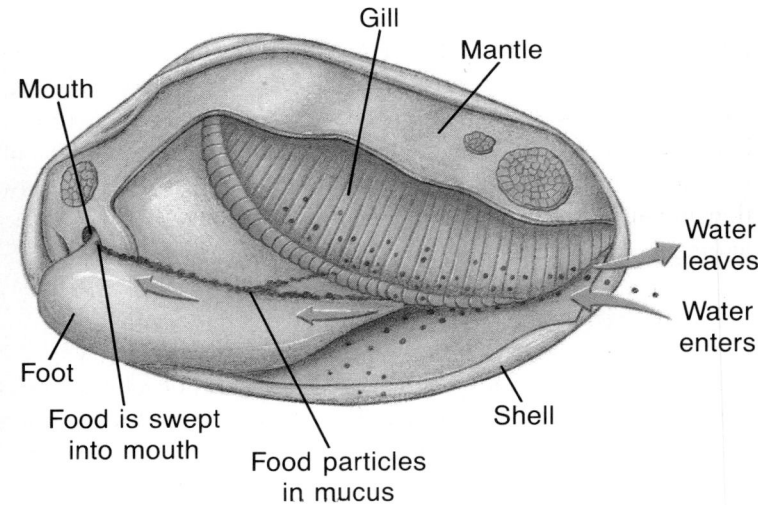

Mouth

Gill

Mantle

Water leaves

Water enters

Foot

Food is swept into mouth

Food particles in mucus

Shell

Figure 20-11 **Like sponges and some whales, clams are also filter feeders. The diagram shows a clam with part of the upper mantle cut away. You can see how a clam collects food particles on its gills and passes them to the mouth. Its digestive system consists of a stomach and intestine, which are located beneath the structures you see in the diagram. Why do you think that filter feeding is such a successful adaptation for many animals in an aquatic environment?**

Other Filter Feeders   Filter-feeding mollusks such as clams, oysters, and mussels are much more complex than sponges and have true digestive systems. They have sieve-like gill structures that strain out food particles from the water that passes through them. These particles become embedded in a layer of mucus, which is continuously swept into the mouth by cilia. From there, food passes through the esophagus, stomach, and intestine. Digestion in the stomach and intestine is both extracellular and intracellular.

Filter feeding has evolved as an efficient feeding strategy for larger animals that live on tiny organisms scattered through a large volume of water. You have read in the opening to this chapter that baleen whales are adapted for filter feeding. It may surprise you to learn that many other vertebrates are filter feeders. Ducks and flamingoes use their bills to strain food from the water. Tadpoles and some types of fish remain stationary while sucking in water and filtering tiny food particles from it. The evolution of different structures that perform similar functions —sieve-like gills of mollusks and the baleen of whales, for example— illustrates convergent evolution.

Section Review

Understanding Concepts
1. What is the adaptive advantage of a digestive system with two openings?
2. How are the length and inner surface area of the intestine important in digestion?

3. A barnacle is a sessile crustacean. How do you think it obtains food?
4. ***Skill Review—Sequencing:*** List the series of structures through which food passes in an earthworm, and describe the function of each of those structures. For more help,

refer to Organizing Information in the **Skill Handbook**.
5. ***Thinking Critically:*** Do you think the mouth and throat structures of chunk feeders are very similar to those of closely related species that filter feed? Explain.

20.3 Nutrition and Digestion in Humans

Do you eat the same foods every day? Maintaining the health of your body requires many different nutrients, and varying your diet is the best way to get them. Your digestive system is adapted for the thorough digestion and absorption of many types of food.

Human Nutrition

Humans are heterotrophs, so our food includes complex molecules that must be digested and simple substances that don't require digestion. The complex nutrients are carbohydrates, fats, and proteins. The simple nutrients are vitamins, minerals, and water. It is important not only to consume foods containing all these nutrients, but also to consume them in the right proportions. In other words, we all need to eat a balanced diet.

A Balanced Diet Nutrition is the study of nutrients and how they are used by the body. For many years, nutritionists recommended that a balanced diet include several portions of food from each of four basic food groups—breads and cereals, fruits and vegetables, eggs and dairy products, and meats. More recent guidelines refer to five basic food groups and include suggestions for the number of servings from each group that a healthy individual should consume every day. These new guidelines are represented by the food pyramid shown in Figure 20-12.

Objectives

Explain the importance of a balanced diet.

Identify physical and chemical aspects of digestion in each part of the human digestive system.

Compare the absorption and transport of sugars and amino acids with that of fatty acids and glycerol.

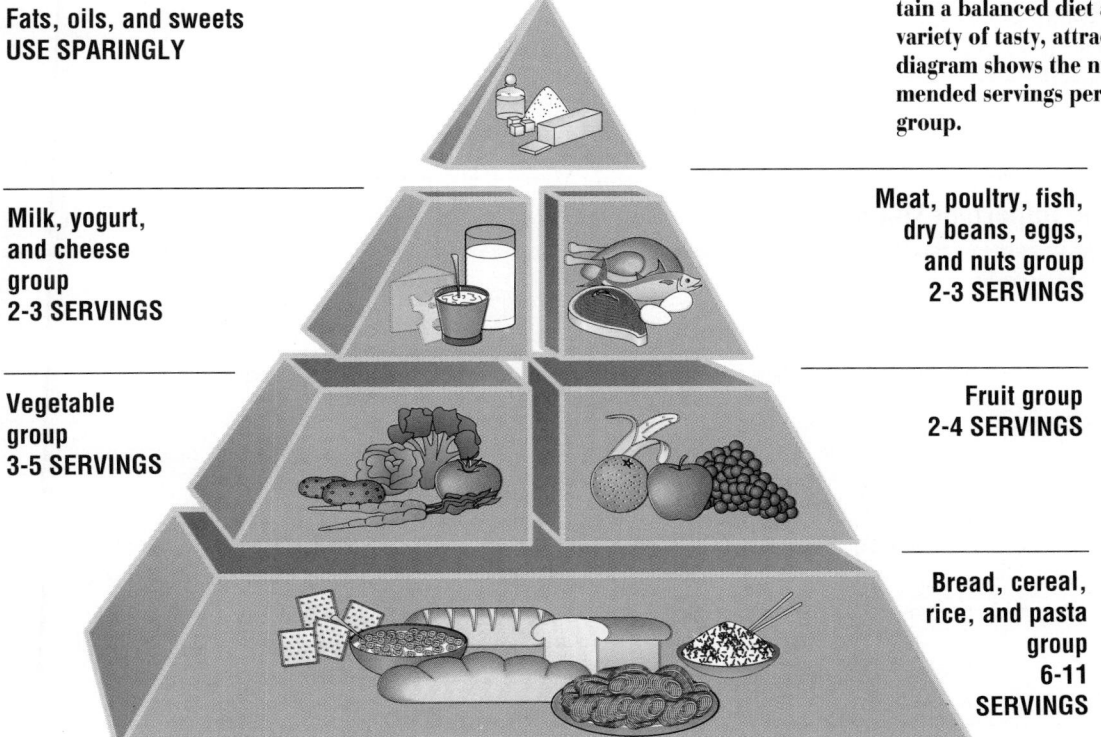

Figure 20-12 You can see from this food pyramid that it's easy to maintain a balanced diet and still eat a great variety of tasty, attractive foods. The diagram shows the number of recommended servings per day of each food group.

Fats, oils, and sweets USE SPARINGLY

Milk, yogurt, and cheese group 2-3 SERVINGS

Meat, poultry, fish, dry beans, eggs, and nuts group 2-3 SERVINGS

Vegetable group 3-5 SERVINGS

Fruit group 2-4 SERVINGS

Bread, cereal, rice, and pasta group 6-11 SERVINGS

Eating a balanced diet provides your body with a daily supply of all the essential nutrients. If one or more nutrients are missing from the diet for a period of time, malnutrition can result. For example, inadequate protein in the diets of growing children can lead to weight loss, liver damage, and anemia. Table 20.1 lists some of the minerals that are important to health. Table 20.2 lists symptoms of some vitamin deficiencies.

Table 20.1 **S**ome Important Minerals

Mineral	Sources	Function
Calcium	Milk, cheese, meats, vegetables, whole-grain cereals	Blood clotting, muscle contraction, bone and tooth formation
Iron	Liver, leafy vegetables, meats, raisins	Production of hemoglobin, part of molecules used in respiration
Iodine	Seafood, iodized table salt	Part of hormone that controls rate of metabolism
Magnesium	Leafy vegetables, meats, potatoes	Proper functioning of some enzymes
Phosphorus	Milk, meats, eggs, vegetables	Bone and tooth formation, part of ATP and nucleic acids
Potassium	Vegetables, bananas	Conduction of nerve impulses
Sodium	Table salt, vegetables	Conduction of nerve impulses, maintaining osmotic balance

Table 20.2 **S**ome Important Vitamins

Vitamin	Sources	Deficiency Symptoms
A	Green and yellow vegetables, eggs, fruits, liver	Night blindness, dry skin
B-1	Whole-grain cereals, liver, other meats, nuts, most vegetables	Weak muscles, paralysis (beriberi)
B-2	Milk, cheese, eggs, liver, whole-grain cereals	Poor vision, sores in and around mouth
B-12	Liver, other meats, eggs, milk products	Anemia
C	Citrus fruits, tomatoes, some vegetables	Loose teeth, bleeding gums, swollen joints, poor muscle development (scurvy)
D	Milk, egg yolk, fish liver oil, also made in skin	Soft bones, deformed skeleton (rickets)
K	Green vegetables, liver, also made by intestinal bacteria	Slow blood clotting

Figure 20-13 You might enjoy eating a meal like this one. The problem, though, is that you would be eating large quantities of fats and sugar and relatively small quantities of proteins, vitamins, and minerals. In other words, the meal is high in Calories, but low in nutrition.

Diet and Obesity   Are you fond of potato chips, candy bars, cookies, and soda? These foods fall into the fats, oils, and sweets portion of the food pyramid. They aren't harmful if eaten in moderation, but including too many of them in your diet can contribute to health problems. Often called junk foods, they contain large proportions of sugar and/or fat, but very little protein and few vitamins and minerals. It's true that sugar and especially fat are good energy sources for the body. However, when we try to make a meal of junk foods, we take in much more sugar and fat than we would in a balanced diet. A diet containing a lot of junk food can lead not only to vitamin deficiencies and malnutrition, but also to obesity. Obesity is a condition of being overweight. It results when the body takes in more food than it needs. The extra food that isn't broken down to release energy is stored in the body as fat. Obesity usually can be avoided by eating a balanced diet with moderate portions and getting plenty of exercise.

Figure 20-14 Vegetarians can obtain all essential amino acids by eating certain combinations of foods such as corn (used to make tortillas) and beans (in the filling of refried beans).

Vegetarian Diet   Your cells must have all 20 amino acids available if they are to make the many proteins needed to build enzymes, muscles, connective tissue, skin, and other tissues. Human cells can manufacture 12 of the 20 amino acids from other nutrients. But the remaining eight amino acids cannot be manufactured in the body. Thus, they are often called the essential amino acids and must be supplied in food. Meat and dairy products supply our major sources of protein and all of the essential amino acids.

For a variety of reasons, many people decide to become vegetarians—they choose not to eat meat. Some vegetarians further limit their diet by eating only plant foods and avoiding seafood, eggs, and dairy products as well as meat. But, while many vegetables, nuts, and grains are sources of protein, they are not as protein-rich as animal products. How, then, do vegetarians maintain a balanced diet?

Individual vegetables and grains supply different combinations of essential amino acids, but none contains all of them. Therefore, a person on a vegetarian diet must include a variety of cereals, grains, peas, beans, and nuts that, taken together, contain proteins having the eight essential amino acids.

The Beginning of Digestion

Digestion begins when you put food in your mouth. In the mouth, both physical and chemical processes begin to break down the food. Teeth physically grind and tear the food into smaller pieces. This process makes the food particles small enough to swallow and increases their surface area. When food is chewed well, the enzymes are better able to begin the chemical process of digestion. Another important function of the mouth is to evaluate the quality of food by sensing taste and texture. This evaluation allows us to spit out food that may be spoiled, contain disagreeable substances, or contain foreign matter that could be harmful.

Salivary Glands   Several structures along the length of the digestive tract aid in digestion. In and near the mouth are three pairs of **salivary glands,** which secrete saliva into the mouth. The taste, smell, or sometimes even the thought of food causes nerves to stimulate the glands and increase their secretion of saliva. Like other structures that secrete fluids important in digestion, the salivary glands secrete saliva into the mouth through small tubes called ducts.

Saliva contains mucus, which moistens and lubricates food, thus making it easier to swallow. Saliva also contains an enzyme that begins the breakdown of starch into molecules of maltose, a disaccharide. Chemical digestion of fats and proteins does not take place in the mouth.

G L O B A L   C O N N E C T I O N

Salt of the Earth

"Here it comes!" shouted a merchant in the ancient city of Timbuktu. Squinting against the desert sun, she spotted the huge caravan, with its 2000 camels and hundreds of guards, weaving slowly across the Sahara. The bulky, 300-pound load on each camel's back represented wealth to finance schools, libraries, and other public services in this affluent west African city. The caravan's cargo was not gold or glittering jewels; it was sodium chloride—everyday salt.

In ancient times, salt was in great demand because it is an essential nutrient, a seasoning, and a food preservative. The problem was that salt was available in large quantities only at a few sites. Also, in those times, shipping salt over great distances was expensive, risky, and often took many months. All across Africa, Europe, and Asia, ancient salt routes were established to trade the precious commodity. The city of London, among other cities, began as a stop along a trade route for salt. In some places, salt was even used as money. Roman soldiers received part of their pay in salt, a practice that was the origin of the English word "salary."

Salt also played an important part in religious life. In many early societies along the Mediterranean, offerings of salt were made to the gods. On the North American continent, the Aztecs honored a salt goddess, the Navajos prayed to Salt Woman, and the Hopis' war god was Salt Man.

Because salt was widely used as a food preservative, it came to symbolize permanence in many societies. An Arab saying, "There is salt between us," implies a lasting trust. The Hebrew expression, "To eat the salt of the palace," means to think highly of someone. When English-speakers call a person "the salt of the earth," they mean that he or she possesses admirable basic qualities.

In contrast, salt has often been a cause of dissension. During the first century A.D., a spear battle between

The Esophagus   Next, chewed, moistened, slightly digested food is swallowed through the pharynx and pushed into the long, slender esophagus. Digestive fluids are not added to the food while it is in the esophagus. Food is moved down the esophagus and through the rest of the digestive tract by **peristalsis,** the alternating contraction and relaxation of muscles along the digestive tract. First, one area of the esophagus contracts, squeezing the food from above and pushing it down. Then the muscle in that first area relaxes, and the next area contracts, again squeezing the food from above and moving it farther along. Alternating contractions and relaxations continue to push the food into the stomach. Similar contractions move food through the remainder of the digestive tract.

The Stomach

The **stomach** is a large, hollow organ in the digestive tract. The stomach performs three main functions. (1) The stomach's muscular walls provide a churning action that continues the physical breakup of food particles. (2) The stomach secretes acid and certain enzymes that begin breaking up protein. (3) The stomach acts as a storage area for food, which allows food to be released slowly into the intestine.

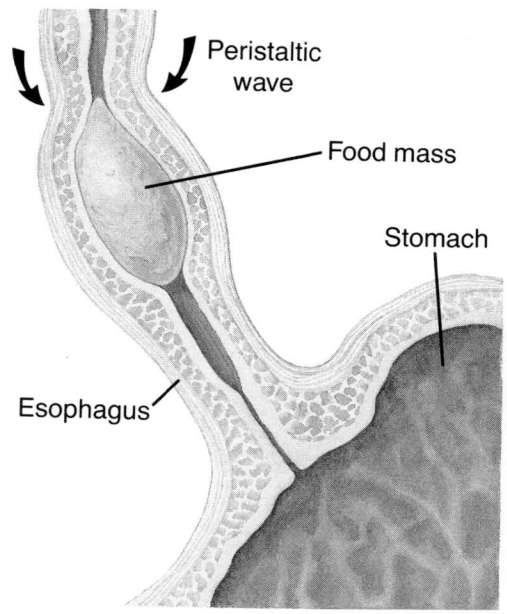

Figure 20-15   When you swallow food, waves of muscular contraction, called peristalsis, push the food through your esophagus and into the stomach.

---

two German groups left thousands of dead warriors beside the contested prize, a salty stream. In France, the peasants' revolt against a heavy tax on salt contributed to the French Revolution that began in 1789. Similarly, Mahatma Gandhi of India led a "Salt March" in 1930 to protest the tax on salt that the English imposed on his country.

Victory and defeat in wartime has also been influenced by salt. The army of the French general Napoleon lost contact with its supply wagons during the 1812 invasion of Russia. Deprived of salt, the soldiers became sick and their wounds would not heal. During the American Civil War, the South ran low on salt, and the price of a pound of salt shot from one-quarter cent to five dollars. In more modern times, World War II fighter pilots flying over China carried salt in their emergency kits. If they were shot down in rural areas, the pilots could use the salt in exchange for help from local people.

Salt has even been a sign of social rank. At royal banquets in past centuries, guests eagerly surveyed where their assigned seats were in relation to the salt on the table. Those seated "below the salt" realized that they were unlikely to win favors from the king or queen.

Because salt is so vital to human health, it is fortunate that today the mineral is plentiful and cheap in most parts of the world.

1. **Connection to Literature:** From what you've learned about salt, tell what you think are the meanings of these sayings.
   (a) That worker isn't worth his salt.
   (b) I've salted away my allowance instead of spending it.
   (c) Wit is the salt of conversation, not the food.
      —William Hazlitt
2. **Solve It:** When humans lived a hunting, gathering, and herding way of life, their diet contained a lot of milk and roasted meat. When societies turned to farming, they began to need salt for good health. How can you explain this observation?

Digestion in the Stomach Food enters the stomach through a ring of muscle that is normally closed tight. This muscle prevents food from moving back up into the esophagus. As the food moves into the stomach, it is moistened further by mucus that coats the inner lining of the stomach. The presence of proteins in the food stimulates nerves that signal the gastric glands lining the stomach wall to release gastric juice. Gastric juice contains hydrochloric acid and an enzyme that is in an inactive form until it comes into contact with the acid in the stomach. The hydrochloric acid gives the stomach fluid an acidic pH of about 2. The enzyme begins protein digestion by breaking some of the bonds between amino acids. However, because the enzyme can't break the bonds between all amino acids, it cannot complete protein digestion. Instead, the proteins are broken down into smaller polypeptide molecules.

As polypeptides are formed, their presence stimulates cells in the stomach lining to release a hormone into the bloodstream. Although the hormone travels throughout the body, it affects only the gastric glands, signaling them to release more gastric juice. As a result, large amounts of gastric juice are secreted only when protein is present in the stomach.

Digestion is also aided by a churning motion that constantly mixes the food in the stomach. This action continues to break up food and helps bring the surfaces of all food particles into contact with the enzymes. When this stage of digestion is complete, the food is in the form of an acidic liquid mixture known as chyme.

The stomach plays yet another important role. Like the crop of an earthworm, it stores food. Adaptations for food storage are important because they enable an animal to eat periodically, rather than having to eat almost continuously.

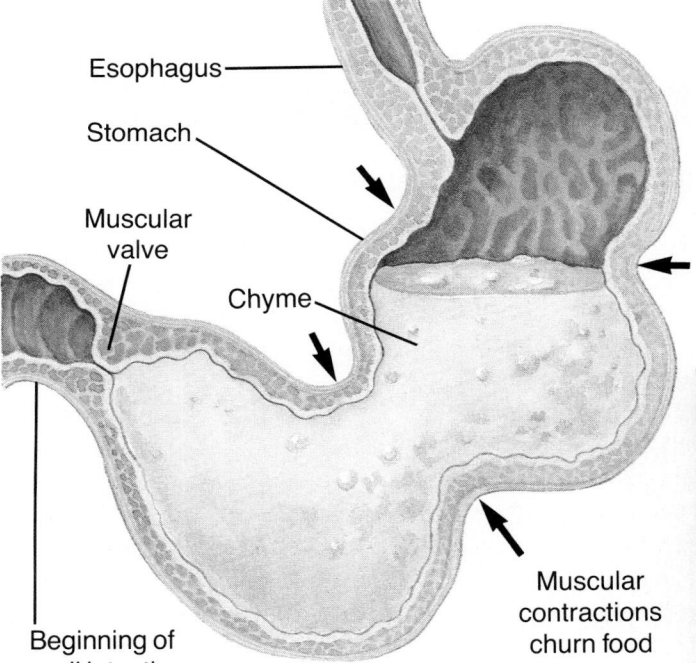

Esophagus

Stomach

Muscular valve

Chyme

Beginning of small intestine

Muscular contractions churn food

Figure 20-16 After food enters the stomach, acid is secreted along with enzymes that begin protein digestion, and muscular contractions churn the food. The stomach also acts as a storage container, releasing food into the intestine at a slow rate for maximum efficiency in digestion and absorption.

The Small Intestine

Another muscular ring regulates the passage of chyme from the stomach into the longest part of the digestive tract, the small intestine. It is in the **small intestine** that the chemical digestion of food is completed and that most of the nutrient molecules are absorbed. The small intestine in an adult human is about seven meters long. The length of the small intestine helps ensure that the food stays in the organ long enough to be completely digested and absorbed.

The Gallbladder   Two organs that are not part of the digestive tract secrete important digestive fluids into the small intestine. The first of these organs is the **gallbladder.** The liver produces **bile,** which is stored in the small, saclike gallbladder. As soon as proteins or fats are detected in the intestine, certain cells of the intestine release a hormone that stimulates the gallbladder to empty the stored bile into the small intestine. Bile moves from the gallbladder into the small intestine via the common bile duct as shown in Figure 20-17.

Bile is made up of several substances, but only bile salts are important in digestion. Bile salts are a mixture of soaplike substances that the liver makes from cholesterol. These bile salts break up fats into small globules. This process is called emulsification and is similar to what takes place when you use detergent to wash fats and oils from dirty dishes. In the intestine, emulsification increases the surface area upon which fat-digesting enzymes can operate. The bile salts also aid in the absorption of digested fats.

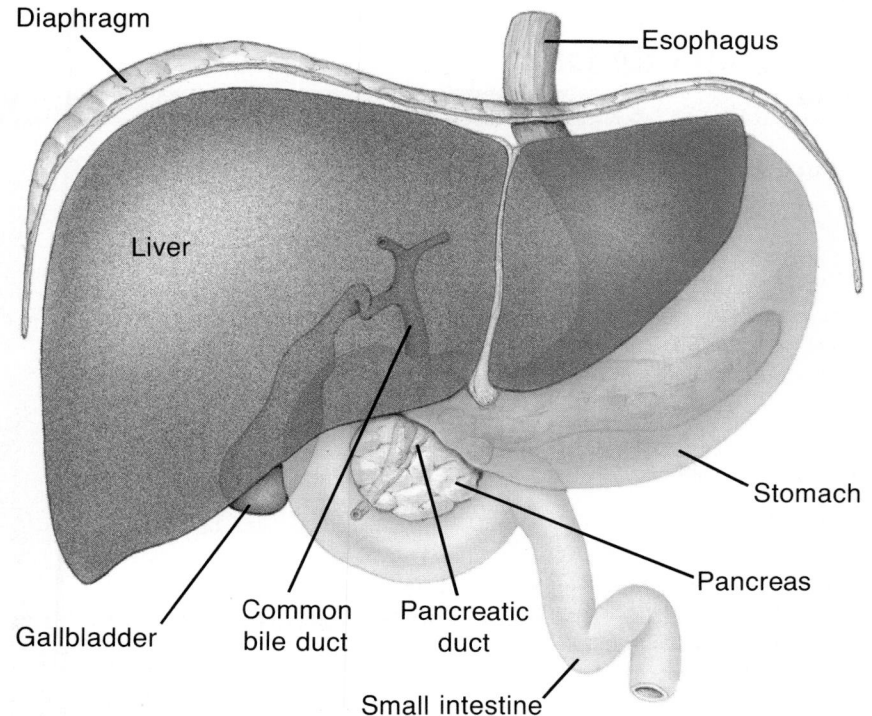

Diaphragm

Esophagus

Liver

Stomach

Pancreas

Gallbladder

Common bile duct

Pancreatic duct

Small intestine

Figure 20-17   When food passes into the small intestine, digestion is continued by enzymes from the pancreas. Bile, which is produced by the liver and stored in the gallbladder, is released. Bile salts break up fat globules so that they can be attacked more efficiently by enzymes. Further along, enzymes secreted by the small intestine itself complete digestion. Sometimes people must undergo surgery to remove most of the small intestine because of disease or injury. What kinds of foods would these people have to eat after such surgery?

Minilab 2

Which fruits keep gelatin from jelling?

Gelatin that you use to make desserts is partially composed of protein. Certain raw fruits contain an enzyme that digests protein. When these fruits are added to gelatin, it will no longer jell when cooled. Fill five test tubes with a warm (not hot) solution of gelatin. Label them 1 through 5. Add the following: 1 = nothing, 2 = fresh mashed pineapple, 3 = fresh mashed apple, 4 = fresh mashed kiwi fruit, 5 = fresh mashed orange. Allow all tubes to cool overnight. Observe whether each tube did or did not jell.

Analysis *Which fruits contained a protein-digesting enzyme? How were you able to tell? If this same enzyme were present in your body, what organs might produce it?*

The Pancreas The second organ that secretes digestive fluid into the small intestine is the **pancreas,** which lies behind and partly below the stomach. It secretes enzyme-containing pancreatic juices into a duct that merges with the common bile duct. The pancreas releases its juices in response to the same hormonal signal that causes the release of bile from the gallbladder.

Presence of the acidic chyme in the small intestine causes the secretion of yet another hormone from cells of the intestine. This hormone causes the pancreas to release sodium hydrogen carbonate along with its enzymes. Sodium hydrogen carbonate is the same substance as baking soda and helps neutralize the acidic chyme. Bile also contains basic substances and, together with sodium hydrogen carbonate, raises the pH of the chyme to about 8. The pancreatic enzymes work most effectively in this slightly basic environment. Over-the-counter remedies used to relieve acid indigestion and heartburn contain either sodium hydrogen carbonate or other basic substances such as magnesium hydroxide. Heartburn is a painful condition that occurs when acidic gastric juice from the stomach enters and irritates the esophagus. The basic substances in the medicine neutralize stomach acid, thus relieving the symptoms of heartburn.

One of the pancreatic enzymes is similar to the salivary enzyme that converts starch to maltose. A second pancreatic enzyme helps convert fats to fatty acids and glycerol. A third group of enzymes acts on the polypeptides that were produced in the stomach, breaking them down into smaller polypeptides. Several enzymes of this third type are needed because each one serves to break only the bonds between certain amino acids. A fourth type of pancreatic enzyme converts nucleic acids to nucleotides.

Thinking Lab

DRAW A CONCLUSION

Why are Michael's data different?

Background
Salivary glands form an enzyme that digests starch into disaccharides. The amount of enzyme present in saliva varies from person to person. A deep blue color appears when iodine is added to starch. The color becomes progressively lighter as the amount of starch decreases. A number system from 10-0 can be used to judge color intensity and starch amounts. Let 10 = deep blue and 0 = no blue color.

You Try It
Each of six students received a tube containing 10 mL of a starch suspension. They then added 5 drops of their saliva and stirred the mixture thoroughly. After 10 minutes, each student added a drop of iodine to the tube and mixed thoroughly. The students' results are shown in Table A.

Draw a conclusion to explain Michael's results. Michael correctly repeated the experiment five more times and recorded the same results each time. Draw another conclusion for his five similar results.

Table A

Student	*Color Intensity*
Shawana	3
Talyn	1
Jay	8
Maria	2
Larry	1
Michael	10
Control	10

Final Steps in Digestion   At this point, the food in the small intestine is being acted upon by bile and pancreatic enzymes. The intestine now contains both simple nutrients—vitamins, mineral ions, and water—and complex nutrients. Of the complex nutrients, only the fats have been fully digested. They have been broken down into fatty acid and glycerol molecules. Proteins and carbohydrates have been only partially digested. So far, proteins have been broken down into small polypeptides, nucleic acids have been broken down into nucleotides, and carbohydrates exist in the form of the disaccharides maltose, sucrose, and lactose.

Protein and nucleic acid digestion is completed by enzymes secreted in the intestinal walls. A group of these enzymes converts nucleotides into their component sugars, bases, and phosphates. Other enzymes break down polypeptides into dipeptides and then into amino acids.

Carbohydrate digestion is completed by enzymes that convert disaccharides to monosaccharides. These enzymes are located in the membranes of cells lining the intestine. By the time digestion in the small intestine is complete, all complex food molecules have been converted to smaller molecules that can pass across cell membranes. A summary of the process of human digestion is shown in Table 20.3. The right column in the table shows the overall enzyme reactions that occur in digestion. The names of the enzymes involved are printed over the arrows.

Table 20.3   Summary of Digestion in Humans

Digestive Organ	Major Physical Actions	Major Chemical Actions
Mouth	Chewing Grinding Moistening	starch $\xrightarrow{\text{salivary amylase}}$ maltose
Esophagus	Moistening, Peristalsis	
Stomach	Moistening Churning Peristalsis Some water absorption	proteins $\xrightarrow{\text{pepsin}}$ polypeptides
Small intestine	Peristalsis Fat emulsification Most nutrient absorption Some water absorption	starch $\xrightarrow{\text{pancreatic amylase}}$ disaccharides proteins $\xrightarrow{\text{proteases}}$ polypeptides polypeptides $\xrightarrow{\text{peptidases}}$ amino acids fats $\xrightarrow{\text{lipases}}$ fatty acids and glycerol disaccharides $\xrightarrow{\text{maltase, lactase, sucrase}}$ monosaccharides
Large intestine	Most water absorption Peristalsis Waste elimination	some vitamins synthesized

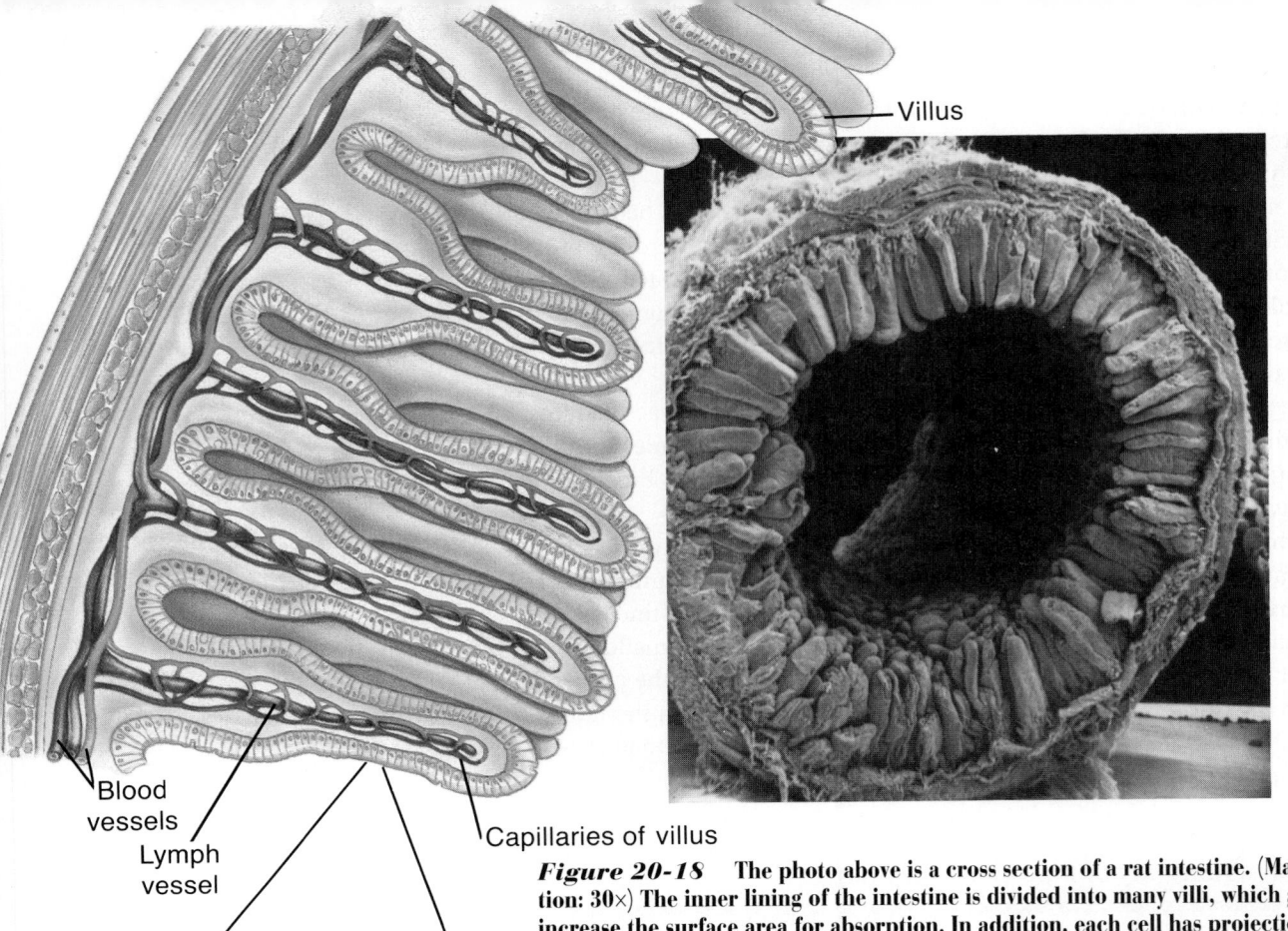

Villus

Blood vessels

Lymph vessel

Capillaries of villus

Figure 20-18 The photo above is a cross section of a rat intestine. (Magnification: 30×) The inner lining of the intestine is divided into many villi, which greatly increase the surface area for absorption. In addition, each cell has projections called microvilli as shown in the photo below left. (Magnification: 6000×) Enzymes that break disaccharides into monosaccharides are located in the microvilli.

Absorption of Nutrients   The lining of the small intestine is not smooth. If it were, the surface area available for absorbing digested food would be too small to complete the job. The lining is folded, and fingerlike projections called **villi** extend inward from each of the many folds. The membranes of the villi cells that face the intestine's interior are also highly folded, forming microvilli, as shown in Figure 20-18. The villi and microvilli provide a very large surface area inside the intestine where absorption of nutrients can take place. The surface area inside the human intestine has been estimated as 300 m^2, which is roughly the area of a tennis court.

Transport of Nutrients

Before nutrients can reach all body cells, they must first enter villi cells. Then they are absorbed by the blood and transported through the body. Passage into villi cells and from there into the blood involves both diffusion and active transport. Within the villi are many tiny blood vessels called capillaries. Monosaccharides, dipeptides, amino acids, products of nucleotide digestion, minerals, small fatty acids, and vitamins enter the blood flowing through the capillaries. The blood, laden with nutrients, then moves on to the vein that carries blood to the liver.

Glucose and Amino Acids   The liver has the important task of monitoring some nutrients as they leave the small intestine. Under the direction of hormones, the liver controls the levels of these nutrients in the body. For example, it is important that the level of glucose in the blood be kept fairly constant. If all the glucose produced during digestion of a meal remained in the bloodstream, the level of glucose in the blood would increase dramatically. To prevent this, the liver converts excess glucose to the polysaccharide glycogen, which is stored for later use. Thus, the blood that leaves the liver and transports nutrients to the rest of the cells in the body has the correct glucose level. As time passes between meals, the level of glucose in the blood gradually falls. To keep the level constant, the liver gradually converts the stored glycogen back into glucose and releases it into the bloodstream as needed.

The liver also regulates the levels of amino acids in blood that passes through it. Excess amino acids, unlike excess glucose, can be stored only for a short time. As a result, most are converted to other organic molecules such as carbohydrates or fats. The body's inability to store amino acids is the primary reason why it is important to eat a diet that provides a constant intake of proteins containing all the essential amino acids.

Fatty Acids and Glycerol   Large fatty acid molecules and glycerol do not enter the blood directly. After they pass from the intestine into villi cells, they are recombined to form molecules of fat, which are then coated with proteins. These particles are absorbed into small vessels in the villi that are part of the lymphatic system, which will be described in more detail in Chapter 22.

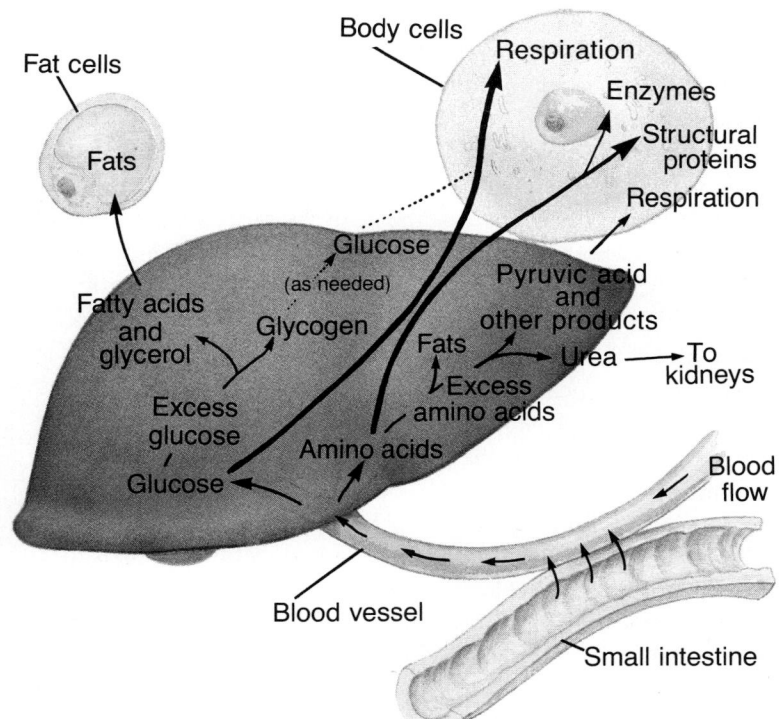

Figure 20-19   This diagram shows how the liver handles the products of carbohydrate and protein digestion. Notice that excess glucose from carbohydrate digestion is stored as glycogen. Any excess beyond what can be stored as glycogen is converted to fats and stored in fat cells. Amino acids, on the other hand, cannot be stored in reserve. Excess amino acids are converted either to products that can enter respiration or to fats. In either case, urea is formed as a waste product and must be eliminated by the kidneys.

Cholesterol molecules are coated with protein to form molecules known as LDL complexes. These complexes also enter the lymphatic vessels. These and other substances that travel through lymph vessels are eventually emptied into the bloodstream. Fat and cholesterol molecules must be coated with proteins because they are not soluble in water. Insoluble particles cannot be transported in the watery liquid part of the blood. In the blood, the protein-coated fats are slowly broken apart and delivered to cells for use as a source of energy. If they are not needed for energy, the fats are stored in fat cells. The LDL complexes remain intact. The liver absorbs the LDLs and regulates blood cholesterol by storing or releasing it.

Absorption of Nutrients by Cells   When digested nutrients reach the bloodstream, they are in a form that can enter your cells. Some nutrients, such as fatty acids and glycerol, diffuse directly through cell membranes. Others, such as glucose, amino acids, and mineral ions, enter cells by means of facilitated diffusion. LDL complexes are brought into cells by receptor-aided endocytosis. Glucose and some fats are used in cellular respiration, which releases energy for the cell's many functions. Other nutrients are used for growth. For example, amino acids that were linked to form proteins in the beef, chicken, fish, or beans you ate for dinner can now be joined in different sequences to form human proteins. Fats and cholesterol are used to build new membranes, and certain cells convert cholesterol to hormones.

HISTORY CONNECTION

Lillie Rosa Minoka-Hill

A squawking chicken, a bushel of newly dug potatoes, or a bucket of foamy and still-warm milk—that was the kind of payment doctors in rural America often received in the early 1900s. Dr. Minoka-Hill, a physician in the small town of Oneida, Wisconsin, accepted these offerings graciously. She understood her patients' financial situations and explained, "If I charged too much, I wouldn't have a very good chance of going to heaven."

Dr. Minoka-Hill (1876-1952) was a Native American born on a Mohawk reservation in New York state. After her mother's death, she stayed on the reservation with her maternal relatives until she was old enough to go to school. Then her Quaker father, a doctor, brought her to Philadelphia, where he taught her about her Mohawk heritage and sent her to a Quaker boarding school.

She continued her education at the Woman's Medical College in Philadelphia. After graduation, she helped penniless immigrant women at a clinic associated with the medical school.

When she married Charles Hill, a member of the Oneida tribe, and moved to Wisconsin, Dr. Minoka-Hill thought that her days of practicing medicine were over. There were too many other chores on the land she and her husband farmed. However, she found time to visit the nearby Oneida reservation and talk with the people living there. They taught her about traditional herbal remedies for illness and injuries. Soon, neighbors began to come to Dr. Minoka-Hill for her combination of traditional and modern medical knowledge.

Charles Hill died in 1916, leaving his wife with only a debt-ridden farm and a few farm animals to support six young children. Dr. Minoka-Hill ignored the advice of friends who urged her to move

The Large Intestine

After the small intestine has absorbed most of the nutrients from digested food, undigestible solids and some water remain. These materials pass from the small intestine into the **large intestine** or **colon.** This organ is composed of three sections that loop around and over the coiled small intestine. Slightly below the point at which the small and large intestines join is the appendix, a vestigial structure.

Absorption of Water   During digestion, a lot of water has been added to the food by secretions from the mouth, stomach, and small intestine. A major function of the colon is to reclaim that water by absorption. In this process, water and dissolved minerals pass from the hollow interior of the colon into the bloodstream of the colon walls. Absorption of water causes the remaining material to become a mass of damp solids, called feces. These solids are stored in the rectum and pass out of the body through the anus. Some excess minerals, such as iron and calcium, may pass from the bloodstream into the colon to be eliminated along with feces.

The interior of the human colon provides an ideal habitat for many species of bacteria. These bacteria have a mutualistic relationship with humans. The wastes of digestion provide the bacteria with food, and the bacteria make the vitamins B-12 and K along with some amino acids. These bacteria also inhibit the growth of disease-causing bacteria.

back to New York. She said, "While in college, I resolved to spend some time and effort to help needy Indians. In Wisconsin I found my work."

From seven in the morning until ten at night, Dr. Minoka-Hill welcomed patients to what she called her "kitchen-clinic." She traveled long distances to treat patients or deliver babies, sometimes taking her younger children along. She also spent much of her time teaching the Oneida people about nutrition and good eating habits in an effort to relieve their constant problem of malnutrition. Because of her dedication, the Oneida adopted the doctor into their tribe and gave her an Oneida name, You-da-gent, which means "she who serves."

Dr. Minoka-Hill died in 1952. A memorial erected by her friends has this simple inscription: "I was sick and you visited me."

1. **Explore Further:** What "home remedies" are used by people in your family or community?
2. **Connection to Health:** Why do you think today's recent immigrants to the United States might find it difficult to get health care?

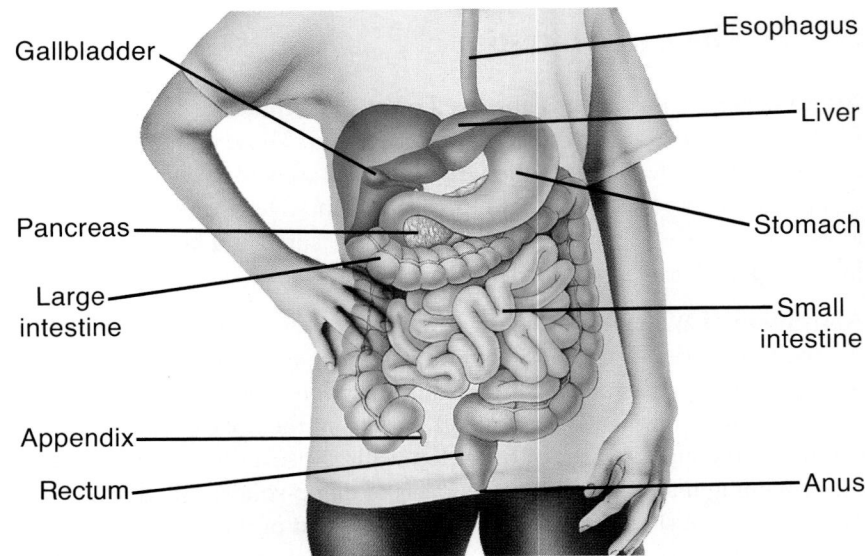

Figure 20-20 Shown here are the major organs of the digestive system. Notice how the shape of the large intestine adapts it to collect undigested waste, absorb water from the waste, and compact the waste. In this diagram, the front of the liver is raised slightly so that you can see the gallbladder underneath. The large intestine is usually described as having three parts—ascending, descending, and transverse. Which part do you think goes with each term? Why?

Gallbladder

Pancreas

Large intestine

Appendix

Rectum

Esophagus

Liver

Stomach

Small intestine

Anus

LOOKING AHEAD

YOU HAVE LEARNED in this chapter that autotrophs take in simple inorganic materials that they can use in photosynthesis to produce food molecules. Heterotrophs digest complex molecules in their food into simpler molecules. The simple inorganic nutrients and the products of photosynthesis of autotrophs must be distributed to all of the organism's cells. The products of digestion in heterotrophs also must be distributed. This distribution depends upon some type of transport mechanism. In the next chapter, you will study the function of transport, not only of nutrients, but also of other substances important to life. You will also learn about how part of the transport system of animals functions in other vital ways.

Section Review

Understanding Concepts

1. Suppose a person always ate a breakfast that included buttered toast, two eggs, and bacon. How would eating only those foods be likely to affect the person's health? How could this person improve his or her diet?

2. How can the human digestive tract be compared with an assembly line?

3. How do physical and chemical processes of digestion work together in your mouth, your stomach, and your small intestine?

4. *Skill Review—Classifying:* Think about what you have eaten in the last two days. Use the information provided in Figure 20-12 to arrange the various foods by food groups. Is your diet of the last two

days a balanced one? For more help, refer to Organizing Information in the **Skill Handbook**.

5. *Thinking Critically:* What is the advantage of having the same starch-digesting enzyme in both the mouth and the small intestine?

Health Connection

Does cholesterol count?

Suppose that in the grocery store, you see two one-pound jars of peanut butter, Dippy's brand and Duffy's brand. The price tags are identical, and about the only difference you can spot is the claim printed in bright yellow letters on the Dippy's jar: *Now! Cholesterol-Free!* Which jar will you put into your shopping cart?

If you choose Dippy's because of the label, the manufacturer's strategy has worked. Of course, Dippy's Peanut Butter contains no cholesterol. But, neither does Duffy's or any other brand of peanut butter. Only foods containing animal products are likely to contain this fat-like, soft, waxy substance. Peanut butter contains no animal products.

Advertisers' claims may have persuaded you that cholesterol is always a villain, but that idea is misleading, too. As you learned in Chapter 4, cholesterol is an important part of the membranes of cells. In your body, cholesterol is used to make several types of hormones, the chemical messengers of the body. Problems occur when your body has too much cholesterol.

Cholesterol is transported in two forms, low-density lipoproteins (LDLs) and high-density lipoproteins (HDLs). Although it's not entirely accurate, the label of "good" has been attached to HDLs

because they seem to act something like a vacuum cleaner, taking cholesterol from cells and delivering it to the liver. In contrast, LDLs tend to deposit cholesterol on the inside lining of arteries. These deposits are referred to as *plaque.* As layers of this plaque thicken, they can block the flow of blood in much the same way that mineral deposits inside water pipes can restrict water flow.

When blood flow is blocked, a heart attack or stroke can result.

Eating saturated fats tends to increase a person's cholesterol. As a result, the National Heart, Lung, and Blood Institute recommended in 1990 that everyone eat a diet with only 10% of daily Calories coming from saturated fats. Instead, people are advised to replace saturated fats in the diet with unsaturated fats. Vegetable oils such as corn oil, olive oil, and peanut oil are typically low in saturated fats.

However, "vegetable oil" printed on a label is no guarantee that the food contains no saturated fats. The so-called tropical oils such as palm oil and coconut oil can properly be called vegetable oils. However, these tropical oils usually contain saturated fatty acids and should only be used in moderation, just as with animal fats.

Most of the earliest findings about cholesterol were based on a study of 362 000 middle-aged men. Some scientists have criticized the application of these results to women and people of other ages. Currently, research is taking place to determine the relationship between cholesterol and arterial disease in diverse groups of people.

1. **Connection to Economics:** People living in wealthy countries tend to have more cholesterol in their blood than people in less affluent nations. What might account for the difference?

2. **Connection to Chemistry:** Chemists are synthesizing and testing fake fats, substances that can be used in food preparation because their physical properties closely resemble real fats. However, fake fats pass through the digestive system without being absorbed. What advantages and disadvantages could result from using fake fat in your diet?

Issues

Are liquid diets a losing proposition?

How much are Americans willing to spend to become thin? Apparently, quite a lot, because the U.S. diet industry takes in more than $30 billion a year. Displaying the typical American trait of trying to accomplish things quickly, the fastest-growing group of dieters are

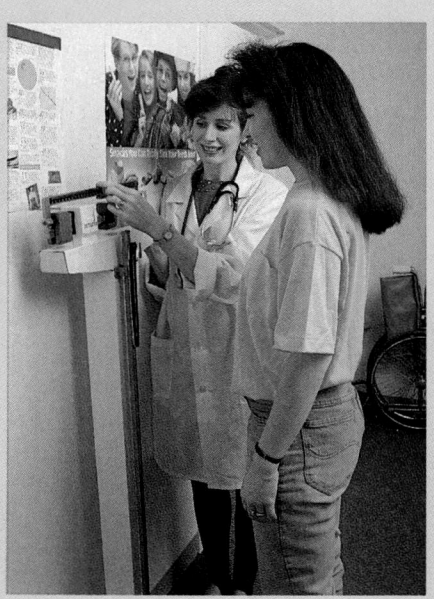

those on liquid diets that substitute high-protein "milkshakes" for regular meals.

Winning with Liquid Diets
Liquid diets have been around since the late 1970s. At that time, some were considered dangerous and were linked to more than 50 deaths, mostly from heart problems. Since 1984, the U.S. Food and Drug Administration has required all very-low-calorie protein diets to carry a warning that they may cause serious illness and should be medically supervised. Today's liquid-protein mixes contain more and better quality protein than those of the 1970s. In addition, carbohydrates, vitamins, and minerals are now included.

For those patients who are truly obese, medically monitored liquid-protein weight-loss diets can be life-saving. Most programs run by hospitals or through doctors' offices accept only adult patients who are 20 to 50 percent over their ideal weights. These people are at risk for serious heart problems, stroke, high blood pressure, gallbladder disease, diabetes, and other life-threatening medical problems if they don't lose weight.

The psychological benefits of quick weight loss for people with many pounds to lose must also be considered. Ordinary, slow-paced diets, with their losses of only a pound or two a week, may be too discouraging for very obese people.

The most regimented liquid-protein diet plan replaces food entirely for weeks at a time. In the first stage of the program, dieters consume protein shakes that provide 400 to 800 Calories a day. Weight loss averages three to ten pounds a week. Dieters must have weekly checks of blood pressure, heart function, urine content, and electrolyte levels.

Following the initial stage, the "refeeding" portion of the diet begins. The protein shakes are gradually replaced with sensible meals that are low in fat.

When patients reach their ideal weight, the third phase of the program, maintenance, begins. During this phase, patients practice eating a moderate, balanced diet on their own to keep the lost pounds from returning.

Dieters who decide to use liquid diets have several plans from which to choose. Some programs involve the supervision of only a doctor, while others promote a team approach that includes a doctor, dietitian, exercise physiologist, and psychologist.

Losing with Liquid Diets   It's hard to pick up a magazine without finding articles with titles such as "One-Day Quick-Fix Diet" or "Holiday Weight-Watching." The magazines' publishers know their audience well, because about one-third of American women and one-fourth of American men are trying to lose weight.

Television, too, makes people aware of weight-loss programs. Fans of a popular TV talk show cheered when the formerly plump hostess of the program modeled her size-10 jeans in front of 20 million viewers and touted the wonders of a liquid-protein diet. Unfortunately, in the following months, she regained many of her banished 67 pounds.

Why do liquid diets often fail? In most cases, it is because the people who have lost weight on these diets have not changed their lifestyles. When they return to eating regular meals, they often cannot eat sensibly and maintain an exercise program. They discover the harsh truth that there is no magic in a liquid-protein diet itself and that losing weight means limiting caloric intake, no matter what the source of those Calories is.

The big selling point for liquid-protein diets is how quickly they

work. However, this type of rapid weight loss can have bad effects on the body's chemistry. When fueled with few Calories, the body is fooled into thinking it is in a state of starvation. Then, to save energy, the thyroid and sympathetic nervous systems slow down the body's metabolic rate. Because the body has to decide where to spend its limited energy, people on rapid weight-loss diets often complain about dry skin, bad breath, and hair loss and of being cold, tired, or dizzy.

Scientists think that after repeated diet cycles, the body stays in this semi-starvation mode. As a result, the repeat dieter's metabolism slows with each diet cycle, and it becomes harder to lose weight or to maintain a weight loss. This effect is sometimes dubbed the "yo-yo syndrome."

Are liquid-protein diets the miracle weight-loss method of the future? Probably not. Although the diets help with initial weight loss, most dieters fail to maintain that loss. In fact, 90 percent of dieters who lose weight regain it within three years, and half gain even more.

Understanding the Issue

1. Many liquid diets are sold in drug stores and grocery stores. What steps would you recommend that a person take before starting one of these programs? What criteria do you think should be used in choosing a program?

2. How would you advise someone who is ten pounds overweight and is considering a liquid weight-loss program?

3. The medically supervised team approach to dieting seems to be more successful for maintaining weight loss than are personal diet plans. How can you account for this difference?

4. Many teenagers have an inaccurate view of their bodies and consider themselves fat even when they are not. How do you think they could get a more realistic picture of themselves?

5. While on a liquid-protein diet, people can just mix a packet of powder with water at mealtime. Why do you think this approach appeals to many dieters?

Readings

- "The Losing Formula." *Newsweek,* April 30, 1990, p. 52.
- "Inside America's Hottest Diet Programs: Losing BIG with Optifast." *Prevention,* January 1990, p. 52.

Summary

Organisms must take in either complex molecules or the simple materials to make them. Autotrophs make complex organic molecules by using the products of photosynthesis, plus other nutrients absorbed from the surroundings. However, heterotrophs must get complex molecules that are already made.

Digestion, which may be intracellular or extracellular, breaks up complex organic molecules into smaller ones that can pass through cell membranes.

Animals may be either filter feeders or chunk feeders, and the digestive system of either type may be simple or complex. Digestion is more efficient in animals having two body openings than in animals in which food and wastes must pass through the same opening.

Humans should eat a balanced diet that includes sufficient amounts of food from all five food groups. Digestion in humans and many other animals is a complex, step-by-step process involving both physical breakdown of food chunks and chemical breakdown of complex molecules. Nutrients are absorbed by the capillaries or lymphatic vessels in villi and transported to all body cells.

Language of Biology

Write a sentence that shows your understanding of each of the following terms.

anus	intracellular digestion
bile	large intestine
colon	nutrient
crop	pancreas
digestion	peristalsis
esophagus	pharynx
extracellular digestion	salivary gland
gallbladder	small intestine
gastrovascular cavity	stomach
gizzard	villi

Understanding Concepts

1. Plant food can be purchased at garden shops. Does this product really contain food? What does it contain?
2. Would you expect digestion in filter feeders to be mainly intracellular or extracellular? Explain.
3. Digestion in protozoans is intracellular. In what way, however, does it resemble extracellular digestion?
4. What nutrients would a chimpanzee require that an oak tree would not?
5. In which animal, a flatworm or a segmented worm, would digestion be more efficient? Why?
6. A certain aquatic animal is slow moving, is enclosed by shells, and has gills. What is the probable diet of this animal and how does it obtain its food?
7. A cheeseburger on a bun consists of many substances including starch, fats, cholesterol, protein, minerals, and vitamins. Which of these, if eaten in excess, could lead to health problems and what are those problems?
8. A person has just begun a meal of foods containing starches, fats, and proteins. Describe the state of that food, physically and chemically, by the time it (a) leaves the mouth; (b) leaves the stomach; (c) has been digested in the interior of the small intestine; and (d) has moved through the colon.

9. In terms of events in the liver, explain why vegetarians should eat vegetables and grains that contain a variety of proteins.

Applying Concepts

10. Some people develop stomach ulcers, which are sores in the stomach wall. In severe cases, the ulcer does not respond to medication or diet and the part of the stomach containing the ulcer is removed surgically. How do you think a person's eating habits might change after such surgery?
11. Even if a person's entire stomach is removed, protein digestion can occur normally. Can you suggest a reason for this fact?
12. After removal of the gallbladder, a patient is told to stick to a low-fat diet. Why?
13. The inner surface of a shark's intestine is coiled like a carpenter's drill, and the food must move through the intestine in a spiral fashion. What is the adaptive advantage of such a structure?
14. Normally, materials pass through the colon at a relatively slow rate. What condition results from too rapid a passage? How might that condition, if prolonged, become dangerous?
15. How would the relative lengths of the small intestine of a lion (a meat eater) and a zebra (a plant eater) compare? How do you suppose the relative length of a human's small intestine compares to the intestines of these two animals?
16. It is thought that the evolution of a digestive system that allowed animals to consume a large meal and then digest it gradually was a major factor that allowed animals to become highly mobile. Explain how this could be true.
17. Hospitalized people must sometimes be fed by means of solutions that flow directly into the bloodstream. What might these solutions contain? Why is it difficult to maintain nutrition over time by this means?
18. **Global Connection:** Why do you think salt is cheap and plentiful today instead of expensive and scarce as it was in ancient times?
19. **Health Connection:** Suppose you saw a muffin mix that advertised, "Contains no cholesterol or animal fats!" Do you think such a product would be safe for a person who needs to reduce blood cholesterol? Explain your answer.

Lab Interpretation

20. **Investigation:** Suppose you conducted this investigation by placing the starch suspension and enzyme in an open container and filling the dialysis tubing with plain water. Would the results be different? Explain what you would observe.
21. **Thinking Lab:** Explain how it is possible for a person to survive if the person forms no salivary amylase enzyme.
22. **Minilab 2:** A student conducts Minilab 2 to compare the effects of cooked, fresh, and frozen fruits on gelatin. Write a conclusion based on the data shown in the following table.

Condition after cooling

Fruit	Fresh	Frozen	Cooked
Pineapple	Not Jelled	Not Jelled	Jelled
Orange	Jelled	Jelled	Jelled

Connecting Ideas

23. A protein deficiency can be especially harmful early in life. Why?
24. Explain how an increase of blood-glucose level above normal would result in osmotic imbalance between the blood and body cells. What would result from such an imbalance?

CHAPTER
21

TRANSPORT

"OH, WHAT big roots you have!"

"The better to nourish myself with, my dear."

This revision of some dialogue from "Little Red Riding Hood" illustrates a point—people don't realize how much of a plant is underground until they see the exposed roots of a fallen tree or of a tree on a riverbank as shown here. The exposed roots of this tree mean that the tree will soon die. Its fragile root system has been extensively damaged. Fragile? Yes. Even though many of the tree's roots are as big around as your arm, about 90 percent of all absorption of water and minerals from the soil is done by tiny root hairs.

Root hairs grow near the root tips and are composed of just one cell. They worm their way between soil particles and help to absorb water and nutrients for the plant. From within the roots, the water and minerals spire upward through columns of tubes to every part of the plant. Another set of tubes transports sugars and other nutrients from the leaves, around the plant, and back to the roots for growth. This tree's roots have been exposed by water erosion, but much of the root system still remains hidden. Such extensive damage to a plant's transport system greatly reduces its capacity for absorption and often results in death.

Root systems may be huge. A single rye grass plant has been found to have almost 14 million roots after only four months of growth. The total surface area of the root hairs was more than half the size of a basketball court.

Animals, too, can have a system of tubes in which water, nutrients, and other substances move throughout their bodies. Unlike plants, though, the nutrients in animals are circulated all together throughout their bodies. In this chapter, you will read about how the transport systems in plants and animals are vital to survival of these organisms.

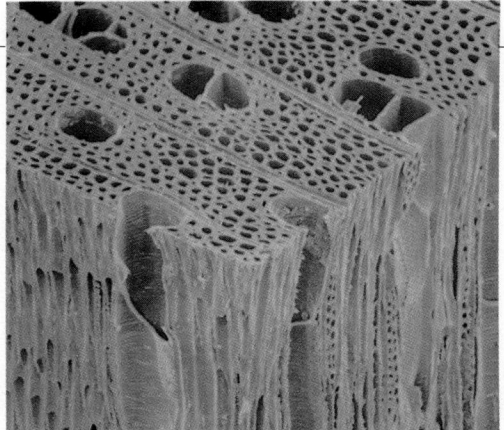

How are these plant cells, which connect roots to leaves and are stacked end to end to form long tubes, like drinking straws? (Magnification: 300×)

Chapter Preview

Objectives

Determine why simple organisms do not have a transport system.

Compare the transport of water and minerals with transport of organic molecules in a vascular plant.

21.1 Nutrient Transport in Plants

Imagine a busy factory. New supplies continually arrive at the factory and are unloaded. These supplies—raw materials used in making products or fuel to provide the energy to make them—then must be transported to the areas within the factory where they are needed. Once made, the products are sent to different areas of the factory for packaging and storage. Eventually, they are shipped out. As the raw materials and fuel are used, wastes are produced. These wastes somehow must be removed from the factory. Like factories, organisms must take in raw materials from the outside world, and those materials must be transported to cells to supply life's processes. Products and wastes made by cells then must be transported away and removed from the organism.

Nutrient Supplies for All Cells

Every cell is surrounded by a liquid environment from which needed materials enter the cell and into which wastes are dumped. The liquid surrounding unicellular organisms is the external environment in which they live. These organisms take in needed materials such as gases, ions, and food directly from their external environment and pour their wastes back out. Distribution of materials within the cells often occurs by diffusion. However, diffusion within a cell is effective only over short distances. Movement of substances within cells is aided in many cells by the streaming of cytoplasm. There is a gentle but steady flow of cytoplasm within the cell. Streaming also helps in the distribution of larger substances, as shown in the amoeba in Figure 21-1.

Figure 21-1 The internal environment of this hydra (left) is in direct contact with the watery external environment. The cells of this simple multicellular organism and unicellular organisms such as the amoeba (right) obtain their nutrients directly from the surrounding water.

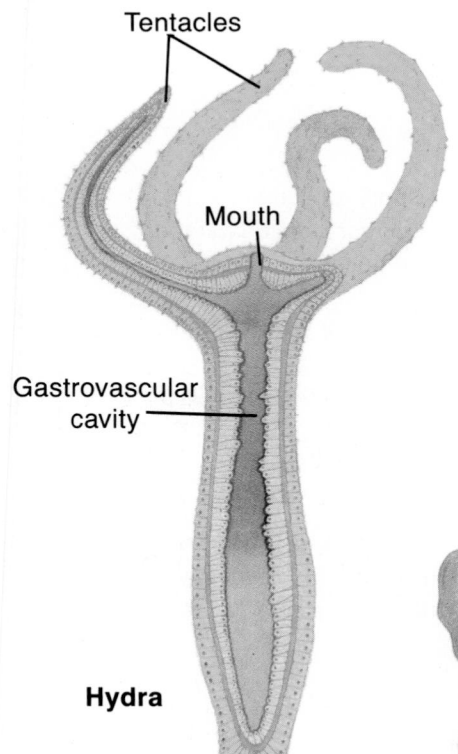

Tentacles

Mouth

Gastrovascular cavity

Hydra

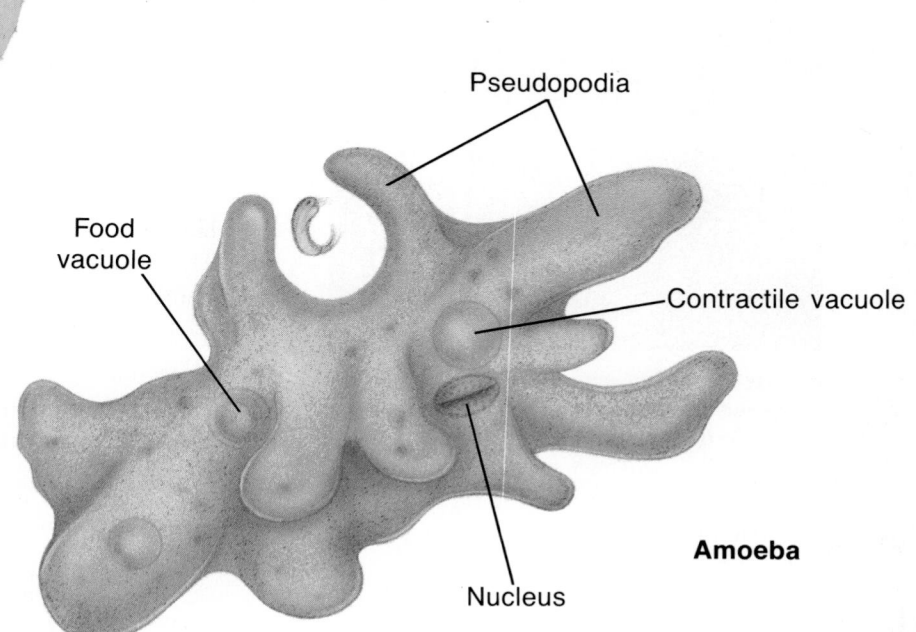

Pseudopodia

Food vacuole

Contractile vacuole

Nucleus

Amoeba

The cells of multicellular organisms such as most algae, fungi, nonvascular plants, sponges, cnidarians, and flatworms also interact directly with their external environment. Some of these organisms, such as kelp seaweed, may be large in terms of length, but they all are relatively thin, consisting of just a few layers of cells. Thus, each cell is in direct contact with or very close to the external environment. Gases, water, ions, and wastes can be exchanged between each cell and the environment. Most cells of organisms with little tissue differentiation act independently of one another in exchanging materials with their environment or can easily get materials by diffusion from other cells. There is no organized transport system in these organisms.

On the other hand, in plants and animals that have more complex tissue organization, most cells are too far away from the external environment for direct exchange of materials. Diffusion within an organism is ineffective over several thicknesses of cells. In these complex organisms, materials enter and leave only at certain points and then are carried between the cells and these points by some kind of fluid. The fluid is part of the organism's transport system, a link between the organism's cells and the external environment. Because exchange of materials occurs between cells and the transport fluid, the fluid can be called the internal environment of the organism. Evolution of an internal environment in complex organisms provided each of their cells access to an environment much like the external environment of their unicellular or simpler multicellular ancestors.

Figure 21-2 The cells of many multicellular protists, such as seaweed (left), are in direct contact with water that contains their required nutrients. A giraffe is an unusual and very complex organism. How can all the cells of a giraffe obtain the nutrients they need to survive?

Transport Tissues in Vascular Plants

One group of organisms that uses a transport system is the vascular plants, so named because they contain vascular tissues that transport water, minerals, and food in the form of sugars. These vascular tissues are of two types: xylem tissue and phloem tissue.

Xylem and Phloem   Water and dissolved mineral ions are transported upward from roots to leaves within a series of tubular cells that make up **xylem** tissue. In the roots of a young plant, xylem cells are at first short and narrow, but as the plants mature the xylem cells elongate, the end walls break down, die, and the cells then form hollow tubes consisting only of the cell walls. The cells are linked end to end, forming long pipelines called **vessels.** Xylem vessels extend from near the tips of the roots up into the rest of the plant. The water transported in xylem vessels helps to keep the plant from wilting and is important in the process of photosynthesis in the leaves. Some of the sugars produced during photosynthesis are transported within the plant by cells of **phloem** tissue. Phloem is another part of the vascular plant's transport system. As in xylem, phloem is composed of cylindrical cells joined end to end to form pipelines. Unlike xylem, however, phloem cells are not dead. They have cytoplasm but lose their nuclei as they age. Their life processes are controlled by smaller companion cells that retain their nuclei. The end wall of each phloem cell is perforated, like a sieve. Therefore, a pipeline of sieve cells is called a **sieve tube.** The tissues of xylem and phloem form vascular bundles throughout the vascular plant.

Figure 21-3   The strands of vascular tissue in a plant supply all the different organs with required nutrients. How does the arrangement of xylem tissue change as it moves up into the stem?

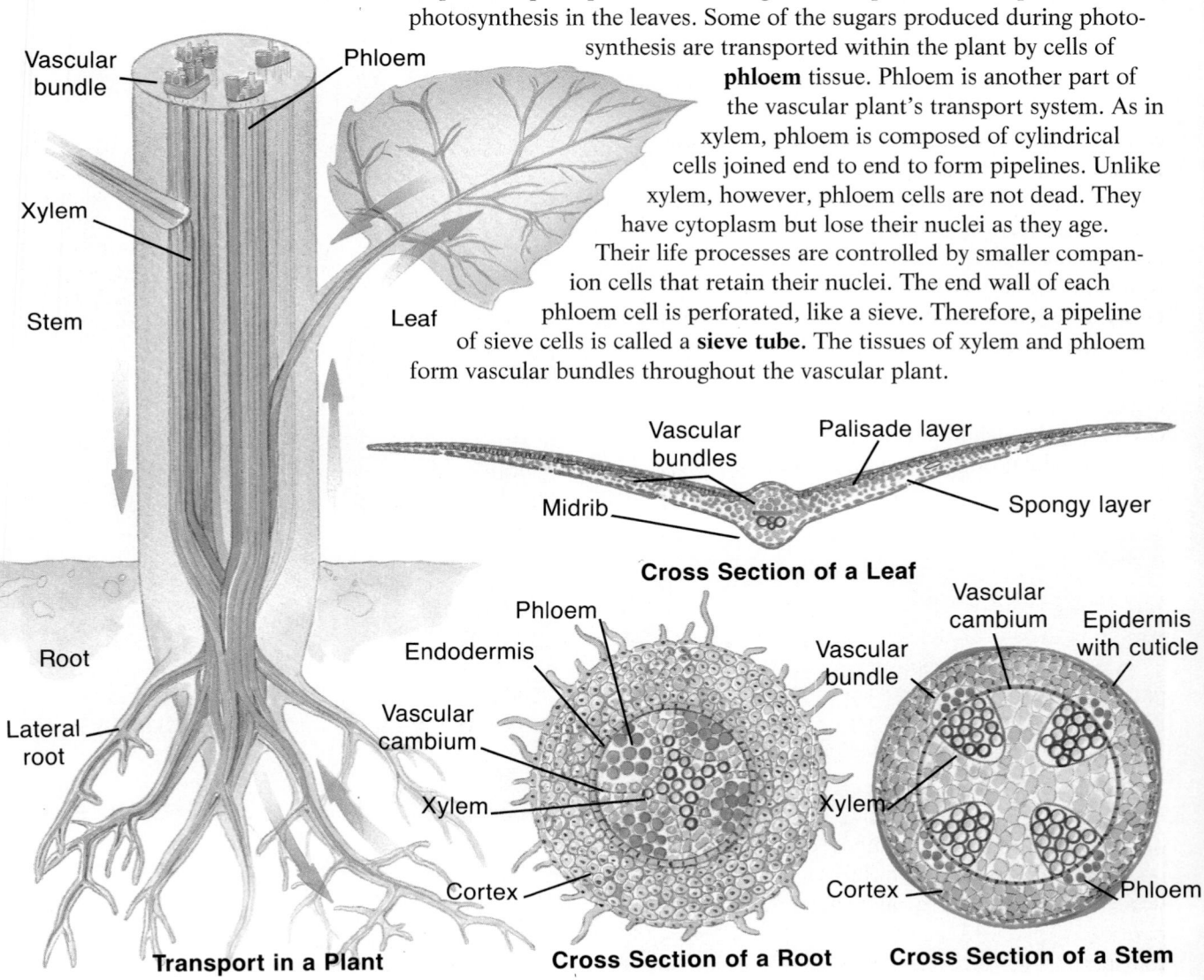

Vascular bundle

Phloem

Xylem

Stem

Leaf

Root

Lateral root

Transport in a Plant

Vascular bundles

Midrib

Palisade layer

Spongy layer

Cross Section of a Leaf

Phloem

Endodermis

Vascular cambium

Xylem

Cortex

Cross Section of a Root

Vascular cambium

Epidermis with cuticle

Vascular bundle

Xylem

Cortex

Phloem

Cross Section of a Stem

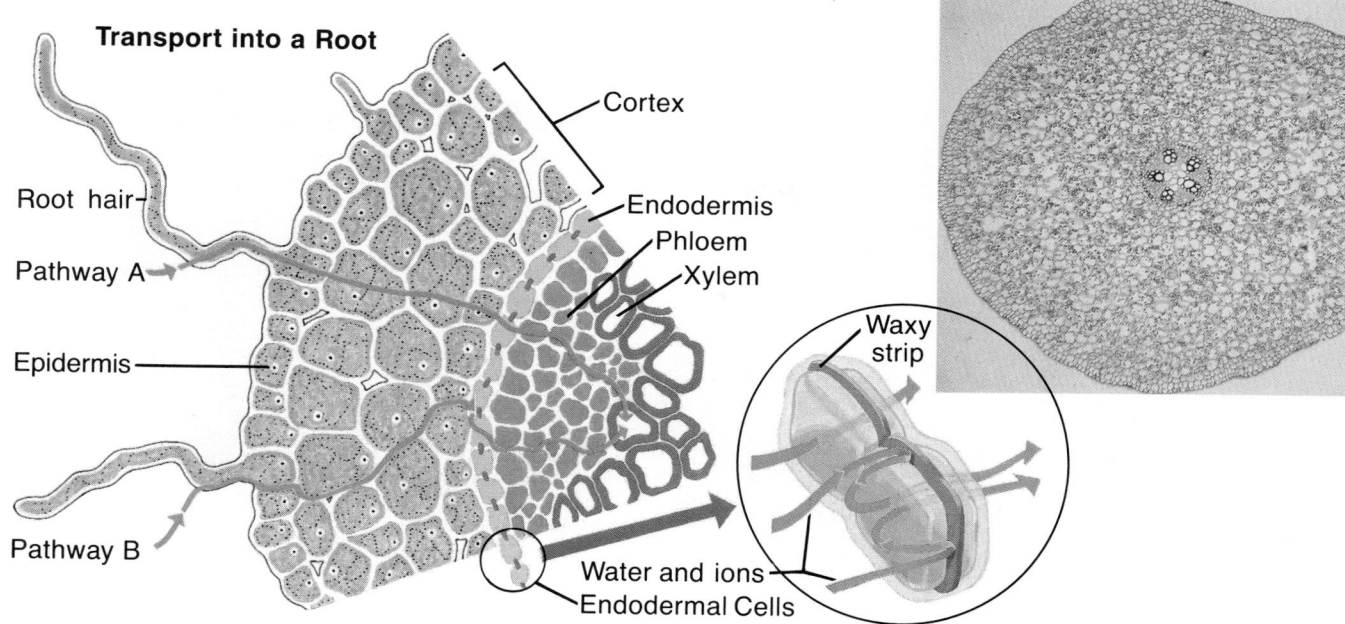

Transport into a Root

Cortex
Endodermis
Phloem
Xylem
Root hair
Pathway A
Epidermis
Pathway B
Waxy strip
Water and ions
Endodermal Cells

Figure 21-4   The cross section of a root (right) shows that the xylem tissue forms a central core surrounded by a single layer of cells called the endodermis. Follow the pathways A and B of water molecules into a root. What happens when they reach the waterproofed cells that make up the endodermis?

Root Tissues   Water and minerals enter a plant via the roots, which are also important in anchoring the plant and in storage of food in the form of starch. Figure 21-4 shows a cross section of a typical root of a flowering plant such as a sunflower. Notice that the tissues are arranged somewhat in circular layers. Notice also the numerous **root hairs,** which are outgrowths of the outer layer of cells, or epidermis, of the root. The hundreds of tiny root hairs together provide a very large surface area for absorbing water and dissolved ions from the film of moisture around soil particles. Entrance of water into the epidermal cells by osmosis results in a greater concentration of water molecules there than in the layer of loosely packed cells that form the next inner tissue layer. Thus, water moves by osmosis into this inner tissue, called the cortex. In addition to being involved in transport, cells of the cortex are food storage areas. For example, when you eat carrot sticks you are eating the cortex with stored carbohydrates of a carrot root.

Endodermis: The Plant's Gatekeeper   Some water and dissolved ions pass from the soil all the way through the root cortex without actually entering the cells. Instead, they pass *between* the cells of the cortex. This uncontrolled flow is stopped short by the ring of endodermis. The endodermis is a layer of cells one cell thick that completely surrounds the central core of the root. The radial and end walls of endodermal cells are surrounded by a strip of wax. These strips form a waterproof seal between the outer and inner layers of the root. Water and ions can no longer pass between cells. Instead, they are forced to enter the endodermal cells.

Although water readily diffuses across the cell membranes, ions in the water cannot cross these membranes by simple diffusion. Rather, ions pass into the cells of the endodermis by facilitated diffusion or by active transport. From the cortex, water and ions pass through the endodermis

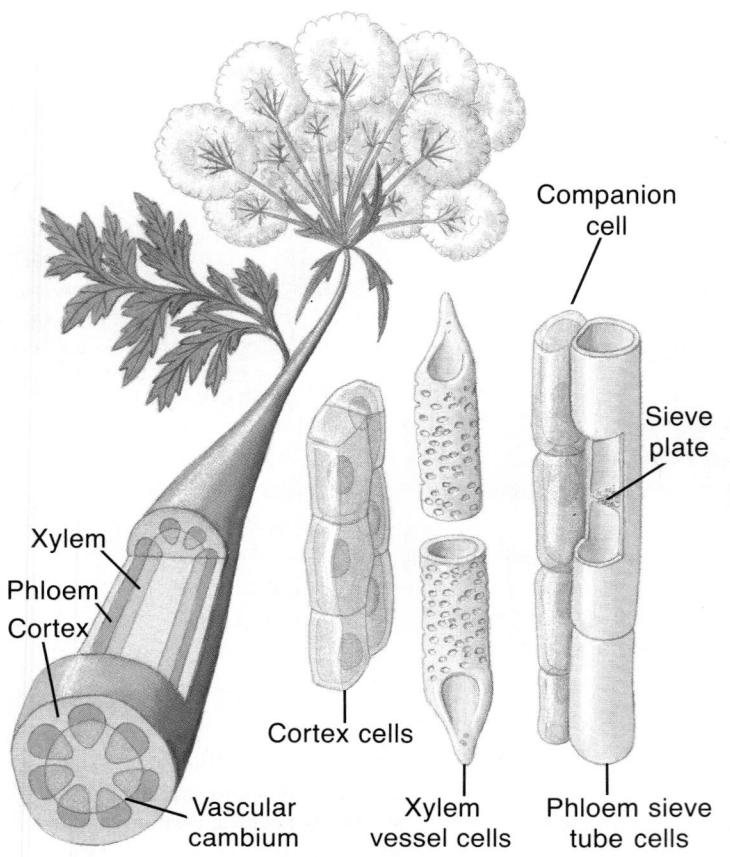

Companion cell

Sieve plate

Xylem
Phloem
Cortex

Cortex cells

Vascular cambium

Xylem vessel cells

Phloem sieve tube cells

Bark
Phloem
Xylem

Figure 21-5   In the stems of many flowering plants such as this Queen Anne's lace (left), the vascular tissues are arranged in a circle of discrete bundles. In stems of plants such as lilies, palms, or grasses, the vascular bundles are scattered throughout the stem. In woody stems (right), new layers of xylem and phloem form complete cylinders of woody tissue surrounded by a protective bark.

and into the xylem in the central core of vascular tissues. Only substances that can cross the cells' plasma membranes can get into the inner column of vascular tissue and up into the plant. At the same time, the endodermis prevents a backflow of ions. Thus, the membranes of the endodermal cells regulate the passage of ions by active transport into the xylem. The higher ion concentration in the central core of vascular tissue is a major driving force of water flow up a plant. In other words, the endodermis acts as a checkpoint or gatekeeper to control substances entering the plant. Once in the xylem, the water and some of the ions move up the stem of the plant to the leaves for use in photosynthesis and other cell processes. Other ions remain in the root cells and are used for their cell processes.

Stem and Leaf Tissues   From the roots, the water and ions move up the xylem vessels and into the stem. Figure 21-5 shows a cross section of a herbaceous plant stem. Note the position of xylem and phloem in the vascular bundles. These bundles are continuous with the core of vascular tissues in the roots. Water and ions that have entered the xylem of roots now pass up the xylem of stems and into the vascular bundles of the leaves. In many plants, xylem and phloem cells develop on either side of a ring of meristematic tissue called vascular cambium. This vascular cambium is an active region of cell division. In plants that live longer than one year, the cells that are produced in the vascular cambium each year develop either into new xylem or new phloem cells; xylem cells form toward the inside of the stem, and phloem cells form toward the outside. The vascular cambium links each bundle of xylem and phloem to form a cylinder of vascular tissue. Addition of new xylem and phloem thickens the stem. A woody stem is usually formed by the buildup of xylem tissue and by the formation of bark, a thinner outer layer of corklike cells that includes and protects the phloem (Figure 21-5).

From the stem, the water and ions enter the xylem of the leaves. Located throughout a leaf are vascular bundles, which cause the appearance of veins on the leaf's surface. Veins in a leaf are continuous with the vascular bundles of the stem. As water enters a leaf, it moves from the xylem in the veins into layers of leaf cells that make up the palisade tissue and spongy tissue. Cells in the palisade layer are long and narrow, packed closely together, and contain chloroplasts. They are lined up side by side like notes on a keyboard (Figure 21-6). Cells in the spongy layer also contain chloroplasts. The spongy cells are round and loosely packed and have many air spaces between them, like a sponge. It is in these two layers that most of a plant's photosynthesis occurs. The epidermis is the outer layer of cells of a plant. The epidermis of the leaf protects the cells inside and is coated by a cuticle, a layer of waxy secretion from the epidermal cells that prevents evaporation.

You know that carbon dioxide is needed for photosynthesis. It enters through structures with openings, called **stomata** (*sing.,* stoma), in the epidermis and passes to the air spaces between the palisade and spongy cells. It then dissolves in the watery film around those cells and diffuses into them and on into the chloroplasts. Oxygen produced during photosynthesis passes out of the leaf through the pores in the stomata. Oxygen also enters the leaf for use in respiration by the plant. In this process, carbon dioxide is released by the plant via the stomata.

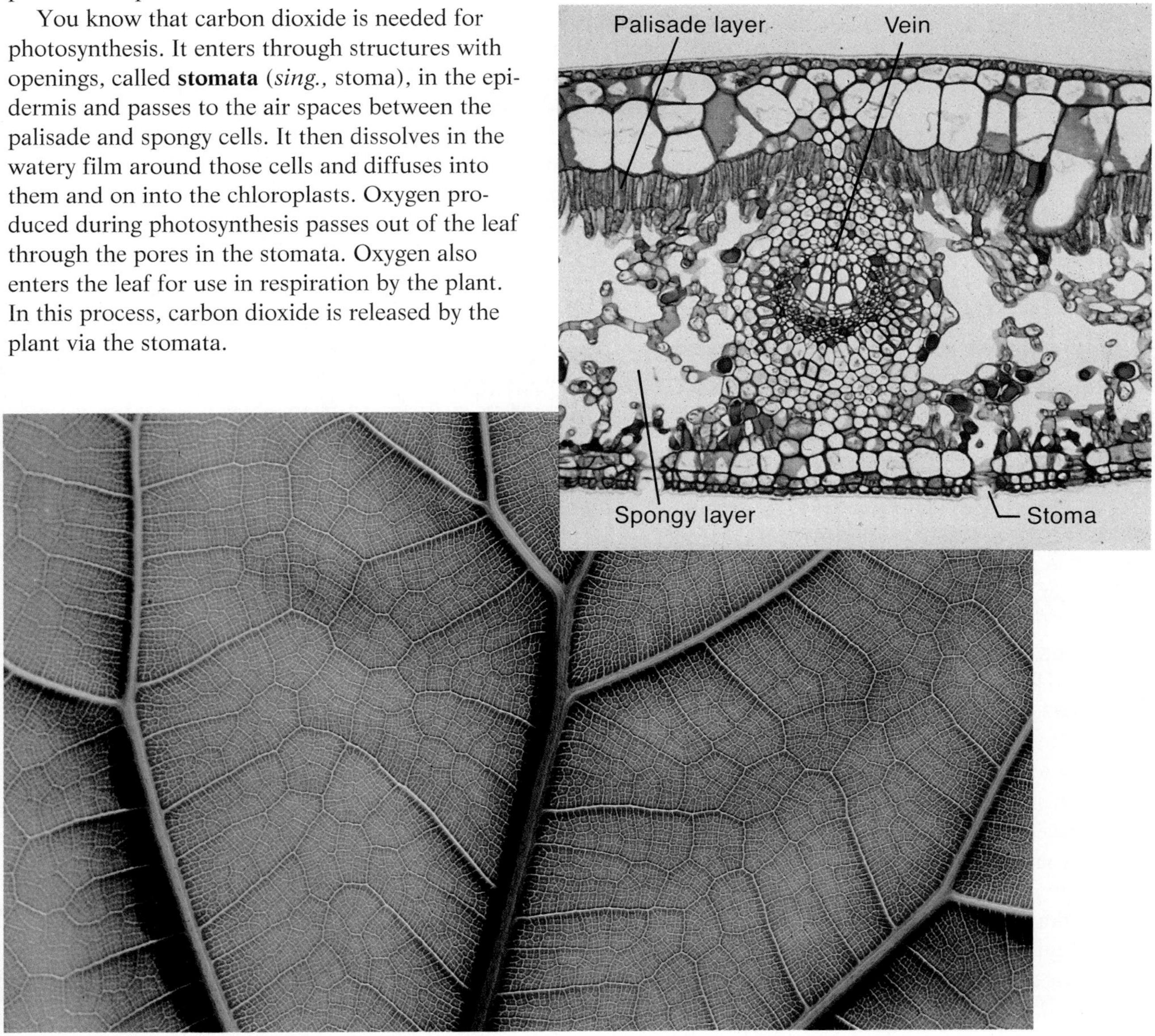

Figure 21-6 The veins of a leaf are clearly visible (left). However, a cross section of a leaf (right) will provide a better understanding of the function of this complex plant organ. Notice the stomata in the epidermis where gases are exchanged and the compact palisade layer where the energy of sunlight is captured.

Water Movement in a Vascular Plant

To get to the leaves, water from the roots sometimes must travel 100 meters or more. How can this be done? Plants do not contain a pump to push the water upward.

Properties of Water In fact, the water is not pushed from below, but rather is pulled from above. For such a pull to occur, water must form unbroken columns from roots to leaves, as the lead in a pencil extends from point to eraser. Two properties of water combine to produce such continuous columns. Because of the polarity of water molecules (Chapter 3), they tend to cling to one another like magnets; the positive end of one molecule attracts the negative end of another. This attraction among the same kind of molecules is known as cohesion. You can see cohesion of water molecules when you turn on a faucet. The water flows out in a solid, steady stream, rather than sprinkling out like separate grains of sand. Water molecules also are attracted to the inner walls of xylem vessels, a property known as adhesion. You can see adhesion of water molecules when you step out of a shower—water droplets cling to your skin. Remember that xylem cells are microscopic. In the very narrow pipelines that xylem cells form, the combination of cohesion and adhesion gives a column of water properties more like those of a solid than of a liquid. The

A R T C O N N E C T I O N

Is a root more than a root?

Diego Rivera returned to Mexico in 1921, after the revolution of 1910-1917 had brought about land reform and a more democratic constitution. He had been in Europe since 1907 painting and studying, but once back in Mexico, art became a way for him to communicate the history of Mexico's political struggles against foreign powers and the ruling classes. Rivera is best known for the murals he painted both in Mexico and the United States illustrating the social conditions and the heritage of the

column of water remains continuous, in complete contact with the walls of the xylem vessels, and could be pictured as a long but very fine thread, extending from roots to leaves.

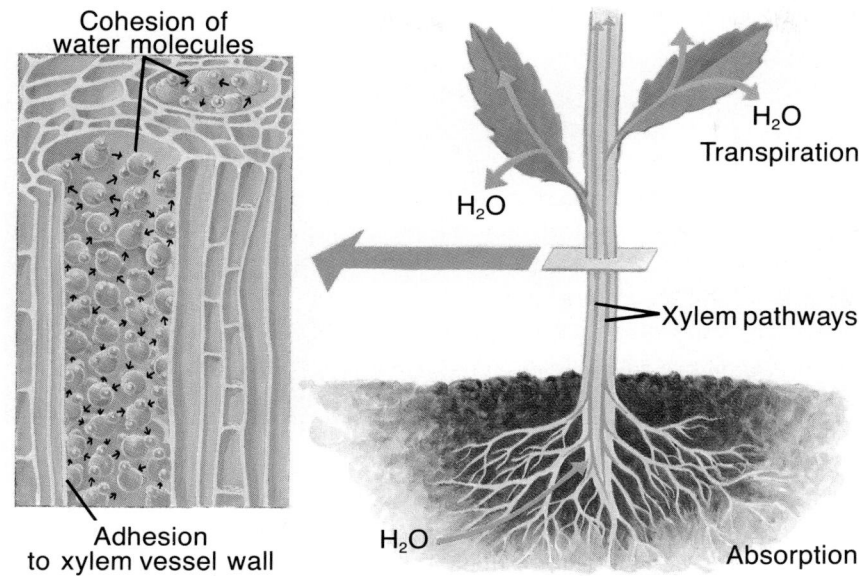

Cohesion of water molecules

Adhesion to xylem vessel wall

H_2O Transpiration

H_2O

Xylem pathways

H_2O

Absorption

Figure 21-7 When water moves up the xylem vessels in a plant, the molecules stick together by cohesion and to the walls of xylem vessels by adhesion.

Mexican people. Rivera and his fellow muralists helped to give Mexico contemporary art based on native tradition.

Roots The geography of Mexico is part of that native heritage, and many of Rivera's paintings and murals show the tie between the people and the land. "Roots," painted in 1937, is one of six watercolors Rivera did of Mexican plants. The "roots" of this ancient-looking plant appear to be more like rhizomes, which are underground storage organs modified from stems. In Rivera's painting, the eye is immediately drawn to the rhizomatous "root" system that seems to be pushing its way up through massive

rocks. There appears to be little or no soil or water anywhere. The rhizomes look old and gnarled, as if their push to the surface has been long and difficult. At the tips of the giant rhizomes, thin, flowerlike shoots of yellow and green burst out, almost like firecrackers. They alone seem to be young and healthy.

The Struggle for Survival
Knowing Rivera's use of art to tell a story, one wonders whether the rocks perhaps represent the government and the ruling class of Mexico, and the "roots" the long struggle of the Mexican people to overthrow them. In some of Rivera's other paintings, trees and plants take on clearly animal or

human shapes. Some of the "roots" in this picture look dead, perhaps representing the people who died in the struggle. The flowerlike shoots seem to symbolize the people who have finally won the struggle and are just beginning to claim their own land. What at first looks like a bleak, gray scene becomes a sign of hope for the future.

1. **Connection to the Environment:** How might this picture tell a story about the environment?
2. **Writing About Biology:** Describe how desert plants might be adapted for survival.
3. **Explore Further:** Compare the tissues of roots and rhizomes.

Figure 21-8 A common sight over a forest on an early summer morning is the rising of a cloud of water vapor. This is one sign that the forest is alive. Most of the water taken up by plants is given off by transpiration from stomata in the leaves.

Transpiration What causes the pull? Water is lost constantly through the pores in stomata in a process called **transpiration.** As water molecules exit the leaf tissues through stomata, they are replaced by water molecules from the xylem vessels in the veins of the leaf. Because of cohesion, this movement of water molecules creates a tension, or pull, on the rest of the water column up the stem. This pull can be thought of as a stretching effect that extends all the way to the roots. As the water in the root xylem is pulled up, more water moves into the xylem from the surrounding root cells and ultimately from the soil. It's like a magician pulling an unending string of scarves from a sleeve. According to this model, known as the transpiration-cohesion theory, it is important that no air bubbles form in a xylem vessel. Such bubbles would break the continuous column of water. Obviously, movement of water columns from roots to leaves is work. What is the source of energy for this work?

Stomata: The Pores of a Plant

Transpiration is important in the upward transport of water and dissolved ions. However, excess water loss from the leaves could cause a plant to dehydrate and die. Each stoma in a leaf is a pore surrounded by **guard cells** that regulate the pore's size. Guard cells prevent a plant from drying out. During the day, the pores of the stomata are wide open, allowing an ample supply of carbon dioxide to enter the leaf for photosynthesis. But much water escapes from these stomata also during the day. At night, however, the light reactions of photosynthesis do not occur, and large supplies of carbon dioxide are not needed. At that time stomata are partially closed, decreasing the amount of transpiration and conserving water. Stomata do not close up completely because respiration continues both day and night. The stomata still allow the exchange of oxygen and carbon dioxide for this essential life process.

How do guard cells control opening and closing of stomata? When guard cells take in water by osmosis, they swell. The cell walls of guard cells have rings of cellulose microfibrils (Figure 21-9) similar to the belts in radial tires. When water enters the guard cells, the rings of microfibrils prevent the cells from swelling in width. They can only swell in length and thickness. Because the guard cells are joined at each end, they bow apart when they swell. They take on a balloonlike crescent shape. Thus, the size of the stoma between them increases. At night, or when there is a water shortage, water moves by osmosis from the guard cells into neighboring epidermal cells. The guard cells collapse and lose their crescent shape, and the size of the stoma decreases.

What determines when guard cells swell or collapse? As day breaks, light stimulates photosynthesis in the guard cell chloroplasts, which makes ATP available for the active transport of potassium ions from neighboring cells. This uptake of ions causes water to enter by osmosis and the guard cells swell. At night, when potassium ions leave the guard cells, water is lost by osmosis and the stoma closes.

Transport in Phloem Tissue

From the leaf, the sugars made during photosynthesis are converted to sucrose, a more soluble form of sugar, which then meanders through the phloem to all parts of the plant. You know that the sieve cells of phloem pipelines, which transport sugars, are not dead and that they contain cytoplasm. The sieve tubes, therefore, are not comprised of continuous hollow pipelines as in the xylem. The side walls of phloem sieve cells have small holes, and strands of cytoplasm pass horizontally between them and their companion cells. Ions and amino acids are known to pass into sieve tubes via these strands of cytoplasm. Phloem tissue is a living tissue.

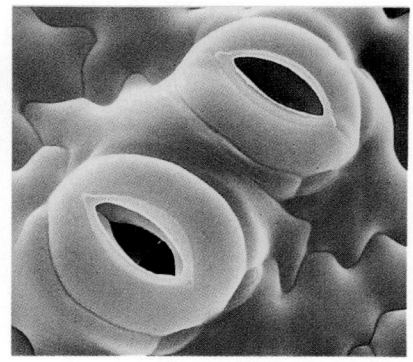

Figure 21-9 The stomata in a leaf, as seen in this electron micrograph (magnification: 257×), are the doorways between a plant's internal and external environments (right). When light stimulates the movement of potassium ions (K^+) into the guard cells (left) the cells take up water and swell, and the stoma opens. The stoma closes later when daylight fades and the movement of potassium ions is reduced.

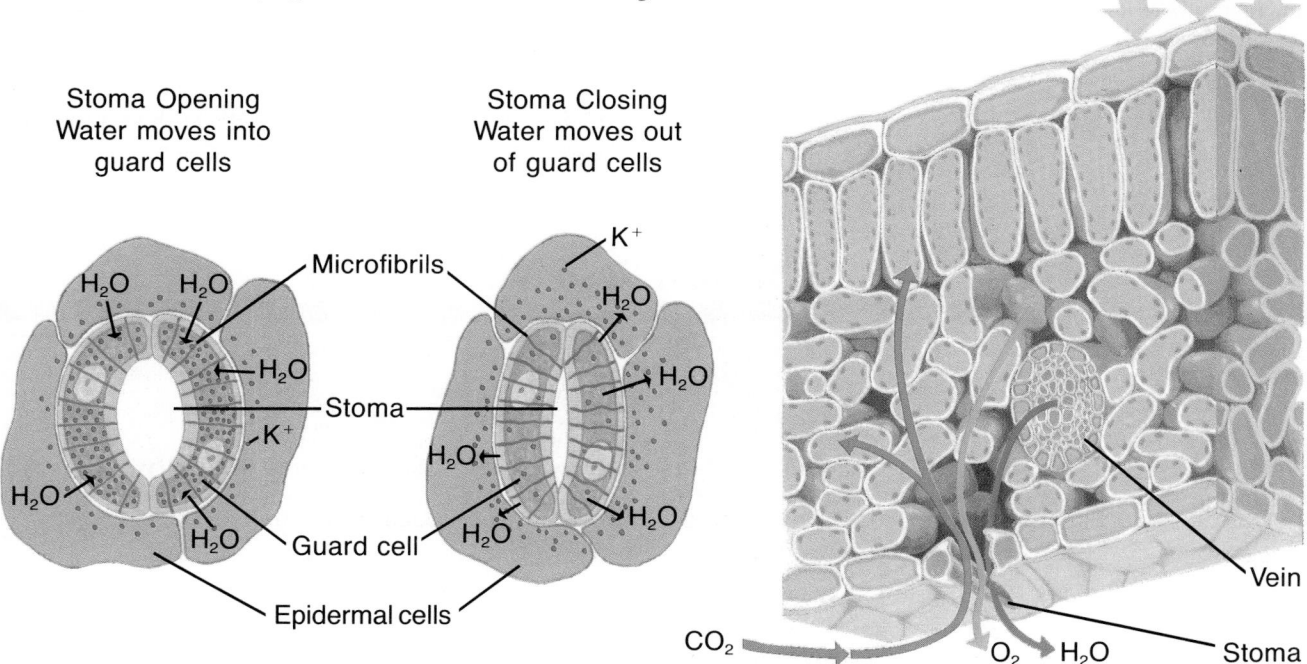

21.1 NUTRIENT TRANSPORT IN PLANTS   **587**

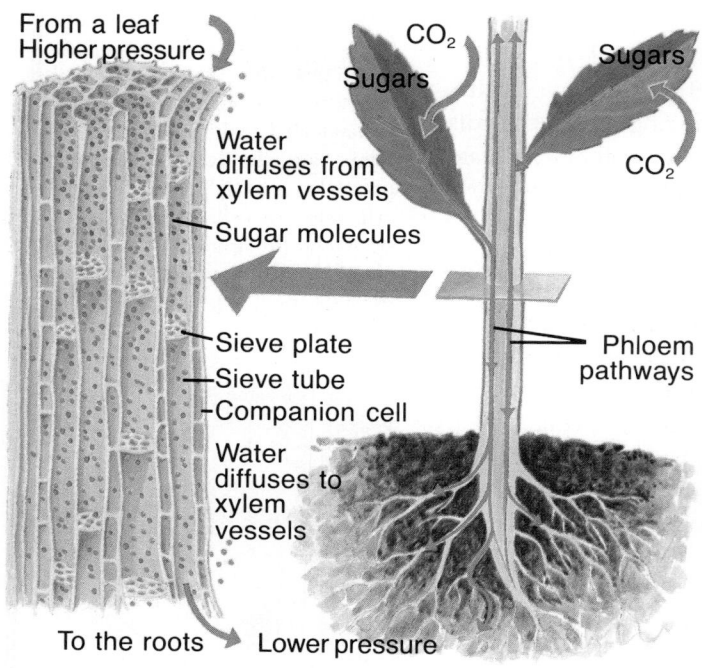

From a leaf
Higher pressure

CO₂
Sugars

Sugars

CO₂

Water
diffuses from
xylem vessels

Sugar molecules

Phloem
pathways

Sieve plate
Sieve tube
Companion cell

Water
diffuses to
xylem
vessels

To the roots   Lower pressure

Figure 21-10   **The pressure flow (from the phloem in the leaves down sieve tubes in the stem to the roots) is due to a higher concentration of sugar in the leaves. Sugar molecules enter companion cells by active transport. The sugar then passes into the sieve tubes across the interconnecting strands of cytoplasm and on down to the roots, where it is stored as starch or used for growth and metabolism.**

The dissolved sucrose and other nutrients made in the cells pass from the palisade and spongy tissues of the leaf into the sieve tubes in the leaf's veins. This increase in concentration of solutes in the sieve cells causes water to enter by osmosis. The sucrose and nutrients are carried by this water downward to the roots to be used in cell processes as an energy source or converted to starch for storage. Sucrose also can be transported upward, where it could be stored in fruits or used as an energy source for the development of new leaf or flower buds.

How does phloem transport work? One model suggests that transport in phloem is a result of pressure. The entrance of water from the surrounding cells into the sieve cells increases the pressure inside the cells at the top of the sieve tubes. Because of this pressure, the sucrose, nutrients, and water are forced out through the sieve end and into the next lower sieve cell, where pressure is lower. The pressure differences in sieve cells from leaves to roots may account for this downward movement. As the solution works its way down to the roots, some of the sucrose, nutrients, and water leaves the sieve cells to nourish the developing tissues of the roots or to be held for storage. Thus, pressure remains lower in the roots. This model of movement in phloem tissue is called the pressure-flow hypothesis.

The pressure-flow hypothesis also explains the upward movement of organic molecules. For example, in spring, starch stored in the roots is converted to sucrose, which then enters sieve cells. Water enters the sieve cells from other root cells by osmosis, increasing the pressure within them. In this case, the higher pressure is in the sieve cells at the bottom of the phloem pipeline. As a result, the sucrose is forced upward.

Section Review

Understanding Concepts

1. Explain why humans must have a transport system, whereas a planarian can exist without one.
2. How are the structures of xylem and phloem cells adapted to their functions?
3. Explain how the various parts of a leaf contribute to its ability to carry out photosynthesis.
4. ***Skill Review—Recognizing Cause and Effect:*** Sieve tubes form a circle of tissue within the bark of a woody stem. What would be the result of removing a strip of bark all the way around such a stem? For more help, see Thinking Critically in the **Skill Handbook.**
5. ***Thinking Critically:*** How can it be explained that organic molecules can be moving both upward and downward in the same kind of vascular tissue at the same time?

21.2 Transport in Animals

Vascular plants, as you have read, obtain their nutrients by a system of two kinds of tissues that make up an internal transport system. Most animals, too, have a transport system to move materials to and from body cells. As in plants, each cell requires a supply of oxygen and food molecules, and each cell must get rid of waste products. In all animals that have a transport system, two parts are essential: (1) circulating fluid, and (2) one or more hearts to pump the blood. All of the materials moved by a transport system travel together in the same fluid, which circulates over and over again from the cells, sometimes through vessels, in and out of the heart, and back to the cells. Thus, the transport system of animals is a *circulatory* system.

Closed Circulatory System—The Earthworm

Segmented worms are the least complex animals to have a true circulatory system. The system is closed; that is, the blood is carried within blood vessels. It does not slosh around freely in the worm's body cavity. This kind of animal transport system, in which the blood is enclosed at all times within vessels, is called a closed circulatory system.

Although simple in structure, the worm's circulatory system permits rapid and efficient exchange of materials. In an earthworm, the blood flows through two major blood vessels—the dorsal blood vessel and ventral blood vessel, shown in Figure 21-11. These two vessels are connected near the anterior end of the worm by a series of five pairs of pumps. These tubelike pumps, known as aortic arches, function as hearts. The hearts, like all hearts, contract rhythmically and pump the blood along. The elastic dorsal vessel also helps squeeze the blood. Under these pressures, the blood flows away from the hearts through the ventral vessel. It returns to the hearts from the body tissues and through the dorsal vessel. Valves in the hearts and in the dorsal vessel, like one-way doors, keep the blood flowing in the right direction.

Objectives

Distinguish between open and closed circulatory systems.

Sequence the flow of blood in the circulation of birds and mammals.

Discuss the various means by which materials are exchanged between the blood and the body cells.

Figure 21-11 Notice how the dorsal (back) and ventral (belly) blood vessels of an earthworm are connected by the five pairs of hearts at the anterior or head end. How do nutrients enter the blood of an earthworm?

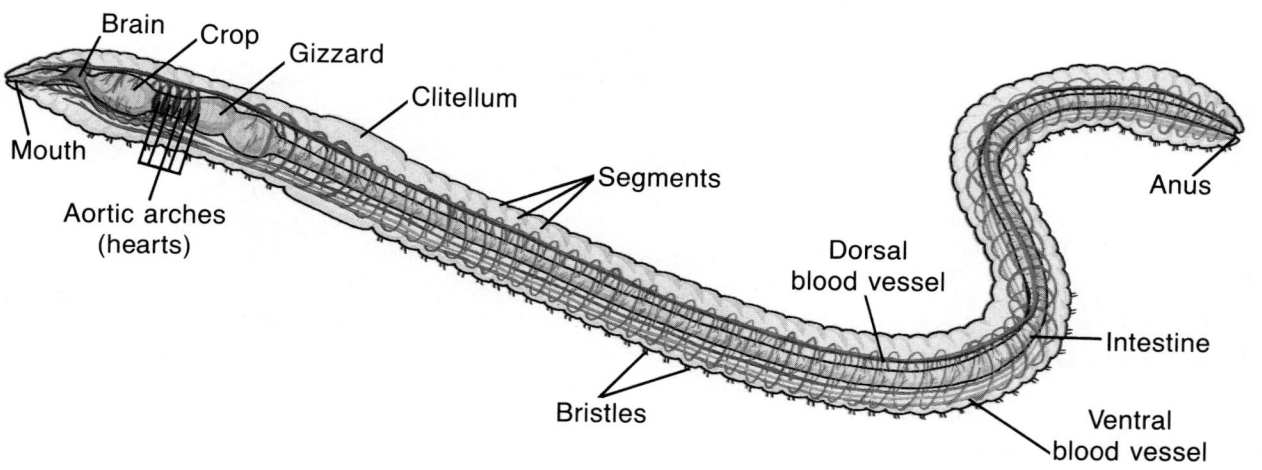

Brain
Crop
Gizzard
Clitellum
Mouth
Aortic arches (hearts)
Segments
Dorsal blood vessel
Anus
Intestine
Bristles
Ventral blood vessel

The blood flows from the ventral blood vessel through smaller and smaller branches of the vessel to the worm's tissues. Exchange of materials between blood and cells takes place through the smallest blood vessels, called **capillaries.** Capillaries have walls only one cell thick, so materials can pass easily in and out of them. The blood picks up digested food molecules from within capillaries in the intestine and picks up oxygen from within capillaries in the skin. As the blood in the capillaries passes by cells, food and oxygen molecules in the blood pass through the capillary wall to the cells. Wastes pass from the cells, through the capillary wall, to the blood. The capillaries then converge into larger and larger branches of blood vessel until they eventually empty into the dorsal blood vessel. From there, the blood flows back to the hearts, and the circuit continues. Capillaries are so numerous that no cell is far from the blood. Thus, each cell is linked by means of blood, the animal's internal environment, to the external environment.

Open Circulatory System—The Grasshopper

A different kind of circulation later evolved among arthropods. For example, in a grasshopper, blood is pumped toward the head by a dorsal, segmented heart, into the **aorta,** the only blood vessel in its body, Figure 21-12. From the aorta, the blood empties into the insect's body cavity, where it moves into spaces surrounding the internal organs. While in these spaces, the blood bathes the cells directly and exchange of materials occurs. Muscular movements of the grasshopper circulate the blood within the spaces. Eventually, the blood collects in the space surrounding the heart and passes into the heart through tiny openings. This system, in which blood does not always move within vessels, is known as an open circulatory system.

Figure 21-12 The dorsal blood vessel is the only blood vessel in an insect such as this grasshopper. The tubular heart has slits through which blood enters and is then pumped forward along the aorta to the head and out into the body cavity. The direction of blood flow around the organs is controlled by partitions of tissues.

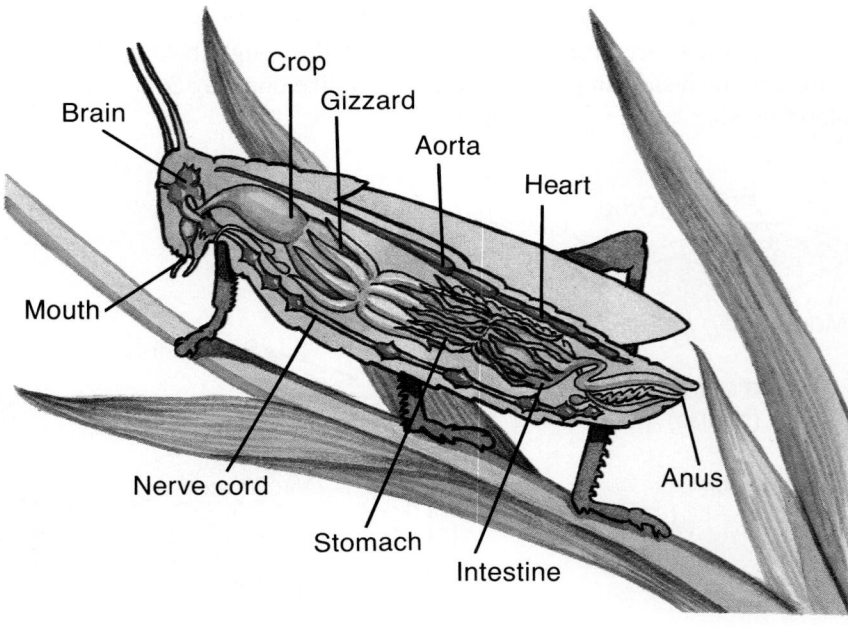

Crop
Brain
Gizzard
Aorta
Heart
Mouth
Nerve cord
Stomach
Intestine
Anus

An open circulatory system may seem less efficient than a closed system because movement of blood around the body is less rapid and under less pressure. Insects are active animals with high energy demands. How can the open system of an insect provide a sufficient supply of oxygen for cellular respiration? The answer is that it doesn't. Oxygen is not transported by the blood at all. It is transported very quickly to the cells by a separate system of tubes, as you will learn in the next chapter. Food and wastes, though, are carried efficiently by the blood.

Other arthropods, such as spiders and crayfish, have open circulatory systems in which the blood does indeed carry oxygen and carbon dioxide. In crayfish, the gills have a rich supply of blood. Within the gills, oxygen from the water diffuses into the blood, and carbon dioxide in the blood is expelled into the water.

Circulation in Vertebrates

Vertebrates have a closed circulatory system made up of a single heart and three types of blood vessels. The heart has two kinds of chambers. An **atrium** is a thin-walled chamber—a kind of receiving room in which incoming blood from the body is collected. A **ventricle** is a larger, more muscular, thicker-walled chamber. It pumps blood to the body. The three types of blood vessels differ in structure and function, as seen in Figure 21-13. Arteries and veins are thick-walled, large-diameter vessels composed of three layers of cells. **Arteries** transport blood away from the heart, and **veins** transport blood back to the heart. Veins contain one-way valves that aid in the return flow of blood to the heart. Capillaries, as you have already read, consist of just one layer of cells. The function of these very narrow vessels is exchange of materials.

Figure 21-13 Arteries (left, 440×) carry blood away from the heart to supply the body's tissues with a network of thin-walled capillaries (center, 28 000×). Veins (right, 16×) carry blood back to the heart. One-way valves (lower right) keep blood from flowing back.

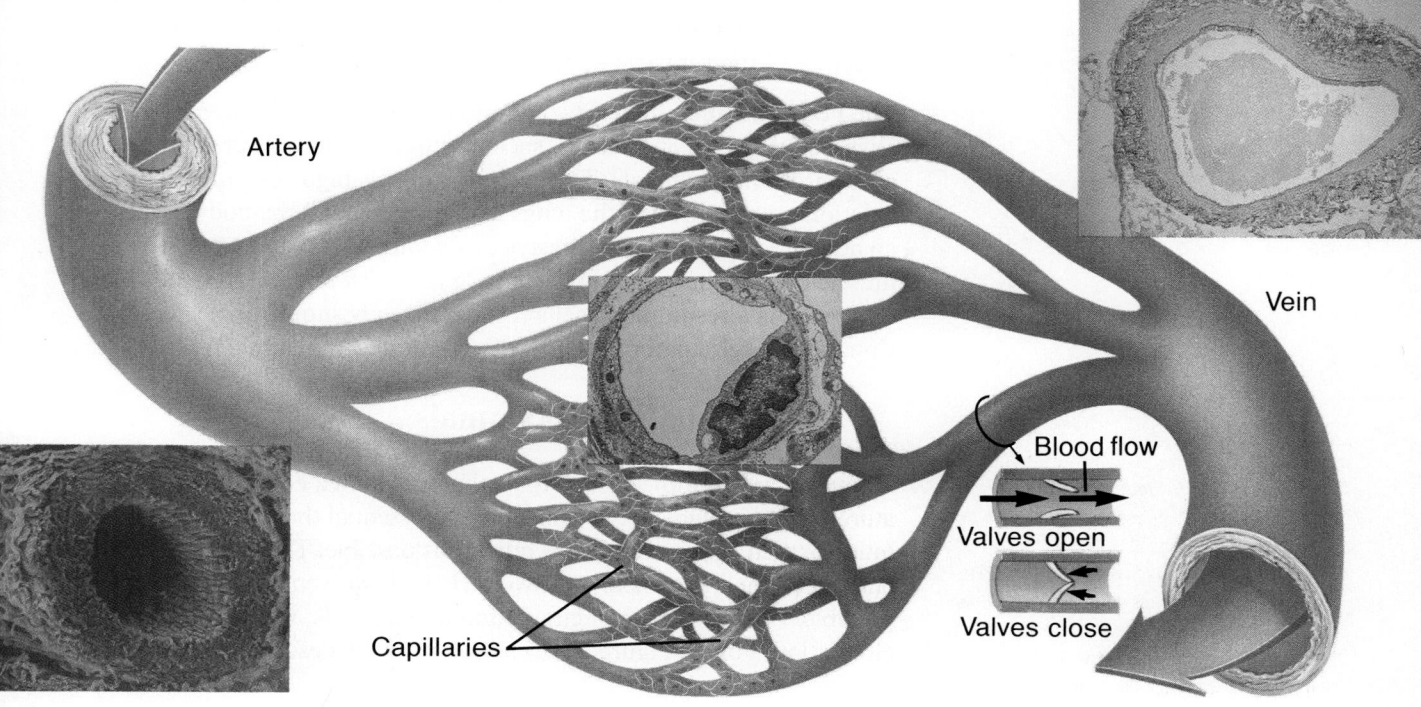

Artery

Vein

Blood flow

Valves open

Valves close

Capillaries

Minilab 1

What affects pulse rate?

Take your resting pulse rate by counting the pulse on your wrist for 20 seconds. Multiply by three to find your pulse rate for one minute. Repeat this procedure for four trials. Find the average. To find your after-exercise pulse rate, do 20 deep-knee bends in 40 seconds. Immediately take your pulse and find the average rate as before. To find the effect of increased weight on pulse rate, first wait about 10 minutes to be sure your pulse has returned to normal. Put 10 lbs. (4.5 kg) of books in a backpack on your back. Do knee bends as before, then find your new average pulse rate. **Analysis** *What is the effect of exercise on pulse rate? Explain. What is the effect of increased weight on your pulse rate? Explain.*

A Heart with Two Chambers   As vertebrates evolved, they developed a greater body complexity and a more active lifestyle from life in the water to life on land. These changes made necessary a more efficient gas exchange with the environment, and a high-pressure circulatory system that supplies all the cells of the animals with necessary nutrients and removes the waste products.

The vertebrate heart is one organ that shows a range of adaptations to different vertebrate lifestyles. For example, fish have a heart with two chambers, an atrium and a ventricle. The atrium receives blood, and the ventricle forces blood out to the gills. The blood flow back from the gills is much slower than to the gills. Due to friction, the pressure of the blood is reduced as it passes through narrower and narrower vessels. Blood pressure in a fish drops as blood passes through the gill capillaries. The blood moves sluggishly from the gills to other parts of the body, and cells do not receive oxygen quickly. The low pressure of fish circulation is adapted to the lower energy demands of these ectotherms, animals whose body temperature changes with the temperature of the environment.

The Three-chambered Heart   Amphibians have three heart chambers—two atria and one ventricle. One atrium receives blood directly from the lungs and skin; another receives blood from the body tissues. Blood from both atria is then pumped into the ventricle. The oxygen-rich blood from the lungs and skin is then pumped to the body systems, and the oxygen-poor blood is pumped back to the lungs and skin where gas exchange takes place. Although the blood pressure falls as it passes through lung and skin capillaries, the blood returns to the heart before being pumped to the rest of the body. Therefore, the circulation is repressurized for its journey to body tissues. Amphibians are also ectotherms, and their circulatory system is an adaptation to their lifestyle. There is some mixing of oxygen-rich and oxygen-poor blood in the heart, although the internal structure of the ventricle keeps this to a minimum.

The Four-chambered Heart   In reptiles, some animals have a three-chambered system and others, such as alligators, have four chambers, just as in birds and mammals. The evolution of this high-pressure, double circulation—once through the lungs and once through the body tissues—is an adaptation to the active lifestyles of complex vertebrates. A wall of muscle divides the ventricle keeping oxygen-rich and oxygen-poor blood separate. As a result, body cells receive a steady supply of oxygen. Compare the circulatory systems of vertebrates in Figure 21-14.

Circulation in Mammals

Mammals are endotherms, animals that maintain a constant body temperature and have high energy demands. It is essential that their cells receive as much oxygen as possible and as quickly as possible. The evolution of a four-chambered heart consisting of two atria and two ventricles has made this possible. As an example of circulation in mammals, we will study human circulation. Follow Figure 21-15 as you read the description that follows.

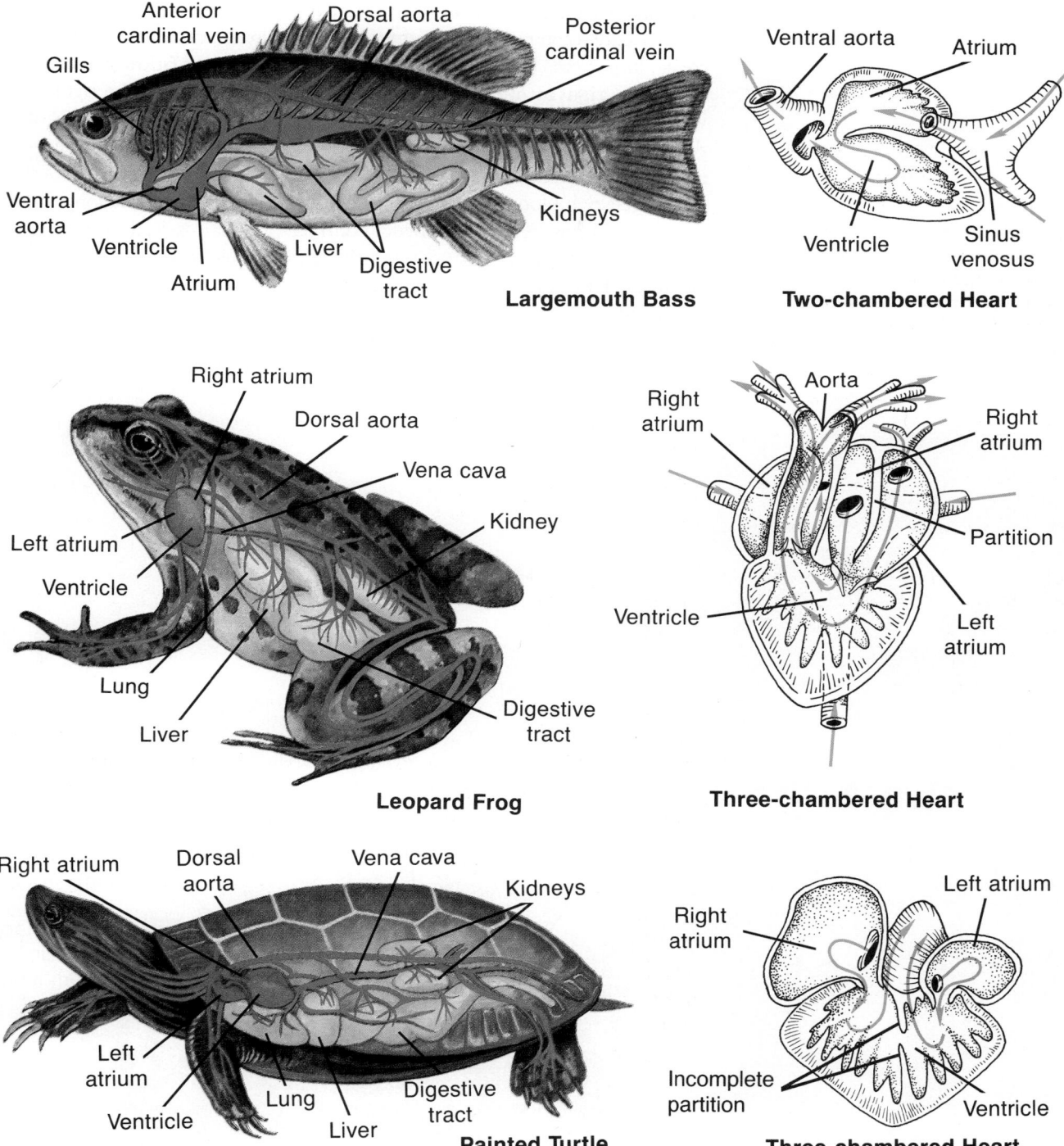

Largemouth Bass

- Anterior cardinal vein
- Gills
- Dorsal aorta
- Posterior cardinal vein
- Ventral aorta
- Ventricle
- Atrium
- Liver
- Digestive tract
- Kidneys

Two-chambered Heart

- Ventral aorta
- Atrium
- Ventricle
- Sinus venosus

Leopard Frog

- Right atrium
- Dorsal aorta
- Vena cava
- Kidney
- Left atrium
- Ventricle
- Lung
- Liver
- Digestive tract

Three-chambered Heart

- Right atrium
- Aorta
- Right atrium
- Partition
- Ventricle
- Left atrium

Painted Turtle

- Right atrium
- Dorsal aorta
- Vena cava
- Kidneys
- Left atrium
- Ventricle
- Lung
- Liver
- Digestive tract

Three-chambered Heart

- Right atrium
- Left atrium
- Incomplete partition
- Ventricle

Figure 21-14 The circulatory systems of vertebrates show a trend from the two-chambered heart of a fish (top) to the three-chambered heart of amphibians (center) and reptiles (bottom).

Human Circulatory System   Blood low in oxygen and high in carbon dioxide enters the right atrium through the superior and inferior vena cavae, large veins leading into the heart from above and below. Contraction of the right atrium forces blood into the right ventricle. The right ventricle contracts, forcing blood into the pulmonary artery. Branches of this artery lead to capillaries in each lung where blood picks up oxygen and carbon dioxide is given off.

Oxygen-rich blood from the lungs courses to the left atrium along four pulmonary veins. As the left atrium contracts, blood is forced into the left ventricle. A powerful contraction of the left ventricle forces blood into the aorta, the largest blood vessel in the body. The aorta forms an arch, like the handle of a cane, that bends down behind the heart. Smaller arteries branching from the aorta direct blood to all parts of the body. The arteries branch and rebranch to form capillaries in which the blood gives up

Figure 21-15   Some activities require large supplies of oxygen to the muscles. Olympic athletes learn to understand efficient use of their body's circulatory system as they strive to reach their greatest physical potential.

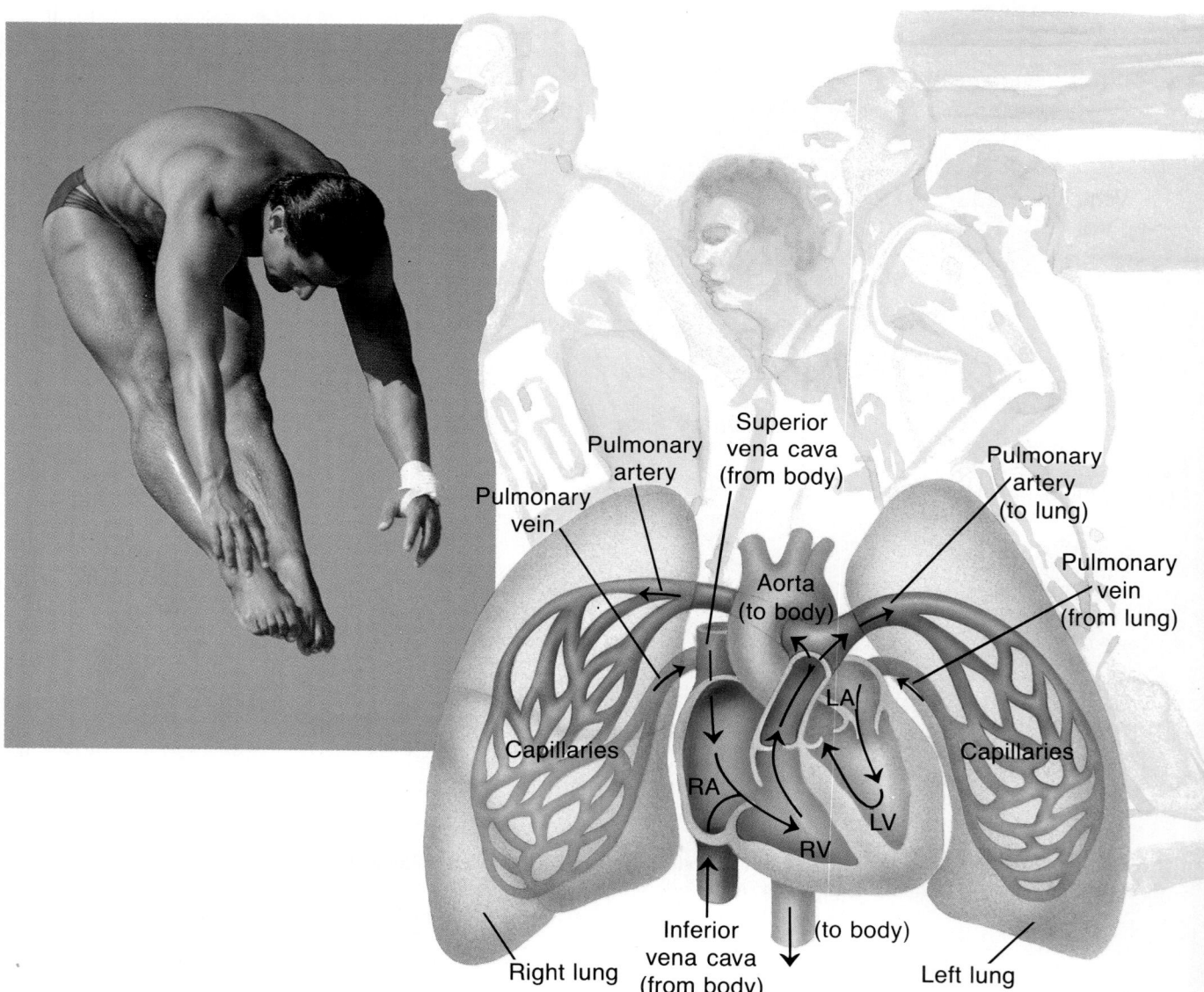

oxygen to body cells and receives carbon dioxide from these cells. In turn, these capillaries merge to form larger and larger veins that eventually form the superior and inferior vena cavae that return blood to the right atrium.

One-way valves between the atria and ventricles and at the base of the arteries leading from the ventricles allow blood to flow in only one direction. In the heart, valves are found between the right atrium and right ventricle, and between the left atrium and left ventricle. Valves also are located at the entrances of the pulmonary artery and the aorta. Find these structures in the heart in Figure 21-14. Why do you think that blood must flow in only one direction?

All blood travels through the heart twice during each circulation of the body. Oxygen-poor blood travels through the right side of the heart and to the lungs, where it picks up oxygen. The oxygen-rich blood travels through the left side of the heart, where it is forced out under great pressure to the body. Presence of two distinct ventricles ensures no mixing of oxygen-rich and oxygen-poor blood.

Blood Pressure As blood is pumped, it exerts a pressure on the walls of blood vessels. In arteries, the pressure causes the walls to bulge outward and then bounce back. You can feel this as a pulse in arteries near the surface of your skin. The surface area of blood is reduced on its journey along this passageway of narrowing vessels. The greatest loss of pressure occurs as blood travels through capillaries. Thus, when blood enters veins on its way back to the heart, the pressure is quite low. How does it get back to the heart, especially from the lower parts of the body? The one-way valves in the veins are important in this return flow of blood. Contraction of skeletal muscles squeezes the veins, forcing blood to move through the valves, which open only toward the heart. When muscles relax, the valves snap shut, preventing blood from flowing down and away from the heart. By the time blood enters the inferior vena cava, the pressure is very low. Breathing creates a vacuum effect in the chest cavity that helps draw the blood from the veins to the heart.

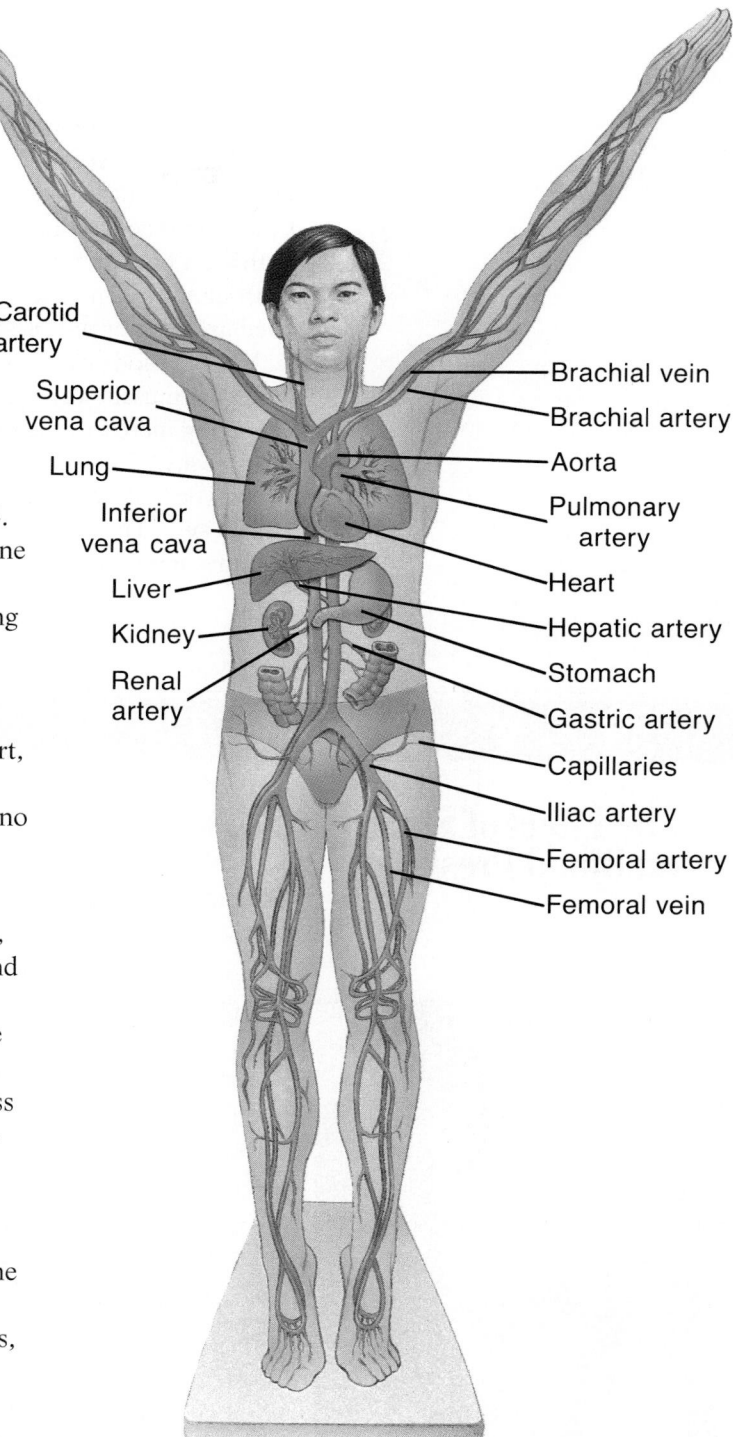

Figure 21-16 The human circulatory system comprises blood circulation to and from the lungs, and circulation of blood to and from all the other organs, tissues, and cells of the body. The thousands of arteries and veins are linked by capillaries within the tissues and form a closed circulatory system.

The Heartbeat

The source of each heartbeat is from within the heart itself. Each beat of the heart originates in a small bundle of tissue in the right atrium. This bundle of tissue is called the **sinoatrial node,** or **S-A node,** and also is known as the pacemaker. The S-A node acts like a kind of spark plug. A flow of ions acts as the electrical impulse that causes the muscle fibers of both atria to contract. The contraction forces blood from the atria into the ventricles. The impulse then reaches another small bundle, the **atrioventricular node,** or **A-V node,** located between the atria and the ventricles.

The A-V node causes an impulse that sweeps outward and downward along the muscle walls of the ventricles, which makes them contract. The squeezing increases the pressure of the blood. This pressure closes the heart's valves and opens the valves of the arteries that lead away from the heart. The "lub-dup" of a heartbeat is the sound of these valves opening and closing. The heart of an adult human at rest beats on average about 70 times per minute.

I N V E S T I G A T I O N

The Effect of Stress on Blood Pressure

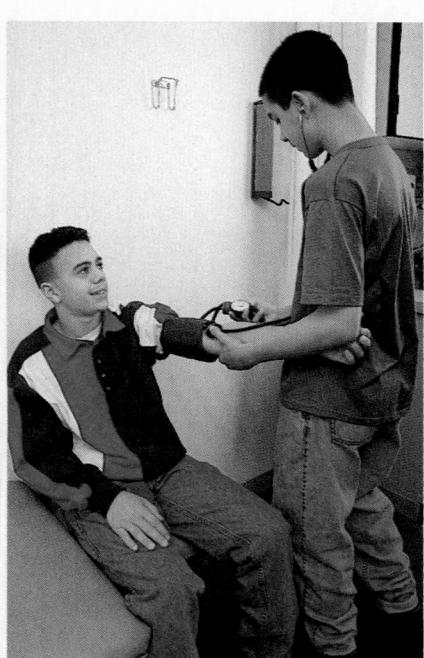

Your teacher hands out the big test. Your hands feel cold and clammy. Your mouth is dry, and the back of your neck is tight and stiff. You are experiencing stress. Stress has many physiological effects on the body, with wide variations from one person to another. Anxiety, which is stress over a long period of time, can cause a rise in blood pressure. This higher blood pressure might place a person at greater risk for stroke or heart attack.

The average normal blood pressure is around 120 over 80. Each heartbeat consists of two phases. The forceful, muscular contraction of the ventricles is called the systole. The top or first number in a blood pressure reading is the systolic pressure. Systole is followed by a short period called diastole, during which time the ventricles fill back up with blood. The second or bottom reading is the diastolic pressure. In this lab, you will test your blood pressure before and after a stressful event.

Problem

What effect does stress have on blood pressure?

Safety Precautions

Do not pump the blood pressure cuff beyond 180. Do not leave the cuff inflated any longer than necessary.

Hypothesis

Make a hypothesis about how your blood pressure will change following a stressful experience. Explain your reasons for making this hypothesis.

Control of the heartbeat is an example of homeostasis, which, as you learned in Chapter 1, is the maintenance of a constant internal environment despite external change. Blood is part of the internal environment of humans. The blood must deliver and remove materials from body cells under varying conditions. At rest, a heartbeat rate of 70 beats per minute is adequate, but during heavy exercise, cells require more oxygen and glucose and produce more wastes. The heartbeat rate must increase so that the exchanges between blood and cells occur more quickly. Once exercise stops, the heartbeat rate gradually slows and soon returns to the resting rate. Heartbeat rate is under the control of part of the nervous system, which will be discussed in Chapter 25.

Exchange of Materials

Capillaries are the sites of exchange. However, there are several methods by which different materials are exchanged between the body cells and the blood in the capillaries. One method is diffusion, the means by which

Experimental Plan

1. In your group, make a list of possible ways you might test your hypothesis.
2. Agree on one idea from your group's list that could be investigated in the classroom. Be sure that your subjects will be treated humanely.
3. Design an experiment that will test one variable at a time. Plan to collect quantitative data.
4. Following the style of a recipe, write a numbered list of directions that anyone could follow.
5. Make a list of materials and the quantities you will need.

Checking the Plan

Discuss the following points with other group members to decide the final procedures for your experiment.
1. What variables will need to be controlled?
2. What is your control?
3. What procedures or props will you use to cause stress?
4. What will you measure or count?
5. How long will you apply stress?
6. How many trials will you run?
7. Have you designed and made a table for collecting data?

Make sure your teacher has approved your experimental plan before you proceed further.

Data and Observations

Carry out your experiment, make your measurements, and complete your data table. Make a graph of your results.

Analysis and Conclusions

1. What is your measured systolic pressure before stress? What does this number represent?
2. What is your measured diastolic pressure before stress? What does this number represent?
3. How did stress alter your blood pressure?
4. How is this change beneficial?
5. If the source of stress did not alter your blood pressure, why do you think it didn't?
6. If you were subjected to a continual high level of stress, how might this stress level affect your health?
7. What could you do to alleviate some stress in your life?

Going Further

Based on this lab experience, design another experiment that would help explain the effects of different kinds of stress in everyday life. Assume that you will have unlimited resources and will not be conducting this lab in class.

Minilab 2

How do you interpret an EKG?

Examine the graph below of a normal human heartbeat. This is called an electrocardiogram or EKG. By placing electrodes on the body, the pattern of electrical messages stimulating the heart muscle can be recorded. Segment T-P in the graph represents the heart at rest; P-Q represents atrial contraction; Q-T represents ventricles contracted. Study the samples of actual EKGs provided by your teacher. Label them with the appropriate letters.

Analysis *Examine the normal EKG. Compare the times spent by the heart in each segment—at rest, atrial contraction, and ventricle contraction. Were the sample EKGs normal or abnormal? Explain.*

carbon dioxide and oxygen can pass between the blood and the body cells. For example, oxygen diffuses from blood in capillaries to muscle cells. Together, the great number of capillaries provide an enormous surface area for exchange. Because these tiny vessels branch and rebranch many times, no cell is far from a capillary. These two features promote ready exchange of materials between cells and blood. In a second method, dissolved materials are transported across the cells that make up the capillary wall in small vesicles. The vesicles form on one side of a capillary cell by endocytosis, and the material is deposited on the other side by exocytosis.

Tissue Fluid   A third method for exchange involves seepage of fluid through the spaces between capillary cells. As blood passes through the capillaries, pressure is exerted on the capillary walls. The pressure is greatest at the arterial end of the capillary. This pressure forces some water and small dissolved particles out of the capillaries and into the tissues. The water and particles are called tissue fluid. Large molecules such as proteins, as well as blood cells, remain in the capillaries. At the venous end of the capillary, blood pressure is lower. Some tissue fluid re-enters that end of the capillary by osmosis. Osmosis occurs in this direction because there is a larger concentration of protein and a smaller concentration of water inside the capillaries than in the surrounding tissue fluid. The tendency of fluid to move into the capillary by osmosis at the venous end of the capillary is greater than the tendency of fluid to move out due to blood pressure.

Tissue fluid is important in the exchange of materials to and from cells. Materials such as fatty acids and sodium ions enter cells from tissue fluid. Also, wastes, such as urea, and products produced by cells, such as hormones, enter the tissue fluid. Thus, tissue fluid is part of the internal environment. When you scrape your knuckle, the sticky fluid (not the red blood) that collects is tissue fluid.

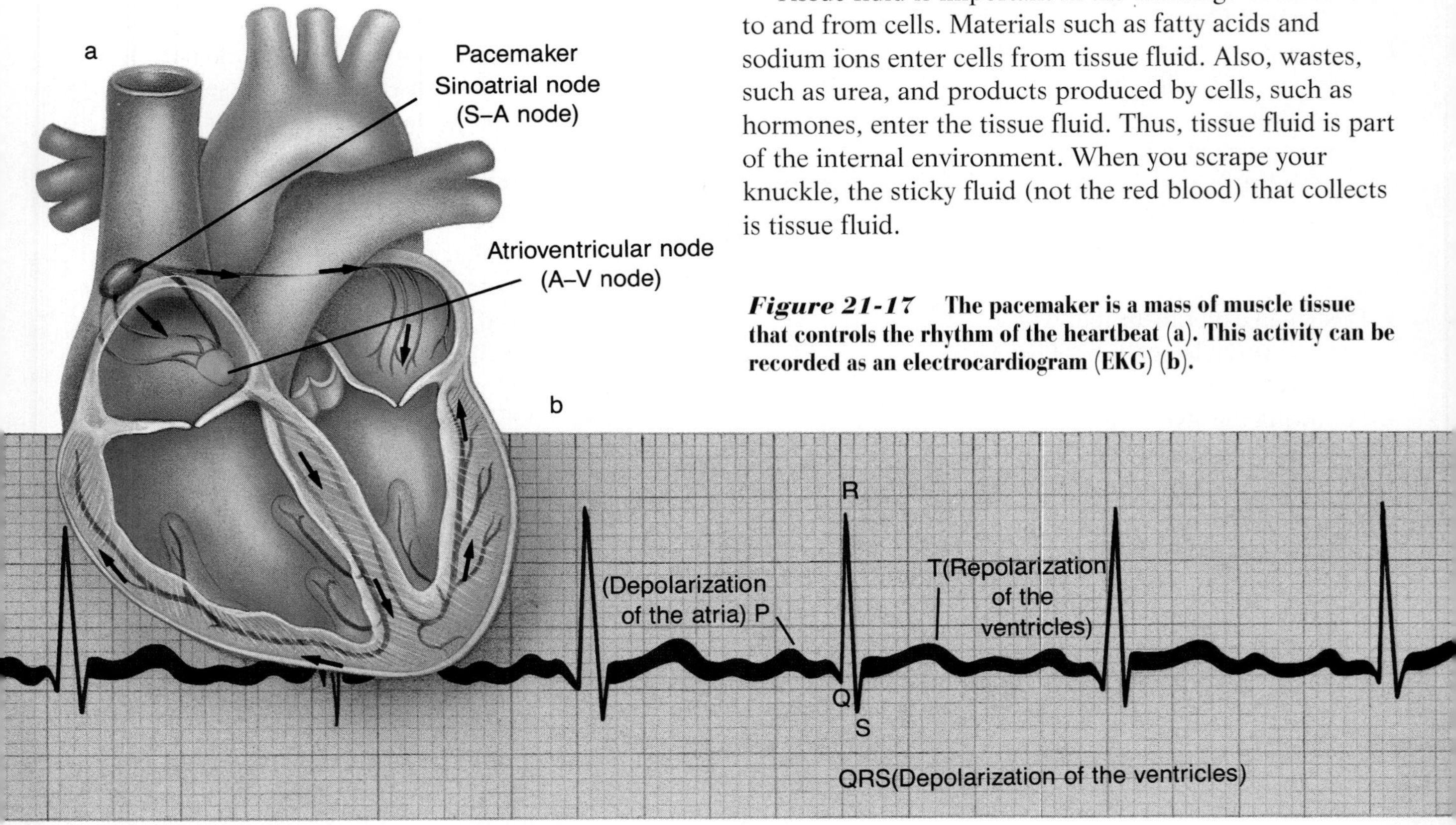

Figure 21-17   **The pacemaker is a mass of muscle tissue that controls the rhythm of the heartbeat (a). This activity can be recorded as an electrocardiogram (EKG) (b).**

The Lymph Network Not all tissue fluid that leaves a capillary returns to it by osmosis. Some of the fluid enters a network of small vessels, called lymph vessels, that are closed at one end. The fluid that collects in lymph vessels is then called lymph fluid. There is no pump for lymph circulation, but random body movements force the lymph toward a major collecting duct located near the heart. This duct empties into a large vein. In this way, the liquid portion of blood lost from within the capillaries is restored to the blood.

Scattered throughout the lymphatic system are **lymph nodes,** small, spongy capsules—more than 100 of them—that appear as bumps along the lymph vessels.

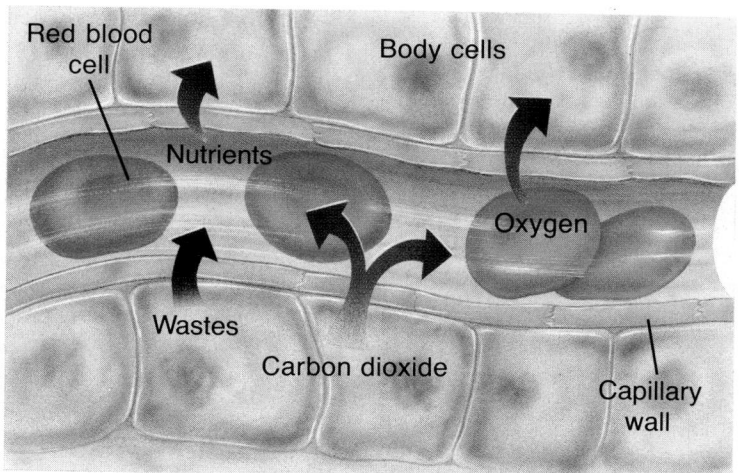

Figure 21-18 Exchange of carbon dioxide and oxygen between the red blood cells and body cells occurs in the thin-walled capillaries.

HISTORY CONNECTION

A Pioneer in Medicine

Today, surgeons can replace diseased heart valves, clear blocked arteries or put in a new artery to bypass the blocked one, implant a pacemaker to make the heart beat regularly, and even transplant a heart from one person to another. Until the middle of this century, however, such heart operations were unheard of.

Dr. Myra Logan was one of the pioneers of heart surgery. Not only was she one of the first surgeons to operate on the heart, she was the first woman ever to do so. In 1951, she became the first African American woman to be elected to the American College of Surgeons, a prominent medical honor society. When Dr. Logan graduated from medical school in 1929, however, a surgeon did all kinds of operations.

During her internship and residency at Harlem Hospital, she did everything from delivering babies to closing stab wounds to the heart. Today, heart surgery is performed by specialists called cardiovascular surgeons.

In addition to performing heart surgery, Dr. Logan continued to perform many other kinds of operations, especially on children. She was one of the first doctors to practice group health maintenance, in which several doctors with different specialties work together. In the 1960s, she conducted an important study on breast tumors and improved the use of X rays to detect breast cancer in women. Dr. Logan was a pioneer not only as a successful female African American, but also as an outstanding medical researcher and a surgeon who served the Harlem community for more than 35 years.

1. **Explore Further:** Breast cancer is the third leading cause of death from cancer. How can a breast X ray, called a mammogram, help prevent breast cancer?
2. **Writing About Biology:** Explain why heart transplant operations are not always successful.

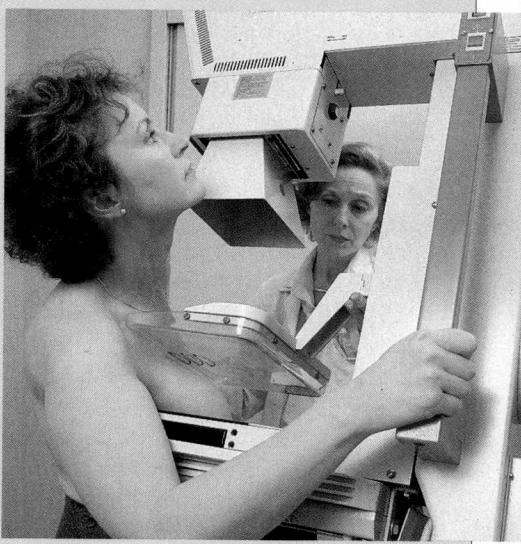

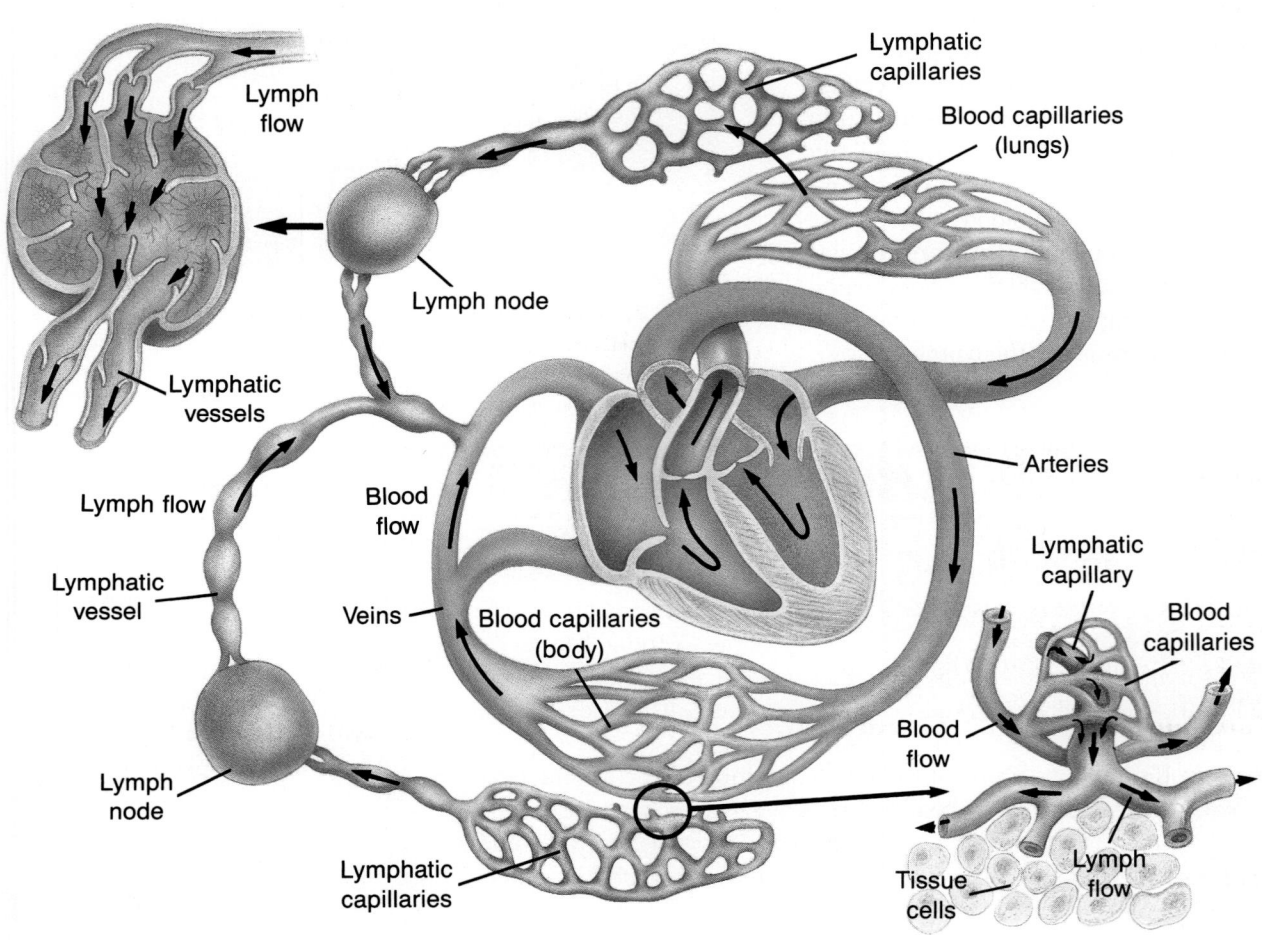

Figure 21-19   Fluid lost from the capillaries to the lymphatic system is later returned to the blood via a duct near the heart.

Destruction of bacteria occurs mainly within the lymph nodes. You will read more about the lymphatic system and its role in immunity in Chapter 23. Your tonsils and adenoids act in much the same way as your lymph nodes. Have you ever experienced swelling in the lymph nodes in your neck? What does such swelling indicate?

Lymph flow

Lymph node

Lymphatic vessels

Lymph flow

Lymphatic vessel

Lymph node

Lymphatic capillaries

Blood flow

Veins

Blood capillaries (body)

Lymphatic capillaries

Blood capillaries (lungs)

Arteries

Lymphatic capillary

Blood capillaries

Blood flow

Tissue cells

Lymph flow

Section Review

Understanding Concepts

1. Why is circulation more efficient in an animal with a closed circulatory system than in an animal with an open circulatory system?

2. For each pair of fish circulatory structures, indicate where the amount of oxygen is higher: ventral aorta—dorsal aorta; ventral aorta—gill capillaries; dorsal aorta—atrium; atrium—ventricle.

3. How is circulation in an amphibian more efficient than in a fish?

4. **Skill Review—Sequencing:** Sequence all the structures through which blood passes as it circulates from your heart to your left foot and back again. For more help, see Organizing Information in the **Skill Handbook.**

5. **Thinking Critically:** Blood flows most slowly in capillaries. How is that rate of flow adaptive?

21.3 The Blood

As your internal environment, your blood has many functions. It transports nutrients such as glucose, fats, amino acids, ions, and vitamins to your body cells. It delivers oxygen, removes carbon dioxide, and carries hormones from the glands where they are formed to the tissues or organs they regulate. Blood passing through your liver picks up a nitrogenous waste called urea, which is carried via the bloodstream to the kidneys, where it is removed from the body. Blood also has a major role in protecting the body from infection.

Gas Transport

A microscope reveals that a drop of human blood is a pale yellow fluid, called plasma, that contains a variety of cell-like structures as shown in Table 21.1 below. These structures differ from most other body cells and include red blood cells, white blood cells, and platelets, which are cell fragments. All are produced from central bone tissue called marrow—a mesh of delicate connective tissue and blood vessels. Blood cells are formed in the marrow of long bones, such as the thighbone, and flat bones, such as ribs and the breastbone. The characteristics of blood components are outlined in Table 21.1.

Objectives

Compare the structure and function of components of the blood.

Define the roles of white blood cells and platelets.

Predict the outcomes of mixing the four blood types of humans.

Table 21.1 Blood Components

Component	Characteristics
Red blood cells	biconcave, disk-shaped cells with no nuclei; contain hemoglobin, transport oxygen and some carbon dioxide
White blood cells	five types; shape is round with a nucleus; some fight infection; some control immunity
Platelets	cell fragments that liberate substances necessary for clotting blood
Proteins	some are involved with clotting activities; some are antibodies
Hormones	chemicals secreted by glands; control and coordinate the activities of the body
Urea	waste product formed by liver; transported to kidneys, where it is filtered out of the body
Glucose, amino acids, fats, vitamins, minerals, lipids	nutrients transported to all cells and tissues; necessary for metabolism

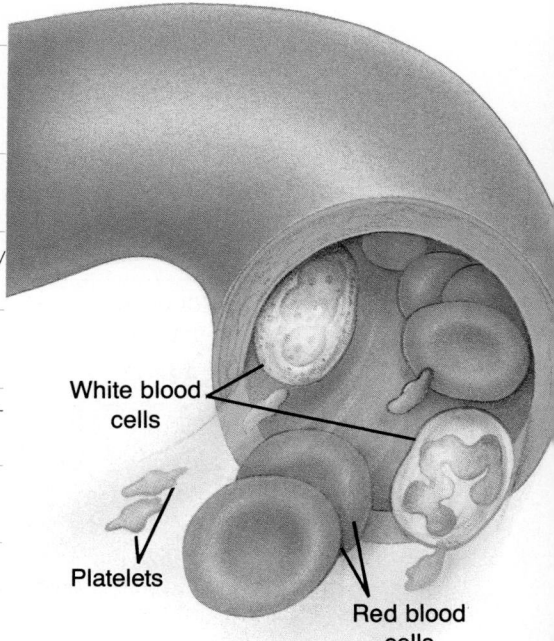

White blood cells

Platelets

Red blood cells

The Work of Red Blood Cells　The most numerous of all kinds of blood cells are the **red blood cells.** About 5 million red blood cells are present in a cubic millimeter of blood. Disk-shaped and concave on each side, these blood elements lose their nuclei as they age. Red blood cells carry an iron-containing protein molecule—hemoglobin—that functions in the transport of oxygen. In the lungs, oxygen combines with hemoglobin to form the compound oxyhemoglobin, which makes the blood bright red. Oxyhemoglobin is then transported to the heart and the body cells.

In the capillaries, oxyhemoglobin breaks down into hemoglobin and oxygen. The liberated oxygen then diffuses from the red blood cells to the body cells. As oxygen leaves the blood, the blood turns a dark red, or maroon. At the same time, carbon dioxide diffuses from body cells into the capillaries. Some of the carbon dioxide forms ions in the plasma. Some combines with hemoglobin. The remaining carbon dioxide dissolves in the liquid within the red blood cells. The blood then returns to the lungs, where the carbon dioxide is released and oxygen is picked up.

Thinking Lab

INTERPRET THE DATA

What are the effects of iron deficiency?

Background

Have you ever heard the expression "tired blood"? This expression refers to iron deficiency in blood. Iron is an important element of the hemoglobin molecule. Without enough iron, the body cannot make adequate amounts of hemoglobin, which is the molecule that carries oxygen to body cells. Iron deficiency also interferes with the exchange of oxygen and carbon dioxide in muscle cells.

The Harvard step test is designed to test physical ability. The subject steps up and down from a platform for a specified time, after which the pulse is measured. This pulse rate after exercise can be correlated to the body's ability to do physical work.

You Try It

Examine Graph A below, which demonstrates the results of the Harvard step test on a small population of people. **Interpret** the results. Look at Graph B, which compares available iron content of different foods, and describe a meal that might help someone with an iron deficiency.

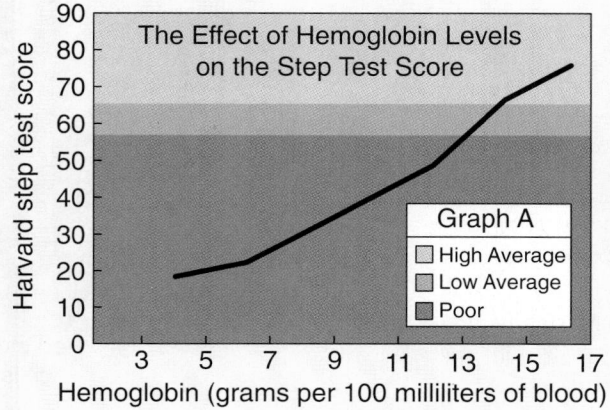

The Effect of Hemoglobin Levels on the Step Test Score

Graph A
- High Average
- Low Average
- Poor

Harvard step test score (y-axis: 0–90)
Hemoglobin (grams per 100 milliliters of blood) (x-axis: 3, 5, 7, 9, 11, 13, 15, 17)

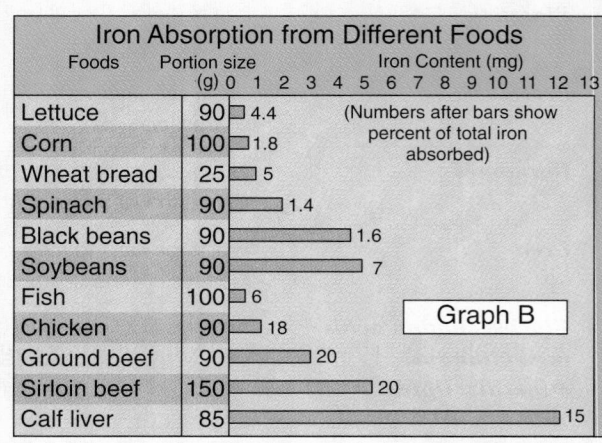

Iron Absorption from Different Foods

(Numbers after bars show percent of total iron absorbed)

Graph B

Foods	Portion size (g)	Iron Content (mg)
Lettuce	90	4.4
Corn	100	1.8
Wheat bread	25	5
Spinach	90	1.4
Black beans	90	1.6
Soybeans	90	7
Fish	100	6
Chicken	90	18
Ground beef	90	20
Sirloin beef	150	20
Calf liver	85	15

Anemia   Red blood cells live for about three months. After they die, they are filtered from the blood by the liver and another abdominal organ, the spleen, which also acts as a reservoir for blood. Normally, new red blood cells are produced in the bone marrow at the same rate as old ones die. However, if more are destroyed than are produced, a type of anemia may result. Anemia also may be caused by lack of iron or hemoglobin in blood or by a lowering of blood volume following a serious accident. Foods rich in iron or injections of iron-containing drugs may help correct certain forms of anemia. Do you recall the cause and symptoms of sickle-cell anemia?

Healing and Protection Against Disease

Blood has other functions besides transport. Among the functions are repair of wounds and defense against disease.

Healing Wounds   Most people assume that a cut will heal quickly. But a cut blood vessel would be fatal if the flow of blood did not stop. Cuts are repaired by the clotting of blood, which depends on the blood components called **platelets.** Platelets are cell fragments that arise from the breaking apart of very large cells produced in the bone marrow. The colorless, sticky, disk-shaped fragments lack nuclei and live for only about ten days. A cubic millimeter of blood contains up to 400 000 platelets.

When an injury occurs, platelets in the spilled blood stick to the edges of the wound and clump together to plug the gap. Contact with the jagged edges of injured tissue causes the platelets to burst open and release a protein substance. The injured tissue also releases chemicals that trigger a set of reactions in the blood. The chain reaction of a protein in the blood activating another protein continues until a protein called fibrin is produced. Fibrin is insoluble, so it settles out of the blood as a tangled web of threads, as seen in Figure 21-20. As a spiderweb traps insects, this web traps other blood parts and forms a blood clot that seals the wound, preventing excessive loss of blood and protecting against the entry of microorganisms. As the wound heals, the clotted material dries and forms a hard scab that protects the tissues from infections. Do you remember in what genetic disease the blood fails to clot properly?

Protecting Against Disease   The clotted blood protects not only against cuts but also against disease-causing viruses, microorganisms, and some parasites. Protection against these foreign invaders involves another kind of blood component, the **white blood cells.**

There are five types of white blood cells. They differ from the other components of blood in that they each have a nucleus. Unlike the red blood cells, they are colorless. White blood cells play essential roles in the body's immune system, which you will study in Chapter 23. Normally, a cubic millimeter of blood contains up to 10 000 white blood cells. During an infection, however, the number may rise to 25 000 or more as the body mobilizes its defenses against invading organisms.

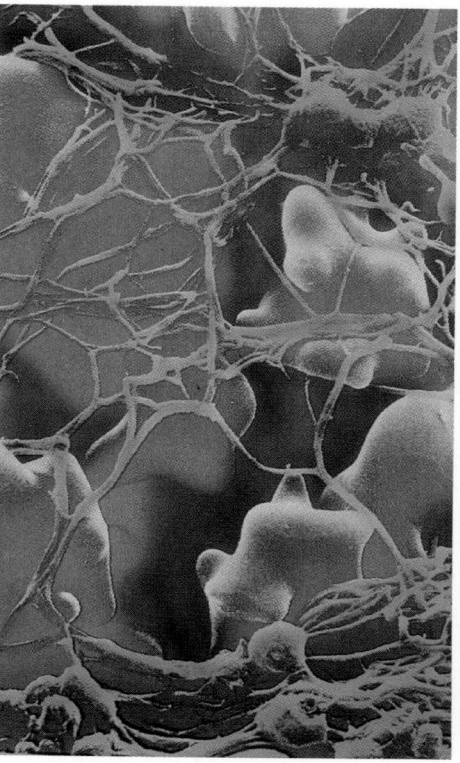

Figure 21-20   **When exposed to air, blood will congeal and solidify rapidly. Chemicals from platelets react with protein in the plasma to form thrombin. Thrombin precipitates the protein called fibrin, which forms a tangled web and seals the wound. What do you think causes the medical condition called thrombosis?**

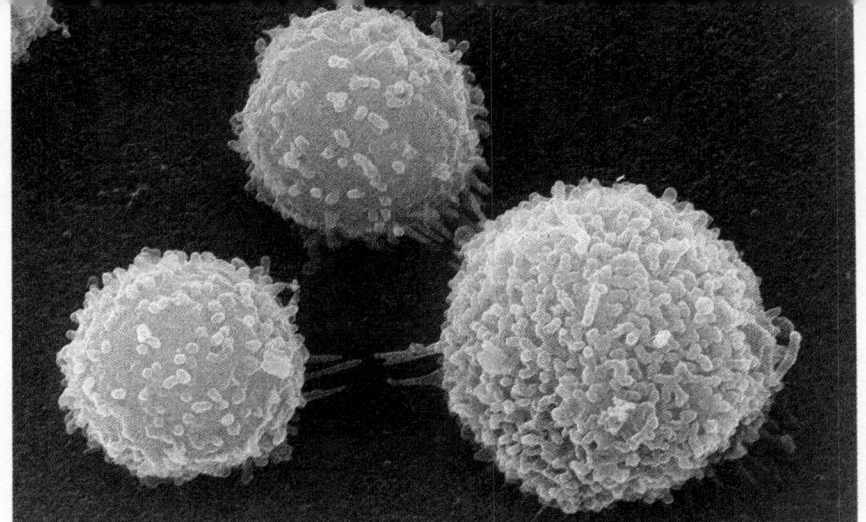

Figure 21-21 White blood cells are formed within lymph nodes and bone marrow and patrol the body within lymph tissue and blood to guard against foreign invaders.

Blood Types

Blood types, as you read in Chapter 8, are genetically determined by three alleles. The four blood types are A, B, AB, and O. Differences in blood type are due to the presence or absence of proteins, called **antigens,** on the membranes of the red blood cells. Antigens are normally foreign substances that stimulate an immune response in the blood. Type A blood has A antigens, and type B blood has B antigens. Type AB blood has both A and B antigens, and type O blood contains no antigens.

The liquid part of blood—plasma—may contain other proteins whose shapes match those of the different antigens like matching pieces of a jigsaw puzzle. These proteins are called **antibodies.** Antibodies that attack type A antigens are labeled anti-A antibodies. Antibodies that attack type B antigens are called anti-B antibodies. When the antibodies in the plasma are mixed with matching antigens on the red blood cells, the cells clump together. Any antigen not present in a blood type will be recognized as a foreign substance if it is introduced by transfusion.

If you had to receive a transfusion of blood, your doctor would be concerned mainly about the effect of your plasma with its antibodies on the donor's red cells, because your antibodies by locking onto their matching antigens can cause clumping of the transfused donor cells. The antibodies

Table 21.2 **B**lood Types and Transfusion Possibilities

Type	Antigens	Antibodies	Can Receive	Can Donate to
A	A	Anti-B	O, A	A, AB
B	B	Anti-A	O, B	B, AB
AB	A, B	none	all	AB
O	none	Anti-A Anti-B	O	all

in the donor's plasma will not affect your blood cells as much because they will be greatly diluted by your own plasma. This is the reason why type O blood, which has no antigens but has both anti-A and anti-B antibodies, can be transfused into all persons no matter what their blood type. People with type A blood have type anti-B antibodies. If type B blood is mistakenly transfused into a person with type A blood, clumping in the recipient's blood will occur. Such clumping may plug up blood vessels, causing the recipient's death. The reverse is also true; that is, type B blood has type anti-A antibodies. Clumping of the recipient's blood will occur if type A blood is given to a person with type B blood. People with type AB blood have neither type of antibody. Transfusion possibilities are summarized in Table 21.2.

Because people with type AB blood have no antibodies, they can, therefore, receive any blood type. Such people are known as universal recipients. People with type O blood are known as universal donors because their red blood cells have no antigens. If no foreign antigens are present, clumping will not occur. Thus, when transfusions are performed, it is important that no foreign antigens be introduced.

LOOKING AHEAD

As you have read, transport within organisms is accomplished in a variety of ways. Cells must take in nutrients and oxygen for life processes, and waste by-products from these processes must be removed. What prevents so much material from entering the cells that they burst? What prevents so much material from leaving the cells that they collapse? And once the wastes leave a cell, how are they removed from the organism? In the next chapter, you will explore how osmotic balance is maintained, how gases are exchanged, and how cell wastes are removed from the body.

Section Review

Understanding Concepts

1. Why do you think the number of red blood cells is so much greater than the number of other blood components?

2. Sometimes a blood clot forms on the inner wall of a blood vessel and then breaks free. How might that event be dangerous?

3. Explain what will happen if a person with type O blood receives a transfusion of type B blood.

4. **Skill Review—Making Tables:** Make a table comparing the components of blood in terms of their structure, number, and function. For more help, see Organizing Information in the **Skill Handbook.**

5. **Thinking Critically:** Why are people with certain immune system problems sometimes given a bone marrow transplant?

Biotechnology

Looking Inside Your Body

A new way of operating without making large cuts into your body is no longer science fiction. Already, doctors can look inside organs such as your heart by making a tiny cut and threading optical fibers through your blood vessels.

Viewing with a Fiberscope   The first fiberscope was tested in 1957. A fiberscope has two bundles of optical fibers made of very pure glass called silica. One bundle transmits light to the body tissue being examined. The other bundle sends back an image to an eyepiece, camera, or TV screen. Each bundle contains about 10 000 optical fibers and measures less than 1 mm in diameter. The two bundles are wrapped in a flexible catheter thin enough to fit into the blood vessels.

A fiberscope lets doctors observe your body at work. They can watch your heart valves open and close and spot cholesterol in your arteries. A light reflected from the tip of the catheter can tell them how much oxygen is in your blood and how fast the blood is flowing. By adding special sensors called optrodes, doctors can measure the pressure in your arteries and the pH of your blood. They can also peer into your lungs, your stomach and intestines, and other organs.

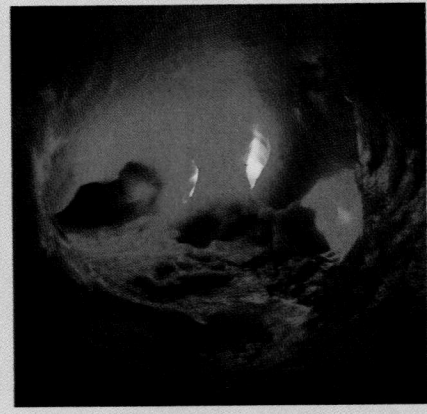

Operating with an Endoscope   A fiberscope is often part of a larger instrument called an endoscope. Endoscopes contain other fibers through which doctors can withdraw fluids or perform surgery. An entire endoscope can measure as little as 2.5 mm in diameter—about a third of the diameter of a pencil.

Endoscopes are used in laser surgery. Laser light is a form of radiant energy. Low-power lasers cause blood to coagulate by gentle heating, while high-power lasers boil tissue away. During laser surgery, high concentrations of laser energy are beamed through optical fibers made of quartz or crystals onto the tissue to be treated.

Cleaning the Arteries   Laser surgery offers a new way to treat heart disease. Blood clots and plaque in the arteries reduce blood circulation to the heart and brain. If circulation is cut off completely, a heart attack or stroke occurs. Until recently, open heart surgery was the only way to open clogged arteries. Now doctors are testing a procedure called laser angioplasty. A laser beam is transmitted through an optical fiber to a metal tip, which heats up and melts the plaque or blood clot.

1. **Issue:** Why is laser angioplasty not the perfect solution to atherosclerosis?

2. **Writing About Biology:** Describe other technologies that doctors use to view what is going on in your body.

More than twice as many Americans die from diseases of the heart and blood vessels than from any other cause. The risk of heart disease rises as you get older, but the early stages can start in your teens. The culprit is atherosclerosis, a buildup of fat deposits on the artery walls.

Clogged Arteries Your liver removes fats called cholesterol from your blood, but if there is too much cholesterol for your liver to handle, the excess stays in your blood. Some of it seeps into the lining of your arteries. White blood cells come along, gobble up the cholesterol, and then swell and get trapped in the artery wall. A small mass of these bloated white cells causes a fatty streak.

As more and more bloated white blood cells clump together, they create a bulge on the inside of the artery and damage the lining of the artery wall. Other cells move in to repair the damage, but they end up turning the bulge into a buildup of fats and dead cells, called plaque. The artery wall becomes hard. Less blood can flow through the artery, so your heart doesn't get as much oxygen. You are now at risk for a heart attack or stroke.

Hypertension Some people are more prone to atherosclerosis, but there are things you can do now to help reduce your risks. High blood pressure, called hypertension, is thought to be the biggest risk factor in heart disease. An average normal blood pressure is around 120 over

80. Diastolic blood pressure (the bottom number) higher than 90 signals hypertension. High blood pressure causes the walls of the arteries to become thick and hard, although exactly how is not fully known. You can greatly reduce your risks of developing hypertension by keeping your weight down, eating a low-salt diet, and getting regular exercise.

High Cholesterol Research studies show that high cholesterol is directly linked to heart disease. High cholesterol leads to a buildup of plaque in your arteries. The best way to keep your cholesterol low is to eat low-fat foods. That means cutting down on hamburgers, milk shakes, potato and corn chips, fried foods, ice cream, and cheese.

Smoking Smoking is another major risk factor in heart disease. More than 170 000 people die each year from the damage cigarette smoke causes to their arteries. Smokers are two to four times more likely to have heart attacks than nonsmokers, and their heart attacks are more likely to end in death. Researchers have found that cigarette smoke increases the buildup of plaque in your arteries. Ten to 15 years after you stop smoking, your risk of dying from a heart attack approaches the same level as that of a nonsmoker.

Control of Risk Factors If you don't smoke and don't have high blood pressure or high cholesterol now, are you home free? Not totally. Several other factors play an indirect role in atherosclerosis.

Obesity is one. Extra pounds strain your heart muscles and often cause your blood pressure and cholesterol to go up. Obesity can also cause the breakdown of pancreatic tissues resulting in adult diabetes. Diabetes raises cholesterol levels. Lack of exercise is another risk factor. Exercise lowers blood pressure and cholesterol and strengthens your heart muscles.

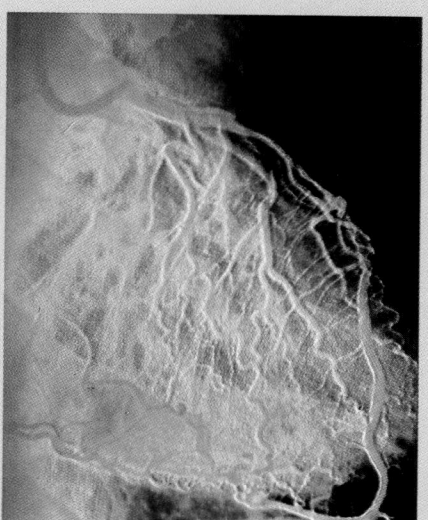

You can't absolutely prevent heart disease, but exercising, eating a low-fat, low-salt diet, and not smoking can greatly reduce your risks. You'll feel better and look better, too.

1. **Writing About Biology:** Describe some of the risk factors of heart disease that you can't control.
2. **Explore Further:** Why is taking aspirin thought to help prevent atherosclerosis?

Summary

Cells take needed materials from and release wastes into their external environment. Unicellular organisms and cells of many simple multicellular organisms can directly exchange materials with their external environment. However, more complex organisms cannot do so. They have transport systems in which internal fluids carry materials to and from their cells, and which link them to the external environment.

In vascular plants, water and dissolved minerals are pulled upward in xylem vessels as a result of transpiration. Due to pressure differences, organic molecules made by these plants move either up or down within phloem sieve cells.

In animals that have a transport system, blood circulates around the body, returning to the heart, which has a pumping action that keeps the blood moving. Circulatory systems may be open or closed.

Humans have a four-chambered heart that rapidly moves fully oxygenated blood to body cells. Besides gas transport, the blood cells also function in clotting and defense against infectious disease. Different blood types are due to presence or absence of antigens on red blood cell membranes.

Language of Biology

Write a sentence that shows your understanding of each of the following terms.

antibody	red blood cell
antigen	root hair
aorta	sieve tube
artery	sinoatrial (S-A) node
atrioventricular (A-V) node	stoma
atrium	transpiration
capillary	vein
guard cell	ventricle
lymph node	vessel
phloem	white blood cell
platelet	xylem

Understanding Concepts

1. Trace the path of water from soil to the leaves of a vascular plant. Explain how that water moves into and up through the plant.
2. Trace the movement of glucose produced in a leaf to a bud on a higher branch and explain how the sugar is transported.
3. In general, how is the transport system of an animal different from that of a vascular plant?
4. Suppose you were viewing a cross section of vascular tissue from a root, stem, or leaf. How would you distinguish between xylem and phloem?

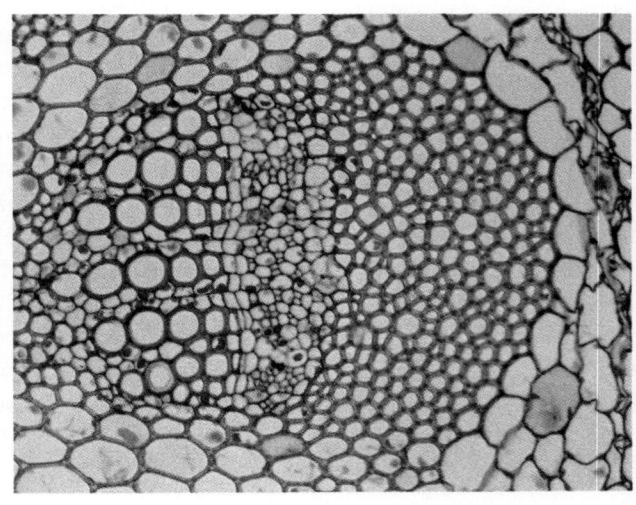

5. Identify the two adaptations in the circulatory system of humans that enable fully oxygenated blood to be quickly delivered to body cells.

6. Why are the walls of atria much thinner than those of ventricles?

7. Explain the origin and importance of tissue fluid.

8. How is your blood adapted to preventing blood loss and infection after the vessels are damaged?

Applying Concepts

9. Describe and explain the effects of a birth defect in which the infant had a hole between either the left and right atria or the left and right ventricles.

10. What blood type could be given safely to a person in an emergency? Explain.

11. Compare and explain the size of a stoma on a humid day and a dry day.

12. Can you suggest a reason why a soldier standing at attention for a long period of time might faint?

13. A roundworm, *Filaria,* is a parasite that lives in lymph vessels. Suppose many of these worms were in the lymph vessels of a person's leg. Describe and explain how the leg would be affected.

14. Oxyhemoglobin releases more oxygen at higher temperatures than at lower ones. How is that property important to muscle tissue?

15. Describe and explain the probable location of stomata on a floating lily pad.

16. An artificial blood that carries oxygen has been developed. Explain why this blood could not replace normal blood.

17. **Biotechnology:** How can a laser be used to view tissues and also to cut tissues?

18. **Health Connection:** How is plaque in an artery different from plaque on your teeth?

Lab Interpretation

19. **Investigation:** You go for a checkup by your doctor, who tells you that your blood pressure has gone up. What are some possible causes of this increase?

20. **Thinking Lab:** The following are Harvard step test measurements of adults from various countries. Which countries might have adults with a diet low in red meat? Explain.

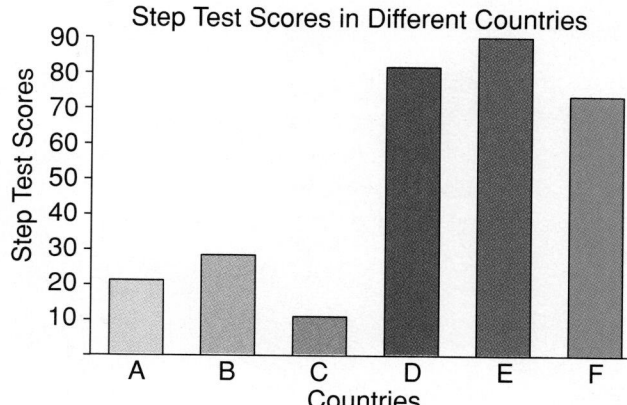

Step Test Scores in Different Countries

21. **Minilab 2:** Examine the abnormal EKG shown here. What part of the heart might be affected? Explain.

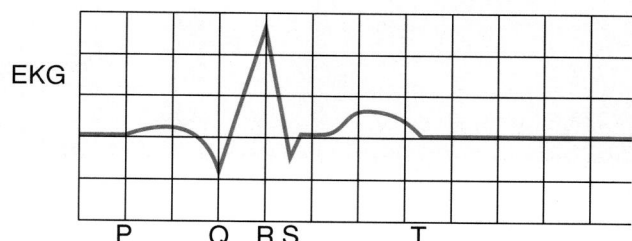

Connecting Ideas

22. In a human fetus, most blood passes directly from the right atrium to the left atrium, and some blood passes through a duct from the pulmonary artery to the aorta. As a result, only a small amount of blood passes to the lungs. Explain the reason for this circulatory pattern.

23. Trace the pathway of blood in a human from the heart to the small intestine and back again. What materials are exchanged along the way and where does the exchange occur?

CHAPTER
22

GAS EXCHANGE AND EXCRETION

ARE THESE FISH puzzled by the odd equipment this diver is wearing? It's not likely, because they have probably never seen a human without it. What is the purpose of all that equipment? If the diver could talk with that thing in his mouth, he would explain that the mouthpiece is the part of the scuba equipment that controls the volume of air leaving the air tanks and entering the lungs. The deeper the diver goes, the greater the water pressure will be and the more air the lungs will require. Increased air pressure in the lungs helps equalize the pressure inside and outside the body.

Why don't fish release bubbles under water the way a human diver does? Don't they both have to breathe? Of course they do. But the fish is adapted to take oxygen from the water and return carbon dioxide to the water. Humans are adapted to take oxygen from the air and return carbon dioxide to the air. A human diver, therefore, has to carry air along in tanks in order to breathe under water, and the water bubbles as the diver exhales carbon dioxide from his lungs. The only way a fish could survive away from its underwater home would be to take along a water tank—one that is built in, not one that it carries on its back. Before you finish this chapter, you will understand why fish and people breathe differently. You will also see how organisms are adapted to getting rid of bodily wastes and maintaining osmotic balance in different environments.

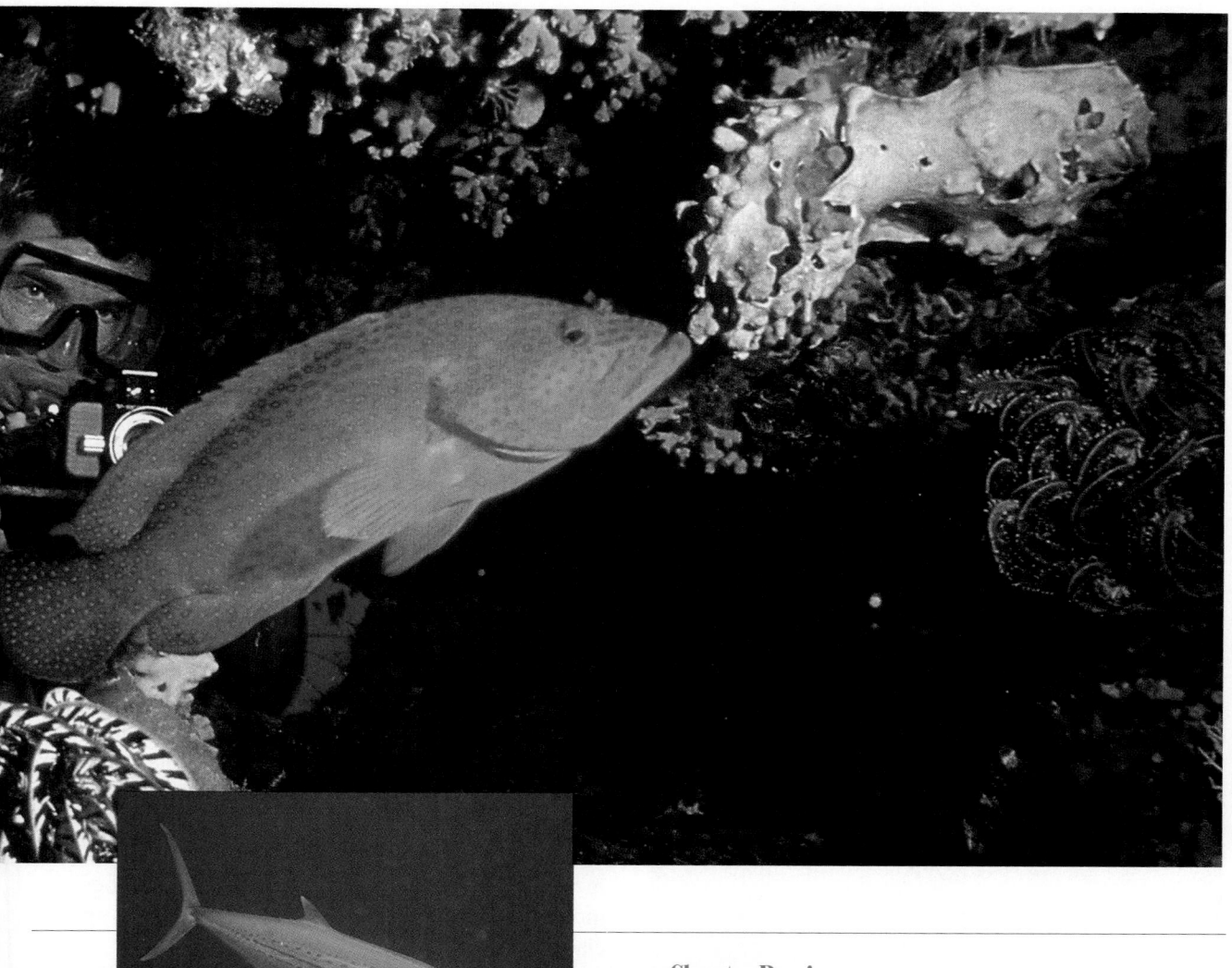

Mackerel live in the Atlantic Ocean. They swim with their mouths open. This forces water to flow into their mouths and out the gills. What would happen if the mackerel in the photo were placed in an aquarium with limited space? Why?

Chapter Preview

Objectives

Discuss how gases are exchanged in vascular plants.

Compare respiration in segmented worms and bony fish.

Describe respiration in humans.

Almost every organism on Earth lives in close relationship with two gases—carbon dioxide and oxygen. Most autotrophs take carbon dioxide from their environment and use it as one of the raw materials for photosynthesis. They release oxygen as a by-product of photosynthesis. In addition, nearly all autotrophs and all heterotrophs require oxygen for cellular respiration. As they carry out the respiration process, organisms take in oxygen from their physical surroundings and return carbon dioxide in a continuous process. This general back-and-forth movement of carbon dioxide and oxygen between organisms and their environment is known as gas exchange. All organisms, from the simplest to the most complex, meet certain requirements in order to accommodate gas exchange.

Simple Organisms vs. Complex Organisms

In order for a gas to pass into or out of a cell, it must be dissolved in a liquid. Think for a moment about where unicellular and simpler multicellular organisms are found. They live either in water, in damp areas on land, or within the moist environment of a host. Every cell among these organisms lives in direct or close contact with a damp environment. Each cell exchanges gases directly with the liquid that surrounds it.

Among larger organisms, gas exchange cannot occur as readily as among simple life forms. Many do not live in watery environments. And even if they do, most organisms' bodies are too thick for each cell to exchange gases directly with the external environment. To accommodate gas exchange among the more complex forms of life, certain body areas specialized in gas exchange have evolved.

Body tissues in gas exchange areas are moist, which takes care of the requirement that gases be dissolved in liquid. Gas exchange areas also have large surfaces to ensure that the volume of gases entering and leaving the organism is sufficient to meet its

Figure 22-1 Among organisms, there is a wide variety of structures designed for gas exchange. Most land organisms such as humans have protected internal respiratory structures. Small aquatic organisms exchange gases directly with their environment.

needs. Tissue membranes that make up the gas exchange surfaces are usually thin, so that gases can easily cross over. Because these tissues are so delicate, the bodies of most large, complex animals have evolved structures that shield the gas exchange area from possible injury. Finally, many complex organisms have further specialized structures that work together as a transport system. The job of the transport system is to carry the gases back and forth between the exchange area and other body cells.

Gas Exchange in Plants

Most plants could be considered large organisms, especially some vascular plants like trees. Yet plants do not have a specialized gas transport system. There is no need for one. Gas exchange occurs at many different places within plant structures.

Photosynthesis may be the first process in which gas exchange occurs that you think about when it comes to plants. Leaves of plants do take in carbon dioxide and give off oxygen during photosynthesis. But leaves, and other parts of plants, also do the reverse. They take in oxygen and give off carbon dioxide during cellular respiration. As in all organisms, cellular respiration occurs in plants at all times, in sunlight and in darkness. Photosynthesis, on the other hand, occurs only in sunlight.

Gas Exchange in Roots Plant roots extract some mineral nutrients from water by active transport. Roots get the energy they need for active transport and for other processes from cellular respiration. The oxygen needed for cellular respiration is in the air spaces between soil particles. This oxygen dissolves in the moisture of the soil and then diffuses into the root hairs. Further diffusion distributes the oxygen to other root cells, such as the cortex. As cellular respiration occurs within the root cells, carbon dioxide is produced. The carbon dioxide diffuses from the root cells to the root hairs and then to the soil.

Figure 22-2 **Plants obtain oxygen from water they absorb through their roots. Root hairs increase the surface area of the root system, facilitating gas exchange.**

Gas Exchange in Stems   Most of a tree's stem, which we usually call the trunk, consists of dead xylem cells. That's what wood is. But some cells, namely those in the bark region, are alive and so need oxygen for respiration. Although a tough barricade made up of dead cells and a waxy substance surrounds the stem to protect it and prevent evaporation, oxygen can still get in through tiny openings called lenticels. After entering the stem via these lenticels, oxygen travels by diffusion through tiny air spaces and then diffuses across the moist membranes of nearby cells. Carbon dioxide travels and finally exits in the opposite direction.

Gas Exchange in Leaves   In the discussion of photosynthesis in Chapter 21, you read how carbon dioxide and oxygen enter and exit a leaf through stomata. The actual exchange of gases takes place in the cells of the palisade and spongy layers, headquarters for photosynthesis. A thin film of moisture surrounding the palisade and mesophyll cells dissolves the gases so they can pass in and out of the cell.

Palisade and spongy cell layers are also responsible for a lot of the leaf's cellular respiration, which occurs at the same time that photosynthesis occurs. The cells can use oxygen produced by photosynthesis for cellular respiration, and carbon dioxide released during cellular respiration for photosynthesis. Some of the oxygen produced by photosynthesis is used in cell respiration. Additional oxygen enters the leaf through stomata; carbon dioxide departs through the same openings.

Figure 22-3   Gases can be exchanged in a plant stem through tiny openings called lenticels. Plants also exchange gases through the cells of the palisade layer and spongy layer of the leaf. The stomata also allow gases to be exchanged in a leaf.

Animal Respiratory Systems

Cellular respiration is vital for animals just as it is for plants. In contrast to plants, however, complex animals require a transport system to move around the gases involved in cellular respiration. Complex animals have evolved complete respiratory systems with the sole purpose of acquiring oxygen from the external environment, distributing it throughout the body, and discharging carbon dioxide. The work of animal respiratory systems is simply referred to as **respiration.**

A gas exchange area, where specialized cells exchange gases with the external environment, is part of an animal's total respiratory system. Commonly, but not always, blood is the transport vehicle. Certain blood cells, traveling the routes of the blood vessels, carry oxygen to all body cells. Oxygen diffuses into the cells; carbon dioxide diffuses out of the cells. The carbon dioxide is carried back to the exchange area and then is released into the external environment. Because of the requirement that the exchange area be moist, the type of respiratory system an animal has is related to the environment in which it lives. Animals that live in watery environments are adapted to obtain oxygen from water. Those that live in dry air environments are adapted to obtain oxygen from the atmosphere.

Respiration in Segmented Worms

The earthworm lives in moist soil. In this damp environment, it might seem appropriate that the earthworm's gas exchange area be located somewhere close to its outer surfaces. And such is the case. The entire skin area of the earthworm, in fact, is its gas exchange area. Thus, the earthworm's respiratory solution meets the requirement that the exchange area be large as well as moist.

Mucus secreted by the worm, as well as water in the soil, maintains a film of moisture all over the earthworm's skin. Oxygen from water in the soil dissolves in the film, diffuses across the thin skin, and enters capillaries. The blood transports the oxygen to all body cells, exchanges it for carbon dioxide, and transports the carbon dioxide to the skin. From the skin, the carbon dioxide diffuses into the soil.

Figure 22-4 Some species of segmented worms, such as this marine fireworm, have parapodia—thin, paddle-like, bristled extensions of the body where gas exchange occurs. How do you think the worm benefits from having these specialized structures for gas exchange?

Respiration in Bony Fish

Like earthworms, aquatic animals such as mollusks, certain aquatic arthropods, echinoderms, and fish spend their lives surrounded by liquid. Aquatic animals, too, have gas exchange areas located near body surfaces. Their exchange areas, however, are not spread over their entire outer body surfaces. Instead, their exchange areas are concentrated in organs called **gills.** Gills vary in appearance among different aquatic animals, but most often they have a delicate, feathery structure.

INVESTIGATION

Carbon Dioxide Levels

Is there a difference in the chemical makeup of air inhaled versus air exhaled? A more specific question might be, does inhaled air contain more, the same, or less carbon dioxide than air exhaled? To help answer this question, a chemical called bromothymol blue can be used to detect carbon dioxide. This chemical, as its name suggests, is normally blue. However, if carbon dioxide gas is added, bromothymol blue changes color progressively from blue to aqua to green to yellow as the amount of carbon dioxide increases.

The two pieces of equipment supplied by your teacher allow you to pass either your inhaled or exhaled air through the bromothymol blue solution. Any differences in the concentration of carbon dioxide gas in inhaled versus exhaled air can be observed.

Working in groups of two, form a hypothesis that could be tested.

Problem
Is there a difference in carbon dioxide concentration between inhaled and exhaled air?

Hypothesis
What is your group's hypothesis?

Safety Precautions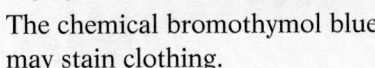
The chemical bromothymol blue may stain clothing.

Experimental Plan
1. Decide on a way to test your group's hypothesis. Keep the available materials provided by your teacher in mind as you plan your procedures.
2. Remember to test only one variable at a time and use suitable controls.
3. Record your procedure and list materials that you will need.

Checking the Plan
1. What controls will be used and what variables will have to be controlled?
2. What data will you collect and how will it be recorded?
3. How will you determine if there is a difference in carbon dioxide amounts between inhaled and exhaled air?

In bony fish, oxygen-rich water enters through the mouth and passes back to the throat or pharynx. From the pharynx, the water travels to the gills. As water is taken into the mouth, the bony gill cover closes. Contraction of the pharynx forces water over the gills, the gill cover opens, and water is expelled. You have probably seen the gill cover opening and closing if you have observed a bony fish in an aquarium.

Each arch of the gill has rows of feathery gill filaments. The surface area of these filaments is further increased by tiny extensions called lamellae that protrude from the filaments. The gill filaments and lamellae house

4. How many trials will your experiment need?
5. Assign roles for this investigation.

Make sure your teacher has approved your experimental plan before you proceed further.

Data and Observations

Carry out your experiment. Prepare a data chart to record your findings.

Analysis and Conclusions

1. What is the role of bromothymol blue in this experiment?
2. What type of control was used?
3. What are the variables?

4. Two terms are used to describe data, *quantitative* and *qualitative*. Quantitative refers to data presented with numbers (quantities). Qualitative data usually is presented with a general description that uses words. An example might be: thick, moderate, or thin (quality).
   (a) Was your experiment qualitative or quantitative? Explain why.
   (b) Suggest a way that the experiment might be changed to make the data the opposite of (a).

5. Explain how your experimental findings tend to confirm what you already know about the processes of cellular respiration and external respiration.
6. Write a conclusion to this investigation.

Going Further

Based on this lab experience, design an experiment that would help to answer another question that arose from your work. Assume that you will have limited resources and will be conducting the lab in class. Include this in your conclusion.

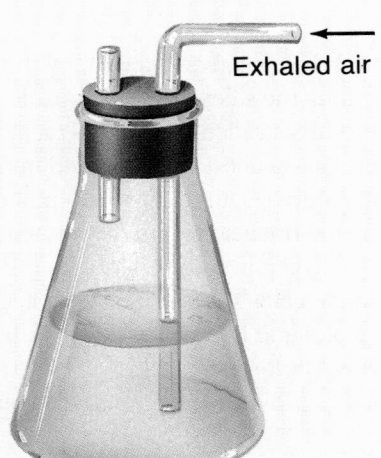

a

Exhaled air

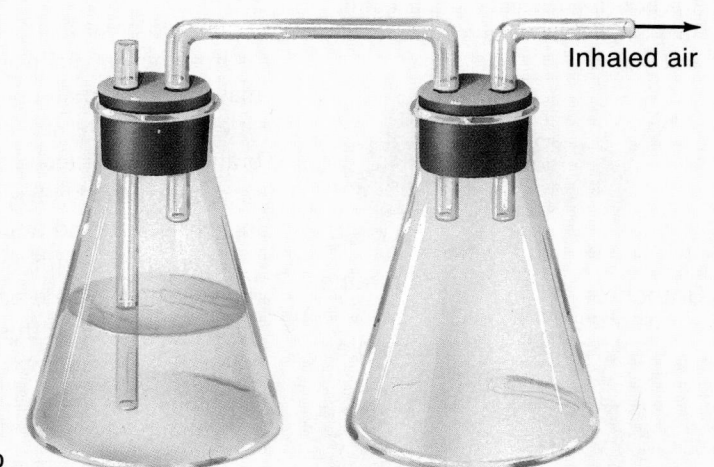

Inhaled air

b

a broad network of capillaries, which provides a very large surface area for gas exchange. The gills lie very close to the heart of the fish. Blood low in oxygen leaves the heart and travels to the gill capillaries. As water passes over the gills, oxygen dissolved in the water diffuses into blood in the capillaries, and carbon dioxide diffuses from blood to the water. Oxygen-rich blood is transported from the gills to all parts of the body, where exchange of oxygen and carbon dioxide takes place.

Figure 22-5   In bony fish, the gills are covered by a bony flap called an operculum. The operculum of this fish (a) is raised, exposing the gills. As water passes over the gill filaments, oxygen enters the blood in the capillaries and is circulated to the body of the fish (b,c). How do you think CO_2 is eliminated from the blood?

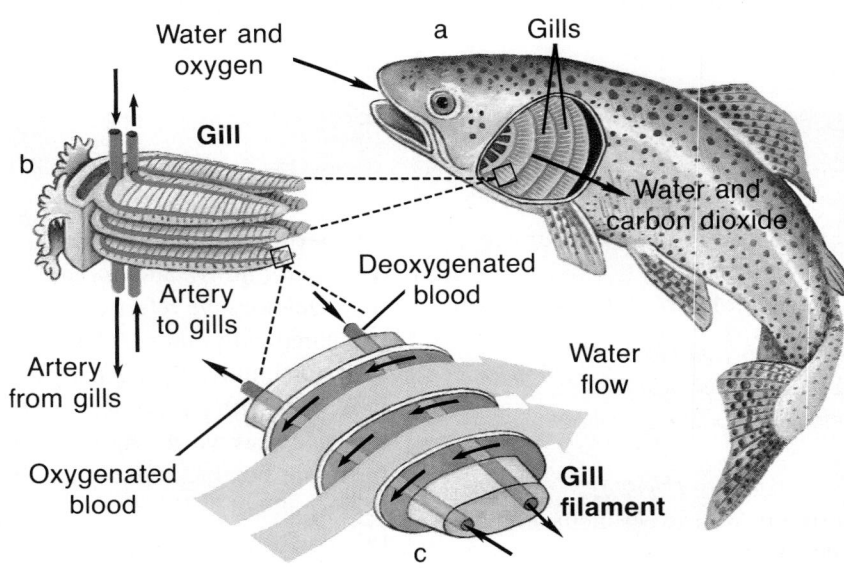

Respiration in Insects

Figure 22-6   A grasshopper has a specialized respiratory system composed of spiracles and tracheae. The tracheae branch and rebranch until they end in tiny air sacs called tracheoles.

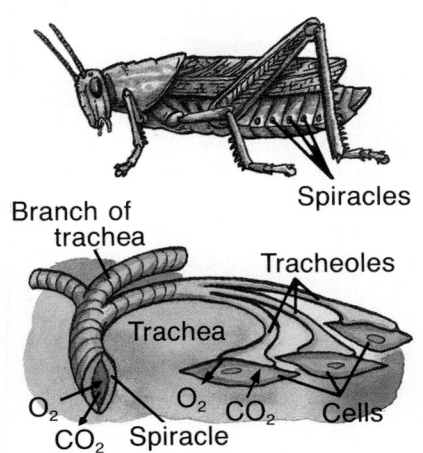

Most land animals live in an environment that exposes them to dry air. They cannot undergo gas exchange on their body surfaces because there is not enough moisture to dissolve the needed gases. Instead, dry air is drawn into moistened areas inside their bodies.

In some land arthropods (insects and their relatives), air enters the animals through small openings located along the sides of the body called spiracles. These external openings lead into tubes called **tracheae.** Tracheae branch and rebranch inside an animal. The smallest branches end in tiny air sacs, called **tracheoles,** that contain moisture (Figure 22-6). External openings, tracheae, and tracheoles make up the tracheal respiratory system.

In an animal with a tracheal system, oxygen in the air dissolves inside the moist tracheoles and passes directly to the cells by diffusion. Carbon dioxide from the cells diffuses to the tracheoles and is expelled through the external side openings. The system works because no cell is very far from a tracheole. The large number of tracheoles provides a large surface area for gas exchange.

Blood does not transport gases in a tracheal system. Gases travel via simple diffusion or, in large insects, via muscular contraction. The blood's main function is limited to the transport of nutrients and wastes.

Respiration in Humans

As a human, you live on the land in a dry air environment. Does this mean that you have a tracheal respiratory system, like an insect's? No, but you do have a tube called a trachea, which conducts air from the outside into an internal gas exchange area. That gas exchange area is composed of two organs called **lungs.** Your lungs have many tiny, moist air sacs similar to those in a tracheal system. But unlike the insect's, your respiratory system relies on circulating blood to transport gases between body cells and the gas exchange area. All other mammals, as well as reptiles, birds, and most adult amphibians, have respiratory systems that have lungs and use blood to transport gases.

From Atmosphere to Air Sac   You know your body well enough to sense that your lungs are located deep inside your body structure. This is a safe place for them, considering that gas exchange surfaces are always thin and delicate. But how does oxygen reach the lungs?

LEISURE CONNECTION

Scuba Diving: Exploring Below the Surface

People have always wanted to explore places that are beyond the physical abilities of their bodies to reach. For example, the ancient Greeks seemed eager to explore both the heavens and the oceans. One myth tells of Icarus and Daedalus flying with wings of wax. The ancient philosopher Aristotle is credited with describing a crude diving bell that allowed a diver to work under water. It wasn't until this century, however, that people have been able to explore both places. Just as the Wright brothers opened up the sky with their invention of the airplane, Jacques Cousteau and Emile Gagnan opened up the oceans to exploration with their invention of scuba, or *Self-Contained Underwater Breathing Apparatus.*

Several factors inhibit a person's ability to work under water. The most obvious is that while fish have gills to remove oxygen from water, your lungs function only on land. The second factor involves the amount of pressure exerted on your body. Your body is adapted to function at sea level. At sea level, the mass of air in the atmosphere presses on your body at a force of 14.7 psi (pounds per square inch), or 1 atmosphere. You inhale and exhale at a normal rate and without much effort. As you descend into the ocean, the mass of the water increases the pressure on your body. The pressure doubles to 2 atmospheres at a depth of 10 meters and is 3 atmospheres at a depth of 20 meters. On land, an increase in pressure like this would make it impossible for you to inhale

and inflate your lungs. Scuba equipment provides a way for you to breathe under the increased pressure. A tank of compressed air and a demand regulator make up the scuba equipment. With this equipment, air is supplied from the tank to your lungs, at the same pressure as the surrounding water, when you inhale. As you exhale, air is released through the regulator into the water. You can breathe easily as you explore the wondrous habitat of the ocean organisms.

1. **Find Out:** Why do divers have to take time to "decompress" when they go back up to the surface after a deep dive?
2. **Connection to History:** How can scuba gear be used to find out about how Spanish explorers lived in the newly discovered lands of North and South America?

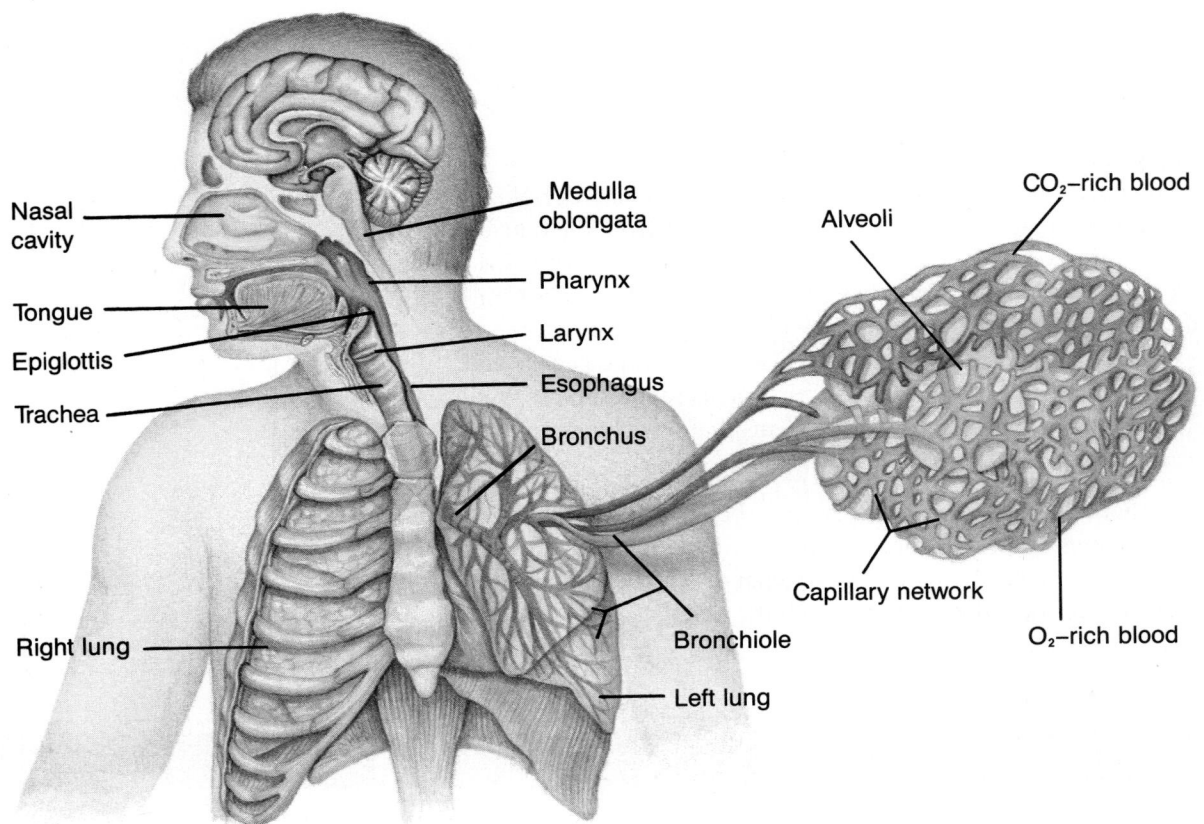

Nasal cavity

Tongue

Epiglottis

Trachea

Right lung

Medulla oblongata

Pharynx

Larynx

Esophagus

Bronchus

Bronchiole

Left lung

Alveoli

CO_2–rich blood

Capillary network

O_2–rich blood

Figure 22-7 The human respiratory system consists of air passageways and the lungs. The bronchioles end in clustered air sacs called alveoli. Diffusion of oxygen and carbon dioxide into and out of the lungs occurs at the alveoli.

When you inhale, oxygen (as part of the air) enters the nose or mouth, passes to the pharynx, and then moves downward into the trachea, also called the windpipe. The opening of the trachea is protected by a structure called the epiglottis. During breathing, the trachea is open. When a person swallows food, the epiglottis closes over the top of the trachea so food goes down only the esophagus. Have you ever choked from talking while eating? Choking occurs when the epiglottis fails to close properly and food starts into the trachea.

Branching from the trachea are two large tubes called **bronchi.** The bronchi branch and rebranch into many smaller bronchial tubes that end as numerous small, moist sacs called **alveoli.** Gas exchange occurs in the alveoli. Like the tracheoles in insects and gill filaments in fish, groups of alveoli greatly increase the surface area of the lungs for gas exchange. Thus, the human body is assured an adequate supply of oxygen.

Blood as Transporter of Gases Oxygen in the bloodstream travels exclusively in the red blood cells. Red blood cells, you may remember from Chapter 21, contain hemoglobin—an iron-rich protein that can combine with and carry oxygen. Carbon dioxide, on the other hand, is not so limited in its travel arrangements. It combines with hemoglobin, with liquid in the red blood cells, and with the watery portion of blood.

Blood rich in carbon dioxide reaches the alveoli via capillaries that branch from the pulmonary arteries. In the alveoli, the level of carbon dioxide is low and the level of oxygen is high. Under these conditions, the carbon dioxide diffuses from the blood into the alveoli and is exhaled. At the same time, oxygen diffuses from the alveoli into the blood and there combines with hemoglobin to form oxyhemoglobin. The oxygen-rich blood returns via the pulmonary vein to the left side of the heart. From there, it is pumped throughout the body.

At the body tissues, cells are rich in carbon dioxide and poor in oxygen. Under these conditions, oxyhemoglobin readily breaks down into hemoglobin and oxygen, and the oxygen diffuses from the blood to the tissues. Carbon dioxide diffuses from the tissues to the blood and is carried back toward the heart and lungs.

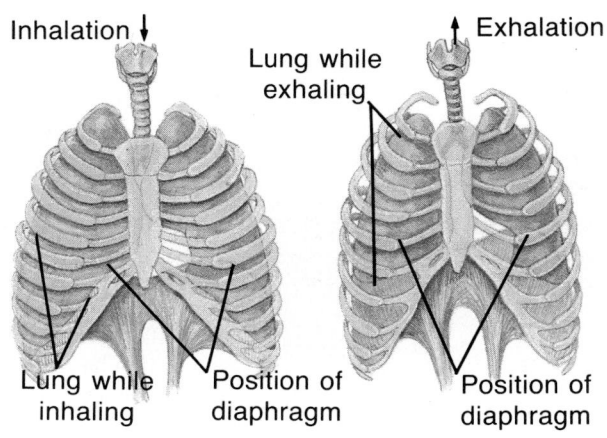

Inhalation ↓   Lung while exhaling   ↑ Exhalation

Lung while inhaling   Position of diaphragm   Position of diaphragm

Figure 22-8 During inhalation, rib muscles contract, lifting the ribs apart, and the diaphragm flattens. This increases the size of the chest cavity, allowing air to enter the lungs. During exhalation, the rib muscles relax, the diaphragm moves up, and the size of the chest cavity decreases.

Control of Breathing

What happens when you inhale? Air is drawn from the environment into your lungs. When you exhale, air is pushed out of your body into the environment. This movement of air is called breathing.

Breathing requires action inside the body. Taking in air (inhalation), requires the action of the rib muscles and a large flat muscle called the

GLOBAL CONNECTION

How high can we live?

Imagine a hot summer morning in the California desert. To escape the oppressive heat, you ride the aerial tramway to the top of Mt. San Jacinto. On the way up, you watch the rugged desert landscape fall far below. At the station on top of the mountain, you see squirrels and chipmunks scampering among pine trees. The air is refreshing and you look forward to a day of hiking. However, a few hours later you begin to feel a little dizzy. You're breathing faster and you can't walk very quickly. Your symptoms are caused by the high altitude of Mt. San Jacinto, which has a summit more than 2000 meters above sea level.

High altitude sickness, or hypoxia, often strikes skiers, hot air balloonists, mountain climbers, and people who visit places like Mexico City, the Himalayas, and the Andes. Hypoxia is caused by a reduction in the amount of oxygen that enters a person's body. While the atmosphere has the same percentage of oxygen at all altitudes, the air is less dense and the air pressure is lower at higher altitudes; thus, less oxygen can move into the lungs, the blood, and the individual body cells. A person suffering from hypoxia will try to compensate for the oxygen deficiency by hyperventilating, or breathing more quickly and more deeply than normal.

People who live at high altitudes are adapted to the lower air pressure. For example, the native

Aymara and Quechua Indians that live in the Andes mountains have barrel-shaped chests, strong diaphragms, and large lungs that have a greater capacity than the lungs of people who live at lower altitudes. The Aymara and Quechua also have more capillaries around their alveoli, larger hearts, and more red blood cells. All of these adaptations help deliver oxygen to their body cells and tissues.

1. **Find Out:** How can the effects of high altitude sickness be reduced or avoided?
2. **Explore Further:** Modern jets fly at altitudes of 10 000 to 13 000 meters. Why don't passengers have difficulty breathing?

How much air do you inhale in one minute?

An average adult breathes in 500 mL of air each time he or she inhales but only 350 mL reach the alveoli. The remaining 150 mL is called dead air. It fills your nasal chamber, trachea, and bronchial tree. Determine the number of times in one minute that you inhale. Calculate the total volume of air in one minute: (a) that you breathe in, and (b) that comes in contact with your alveoli.

Analysis *How might the concentration of oxygen and carbon dioxide gas vary with: (a) The 350 mL of air inhaled compared to air exhaled? Why? (b) The 150 mL of dead air inhaled compared to dead air exhaled? Why?*

diaphragm, which stretches across the base of the rib cage. The diaphragm and the rib muscles contract at the same time. As a result, the diaphragm muscle pulls downward and the ribs move upward and outward. At this point, the size and volume of the chest cavity have expanded. Air pressure inside the cavity is less than air pressure in the atmosphere. Air rushes into the trachea and the lungs to correct the imbalance. About 500 mL of air are brought into the lungs at a time. Now the muscles relax. The diaphragm pushes upward and rib muscles allow the rib cage to relax, thus decreasing the size of the chest cavity. Inside air pressure becomes greater than air pressure in the atmosphere, so the air rushes out. This is called exhalation.

When you think about it, you can control the action of the diaphragm and rib muscles. But you don't have to worry about remembering to breathe. Specialized nerve cells, which are scattered throughout certain arteries and concentrated in the medulla, control your breathing for you. The medulla is the part of the brain that contains the breathing control center. The nerve cells along the arteries send impulses to the medulla, keeping it informed about the levels of carbon dioxide in the blood. If the level of carbon dioxide gets too high, the medulla sends impulses of its own along nerves leading to the diaphragm and rib muscles. These nerves stimulate the muscles to contract more frequently and so increase the breathing rate. The faster a person breathes, of course, the greater the amount of carbon dioxide leaving the blood and the greater the amount of oxygen entering the blood. Too much carbon dioxide can build up acidity in the bloodstream and cause illness, even death in extreme cases. Too little oxygen inhibits the work of the cells.

When you exercise, you are asking the cells to work harder to give you the energy you need. Thus, you raise the level of carbon dioxide and lower the level of oxygen in the blood. Your breathing rate increases to adjust the rate of gas exchange to the needs of the cells. Smoking or breathing heavily polluted air can raise levels of carbon dioxide in the blood. The respiratory system may not be able to adjust to these conditions, and illnesses such as chronic coughing, persistent respiratory infections, and eventually emphysema and cancer could result.

Section Review

Understanding Concepts

1. Explain why gases need not be carried by a specialized transport system in plants.
2. How are the respiratory systems of an earthworm and a bony fish similar? Why?
3. Explain how the respiratory system of a grasshopper is adapted with respect to each of the following: maintaining a moist gas exchange area; having a gas exchange area with a large surface; protecting the gas exchange area; and transporting gases between body cells and the gas exchange area.
4. **Skill Review—Sequencing:** Trace the path of oxygen in a human from the point of air intake to body cells. For more help, refer to Organizing Information in the **Skill Handbook.**
5. **Thinking Critically:** Why is it not possible for you to hold your breath for very long?

22.2 Excretion and Maintenance of Osmotic Balance

Objectives

Discuss why many organisms must have adaptations for maintaining osmotic balance.

Describe maintenance of osmotic balance and the process of excretion in simple freshwater organisms.

Discuss the function of excretion and maintenance of osmotic balance in animals having circulatory systems.

If you light a fire using several wood logs in a fireplace, you will have ashes to clear away after the fire has died. The ashes are wastes left after the wood has burned. Your body cells also produce wastes as they metabolize nutrients. So do the body cells of all animals and protozoans. Some of these wastes are potentially toxic and must be removed. Humans and many other animals have specialized systems that clear away the unwanted wastes.

Adaptations that dispose of cellular wastes often serve double duty. That is, they may also be responsible for maintaining osmotic balance. A cell is bathed in liquids, both internally and externally. The liquids hold dissolved particles. If the liquids inside and outside a cell do not contain about the same concentration of dissolved particles, the cell is not in osmotic balance. The kinds of adaptations animals may or may not have for disposing of wastes and maintaining osmotic balance are closely related to the environments in which they live.

Aquatic Environments

The watery environment of simple organisms influences the way they remove cellular wastes as much as it affects the way they exchange carbon dioxide and oxygen in respiration. It also determines the way they maintain osmotic balance.

Excretion Metabolism of nutrients in animals and protozoans produces waste products. These waste products include carbon dioxide gas, excess water, and nitrogen wastes. Nitrogenous chemicals are released during the cellular breakdown of proteins to produce energy or to make carbohydrates or fats. The nitrogen-containing waste product is ammonia. Ammonia is a highly toxic molecule and must be either excreted directly or converted to a less toxic waste such as urea or uric acid. Figure 22-9 shows the molecules of ammonia, urea, and uric acid. The production of uric acid and urea are two important adaptations of land animals that enable them to excrete nitrogenous wastes with a minimum of water loss.

A liquid environment facilitates the removal of ammonia. Since ammonia dissolves readily, it simply diffuses from the organism into the external environment. Land animals and complex aquatic animals cannot rely on such direct methods for removal of wastes. But whatever the methods or steps required, removal of toxic nitrogenous wastes is called **excretion**.

Figure 22-9 **Birds excrete uric acid as a pastelike substance that contains almost no water. How is the excretion of uric acid a beneficial adaptation to animals that live in dry environments?**

Figure 22-10 Many marine organisms, such as this scarlet reef lobster, are able to maintain osmotic balance without much trouble. To excrete nitrogenous wastes, most often in the form of ammonia, they have simple excretory organs located in the head.

Osmotic Balance and Sea Water Among simple organisms, excretion and osmotic balance are independent of each other. Life is thought to have first evolved in the sea, a saltwater environment. Both the cytoplasm of those first cells and the surrounding sea water doubtless had similar concentrations of dissolved particles, such as salts. Thus, the first organisms would have been in osmotic balance with their environment. Descendants of these early forms of life that remained in the sea continued to maintain that balance. Organisms such as today's seaweeds, sponges, and jellyfish have cells directly bathed by ocean water; their cells are in osmotic balance with that water. More complex organisms, such as lobsters or clams, have internal tissue fluids and blood that bathe their cells. The concentration of dissolved particles in the tissue fluid and blood is close to the concentration of dissolved salts in ocean water. Thus, these animals, too, are able to maintain osmotic balance without any special adaptation.

Osmotic Balance in Fresh Water During the course of evolution, many organisms moved to environments other than sea water. Consider for a moment a single-cell organism bathed by fresh water. The surrounding water has a much higher concentration of water and a much lower concentration of salts than the cytoplasm of the cell. As a result, there is a net movement of water into the cell. If there were no way to counteract this inward flow of water, the cell would eventually burst. Multicellular organisms in fresh water face this situation many times over. They can survive in such an environment only if they have adaptations that allow them to maintain osmotic balance between their cells and the surrounding fluid.

Thinking Lab

INTERPRET THE DATA

What may cause "fish kills" in the summer?

Background

In late summer, fish may die off in small lakes as water temperatures rise. Are the fish boiling? This is not likely, as the water doesn't get that hot. What else could be affecting them?

You Try It

Examine the data in the table. It shows the amount of dissolved oxygen available at different water temperatures. The amount of oxygen available to fish is in units marked ppm or "parts per million." A reading of 8 ppm means 8 parts oxygen for 1 million parts of water. The higher the ppm reading, the more oxygen there is. Prepare a graph and then **interpret the data.**

Water temperature	Amount of dissolved oxygen
30°C	7.8 ppm
20°C	9.0 ppm
10°C	10.5 ppm
0°C	14.1 ppm

Why do fish require oxygen? What happens to dissolved oxygen as water temperatures rise? When they fall? Explain why fish may die off when water temperatures rise.

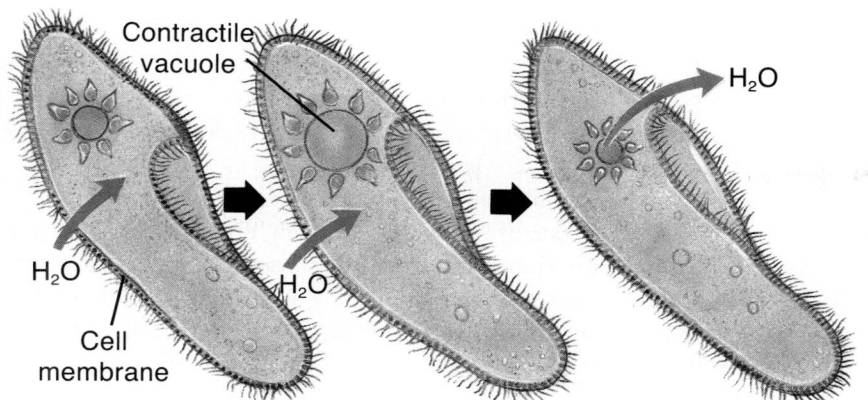

Contractile
vacuole

Cell
membrane

H₂O

H₂O

H₂O

Figure 22-11 *Paramecium* maintain osmotic balance by means of contractile vacuoles. Water moves into these vacuoles, which expel the excess through the cell membrane into the environment. What would happen to this organism if the contractile vacuoles failed to function?

A single-cell *Paramecium* is adapted to maintain osmotic balance in fresh water. This protozoan has contractile vacuoles (Chapter 5) that remove excess water. The net movement of water into the cell fills the canals that surround these vacuoles. When a vacuole is full, its membrane contracts, forcing the water from the vacuole and out through the cell membrane. ATP energy is needed for the contraction of the vacuole.

A planarian is a freshwater animal that has a special system to maintain osmotic balance. A planarian has no circulatory system to transport wastes and excess water to a collecting site. Instead, a system of excretory canals and flame cells is found throughout the animal. Each flame cell has several flagella, the movement of which reminded scientists of candle flames—thus the name *flame cells*. Excess water and some wastes are drawn into the flame cells, where movement of the flagella sets up a current. This current moves the water into and along the excretory canals. At certain places, the canals branch into excretory ducts that open as pores on the surface of the planarian. Water is given off through these pores.

Most of the nitrogenous wastes of a planarian do not enter the excretory canals. They pass from cells to the digestive system and are excreted through the mouth. Some wastes may diffuse directly from cells to the water.

Because their chemical reactions differ from those of protozoans and animals, plants do not build up dangerous nitrogenous wastes. Thus, they have no mechanisms for excretion. However, the fluid that bathes their cells is fresh water, and they must have a means of maintaining osmotic balance. You learned how they do this in Chapter 4. Some water enters plant cells by osmosis, but the cell wall prevents much swelling from occurring. The presence of the cell wall prevents the bursting of cells that would otherwise occur.

Figure 22-12 Flagella in the flame cells of *Planaria* draw excess water from the body. The water collects in ducts and moves out of the body through pores.

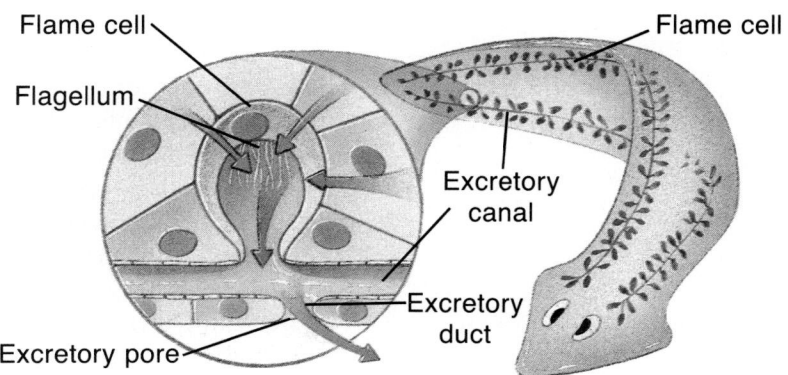

Flame cell

Flagellum

Flame cell

Excretory
canal

Excretory
duct

Excretory pore

Land Environments

Animals living on land are surrounded by dry air. Thus, they cannot diffuse dissolved ammonia into water as some aquatic organisms do. In these animals, the ammonia produced must be changed to another form—a substance that can safely build up in body fluids until it is excreted.

Because land animals occupy a dry environment, conservation of water is crucial. Thus, it is important that they do not lose a great deal of water as they excrete their wastes. In these animals, wastes are picked up and carried by blood or tissue fluid. Their excretory systems not only rid the animals of nitrogenous wastes, but are also a factor in maintaining osmotic balance.

The Earthworm Fluid in the body cavity of the earthworm contains both useful substances and wastes. The wastes take two forms: ammonia and another much less toxic compound called urea. Because the earthworm lives in the land rather than on it, its environment contains soil moisture. Thus, the earthworm diffuses ammonia wastes into the soil. But the urea is excreted through its excretory system.

Figure 22-13 shows the excretory system of an earthworm. Pairs of excretory organs, **nephridia,** are found in almost every segment. Body fluid is drawn into the funnel-shaped entrance of a nephridium by the action of cilia and muscles. As the body fluid flows through the long, narrow tubule of the nephridium, the useful materials such as water, food molecules, and ions are taken in by certain cells of the tubule. These materials then pass to surrounding capillaries and can be recirculated. The nitrogen wastes and a small amount of water remain in the nephridium and are eventually excreted to the outside.

Nephridia in the earthworm act as filters, removing wastes and returning usable substances to the circulatory system. Excretory systems, likewise based on filtration and the return of usable substances to the circulatory system, are found in most complex animals.

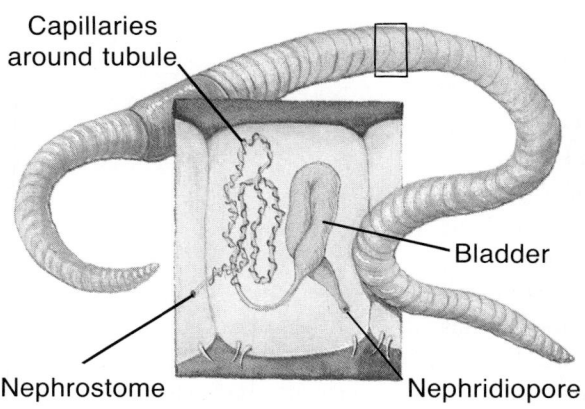

Capillaries around tubule

Bladder

Nephrostome

Nephridiopore

Figure 22-13 In an earthworm, wastes are separated in the nephridia and excreted through nephridiopores. Certain materials are reabsorbed by tubules and passed to capillaries for recirculation.

The Grasshopper True land animals, such as insects, cannot excrete ammonia and also must conserve water in their body fluids. The ammonia their cells produce is converted to a less toxic compound called uric acid.

Uric acid takes the form of insoluble crystals. The insolubility of uric acid is important for two reasons. First, because it does not dissolve in water, it does not enter into chemical reactions within fluids and so can be concentrated in the body without interfering with cellular metabolism. Second, it does not require excretion of water, so water is conserved. Uric acid's insolubility is important not only in the adult stage of life, but also for the embryo. Insects develop in a shelled egg. That means their wastes must be stored safely until they hatch out of the egg. Uric acid meets that requirement. Other embryos that develop in shelled eggs—reptiles, birds, and some land snails—also produce uric acid.

Attached to the intestine of a grasshopper, as seen in Figure 22-14, is a group of stringlike excretory structures. The free ends of these structures, known as **Malpighian tubules,** lie among the cavities of the body. Blood enters and passes through these tubules. As it does so, needed materials and most of the water are resorbed, meaning taken back, into the body, usually through osmosis and active transport. The uric acid and remaining water enter the intestine, where the rest of the water is resorbed. The solid uric acid crystals are excreted from the anus as undigested food is expelled.

Filtration and Excretion in Humans

Like other animals, humans have an excretory system that filters out nitrogen wastes and helps maintain osmotic balance. The main excretory organs are two **kidneys,** fist-sized, bean-shaped structures that lie near the dorsal abdominal wall. Each kidney consists of an outer part, an inner layer, and a center hollow cavity, as seen in Figure 22-15.

Nephron Filtration Centers   Inside each kidney are about a million tiny units called **nephrons.** Nephrons are similar in function to the nephridia of an earthworm. Each nephron is made up of a cup-shaped **Bowman's capsule,** which narrows into a long, coiled tubule. In the center of each Bowman's capsule is a mass of capillaries called the **glomerulus.**

A glomerulus forms from a small artery that branches from one of the two renal arteries. The renal arteries direct blood from the aorta into the kidneys. Another small artery loops away from the glomerulus, outside the wide entrance to the Bowman's capsule. This artery, too, divides into a set of capillaries, which extend to surround the outside of the tubule. The capillaries merge to form small veins. These veins merge to form one of the two renal veins, which return purified blood to general circulation.

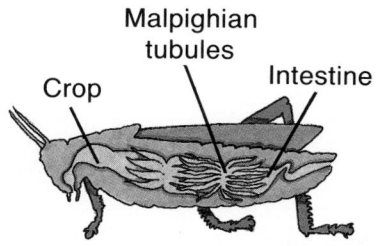

Figure 22-14   In a grasshopper, Malpighian tubules absorb uric acid and water from blood. Water is reabsorbed, but uric acid is passed to the intestines and excreted through the anus.

Figure 22-15   A human kidney consists of three distinct areas. The outer layer or cortex, the inner layer or medulla, and the inner collecting area, the pelvis. A detailed diagram of an individual nephron is shown on the right.

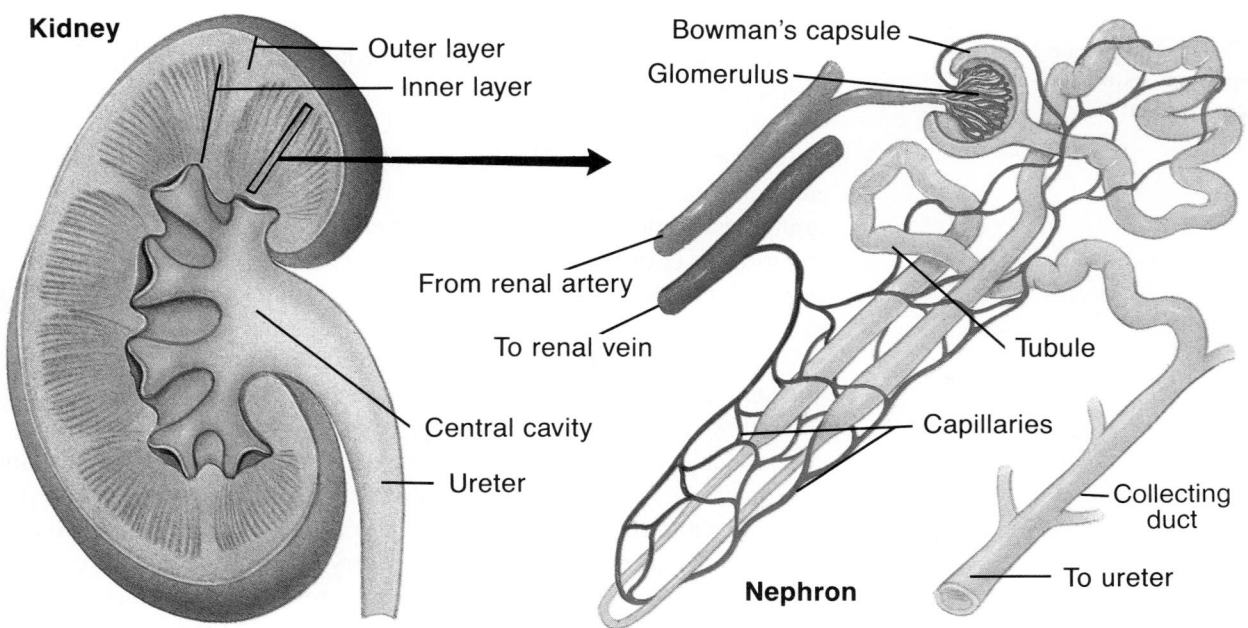

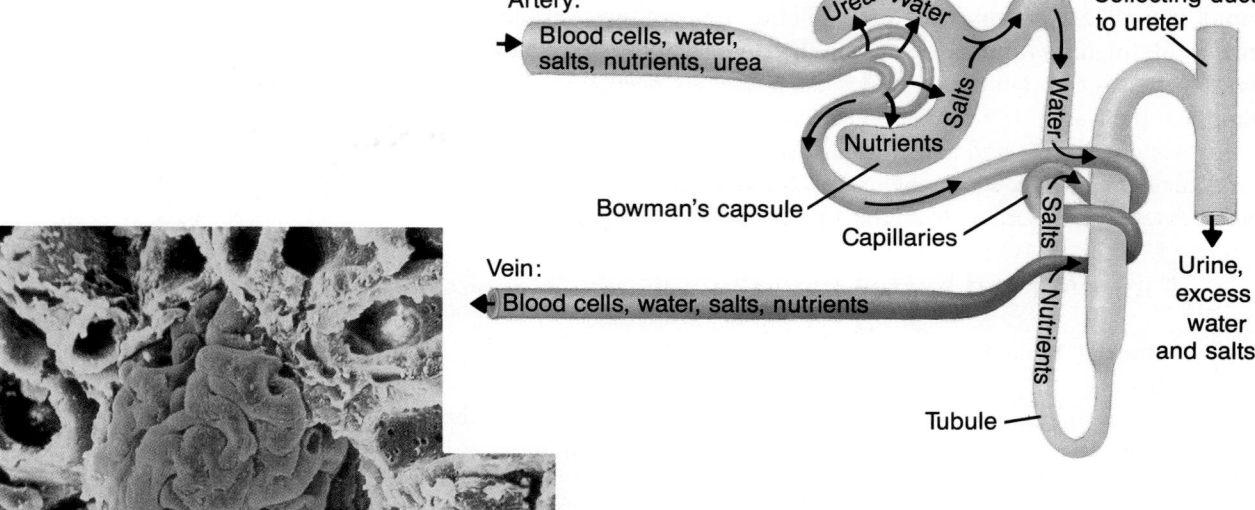

Artery:
Blood cells, water, salts, nutrients, urea

Urea   Water

Collecting duct to ureter

Salts

Nutrients

Water

Bowman's capsule

Capillaries

Salts

Vein:
Blood cells, water, salts, nutrients

Nutrients

Urine, excess water and salts

Tubule

Figure 22-16 **This single nephron is one of approximately one million functional units of a kidney. The arrows show the movement of materials through the nephron. Water, nutrients, and salts are returned to circulation. Excess water and salts, as well as urea, form urine and are excreted. The glomerulus of one nephron is shown at the left. Magnification: 300×**

The Filtering Process   Blood entering a kidney contains needed materials being transported to body cells, plus urea and excess salts. The blood acquires urea when it passes through the liver, which converts ammonia to urea. Urea is less toxic than ammonia and also conserves water because it requires less water for excretion.

From a renal artery, blood is channeled into smaller arteries that lead to each nephron and its glomerulus. The first step in the filtering process takes place as blood pressure forces most of the water and dissolved particles out of the glomerulus and into the cup portion of the Bowman's capsule. Large proteins and the formed elements of blood (red and white blood cells, platelets) do not normally enter the capsule. They remain in the capillaries of the glomerulus.

The continuous flow of blood into the kidneys forces the liquid out of the cup portion of the Bowman's capsule and into the tubule. Here is where the actual filtering takes place. Certain cells lining the tubule use active transport to transfer needed materials such as food molecules, ions, and hormones out of the liquid in the tubule and into the surrounding capillaries. Some of the water also travels, by osmosis, out of the tubule and into the entwined capillaries. The blood, now reassembled and cleansed, makes its way out of the kidneys through a renal vein and back to general circulation.

The liquid left to move into the collecting duct contains urea, excess salts, and a great deal of the water originally present in the blood that entered the nephron. As this liquid moves down through the collecting duct, much of the remaining water moves by osmosis out of the duct and into the tissue fluid of the kidney. Movement of water in this direction occurs because the surrounding tissue fluid is less watery than the liquid in the duct. This resorption of water is very efficient. In fact, 99 percent of the water passing through the kidneys is resorbed! The amount of water lost through excretion is replaced by water in food and beverages.

Excretion of Urine What remains in the collecting duct—urea, salts, and a small amount of water—is called **urine.** Slowly but constantly, urine passes from the collecting ducts in the outer layers of a kidney to the center cavity. Leading from the center of each kidney is a tube called the **ureter,** which conveys the urine to a muscular storage sac, the **urinary bladder.** When the bladder becomes filled, a muscular valve relaxes, and urine is excreted through the **urethra** (Chapter 18). Table 22.1 compares the contents of urine with the contents of blood.

Kidney Failure When a person's kidneys fail to function, the body has a big problem. The kidneys may fail to work properly for a number of reasons. Bacteria may attack the kidneys and do damage. Kidneys may also be damaged by environmental toxins such as mercury and lead. When enough of a person's nephrons are affected, the kidneys fail to filter the blood effectively. Persons whose kidneys have failed must have their blood filtered by an artificial kidney machine in a process called dialysis. During dialysis, blood is pumped from the person's artery through tubing that is bathed in a salt solution similar to plasma. Waste materials diffuse from the tube containing blood and are washed away by the salt solution. The cleaned blood left behind is returned to a vein. Dialysis sessions may last from five to ten hours every two to three days.

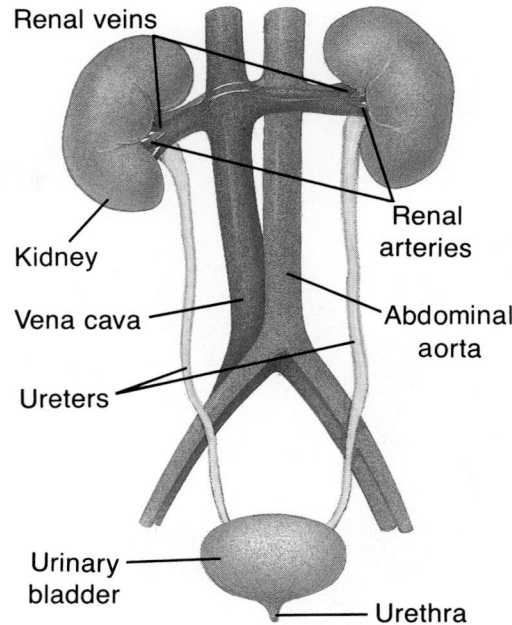

Figure 22-17 The human urinary system is made up of a pair of kidneys, ureters, a urinary bladder, and a urethra.

Table 22.1 **Composition of Urine and Blood**

Substance	In Blood	In Urine
Urea	small amount	large amount
Water	large amount	small amount
Salts	small traces	moderate traces
Glucose	present	usually absent
Amino Acids	present	usually absent
Proteins	present	usually absent
Blood Cells	present	usually absent

Control of Osmotic Balance

The removal of excess water and salts by the kidneys is as crucial as the removal of urea. The amount of water and salts removed—or not removed—affects the composition of blood and tissue fluid. These liquids must contain appropriate amounts of salts and water in order to help maintain osmotic balance between internal fluids and body cells.

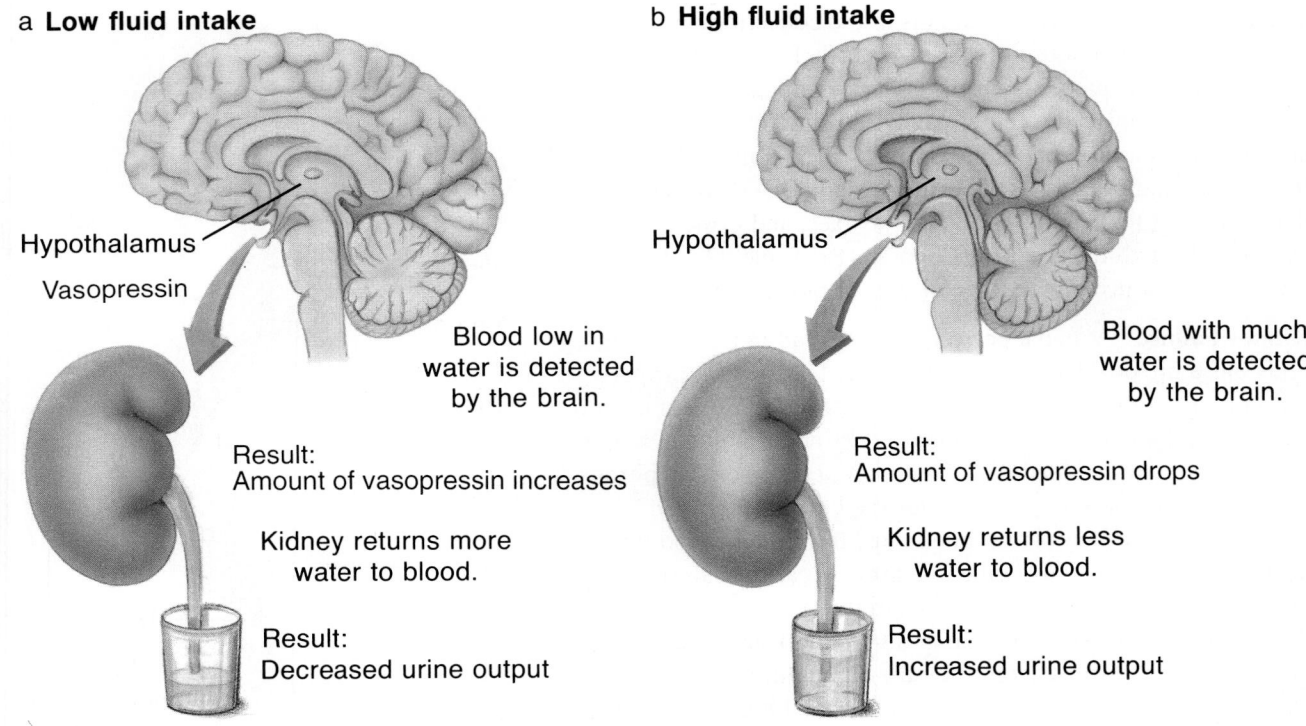

a **Low fluid intake**

Hypothalamus

Vasopressin

Blood low in
water is detected
by the brain.

Result:
Amount of vasopressin increases

Kidney returns more
water to blood.

Result:
Decreased urine output

b **High fluid intake**

Hypothalamus

Blood with much
water is detected
by the brain.

Result:
Amount of vasopressin drops

Kidney returns less
water to blood.

Result:
Increased urine output

Figure 22-18 **The amount of
urine produced by the kidneys is a
result of an interaction between body
fluid levels and vasopressin.**

You have learned that almost all the water that enters the kidneys is
conserved. Exactly how much water is conserved or excreted in urine can
vary, however. For example, a person may need to conserve even more
water than usual if much has been lost because of heavy perspiration or
illness. On the other hand, removal of more water than usual may be necessary for a person who has consumed a large volume of liquid.

The kidneys respond to a variety of regulatory substances called hormones, secreted by various parts of the body, to monitor the amount of
water and salts excreted. The amount of water excreted in urine is controlled by the hormone **vasopressin.** Vasopressin causes the collecting
ducts of the kidneys to resorb more water. Thus, a higher level of vasopressin will result in more water resorption and a more concentrated (less
watery) urine. When too much water is present in the blood, the opposite
occurs. The level of vasopressin is reduced, a lesser amount of water is
resorbed in the collecting ducts, and a more dilute (watery) urine is produced.

Other hormones play a role in determining the amount of certain salts
excreted in urine. Control over the amount of water and salts in the body
is an important means of homeostasis. Homeostasis refers to the ability of
an organism to maintain a steady state of internal operation regardless of
changes in the external environment. By regulating the salt and water content in the blood and tissue fluids, the kidneys help maintain a constant
internal environment for cells regardless of changes in diet, activity, or
other factors that affect the body.

Kidneys in Other Vertebrates

Over the course of evolution, freshwater fish were the first vertebrates with kidneys. The structure and function of the kidneys were modified, however, as later vertebrates began occupying different environments. Bony fish, for example, all have kidneys, but the role of the kidneys in both excretion and maintaining osmotic balance is more vital for freshwater fish than for saltwater fish.

Bony Fish   Body fluids of bony fish are not in osmotic balance with the water in which they live. Most of the surface of a fish is covered with scales, which water and salts cannot cross. Water and salts may enter or leave the fish at the gills, however, causing an imbalance between internal fluids and the cells.

Body fluids of freshwater bony fish are more concentrated than the surrounding water. Thus, these fish tend to gain water and lose salts through the gill membranes. Three factors compensate for these conditions. First, the kidneys of freshwater fish resorb most salts but very little water. As a result, the urine produced is dilute, and large amounts of urine are excreted. Second, a freshwater fish drinks very little water, so no additional water builds up in its fluids. Finally, salts are actively transported from water into the gills. Together, these adaptations enable the animal to conserve salts, rid itself of excess water, and excrete wastes.

You might expect the internal fluids of marine bony fish to be in osmotic balance with sea water. But marine fish descended from freshwater fish, not saltwater organisms. Thus, the salt content of their body fluids is less concentrated than that of the water in which they live.

Figure 22-19   **Freshwater and marine bony fish have evolved different adaptations for maintaining osmotic balance.**

Salts   Water

Do not drink water

Salts

Freshwater Fish   Large amounts of dilute urine

Water

Salts

Drinks

seawater

Ammonia   Salts

Saltwater Fish   Small amount of urine

Figure 22-20 The kangaroo rat, a desert animal, excretes highly concentrated urine and, thus, conserves water. This animal obtains enough water from food and its metabolic reactions to survive without drinking water.

The gills play a major role in the adaptation of marine bony fish to their saltwater environment. Some of the water lost from their body fluids is restored by drinking large amounts of sea water. However, this water contains salts, so even more salts build up in the fluids. This excess salt is removed by active transport across the gill tissue to the water. Also, most nitrogenous wastes in these fish are excreted from the gills in the form of ammonia. Because very little water enters the kidneys, even more water is conserved. The end result is that the kidneys play only a minor role in both excretion and osmotic balance.

Land Animals Vertebrates living on land also exhibit variety in the structure, function, and efficiency of their kidneys. Amphibians and mammals, for instance, need to drink a certain amount of water to make up for water loss through their kidneys. Birds and reptiles conserve water by excreting uric acid and so resorb almost all the water passing through their kidneys. The tiny kangaroo rat, a desert dweller, can survive without drinking any water at all. Its kidney tubules resorb most of the water that passes through them. So efficient are its kidneys that a kangaroo rat can survive without drinking water. Kangaroo rats obtain enough water from the food they eat and as a by-product of metabolism.

LOOKING AHEAD

YOU HAVE OBSERVED that organisms are closely involved with the external environment for many vital functions including gas exchange, excretion, and maintaining osmotic balance. While the environment is a source of air, water, and food, it is also a potential source of danger. Any number of viruses and microorganisms can contaminate life-sustaining materials in the environment. In the next chapter, you will learn about some of the diseases these tiny organisms can spread. You will also study the means by which the body defends itself against the agents of disease.

Section Review

Understanding Concepts

1. In terms of osmotic balance, what common problem is faced by a tomato plant and an amoeba?
2. In general, how are the excretory systems of an earthworm and a grasshopper similar?
3. Could a marine bony fish survive in fresh water? Explain.

4. **Skill Review—Sequencing:** Trace the path of blood from a renal artery to a renal vein, indicating how the various structures through which blood passes function in excretion and maintenance of osmotic balance. For more help, refer to Organizing Information in the **Skill Handbook**.

5. **Thinking Critically:** A urinalysis is a common test from which much can be learned about a patient's health. A urinalysis reveals the presence of proteins in the urine. Is the patient healthy? Explain. What part of the excretory system is not functioning properly?

BIOLOGY, TECHNOLOGY, AND SOCIETY

Health Connection

Let's Clear the Air

By now, most people know the serious damage smoking does to a person's health. A smoker is twice as likely to die from heart disease, ten times more likely to have mouth or throat cancer, and 25 times more likely to develop lung cancer than a non-smoker. Bronchitis, an inflammation of the air tubes leading to the lungs, is more common and harder to cure in smokers. Many long-term, heavy smokers contract emphysema, a disease that causes the flexible air sacs of healthy lungs to break down, reducing the amount of oxygen the body can absorb. This makes breathing extremely difficult, and eventually restricts a patient's activities to the bare minimum.

All of this information has little relevance for the average young person who lights up a cigarette. The dire consequences of smoking lie not in the present but in the dim future. The cold statistics don't translate into a meaningful, real-life situation. Maybe it would help to think about the risks of smoking in other ways. Suppose you attended a

large high school of 1000 students. All of the students smoked. Cold statistics show that on average, one of the 1000 will be murdered, six will die in traffic accidents, and 250

(or one out of four) of your fellow classmates will die because of a smoking-related disease before reaching the age of 65.

CAREER CONNECTION

A **respiratory therapist** assists patients who have cardiorespiratory problems arising from illnesses, heart attacks, or strokes. Respiratory therapists operate and monitor equipment that patients may need to support their breathing. They also teach breathing exercises.

While this statistic may be more meaningful, it doesn't show how smoking might influence your immediate life. Perhaps the following situation will show how smoking can make an unattractive impression on someone you might want to meet.

What a week! Three tests, an essay, and a science project due at the same time. By Friday, you're looking forward to a movie with some friends. As you enter the theater, your eyes settle on a stranger buying a soft drink at the concession stand. Wow! Even from a distance you know that you really want to meet this person. Taking a deep breath, you boldly step up, tap the shoulder, and wait as the stranger slowly turns around.

Suddenly, anticipation turns to disappointment. The stale odor of cigarette smoke drifts from the stranger's hair and clothes to your nose. Red-rimmed eyes peer at you. Yellowed teeth and the unmistakable odor of smoker's breath flash your way as the stranger searches for a breath mint and brushes away some ashes that have just burned a hole in what looks like a new sweater. Unable to hide a look of disappointment, you mumble an apology of mistaken identity and walk away.

1. **Issue:** Smokers usually have more health problems than non-smokers. They often suffer from frequent respiratory infections such as colds, bronchitis, and pneumonia. These illnesses cause many days of missed work. Because of the rising cost of health insurance, some companies are refusing to hire smokers, and are even firing some present employees who smoke only in their own homes. How much right does a company have to regulate what a person does within his or her home?

2. **Writing About Biology:** Explain how smoking can make your skin look older than your actual age.

Biotechnology

A Treatment for Respiratory Distress Syndrome

The birth of a baby is often a time of great joy. However, sometimes the joy is erased by fear and apprehension when the newborn arrives prematurely. Premature babies may be affected by a disorder called respiratory distress syndrome, or RDS. The lungs of babies with RDS are not flexible and elastic enough to stay inflated. Until recently, babies with this disorder died when the air sacs of their lungs collapsed.

Doctors know that fully developed lungs secrete a natural mucus-like substance that is a combination of fats and proteins. The substance is called a surface-active agent or surfactant. Surfactants lower the surface tension of liquids. They act as lubricants to prevent tissues from sticking together. The natural surfactant produced in healthy lungs lowers the surface tension of fluids found in the air sacs and enables the lungs to remain inflated. Most babies do not produce enough surfactant to prevent the lungs from collapsing until they are at least 32 weeks gestational age. Without this substance, babies born prematurely are unable to breathe independently.

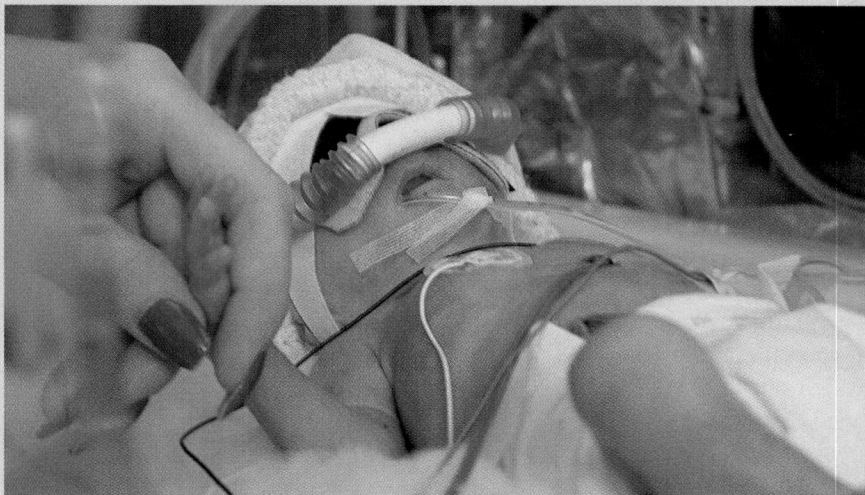

Thus, the most effective way of preventing RDS is to prevent premature births.

While prevention is the best method, doctors now have an additional method of treating the disorder. Since the 1980s, surfactant-replacement therapy has saved the lives and reduced the suffering of babies with RDS. Synthetic surfactant is mixed with sterile water and administered directly into the windpipe of an infant suffering from RDS. Almost half of the infants improve within hours. After a short time, an affected baby's lungs will begin manufacturing surfactant by themselves. Based on the results of clinical trials, the Food and Drug Administration has now approved for use a surfactant that is produced from calf lung tissues and a synthetically produced surfactant. Continued research on the disorder should provide for even better treatment in the future.

1. **Connection to Health:** What measures can be taken to reduce the number of premature births in the United States?

2. **Find Out:** What common household products are surfactants?

Asbestos Hazard

The scene looks like something out of a science fiction movie. Workers wearing gloves, special hooded suits, and face masks enter the furnace room of an old basement. The area around the furnace has been sealed off so that air can't escape and contaminate the outside. What's the danger? It isn't a new form of virus or a radiation leak. These workers are about to remove a furnace that had been insulated with asbestos.

Asbestos has been used for thousands of years. Romans used it for wicks in oil-burning lamps, and Egyptians clothed their mummies in garments made with asbestos fibers. More recently, asbestos has been mixed with paint, cement, plastic, rubber, and asphalt to produce a variety of products used by industries and in construction. Asbestos adds texture to paint. When mixed with cement or wall and ceiling materials, it insulates pipes and buildings. It even helps soundproof rooms. The brake linings of new cars are still made from asbestos materials, and firefighters wear asbestos suits so they can work in burning buildings without fear of being burned.

To understand the dangers of asbestos, you must know about the properties of this material. Asbestos is a fibrous, or flexible mineral found in serpentine and amphibole rocks. Asbestos fibers from amphibole rocks are blue or brown in color, brittle, straight, and needle-like. When people inhale the fibers of blue and brown asbestos, these fibers become lodged in the lungs.

Over time, the lungs become scarred, breathing becomes difficult, and a disease called asbestosis develops. In some cases, the lodged fibers cause lung cancer to develop. Asbestos fibers have even contaminated the water supply of older homes. This happens when fibers from insulating materials around the water tank or pipes crumble and fall into the tank. The fibers may then become lodged in the stomach or intestines of people who drink the water. Doctors suspect that some cancers of the gastrointestinal system have been caused by the fibers. Some cancers of the larynx, pancreas, kidneys, ovaries, and lymph glands may also be traced to asbestos. Because of these health problems, the use of blue and brown asbestos has been banned. Possibly because of its spiral shape, the fibers of white asbestos are not as dangerous as blue and brown asbestos. Although these fibers may also be inhaled or ingested, they don't remain lodged in the tissues. The use of white asbestos is greatly restricted, and safer alternatives are used whenever possible.

An immediate danger now involves removing asbestos materials installed in schools, homes, and office buildings before the dangers of the mineral were identified. As these materials get older, they deteriorate and crumble apart, releasing the fibers into the air and water. Only qualified workers using special equipment and following carefully developed precautions should remove asbestos. These activities may remind you of a movie scene, but the results help provide a greater chance of good health for those people living and working in the building.

Understanding the Issue

1. Should old asbestos insulation be removed even though it hasn't begun to break apart or crumble?
2. What new products are being used to replace asbestos?

Readings

- Mathews, J. "To Yank or Not to Yank?" *Newsweek,* April 13, 1992, p. 59.
- Stone, R. "No Meeting of the Minds on Asbestos." *Science,* Nov. 15, 1991, pp. 928-931.

Summary

Simple organisms have no specialized structures for gas exchange; each of their cells carries out that function independently. Gas exchange areas in plants are located in stems, leaves, and roots, and there is no transport of gases from one part of the plant to another.

Location of respiratory exchange surfaces of animals is related to the environment in which they live. In aquatic animals, the exchange area is near or projecting from the surface, where it is kept moist by the surrounding water. In land animals, the exchange surface must be deep within the body where the delicate tissues can be surrounded by moisture. Humans exchange gases at the lungs, and gases are transported between lungs and body cells by blood.

Terrestrial and freshwater organisms as well as some marine forms must have a means of maintaining an osmotic balance between their cells and the fluid that bathes those cells. Protozoans and animals excrete various forms of nitrogenous wastes. In animals with a circulatory system, the excretory system also functions in maintaining osmotic balance. Kidneys are the excretory organs of humans and other vertebrates.

Language of Biology

Write a sentence that shows your understanding of each of the following terms.

alveolus	nephridium
Bowman's capsule	nephron
bronchus	respiration
diaphragm	trachea
excretion	tracheole
gill	ureter
glomerulus	urethra
kidney	urinary bladder
lung	urine
Malpighian tubule	vasopressin

Understanding Concepts

1. Why is the phrase *gas exchange* suitable when talking about gases moving into and out of all kinds of organisms?
2. What do the terms *respiration* and *cellular respiration* have in common?
3. How is the human respiratory system adapted for each of the following: a large gas exchange surface; protection of the exchange surface; a moist environment for the exchange surface; transport of gases between body cells and the exchange surface?
4. How is a bony fish adapted for each of the features or functions listed in question 3?
5. Describe inhalation, gas exchange, and exhalation in a human. List all the structures of the respiratory system and their functions, in order, for each process.
6. Why have most marine organisms not evolved a means of regulating their body fluids?
7. Why are excretion and maintenance of osmotic balance linked together in animals having a circulatory system?
8. Many people add plants to their freshwater aquaria. How can the plants maintain osmotic balance in that environment?
9. Could a seaweed survive in a freshwater aquarium? Explain.

10. Explain why the amount of water and salts excreted by a human is not always the same. How are the amounts that are excreted controlled?

Applying Concepts

11. Shore crabs can exist in sea water, but can also live in brackish water, water that contains a mixture of salt water and fresh water. In which environment are their internal fluids in osmotic balance with the water? In general, what adaptation must they have to be able to live also in the other environment?
12. Why must the space shuttle contain equipment that removes carbon dioxide from the cabin?
13. A person is caught in an avalanche of snow. The snow is tightly packed around the person up to the neck. Help is three days away. Will the person be able to survive that long? Explain.
14. A person is adrift on a raft in the ocean and is very thirsty. Will drinking sea water help? Explain.
15. On a relative basis, do you think that the tubule and collecting duct of a kangaroo rat nephron are longer or shorter than those of a human nephron? Why?
16. You are observing a paramecium under high power of a microscope and measuring the rate at which its contractile vacuole is contracting. How will that rate change if you slowly add more and more salt to the water? Explain.
17. Why do marine birds have special glands that excrete salt?
18. **Health Connection:** Cigarette smoking is hazardous to a person's health, but becomes much more hazardous when the individual is exposed to other air pollutants as well. The two pollutants may work together to create an even more toxic situation. What other pollutants might a smoker encounter daily that may cause more lung damage than the smoke alone?
19. **Leisure Connection:** High altitudes and low air pressure create problems for mountain climbers. Would scuba equipment be beneficial for these people? How might it be adapted?

Lab Interpretation

20. **Investigation:** A student group ended up with the following data: inhaled air flask = bright blue, exhaled air flask = bright blue. Suggest a possible reason for their results.
21. **Minilab 2:** You are provided with the following data: breathing rate = 10 times per minute, volume of air inhaled = 400 mL, dead air space = 100 mL, inhaled air is 20% oxygen.
    (a) Calculate the volume of air reaching the alveoli in one hour.
    (b) Calculate the volume of oxygen reaching the alveoli in one minute.
22. **Thinking Lab:** Water gets its dissolved oxygen from the air that mixes into it. How might it be possible for fish to survive in bodies of water that are covered during the winter by a layer of ice?

Connecting Ideas

23. Based on what you learned about digestion in humans, how does the liver play a role in maintaining osmotic balance? The colon?
24. Compare the amounts of urea and oxygen present in blood in the right atrium with the amounts present in the left ventricle.

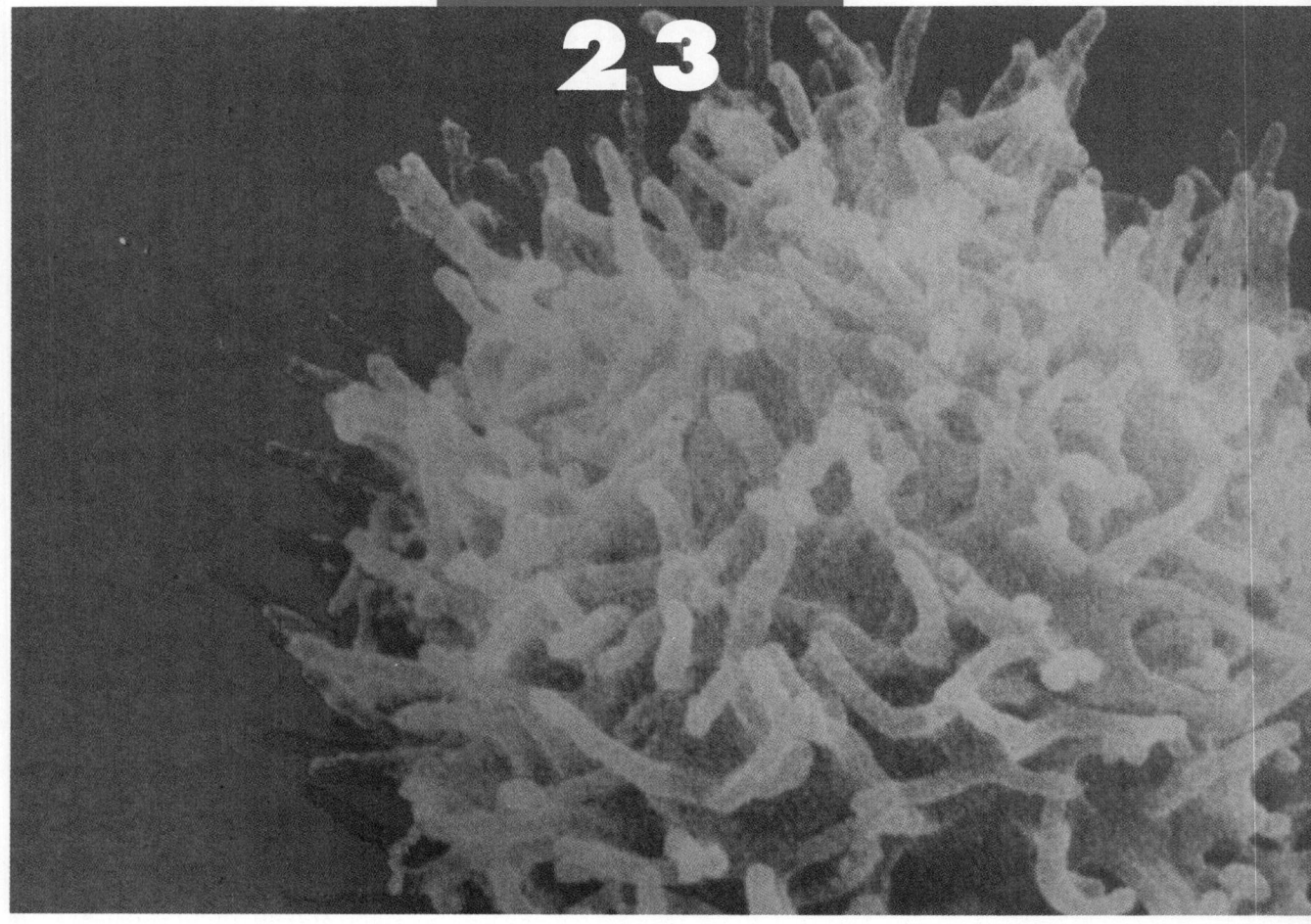

IMMUNE SYSTEM

YOU MAY NOT realize it, but at this very moment, a battle might be raging inside your body. Are you sniffling? Do you have a scratchy throat? These symptoms may mean you are fighting off an infection. Within your body, white blood cells like the one shown in the photograph are busy consuming the bacteria or viruses that are causing the disease. At first, the invaders reproduce rapidly and seem likely to overwhelm your defenses. But then your body rallies. Thousands of white blood cells go on the counterattack. The battle rages on. At times, the outcome may appear to be in doubt. But if you recover, you can be sure your body has won the war.

The response of the body to attack by outside agents is a remarkable process to experience, whether it is going on inside you or you are viewing it under a microscope. It involves thousands of troops as well as time and energy. Because of it, some infections—such as colds—are of no permanent concern. Others—such as pneumonia—are more serious but still likely to be conquered by most people. Our knowledge of the body's ability to fight disease has helped us prevent diseases such as polio and eliminate smallpox. Perhaps someday that knowledge will extend to a cure for the disease called AIDS.

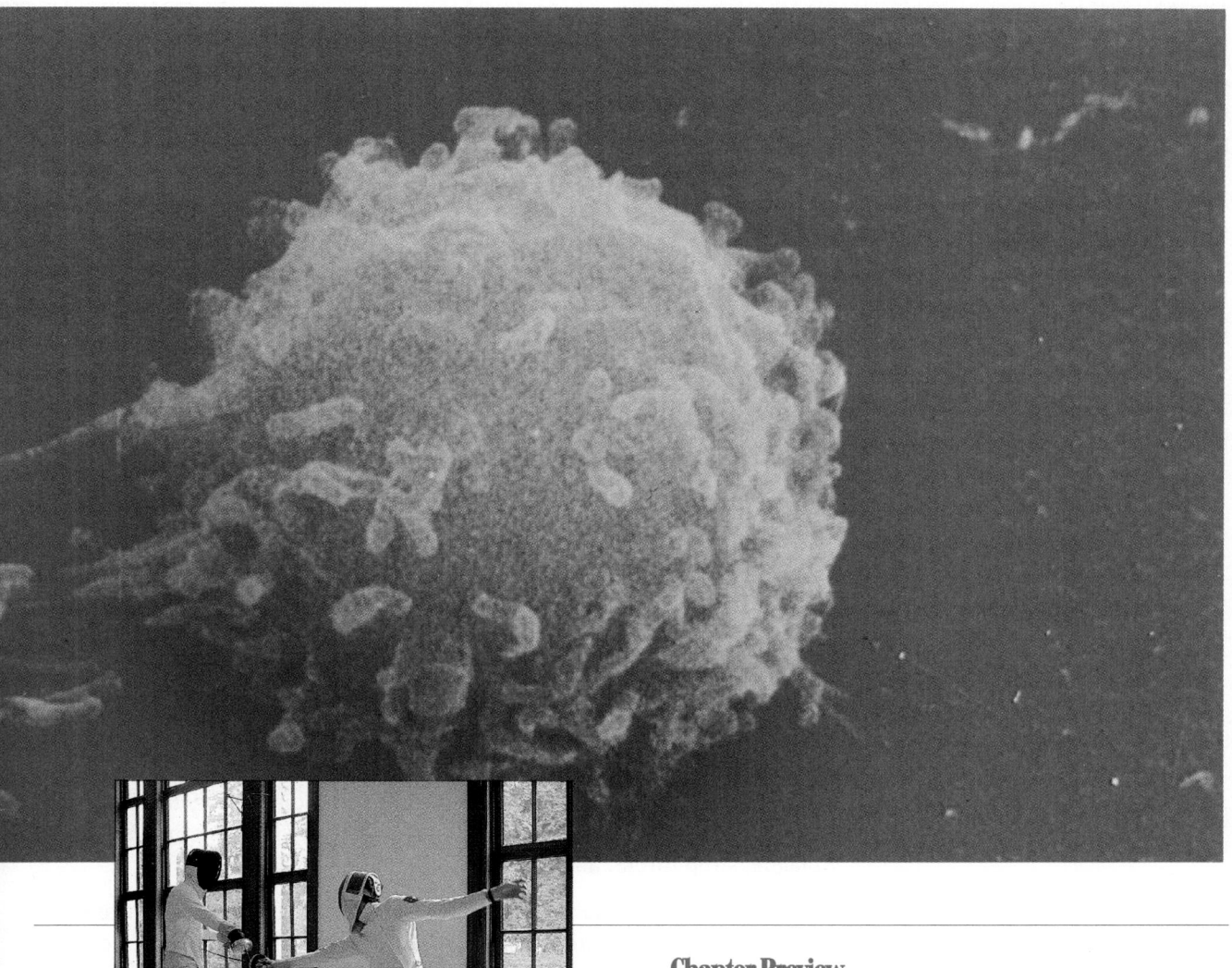

These fencers are fighting a mock battle in which no one will be hurt. Does this type of battle go on inside your body?

Chapter Preview

639

23.1 Bacterial and Viral Agents of Disease

When is the last time you had a cold? Not only were you sneezing and coughing, but you also probably had a sore throat and a fever. Did you ever suffer from an intestinal flu? Have you ever had strep throat? Do you know why you were vaccinated against measles, mumps, and polio? All these illnesses are caused by either bacteria or viruses, unseen enemies that can enter human tissues. There, they may cause discomfort, disease, and sometimes even death. Not all microorganisms cause disease. In fact, many are important in nature or useful to humans.

Microorganisms that do cause disease are called **pathogens.** Besides bacteria and viruses, simple organisms such as fungi and protozoa may also be pathogens. Many diseases caused by pathogens—including the common cold, the flu, and strep throat—are spread from one person to another and are known as infectious diseases.

Bacteria and Disease

The relationship between bacteria and certain diseases was not well understood until the middle of the 19th century. At that time, Louis Pasteur proposed the idea that bacteria can cause disease. About the same time Pasteur developed this germ theory of disease, a standard method of identifying bacterial pathogens was devised. Since then, many bacteria have been found to be responsible for infectious diseases in animals. Among humans, bacterial diseases include bacterial pneumonia, bubonic plague, strep throat, and meningitis.

Transmission of Bacterial Diseases   Bacteria are adapted to be transmitted from one host to another in a variety of ways. Often, their mode of transmission is related to the symptoms they cause. For example, some bacterial pathogens are airborne. That is, the bacteria are carried through the air on small liquid droplets produced when the infected person coughs or sneezes. The bacteria responsible for strep throat are transmitted in this way.

Figure 23-1   Bacteria that cause diseases include various species of *Streptococcus*, left; *Vibrio cholerae*, the organism that causes cholera, center; and *Yersinia pestis*, right, the bacteria that cause bubonic plague. Magnifications: 40 000× (left), 6500× (center), 1200× (right)

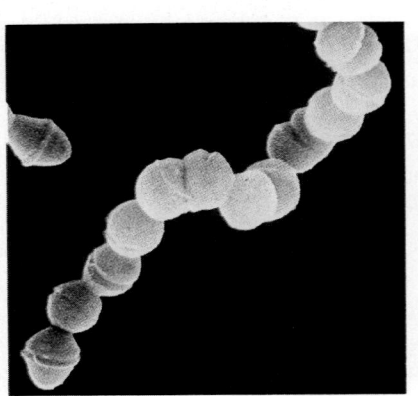
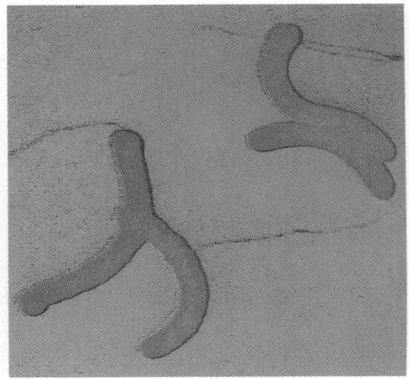
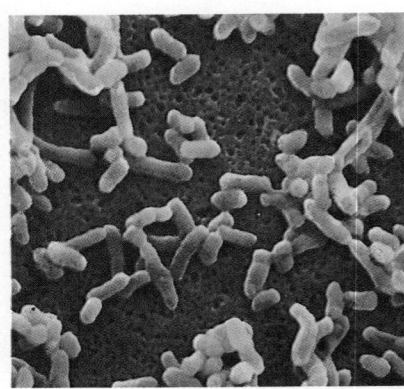

Figure 23-2 When you sneeze, bacteria carried on tiny water droplets spread out in all directions. Does sneezing spread pathogens?

Other bacterial pathogens are carried in water that has not been treated, or purified. Waterborne bacteria are spread when a person drinks water or eats food that once lived in water contaminated by human wastes. The bacteria responsible for cholera are transmitted in this way. When a person contracts cholera, the bacteria cause the intestine to secrete excessive amounts of fluids. This constant secretion results in diarrhea. The diarrhea, if untreated, can quickly lead to dehydration, decreased blood pressure, kidney failure, and then death. Bacteria may reproduce within a host, but if the host dies before the bacteria find another host, the bacteria will die too. However, even if a person dies of cholera, the bacteria responsible for the disease can be spread because they are present in the fluids lost by the dying host. If those fluids enter water supplies from which people drink or obtain food, the bacteria may be transmitted to others. An outbreak of cholera that began in Peru in early 1991 spread along the entire Pacific coast of South America and east to Brazil by 1992, infecting more than 300 000 people.

Still other bacterial pathogens are transmitted by direct contact. The bacteria are spread when one person touches an object or a living thing on which the bacteria are located. Sexually transmitted diseases caused by bacteria—syphilis and gonorrhea—are spread by sexual contact. Some other bacterial diseases are transmitted by arthropods. For example, bubonic plague is spread by fleas, epidemic typhus fever is transmitted by lice, and Lyme disease is carried by ticks. The bacteria may enter the body when the arthropod bites or pierces the skin.

Causes of Bacterial Diseases When bacteria enter the body, they can cause disease in several ways. One way many bacteria cause disease is by producing poisonous chemicals called toxins.

Exotoxins are released by living bacteria. They travel through the bloodstream and affect particular tissues or act locally. For example, one species of bacterium secretes a toxin that stimulates certain nerves, causing some muscles to remain contracted. The result is tetanus, a condition that can be very painful and may even result in death. The bacterium that causes diphtheria secretes a toxin that may damage the heart, kidneys, and nerves.

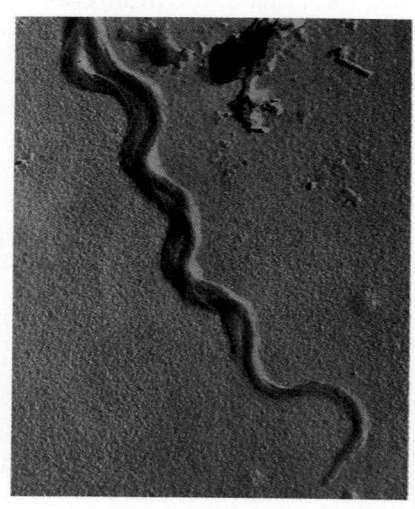

Figure 23-3 Syphilis is a sexually transmitted disease caused by the bacterium *Treponema pallidum.* Magnification: 10 000×

Minilab 1

How is disease spread?

Your teacher will give you 2/3 of a cup of clear liquid. Ten percent of the class will have "diseased" liquid and 90 percent will have water. Pour 1/4 of your liquid into your empty cup and mark it "control." Exchange your original liquid with another student by pouring half of your remaining liquid into his or her liquid, then taking half back. Do this two more times with a different student each time. Add 1 to 2 drops of phenolphthalein to both of your cups. CAUTION: *Be careful using phenolphthalein. It can stain clothing.* The dark pink liquid indicates which cups have the "disease."

Analysis *What is the percentage of students who had the "disease" at the end of this lab? How does this lab simulate the spread of an infectious agent?*

Endotoxins are released by dead bacteria. These toxins are associated with bacterial membranes until the bacteria die and break open. Endotoxins are released when the bacterial cells rupture. Endotoxins are the cause of tuberculosis and bubonic plague.

How do toxins produce their effects? The answer is not always known. But in the case of cholera, the bacterium exerts its deadly effects by means of a toxin whose action *is* understood. The toxin alters a protein in the membrane of intestinal cells, causing negatively charged chloride ions to continually leave the cells. Positively charged sodium ions, attracted to the chloride ions, follow them out. As a result of the loss of these ions, the cells lose water by osmosis. The effect is like opening a gate in a dam, and the constant loss of water results in the symptom of diarrhea.

Disease symptoms do not always result from the action of toxins. In some cases, symptoms result from the host's reaction to the invasion of the bacteria. Bacterial pneumonia is an example. The presence of the bacteria results in the loss of a great deal of fluid from cells. This fluid accumulates in the lungs, interfering with breathing.

Viruses and Disease

Since viruses were first isolated in 1935, a large number have been discovered and studied. Many viruses cause diseases and are transmitted in the same ways as bacterial pathogens. Human diseases caused by viruses include the common cold, polio, the flu, measles, mumps, rabies, and AIDS. One form of a herpes virus causes a sexually transmitted disease. Other herpes viruses cause infectious mononucleosis and cold sores.

Some viruses that infect humans also attack other animals. For example, the virus that causes encephalitis, an inflammation of the brain, invades horses as well as humans. In addition, viruses that cause disease in both humans and other animals can be spread among the different host species. Rabies is such a disease. It can be spread among animals such as raccoons, dogs, and humans when a rabid animal bites another animal.

The symptoms of viral diseases vary with the virus and the tissues under attack. Recall that as viruses reproduce, they often destroy their host cells. For example, polio viruses attack and damage or destroy nerves in the brain and spinal cord, causing paralysis or even death.

Figure 23-4 **Viruses contain either DNA or RNA, but not both. The RNA viruses shown here cause poliomyelitis (left) and influenza (center). The smallpox virus (right) contains DNA. Magnifications: 4000× (left), 300 000× (center), 65 000× (right)**

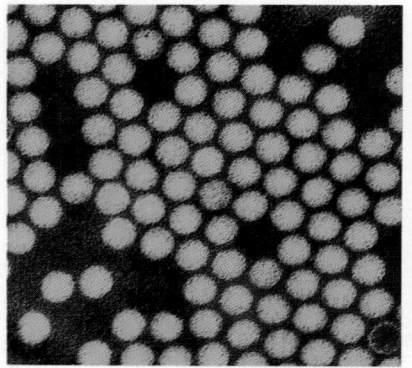
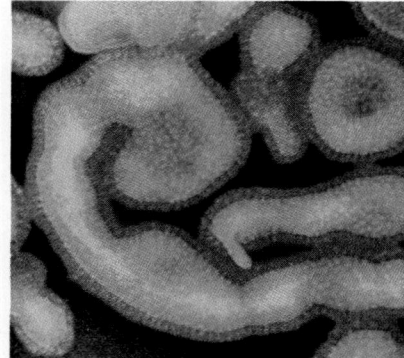
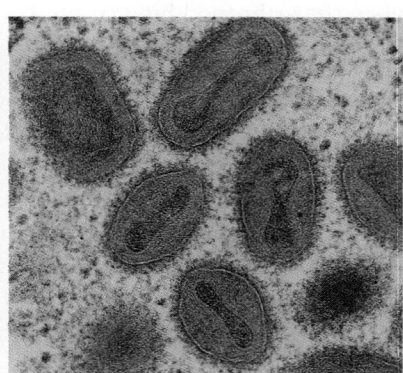

The Evolution of Disease

How did the pathogens that infect humans come to do so? Many human pathogens may be descendants of pathogens that invade other animal hosts. Most bacteria and viruses that pass from other animals to humans cannot survive and cause disease in humans. However, some of these pathogens may have evolved over time, increasing their chances of survival. For example, the virus that causes smallpox in humans is thought to have evolved from a virus that causes a disease in cattle called cowpox. Scientists hypothesize that the virus responsible for AIDS may have descended from a usually harmless virus present in the African green monkey. In addition to infecting new hosts, some pathogens also evolved different ways of being transmitted from one host to another.

Figure 23-5 Pathogens that affect humans may have evolved from bacteria and viruses that are found in animals like the African green monkey, cow, and red fox.

Section Review

Understanding Concepts

1. Malaria is caused by a protozoan and is transmitted by a particular species of mosquito. How is this protozoan transmitted? Explain why malaria is considered an infectious disease.

2. A person is infected with the bacterium that causes cholera. How does the transmission of the bacterium relate to how it causes the disease?

3. Given the symptoms of a common cold, in what kinds of cells must cold viruses reproduce?

4. **Skill Review—Designing an Experiment:** How would you set up an experiment to test the hypothesis that a certain bacterium causes a disease? For more help, see Practicing Scientific Methods in the **Skill Handbook.**

5. **Thinking Critically:** What would be a logical plan of attack for preventing the spread of waterborne diseases? Of those spread by insects?

Discuss nonspecific lines of defense
against pathogens.

Describe how the immune system
defends against pathogens present in
blood and tissue fluid.

Demonstrate how the immune system
functions in destruction of infected,
abnormal, or foreign cells.

23.2 Defending Against Pathogens

The environment that you and other humans inhabit includes a great number and a great diversity of potential pathogens. Those pathogens, however, can cause disease only if they gain entry into your body. If they do, an army of specialized cells and tissues is present and ready to respond to the invasion. Working together, these cells and tissues become mobilized to identify, seek out, and destroy the pathogens. While the battle is raging, you feel sick. The symptoms you feel are the result not only of the effect of the pathogens, but also of your body's defense against the pathogens.

Your body possesses several lines of defense against diseases. Some are nonspecific in their response; that is, they attempt to take on all potential invaders. Others target specific pathogens.

Nonspecific Defense Mechanisms

The first line of defenders to meet invading pathogens are nonspecific. They stand guard at or near the entryways to your body.

Skin and Membranes   Your first line of defense against pathogens is your skin and the membranes of body openings. Your outer layer of skin is made up of dead cells that microorganisms cannot penetrate. Only if your skin is broken, as when a cut occurs, can microscopic pathogens enter.

Membranes that line openings in your body, such as your nostrils and mouth, are more vulnerable to entry of pathogens. Microorganisms can easily enter through your nose, for example. However, several adaptations prevent pathogens from gaining an upper hand. For example, hairs in your nose filter the air, while mucus in the lining of your nose traps and destroys microorganisms. Saliva in your mouth and tears in your eyes both contain substances that kill microorganisms. An enzyme in tears breaks down the cell walls of bacteria that enter the tear ducts of your eyes. The lining of your trachea and bronchi consists of cells that bear many of the tiny hairlike structures called cilia. These cilia wave as they move mucus from your lungs to your mouth. As air moves toward your lungs, potential pathogens are trapped in the mucus. Action of the cilia and mucus together prevents respiratory infection.

You have still other defense mechanisms against infectious diseases. Acids and enzymes in your body can kill many pathogens that enter body openings. Your stomach and other digestive organs produce a variety of chemicals that destroy pathogens. Bacteria present in your lower intestine also attack pathogens.

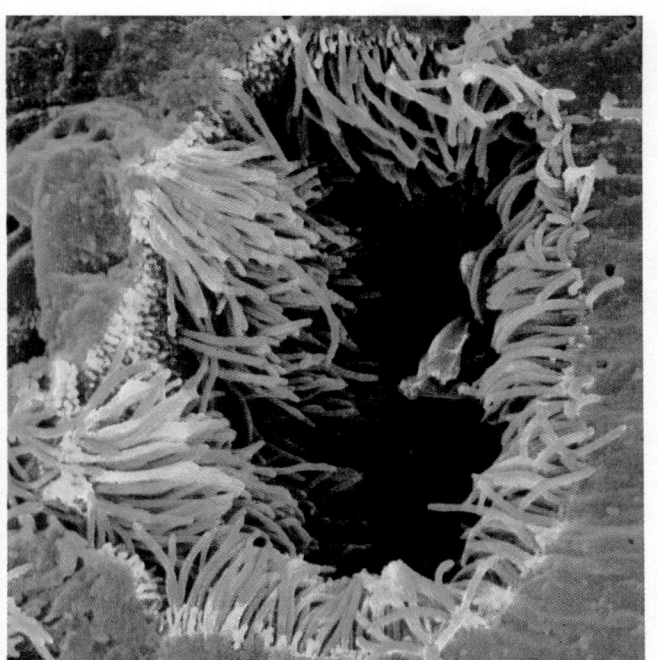

Figure 23-6   **Cilia lining your windpipe trap foreign particles to prevent them from entering your lungs. Magnification: 3570×**

The Inflammatory Response   If pathogens get by the first line of defense, a second line begins to function. Suppose you cut your finger, and pathogens enter your body. Within a short time, you notice that the cut area has become swollen, hot, painful, and red. Specialized white blood cells in the area have released a chemical called histamine. Histamine increases the flow of blood to the wound. It also increases the permeability of capillaries, allowing fluid and a variety of other white blood cells to escape into the surrounding tissue. Some of the white blood cells, called **macrophages,** are scavengers. Like amoebas, they engulf and digest pathogens that entered the cut area. The increased blood flow and accumulation of fluid are the reasons that the cut area becomes inflamed.

The temperature in the cut area often rises. This response is still another adaptation, one that interferes with the reproduction of the pathogens. Together, these various events are known as the inflammatory response.

Besides these localized events, inflammation can also cause responses throughout your body. Some injuries result in production of large numbers of white blood cells. Physicians use the relative numbers of white blood cells in a sample of your blood to help them make a diagnosis of illness. Another such response is fever. A fever is caused by a protein released by white blood cells during an inflammatory response.

Over time, these nonspecific defense mechanisms destroy most pathogens, and the infected tissue returns to normal. If a cut results in a particularly bad infection, pus may form. Pus is tissue fluid containing macrophages, other white blood cells, and dead pathogens.

Figure 23-7   **When your skin is cut, the inflammatory response rids the injured area of pathogens. Sometimes a bad infection results in the formation of pus, a combination of tissue fluid, white blood cells, and dead pathogens.**

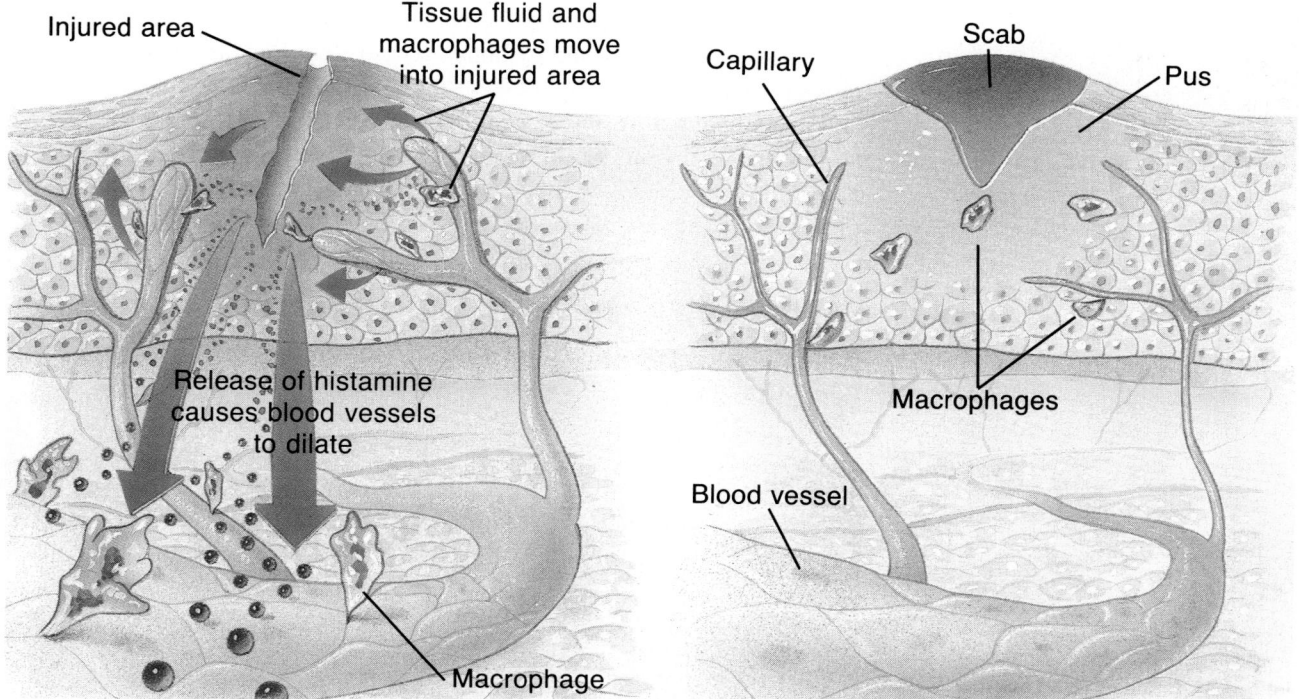

Injured area

Tissue fluid and macrophages move into injured area

Release of histamine causes blood vessels to dilate

Macrophage

Capillary

Scab

Pus

Macrophages

Blood vessel

Minilab 2

Do mouthwashes kill bacteria?

On the bottom of a petri dish prepared with sterile nutrient agar, draw a line that divides the dish in half. With a sterile toothpick, gently scrape the side of a back tooth. **CAUTION:** *Do not scrape hard enough to cause the gum to bleed.* Being careful not to cut into the surface of the agar, gently run the toothpick in an *S* shape on the surface of 1/2 of the agar. Mark C for control on the bottom of that half of the dish. Rinse your mouth with your favorite mouthwash or that supplied by your teacher. Follow the same procedure as above with another sterile toothpick. Mark an M for mouthwash on that side of the bottom of your dish. Put on the lid and seal your dish all the way around with masking tape. Incubate in a warm place for three to four days. Without opening the dish, count the bacterial colonies on each side of the dish. **CAUTION:** *Handle petri dishes with care. Be sure to wash your hands after handling petri dishes.* Put your results on the board. Make a graph of class results.

Analysis *According to class results, which section of the dish had the most bacteria colonies? According to class results, does mouthwash kill bacteria? Explain.*

Specific Defense Mechanisms

Of course, you're well aware that the adaptations you just read about are not foolproof. Pathogens can sometimes escape the first and second lines of defense. Your last cold or strep throat infection will remind you of this fact. You know, too, that you survived that cold or throat infection. You regained your health as a result of a third line of defense, called the immune response. The immune response is a highly specific means of defense. The response involves recognition and destruction of a particular pathogen. In the case of your cold or strep throat, for example, it involves recognizing and destroying the virus that causes the cold or the bacterium that causes strep throat.

The Immune Response The ability to recognize a particular pathogen occurs early in development, as an embryo learns to detect the difference between self and non-self. That is, it becomes able to distinguish between its own chemicals and those foreign to it. In general, any foreign chemical is called an antigen. An antigen may be a protein in the membrane of a bacterium or coat of a virus. It may also be a chemical secreted by a pathogen, such as a toxin. Strep throat bacteria are distinguished from all other bacteria because of their unique antigens. Similarly, cold viruses have antigens different from antigens of other viruses. A single bacterial cell's surface may have a number of different antigens. Antigens are chemical identification markers.

Lymphocytes How does your body use antigens to defend itself against infection? The immune response is carried out by the interaction of different white blood cells. Among them are cells known as **lymphocytes,** which recognize antigens and then begin to destroy specific pathogens. Lymphocytes are produced by bone marrow, the tissue in the hollow parts of some bones. Some lymphocytes mature within the bone marrow itself and are called **B cells.** Other lymphocytes pass from the bone marrow and mature in the **thymus gland,** which is located beneath the breastbone above the heart. Such lymphocytes are called **T cells** because they are processed in the thymus.

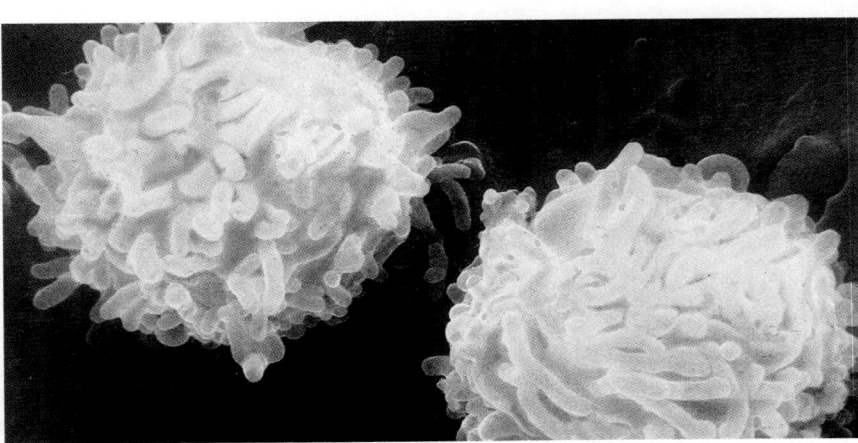

Figure 23-8 **Lymphocytes are one type of white blood cell. They may be B cells or T cells. Magnification: 10 000×**

Macrophages   Macrophages, which are involved in nonspecific defense, are also involved in specific immune responses. Once mature, some of these cells take up residence in a variety of structures around your body—the lymph vessels and lymph nodes, the spleen, and the tonsils. These structures, along with the bone marrow and thymus, form the human immune system, which you can see in Figure 23-10.

Lymph   You learned in Chapter 21 that lymph vessels carry fluid from the microscopic spaces around cells in body tissue back to your blood-stream. This tissue fluid contains lymphocytes and macrophages. Microorganisms, as well as foreign particles, are also present in tissue fluid. Located among the lymph vessels are lymph nodes through which the fluid—at this point called lymph—passes. Lymph nodes are scattered throughout your body, but they are most prominent in your neck, groin, and armpits. The lymph nodes act as filters, as macrophages within them trap and destroy some pathogens as well as debris present in the lymph fluid. B cells in the lymph nodes recognize particular pathogens and begin to kill them.

Not all pathogens, though, are present in lymph. Some are in the blood. These pathogens are detected, and some are destroyed as the blood passes through the spleen. Other pathogens are present in the air as it enters your body through your nose or mouth. These pathogens may be trapped by your tonsils.

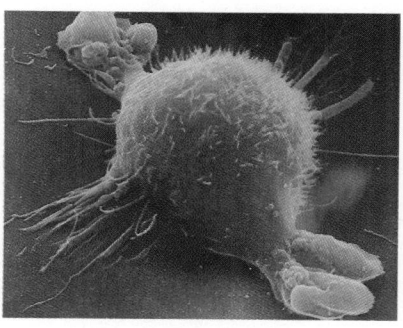

Figure 23-9   **Macrophages are white blood cells that engulf and digest pathogens. Magnification: 9000×**

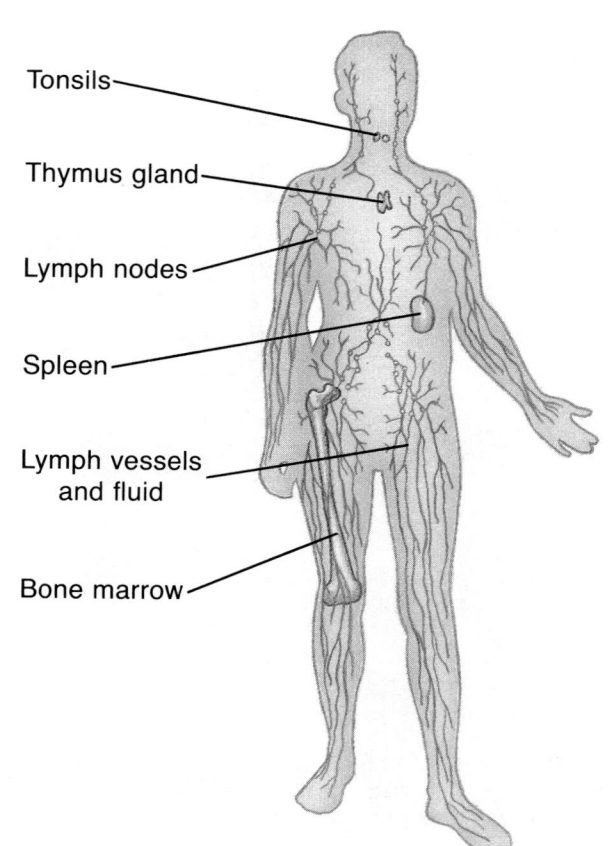

Tonsils
Thymus gland
Lymph nodes
Spleen
Lymph vessels and fluid
Bone marrow

Figure 23-10   **The human immune system includes the tonsils, thymus gland, spleen, lymph vessels and fluid, lymph nodes, and bone marrow.**

B Cells and Antibody Formation

The lymphocytes called B cells help defend against bacteria, viruses, and foreign chemicals that may be present in blood or tissue fluid. There are easily millions of possible antigens that may be detected. How can B cells recognize all of them?

Antibodies   B cells have protein molecules called antibodies embedded in their outer membranes. An antibody molecule is Y-shaped, composed of two smaller (sometimes called light) polypeptide chains bound to two larger (sometimes called heavy) polypeptide chains. Each kind of B cell bears a slightly different kind of antibody on its surface. Recognition occurs when the shape of an antibody molecule on a B cell matches the shape of part of an antigen molecule on a pathogen.

How can there be so many differently shaped antibody molecules? Look at Figure 23-11. You can see that certain areas of both the light and heavy regions are constant. That is, they have the same sequence of amino acids. But other regions of the light and heavy regions that form the two branches of the Y vary from one antibody molecule to another. In other words, they have different sequences of amino acids. The variations in amino acid sequence determine the shape of these variable regions. Each kind of antibody molecule has variable regions with a distinctive shape. In addition, each shape matches part of a specific antigen molecule in much the same way that the active site of an enzyme matches a portion of its substrate (Chapter 3). Because the two branches of the Y are the same as one another, one antibody molecule can bind to two antigen molecules.

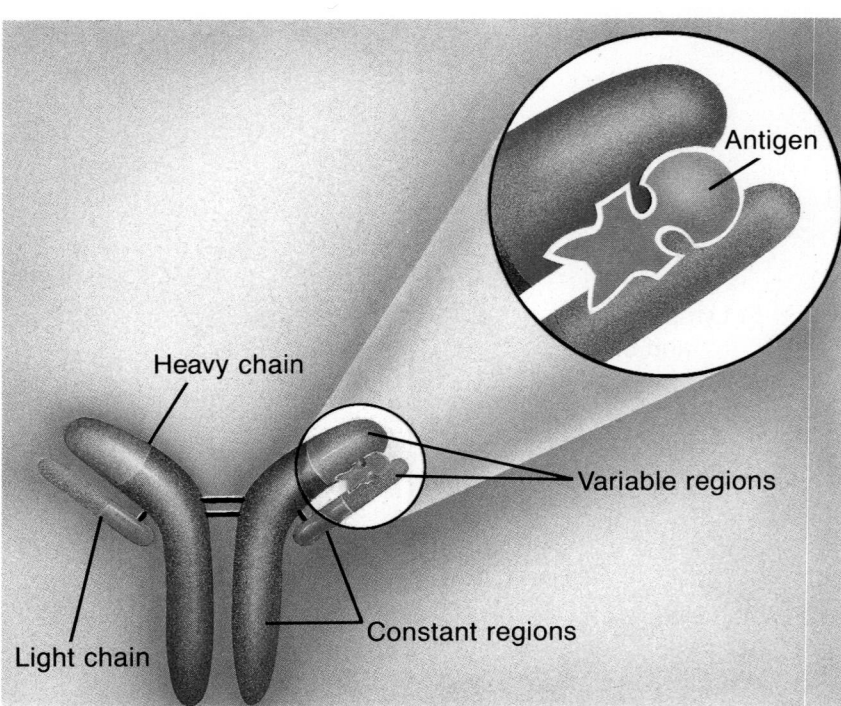

Figure 23-11   A Y-shaped antibody consists of light and heavy polypeptide chains. The tips of the Y are variable regions. The variable regions are so specific that each antibody can bind with just one type of antigen.

B Cells in Action   Look at an example of how B cells work. Suppose a person with strep throat coughs, and you inhale some of the bacteria that cause this illness. The bacteria escape your first and second lines of defense and begin to reproduce within your throat tissues. Some of the bacteria enter lymph fluid, traveling with it to lymph nodes in your neck. In the nodes are B cells whose antibodies match the antigens on the bacteria. Because they match, the antibodies and antigens become linked.

Once the antibodies of a B cell are linked to antigens, the B cell becomes activated. It begins to enlarge. The large B cell divides many times to form two types of cells—plasma cells and memory cells. You will learn about memory cells later in this chapter. **Plasma cells** produce thousands of antibody molecules per second. These antibodies, identical to the ones in the membrane of the original B cell, are released into the bloodstream and lymphatic system. There they move to the site of the invasion. When the free antibodies locate their matching antigens, they bind to them. Because each antibody joins to two antigens, one antibody molecule may bind to antigens located on two different bacteria. You can see this clumping together of antibodies and bacteria in Figure 23-12. Recall from Chapter 21 that a similar clumping occurs when an improper blood transfusion is given.

Once B cell antibodies and antigens have matched and clumping has occurred, the bacteria may be destroyed. Macrophages may engulf the bacteria by phagocytosis and digest them. This process occurs within the lymph nodes, which become swollen and tender during an infection such as strep throat. In other cases, a set of blood proteins becomes activated, forming a protein that creates an opening in the bacterium. Water enters the opening by osmosis, causing the bacterium to burst.

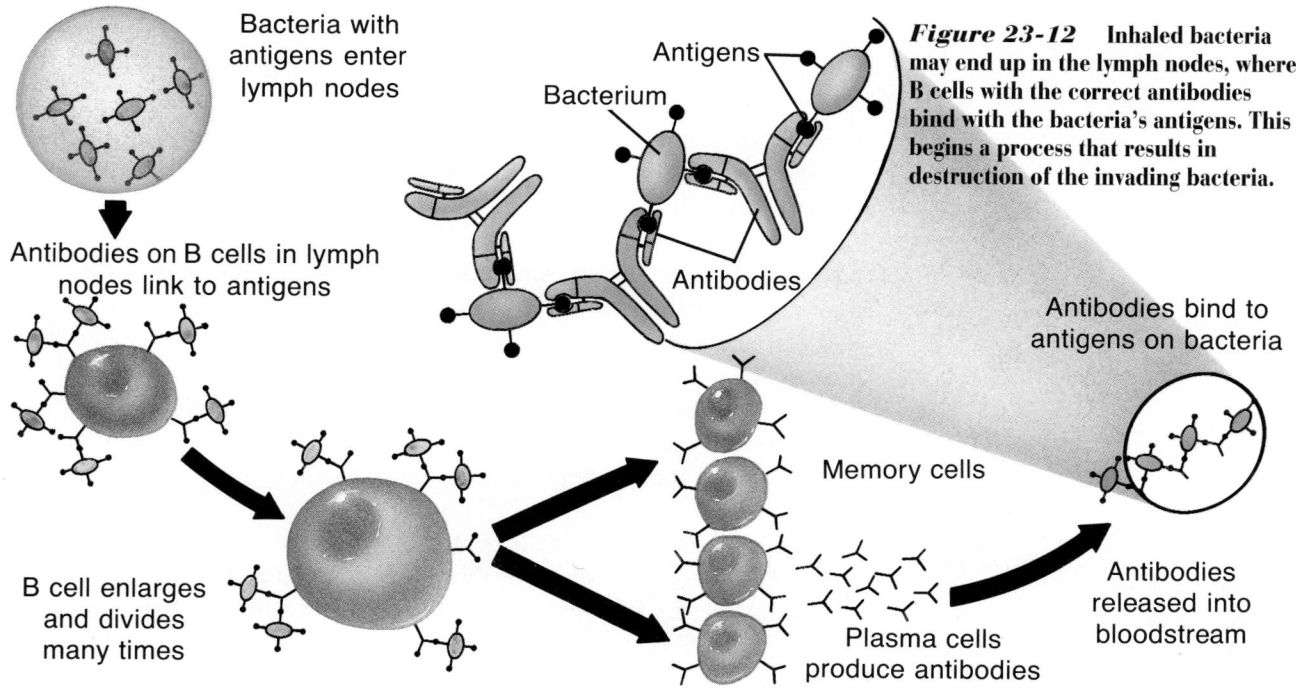

Bacteria with antigens enter lymph nodes

Antibodies on B cells in lymph nodes link to antigens

B cell enlarges and divides many times

Antigens

Bacterium

Antibodies

Figure 23-12   Inhaled bacteria may end up in the lymph nodes, where B cells with the correct antibodies bind with the bacteria's antigens. This begins a process that results in destruction of the invading bacteria.

Antibodies bind to antigens on bacteria

Memory cells

Plasma cells produce antibodies

Antibodies released into bloodstream

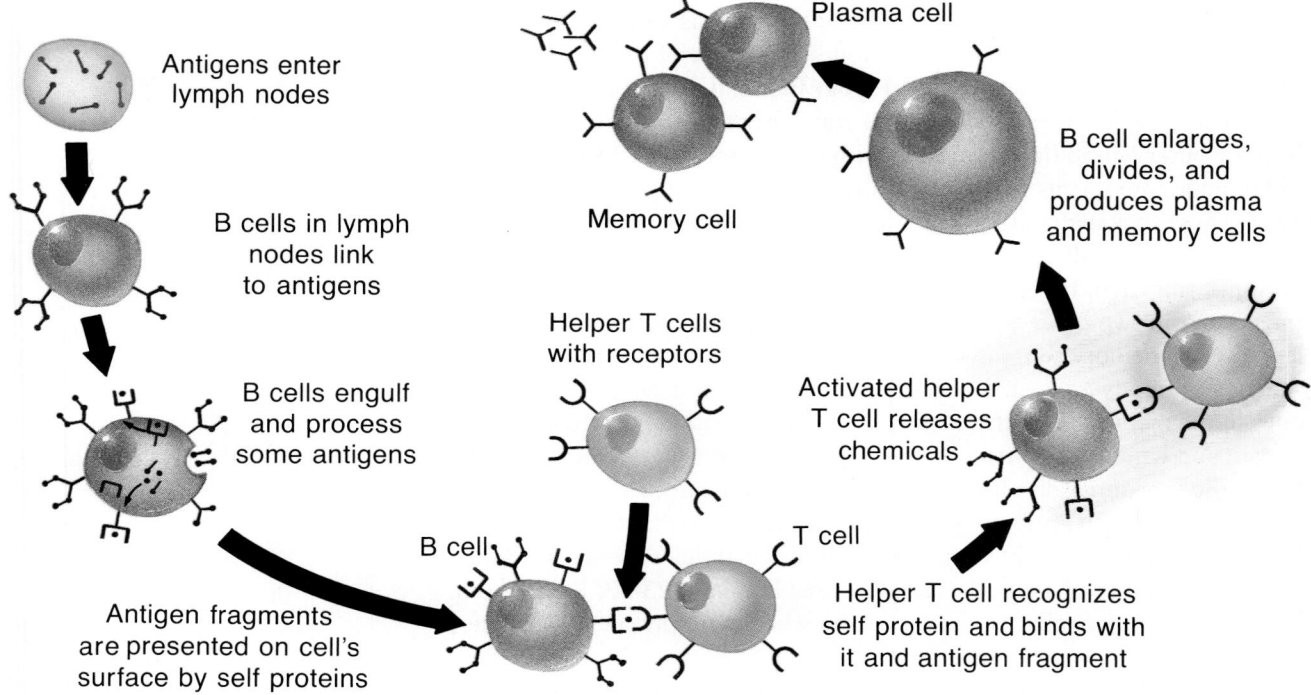

Antigens enter
lymph nodes

B cells in lymph
nodes link
to antigens

B cells engulf
and process
some antigens

Antigen fragments
are presented on cell's
surface by self proteins

B cell

T cell

Helper T cells
with receptors

Plasma cell

Memory cell

B cell enlarges,
divides, and
produces plasma
and memory cells

Activated helper
T cell releases
chemicals

Helper T cell recognizes
self protein and binds with
it and antigen fragment

Figure 23-13 **Helper T cells have
receptors that recognize self proteins
on B cells. When the self proteins have
antigen fragments, helper T cells bind
with them and become activated.**

Regulation of B Cells by T Cells   Can you see how the formation of antibodies depends on the activation of B cells? In turn, the growth and division of activated B cells depend on their binding with certain T cells.
These special cells are called **helper T cells.** In order to bind with a
B cell—or, for that matter, to affect any cell—a T cell must first recognize the other cell. How does a T cell recognize a B cell?

Embedded in the membrane of all T cells are **receptors,** proteins similar to the antibodies of B cells. Receptors recognize certain self proteins
that are present in the membranes of other cells in the body. However,
they will bind to a self protein only if that protein is linked to an antigen.
If any T cell encounters another cell displaying a self protein and an antigen that matches its receptor, the receptor will bind to the self protein
and antigen, and the T cell will become activated.

For a helper T cell to bind with a B cell, therefore, the B cell must first
be joined with an antigen. After a B cell's antibodies have linked with
their matching antigens, some of those antigens are brought into the
B cell by endocytosis. There, the antigens are broken up. A fragment of
the original antigen is linked to a self protein inside the B cell. Once
linked, the fragment and self protein are shuttled back to the B cell's
membrane. The helper T cell's receptor recognizes the self protein and
binds to it and the antigen fragment on the B cell, and the T cell becomes
activated.

Helper T cells can also be activated by binding to macrophages.
Macrophages break up antigens they have engulfed and, like B cells, send
fragments linked with self proteins back to the membrane. Once activated,
the helper T cell releases chemicals that stimulate the B cell to divide and
form plasma cells. These events are summarized in Figure 23-13.

The chemicals released by activated helper T cells also attract macrophages and activate other kinds of T cells. **Suppressor T cells** counteract the effect of helper T cells by releasing chemicals that inhibit activity of B cells and macrophages. The number of helper T cells decreases and the number of suppressor T cells increases as an infection is brought under control.

Allergies   Do you suffer from hay fever or know someone who does? Hay fever is an immune response to an antigen that does not cause a reaction in most people. As such, it is an allergy. In hay fever, pollen acts as an antigen against which plasma cells begin producing antibodies. Some of the antibodies cause release of histamine. The continued high level of histamine accounts for the sneezing and runny noses of hay fever victims. Allergies are thought to result when suppressor T cells fail to balance the effects of the plasma cells, antibodies, and histamine. Other allergies can be caused by certain foods. Hives, reddened swellings on skin similar to a rash, are a common indication of such allergies.

Thinking Lab

DRAW A CONCLUSION

How do helper T cells act?

Background
How do you fight off a cold, "strep" throat, or cancer? Your immune system protects you from harmful microorganisms. Recently, scientists have learned how different white blood cells function as part of your immune system. B cells, white blood cells made in the bone marrow, make receptors called antibodies that can deactivate toxins produced by bacteria. Helper T cells,

made in the thymus gland, get information from another white blood cell, a macrophage, about the type of toxin present in the body. The helper T cell, activated by the macrophage, gives off a substance which stimulates B cells.

You Try It
Study the summary diagram of recent discoveries of interactions of B cells, helper T cells, and macrophages.

Draw conclusions about the diagram by answering the following questions.
1. How do macrophages get information about bacterial toxins?
2. What do B cells do in the presence of interleukin?
3. Where does interleukin come from?
4. What happens to the antibodies produced by B cells?

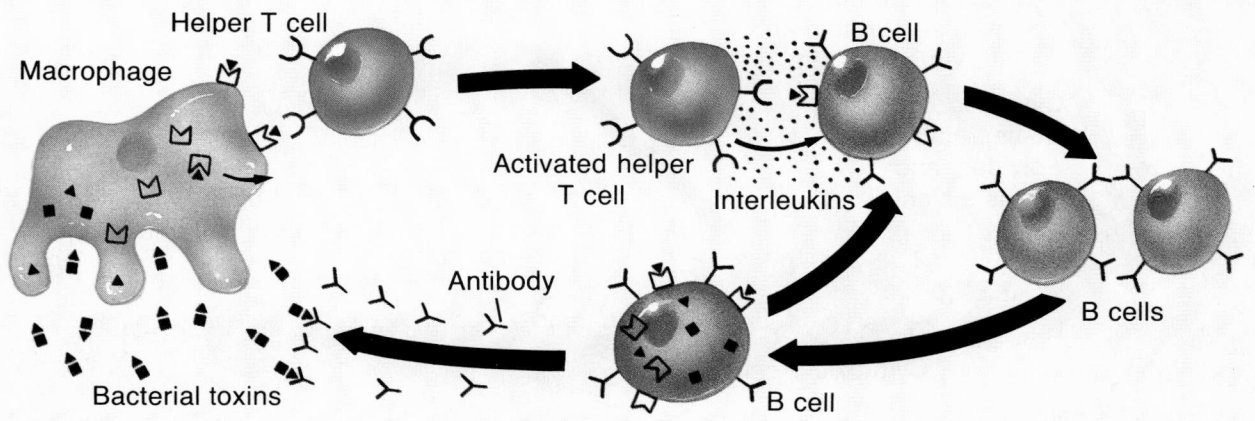

T Cells and Cellular Immunity

You've just read how B cell stimulation leads to the production of antibodies that detect and destroy pathogens present in blood and tissue fluid. You've also seen how different T cells are involved in that process. Yet T cells function in ways other than regulating the activity of B cells and macrophages. They also interact directly with cells infected by viruses and with foreign cells.

One kind of T cell defends against body cells in which viruses are reproducing. This cell, called a **cytotoxic T cell,** acts to destroy the infected cell. But like any other T cell, to be activated it must first bind with a self protein linked to a foreign antigen. The viral mRNA within the infected cell directs the production of proteins needed to make new virus coats. Those proteins are foreign to the person's body and therefore serve as antigens. Some of those antigens move to the infected cell's membrane and become attached to a self protein. At this point, a receptor on a cytotoxic T cell recognizes the antigen linked to the self protein on the infected cell's membrane and binds with it. After binding to the infected cell, the cytotoxic T cell becomes activated. It then forms a clone of cells that destroy the infected cells. It also releases chemicals that attract macrophages. Some cytotoxic T cells secrete powerful chemicals that kill the infected cells. Helper and suppressor T cells help regulate the activity of cytotoxic T cells. How is a cytotoxic T cell similar to a helper T cell?

Figure 23-14 **Cytotoxic T cells kill infected body cells indirectly by attracting macrophages. Cytotoxic T cells also kill infected cells directly by releasing proteins that disrupt the cells' plasma membranes.**

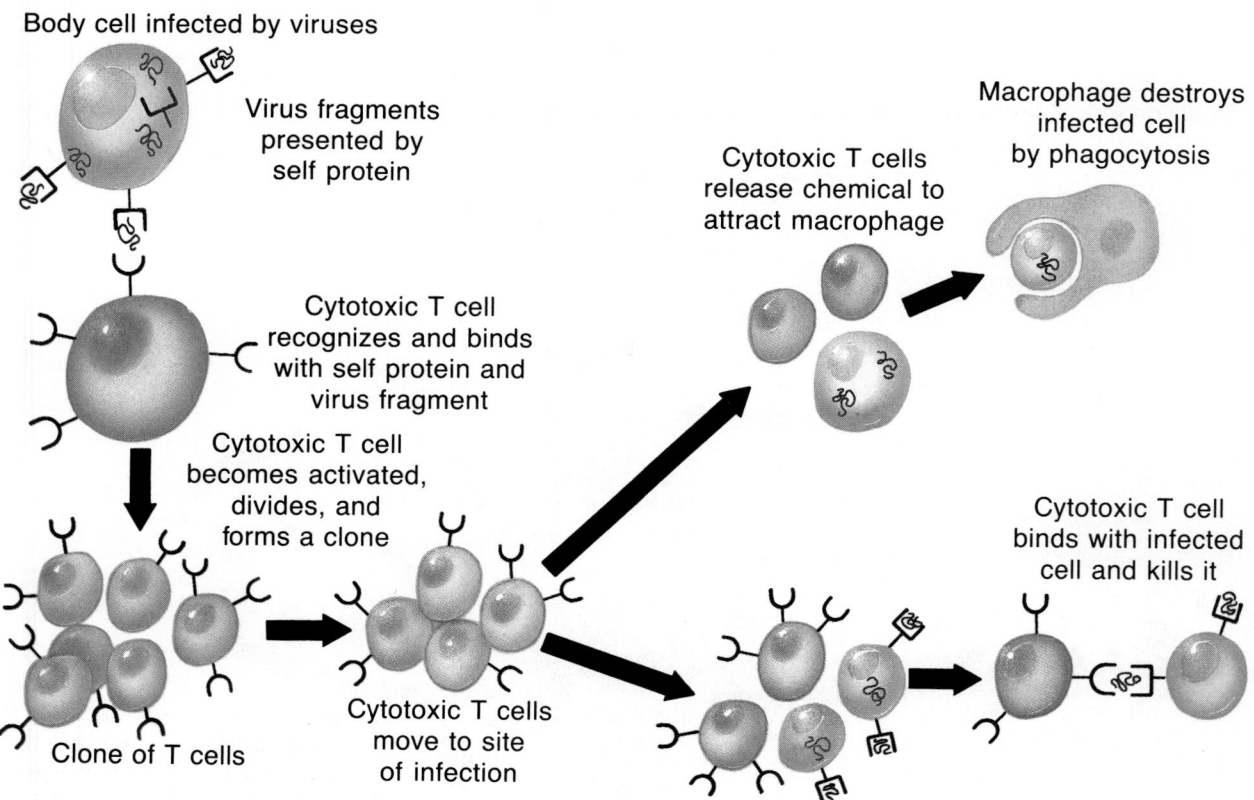

Body cell infected by viruses

Virus fragments presented by self protein

Cytotoxic T cells release chemical to attract macrophage

Macrophage destroys infected cell by phagocytosis

Cytotoxic T cell recognizes and binds with self protein and virus fragment

Cytotoxic T cell becomes activated, divides, and forms a clone

Clone of T cells

Cytotoxic T cells move to site of infection

Cytotoxic T cell binds with infected cell and kills it

Why You Feel Sick

Some of the chemicals that act to signal and regulate various parts of the immune system are also responsible for making a person feel sick. You've already learned about histamine and its effects. When you are fighting an infection, macrophages release another chemical that resets your body's thermostat to a higher setting, resulting first in chills and then in fever. Though fever may make you feel uncomfortable, it is an adaptation. A slightly higher-than-normal body temperature interferes with the ability of viruses and bacteria to reproduce. It also increases the production of antibodies and replication of T cells. The same chemical that increases your body temperature also causes an increase in the production of certain hormones that are secreted during periods of stress. High levels of these hormones inhibit reactions that store fat or convert glucose to glycogen. Instead, fats and sugars are promptly delivered to your body tissues, where they provide the energy needed to fuel the immune response. In addition, the hormones tend to decrease your appetite and cause sleepiness. They also make a sick person more susceptible to pain, which explains why many illnesses are accompanied by headaches and sore joints. Aspirin is effective in combating fever and pain associated with many illnesses because it interferes with chemical changes that lead to those symptoms.

Defense Against Cancer

Cytotoxic T cells defend not only against infectious diseases but also against cancer. Recall from Chapter 10 that cancers are diseases in which cells grow and reproduce uncontrollably. In addition, the cells are abnormal, resembling embryonic cells rather than differentiated cells. They also differ in shape from normal cells, have unique antigens on their surfaces, and have self antigens. Cytotoxic T cells can bind to these cancer cells and destroy them. It is likely that many cells become cancerous during a person's life, but they are destroyed by the immune system before they can multiply and cause damage. A person may develop cancer when his or her immune system fails to perform its normal function of detecting and destroying cancer cells.

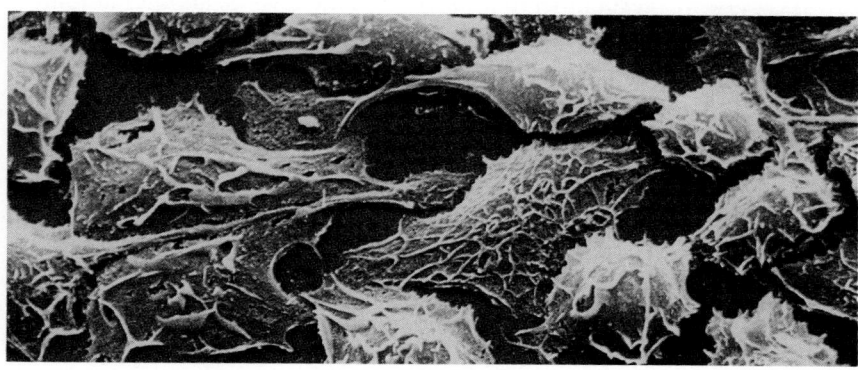

Figure 23-15 Cancer cells like these are usually killed by cytotoxic T cells. What happens if cancer cells are not destroyed? Magnification: 2000×

Rejection of Transplants

Organ transplants are considered by many to be a miracle of modern medicine. Replacement of a failed heart or liver offers a person a promise of survival when no other medical technology can do so. However, the person's immune system treats the transplanted organ as it has evolved to do—as foreign matter to be destroyed. The foreign tissue will have some self proteins in common with the person into which it has been transplanted. However, it will also have self proteins of its own, which are foreign to the transplant recipient. Cytotoxic T cell receptors will bind to both proteins, and ultimately the T cells will destroy the tissue. Because the T cell receptors can bind only to both a self protein and an antigen, the closer the match between donor and recipient tissues, the less the chance for rejection. In other words, if the transplanted tissue has many proteins like the self proteins of the recipient, fewer T cells will be able to attack.

To combat the rejection of transplanted tissue, patients are usually given drugs that dampen the immune response. The problem with many of these drugs, though, is that the patient then becomes susceptible to other diseases. Transplant recipients, for example, have an increased risk of cancer. However, a drug called cyclosporin affects only the T cells that fight the transplant. Thus, the other cells of the immune system are left to defend against other antigens.

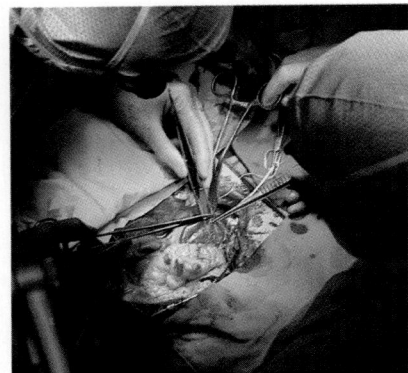

Figure 23-16 **Heart transplants are no longer rare in the United States. What other organs are routinely transplanted?**

Autoimmune Diseases

Recall that an embryo becomes able to distinguish self from non-self early in its development. This distinction is usually retained throughout life. However, for reasons we still don't know, some people's immune systems turn against them. For these people, certain of their own tissues are attacked by antibodies or by cytotoxic T cells as if they were transplanted tissues. This abnormal response of the immune system to part of a person's own body may cause a variety of **autoimmune diseases.**

Examples of autoimmune diseases include myasthenia gravis and multiple sclerosis, diseases affecting the nervous system. In myasthenia gravis, antibodies attack the connections between nerves and muscles. In multiple sclerosis, T cells attack the outer covering around certain nerve cells. Rheumatic fever is another autoimmune disease in which antibodies against certain bacteria react with heart and skeletal muscles. Juvenile onset diabetes mellitus is caused by T cells that attack certain pancreas cells. In this case, the body cells may be infected with a virus.

Treating Autoimmune Diseases Scientists are working to find ways of helping people with autoimmune diseases. However, their work is hampered by the fact that they do not fully understand what makes the immune system turn against the body of which it is a part.

Figure 23-17 **In the autoimmune disease called rheumatoid arthritis, the membranes around joints are attacked by the immune system. Rheumatoid arthritis is extremely painful and often leads to crippling like this.**

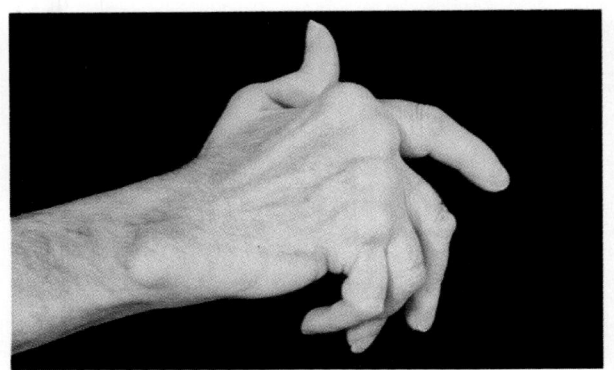

One possible method of treatment is to introduce toxic chemicals that bind to and kill the T cells that cause the disease. This cure might be effected by producing an antibody that matches the T cell's receptor and then linking the toxic chemical to that antibody. The antibodies, attached to the toxic chemicals, would be injected into the body. Because the antibody would match only with receptors of a particular T cell, only that kind of T cell would be destroyed by the toxic chemical. It may also be possible to inject substances that would bind to the self proteins of the body cells under attack. The substances, once attached to the self proteins, would prevent the T cell receptors from binding with them.

Monoclonal Antibodies   You've just read how particular antibodies might be made and used to treat autoimmune diseases. A means of producing pure antibodies has been available since the 1970s. These antibodies are not produced in a test tube. Rather, they are produced by mice.

To produce a specific antibody, a mouse is injected with the proper antigen. Lymphocytes in the mouse's spleen respond by making antibodies against the antigen in the injection. Each lymphocyte makes a specific antibody. The lymphocytes are then removed from the spleen and fused with certain cancer cells, which divide repeatedly in a culture medium in the laboratory. Each fused cell is cultured and forms a pure clone. The clone consists of cells that make only one kind of antibody, so the antibodies produced are called **monoclonal antibodies.**

After the antibody produced by each clone is identified, the clones are grown in culture. Even greater quantities of pure antibody can be obtained by injecting a mouse with the fused cells. The cells form a tumor that produces a large amount of the antibody. The antibody is later filtered from the body fluids of the mouse. These techniques are pictured in Figure 23-18.

Figure 23-18   **Monoclonal antibodies can be produced in the laboratory and might be used to fight autoimmune diseases.**

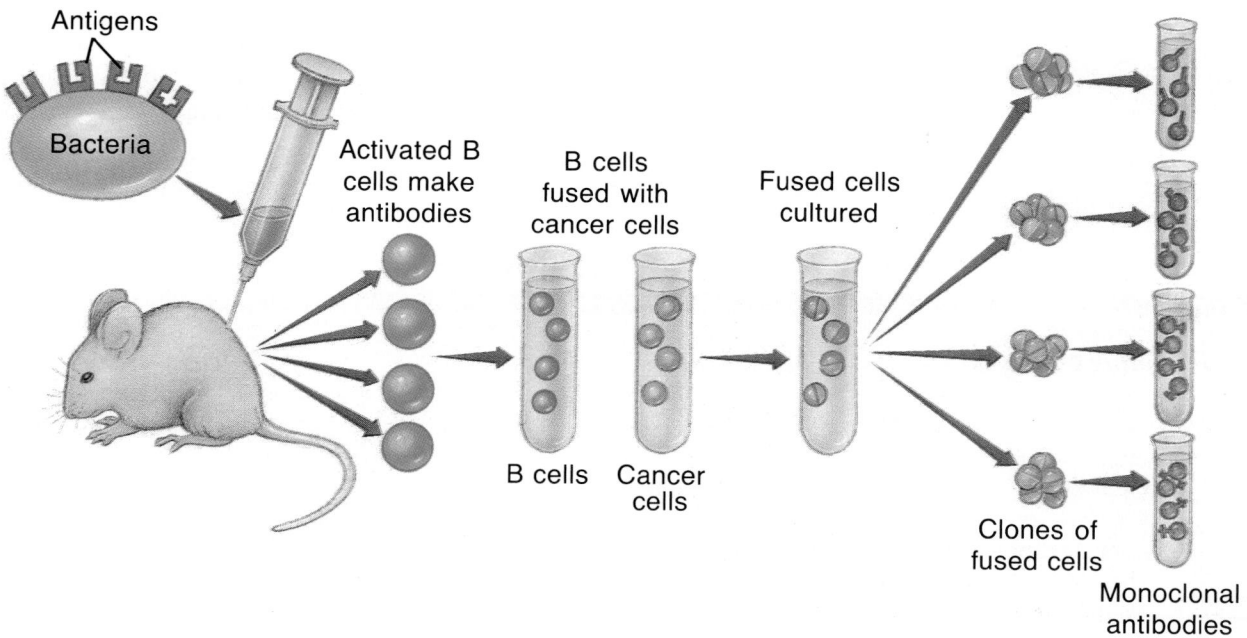

Antigens

Bacteria

Activated B cells make antibodies

B cells fused with cancer cells

Fused cells cultured

B cells   Cancer cells

Clones of fused cells

Monoclonal antibodies

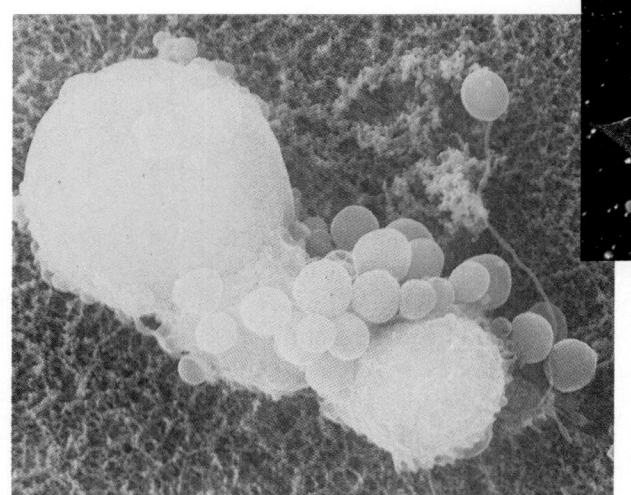

Figure 23-19 Cancer cells in humans vary in size and shape. The photograph on the right shows cancer cells from a brain tumor. In the photograph on the left, a cytotoxic T cell is attacking a cancer cell. Magnification: 500× (right); 2500× (left)

Monoclonal antibodies have taught us a great deal about how the immune system operates. They are also used in home pregnancy tests. In such tests, antibodies react with a chemical present only in the urine of a pregnant woman. Combining the antibodies with the chemical in the urine results in a visible change that the woman can easily see.

In the future, monoclonal antibodies might also be used to detect and treat cancer. The antibodies would be designed to recognize abnormal antigens present on cancer cells. Attaching a poison to the antibodies would lead to the destruction of the cancer cells. After the antibodies bind to the antigens on the cancer cells, the poison would enter and kill the cells. This technique has already proved successful in killing certain cancer cells grown in culture.

Section Review

Understanding Concepts

1. You get a splinter in your foot while walking barefoot. The next day, the area around the splinter is swollen, sore, and red. Explain why.
2. How do two antibodies that recognize two different antigens differ from one another? How are they the same?
3. Explain why skin grafts might be rejected.
4. **Skill Review—Sequencing:** Sequence the steps by which the immune system would identify and destroy a bacterium present in your tissue fluid. For more help, see Organizing Information in the **Skill Handbook.**
5. **Thinking Critically:** Explain why burn victims are susceptible to infection.

23.3 Immunity and Prevention of Disease

Did you have chicken pox as a child? If you did, you might recall having a fever and certainly will remember the rash that appeared all over your body. You also know that you will probably not get chicken pox again. Why are you protected from a second bout of this disease? Why do you get some other infections over and over again? How can many diseases be prevented from occurring in the first place? Why can medicines be used to treat some diseases but not others? The answers to all those questions concern whether and how your body develops immunity to diseases.

Active Immunity

The first time you are exposed to a particular antigen, your immune system springs into action, producing clones of plasma cells or cytotoxic T cells that help stop the infection. However, it takes several days for your immune system to mobilize its forces against the pathogens. During that time, you feel sick. Gradually your symptoms disappear, and your immune system wipes out the enemy. In many cases, the next time you are exposed to the same antigen, your immune system responds much more quickly and you don't become sick at all. Why?

During your first exposure to an antigen, when clones of plasma cells or cytotoxic T cells are being produced, clones of **memory cells** are also produced. Those memory cells remain in your body long after an infection is over. Some live for months or years; others last your lifetime. Thus, if a pathogen reappears, your body can react very quickly, bypassing the early steps of the immune response. The antigens of the pathogen are recognized very quickly by either antibodies or cytotoxic T cells and destroyed before the pathogen can cause symptoms. In such a case, you have become immune to the pathogen. When memory cells enable you to resist a disease, you have what is known as **active immunity.**

Objectives

Distinguish between active and passive immunity.

Discuss the use of drugs in treating bacterial and viral diseases.

Relate the life cycle of HIV to the symptoms of AIDS.

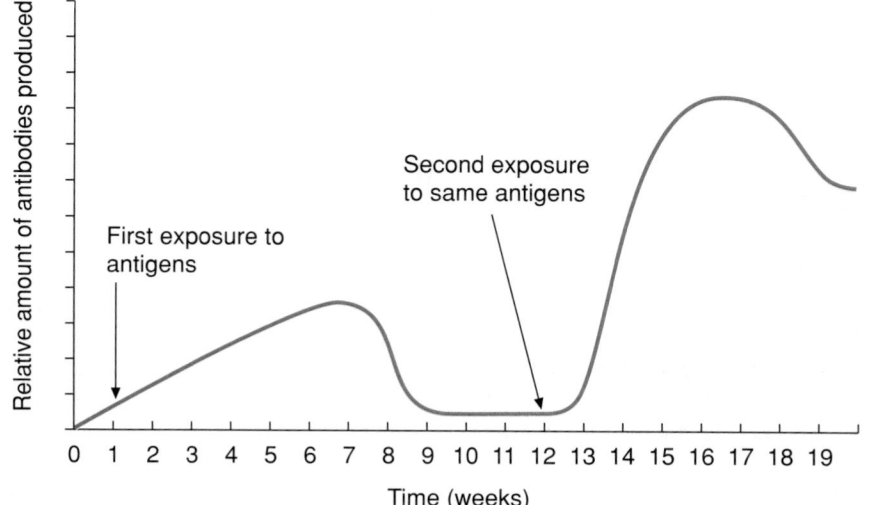

First exposure to antigens

Second exposure to same antigens

Relative amount of antibodies produced

Time (weeks)

Figure 23-20 **Memory B cells have antibodies for antigens that have attacked your body in the past. Memory cytotoxic T cells have receptors. Both types of memory cells allow the immune system to respond quickly to new infections of the same antigen.**

Use of Vaccines   The principle of active immunity may be used to prevent certain diseases from occurring. Solutions prepared from weakened or dead microorganisms, viruses, or toxins are called **vaccines.** Vaccines injected into a healthy person do not normally cause the disease, but they do cause antibodies or cytotoxic T cells to form. When you are vaccinated, you may suffer mild discomfort, but the antibodies or cytotoxic T cells will protect you from the more potent disease organisms or toxins if you are later exposed to them.

Vaccines have been developed for such diseases as polio, measles, mumps, some forms of influenza, tetanus, diphtheria, and bacterial pneumonia. Why is there no vaccination for the common cold? Why, for that matter, do you keep getting colds over and over again?

The reason is that there is no single cold virus. In fact, there are 100 different forms of one kind of cold virus alone. Moreover, these viruses mutate. Mutations result in different alleles, which result, in turn, in different protein antigens. Therefore, one generation of cold virus may have different antigens from its ancestors. Yet one region of the virus's protein coat must remain the same. It is the part of the coat that matches with a receptor on the membrane of cells in the nasal lining. Though different, at least half of all cold viruses attach to the same receptor on nose cell

I N V E S T I G A T I O N

Antibiotic Properties of Plants and Other Substances

Medicine men and women of the past used a variety of plants for healing. Scientists have studied the medicinal effects of a variety of these plants and find that in some cases there is a medicinal effect. However, some folklore remedies have no merit; in fact, some can be harmful. In this lab, you will test the antibiotic properties of plants and other substances you may have heard have healing effects.

Problem

Do plants and other substances have antibiotic properties?

Safety Precautions

Take appropriate precautions when using a Bunsen burner. Wear goggles. Wash hands at the end of this lab. Do not use poisonous plants or substances. Even though *E. coli* is not pathogenic under normal circumstances, it is wise to treat your cultures with caution.

Hypothesis

Make a hypothesis about what plants or other substances might have antibiotic properties. Explain your reasons for making this hypothesis.

Experimental Plan

1. As a group, make a list of possible ways you might test your hypothesis using the materials your teacher has made available or materials you can bring from home.

membranes. Based on that knowledge, scientists are attempting to produce antibodies that would link to the receptors, preventing the viruses from attaching. The antibodies might be given in the form of drops or a spray. The drug would not cure a cold, but it would prevent one. Vaccinations in the future may be sprayed into your nose!

Synthetic Vaccines Vaccines made from weakened bacteria, viruses, or toxins often take a long time to produce and are usually expensive to develop. They also may cause side effects. Recent advances have made possible the production of synthetic viral vaccines. The idea behind these vaccines is that an entire virus is not necessary. Only a specific antigen, a part of a protein in the virus's coat, is recognized by the host. Synthetic vaccines contain only these recognizable antigens. Another advantage of synthetic vaccines is the possibility of producing a single vaccine that would provide immunity against several viral diseases.

Figure 23-21 **Vaccination programs protected children against poliomyelitis in the 1950s.**

2. Agree on one idea from your group's list that could be investigated in the classroom.
3. Design an experiment that will test one variable at a time. Plan to collect quantitative data.
4. Following the style of a recipe, write a numbered list of directions that anyone could follow.
5. Make a list of materials and the quantities you will need.

Checking the Plan

Discuss the following points with other group members to decide the final procedures for your experiment.
1. What variables will need to be controlled?
2. What is your control?
3. What will you measure or count?
4. How will you make your counts?

5. Have you designed and made a table for collecting data?

Make sure your teacher has approved your experimental plan before you proceed further.

Data and Observations

Carry out your experiment; make your measurements; and complete your data table. Make a graph of your results.

Analysis and Conclusions

1. According to your results, which of the plants or substances that you tested have antibiotic properties? Explain.
2. Were your results supported by your hypothesis? Explain.
3. If your hypothesis was not supported, what types of experimental error could be responsible?

4. Consult with another group that tested the same plants or substances. Were their results the same as yours? Explain.
5. Of the plants or substances you tested, which ones had the most antibiotic effect? How could you tell?
6. Why is it important that new antibiotics be developed?

Going Further

Based on this lab experience, design an experiment that would help to answer another question that arose from your work. Assume that you will have unlimited resources and will not be conducting this lab in class.

Figure 23-22 What kind of immunity is passed from the mother to a nursing baby?

Passive Immunity

When a person becomes infected, the body often cannot produce antibodies quickly enough to fight the disease. In the case of some diseases—for example, tetanus—this slow response could be fatal. However, a person might survive the disease by receiving antibodies against the disease made by another animal. The antibodies are extracted from the animal's blood. A solution of antibodies and blood serum is called an **antiserum.** Injection of an antiserum provides borrowed antibodies to fight the disease. Protection in this case results in what is called **passive immunity** because the antibodies are not being made by the person exposed to the pathogen.

Passive immunity works more quickly than active immunity, but its effects last for a shorter time. Human infants have passive immunity at birth because they have some antibodies that passed across the placenta from the mother. Borrowed antibodies may later be passed to an infant in the mother's milk. Passive immunity protects an infant during its early life when its own immune system is just developing and when it is exposed to a variety of pathogens for the first time.

Treatment of Disease

You know that a person is sick while his or her immune system is building its response to a pathogen. If the infection is bacterial, the bacteria are reproducing rapidly during this time. The length of an illness caused by some bacterial pathogens can be decreased by antibiotic medicines, which reduce the rate of bacterial reproduction. Antibiotics are produced by many kinds of simple organisms and destroy bacteria in a variety of ways. Penicillin, for example, interferes with the production of the cell walls of some bacteria.

Viral Disease Treatment   Antibiotics cannot be used against viruses because viruses are not living organisms. Viral diseases must be treated differently from bacterial diseases. Drugs that disrupt the life cycle of a virus cannot be used if they also destroy the infected host cell. Some drugs used to treat viral diseases prevent the viral nucleic acid from replicating. Others interfere with transcription or translation. Still others might work by preventing viruses from entering host cells, as with the possible cold virus drug discussed earlier.

Interferon   Protection from a first attack by viruses is not the result of the immune system. Instead, a protein is released by the infected host cell. The protein, called **interferon,** travels to and enters a healthy cell, where it prevents viral reproduction. Several kinds of interferons have been discovered. Some work by activating an enzyme that breaks down viral mRNA. Others prevent translation of viral mRNA or stop the packaging and release of new viruses.

When interferon was discovered, it was believed that it might be used in antiviral drugs. Using recombinant DNA techniques, researchers have produced large quantities of interferon. Tests have been conducted to determine whether interferon can be used to prevent or to treat a variety of viral diseases or certain forms of cancer. Although the results of many tests have not been promising, interferon has been shown to be effective against the virus that causes hepatitis B, a disease that can lead to liver cancer and other liver problems. A recent study showed that injections of interferon prevented the virus from destroying the liver in about half the patients tested; it cured about one in ten. Considering that 300 million people worldwide suffer from hepatitis B, this study is particularly encouraging.

AIDS

With autoimmune diseases, the body turns against itself; the immune system responds to body tissues as if they were foreign. When a person has AIDS (acquired immune deficiency syndrome), his or her immune system is slowly destroyed. As this destruction occurs, the person suffering from AIDS is not able to fight off pathogens or abnormal cells that would normally be attacked by the immune system of a healthy person. The destruction of the immune system is so overwhelming that AIDS is always a fatal disease.

Transmission   AIDS is caused by a retrovirus (Chapter 15) named human immunodeficiency virus, or HIV. In infected persons, HIV is present in body fluids such as blood and semen. Infection occurs when the virus enters the bloodstream through a cut or tear in the tissues.

HIV can be transmitted in four known ways, all of which involve the exchange of body fluids: (1) Transmission can occur during sexual intercourse in semen or vaginal secretions. (2) Sharing needles during intravenous drug use is another mode of transmission. A small amount of blood can remain on a needle used by a person infected with HIV. If this needle is used by a second person, the virus present in the blood can be injected into the second person's blood. (3) Many cases of infection with the virus have resulted from transfusions of blood or blood products that contained HIV. However, since 1985, this means of transmission has declined as a result of a test that screens blood for the presence of HIV antibodies. Currently, an extremely small chance exists that HIV can be spread through transfusions. (4) Finally, the virus can be transmitted from the blood of an infected female to a fetus during pregnancy. The virus passes to the fetus through the placenta.

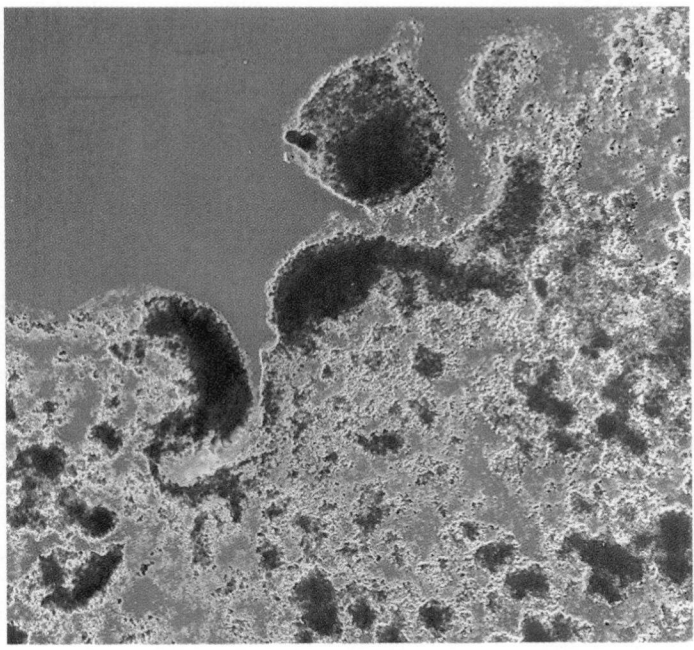

Figure 23-23   **The AIDS virus consists of an outer covering taken from the cells it infects and an inner protein capsule that contains RNA and an enzyme. In this photograph, the AIDS virus is the small red sphere near the top. Magnification: 32 400×**

Infection and Onset of Symptoms   After entering the blood, HIV attacks and destroys helper T cells. HIV can recognize those cells because the proteins projecting from the virus match with receptors on the cells' membranes. Eventually, the viruses destroy the helper T cells as they leave them and move on to infect other helper T cells.

Because helper T cells trigger the production of antibodies by B cells and also stimulate cytotoxic T cells, their destruction causes the immune system to begin to break down. In addition, the activities of macrophages and suppressor T cells are affected. As HIV reproduction begins, so, too, do early symptoms of AIDS. Weight loss, fatigue, and swollen lymph nodes are common. Although antibodies against HIV are produced, they are not made quickly enough, and the immune system gradually becomes overwhelmed. The infected person is then prone to a variety of diseases that normally would be fought off by the immune system.

Three major diseases associated with AIDS are a rare type of pneumonia, a rare form of cancer that appears as spots on the skin or in the mouth, and tuberculosis. In the later stages of AIDS, there may also be a loss of normal mental functions as HIV invades and destroys nerve cells. Eventually, the patient dies from one of these diseases or from heart failure. At this time, it seems that anyone infected with HIV will eventually die from AIDS. The time from infection to onset of symptoms may be as long as ten years, but most cases develop more rapidly.

PSYCHOLOGY CONNECTION

How Attitudes Affect Health

● ●

Do people who are depressed get sick more often? That is the question that researchers at The Ohio State University have been studying. What is the effect of the constant stress of caring for people with Alzheimer's disease on the caregivers? Earlier studies indicate that elderly caregivers were more depressed and had twice as many colds and bouts with the flu as a control group of similar age who weren't caregivers.

It has long been established that stress and anxiety promote heart disease and other maladies, but this study seeks to establish a connection to a specific response by the immune system. One hypothesis to be tested is that stress affects immunity because it causes the release of hormones. Scientists hypothesize that these hormones either suppress or interfere with the immune system.

The study of elderly caregivers and a control group of similar age who are not caregivers involves inoculation with an influenza vaccine. After inoculation, the response of the immune systems of the test subjects will be monitored. The data should yield valuable information on how stress affects parts of the immune system and lead to further discoveries about how to lead a healthy life.

▬▬▬▬

1. **Connection to Nutrition:** Many people under stress have been advised to adopt an ACE diet. This diet is named after three vitamins featured in the cuisine—A, C, and E. Find out why these particular vitamins are recommended and what their effects are.

2. **Solve the Problem:** One problem with the study in the story is that the study population (caregivers) is more isolated than the control group. That is, they tend to remain at home with their ill loved one and do not get out for other activities. Thus, they would be less likely to be exposed to disease than the controls. How have the researchers compensated for this difference?

Prevention and Treatment Clearly, AIDS is a major health problem. To stop the spread of infection by HIV, high-risk behavior associated with transmission of the virus must be avoided. Needles used to inject drugs must not be shared. People must exercise good judgment about their sexual activities and choice of partners. A person infected with HIV may have no symptoms and can unknowingly infect others. Remember that once infected, a person will die from AIDS.

Ideally, AIDS could be prevented if a vaccine against HIV could be produced. A major problem in developing such a vaccine is the fact that the structure of some of the HIV proteins varies greatly, and the proteins are subject to constant mutation. There can be no single vaccine to prevent AIDS. In addition, there have been no animal models in which to study AIDS and test vaccines as they are developed.

Another direction of research is the development of drugs to relieve symptoms or destroy the virus. To date, only two drugs have been approved for use in all patients, AZT and DDI. Although the drugs have been effective in slowing the progress of AIDS symptoms, they do have toxic side effects. New therapies using these drugs have been encouraging.

LOOKING AHEAD

THIS CHAPTER concludes the unit in which you have studied several systems—reproduction, development, digestion, excretion, transport, immunity—that carry out functions. Systems must have controls so that the various tissues and organs operate smoothly under a variety of conditions. One means of control is chemical, which you will study in the next chapter. You will learn about the responses caused by a variety of chemicals and how levels of chemicals change according to the needs of the organism.

Section Review

Understanding Concepts

1. Explain why you would most likely not get measles a second time.
2. How does reproduction of HIV lead to the symptoms of AIDS?
3. Many antibiotics can be used to kill a broad range of bacteria. Why do you think discovery of interferon caused biologists to hope that it could be used against a broad range of viruses?

4. **Skill Review—Comparing and Contrasting:** One person has been vaccinated against tetanus. Another person has received an antiserum to treat tetanus. Compare the degree and type of immunity to tetanus in the two people three years later. Explain. For more help, see Thinking Critically in the **Skill Handbook**.

5. **Thinking Critically:** Suppose a drug were developed that interfered with synthesis of proteins needed for reproduction of a virus causing a particular disease. Why might such a drug not be able to be used safely?

BIOLOGY, TECHNOLOGY, AND SOCIETY

Biotechnology

Defending Against Thousands of Antigens

There are thousands of different disease-causing antigens that could make a person sick. They attack your body every day. But very few cause illness even after they penetrate your defenses. How can this be? How can the body create specific antibodies to fight off thousands of different antigens?

Structure of Antibodies Antibodies consist of two long polypeptide chains and two short ones. Each of these chains has a portion that carries the "self" identifier. This identifier allows the antibody to do its work without being mistaken for an antigen and inviting attack by other lymphocytes. But the other portion of the antibody can respond to thousands of different antigens. How does this happen?

Tonegawa's Discovery Japanese scientist Susumu Tonegawa discovered that the DNA for the parts of the chains that vary consists of more than 400 *variable* gene sequences, about 12 different *diversity* sequences, and four *joining* sequences. Different combinations of these sequences can produce thousands of different antibodies. Here's how.

If *variable* sequence number one (V_1) is combined with *diversity* sequence number one (D_1) and

joining sequence number one (J_1), you get a specific antibody ($V_1D_1J_1$). However, if you substitute *variable* sequence number two (V_2), you get an entirely different antibody ($V_2D_1J_1$).

The immune system has adapted to the variability of antigens by combining and recombining a small

CAREER CONNECTION

Epidemiologists study diseases that spread rapidly through populations. Most modern epidemiologists are doctors (MD) and are experts at gathering and interpreting disease statistics. Almost all practitioners work for governmental agencies or universities.

number of gene sequences into a large number of possible responses. The number of possible combinations of gene sequences that code for antibody formation has been estimated at more than 18 billion. Knowledge of how combinations code for millions of antibodies not only helps scientists understand how the cells of the immune system do their work but also opens up new areas of research on how genes might be manipulated to assist the body in fighting disease.

1. **Issue:** Tonegawa's studies were pure research—they uncovered new knowledge but had little immediate practical application. Some people argue that in today's world, where war, famine, and disease are rampant, we should not be spending our resources on efforts that do not directly help to solve these problems. Write an essay defending or opposing pure research in science.

2. **Mathematics Connection:** If there are at least 400 *variable,* 12 *diversity,* and 4 *joining* sequences, how many different antibodies can be made?

Organ Transplants—Who Decides?

For many years, only organs from an identical twin or close relative could be used for transplant. But now anti-rejection drugs that either fool the immune system or block its action have been developed. Further development of these drugs may lead to the day when transplanted organs are not rejected, even if the donor is not a close relative. However, there are more people needing transplants than there are organ donors.

Sources of Organs   Most organs used for transplantation today are from accident victims. These organs come from individuals who are brain dead, but whose organs continue to function. Some potential donors have been kept "alive" by machines. When their families have agreed to donate their organs for transplantation, the machines are turned off, and their organs are saved and implanted in recipients who need them. In order to make this process as efficient and fair as possible, a computerized network called the United Network for Organ Sharing (UNOS) has been set up by the more than 140 organ transplant centers in the United States. Each network member maintains information on patients who need organs and where they are located. The transplant centers also maintain records of the donor's major histocompatibility complex (MHC), blood type, and other data. MHC is a measurement of the kinds of antigens produced by the organ and includes four important antigens. For a transplant to occur, there has to be at least one matching antigen between donor and recipient. A perfect match of four antigens would reduce the chances of rejection to a minimum.

Choosing a Recipient   Because there are so many more potential recipients than donors, transplant centers maintain waiting lists. When an organ becomes available, the recipient with the best match of antigens usually gets it. Sometimes, however, the organ will go to a recipient who is not as close a match, but desperately ill, especially if the best match has a good chance of surviving the wait for another organ. Other factors also influence the choice of recipient. Doctors try to match the weight of donor to recipient, for example.

Who Makes the Decision?
Today, the decision as to who receives particular donated organs is made by doctors, usually part of a committee at a hospital that does transplants. But some people feel that this should not be just a medical decision. What happens when one organ is available that matches two needy patients? How do doctors choose between two heart patients—say a young boy and a young woman who has just had her first baby? Obviously, these are tough, if not impossible, decisions to make.

Would these decisions be easier if more organs were available for donation? The answer is probably yes, but it is not easy to increase the supply. People who have just lost a loved one in an accident are not thinking about organ donation. By the time they do, it may be too late to remove healthy organs.

Understanding the Issue
1. Outline the issues surrounding organ transplants. List opposing viewpoints on each of the major issues, adding your own facts and opinions. Use the outline to make a balanced presentation on the subject to your class or other group.

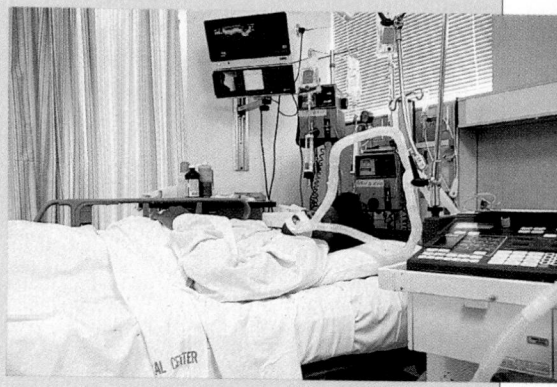

2. One major problem with organ transplants is the cost. Not only does an operation cost hundreds of thousands of dollars, but drugs to prevent rejection must be taken for the rest of the recipient's life. Find out how much an organ transplant operation costs and how these operations are usually paid for, and give a report to your class.

Readings
- Pekkanen, John. *Donor.* Boston, MA: Little Brown, 1986.
- Chandra, Prakash. "Kidneys for Sale," *World Press Review,* February 1991, p. 53.

Summary

Pathogens are adapted so that they can be spread from one host to another. Many bacteria and viruses are pathogens spread through air, water, or direct contact. Bacteria often cause disease by producing toxins. Viral diseases occur as viruses reproduce within and destroy their host cells.

Your skin, membranes, and certain body chemicals form a first line of defense against pathogens. The inflammatory response is a second line of defense. Both these lines of defense are nonspecific.

The immune response is a specific defense that destroys pathogens in the blood or tissue fluid and rids the body of infected, abnormal, or foreign cells. Allergies and autoimmune diseases are conditions in which the immune system either overreacts to antigens or destroys the body's own tissues.

Active immunity may result from exposure to a disease or from a vaccination. Passive immunity makes use of antibodies produced by another organism. Antibiotics can be used to treat many bacterial diseases, but few antiviral drugs have been developed.

AIDS, caused by HIV, gradually destroys the immune system, leading eventually to death.

Language of Biology

Write a sentence that shows your understanding of each of the following terms.

active immunity	memory cell
antiserum	monoclonal antibody
autoimmune disease	passive immunity
B cell	pathogen
cytotoxic T cell	plasma cell
endotoxin	receptor
exotoxin	suppressor T cell
helper T cell	T cell
interferon	thymus gland
lymphocyte	vaccine
macrophage	

Understanding Concepts

1. Outline the events of a cold and the immune system's response to it from the onset of infection until the infection is over.
2. The first organ transplants ever done were kidney transplants. A kidney of a healthy person was removed and donated to an identical twin, whose two kidneys were failing. Why would a transplant between identical twins not result in rejection?
3. Tissue from one mouse is transplanted to a mouse fetus very early in its development. Why is the tissue not rejected by the fetus?
4. Would the fetus in question 3 reject a transplant from the same donor mouse after it is born? Explain.
5. Does aspirin cure you of the flu? Explain. Why do you take aspirin when you have the flu?
6. Should you take penicillin when you have the flu? Explain.
7. Some people, when exposed to dust, begin to have itchy eyes and sneeze repeatedly. Why?
8. Botulism is a disease that may result from eating improperly canned or preserved food. It is caused by a powerful toxin produced by bacteria in the food and may quickly lead to paralysis and death. How might a person who begins to have symptoms of botulism be treated?
9. In terms of AIDS, why is sexual contact between two people who do not know each other's sexual history a risky behavior?
10. What line of defense is not protecting a person when HIV is transmitted by sexual intercourse? By sharing needles?

Applying Concepts

11. Smoking can inactivate the cilia lining the trachea. Why are long-time heavy smokers susceptible to respiratory infections?
12. Many vaccinations require booster shots a few years later. Why do you think they are necessary?
13. Why does a physician usually examine the lymph nodes during a checkup?
14. What do you think might be an effective treatment to prevent death from cholera? Why?
15. Explain why preventing a cold might be a more realistic goal than developing a cure for the cold.
16. Can you suggest a reason that some people are advised to have a flu vaccination each year even though they have had the flu and other vaccinations before?
17. Why are bone marrow transplants sometimes suggested for patients with certain immune system deficiencies?
18. Could monoclonal antibodies be useful in terms of passive immunity? Explain.
19. **Psychology Connection:** Find out some of the symptoms of Alzheimer's disease, then answer the following question. Why would caring for a person who had Alzheimer's disease cause stress to the caregiver?
20. **Issues:** Research on transplantation of organs is often done on animals before attempted with human subjects. Do you think this practice should continue? Defend your answer.

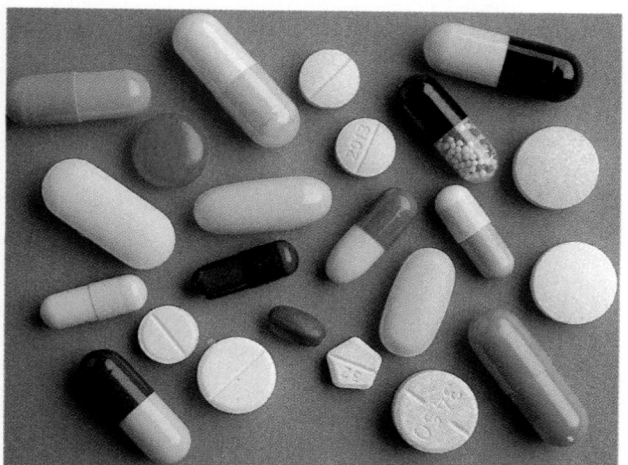

Lab Interpretation

21. **Investigation:** Study the following zones of inhibition produced by a variety of plants in a petri dish culture of a specific bacterium.

    Plant A   1 mm
    Plant B   4 mm
    Plant C   1 cm
    Plant D   3 mm

    Which plant would you continue to investigate as a possibility for producing a new antibiotic? Why?
22. **Thinking Lab:** What would happen if the antibodies made by B cells were not given off into the blood stream, but remained attached to the B cell? Be specific in your answer.
23. **Minilab 2:** The following graph shows the results of an experiment to test how effective brushing teeth with toothpaste is in combating mouth bacteria. Explain these results.

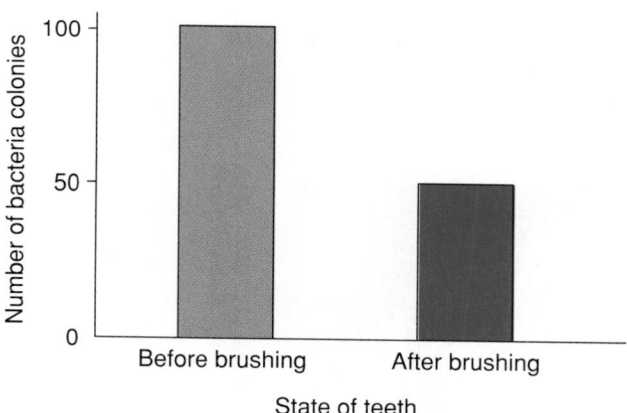

Connecting Ideas

24. Explain how many bacteria become resistant to certain antibiotics over time.
25. Would you expect that a woman's immune system is somehow different when she is pregnant? Explain.

Controlling Living Systems

Sometimes people who have lost an arm or a leg still feel pain in the lost limb. How can that happen? Even though there is no muscle to be stimulated, nerve impulses are still being sent to the brain. Nerve pathways that existed before the limb was lost still exist.

Octopuses change color depending on their state of arousal. At rest, an octopus matches its surroundings and is a mottled gray-brown color. To signal alarm, an octopus can suddenly exhibit strong contrasting light and dark bars or spots. When engaged in a visual battle with another octopus, it changes colors so rapidly that waves of color appear to wash over it.

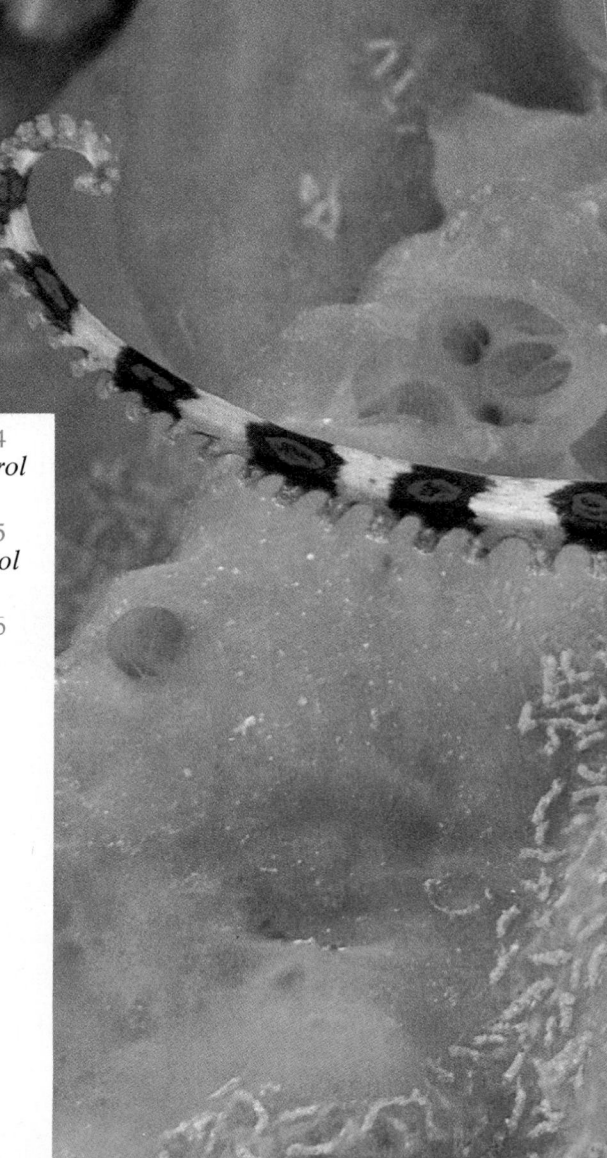

What color is an octopus? If you answered white, brown, gray, black, red, yellow, or orange-red, you would be correct. Octopuses and their relatives, the squid and cuttlefish, are members of a group of invertebrate animals that have complex brains and highly developed central nervous systems. Octopuses can completely change color in less than half a second. Color change in an octopus is controlled by nerve impulses from the brain to pigment cells in the animal's skin.

What does an octopus have in common with a person who has lost an arm or a leg? People who have lost a limb are often fitted with an artificial arm or leg called a prosthesis. Muscle contractions in the remaining portions of a person's arm or leg generate an electric current that activates electrodes implanted in the prosthesis. Color change in an octopus and operation of an artificial arm are both a result of nervous control. The nervous system is one of the living systems you will be studying in this unit.

24

CHEMICAL CONTROL

PICTURE YOURSELF in a beautiful green garden on a warm summer day. The sounds of insects fill the air with peaceful background buzz, flowers are in full bloom, and the graceful leaves of small plants and trees seem to stretch outward, as if to grab every ray of sunlight. Farther along, you notice young fruits and berries starting to appear on small branches, and slender vines twining around poles and fences. Have you ever wondered why some vines wind around poles or why leaves seem to turn to face the sun? If you're a sharp observer, you'll probably wonder whether plants also sense the world around them. Do plants, like animals, respond to the environment?

Unlike animals, plants do not possess nervous systems that allow them to respond quickly to a stimulus. But plants, like all other living organisms, react to changes in their environments. The adjustments plants make are usually slow, and it often takes an entire summer to witness these dramatic changes. In this chapter, you'll learn how plants respond slowly to stimuli through chemical, rather than nervous, control. Animals have the ability to respond to changes both slowly and quickly, and in this chapter, you'll also find out how a plant's chemical control system compares to your own.

Growth patterns in both plants and humans are under chemical control, but the chemicals differ greatly.

Chapter Preview

Objectives

Summarize the experiments that led to the discovery of auxins.

Discuss the role of auxins in the growth and development of plants.

Describe the effects of other plant hormones and how they interact in control of plant growth and development.

24.1 Chemical Control in Plants

Picture a tree near your home or school. Each year that tree gets taller and the size of its trunk increases. In spring, new branches appear and leaves develop on them. Flowers begin to form, pollination occurs, and new seeds are produced. Eventually, leaves are shed, either constantly or once each year. Beneath the surface of the soil, there are similar changes. New roots develop, enabling the tree to absorb necessary nutrients. What controls these changes? In this section, you'll learn that all of these changes in a plant are under control of chemicals that interact to guide growth and development.

Phototropism and the Discovery of Auxins

Although changes such as flowering and leaf development are usually not observable in short periods of time, you know from experience that plants are not inactive. In fact, like yourself and other organisms, plants respond to changes in the environment. They respond to factors such as light, water, and the force of gravity. Such factors, which lead to a reaction or change in an organism, are called **stimuli.** Like other living organisms, you respond to stimuli constantly. In school, when a bell rings, what do you do? You move to another class, right? In this case, the ringing bell is a stimulus; it makes you react.

Phototropism　　Biologists have studied for many years how plants respond to stimuli. One example of such a response is probably familiar to you. Have you ever noticed that a houseplant on a windowsill grows toward the light? In this case, light from one side is a stimulus, and the plant's response is to grow toward the light.

Figure 24-1　　**Plants seem to grow naturally toward light. This tendency is called phototropism and is caused by substances produced within the plant.**

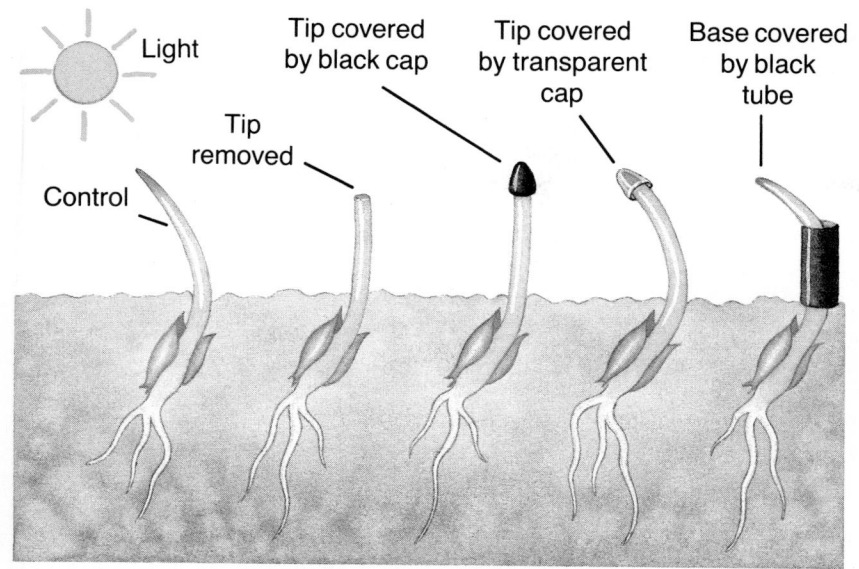

Why do plants grow toward light? It happens because there is unequal growth stimulation on opposite sides of the plant stem. Any plant response caused by unequal stimulation is called a **tropism.** Growth of a plant part toward a stimulus is called positive tropism, and growth away from the stimulus is called negative tropism. A growth response to light is called phototropism. The stem of a plant is said to show positive phototropism because it bends toward the light. Positive phototropism is adaptive for plants because it causes them to grow toward regions where light is more abundant.

Charles Darwin and his son Francis began investigating the causes of positive phototropism in stems more than 100 years ago. They worked with very young seedlings of grass plants, conducting several experiments in which the seedlings were exposed to light from one side. In one experiment, they covered the tips of one group of seedlings with black caps. The seedlings did not curve, but instead grew straight upward. Then, the scientists covered the tips of another group of seedlings with transparent caps. Still another group was left uncovered as a control. Both the plants with transparent caps and the uncovered controls curved toward the light as usual.

The Darwins also showed that enclosing the base of the young stem with a black tube did not prevent the plant from curving toward light. However, when the tip of a stem was removed, the stem failed to grow toward the light. Results of the various experiments, shown in Figure 24-2, all led to the same conclusion. The tip of the stem somehow detects the presence of light and controls the growth of the stem below. How does this control occur?

The Discovery of Auxins Over the years, through much experimentation with young plants, understanding of this plant growth response increased. Many scientists thought that a chemical secreted by the tip of the stem causes plants to grow toward light. In 1926, a Dutch scientist,

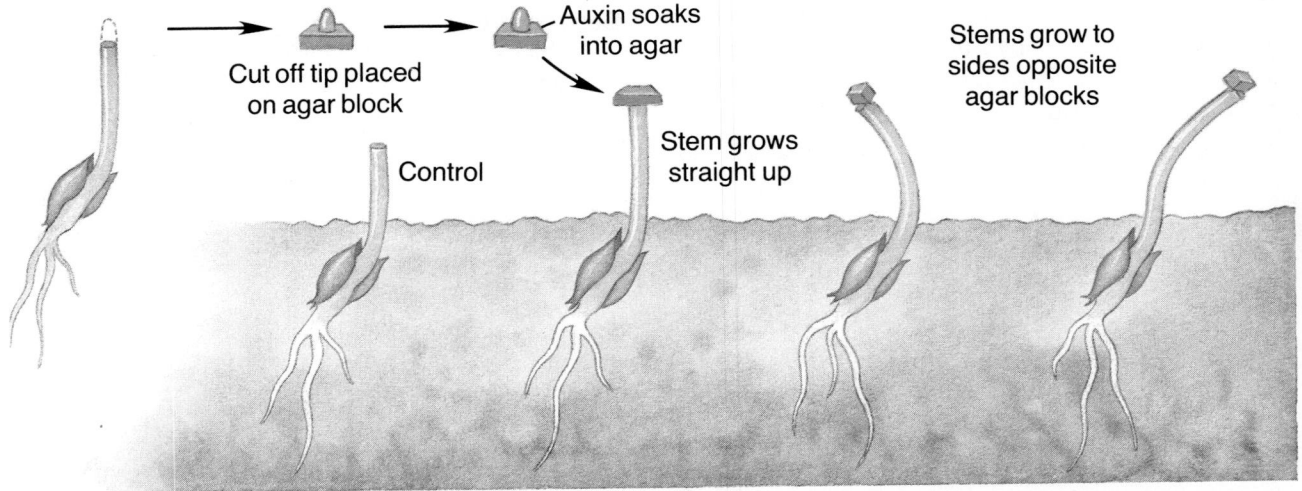

Cut off tip placed on agar block

Auxin soaks into agar

Control

Stem grows straight up

Stems grow to sides opposite agar blocks

Figure 24-3 Went used agar to absorb chemicals from the tips of seedlings. He found that a chemical in the agar blocks would cause growth in seedlings whose tips had been removed. When the block was placed off center, the stem curved toward the side opposite the side on which the agar block had been placed. What do you think Went did to prove that the effects were not due to the agar itself?

Frits Went, carried out a famous experiment. He removed tips from many young stems and placed them on agar, a gelatin-like substance that permits chemicals to diffuse through it.

Went thought that if growth-control chemicals were produced in the tips, then those chemicals would diffuse into the agar. After the tips had rested on the agar for about an hour, he cut the agar into small squares. Next, he placed the squares on the cut ends of the stems from which the tips had been removed. He found that the stems continued to grow, even in darkness. In fact, Went found that the direction in which a stem grew depended on the position of the agar square. When he placed the agar square on the center of the cut end of the stem, the stem grew straight upward. When the agar square was placed only on one side of the cut end, the stem grew in the direction of the opposite side, as you can see in Figure 24-3. As a control, Went used squares of agar on which no tips had been placed. When these agar squares were placed on cut stems, the stems did not grow, regardless of where the squares were placed. From these results, Went concluded that a chemical produced in the tips of stems controls plant growth. Went named the chemical an **auxin,** from the Greek word *auxein* meaning "to grow."

How Auxins Work   Today, we can partially explain how auxins cause stems to grow toward light. A major effect of auxins in some tissues of plants is to cause cells to elongate rapidly. Light coming from one side causes auxin molecules in the tip to move toward the darker side. Thus, auxin becomes more concentrated on the shaded side. By active transport, the auxin moves downward, causing cells on the shaded side to elongate more quickly than cells on the lighted side. Because of this unequal growth in cells, the stem seems to bend toward the light. In stems exposed equally to light from all sides, auxin concentration is the same in cells throughout the stem. In this case, the stem simply grows upward and does not curve. Figure 24-4 summarizes these findings. It's interesting to note that after more than 100 years of research, scientists still do not completely understand how auxins act.

Auxins, as well as other chemical regulators in plants, are examples of hormones. A hormone is a chemical that is produced in one tissue of an organism and then is transported to another tissue where it causes a response. The tissue in which the hormone causes a response is often called a target tissue. How does a hormone cause a response in the cells of a target tissue? It does so in one of two ways. One way is to trigger certain chemical reactions within the cell. The other way is to activate a gene, causing the cell to synthesize a certain protein. In the case of auxin, scientists have found that it lowers the pH in stem cell walls, making them more acidic. This action causes an enzyme in the cell walls to separate cellulose molecules from one another allowing cell elongation to occur more easily.

Other Effects of Auxins

Plant stems grow toward light but grow away from the force of gravity. A growth response to the force of gravity is called geotropism. The stem of a plant placed on its side will respond by growing upward, even in the dark. Because the growth response of a stem is opposite to the direction of the force of gravity, it is called negative geotropism.

Negative geotropism of stems also occurs because of auxins. When a stem is placed on its side, more auxin collects in cells on the stem's lower side. As a result, the cells on the lower side elongate more rapidly than those of the upper side, and the stem grows upward. In contrast to stems, roots grow in the direction of gravitational force—in other words, downward. This response is an example of positive geotropism, and auxin also may play a major part in this process. Increases in auxin concentration inhibit root growth. If a root is placed sideways, auxin collects along its lower side.

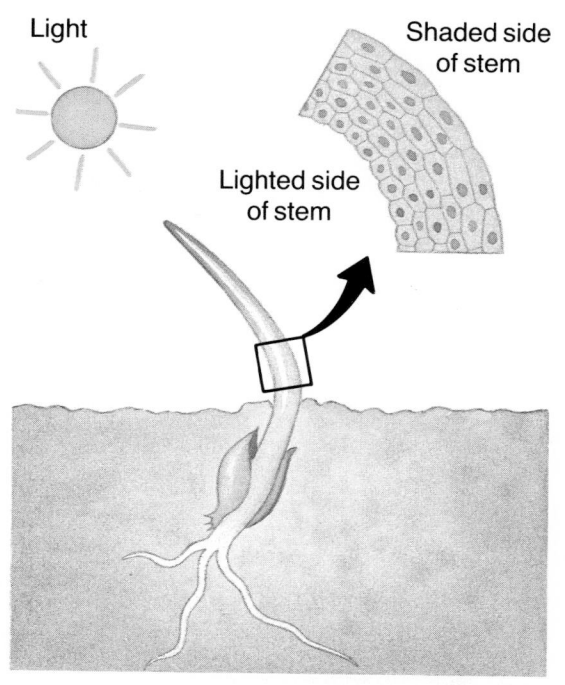

Figure 24-4 **Stems grow toward light because auxin collects in the cells on the shaded side of the stem, stimulating elongation of these shaded cells. As a result, the stem curves toward the light.**

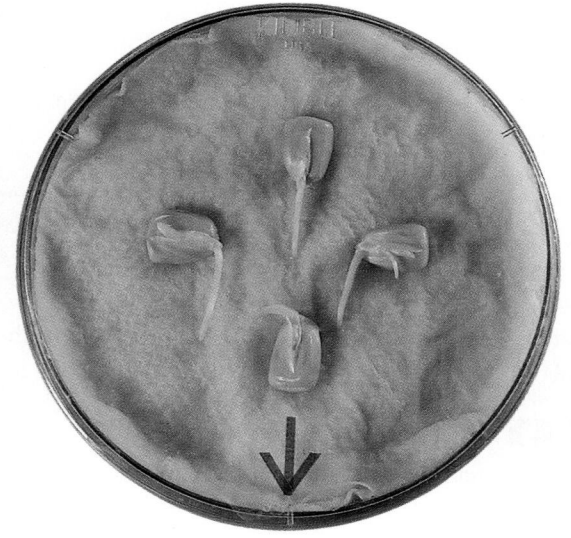

Figure 24-5 **Plant roots grow downward in the direction of the force of gravity. This phenomenon is an example of positive geotropism. On the other hand, stems exhibit negative geotropism because they grow away from the force of gravity. The corn grains you see here were germinated with the dish standing vertically. The arrow points in the direction of gravitational force.**

Cell growth is inhibited on that side while the cells along the upper side continue to elongate. As a result, the root turns downward. Scientists still cannot fully explain the details of positive geotropism in roots.

How Auxin Affects the Shape of a Plant   Auxins are also responsible for actions other than phototropism and geotropism. As an example, let's look at the overall growth pattern of trees. You have probably noticed that many trees grow into cone shapes. This appearance results from the fact that branches near the bottom of the trunk are much longer than those near the top. The actively growing bud at the tip (apex) of a stem is known as the apical meristem. This growing tissue produces auxin, which travels downward through the stem. The presence of auxin in the stem inhibits the growth of buds, including the tips of branches, along the sides of the stem.

Buds along the stem of a plant are known as lateral buds. In the case of a tree, the stem is the tree's trunk. Buds and branches nearest the top of the stem are inhibited the most, because the concentration of auxin is greater there. Those farther down the stem are inhibited less because they are farther from the apical meristem, and thus the concentration of auxin

I N V E S T I G A T I O N

Where to Apply Gibberellin

Gibberellin is a plant hormone that stimulates cell elongation. The result is that some plants tend to grow very tall when this hormone is present in the plant. If you could apply gibberellin to an experimental plant, where do you suppose would be the best place to apply it?

Your task is to determine if the site where the gibberellin is applied causes any difference in growth stimulation. Two different stimulation sites will be tested. One site is the roots. To test the effectiveness of this site, you will add the chemical to the water in which a bean plant is growing. The second site is the growing tips of stems, which you will test by applying gibberellin paste directly to them. Both the paste and liquid have identical strengths of 500 ppm (parts per million). This means that 500 parts gibberellin are present for every 1 million parts of mixture.

Problems
Will gibberellin taken in through a bean's roots influence growth the same as when it is applied to the bean stem's growing tips?

Safety Precautions
Wear aprons and goggles to protect your eyes and clothing.

Hypothesis
Form a hypothesis that could be tested using gibberellin on bean plants. Record your group's hypothesis.

Experimental Plan
1. Decide on a way to test your group's hypothesis. Keep the

is lower. This inhibition of lateral bud development by auxins is called **apical dominance.**

Gardeners put this knowledge about apical dominance to practical use. Many gardeners produce bushy plants by cutting off the apical buds. If the apical bud is removed, auxin production ceases. The lateral buds then develop into many longer branches because they are no longer inhibited by high concentrations of auxin. As a result, the plant becomes bushier instead of growing taller. Contrast the two photos in Figure 24-6.

Figure 24-6   The cone-shaped tree on the left shows apical dominance. Flower growers pinch the tips of plants to prevent apical dominance.

available materials provided by your teacher in mind as you plan your procedure.
2. Remember to test only one variable at a time and use suitable controls. A paste like the one with gibberellin is also available with no gibberellin added.
3. Record your procedure and list the materials and amounts that you will need.

Checking the Plan

1. What controls will be used and what factors will have to be controlled?
2. How long will your experiment run and how many plants will you use?
3. What data will you collect and how will it be recorded? Prepare a data table to record your findings.
4. Assign roles for this investigation.

Make sure your teacher has approved your experimental plan before you proceed further.

Data and Observations

Carry out your experiment. Record your findings in your data table. Show your results by plotting graphs of your data.

Analysis and Conclusions

1. Make a statement comparing the plants treated with gibberellin paste to those treated with paste containing no gibberellin.
2. Make a statement comparing the plants treated with gibberellin solution to those treated with plain water.
3. Make a statement comparing the paste-treated plants to the liquid-treated plants.
4. Do your experimental and control data support your hypothesis?

5. Do your experimental paste and experimental liquid data support your hypothesis? Explain your answer.
6. How does the gibberellin added to water enter a bean plant? How might this affect distribution of gibberellin compared to the distribution of gibberellin applied as a paste?
7. Write a conclusion to the experiment. Make sure your conclusion is based entirely on the results you observed.

Going Further

Based on this lab experience, design an experiment that would help answer another question that arose from your work. Assume that you will have limited resources and will be conducting the lab in class.

Minilab 1

Is light a factor in geotropism?

Line the bottoms of two petri dishes with four layers of paper toweling cut to fit the round shape. Wet the toweling. Add ten radish seeds to each dish and spread them out evenly. Sketch their appearance. Keep dishes moist and flat for four days. On day five, record the position of all young stems. Stand both dishes up on their edges. Next, label the dishes A and B. Tape the covers on. Cover dish A with foil. After 24 hours, record the position of stems. **Analysis** *How does direction of stem growth compare for plants in the light versus those in the dark? Were the plants responding to light, gravity, or both? How do you know?*

Effects of Auxin on Fruit and Leaves Development of fruit also depends on auxins. You learned in Chapter 18 that a fruit develops from the ovary of a flower. When pollination occurs, pollen grains produce an auxin that causes the ovary to begin growing. As seeds develop in the fruit, the seeds release auxin, which continues to stimulate the formation of the fruit. The high level of auxin that leads to development of fruit also prevents the dropping of fruit from the plant. The dropping of fruit or leaves from a plant is called **abscission.** However, the level of auxin declines as the fruit matures and a layer of cells, the abscission layer, forms across the stalk that holds the fruit. Because the cells of the abscission layer are weak, the stalk breaks and the fruit falls from the plant.

High concentration of auxin also inhibits leaf abscission. Deciduous plants are plants that lose their leaves at the end of the growing season. In these plants, auxin concentration declines in autumn. As in fruits, an abscission layer forms across the stalk of the leaf. Eventually the leaf falls, usually because of wind or rain.

Auxin also stimulates cell reproduction within some plant tissues. For example, in spring, auxin produced by buds causes the development of new xylem and phloem tissue. It is also known that application of auxins to stem cuttings can cause the formation of roots from those cuttings.

Gibberellins

A second type of hormone was discovered in Japan, where rice farmers had long noticed that many of their plants grew unusually tall. Because these tall plants usually fell over and did not live to maturity, they were long ignored. But in 1926, scientists found that the tall plants were being

Figure 24-7 Application of gibberellin to some plants causes them to grow much taller than normal. Compare the gibberellin-treated mustard plants on the right of the photo with the normal plants on the left.

affected by a parasitic fungus of the genus *Gibberella*. It was later found that the fungus produces a hormone, named **gibberellin,** that is responsible for the unusual growth of the rice plants. Many different gibberellins have since been discovered, both in fungi and in plants.

Action of Gibberellins Gibberellins, like auxins, help to control cell growth but they do so in a way unlike auxins. Often, gibberellins and auxins contribute to the same response, as in the case of cell elongation. In other cases, gibberellins and auxins counter the effects of one another, as in control of leaf abscission. Gibberellins also increase the rate of germination of many seeds and the development of buds. They may also induce flowering and fruit formation in some species.

You might think that application of gibberellins could produce bigger food plants. However, gibberellins applied artificially often cause abnormalities. They have been successfully used, though, to produce larger seedless grapes and more tender stalks of celery.

Both auxins and gibberellins are hormones that promote growth and development of plants. Their effects are also balanced by a group of plant hormones called inhibitors. Many inhibitors promote dormancy in seeds and buds. Dormancy is important because it prevents development during unfavorable conditions, such as when not enough water is available for seeds to germinate or when cold weather would damage new leaves or flowers. Concentration of inhibitors eventually declines as they are washed away by water, destroyed by cold, or as they simply break down with time. When that happens, hormones such as gibberellins, which promote development, exert their effects. Other hormones work along with auxins, gibberellins, and inhibitors to regulate plant growth. Cytokinins, for example, stimulate lateral bud development, whereas auxin inhibits it. Ethylene, a gas, promotes the formation of an abscission layer and causes some fruit to ripen.

Figure 24-8 Seedless grapes treated with gibberellin develop into larger, juicier fruit than untreated grapes.

Section Review

Understanding Concepts

1. At the time of Went's experiments, it was thought that the tip of a stem "informs" the rest of the stem by some electrical means. How did Went's work show that it is a chemical that passes from the tip to the rest of the stem and not an electric current?

2. Using information from this section, explain why many orchard keepers spray developing fruits with auxin.

3. Give two examples of how different plant hormones interact to control growth and development.

4. ***Skill Review—Designing an Experiment:*** How would you conduct an experiment to test the idea that auxin prevents abscission of leaves? For more help, see Practicing Scientific Methods in the **Skill Handbook.**

5. ***Thinking Critically:*** Review how auxin affects the shape of a plant. How is cone-shaped growth in a plant an adaptive advantage? Hint: Think about photosynthesis.

Explain the role of insulin in regulation of blood sugar level.

Describe the operation of feedback mechanisms that control levels of some hormones.

Discuss the interaction of hormones in control of the menstrual cycle.

Explain how hormones cause responses in their target tissues.

24.2 Chemical Control in Animals

Like plants, animals also respond to stimuli. The stimuli come from outside or inside the body—a sudden drop in temperature, a baseball speeding toward you, hunger pangs, or a rise in the level of carbon dioxide in the blood. In plants, hormones regulate growth and development. There are some hormones that control these functions in animals. However, many other hormones regulate responses to stimuli inside and outside the body in order to maintain homeostasis. In this section, we will focus mostly on the hormones that maintain this condition.

Endocrine Control

In the next chapter, you will learn how the nervous system in animals quickly responds to stimuli by sending electrical "messages" along nerves to various parts of the body. The chemical control system in animals, on the other hand, is usually a much slower system because most hormones must travel through the bloodstream to reach their target tissues. Like plant hormones, though, their effects on the organism may last hours, days, weeks, or even years.

Glands   In animals, chemical control is regulated by several glands located throughout the body. Glands are organs that are specialized for secreting substances needed by the organism. Some glands, such as those in the digestive system, produce substances that are needed quickly. These glands secrete products through small ducts and are known as exocrine glands. Other exocrine glands include sweat glands and milk glands. The glands you will study in this section secrete hormones directly into the bloodstream without the use of ducts. Such glands are the **endocrine glands,** which are sometimes called ductless glands.

Figure 24-9 **Chemical control isn't limited to plants and humans. Chemical substances play important roles in controlling the development of a polyphemus moth from egg to larva to pupa to adult.**

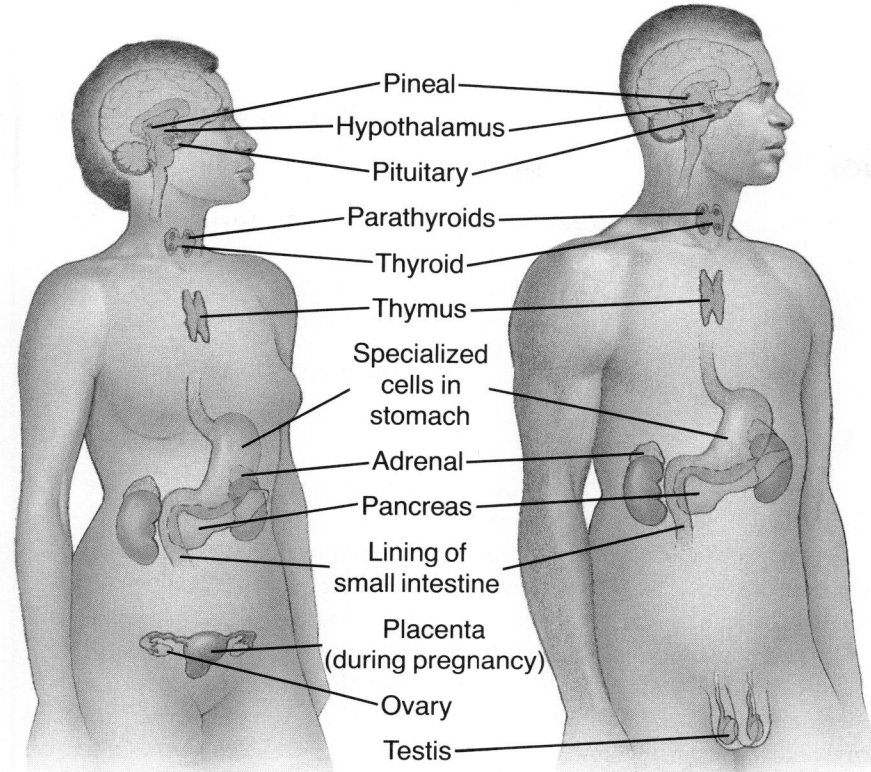

Pineal
Hypothalamus
Pituitary
Parathyroids
Thyroid
Thymus
Specialized cells in stomach
Adrenal
Pancreas
Lining of small intestine
Placenta (during pregnancy)
Ovary
Testis

Figure 24-10   This illustration shows the locations of the major endocrine glands. The small intestine and stomach are listed because certain of their cells produce hormones that affect digestion.

Endocrine glands are found in the brain, chest, abdomen, and pelvis in vertebrate animals. Figure 24-10 shows the names and locations of the major endocrine glands in humans.

Animal Hormones   Animal hormones, like auxins, are recognized only by specific target tissues. In general, hormones either stimulate a target tissue and increase its activities, or inhibit a target tissue and decrease its activities. It's important to note that the target tissue determines the response to a hormone, not the hormone itself. Like auxins, animal hormones also affect the target tissues by affecting chemical reactions. In animals, hormones regulate growth, development, and reproduction as well as maintain homeostasis.

In terms of chemical structure, there are two major types of hormones. Some hormones, such as insulin, are composed of chains of amino acids and are known as protein-type hormones. Others, such as testosterone and estrogen, are called steroid hormones. The chemical structures of the steroid hormones are closely related to cholesterol.

Negative Feedback   How do hormones work? In general, endocrine glands don't secrete hormones into the bloodstream at a constant rate. Rather, the rate of hormone secretion is determined by the needs of the animal at the particular time. Often, the message that causes a gland to speed up or slow down comes from the nervous system responding to some stimulus. In most cases, however, messages come in the form of other hormones.

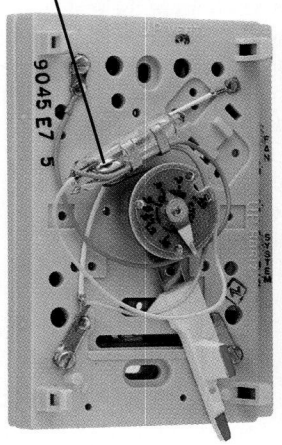
Mercury switch in the "On" position

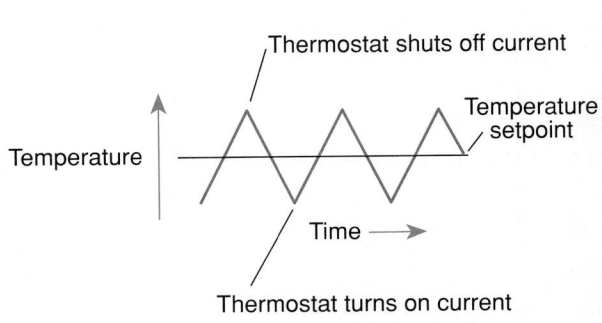
Thermostat shuts off current

Temperature setpoint

Temperature

Time →

Thermostat turns on current

Figure 24-11 **A furnace thermostat is a device that operates on the principle of negative feedback. It responds to temperature and switches heat on and off to keep the temperature within a degree or two of the set point. Negative feedback in humans is similar except that changes aren't as quick as with an on-off switch in a thermostat. Why would feedback responses in humans be slower than with a thermostat and furnace?**

Many body processes are maintained at an appropriate level through the mechanism of **negative feedback.** A familiar example of a negative feedback mechanism is an oven thermostat, which is used to maintain an oven at a constant temperature. In an electric oven, when the electric element raises the oven temperature just above the set temperature, the thermostat shuts off the current to the element. Then, the oven cools down to just below the setting, and the thermostat turns on the current again. You know also that you can set a thermostat to a higher or lower setting as needed. Likewise, negative feedback by the endocrine system maintains body processes at the levels required by the body. Those levels can vary according to the conditions of the body. Those conditions might include sleeping, working, fleeing danger, growing, fighting off an infection, and pregnancy.

In the body, negative feedback works like this. Suppose the concentration of a particular substance, such as thyroid hormone in the blood, increases above the level required by the body. Then, negative feedback acts to decrease it. On the other hand, if the concentration becomes too low, negative feedback will act to increase the level. Now let's look at some specific examples of hormonal control in humans.

Control of Blood Glucose Level

When you digest a meal, carbohydrates are broken down into glucose. As a result, a high concentration of glucose begins to build in the small intestine. In Chapter 20, you learned that after glucose enters the bloodstream, it is transported to the liver, where some of it is stored as glycogen. When glucose is needed by the body, some of the stored glycogen is broken down and the resulting glucose is released back into the bloodstream. This storage and release of glucose from the liver helps maintain the concentration of blood sugar at a level appropriate to the body's activities. Glucose is also stored as glycogen in skeletal muscle, the muscle attached to bones. Excess glucose is also transported into fat cells, where it is converted to fats.

Action of Insulin The uptake of glucose by the liver, skeletal muscle, and fat cells is under control of the hormone insulin. Insulin is produced by a group of cells located in the pancreas. After a carbohydrate-rich meal, the increased level of glucose in the blood is detected and these pancreatic cells respond by increasing their output of insulin. The net effect is the storage of blood glucose as described previously. The cells in the pancreas detect the reduced glucose level. As a result, the pancreas decreases its output of insulin, thus ensuring that some glucose remains in the blood rather than in storage.

Diabetes Mellitus If the level of blood sugar is not regulated properly, such that a person has too much glucose in the blood, a variety of problems will arise. Such a person suffers from the disease diabetes mellitus, or sugar diabetes. In Chapter 22, you learned that one function of the kidneys is to conserve important substances including glucose by resorbing them into the blood. However, in diabetes, so much glucose is in the blood that not all of it can be resorbed, and glucose passes into the urine. Also, because there is an increased concentration of glucose in the blood, water and salts move from body cells into the blood because of osmosis and diffusion. This condition also forces the kidneys to work harder as they act to remove the excess water and salts in the bloodstream. As a result, thirst and frequent urination are common symptoms of untreated diabetes mellitus. A series of complications resulting from untreated diabetes can depress brain function, leading to a condition called diabetic coma.

What is the cause of this disease? You have learned that insulin enables glucose to enter body cells from the blood. In one form of diabetes, lack of insulin prevents glucose from entering cells. Thus, glucose levels in the blood remain high. It's as if the shelves are full of groceries, but there is no way to get at all the food. Thus, even though a person with

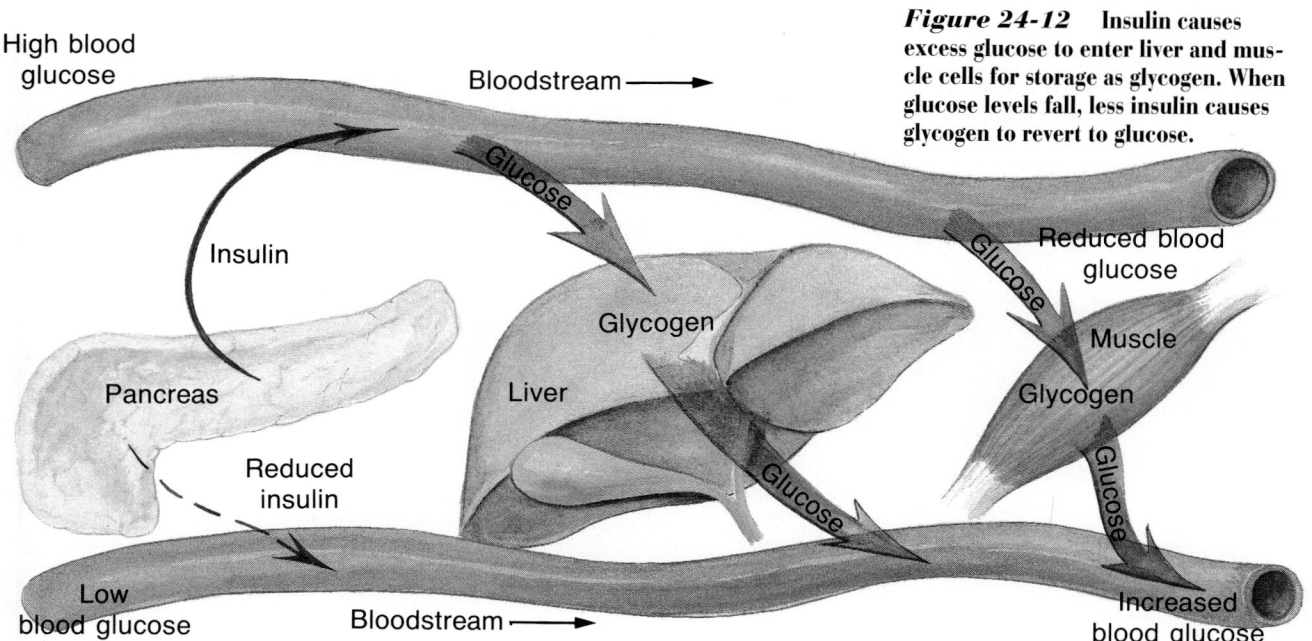

Figure 24-12 Insulin causes excess glucose to enter liver and muscle cells for storage as glycogen. When glucose levels fall, less insulin causes glycogen to revert to glucose.

High blood glucose

Bloodstream →

Glucose

Insulin

Reduced blood glucose

Glucose

Glycogen

Muscle

Pancreas

Liver

Glycogen

Reduced insulin

Glucose

Glucose

Low blood glucose

Bloodstream →

Increased blood glucose

untreated diabetes may have recently eaten a meal, he or she may feel fatigued and hungry. Long-term complications of untreated diabetes include strokes, blindness, kidney failure, and serious circulatory problems.

There are two types of diabetes. In one type, glucose cannot enter cells because cells in the pancreas do not produce enough insulin. This form of diabetes usually develops in people under 20 years of age and is known as type I diabetes. Type I diabetes is usually controlled by injections of insulin to maintain needed levels.

Type II diabetes occurs because muscle and fat cells that normally take up glucose after a meal do not carry out that function adequately. It is much more common and occurs most frequently in overweight people over 40 years old. Diet and exercise have been shown to help reduce blood glucose in these cases. However, as the disease progresses, many people must rely on insulin injections.

Insulin and Uptake of Glucose   If the blood of people with untreated diabetes is so rich in glucose, why can't glucose pass from blood to body cells? Recent research has revealed a great deal about how glucose is transported across cell membranes. The transport relies on proteins called transporters in cell membranes. It is thought that transporters act to form pores that extend through the cell membrane. Figure 24-13 shows the series of steps by which glucose is thought to pass through the transporter and into a cell's interior. Once a glucose molecule enters a pore, the pore changes shape, closing behind the glucose molecule. When the molecule is in the cell, the pore takes its original shape. Scientists have found that the rate at which this transport can occur is greatly influenced by presence of insulin in the bloodstream.

Scientists have worked out a partial explanation of how insulin influences absorption of glucose into cells. In experiments with fat and muscle cells, scientists have found that some transporters are in the plasma membrane, but many more are on the surface of vesicles inside the cell. Evidence indicates that insulin links with a receptor present in the membrane. This linking sets off a chain of events that causes vesicles that have transporters to be brought from inside the cell and to fuse with the

Figure 24-13   **Glucose enters cells through the action of transport proteins in the cell membrane. How would the number of transport proteins in the plasma membrane affect the rate at which glucose enters a cell?**

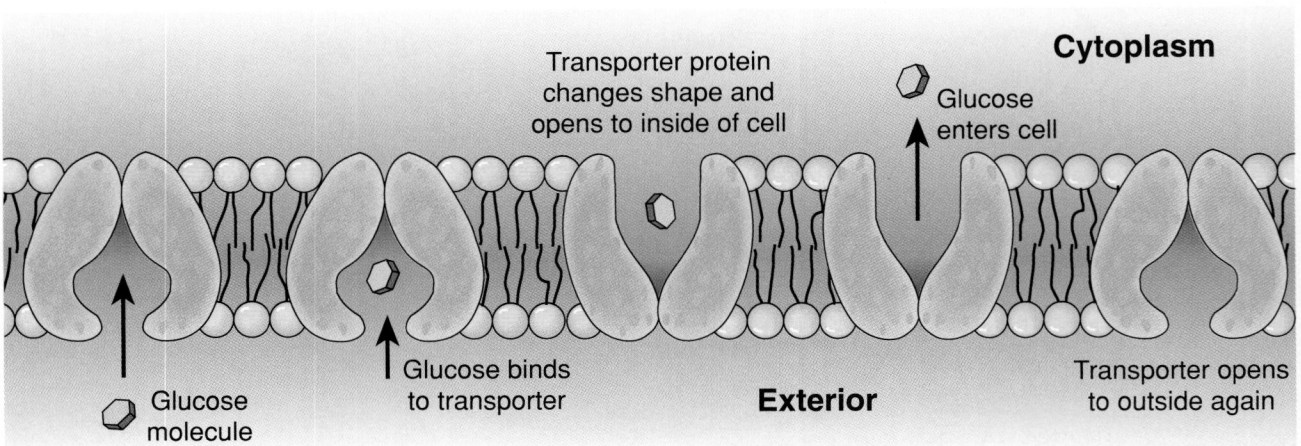

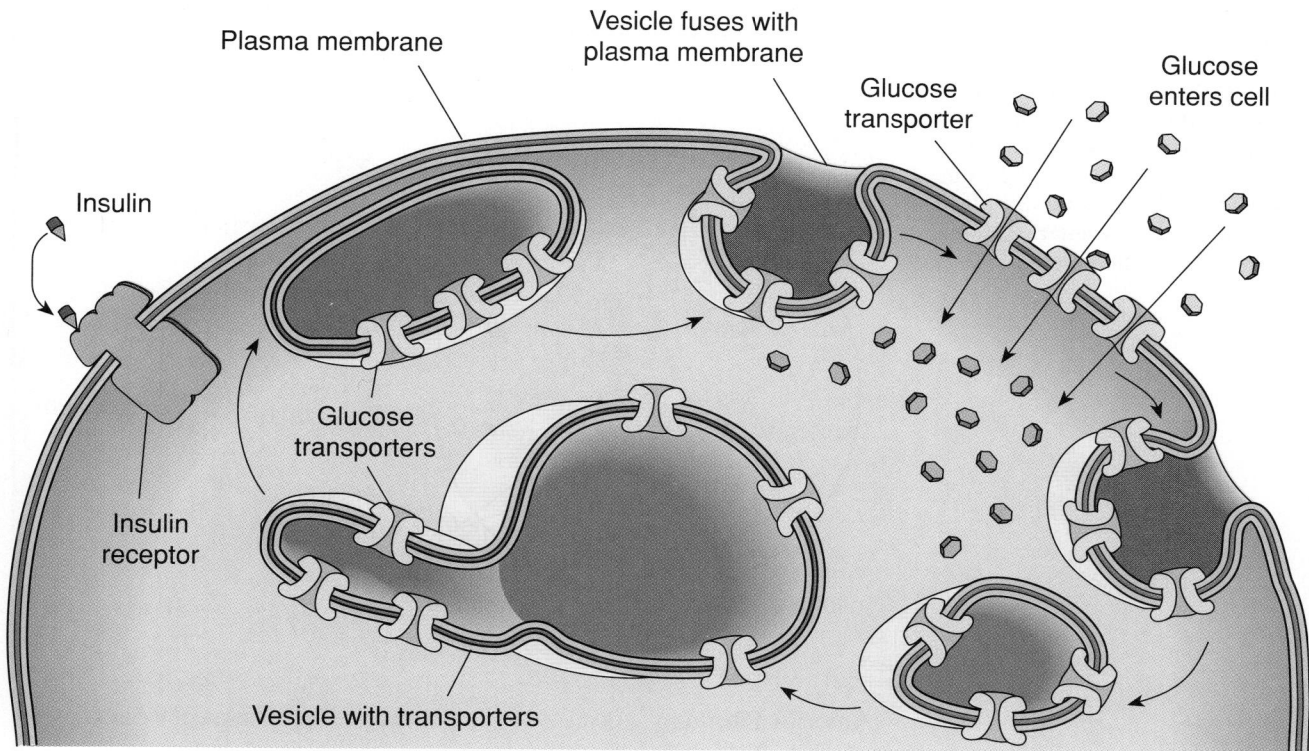

Plasma membrane

Vesicle fuses with plasma membrane

Glucose transporter

Glucose enters cell

Insulin

Insulin receptor

Glucose transporters

Vesicle with transporters

plasma membrane. The cell can then take up glucose more quickly because of the increased number of transporters. Research continues to lead to more information about this process.

Interaction of Hormones   In the first section of this chapter, you learned that some plant hormones interact to control growth. Hormone interaction occurs in animals as well. For example, glucagon is another hormone secreted by cells in the pancreas. Instead of promoting the uptake of glucose by cells, it has nearly the opposite effect. Glucagon increases blood sugar by causing glycogen in the liver to break down and release glucose into the bloodstream. In fact, there are at least five other hormones that interact to maintain blood glucose at a proper level. You will see several examples of hormone interaction in the remainder of the chapter.

Figure 24-14   **Scientists believe that insulin helps glucose enter cells by a process that brings more transport proteins to the plasma membrane from inside the cell. When insulin binds to a receptor, a chain of events causes vesicles containing transport proteins to fuse with the cell's plasma membrane, providing more transporters to bring glucose into the cell.**

The Pituitary Gland

In order to maintain homeostasis, levels of hormones must change in response to different stimuli. Secretion of hormones from many endocrine glands is under control of the **pituitary gland.** The pituitary gland is a three-lobed structure located at the base of the brain. The two major lobes are known as the anterior pituitary and posterior pituitary. Working with a nearby brain structure, the **hypothalamus,** the pituitary releases hormones of its own. The targets of many of these hormones are other endocrine glands around the body, and they cause the glands to either increase or decrease activity.

Figure 24-15 The pituitary gland is embedded in bone below the hypothalamus of the brain. The hypothalamus senses the levels of various substances in the blood. As a result, it secretes (or stops secreting) releasing factors. These factors enter the bloodstream and cause the anterior pituitary to release its own hormones.

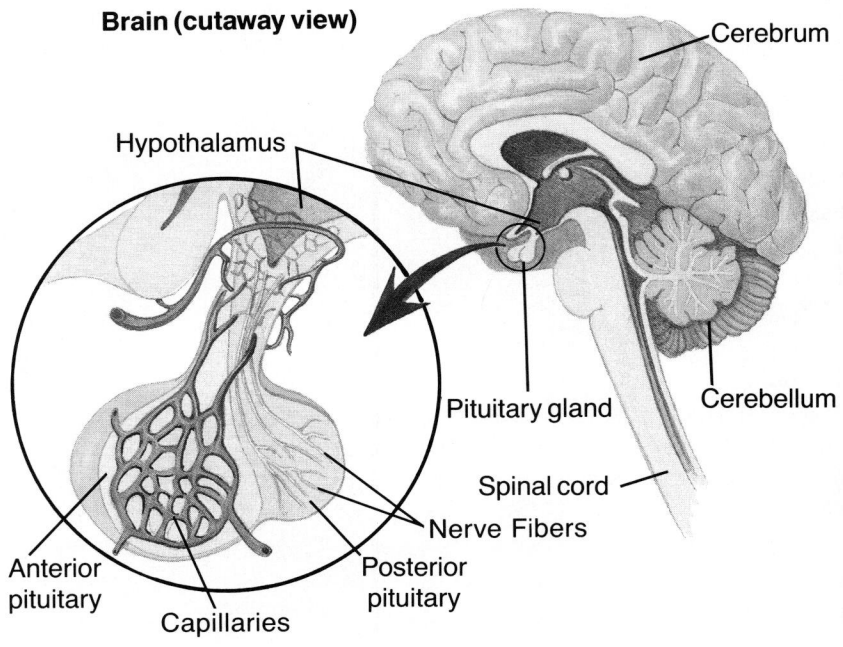

Brain (cutaway view)

Figure 24-16 This diagram shows how negative feedback by hormones keeps the amount of thyroxine at a level suitable to the body's needs.

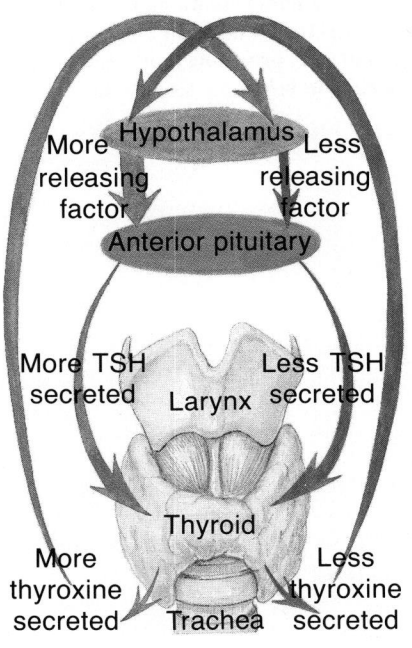

Anterior Pituitary As an example of negative feedback, consider control of the level of thyroxine, a hormone produced by the thyroid gland. Thyroxine increases the rate of metabolism in cells, and the level of thyroxine in the blood must be maintained at a level that meets the needs of the body. Output of thyroxine by the thyroid is controlled by a hormone from the anterior pituitary called thyroid stimulating hormone, usually referred to as TSH. An increase in level of TSH stimulates the thyroid to release more thyroxine and vice versa. How do the two interact and what signals the pituitary to release more or less TSH?

As the level of thyroxine in the blood falls below the proper point, that lower level is detected by the hypothalamus as blood passes through it. The hypothalamus secretes a hormone of its own, called TSH releasing factor, which enters the bloodstream. As the releasing factor reaches the anterior pituitary, the anterior pituitary responds to the stimulus by secreting more TSH. TSH enters the bloodstream and, upon reaching the thyroid, stimulates it to secrete more thyroxine. With time, the thyroxine level rises above the "set point," and this level is also detected by the hypothalamus, which stops secreting TSH releasing factor. The anterior pituitary then decreases TSH output, and the level of thyroxine in the blood begins to fall.

This interaction of hypothalamus, anterior pituitary, and thyroid is another example of negative feedback. Information—in this case high thyroxine level in blood—is fed back as a drop in releasing factor to the structure responsible for controlling that level. The structure, the anterior pituitary, responds accordingly by shutting off production of the stimulating hormone. The fact that the hypothalamus, a part of the nervous system, plays a role in chemical control shows a close link between nervous and endocrine systems.

The Posterior Pituitary   The posterior pituitary secretes two hormones, which are actually made in the hypothalamus. Neither of these hormones controls other endocrine glands. One of them, oxytocin, stimulates contraction of muscles of the uterus, leading to labor and childbirth. The other is vasopressin, which helps maintain the body's water balance by regulating the amount of water lost in urine. Unlike secretions by the anterior pituitary, secretion of hormones from the posterior pituitary is not controlled by other hormones, but by nervous stimulation from the hypothalamus. Still, the control is by feedback.

Hormone Action

How do hormones work to cause a response by a target tissue? In Section 24.1, you learned that hormones can work in one of two ways. In one way, hormones cause a response by triggering chemical reactions within cells of target tissue. In general, protein-type hormones work this way. Steroid hormones work in another way. They activate genes within cells of target tissue. In either case, the hormone must first recognize its target tissue before a response can occur.

HEALTH CONNECTION

Choh Hao Li

• •

Choh Hao Li dedicated his scientific career to the study of the hormones of the pituitary gland. He and his research teammates were either the first or among the first to isolate and identify the eight hormones produced by the anterior pituitary.

Born in China in 1913, Li studied and then taught chemistry at Nanking University. He earned a Ph.D. in the United States at the University of California at Berkeley, and stayed on afterwards as a faculty member. In 1950, he became director of the university's Hormone Research Laboratories. Five years later, he was granted U.S. citizenship.

Among Dr. Li's most famous discoveries is adrenocorticotropic hormone (ACTH), a protein hormone from the anterior pituitary. ACTH stimulates the outer layer of the adrenal gland, called the adrenal cortex, to produce corticoids, also known as corticosteroids. Corticoids are hormones the body needs in order to use nutrients properly and maintain the right balance of salts and water in body fluids. They also affect inflammation and immune response. Li worked out the structure of ACTH and showed that only portions of the hormone's molecule were necessary for it to be effective in the body.

It was also Dr. Li who isolated and discovered the structure of human growth hormone (HGH), also called somatotropin. His work with pituitary growth hormones proved that the hormones work only for the species that produces them. A dog's growth hormone, for example, will help a dog grow, but it will do nothing for a cat or any other non-canine animal.

In 1970, he succeeded in synthesizing HGH. With its 256 amino acid chains, it was the largest protein molecule ever synthesized up to that time. Li's work paved the way for research showing that doctors can use HGH to help children lacking the hormone to grow normally.

━━━━━

1. **Solve the Problem:** Many scientists are concerned about the possible misuse of HGH. In what ways do you think this hormone might be misused?
2. **Explore Further:** ACTH is a protein-type hormone. How do you think it exerts its effect on the cells of the adrenal cortex?

Protein-type Hormones Protein-type hormones locate their target tissue by matching with receptor proteins in the plasma membranes of the cells of their target tissues. Because each hormone can link with only a certain receptor, the target tissue is easily recognized. As a specific example, adrenaline molecules can dock with adrenaline receptors located on certain liver cells and on cells of skeletal muscle. One action of adrenaline is to cause conversion of glycogen to glucose in those target cells.

However, adrenaline does not cause a response directly. After an adrenaline molecule links to its receptor, the receptor activates an enzyme in the cell membrane. This enzyme converts one substance to another substance called a second messenger, which causes a series of reactions. These reactions activate an enzyme that converts glycogen to glucose. In just a few seconds, one adrenaline molecule can lead to production of 10 billion molecules of glucose. This model has become known as the **second messenger hypothesis.** In this case, adrenaline is the first messenger, but it never enters the target cell. It is the second messenger that actually causes the response.

Steroid Hormones Steroid hormones actually pass through the membranes of target cells. Once inside the cell, the hormones combine with receptors in the cytoplasm. No second messengers are involved. Instead, the steroid-receptor complex moves into the nucleus, where it activates a gene. When the gene becomes active, it is transcribed and translated. The protein formed then changes the cell's chemistry, leading to a response.

So far, you've looked at some human hormones, how they function, and a relatively simple example of negative feedback. Now, let's look at how hormones interact to regulate a more complex body process.

Figure 24-17 **The diagram on the left illustrates the action of adrenaline, a typical protein-type hormone. When adrenaline binds to its receptor, a second messenger is produced. This second messenger initiates a series of reactions that finally activates an enzyme that converts stored glycogen to glucose. In contrast, the right diagram illustrates the action of a steroid hormone, which enters the target cell and binds with a receptor. Together they enter the nucleus and cause a gene to be transcribed. Translation of the resulting mRNA produces the needed protein.**

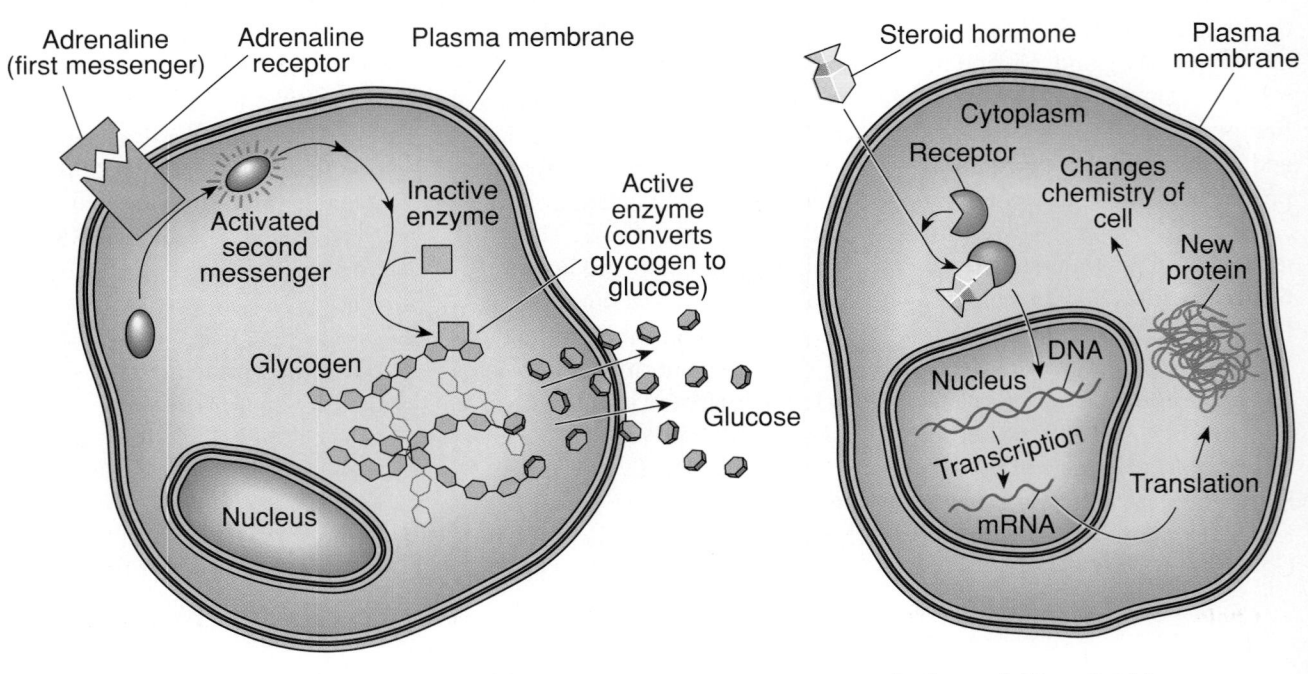

Action of Protein-type Hormone

Action of Steroid Hormone

Table 24.1 **S**ome Human Hormones and Their Main Functions

Hormone	Sources	Location of Source	Functions
Thyroxine	Thyroid gland	Neck	Regulates metabolic rate of body cells
Insulin	Islets of Langerhans (beta cells)	Pancreas	Decreases level of sugar in blood
Glucagon	Islets of Langerhans (alpha cells)	Pancreas	Increases level of sugar in blood
Adrenaline	Adrenal medulla	On kidneys	Prepares body to cope with stress; converts glycogen to glucose
Cortisone	Adrenal cortex	On kidneys	Prevents inflammation Increases level of sugar in blood
Testosterone	Testes	In scrotum	Causes secondary sexual characteristics in males; stimulates sperm production
Estrogen	Ovaries	In abdomen	Causes development of secondary sexual characteristics in females Prepares uterus for pregnancy
Progesterone	Corpus luteum	In abdomen	Maintains uterus during pregnancy
FSH	Anterior pituitary	Base of brain	Causes maturation of egg in females Stimulates sperm production in males
LH	Anterior pituitary	Base of brain	Causes ovulation in females Causes release of testosterone in males
TSH	Anterior pituitary	Base of brain	Stimulates thyroxine production
HGH	Anterior pituitary	Base of brain	Regulates growth of body
ACTH	Anterior pituitary	Base of brain	Stimulates release of adrenal cortex hormones
Oxytocin	Posterior pituitary	Base of brain	Stimulates contraction of uterine muscles
Vasopressin	Posterior pituitary	Base of brain	Regulates water balance by kidneys
Gastrin	Stomach cells	Stomach	Stimulates release of gastric juice
Secretin	Intestinal cells	Intestine	Stimulates release of sodium hydrogen carbonate from pancreas and bile from gallbladder
Cholecystokinin	Intestinal cells	Intestine	Stimulates release of pancreatic enzymes and bile

Control of the Menstrual Cycle

Control of the menstrual cycle involves several different hormones and illustrates the principle of negative feedback very well. You learned in Chapter 18 that the menstrual cycle lasts an average of 28 days. During that time, an egg matures within the follicle of an ovary, and the uterine tissue becomes spongier and enriched with blood. Because the menstrual cycle is a cycle, we could begin studying it at any point. For convenience, let's begin with menstruation, which is marked by loss of blood and some uterine tissue through the vagina and lasts three to five days.

The Follicle Stage   As menstruation is proceeding, a new egg is already beginning to mature within a follicle of the ovary. This event marks the beginning of the **follicle stage,** which lasts for about ten days. Maturity of an egg is controlled by the anterior pituitary gland, which releases follicle-stimulating hormone, or FSH. Secretion of luteinizing hormone, LH, from the anterior pituitary is also increased. As the follicle develops, it also secretes a hormone, estrogen, which stimulates the tissues of the uterus to thicken and the blood supply to increase. This action begins to prepare the uterus for possible pregnancy.

The Corpus Luteum   As the level of estrogen in the blood increases, it is detected by the hypothalamus. In a feedback process, the hypothalamus responds by causing the pituitary to decrease FSH production and increase output of LH. The increased concentration of LH in the blood leads to ovulation, the bursting of a follicle and release of an egg into the oviduct.

Thinking Lab

DRAW A CONCLUSION

Which hormones control egg release or uterine thickening?

Background

The menstrual cycle consists of a number of events that follow a recurring pattern. This pattern repeats because of the interaction of hormones. Two events of this cyclic pattern are (1) release of an egg from an ovary and (2) the buildup of the uterine lining.

You Try It

Use the table to predict which hormone(s) would be associated with the two events described above. Assume that the cycle is one of average length (28 days) and the egg is not fertilized.

Day of Menstrual Cycle	Concentration of Hormones		
	A	B	C
1	12	5	10
5	14	5	14
9	14	5	13
13	70	10	20
17	12	60	9
21	12	150	8
25	8	100	8
1	12	5	10

Draw a conclusion by answering the following questions. Which hormone(s) may be involved with the: (a) thickening of the uterine lining and (b) release of an egg from the ovary? Explain your answers.

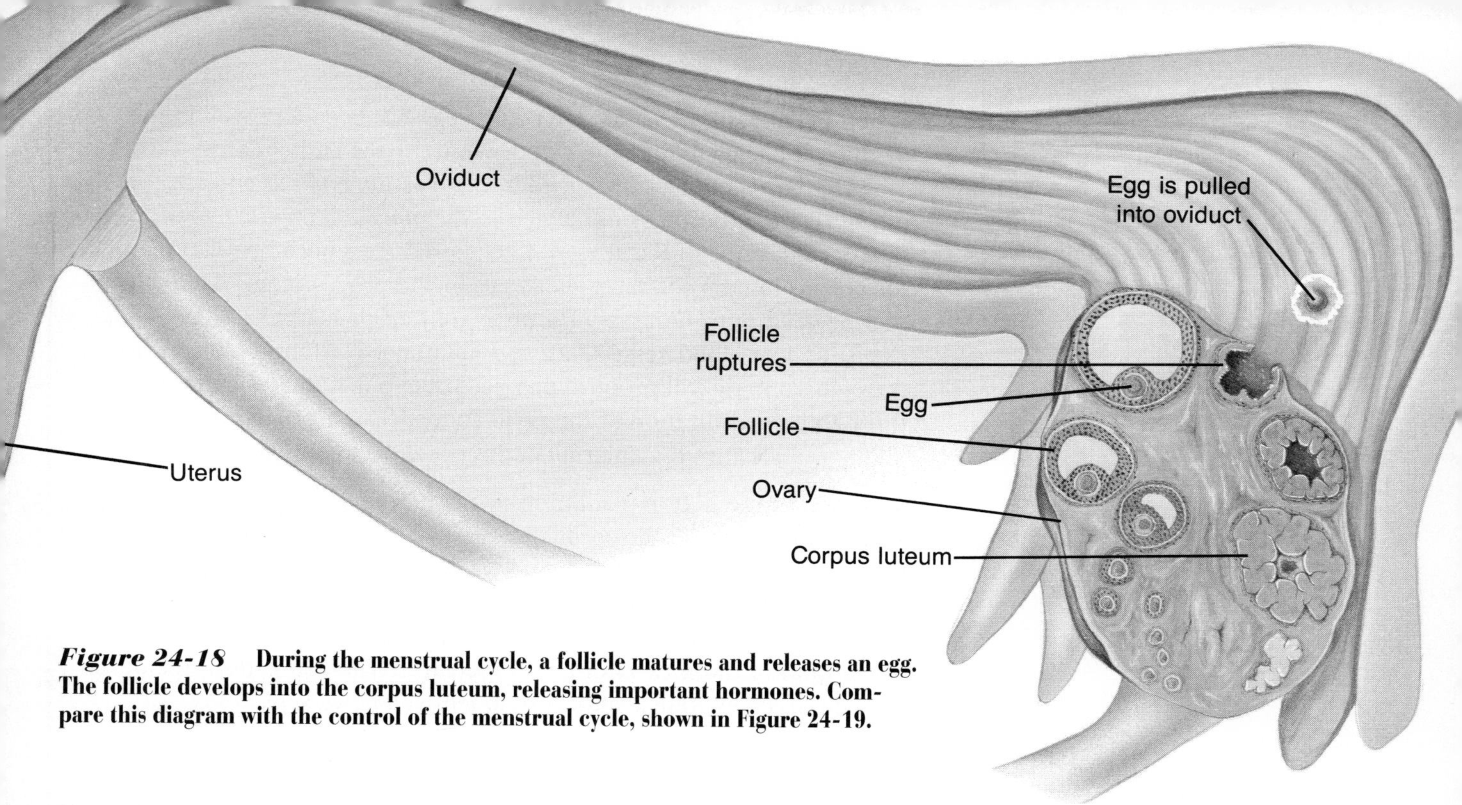

Oviduct

Egg is pulled
into oviduct

Follicle
ruptures

Egg

Follicle

Ovary

Corpus luteum

Uterus

Figure 24-18 During the menstrual cycle, a follicle matures and releases an egg. The follicle develops into the corpus luteum, releasing important hormones. Compare this diagram with the control of the menstrual cycle, shown in Figure 24-19.

This release is called ovulation, which usually occurs about midway through the cycle.

LH also causes the ruptured follicle to be converted to a yellowish body, the **corpus luteum.** Under the influence of LH, the corpus luteum secretes some estrogen and the hormone progesterone. Progesterone further prepares the uterus for pregnancy. The high concentration of progesterone in the blood is again detected, and the pituitary reduces output of FSH and LH. This stage of the cycle, which lasts about two weeks, is called the **corpus luteum stage.**

Figure 24-19 This diagram of the menstrual cycle shows the relationship among the levels of pituitary hormones, levels of ovarian hormones, events in the ovary, and the condition of the uterine lining.

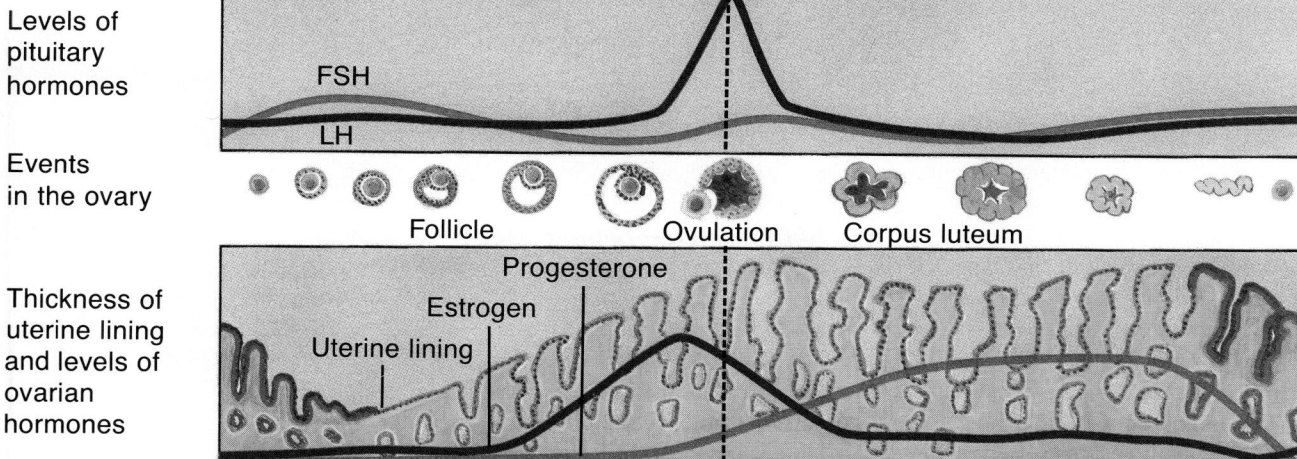

Levels of
pituitary
hormones

FSH

LH

Events
in the ovary

Follicle

Ovulation

Corpus luteum

Thickness of
uterine lining
and levels of
ovarian
hormones

Progesterone

Estrogen

Uterine lining

Time (days)

| 2 | 4 | 6 | 8 | 10 | 12 | 14 | 16 | 18 | 20 | 22 | 24 | 26 | 28 | 2 |

Flow stage Follicular stage Luteal stage Flow

When Fertilization Occurs   When an egg has been fertilized in the oviduct, it begins to undergo cell division. By the end of the first week of growth, the zygote has developed into a blastula and has entered the uterus. Here, it becomes implanted in the uterine lining. When implantation occurs, some of the embryo's cells produce a hormone that maintains the corpus luteum so that it keeps producing progesterone. Thus, the uterine conditions necessary for pregnancy, such as a thick lining and a rich supply of blood vessels, are also maintained. After about five weeks, the embryo itself will produce progesterone to maintain the uterus throughout the duration of the pregnancy. Because progesterone inhibits FSH and LH output, menstrual cycles cease during pregnancy.

Menstrual Flow   If implantation of an embryo does not occur during the corpus luteum stage, the corpus luteum begins to break down. As it does so, progesterone level declines and the uterine lining cannot be maintained. The uterine lining begins to break down, and menstruation occurs once more. In addition, production of FSH and LH is no longer inhibited, and output of those two hormones increases. The net result is the development of a new follicle and, consequently, a new cycle.

LOOKING AHEAD

ORGANISMS RESPOND to stimuli by means of chemical control—hormones. Chemical response to stimuli is slow, but in Chapter 25 you will examine a faster means of response—nervous control. You will learn how stimuli are detected and how those stimuli set off impulses that are transmitted from one cell to another and to the structures that respond to the stimuli. 🐚

Section Review

Understanding Concepts

1. Explain why the blood glucose level is often high in people with untreated diabetes.
2. Ecdysone, a steroid, stimulates metamorphosis in insects. Describe how this hormone causes a response in target tissue.
3. How does the follicle stage of the menstrual cycle illustrate the idea that some hormones work together to cause a response? Give examples of negative feedback at work in control of the menstrual cycle.

4. **Skill Review—Sequencing:** Output of cortisone from the adrenal gland is under control of ACTH, an anterior pituitary hormone. Sequence the series of events that would lead to a decrease in output of cortisone. For more help, see Organizing Information in the **Skill Handbook**.

5. **Thinking Critically:** People who do not produce enough of the hormone vasopressin suffer from the disease diabetes insipidus. Consider what their symptoms might be and why the term *diabetes* applies to their illness, as well as to people with diabetes mellitus.

Considering the Environment

The Pesticide Tree

Since the dawn of agriculture, farmers have been at war with insects seeking to steal crops for their own food. For the past several decades, humans have been winning the war on insect pests, thanks in great part to chemical insecticides. But there's been growing concern about the environmental damage that some of these insecticides may cause. As a result, there's been an increasing demand for safer ways to control insect pests.

Recently, a lot of scientific interest has been focused on the neem tree, a plant long known in India for its insect-fighting properties. Of particular interest is the oil from

neem seeds. Neem oil shows insecticidal properties that are effective against a wide range of insect pests. One component of neem oil is a compound called azadirachtin, which fatally disrupts insect growth.

How Neem Oil Acts   One reason azadirachtin is so effective is that it has a molecular structure very similar to ecdysone. In some insects, it bonds to ecdysone receptors, blocking the insect's natural hormone and preventing it from acting. Molting stops and the larva dies.

Safety of Neem Oil   Despite its toxicity to many insect pests, azadirachtin has so far shown no ill effects on most other animals. Laboratory tests indicate it to be safe for birds and rats, although it might pose a threat to some fish. It seems to offer no danger to humans or pets.

Many beneficial insects, such as those that prey on pests, have also survived well among azadirachtin-treated test plants. The reason seems to be that plants absorb the compound. Those insects that feed on the plant die, while those that don't remain safe.

Another advantage of neem-based insecticides is that they rapidly degrade, losing all toxicity. That may mean extra work for

farmers since they might have to treat crops several times each growing season. However, it greatly reduces any risk of buildup in the soil or groundwater.

1. **Explore Further:** Do library research to find out about other insecticides that have been derived from plants. Against what insects are they effective?
2. **Connection to Economics:** Many substances that are useful to humans can be found in plants and other organisms. Assume that a substance from plants found only in a desert environment is discovered to be very active in treating heart disease in humans. It is complex and cannot be manufactured. What do you think would be the consequences of such a discovery, both economic and ecological?

Biotechnology

Insulin Without the Sting

Nobody enjoys taking shots, but people who suffer from juvenile diabetes mellitus have learned to live with them, because the alternative may be not living at all. The name "juvenile diabetes" does not mean that everyone who has it is a child. Rather, it's called "juvenile" because its onset is usually during childhood.

Causes of Diabetes People who have diabetes mellitus either don't produce enough insulin in their bodies or can't efficiently use the insulin they do produce. The form of diabetes mellitus known as adult diabetes can often be controlled with careful diet and exercise. The form called juvenile diabetes, however, usually cannot. Those with juvenile diabetes need to take regular doses of insulin to keep their blood sugar under control.

Unfortunately, insulin can't be taken by mouth. It's a simple protein—exactly the sort of material your stomach easily digests. Insulin would be quickly broken

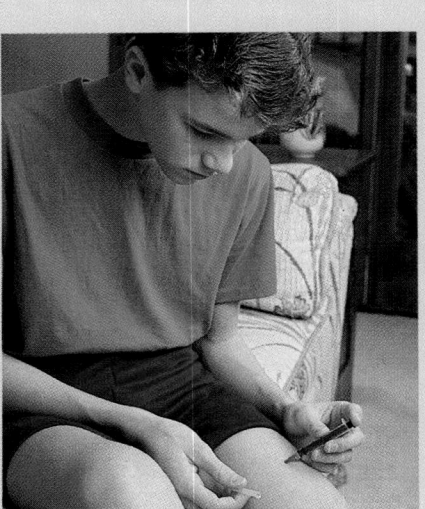

down by digestive enzymes and would never make it into the bloodstream. Diabetics in need of insulin have to take it by frequent injections.

A New Way to Take Insulin The good news is that the need for injections may soon disappear. New coatings are being developed that go beyond allowing insulin in tablet form to survive the digestive system. These coatings actually use the digestive process to release the insulin at the right time and place.

The new coatings are made of substances called proteinoids, combinations of amino acids that aren't quite normal proteins and so resist digestion. The unusual thing about proteinoids is that they shape themselves into tiny bubble-like spheres when placed in water. Put insulin, or any other organic material, in the water with them and some will be caught up inside the spheres. Since the spheres are so small that millions would fit in a single tablet, they're referred to as microspheres.

The molecules in proteinoids can be treated so that they have electrical charges. Some become positively charged and others negatively charged. By selecting ingredients carefully, it's possible to make microspheres with an abundance of either positively or negatively charged molecules on the outside.

How Microspheres Avoid Digestion To carry insulin, it's best to have microspheres with a shell of positive charges. Because the stomach contains acid, there is a high concentration of positive

hydrogen ions. Because like charges repel each other, the positive charge on the outside of the microsphere would deflect attacks by digestive acid. The microspheres would then carry their cargo of insulin safely into the small intestine where they are absorbed without having been broken down.

In the bloodstream, the microspheres are in a very different chemical environment. Because opposite charges attract, the negative ions would attract the microspheres and begin breaking them down. Soon, the microspheres would rupture, releasing insulin directly into the bloodstream.

Other Roles for Microspheres Proteinoid microspheres aren't limited to carrying insulin. They could transport just about any type of medication into the bloodstream— including vaccines. By carefully selecting the proteinoids to be used, pharmaceutical companies can engineer microspheres to release each medicine at its most beneficial point.

1. **Issue:** When encapsulated insulin finally becomes available, it is likely to be expensive. What factors might offset this expense for people who must take frequent insulin injections? Can you think of any health advantages that encapsulated insulin might have over injected insulin?
2. **Connection to Chemistry:** What causes an ion to have a positive or a negative electrical charge?

Steroids and Athletes

There's an old doctor's joke about the hospitalized patient whose condition kept improving day by day until he died. He died of too much improvement. That joke has become a tragic reality for some athletes and for some who merely want to look like athletes. The people are those who have "improved" their muscle size and strength using drugs called anabolic steroids.

The Action of Anabolic Steroids

Anabolic steroids are synthetic chemicals similar to testosterone, which is normally produced in the testes of the male. Although both males and females have testosterone, males have greater amounts.

Because anabolic steroids mimic testosterone, some obvious side effects occur in female users. These effects include deeper voices, beard growth, and male pattern baldness.

Males, on the other hand, do not grow increasingly "masculine" in appearance except for larger muscles and an increase in facial and body hair. The testosterone-like action of the steroids tends to shut down natural testosterone production. As a result, male users' testicles may permanently shrink and their breasts may enlarge.

Side Effects of Anabolic Steroids

Anabolic steroids can greatly reduce the body's level of HDLs, substances necessary for a healthy heart. As a result, many young users have died from sudden heart attacks. Long-term use of the drugs has been linked to liver damage, kidney disease, strokes, immune system disorders, and cancer.

Teenage users face even more dangers. Teenage bodies are not yet fully formed, and some parts, such as the skeletal and reproductive systems, are still developing. Anabolic steroid use can bring that development to a premature and permanent halt.

Can They Be Used Safely?

Still, many who bulk up their muscles with anabolic steroids are convinced that the drugs are safe if used wisely and carefully. A number of doctors say that this might be true, although they stress that it's by no means certain. Furthermore, medical experts insist that wise and careful use requires a high level of medical supervision. Very few users have such guidance, and the temptation to overuse is very great.

Nor are many likely to get the medical help required because the use of anabolic steroids without a doctor's prescription is illegal. In addition, all anabolic steroid use is against the rules of most professional and amateur sports. It's considered cheating.

Psychological Effects

Anabolic steroids affect the brain as well as the body. The drugs tend to give users feelings of extreme self-confidence while making them more aggressive.

These psychological effects can be drastic. Many users have undergone major personality changes. Steroid use has been linked with cases of violent behavior, and one psychiatric study showed that 14 out of 41 anabolic steroid users displayed signs of mental illness.

1. **Understanding the Issue:** In some countries, sports organizations and schools require athletes to take random tests for steroid use. Do you think random testing of athletes for steroid use should be carried out? Do you believe it's an invasion of privacy? Explain your position.

2. **Explore Further:** Some athletes feel that anything they can do to improve their performance should be legal and that steroid use is as fair as having a proper diet. Others think that strength from steroids is an artificial boost and should be illegal. Do you think a steroid advantage is fair or unfair? Why?

Readings

- "Pumped Up." *U.S. News and World Report,* June 1, 1992, pp. 54-63.
- "Roid Rage." *American Health,* May, 1991, pp. 60-65.

Summary

Plants respond to stimuli by means of hormones that interact to control growth and development. Hormones control such activities as tropisms, seed germination, cell division, fruit formation, and abscission of fruit and leaves. A plant's overall pattern of growth and development depends upon a balance among those hormones that promote these activities and those that inhibit them.

Hormones are one means of controlling responses in animals. Some animal hormones control growth and development. Others play a role in maintaining homeostasis—for example, the hormones that regulate blood sugar level. As in plants, hormones may work together to cause a response or in opposition to one another. Levels of many hormones in the blood are regulated by negative feedback mechanisms such as those that occur during the menstrual cycle.

As in plants, hormones in animals cause a response in one of two ways, each of which results in a chemical change within target tissues. A hormone may cause such a change by directly activating an enzyme within a target cell or by indirectly activating a gene.

Language of Biology

Write a sentence that shows your understanding of each of the following terms.

abscission
apical dominance
auxin
corpus luteum
corpus luteum stage
endocrine gland
follicle stage
gibberellin

hypothalamus
negative feedback
pituitary gland
second messenger
  hypothesis
stimulus
tropism

Understanding Concepts

1. In the Darwins' experiment, what was the purpose of the stem with the black tube?
2. A plant is placed on its side in the dark. Explain the growth of the plant's stem and roots in terms of stimuli and the action of hormones.
3. In terms of plant growth and development, discuss an example that shows that it is the target tissue, not the hormone, that determines a response.
4. Some weed killers are chemicals with powerful auxin-like activity. Suggest a way that these chemicals kill a plant. Why do you think that auxin-like weed killers would not kill plants during cold or very dry weather?
5. Drugs called diuretics cause the body to eliminate more water. Which hormone in Table 24.1 would be most affected by diuretics?

6. Explain the high blood sugar level of a person with type I diabetes in terms of insulin's effect on body cells.
7. Explain why a female does not usually ovulate during the corpus luteum stage.
8. You have probably heard of the idea that adrenaline can result in a burst of energy. Explain why this is true in terms of how adrenaline causes a response in liver and skeletal muscle cells.
9. Glucagon, a protein-type hormone, causes the same response in liver cells as adrenaline. How can it be explained that two different hormones can cause the same response in the same target cell? Can you suggest a reason why adrenaline, but not glucagon, can cause that response in skeletal muscle cells?

10. Insulin causes a response in liver cells opposite to that caused by adrenaline and glucagon. It exerts its effect by a second messenger. How can these opposite responses be explained if all three hormones involve a second messenger?

Applying Concepts

11. How would you design an experiment to show that it is auxin secreted from an apical bud that controls apical dominance?
12. Oxytocin and vasopressin are protein-type hormones composed of eight amino acids, only two of which differ. How can the different responses to these two hormones be explained?
13. Explain why unfertilized flowers fall from a plant much more quickly than fertilized flowers.
14. Explain why auxins are used to cause seedless varieties of some plants to develop fruit.
15. Suggest an explanation for the fact that a given protein-type hormone can cause two different responses in the same target cell.
16. Using Table 24.1 as a guide, identify the hormone(s) that would be associated with each of the following: (a) increased heartbeat rate and blood pressure; (b) loss of weight; (c) infertility in males; (d) infertility in females; (e) failure to maintain a balance of osmotic pressure between cells and surrounding fluid or blood.
17. In the menstrual cycle, when the corpus luteum breaks down, the uterine lining begins to break down also. Why does the corpus luteum break down if fertilization does not occur?
18. People with insulin-treated diabetes run some risk of a condition called insulin shock, which usually results from taking too much insulin. Why would this condition affect the body adversely? Why do some diabetics carry candy to eat in case insulin shock sets in?
19. **Health Connection:** Why do you think it's difficult to synthesize human growth hormone and similar substances in the laboratory?
20. **Biotechnology:** There are medicine capsules available that will pass through the stomach without being digested. Why not just fill these capsules with insulin?

Lab Interpretation

21. **Investigation:** You have tested the effect of various concentrations of gibberellin applied as a solution to the roots of bean plants. The data you obtained are shown in the following graph. Write a conclusion to the experiment.

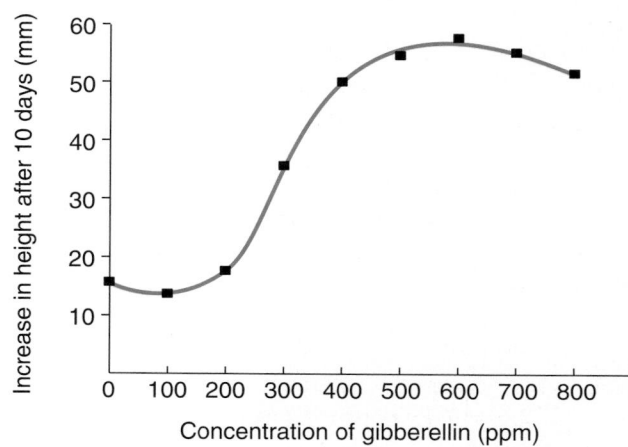

22. **Minilab 1:** After the activity is finished, what steps would be needed to return the plant stems back to their original vertical position? Explain why your idea would work.
23. **Minilab 2:** Which tissue found in the pancreas should have ducts leading from it? Which should not have ducts leading from them? Explain your answers.
24. **Thinking Lab:** Using what you learned in the text, identify hormones A, B, and C.

Connecting Ideas

25. Explain the following events in plant development in terms of chemical control: (a) dormancy of the seed after the embryo begins its development; (b) breaking of dormancy and onset of germination; (c) lengthening of cells in the zone of elongation of the root; (d) growth of the root downwards even if the seed begins germinating with the radicle growing horizontally; (e) the different lengths of branches.
26. Why must insulin be taken by injection rather than orally?

CHAPTER

2 5

NERVOUS CONTROL

A HAWK SOARS majestically through the air, its sharp eyes surveying a meadow hundreds of feet below. On the ground, small animals carry on with their activities, oblivious to the hawk above. The instant the hawk catches sight of its prey it swoops down in a steep dive and snatches the unsuspecting animal in its talons. Hawks and other birds of prey have such a keen sense of sight that they can hunt from a distance and take their quarry by surprise.

When a hawk is hunting, light rays entering its eyes are transformed into electrical impulses that travel to its brain. The brain processes the information and transmits impulses to the bird's muscles that send it into a dive.

How do light rays become electrical impulses? How do these impulses cause muscle action? How does this happen quickly enough for a hawk to catch its prey? In this chapter, you will learn how animals and other organisms sense their environment, and how those sensations are transformed into responses. You will discover that nerves enable organisms to respond to stimuli very quickly. And you will learn how the most highly developed nervous system in the animal kingdom—the human nervous system—operates.

A hawk's eyes are able to focus on two objects at the same time. What advantage is this to the hawk?

Chapter Preview

Objectives

Diagram the transmission of an impulse along a neuron.

Explain why an impulse is an all-or-none response.

Sequence the events in the transmission of an impulse from one neuron to another, and describe the different ways the neuron may respond.

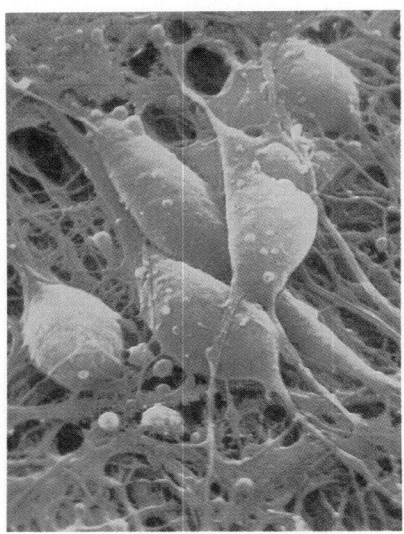

Figure 25-1 **This electron micrograph shows one of the body's billions of neurons.**

25.1 Conduction of Impulses

You learned in the last chapter that many of an organism's responses to stimuli are controlled chemically by hormones. Hormonal control is usually involved in responses that occur slowly, such as positive phototropism in plants, metamorphosis in insects, or growth in humans. But what about responses that must take place within a fraction of a second, like jerking your hand away from a hot stove or dodging out of the path of a speeding car? Rapid responses such as these are controlled by the nervous system. The nervous system is responsible for receiving information from internal and external stimuli and responding to that information. While bacteria, protists, and some plants are capable of nervous response, animals are the only organisms that possess true nervous systems.

Nervous Response

What kinds of cells, tissues, and organs are involved in your response to a hot stove or a speeding car? Four requirements must be met for a nervous response to occur. First, there must be a means of detecting a stimulus. In most animals, stimuli are detected by sensory receptors, which may be either individual nerve cells, such as touch-sensitive cells in your skin, or nerve cells that form part of a sense organ, such as the eye or ear. These cells transform sensory input into electrical impulses.

Second, there must be a way for the impulses to be transmitted to other parts of the body. Impulses are transmitted along a network of **neurons**—cells that are specialized for conducting nerve impulses. Third, the impulses must be interpreted and analyzed. This process takes place in a coordination center, such as the brain or spinal cord of vertebrates.

Fourth, an appropriate response must be carried out by an effector. An effector is usually a muscle or gland. For example, the muscles of your arm are the effectors that pull your hand away from the stove. Your adrenal glands are effectors that participate in your response to a speeding car by secreting adrenalin, which increases your metabolic rate and helps you respond more quickly to the emergency.

Nerve Tissue

Nerve tissue is composed of specialized cells that transmit nerve impulses through the body. Vertebrates and the more complex invertebrates, such as mollusks, annelids, and arthropods, have nerve tissue that is made up of different types of neurons. Each type differs in its structure and in the direction in which it carries impulses.

Sensory Neurons   In most animals, **sensory neurons** transmit incoming impulses from receptors in sense organs to a coordination center, where the impulses are interpreted. For example, a sensory neuron picks up impulses from touch-sensitive receptors in your fingertips and transmits those impulses to your brain or spinal cord for interpretation and processing.

Motor Neurons Once the incoming impulses have been analyzed, the brain or nerve cord will send out a response. **Motor neurons** transmit these outgoing impulses to the effectors. When your brain or spinal cord receives impulses from a receptor that is detecting too much heat, impulses are sent along the motor neurons to your muscles, which pull your hand away from the stove.

Interneurons In the brains and nerve cords of vertebrates, most impulses pass from sensory neurons to motor neurons through **interneurons.** In some simple animals, interneurons are not present; impulses pass directly from a sensory neuron to a motor neuron. In some simple multicellular organisms, such as *Hydra,* responses may be carried out without either interneurons or motor neurons. In these organisms, some sensory neurons detect a stimulus and conduct the impulse directly to the effector.

Nerves All neurons are cells that are specialized for transmitting impulses from one part of the body to another. Figure 25-3 shows the structure of a vertebrate motor neuron. It looks like a cell that has been drawn out into a long, thin fiber or wire. When many neurons are grouped together, they form a **nerve.** A nerve is like an electric cord composed of many individual wires.

Neuron Structure

The portion of a neuron that most resembles other types of cells is the cell body, which contains a nucleus and cytoplasm. On one side of the cell body is the narrow **axon.** The end of the axon is subdivided into many filaments, forming the end brush.

The other side of the cell body, opposite the axon, is often highly branched into fibers called **dendrites.** Axons transmit impulses *away* from the cell body. Impulses are conducted *toward* the cell body from the dendrites. Thus, an impulse flows into a neuron through its dendrites. Within the neuron, the impulse flows from dendrites to cell body to axon. The impulse is then transmitted from the end brush of the axon to the dendrites, axon, or cell body of another neuron or to an effector.

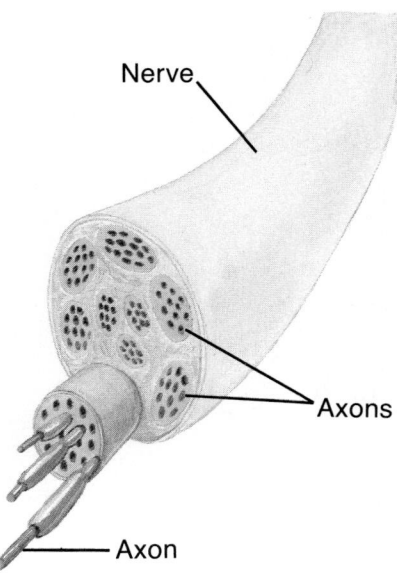

Figure 25-2 A nerve is made up of many neuron axons held together by connective tissue. What is the comparison between a nerve and a telephone cable?

Figure 25-3 A typical motor neuron is composed of dendrites, a cell body, an axon, and an end brush. Impulses are transmitted from the dendrites to the cell body. The axon carries impulses away from the cell body to the end brush and on to the dendrites of the next neuron.

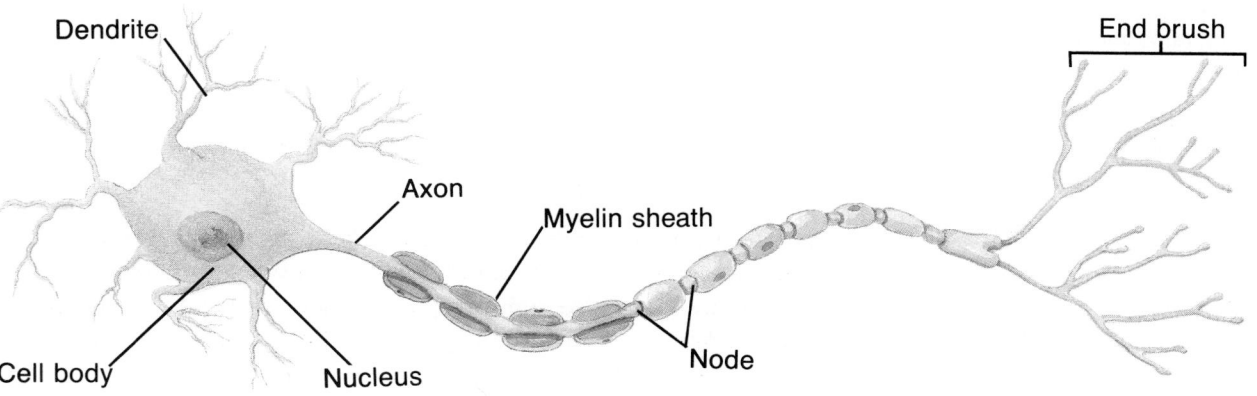

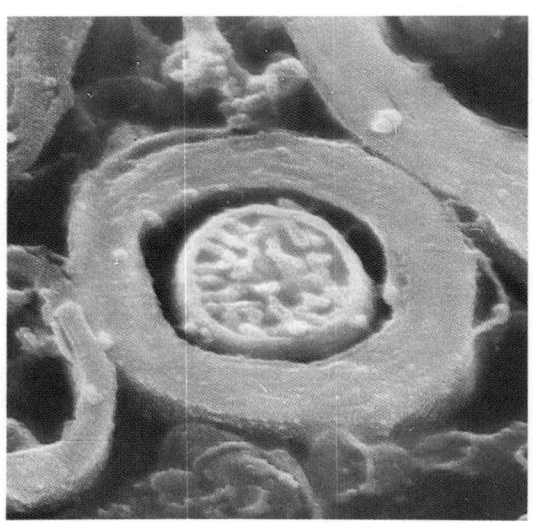

Myelin Sheath   In vertebrates, the axon of a neuron often has an outer layer of material called the myelin sheath. This sheath is composed of an insulating lipid material. A myelin sheath allows a nerve impulse to move very rapidly. In a motor neuron, the myelin sheath consists of many layers that wrap around the axon. These layers, made of a Schwann cell wrapping around and around, form a type of nerve tissue that helps provide nourishment to the axon and aids in axon regeneration.

Between Schwann cells are points where the axon is left uncovered. These points are called nodes. Both the myelin sheath and the nodes are important in determining the speed with which nerve impulses are conducted. An impulse is able to move from node to node instead of moving continuously along the membrane. Neurons lacking these features transmit impulses much more slowly than neurons that have a myelin sheath and its associated nodes. Some invertebrates such as insects and some crustaceans have a sheath of a myelin-like substance and nodes. This helps transmit impulses more rapidly but not as fast as vertebrates. In some invertebrates, axons are not myelinated but are of a large diameter. Axons with a large diameter transmit impulses much more quickly than those with small diameters. A squid neuron has a giant axon that transmits impulses ten times faster than an ordinary size fiber. Earthworms also have these giant axons that allow rapid movement.

Figure 25-4   **The axons of most neurons are wrapped in a myelin sheath composed of a Schwann cell. Magnification: 10 000×**

HISTORY CONNECTION

Dr. Humberto Fernandez-Moran

Scientists have been studying the human body and its systems for hundreds of years, yet one of the most difficult systems for them to understand was the nervous system. Only in this century, especially in the last 50 years, have real strides been made in effectively studying and understanding the workings of this body system.

One important person who helped unlock the mysteries of the nervous system is the Latin-American physician, biophysicist, and cell biologist, Humberto Fernandez-Moran. Born in Venezuela, Dr. Fernandez-Moran has studied and worked in Latin America, Europe, and the United States. Among his many accomplishments is the use of a superconducting objective lens in electron microscopes. He also discovered that the fatty substance, called the myelin sheath, surrounding some axons can be compared to a semiconductor. This fatty material aids in the passage of electrical charges and increases the speed at which impulses travel from axons.

For his contributions to our knowledge and his inventions, Dr. Fernandez-Moran has received numerous awards in Latin America, Europe, and the United States. The 1967 John Scott Award was presented to him for his invention of the diamond knife. This instrument allows very thin samples of cells to be sliced and used for microscopic studies. Today, Dr. Fernandez-Moran continues his work as professor of biophysics at the University of Chicago.

1. **Find Out:** What is the advantage of using a superconducting objective lens?
2. **Writing About Biology:** Explain why it might be more difficult to find out about the nervous system than other body systems.

Neuron Length and Function Neuron length varies considerably. While many neurons are comparatively short, others are long, thin, and delicate. For example, a single motor neuron may extend from your spinal cord all the way to your foot, which is a distance of about a meter. Interneurons in the brain and spinal cord are very short and densely packed together.

Why would it be useful for a motor neuron in your leg to be so much longer than an interneuron in your brain? It makes sense when you realize that the motor neuron's function is to transmit impulses from your spinal cord to a muscle in your foot, while the interneuron's function is to interpret impulses received from a nearby sensory neuron and transmit a response to a nearby motor neuron. Regardless of these differences, all neurons conduct nerve impulses in about the same way.

Neuron Function: Resting Potential

A neuron at rest is polarized. In other words, the fluids outside the neuron's cell membrane have an electrical charge that is different from the electrical charge inside the cell. This difference in electrical charge is a form of potential energy that is released when the neuron receives a stimulus.

What are the factors that cause polarization? One factor has to do with the concentration of sodium ions (Na^+) and potassium ions (K^+) inside and outside the cell. Outside the cell, the concentration of sodium ions is high while that of potassium is low. Inside the cell, the concentration of potassium ions is greater than that of sodium ions. As a result, the outside of the cell has a greater positive charge than the inside. How is this unequal distribution of sodium ions achieved? The neuron possesses a membrane protein, known as the sodium-potassium pump, which removes sodium ions from the cell by active transport.

Other factors also play a role in polarization. The cell membrane is slightly permeable to potassium, so potassium ions can diffuse out of the cell. The loss of potassium ions by diffusion adds to the more positive charge outside the cell membrane. Also, there are many large, negatively charged ions inside the cell. These negative ions cannot cross the membrane so they contribute to the less positive charge inside the cell. The factors that cause the fluids outside the neuron to be more positively charged than the fluids inside the cell are shown in Figure 25-6.

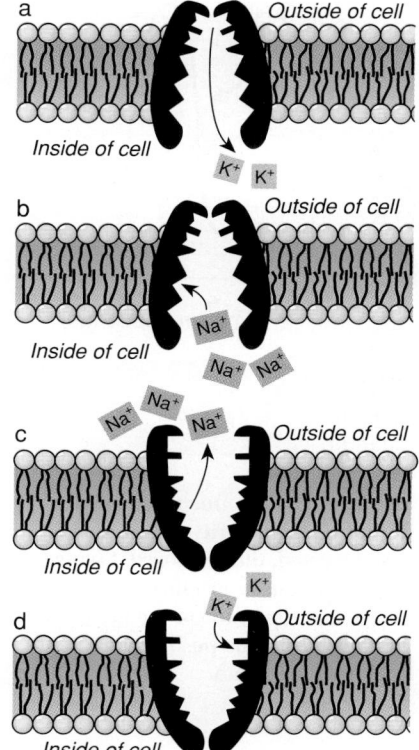

Figure 25-5 The sodium-potassium pump is made of proteins embedded in the cell membrane. These proteins are able to move Na^+ and K^+ ions in and out of a cell by active transport. When the pump is open to the inside of the cell (a, b), K^+ no longer binds and Na^+ sites become active. When the pump is open to the outside of the cell, Na^+ no longer binds and K^+ sites become active (c, d).

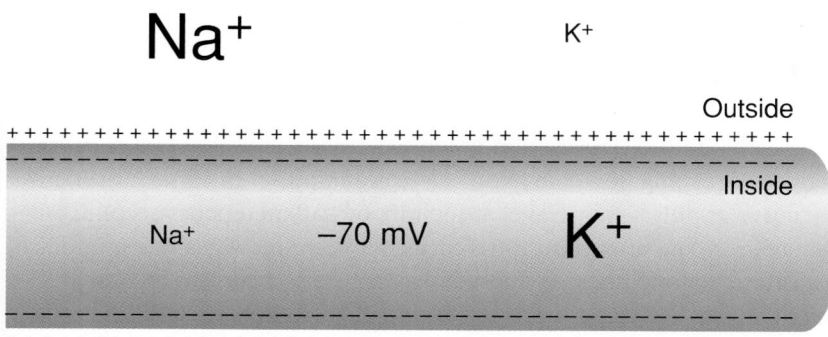

Figure 25-6 The unequal distribution of ions inside and outside its cell membrane polarizes a neuron. This is known as resting potential. Although the inside of a neuron contains positive ions, it can be considered as being negative compared to the outside.

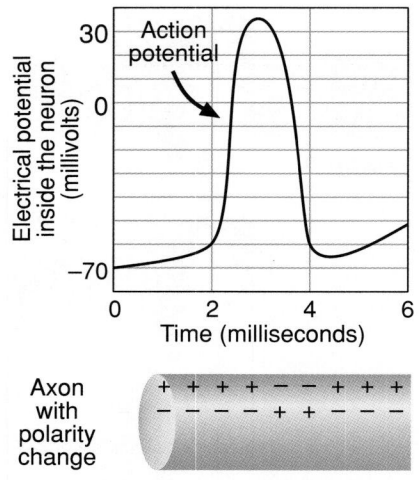

Axon
with
polarity
change

Figure 25-7   **During an action potential, the polarity of a neuron is first reversed, then restored. Changes in polarity can be graphed. The graph shows that the inside of the neuron is first negative, then positive, and then negative once again.**

The difference in electrical charge between the inside and outside of a resting neuron is called **resting potential.** The word *potential,* as used here, means voltage, which is a measure of electrical energy. You can think of the electrical charges on either side of the cell membrane as the two poles of a battery. Although all cells have an electrical potential across their membranes, only neurons use this potential energy to transmit nerve impulses.

Neuron Function: Action Potential

When a neuron is excited by a stimulus of sufficient strength such as light, pressure, chemicals, or an electric current, changes in the resting potential take place. A stimulus alters the permeability of the cell membrane. The change in permeability is what allows impulses to move along neurons, carrying messages throughout the nervous system. Let's see how this process works.

Sodium and Potassium Gates   First, the membrane becomes highly permeable to sodium ions. Recall from Chapter 4 that membrane proteins form openings, or gates, that allow certain ions and molecules into and out of the cell. When a neuron is excited by a stimulus, gates permeable to sodium ions are opened and the ions rush into the cell. The rapid movement of sodium ions into the cell is a result of the greater concentration of sodium ions outside the cell and the difference in electrical charge. The positively charged sodium ions are attracted to the less positive charge inside the cell. The inward rush of sodium ions reverses the polarity of the cell. Now the interior of the cell has a greater positive charge than the exterior.

Once this reversal of polarity has taken place, the sodium gates close and potassium gates open. Potassium ions rush out of the cell for the same reasons sodium ions rushed in: there is a higher concentration of potassium ions inside the cell, and the positive potassium ions are attracted to the less positive charge outside the cell. Polarity is again reversed so that, once more, there is a greater positive charge outside the cell than inside it.

Action Potential   The flow of ions and the changes in polarity caused by a stimulus create an electrical current called the **action potential.** The action potential, which can be detected by equipment that measures electrical current, shows that an impulse is being sent along a neuron.

An action potential in one region of a neuron causes a change of polarity in the next region. In this way, action potentials, which may be thought of as currents, move along the entire length of the axon. A neuron can send 250 to 2500 action potentials per second. A series of action potentials sweeping down an axon is a nerve impulse. In some large neurons, nerve impulses may travel at the rate of 130 meters per second.

There are some important differences between the way a nerve impulse is transmitted along neurons and the way an electric current moves through a wire. An electric current travels close to the speed of light and

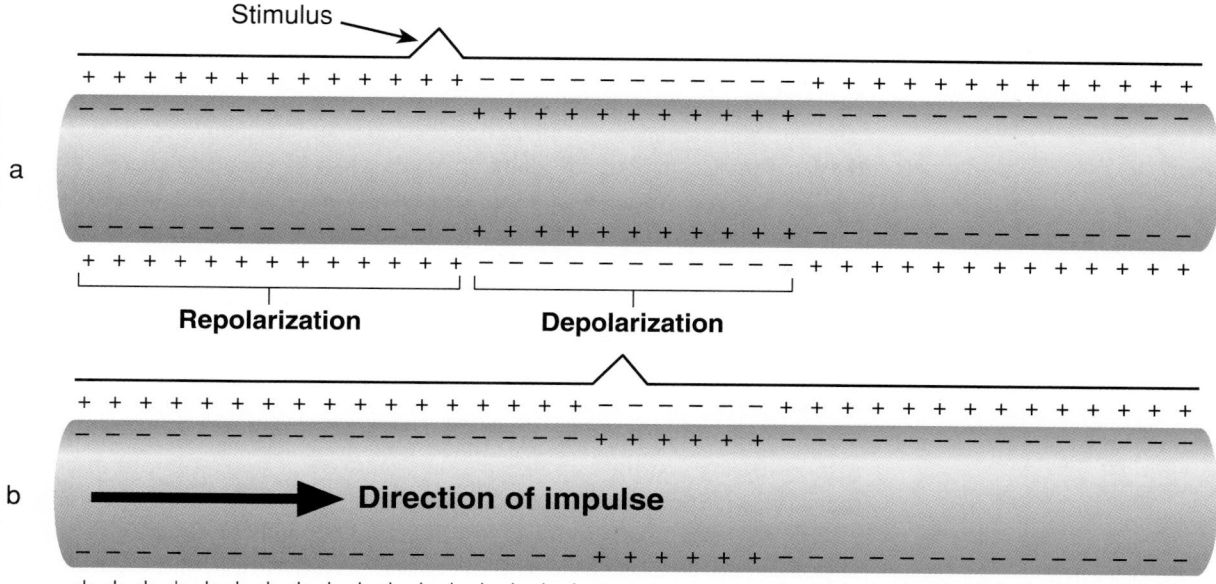

Repolarization   Depolarization

Direction of impulse

consists of electrons flowing smoothly along a wire. A current in a nerve impulse is caused by the flow of ions across the cell membrane. Thus, a nerve impulse is an electrochemical reaction that behaves somewhat like a fuse. Ignited at one end, a fuse will burn continually and very rapidly. Like a nerve impulse, this action can move in only one direction. Unlike the fuse, the nerve impulse can restore itself.

Restoring Resting Potential   After an action potential has swept along a neuron, some of the sodium ions that were originally outside the cell membrane are now inside it. Similarly, some potassium ions have moved from inside the cell into the fluids surrounding it. However, because the number of sodium ions that were originally outside the cell is very large, there are still more sodium ions outside the cell than inside it. Similarly, there are still more potassium ions inside the cell than outside it.

But as action potentials continue moving along the neuron, a point is reached when the balance of sodium and potassium ions inside and outside the cell becomes equalized. It is at this point that the sodium-potassium pump operates to once again create a resting potential across the cell membrane. Sodium ions once again move into the cell and potassium ions move out in a ratio of approximately three to two. Each pump can move several hundred ions per second, and a single neuron has about 1 million pumps. So, even though the neuron cannot carry a nerve impulse while the sodium-potassium pumps are restoring the resting potential, there is little effect on the nervous system. Resting potential can be restored in about 1.0 to 1.5 milliseconds.

Figure 25-8   Repolarization occurs as K$^+$ flows out of the cell, restoring resting potential (a). Transmission of an impulse occurs as the action potential (polarity change) moves along an axon (b).

Figure 25-9   The burning of this fuse resembles the way an impulse travels down an axon. Unlike the fuse, however, the nerve impulse can be restored.

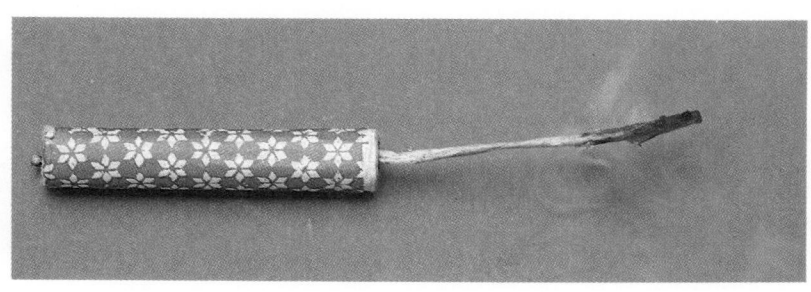

Minilab 1

How sensitive is skin to the sense of touch?

Tape two toothpicks to six separate pieces of masking tape the following distances apart: 2, 4, 6, 8, 10, 12 mm. Allow a small amount of the toothpick to extend beyond the edge of the tapes. Working with a partner, test the fingertip, palm, back of hand, inside forearm, and back of neck.

Be sure to apply the same pressure each time. **CAUTION:** *Be sure not to press too hard with the toothpicks.* The subject should let the tester know when two points are felt and when one point is felt. Conduct four trials in each location. Make a graph of the average for each location.

Analysis *Which location was most sensitive to touch? How could you tell? Which location was least sensitive to touch? How could you tell? How might the distribution of receptors for touch be a reflection of the functions of those body parts?*

All-or-none Response

Not all stimuli cause action potentials (impulses). A weak stimulus might generate some changes in the dendrites and cell body neuron, but unless the depolarization is of sufficient strength when it reaches the axon, there will be no nerve impulse. A stimulus must have a certain intensity to change polarity in the beginning of the axon and start an impulse. This level of intensity is called the threshold.

A stimulus stronger than threshold level will produce the same depolarization of the axon and thus the same impulse as a stimulus just at the threshold level. No matter how strong the stimulus is, as long as it is at or above the threshold level, the nerve impulse it causes will travel at the same speed and in the same way. Thus, a nerve impulse is an all-or-none response, meaning the impulse is either sent or not sent. If impulses are the same whether the stimulus is at or above threshold, how do you detect different strengths of stimuli? This detection depends on the number of impulses carried along a neuron in a given period of time and the number of neurons carrying the impulse. For example, you can tell whether the metal handle of a frying pan is warm or burning hot because the stimulus of a hot handle will excite more neurons, and cause those neurons to send impulses more frequently, than the stimulus of a warm handle.

All nerve impulses are essentially the same. If this is so, how do you interpret *what* the stimulus is? How do you know whether you are sensing light or sound or touch? The answer is that your interpretation and response to a stimulus depend on the pathway of the nerve impulse. For example, when a touch-sensitive receptor in your skin is stimulated, the nerve impulse travels to a specific area of your brain or spinal cord that interprets nerve impulses coming from touch receptors. Impulses transmitted by receptors in your eyes or ears travel to areas of your brain that determine your response to light or sound. Impulses traveling along a particular nerve pathway will always go to the same area of the brain or spinal cord.

Figure 25-10 **A "strong enough" push is needed to begin these dominos falling. The same is true for a nerve impulse. A stimulus must be of a certain intensity to begin an impulse. This level of intensity is called the *threshold*.**

Transmitting Impulses from Neuron to Neuron

So far, you have learned how an action potential, or nerve impulse, moves along the axon of a single neuron. Virtually all nerve impulses must travel through many neurons before reaching their target. But there is a very small, fluid-filled space, or **synapse,** between each neuron along a nerve pathway. How does a neuron transmit across a synapse?

Remember, an action potential can travel through a neuron in one direction only—through the dendrites, into the cell body, and then along the axon. It follows that the impulse travels across a synapse in one direction only—from the axon of one neuron, across the synapse, to the dendrites of the next neuron. Since the end of the axon is so highly branched, a single neuron can send impulses across the synapse to the dendrites of many neurons.

Neurotransmitters   Located in the cell membrane of the end brush of the axon are membrane proteins that act as calcium ion gates. When the neuron is at rest, the concentration of calcium ions outside the membrane is much greater than inside it. As an action potential sweeps toward the end brushes of the axon, the calcium gates open, permitting an inward rush of calcium ions. Once inside the end brushes of the axon, the calcium ions stimulate vesicles in the cytoplasm to move toward the membrane. These vesicles contain **neurotransmitters,** which are chemicals that transmit the nerve impulse across the synapse. The vesicles fuse with the

Figure 25-11   At a synapse, neurotransmitters pour into specific sites on nearby dendrites. Compare the diagram and the electron micrograph. Find the synaptic vesicles in the micrograph.

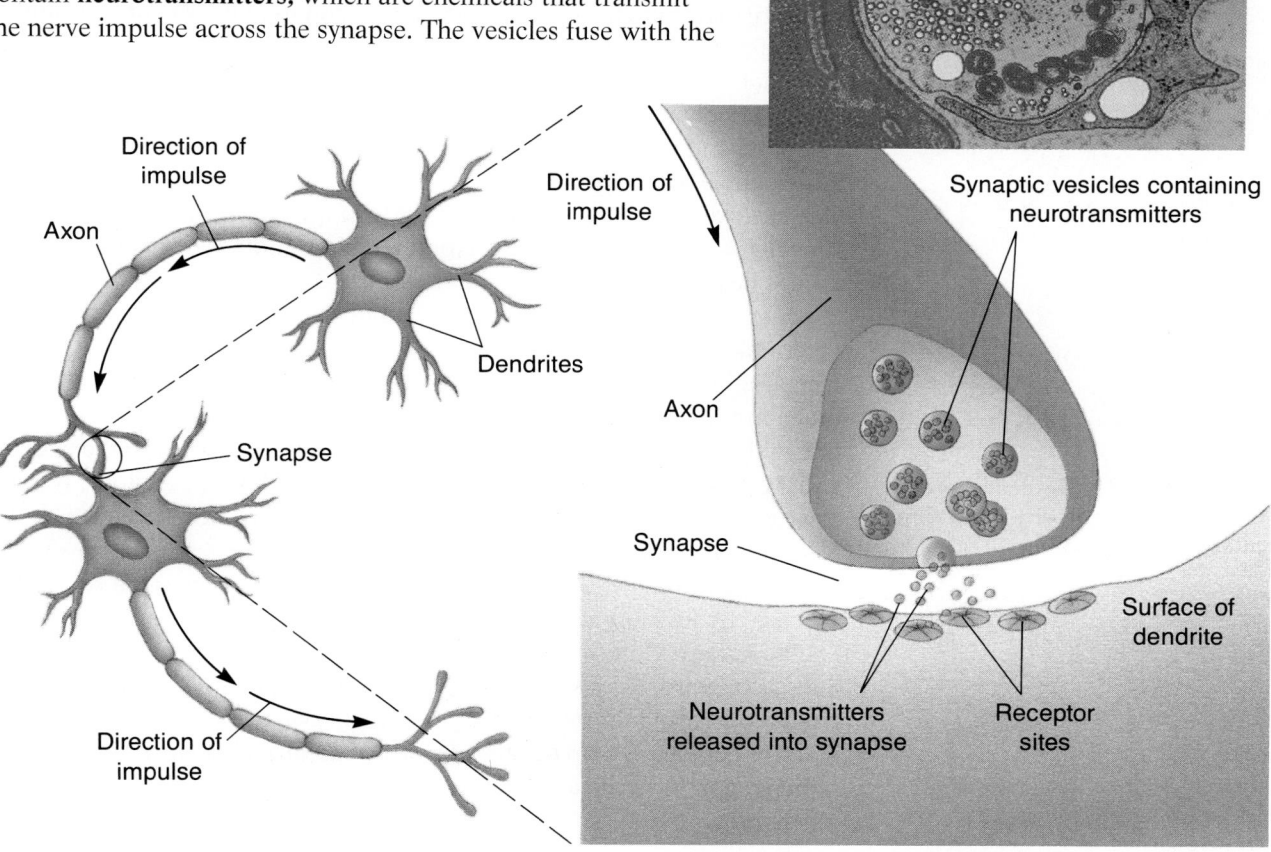

axon membrane, and the neurotransmitters are released into the synapse by exocytosis. The neurotransmitters diffuse across the fluid-filled synapse and combine with receptor molecules embedded in the membrane of the dendrites of neighboring neurons.

Excitation   There are many types of neurotransmitters and many types of receptor molecules. Some neurotransmitters, when combined with certain receptor molecules on the dendrites, will cause the dendrite membrane to become more permeable to sodium. When this happens, sodium ions rush into the neuron and an action potential is started. In this case, the neurotransmitter is said to excite the neuron.

Inhibition   Other combinations of neurotransmitters and receptor molecules may have the opposite effect. In this case, the combination may cause the dendrite membrane to become more permeable to negatively charged chloride ions, which enter the cell and reinforce its resting potential. The polarization of the cell increases (the interior becomes even less positive than the exterior), which prevents an action potential from starting. Some neurotransmitters produce the same inhibitory effect by inducing the dendrite membrane to release positively charged potassium ions.

Determining the Effect of a Neurotransmitter   Recall that a single axon may pass impulses to the dendrites of many neurons. In addition, the dendrites of a single neuron may receive neurotransmitters from many axons. Some neurotransmitters will excite the target neuron, while others will inhibit it. The net effect of these excitatory and inhibitory neurotransmitters determines whether or not the target neuron will carry an action potential.

The effect of a particular transmitter on a neuron depends on the receptors and membrane proteins in the dendrites of the target neuron. For example, the transmitter acetylcholine excites the neurons that activate skeletal muscle and muscles of the digestive tract, but inhibits neurons that activate heart muscle. Noradrenaline, another neurotransmitter, excites heart muscle, inhibits muscles of the digestive tract, and has no effect at all on skeletal muscle. It is the target, not the transmitter, that determines the response. Do you see that nerves ultimately cause a response by means of chemicals? Many of those chemicals, such as noradrenaline, are very similar to hormones. Not only that, but some transmitters within the brain, are, in fact, hormones such as vasopressin, oxytocin, and ACTH. As transmitters, they have evolved different functions. It is clear that endocrine and nervous control systems are closely related. Nerves can be thought of as structures specialized for delivering chemicals quickly to target organs of animals, thus ensuring the rapid responses animals need to survive.

The hormones of the endocrine system for the most part control slow, long-lasting processes such as growth and development. The nervous system is specialized for much faster movement of chemicals through the body. A nervous system of this type is needed to enable animals to make rapid responses necessary for survival, such as fleeing from danger or capturing prey.

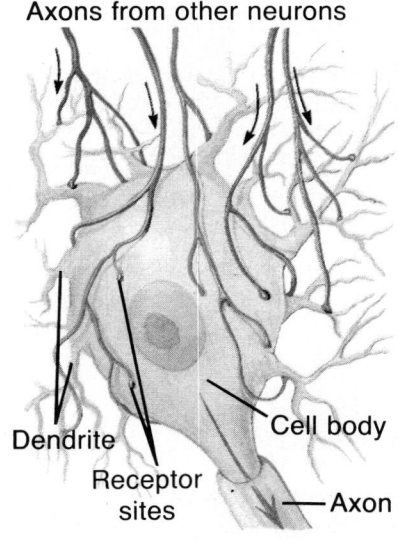

Axons from other neurons

Dendrite

Receptor sites

Cell body

Axon

Figure 25-12   **The dendrites of a single neuron may receive neurotransmitters from many axons.**

Suppression of Neurotransmitter   Soon after a neurotransmitter is released from the axon, it is suppressed so that the target neuron does not continue to respond. For example, after acetylcholine has combined with the receptor proteins of a dendrite membrane, it is broken down by an enzyme into two smaller molecules. Neither of these smaller molecules can excite or inhibit the neuron. Noradrenaline is not chemically changed by the target neuron. Instead, it is recaptured by the axon that released it.

Effects of Drugs   Many drugs, both legal and illegal, exert their effects at the synapse. Depressants such as hypnotics and anesthetics are drugs that reduce nerve transmission. This reduction in transmission can result in impaired coordination, slowed reaction time, and impaired judgment. Some depressants have molecular structures that are very similar to those of neurotransmitters that occur naturally in the body. These depressants combine with receptors in the cell membranes of dendrites, blocking the natural neurotransmitters and reducing the number of impulses transmitted along the neuron. Other depressants may prevent the production of neurotransmitters or prevent their release from the vesicles of the axon.

Stimulants such as caffeine increase the activity of the nervous system. This causes heartbeat rate to increase, an increase in blood pressure, and an increase in mental alertness. Some stimulants combine with receptors in the dendrite and excite the neuron by causing action potentials. This occurs in the same way as the transmitters that normally link with the receptors. Other stimulants prevent the recapture or destruction of neurotransmitters or cause the excessive release of neurotransmitters.

Figure 25-13   **Many people use the caffeine in coffee as a stimulant in the morning. What other food products contain caffeine?**

Section Review

Understanding Concepts

1. Suppose you accidentally stick your finger with a pin. Explain how a neuron in your finger responds by conducting an impulse.

2. In a lab, an axon is exposed to 1 volt of electricity, then 3 volts, and then 5 volts. The axon does not carry an impulse until stimulated by 5 volts. Why not? How will stimulating the axon with 10 volts affect the axon? Explain.

3. An interneuron in the brain synapses with hundreds of sensory neurons. What factor determines whether or not that interneuron will carry an action potential at any given time?

4. ***Skill Review—Recognizing Cause and Effect:*** Novocaine interferes with operation of the sodium-potassium pump. Explain why Novocaine causes numbness. For more help, see Thinking Critically in the **Skill Handbook.**

5. ***Thinking Critically:*** Suppose your nervous system were rewired so that the nerve from your eye led to the area of your brain to which the nerve from the ear is supposed to lead and vice versa. How would you respond to light and sound? Explain.

Objectives

Compare the nervous systems of invertebrates and vertebrates.

Explain the functions of the parts of the central nervous system of humans.

Describe examples of reflexes and how those responses are carried out by the nervous system in humans.

Explain the functioning of the human ear and eye.

25.2 Nervous Systems

Although animals (except sponges) are the only living organisms that have nervous systems, all cells possess the basic elements of nervous control. Even unicellular organisms are capable of detecting stimuli. They can transmit signals from one part of the cell to another, analyze those signals, and respond accordingly. As animals evolved, certain cells became specialized for the functions involved in the rapid responses that are crucial to survival. Those specialized cells gradually evolved into the nervous systems of animals. As more complex animals evolved, their nervous systems also became more complex. Today, the nervous systems of complex organisms, such as humans and other vertebrates, exert a great deal of control over behavior and permit a large variety of responses to a given stimulus.

Unicellular Organisms

Many bacteria can sense the presence of chemicals in their environment. They move toward useful substances and away from harmful ones. The movement of an organism toward or away from a stimulus is called a taxis. It is thought that the bacteria detect stimuli by means of receptor proteins in their cell membranes and cell walls. Stimulation of these receptors leads to production of chemicals within the cell. After these chemicals are "analyzed" by the cell, an electrical signal travels through the cytoplasm to effectors, such as the flagella. The flagella respond to the signal by beating in a direction that propels the cell toward or away from the stimulus.

Some simple eukaryotes have cell parts that are specialized for responding to the environment. For example, *Euglena* has a light-sensitive structure under the eye-spot or "stigma." The pigment in the stigma

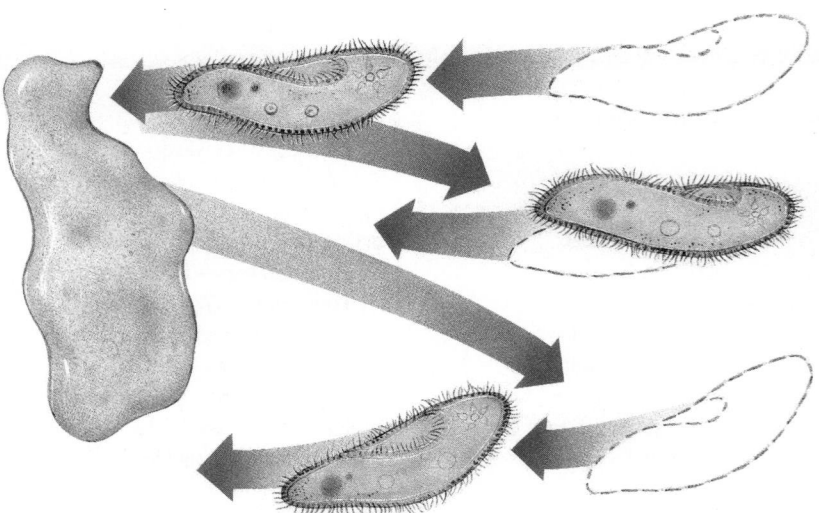

Figure 25-14 Paramecium responds to obstacles in its path by turning and moving forward again. The organism continues responding this way until it has cleared the obstacle.

allows this protozoan to detect the direction of the light source. This allows *Euglena,* a photosynthetic protist, to swim in the direction of light.

Paramecium can respond to obstacles in its path. If *Paramecium* comes into contact with an object, it will reverse the beating of its cilia and back away. Then it turns at an angle of about 30° and moves forward again. The organism continues to respond in this way until it passes the object (Figure 25-14). Responses in *Paramecium* seem mainly to be under electrical control.

Invertebrates

The cnidarians are the simplest phylum of animals that possess a nervous system. Recall that cnidarians are radially symmetrical. They have no front or back, and no left or right sides. Those that can move do not travel in a definite direction. There is no center of nervous control in these animals. Instead, many short neurons are scattered throughout the animal, forming what is called a nerve net. Figure 25-15 shows the nerve net of *Hydra.* All its neurons are short and similar in structure, and there are no definite nerve pathways. Because the neurons are short, an impulse must cross many synapses, and the impulse often dies out before being transmitted very far. As a result, most responses are localized, involving only one part of the body such as a single tentacle. Strong stimuli, though, may lead to a response of the entire organism. For example, *Hydra* may respond to a strong disturbance by contracting into a ball. Unlike the synapses described earlier, those of cnidarians send messages in both directions. A light touch to one tentacle of a *Hydra* causes the tentacles on either side to respond at the same time.

Flatworms   Flatworms, unlike cnidarians, have bilateral symmetry. Like most animals, they have anterior and posterior ends and a definite head-first direction of movement. Animals with these characteristics have evolved a nervous system that is especially suited for movement in one direction. Animals with bilateral symmetry have nervous systems with a control center, or centers, and specific nerve pathways.

A planarian is a flatworm whose nervous system shows the early evolution of such systems. At its anterior end are two bundles of neuron cell bodies called ganglia, which are like the brains of more complex animals. Extending from the ganglia are two ventral nerve cords that run the length of the body. The cords consist of the axons of the neurons. Connecting nerves join the two main cords. Smaller neurons branch from the cords and act much like a nerve net. The planarian's primitive brain does not exert much control.

Other Invertebrates   Segmented worms, such as the earthworm, have two ventral nerve cords, like planarians. However, in each segment of the worm there is a pair of ganglia along the nerve cord. These ganglia act as control centers. The brain, which is located on the dorsal side of the head, is really just a slightly larger ganglion. It exerts a bit more control than do the other ganglia. Arthropods are segmented, too, but they have fewer

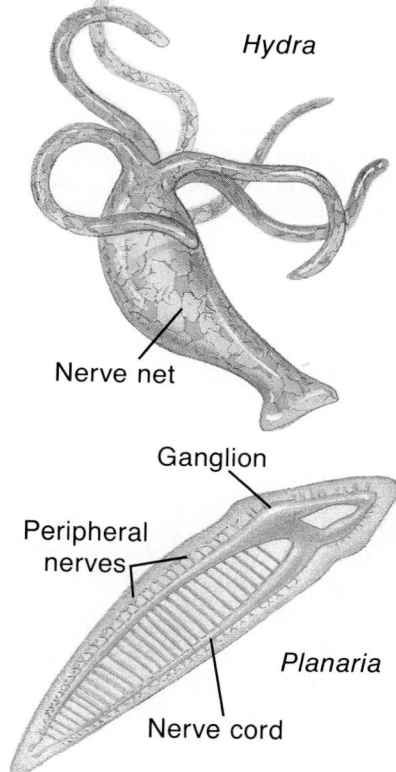

Hydra

Nerve net

Ganglion

Peripheral nerves

Planaria

Nerve cord

Figure 25-15 ***Hydra*** **(a) has a nerve net nervous system. *Planaria* (b) has a ladder-type nervous system with the beginnings of central control in the ganglia of the head region.**

ganglia and a somewhat more prominent brain that exerts a greater degree of control than the earthworm's brain. Like other bilateral invertebrates, arthropods have a dorsal brain and ventral nerve cords.

Definite evolutionary trends can be seen when you examine the nervous systems of bilaterally symmetrical animals. As more complex animal phyla evolved, the brain began to exert more control, and the nerve cords gradually lost many of their control functions. Location of the major control center, the brain, at the animal's anterior end allowed organisms to adapt more easily to the environment. As an animal moves, its anterior end is the first part to encounter stimuli from the environment. As sense organs—such as eyes, antennae, and hearing structures—evolved, they became more concentrated at the anterior end. It makes sense that sensory organs, which gather information, and the brain, which analyzes the information, should be located near each other.

Vertebrates

Evolutionary trends observed among invertebrates continued as the vertebrate nervous system evolved. The nervous systems of bilaterally symmetrical animals are divided into two main parts. The **central nervous system,** or CNS, consists of a brain and one or more nerve cords, which serve as control centers. The **peripheral nervous system** consists of sensory and motor neurons connected to the CNS. Information about the animal's environment is gathered by receptors and brought to the CNS by sensory neurons. That information is analyzed by means of interneurons in the CNS, and impulses are then carried to appropriate effectors by motor neurons of the peripheral nervous system. Vertebrates have an anterior brain, which serves as the major center for coordinating nervous response, and much of the sensory equipment is located in the head. Although ganglia are located along the length of the single, dorsal spinal cord, they do not control complex responses. The spinal cord is an extension of the vertebrate brain; together, the brain and spinal cord make up the CNS.

The peripheral nervous system consists of nerves leading to and from the brain and spinal cord. In humans, the peripheral nervous system is composed of 12 pairs of cranial nerves attached to the brain, and 31 pairs of spinal nerves attached to the spinal cord.

Vertebrate brains are made up of three general areas—forebrain, midbrain, and hindbrain. The structure of each area and the functions it performs differ from one kind of vertebrate to another. We'll study the human nervous system as a model of the vertebrate nervous system as it exists in the most complex vertebrate.

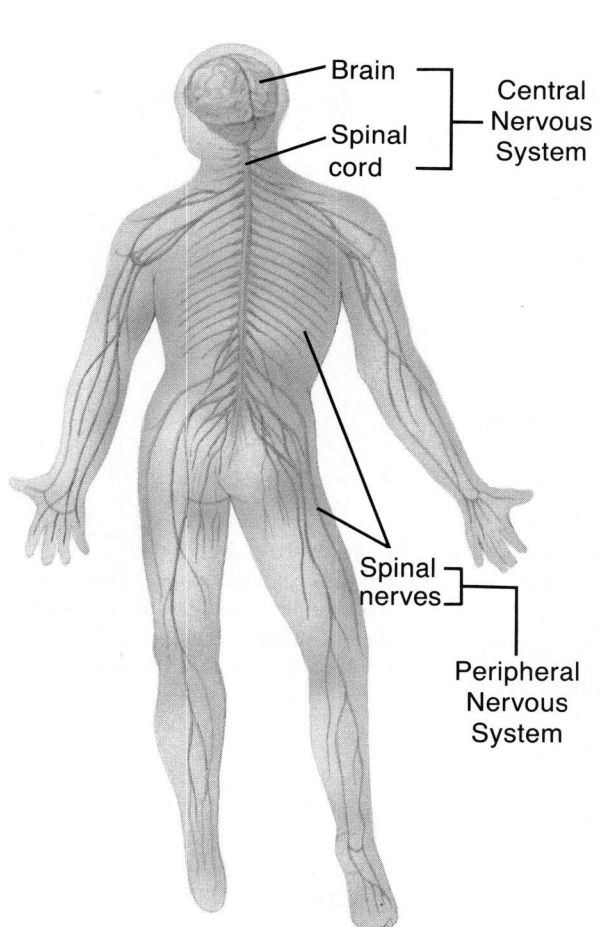

Figure 25-16 **The brain and spinal cord make up the central nervous system (CNS). The peripheral nervous system (PNS) is made of sensory and motor neurons connected to the CNS.**

Brain
Spinal cord
Central Nervous System

Spinal nerves
Peripheral Nervous System

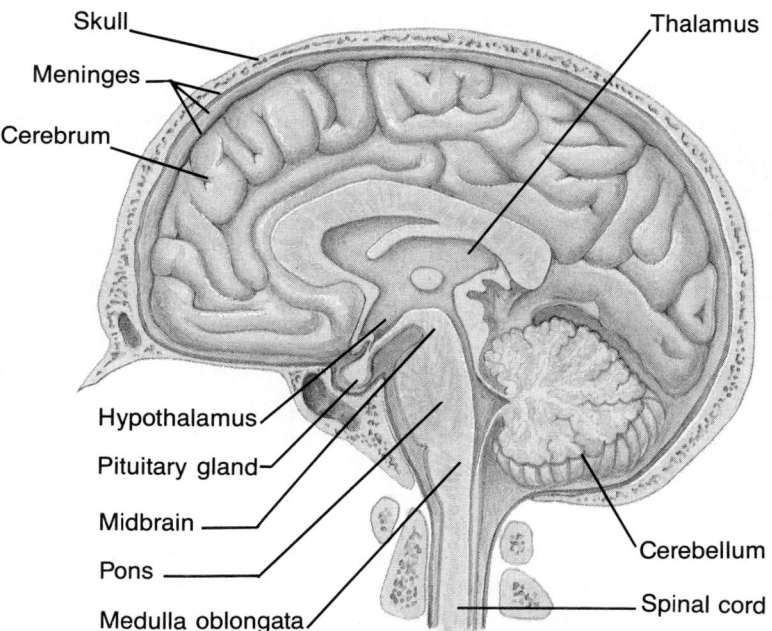

Skull

Meninges

Cerebrum

Thalamus

Hypothalamus

Pituitary gland

Midbrain

Pons

Medulla oblongata

Cerebellum

Spinal cord

Figure 25-17 The human brain is composed of three major regions—the forebrain, or cerebrum; the midbrain; and the hindbrain, made up of the medulla oblongata and the cerebellum.

Central Nervous System of Humans

The human brain is the most complex structure known to humans. It is a soft, wrinkled mass composed of billions of neurons. A human brain is protected by the skull and three layers of membranes called the meninges. These membranes protect and nourish both the spinal cord and brain. The space between the inner two meninges is filled with cerebrospinal fluid. Cerebrospinal fluid also fills the ventricles (cavities) of the brain. Capillaries in the walls of the ventricles exchange nutrients and waste materials between the blood and the cerebrospinal fluid. The fluid, in turn, exchanges these materials with the brain cells. This clear fluid cushions the brain against mild shock, though a severe shock, such as a blow to the head, can sometimes damage the brain. For example, a concussion, a brain injury resulting from a severe blow to the head, temporarily affects the brain's ability to function and can cause dizziness or unconsciousness.

The Forebrain The largest part of the forebrain is the **cerebrum,** which consists of two hemispheres. Each hemisphere is made up of four lobes, Figure 25-18. The surface of the cerebrum is called the cerebral cortex and is composed of neuron cell bodies that are collectively known as gray matter. These cells are gray in color because they are not covered with a myelin sheath. The interior of the cerebrum is made up of white, myelin-sheathed axons.

The cerebral cortex is highly folded. These folds provide more surface area, and therefore more space for neuron cell bodies, than a smooth surface would. The large number of neuron cell bodies in the human brain is

Figure 25-18 The cerebrum is divided into two hemispheres, each consisting of four lobes. These lobes are named after the bones of the skull that protect them.

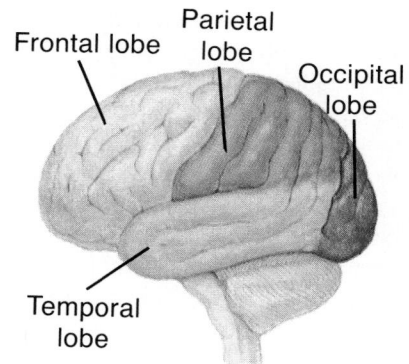

Frontal lobe

Parietal lobe

Occipital lobe

Temporal lobe

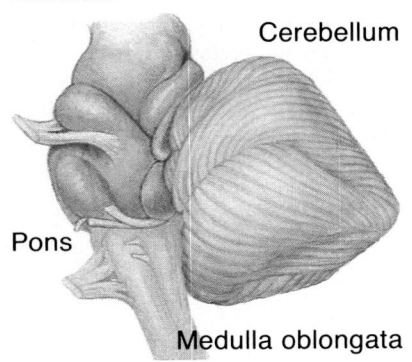

Figure 25-19 **The medulla oblongata and the pons make up the brainstem. The brainstem connects the brain and the spinal cord.**

important to our capacity for complex behavior and intelligence. Many human abilities are related to our very large cerebrum; for example, speech, reasoning, emotions, and personality are all functions controlled by this area of the brain. The cerebrum is also where sensory inputs are interpreted and motor impulses are started.

Located under the cerebrum are the thalamus and hypothalamus. The thalamus aids in sorting sensory information. The hypothalamus, which is a small structure just under the thalamus, controls hunger, body temperature, aggression, and other aspects of metabolism and behavior. It also plays an important role in endocrine control.

Midbrain The midbrain is an important and prominent structure in fish and amphibians. But in humans and other mammals, it mainly functions as a station for message exchange between the forebrain and hindbrain. The midbrain also is involved in some sight, hearing, and orientation responses.

Hindbrain: The Cerebellum The **cerebellum,** a motor area of the brain, lies at the back of the head, below the cerebrum. Like the cerebrum, the cerebellum is highly folded; its outer surface is made up of gray matter, and its interior is made up of white matter. The cerebellum coordinates

Thinking Lab

DRAW A CONCLUSION

Can brain cells regenerate?

Background

Human brains do not have the ability to replace nerve cells that have been damaged through disease or injury. A team of scientists recently decided to try to grow brain cells in a laboratory. They obtained brain tissue from an infant with a rare brain disease. One side of this infant's brain was growing at an extremely rapid rate. The researchers began growing these cells from the cerebral cortex of the child. Growing human cells in laboratory dishes is very difficult. They must be given precisely the same amounts and kinds of nutrients and growth hormones that they have in the body. Surprisingly, some of the brain cells in the lab dishes began to grow. The scientists had to decide if they were neuron or other types of neuron tissue called glial

cells—cells that provide support to the brain. Neurons have a pyramid-shaped cell body and long, thin tendrils that stretch out from the cell body. The support glial cells are flat and many-sided with no tendrils, and these cells have the ability to divide.

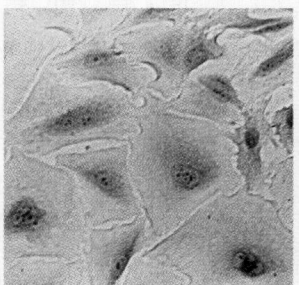

A

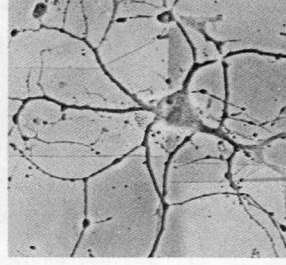

B

You Try It

Draw conclusions by examining the microscopic views of the brain cells in this study. Figure A shows brain cells before treatment with nutrients and hormones. Figure B shows brain cells after treatment. (1) If you were a scientist

working on this study, what hypothesis would you make after observing these cells? (2) Did the brain cells reproduce forming new nerve cells? How could you tell? (3) Remember that the cells were originally taken from the brain of a child with a brain disease. How would this fact affect your conclusion?

impulses sent out from the cerebrum. It constantly receives sensory impulses from certain receptors in muscles, tendons, and joints. If the sensory information and cerebral impulses do not agree, the cerebellum sends messages to the cerebrum and the cerebral impulses change. The cerebellum also controls posture and balance and maintains muscle tone. Extending down from the center portion of the brain is the **medulla oblongata,** which controls many responses of internal organs, such as breathing rate, heartbeat rate, peristalsis, and some gland secretions. The medulla oblongata and the pons make up the brainstem. The pons links the spinal cord, medulla, and the midbrain. It is composed of a bundle of fibers that transmit impulses from one side of the cerebellum to the other. The pons, together with the respiratory center in the medulla, helps to control respiration.

Spinal Cord Extending down from the medulla oblongata is the long spinal cord, which has a fluid-filled central canal. The spinal cord is surrounded and protected by an extension of the three-layered meninges that cover the brain and by a series of bony vertebrae. The outer surface of the spinal cord is composed of white matter (myelin-sheathed axons), and its interior is made up of gray matter (unsheathed cell bodies), an arrangement opposite of that in the brain (Figure 25-20).

In humans, the spinal cord connects the brain with the peripheral nervous system. The spinal nerves, which are part of the peripheral system, serve as the communication link between the spinal cord and the rest of the body. Spinal nerves contain both sensory and motor neurons, which enter and leave the spinal cord through nerve roots. Impulses reach the spinal cord from receptors via sensory neurons and then travel to interneurons in the spinal cord. The impulses may be transferred to the brain for interpretation or may be passed directly to motor neurons in the spinal cord. Impulses travel to effectors by motor neurons of the peripheral nervous system.

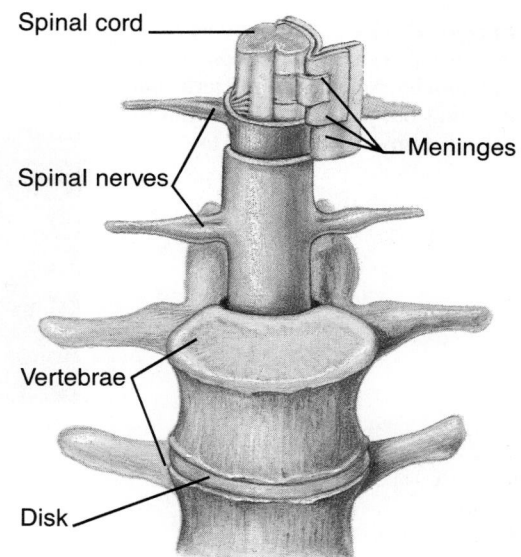

Figure 25-20 The spinal cord is protected by vertebrae and discs made of cartilage. Even with this protection, many thousands of people incur spinal injuries every year. Many happen during sports activities such as diving and skiing.

The Sensory-Somatic System of Humans

The sensory and motor neurons of the spinal and cranial nerves are part of the human's peripheral nervous system. Some sensory neurons carry impulses from receptors to the CNS and are linked to motor neurons that carry impulses to skeletal muscles. They are part of the peripheral nervous system known as the **sensory-somatic system.** Responses that are under conscious control—voluntary responses—are part of this system. The sensory-somatic system is also responsible for controlling certain responses not under conscious control—involuntary or automatic responses to stimuli. Such responses are called **reflexes.** Reflexes involve the CNS but require no conscious control or decision making because they are determined by fixed pathways of the nervous system. We are aware of reflexes controlled by the sensory-somatic system.

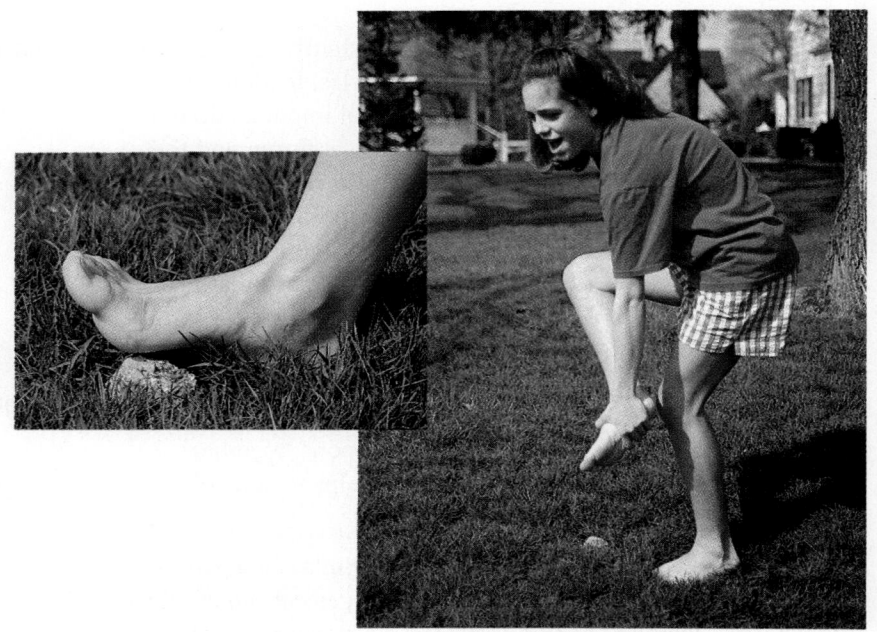

Figure 25-21 The response to stepping on a sharp object takes nearly 100 muscles. Not only do you jerk your foot away but you extend your arms to balance yourself on one foot, too.

The Effect of Stimulants on *Daphnia*

How many people do you know who start each morning with a cup of coffee? What do you do when you're feeling sleepy in mid-afternoon? You might drink a cola beverage with caffeine. Caffeine is a drug. It is a stimulant that is addictive and can produce withdrawal symptoms such as headaches when limited or eliminated from a person's diet. Caffeine is contained in many carbonated soft drinks, coffee, tea, chocolate, and many over-the-counter medications. It can produce such physical symptoms as ulcers, heart irregularities, and irritability. Recently, it has become a suspect in the cause of fetal development problems. In this lab, you will determine the effect of caffeine on the crustacean *Daphnia*.

Problem

What is the effect of a stimulant on *Daphnia*?

Safety Precautions

Do not drink any of the beverages used during this lab. Treat invertebrates used in a humane fashion at all times.

Hypothesis

Make a hypothesis about how caffeine will affect *Daphnia*. Explain your reasons for making this hypothesis.

Experimental Plan

1. As a group, make a list of possible ways you might test your hypothesis using the materials your teacher has made available.

Reflex—Initial Response   In some reflex actions such as blinking, the fixed pathway runs through the brain. For other reflexes, the fixed pathway runs through the spinal cord. For example, consider your response to stepping on a sharp rock while walking barefoot. Before you are aware of what is happening, certain muscles in your leg have contracted, lifting your foot from the ground. What is the nerve pathway of this response?

Receptors in the skin of your foot are activated by the pressure of the rock and the pain it causes. Each receptor stimulates a sensory neuron and causes an impulse to travel to the spinal cord via one of the spinal nerves. Each impulse crosses a synapse to an interneuron.

The impulse travels along the interneuron and across another synapse to a motor neuron, which leaves the spinal cord. As the impulse passes out of the end branches of the motor neuron, it crosses several synapses and activates several muscle cell effectors. Acetylcholine is the neurotransmitter that carries the impulses across these synapses and causes the muscles to contract. Once the reflex action has taken place, the motor neuron stops carrying impulses, acetylcholine is no longer released, the muscles relax, and your foot returns to the ground.

2. Agree on one idea from your group's list that could be investigated in the classroom.
3. Design an experiment that will test one variable at a time. Plan to collect quantitative data.
4. Following the style of a recipe, write a numbered list of directions that anyone could follow.
5. Make a list of materials and the quantities you will need.

Checking the Plan

Discuss the following points with other group members to decide the final procedures for your experiment.

1. What variables will need to be controlled?
2. What is your control?
3. What will you measure or count?
4. How many *Daphnia* will you examine?
5. How many trials will you run?

6. Have you designed and made a table for collecting data?

Make sure your teacher has approved your experimental plan before you proceed further.

Data and Observations

Carry out your experiment; collect your data and complete your data table. Make a graph of your results.

Analysis and Conclusions

1. According to your results, how does caffeine affect *Daphnia*?
2. What is your experimental evidence for your conclusion in question 1?
3. Was your hypothesis supported by your data? Explain.
4. If your hypothesis was not supported, what type of experimental error could be responsible?

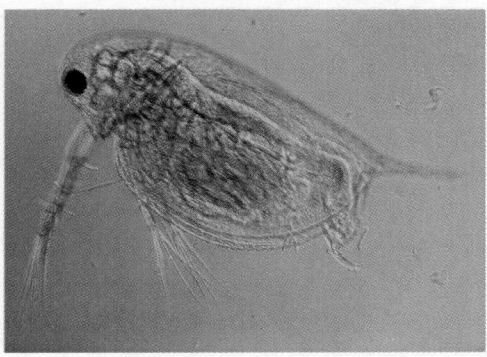

5. Consult with another group that used a different procedure to examine this problem. Were their conclusions the same as yours? Why or why not?

Going Further

Based on this lab experience, design an experiment that would help to answer another question that arose from your work. Assume that you will have unlimited resources and will not be conducting this lab in class.

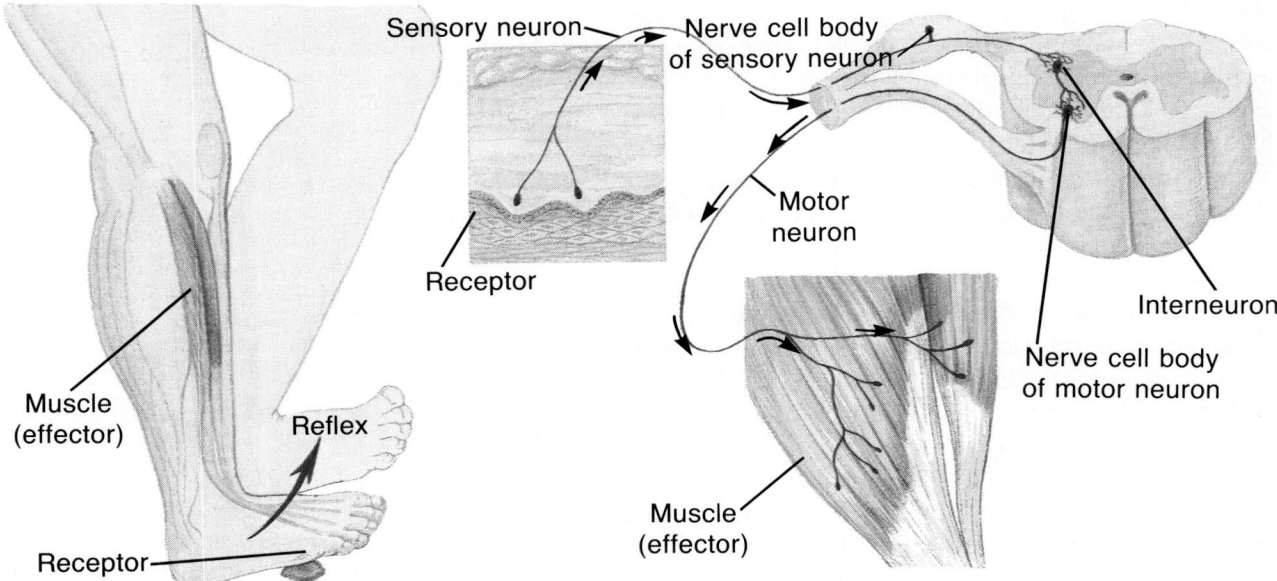

Labels in figure:
Sensory neuron — Nerve cell body of sensory neuron
Receptor
Motor neuron
Muscle (effector)
Reflex
Receptor
Muscle (effector)
Interneuron
Nerve cell body of motor neuron

Figure 25-22 Reflexes operate with such speed that you can move away from a potentially dangerous situation quickly—hopefully without much injury.

The path a nerve impulse takes during a reflex action is called a **reflex arc.** A reflex arc often includes five parts—receptor, sensory neuron, interneuron, motor neuron, and effector.

Reflex—Secondary Response   A reflex action takes place within a fraction of a second. Since your response to a sharp rock underfoot does not involve your brain, how do you become aware of what is going on? Some of the spinal cord interneurons that receive sensory impulses connect with other neuron pathways leading to the brain. The brain becomes aware of what has happened. At this point, secondary responses occur. A loud "ouch" is a common secondary response to stepping on a sharp rock.

The Autonomic System of Humans

Many reflexes involve effectors other than skeletal muscle, including glands and muscles of the heart and digestive tract. You usually remain unaware of these reflexes, which are often responses to internal stimuli. These responses—such as the beating of your heart, or the peristaltic motions of your intestines—are under the control of the **autonomic system.** The autonomic system is made up of two groups of nerves that serve to check and balance each other. These two groups are called the sympathetic and parasympathetic nerves.

Sympathetic Nerves   Imagine you're out riding your bicycle on a pleasant Saturday afternoon. As you pedal up a long, steep hill, your breathing gets faster and deeper, and you can tell by the pounding in your temples how fast your heart is pumping. You are sure it's going to take all your strength to get up this hill without stopping. Suddenly, a car comes speeding out of a parking lot right in front of you. You see it just in time to put on an extra burst of speed and swerve out of its path.

Sympathetic nerves speed up the activity of the heart and lungs during emergencies or strenuous exercise. A high rate of cellular respiration increases the level of carbon dioxide in your blood. Receptors in your aorta and other major arteries detect the higher carbon dioxide level and send impulses to the part of your medulla oblongata that controls heartbeat rate. This region of your brain is also receiving signals from the right atrium of your heart that indicate this chamber is receiving an increased volume of blood. In response to these signals, your brain sends impulses along the spinal cord and spinal nerves that release the neurotransmitter noradrenaline, which stimulates heart muscle contractions and increases your heartbeat rate.

The fear that you feel when you see the speeding car in your path also stimulates the acceleration center of your medulla oblongata. As a result of all this sympathetic nerve activity, you get an unusual burst of energy that enables you to not only pedal up the hill, but also to get out of the way of that car.

Parasympathetic Nerves Once you've escaped a potential collision and reached the top of the hill, your heart no longer needs to work so hard. In fact, a too-rapid heartbeat can damage blood vessels. Your sympathetic nerves are capable only of speeding up your heartbeat rate. Slowing it back down to normal is the job of the parasympathetic nerves.

When your heartbeat rate increases, so does your blood pressure because the blood vessels have constricted. The increase in blood pressure is detected by receptors located in the same blood vessels that have receptors to detect high carbon dioxide levels. The blood pressure receptors send impulses to the inhibition center of the medulla oblongata, which sends impulses that cause the release of acetylcholine, a neurotransmitter that inhibits heart muscle contractions.

The responses of most of your internal organs are regulated by the two parts of the autonomic nervous system. For example, sympathetic nerves cause the pupils of your eyes to become larger in response to low light levels, and parasympathetic nerves reduce the size of the pupils when light levels are high. In other words, the sympathetic system copes with stress or emergencies and the parasympathetic system does the opposite.

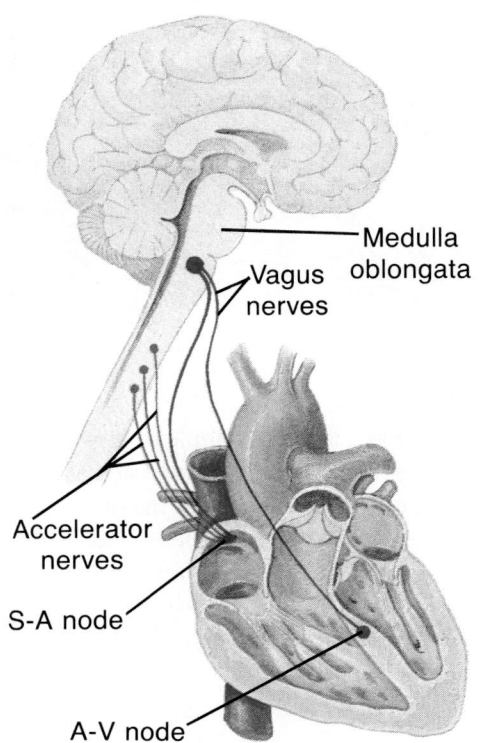

Medulla oblongata

Vagus nerves

Accelerator nerves

S-A node

A-V node

Figure 25-23 **Heartbeat rate is controlled by nerves and the medulla oblongata. The medulla sends messages by way of nerves to the heart to accelerate it or to slow it down.**

The Senses

Animals are constantly bombarded by stimuli from their environment. Detecting and reacting to stimuli are important for an animal's survival. Humans have adaptations to sense a variety of stimuli. Whether stimuli are detected by a simple nerve ending or special cells of the ear, eye, or other complex organs, sensory information causes action potentials in neurons. Each receptor or sense organ sends impulses to particular parts of the brain where they are interpreted. We have the ability to then act on the information that has been provided.

There are five major human senses—vision, hearing, smell, taste, and touch. Touch is a general term that includes several related senses such as pain, pressure, heat, and cold. Each of these "touch" senses is detected by separate skin receptors. Each kind of receptor detects only one sense. The number and location of touch receptors around the body varies. For example, there are more cold receptors than heat receptors, and there are more touch receptors in the fingertips than in the legs.

The senses of taste and smell are not essential to the survival of humans, but for many, many animals, they are the most important senses. For nocturnal animals and very small mammals, the sense of smell provides the most information about what is happening in their environments. Can you think of a reason smell is not more highly evolved in humans?

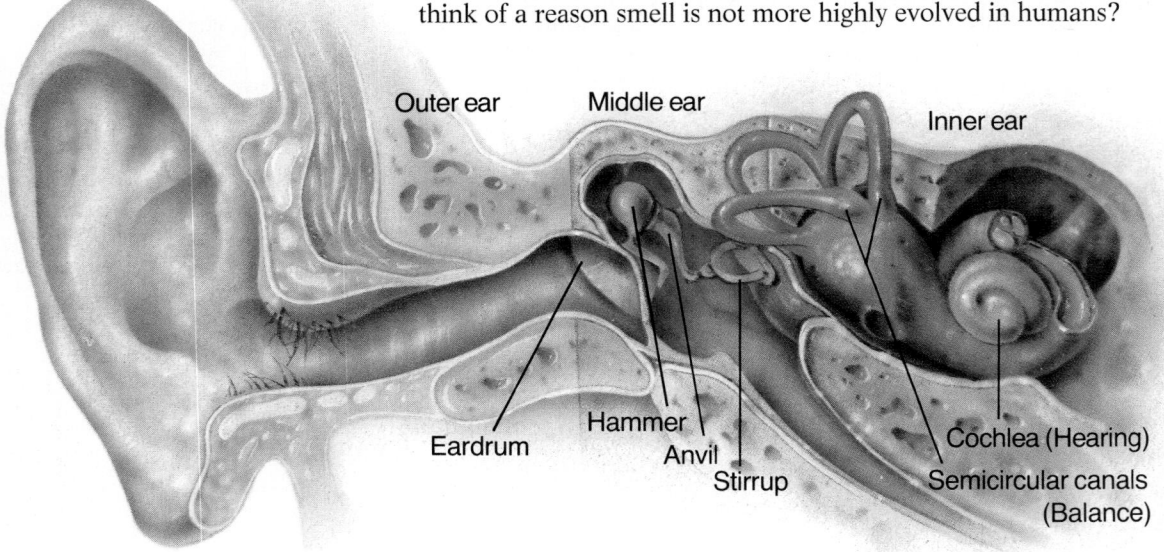

Outer ear Middle ear Inner ear

Eardrum Hammer Anvil Stirrup Cochlea (Hearing) Semicircular canals (Balance)

Figure 25-24 **The human ear is a complex sense organ that converts sound waves into vibrations. These vibrations ultimately trigger action potentials in the neurons of the auditory nerve, which ultimately connect with neurons in the auditory center of the brain.**

The Ear As an example of how a sense organ works, consider the ear. The ear is made up of three areas—outer ear, middle ear, and inner ear. The outer ear consists of the outer flap, the auditory (hearing) canal, and the eardrum. The middle ear begins on the inner side of the eardrum and consists of a chain of three small bones commonly called the hammer, anvil, and stirrup. The stirrup contacts a second membrane called the oval window. Also in the middle ear is the eustachian tube, a canal that connects the middle ear to the nasal passages and throat. The eustachian tube contains air and maintains equal air pressure between the middle and outer ear. This is unfortunately a prime breeding ground for infectious microorganisms that enter the body through the mouth or nose. The inner ear contains two sets of fluid-filled canals. One set, the semicircular canals, controls the sense of motion and is not involved in hearing. The other fluid-filled canal is a coiled structure called the cochlea, which is involved in hearing. Part of the cochlea is in contact with the oval window. The cochlea contains sensitive hair cells that are connected to fibers of the auditory nerve.

Detecting Sound   How does hearing occur? Sound waves enter the outer ear, pass through the auditory canal, and cause the eardrum to begin vibrating. The vibrations are picked up by the middle ear bones, which amplify the vibrations and transfer them to the oval window. The oval window then vibrates, causing the fluid in the cochlea to move. Movement of the fluid within the cochlea causes the sensitive hairs to bend, and to cause an action potential along axons of the auditory nerve.

Transmitting and Interpreting Sound Impulses   There are thousands of sensitive hairs inside the fluid-filled cochlea. When each hair bends, even slightly, in response to a vibration, it creates an action potential in the auditory nerve fiber to which it is attached. These impulses travel to the auditory cortex located on each side of the cerebrum.

How does the brain interpret these impulses? How can you tell the difference between a bass guitar and a bird's song? The hairs in different regions of the cochlea respond to different sound frequencies. Each hair is connected via a neuron pathway to a specific location in the auditory cortex. The frequency of a sound vibration determines which of the hair cells will be stimulated. The loudness of the sound determines how many times the hair cell will be stimulated. What you hear depends on the pattern of hair cells stimulated into movement by the sound. You respond, perhaps by turning down the volume of a too-loud stereo or jumping up to dance to a favorite tune.

Figure 25-25   The eardrum or tympanic membrane (a) begins sending vibrations through the rest of the ear structures. The hammer, anvil, and stirrup are located in the middle ear (b). The semicircular canals are located in the inner ear (c).

a

b

c

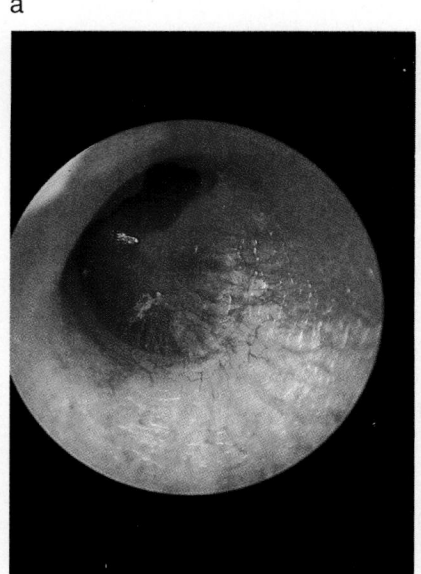
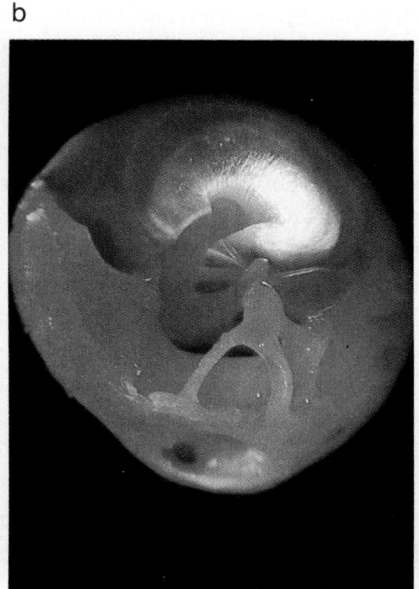

Figure 25-26 The eye is a sensory structure composed of three layers of tissue that enclose a fluid-filled space. The innermost layer, the retina, contains rods and cones, the photoreceptor cells. These cells transmit impulses to the brain, where they are interpreted.

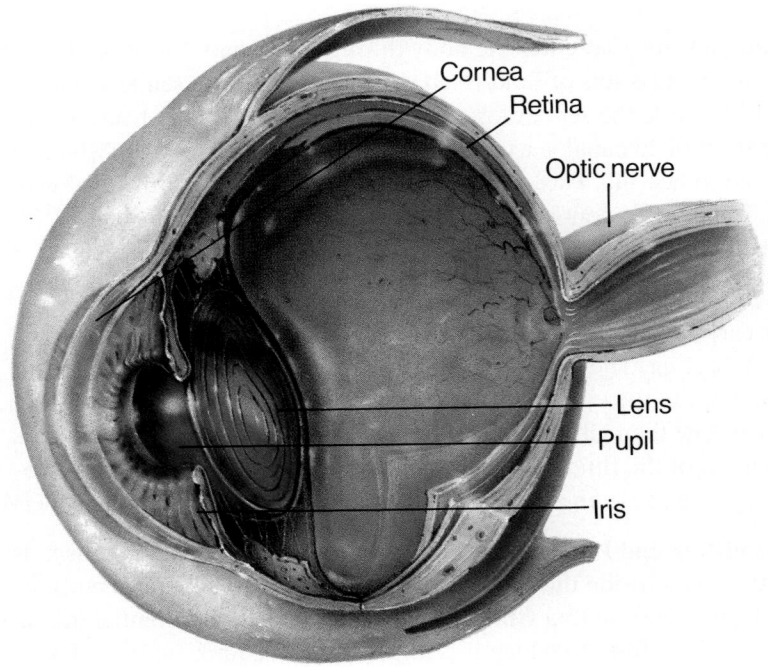

Cornea
Retina
Optic nerve
Lens
Pupil
Iris

The Eye   Ability to detect light, an important source of information about the environment, has evolved among a wide variety of organisms. *Euglena,* a unicellular protist, and planarians, flatworms, both have structures that enable them to distinguish between light and dark. Light detection pigments are also present in plants to aid in phototropism and flowering. Only in certain animals, however, have structures evolved that enable the organism to form an image. Image-forming structures, eyes, are characteristic of arthropods, some mollusks, and vertebrates.

Humans (and other vertebrates) have what is known as a camera eye, so named because it works much like a camera. Rays of light reflected from an object enter the front of the eye, pass through a lens, and are focused on the back of the eye much as light passing through a camera lens is focused on film at the rear of the camera.

Figure 25-26 shows the structure of the human eye. Notice that only the front of the eye is exposed. The rest of the eye is protected by the skull. Surrounding most of the human eye is a tough layer of connective tissue, the sclera, which you know as the white of the eye. At the front of the eye, the transparent sclera bulges outward and is known as the **cornea.** The cornea admits light into the eye. Beneath the sclera is a dark layer, the choroid, which contains many blood vessels that nourish the eye. The dark color of the choroid reduces reflection within the eye. Toward the front of the eye, the choroid extends upward, forming a ring of cells known as the iris. The iris contains pigments, which give the eye its color. The iris contains sets of muscles that control the size of the opening in its center, the pupil. When one set of iris muscles is contracted, the pupil is dilated, admitting more light. When the other set of muscles is contracted, the size of the pupil is reduced and less light can enter.

Light passing through the cornea travels through a watery fluid, the aqueous humor, and then through the pupil. From there, the light rays pass through the transparent **lens** of the eye. The lens is an elastic ball that bends rays of light. The shape of the lens is changed by the action of very small muscles. Such changes in shape focus the rays of light.

Light focused by the lens strikes the innermost layer of the eye. That layer, the "film" of the camera eye, is called the **retina.** The retina contains two types of photoreceptor cells that are stimulated by the rays of light striking them. **Rod cells** are most sensitive and permit vision even in dim light. However, they do not function in color vision. Distinguishing colors is the function of the **cone cells,** most of which are located near the center of the retina. Cone cells also contribute to sharper images.

Light rays falling on rods and cones cause changes in pigment molecules present in those cells. Those changes lead to action potentials in the rods and cones. Impulses cross synapses to other retinal nerve cells, and eventually are carried along axons that form the optic nerve, which leads to vision centers in the cerebrum of the brain. Because of the way that light rays behave as they pass through a structure such as the lens of the eye, the image formed on the retina is upside down. Our brain, however, interprets images as being right side up, and that is the way we see them.

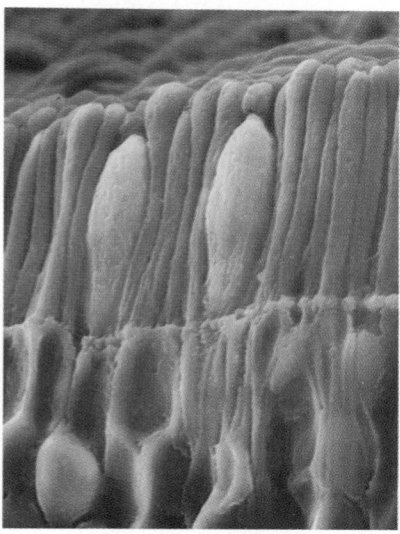

Figure 25-27 **This electron micrograph shows the rods and cones of the human eye. Magnification: 10 000×**

LOOKING AHEAD

In THIS CHAPTER, you have studied neuron structure and function and the nervous pathways involved in reflex responses, some of which depend upon skeletal muscle. In the next chapter, you will learn how the skeleton and muscles interact to provide support and locomotion, and will study the mechanism by which skeletal muscle causes movement. You will also examine various ways in which energy is made available for the activities of skeletal muscle.

Section Review

Understanding Concepts

1. Compare and contrast the nervous systems of invertebrates and vertebrates. Concentrate on major differences.
2. Suppose a person suffered brain damage as a result of an accident. What part of the brain was injured if the symptoms of the injury were (a) loss of speech, (b) emotional problems; (c) loss of control of internal organs; or (d) loss of balance and coordination?
3. What is the difference between a reflex arc and a reflex act? Name the components of a typical reflex arc.
4. **Skill Review—Sequencing:** Sequence the nervous pathway structures involved in the response made to touching a hot stove. For more help, see Organizing Information in the **Skill Handbook.**
5. **Thinking Critically:** Why do the ears of a person going up or coming down in an airplane often feel stuffed up?

BIOLOGY, TECHNOLOGY, AND SOCIETY

Biotechnology

Fighting Noise Pollution

When you think of pollution, usually water and air pollution quickly come to mind. Generally, people don't think of sound as a polluting agent. Yet sound can cause people to become nervous, irritable, or tired. It can make people nauseous and damage their hearing. Loud

sounds such as fireworks and the roar of jet engines would seem the obvious cause of these health problems, yet scientists and doctors know that the sound doesn't have to be loud to cause problems. It only has to be constant or prolonged. Scientists who work to control sound or noise pollution are in the field called environmental acoustics.

Acoustical engineers know that noise pollution can be controlled in several ways. For example, people who work around loud sounds, such as the operators of printing presses and rock musicians, should wear earplugs or muffs to deaden the constant loud noise. Architects and acoustical engineers can design buildings with rooms that absorb the sounds of appliances and office machinery. Special acoustic tiles that trap sound are placed on walls and ceilings. Carpets cover the floor. Draperies are used as window treatments. Even fabric is used on office walls to absorb sounds.

Recently, technology has provided an additional means to control noise pollution. Scientists know that if an identical tone is produced by two different sources, the sound waves may cancel out each other.

CAREER CONNECTION

Audiologists work with both children and adults who have hearing disorders. They identify and evaluate the severity of the hearing disorder as well as provide a therapy program to treat the problem. Most states require audiologists to have a master's degree in audiology and to meet special licensing requirements.

No sound is heard. This phenomenon is called *interference.* They have used this knowledge to build active noise control (ANC) devices. Sounds produced by active noise control devices match the frequency and amplitude of the polluting sound; however, the two sounds vibrate in opposite directions. This results in interference, and as a result, neither sound is heard. Think of the improvement that ANC devices can make on annoying machines such as air conditioners and heaters. Such devices can even make a whole office a quieter place to work.

1. **Find Out:** Two sound waves that cancel each other out are an example of destructive interference. Find out how destructive interference is related to constructive interference and the production of musical beats.
2. **Issue:** Suppose you work after school in an office. The copying machine is located next to your desk. The machine doesn't make a loud sound, but after sitting next to it for an hour, you find yourself becoming annoyed by its sound. Is the sound produced by the machine an example of noise pollution? Explain.

Alzheimer's: The Disease that Takes Everything

Jackson walked up the front steps of his grandmother's house. It had been months since he was last there and he was eager to see her. His excitement turned to dismay as he looked around the familiar house. His grandmother had always been a meticulous housekeeper. She took pride in keeping herself well-groomed as well. But as Jackson looked around the living room, he saw that the once-tidy room was now cluttered and unkept. In fact, Grandma Sweetwell herself looked unkept. Even more disturbing, his grandmother did not seem to recognize him. What had happened? Something was quite wrong. Jackson's grandmother was the victim of a progressive, degenerative, terminal disease that attacks the brains of approximately 5 percent of people over the age of 65. This disease, called Alzheimer's disease, impairs the memory, thinking ability, and behavior of its victims. In the final stages, its victims are bedridden and completely helpless. Today, Alzheimer's is the fourth leading cause of death in the elderly population following heart disease, cancer, and strokes. As Jackson was to find out, coping with the mental and physical degeneration of the victim is only part of the problems facing Alzheimer's patients and their families. The other part deals with the financial burdens incurred.

At first, Jackson's parents were able to care for Grandma Sweetwell. They had her move into their house. As the disease pro-

gressed, they took turns dressing her and caring for her needs. They watched to be sure she didn't wander out of the house, hurt herself, or damage the house. But as the disease progressed, she needed more care. Doctors suggested a nursing home; however, the cost of one month in a home was more than $2000. Jackson's parents couldn't afford that. Grandma Sweetwell's health insurance paid for hospital and doctor bills, but it didn't pay for long-term care in a nursing home. The federal program, Medicare, paid for "skilled care" in a nursing home; but Jackson's grandmother didn't need that. She needed custodial care. The program Medicaid would cover custodial care but she had a savings account, a small monthly pension, and owned her own house. She had too many assets to qualify for Medicaid.

Jackson's parents were forced to go to the courts to get authority to

sell Grandma Sweetwell's house and its contents. They used the money from the sale and her savings to pay for the nursing home. Once all of these assets were gone, she qualified for government assistance. Jackson's grandmother died nine months after getting government assistance. Alzheimer's had stripped away her mental and physical abilities, her dignity, and just about everything else that she owned.

Understanding the Issue
1. How do you think Jackson's parents felt about having to sell all of Grandma Sweetwell's possessions and using all of her savings before she could qualify for government assistance?
2. The cost of health care is a crisis in the United States. The number of senior citizens is increasing. How will this increased population impact the cost of health care? What suggestions might you make to provide the necessary care for this segment of the population, yet still keep down the costs?

Readings
- Sandstrom, B. "Difficult decisions: A family deals with Alzheimer's disease." *Aging*, no. 363-364: 22-3, 1992.
- Stoneman, D. "The other side of Alzheimer's." *Health*, May/June 1992.

Summary

With the exception of sponges, animals have nervous systems that control responses to many stimuli. Nervous control requires detection of stimuli, transmission of impulses, analysis of impulses, and an appropriate response to the stimuli.

Transmission of impulses in animals depends upon reversals of polarity along neurons. Those reversals result from changes in membrane permeability to different ions, which cause action potentials, or impulses. Impulses cross synapses by means of chemical transmitters, which may either excite or inhibit the target cell. Analysis of impulses is made on the basis of the number of excitatory and inhibitory effects of transmitters at any given time.

Nervous systems in animals became more complex as more complex animals evolved. Most animals have a central nervous system that analyzes information from sensory neurons, and in response sends impulses along motor neurons to effectors. In humans, many responses are voluntary, but others are involuntary reflexes. The sense perceived depends on the region of the brain receiving the sensory message. The strength of the sensation is reflected in the frequency of nerve impulse in the sensory nerves.

Language of Biology

Write a sentence that shows your understanding of each of the following terms.

action potential	nerve
autonomic system	neuron
axon	neurotransmitter
central nervous system	peripheral nervous
cerebellum	system
cerebrum	reflex
cone cells	reflex arc
cornea	resting potential
dendrite	retina
interneuron	rod cells
lens	sensory neuron
medulla oblongata	sensory-somatic system
motor neuron	synapse

Understanding Concepts

1. A neuron stimulated in a laboratory will carry an impulse in either direction away from the point of stimulation. Why don't impulses travel in two directions along your neurons?
2. Can you suggest a reason why noradrenaline has no effect on skeletal muscle?
3. When does the sodium-potassium pump go into action? Why must it do so?
4. Cite two examples from this chapter that illustrate the close relationship between the endocrine and nervous systems.
5. What would be the effect on skeletal muscle if the enzyme that breaks down acetylcholine were not present?
6. Explain several ways in which drugs affect the nervous system.
7. Muscles controlling peristalsis are inhibited during an emergency or stress situation. Afterwards, peristalsis again occurs normally. How is regulation of peristalsis controlled? How is such regulation adaptive?
8. How are the nervous systems of a hydra and a planarian related to the body plans of those animals?
9. Relate the concepts of action potentials and crossing of synapses to a reflex arc of your choosing.
10. During an outer ear infection, the auditory canal may become inflamed and swollen. Explain why hearing is impaired during such an infection.

Applying Concepts

11. Nicotine mimics the effect of acetylcholine. Is nicotine a stimulant or depressant? Explain.
12. Other than arthropods, can you think of invertebrates in which the brain is complex and exerts much of the control over behavior? Explain.
13. How would a drug that occupies sodium-potassium pumps affect transmission of impulses? Explain.
14. In the cerebrum, the areas that receive impulses from sensory neurons vary in size. For example, the area receiving impulses that originate in the fingers is larger than the area receiving impulses that originate in the upper arm. Explain the reason for the differences.
15. Brains that exert greater control over responses have relatively more interneurons than brains that do not exert so much control. Can you explain that fact?
16. Muscle cells, like neurons, have resting potentials and action potentials. They contract in response to an action potential. Explain how transmitters released from sympathetic and parasympathetic motor neurons can have opposite effects on the same muscle tissue.
17. Based on your own experience, what kinds of responses, in general, are reflexes? How is it adaptive that such responses occur automatically rather than as a result of conscious effort?
18. **Biotechnology:** Explain how sounds can damage your ears and cause hearing loss.
19. **Issues:** What part of the brain does Alzheimer's affect? Explain.

Lab Interpretation

20. **Thinking Lab:** Oncogenes are genes that produce cancerous cells—cells that multiply very rapidly. Adult brain neurons ordinarily do not reproduce once they have been destroyed by accident or disease. Can you think of a way oncogenes might be used to solve the problem of lack of brain cell division?

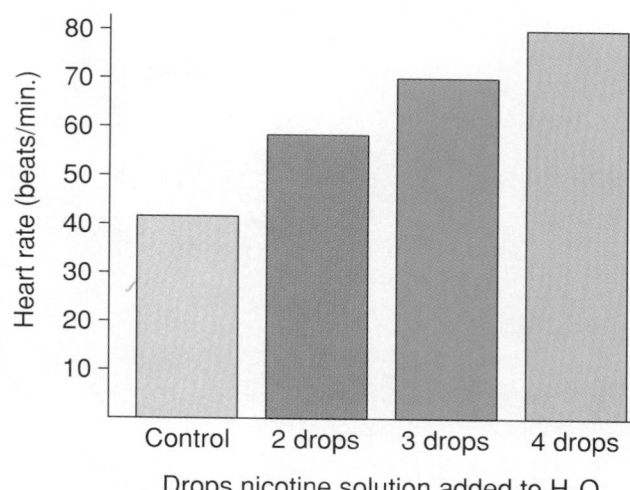

Effects of nicotine on *Daphnia*

21. **Investigate:** A student conducts an experiment to determine the effect of nicotine on *Daphnia*. Examine the results in the graph and tell how nicotine affects *Daphnia*.
22. **Minilab 2:** Reaction time may be affected by a variety of circumstances. Make a hypothesis about how reaction time would be affected by the following: alcohol, lack of sleep, after a meal, after eating a chocolate bar.

Connecting Ideas

23. Review the means by which breathing rate is controlled. Is that control by the sensory-somatic or autonomic system? Explain.
24. How do hearing and sight illustrate the principle of energy transformation?

26

MOVEMENT

LIKE HUNDREDS of tiny oars, the many cilia of *Paramecium* propel it through the water. Powered by contraction of muscles in its hind legs, a frog leaps into the air. Muscles also control the burrowing of an earthworm, the graceful flight of a butterfly, the sidewinding movement of a snake, the creeping of a snail, and the jumping and tumbling of a cheerleader. Cilia and muscles are examples of effectors. After stimulation by a motor nerve, effectors such as muscles respond.

Muscle contraction may contribute to locomotion, the movement of an entire organism, as in the examples above. Effectors also may contribute to the movement of individual parts of an organism, like the tentacle of a hydra, the mouthparts of a grasshopper, or the fingers of a violin player. Movement, whether of the organism or one or more of its parts, is a characteristic of animals and animal-like organisms. It is related to their heterotrophic way of life, in which they must secure food and avoid becoming food for some other organisms.

Predators like the wolf are adapted to capturing prey on the run. How are the wolf and the snake alike? What features enable both organisms to capture their food?

Chapter Preview

Objectives

Compare the structures and functions of exoskeletons and endoskeletons.

Discuss the development of bone and its structure.

Demonstrate various ways in which different bones form joints.

26.1 Skeletal Systems

Most animals have the ability to move at some stage of their lives. Even if an animal does not move, it requires some kind of support. Without support, an animal would collapse from its own body weight. Many aquatic invertebrates such as jellyfish are supported by the water in which they live. Movements of these animals, though, are most often slow and sluggish.

The Exoskeleton

Think about the most active animals you know. You probably thought of animals belonging to two groups—arthropods and vertebrates. These animals, whether living in water or on land, have muscles that function in both movement and support. They also have an additional means of support—a skeleton. The muscles and skeleton work together, providing an efficient means of locomotion and movement of individual body parts. The skeletons of arthropods and vertebrates also protect the organism and serve as holding areas for needed minerals. Invertebrates such as sponges, sea stars (starfish), and clams do not have skeletons, although they do have mineral deposits in their tissues. These deposits contribute to support and protection of the organism, but not to movement.

Figure 26-1 **Arthropods, left, and vertebrates, right, have skeletons and muscles that work together to create movement. What types of organisms have no skeletons?**

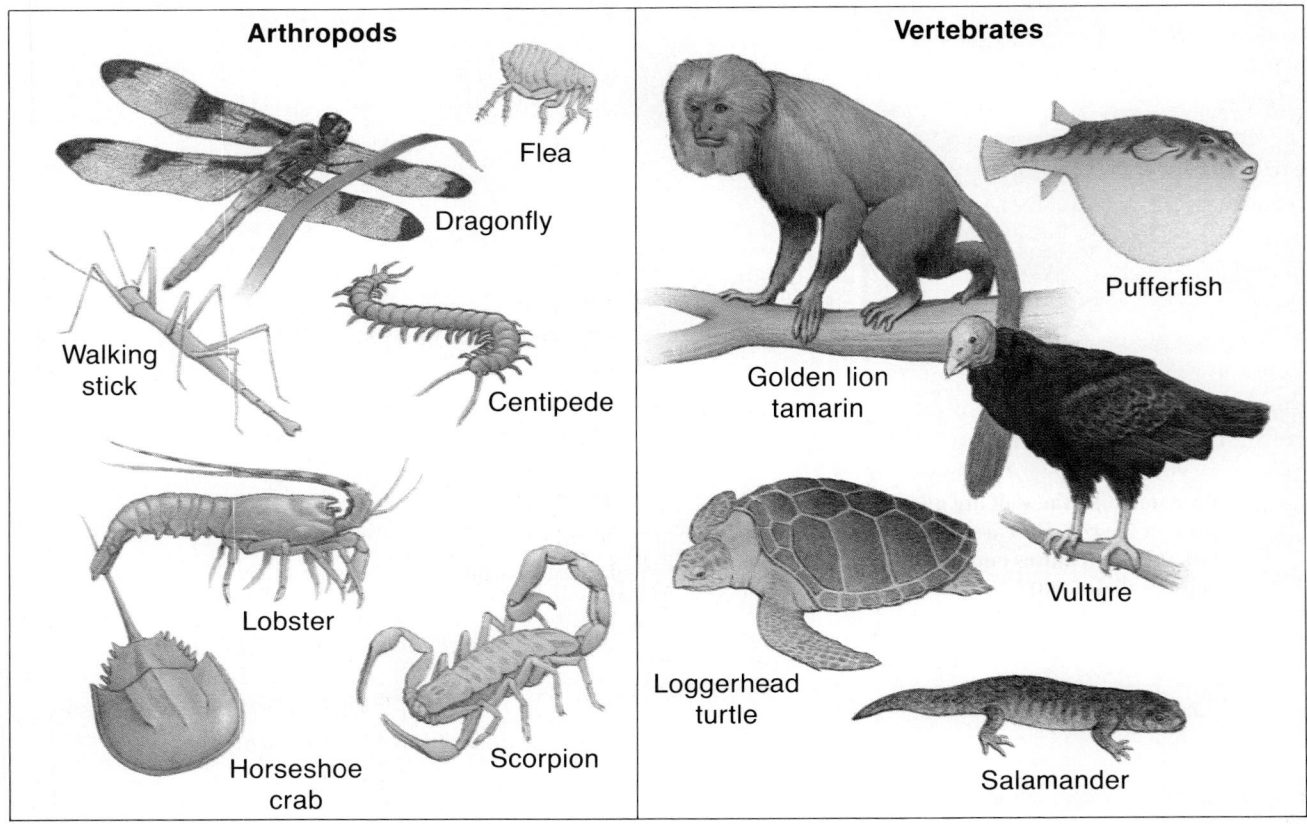

Arthropods

Flea

Dragonfly

Walking stick

Centipede

Lobster

Horseshoe crab

Scorpion

Vertebrates

Pufferfish

Golden lion tamarin

Vulture

Loggerhead turtle

Salamander

Composition of Exoskeletons   Recall from Chapter 17 that the skeleton of arthropods is external and called an exoskeleton. Exoskeletons are composed of proteins and a specialized carbohydrate material called chitin. The outer surface of many arthropods is waxy. The waxy layer protects against water loss in terrestrial arthropods. Parts of the exoskeleton may also extend inwards. In arthropods, such as insects, the muscles are attached to these inward sections of the exoskeleton. The exoskeleton of insects also forms the lining of the tubes through which air enters the body, the tracheae, thus supporting them and keeping them open.

An exoskeleton is a tough, rigid structure that is produced by the epidermal cells. The exoskeleton is divided into segments connected by soft, flexible **joints.** Joints are located between body segments, at the base of appendages, and between segments of appendages. Without joints, movement of body parts and of the entire animal would be impossible. Though protected, an arthropod would be a prisoner in its own armor.

Figure 26-2   In this photograph of a praying mantis, you can see why insects are arthropods. The word *arthropod* means *jointed leg.*

Disadvantages of Exoskeletons   An exoskeleton has certain disadvantages. It limits the size and movement of an arthropod. Because the heavy exoskeleton needed to support a large arthropod would seriously limit movement on land, most adult terrestrial arthropods are small. However, some arthropods that live in water are able to grow much larger than those on land. A giant crab that lives in the ocean near Japan can be as large as 3.6 m across!

Growth is also limited in arthropods. Once the exoskeleton is made and it hardens, its size cannot be changed. Growth only takes place because molting, the shedding of the exoskeleton, occurs. When an arthropod sheds its exoskeleton, the new exoskeleton underneath is stretched by increased blood pressure, and by the intake of air or water by the organism. This increases the size of the animal before the new exoskeleton hardens. Molting may occur many times before an arthropod reaches its final adult size. Some spiders molt 10 times before they reach sexual maturity and molting stops. After each molt, the new skeleton is soft, making an arthropod easy prey for predators. For example, soft-shell crabs are merely crabs that have been caught after a molt but before the exoskeleton has hardened. The animal has neither its armorlike protection nor the ability to move rapidly. Land arthropods lose water when molting. Would you expect to find terrestrial arthropods hiding under rocks just after they have molted? Why?

Figure 26-3   Arthropods must shed their exoskeletons in order to increase in size.

Advantages of Exoskeletons Despite the limitations on size and growth, the exoskeleton has certain advantages. An exoskeleton protects animals from predators and from physical harm. In terrestrial arthropods, the exoskeleton guards against water loss. The exoskeleton also allows movement and supports the animal. Many scientists credit the evolution of the exoskeleton as the reason for the amazing versatility of the arthropods. There are more species of arthropods known than are found in all other animal phyla combined. This great diversity of arthropods is due in part to the exoskeleton. In fact, the presence of an exoskeleton may have been critical to the success of arthropods on land.

The Endoskeleton

In vertebrates, the skeleton lies within the soft tissues of the body rather than outside the body. Recall that such an internal skeleton is called an endoskeleton. Although an endoskeleton and an exoskeleton function similarly, they differ in development and structure and are not homologous. Vertebrates and arthropods (which are invertebrates) evolved from different ancestors.

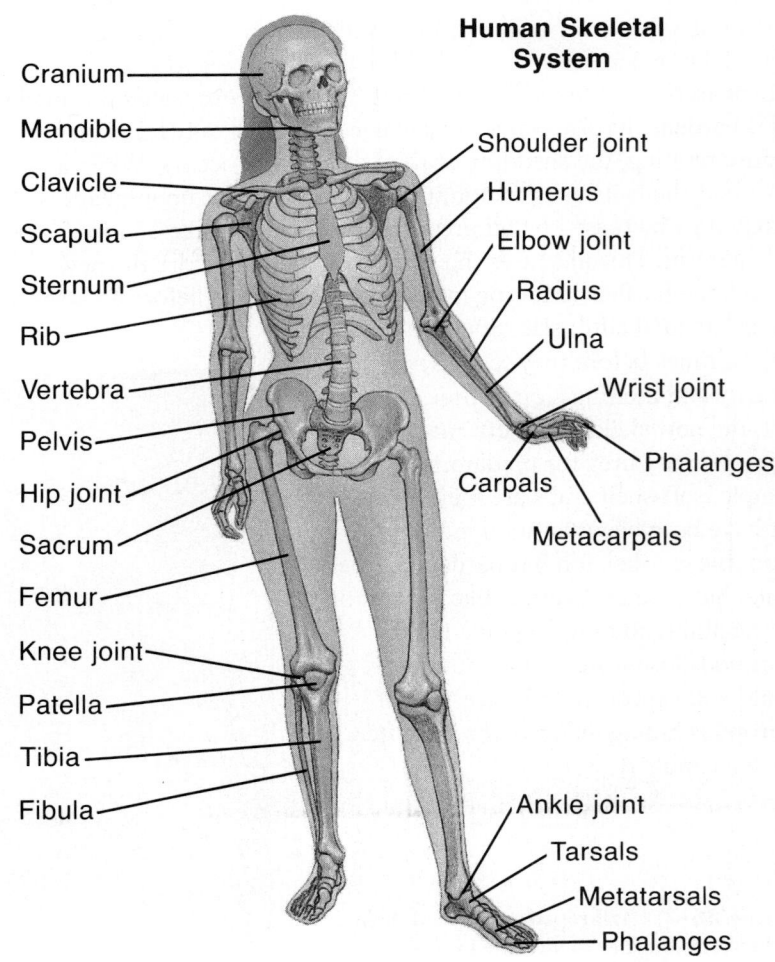

Human Skeletal System

Cranium — Mandible — Clavicle — Scapula — Sternum — Rib — Vertebra — Pelvis — Hip joint — Sacrum — Femur — Knee joint — Patella — Tibia — Fibula

Shoulder joint — Humerus — Elbow joint — Radius — Ulna — Wrist joint — Phalanges — Carpals — Metacarpals — Ankle joint — Tarsals — Metatarsals — Phalanges

Figure 26-4 **The human skeletal system includes 206 bones. Why do you think an endoskeleton has so many parts?**

Functions of Endoskeletons Unlike an exoskeleton, which covers the entire animal in one continuous unit, an endoskeleton is subdivided into many distinct parts. These individual parts, called bones, are specialized for certain functions and types of movements.

Some bones function in protection. Skull bones, ribs, and the breastbone protect the brain, lungs, and heart. The column of vertebrae called a backbone forms a hollow tube to protect the spinal cord.

In humans, the vertebrae also function in another way. The shape of the human vertebral column is unique. Humans have upright posture and locomotion. The "s" curve of the vertebral column aids in distributing and supporting body weight, thus making upright locomotion possible.

Bones also function in support. The long, dense thighbones support the weight of the body and aid the speed and strength of movement. Some bones, such as the delicate finger bones, are adapted for precise, small movements. Many bones function in several ways. For example, the ribs not only provide protection, but also aid breathing.

Advantages of Endoskeletons Internal skeletons protect the internal organs of the body but give little external protection, except the skull, which protects the brain. Some animals with endoskeletons also have skeletal adaptations that give more external protection. For example, turtles are covered by a shell made of bony plates. The plates are made of bone covered by scales. The plates are fused with the ribs of the endoskeleton but do not originate from the ribs in their development.

Limited size and growth are the main disadvantages of an exoskeleton, but size and growth are not as limited in animals with endoskeletons. Because bones are alive and can grow, there is no loss of protection or support as the animal grows. Animals with endoskeletons do have an upper size limit, but as bones are stronger than exoskeletons, they can support greater mass. Muscles of animals with endoskeletons are larger, too, thus providing additional support.

Figure 26-5 Like the exoskeleton of a crab, a turtle's shell protects it from predators. Is a turtle's shell homologous to the shell of a crab?

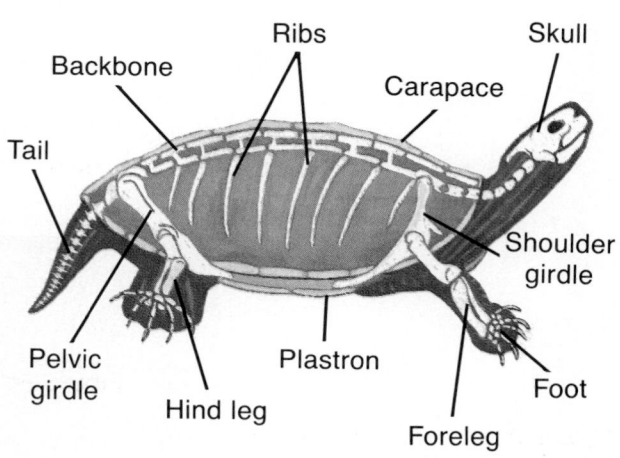

Turtle Skeleton

Development of Bone

Because bone is made up of minerals and is hard, many people think that it is not living material. But a bone in a living animal is a living tissue.

Ossification   In a vertebrate fetus, most of the skeleton is made up of **cartilage,** a tough, flexible tissue that has no minerals. As the fetus grows, living cells called osteoblasts slowly replace cartilage cells and ossification begins. **Ossification** is the replacement of cartilage with bone by the activity of osteoblasts and addition of minerals such as calcium compounds. By the time an animal is born, many of the bones have been at least partly ossified. Some bones are not formed by ossification of cartilage. Bones of the face and skull, for example, are formed as osteoblasts fill in membranes. In humans, ossification is not completed until about age 25. During this period, the stresses of physical activity result in the strengthening of bone tissue. This process continues until about age 40, when the activity of osteoblasts slows and bones become more brittle.

Materials Used in Ossification   Calcium compounds must be present for ossification to take place. Osteoblasts do not make these minerals, but must take them from the blood and deposit them in the bone. Because this process begins in the fetus, the minerals are first obtained from the mother's blood through the placenta. Thus, it is important that the mother eat a well-balanced diet, taking in the proper amounts of minerals and other nutrients.

Because bones grow after the birth of an animal, ossification continues and calcium compounds are still needed. Foods rich in calcium include milk, cheese, meats, vegetables, and whole grain cereals. Vitamin D is needed for proper bone development because it is necessary for calcium absorption from blood. Vitamin D is present not only in milk, but also in egg yolk, fish oils, and liver. It is also made in the skin in the presence of sunlight. Rickets, a disease that results in severe bone deformities, is caused by inadequate amounts of vitamin D.

Not all cartilage is replaced by bone. Some cartilage remains at the ends of bones to provide smooth joint surfaces. In some vertebrates, such as sharks, the whole skeleton remains cartilage throughout the life of the animal. However, in most vertebrates, only a few parts remain as cartilage. Where in the human body is cartilage found? How is that cartilage important?

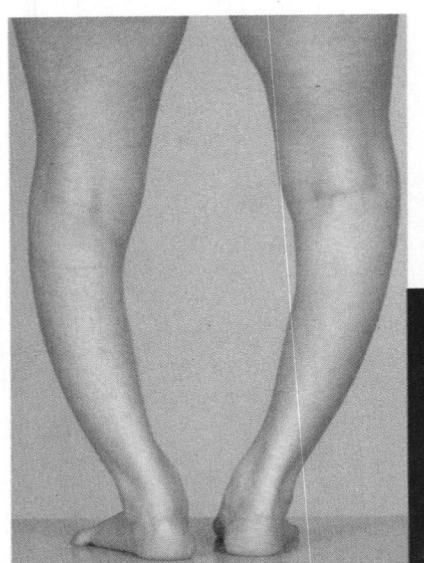

Figure 26-6   **A child with rickets, left, is unable to form bone properly due to a lack of vitamin D. Sharks, right, are fish that do not undergo complete ossification.**

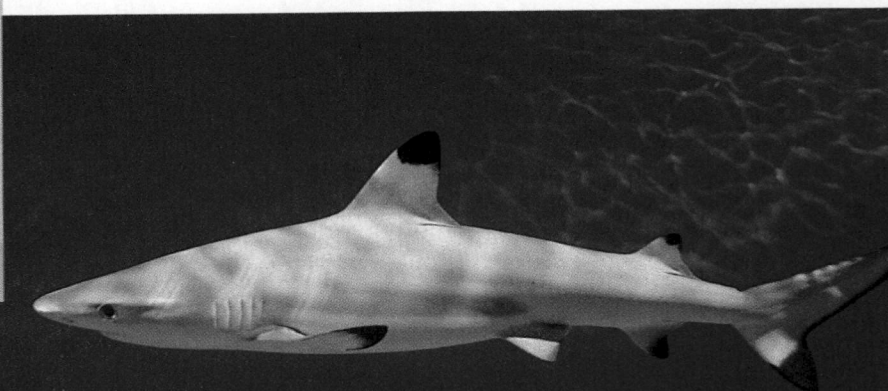

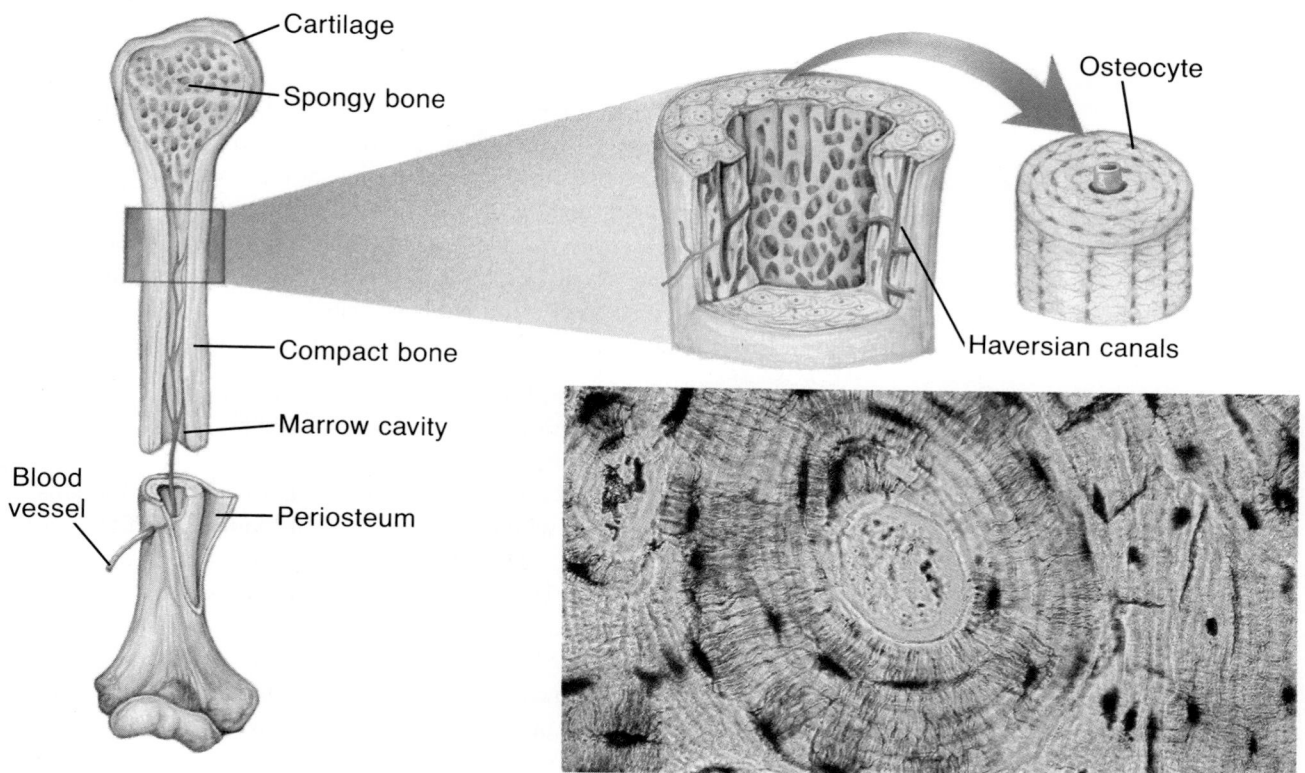

Figure labels:
- Cartilage
- Spongy bone
- Osteocyte
- Compact bone
- Haversian canals
- Marrow cavity
- Blood vessel
- Periosteum

Structure of Bone

Have you ever broken a bone? Did your bone bleed? Bleeding is not surprising when you have broken a bone because bones are living tissue.

Types of Bone Tissue   All bones are enclosed by an outer membrane called the periosteum. The periosteum supplies blood vessels and nerves to the bone, and contains cells needed by the bone for growth and repair. Most bones are made up of two types of bone tissue: compact bone and spongy bone. The shafts of long bones, such as the femur, are composed of hard, compact bone. The ends of these bones are composed of spongy bone underneath a layer of compact bone. Compact bone is made of alternating layers of a protein called collagen and minerals, and contains many tiny channels called **Haversian canals.** Haversian canals carry blood vessels and nerves that supply osteocytes in the bone tissue. **Osteocytes** are mature osteoblasts found in most formed bone tissue. Spongy bone is composed of a system of cavities and intersecting plates. Spongy bone contains blood vessels and marrow, a tissue that produces blood cells and stores fat.

Bone Marrow   Marrow is found in the central hollows of the sternum, ribs, pelvis, vertebrae, skull, and the long bones of the arms and legs. Red marrow produces red blood cells, some white blood cells, and platelets. Yellow marrow stores excess fat. Marrow also functions in formation of bone cells. Bones without marrow—such as the small bones of the wrist and ankle—are spongy bone.

Figure 26-7   There are three types of bone cells—osteoblasts, osteoclasts, and osteocytes. Osteoblasts form new bone, whereas osteoclasts break down bone. Osteocytes, as seen in the diagram, are found in most formed bone tissue. What kind of bone cells would you expect to see at the site of a broken bone? Magnification: 400×

Joints in Vertebrates

Every time you open a door, you are using a type of joint. The door is joined to the door frame with a joint called a hinge. In vertebrates, joints are the points where bones connect with one another. Most joints are movable and are held together by muscles and by connective tissue called **ligaments.** Ligaments are made of collagen and some elastic fibers. The ends of bones are covered by a layer of cartilage. Between cartilage and the ends of the bones are fluid-filled bags called bursas. Bursas cushion the ends of the bones and prevent them from rubbing against each other.

Types of Joints   There are four main types of joints in vertebrates: ball-and-socket joints, hinge joints, pivot joints, and gliding joints. You can see the types of joints in Figure 26-8.

The most flexible of all joints is the ball-and-socket joint. In humans, this type of joint is found where the upper end of the femur (thighbone) joins the pelvis (hipbone). The end of the femur is rounded into a knob that fits into a depression in the pelvis. The femur can rotate, as well as move from front to back or side to side. Are there other ball-and-socket joints in the human body? If so, where?

Hinge joints exist in many places within the human body, such as the fingers. Like a hinge on a door, hinge joints can move in only one direction. A door can only be opened or closed; your fingers can only be bent toward your palm or straightened. They cannot be bent in another direction. Hinge joints are also found in the elbows, knees, and toes.

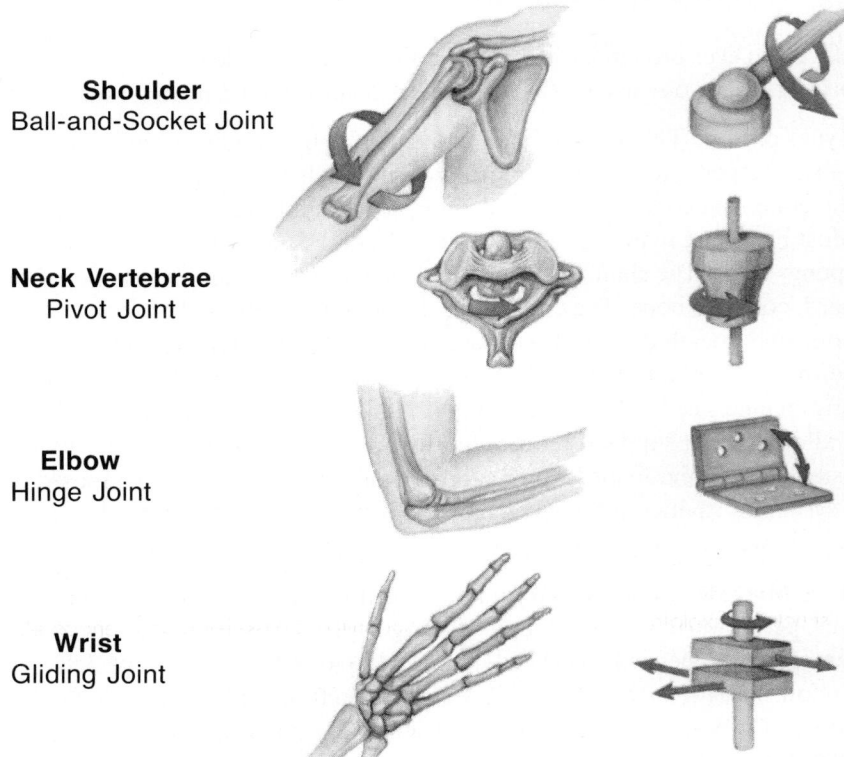

Shoulder
Ball-and-Socket Joint

Neck Vertebrae
Pivot Joint

Elbow
Hinge Joint

Wrist
Gliding Joint

Figure 26-8   **The four main types of joints in vertebrates are the ball-and-socket, pivot, hinge, and gliding joints. What kind of joint is found in your ankle?**

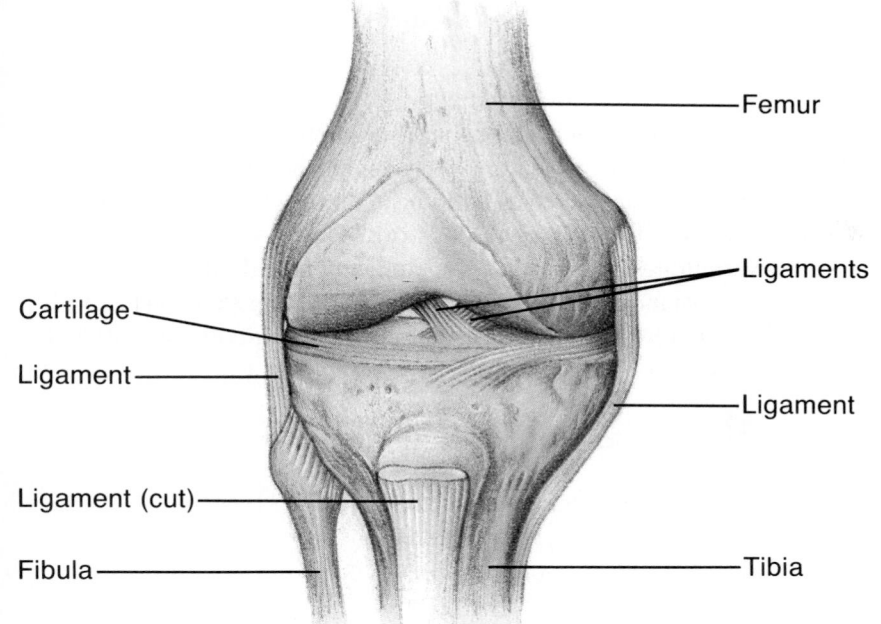

Femur

Ligaments

Cartilage

Ligament

Ligament

Ligament (cut)

Fibula

Tibia

Figure 26-9 This diagram of the right knee with the knee cap removed shows how the femur is attached to the lower leg bones by ligaments. Most knee injuries are the result of torn ligaments. Can you see why ligaments would be torn if a hinge joint like the knee was hit from the side?

Rotation can occur in a pivot joint. One example of a pivot joint is where the bones of the lower arm connect near the elbow. This joint allows the radius to rotate over the ulna when the hand is turned over. The cranium, or skull, is connected to the spine by a pivot joint. With this joint, the head can turn, as well as move up and down.

Vertebrae are linked by gliding joints. In gliding joints, the bones move easily over one another in a back-and-forth manner. Gliding joints aid the flexibility of the backbone and are found in wrists and ankles.

Fixed Joints Some joints in vertebrates do not move. In adult humans, the individual bones of the skull are held together by fixed joints. In fixed joints, the bones are fused; these joints are not connected by ligaments. In babies and children, the skull bones are not fused together, but rather have soft spots between them. Why do you think this is so?

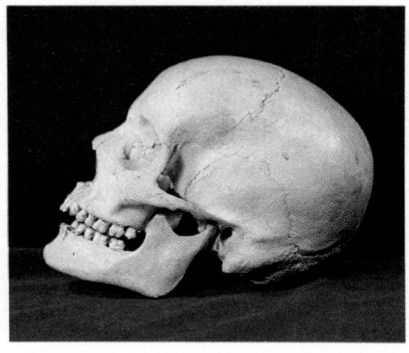

Figure 26-10 The bones of the skull meet in fixed joints. Where are the fixed joints on this human skull?

Section Review

Understanding Concepts
1. Think about an arthropod, such as an ant, and a vertebrate, such as a lizard. What functions do the skeletons of these animals have in common? Do they have a common structure? Explain.
2. Why do you think spongy bone has an outer layer of compact bone?

3. Think about the possible movements of your lower arm and lower leg. How do the movements differ and why?
4. **Skill Review—Comparing and Contrasting:** Compare and contrast exoskeletons and endoskeletons in terms of how they determine growth and size of the animals that

have them. For more help, see Thinking Critically in the **Skill Handbook.**
5. **Thinking Critically:** Why are the largest arthropods, such as lobsters, found living in water?

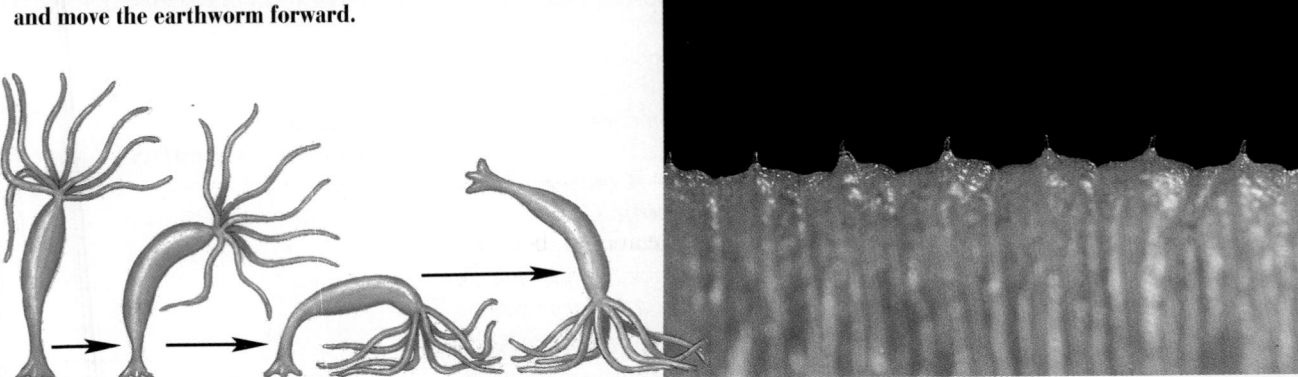

Objectives

Describe examples of movement in animals lacking skeletons.

Distinguish among the three types of vertebrate muscles and their characteristics.

Relate the events of muscle contraction to the ways in which energy is made available for that contraction.

26.2 Effectors and Their Actions

Locomotion and movement of body parts are important in many ways: obtaining food, avoiding predators, reproducing, and reacting to a variety of environmental factors such as light, heat, and chemicals. How do the effectors that contribute to movement work? How do muscles of arthropods and vertebrates attach to the skeleton, and how do the two work together to cause movement? How do animals lacking skeletons move by muscles alone? How do unicellular organisms lacking both muscles and skeletons move?

Movement Without a Skeleton

Movement in animals without a skeleton is limited. How do invertebrates move? Recall from Chapter 17 that cnidarians exist in two forms, polyp and medusa. Let's see how some cnidarians move.

Hydra is most often a sessile polyp, but it can move by somersaulting, as seen in Figure 26-11. Other cnidarians spend most of their life cycles as medusas, such as jellyfish. Jellyfish float freely in the water or move by contracting their muscle cells, forcing water from under the bell-shaped medusa to create a jet propulsion effect. The jellyfish moves in the direction opposite to that of the expelled water.

Earthworms are adapted for locomotion in soil. Their movement is controlled by two sets of muscles. As one set of muscles contracts to shorten the body, the other set then contracts, resulting in a lengthening of the body. Interaction of these two sets of muscles causes movement. In addition, bristles called setae extend backward from each segment of an earthworm. With setae, a worm "grasps" the soil, resulting in smoother movement.

Other means of locomotion occur in mollusks and echinoderms. Clams have a muscular foot that can be extended between the shells. To move, a clam extends the foot and anchors it in the sand. The muscles of the foot contract at the rear, lifting and coming down again. The rest of the foot moves forward from the rear in a wavelike series of movements.

Figure 26-11 **Invertebrates have evolved unique ways to move. *Hydra*, left, somersaults from place to place. The setae of earthworms, right, grip the soil so that muscles can contract and move the earthworm forward.**

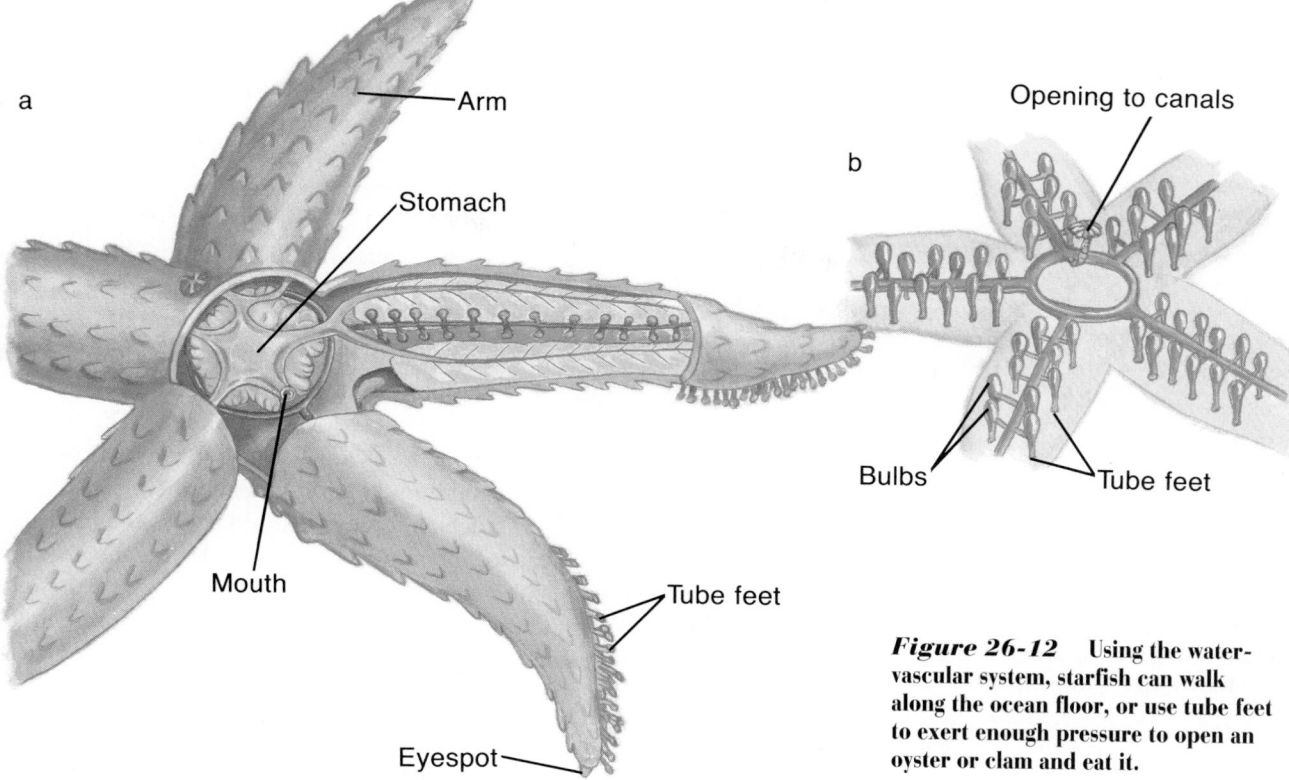

a

Arm

Stomach

Mouth

Tube feet

Eyespot

b

Opening to canals

Bulbs

Tube feet

Figure 26-12 Using the water-vascular system, starfish can walk along the ocean floor, or use tube feet to exert enough pressure to open an oyster or clam and eat it.

In starfish, movement is controlled by a unique **water-vascular system,** pictured in Figure 26-12. Water is drawn into the animal through a small opening in the central disc and is passed to canals running along each "arm." Along each canal are hollow, muscular tube feet that are found on the underside of the starfish. The bottom part of each tube foot flattens out like a sucker. The upper part of the tube foot is a bulb-like structure. The bulb contains water that comes from the canal. When the bulbs contract, water is forced into the tube feet and they lengthen. If pressed against a surface, the tube feet stick on like small suction cups. When muscles of the bulbs relax, water pushes back up through the canals and the tube feet shorten, releasing the suction. Each tube foot works independently of the others. A starfish can creep along rocks by alternately pushing out and pulling in its tube feet.

Movement: Vertebrates and Arthropods

Because they have both muscles and a skeleton, vertebrates are capable of a great variety of movement. Muscles are attached to the skeleton. As muscles contract, they pull on the skeletal system. The pull results in movement. In vertebrates, most muscles are attached to bones by tough connective tissues called **tendons.** Muscles are attached to bones at two different sites. During contraction, one end of the muscle and the bone to which it is attached do not move. This attachment site is called the point of origin. The other end of the muscle and the bone to which it is attached move when the muscle contracts. This attachment site is called the point of insertion. Attachment of a muscle to two sites is necessary for movement.

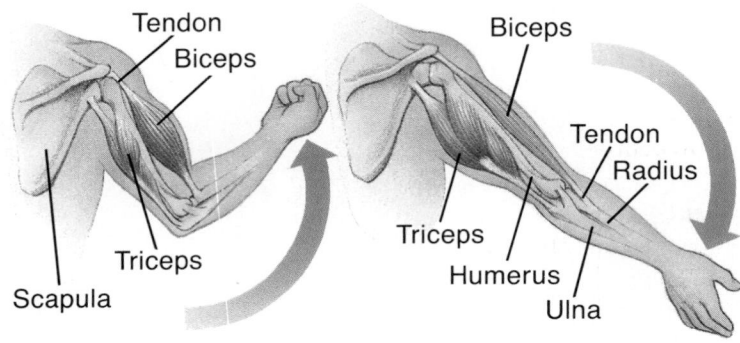

Figure 26-13 To bend your arm, the biceps muscle contracts while the triceps muscle relaxes (left). Your arm straightens when the biceps relaxes and the triceps contracts (right).

Paired Muscles Muscles that cause loco- motion work in pairs (or at least in two opposing groups) in arthropods and verte- brates. For example, in humans the biceps and triceps muscles work opposite each other to cause motion. Examine Figure 26-13. The biceps originates on the scapula (shoulder blade). The other end of the biceps is attached to the radius (lower arm), the point of insertion. The radius is the smaller of the lower arm bones. The biceps is called a **flexor** because it causes flexing (bending) at a joint. When the biceps con- tracts, a force is exerted across the hinge joint of the elbow, which causes the arm to bend. In order for the arm to bend, the triceps must relax.

To straighten the arm, the process is reversed. The triceps muscle links the back of the humerus and scapula with the ulna, the other lower arm bone. As the triceps contracts, the ulna is pulled downward, the biceps relaxes, and the arm straightens. The triceps is called an **extensor** because

MATH CONNECTION

How fast do animals move?

In the United States, most humans don't often need to think about how fast they can move on foot. If they're in a hurry, they ride bicy- cles, drive cars, catch buses, or even take jet planes. Some animals, though, would break the speed laws if they ran on the road at top speed. For example, a cheetah can sprint at 113 kilometers (70 miles) per hour!

Locomotion Speed in Animals
Without transportation, how do humans compare in speed of loco- motion to other animals? Which animals can move faster than the fastest humans? Of the animals that

are faster than humans, some swim, some fly, and others run. For exam- ple, cheetahs, horses, blue whales, and vultures all travel faster than humans do.

Some animals can travel only through water. Many are amphibi- ous and can swim or travel on land. Birds are the most versatile travel- ers. Many of them can swim, fly, and walk.

Locomotion Systems Locomotion systems in vertebrates use movable bones that interact with attached muscles. Most vertebrates use legs, arms, wings, tails, or fins to propel themselves, but snakes move even without limbs.

Invertebrates use a variety of specialized structures, systems, and interactions to move. They may use waving hairs or spines, for example.

Some, like vertebrates, use legs, wings, fins, or tails.

The speeds listed in the follow- ing table are the fastest speeds ever recorded for given animals. An ani- mal can move only a short distance at its top speed. The average speed for an animal is lower than its top speed.

The animals that can move body parts fastest are not necessarily the ones that cover the greatest dis- tance in the shortest time. For example, a hummingbird's wings move much faster than a duck hawk's do, but a duck hawk soars through the air more than twice as fast as a hummingbird. An ant's legs move rapidly, but you can see from the chart that a turtle can travel a given distance in less time.

it causes the extension (straightening) of a joint. What are the points of origin and insertion of the triceps?

Note that the bending of a joint is always due to contraction of one of the muscles of the pair. In other words, both flexing and extending are due to contraction. In order for one muscle of a pair to contract, the other must relax.

Besides bending or straightening a joint, paired muscles are important in another way. No muscle is ever completely relaxed; both muscles of a pair are always slightly contracted whether or not a movement is taking place. This condition provides enough contraction of the muscles to help support the body. Called muscle tone, this condition also keeps the muscles ready for quick contractions.

Arthropod Movement   In arthropods, muscles are attached to the inside surface of the exoskeleton. Contraction of one muscle exerts a pull on the exoskeleton that results in the bending of a joint and a movement. When this muscle relaxes and the opposing muscle contracts, an opposite movement occurs.

Note that the paired muscles of arthropods function as flexors and extensors. Although arthropods and vertebrates are not closely related, they have evolved a similar type of muscular control of skeletal movements.

In general, large animals move faster than smaller ones, but many exceptions exist. For example, cheetahs are smaller than horses or elephants, but much faster. The cheetah has an unusually long stride for its size because of a flexible spine that lets its hind legs reach ahead of its front legs as it runs. Its structure and the interaction of its parts enable it to accelerate from standing still faster than most automobiles.

---

1. **Solve the Problem:** At the speeds provided in the chart, how many meters could a squirrel travel while an ant was traveling one meter?
2. **Explore Further:** Examine speed records for running and swimming. Why can people run faster than they can swim?

Speeds of Animals

Mammals		Reptiles	
cheetah	26.7 m/s	lizard	6.7 m/s
racehorse	19.1 m/s	turtle	2.0 m/s
blue whale	18.0 m/s	**Amphibians**	
dog	16.0 m/s	frog	1.5 m/s
human	11.0 m/s	**Fish**	
cat	10.0 m/s	tuna	20.0 m/s
rabbit	8.0 m/s	flying fish	10.0 m/s
squirrel	2.0 m/s	salmon	3.0 m/s
Birds		**Invertebrates**	
vulture	17.0 m/s	locust	4.5 m/s
ostrich (running)	9.5 m/s	dragonfly	3.0 m/s
penguin (swimming)	3.5 m/s	ant	0.03 m/s
duck (swimming)	0.7 m/s	snail	0.0025 m/s

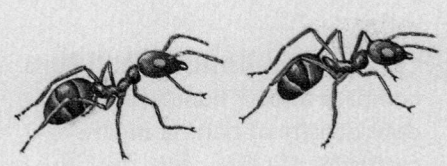

Types of Vertebrate Muscles

Have you ever felt nervous before riding a roller coaster at an amusement park? Perhaps you took a deep breath to try to control all the butterflies fluttering in your stomach. Did it work? In vertebrates, there are some muscles that the animal cannot control, like the muscles of your stomach or heart. Other muscles can be controlled by your actions. Why is there a need for different kinds of muscles?

Striated Muscle   Three types of muscle tissue are present in vertebrates. Of these, the most important in terms of locomotion is **striated muscle** (also known as skeletal muscle). Striated muscles are so named because of their striped appearance when viewed with a microscope. Striated muscle

I N V E S T I G A T I O N

Bone and Muscle Density

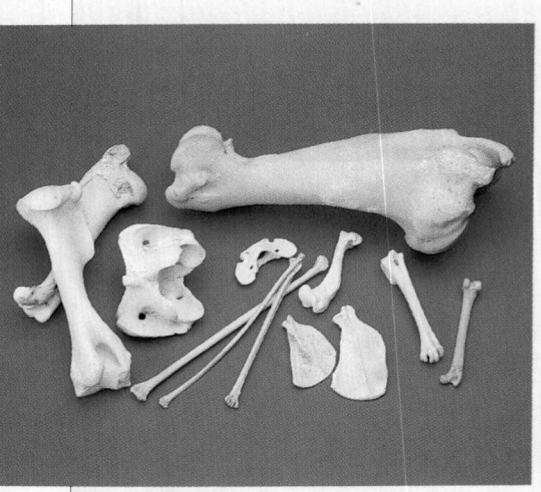

The term *density* refers to compactness. It is an indication of how much mass is packed into a specific space. For example, the density of a pillowcase filled with feathers is very low, whereas the density of the same pillowcase filled with sand would be much higher. One can express density in a mathematical formula as follows:

$$\text{Density} = \frac{\text{Mass}}{\text{Volume}}$$

How might one go about measuring density of an object? Try these steps to find the density of a stack of ten pennies:

1. Use a balance to find and record their mass in grams.
2. Find the volume in mL of the stack using water displacement.
3. Divide volume into mass. This gives you density in g/mL.

Repeat the entire procedure for a stack of five pennies. How do the densities compare for five or ten pennies? How should they compare if you were finding the density of the same coin type in both trials?

Now that you know how to find density, this technique will be used to solve several problems. How might the densities of muscle and bone compare? How might densities of bone and cartilage compare? How might densities of bones of different animals compare? Work with a group assigned by your teacher. Form hypotheses that could be tested using bone, cartilage, and muscle.

Problems

(1) How does density of bird bone compare to beef bone? (2) How does density of bone compare to cartilage? (3) How does density of bone compare to muscle? (4) How does density of an exoskeleton compare to that of an endoskeleton?

consists of long cells, each called a **fiber,** that have many nuclei. Fibers have many nuclei because they are formed by the fusion of many embryonic cells during the development of a fetus. Each fiber is surrounded by connective tissue, and the fibers and connective tissue together make up a muscle. Striated muscles are attached to the bones. They are voluntary muscles because they can be controlled at will.

Each skeletal muscle fiber is stimulated by just one nerve. Your brain sends impulses to motor neurons in your right arm. The motor neurons in turn stimulate several muscle fibers in your right arm muscle. These nerve impulses cause the muscle fibers to respond (contract). Muscle contraction causes your arm to bend at the elbow joint. When the impulses stop, the muscle fibers relax and your arm can straighten. Remember that your arm straightens because it is pulled by the contraction of the opposing muscle.

Hypothesis

What are your group's four hypotheses?

Experimental Plan

1. Decide on a way to test your group's four hypotheses. Keep the available materials provided by your teacher in mind as you plan your procedure.
2. Record your procedure and list materials and amounts that you will need.

Checking the Plan

Discuss the following points with other group members to decide the final procedure for your experiment.

1. What data will you collect and how will it be recorded?
2. What variables will have to be controlled?
3. Assign roles for this investigation.

Make sure your teacher has approved your experimental plan before you proceed further.

Data and Observations

Carry out your experiment. Prepare a data chart to record your findings.

Analysis and Conclusions

1. State each hypothesis separately. Explain how your data either support or reject each hypothesis.
2. (a) What advantage to birds is having a bone density that is different than that of a cow?
   (b) What advantage to a cow is having a bone density that is different than that of a bird?
   (c) What advantage to animals is having an exoskeleton density that is different than that of animals having an endoskeleton?
3. Offer an explanation as to why the densities of bone differ from cartilage.
4. What advantage to animals is having bone density that differs from muscle?
5. The student group working next to you had bone samples that were much larger than yours. Yet, your data and theirs were almost identical. Explain how

this is possible. Offer several reasons why your data may have differed slightly.

Going Further

Based on this lab experience, design another experiment that would help to answer questions that arose from your work. Assume that you will have limited resources and will be conducting this lab in class. Include this as part of your report.

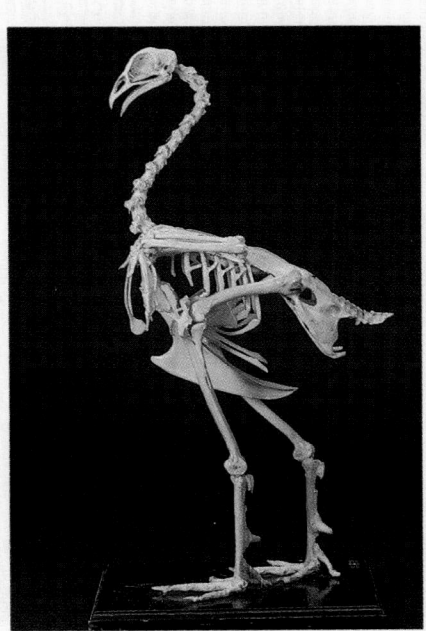

Smooth Muscle As well as muscles for locomotion, vertebrates have muscles for other functions. Nonstriated or **smooth muscle** moves many of the internal parts of the body. Smooth muscle is made up of sheets of distinct, tapered cells, each with a nucleus. Smooth muscles are called involuntary muscles because an animal cannot control the activities of these muscles at will. Smooth muscles make up the walls of the hollow organs of the body, such as those of the digestive tract. Blood vessels also have a layer of smooth muscle. Contraction and relaxation of these muscles can regulate the activity of the organs or the diameter of vessels. Peristalsis is caused by smooth muscle contractions.

Structures of which smooth muscles are a part are controlled by a pair of nerves. These nerves belong to the autonomic nervous system. One nerve causes contraction; the other inhibits it. Smooth muscles contract less forcefully and somewhat more slowly than skeletal muscles and are not involved in locomotion.

Cardiac Muscle The heart is composed of individual muscle cells. The cells are branched and form networks which wrap around the chambers of the heart. Heart muscle is similar to smooth muscle in that it is an involuntary muscle and it is controlled by two nerves. Heart muscle looks much like striated muscle in that both show striations, but heart muscle is different from striated muscle because it contracts rhythmically. The contraction of heart muscle cells forces blood out of the heart chambers and into arteries. Heart muscle is known as **cardiac muscle.** All three types of vertebrate muscle are shown in Figure 26-14.

Figure 26-14 **Three types of muscle in vertebrates include striated, smooth, and cardiac muscles. Striated muscles, left, are muscles that attach to bones. (Magnification: 250×) Smooth muscles line internal organs and blood vessels. (Magnification: 100×) Cardiac muscles, right, are found only in the heart. (Magnification: 140×)**

TYPES OF MUSCLE

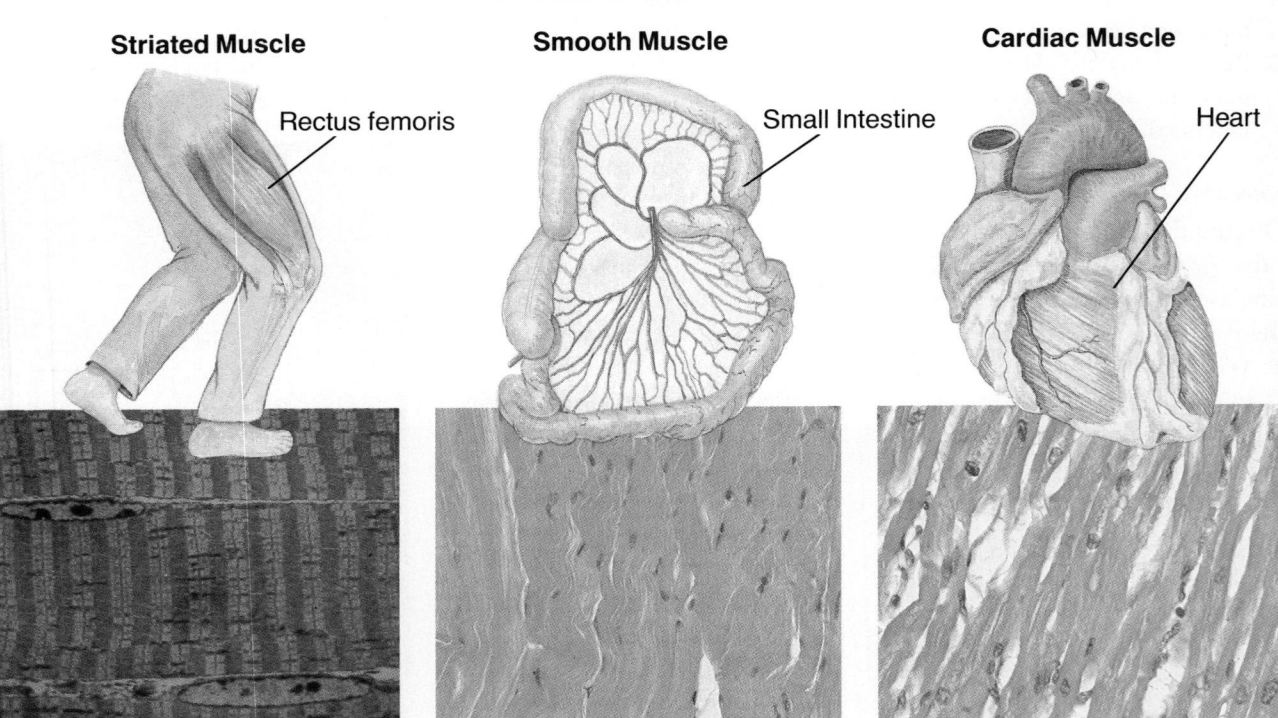

Striated Muscle — Rectus femoris

Smooth Muscle — Small Intestine

Cardiac Muscle — Heart

Contraction of Vertebrate Skeletal Muscle

The fibers of a skeletal muscle are closely packed together. Each cylindrical fiber is made up of smaller units called **fibrils.** Fibrils consist of many microfilaments. You learned in Chapter 5 that microfilaments are composed of the proteins actin and myosin. In vertebrate muscle cells, microfilaments are more simply called filaments. The arrangement of filaments gives striated muscle its striped appearance. There are two kinds of filaments, thick and thin. The thick filaments are made of the protein myosin, and the thin ones are made mainly of the protein actin. Along the thick filaments are knoblike parts of myosin molecules that bridge (link) the two kinds of filaments together during contraction. The thin filaments are anchored to vertical bands called Z lines. The part of a fibril from one Z line to the next is called a **sarcomere.**

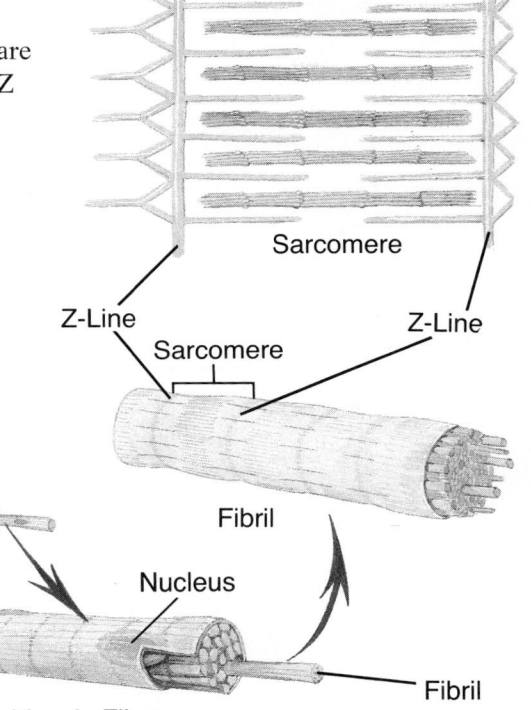

Figure 26-15 Muscles consist of bundles of fibers that are made up of smaller fibrils. Fibrils consist of filaments of the proteins actin and myosin.

Thick filament (myosin)

Thin filament (actin)

Sarcomere

Z-Line

Sarcomere

Z-Line

Fibril

Nucleus

Fibril

Muscle

Bundles of Muscle Fibers

Muscle Fiber

Thinking Lab

DRAW A CONCLUSION

What kind of muscle is being used?

Background
The three muscle types each have their own characteristics. For example, smooth and cardiac muscle are involuntary, while skeletal muscle is voluntary. Smooth muscle does not fatigue easily, but skeletal muscle does. Cardiac muscle is only found in the heart.

You Try It
Chewing food with jaw muscles must be voluntary. Therefore, jaw muscles must be skeletal. Swallowing and the passage of food down your esophagus also depend on muscles. Using your own experience with swallowing and what happens after you swallow, **draw**

a conclusion and offer support for your answer for: (a) the type of muscle found toward the back of your mouth; and (b) the type of muscle found in your esophagus.

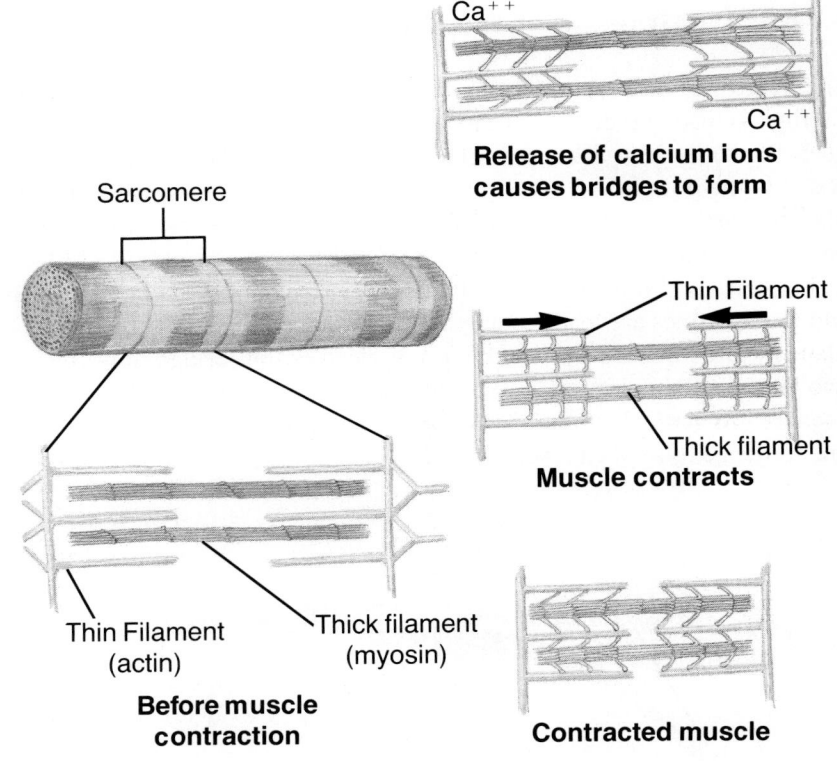

Figure 26-16 The sliding filament hypothesis suggests that actin and myosin filaments slide over one another to shorten the fibrils during muscle contraction.

Ca⁺⁺

Ca⁺⁺

Release of calcium ions causes bridges to form

Sarcomere

Thin Filament

Thick filament

Muscle contracts

Thin Filament (actin)

Thick filament (myosin)

Before muscle contraction

Contracted muscle

Sliding Filament Hypothesis Electron microscope studies show that Z lines of each sarcomere are closer together when a muscle is contracted than when a muscle is relaxed. Such observations as well as experiments on striated muscle have led to the **sliding filament hypothesis,** a model that explains contraction.

Motor nerves leading to a skeletal muscle branch many times so that a number of individual muscle fibers are stimulated. Acetylcholine is released by the motor neuron and diffuses across the synapse to the muscle fiber. The membranes of muscle fibers, like neurons, are polarized. The acetylcholine causes depolarization of the membranes of the fibers. This action potential causes a release of calcium ions stored in special membranes inside the fiber. Calcium ions are needed to allow the ends of the myosin molecules of the thick filaments to form bridges with the actin molecules of the thin filaments. Bridges form as the ends of the myosin molecules "hook" into grooves along the actin molecules. Using energy of ATP, the bridges bend, causing thin filaments to slide past thick filaments. Myosin detaches from actin and combines with another site to repeat the process. As a result of this repeated pattern, the thin filaments slide toward and over one another. The chemical energy of ATP is thus transformed to the mechanical energy of sliding filaments. Note that the filaments themselves do not shorten, but the filaments' sliding over each other causes the muscle fiber to shorten. After the nerve impulse is over, the calcium ions are once again stored. Bridges cannot form and the muscle relaxes. These events are diagrammed in Figure 26-16.

All-or-none Response   Once an impulse reaches a muscle fiber, there is an all-or-none response. That is, either the muscle fiber contracts fully or it does not contract at all; there is no partial contraction for a given fiber. Also, all contractions are of the same intensity. How, then, can a muscle contract to a greater or lesser degree? The number of fibers that contract at one time determines the strength of the contraction of the whole muscle. The greater the number of fibers that contract, the greater the contraction of the muscle as a whole.

The all-or-none response can also explain muscle tone. The nervous system constantly sends out messages that you are not aware of. These messages keep some fibers in each muscle relaxed and some fibers contracted. A muscle fiber must undergo a short recovery period after it contracts before it can contract again. The nerves constantly cause different fibers to contract so that while some fibers are contracting, others are recovering. Thus, some fibers will always be able to contract, and the muscle as a whole is always prepared for a complete contraction. Why is constant readiness of a muscle for contraction important?

Energy for Skeletal Muscle Contraction

Contraction of muscles is biological work requiring ATP energy. The energy needed for muscle contraction is made available in several ways, depending on muscle activity.

Aerobic Respiration Provides ATP   During rest or very light activity, ATP is produced in muscle cells by aerobic respiration. Some of the ATP produced this way is immediately used for muscle contraction. The remaining ATP is stored in the muscle cells either as ATP itself or as another compound, creatine phosphate (CP), that can be converted quickly to ATP. The CP is formed when molecules of ATP donate phosphate groups to molecules of a compound called creatine. As muscles become more active, the energy stored in CP can be used. CP is changed to creatine, and the phosphate released combines with ADP to form ATP. The energy in the ATP can then be used for muscle contraction.

ATP production

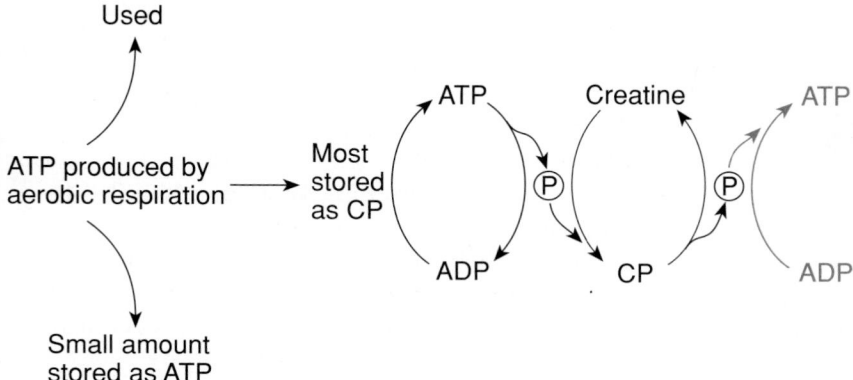

Figure 26-17   **The source of energy for muscle contraction is ATP.**

Need for Lactic Acid Fermentation   During prolonged or heavy exercise, the supplies of stored energy (both ATP and CP) are quickly used. Even though the heartbeat and breathing rates are increased during heavy exercise, not enough oxygen is delivered to muscle cells for aerobic respiration to continue. Under these conditions, the muscle cells switch to lactic acid fermentation. During lactic acid fermentation, glycogen stored in muscles is converted to lactic acid, and ATP is produced in the process. The ATP provides energy for muscle contraction. During this process, the muscles are said to acquire an oxygen debt. **Oxygen debt** occurs when the cell is obtaining ATP energy without going through normal, aerobic respiration. The cell is "borrowing" against future oxygen intake.

After strenuous activity, rapid breathing occurs, increasing the supply of oxygen to cells. With an increase in oxygen, aerobic respiration is again possible. Blood circulation removes the lactic acid from your muscles and transports it to the liver. Using ATP produced by the oxidation of some lactic acid, the liver converts the remaining lactic acid to glycogen. The oxygen debt is "paid" when all the lactic acid is removed.

LEISURE CONNECTION

Exercise for Muscle Fitness

How many hours of television do you watch in a week? How much of your leisure time is spent sitting

around? Outside school, exercise for some young people consists of getting up for a snack. They don't even need to get up to change the

channel on the television set. As a result, many young people are overweight and out of shape.

In a 1985 National Adolescent Health Survey, of the students polled in grades eight and nine, 41 percent exercised 20 minutes or more five times per week. Almost 16 percent didn't exercise, or did so only once per week.

Why You Need to Exercise
Your muscles need exercise to stay strong and flexible. Muscles that are toned resist injury from strenuous or unexpected activity.

Exercise has other benefits as well. Like other muscles, the muscles in the heart need regular exercise to stay strong. A strong heart muscle will pump more blood per heartbeat to your body's cells. The more blood pumped with each heartbeat,

the greater the amount of oxygen provided to body cells.

The most healthful exercises are aerobic, the kind that keep your heart beating faster and your lungs drawing more air for an extended period of time. Aerobic exercise will strengthen your heart and lungs, open your arteries, and keep these systems working and interacting efficiently. Thirty minutes of continuous aerobic activity every other day is recommended.

Exercise can have many benefits. It helps reduce blood cholesterol levels and weight. The bones of people who exercise are stronger. In fact, as people age, they tend to lose calcium from their bones from lack of exercise.

Exercise for Mental Health   In addition to physical health, exercise enhances mental and emotional health. It also boosts energy and releases tension, anger, and

Ciliary and Flagellar Motion

Some organisms or specialized cells can move even though they lack both muscles and skeletons. Many unicellular eukaryotes have cilia or flagella composed of microtubules. These effectors may be used not only to propel cells, but also to move particles past cells. In a paramecium, the cilia do both, moving the protozoan through the water and setting up a current that draws food into its oral groove.

Microtubules in cilia and flagella of eukaryotes are arranged in nine pairs with a pair in the middle. Figure 26-18 shows the arrangement of the microtubules, although it is not detailed enough to show that the outer pairs of microtubules are connected to one another. The connections are projections, or arms, made of a protein. Each outer pair is also connected to the pair in the center by structures resembling the spokes of a wheel.

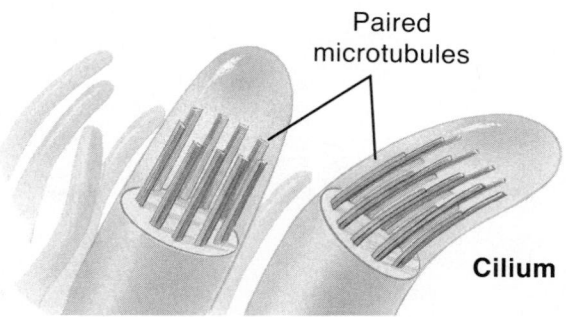

Paired microtubules

Cilium

Figure 26-18   The cilia and flagella bend when one tubule of each pair slides past the other, similar to the sliding found in muscle filaments.

stress. Studies have shown that exercise is helpful in fighting depression. Regular aerobic exercise can result in elevated self-confidence, personal responsibility, and a sense of well-being. By contrast, inactivity tends to make your mind and body sluggish.

If you develop the habit of exercising regularly now, you are more likely to incorporate exercise into your life as an adult. Exercise needn't be a chore if you choose an activity you enjoy. Skiing, soccer, rowing, hiking, swimming, bicycling, basketball, dancing, and many competitive sports are all excellent forms of exercise. For some people, such activities are the best part of a weekend. Many are social activities that help them make new friends and enjoy the ones they have. What television program or video game could provide so many benefits and so much fun?

1. **Explore Further:** Examine some studies comparing various types of exercise. Which are best for weight control?

2. **Explore Further:** Which types of exercises are most effective for building muscle mass and strength?

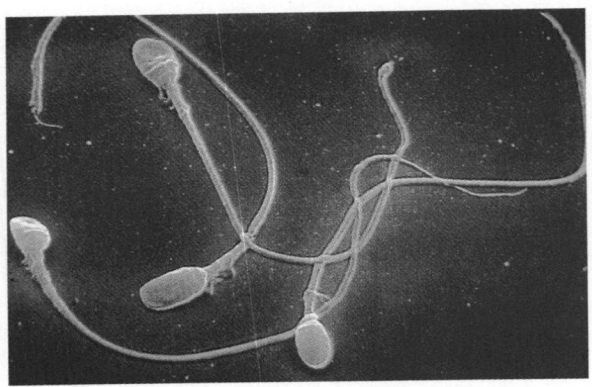

Figure 26-19 Each human sperm moves by means of a flagellum. Magnification: 5000×

It is thought that the outer pairs of microtubules slide past one another in much the same way as the filaments of a muscle fiber do. According to this model, the arms that connect the outer pairs of microtubules act in much the same way as the bridges that connect thick and thin muscle filaments. The energy of ATP is necessary for the arms to form these connections between microtubule pairs and to cause them to move. As the pairs of outer microtubules slide over one another, the cilium or flagellum bends, because all the microtubules are interconnected. Sliding of different pairs of microtubules results in movement such as that which occurs as an oar moves through the water.

Locomotion of unicellular eukaryotes thus has similarities to locomotion of vertebrates. However, the two systems of locomotion—bending of cilia and the mechanics of muscle contraction—evolved independently, another good example of convergent evolution.

LOOKING AHEAD

THIS CHAPTER completes a unit in which you have studied how chemical control, nervous control, and effectors enable organisms to respond to stimuli. Many of those stimuli are present in the external environment. In the next chapter, you will begin to explore the relationship between organisms and their environment from another point of view. You will study how various factors limit the growth of populations of organisms. Those factors include other organisms as well as physical factors of the environment. ❧

Section Review

Understanding Concepts

1. As an earthworm moves, part of its body seems to get shorter, and then the body gets longer again. How are these changes in body length explained?

2. You are observing a section of muscle tissue with a microscope. You have difficulty distinguishing separate cells but notice that the muscle has a branched appearance. Is the muscle tissue you are observing voluntary or involuntary? Explain.

3. You are sitting watching TV, but some of your skeletal muscles are moving. By what process is energy made available for that movement? How, specifically, does the ATP contribute to contraction of your muscles?

4. **Skill Review—Designing an Experiment:** Assuming you had the necessary equipment, how would you test the idea that a muscle contraction is an all-or-none response? For more help, see Practicing Scientific Methods in the **Skill Handbook.**

5. **Thinking Critically:** What effect would a chemical that interferes with ATP production have on cilia and flagella? Discuss the role of ATP in explaining your answer.

BIOLOGY, TECHNOLOGY, AND SOCIETY

Biotechnology

Bionics and Artificial Limbs

Most people take locomotion for granted. A person with full use of all body parts can scarcely imagine the impact of paralysis or the loss of a limb. In fact, to most wild animals, the loss of a limb is so disabling that the animal is unable to survive.

The First Substitute Limbs
The development of human societies has enabled humans to survive the loss of limbs. According to archaeologists, humans were caring for the physically challenged among them as early as 250 000 years ago. More than 2000 years ago, humans began using substitutes for lost legs, arms, and hands. The earliest known artificial leg, made of metal plates around a wooden core, dates from 300 B.C. By 1990, human technology had evolved to the point where some quadriplegics could thread needles, and individuals with artificial legs could walk up and down stairs, jump, ski, and run marathons.

Artificial Limb Replacement
Modern surgeons can amputate limbs, leaving enough muscle structure in place to control an artificial replacement. If enough stump and muscle remain after amputation of a hand or an arm, replacements are

usually operated by voluntary muscle control. An artificial hand, for example, may be held closed with a built-in spring. Straps connected to the opposite shoulder pull the fingers and thumb apart.

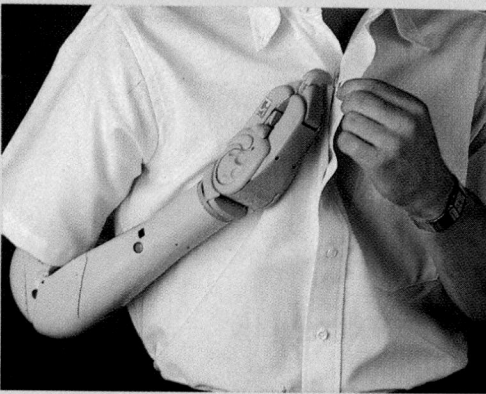

Designing an Artificial Limb   To design an artificial limb, a prosthetist needs to examine the function of the natural limb and how it contributes to locomotion. For example, a living foot extends and flexes, bends at the ball for walking or running, and acts as a shock absorber, bearing the entire weight of the body. Some modern artificial feet are designed to flex at the ball like natural feet. They contain springs to simulate the action of the sole of the foot in walking.

Stimulating Muscles by Electrodes
Locomotion requires more than the

presence of the necessary limbs, muscles, and joints. The muscles must receive stimulation from the nerves. Injuries to the nerves or spinal column can leave victims unable to control the muscles that move their limbs. However, researchers learned to use electrodes to stimulate the muscles in paralyzed limbs. Muscles can be made to contract or relax to control the limbs. Using these electrodes, patients have learned to walk with paralyzed legs. Some people with paralyzed arms can even write.

Nothing can fully replace a functioning limb. However, modern researchers constantly seek ways of improving survival and quality of life for those who have lost use of limbs.

1. **Writing About Biology:** Experiment with your leg, knee, ankle, or hand. Observe the body part while performing several everyday functions. In order to replace that part fully, what would an artificial body part need to do?
2. **Explore Further:** Research the latest developments in one of these areas: electrical stimulation of paralyzed limbs, artificial hands, artificial feet.

CHAPTER 26 MOVEMENT **751**

A Day in the Life of a Physically Challenged Person

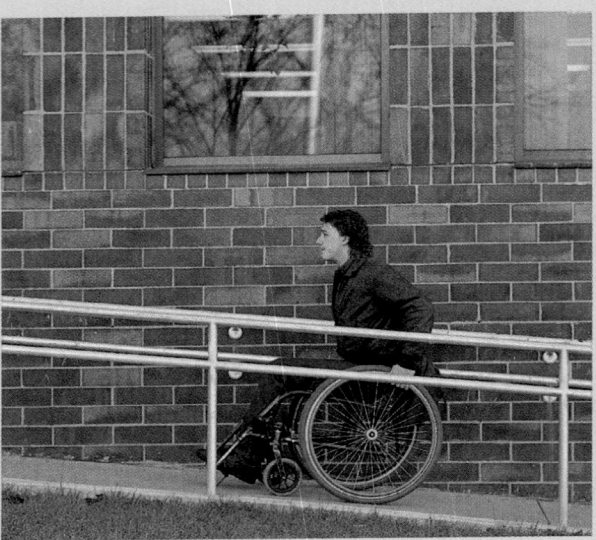

"One of the biggest differences is that everything takes a lot longer," said former police officer Chris Barrett, describing changes in his life since a bullet paralyzed him from the waist down. It took just a fraction of a second for the bullet to damage his spine and destroy the systems he used for locomotion and everyday activity. However, the pace of recovery was laborious and slow as he trained his body parts to interact in new ways. For example, since his injury it takes him six times as long to get up and prepare for work.

Getting Out of Bed First, he needs to get out of bed. With his back to the wheelchair beside the bed, he uses his arms to push himself to a sitting position. Then he uses his arms to pull his torso onto the seat. Next he uses his arms to lift his legs onto the footrest of his wheelchair. When the legs and stomach muscles can't interact with the arms, head, and torso, all these maneuvers require abundant energy and tremendous strength in the arms. Fortunately, Barrett was an athlete before his injury. Injured people in poor shape have much more trouble developing the necessary strength.

Upon arising, Barrett navigates the doorways and hallways to the kitchen. In the kitchen, most cabinets are too high or too low for him to reach from a wheelchair. He uses a set of tongs to grab the food and dishes he needs. The sink is too high, but by placing his wheelchair beside it, he is able to reach up and to the side to use it. "Some disabled people who live alone have special cabinets and counter tops that can be raised or lowered at the touch of a button. Even the sink moves up and down," says Barrett.

How to Use the Bathroom
Wheeling to the bathroom, Barrett is grateful that the family rebuilt their home, widening doorways and hallways. After Barrett's injury, he had to maneuver his wheelchair through narrow doorways and hallways, and cross raised thresholds when going from one room to another.

The bathroom presented problems, too. When Barrett was injured, his bathroom was small and the door opened inward.

Imagine trying to wheel a chair into a small bathroom and close the door behind you, especially if the door opens inward.

Unfortunately, as Barrett points out, many physically challenged people are renters and can't remodel their homes. Those who don't have much money must choose housing they can afford and can't shop around for homes where they can easily navigate. They must wrestle with thresholds, narrow doors and hallways, and cramped bathrooms without space to maneuver.

Obstacles in Public Even if physically challenged people can afford to design their homes to accommodate their challenges, they have no control over public streets and buildings, friends' homes, or the work place.

Because the front of a wheelchair extends about two feet in front of the torso, it can't be pulled close to tables and desks. As a result, there are many restaurants that Barrett can't use. Transferring to a chair or booth doesn't work well for him. Unless he supports himself with his hands, he tends to fall over if he leans too far forward or to the side.

Challenges at the Work Place Barrett faces additional challenges at work. For example, he must go through heavy, self-closing doors. Fortunately, the doors open with levers. Barrett works up some speed with his wheelchair and runs into the door. If he depresses the lever fast enough, the momentum opens the door. Then he holds it open with his footrest while he struggles to wheel the chair through.

Turning a knob, pushing a heavy door open, and holding it open while wheeling a chair through is more than many physically challenged people can manage. "Levers are a big improvement," says Barrett. "Some quadriplegics are accompanied by dogs, and the dogs can trip the levers to open these doors." Quadriplegics have injuries high on the spinal cord and can't control their arms and legs.

The two-story building where Barrett works doesn't have elevators, and Barrett has duties on both floors. Fortunately, the building has a two-story garage with entrances on both levels. "It takes a lot of planning," says Barrett. First, he completes all his duties on the top floor. Then he goes outside, gets into his car, and drives to the ground floor.

Barrett doesn't want to make the drive more often than necessary. Getting in and out of the car and collapsing or opening his wheelchair takes time, energy, effort, and careful maneuvering. In Barrett's car, all controls are operated by

hand. "There's a quadriplegic in town who drives his car with his mouth," says Barrett. "Plenty of people have it harder than I do."

Laws Help the Physically Challenged Laws now require public buildings to provide automatic doors, and ramps for people who can't use stairs. Hallways, aisles, and doorways must be wide enough so that people with wheelchairs can use them. Rest rooms must be constructed so that people in wheelchairs can get in and out of them. Handrails are provided beside toilets and in some showers and bathtubs.

For the sake of people who find locomotion difficult, handicapped parking spaces are required in public parking lots. Such spaces are located near the entrances and have safety areas where people can get in and out of wheelchairs. In many cities, physically challenged citizens may have blue-curbed handicapped parking spaces designated outside their homes.

In addition to wheelchair-friendly construction, Barrett reports that he has found workers in public and commercial buildings very helpful. For example, in restaurants with steps but no eleva-

tors, staff members willingly take him through the kitchen or carry him up steps. However, like most physically challenged people, Barrett wants to be as independent as possible. What he appreciates most are the facilities that allow him to be independent in public places.

Fortunately, technology is constantly creating new devices to enable the physically challenged to be more independent. Many devices, for example, can be activated by voice, breath, or head movements. Researchers continuously seek to improve not only locomotion, but the quality of life and the range of possible activity for the physically challenged.

1. **Writing About Biology:** Examine the doors to your school and classroom. Describe what a person in a wheelchair would need to do to enter the school, travel to the classroom, enter the room, and reach your desk.

2. **Explore Further:** Examine ten public or commercial buildings in your community. How many are designed to accommodate the physically challenged? What challenges would the other buildings present for someone in a wheelchair or walker?

Readings

- Hertz, Sue. "The User-Friendly Home." *House Beautiful*, Nov. 1992, p. 90.
- Kridler, Charles and R.K. Stewart. "Access for the Disabled." *Progressive Architecture,* Sept. 1992, p. 45.

Summary

Effectors such as cilia or muscles are structures by which organisms respond to stimuli. Muscles not only aid in movement, but also provide support. Animals having skeletons, arthropods and vertebrates, are capable of the most efficient movement, and their skeletons also provide support and protection.

Arthropods have a nonliving exoskeleton that limits their size and growth patterns. Vertebrates, having an endoskeleton, are not so severely limited. Most vertebrates have a skeleton composed of bone, which replaces cartilage during development. The individual bones are joined in different ways, providing a great variety of movements.

Much movement in arthropods and vertebrates depends upon interaction of muscles and skeleton. Flexors and extensors work together to move parts of the body. Smooth muscle and cardiac muscle also cause movement, but do not interact with the skeleton. Energy for muscular movement is provided by ATP, which also fuels the movement of cilia and flagella.

Language of Biology

Write a sentence that shows your understanding of each of the following terms.

cardiac muscle	ossification
cartilage	osteocyte
extensor	oxygen debt
fiber	sarcomere
fibril	sliding filament hypothesis
flexor	smooth muscle
Haversian canal	striated muscle
joint	tendon
ligament	water-vascular system

Understanding Concepts

1. Sharks have an endoskeleton composed entirely of cartilage. Why are there no terrestrial vertebrates with such a skeleton?
2. Do you think that it is a coincidence that the largest vertebrates, whales, and the largest arthropods, such as lobsters, both inhabit water? Explain.
3. Can you offer an explanation of why there are so many knee injuries in athletics?
4. How can you tell if a striated muscle is skeletal muscle or heart muscle when viewed under a microscope?
5. Why can't a starfish move as rapidly and smoothly as an earthworm?
6. Have you heard of someone suffering from a torn ligament? How would such an injury affect a person?
7. What effect would a cut or injured tendon have?
8. Explain the series of events beginning with an impulse from a motor neuron and ending with contraction of a flexor in your lower leg.
9. Suppose you were running a race. Toward the end of the race, you are gasping for breath. How do you manage to keep running a little bit more?
10. How is the movement of cilia and flagella similar to the contraction of muscle?

Applying Concepts

11. The graph below indicates growth of a vertebrate over time. How would you graph the growth of an arthropod? Explain.

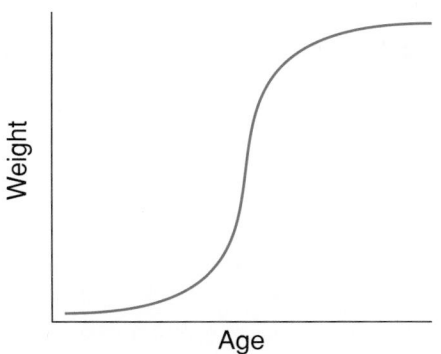

Age

12. Which bone(s) of the human body would you expect to be the strongest? Explain.
13. How would ATP be provided when a person is lifting a heavy object for a very long time? Explain.
14. How is the slow contraction of smooth muscle adaptive?
15. Why can skeletal muscles be controlled by just one nerve rather than two, like smooth or cardiac muscle?
16. Parathyroid hormone acts to stimulate release of calcium from bones to blood. Calcitonin has an opposite effect. What symptoms might arise if the levels of these hormones were not in normal balance?
17. **Biotechnology:** Why would the loss of a leg be disabling to an animal such as a rabbit?
18. **Issues:** Walk around your home and make an inventory of areas that you think would be difficult to maneuver in if you were in a wheelchair. Look carefully at both the kitchen and the bathroom. How could you remodel these areas if you had no money for major changes?

Lab Interpretation

19. **Investigation:** The following data are provided for your use. Determine which sample is bone and which is cartilage. Explain your answers.

Sample	Mass	Volume
A	6.2 g	8 mL
B	88.4 g	52 mL

20. **Thinking Lab:** Explain why you cannot stop vomiting once the process begins.
21. **Minilab 2:** Suggest two ways in which you could delay the onset of fatigue with each succeeding trial if doing an experiment similar to the one you did with the book.

Connecting Ideas

22. How is it possible for mollusks such as squid and octopuses to move much more rapidly than most other mollusks? How is that movement related to their food-getting abilities?
23. Explain why skeletal muscles and cilia and flagella illustrate convergence rather than homology.

Interactions in the Environment

How is a city like any other environment? Organisms that live in a city interact with other living and nonliving things. Dandelions sprout in sidewalk cracks, and robins pull worms out of the soil in vacant lots. A city, like other environmental systems, has a life of its own.

Grasslands, such as this one in Africa, are characterized by the lack of trees and the presence of various types of grasses. Large herds of grass-eating mammals are generally found on grasslands, but the species vary depending upon where the grassland is located. Wind, rainfall, and temperature also influence the formation of grasslands, as does soil type. Living and nonliving things interacting help to define a grassland.

What organisms can you see in a grassland? Shrubs, grasses, giraffes, zebras, and wildebeests are residents of this African grassland. Waterholes harbor protists and algae, and even fingerling fish. Nonliving factors also influence what goes on in the grassland. The amount of sunlight, rainfall, and wind intensity are all important to the life of the grassland.

What do you see when you look at a photograph of a city? You see buildings, vehicles, and people; but, a city also has a natural environment just like a grassland.

Pigeons are found in many cities, but so are peregrine falcons that nest on tall skyscrapers. Cities near the ocean have gulls that wheel overhead. Look closely at the trees in a city park, and you will see rows of holes made by woodpeckers looking for an insect meal. The next time you visit a city, look for signs of living things. There are complex relationships among the organisms in a city, just like there are in a grassland. Interactions among living and nonliving things are the focus of this final unit.

POPULATION BIOLOGY

THERE'S A PLACE for everything—food, nursery, work places—and everything is in its place inside a beehive. Each individual has a role, which it performs faithfully. All the work gets done inside and outside the hive. Everybody eats and nobody fights. Sounds like the perfect success story, doesn't it? And it is, until success fills the hive with too many bees. But not to worry; the bees have an adaptive mechanism for handling the situation.

It's time to move—for the bees, as well as for anyone who might be in the path of a swarm such as the one shown here. If a hive becomes too crowded, a new queen bee is produced. The old queen gathers together a group of bees, and they move away to start a new hive. Does it matter how far away the bees move from their old home? Will other species of organisms living in the area of the new home site compete with the bees for food? In this chapter, you will think about questions such as these. You will also see how whole populations of organisms can reproduce successfully without overcrowding the planet.

How many family members would it take to create a crowd in your home? For bees in this hive, the number is in the tens of thousands. But even bees have their limits.

Chapter Preview

Objectives

Examine the potential of organisms to reproduce in great numbers.

Compare the growth of populations when limited by density-dependent or by density-independent factors.

27.1 Growth of Populations

People who live in climate zones with four seasons expect new populations of living things to arrive every spring. Perhaps you, too, have watched dandelions sprout up in lawns or vacant lots in springtime. At first, there are just a few here and there. After a week or so, there may be more than you can easily count. As the summer months progress, you may see numerous grasshoppers and houseflies, or feel the sting of mosquitoes. At the same time you are observing—or experiencing—the dandelions and insects, you know that their days are limited. Their populations may grow rapidly for a while, but they will not continue to grow endlessly. They will, in fact, all disappear with a new change of season. Scientists of Charles Darwin's day took careful note of the fact that no population continues to grow unchecked. This realization helped Darwin form his ideas about natural selection.

Biotic Potential

How big is a population? The answer depends partly on where the organisms live. Suppose you discover a pond where 40 frogs are living. The frog population of the pond at this time is 40. Suppose you can count 75 cattails growing along the edges of the pond. The cattail population of the pond is 75. The frogs make up one population at the pond; the cattails make up another. There probably are populations of other kinds of organisms at the pond as well. A population is a group of organisms belonging to the same species and living within a certain area, such as the pond. The number of organisms in each group is the **population size.** As the organisms reproduce, the population size increases.

Look at the growth curve, shaped like the letter J, in Figure 27-2. It shows that under ideal conditions—unlimited food, absence of disease, lack of predators, favorable temperatures, and so on—the size of a population would increase indefinitely. Increase in the size of a population with time is called **population growth.**

Figure 27-1 **Dandelions seem to sprout up everywhere in the springtime. They produce huge numbers of seeds that are carried by air currents (right).**

Under ideal conditions, many species of bacteria reproduce by binary fission and double their population size every 20 minutes. If no environmental factors limited their survival, in 36 hours one bacterium could give rise to enough descendants to cover Earth with a layer of bacteria 30 cm deep! Even organisms that reproduce more slowly could produce many offspring under ideal conditions. Darwin estimated that, starting with one pair of elephants, there would be more than 19 million elephants in only 750 years. This highest rate of reproduction under ideal conditions is called a population's **biotic potential.**

Do most organisms reach their biotic potential? Obviously, they do not. If they did, Earth would be overrun with elephants, bacteria, cattails, grasshoppers, and other organisms. **Limiting factors,** which are circumstances that keep organisms from reaching their biotic potential, prevent such chaos. For many organisms, such as grasshoppers and mosquitoes, unfavorable temperatures that come with seasonal changes are limiting factors. Disease, predators, and lack of food or space are other common limiting factors.

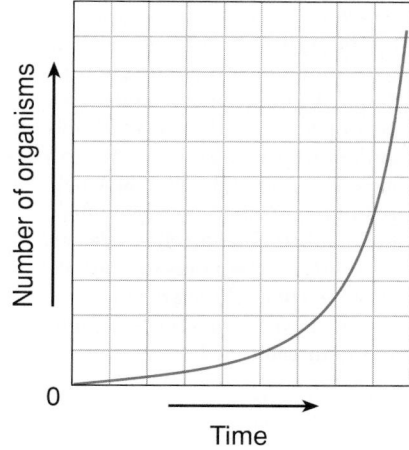

Figure 27-2 **Under ideal conditions, a population's size would continue to increase indefinitely. This increase can be shown graphically and resembles the letter *J*.**

Population Growth Curves

How do we know how populations grow and respond to their environment? Growth of populations can best be studied in the laboratory because conditions can be controlled. To study the population growth of yeast cells, a culture medium can be inoculated with a few yeast cells. Counting the yeast cells at regular time intervals would give data much like those shown in Table 27.1. Note that the size of the population increases slowly at first. This period of slow growth in population size is followed by a period of rapid

Table 27.1 Growth of a Yeast Population

Population Age (hours)	Number of Cells/mL	Increase in Cell Number/mL
0	10	
2	29	19
4	71	42
6	175	104
8	351	176
10	513	162
12	595	82
14	641	46
16	656	15
18	662	6

Figure 27-3 A yeast population can be grown under controlled laboratory conditions. When the growth is graphed, an S-shaped population growth curve results. Growth levels off at the carrying capacity of the environment. The liquid medium in which the yeast are grown becomes cloudier as the number of yeast cells increases. Each tube corresponds to a population at the portion of the growth curve shown above it. Data for this graph are shown in Table 27.1.

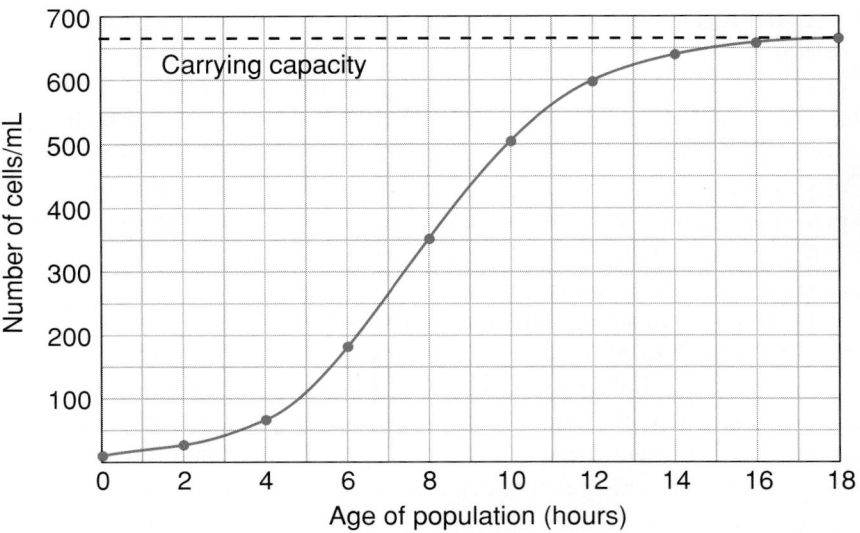

growth, followed by another period of slow growth. Eventually, the population stops increasing altogether. The population size remains stable after that point. Limiting factors such as available food supply or the buildup of wastes have determined that the yeast population will grow no larger.

If you were to graph the changes in the size of the yeast population, you would get an S-shaped curve, as in Figure 27-3. An S-shaped curve is also known as a **population growth curve.** The low end of the curve, on the left, indicates the early slow growth. The steep rise in the curve illustrates the period of rapid growth over a comparatively short span of time. The high end, on the right, indicates the final, stable population size. At this point, the population has stopped increasing in size.

The growth of animal populations shows a similar pattern. Imagine several families of raccoons newly relocated to a woodland where there were no raccoons before. If you could keep count of the raccoons over several years, you would find that the size of the raccoon population grows in much the same way as that of the yeast population. Eventually, the population would stop increasing in size and then would remain stable. Limiting factors such as available food supply, space, or predators have determined that the raccoon population will grow no larger.

If you were to make similar studies of deer, mice, snakes, different species of trees, shrubs, or other complex organisms in the woodland, your population growth curves would also follow an S-shaped pattern. The limiting factors might vary from population to population, but eventually they would check each population's growth.

Population Growth Rate Curve   You could also use graphs to show how fast the yeast population and other populations grow. Such a graph, as seen in Figure 27-4, is called a **population growth rate curve.** It demonstrates that, up to a point, the size of a population increases more and more rapidly. Then the rate of increase slows down. The *number* of organisms still increases, but the *rate* at which new organisms are added to the population decreases. When the population growth levels off, the population growth rate approaches zero. To picture this another way, think about the acceleration of a car. As the car pulls away from a stop sign, it moves slowly at first. Then the speed rapidly increases, and more distance is covered in less and less time. As the car approaches the speed limit, its speed continues to increase, but not as quickly as before. The car continues to move; eventually, its speed no longer increases. Instead, it remains steady at the speed limit.

Carrying Capacity   When a population arrives at the point where its size is no longer increasing, it has reached the carrying capacity of the environment. The **carrying capacity** is the greatest number of individuals in a certain population that a given environment—such as a pond or a woodland—is capable of supporting under a specific set of conditions. The carrying capacity of an environment can vary with the time of year as conditions change. For example, as the weather becomes colder in winter, food supply may diminish and the environment will not be able to support as many organisms. For plant populations, the carrying capacity may decrease in summer as water supply or access to sunlight becomes limited. At the carrying capacity, the number of organisms born or produced in a given period of time, the **birthrate,** balances the number of organisms that die during that time, the **death rate.** At this point, the size of the population remains fairly stable.

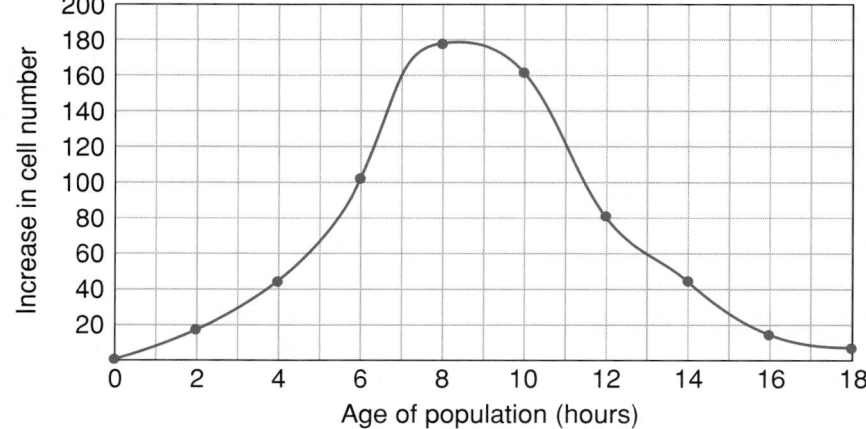

Figure 27-4   **The growth rate of a population can be graphed. This growth rate curve is for the same yeast population whose growth is graphed in Figure 27-3. The growth rate data are shown in the right-hand column of Table 27.1.**

Density-dependent Limiting Factors

Suppose you wanted to predict population growth among a group of organisms, perhaps among the frogs in a pond. You would need to know as much as possible about the limiting factors of the frogs' environment. Limiting factors for organisms whose growth pattern takes an S-shaped curve, as is the case with frogs, are almost always density dependent. This means that the influence of any limiting factor varies with the population density. **Population density** refers to the size of a population that occupies a given area at any given point in time. A small population of frogs living in a given pond has a low population density. A large population living in the same pond has a high population density.

The growth of many populations decreases as density increases. Thus, whatever limiting factor is acting on the population—possibly a shortage in the food supply—is tied to population density. The greater the density becomes, the more vigorously the limiting factor slows down population

I N V E S T I G A T I O N

Limiting Factors for Duckweed Growth

During the summer, you may have noticed green mats of plants on the surface of ponds or in areas of slowly moving streams. Some of this green material was probably algae, but some may have been duckweed—the smallest flowering plant in the world. Its tiny, round "leaves" are actually the stems of the plant. They can be as small as 1 mm in diameter. Duckweed reproduces asexually with the daughter plants clinging to the parent plant. As growth progresses, the plant bodies separate, producing new individuals.

Conditions for growth must be optimal if plants are to grow well. For example, water temperature and pH must be in a narrow range that the plants can tolerate, light must be available, and the water must contain sufficient minerals. If any of these factors are limiting, plants will not grow well.

Duckweed

Problem

What are the limiting factors for duckweed growth?

Safety Precautions

Be sure to wash your hands at the end of each lab session in which you count or handle duckweed.

Hypothesis

Make a hypothesis about what might be a limiting factor for duckweed growth. Explain your reasons for making this hypothesis.

Experimental Plan

1. As a group, make a list of possible ways you might test your hypothesis using the materials your teacher has made available.

growth. Gradually, the population density approaches the carrying capacity of the environment. After the density reaches the carrying capacity, birthrate and death rate become about equal; population growth levels off.

Density-independent Limiting Factors

Organisms such as the ladybugs shown in Figure 27-5 show a different pattern of population growth. The size of the population does not level off in a smooth S shape. Rather, the curve showing the growth of these organisms begins in a way that resembles the J-shaped curve in Figure 27-2. First there is a rapid increase in population size, also similar to the early stages of S-shaped growth but occurring at a higher rate. But then the population crashes, which means that it rapidly declines rather than levels off, and the population size never reaches the carrying capacity of the environment. The factor limiting the growth of the ladybug

Figure 27-5 **This dense population of ladybugs probably will never reach the carrying capacity of the environment. Why might this be so?**

2. Agree on one idea from your group's list that could be investigated in the classroom.
3. Design an experiment that will test one variable at a time. Plan to collect quantitative data.
4. Following the style of a recipe, write a numbered list of directions that anyone could follow.
5. Make a list of materials and the quantities you will need.

Checking the Plan

Discuss the following points with other group members to decide on the final procedures for your experiment.
1. What variables will need to be controlled?
2. What is your control?
3. What will you measure or count?
4. How many setups will you make?
5. How many groups of duckweed plants will you start with? How many plants will be in a group?

6. How will you raise the plants? What kinds of containers will you keep them in? Will you supply any nutrients?
7. How long will your experiment run?
8. Have you designed and made a table for collecting data?

Make sure your teacher has approved your experimental plan before you proceed further.

Data and Observations

Carry out your experiment, make your measurements, and complete your data table. Make a graph of your results.

Analysis and Conclusions

1. Under what conditions did your duckweed reproduce faster?
2. What was your experimental evidence for this conclusion?
3. Was your hypothesis supported by your data?

4. If your hypothesis was not supported, what types of experimental error could be responsible?
5. Consult with another group that used a different procedure to examine this problem. What are its conclusions?
6. What does your experimental evidence tell you about limiting factors for duckweed growth?
7. What other limiting factors might affect duckweed growth?

Going Further

Based on this lab experience, design an experiment that would help to answer a question that arose from your work. Assume that you will have unlimited resources and will not be conducting this lab in class.

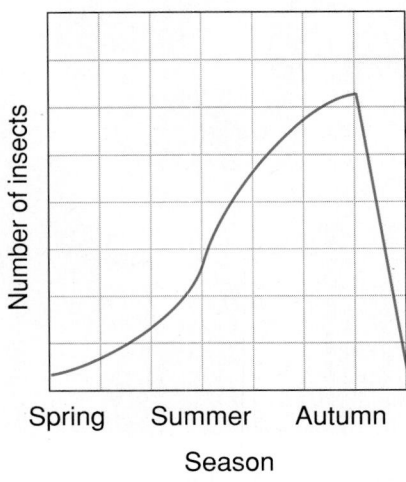

Figure 27-6 An insect population in a temperate climate, such as the ladybugs shown on the previous page, increases each spring and summer and dies rapidly with the freezing weather of late fall. Temperature is therefore a density-independent limiting factor.

population is, thus, not related to population density and is said to be density independent. Density-independent factors, such as temperature and oxygen supply, are often related to physical aspects of the environment.

Temperature is a physical factor that frequently limits the size of insect populations. In temperate zones, the population size begins to grow slowly in late spring and then more rapidly through the summer. Then, when cold weather begins in late autumn, population size declines rapidly as adults die. The environment could support further population growth, but the population declines before that point is reached. However, the population is not destroyed by the weather. Before they die, the insects deposit their fertilized eggs in a weather-resistant area. The following spring, the young insects emerge and begin another population explosion.

Oxygen can be a physical limiting factor in aquatic communities. Oxygen is needed for cellular respiration by almost all organisms. If the oxygen level drops too low, many animals that have a high demand for oxygen cannot survive. Decline in oxygen content of water can be caused by the decomposition of polluting organic wastes or by an increase in water temperature. More oxygen can dissolve in cold water than in warm water. Sometimes, hot water is released into lakes or rivers from factories or power plants. The hot water raises the temperature of the water to which it is added, causing the oxygen content to be lowered. Lack of oxygen becomes a limiting factor. As fish and other organisms die, their overall populations decline.

Section Review

Understanding Concepts

1. Suppose you were to observe the growth of a population of amoebas in a very large container. You provide unlimited numbers of smaller microorganisms for food, the water contains plenty of dissolved oxygen, and all other conditions are favorable for growth. Describe the growth of such a population.

2. Now suppose you repeat the experiment in question 1, but this time you limit the amount of food organisms. Assuming all other factors are the same as in the first experiment, describe the growth of this population of amoebas.

3. Suppose you were conducting the experiment as in question 1, but after several weeks you add a chemical that the amoebas cannot

tolerate. Describe the growth of the population up to and after the addition of the chemical.

4. **Skill Review—Making and Using Graphs:** The following table shows the growth of a population of unicellular organisms.

Time (hours)	Number of Organisms/mL
0	17
4	41
8	89
12	202
16	418
20	510
24	530
28	537
32	542
36	544

Prepare a graph of these data. What kind of factor is limiting the population's growth? Explain. For more help, refer to Organizing Information in the **Skill Handbook**.

5. **Thinking Critically:** Which kinds of populations, those limited by density-dependent or by density-independent factors, are more likely to be made up of individuals that reproduce quickly and produce many offspring? Why?

27.2 Regulation of Population Size

You have learned that density-independent limiting factors, such as climate and seasonal temperature change, have the same effect on a population regardless of its size. A large population of houseflies will die in the autumn cold just as surely as a small one will. In contrast, the effect of density-dependent limiting factors varies with the size of the population. The larger the population size, the more intense a density-dependent factor will be.

While density-independent limiting factors tend to be rooted in the physical environment, density-dependent factors most often involve interactions among organisms, and so they are biological. A population may be limited by availability of food, or by other organisms that feed or prey upon it or cause disease. Members of a population may compete with one another for some important resource in the environment. Organisms belonging to different populations may also compete. In any case, populations responding to density-dependent factors can attain a maximum population size that matches the carrying capacity of the environment.

Predation and Food

Almost all organisms are food for other organisms. **Predation,** the feeding of one organism on another, can be a factor limiting the size of a population. Often it is hard to tell, however, if predation is a limiting factor. For example, the Canadian lynx preys on the snowshoe hare. At about ten-year intervals, the sizes of both populations rise and fall almost simultaneously. Does this mean that the lynx regularly overkill their prey? It has been observed that snowshoe hare populations on islands where the lynx does not live show similar cycles. Perhaps the hare's food supply or another limiting factor periodically increases and decreases. If so, both lynx and hare populations would increase and decrease with it. Or, perhaps the lynx

Objectives

Predict how predation and food supply may interact to limit two populations.

Discuss the roles of parasitism and disease as density-dependent limiting factors.

Distinguish between interspecific and intraspecific competition, and identify ways in which both may be reduced.

Analyze the growth of the human population.

Figure 27-7 As the prey population rises and falls in number, so does the predator population, but with a slight lag. These data come from records of hare and lynx pelts sold by trappers to the Hudson Bay Company in northern Canada.

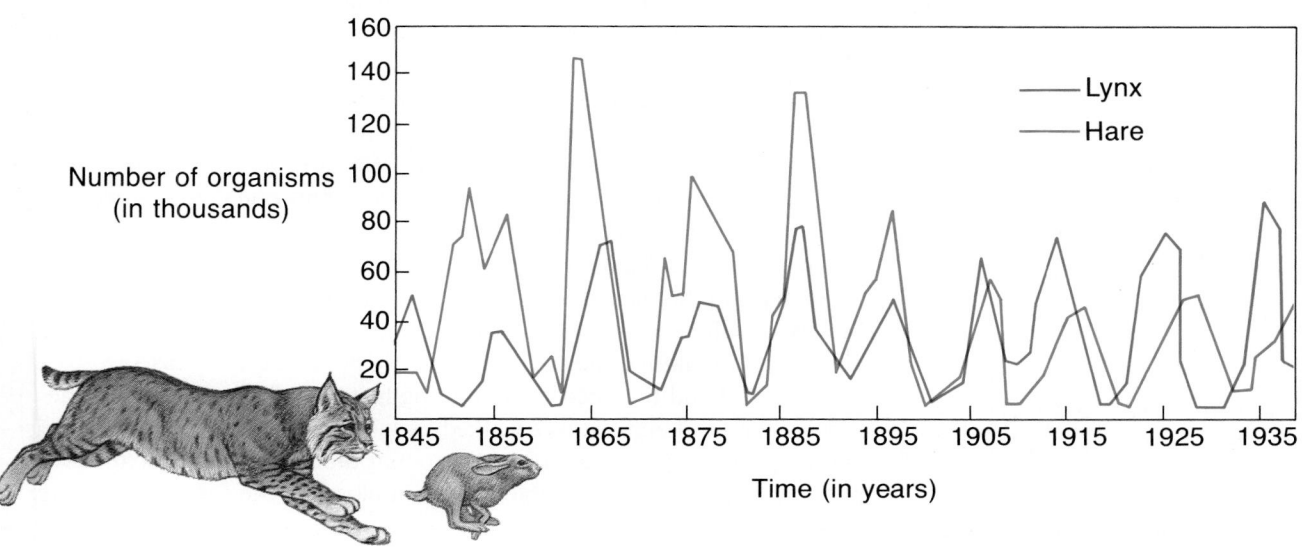

Number of organisms (in thousands)

Time (in years)

Figure 27-8 **Eagles are predators of fish. What would you expect to happen to a fish population in this river if the eagle population declined?**

undergoes a regular cycle, independent of the hares but affecting them. The situation is complex, yet it is clear that the size of most populations is at least partly checked by predators.

Predation as a Density-dependent Factor How may predation be density dependent? Consider a prey population such as rabbits. The greater the number of rabbits in a given area, the more likely it is that a predator, perhaps a coyote, will find and kill one. As the density of the prey population begins to fall as a result of predation, fewer prey are killed.

Most often, a balance is reached in a predator-prey relationship. The size of one population affects the size of the other. An increase in size of the prey population leads to an increase in size of the predator population because more food is available for the predators. Later, the size of the prey population begins to fall. Soon, because less food is available, fewer predators survive, and the size of the predator population falls. Thus, with fewer predators in its environment, the prey population will rise again.

The predator-prey interaction creates a cycle of population increase and decrease in both populations. Notice in Figure 27-7 on the previous page that changes in the size of the predator population follow those in the prey population. In a typical predator-prey relationship, predation is the limiting factor for the prey population, and the availability of food is the limiting factor for the predator population.

Predation as Beneficial Predation is healthy for a prey population. Very often, the prey caught and killed are those that are very young, very old, or less capable of coping with the environment. Thus, the healthiest, best-adapted individuals are most likely to survive, reproduce, and pass their genes on to the next generation. In this sense, predation is a form of natural selection that benefits the population as a whole.

Predation may also maintain prey population size very near the carrying capacity while other factors would not. Suppose, for example, predators were removed from the environment. At first, in the absence of predators, the size of the prey population would probably increase, exceeding the carrying capacity. However, food might then become a limiting factor. The unusually large number of prey organisms might over-browse, eating plants more quickly than the plants could reproduce. Eventually, almost all of the animals would starve to death, and the population size would decline to a point well below the carrying capacity.

Such a case happened in a moose population on an island in Lake Superior. Before wolves lived on the island, many moose starved. In the absence of predators, the size of the moose population increased unchecked. Food then became a limiting factor, and the majority of the moose starved. Later, wolves walked across the frozen lake to the island and began to prey on the moose. Since that time, the size of the moose population has remained stable at a point closer to the carrying capacity.

Parasitism

When you think of a predator stalking prey, you probably think of a larger or stronger animal, such as a lion, attacking a smaller or gentler animal, such as a zebra. There are many instances, however, in which smaller— much smaller—organisms attack organisms much larger than they are. These kinds of harmful organisms are called parasites. Parasites include such organisms as fleas and leeches, which live on another organism called the host. Parasites also include many organisms, such as tapeworms and flukes, that live inside host organisms.

Parasitism can be a limiting factor for population size. Almost all organisms have some parasites. Parasites usually harm the host but typically do not kill it, because the death of the host would result in the death of the parasites. However, if a host has too many parasites, the host might die. Dogs, for example, may die from the effects of harboring large numbers of heartworms.

Parasitism is a density-dependent factor. The possibility that different kinds of parasites will be passed from one organism to the next is greater in a dense host population. As the rate of transfer increases, the number and kinds of parasites in each host also increase. Too many parasites on one host may cause death by interfering with the host's nutrition or immune responses. The parasites may also reduce fertility of the host. As a result, the size and density of the host population decreases. Host organisms are then more widely spaced, and transfer of parasites is less frequent. Because hosts have fewer parasites, the host population can survive and increase in number again. Thus, the cycle continues.

Figure 27-9 The head of a tapeworm is adapted with hooks and suckers for grasping the inner lining of the host's intestine. Magnification: 80×

Thinking Lab

DRAW A CONCLUSION

Why do newts survive their parasites?

Background
Why is it that parasites haven't killed everything on the planet? This question can be answered in part by examining the way in which parasites are transmitted from one host to another. If a parasite kills its host, it must have a method of transportation for moving to a new host. A biting fly or mosquito would serve this purpose. If a parasite has no way to get to another host except by contact with that host, then the parasite must not kill the first host. The host must

be permitted to live and associate with other, potential host organisms.

Red-spotted newts often are infected with a parasite that is transmitted by the bite of a leech. Therefore, infection can be transmitted only in ponds, the leech's habitat. At the end of their first summer, young newts leave the pond and spend about six years maturing before they return to the pond to reproduce. When the newts are analyzed for parasites, it is found that they do not suffer from any adverse effects—even

when their blood contains a million parasites per milliliter of blood. This number is about the same as the number of red blood cells in the blood of a newt!

You Try It
In general, parasites that kill their hosts have intermediate hosts such as insects or leeches. Why are the newts, with a leech as intermediate host, not killed by their parasites? Analyze the information and **draw a conclusion.** Hint: Think about the newt's life cycle.

Disease and Populations

Population density is closely related to the spread of certain diseases. For example, malaria is a disease caused by the sporozoan protist *Plasmodium*, which is carried by female *Anopheles* mosquitoes and transferred to humans by the bites of these mosquitoes. Malaria is a density-dependent factor because the denser the human population, the greater the chance of the disease being passed to others. The spread of malaria is also dependent on the density of the mosquito population. Without these mosquitoes, malaria could not be transmitted, even if people stood cheek-to-cheek.

Disease can control the size of animal populations. In a dense population, disease can spread rapidly and kill many organisms in a matter of days. Sometimes people use this fact as a way to control oversized animal populations. For example, rabbits are not native to Australia; they were brought there from Europe by humans. The result was a disaster. Because there were no effective native predators or competitors at the time, the size of the rabbit population increased rapidly. Soon rabbits began competing with sheep for food. In order to reduce the rabbit population and preserve the sheep population, biologists injected some of the rabbits with a virus from South America. The virus was transferred directly from rabbit to rabbit, and rapidly killed large numbers of the animals because most had no immunity to the disease. As the number and density of rabbits decreased, the rate of disease transfer also decreased. A few rabbits were resistant enough to the virus to survive the disease. Because resistance is an inherited trait, the rabbit population became more resistant through natural selection. At the same time, there was selection for less virulent viruses, which had greater success at being transmitted by hosts that survived. Today, Australia's rabbit population is stable.

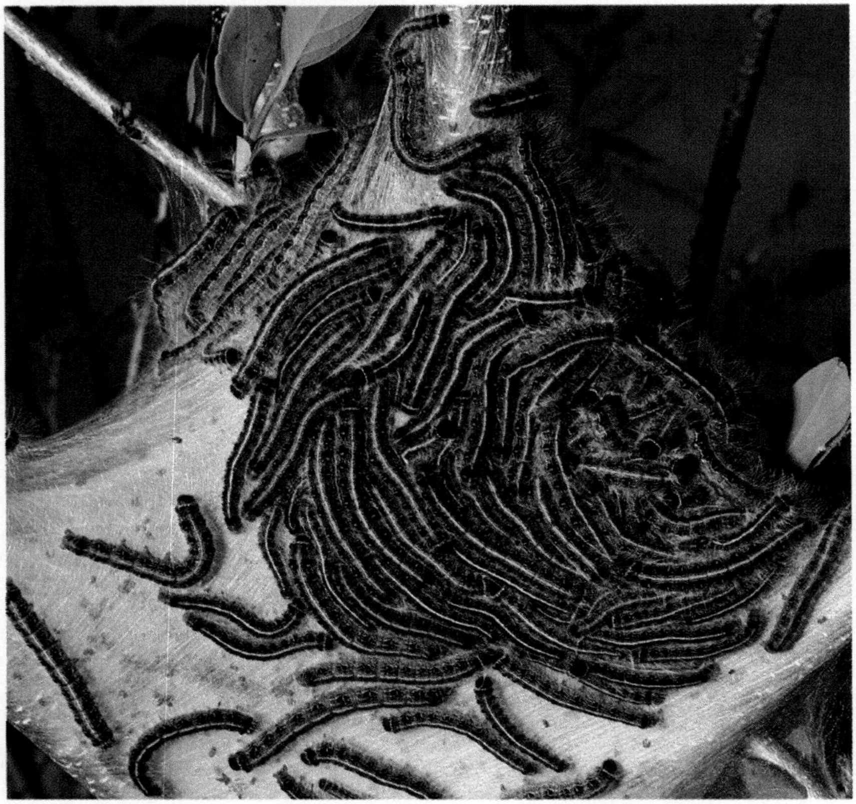

Killing pests by using disease-causing organisms must be carefully controlled. Care must be taken that only the pest population is affected. If other organisms are killed, the balance of nature could be upset, causing even more harm.

Figure 27-10 **Large populations of caterpillars, such as these tent caterpillars, can be controlled by spraying them with bacteria that cause the caterpillars to become sick and die. Other types of organisms, such as birds and humans, are not affected by these bacteria.**

Interspecific Competition

As you have perhaps observed, a number of different species usually occupy the same area at the same time. How many species of birds, for instance, or rodents such as squirrels or field mice, can you spot in your locality? Now suppose a food source shared by several populations in a given area becomes limited. What happens? The populations of different organisms compete for it, and the competition itself may become a limiting factor. Competition among populations of different species is called **interspecific competition.** As a limiting factor, it is density dependent.

Extinction When two or more species living in the same environment compete for the same resources, one of several different events may occur. One possibility is that one of the species will die out and become extinct in that environment. This event was demonstrated in a classic lab experiment using two species of paramecia. First, *Paramecium caudatum* and *Paramecium aurelia* were grown in two separate culture dishes. Food was kept at a constant level in both cultures by adding a species of bacterium that both paramecium species use for food. Wastes were removed regularly so they did not accumulate. The growth of each *Paramecium* population was measured and recorded, Figure 27-11 a and b. Then the two species of paramecia were placed in the same culture dish with a constant limiting supply of the same kind of bacteria as the food source. Wastes were removed by washing the cultures. This was done to ensure that competition for the fixed amount of food was the limiting factor. In the competition that followed, *Paramecium aurelia* proved better adapted to rapid growth and so trapped a larger share of the limited food supply. In the end, *Paramecium aurelia* survived, but *Paramecium caudatum* died.

Figure 27-11 **Population growth curves show that *Paramecium aurelia* (a) grows to a higher density than *Paramecium caudatum* (b) before leveling off. In this experiment, when the two species are grown together, interspecific competition occurs (c). *P. aurelia* survives while *P. caudatum* dies.**

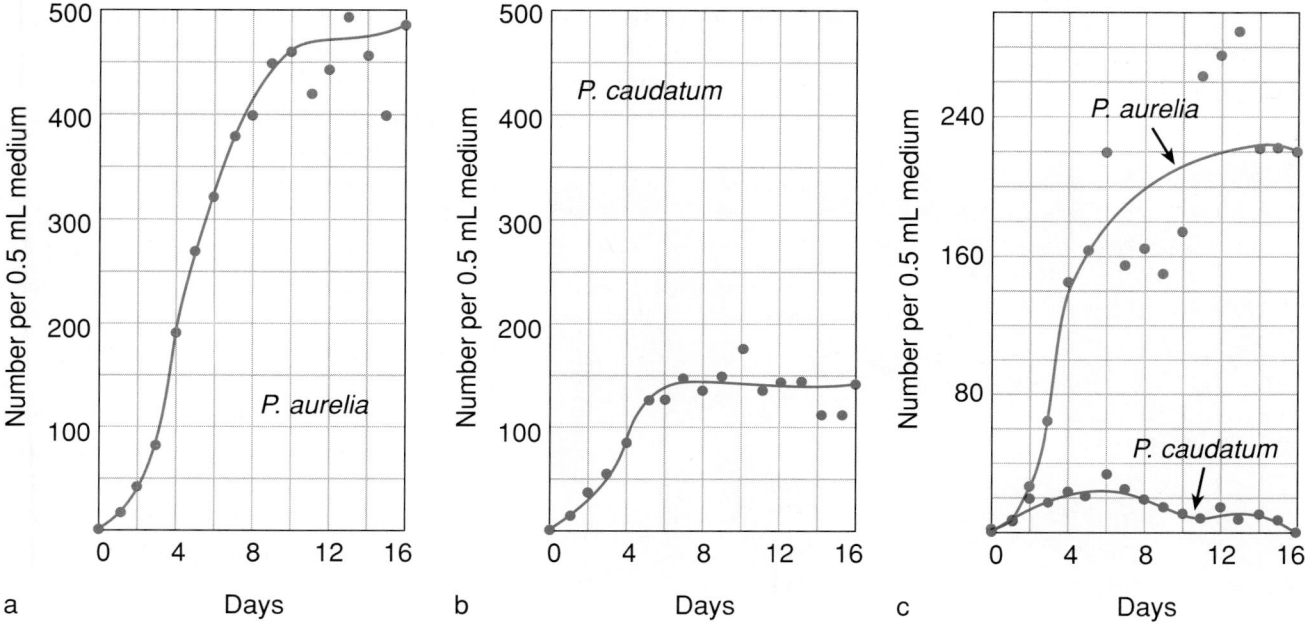

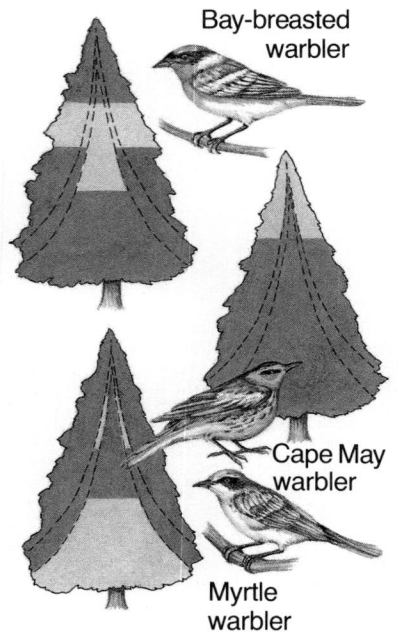

Bay-breasted warbler

Cape May warbler

Myrtle warbler

Figure 27-12 **Several species of warblers are able to feed and nest in the same kind of spruce tree without competing because they live in different parts of the tree.**

Figure 27-13 **Animals disperse seeds to distant locations, where the new plants will not compete with the parent plant. Can you think of an adaptation that would help the seed cling to the animal's fur?**

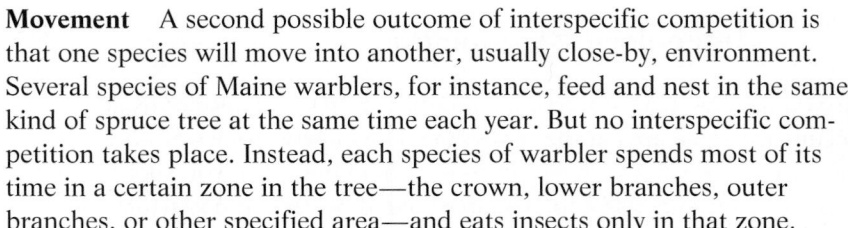

Movement   A second possible outcome of interspecific competition is that one species will move into another, usually close-by, environment. Several species of Maine warblers, for instance, feed and nest in the same kind of spruce tree at the same time each year. But no interspecific competition takes place. Instead, each species of warbler spends most of its time in a certain zone in the tree—the crown, lower branches, outer branches, or other specified area—and eats insects only in that zone.

Adaptation   Third, interspecific competition may spur rapid evolution in which species acquire different traits. As a result, the populations no longer compete with each other. You may recall that several of Darwin's finches live on the ground and eat seeds. But the size and shape of the different birds' beaks provide each species with an adaptation that allows its members to feed on seeds of different sizes and shapes. The insect-eating finches also have different kinds of beaks—ones that allow the birds to pick out insects from different kinds of places. Hence, interspecific competition loses its limiting effect.

Intraspecific Competition

Often, competition occurs between members of the same species. This competition is called **intraspecific competition.** It is more severe than interspecific competition because all members of the species have the same requirements. Intraspecific competition also is density dependent. The greater the population density, the greater the chance for competition because there are more contacts among individual organisms. Some degree of competition is desirable for a species because, through natural selection, the best adapted individuals survive to reproduce. However, too much competition is not desirable because it could lead to a drastic decline in population size.

Life Cycles and Life Spans   Excess intraspecific competition is avoided in many ways. Life cycles may be such that competition is avoided within a species. For example, adult frogs do not compete with tadpoles because the habitats and foods of frogs and tadpoles are different. Life spans also may reduce competition. In many animal species, such as insects, adults die shortly after the young are produced. Thus, old and young do not compete for food, space, or other factors. In animal species in which the parent does not die after the young are produced, parental care is common. This avoids competition because the offspring depend on the parents. In plants and fungi, dispersal of seeds or spores to distant locations by wind and animals also prevents competition among members of the same species.

Dominance and Social Hierarchy   Social behavior can reduce intraspecific conflicts. Consider, for example, how levels of authority reduce conflicts and preserve order in certain human groups, such as the military or a business. A designated chain of command, based on dominance, eliminates confusion and promotes group efficiency. Behavior patterns based on dominance and chains of command are characteristic of some animal populations. Such a pattern is known as a pecking order or **social hierarchy.**

The term *pecking order* grew out of studies of dominance relationships among chickens. A society of chickens forms a hierarchy by pecking at each other. Two chickens approach each other and may fight, or one may peacefully give way to the other, Figure 27-14. In either case, a dominance relationship results between these two chickens.

Dominance leads to a system of authority relationships spread throughout a society. Contacts similar to the one between the two chickens just described occur throughout a flock, and a total hierarchy emerges. One chicken becomes dominant to all the others. Another chicken may be dominant to all but the first one, and so on. Once the order is decided, it remains constant unless new chickens join the flock. Then the entire pecking order must be reestablished. Sometimes a chicken with lower status challenges a more dominant one, but such encounters rarely change the pecking order. This type of social hierarchy occurs in many groups of animals such as birds, some reptiles, and mammals, particularly wolves, dogs, and a variety of primates.

A hierarchy promotes survival in several ways. The overall result of dominance is increased order and reduced aggression throughout the society. After the first pecking is over, the less dominant individual accepts its status, and there is less fighting over food and mates. If conditions are such that not every individual can eat and mate, then at least the more dominant ones will be able to do so. These members are usually the better

Figure 27-14   **Chickens form a pecking order, a social hierarchy in which more dominant individuals peck at less dominant individuals.**

Figure 27-15   **Some animals engage in aggressive behaviors in order to establish a social hierarchy. Once established, the hierarchy reduces conflict.**

Minilab 2

How does crowding affect plant growth?

Radishes are your favorite garden vegetable. You want to plant enough so that you will have radishes for many months of salads, but you have only a small garden patch. If you plant the radish seeds very close together, will you get as many good-quality radishes as you will if you follow package directions? In one cup of potting soil, plant five radish seeds according to package directions. In another cup with the same amount and type of soil, plant 30 seeds. Keep the soil moist and place the cups in a well-lighted area. Measure growth at the end of one week and at the end of two weeks.

Analysis   *What is the effect of crowding on the growth of radish seeds? Speculate about why crowding is a limiting factor for plant growth.*

adapted. Thus, a social hierarchy is good insurance that the most adaptive genes are the ones passed on. A social hierarchy also prevents overbreeding because the most dominant individuals may prevent the lower individuals from mating. In any event, the pecking order reduces competition among the members of a population. Therefore, it is adaptive to the total group of animals and promotes group survival.

Role Separation   Competition is also reduced in societies in which members have definite roles. In insect societies such as those of bees, ants, and termites, roles are both genetically and physiologically determined. Usually, there is one queen that produces all new members of the society. Other members may have roles as workers, soldiers, or foragers. Among termites, the relative densities of workers, soldiers, and foragers seem to determine what role an individual will play. Spreading out the work reduces conflict and competition among the members of the society.

Behavioral and Physiological Changes   Overcrowding can lead to behavioral and physiological changes among members of the same species. If rats are overcrowded and cannot escape, they may become increasingly aggressive to the point where more deaths than usual occur. They may build poorly constructed nests or none at all. Hormonal changes resulting from stress may reduce litter size. In some cases, no young are produced. In other cases, the young are abandoned or eaten by their parents.

As a result of these behavioral and physiological changes, birthrate decreases and death rate increases. The outcome is a reduction in the size of the population to a number that the environment can support.

Emigration   Another solution to overcrowding in some populations is emigration, the moving out of an area. You read in the chapter opener that bees swarm to new areas when the hive becomes overcrowded. Not all the bees in the hive emigrate, however. Some remain in their original environment with a new queen and are able to survive under conditions of reduced population size and density.

Figure 27-16   **Howler monkeys establish and protect their territory by howling loudly when an intruder approaches. This practice ensures that each group of monkeys will have sufficient living space.**

Territoriality Intraspecific competition among some animals may be held in check as animals occupy and defend specific territories. This behavior, known as **territoriality,** was first observed in birds. Among many bird species, the song of the male announces that he has established his territory. The song tells other males to stay away and informs females of the species that a male has staked out his territory. Howler monkeys, another territorial species, travel in clans. They establish and then defend their territory against rival howler clans by howling and bluffing attacks against the rivals. Rarely is anything more serious needed to preserve a clan's territory.

Like a social hierarchy, possession of territories reduces conflicts within a species. It spreads the members of a population over a large area, thus ensuring a better food supply for all members. In addition, it is usually the strongest, largest, most aggressive, and otherwise best-adapted members of the species that win a territory and mate. Thus, their genes are the ones most likely to be passed on.

The Human Population—Without Limits?

If you examine Figure 27-18, you will see that the human population grew very slowly at first. Even after the year A.D. 1, it took 1600 years for the number of humans to double. After the year 1650, however, the rate of doubling rapidly increased. About this time, the rapid development of science, technology, and industrialization led to an explosion of the population. The population doubled between the years 1650 and 1850, then doubled again by 1930. By 1990, the world's population was estimated at 5.3 billion. Projections suggest that it will be 6.3 billion by the end of this century and 10.8 billion by the year 2050. Thus, the current population will double in roughly 50 years. The graph of human population growth closely resembles the graph of population growth under ideal conditions—the J-shaped curve you saw in Figure 27-2. Such a pattern, remember, occurs in the absence of limiting factors.

Figure 27-17 **Much of the world's human population is crowded together in cities.**

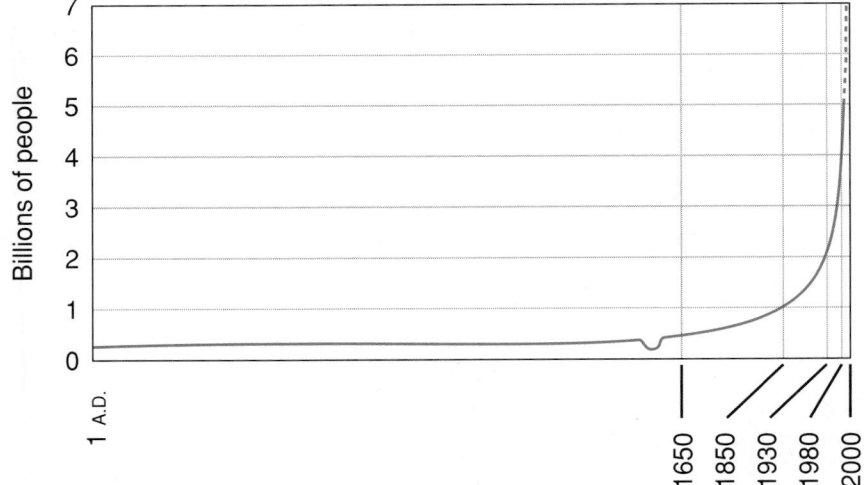

Figure 27-18 **Worldwide, human population growth has risen dramatically since the middle of the nineteenth century. The dotted portion of the curve represents projected future growth. Note the dip in the curve, due to the death of 75 million Europeans by bubonic plague between 1347 and 1351.**

Declining Growth Rate   Percentages are often used to illustrate the population growth rate pattern. To arrive at such percentages, think of the birthrate as the number of births each year for every 1000 persons. The death rate is the number of deaths each year for every 1000 persons. To obtain a growth rate, figure the difference between the two numbers and express it as a percentage of 1000. For example, think of a population where the birthrate is 30. The death rate is 10. The difference between the two figures is 20. The population growth rate, then, is 2 percent because 20 is 2 percent of 1000. A growth rate of 2 percent is high. At that rate, a population would double its size about every 35 years.

Given the predictions for human population growth in the next 50 years, you might conclude that the rate of growth will continue to increase. That is not the case, however. Since the 1970s, the worldwide growth rate has actually been decreasing. In 1989, the rate of growth was 1.8 percent, and it is expected to fall to 1.3 percent by 2025. Despite lowered growth rates, the human population will continue to increase in size. It will just grow more slowly than in recent centuries.

Uneven Growth Rates Worldwide   Declines in population growth rates are expected to occur in almost all countries of the world. But the rates will vary. The steepest declines are expected among developed countries—those that are highly industrialized and able to provide healthful, comfortable living conditions for their people. Such countries can expect their growth rates to drop from a present rate of about 0.5 percent to about 0.3 percent. Some industrialized countries, such as Germany and

LITERATURE CONNECTION

Rafael Jesús González
. . . .

Needing to borrow the sharp eyes of youth to help her make her embroidered flowers and birds come to colorful life, the old woman always sought out Rafael. The young man, a pre-med college student, worked in a five-and-dime store on the Mexico-Texas border. The pair's frequent consultation over the rainbow-hued display of thread drew them into a special friendship. Then, one day the old woman stopped coming to the store. Missing her visits, Rafael wrote this poem, *To an Old Woman*, to honor his older friend.

To an Old Woman
Come mother—
  your rebozo trails a black web
  and your hem catches on your
  heels;
you lean the burden of your years
on shaky cane and palsied hand
pushes
  sweat-grimed pennies on the
  counter.
Can you still see, old woman,
the darting color-trailed needle of
your trade?

The flowers you embroider
with three-for-a-dime threads
cannot fade as quickly as the leaves
of time.
  What things do you remember?
Your mouth seems to be forever
tasting
the residue of nectar-hearted years.
  Where are the sons you bore?
Do they speak only English now
and say they're Spanish?
  One day I know you will not
  come
  and ask for me to pick
  the colors you can no longer see.
I know I'll wait in vain
  for your toothless benediction.

POPULATION BIOLOGY

Figure 27-19 Many people in developing countries struggle to grow enough food to feed their families. In developed countries, however, surplus food is grown by modern agricultural practices.

Denmark, have already begun to experience declines in not only growth rate, but also population size. The rest of Europe and Japan will probably see the same kinds of decreases over the next 50 years.

The outlook is much different in the world's developing countries. Developing countries are those in which industrialization is only in its early stages. Many of their people, a large portion of whom try to grow their own food on small farms, have trouble meeting their needs. In these countries, the present growth rate of about 1.9 percent is expected to

I'll look into the dusty street made cool by pigeons' wings until a dirty child will nudge me and say:

"Señor, how much ees thees?"

As he grew up on the Mexico-Texas border, Rafael Jesús González (1935-   ) was surrounded by older family members who were highly valued and respected. Moving further into the United States, he was saddened to find such a division between age groups. He expressed this contrast in his poem. The last line of the poem completes the contrast of the values of the old woman and the more materialistic values of the young.

Today, Rafael wears a denim jacket with a beautiful embroidered flower, heart, and yin-yang symbol. It's a piece of art that he created. He says that he certainly couldn't have taken up an embroidery needle when he was young because friends his age would have teased him for doing what they considered women's work. "Hopefully, when we get older, we can see through nonsense and can understand that some strictures make no sense," he explains. "I would also hope that old age brings increased patience, a more life-accepting set of values, and the ability to love more wisely and fully."

━━━━

1. **Issue:** What activities in American culture seem to be restricted by gender or age? Why do you think this has happened?
2. **Explore Further:** Conduct an informal opinion survey in your class, school, or neighborhood. Ask how people define *old*.

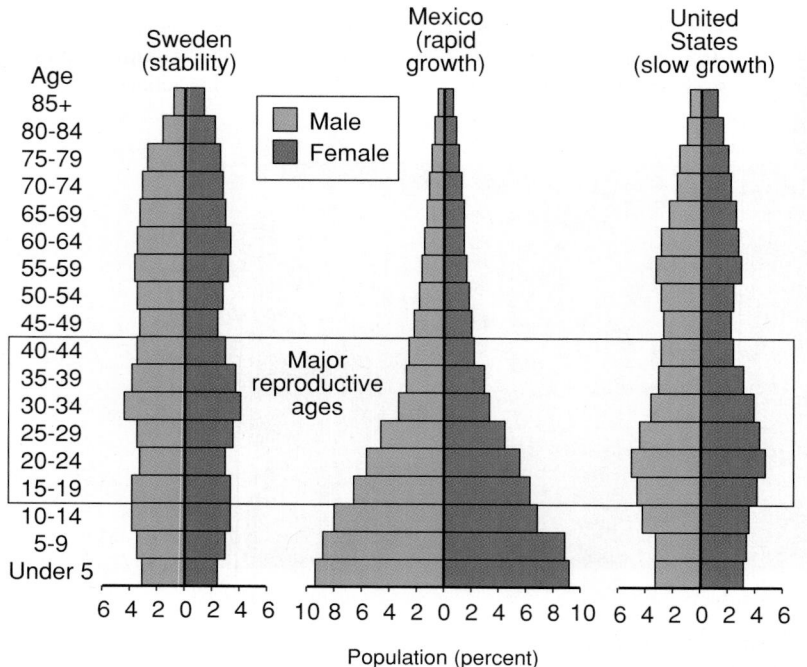

Age

| Sweden (stability) | Mexico (rapid growth) | United States (slow growth) |

85+
80-84
75-79
70-74
65-69
60-64
55-59
50-54
45-49
40-44
35-39
30-34
25-29
20-24
15-19
10-14
5-9
Under 5

Male
Female

Major reproductive ages

6 4 2 0 2 4 6   10 8 6 4 2 0 2 4 6 8 10   6 4 2 0 2 4 6

Population (percent)

Figure 27-20 **The proportion of females of reproductive age in a population has a significant impact on the future growth of that population.**

drop to about 1.5 percent. Even 1.5 percent is still a substantial rate of growth. Two factors account for it. First, birthrates in developing countries are high, as much as four times higher than in developed countries. Second, a very large segment of the total population in developing countries is of reproductive age, Figure 27-20. Worldwide, the increases in population sizes of developing countries will more than make up for the decreases in developed countries. The total world population, therefore, will continue to grow.

Another way to study population growth is to look at the number of births among women of reproductive age. For example, the birthrate in the United States has dropped from a peak of nearly 3.8 births per woman in the late 1950s to 1.84 by the late 1980s. However, the size of the population still increased. A rate of 2.1 births per woman for many generations is necessary to attain **zero population growth,** a condition in which the birthrate equals the death rate and the rate of population growth equals zero. Because a large proportion of the United States population is now of reproductive age, even a birthrate of 1.84 births per woman—well below 2.1—would continue to cause significant increases in population size in the years ahead. According to U.S. Census figures, the United States population was 248.7 million in 1990. It is projected to grow to 268.1 million by the year 2000 and to 312.7 million by 2025. At this rate, the population in the United States will double in fewer than 100 years.

Future of the Population Explosion

What has caused the human population explosion—especially in the last several hundred years? It has not happened because the birthrate steadily increased. You learned in Chapter 13 that agriculture and technology are unique attributes of the human population. You also saw that recent human progress stems not from biological evolution, but from cultural evolution. Cultural advances have greatly improved the quality of life among many individuals. They have also greatly impacted the growth of the human population, mainly by reducing the death rate. Consider the importance of sanitation, for instance. Bubonic plague is a bacterial disease carried by fleas that hitchhike on rats. In the 14th century, one-fourth of the population of Europe died of that disease. Later, improved sanitation eliminated most rats, which reduced people's contact with fleas and, therefore, the incidence of the plague.

So far, cultural evolution has placed the human population beyond the environmental factors that limit all other populations. But the human population cannot continue to expand indefinitely. It, too, is subject to a point at which the environment cannot support additional growth. There is also the chance that a density-independent factor could take effect. Based upon what you've learned, what might happen if the human population reaches carrying capacity? What could happen if a density-independent factor, such as an earthquake or volcanic eruption, intervenes?

Most natural checks of population growth cause an increase in the death rate. One alternative for controlling population growth of humans would be to decrease the birthrate. This can be accomplished as part of family planning methods. Biologists and physicians have developed several artificial ways of preventing pregnancy based on the biological aspects of reproduction. Natural methods that prevent pregnancy have also been studied.

Many countries around the world educate their citizens about family planning. Which, if any, family planning method is used depends on physical, emotional, religious, and moral factors. The results of this education may be a factor in the recently declining growth rates in some countries.

Figure 27-21 Human populations are affected by density-independent factors such as earthquakes.

LOOKING AHEAD

THE GROWTH OF populations is affected by environmental factors, both living and nonliving. In the next chapter, you will look at the bigger picture as you explore factors that influence the functioning of entire communities. You will study ways in which different populations interact with one another, how populations respond to the availability of energy, and how materials are continually made available to organisms.

Section Review

Understanding Concepts

1. A population of insects is fed upon by spiders. What change in the insect population will cause a decline in the size of the spider population? Why? What will then happen to the size of the insect population? Why? What is the limiting factor for each population?

2. Explain why clearing land and planting a large area with a single crop makes that crop subject to destruction by parasites.

3. Why did the human population size not begin to increase rapidly until after 1650?

4. **Skill Review—Comparing and Contrasting:** Caterpillars of two different species of insects both feed on leaves. Members of a population of birds compete for the same nesting sites. Compare the two types of competition and mechanisms for avoiding each type of competition. For more help, refer to Thinking Critically in the **Skill Handbook.**

5. **Thinking Critically:** What method might prove most effective in preventing the spread of malaria?

BIOLOGY, TECHNOLOGY, AND SOCIETY

Issues

Old Versus Young in America?

Take a look at your parents. They don't look much like babies, do they? But to demographers, scientists who study population trends, your parents will always be "baby boomers." That's the name tag that has been pinned on all people born in the United States between 1947 and 1961. In those years, the number of babies born seemed to explode with a resounding boom.

An Aging Population   The baby boomers make a "bulge" in the population of the United States. When they started school, they crowded the classrooms. Entering the work force, they composed a group to which marketers eagerly geared their products. And, of course, baby boomers will grow old together. The number of Americans age 65 and over is expected to leap beginning in the year 2010.

Better health care and nutrition account for some of the growth in the over-65 category. In 1900, the average American could expect to live to 47. Today, the average American will quite likely have a 75th birthday party. By the time the baby boomers reach old age, they may have an even longer life span.

With their professional responsibilities accomplished and their children grown, older Americans have

more leisure time to develop their minds and their imaginations. Although modern technology has caused enormous changes in lifestyles, most older Americans have been able to keep up through education, both formal and informal. The number of people over 65 who take classes in everything from computer science to painting has never been higher.

Many older Americans disregard the traditional retirement age of 65 and continue to work. They may continue their careers or even enter entirely new job areas.

Many older Americans are the backbone of volunteer organizations. These people are generous in giving their time to run soup kitchens, help people learn to read,

provide child care and hospital services, and so much more.

Problems of an Aging Population
An aging population could bring some difficult problems for the United States. Although today's older Americans are the healthiest in history, they do inevitably require more health care. Can America afford health care for both ends of the age spectrum?

Another area of concern is the Social Security system. Older Americans paid into the system during their working years and deserve to draw the benefits after they retire. However, that system depends on a balance between working people and retired people. In 1985, that ratio was 3.4 to 1; by 2030, it is projected to decline to 2 to 1. That means there will be fewer people in the work force supporting more people outside it.

Even in everyday life, an aging population may require some accommodation. For example, traffic signals might have to stay green longer for older pedestrians, and the soundtracks of movies may need to be turned up.

Clearly, changes will have to be made to accommodate our aging population. Older Americans are too valuable a resource to be ignored.

Understanding the Issue

1. What do you see as advantages and disadvantages of old age?
2. Find out about Americans who have made important contributions in their older years.

Readings

- Ames, K. "Cheaper by the Dozen." *Newsweek*, Sept. 14, 1992, pp. 52-53.
- Baltes, Paul B. "Wise, and Otherwise." *Natural History*, Feb. 1992, pp. 50-51.

Considering the Environment

Pile It On!

The Great Wall of China snakes across 2400 kilometers of hilly terrain. It averages about 8 meters in height and 4 meters in width. The Wall has been for centuries the largest human construction on Earth. However, it may lose that distinction in the next few years. Every day in New York, another structure grows a little larger. It's not a wall or a beautiful building—it's a garbage dump! The 3000-acre Fresh Kills landfill will soon rise to a height of almost 170 meters, making it the tallest "mountain" on the east coast of the United States.

The Fresh Kills landfill is only one of 7300 dumps in the United States, about half of which are nearing capacity. As the human population continues to grow, city officials are asking where they are going to put all the trash.

Mulch   There is a large component of landfills that many people can reuse directly. That's the material that consists of yard wastes—leaves, twigs, and lawn clippings. When people take care of their yards, they often throw these materials into the garbage can. Then they spend money on mulch to control weeds and retain soil moisture. And what is mulch made of? Leaves, twigs, and lawn clippings—the same

materials they just discarded! Some people have added chippers and shredders to their yard-care equipment to make their own mulch.

Compost   Another resource that you can save from landfills and turn into garden "gold," or compost, is organic kitchen scraps. You don't want all of it; meat, bones, or anything oily won't do. Fruit and vegetable peelings, coffee grounds, teabags, and eggshells are good materials for compost. So are small materials from your yard, such as leaves and grass clippings.

You can buy or make a container for your compost pile, or you can just start a pile in an out-of-the-way spot. To begin, make a pile of kitchen and garden leftovers. When the pile is about one meter high and equally wide, buy some nitrogen fertilizer, which will speed the

breakdown of your compost. Add a handful or two of fertilizer, sprinkle the compost pile with water, and cover it with black plastic anchored with rocks.

Wait about a week. Then remove the cover and stir up the pile. You'll probably find that the middle is warm. That's because the materials are decomposing. Add a little more water, just enough so that the material is as damp as a squeezed-out sponge. Put the cover back on, and repeat the process every week. In a month or two, the compost will look dark and crumbly. Then you can dig it into your garden plot. You will have done two good deeds at once—improved the soil and saved space in a landfill.

1. **Explore Further:** Where does your garbage go? Call a city official and find out. Try to learn how much trash is dumped in your local landfill.
2. **Solve the Problem:** Scientists have discovered 20-year-old newspapers that are still readable and five-year-old hotdogs that are recognizable although certainly not appetizing. Why do you think these items haven't decomposed the way material in a compost pile does?

Summary

Under ideal conditions, a population would continue to expand indefinitely. However, environmental factors limit the growth of populations. Some of these factors may be related to the density of a population. The size of the population levels off at the carrying capacity of the environment. The size of populations limited by density-independent factors, often involving a sudden environmental change, never reaches the carrying capacity.

Density-dependent factors include availability of food, predation, disease and parasitism, and competition within or between species. Each of these factors becomes more severe as the number of organisms in the population increases.

Growth of the human population is unlike that of any other population. Advances due to cultural evolution have allowed humans to escape the effect of limiting factors, resulting in a decrease in the death rate and contributing to a population explosion. Even though the growth rate of the human population has recently begun a slight decline, the population size will continue to increase in the future because there are so many people of reproductive age. Decreasing the birthrate seems to be the most logical and least drastic means of reversing this trend.

Language of Biology

Write a sentence that shows your understanding of each of the following terms.

biotic potential	population growth
birthrate	population growth curve
carrying capacity	population growth rate
death rate	curve
interspecific competition	population size
intraspecific competition	predation
limiting factor	social hierarchy
parasitism	territoriality
population density	zero population growth

Understanding Concepts

1. Explain how an S-shaped curve and a population growth rate curve illustrate the same ideas in different ways.
2. Explain why, in terms of population density, population growth slows more and more as population size approaches the carrying capacity.
3. Why is there a similarity between the early stages of population growth that is limited by density-dependent factors and by density-independent factors?
4. Prior to development of the measles vaccine, that disease spread rapidly among young schoolchildren. Why?
5. Can legalized hunting be "good" for a population of animals? Explain.
6. Why do farmers find it necessary to spray their crops with pesticides? What might happen if they didn't? Why?
7. Will two species of amoebas with identical growth requirements both survive if they are placed in the same container in a lab? Explain.
8. Although some adult animals are sessile, their larvae are motile. How are motile larvae important with respect to competition?
9. Life expectancy in developed countries is far greater than in developing countries. Why is population growth greater in developing countries if people do not live as long?

10. Reduction in death rate among humans has been especially important with respect to infant mortality. Many more infants survive today, at least in developed countries, than in the past. How has reduction of infant mortality played a role in population growth?

Applying Concepts

11. Two species of flour beetles are introduced into a single jar of flour. What will happen? Why?

12. Two species of flour beetles are placed in a single container of flour to which some glass tubing has been added. One species is found to occupy the glass tubing and both species survive. What idea do these results show?

13. Assuming that a human female has a reproductive life span of 30 years, how many offspring could a couple hypothetically produce during that time? Assume one birth per year, with no multiple births. If half of all the offspring were females, how many children could be produced in the next generation? What do your answers indicate about the biotic potential of humans?

14. Choose a particular example of technology and explain how it has contributed to growth of the human population.

15. Suppose a few mice had invaded a new habitat and their population size had begun to grow. A flood then swept over the habitat of the mice. What kind of limiting factor is the flood? Why?

16. What kind of factor might act to limit the growth of a population of dandelions in a lawn? Explain.

17. What factor probably explains why desert plants of the same species are widely spaced? Explain.

18. Look again at the S-shaped curve in Figure 27-3. Suppose the graph represented a population of fish regularly caught by humans and that at a given point in time, the population was at the carrying capacity. How far should the fish population be allowed to fall in order to ensure that the population will once again grow to maximum size as quickly as possible? Explain.

19. In India, the number of people between 15 and 40 is more than twice the number of people of those ages in Sweden. Predict the future growth of the population of each of those countries.

20. **Issues:** What do you think caused the boom in the number of babies born beginning in 1947?

21. **Considering the Environment:** What makes compost such good fertilizer?

Lab Interpretation

22. **Investigation:** Growth of a bacterial species under different pH conditions is shown below. Examine the graph and determine if pH is a limiting factor for growth of this bacterium. Explain.

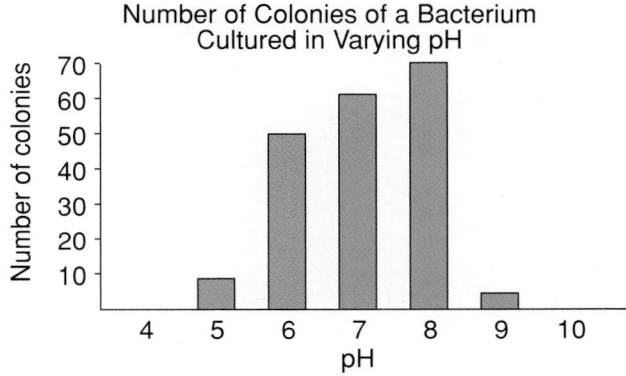

Number of Colonies of a Bacterium Cultured in Varying pH

23. **Minilab 1:** You examine remote sensing photos of the same farmland today and 10 years ago. In the most current photos, the crop fields are larger and contain only one type of crop. Ten years ago, the fields were smaller and had a variety of crops. Discuss the pros and cons of the conversion to larger fields and one type of crop.

24. **Minilab 2:** Might the results of this experiment have been different if two species of plant seeds had been planted together? Explain your answer.

Connecting Ideas

25. Explain how natural selection might lead to evolution of different traits among members of two similar species living in the same environment.

26. How can you explain the fact that many different mammals limited by density-dependent factors share the following characteristics: They usually live in stable environments, they reproduce only a few offspring at a time and do so slowly, and their offspring mature slowly and live a long time?

ECOSYSTEMS

HAVE YOU EVER wondered what those colorful, fuzzy patches of blue, red, brown, or green growing on damp trees and rocks are? What about the small, gray, shrublike organisms that sprout from the soil in all types of weather? Such organisms are called lichens, and they certainly rank among the most unusual of all living things. Lichens defy insertion into any general classification because they are actually dual organisms, an alga and a fungus. The typical lichen consists of a sac fungus and either a photosynthetic bacterium or a green alga, and their relationship represents one of the more classic examples of mutualism in nature. In this relationship, fungi provide crucial protection from the elements, while the photosynthetic bacteria or algae manufacture food that is used by both organisms. Together, the two components form a compact and highly efficient unit that may occupy even the most inhospitable habitats.

In nature, living organisms interact in a variety of ways. Sometimes, these relationships are surprising, as in the case of lichens. But more often, interactions between organisms are much simpler. In this chapter, you'll learn how groups of organisms interact and depend on each other for survival. You'll also consider how organisms are affected by physical factors in the environment.

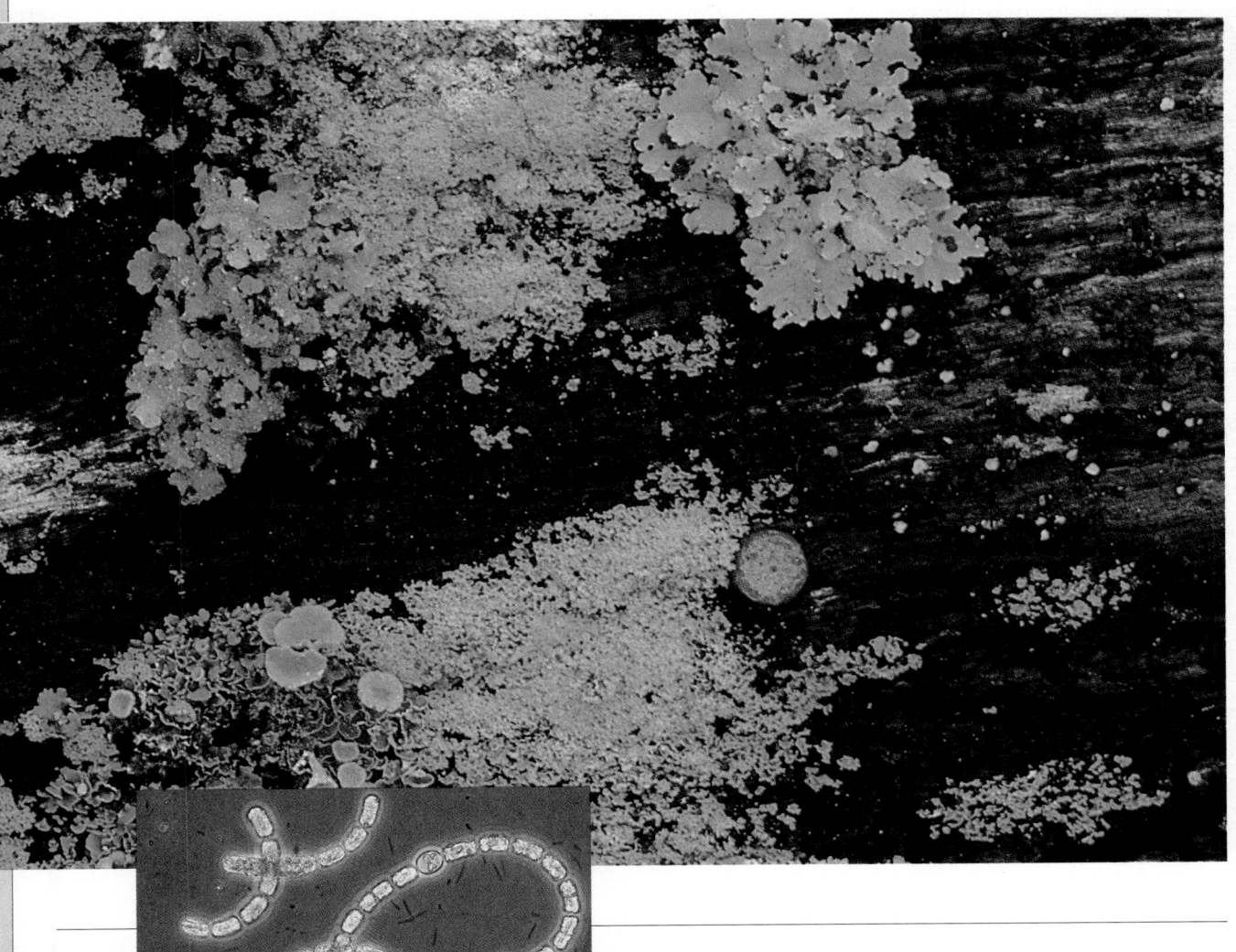

In communities, organisms depend on and interact with each other in ways similar to the way this alga depends on and interacts with the fungus. In what ways do the members of the community you live in interact like organisms in these environments?

Chapter Preview

Contents

Objectives

Trace the passage of energy from producers to consumers in an ecosystem.

Relate the flow of energy through an ecosystem to the pyramid of energy.

Analyze the importance of recycling of materials in an ecosystem and **describe** how such recycling occurs.

Distinguish among parasitism, commensalism, and mutualism.

28.1 Interactions Among Organisms

Can you picture a lake and think about the many living things that inhabit such a community? Around the edge live a variety of plants—grasses, cattails, and ferns, perhaps. The water itself is home to a great diversity of invertebrates such as worms, snails, hydras, and insect larvae. Vertebrates include a variety of fish, tadpoles and adult frogs, snakes, turtles, and maybe ducks or coots. Unseen are the many microscopic organisms present: bacteria and fungi, algae, and tiny crustaceans. In what ways do these many organisms interact with one another? How does each organism obtain the materials and energy it needs to remain alive? Do the organisms depend upon one another in ways other than as food sources? Could the community continue to exist if the diversity of life forms were altered in some way? Knowing answers to questions such as these will be important to you when you must make informed decisions concerning the environment.

Trophic Levels

Just as the homes, schools, and businesses in your town make up a community, populations of organisms living together in the same area make up a community. Within any community, the organisms interact with members of their own species and with those of other species. Such relationships between or among different organisms are known as **biotic factors** of the environment. These organisms interact with their physical environment as well. Interaction of a community with its environment is an ecological system or **ecosystem.**

Figure 28-1 This pond is an example of an ecological system or ecosystem. The microscopic plants and animals (left), pintail duck (top right), and sunfish (bottom right) all contribute to the balance of the pond ecosystem.

The interactions between a community of organisms and the environment are give-and-take relationships. Organisms take from their environment materials and energy needed for life, and, as they are used, they are transferred back to the environment. Because there is a tight relationship between a community of organisms and the environment, there is a delicate balance among the many forms of life in a healthy ecosystem.

How do organisms interact with each other? Perhaps the most obvious way in which different populations of organisms interact is by feeding upon one another. Because food is a source of energy, availability of food is a limiting factor for many animal populations. You are well aware by now that all organisms require energy in order to carry out their many functions. In almost all ecosystems, the original source of that energy is sunlight. The light can be used directly only by autotrophs such as grasses, trees, or algae. In photosynthesis, autotrophic organisms trap light energy and change it to chemical energy stored in the form of sugars and other organic compounds.

Because only photosynthetic autotrophs can trap and change light energy to chemical energy, all other organisms depend on them for their energy sources. Energy stored in energy-rich organic molecules is transferred from one organism to another as organisms feed upon each other within an ecosystem. In this way, each organism represents a feeding step, or **trophic level,** in the passage of energy and materials.

All organisms within an ecosystem have functions in that ecosystem. An organism's **niche** describes how an organism fits into an ecosystem. Where an organism lives, what it eats and what eats it, and how it interacts with all the biotic and abiotic parts of the ecosystem describe its niche. Since each organism plays a unique role, no two organisms can occupy the same niche in an ecosystem.

Trophic Levels   In order for a community to function, people depend on other people. How do you depend on work done by police? Mail carriers? The same is true for organisms within an ecosystem. In order for an ecosystem to function, organisms also depend upon each other. For an ecosystem to be successful, energy must first pass from producers to consumers. A consumer that feeds directly on a producer is called a **first-order consumer.** First-order consumers are also called **herbivores,** or plant eaters. A cow grazing on grass, a deer browsing on foliage, and a tadpole eating algae are examples of herbivores. Herbivores make up the second trophic level in ecosystems.

Consumers that feed on other consumers are called **carnivores,** or meat eaters. A mouse is preyed upon by snakes. Therefore, a snake is a carnivore and a **second-order consumer.** An owl that eats the snake is a **third-order consumer.** Second-order and third-order consumers make up higher trophic levels.

Decomposers are consumers as well. Protists and fungi are common types of decomposers. Scavengers such as vultures may prey on decomposers as well as feeding on dead animals. Decomposers break down and consume dead organisms and wastes at every trophic level. In the process, they recycle a large part of the broken-down substances back to the producers.

Food Chains   The transfer of materials and potential energy from organism to organism (trophic level to trophic level) forms a series called a food chain. A food chain is represented using arrows that indicate the direction in which materials and energy are transferred from one organism to another. Decomposers can function at any point in a food chain.

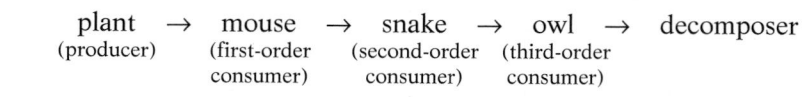

plant $\rightarrow$ mouse $\rightarrow$ snake $\rightarrow$ owl $\rightarrow$ decomposer
(producer)   (first-order   (second-order   (third-order
                consumer)   consumer)   consumer)

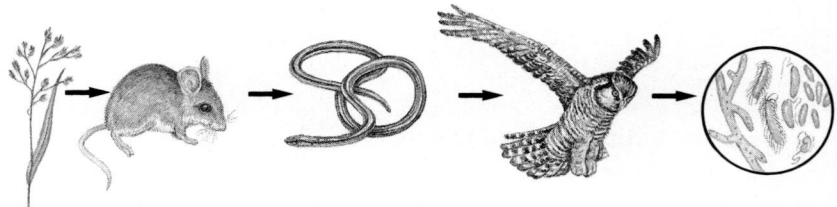

A food chain represents *one* possible route for the transfer of materials and energy through an ecosystem. Many other routes exist. In an ecosystem, many species are found at each trophic level, and one species does not always feed on the same food source. A snake may also eat a lizard or a toad; an owl may feed directly on a mouse. Can you think of any other possible food chains? In addition, some animals are **omnivores,** organisms that eat both plants and animals. Humans are omnivores. All the possible feeding relationships that exist in an ecosystem make up a **food web,** pictured in Figure 28-2.

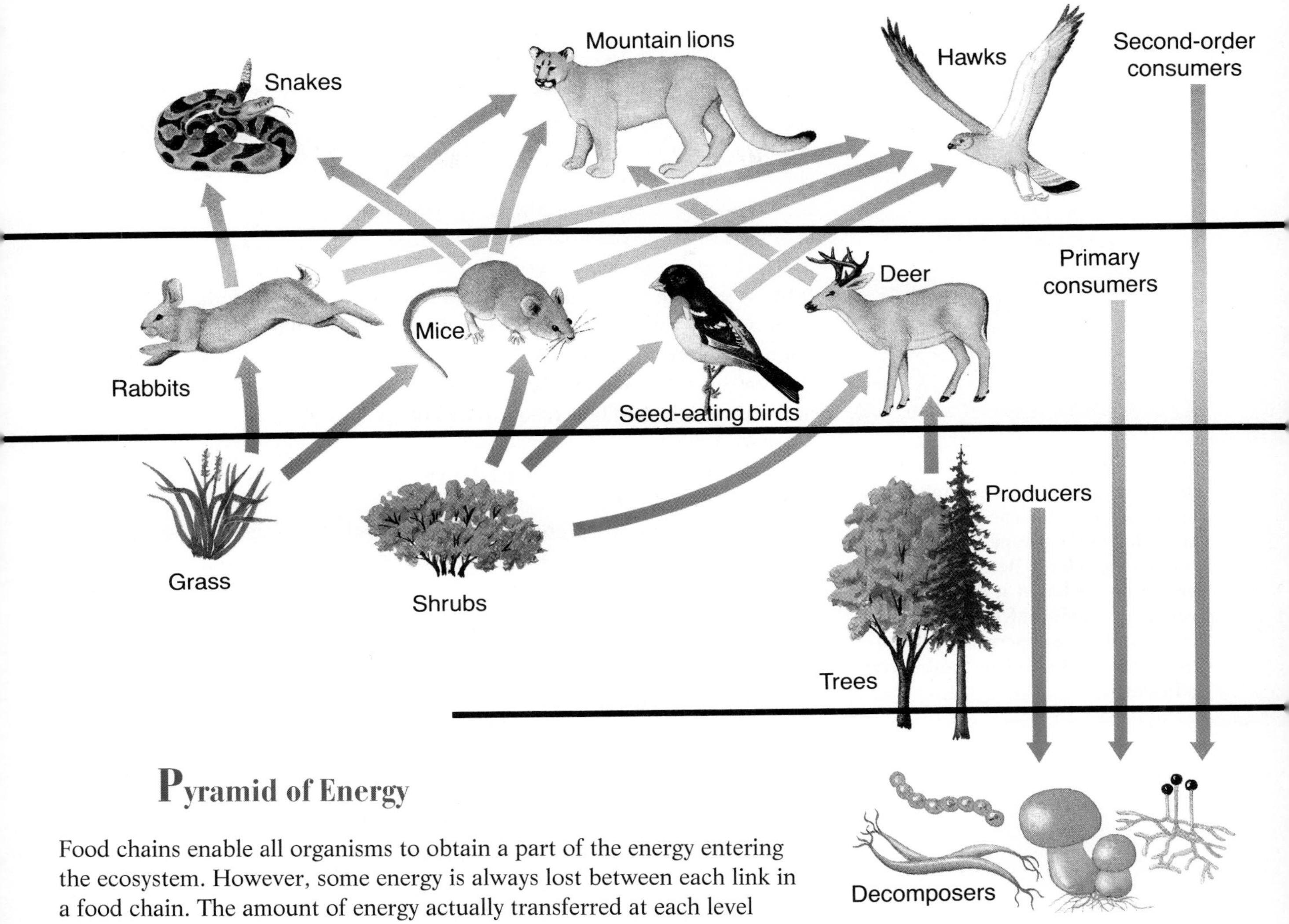

Snakes

Mountain lions

Hawks

Second-order
consumers

Rabbits

Mice

Seed-eating birds

Deer

Primary
consumers

Grass

Shrubs

Trees

Producers

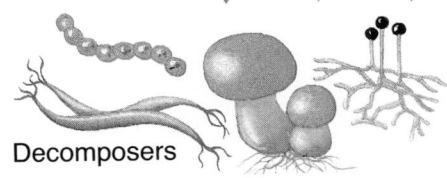

Decomposers

Pyramid of Energy

Food chains enable all organisms to obtain a part of the energy entering the ecosystem. However, some energy is always lost between each link in a food chain. The amount of energy actually transferred at each level depends on the ecosystem. Accurate data are difficult to obtain because of variables such as season, age and size of organisms, and amount of sunlight. A study of a river ecosystem in Silver Springs, Florida, was made by Howard Odum. It showed that only 16 percent of the potential energy in producers was transferred to herbivores. Second-order consumers obtained only 11 percent of the potential energy in herbivores. Only about 5 percent of the energy in second-order consumers was transferred to third-order consumers. The third-order consumers received only about 0.1 percent of the potential energy originally available in the producers. A graph of data such as these has the pyramid shape shown in Figure 28-3. Thus, ecologists refer to the transfer of energy as a **pyramid of energy.**

What happens to the energy originally produced? According to the law of conservation of energy, the energy is not lost—it still exists—but it is in a form that cannot be used again. Remember, some stored energy is used by the producers themselves for their life processes. Some energy is bound up in molecules the consumer cannot digest. More than half of the potential energy present in each food molecule is lost as heat energy during cellular respiration.

Figure 28-2 The food web pictured here describes some of the feeding relationships within an ecosystem. The arrows show the direction of energy flow. In a real ecosystem, many more plants and animals would be involved in the food web.

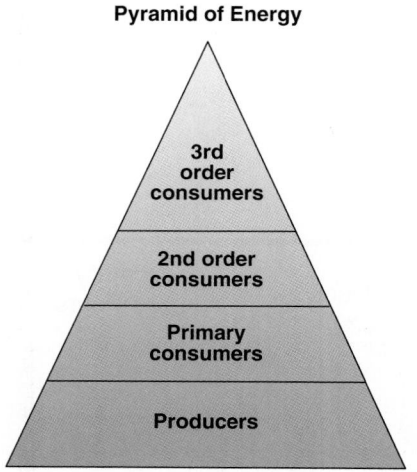

Pyramid of Energy

3rd order consumers

2nd order consumers

Primary consumers

Producers

Figure 28-3 **A pyramid can be drawn to show the energy levels within an ecosystem. Little energy is available at the top of the pyramid because some is lost at each level. What types of organisms are found at the top of the energy pyramid? Which trophic level contains the most energy? Why is eating low on the food chain most energy efficient?**

There are other reasons why the amount of energy available to each level decreases. Some organisms die and are acted on by decomposers. Also, members of higher trophic levels often don't consume every part of an organism. For example, only a small fraction of the organic material in a tree is consumed by herbivores. In addition, not all parts of the material eaten by an organism, such as bones or cellulose, are digested. For all these reasons, each link in the food chain has less potential energy available to it than the previous link.

The pyramid of energy also offers an explanation as to why organisms feed on particular organisms and not others. Lions prey on large animals because they can get more energy from a large animal than from a small animal. In fact, lions might use more energy trying to catch a rodent than they would get by eating the rodent. Therefore, lions hunt and eat zebras, and a single chase might result in an energy "profit" for the lion.

Pyramids of Numbers and Biomass

Loss of energy in a food chain explains several ecological principles. A food chain rarely includes more than four links. So much energy is lost at each trophic level that enough energy seldom remains to support fourth- and fifth-order consumers. The loss of energy also explains why there are usually fewer organisms in each higher trophic level than in the previous level. For example, there are more mice than there are weasels that feed on those mice. This relationship is known as the **pyramid of numbers.** Each trophic level has fewer organisms than the previous level. A pyramid of numbers shows the same general pattern as a pyramid of energy.

The pyramid of numbers does not apply to all food chains, especially where a large organism is fed upon by smaller ones. For example, a single tree might be food for thousands of caterpillars, and a dog might be infested with many parasites.

Decrease in energy along food chains suggests that the total amount of organic matter present at each trophic level should also decrease because energy is required to build organic matter. Research has shown that this

Table 28.1 **Biomass in an Ecosystem**

Trophic Level	Biomass (g/m²)	Percent of Biomass Compared to Previous Level
Producers	809	—
Herbivores	37	4.5
Second-order consumers	11	29.7
Third-order consumers	1.5	7.3

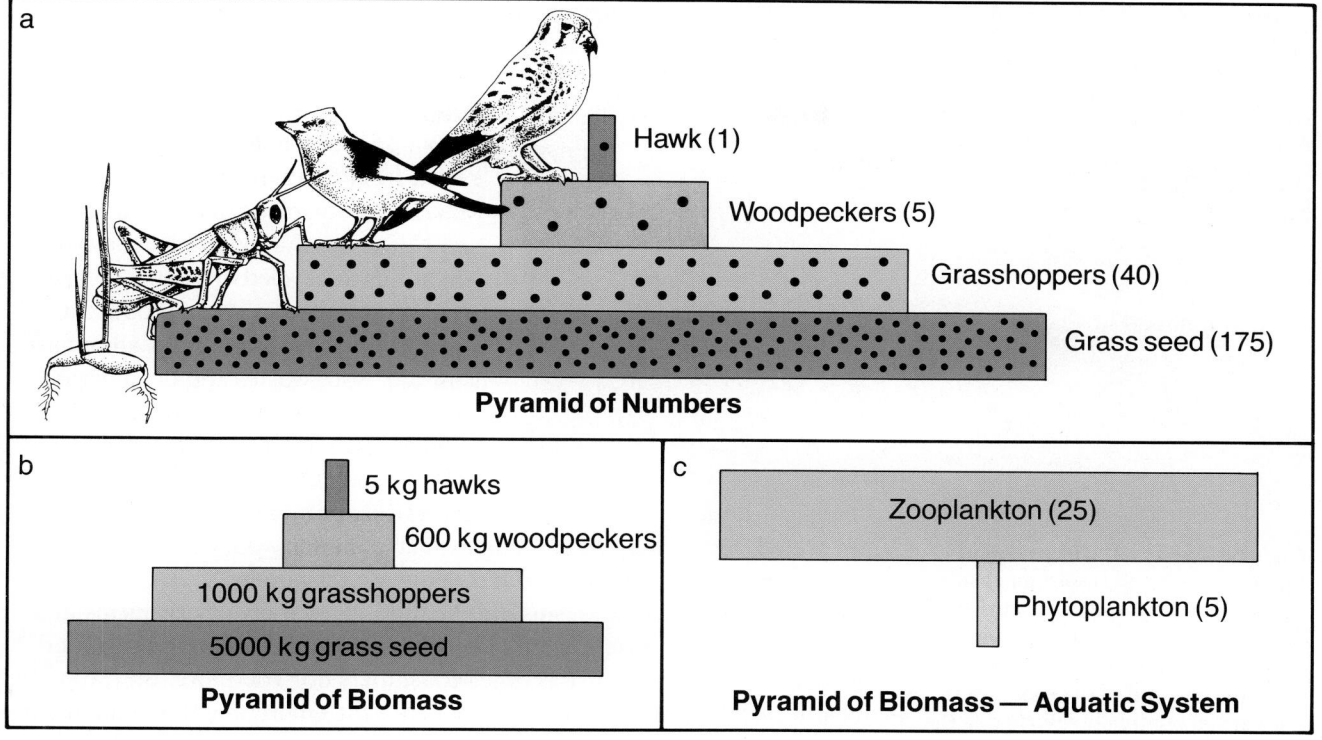

a

Hawk (1)

Woodpeckers (5)

Grasshoppers (40)

Grass seed (175)

Pyramid of Numbers

b

5 kg hawks

600 kg woodpeckers

1000 kg grasshoppers

5000 kg grass seed

Pyramid of Biomass

c

Zooplankton (25)

Phytoplankton (5)

Pyramid of Biomass — Aquatic System

prediction is true in some cases. Examine Table 28.1, which lists biomass data from the Silver Springs ecosystem. These data represent the total amount of dried weight of organic matter, called **biomass,** at different trophic levels. Do you see that there is indeed a decrease in total biomass? If data such as these are graphed, the familiar pyramid shape results, showing a **pyramid of biomass** for most ecosystems on land. The reverse is true, though, for aquatic ecosystems, because the producers are mainly microscopic algae, not large plants or trees. Though the total number of algae is great, their total biomass is quite small at any given time.

Figure 28-4 Pyramids of numbers (a) show how many individuals an ecosystem can support. Pyramids of biomass (b) show how much living matter an ecosystem can support. The inverted pyramid (c) shows that a relatively small biomass can support a larger biomass, such as a whale being supported by microscopic plants and animals.

Cycling of Materials

Although energy is not cycled in an ecosystem, the energy lost is constantly replaced by energy of sunlight. As a result, an ecosystem always has a new supply of energy. However, if loss of materials occurred, an ecosystem could not continue to survive, because the amount of matter is limited. No new matter is being made. Thus, matter must be recycled— the same materials used over and over again—if an ecosystem is to continue to function.

Look at Figure 28-5, which illustrates the carbon cycle. Almost every member of an ecosystem contributes to the cycling of carbon. Producers do so as they carry out both photosynthesis and aerobic respiration. Consumers do so as they undergo aerobic respiration. The products of aerobic respiration, carbon dioxide and water, are cycled to producers, which use those products to make sugars and other organic compounds in

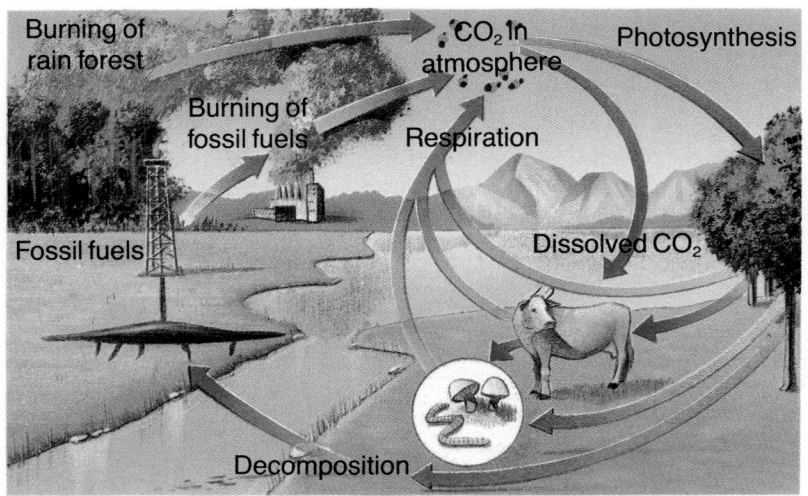

Burning of rain forest

Burning of fossil fuels

Fossil fuels

CO_2 in atmosphere

Photosynthesis

Respiration

Dissolved CO_2

Decomposition

Figure 28-5 The processes of photosynthesis and respiration cause carbon to cycle through the environment. As the diagram shows, some carbon can accumulate for years in the wood of trees and in the soil. How is this carbon eventually returned to the atmosphere?

photosynthesis. Most producers and consumers use those organic compounds, along with oxygen produced during photosynthesis, in aerobic respiration. Thus, carbon is continually cycled from inorganic form, carbon dioxide, to organic form, such as sugars. Note, too, that hydrogen and oxygen atoms are cycled as well.

As organisms live, they produce a variety of wastes. Eventually, all organisms die. Both wastes and dead organisms contain materials. Some are organic, such as carbohydrates, fats, and proteins. Others are inorganic, such as a variety of minerals. These materials are cycled, too. If they weren't, all available matter would one day be locked up in wastes or dead bodies. How are they cycled?

Wastes and dead organisms are the food of a variety of organisms in an ecosystem. Decomposers, such as many bacteria and fungi, use this food in cellular respiration, just as other consumers do. Therefore, they, too, release carbon dioxide and water back into the environment, making them available to producers. Various animals also cycle carbon as they feed upon dead organisms and wastes. Among them are insects, buzzards, lobsters, crabs, and jackals. Such animals are often called scavengers.

Carbon is also recycled in other ways. For example, the burning of fossil fuels such as oil, gasoline, and coal by humans releases large amounts of carbon dioxide into the environment. Carbon dioxide is also released during volcanic eruptions, and when carbonate ions present in rock are slowly changed in chemical reactions.

Many decomposers recycle materials other than carbon as well. Some break down nitrogen-containing organic molecules such as nucleic acids, proteins, and urea, converting them to inorganic ions. Others then convert those ions to other forms, such as nitrates, which are absorbed by producers. Producers use nitrates to manufacture the organic nitrogen-containing molecules mentioned above, which can then be passed along food chains. In addition to nitrogen, decomposers play a role in recycling of other materials, such as phosphorus and sulfur. Availability of these inorganic materials is crucial, and the supply of them may act to limit distribution of organisms and the size of their populations.

Other Biotic Relationships

Many relationships and interactions exist among organisms of different species. Some of these relationships are necessary to the survival of both organisms. Other relationships are of no consequence to either organism. In ecosystems, the most obvious biotic relationships involve feeding, but organisms also interact in other ways.

Commensalism   In some relationships, one organism benefits from a host without aiding or harming the host. Such a relationship is called **commensalism,** and an organism that benefits is called a commensal. For example, in dense tropical rain forests, the tree branches are so thick that the forest floor is dark. As a result, small plants cannot grow on the forest floor. Some small plants, such as orchids, live in the branches of trees and, thus, get enough light. Because of the high humidity in a rain forest, their roots absorb water from the air. Orchids benefit greatly in this relationship because they can receive ample sunlight. Large trees receive no benefit from this relationship, but they also are not harmed in any way.

The bromeliad, a member of the pineapple family, also lives in the branches of rain forest trees. Not only is a bromeliad a commensal, it is also a small ecosystem. The leaves of a bromeliad overlap to form a hollow in which water collects. Many arthropods and even tree frogs reproduce in the water. Some of these frogs live entirely in the treetops.

Animals exhibit commensalism also. A remora is a small fish that attaches itself to the belly of a larger animal, such as a shark. The shark provides a "free ride" for the remora, which feeds from the leftovers of the shark's meals. Are sharks harmed in any way from this relationship?

Mutualism   Often, both organisms in a relationship benefit in some way. As you know, this relationship is called mutualism. For example, bees pollinate plants and at the same time receive food in the form of nectar and pollen. Termites show another classic example of mutualism. Termites are able to obtain energy from wood because their intestines are home to protozoans that digest the cellulose of which the wood is composed. How do

Figure 28-6   **This orchid depends on a tree for a place to live. The tree neither benefits nor is harmed by this relationship.**

Thinking Lab

ANALYZE THE PROCEDURE

How do root nodule bacteria benefit clover plants?

Background
Scientists have known for years that clover roots contain nodules filled with bacteria. They thought that the bacteria were parasites and caused harm to clover. Then, when checking on this "parasite," they found that clover with bacteria grew better than when no bacteria were present.

You Try It
Analyze each experiment listed below. Explain why you think it would or would not be helpful in explaining why or how these bacteria improve clover growth. Explain what information, if

any, each experiment would give you. Describe the errors in experimental method you find. (1) Clover with bacterial nodules was grown in soils having different nutrients available. Clover was observed for differences in growth. (2) Clover was grown with and without nodules in beakers of water. Kinds of chemicals present in the water before and after several weeks were analyzed and compared. (3) Clover nodule bacteria were grown in culture medium at different temperatures. The optimum temperature for their growth was determined.

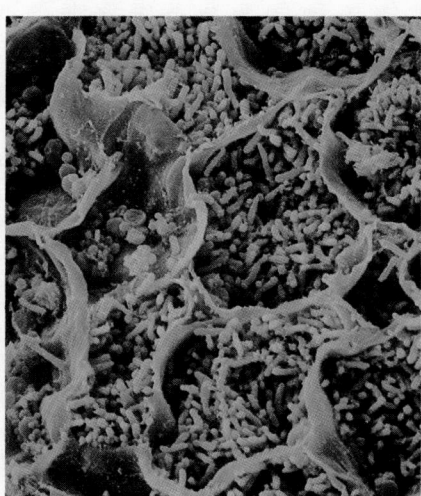

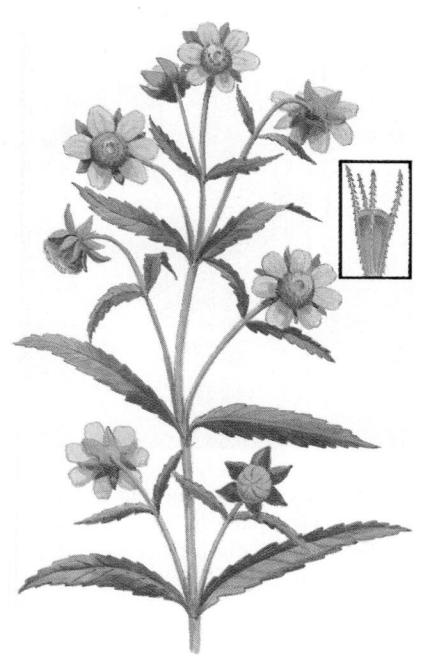

Figure 28-7 **Bur marigold seeds, commonly called sticktights, cling to fur, feathers, or even clothing. This plant depends on other members of the ecosystem to disperse its seeds.**

termites and protozoans benefit from this relationship? Another example of mutualism is that of the yucca plant and yucca moth. Female yucca moths transfer pollen from one yucca flower to another. These moths also deposit eggs in the developing seedpods of yucca plants. In this relationship, moths benefit because the eggs have a source of food when they hatch, and yucca plants benefit because pollination and reproduction of yucca plants is assured.

Cycling of nitrogen depends in part on a mutualistic relationship. The atmosphere consists of about 78 percent nitrogen gas by volume, but most organisms cannot use nitrogen in that form. Many forms of bacteria, though, can convert nitrogen gas to a usable form, a process known as nitrogen fixation. Many of those bacteria live in the roots of certain plants, especially legumes such as peas, clover, and alfalfa, where they break down dead plant material for food. These bacteria reside in the cortex cells of the roots and cause them to become swollen areas called nodules. Within the root cells, the bacteria convert nitrogen gas to ammonia. The plant uses the ammonia to make proteins and other nitrogen compounds. The bacteria benefit by having a place to live and a constant food supply.

Parasitism   Parasitism is another relationship between organisms. In parasitism, one organism, the parasite, is completely dependent at some point in its life cycle upon a host. Unlike mutualism and commensalism, the host is usually harmed in some way.

Besides these specific biotic relationships, many more relationships exist among organisms. Each ecological niche involves many obvious and not-so-obvious relationships among living or once-living organisms. For example, many species of birds build nests in trees. Rotten logs provide a place for some types of insect larvae. And some plants such as the bur marigold and tick trefoil depend on animals for seed dispersal. Think of some of the ways humans depend on other organisms.

Section Review

Understanding Concepts

1. A certain fish feeds upon tiny crustaceans in the ocean. The crustaceans feed upon algae. How do these feeding relationships provide energy to the organisms involved?

2. In a certain ecosystem, owls feed on field mice. Explain two reasons why the total energy present in field mice is not available to the owls.

3. Suppose a polluting chemical entered a lake and that it killed most bacteria but not other life forms. What effect would that chemical have on the lake ecosystem? Why?

4. **Skill Review—Making Tables:** Make a table that compares the effect on both organisms in each of the following relationships: mutualism, parasitism, commensalism. Use a "+" to indicate benefit, a "−" to indicate harm, and a "0" to indicate no effect. For more help, see Organizing Information in the **Skill Handbook.**

5. **Thinking Critically:** Consider a New England forest ecosystem in summer and winter. How would the number of organisms at each trophic level differ during those two seasons? Explain.

28.2 Abiotic Factors of the Environment

In the previous section, we focused on ways in which organisms interact. But organisms are affected not only by other organisms, but also by physical factors of the environment. Let's think again about a lake ecosystem. Why are plants such as grasses and cattails found only around the edges of the ecosystem? Why are the algae located only near the surface? Does the temperature of the water affect the kinds of animals that live there or the activity of those animals? How much oxygen is dissolved in the water, and how does that oxygen supply influence the organisms? Do different animals live at different depths? If so, why? Such questions can be answered by analyzing physical factors of an environment and their effects.

Water

Physical aspects of the environment are important in many ways. They can influence the distribution and range of organisms, as well as their reproduction, feeding, growth, and metabolism. Physical aspects of the environment, known as **abiotic factors,** can also limit the size of a population, as you learned in Chapter 27. Therefore, an entire ecosystem, composed of many different populations, can be affected by abiotic factors. Pinpointing the role of a particular abiotic factor is often difficult because many of those factors interact to affect the makeup and functions of the ecosystem. We'll take a look at a few of these factors separately, but keep in mind that they operate together.

All organisms need water. Water is important as a medium in which many organisms live. The organisms living in a pond or lake ecosystem, or in a river, stream, or ocean, are adapted for living in water and could not live in the dry medium of air. If you consider functions such as gas exchange, excretion, and locomotion, you'll understand why aquatic organisms could not carry out those functions on land.

Water is crucial for organisms in other ways. It is a major component of cytoplasm and of the fluids that bathe cells. It is also the substance in which numerous metabolic reactions take place.

Because water is so important for living organisms, availability of water can affect the distribution of land plants. Why? Plants require water for photosynthesis, and plants such as mosses and ferns

Objectives

Analyze the ways in which abiotic factors of an environment may influence the activities and distribution of organisms.

Predict the effects of changes on the balance of an ecosystem.

Figure 28-8 **Water moves in a cycle between organisms on land, the land itself, and the atmosphere. From where do you think the largest amount of evaporation of water occurs?**

Figure 28-9 **Grasslands cover about 25 percent of Earth's land surface and are North America's largest biome. Grasslands typically receive between 25 and 75 cm of rainfall each year.**

also need at least a thin film of water in which sperm cells can swim to eggs. The amount of water on land is controlled by rainfall. The amount of rainfall, in combination with temperature, determines the type of plant that will be dominant in a given area. For example, forests are common in areas receiving 75 cm or more of rain per year. However, in regions where rainfall is between 25 and 75 cm per year, grasslands are common. Water is scarce in desert regions, and the plants living in such ecosystems are those adapted to obtain and conserve the small amount of water that is available. Thus, water availability is a limiting factor. The types of plants in an area determine, in turn, the animal life in that area. For example, monkeys can live in a tropical forest, but not in a grassland.

You know that water is constantly cycled as organisms carry out cellular respiration. However, that process is just one aspect of water cycling. Examine Figure 28-8 on the previous page. It shows how physical events—evaporation, condensation, and transpiration (water loss from leaves)—are all important components of the water cycle.

ART CONNECTION

Abuelitos Piscando Napolitos

● ●

The painting *Abuelitos Piscando Napolitos* shows the harvest of prickly pear cactus fruits in the southwest. The artist, Carmen Lomas Garza, grew up in Kingsville, Texas, where her mother's family had lived for many generations, working as ranch hands and on the railroads. She inherited her artistic talent from her grandmother, a skilled needle-worker, and from her mother, a self-taught artist. In choosing a career as a professional artist, however, Garza has challenged the traditional roles that many Latin women have followed.

Abuelitos Piscando Napolitos is one of several works in which Garza shows her response to the Latino movement. In painting these pictures, Garza has said that she wanted to portray events that are meaningful to all Mexican-American culture.

The prickly pear cactus has long played an important part in the life of Mexico and the southwest. Fossil evidence indicates that the fruit of the prickly pear cactus was part of the diet of the native Indians of this region as long as 9000 years ago. The Aztecs prized the prickly pear cactus for the red dye that was extracted from the bodies of an insect that lived on the plant. After Cortes invaded the Aztec empire, there was a great demand for the dye in Europe, and prickly pear cac-

tus plants were exported to Spain and the Canary Islands. Today, the fruit of the prickly pear cactus is eaten raw or cooked and made into preserves, pickles, a sweet drink called "miel de tuna," and even a taffylike candy called "queso." The pads can be used as an ingredient in soup.

While most plants cannot survive in a very dry climate, the prickly pear cactus is well adapted to its desert environment and plays an important part in the local ecology. The fleshy pads of the prickly pear and other cacti are actually stems whose spongy tissues retain moisture. During the short rainy season, these stems expand quickly to hold the water absorbed by the plant's shallow but extensive roots. The vascular tissue of the cactus is

Soil

Soil is important in many ways to land organisms. Most plants are anchored in soil, which is also the source of water and minerals the plants require. Soil is the surface upon which many animals move and the home for countless microorganisms and decomposers. Activities of these soil-inhabiting organisms not only help to enrich and aerate the soil, but also play an important role in cycling materials to other organisms in the ecosystem.

Soil Formation   Organisms do more than enrich existing soil. They also play a major role in early steps of soil formation. Soil formation begins with a mechanical process in which weathering (freezing, thawing, and erosion) breaks rock into small mineral particles. Lichens erode away rock surfaces by secreting carbonic acid. Activities of organisms that inhabit soil further modify it. Leaching continues the breakdown of the rock. Leaching is the dissolving of minerals out of the rock or soil by water. Eventually, fine particles such as sand or clay are formed.

By painting Mexican-Americans picking the fruit of the prickly pear cactus, Carmen Garza is portraying something that her culture finds important and beautiful.

continuous from roots into the stem, so that water is easily distributed throughout the plant. The cacti also grow special rain roots within a few hours to a few days after rainfall, which helps them absorb water more rapidly. Most of the water is stored in the cortex and the pith of the stem. The stomata of cacti open mostly at night, when temperatures are lower, so that less water is lost by transpiration. Thus, the cactus retains a large internal store of water—enough to carry it through the long dry season.

Spines on the cactus take the place of leaves and help to discourage predators. Small desert animals, such as the cactus wren, use the stems to protect the entrance to their burrows. Larger animals, however, find the stems very unpleasant because they break off easily. A number of desert animals, including the pack rat, use the prickly pear cactus as a source of food and water.

In these ways, the prickly pear and other cacti have both adapted to their dry environment and become highly useful to the people and animals sharing their habitat.

1. **Explore Further:** In what other ways does the prickly pear cactus play an important role in the ecology of the desert?

2. **Writing About Biology:** Choose another ecological area, such as the coast, tundra, rain forest, tropics, savannah, or high mountains, and research the ways in which the ecology of the area influences the culture of the people who live there.

After the breakdown of rock into soil begins, organic material is added as organisms die. More plants begin to take root in the soil, and then the diversity of animal life both in and on the soil increases. As these plants and animals die, they build up on the surface of the soil and are gradually decomposed. The decayed remains of organisms are called **humus.** Humus is an important part of soil because it contains organic material that enriches the soil. Thus, the remains of once-living organisms provide the materials that living organisms will later use. Humus also increases the soil's capacity for holding the water and air necessary for plant growth. Humus is found in the dark upper layers of soil called topsoil. Removal of topsoil by erosion is a serious agricultural problem.

I N V E S T I G A T I O N

Do soil samples differ in amount of humus present?

Humus is the decomposed organic matter in soil. This means that anything that was once alive (organic matter such as leaves, insects, roots, and dead birds or mammals) that is now in the soil is part of humus. Decomposition takes place as soil bacteria use this organic matter for food and slowly break it down into mineral nutrients.

Do all soils contain the same amount of humus? It might depend on the amount of organic matter originally present in the soil. Which types of soils would have the most or least humus? Would depth of a soil sample make any difference as to amount of humus present? How can you find out?

The following experimental setup can be used to indirectly measure the amount of humus in a sample of soil. Look it over carefully. The amount of gas collected in the graduated cylinder after waiting 10 minutes is an indication of the amount of humus in the soil sample.

Now that you know how to measure soil humus, use the technique to solve several problems. How might soil humus differ if samples are taken from the surface and 30 cm below the surface? Will humus vary in soil samples taken from different areas? Work with a group assigned by your teacher. Form hypotheses that could be tested using soil samples.

Problem

Does humus vary in different soil samples? Does humus vary in the same samples of soil taken from different depths?

Safety Precautions

Be sure to wash your hands carefully after handling soil.

Hypotheses

What are your group's hypotheses?

Experimental Plan

1. Decide on a way to test your group's two hypotheses. Keep the available materials provided by your teacher in mind as you plan your procedure. If possible, you may wish to collect your own soil samples for testing.

2. Record your procedure and list materials and amounts that you will need.

Types of Soil   The relative amounts of minerals present in soil influence the kinds of plant life that can exist. Consider, for example, a sandy soil. There are many air spaces in such soil, but the minerals do not attract water. Water drains rapidly through the soil. Since little water is held in the soil, water is not available for plant life. These facts explain why you see no life on a beach or a scarcity of life in the middle of a desert. While there is no "best" type of soil, some soils support more plant life than others. Clay particles are minerals in soil that retain water. A soil that contains a higher proportion of clay will, therefore, be able to sustain more plant life. Too much clay, however, prevents drainage of water and the air spaces become filled.

Checking the Plan

Discuss the following points with other group members to decide the final procedure for your experiment.

1. What data will you collect and how will it be recorded?
2. What variables will have to be controlled?
3. What control will be used?
4. Assign roles for this investigation.

Make sure your teacher has approved your experimental plan before you proceed further.

Data and Conclusions

Carry out your experiment. Prepare a data chart to record your findings.

Analysis and Conclusions

1. State your hypothesis regarding the amount of humus present in the soil samples taken from different depths. Explain whether your hypothesis supports or rejects the information you now have from the answer to question 1. Use specific experimental data to confirm or reject your hypothesis.

2. State your hypothesis regarding the amount of humus present in different soil samples. Explain whether your hypothesis supports or rejects the information you now have from the answer to question 1. Use specific experimental data to confirm or reject your hypothesis.

3. Explain what you believe may have been the role of adding hydrogen peroxide (H_2O_2) to each soil sample.

4. Describe some of the variables that had to be controlled in this experiment. How were they controlled?

5. Describe the control that was used and how it enabled you to conclude that the gas being released did come from the soil. Did your experi- ment actually show that the gas given off did come directly from the humus alone? Explain how you might have altered your experimental procedure to prove this.

Going Further

Based on this lab experience, design another experiment that would help to answer questions that arose from your work. Assume that you will have limited resources and will be conducting this lab in class. Include this as part of your report.

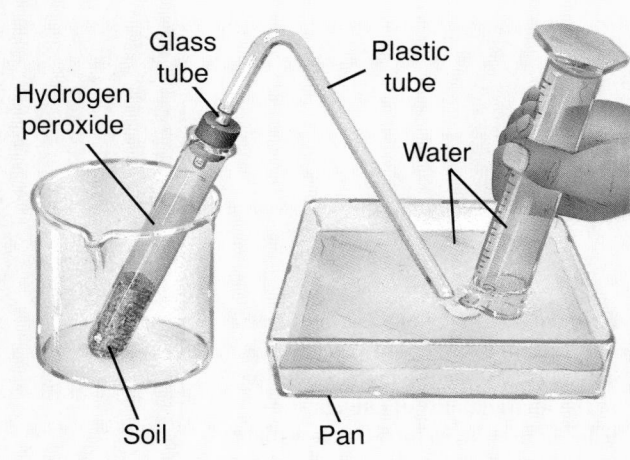

Glass tube

Plastic tube

Hydrogen peroxide

Water

Soil

Pan

Minilab 1

Do plant roots help with soil formation?

Plant roots may chemically aid in soil formation if they give off CO_2 gas into soil water. CO_2 plus water forms carbonic acid, which hastens the change of rock minerals into soil. Do plant roots give off CO_2? Put the root portion of a bean ·plant into a test tube of bromothymol blue (BTB) solution. **CAUTION:** *Be careful handling BTB solution. It will stain clothing.* Be sure all soil is rinsed from the roots. The change in color of BTB indicates the presence of CO_2. Seal the stem from outside air with a clay plug. Prepare a suitable control for this experiment. Note the color of the BTB 24 hours later. A color change means CO_2 has been given off.
Analysis *Explain the results of your experiment. Do plant roots give off CO_2? Why is CO_2 given off?*

Plants have adapted to grow in many different types of soil. Cacti, for example, have adaptations to grow in extremely dry, sandy soils. Water lilies are adapted to a different type of soil. They grow in fresh water, sometimes up to 15 feet deep.

In addition to organic and mineral content, soil fertility is also determined by pH. Some plants such as azaleas, cranberries, and rhododendrons can grow well in acidic soils, but most cannot. Most grow best in neutral or basic soils. Lime, a basic material, is often added to acidic soils to make them neutral. Acid rain, which you will study in Chapter 30, alters the pH of soil and therefore affects plant life.

Studying soil is an excellent way to learn how organisms are affected by the physical environment and how they, in turn, affect that environment. The nature of soil determines the diversity of plants and all other organisms in an ecosystem on land. At the same time, the presence of organisms is crucial to the formation and enrichment of the soil.

Light

As you've learned, light, or radiant energy, is the source of energy for almost all ecosystems. By photosynthesis, light energy is changed to the chemical energy needed by every living system.

As you may have guessed, plant and protist distribution is also affected by the amount of light. Because light cannot penetrate deep into water, most algae live near the surface of aquatic ecosystems. In a forest, trees are exposed to the greatest amount of light through leaf arrangements that are adaptations for maximum light exposure. In temperate forests, other plants are adapted to grow in the dim light of the forest floor. Some of these other plants begin to grow before leaves appear on the trees. In tropical rain forests, many plants, like bromeliads, receive enough light by growing among tree branches.

Figure 28-10 **In a mature forest, only a limited amount of vegetation is able to survive on the forest floor due to the unavailability of enough sunlight. In deciduous forests, many plants begin to grow before the leaves appear on the trees in order to utilize the sunlight available.**

Light is also important to organisms in other ways. Light is required for vision. Some animals see well in dim light, but no animal can see in the complete absence of light. Light is also a factor in the migration of birds and the flowering of certain plants. Light energy is needed to make vitamin D in humans. Can you think of other examples of the importance of light?

Only about two percent of the light striking Earth is used for photosynthesis. Much of it is absorbed by the atmosphere, land, and seas and is converted to heat or thermal energy. Light absorbed by the surfaces on Earth and trapped by atmospheric gases helps maintain its temperature.

The heat produced by this absorption of light also plays a role in maintaining life. For example, heat is needed for the transpiration of water from plants. Heat also causes the evaporation of water from oceans and rivers. Without heat, water could not be cycled. As you'll learn, temperature also plays a more direct role in the survival of organisms.

Temperature

Temperature, along with factors such as wind and amount of rainfall, plays an important role in the distribution of organisms. In general, temperature changes with altitude and latitude; that is, temperature decreases from Earth's surface upward, and from the equator north and south to the poles. Thus, the types of organisms found in Alaska or high on a mountain are different from those living in Florida or in a valley. Would you expect to find polar bears in Florida or orange groves in Alaska? Organisms are adapted to survive within particular average temperature ranges.

Figure 28-11 **Temperature plays an important role in the distribution of organisms. Organisms are adapted to survive within certain temperature ranges: Adelie penguins, Antarctica (top left); bananas, Madagascar (bottom left); saxifrage, New York State (center); Eastern Rosella parrot, Australia (right).**

Temperature varies with time as well as with location. Temperatures differ from day to night and from season to season. Temperature differences also exist between times of sunshine and times of cloudiness. Such changes affect land organisms more than water organisms because air temperatures change more rapidly and to a greater degree than water temperatures. Air temperatures can change as much as 40°C between day and night. In the same time period, the temperature of the surface (top ten meters) of the ocean changes no more than 4°C. Below ten meters, there is little change in ocean temperature. How do you think the differences in temperature at different depths affect the diversity of organisms that live there? In terms of temperature, water is a more stable environment than land, but temperature on land is also an important ecological factor. Bacteria can live within the greatest temperature range. Some bacteria form endospores and can withstand temperatures above the boiling point and far below the freezing point of water (Chapter 15). For this reason, bacteria are found almost everywhere. The spores of fungi and seeds from some plants are adapted to survive temperatures that would kill mature fungi and plants. In this way, many species are able to survive extreme temperatures.

Temperature often starts biological processes. That is, temperature changes act as stimuli that activate chemical control systems of some organisms. For example, peach trees bloom when it is warm, but they must first be exposed to chilling during winter. The trees will not bloom without the chilling. Gypsy moth pupae will not develop into adults unless a freezing temperature has occurred. An adult, though, would die in freezing temperatures. Temperature requirement is a limiting factor that affects distribution of organisms. Do you think peach trees and gypsy moths could reproduce in the tropics?

Figure 28-12 Gypsy moth pupae need freezing temperatures in order to develop into adults.

Temperature and day length can also affect the movement of organisms from place to place. You know that many birds and some mammals migrate in the winter. Animals such as bighorn sheep, caribou, and many birds migrate annually because temperature, length of day, and weather changes in winter create food shortages for these organisms.

Metabolic rate in some animals varies with temperature as well. Examine Figure 28-13. As you see in the graph, up to about 45°C, metabolic rate increases two or three times for every 10° the temperature rises. Remember that metabolism is related to the activity of enzymes. At high temperatures, metabolic reactions cannot occur; thus, most forms of life cannot survive when their body temperatures are above 50°C as enzymes are destroyed above this temperature.

Many animals have behavioral and physiological adaptations that prevent severe changes in body temperature and metabolism as the temperature of the environment changes. Desert lizards, for example, are active only in the early morning when it is cool. They spend the hottest part of the day under rocks where it is cooler. Also, certain species of frog go through dormant periods during the summer. This period of summer dormancy is called **estivation.**

Animals such as birds and mammals maintain constant body temperatures. A human has a body temperature of about 37°C, and a polar bear's body temperature is about 38°C. Why are these temperatures so similar? Look again at Figure 28-13. This temperature range is one at which metabolic reactions operate most efficiently. During severely cold weather, though, this body temperature is difficult to maintain because food is scarce. That's why many small mammals, such as chipmunks and bats, hibernate during the winter. During hibernation, metabolic rate decreases as body temperature decreases. In this way, food is conserved.

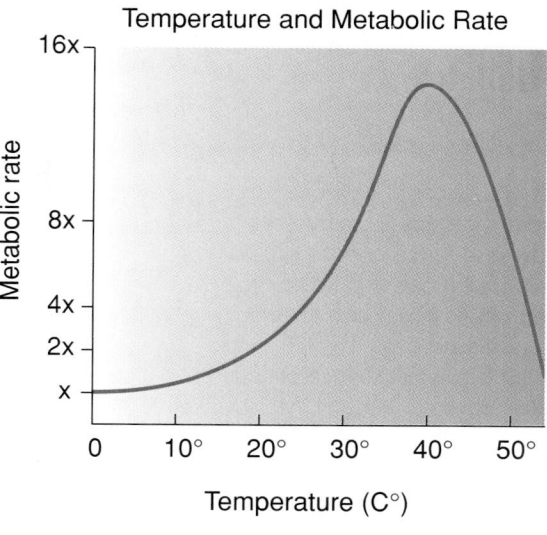

Temperature and Metabolic Rate

Figure 28-13 As this graph shows, metabolic rate is highest at approximately 40°C.

Figure 28-14 Many organisms have behavioral adaptations that allow them to maintain constant body temperatures. This sand-diving lizard cools itself by lifting its feet off the hot sand.

You may be familiar with some of your own adaptations to temperature. For example, humans often perspire when hot. As perspiration occurs, the skin is cooled. When you're hot, arteries in your skin also expand to increase heat loss. Humans also tend to be less active when hot, so less heat is generated by the muscles. When you are cold, you may become more active and even shiver. When this occurs, arteries in the skin contract, conserving heat.

Disrupting the Balance

An ecosystem's balance can be altered temporarily by natural biotic and abiotic factors. Both biotic and abiotic factors play a role in limiting population sizes in such a way that the resources of the environment are not overused. Energy is captured in photosynthesis by green plants and made available to all organisms by means of food webs. Materials are cycled and can be used over and over again. Organisms are adapted to both living and physical factors of the environment and are capable of responding to some changes, such as different seasons or a temporary shortage of water. However, drastic or prolonged changes can affect the delicate balance of an ecosystem.

Natural occurrences such as a flood, earthquake, or volcanic eruption may certainly upset the balance of an ecosystem. Such events cannot be prevented. Other events, though—ones that might not be predicted to cause disturbances—may, in fact, lead to major problems in an ecosystem. Often, those events are a consequence of human activities. The real threat to ecosystems is the frequency, intensity, and uniqueness of human activity.

As an example of how human activities can upset the balance in an ecosystem, let's return once more to our lake community. Phosphates are chemicals present in detergents. Phosphates may enter a lake from sewage treatment plants. Phosphates, as well as nitrates present in fertilizers, may also enter the lake from soil runoff. These phosphates and nitrates are important nutrients for algae. Low levels of these chemicals limit the growth of algal populations. However, with large amounts of these chemicals and in the presence of warm temperatures, the algal populations thrive and grow rapidly in what is called an **algal bloom.** You might think that an increase in the algal population would be a benefit because more food would be available for first-order consumers.

Figure 28-15 **An extreme increase in the population of one organism in an ecosystem can disrupt the balance of the entire ecosystem. In this case, an algal bloom caused this pond to become stagnant.**

At best, though, the benefit is temporary. The algal population may become so large that the algae finally use up all available nutrients and begin to die. The dead algae settle to the bottom where they are decomposed. Large amounts of oxygen are used up as decomposition occurs. As a result, less oxygen is available for cellular respiration, and other organisms begin to die. Death of a large number of organisms disrupts the food web, and continued decomposition uses even more oxygen. Consumers such as oxygen-sensitive insect larvae, mayflies, and fish may die, while other organisms such as bacteria and mosquitoes may thrive. The entire lake may become stagnant as the ecological balance is drastically altered to the new factors.

Analysis of an algal bloom shows how a change in just a single physical factor, amount of a particular chemical, can have serious consequences for an ecosystem. Changes in other factors, both abiotic and biotic, can also affect and disrupt the balance of ecosystems. In Chapter 30, you'll study examples of how human actions contribute to the disruption of ecosystems.

LOOKING AHEAD

ORGANISMS IN an ecosystem interact with each other and are affected by them as well as by physical factors in the environment. In this chapter, you've considered how disruptions caused by biotic and abiotic factors can alter the delicate balance of ecosystems. In the next chapter, you'll discover how ecosystems develop over time and how ecosystems are replaced after being destroyed. You'll also learn how physical factors operate together to determine the various kinds of ecosystems found throughout the world.

Section Review

Understanding Concepts

1. Grasslands are common in Iowa and beech-maple forests are natural ecosystems in New England. Why are the ecosystems different in these two areas?

2. Food chains exist within soil, yet no food is produced there. How can this be explained?

3. In northernmost regions, the ground is permanently frozen except for the very top layer, which thaws in spring and summer. What effect does this physical factor have on plant and animal life in such regions? Explain.

4. **Skill Review—Recognizing Cause and Effect:** In an abandoned field, grasses and shrubs begin to grow. Later, the number of these smaller plants decreases as trees sprout up. How does the presence of trees result in a decrease in the numbers of grasses and shrubs?

For more help, see Thinking Critically in the **Skill Handbook.**

5. **Thinking Critically:** Suppose a pesticide were applied to rid a forest of insect larvae that were hungrily consuming leaves of the trees. Suppose, too, that the pesticide killed natural predators of the insect larvae. The natural predators are food for other animals. Describe some possible consequences of applying the insecticide to this forest ecosystem.

Considering the Environment

Is the lowly lichen smarter than we are?

What plant-like organism can live in both arctic regions and deserts, doesn't need soil to supply its food, and can tell you how bad the air is? Answer: the lichen, a rootless organism that is actually a combination of a fungus and a green alga or

photosynthetic bacterium. Lichens are perhaps the hardiest organisms on Earth. They are largely unaffected by extremes of temperature, and they can lie dormant through long periods of drought. Some lichens are known to live for more

than 4000 years. You have probably seen lichens growing on bare rocks, tree trunks, down the sides of cliffs, or along stone walls. Lichens can survive in these seemingly inhospitable places because they absorb all their nutrients from the air.

Just as most plants are vulnerable to pollutants in the soil and water, lichens are vulnerable to pollutants in the air. They absorb anything the air brings their way, including sulfur dioxide from industrial and urban emissions, heavy metals, and radiation. This makes them good early indicators of air quality. Scientists noted more than a century ago that most lichens don't survive in cities and industrial areas because of the poor air quality. Today, by chemically analyzing lichens or charting their presence or absence in an area, scientists can determine the amount and kind of pollution present in the air.

There are more than 18 000 varieties of lichens, and different varieties are more sensitive to some pollutants than to others. Sulfur dioxide, for example, kills some lichens, causes others merely to sicken, and doesn't affect several types at all. By studying different lichens, scientists have been able to categorize them according to which pollutants they are sensitive to. The presence or absence of these particular varieties in an area then alerts the scientists to the presence of pollutants in the air long before other kinds of plants start to die or people become affected. Sometimes lichens tell a story of air pollution in remote rural areas long before any hint of haze can be detected in the air.

One variety of lichen, *Cladina*, is commonly known as reindeer moss. It grows like a thick, green carpet in arctic regions and is the main source of food for caribou and reindeer in North America and Eurasia. The animals dig through the snow with their hooves to eat the lichen during the long winters. For the Lapps, or Sami as they prefer to be called, semi-nomadic herders who live on the tundra of Scandinavia, the reindeer are the basis of their economy. Each Sami

eats an average of eight to ten reindeer a year. They also make their coats, shoes, and tents from reindeer hides, use the antlers to make knife handles and glue, and sell reindeer meat to markets in the southern cities. Following the migration of reindeer herds has been their way of life for more than 10 000 years.

In the fall of 1986, that way of life was suddenly interrupted because of the lichen's ability to soak up pollution from the air. On April 26, 1986, a huge reactor at the Soviet nuclear power plant in Chernobyl burned, releasing a cloud of radioactive particles. The cloud blew northwest, arriving over Norway and Sweden within two days. Rain in parts of these countries brought radioactive particles, especially cesium-137, to Earth. The *Cladina* absorbed the cesium-137 along with its usual nutrients, and when the reindeer began to feed on the moss in the early fall, they, in turn, absorbed cesium-137. Passed through the food chain to humans, cesium-137 can cause cancer. Fortunately, the Swedish government detected the

radioactive material in the reindeer meat and acted to prevent its sale. However, about 50 000 reindeer had to be destroyed. The government provided commercial feed for some reindeer herds and moved others to Norway, which received less radioactive fallout. This disrupted the traditional Sami economy and way of life and cost the Swedish government millions of dollars. The price of uncontaminated reindeer meat rose, and consumer demand fell as people worried about its safety or could not afford the higher price. Cesium-137 decays slowly, and it may be more than 30 years before the lichen is safe again for grazing.

Lichen such as the reindeer moss on the arctic tundra act as a kind of advance warning system, telling us that we are polluting our world without even knowing it. New technologies now allow scientists to test lichen to detect the presence of

many chemicals that are dangerous to humans, including lead, nitrates, fluorides, and ammonium. The lowly lichens may prove to be one of our best friends.

1. **Explore Further:** The 1986 accident at the Chernobyl nuclear power plant led to the largest release of radioactive material ever from one accident. What were some of the other effects of this disaster?
2. **Writing About Biology:** Find out more about the Sami. Locate where they live on a map. Write a paragraph about the niche these people fill in the ecosystem.

Readings

- Bruemmer, Fred. "In Praise of Lowly Lichen." *International Wildlife,* Vol. 21, No. 6, Nov./Dec. 1991, pp. 30-33.
- Kiester, Edwin, Jr. "Prophets of Gloom." *Discover,* Nov. 1991, pp. 53-56.
- Lofstedt, Ragnar, and Allen White. "Chernobyl: Four Years Later." *Environment,* Vol. 32, No. 3, April 1990, pp. 1-8.

Issues

Can agriculture be climate free?

The Southwest is the driest region of the United States. Average annual rainfall is less than 20 inches. Summers are hot and dry, winters cold and windy, and drought years are frequent. Yet if you drive through the Texas Panhandle, you will see fields of cotton, wheat, grain sorghum, and corn, as well as sugar beets, potatoes, and hay. In all, more than 5 million acres, or one-fourth of the land in northwest Texas, have been turned into productive cropland.

Farming in the Texas High Plains began to boom in the 1950s with introduction of the central-pivot

irrigation, or watering, system. This strange looking structure is made of a vertical pipe, or tower, connected to a horizontal pipe. Both are on wheels so they can rotate. The horizontal pipe, sometimes as long as 1000 feet, circles slowly around the field, about eight feet above ground, spraying water from holes along its length. The tower, usually the height of a house, is connected to a well that pumps water from underground. A thousand gallons of water spew each minute from the biggest of these center-pivot irrigation systems. The force of the water moves the pipe slowly around in a

circle, usually taking about 12 hours to complete a full turn around a quarter-square-mile area.

Beneath the Texas High Plains lies the southern part of the Ogallala aquifer. An aquifer is an underground formation of sand, gravel, and clay saturated with water. The top of the aquifer is called the water table. The Ogallala stretches under six states: Texas, New Mexico, Colorado, Oklahoma, Kansas, and Nebraska. In Texas alone, the Ogallala covers 35 000 square miles and holds an estimated 400 million acre-feet, or more than 1.3 quadrillion gallons of water.

Farmers in Texas started tapping the Ogallala with small wells in the late 1800s. Agriculture expanded in the area during the first half of the 20th century, but the drought in the 1930s wiped out much of the area. It was the invention of better irrigation systems in the 1950s that opened the way for large-scale farming in this semiarid region. Crop production per acre more than doubled, the economy boomed, and more farmers moved into the Southwest. Technology, it seemed, could change the course of nature. Climate-free agriculture it was called.

But now we are realizing that the technology that turned the Texas High Plains into an agricultural gold mine may be leading to its ultimate ruin. Today, more than 150 000 wells dot the Ogallala, most of them driving center-pivot irrigation towers. Together, they are

capable of pumping more than 50 billion gallons of water from the aquifer in one 12-hour period. How does all that water get replaced?

The Ogallala was once called the "land of underground rain" or the "sixth" Great Lake. Everyone assumed that its supply of water was endless. But we now know that the Ogallala is slowly drying up. At its northern end, there is enough rainfall to nearly recharge the Ogallala each year. In the dry High Plains, however, there is very little rainfall. In some places, only a teacupful of water is put back for every gallon that is pumped out. Each year, almost 5 trillion gallons of water are taken from the Ogallala without being replaced. That's almost as much water as flows through the entire Colorado River. And only 60 to 70 percent of that water gets into the soil; the rest is lost to evaporation and runoff. The Ogallala took about 10 million years to fill. At the present rate of water use, it may be emptied in fewer than 100 years.

The first effects are already being felt. As the aquifer is depleted, the water table sinks lower and lower.

It becomes more expensive to pump water to the surface, which in turn makes crops less profitable to grow. Some farmers already have to redrill their wells every year, or connect smaller wells to the central well in order to have enough water power to drive their irrigation systems. Eventually, fields may be left unplanted, and then wind will erode the soil, just as it did during the Dust Bowl of the 1930s.

In parts of the Texas High Plains, more than water is disappearing. When the water table drops, the soil above it compacts. If it compacts enough, the ground will sink and form cracks. This process is called land subsidence. Eventually, these cracks widen and deepen into huge ditches called fissures. Fissures have opened up all across the plains, damaging wells, water pipes, and roads; trapping cattle; and making fields impossible to irrigate.

Thus, the High Plains face a dilemma. While irrigation has turned this and other dry regions of

the southwest, such as California's Imperial Valley, into an agricultural gold mine supplying food to millions of people and livestock, it has also depleted sources of underground water that can never be replaced. Eventually, the Ogallala may run dry. That is a price we may have to pay for climate-free agriculture.

Understanding the Issues

1. What might be some ways to conserve water on the High Plains without giving up agriculture?

2. Do you think it is more important to conserve our groundwater supplies or grow as much food as possible? Why?

Readings

- Barnes-Svarney, Patricia. "Ground Water in Peril." *Earth Science,* Vol. 42, No. 4, Winter 1989, pp. 14-16.

- Glantz, Michael. "An Analog for the Texas High Plains." *Environment,* Vol. 33, No. 5, June 1991, pp. 30-31.

CAREER CONNECTION

Hydrologists study Earth's water system including precipitation, storm patterns, the causes and effects of droughts and floods, the water supply in aquifers, and the exchange of water between the oceans and the atmosphere. They plan soil and water conservation programs, flood control systems, and irrigation and drainage projects for agriculture. They work for the government and in private industry.

Summary

Producers are the breadwinners of an ecosystem. In almost all ecosystems, producers trap the energy of sunlight, changing it to chemical form. All other members of the ecosystem depend directly or indirectly on producers as a source of energy, which they obtain via food webs.

Less and less energy is available to each trophic level in an ecosystem, and, because energy is lost and cannot be recycled, each ecosystem must have a continual input of energy in order to survive. However, materials are constantly recycled and can thus be used over and over again.

In addition to their need for food, organisms depend upon one another in other ways. Specialized relationships include mutualism, commensalism, and parasitism.

Organisms are influenced not only by other organisms, but also by abiotic factors of their environment. Factors such as water, soil, light, and temperature may affect the distribution and activities of organisms and may act to limit population sizes. A stable ecosystem is a delicate balance of biotic and abiotic interactions. Changes in even one of these factors may disrupt that delicate balance.

Language of Biology

Write a sentence that shows your understanding of each of the following terms.

abiotic factor	herbivore
algal bloom	humus
biomass	niche
biotic factor	omnivore
carnivore	pyramid of biomass
commensalism	pyramid of energy
ecosystem	pyramid of numbers
estivation	second-order consumer
first-order consumer	third-order consumer
food web	trophic level

Understanding Concepts

1. Identify a particular food chain of four links in an ecosystem of your choosing. You will use your food chain to answer questions 2-6.
2. Identify the producer and second-order consumer in your food chain. Are there any omnivores? If so, identify them.
3. Compare the total amount of energy present in the herbivores in your food chain with the amount available to the organisms in the next trophic level. Explain several reasons why the amounts are not the same.
4. If you could determine the amount of biomass present in each trophic level, what pattern would you see? Explain the reason for that pattern.
5. Does your food chain show a pyramid of numbers? If so, explain why. If not, explain why.
6. How does each member of your food chain contribute to the cycling of carbon? Where do decomposers "fit" in your food chain? How do they contribute to the cycling of carbon?
7. Why do fertilizers contain nitrates? Could planting of legumes reduce the need for fertilizers? Explain.
8. Certain fungi grow on the roots of conifers. The fungi absorb nutrient-rich fluids from the roots and help draw water and mineral ions from the soil into the roots. Of what type of relationship is this an example?

9. In cacti, the stem is the organ of photosynthesis rather than leaves. Explain this fact in terms of an abiotic factor.
10. Why would you expect to find moss plants growing on or near the base of a tree rather than in an open spot in a forest?
11. How can temperature cause food to become a limiting factor for some animals?
12. Why does a desert snake sun itself on a rock early in the day?
13. How might a lake ecosystem be affected if there were no limits on the number of a certain fish that humans are allowed to catch?

Applying Concepts

14. Even though mammals are endotherms, heat is lost from the surfaces of their bodies. Smaller endotherms lose relatively more heat than do larger ones. How would you expect the relative metabolic rates of small and large endotherms to compare? Why?
15. Allen's rule states that the size of extremities (ears, for example) of animals decreases as temperature decreases. Why are extremities smaller in cold temperatures and larger in warm temperatures?
16. In what ecosystem do you suppose the greatest amount of photosynthesis occurs? Why?
17. A home aquarium can be considered a miniature balanced ecosystem. Explain how balance is maintained in this tiny ecosystem.
18. Explain why more people could be fed if food chains were shortened.
19. How may clearing land for agriculture affect a natural ecosystem?
20. How does destruction of large portions of tropical rain forests affect the carbon cycle?
21. **Issues:** Using more water from an aquifer than is replaced is called overdrafting. Overdrafting is one threat to our groundwater. What are other possible threats?
22. **Considering the Environment:** Why do you think that lichens are particularly sensitive to air pollutants?

Lab Interpretation

23. **Investigation:** The following bar graph depicts the data that a student group presented of their results when measuring gas release from soil treated with hydrogen peroxide. Which bar letter best represents:
    (a) Their control? Explain why.
    (b) A tube in which sand was tested? Explain why.
    (c) A soil sample in which they forgot to add hydrogen peroxide? Explain why.
    (d) The soil sample with the most humus? Explain why.

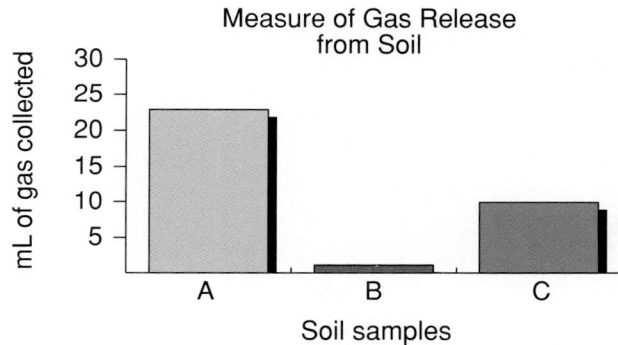

24. **Minilab 1:** Predict the experimental outcome of this same experiment if the bean plant placed in BTB solution had been supplied with light overnight and explain your answer.
25. **Minilab 2:** Predict and explain your observation on lens paper after one week if it had been placed in sterile soil. Note: All life forms in sterile soil have been killed.

Connecting Ideas

26. Cite and explain several structural adaptations that permit endotherms to withstand extreme cold.
27. In what way is a mutualistic relationship between two organisms in an ecosystem similar to the relationship that is thought to have existed between the first mitochondria and chloroplasts and their host cells?

29

ORIGIN AND DISTRIBUTION OF COMMUNITIES

ON AUGUST 27, 1883, what may have been the loudest sound in recorded history was heard when the volcano on the island of Krakatau blew apart. It was heard 4800 km away. The island, which is near Indonesia and was about the size of Manhattan before the blast, was now only one-quarter that size.

When it was all over, the first landing party on what was left of the island found not a tree, shrub, or blade of grass growing. All vegetation was buried under ash and cinders and hardened lava. Within a year, however, grasses and other pioneer plants began to appear in the ashes. As these plants died and decayed, they formed a fertile soil; less hardy plants could then grow there. Airborne spores and seaborne seeds could germinate and begin to grow. Within 50 years, the island was well on its way to renewing itself; it was covered with a thick, young forest.

Like human pioneers, plant pioneers must withstand the harshest conditions—among them, intense sunlight and scarce or infertile soil. The growing conditions are not suitable for most other species, so pioneer plants have the area pretty much to themselves. Few competitors will challenge their existence until the pioneers have done the homesteading—that is, until the pioneers have changed the site enough for others to survive.

As you read this chapter, you will see how pioneer organisms modify their environment. You will follow the sequence of events that occurs as an area changes over time from a barren wasteland to a stable, flourishing community of living things. You will also examine the major kinds of ecosystems that exist on Earth.

Like Davy Crockett and Johnny Appleseed, this plant is a pioneer. It won't tussle with bears or sow apple trees, but it will change its environment so other seeds that land there can sprout and survive.

Chapter Preview

Objectives

Relate the events of a particular ecological succession to the factors that cause succession.

Compare and contrast primary succession and secondary succession.

Apply the ideas of the general trends seen in succession to a particular example of succession.

29.1 Ecological Succession

Have you ever walked along an abandoned railroad track and seen young shrubs sprouting up beside and between the tracks? Have you ever noticed moss growing in cracks in the sidewalk or along a curb—or the grasses growing around a house or building where the land has not been tended? Although you probably haven't thought about it, the shrubs or moss or grasses could be the beginnings of new, natural communities. If these areas were not altered by human activities, each would undergo a series of changes in which different species of organisms would appear, only to be replaced later by others. Given enough time, each area would become a complex and stable community.

Succession and Its Causes

Picture several types of communities—lake, forest, desert. Each of these communities has a history. The diversity of organisms living in each community today is not the same as it was in the past. Biotic and abiotic factors have changed over time. Some communities today may be stable, having undergone many changes over a long period of time. Others may not be quite so stable and will undergo still further change. The series of changes in a community is called **ecological succession.** Succession may occur in areas where no life has previously existed, or it may occur where an established community has been affected by a disaster such as a fire or volcanic eruption, or by human activities such as the clearing of a forest. In either case, succession is marked by differences in organisms and physical factors over time.

What causes ecological succession? A major factor is the effect a particular set of organisms has on the physical environment. Think back, for example, to the effect of organisms on the formation and characteristics of soil. During succession, organisms alter their physical surroundings in a way that makes those surroundings more suitable for other kinds of organisms and less suitable for themselves. As a result, new organisms invade a habitat, and the organisms previously living there cannot compete with them. Thus, many of the earlier organisms gradually disappear.

Physical changes in the environment during succession may result from factors other than the action of organisms. For example, soil from surrounding land may run off into a pond, causing the pond to become shallower and eventually fill in entirely.

Figure 29-1 **When a forest is cleared, either by natural causes or by human intervention, the process of succession soon will follow.**

In any succession, the type of plant life present influences the kinds of animals that can live there. Different stages of succession are characterized by different sets of dominant plant species. Which plant species exist at any one stage depends not only on the physical factors present, but also on the life cycles and dispersal characteristics of the plants. Plants like mosses or grasses tend to appear first. They produce numerous spores or seeds that are dispersed easily by the wind, and their life cycles are short. Larger plants, like trees, produce fewer, larger seeds and have longer growth periods. They appear and become dominant much later.

Primary Succession

How does succession occur in an area for the first time? Imagine the surface of land before any life existed on it at all. You would see bare rocks or sandy beaches devoid of soil. From these barren areas emerged the many types of communities that live on land today. Succession that begins in areas where there is no life is called **primary succession.** Obviously, the first stage of primary succession must begin with very hardy organisms— pioneers—that can grow under adverse conditions. The first stage is called the **pioneer stage.**

Pioneer Stage   Consider a natural forest community in New England. This forest might have begun long ago with lichens growing on bare rock. Lichens produce acids that may corrode rock. Freezing and thawing of water in cracks in the rock then breaks the rock into smaller fragments. As the lichens die, bacteria and other decomposers break them down. Organic materials collect in the cracks of the rock, thus beginning soil formation. Mosses then may anchor in the primitive soil. As the mosses die and decompose, they further enrich the soil, making it possible for more complex plants such as ferns, grasses, and shrubs to appear. As these complex plants flourish and grow, the mosses cannot compete, and they die out. The eventual death and decay of the more complex plants further increase the humus content of the soil, making it thicker and richer.

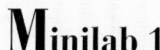

Figure 29-2   **Plants with short life cycles, such as the mosses on the right, are usually among the first plants to grow in a new area. The large trees on the left have longer life cycles and appear much later. They may not mature until 150 years after succession begins.**

Beeches and maples

Pines

Ferns, shrubs,
and grasses

Mosses

Lichens

Soil

Rock

Figure 29-3 The first stage in primary succession of a forest consists of lichens growing on bare rocks. As they die and form soil, larger plants such as mosses, grasses, ferns, and shrubs can grow. Only one stage is present at any time.

Figure 29-4 Plants change their surroundings when they grow, creating microenvironments to which other kinds of plants may be better adapted.

While these changes in plant species are occurring, Figure 29-3, changes in animal species are occurring, too. Animals such as roundworms and earthworms add to the richness of the soil, and insect life becomes abundant above the soil. The insects may help pollinate the plants and speed their population growth. Birds and small mammals may also enter the area.

Microenvironments The kinds of plants and animals present at each stage of succession change not only the soil but also other abiotic and biotic factors. For instance, larger plants cast shadows that cause subtle differences in temperature, moisture, and light in some small parts of the community. Different organisms can live in these slightly different areas, which are known as **microenvironments.** Differences in organisms also influence food chains and energy flow. All of these changes result in the eventual replacement of the organisms present at each stage by different organisms in later stages.

In the New England forest community, pine trees gradually come to replace the ferns, grasses, and shrubs. Pines grow well in direct sunlight, but young pines do not prosper in the shade of older, taller pines. Seedlings of beech and maple trees, however, can tolerate the shade. Gradually, beeches and maples replace the pines by crowding them out.

Climax Community

Beeches and maples are the dominant species in the mature New England forest. The beech-maple forest is the final, stable stage in the development of the community and is known as the **climax community.** The type of climax community that emerges in a given area depends on the interaction of all factors of the environment, both abiotic and biotic. In the southeastern United States, the climax community might be an oak-hickory forest. In the climax community of a North American desert, the creosote bush may be the dominant species.

Because it is stable and complex, a climax community tends to remain the same over a long period of time without further succession. But factors such as changes in climate, disease, a natural disaster, or human interference could be so severe that succession might start over again later, though not necessarily from the pioneer stage.

From Pond to Forest

A pond is a small body of water surrounded by land. It is an ecosystem in which blue-green bacteria and green algae are the main producers. Toward the shore, plants such as cattails and reeds also trap energy from the sun. Consumers consist of ciliates and other protists, insect larvae, worms, crayfish, and fish.

A pond is organized, but it is not a climax community. A pond is not a permanent ecosystem; it undergoes succession. The life span of a pond depends on its size and the conditions of the environment. Consider a pond surrounded by a forest. As the pond ages, sediments collect at its bottom, in much the same way as dust collects under your bed. The shallow portions around the shoreline fill in first. In this marshy soil, simple plants such as sphagnum moss, insectivorous plants, and shrubs take root and help bind the soil. Sediments continue to collect on the bottom of the pond, decreasing its depth. As land takes the place of water, simple plants continue to grow. Meanwhile, the areas around the old shore undergo further succession. Shrubs are replaced by trees, and finally trees become dominant. This pattern of simple plants to shrubs to trees continues until no more water remains, as seen in Figure 29-5. A forest—the climax community—gradually replaces the pond.

Secondary Succession

Succession often begins as a result of natural destruction, such as a fire started by lightning. It also may result from interference by humans, such as logging. In either case, new succession begins when the dominant plants of a plant community are removed. If the land is later left untended, **secondary succession** will occur. Secondary succession occurs more rapidly than primary succession, especially in its early stages, because soil is already formed and a few life forms such as seeds may be present. Thus, the area does not have to "start from scratch."

Figure 29-5 A pond is not a climax community. Over a period of time, a pond becomes filled in with soil and plant material, and eventually is replaced by the surrounding climax community—the forest.

During the early history of the United States, much of the forest land of New England was cleared and used for farming. As people moved westward, many of these farms were abandoned. Secondary succession then began in the abandoned fields. This type of succession is still occurring today in New England (and in other parts of the world). At first, hardy pioneer grasses and weeds appear in the fields. Later, shrubs grow, followed by junipers, poplars, and white pines. Eventually, beeches and maples begin to sprout in the shadow of the other trees. What was once an open field becomes a beech-maple forest.

In this example, the sequence of stages of community development is different from that beginning with bare rock, but the climax community is the same. However, succession in a given area does not necessarily lead to the same climax community each time. The characteristics of a climax community probably result from many chance occurrences including biotic and abiotic environmental factors over long periods of time. In general, though, each climax community is more stable and productive than previous stages of the succession that produced it.

INVESTIGATION

Factors Involved in Grass Seed Germination

Did you ever have a garden or grow seeds in a paper cup? What caused your seeds to grow? Some seeds require a period of cold before they will germinate. Others need to have their seed coats burned by a fire or exposed to charate, which is the remains of burned plants. Some seeds must pass through the digestive system of an animal before they will germinate. Others must be in the light, while still others must have darkness for germination.

Ranchers and farmers use a variety of methods to stimulate growth of the fields of grasses they make into hay. Some fertilize their fields either with organic manure or synthetic fertilizer; some burn their fields before the growing season begins; some irrigate heavily, ensuring adequate water supply.

During various stages of succession, conditions for growth change. In a recently burned meadow, seeds requiring charate, heat, and sunlight may germinate best. Later, as plants grow in this meadow, shade creates different conditions more suited for germination of other kinds of seeds.

Problem
Under what conditions does grass seed germinate fastest?

Safety Precautions
Wash your hands after setting up this lab and after making measurements.

Hypothesis
Make a hypothesis about the conditions under which you think grass seed will germinate fastest. Explain your reasons for making this hypothesis.

Desert communities may not always follow the same sequence. In climates more favorable for plant growth, the plants compete for space; the result is dominance by some plants and elimination of others. In the desert, however, the primary struggle of the plants is for water. Also, there is little plant debris on the ground to help form fertile soil for successional changes. When a forest is destroyed by cutting or burning, new light-loving species spring up, and it may take decades for the forest to regain its original appearance. When a desert community is destroyed, the first plants that spring up are almost always the very same species that were destroyed.

Trends in Succession

Whether primary or secondary, every succession on land has common characteristics. As succession occurs, small plants with short life cycles are replaced by larger plants with longer life cycles. With more producers, the amount of energy trapped during photosynthesis increases. In the

Experimental Plan

1. As a group, make a list of possible ways you might test your hypothesis using the materials your teacher has made available.
2. Agree on one idea that could be investigated in the classroom.
3. Design an experiment that will test one variable at a time. Plan to collect quantitative data.
4. Following the style of a recipe, write a numbered list of directions that anyone could follow.
5. Make a list of materials and the quantities you will need.

Checking the Plan

Discuss the following points with other group members to decide the final procedures for your experiment.

1. What variables will need to be controlled?
2. What is your control?
3. What will you measure or count?

4. How many groups of seeds will you count, measure, or examine? How many seeds will you have in each group?
5. What treatment will each group of seeds receive?
6. Have you designed and made a table for collecting data?

Make sure your teacher has approved your experimental plan before you proceed further.

Data and Observations

Carry out your experiment, make your measurements, and complete your data table. Make a graph of your results.

Analysis and Conclusions

1. According to your results, what conditions promote faster grass seed germination?
2. What was your experimental evidence for your conclusion?

3. Was your hypothesis supported by your data? Explain.
4. If your hypothesis was not supported, what types of experimental error might be responsible?
5. Consult with another group that used a different procedure to examine this problem. What was its conclusion?
6. In what way might farmers or ranchers benefit from the results of your experiment?
7. During what stage of succession after a forest fire would grass grow fastest?
8. Write a brief conclusion to your experiment.

Going Further

Based on this lab experience, design an experiment that would help answer a question that arose from your work. Assume that you will have unlimited resources and will not be conducting this lab in class.

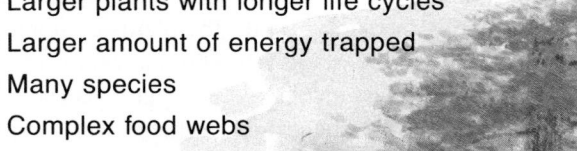

Figure 29-6 Every succession has common characteristics, which are summarized here.

- Small plants with short life cycles
- Small amount of energy trapped
- Few species
- Simple food chains
- Wasted energy

- Larger plants with longer life cycles
- Larger amount of energy trapped
- Many species
- Complex food webs
- Efficient energy use

Early stage

Climax community

early stages of succession, few species are involved, but as time goes on, the diversity of species increases. Food chains are simple at first, consisting mainly of producers, first-order consumers, and decomposers. Later, as the diversity of species increases, more complex food webs develop that include higher trophic levels and a greater variety of decomposers. More complex food webs result in greater use of the energy that enters the community, as less energy is wasted. Recycling of materials also becomes more efficient. Population sizes are kept steady by both biotic and abiotic factors.

All of these trends contribute to production of the climax community, which is complex and stable. At this point, the activities of the organisms are not likely to alter the physical environment in ways that would lead to further change. Thus, the community will remain basically the same unless disturbed by outside influences of nature or the activities of humans.

Section Review

Understanding Concepts

1. Use two examples from this section to explain how organisms alter the physical environment in such a way as to lead to their own elimination.
2. Suppose a meadow community were destroyed by fire. Will succession in that community occur more rapidly than the succession that originally produced the community? Explain.
3. Using the primary succession of a forest community from bare rock as

an example, explain how changes during succession illustrate each of the following trends in succession: (1) increase in species diversity; (2) change in size of dominant plant forms; (3) greater complexity of food webs; and (4) increase in amount of light energy trapped.

4. **Skill Review—Sequencing:** Sequence the steps you might expect to occur in a succession that begins on rock that has formed from lava after a volcanic eruption.

Assume the volcano erupted in a tropical climate. For more help, refer to Organizing Information in the **Skill Handbook.**

5. **Thinking Critically:** You know that a pond will undergo further succession eventually. How do you think the pond community developed in the first place?

29.2 Biomes

Within the United States are several types of communities. In much of the eastern part of the country, forests predominate. In the Midwest, grasslands are common. Deserts are found in the Southwest. These types of communities are not limited to the United States. They are found elsewhere in the world, too—forests in Scandinavia, grasslands in Argentina, and deserts in Northern Africa. Why are certain kinds of communities found only in certain parts of the world? What do those parts of the world have in common? Basically, they share similar climates.

Climate

Climates are a function of many factors. Among them are amount and pattern of rainfall, humidity, and temperature ranges. Such factors are determined by altitude, latitude, wind patterns, and topography (surface characteristics).

Temperature   The sun is important to organisms not only because it provides the light energy for photosynthesis but also because it heats Earth, providing temperatures at which organisms can live. Not all parts of Earth, however, receive the same amount of sunlight. Because of Earth's curvature, sunlight hitting areas at higher latitudes is spread over a wider area than is the sunlight striking near the equator, Figure 29-7a. Put another way, the area of Earth near the equator receives more light energy per unit area than areas north and south of the equator. Because they receive different amounts of light energy, different areas have different temperatures. These temperature differences are a major factor influencing distribution of organisms.

The tilt of Earth and its orbit around the sun influence the angle at which sunlight is received and, thus, the temperature, as seen in Figure 29-7b. For example, during the summer the United States is tilted

Objectives

Explain how physical factors interact to determine the life forms found in different biomes.

Compare the climates and common life forms of the world's major land biomes.

Differentiate among the various zones of which the oceans are composed.

***Figure 29-7**   Sunlight is five times stronger at the equator than at the poles because the surface of Earth is curved (a) and because Earth's axis is tilted (b). Seasons occur because of Earth's tilted axis and its orbit around the sun. December is winter in the Northern Hemisphere because the North Pole is tilted away from the sun. Why is June warm in the Northern Hemisphere?*

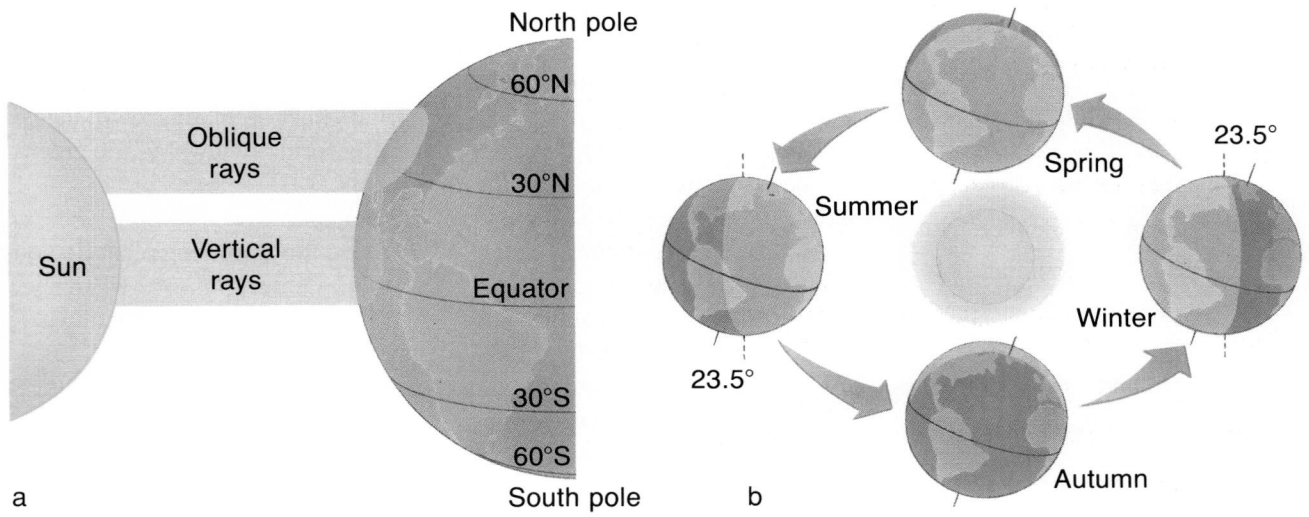

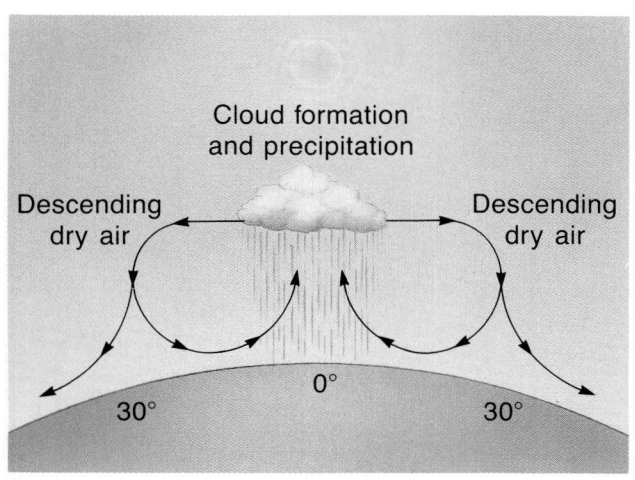

Cloud formation
and precipitation

Descending
dry air

Descending
dry air

30°   0°   30°

Figure 29-8   Deserts are common at latitudes 30° north and south because the descending air is so dry.

closer to the sun than in the winter, so it receives sun more directly in the summer. Less energy is lost as sunlight travels toward the United States in summer. In addition, the United States receives more hours of sunlight. As a result, temperatures are higher in summer than in winter. Seasonal temperature differences are greater at latitudes farther from the equator. You know this by comparing the seasonal differences in temperature in Maine or Michigan with those in southern Florida. Tropical areas, at or close to the equator, receive about the same large amounts of solar energy year round, resulting in a steady, warm temperature and no changing seasons.

Precipitation   Differences in temperature contribute to precipitation patterns. Warm temperature causes air to rise because warm air is less dense than cooler air. Near the equator, where air is warmest, the rising warm air takes with it a great deal of moisture from the oceans, Figure 29-8. As the warm air rises, it begins to cool. Because cooler air cannot hold as much moisture as warmer air, the moisture condenses as rain. Tropical areas receive large amounts of rainfall all year long. The cooled air that rises from the tropics and loses its moisture then spreads to latitudes 30° north and south before descending.

LITERATURE   CONNECTION

Winter Poem
by Nikki Giovanni

once a snowflake fell
on my brow and i loved
it so much and i kissed
it and it was happy and called its
cousins
and brothers and a web
of snow engulfed me then
i reached to love them all
and i squeezed them and they
became
spring rain and i stood perfectly
still and was a flower

Nikki Giovanni (1943-   ) is an African American poet, writer, and lecturer at Queen's College of the City University of New York. She says of poetry and life in general, "I just think things should mean something and I get confused when no meaning is found. We waste too much, we humans, because we refuse to recognize that there is a possibility of order and things making sense and us as a planet doing better."

As you can see, *Winter Poem* evokes not only the season of winter, but the quiet passage of winter into spring. Giovanni is fond of the order expressed in the quotation, "To everything, there is a season and a purpose under heaven."

Winter Poem suggests the flowering of the human spirit in the growth of a new season. It also describes the acceptance and joyful challenge of difficult times as in winter. On another level, the poem describes not only the cycle of seasons, but the cycle of phase change in water from solid to liquid. This, in turn, has an effect on plant growth, causing the spring renewal of the green Earth.

1. **Solve the Problem:** What do you think Giovanni means when she says, "We waste too much, we humans . . ."?
2. **Explore Further:** How does the cycle of seasons affect the life cycle of plants?

When it does descend, it is very dry. This dry air is a factor in the formation of deserts common at these latitudes, such as the Sahara Desert of northern Africa and the vast deserts that cover much of Australia. Changes in the temperature and moisture content of air occur at other latitudes as well. Depending on differences in temperature at those latitudes, different patterns of precipitation and climate result. In general, air moves from the equator north and south toward the poles. As Earth spins on its axis, that spin causes wind patterns in the air moving toward the poles.

Topography Mountain ranges illustrate the importance of topography in determining climate. Mountains can contribute to rainfall patterns. When wind pushes air toward mountains, the air is forced upward along the mountain slopes. As the air rises, it cools, and the moisture it holds condenses. Rain falls, like water being wrung from a wet sponge, on the side from which the wind is blowing (the windward side). The ample rainfall may be sufficient to support a forest community. By the time the cool air descends on the other side of the mountains (the leeward side), it has been wrung dry. Little precipitation falls on this side of the mountains, giving rise to a desert community.

Wind, temperature, altitude, and latitude interact to produce similar climates in large land areas around the world. These large geographic regions are known as **biomes.** Each type of biome is characterized by similar forms of dominant plant life such as grasses, conifers, or cacti.

Figure 29-9 Warm, moist air moving up the mountains cools and forms rain. Deserts form on the leeward side of the mountains as dry air descends to the surface and warms. These conditions exist on either side of the Sierra Nevada range in California.

Figure 29-10 The distribution of the world's biomes is shown on this map. Notice how areas with similar climates have the same biomes.

Ice

Tundra

Taiga

Temperate Forest

Tropical Rain Forest

Grassland

Desert

Tundra

Circling the globe at the northernmost latitudes is a biome having low average temperatures and little annual precipitation. The precipitation that does fall is mostly in the form of snow, in an amount equivalent to about 12 cm of rain per year. This biome, called the **tundra,** includes areas where the water in the soil is frozen during winter. The winter is long in the tundra. There are many months of twilight and darkness, when the sun is below the horizon the entire day. The result is a growing season that is very short. Although the top layer of soil thaws during the short summer season, the lower layer remains permanently frozen. This permanently frozen soil layer is known as **permafrost.** Low temperatures, poor soil, and limited water are the critical factors that determine which plants grow in this biome. Permafrost and the low precipitation level prevent the rooting and growth of forests in the tundra. There may be a few small trees, but they are widely scattered. Thus, the tundra resembles a plain. Lakes and ponds are plentiful.

Producers of the tundra include lichens known as reindeer moss, true mosses, grasses, sedges, and some low, stunted shrubs. During the brief summer season, which is quite cool, a variety of small perennial plants sprout and quickly flower.

Common consumers of a tundra include rodents such as mice, voles, and lemmings. Snowshoe hares, reindeer and caribou, wolves, and foxes live there, as do polar bears in areas near the ocean. Biting flies and mosquitoes are numerous; great swarms of them make life almost unbearable for other animals. Many birds, including ducks and geese, migrate to the tundra in summer to nest. Insects, seeds, and berries are plentiful. For the few months of summer, when the sun is above the horizon the entire day, the unbroken daylight permits nonstop feeding.

The fragile tundra biome is being changed by human activities. Disturbance of the vegetation that insulates the permafrost causes widespread erosion. Footpaths used for one season are still visible decades later. Vehicles have left tracks that have eroded into gullies three meters deep. The search for oil also poses grave threats to the tundra, reducing grazing ranges and disturbing the migratory patterns of animals.

Figure 29-11 **Animals of the tundra are adapted to life in the cold. Musk oxen, caribou, and snowy owls (from top to bottom) are found here. Harsh winters with strong winds, permafrost, short summers, and lack of available water prevent growth of large plants.**

Taiga

South of the tundra are the great coniferous forests that cover large portions of Canada, Alaska, and Eurasia. This biome, called **taiga,** is probably the largest of the world's biomes. The more southern areas of the taiga have a rainfall of about 35 to 40 cm per year. The taiga also has frequent fog and a low evaporation rate that results in very wet conditions. The soil is saturated with water and is acidic. There is no permafrost because the temperature is higher than that in the tundra.

Trees can grow in taiga soil. Once spruce and firs take hold, they further change an area by creating microenvironments. Trees of the taiga provide shade, protection, and homes for many organisms. Trees also cause changes in the soil. Although conifers are prominent, other trees grow there too.

Animal life common in the taiga includes moose, caribou, black bears, wolves, porcupines, weasels, and mink. Crossbills are birds well adapted to the taiga. Because their upper and lower bills overlap like a pair of scissors, Figure 29-12 (right), crossbills are able to break apart pine and spruce cones and feed on the seeds.

Temperate Forest

South of the taiga, where rainfall is greater and summers are longer and warmer, a different kind of forest biome is found. These regions, with a rainfall totaling about 100 cm per year, make up the **temperate forest** biome. As the name suggests, this biome is found in temperate regions of North America, South America, Europe, and Asia. Temperate forests are usually characterized by definite seasons. The growing season lasts at least six months.

Figure 29-12 The taiga, land characterized by coniferous forests, is the largest of the world's biomes. Taiga animals include moose (left), porcupines (center), and crossbills (right).

Layers of Vegetation   Temperate (and other) forests form a series of layers of vegetation from top to bottom. The top layer is the canopy (branches) of the trees. The canopy shades the ground below, much as an awning shades a window. It provides homes for animals such as insects, squirrels, and birds. Most of the food for these animals is produced in the canopy. Next is the shrub layer, consisting of shorter, shade-tolerant bushes. Under this is the forest floor. Light is scarce on the floor, so only plants that can grow well in dim light are present. Insects, other arthropods, and snakes are common on the forest floor. A fallen, rotting log may contain a microcommunity of life. Beneath the ground, in the soil, is the lowest level with many small arthropods, worms, and decomposers.

Each level of the forest has its own set of environmental conditions. Each is a separate microenvironment. Although an organism usually lives in only one level, the organization of the community depends on the interaction of all organisms.

Temperate Deciduous Forest   Most temperate forests in North America and Europe are composed of trees that lose their leaves each autumn. During the winter season, the trees are bare. Such trees are called deciduous, and the biome is known as a temperate deciduous forest. Most of the region east of the Mississippi River is the temperate deciduous forest biome of the United States, and is shown in Figure 29-2 (left) on page 815.

A temperate deciduous forest changes with the seasons. The falling of leaves in the autumn affects the community in many ways. Most important is that the food supply is reduced because food sources such as leaves, seeds, and fruits are less abundant. Also, without leaves, the effects of cold temperatures and wind are more severe. The net effect of these changes is a dormant period during the winter months. The forest floor is covered with dead leaves. Animal life is scarce as hibernation and long winter sleeps are common among the animals. The spring brings a rapid burst of growth as plants on the forest floor bloom and leaves on the trees unfurl.

Some of the major animals of the temperate deciduous forest in the United States are white-tailed deer, black bears, opossums, salamanders, and squirrels. Birds and insects are abundant.

Figure 29-13   Forests form layers of vegetation in which life forms vary. Why might a plant such as a vine be able to grow in the canopy but not in the shrub layer or on the forest floor?

Tropical Rain Forest

A different kind of forest occupies the equatorial regions around the world. These regions comprise the **tropical rain forest** biome. Tropical rain forests are found in Central and South America, central Asia, parts of Australia, and Africa. In a tropical rain forest, the temperature is almost constant at 25°C, and rainfall is heavy. A total of 200 cm or more of rain falls per year. Heavy rainfall and high temperatures produce a humid environment much like a florist's greenhouse, and growth occurs constantly. The tropical rain forest has the most abundant growth and the greatest species diversity of any biome.

Plants Most prominent of the plants in the tropical rain forest are large trees reaching 30 to 40 meters in height. Unlike coniferous and deciduous forests, a tropical rain forest has no dominant species of tree. In fact, hundreds of species may inhabit the same forest, and the few individuals of each species are widely scattered. As in a temperate forest, the branches of the trees form a large canopy, which is a major layer of the forest. Some taller trees may poke out above the canopy. The canopy is so thick that the forest floor may be dark even at noon. Because the forest floor is so dim, plant life there is scarce. A few shrublike plants may survive, but often the soil is bare.

Figure 29-14 Monkeys (center) and bromeliads (bottom) are among the enormous number of species that live in the hot, humid tropical rain forest.

Common in this biome are woody, vinelike plants called lianas. Lianas are rooted in the ground, but their leaves are up in the canopy where they receive sunlight. Also in the canopy are commensals such as orchids and bromeliads. The bromeliads trap and store water in the hollow, cuplike areas where their rosettes of leaves are joined. This water is used by the plant and by the animals that live in the canopy.

Animals Although plants are more evident, animals abound in the tropical rain forest. The forest is alive with arthropods such as beetles, butterflies, and grasshoppers and vertebrates such as snakes, lizards, and tree frogs. Beautifully colored birds including parrots, parakeets, and toucans are common. Most of the mammals in the rain forest live in the trees. Among them are bats, sloths, and monkeys.

Effects of Destruction Many tropical rain forests exist in developing countries of the world. For economic reasons, huge tracts of these forests are being destroyed. Trees are sometimes slashed, dried, and then burned, and crops are planted right in the ashes. Some cleared areas have been planted with grass to feed cattle. However, these cleared areas can be used in this way for only a few years. Because high temperature and humidity cause rapid decay of organisms, little humus accumulates in the thin tropical rain forest soil. Heavy rainfall leaches minerals from the soil. Thus, the soil is poor in nutrients to begin with, and those that are present are quickly used up by the new plants. The land is then abandoned as useless.

Tropical rain forests also have been cleared to provide a source of wood and to make space for new roads and mines. In the process, a great many species have been destroyed. Because the biodiversity in the tropical

Thinking Lab

ANALYZE THE PROCEDURE

How are tropical rain forests studied?

Background
Fifteen places in the world, totaling about half the size of Mexico, grow one-third to one-half of all the world's species of plants. These areas are parts of tropical rain forests that have been declining in area from more than 3.5 billion total acres in the early 1980s to fewer than 2 billion acres today. This map illustrates the 15 "hot spots," the most diverse parts of tropical forests.

Deforestation of 42 million acres per year results in losses of hundreds of species annually as roads are built and logging, ranching, farming, and con-

struction of settlements continue. Scientists agree that something must be done, but they disagree on which of two different research procedures should be used to study the problems of tropical rain forests. One procedure uses lengthy studies of entire forests and examines changes as they occur over long periods of time. Scientists from different parts of the world collect data from the rain forests, then take their data home, analyze it, and submit reports and recommendations for protection and management to the countries involved. In a contrasting proce-

dure, the focus of study is limited to gathering data over a short time in a few selected "hot spots" of biodiversity. Scientists then share their findings, which are used immediately in land-use planning and conservation efforts. Some researchers say that short-term studies cannot give accurate information about long-term ecological effects. These scientists also say that complex ecosystems cannot be studied for a brief period and yield useful data needed for long-term management. Advocates of short-term study say that their findings could be applied immediately and thereby give immediate protection to critical habitats.

You Try It
You are the scientist in charge of making a grant proposal for a research project that will provide data for developing conservation methods in South American tropical rain forests. **Analyze the procedures** that could be used in your research project. Write a short proposal that advocates one of the two procedures described above. Remember that what you write will determine how much money you get for your project.

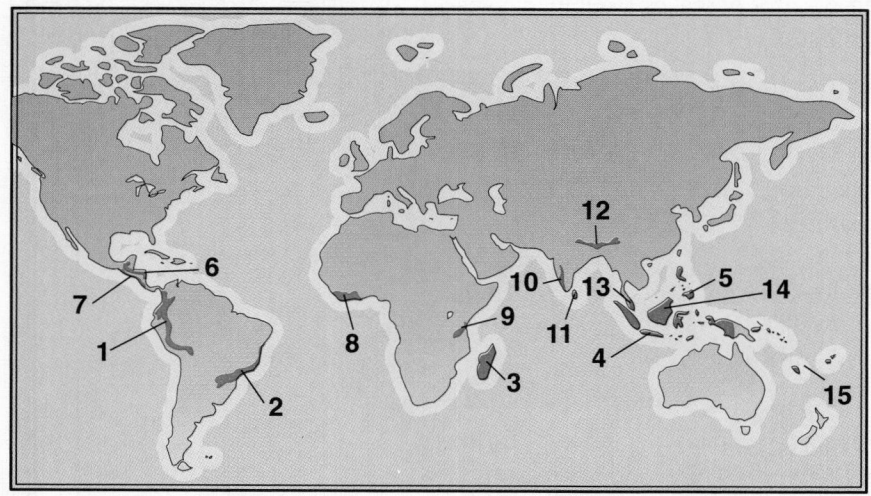

rain forest is the greatest in the world, biologists are concerned that potentially useful organisms (and chemicals produced by those organisms) are being eliminated before they can be discovered. It is also possible that removal of the forests may affect concentrations of both oxygen and carbon dioxide in the atmosphere and contribute to an increase in temperature worldwide. We will explore this idea in Chapter 30. It is clear that succession will not lead to restoration of a climax community like the original. Once destroyed, a tropical rain forest cannot be replaced. Recent efforts by scientists, governments, and environmental groups offer hope that the rate of destruction will be greatly reduced in the years ahead.

Grassland

Some areas in both temperate and tropical regions receive between 25 and 75 cm of unevenly distributed rainfall per year. Grasses are the major species of the climax community in these areas. Grasses also will grow where there is more than 75 cm of rain per year, but they are often crowded out in the shade of trees. Biomes in which grass is predominant have different names in different regions of the world. These names include steppe, prairie, plain, pampa, savannah, and veldt. In some regions where grasses predominate, there are also scattered trees. However, in general, these biomes are all **grasslands.**

Figure 29-15 **Herds of bison (bottom) and colonies of prairie dogs (top) are native to the North American grasslands.**

Plants and Animals   In the United States, grassland is found west of the Mississippi River to the foothills of the Rocky Mountains. Grasses are the natural climax species of this biome. The height of the grasses increases from west to east as average rainfall increases. At one time, grasses close to two meters tall were common.

Wildlife of the grasslands in the United States once included grizzly bears, bighorn sheep, and bison. Today, the wildlife includes antelopes, jackrabbits, gophers, ground squirrels, and meadowlarks. Hunting of ground-nesting birds, such as the prairie chicken, is restricted to prevent their extinction.

Effects of Humans   Grasslands are important to humans because of their use in agriculture. To take advantage of the high fertility of grassland soils, humans have changed this biome. They have replaced the natural grasses with cereal grains such as wheat and corn. Humans have been

Minilab 2

Which biomes are found in the United States?

The Potential Vegetation Map provided by your teacher is a map of the natural vegetation that could grow if humans had not interfered with the natural course of events. Using this map, plot a trip of 2000 miles between two major cities in the United States. Plan stops every 400 miles. Assume that the potential vegetation is growing. Describe what you would see at the beginning and end of your trip and at each of the 400-mile stopping points.

Analysis   *What biomes were you in at the beginning and end of your trip and at each 400-mile stopping point? How have human activities changed the environments where you stopped?*

careless with this biome. Overgrazing and trampling by herds of domestic animals have killed much of the grass. Much of the rich topsoil has been lost as a result of wind and water erosion of plowed soil. Constant plowing and lack of rain change the rich soil into loose dust. In the 1930s, winds picked up the dry soil and carried it for great distances. The south central part of the United States became known as the Dust Bowl, and many farms had to be abandoned. The tragedy of the Dust Bowl was a major factor in starting conservation practices.

Desert

Where rainfall is less than 25 cm per year and evaporation is rapid, there is little plant and animal life. Such areas are desert biomes. **Deserts** are found in the western United States, Africa, India, Asia, South America, and Australia. Depending on the amount of rainfall relative to the average temperature, deserts can be hot, cold, or temperate. The Mojave Desert of southern California is an example of a hot desert. The Northern Great Basin Desert in parts of Nevada, Utah, and California is a cold desert.

Deserts have little plant cover; bare ground is a common sight. However, the endless stretches of barren sand dunes that you probably picture in your mind are not typical of most deserts. The plants that are present are widely spaced due to the scarcity of water, Figure 29-16.

Not only is water scarce, but there is no regular pattern of rainfall. Long periods of drought are not uncommon and may last for many years. When it does rain, much of the water quickly runs off and is not available for use by the plants.

Adaptations of Plants   Organisms in desert biomes have many adaptations to dryness and temperature extremes. In plants, these include reduction of the surface area of leaves, waxy coatings on the leaves and stems, storage of water in fleshy parts, and large, shallow root systems for

Figure 29-16 **Poor soil and lack of water cause desert plants to be widely spaced.**

absorbing water quickly. Cactus leaves have become modified into spines, and the stem of the plant carries out photosynthesis. Some plants mature quickly and produce seeds in the few days when water is available. During droughts, some parts of plants may die, but the remaining parts can produce new growth when water again becomes available. Seeds of some desert plants can remain dormant for 50 years or more, so a species can survive even under the most adverse conditions. Other seeds contain inhibitors that prevent germination until the inhibitor is dissolved by water. Because a large amount of water is needed to dissolve the inhibitor, the plant starts to grow only when water is plentiful.

Adaptations of Animals Animals, too, are adapted to desert life. Conservation of water is important to their survival. Some desert animals obtain water from their metabolism. Water produced as a by-product of the breakdown of foods can be used for other purposes. Also, animals such as kangaroo rats excrete concentrated urine to conserve water.

Due to the sparse vegetation, temperature extremes are common in deserts, and many animals have evolved behavioral adaptations for protection. Small, burrowing rodents are active only in evening or during early morning or late afternoon. By staying in their burrows, they avoid the heat of the day and the cold of the night. Others stay in shady places during the hottest part of the day. Some burrowing animals undergo a period of estivation. As in hibernation, metabolism is slowed down. In this way, an animal survives adverse conditions.

Many desert animals, such as jackrabbits and kangaroo rats, move by jumping, allowing them to move quickly from place to place. Large animals do not usually live in a desert. They would be unable to find enough food, and they are not adapted to drought and temperature extremes.

Ocean

You are familiar with life on land. However, if you examine a globe, you will see that the surface of Earth is more than two-thirds water, mostly in the form of oceans. The kinds and number of organisms living in the oceans far exceed those on land.

Abiotic Factors   Distribution of organisms in the oceans depends mostly on abiotic factors. The most important of these is light. Light intensity decreases with depth because light is both reflected and absorbed by the water. The penetration of light varies with the contents of the water and with latitude, but it generally does not reach a depth greater than 200 meters. The oceans have an average depth of about 4000 meters, so the zone of water in which light is present is relatively small. Producers, which affect the makeup of any ecosystem, live only where light is sufficient for photosynthesis to occur. Thus, light is an important factor in limiting distribution of life in the ocean.

Temperature is also important to life in water. However, temperature does not vary as much in the ocean as on land because bodies of water absorb heat during the warm seasons and hold it during cold seasons. Thus, the sea is a more stable environment than the land. In the middle North Atlantic, for example, water temperature varies only about 10 C° from surface to ocean floor.

Organisms   Ecologists classify aquatic organisms into three major groups. Organisms that float in the water and are carried by ocean currents are called **plankton.** These include many types of unicellular algae, heterotrophic protists, and small multicellular animals. Copepods, which are microscopic crustaceans, are a major form of animal plankton. Jellyfish and their relatives and the larval forms of some other animals are also plankton.

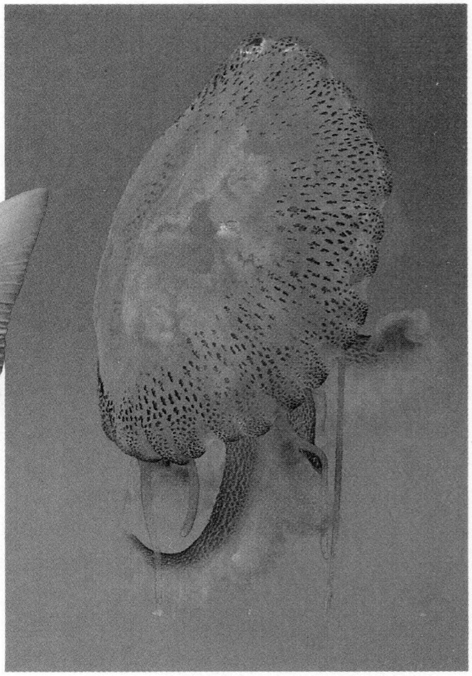

Figure 29-18   **Jellyfish (right) are plankton because they are carried by ocean currents. Fish (left) can swim, and thus are examples of nekton.**

A variety of animals can move freely through the water under their own power. These animals are called **nekton.** Nektonic animals include fish, whales, shrimp, and squid. Other animals live attached to or crawl on the ocean floor. They are called **benthos.** Included in this group are organisms such as barnacles, starfish, sponges, and clams. Benthic organisms are mostly found close to shore or in shallow water.

Zones Different parts of an ocean differ from one another in terms of both physical factors and the organisms present. Rather than referring to these different parts of the ocean as biomes, ecologists usually identify them as zones. Two zones are classified on the basis of presence or absence of light. Other zones are based on depth and the kinds of life present at different depths. These zones overlap one another. For example, the lighted zone is part of the top 200 meters of the entire ocean, regardless of total depth at any point. The different zones of the ocean interact with one another to a greater extent than do distinct biomes on land.

Ecologists often classify two main ocean regions with respect to depth. The area above the continental shelf is known as the **neritic zone.** The region of the deeper waters of the ocean basin is called the **oceanic zone.**

The portion of the neritic zone closest to the shore is the region most subject to change. Organisms in this area, called the **littoral zone,** are influenced by the action of tides, exposure to both water and air, temperature changes, and changes in amount of light and salinity of the water. Algae, both nekton and forms that are attached, are the producers of this zone. Animal life is diverse and includes clams, crabs, and barnacles—animals adapted to the changing conditions.

The littoral zone extends out to a point at which conditions become more stable. Life in this portion of the neritic zone is different. All organisms living there are strictly aquatic. Plankton and nekton populate the surface waters, and benthic animals are plentiful.

Figure 29-19 Barnacles are examples of benthic animals that live attached to surfaces.

Figure 29-20 The ocean is divided into zones based on depth and amount of sunlight penetration.

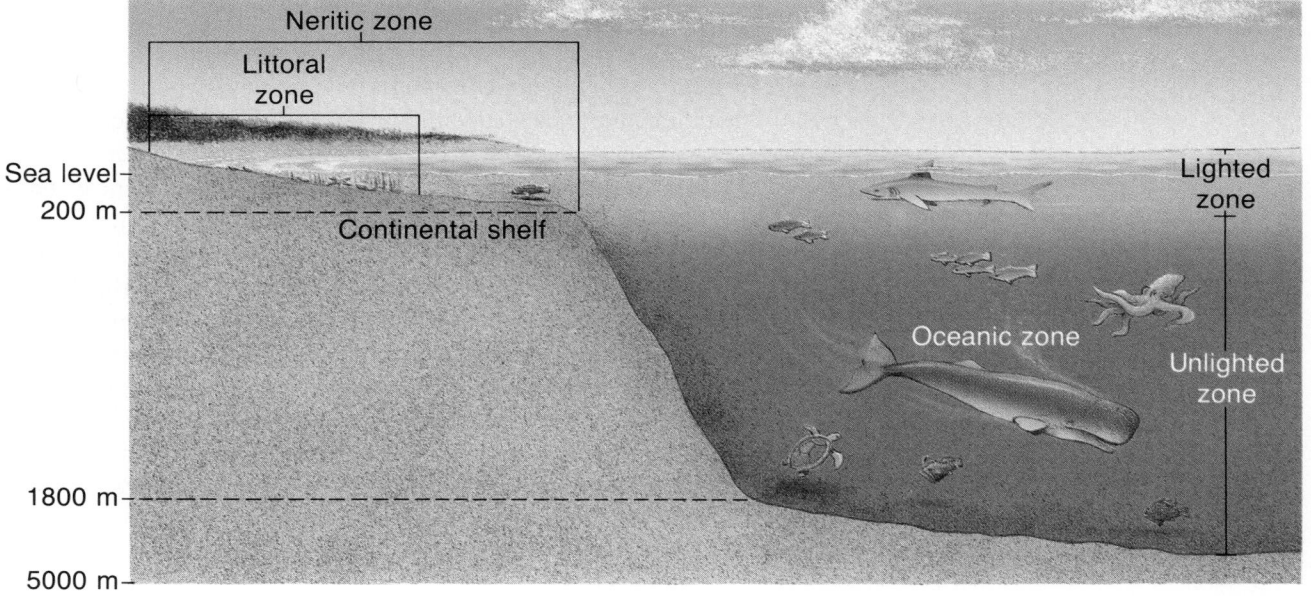

Figure 29-21 **The "headlight" of the angler fish attracts prey.**

In the deepest part of the ocean, the oceanic zone, life in the surface waters is similar to that of the neritic zone. In deeper waters, though, conditions and life forms are quite different. There light is absent, so no photosynthesis occurs. Some nektonic animals, mostly fish, live there. They feed on other fish and on dead organisms that sprinkle down from above. Some organisms that live in deep waters emit light. The light given off by the "lure" of one type of angler fish attracts prey.

Bacteria decompose organic materials that settle onto the ocean floor when organisms die. The organic materials are broken down into simpler molecules such as carbon dioxide, nitrates, and phosphates. These simple molecules must be made available to producers in the ocean. If materials continue to build up on the ocean bottom, a point would be reached where life could no longer continue. An ecosystem can survive only if materials are recycled.

Circulation of Water   Return of materials to the surface of the oceans is affected mostly by water temperature and wind. In cold climates, surface water becomes colder than water below the surface. Because cold water is more dense than warm water, cold surface water sinks and warm bottom water rises, setting up a vertical current. Also, the wind causes horizontal currents that move the surface water. Water below the surface rises to replace the water displaced by the wind. Thus, the ocean water is stirred as if in a huge mixing bowl. Although both of these processes return decomposed materials to the surface, water temperature differences account for most of the return.

Unique Ocean Communities

Most of the ocean floor is like a cold, barren desert. Light does not reach the ocean floor so there can be no photosynthesis. Because of this, it was long thought that few, if any, life forms could exist in this area. However, in recent years, scientists have discovered unique communities of animals living on the ocean floor at depths as great as 3000 meters.

Deep-sea Vents   Many of these unusual communities exist in the Pacific Ocean. The first community was found in 1977 off the Galapagos Islands, and others were later discovered off the coasts of Mexico and Peru. Although the water temperature on most of the ocean's floor is near freezing, each of these Pacific communities is in an area of warm water. The communities are located over areas of molten rock where new ocean floor is forming. Cold seawater seeps into cracks in the new crust of the ocean floor, comes into contact with the molten rock below, and becomes heated. The hot water, which may be as hot as 350°C, gushes up through deep-sea vents—openings that look like small volcanoes. The newly discovered communities thrive in the warm water around the vents.

Animal life never before seen inhabits these communities. Brown mussels and white clams that average 30 cm long live among pillows of cooling lava. Clusters of orange and white crabs, starfish, octopuses, and unusual fish move around the vents. Strangest of all are colonies of giant tube worms, Figure 29-22, some of them as long as three meters, that live around the vents. Many other life forms collected from the sites differ so much from known animals that they have not yet been classified.

Deep-sea Energy Relationships Every community must have an outside energy source and producers that trap the energy to make food. In most communities, light is the energy source, plants or algae are the producers, and carbohydrates are made by photosynthesis. But, there is no light energy in these ocean floor communities. Instead, there is geothermal energy. **Geothermal energy** is heat energy from the interior of Earth. This energy, present in the molten rock, increases the temperature of the water. The very hot water is important because it reacts with minerals in the rocks and the seawater to produce an energy-rich compound—hydrogen sulfide. Chemosynthetic bacteria are the producers of these unique communities. Like other producers, chemosynthetic bacteria make organic molecules. However, these bacteria make carbohydrates by using the energy in hydrogen sulfide. Thus, these bacteria are the basis of the community's food web. Animals such as mussels and fish feed on the bacteria and are, in turn, preyed on by other animals.

Tube worms and clams depend on the bacteria as well, but in a unique way. Instead of feeding on the bacteria, these animals live with them in a mutualistic relationship. The bacteria live in different parts of the tube worm's body. In clams, the bacteria live in the gills. Both animals extract hydrogen sulfide from the water and transport it in their blood to the bacteria. In turn, the bacteria use the chemical energy in hydrogen sulfide to provide food for the animals through chemosynthesis.

In 1986, deep-sea vents were discovered in the Atlantic Ocean. This discovery suggests that hot-spring ocean communities may occur wherever new ocean floor is forming. It also may mean that these vents have a marked effect on both the temperature and mineral content of the oceans. More recently, a freshwater vent was discovered in a lake in Siberia. The water there is not nearly as hot as that around ocean vents, and the community around the vent has fewer animals. It is thought that the community is quite young, and study of it should shed light on how new species colonize the area.

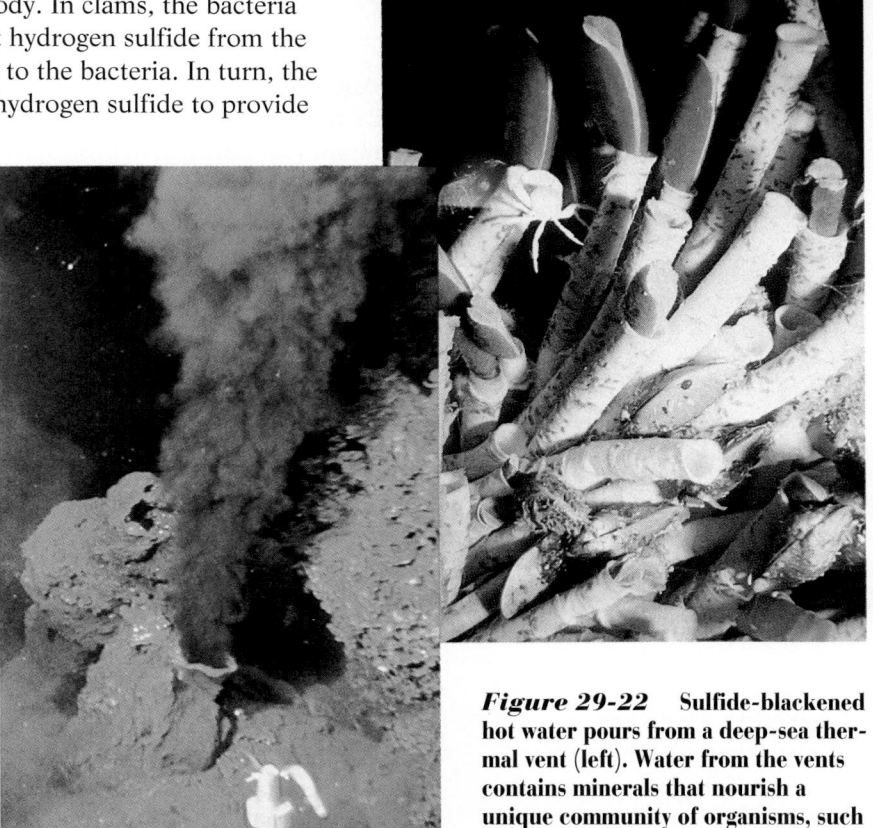

Figure 29-22 Sulfide-blackened hot water pours from a deep-sea thermal vent (left). Water from the vents contains minerals that nourish a unique community of organisms, such as the tube worms shown here (right).

The Biosphere   Just as the many zones of an ocean interact with one another, so, too, do the various biomes on land. There are no sharp divisions between biomes. Rather, one type of biome gradually gives way to another, much as wet strokes of watercolors blend into each other along their edges. All the world's land biomes, as well as ocean and freshwater ecosystems, interact with one another. Together, they make up the highest level of biological organization, the **biosphere.** No one part of the biosphere is isolated from another. Therefore, what happens in one region may affect other regions as well. Understanding of this relationship among the world of life and making sound decisions based on such understanding is crucial to the future of humans and all other life forms.

Figure 29-23   **The biosphere is the thin layer of Earth's surface where life exists.**

LOOKING AHEAD

COMMUNITIES DEVELOP over time as they undergo ecological succession. Biotic and abiotic factors interact to influence the distribution of communities on Earth. What role have humans played in ecological succession? How can we lessen the harm we do to the environment? In the next chapter, you will study ways in which human activities have adversely affected individual organisms, communities, and entire ecosystems, and how many of those activities pose a threat to humans themselves. You will also examine possible remedies for correcting past mistakes and plans for avoiding future ones. ✍

Section Review

Understanding Concepts

1. How can the cold, dry conditions of the tundra biome be explained?
2. Why do you think the taiga receives more precipitation than the tundra?
3. Describe the changes in benthos you would note if you could walk along the ocean bottom from the shoreline to the greatest depths.

4. **Skill Review—Making Tables:** Construct a table comparing the following characteristics in the tundra, taiga, temperate deciduous forest, desert, grassland, and tropical rain forest biomes: average precipitation; temperature range; dominant plant species; and common forms of animal life. For more

help, refer to Organizing Information in the **Skill Handbook.**

5. **Thinking Critically:** Consider the layers of vegetation in a tropical rain forest. Most organisms in that forest inhabit mainly one layer, such as monkeys in the canopy. How is the survival of the monkeys linked to organisms living in different layers?

Considering the Environment
•••••••••••••••••••••

Wetlands

Swamps, bogs, potholes, muskegs, marshes—the names vary by region and location, but they are all lands that are wet for at least part of the year. Historically, wetlands have been places of legends and loathing. These myths, combined with a proven relationship between wetlands and certain mosquito-borne diseases such as malaria and yellow fever, have shaped our attitudes about wetlands.

Two thousand years ago, one of the first large-scale public works programs was the draining of the Pontine Marshes in Italy. This project was started by the Roman emperors as a solution to a problem with malaria and to create prime farmland. In the early history of the United States, the federal government encouraged the draining of swamps by giving 64 million acres of wetlands to those states that would drain them.

But a lot has been learned since then. We have found more effective ways to control disease and have discovered that wetlands are essential to a healthy ecosystem. After centuries of neglect, why are we so concerned about the preservation of wetlands?

Coastal Wetlands   Consider just one type of wetland—the type that borders our coasts. These coastal marshes, swamps, and bogs are the homes of many unique plants and animals. Besides providing humans with food such as cranberries, ducks, geese, and turtles, coastal wetlands filter water before it enters underground reservoirs. These reservoirs are the source of much of the water we use for drinking and washing. Another benefit of coastal wetlands is that they absorb many pollutants in the rivers and streams that feed them, preventing contaminants from entering the ocean. Coastal wetlands also have a direct effect on jobs. Most of the fish harvested by the United States fishing industry live part of their life cycle in coastal marshes. In addition to providing a livelihood for people along the coast, wetlands also provide a protective buffer against the seasonal storms and flooding that threaten coastal communities every year.

Potholes   Other wetlands, such as the potholes in the northern prairies, have a direct effect on wildlife. The potholes serve as the breeding ground for about half the wild ducks in North America. For decades, farmers were encouraged to drain and fill these small ponds and marshes, turning them into farmland. However, when the duck population started to decline, it was realized that the potholes were a valuable resource. Efforts are now underway to preserve them.

Bogs   A fascinating type of wetland with many unique inhabitants is the bog. In the northern part of North America, these diverse communities are called muskegs. Floating on the acidic water of these bogs is a mat of sphagnum moss thick enough to support a person's weight. You may have seen sphagnum moss at garden centers, where it is sold as a soil conditioner. Because it absorbs large amounts of water and has antibiotic properties, it was used as a diaper by Native Americans and as a surgical dressing in World War I.

Bog Life   Bogs support a unique and diverse community of plants and animals. Because the water is acidic and nutrient-poor, most species of plants cannot grow there. However, the ones that do survive among the sphagnum are adapted in unusual ways. A major requirement for plant growth is nitrogen, and bogs have notoriously low levels of nitrogen. But plants such as the Venus's flytrap, pitcher plant, and sundew solve this problem in a distinctive way—they trap and digest insects. Pitcher plants are shaped like a funnel. Along the sides are attractive spots of color and drops of nectar. When the insect enters the funnel, it follows the nectar deep into the plant. However, when the insect tries to leave, it finds that its path is blocked by inward-bending hairs. Exhausted by the struggle to get out, it eventually falls to the water-filled base of the plant and is digested. An even more active bog-based insect hunter is the Venus's flytrap from North and South Carolina. These plants have V-shaped leaves that are fringed with long spines. When an insect lands on the leaf, attracted by the color and nectar on the surface, it brushes against one or more trigger hairs in the center of the leaf; the leaf then closes. In a short time, the insect is digested and the leaf opens again, ready for the next victim.

Because bogs are acidic and lack oxygen, plants do not decay rapidly when they die. Thus, bogs build up thick layers of plant material called peat. Peat, after drying, is used as fuel in many parts of the world. Ireland, for example, uses peat as a major source of energy and, because of overharvesting, has destroyed the majority of its bogs. Fortunately, most North American bogs are untouched, so we have the opportunity to preserve these unique and productive ecosystems.

The Everglades   The Everglades, a huge swamp in Florida, is a study of what can happen if humans aren't careful when they manage wetlands. For 5000 years, a huge wet prairie of plants growing in constantly wet muck prospered on the southern tip of Florida. Fresh water from Lake Okeechobee and the Big Cypress Swamp combined to form a river 50 miles wide, but only inches deep, flowing south. The constantly saturated muck underlying this river supported a lush and varied freshwater ecosystem that was unique in all the world. An added benefit in low-lying Florida was that this constant flow of fresh water replenished the water table and kept salt water from the ocean from migrating inland.

But the winter climate in this area is very comfortable, and increasing numbers of people have moved to Florida to live. In addition, it is a prime area for agriculture. So some of the land was drained and the natural system was replaced with a series of artificial canals and levees. These allow engineers to adjust the water levels and prevent flooding during hurricanes. However, there isn't enough reserve water to keep the Everglades wet during a drought and it dries out. Under these conditions, massive fires have erupted. Because the soil is almost pure organic matter, it can smolder and burn for months, destroying the soil and seeds of the future generations. At the same time, the lack of fresh water allows salt water from the surrounding ocean to infiltrate the water table, making it unsuitable for human or plant consumption.

Now that these relationships are better understood, efforts have been proposed to prevent drying and to preserve the wildlife in the Everglades. These will require billions of dollars and years of time to accomplish. Perhaps the best lesson to be learned is that humans must proceed very carefully when changing what millions of years of natural processes have built.

1. **Connection to Chemistry:** Research the composition and chemistry of swamp gas. Propose ways in which it could be used by humans.
2. **Connection to History:** George Washington was one of the surveyors who laid out the plan to drain the Dismal Swamp. Where is the Dismal Swamp? Find out if the plan was carried out.

Friendly Fires

Fire can be deadly, and it may come as a surprise to you that wildlife biologists sometimes start fires deliberately. Are these biologists literally playing with fire?

Most forest fires that reach the evening news are terrifying. Some, such as the huge lightning-triggered 1988 conflagration in Yellowstone National Park, create fire storms measuring more than 1000°C with tornado-like winds that incinerate everything in their path. Why would anyone deliberately start something that had such a devastating potential?

One reason is that controlled fires can actually be beneficial to a wild community, which needs periodic fires to maintain a healthy environment. If a wild area is protected from fire, the underbrush grows thick, and dead branches build up on the forest floor. This inhibits healthy new growth, and when a fire does occur, it is much larger and hotter than normal. However, if burning is controlled, the size and temperature can be held to levels that encourage recovery.

A fire triggers germination of many seeds, which starts the rebuilding process. For example, the lodgepole pine produces two kinds of cones. One type is like other pine cones and opens in the fall, releasing seeds. However, the other type opens only when it is heated. This ensures that the species will survive a fire. Other species survive by producing tough-coated seeds that can resist fire and germinate only when wet.

Why Not Set Friendly Fires?

Opponents of controlled burns point out that these fires sometimes grow uncontrollably and exceed the ability of humans to manage them. If the wind shifts or something unforeseen happens, there is little that humans can do to reestablish control. In the past, homes and businesses have been lost to controlled fires that got out of control. Opponents ask, Why burn trees that can provide jobs and lumber, when loggers could accomplish the same results? Opponents feel that humans have been given a responsibility to manage these resources, and anything less than our maximum effort to preserve and use them wisely is a breach of trust.

Prairie Fires   Another area where fire is in the natural scheme of life is the prairie. If not burned regularly, woody plants and other invaders start to take over these fertile grasslands. Before it was settled by humans, the prairie was regularly ignited by lightning and other natural causes. Prairie grasses are protected from fire by their deep roots. When fires race across the flat grasslands, the tops of the grasses are burned along with everything else. However, deep in the rich turf, the roots survive—ready to burst forth in growth after the first rain. Other seed plants are not so well adapted. The fire destroys the mature plants and seeds alike, and it may take several decades before a tree can grow again. By that time, other fires may have destroyed any succeeding seed

plants that have begun to grow. Thus, the vast sea of grass that covered the Midwest prospered, in part, because of fire.

Understanding the Issue

1. Recently, the United States Park Service adopted plans that give managers great latitude in deciding whether to actively fight fires ignited by natural causes such as lightning. Unless they threaten human lives or property, many fires have been allowed to burn themselves out as part of a natural process. Discuss the advantages and disadvantages of this policy.

2. Do you agree with opponents of controlled fires when they say that loggers could accomplish the same results as a fire? Why or why not?

Readings

- Farney, Dennis. "The Tallgrass Prairie: Can It Be Saved?" *National Geographic,* January, 1980, pp. 37-61.
- Jeffery, David. "Yellowstone, The Great Fires of 1988." *National Geographic,* February, 1989, pp. 255-273.

Summary

During ecological succession, communities change over time. Succession may begin where no community has existed before or may follow destruction or alteration of a previously existing community.

In each stage of succession, the organisms alter the physical environment, making it more suitable for other organisms and less suitable for themselves. Succession ends with the development of a climax community characterized by stable population sizes, complex food webs, and efficient use of energy and materials.

Physical factors including temperature, rainfall, altitude, latitude, and topography interact to produce major regions of climate called biomes. Each type of biome is home to a similar group of communities.

Major world biomes include tundra, taiga, temperate forests, tropical rain forests, grasslands, and deserts. The ocean consists of several zones that are similar to but more closely interdependent than biomes. All the world's biomes and aquatic ecosystems interact. Together they form the total world of life—the biosphere.

Language of Biology

Write a sentence that shows your understanding of each of the following terms.

benthos	neritic zone
biome	oceanic zone
biosphere	permafrost
climax community	pioneer stage
desert	plankton
ecological succession	primary succession
geothermal energy	secondary succession
grassland	taiga
littoral zone	temperate forest
microenvironment	tropical rain forest
nekton	tundra

Understanding Concepts

1. A meadow exists next to a temperate deciduous forest. Is the meadow undergoing succession? Explain.
2. The banks around a small lake are sloped, and soil slowly but constantly runs off into the lake. What effect will such runoff have on the lake? Explain.
3. Construct a graph in which the *x*-axis represents time and the *y*-axis represents number of species. Use a dotted line to represent plant species and a solid line to represent animal species present during the course of a succession.
4. Why is a climax community not likely to alter the physical environment in such a way as to lead to further succession?
5. Branches of tropical rain forest trees are arranged in a crown near their very tops. Why?
6. Explain how stages of succession would result in the formation of layers of vegetation in a temperate deciduous forest.
7. In what ways does a pattern of layers of vegetation, as it is established, modify the environment?
8. Predict how modification due to the formation of layers of vegetation may influence the kinds of organisms that appear later in the succession.
9. What factor limits life on the floor of a tropical rain forest and in deep ocean waters? How is life in those two places similar? Why?
10. Cite an example that shows how different ocean zones are interdependent.

REVIEW

Applying Concepts

11. How would the rotting of railroad ties on abandoned tracks modify the environment?
12. What might be the results of the modification of the environment caused by the rotting of the railroad ties?
13. Would you say that there is a pattern of layers in an ocean? Explain.
14. Think about the general pattern of biome types that exists from the equator to the poles. Do you think that you would see a similar pattern if you were to climb from the base to the top of a very high mountain? Explain.
15. If you were in the commercial fishing business, would you fish in coastal regions or far out in an ocean? Why?
16. Suppose you are analyzing two soil samples, one from a temperate deciduous forest and one from a tropical rain forest. Which sample would you expect to have a greater concentration of organic material? Why?
17. Compare the length of plant life cycles in a tundra or desert and in a tropical rain forest. How is the difference related to climate?
18. How would you expect differences in plant species to affect differences in bird species during a succession?
19. **Considering the Environment:** Why is sphagnum moss sold in garden supply stores as a soil conditioner?
20. **Issues:** What would most likely be the climax community of a prairie if fires did not occur regularly? Explain your reasoning.

Lab Interpretation

21. **Investigation:** Examine the graph at the top of the next column. What conditions are needed for radish seed germination?

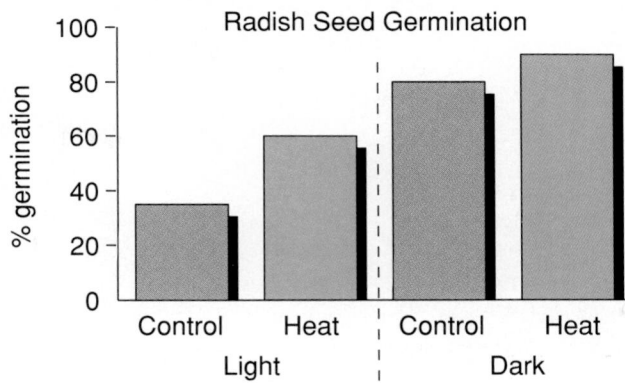

22. **Minilab 1:** The following graph represents percentages of organic matter in four soil samples. Which sample would be best for plant growth? Explain why.

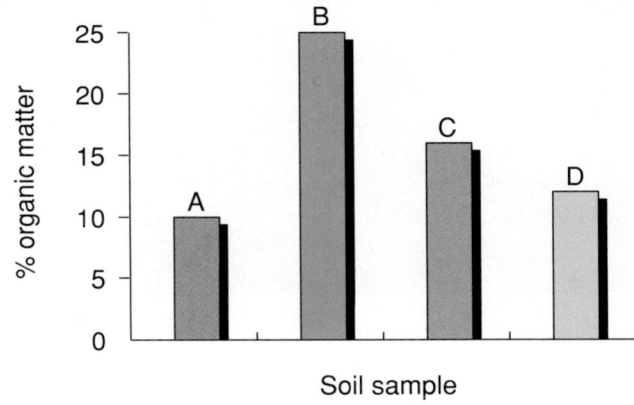

23. **Minilab 2:** Assume that you are a scientist studying the effects of acid rain in the United States. You want to determine its effects on temperate deciduous forests. To what part of the country would you travel to study these forests?

Connecting Ideas

24. Could a succession involve evolution of new species? Hint: Think about the Galapagos Islands. Was succession on the Galapagos Islands primary or secondary?
25. In what ways are the organelles in a cell or the organs in a system like the biomes of the biosphere? Explain.

HUMANS AND THE ENVIRONMENT

RAIN FALLS in this tropical forest almost every day. The tallest trees and vines grow hundreds of feet above the ground and are the first to catch the rain. They bend and sway under the powerful force of torrential downpours. Water runs off one leaf onto another, then another, then along a mossy tree branch. Now flowing gently, the rainwater continues its downward course, dripping softly onto ferns and other delicate undergrowth. Finally, the water soaks into the soil of the forest floor.

What would happen if those torrential rains fell, not on the lush growth of a forest, but on bare ground? When forest, grassland, or other plant cover is destroyed, the soil is left exposed to the ravages of rain and wind. Precious topsoil can be washed or blown away, making it difficult for plants to take root. The forces of erosion can gouge out barren gullies where healthy plant life once thrived. In this chapter, you will learn how the activities of humans can change the environment. You will find out how pesticides have been both helpful and harmful. You will learn the causes of several environmental problems humans face today, including acid rain, global warming, and ozone depletion. And you will discover how conservation of Earth's forests and other natural resources can help us overcome these problems.

What do you think are the biggest environmental problems facing the human population today? What kinds of actions can you and your classmates take to help reduce pollution in your community?

Chapter Preview

Contents

Objectives

Analyze the pros and cons of pesticide use.

Discuss methods that can be used to control pests while also safeguarding the environment.

30.1   The Pesticide Dilemma

Imagine taking a long car trip. Perhaps you're traveling from New York to Los Angeles, or from Chicago to Dallas. Your journey begins downtown, where streets and freeways are crowded with cars, trucks, and buses, and high-rise office buildings tower overhead. As you reach the outskirts of the city, you pass warehouses, factories, railroad yards, and finally, the airport. Next, you travel past row after row of suburban houses and lawns. Then you begin to see herds of dairy and beef cattle. Fields of wheat or corn, and rows of tomatoes, squash, and melons stretch out into the distance. Farther on, you see streams, meadows, a lake, and perhaps some distant mountain peaks.

Sources of Pollution

Without modern advances in technology, this trip wouldn't be possible. The car itself, the gasoline that fuels it, and the highways it travels on are all products of technology. So are factories, power plants, trains, and planes. Today's food crops are grown by agricultural techniques—including the use of pesticides, fertilizers, and farm machinery—that help with almost every task from plowing the ground and planting the seed to harvesting the crop. Technology also provides us with television, artificial lighting, computers, telephones, and many other items we consider indispensable. But are we paying an unexpectedly high price for our modern conveniences?

Figure 30-1   **Technology has provided us with many modern conveniences. Most of them come with a high environmental cost.**

While advances in agriculture and industry have been responsible for enriching and prolonging the lives of humans, they have also created undesirable by-products. The exhaust from your car or the smoke from a city factory may have an effect on the lake you've traveled hundreds of miles to see. The pesticides that protect a corn crop from insect damage also affect the health of birds, fish, and humans that consume the corn. Automobile exhaust, factory emissions, and pesticides are examples of materials that can cause **pollution,** which is the contamination of air, water, or soil by materials released into the environment by the ever-increasing numbers of humans and their activities.

Pest Control and Pollution

Have you ever put a flea collar on a cat or dog, put out poisoned bait for ants or cockroaches, or watched a gardener spray weed killer on a patch of crabgrass? These are examples of pesticide use. A **pesticide** is a chemical poison that is used to kill an unwanted organism. There are many organisms that humans consider pests, including termites that destroy wood, moths that eat woolen fabrics, and aphids that can kill houseplants.

Agricultural Pests   In addition to pests that directly affect humans in and around the home, there are many pests that cause severe damage to agricultural crops, livestock, and forests. For example, weeds compete with and crowd out cultivated plants, causing nearly ten percent of each year's crop losses. Types of fungi known as "rusts" reduce the yield of wheat and other grain crops by millions of bushels each year.

Many agricultural pests were brought to North America from other parts of the world. Among these are the European corn borer, a moth larva that attacks ears of corn and limits the corn growing season in some parts of the country. Another example is the Mediterranean fruit fly, sometimes called the medfly. The larvae of this tiny fly can so severely damage citrus, apple, and other fruit crops that infested fruit cannot be shipped to market for fear the pest will be carried to other regions.

Altering Natural Ecosystems   A field of corn can tolerate a few corn borers without suffering much damage. And a few medflies probably wouldn't destroy all the apples in an orchard. These organisms become pests when their populations grow so large that they cause severe damage and significantly reduce crop yields.

Figure 30-2   **Medfly larvae can damage crops of fruit severely. The pesticide malathion was used widely in California to control the medfly.**

Humans have actually contributed to pest problems by altering natural ecosystems. For example, in nature, several plant species usually grow together. One plant may serve as host to an organism that preys on the pest of a neighboring plant. By planting large areas of land with a single food crop, humans reduce this natural diversity and disturb the relationships among predators and prey, beneficial and pest organisms. As a result, single-crop planting can give pests an opportunity to thrive because fewer of their natural enemies are present.

Humans have also contributed to changes in natural ecosystems by transporting organisms from one part of the world to another. For example, medflies were probably originally introduced into the continental U.S. in fruit smuggled from Hawaii. Since medflies had no natural predators here, their numbers grew unchecked, their population density increased, and they became a pest. Many other pests, including the Japanese beetle and the gypsy moth, have been introduced into the U.S. from other parts of the world.

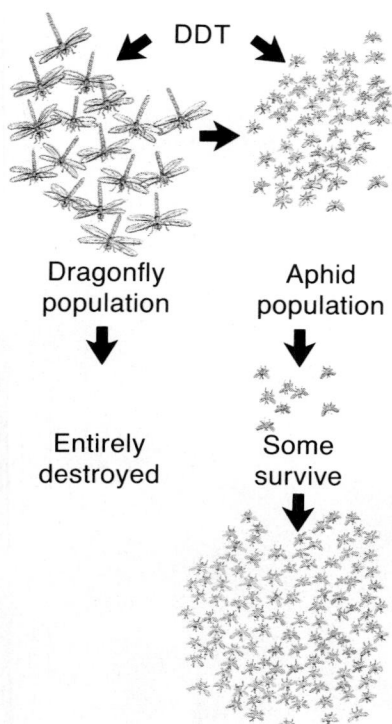

DDT

Dragonfly
population

Aphid
population

↓

↓

Entirely
destroyed

Some
survive

↓

Figure 30-3   **Although pesticides are useful in eliminating pests from crops and allowing us to purchase mold-free and worm-free fruit, they do not come without problems. In some instances, pesticides destroy not only the target pest but another whole population. Dragonflies are natural predators of aphids. Spraying an area with a pesticide to kill aphids may kill all dragonflies, while some aphids may survive. Without the natural predator, the aphid population would grow even larger than it was originally.**

The Benefits of Pesticides   Chemical pesticides have made it possible for humans to successfully replace natural ecosystems with farmland, lawns, golf courses, and gardens. Pesticides have enabled farmers to produce larger quantities of food and have helped reduce the incidence of insect-borne diseases such as malaria, yellow fever, and typhus by reducing the populations of the disease-carrying insects. Because pesticides are effective, easily obtainable, and relatively inexpensive, they have become convenient weapons for farmers, homeowners, and public health workers all over the world. Today, about 2.3 million metric tons of synthetic pesticides are used worldwide each year.

Problems with Pesticides

If pesticides killed only the pests for which they were intended, there would be few problems with their use. Unfortunately, that is rarely the case. When a pesticide is sprayed on a crop, it is estimated that only ten percent, or even less, of the chemical spray actually comes into contact with the insect pests that are being targeted. The remaining 90 percent is taken up by other, often harmless, organisms, as well as soil and water.

Pesticides often destroy not only pest organisms, but also their natural predators. Predators are fewer in number than their prey, and it is possible for a pesticide to completely destroy the predator population. If any pests survive and reproduce, or if they are reintroduced from another area, there may no longer be any natural predators to keep their numbers in check. As a result, the pest problem can become worse than it was originally. Other beneficial organisms, such as honeybees and other insects that help pollinate flowering plants, may also be destroyed by pesticide applications. Entire ecosystems are affected by pesticide use. Sometimes these effects can spread over great distances and last for many years.

DDT   In 1939, a chemical called DDT was found to be extremely effective in controlling insect pests. It quickly gained popularity and was used all over the world to control many kinds of insects from malaria mosquitoes and corn borers to fleas and houseflies. However, DDT has some very serious side effects that were not immediately evident. By 1972, the severity of these side effects had become so alarming that the use of DDT was banned in the U.S. Even though DDT is banned in the U.S., it is still used in large quantities around the world. It is very inexpensive to manufacture, and more DDT than ever before is being sold—mostly to developing countries.

Biological Resistance   During the 1940s, DDT was widely used to kill houseflies. However, some flies were genetically resistant to the chemical. These resistant flies survived and reproduced, passing their DDT-resistant genes on to their offspring. As a result of natural selection, a resistant strain of houseflies rapidly evolved.

DDT is not the only pesticide to which insect pests have developed resistance. There are several species of mosquitoes that transmit malaria; most of these have become resistant to a number of pesticides, including

DDT. This resistance was responsible for a great increase in the number of cases of malaria during the 1970s and 1980s. It is estimated that a total of 500 insect species around the world have become resistant to some of the pesticides that are currently in use. Twenty insect species are thought to be resistant to all currently used pesticides.

This trend toward development of pesticide resistance also has been observed among non-insect pests including rodents, fungi, and weeds. As organisms develop resistance to pesticide chemicals, it becomes necessary to apply the chemicals in stronger concentrations or spend time and money developing more effective, and sometimes more poisonous, compounds.

Figure 30-4 One of the most effective and toxic pesticides ever used to kill insects was DDT. It was developed in the 1940s and used extensively to kill disease-carrying mosquitoes, as well as household pests and insects that destroyed crops. DDT was even used on mattresses in Army barracks to fight bed bugs. Unfortunately, DDT affected many organisms besides pests, disrupting entire ecosystems.

Biological Magnification   Another problem with DDT is that it remains toxic long after it has been released into the environment. A chemical or other substance that cannot be broken down or degraded into harmless, inactive compounds by the metabolism of living organisms is **nonbiodegradable.** A number of nonbiodegradable pesticides that are chemically related to DDT have been developed, including DDD, DDE, dieldrin, and chlordane. DDT breaks down into DDD or DDE in the environment. DDT and its chemical relatives are insoluble in water but can be dissolved in fats and oils. Molecules of these substances remain on plants or in the soil for long periods because they are not easily washed away. Organisms that absorb molecules of DDT, or consume plants or soil contaminated with it, cannot break down the chemical in their bodies. Instead, it is stored in their fatty tissue.

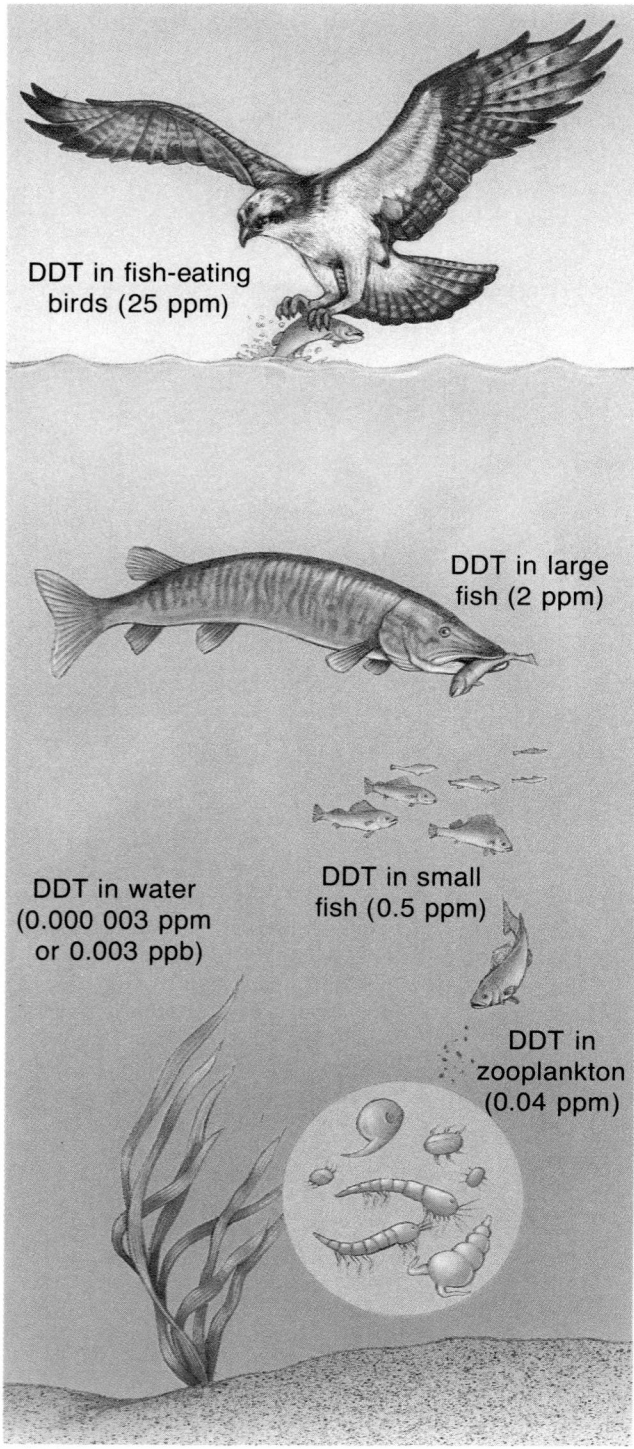

DDT in fish-eating
birds (25 ppm)

DDT in large
fish (2 ppm)

DDT in water
(0.000 003 ppm
or 0.003 ppb)

DDT in small
fish (0.5 ppm)

DDT in
zooplankton
(0.04 ppm)

Figure 30-5 **As top-level carnivores, ospreys can accumulate large concentrations of DDT in their fatty tissues. DDT concentration is measured in parts per million (ppm) or parts per billion (ppb).**

Since DDT isn't metabolized, it is passed along the food chain from one animal to another. Imagine a stream that runs through acre after acre of farmland or forest. DDT is sprayed to control insect pests that are reducing farm yields or damaging valuable timber. Some of the DDT molecules are ingested by the pests, and these insects die. But the DDT in their dead bodies remains toxic. The remainder of the DDT may kill beneficial insects or adhere to leaves or particles of soil. As the dead insects decay, the DDT contained in their bodies is also added to the soil.

With each rainfall, some of these soil particles are washed into the stream. Small amounts of DDT are absorbed or ingested by the organisms that serve as food for fish living in the stream. As time goes by, DDT builds up in the fatty tissues of the fish. Now imagine an osprey or other fish-eating predator that gets most of its food from this stream. It will consume many DDT-contaminated fish over time. This organism will build up a higher and higher concentration of DDT in its fatty tissues. You can see how the concentration of DDT increases as it passes higher up the food chain—the osprey has a much greater concentration of DDT in its fatty tissues than any single fish. This increasing concentration of nonbiodegradable chemicals along a food chain is called **biological magnification.**

In small doses, DDT is not lethal to birds and other animals. But its gradual buildup in the fatty tissues of an animal eventually affects the animal's health. One of the effects of biological magnification of DDT, and one of the discoveries that led to its ban in the U.S., was the thinning of eggshells in birds. Top carnivores—including ospreys, falcons, eagles, and pelicans—with high levels of DDT in their bodies often could not reproduce successfully. Many of these birds produced eggs with shells so thin that they broke long before the embryo had a chance to develop into a hatchling. The populations of these birds decreased so sharply that many biologists feared they would become extinct.

In a study done in 1984, 12 years after DDT had been replaced in this country with other types of pesticides, DDT was still the most common pesticide residue found on fruits and vegetables. Even though it has been banned in the U.S. for more than 20 years, DDT is still being used for pest control in countries that supply the United States with produce.

Biodegradable Pesticides   Another group of pesticides that, unlike DDT, are water soluble and break down quickly, have also been developed. These pesticides—which include parathion, malathion, and phosdrin— have taken the place of DDT and other fat-soluble pesticides and are presently in use. Since they are biodegradable, they do not spread from one animal to another through the food chain. Although these pesticides do not persist in the environment like DDT, their use still poses pollution problems. For example, since they are water soluble, they more easily con- taminate water supplies. Also, these pesticides are actually more toxic than DDT, and are highly poisonous not only to insects but also to verte- brates, including humans. And, like DDT, they kill both pests and their predators, and so disturb natural controls of pest populations. Because they break down so quickly, they are applied frequently. As a result, they are almost always present and, in that sense, are similar to DDT and other pesticides that break down slowly.

Alternatives to Pesticides

Pesticides are not the only answer to pest control problems. Since natural population controls already exist in nature, it makes sense to try to take advantage of them. Some alternative solutions to pest problems involve using natural predators or a pest's own biochemical characteristics to con- trol its population. Other alternatives called **cultural controls** involve devising growing methods that discourage pest growth, or breeding plants that are resistant to pests.

These alternatives are not foolproof—they don't always work quickly enough to keep a pest population from severely damaging a crop. Sometimes pesticides are needed. But if alternative pest control methods are implemented before a pest population gets completely out of hand, they can help reduce the need for pesticides.

Biological Control   Natural predators of agricultural pests include insects, spiders, frogs, and birds. Preserving the natural predators living in an infested area, or introducing predators from another area to help keep a pest population in check, is called **biological control.** The first use of this type of pest control took place in 1888. An insect called the cottony cushion scale had become a severe pest in the cit- rus groves of California. Scale insects exude a sticky, sweet, honeydew material that encourages mold growth, which destroys the fruit. The Australian ladybird beetle, a predator of scale insects, was introduced to control the pest. The beetle saved the citrus industry.

Introducing predators from other parts of the world has proven to be a useful method of pest control. Can you think of some precautions that should be taken before a "foreign" insect is intro- duced into an ecosystem?

Figure 30-6   The Australian lady- bird beetle was introduced to control the cottony cushion scale that was causing great damage to the citrus groves of California.

Figure 30-7 The release of sterile males into an area can help control the numbers of a pest population. When a sterile male mates with a female, no offspring will be produced. What other factors must be taken into consideration for this type of control to be successful?

Autocidal Control In another type of alternative pest control, called **autocidal control,** the insect actually contributes to its own population control. One type of autocidal control involves sterilizing males of a pest species and introducing them into the infested area. Males of the pest species are raised in the laboratory, where they are exposed to radiation that renders them incapable of producing viable sperm. When the sterilized males are released into the infested area, they mate with the females but the eggs are not fertilized and no offspring are produced.

This type of control can work only if a huge number of sterilized males is introduced into the area to make certain that very few of the normal, unsterilized males are able to find females that have not already mated. Research indicates that if there are ten times more sterile males than fertile ones in a population, the entire population can be destroyed within four generations. In California, application of the conventional pesticide malathion, followed by the introduction of sterile males to mate with surviving females, has proved useful in controlling the spread of the medfly.

Integrated Pest Management As we learn more about relationships among organisms, it becomes clear that the problems of pest control are complex, and solutions must be carefully thought out. The most successful, least toxic solution to a problem takes advantage of all available methods of pest control. Several control methods used simultaneously can be much more effective than a single control method used alone. **Integrated pest management** is a system that uses biological, autocidal, and cultural controls along with chemical pesticides to manage the size of pest populations. An integrated pest management system designed to check population growth of the medfly, for example, might include an initial pesticide application followed by trapping with chemical lures and the periodic release of large numbers of sterilized males. The goal of any integrated pest management program is to reduce the numbers of pest organisms to tolerable levels to maximize crop yields at a minimal cost.

Section Review

Understanding Concepts

1. A new insecticide is developed. When first applied to a population of disease-carrying insects, it successfully eliminates most of them. Within several years, however, the insecticide is no longer effective against these insects. How does this case illustrate the benefits and drawbacks of pesticide use?
2. Suppose an insecticide with properties similar to those of DDT accumulates over the years in a lake as a result of runoff from land. What animals are likely to be most severely affected by the insecticide? How might other animals also be affected as a result?
3. Crops are being destroyed mainly by one type of insect. Another kind of insect does not cause much damage because of its small numbers. An insecticide is applied to the crops and kills the natural predators of both insects. What may be the consequences?
4. ***Skill Review—Making Tables:*** Construct a table that compares the pros and cons of pesticide use. For more help, see Organizing Information in the **Skill Handbook.**
5. ***Thinking Critically:*** Explain several ways that biological and autocidal control methods may overcome the disadvantages of using chemical pesticides.

30.2 Pollution of Air and Water

Pesticide pollution is associated primarily with agriculture. You've learned that pesticides can affect the health of virtually all members of an ecosystem including birds, insects, fish, and humans. Pollutants associated with industry, transportation, and other aspects of technology also affect Earth and its inhabitants. Remember the imaginary car trip you took at the beginning of this chapter? Your journey started in the center of a big city, where the air is full of exhaust fumes from automobiles, factory smokestacks, and jet airplane engines. Most pollutants associated with technology are waste materials produced during the manufacture of goods and the burning of fuels.

Air Pollution

Take a moment to think about some of the reasons why humans burn fossil fuels such as oil, coal, natural gas, and gasoline. We burn fuel to heat homes and businesses; to cook food; to run plane, car, train, and bus engines; to produce electricity in power plants; and to drive manufacturing and industrial processes of all kinds. Most of today's air pollution problems result from the burning of fossil fuels. The smoke produced from burning these fuels releases materials into the atmosphere that can either harm living organisms directly or change the environment in ways that are harmful to life. One of the chief causes of air pollution is the burning of gasoline in automobiles. By-products of the combustion of fossil fuels include carbon monoxide, carbon dioxide, sulfur dioxide, nitrogen oxides, lead compounds, and hydrocarbons. All of these by-products are considered pollutants.

Objectives

Discuss the effects of various pollutants that contaminate the air.

Analyze the cause and consequences of destruction of the ozone layer.

Explain the difficulties involved in determining the extent of the greenhouse effect and finding solutions to the problem of global warming.

Describe the sources and effects of pollutants entering ground and surface waters.

Figure 30-8 **The burning of fossil fuels is the cause of most of today's air pollution problems. How many of our modern conveniences are available due to the burning of fossil fuels?**

Carbon Monoxide   Hemoglobin is the component of blood that carries oxygen from the lungs to the rest of the body. When carbon monoxide (CO) is breathed into the lungs, it enters the bloodstream and bonds with hemoglobin at sites usually reserved for oxygen molecules. If hemoglobin's oxygen-carrying sites are taken up with CO, tissues are deprived of the oxygen needed for metabolism. Unfortunately, the bond between CO and hemoglobin is more stable than the bond between oxygen and hemoglobin, so oxygen deprivation occurs even when the concentration of CO in the air is very low.

Nitrogen Oxides   When fossil fuels are burned at high temperatures, nitrogen from the fuel and oxygen from the air combine to form compounds called nitrogen oxides. Nitric oxide (NO) is made up of one nitrogen and one oxygen atom. It can interact with oxygen in the air to form nitrogen dioxide (NO_2), which has two oxygen atoms. Nitrogen dioxide is a brownish-colored gas that can kill lung cells and cause fluid to build up in the lungs.

In the presence of sunlight, NO and NO_2 combine with hydrocarbons (also produced by combustion) to form carbon compounds that irritate the eyes and stimulate the production of tears. Nitrogen dioxide also contributes to the formation of ozone. The oxygen molecules we breathe are made up of two atoms of oxygen; ozone is composed of three atoms of oxygen bound together. Ozone and eye-irritating carbon compounds contribute to photochemical smog—that is, smog produced by the interaction of sunlight and air pollutants. This type of smog, which often surrounds large cities such as Los Angeles, Denver, and Washington, DC, is damaging to lung tissue and can be extremely dangerous to people who already suffer from respiratory illnesses like asthma or emphysema.

One of the major difficulties in analyzing and solving air pollution problems stems from the tendency of many pollutants to interact with one another and form compounds that may be even more dangerous than the original substances. Examples of this phenomenon include the interaction of nitrogen oxides with hydrocarbons and sunlight to form components of photochemical smog.

Figure 30-9   Photochemical smog, the smog that surrounds cities, is produced by the interaction of chemicals released from the burning of fossil fuels and sunlight.

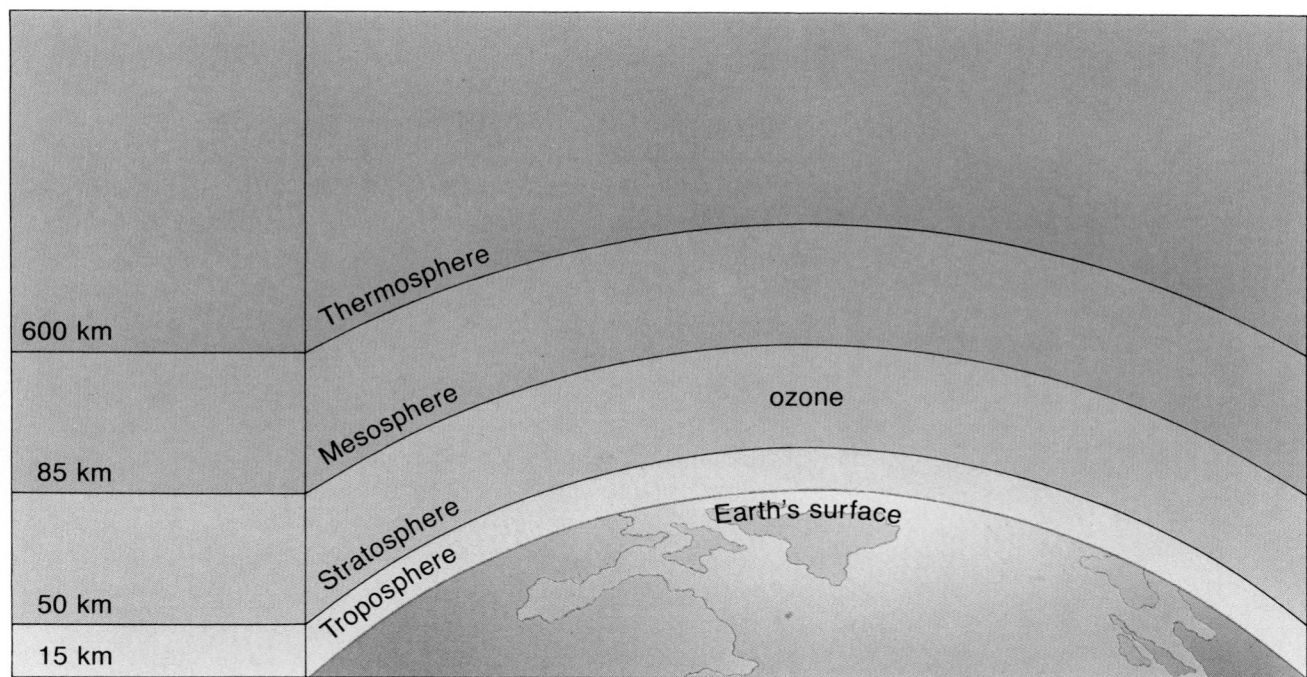

600 km	Thermosphere
85 km	Mesosphere
50 km	Stratosphere
15 km	Troposphere

ozone

Earth's surface

The Ozone Hole

The atmosphere extends several miles above Earth's surface and is made up of four layers, as shown in Figure 30-10. The layer closest to the surface, the troposphere, is about seven miles thick and contains the air we breathe. Most weather events take place in the troposphere, and most air pollution is confined to this layer. The next layer is the stratosphere, which extends to about 30 miles from Earth's surface.

The Ozone Layer   At the top of the stratosphere, just before the mesosphere begins, is a layer of ozone gas. While ozone in the troposphere can be dangerous and is considered an air pollutant, the ozone layer in the stratosphere is important to the health of all organisms living on Earth. This ozone layer absorbs almost all of the ultraviolet (UV) radiation that reaches Earth from the sun. UV radiation can be extremely damaging. The small amount that does penetrate the ozone layer is responsible for sunburn, can cause skin cancer, and can cause mutations in the genetic material of cells exposed to it. It can reduce crop yields and even kill small organisms such as plankton. If the ozone layer did not exist, it is doubtful that life could survive on the surface of Earth.

During the 1980s, it became evident that protective ozone is being lost from the stratosphere. The ozone layer above Antarctica becomes thinner during every Antarctic spring. This thinning has become known as the "ozone hole." In some years, ozone in this region has decreased by as much as 50 percent. The apparent loss of ozone is not limited to the Antarctic; it is just most severe there. Decreases as great as 10 percent have been detected over parts of Europe and North America.

Figure 30-10   Ozone, a layer of gas located in the stratosphere, absorbs almost all the UV radiation that reaches Earth from the sun. Many scientists think that the ozone layer is thinning due to the release of CFCs into the atmosphere.

Figure 30-11 CFCs work as coolants in appliances such as refrigerators and air conditioners. How are the CFCs in these appliances lost to the atmosphere?

CFCs What is causing the thinning of the ozone layer? Most of the blame has been placed on air pollutants called **chlorofluorocarbons,** which are also known as **CFCs.** These chemicals are used as coolants in refrigerators and air conditioners, as foaming agents in the manufacture of Styrofoam, and as propellants in spray cans. CFCs contain chlorine. Scientists have found that as amounts of chlorine in the atmosphere increase, the amount of ozone decreases. In the stratosphere, CFCs destroy ozone by reacting with it to form chlorine gas and oxygen.

What can humans do to stop this loss of ozone? In 1987, governments of the world's industrial nations became concerned enough to sign an agreement to reduce the production of CFCs. In 1990, the agreement was revised. This time, it called for complete elimination of CFCs, worldwide, by the year 2000. What about the CFCs already present in refrigerators and air conditioners? Removing the coolants from discarded appliances and recycling them instead of just releasing them into the atmosphere can help. The efforts of government and industry will probably not be able to prevent CFC levels from continuing to increase, perhaps by as much as 20 percent, by the end of the 20th century. The effect of this increase on the ozone layer remains to be seen.

CFCs can remain unchanged in the troposphere for as long as 70 to 150 years, rising into the stratosphere very slowly. It is believed that CFCs already released at ground level will continue to rise into the stratosphere for many more years. As a result, ozone depletion would likely continue to worsen even if we stopped all CFC use immediately.

Acid Precipitation

Water, in the form of water vapor, enters the atmosphere by the processes of evaporation, transpiration, and respiration. Water returns to Earth's surface in the form of rain, snow, and other types of precipitation. While in the atmosphere, water molecules may come into contact with air pollutants. Analysis of rain and snow falling over many parts of Earth today illustrates how pollutants can react with other compounds to cause unforeseen problems.

In the atmosphere, carbon dioxide gas dissolves into droplets of water to form weak carbonic acid. As a result, rain is normally acidic, with a pH of about 5.6 to 5.7. Release of sulfur dioxide and nitrogen oxides into the atmosphere can further reduce the pH of rain and snow. In the presence of sunlight, these pollutants react with water and/or oxygen in the air to form sulfuric acid and nitric acid. In some urban and heavily industrialized areas, the amount of these pollutants released into the air is so great that the pH of rain or snow may be as low as 3.0, which is about the same as the pH of vinegar. Precipitation with a pH below 5.6 is called **acid precipitation.** Dew, fog, and even airborne dust can also become acidic as a result of air pollution.

The region of the United States that has suffered the most damage from acid precipitation is the northeast, where the average pH of rain and snow is between 4.0 and 4.5. However, most of the sulfur dioxide pollution that causes this acid precipitation does not originate in the northeastern part of the country. It comes from coal-burning power plants located as far west as Illinois, hundreds of miles away. Prevailing winds move the acids toward the east. Emissions from factories and automobiles in the northeast compound the problem. As you can see, air pollution events in one part of the biosphere can affect other regions.

Effects of Acid Rain   Precipitation that is only slightly more acidic than normal may actually increase the yield of some crops, especially those grown in fields that receive little or no fertilizer. Depending on the nutrient balance of the soil, tomatoes, green peppers, and strawberry plants may benefit from the nitrogen and sulfur added to the soil by acid rain. However, as acidity increases (and pH decreases), nutrients that are easily dissolved in acid, such as calcium and potassium, are washed out of the soil. As these nutrients are washed away, the soil becomes less fertile and soil microorganisms become less numerous. Decomposition and the cycling of essential nutrients slows down so that fewer nutrients are available to plants. This loss of nutrients can lead to the death of trees, especially conifers.

Acid rain and snow can damage or destroy leaves and other plant tissues, reduce photosynthesis and nitrogen fixation, and limit the types of plants that can grow in the affected area. Acid precipitation can also have severe effects on lake ecosystems. Acid precipitation falling into a lake, or entering it as runoff from streams or neighboring slopes, causes the pH of the lake water to fall below normal. This acidity may be great enough to cause the death of plankton; destroy the eggs of fish, salamanders, and frogs; and even kill adult fish. Lake food chains can be destroyed because

Figure 30-12   **Many trees in the mountains of New England are dying at a high rate due to acid rain. Acid rain may damage plant tissues and interfere with photosynthesis and nitrogen fixation.**

many organisms disappear as a direct result of acid rain or because their food supply is gone. Low pH can reduce the availability of nutrients the same way nutrient availability is reduced in soils—microorganisms become less numerous, slowing decomposition and the recycling of nutrients.

Low pH also increases the solubility of dangerous metals such as mercury, aluminum, and lead. Under normal circumstances, molecules of these metals cannot dissolve in water. They remain buried in the sediments at the bottom of the lake and pose little hazard to life. But once they become dissolved in lake water, they can be taken up by plants, fish, and other organisms.

Acid rain also has a direct effect on human health. It can reduce the pH of dust particles in the air. A buildup of these particles in the lungs could become a factor in the onset of diseases such as bronchitis, asthma, and emphysema.

INVESTIGATION

Effects of Pollutants on Seed Germination and Plant Growth

Pollution of land and water by everyday common chemicals is a concern of everyone. Barely a day goes by without the news of some chemical that has leaked from a truck, train, or storage site into the ground or water supply. How much damage is being done to life from these kinds of chemical spills? It may be difficult for you to evaluate the damage on animals directly. However, you could experiment with seeds to determine if these chemicals do have a harmful effect on their germination and growth.

What might be the effect on germination if seeds were exposed to such chemicals as oil, cleaning detergents, insecticides, and excess fertilizers? You may wish to use chemicals other than those suggested as you plan your investigation. What might be the effect on growth if young plants were exposed through their roots to such chemicals as those listed above? You may wish to use chemicals other than those suggested as you plan your investigation.

Work with a group assigned by your teacher. Form hypotheses that could be tested using the chemicals suggested or those of your choice.

Problem

(1) How might seed germination be affected by soaking seeds in oil, cleaning detergent, insecticides, and excess fertilizers? (2) How might growth of young plants be affected by having their roots exposed to oil, cleaning detergent, insecticides, and excess fertilizers?

Hypotheses

What are your group's two hypotheses?

Safety Precautions

Be careful when handling oil, insecticides, and fertilizer.

Possible Solutions   Acid rain, along with ozone depletion, is yet another pollution problem resulting from the burning of fossil fuels. The burning of coal is an especially important source of acid precipitation because most coal contains fairly large amounts of sulfur. Much of the electricity we use is produced by coal-burning power plants. In 1992, 17 million tons of sulfur dioxide were released by coal-fired power plants. Reducing these sulfur dioxide emissions could help reduce the acid rain problem. Regulations that are part of the Clean Air Act of 1990 are intended to do just that. According to these regulations, by the year 2010 most power plants will have to reduce sulfur dioxide emissions by half. How can this goal be accomplished? Emissions can be reduced by burning low-sulfur coal, producing less electricity, by installing scrubbers in coal-burning power plants to filter pollutants from emissions, and by using alternative fuels such as solar or hydroelectric power.

Experimental Plan

1. Decide on a way to test your group's two hypotheses. Keep the available materials provided by your teacher in mind as you plan your procedure.
2. Record your procedures and list materials and amounts that you will need.

Checking the Plan

Discuss the following with other group members to decide the final procedure for your experiment.
1. What data will you collect and how will it be recorded?
2. What will be your controls and what variables will have to be controlled?
3. Assign individual roles for this investigation.

Make sure your teacher has approved your experimental plan before you proceed further.

Data and Observations

Carry out your experiment. Prepare a data chart to record your findings.

Analysis and Conclusions

1. State your hypothesis that relates to the effect of chemicals on seed germination. Use specific data from your experiment to support or reject your hypothesis.
2. Might you be able to conclude from your data that:
   (a) different chemical pollutants will have the same effect on the type of seeds used in your investigation? Explain.
   (b) those chemical pollutants used in your investigation will have the same effect on different seed types? Explain.
3. State your hypothesis that relates to the effect of chemicals on plant growth. Use specific data from your experiment to support or reject your hypothesis.
4. Might you be able to conclude from your data that:
   (a) different chemical pollutants will have the same effect on the type of plants used in your investigation? Explain.
   (b) those chemical pollutants used in your investigation will have the same effect on different plant types? Explain.
5. Explain the advantage of using plants rather than animals in this investigation. Might you be able to conclude from your data that those chemicals used in your investigation will have the same effects on animals as they did on plants? Explain.
6. Design and describe an experimental procedure that you could use to test the effects of chemicals such as lead or mercury on: (a) seed germination and (b) young plant growth.

Going Further

Based on this lab experience, design another experiment that would help to answer questions that arose from your work. Assume that you will have limited resources and will be conducting this lab in class. Include this as part of your report.

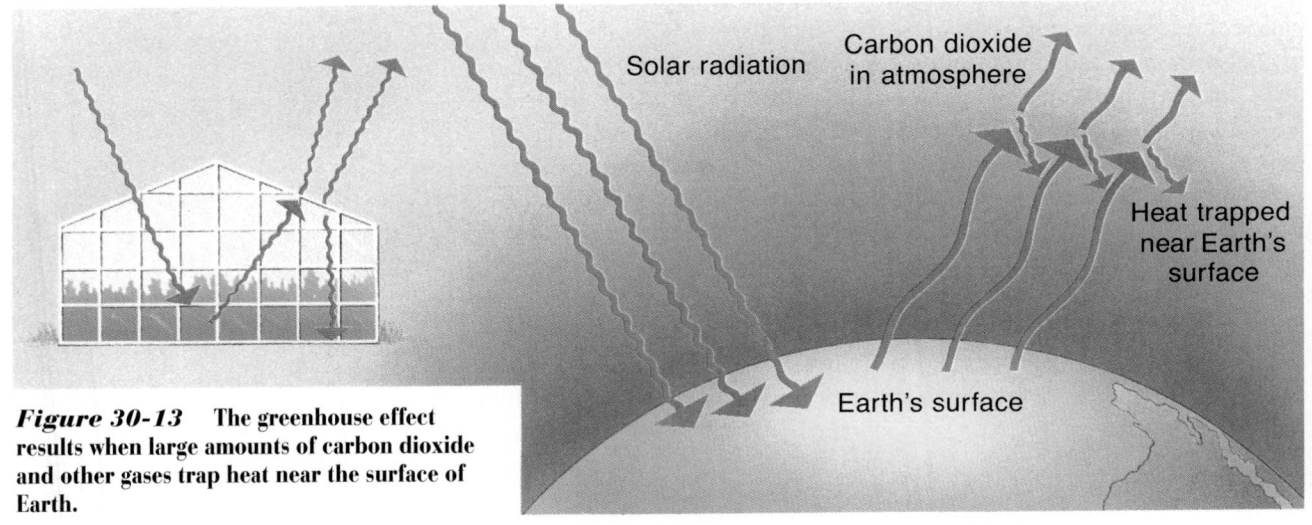

Figure 30-13 The greenhouse effect results when large amounts of carbon dioxide and other gases trap heat near the surface of Earth.

Labels within figure: Solar radiation; Carbon dioxide in atmosphere; Heat trapped near Earth's surface; Earth's surface

Global Warming—Is There a Problem?

You may have heard a great deal of talk in recent years about possible changes in Earth's climate. Many scientists are concerned that average temperatures are slowly getting warmer, a change often referred to as global warming. This problem, too, is related to the burning of fossil fuels, which releases large amounts of carbon dioxide into the atmosphere. Over the past 125 years, as humans have burned more and more fossil fuels for heat, transportation, and industry, the amount of carbon dioxide in the atmosphere has increased by at least 11 percent.

Carbon dioxide levels are also affected by the amount of forest land on Earth's surface. Do you remember the chemical reaction that takes place during photosynthesis? Because trees and other green plants use carbon dioxide, water, and sunlight to manufacture sugars, they help remove carbon dioxide from the atmosphere. But as the global human population has increased, so has the pressure for people to clear forest land for agriculture. In some areas, particularly the tropical countries of South America, millions of acres of forest are burned each year to make way for food crops and livestock pasture. Not only are the forests disappearing, they are being burned away, which adds even more carbon dioxide to the atmosphere.

Greenhouse Effect   What does atmospheric carbon dioxide have to do with temperature? Sunlight passes through the atmosphere to Earth's surface. Part of the heat from sunlight is absorbed by objects and organisms, but much of it is radiated back into the atmosphere. Some of this reradiated heat passes all the way through the atmosphere and escapes into space. The rest bounces off certain atmospheric gases, such as carbon dioxide and water vapor, and returns to Earth's surface. As the amount of carbon dioxide in the atmosphere grows larger, the amount of heat reflected back to Earth instead of escaping into space may also increase.

Carbon dioxide traps heat near Earth's surface in the same way that glass traps heat inside a greenhouse. The increase in temperatures that could result from a buildup of heat-reflecting atmospheric gases is known as the **greenhouse effect.** The heat-reflecting gases that contribute to global warming are called greenhouse gases. They include carbon dioxide, methane (natural gas), nitrogen oxides, and CFCs.

Possible Effects Several groups of scientists have used computer modeling to try to predict future temperatures. Some models suggest that global temperatures will increase only one degree Celsius by the year 2050. Others suggest an increase of up to five degrees Celsius. That may not sound like much, but even a small change in temperature can have a serious effect on the global climate. For example, during the last Ice Age, when glaciers covered most of North America, the average temperature was only five degrees Celsius colder than it is today.

What might happen if the temperature were to rise five degrees during the next half century? The polar ice caps might begin to melt, causing sea levels to rise and possibly submerging some populated coastal areas. The temperate zone could become subtropical, and more northerly regions could become temperate. Such changes would alter the distribution of plants and other life forms, creating major changes in ecosystems around the world. For example, crops like wheat and corn, which are presently grown in abundance in the United States, would no longer thrive here. You might think they could be grown in areas that were formerly cooler, such as Canada, but that is not the case. Even though the weather might be favorable, soil conditions in Canada are not suitable for these crops.

Rising temperatures might change wind and weather patterns, converting grasslands to desert or tundra to taiga. Warmer weather could lead to the spread of diseases now limited to the tropics and might favor the reproduction of crop-destroying insects.

Figure 30-14 **Global warming may result in a decrease in the amount of rainfall different regions of the world receive. The map shows the possible percent decrease in the amounts of rainfall in areas of the United States by 2075 if atmospheric levels of carbon dioxide continue to rise at present rates. Another possible effect of global warming is rising ocean levels due to the melting of polar ice caps. This could result in serious consequences for coastal cities.**

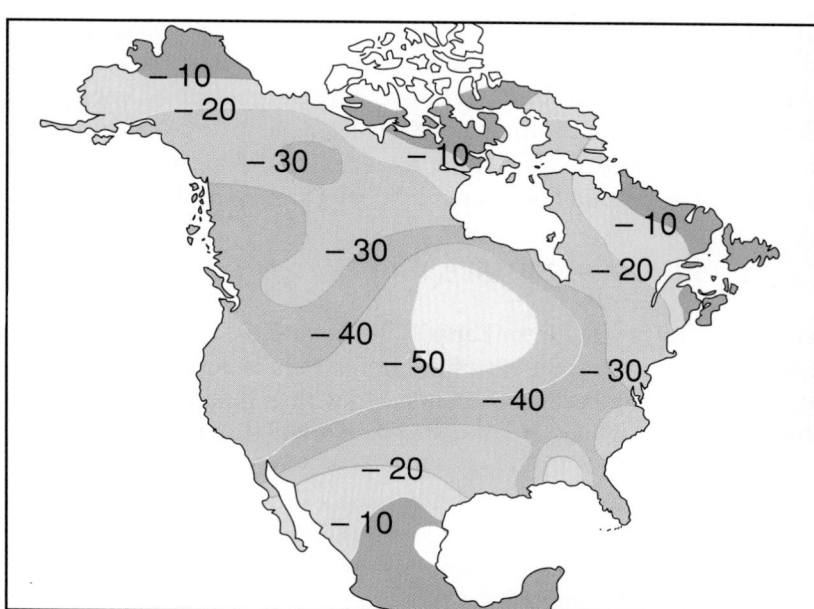

Figure 30-15   **One way to reduce the burning of fossil fuels is to find alternate methods of transportation.**

Possible Solutions   These are frightening possibilities. Could they really happen? Many ideas for reducing the potential for global warming have been proposed. For example, we could make more use of alternative energy sources, such as solar and wind energy, which do not require the burning of fuels. Reducing the use of fossil fuels would, of course, also reduce the output of many pollutants. We could stop removing vegetation, such as tropical rain forests, and start growing new forests. In addition, we could make use of our knowledge of genetics to develop crops capable of tolerating warmer temperatures and a variety of insect pests.

The cost of making these changes would be great, and people would have to endure many alterations in lifestyle. Energy costs would rise. We might have to dramatically reduce the number of car trips we take and severely curtail our use of electricity. As you can see, dealing with environmental problems involves politics and economics as well as science. The issue of global warming may be the most difficult ecological dilemma of our time, yet we are still uncertain of the extent of the problem. This uncertainty is a prime example of how science cannot always provide an immediate solution to a problem. Scientists have not yet learned enough to predict how severe global warming may be, let alone provide safeguards against it. Furthermore, there is still a great deal of disagreement on this issue—some scientists do not foresee a global warming problem at all.

Pollution of Surface Waters

Water is constantly cycled from Earth's surface into the atmosphere and back again. Water vapor that enters the atmosphere is pure, but once in the atmosphere, it can become contaminated with pollutants as in the formation of acid rain. Water on and under the ground can also become contaminated. Contaminants include not only pesticides, oils, detergents, and other substances released by humans, but also biological materials such as decaying plants and animals or animal wastes. These pollutants can enter the water directly from sewers, drainpipes, or runoff from land, or indirectly, picked up as the water flows through polluted soil.

Most of the water on Earth is salt water; less than 0.1 percent is fresh. Organisms that dwell on land need a constant supply of fresh water, as do the organisms that live in lakes, ponds, and streams. Pollution of this tiny percentage of Earth's water can present a serious threat to life. Pollution of fresh water can also affect marine organisms because almost all fresh water eventually finds its way to the ocean. This is another example of how ecosystems interact with one another. Let's take a closer look at some of the ways in which fresh water can become contaminated.

Biological Wastes   In most communities, sewage treatment plants ensure that human wastes, pathogens, and other contaminants are removed from water before it is released into local waterways. But in areas where there is no treatment of sewage, or where treatment systems fail, these wastes find their way into bodies of water, where they are broken down by bacteria and other organisms. If the accumulation of waste materials is large, the decay process may use up all of the oxygen dissolved in the water. Fish and other water dwellers cannot survive without oxygen, and the water may become stagnant and lifeless.

GLOBAL CONNECTION

Michiko Ishimure

Michiko Ishimure became familiar with one of the worst effects of water pollution long before anyone knew its cause. In her Japanese fishing village, it was known as the "cat's dance disease." Cats inflicted with the malady would go mad and drown themselves. After a while, the problem started showing up in humans, who would become crippled and die.

Ishimure became involved when university studies showed that the disease was a neurological disorder caused by mercury poisoning. The source was a chemical plant that had been dumping mercury-carrying waste into the bay. For years, mercury had been thought to be harmless in seawater because it didn't dissolve. It wasn't known that chemical reactions between mercury and seawater would allow the poisonous metal to enter the food chain.

Ishimure helped expose the toxic waste problem to the world by writing books about the people who suffered from the poisoning. She took a leading role in efforts to limit the wastes that industry was allowed to dump.

Yet, while she was defending her village from the immediate threat of the poisons being dumped into its waters, her fellow villagers wouldn't support her. Instead, they made her an outcast. Too many of the villagers depended upon the chemical plants for jobs. To them, an attack on the plant was a threat to their livelihood.

Yet, her books helped make the Japanese government aware of the toxic waste program. They greatly contributed to the passing of pollution control legislation in Japan and around the world.

1. **Explore Further:** Research the famous Love Canal toxic dump site. What happened there? Who were the key people in uncovering the problem?

2. **Writing About Biology:** Write several paragraphs describing human responsibility to control toxic waste worldwide.

Technological Wastes   In technological societies like ours, most water pollutants are by-products of industrial and agricultural activities. These pollutants range from poisonous substances such as pesticides and potentially toxic metals, to biodegradable or inert materials like glass, paper, and chunks of concrete.

PCBs (polychlorinated biphenyls) are a type of industrial pollutant once found in chemical plant wastes and in large electrical transformers, such as those located at the top of power line poles. Their use was banned in 1979. Because they are nonbiodegradable, PCBs remain in the environment. They also pose a health threat to humans. PCBs affect the nervous and muscular systems, and experiments suggest that they may cause cancer and birth defects in animals. Humans come into contact with PCBs through contaminated water, through contact with contaminated soil, or by ingesting PCB-contaminated fish. Most of this contact comes from the improper disposal of waste containing PCBs.

Ocean Pollution   The water in a small stream flows into a larger stream, then into a river, which empties into the ocean. Any pollutants carried in this water will affect the waterways along the coast. These areas of the ocean support the greatest diversity of marine life. They serve as nursery grounds for many species of fish and for marine mammals such as seals and sea lions. Many water birds feed and raise their young there. In regions where a number of rivers and streams flow into the same bay or coastal area, pollutants can become highly concentrated, with severe consequences to marine life.

You have probably heard about oil spills that occur in ocean waters. Sometimes huge quantities of oil are lost from giant tankers as a result of accidents at sea. Smaller quantities leak from offshore oil-drilling rigs or from boat motors. It is estimated that about 0.1 percent of the oil produced each year finds its way into the world's oceans. Some of this oil evaporates and some is broken down by bacteria, but these are very slow processes. Before it breaks down, the oil can be lethal to plankton, fish larvae, crustaceans, birds, and many marine mammals.

Figure 30-16   Oil spills are very costly to marine and coastal ecosystems. After a major oil spill, it can take up to ten years for organisms to recover.

Pollution of Groundwater

Fresh water is present not only in surface waters but also underground in reservoirs called aquifers. Underground water supplies can become contaminated by pollutants washed down through layers of soil. Some pollutants leach into groundwater from landfills where hazardous wastes have been stored. Pollutants also enter groundwater from buried gasoline tanks, septic systems, municipal and industrial landfills, agriculture, and from illegally dumped materials.

Some of the pollutants that have entered groundwater supplies include PCBs, pesticides, and a variety of other organic chemicals. These contaminants most often reach groundwater due to improper disposal. A chemical called TCDD, a dioxin, a by-product of the manufacture of some herbicides, has contaminated groundwater supplies. TCDD is toxic to some animals, but its long-term effects on humans are not yet known. Radioactive wastes, which come from spent nuclear power plant fuel, the manufacture of atomic devices, and from medical waste materials, also pose a threat to groundwater. These materials are usually stored in containers, but leakage sometimes occurs. Radiation from some of these isotopes can be lethal, even in very small amounts.

Pollution of an underground aquifer poses a grave threat to people who depend on it as their primary water supply. It is difficult and expensive to clean large quantities of contaminated groundwater. In many parts of the United States, wells have been shut down and people have been forced to use bottled water. Efforts are being made to clean up existing hazardous waste dumps and to develop safer methods of storage and containment of hazardous wastes to prevent further groundwater contamination.

In examining the effects of pesticides, air and water pollution, acid rain, ozone depletion, and global warming, you have come to see how modern technology contributes to environmental problems. Finding solutions will require contributions from individuals, industries, governments, and scientists. Cooperation is crucial as we explore ways to reverse damage already done to our environment and to prevent future damage.

Figure 30-17 Aquifers become polluted by the improper disposal of hazardous wastes. Contaminants leak into the ground and ultimately into the groundwater from improper storage containers, buried gasoline tanks, and from illegally dumped materials.

Section Review

Understanding Concepts

1. How is the production of the electricity used by many people related to acid precipitation?
2. Landfills may contain old Styrofoam containers, refrigerators, and spray cans. In what way are these materials dangerous to the environment?

3. How does dealing with the issue of global warming illustrate the roles of science, government, and economics in solving environmental problems?
4. **Skill Review—Designing an Experiment:** How would you conduct an experiment, using a

terrarium, to test the adverse effects of acid precipitation? For more help, see Practicing Scientific Methods in the Skill Handbook.
5. **Thinking Critically:** Explain how old motor oil dumped in a vacant lot could affect life in the ocean.

Objectives

Relate availability of food resources to the resource of soil.

Analyze the need for developing non-fossil fuels.

Discuss the importance of preserving forests and wildlife.

30.3 Conservation of Resources

For humans—and all other organisms—the environment is the only source for all materials needed for survival. Humans depend on other organisms as sources of food, clothing, tools, shelter, medicines, and fuel. Humans mine minerals from Earth and the oceans. The development of modern agriculture and technology has enabled humans to extract more and more resources from the environment. How long can we continue to depend on these natural resources? At some point, will we begin to run out of some of the materials we need? How does our use of these resources affect other organisms, and our own future?

Food

In a broad sense, resources are either renewable or nonrenewable. **Renewable resources** can be replaced; **nonrenewable resources** such as minerals and fossil fuels cannot. One of the most important of all resources is food. Food provides the energy needed for life. Because crops can be replanted and livestock can be bred, food is a renewable resource. However, availability of food varies in different parts of the world. Food shortages exist. The problem is most severe in developing countries, where the combination of dense populations, nonproductive farming methods, and poverty results in poor food supplies. Fuel for harvesting crops is expensive and difficult to obtain, limiting food production even more. Prolonged drought in some areas has led to widespread famine. How can food shortages, famine, and starvation be prevented?

Figure 30-18 **Selective breeding practices have resulted in strains of wheat that produce more grain per plant than other varieties. How is this technology beneficial to the environment as well as to famine stricken countries?**

Genetic Engineering Using selective breeding, geneticists have produced new strains of plants, such as rice and wheat, that produce more grain per plant than other varieties. The new strains also respond well to chemical fertilizers. An acre of land planted with these new strains can produce food for more people than the same acre planted with other varieties.

Geneticists have also begun developing a strain of corn that is richer in protein than other types of corn. This protein-rich strain is not yet ready for use because it has several undesirable traits that must be bred out of it, including soft kernels and lack of resistance to pests.

Concentrated Protein One of the major health problems people experience during food shortages is protein deficiency. Protein is very important for normal growth and development. Growing children who don't get enough protein experience severe weight loss, anemia, and liver damage. A possible solution to protein shortage is to make the best possible use of known protein-rich plant foods and to find new sources of protein.

The seeds of many types of plants, especially beans, are good sources of protein. Soybean plants resist drought, thrive in a variety of soils, and can be grown for use as a source of protein in many areas of the world. The fresh seeds can be cooked and eaten, or they can be dried and stored for later use. They can also be ground to make flour or meal. The oil extracted from soybeans can be used to make a concentrated protein to be eaten as a meat replacement, or added to meat for a protein-rich meal.

Soil

Growing food crops on land requires fertile soil. In uncultivated areas, the organic nutrients plants remove from the soil are returned with the decay of the plants or the animals that ate the plants. When humans grow food crops, the plants are taken away from the area. If the soil nutrients are not replaced, the fertility of the soil is reduced and its ability to support crop growth diminishes. This process is called **soil depletion.** Fertilizers, both organic and synthetic, are used in an effort to replace the nutrients that are removed.

Crop Rotation Another method of reducing soil depletion involves alternating the crops grown on a plot of land. In this process, known as **crop rotation,** a soil-enriching plant such as a legume is grown one year; the next year, a soil-depleting crop such as corn or wheat is grown. The legumes return nitrogen and other nutrients to the soil, making it more fertile. As you already know, crop rotation also helps reduce pest problems by removing a pest's host plant.

Erosion Soil is usually considered a renewable resource because it is constantly being formed, primarily by the decay of organisms living on and in it. However, soil resources can be lost as a result of **erosion,** a natural process in which soil is removed by the action of rain and wind. Under the cover of plants, soil is held in place. However, careless farming methods can increase the erosion rate, as happened in the dust bowl, described in Chapter 29. Harvesting and plowing expose topsoil by removing the plant cover. Without rooted plants holding the soil in place, it is easier for wind and water to carry the soil away.

There are several farming methods that can be used to reduce erosion. Many of these involve plowing techniques that reduce erosion caused by the flow of rainwater. They include contour plowing, in which plowed furrows follow the shape of the land, and terracing, which involves creating level banks of land on a slope to slow the downhill flow of water.

Figure 30-19 Rows of taller plants, such as corn planted in strips between rows of small grain plants, can protect the smaller plants from the effects of the wind. Strip-cropping is used to conserve soil in fields that are subject to erosion by wind or water.

Minimum tillage is another method in which fields are not plowed after a crop is harvested. The stalks are kept in place. This can decrease erosion by as much as 90 percent for some crop fields. Wind erosion can also be reduced by planting trees as windbreaks and by plowing at right angles to prevailing winds.

Fuels

The fossil fuels and other petroleum products we rely on today took millions of years to form. Estimates are that humans may use up known petroleum deposits in as little as 645 years at current rates of consumption. New sources of oil and natural gas are constantly being sought. Offshore drilling may provide new supplies, and some shale formations may provide a new source of oil. But petroleum is not a renewable resource. Eventually, the demand is likely to exceed the supply. Humans can help preserve this resource by replacing fossil fuels with other sources of energy.

Figure 30-20 Solar heating systems are available to heat and cool homes. Active solar systems have tracking systems that actually allow solar collecting panels to track the sun across the sky. What are some major limitations to the use of solar energy?

Nuclear Power The United States and other countries are already making use of nuclear power as a source of energy. There are several types of nuclear power plants. In one type, the nuclear reactor contains radioactive isotopes of uranium. These isotopes are split apart in a reaction called nuclear fission. The energy released by nuclear fission is used to generate electricity.

Only a small part of the electricity generated in the United States is produced by nuclear power. While this type of power plant conserves fossil fuels, the radioactive isotopes of uranium used as fuel are also a nonrenewable resource. Radioactive wastes produced by nuclear reactors pose a difficult disposal problem. And the water used to cool the reactors may cause thermal pollution of rivers and lakes. In a properly functioning reactor, radiation does not escape into the environment during fission. However, in the event of a large accident, some radiation could be released.

Solar Energy The energy contained in sunlight is a plentiful, renewable resource. Our use of solar energy is slowly increasing. Solar-heated houses and other buildings are built of heat-absorbing materials and are oriented to catch sunlight during the winter months. You may have seen solar panels installed on the roofs of homes in some parts of the country. These panels collect solar energy for use in heating water. Solar cells have been developed to operate small devices such as calculators and watches. Larger solar cells are being used to power a factory in Saudi Arabia that makes fresh water from salt water, to power satellites, and to power portable field telephones for the military. As the cost of solar cells decreases, more uses for them will undoubtedly be found.

Other Energy Sources   Still other sources of energy are being investigated for use in the future. Possible sources of untapped energy include hydrogen gas, wind, geothermal energy, and tidal power. These energy sources have two advantages—they are renewable, and they produce little or no pollution.

Wildlife and Forests

There have been many cases of human interference causing the extinction of plant and animal species. The dodo bird—a flightless, turkey-sized bird that nested on the ground—was wiped out when explorers brought dogs and pigs to its native island. The bird could not protect its eggs and young from being trampled or eaten by these new predators. Hunters were responsible for the extinction of the passenger pigeon and have almost wiped out other species, such as the tiger. When an organism becomes extinct, the natural balance of the ecosystem is irreversibly altered. It is important to realize that an extinct species will never again live on Earth—once it is gone, it is gone forever. Today, the major cause of extinction is the destruction of natural habitats. It is estimated that humans are causing the extinction of about one species each day.

Figure 30-21   **Manatees and the spotted owl are two animals that are facing extinction due to human interference. For these animals, the major threat is habitat destruction.**

Wildlife Conservation   Efforts to conserve wildlife have included setting up wildlife preserves and imposing limits on hunting and fishing of some species. Wildlife preserves provide a means of making sure the habitat an endangered species needs for survival is not destroyed. Populations of many nearly extinct animals have been increasing as a result of wildlife conservation efforts in the United States. Whooping cranes, bison, sea otters, wild turkeys, and white-tailed deer have all benefited from these efforts. However, animals such as the California condor and the black-footed ferret are still in danger of disappearing.

Forest Conservation   Scientists have estimated that approximately 30 to 50 percent of the world's forests have been destroyed for agriculture, fuel, and commercial use. Before Europeans arrived in North America, forests covered most of the eastern and western regions of the continent. Settlers cleared these forests for farmland, and the timber was burned for fuel or used for building. By the early 1900s, when it was becoming evident that too much of this natural resource was being destroyed, forest conservation efforts began.

Many of today's forest conservation efforts involve harvesting techniques used by the timber industry. To satisfy the demand for lumber, many timber harvesting companies are "farming" forest land by planting trees, both to replace those that are cut and to produce timber for future harvests. They are also planting private tree farms rather than depending on harvesting and replanting public forest lands. Clear cutting—a process of cutting all the trees in an area at one time—is now being replaced with other methods. For example, blocks of trees are sometimes left to reseed nearby cut areas and to maintain part of the complex forest community. Selective cutting and improvement cutting are also being used. In selective cutting, only mature trees are harvested, while young trees are left to mature. In improvement cutting, old, crooked, or diseased trees that will never be useful for timber are removed to make room for more valuable trees to grow.

Cause for Optimism

Pollution of the environment, overconsumption of natural resources, and an expanding human population are real and important problems facing humans today. For too many years, people ignored or failed to recognize the consequences of their actions with respect to the environment. As we improve our understanding of the consequences of our actions, we can make better, more informed choices.

Thinking Lab

DRAW A CONCLUSION

How much and what do we throw away each day?

Background

The graph shows the approximate mass percent of solid waste thrown away each day by each person in the U.S.

You Try It

Draw the following conclusions.

Calculate the mass of solid waste generated in the U.S. each day. If starting a recycling program to reduce solid waste being buried in landfills: (a) What material should be recycled first? Why? (b) What should be recycled last? Why? Does your community recycling program tend to agree with your last answers? Why are they good environmental choices? Explain why it may be a poor choice to begin a recycling program for leather and rubber.

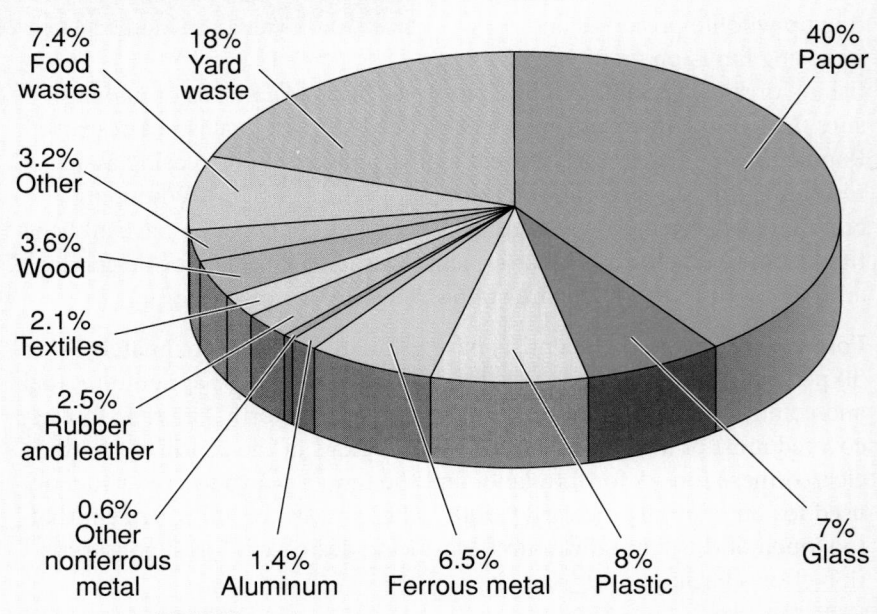

7.4% Food wastes
18% Yard waste
3.2% Other
3.6% Wood
2.1% Textiles
2.5% Rubber and leather
0.6% Other nonferrous metal
1.4% Aluminum
6.5% Ferrous metal
8% Plastic
7% Glass
40% Paper

With careful planning and cooperation, we can reverse many existing problems and avoid creating new ones. Using human reason and intelligence, by balancing scientific facts against sound economic realities, we can take steps to maintain the health of the environment that supports all life on Earth.

LOOKING AHEAD

YOU HAVE NOW completed your course in biology. There are no more chapters to read. You've learned a great deal about life at all levels, from cell to biosphere. You know that all organisms use energy to carry out similar functions and that the instructions for being alive are coded in molecules of DNA. You know now that DNA is copied and transmitted to offspring, and you understand why organisms have the traits they do. You've learned that organisms evolve over time, adapting to environmental changes, and that evolution explains both the similarities and differences among living things. You understand that organisms live together in complex communities, and that they interact with one another and with the physical aspects of their environment. In this chapter, you learned how human activities have caused harm to the environment and some steps that need to be taken to correct and prevent environmental problems. You probably won't remember all of the details you've learned this year. We hope, though, that you will remember and make use of the major ideas. The knowledge you've gained of biology will enable you to make sound decisions concerning your personal life, the lives of fellow humans, and all other living systems. ❧

Section Review

Understanding Concepts

1. A farmer has many acres of land. Each year, he sets several acres aside, planting no crops on that land. He varies the unplanted acres each year. Why?

2. The impact of humans on the environment can be described by the following equation: Impact = Population × Affluence × Techology. Explain how this equation describes the pollution problems in the U.S. and in a developing country. (Hint: You may wish to review Chapter 27.)

3. Some people argue, "What is so important about saving an endangered species?" What would you tell them?

4. **Skill Review—Making Models:** Use mounds of soil to demonstrate the principles of contour plowing and terracing. How can you demonstrate that these practices reduce erosion? For more help, see Representing and Applying Data in the **Skill Handbook.**

5. **Thinking Critically:** Irrigation can make dry land suitable for growing crops. In most irrigation methods, water is "sprayed" over the land. As this occurs, salt accumulates in the soil. The salt may have been present in the water added or it may have come from below as the salty water table rose. What would be the long-range effect of this irrigation procedure?

BIOLOGY, TECHNOLOGY, AND SOCIETY

Considering the Environment
New Uses for Waste Material

Every year, the average American creates approximately 2000 pounds of trash. We're running out of places to put it. So much garbage has already been dumped

into New York's Fresh Kills landfill on Staten Island that it has become the largest human-made object in the world, surpassing the Great Wall of China.

Across the country, existing landfills are filling up but few new ones are opening. Inexpensive land suitable for landfills is becoming rare and few communities are willing to have a landfill nearby.

Approximately 40 percent, by weight, of the trash in landfills is paper. It's paper's role in the landfill crunch that has been behind the growing demand for paper recycling.

During the past decade, paper recycling technology has improved to the point that the only thing holding it back is customer demand. As of 1992, far more paper has been turned in for recycling than the paper companies could accept. They could make only so much recycled paper because they couldn't sell any more. As increasing numbers of people use recycled paper, less and less paper will end up in landfills.

In contrast to paper, plastic accounts for only 8 percent of landfill trash by weight. Although this is true, it makes up approximately 20 percent by volume. That makes plastic a major part of the landfill problem, because volume determines when a landfill is full.

Unlike paper, plastic has been notoriously hard to recycle. One major problem is that there are hundreds of different types of plastic and they usually cannot be mixed. You can melt down soda bottles for reuse easily enough. But, if a single bleach bottle made of a

different type of plastic were accidentally melted down with them, all the plastic would be ruined.

To keep the many types of plastic apart, all plastic has to be hand sorted. The type of plastic is indicated on each container by a number. The type of plastic can be determined easily enough, but hand sorting is an expensive process.

Furthermore, plastic is so inexpensive to manufacture from scratch that recycling has always been more costly. Until the landfill crunch created a demand, few plastic buyers were willing to pay extra for recycled supplies.

Fortunately, new techniques and new products have combined with the growing demand for recycled products to make plastic recycling practical. Among the techniques

CAREER CONNECTION

Chemical engineers develop the manufacturing processes, equipment, and plants that convert raw materials into commercial products. To do their job, they must be skilled in mathematics, chemistry, and physics. A variety of industries require chemical engineers, including plastics, cosmetics, pharmaceuticals, soap, fuels, explosives, and food products.

being pioneered are several processes that completely disassemble plastics to their original chemical components. The components can then be remixed to make fresh plastics or be used for other purposes. For example, since plastics are petroleum-based, scrap plastic can be a source of fuel and fuel additives.

Recycling plastic into fuel may keep it out of the landfills, but it isn't likely to win the hearts of many environmentalists. One use and the fuel is gone in a burst of exhaust fumes. Maybe that's one reason recycled plastic has generated a lot of interest as a construction material.

Plastic lumber has earned itself a place in the construction world. It has proven particularly good for park benches, decks, and other outdoor uses. Plastic makes good lumber for the same reason it makes bad landfill material—it doesn't degrade easily. Plastic is waterproof and rot proof. Insects don't eat it. And it requires no preservatives, as wood does.

Unfortunately, plastic lumber looks like plastic, not like lumber. But another new building material combines recycled plastic with recycled wood to give a more natural appearance. Fibers from scrap wood are encapsulated in the plastic and molded into the desired shape. One of the first uses of this material is in exterior doorways. The material is used at the bottom of the door jamb because it seals well against the weather and resists rot due to moisture.

Recycled plastic is showing up in other parts of new homes and commercial buildings. Plastic foam is being processed into insulation board, while discarded plastic computer housings have been reincarnated as roofing panels.

Discarded tires have created a problem very similar to that caused by plastic. About 242 million tires are tossed away in the United States every year, most of them ending up in landfills, illegal dumping sites, or near-useless stockpiles. One recent Environmental Protection Agency estimate is that the U.S. has a stockpile of more than 2 billion old tires. These old tires are also finding new uses.

The development of rubberized asphalt is putting old tires back on the road. This form of asphalt, made with rubber from discarded tires, has made a name for itself as road repair material. While it costs twice as much as normal asphalt, it also lasts twice as long. What's more, it's less expensive to use and maintain.

One experimental technique that holds much promise involves grinding old tires into small particles and bonding them with other substances to produce new composite materials. The new materials may be used for such things as shoe soles, hoses, and conveyer belts.

Still another technique mixes granules of recycled tire rubber with resins to produce an outdoor flooring material. It's particularly good for coating concrete. It has been used on swimming pool decks at the Disneyland Hotel and the University of California at Los Angeles and to coat the eroding top of a hydroelectric dam.

Another aquatic use for old tires is under water. They've been used to help construct artificial reefs. It takes only days before barnacles and other reef-building organisms start growing on the tires and connecting material. Eventually, reefs they produce become havens for fish and scuba divers.

None of the innovations we've discussed here come close to solving the enormity of the United States' landfill problem. However, they show what's possible when the will to recycle is combined with effort and ingenuity.

1. **Find Out:** Plastic lumber and rubberized asphalt have been used in many city and county government projects. Has your community used these or any similar materials? How?
2. **Issue:** Should city governments require citizens to separate and recycle their trash? Defend your answer.

Summary

Modern technology and agricultural practices have led to pollution of the environment.

Pesticides are an example of chemicals that have positive value but that also have adverse effects including destruction of natural predators of pests and harm to non-pest organisms as well. Use of less toxic, quickly biodegradable pesticides coupled with biological, autocidal, and cultural means of pest control is a reasonable solution to the problem.

By-products of technology often become pollutants of air and water. Individually, many of these pollutants threaten the health and well-being of humans, other organisms, and entire ecosystems. Many of them interact with each other or other chemicals, giving rise to additional problems such as smog, acid precipitation, and destruction of the ozone layer.

Reducing pollution and preventing further harm to the environment involves a cooperative effort among scientists, governments, and individuals.

Not only have humans added pollutants to the environment, they have also depleted it of many resources. Development of new food and fuel sources and conservation of soil, forests, and wildlife are crucial goals if our planet is to continue to support the lives of humans and all other organisms.

Language of Biology

Write a sentence that shows your understanding of each of the following terms.

acid precipitation
autocidal control
biological control
biological magnification
chlorofluorocarbon (CFC)
crop rotation
cultural control
erosion
greenhouse effect

integrated pest
   management
nonbiodegradable
nonrenewable
   resource
pesticide
pollution
renewable resource
soil depletion

Understanding Concepts

1. You are a grocer whose income depends partly on the sale of fresh fruits and vegetables. Are you for or against the use of pesticides? Why?

2. Suppose you are a member of your local town council. Larvae of a certain insect are devouring the leaves on the trees along Main Street. What method(s) would you suggest to your fellow council members to preserve the trees? How would you convince them?

3. More than ten years ago, residents of Times Beach, Missouri, were ordered to evacuate the town permanently. This order was to protect residents against the dangers of dioxin that contaminated roadways in the area. Did scientists act hastily? Did they overestimate the risks of exposure to dioxin? Should they have waited to evacuate the area until further analysis and testing of the hazards of dioxin? How does this story affect your thinking about how we should respond to pollutants and environmental problems in general? Is it better to be safe than sorry, or do you favor a "wait-and-see" approach?

4. Why would it be wise for you to limit your sunbathing?

5. Choose two examples of pollutants discussed in this chapter and explain how solutions to the problems of decreasing those pollutants involve economic factors. If you were the president of a company, how would you feel if regulations forced you to spend more money to counteract pollution? How would you make up your losses? How might your customers react?

6. If you suffered from asthma, how would you feel about regulations like the ones that call for reduced output of sulfur dioxide emissions? Would you be willing to pay a higher electric bill?

7. It's a bright, sunny, though cold day in winter and the heat is on. Why are the rooms on the sunny side of the school warmer than those on the other side?

Applying Concepts

8. Why must offshore oil-drilling procedures be well controlled?

9. Outline some ways in which an energy shortage would affect a technological society.

10. Suppose today a particular area of the United States is not heavily industrialized and is relatively free of pollution. Suppose, too, that the area remains industry-free. Is it safe from pollution in the future? Explain.

11. How can application of pesticides to a field have an effect on large fish in a lake?

12. Which animals, insects or birds, are likely to develop a resistance to a pesticide more quickly? Why?

13. **Considering the Environment:** Explain how recycling reduces pollution problems and helps preserve resources.

14. **Global Connection:** Why is our knowledge of the effects of toxic chemicals on humans so limited?

Lab Interpretation

15. **Thinking Lab:** A recycling program for waste wood is being considered by your community. How would you advise them? Explain why.

16. **Investigate:** The following data were gathered by a student using different detergent concentrations on bean seed germination.

Concentration of detergent	Number of seeds used	# germinated after 5 days
0	100	98
1	100	97
2.5	100	90
5	100	63
10	100	14
20	100	0

(a) At what range may the maximum concentration of detergent be found for a 50% germination rate?

(b) Explain how this data may be used to predict germination of radish seeds in different concentrations of insecticide.

17. **Minilab 2:** The following data gives the pH values for samples of rainwater:

Sample	pH
A	4
B	3
C	5
D	6

(a) Which sample(s) are acid?

(b) Which sample is the strongest acid?

(c) Which sample may contain the fewest air pollutants?

(d) Which sample may contain the most air pollutants?

Connecting Ideas

18. Suppose global warming occurs to an extent such that temperatures in the midwest become more like those of subtropical areas, and farmlands are abandoned. What would happen to those abandoned lands if left undisturbed?

19. Review the graph of human population growth in Chapter 27. Do you see a correlation between that growth and the onset of major environmental problems? What does that suggest to you about another important factor in reducing environmental concerns?

Skill Handbook

Organizing Information

Thinking Critically

Practicing Scientific Methods

Organizing Information

Classifying

You may not realize it, but you impose order on the world around you. If your shirts hang in the closet together, your socks take up a corner of a dresser drawer, or your favorite CDs are stacked in groups according to recording artist, you have used the skill of classifying.

Classifying is grouping objects or events into groups based on common features. When classifying, you first make careful observations of the group of items to be classified. Select one feature that is shared

by some items in the group but not others. Place the items that share this feature in a subgroup. Place the remaining items in a second subgroup. Ideally, the items in the second subgroup will have some feature in common with one another. After you decide on the first feature that divides the items into subgroups, examine the items for other features and further divide each subgroup into smaller and smaller groups until the items have no features in common.

How would you classify a collection of CDs? Classify the CDs based on observable features. You

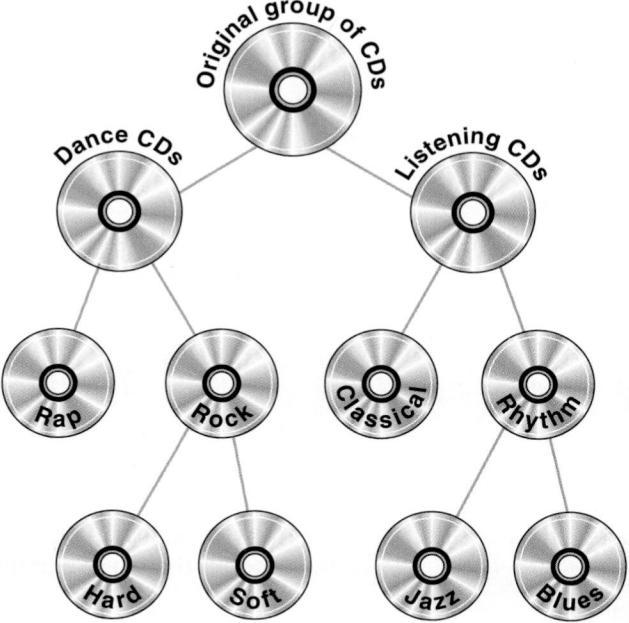

might classify CDs you like to dance to in one subgroup and CDs you like to listen to in another. The CDs you like to dance to could be subdivided into a rap subgroup and a rock subgroup. Note that for each feature selected, each CD only fits into one subgroup. For example, you wouldn't place a CD into both rap and rock categories. Keep selecting features until all the CDs are classified. The chart shows one method of classification.

Remember, when you classify, you are grouping objects or events for a purpose.

Sequencing

A sequence is an arrangement of things or events in a particular order. A common sequence with which you are familiar is students sitting in alphabetical order. Think also about baking chocolate chip cookies. Certain steps have to be followed in order for the cookies to taste good.

When you are asked to sequence things or events, you must first identify what comes first. You then decide what should come second. Continue to choose things or events until they are all in order. Then, go back over the sequence to make sure each thing or event logically leads to the next.

Suppose you wanted to watch a movie that just came out on videotape. What sequence of events would you have to follow to watch the movie? You would first turn the television set to Channel 3 or 4. You would then turn the videotape player on and insert the tape. Once the tape has started playing, you would adjust the sound and picture. Then, when the movie is over, you would rewind the tape and return it to the store. What would happen if you did things out of sequence, such as adjusting the sound before putting in the tape?

Concept Mapping

If you were taking an automobile trip, you would probably take along a road map. The road map shows your location, your destination, and other places along the way. By examining the map, you can understand where you are in relation to other locations on the map.

A concept map is similar to a road map. But, a concept map shows the relationship among ideas (or concepts) rather than places. A concept map is a diagram that visually shows how concepts are related. Because the concept map shows the relationships among ideas, it can clarify the meaning of ideas and terms and help you to understand better what you are studying.

Look at the construction of a concept map called a **network tree.** Notice how some words are circled while others are written on connecting lines. The circled words are science concepts. The lines in the map show related concepts, and the words written on the lines describe relationships between the concepts. A network tree can also show more complex relationships between the concepts. For example, a line labeled "affected by" could be drawn from plants and animals to chemistry, because chemical processes occur in plants and animals.

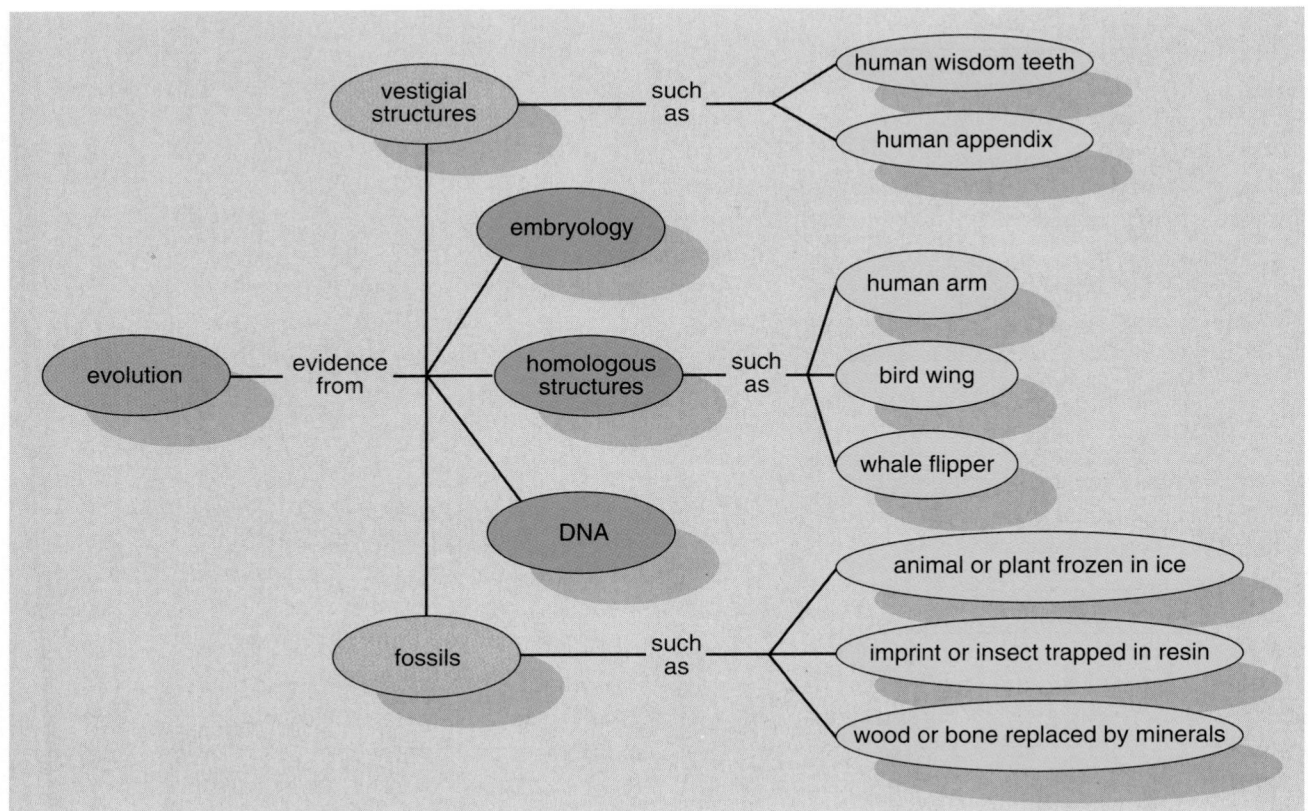

When you are asked to construct a network tree, state the topic and select the major concepts. Find related concepts and put them in order from general to specific. Branch the related concepts from the major concept and describe the relationships on the lines. Continue to write the more specific concepts. Write the relationships between the concepts on the lines until all concepts are mapped. Examine the concept map for relationships that cross branches, and add them to the concept map.

An **events chain** is another type of concept map. An events chain map is used to describe ideas in order. In science, an events chain can be used to describe a sequence of events, the steps in a procedure, or the stages of a process.

Initiating event:
| Mother asks you to wash dishes. |

↓

Event 2:
| You clear the table. |

↓

Event 3:
| You wash the dishes in soapy water. |

↓

Event 4:
| You rinse the dishes in hot water. |

↓

Event 5:
| You dry the dishes. |

↓

Final outcome:
| You put the dishes away. |

When making an events chain, you first must find the one event that starts the chain. This event is called the initiating event. You then find the next event in the chain and continue until you reach an outcome. Suppose your mother asked you to wash the dinner dishes. An events chain map might look like the one in the left column. Notice that connecting words may not be necessary.

A **cycle concept map** is a special type of events chain map. In a cycle concept map, the series of events do not produce a final outcome. The last event in the chain relates back to the initiating event.

As in the events chain map, you first decide on an initiating event and then list each important event in order. Since there is no outcome and the last event relates back to the initiating event, the cycle repeats itself. Look at the cycle map of insect metamorphosis.

Complete Insect Metamorphosis

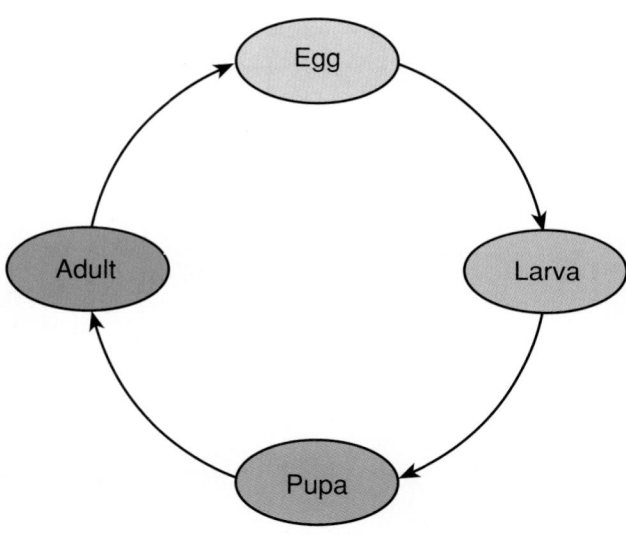

There is usually not one correct way to create a concept map. As you are constructing a map, you may discover other ways to construct the map that show the relationships between concepts better. If you do discover what you think is a better way to create a concept map, do not hesitate to change it.

Concept maps are useful in understanding the ideas you have read about. As you construct a map, you are constructing knowledge and learning. Once concept maps are constructed, you can use them again to review and study and to test your knowledge.

Making and Using Tables

Browse through your textbook, and you will notice many tables both in the text and in the labs. The tables in the text arrange information in such a way that it is easier for you to understand. Also, many labs in your text have tables to complete as you do the lab. Lab tables will help you organize the data you collect during the lab so that it can be interpreted more easily.

Most tables have a title telling you what is being presented. The table itself is divided into columns and rows. The column titles list items to be compared. The row headings list the specific characteristics being compared among those items. Within the grid of the table, the collected data are recorded. Look at the following table and then study the questions that follow it.

Effect of Exercise on Heartbeat Rate

Pulse taken	Heartbeat rate individual	class average
at rest	73	72
after exercise	110	112
1 minute after exercise	94	90
5 minutes after exercise	76	75

What is the title of this table? The title is "Effect of Exercise on Heartbeat Rate." What items are being compared? The heartbeat rate for an individual and the class average are being compared at rest and for several durations after exercise.

What is the average heartbeat rate of the class 1 minute after exercise? To find the answer, you must locate the column labeled "class average" and the row "1 minute after exercise." The data contained in the box where the column and row intersect is the answer. Whose heartbeat rate was 110 after exercise? If you answered the individual's, you have an understanding of how to use a table.

Making and Using Graphs

After scientists organize data in tables, they often display the data in graphs. A graph is a diagram that shows a comparison between variables. Since graphs show a picture of collected data, they make interpretation and analysis of the data easier. The three basic types of graphs used in science are the line graph, bar graph, and pie graph.

A line graph is used to show the relationship between two variables. The variables being compared go on two axes of the graph. The independent variable always goes on the horizontal axis, called the *x*-axis. The independent variable is the condition that is manipulated. The dependent variable always goes on the vertical axis or *y*-axis. The dependent variable is any change that results from manipulating the independent variable.

Suppose a school started a peer study program with a class of students to see how it affected their science grades.

Average Grades of Students in Study Program

Grading Period	Average Science Grade
first	81
second	85
third	86
fourth	89

You could make a graph of the grades of students in the program over a period of time. The grading period is the independent variable and should be placed on the *x*-axis of your graph. Instead of four grading periods, we could look at average grades for the week or month or year. In this way, we would be manipulating the independent variable. The average grade of the students in the program is the dependent variable and would go on the *y*-axis.

After drawing your axes, you would label each axis with a scale. The *x*-axis simply lists the grading periods. To make a scale of grades on the *y*-axis, you must look at the data values. Since the lowest grade was 81 and the highest was 89, you know that you will have to start numbering at least at 81 and go through 89. You decide to start numbering at 80 and number by twos through 90.

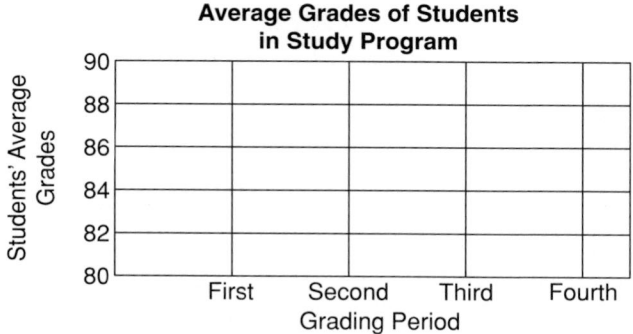

You next must plot the data points. The first pair of data you want to plot is the first grading period and 81. Locate "First" on the *x*-axis and 81 on the *y*-axis. Where an imaginary vertical line from the *x*-axis and an imaginary horizontal line from the *y*-axis would meet, place the first data point. Place the other data points the same way. After all the points are plotted, connect them with a smooth line.

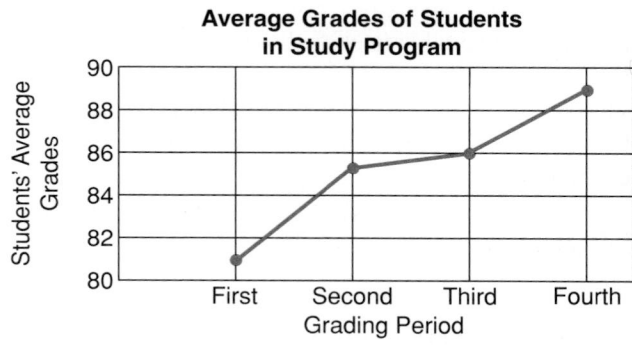

What if you wanted to compare the average grades of the class in the study group with the grades of another class? The data of the other class can be plotted on the same graph to make the comparison. You must include a key with two different lines, each indicating a different set of data.

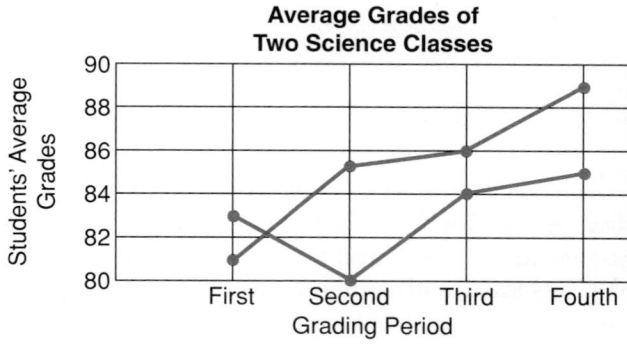

KEY: Class of study students —— Regular class ——

Bar graphs are similar to line graphs, except they are used to show comparisons between data or to display data that does not continuously change. In a bar graph, thick bars show the relationships between data rather than data points.

To make a bar graph, set up the x-axis and y-axis as you did for the line graph. The data is plotted by drawing thick bars from the x-axis up to an imaginary point where the y-axis would intersect the bar if it were extended.

Look at the bar graph comparing the wing vibration rates for different insects. The independent variable is the type of insect, and the dependent variable is the number of wing vibrations per second. The number of wing vibrations for different insects is being compared.

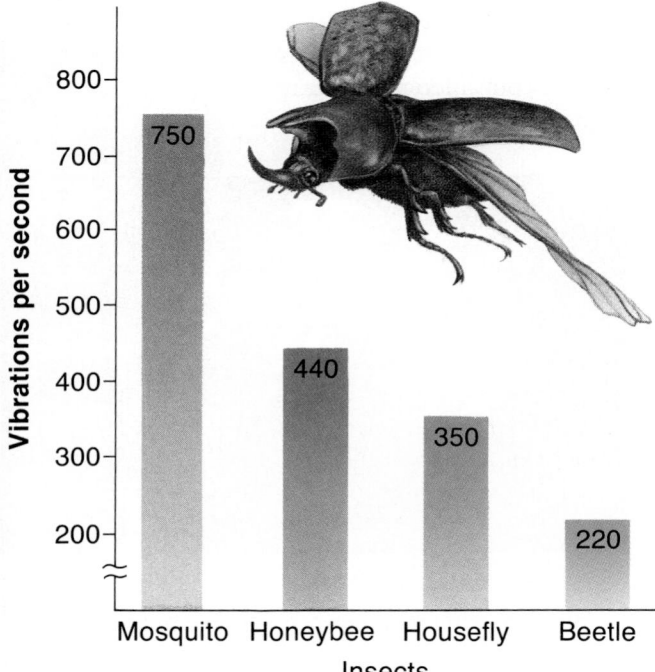

A pie graph uses a circle divided into sections to display data. Each section represents part of the whole. When all the sections are placed together, they equal 100 percent of the whole.

Suppose you wanted to make a pie graph to show the number of seeds that germinated in a package. You would have to determine the total number of seeds and the number of seeds that germinated out of the total. You count the seeds and find that there are 143 seeds in the package. Therefore, the whole pie will represent this amount.

You plant the seeds and determine that 129 seeds germinate. The group of seeds that germinated will make up one section of the pie graph, and the group of seeds that did not germinate will make up another section.

To find out how much of the pie each section should take, you must divide the number of seeds in each section by the total number of seeds. You then multiply your answer by 360, the number of degrees in a circle. Round your answer to the nearest whole number. The number of seeds that germinated would be determined as follows:

$$\frac{143}{129} \times 360 = 324.75 \text{ or } 325 \text{ degrees}$$

To plot this data on the pie graph, you need a compass and a protractor. Use the compass to draw a circle. Then, draw a straight line from the center to the edge of the circle. Place your protractor on this line and use it to mark a point on the edge of the circle at 325 degrees. Connect this point with a straight line to the center of the circle. This is the section for the group of seeds that germinated. The other section represents the group of seeds that did not germinate. Complete the graph by labeling the sections of your graph and giving the graph a title.

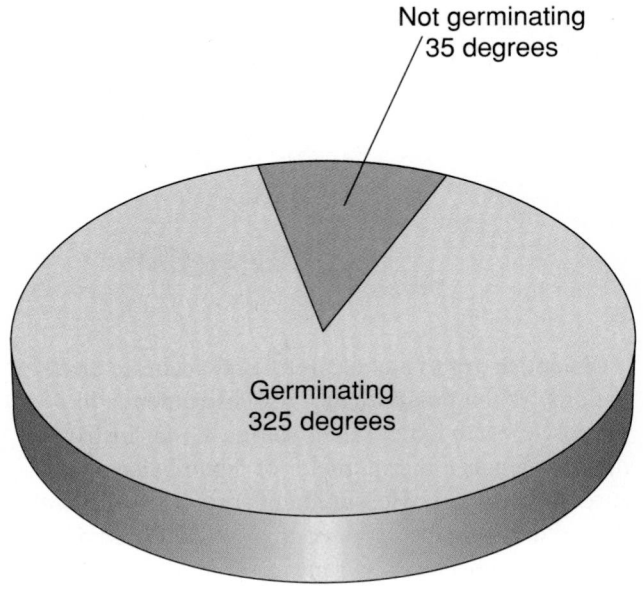

Thinking Critically

Observing and Inferring

Imagine that you have just finished a volleyball game with your friends. At home, you open the refrigerator and see a jug of orange juice at the back of the top shelf. The jug feels cold as you grasp it. "Ah, just what I need," you think. When you drink the juice, you smell the oranges and enjoy the tart taste in your mouth.

As you imagined yourself in the story, you used your senses to make observations. You used your sense of sight to find the jug in the refrigerator, your sense of touch to feel the coldness of the jug, your sense of hearing to listen as the liquid filled the glass, and your sense of smell and taste to enjoy the odor and tartness of the juice. The basis of all scientific investigation is observation.

Scientists try to make careful and accurate observations. When possible, they use instruments, like microscopes, to extend their senses. Other instruments, such as a thermometer or a pan balance, measure observations. Measurements provide numerical data, a concrete means of comparing collected data that can be checked and repeated.

When you make observations in science, you may find it helpful first to examine the entire object or situation. Then, look carefully for details using your sense of sight. Write down everything you see before using another sense to make additional observations.

Scientists often use their observations to make inferences. An inference is an attempt to explain or interpret observations or to determine what caused what you observed. For example, if you observed a CLOSED sign in a store window around noon, you might infer that the owner is taking a lunch break. But, perhaps the owner has a doctor's appointment or has a business meeting in another town. The only way to be sure your inference is correct is to investigate further.

When making an inference, be certain to make accurate observations and to record them carefully. Collect all the information you can. Then, based on everything you know, try to explain or interpret what you observed. If possible, investigate further to determine if your inference is correct.

Comparing and Contrasting

Observations can be analyzed and then organized by noting the similarities and differences between two or more objects or situations. When you examine objects or situations to determine similarities, you are comparing. Contrasting is looking at similar objects or situations for differences.

Suppose you were asked to compare and contrast a grasshopper and a dragonfly. You would start by making your observations. You then divide a piece of paper into two columns. List ways the insects are similar in one column and ways they are different in the other. After completing your lists, you report your findings in a table or in a graph.

Similarities you might point out are that both have three body parts, two pairs of wings, and chewing mouthparts. Differences might include large hind legs on the grasshopper, small legs on the dragonfly; wings held close to the body in the grasshopper, wings held outspread in the dragonfly.

Recognizing Cause and Effect

Have you ever observed something happen and then tried to figure out why or how it came about? If so, you have observed an event and inferred a reason for the event. The event or result of action is an effect, and the reason for the event is the cause.

Suppose that every time your teacher fed fish in a classroom aquarium, she tapped the food container on the edge. Then, one day she tapped the edge of the aquarium to make a point about an ecology lesson. You observe the fish swim to the surface of the aquarium to feed.

What is the effect and what would you infer would be the cause? The effect is the fish swimming to the surface of the aquarium. You might infer the cause to be the teacher tapping on the edge of the aquarium. In determining cause and effect, you have made a logical inference based on careful observations.

Perhaps the fish swam to the surface because they reacted to the teacher's waving hand or for some other reason. When scientists are unsure of the cause for a certain event, they often design controlled experiments to determine what caused their observations. Although you have made a sound judgment, you would have to perform an experiment to be certain that it was the tapping that caused the effect you observed.

Measuring in SI

You are probably familiar with the metric system of measurement. The metric system is a uniform system of measurement developed in 1795 by a group of scientists. The development of the metric system helped scientists avoid problems with different units of measurement by providing an international standard of comparison for measurements. A modern form of the metric system called the International System, or SI, was adopted for worldwide use in 1960.

SKILL HANDBOOK

You will find that your text uses metric units in all its measurements. In the labs you will be doing, you will use the metric system of measurement.

The metric system is easy to use because it has a systematic naming of units and a decimal base. For example, meter is the base unit for measuring length, gram for measuring mass, and liter for measuring volume. Unit sizes vary by multiples of ten. When changing from smaller units to larger, you divide by a multiple of ten. When changing from larger units to smaller, you multiply by a multiple of ten. Prefixes are used to name larger and smaller units. Look at the following table for some common metric prefixes and their meanings.

Metric Prefixes

Prefix	Symbol	Meaning
kilo-	k	1000 thousand
hecto-	h	100 hundred
deka-	da	10 ten
deci-	d	0.1 tenth
centi-	c	0.01 hundredth
milli-	m	0.001 thousandth

Do you see how the prefix *kilo-* attached to the unit *gram* is *kilogram* or 1000 grams, or the prefix *deci-* attached to the unit *meter* is *decimeter* or one tenth (0.1) of a meter?

You have probably measured distance many times. The meter is the SI unit used to measure distance. To visualize the length of a meter, think of a baseball bat, which is about one meter long. When measuring smaller distances, the meter is divided into smaller units called centimeters and millimeters. A centimeter is one hundredth (0.01) of a meter, which is about the width of the fingernail on your index finger. A millimeter is one thousandth of a meter (0.001), about the thickness of a dime. The photograph at the top of the page shows this comparison.

Most metric rulers have lines indicating centimeters and millimeters. The centimeter lines are the longer numbered lines, and the shorter lines between the centimeter lines are millimeter lines.

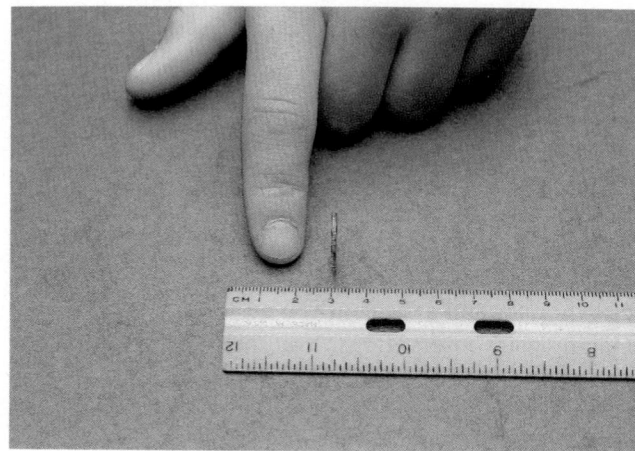

When using a metric ruler, you must first decide on a unit of measurement. You then line up the 0 centimeter mark with the end of the object being measured, and read the number of the unit where the object ends.

Units of length are also used to measure surface area. The standard unit of area is the square meter (m^2), or a square one meter long on each side. Similarly, a square centimeter (cm^2) is a square one centimeter long on each side. Surface area is determined by multiplying the number of units in length times the number of units in width.

The volume of rectangular solids is also calculated using units of length. The cubic meter (m^3) is the standard SI unit of volume. A cubic meter is a cube one meter on a side. You can determine the volume of rectangular solids by multiplying length times width times height.

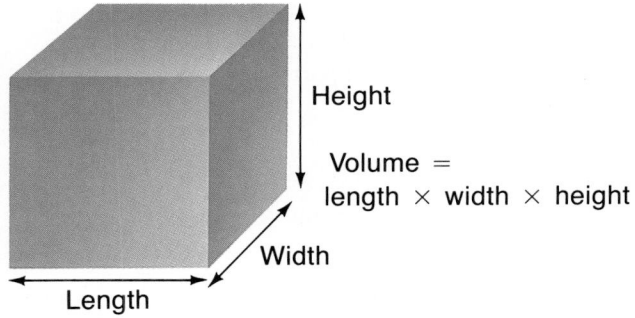

Height

Volume = length × width × height

Width

Length

Liquid volume is measured using a unit called a liter. A liter has the volume of 1000 cubic centimeters. Because the prefix *milli-* means thousandth (0.001), a milliliter equals one cubic centimeter. One milliliter of liquid would completely fill a cube measuring one centimeter on each side.

During biology labs, you will measure liquids using beakers and graduated cylinders marked in milliliters. A graduated cylinder is a tall cylindrical container marked with lines from bottom to top. Each graduation represents one milliliter. You will likely use a beam balance when you want to find the masses of

objects. Often you will measure mass in grams. On one side of the beam balance is a pan and on the other side is a set of beams. Each beam has an object of a known mass called a rider that slides on the beam.

Before you find the mass of an object, you must set the balance to zero by sliding all the riders back to the zero point. Check the pointer to make sure it swings an equal distance above and below the zero point on the scale. If the swing is unequal, find and turn the adjusting screw until you have an equal swing.

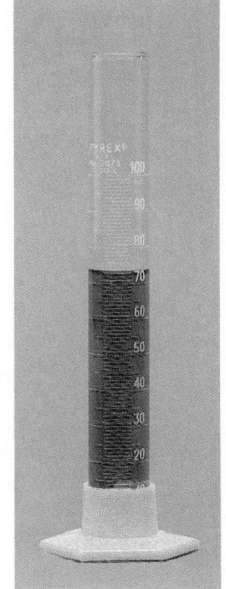

You are now ready to use the balance to find the mass of the object. Place the object on the pan. Slide the rider with the largest mass along the beams until the pointer drops below the zero point. Then move it back one notch. Repeat the process on each beam until the pointer swings an equal distance above and below the zero point. Read the masses indicated on the beams. The sum of the masses is the mass of the object.

Never place a hot object or pour chemicals directly on the pan. Instead, find the mass of a clean container, such as a beaker or a glass jar. Place the dry or liquid chemicals you want to measure into the container. Next, find the combined mass of the container and the chemicals. Calculate the mass of the chemicals by subtracting the mass of the empty container from the combined mass.

Interpreting Scientific Illustrations

Illustrations are included in your science textbook to help you understand, interpret, and remember what you read. Whenever you encounter an illustration, examine it carefully and read the caption. The caption is a brief comment that explains or identifies the illustrations.

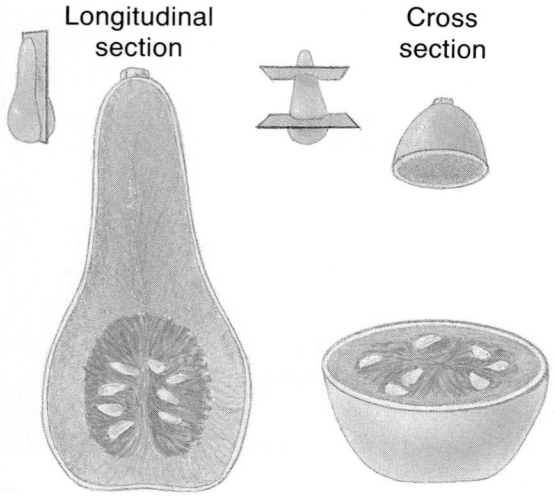

Longitudinal section

Cross section

Other organisms and objects have radial symmetry. Radial symmetry is the arrangement of similar parts around a central point. The anemone in the figure below has radial symmetry. See how it can be divided anywhere through the center into similar parts.

Some organisms and objects cannot be divided into two similar parts. If an organism or object cannot be divided, it is asymmetrical. Study the sponge. Regardless of how you try to divide a sponge, you cannot divide it into two parts that look alike.

Some illustrations are designed to show you how the internal parts of a structure are arranged. Look at the illustrations of a squash. The squash has been cut lengthwise so that it shows a section that runs along the length of the squash. This type of illustration is called a longitudinal section. Cutting the squash crosswise at right angles to the length produces a cross section.

Symmetry refers to a similarity or likeness of parts. Many organisms and objects have symmetry. When something can be divided into two similar parts lengthwise, it has bilateral symmetry. Look at the illustration of the butterfly. The right side of the butterfly looks very similar to the left side. It has bilateral symmetry.

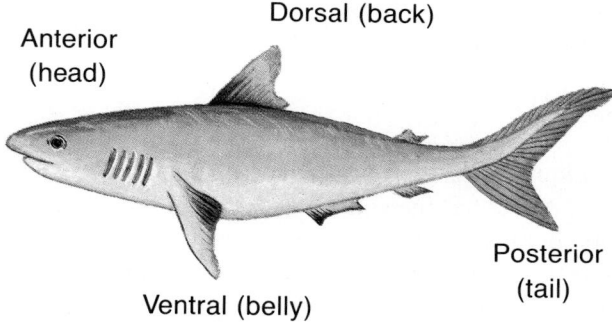

Anterior (head)

Dorsal (back)

Posterior (tail)

Ventral (belly)

In your reading and examination of the illustrations, you will sometimes see terms that refer to the orientation of an organism. The word *dorsal* refers to the upper side or back of an animal. *Ventral* refers to the lower side or belly of the animal. The illustration of the shark has both dorsal and ventral sides.

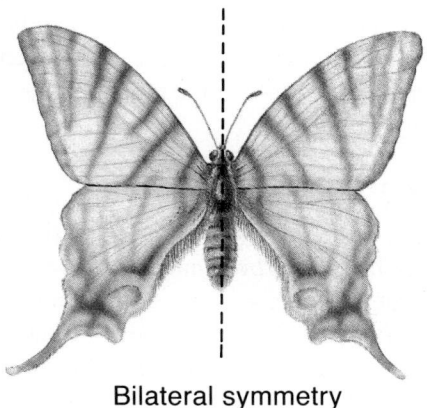

Bilateral symmetry

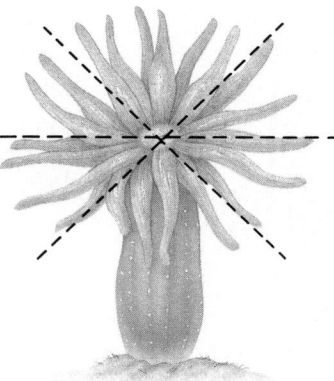

Radial symmetry

Asymmetry

SKILL HANDBOOK

Practicing Scientific Methods

You might say that the work of a scientist is to solve problems. But when you decide how to dress on a particular day, you are doing problem solving, too. You may observe what the weather looks like through a window. You may go outside and see if what you are wearing is warm or cool enough.

Scientists generally use experiments to solve problems and answer questions. An experiment is a method of solving a problem in which scientists use an organized process to attempt to answer a question.

Experimentation involves defining a problem and formulating and testing hypotheses, or proposed solutions, to the problem. Each proposed solution is tested during an experiment, which includes making careful observations and collecting data. After analysis of the collected data, a conclusion is formed and compared to the hypothesis.

Observing

You observe all the time. Anytime you smell wood burning, touch a pet, see lightning, taste food, or hear your favorite music, you are observing. Observation gives you information about events or things. Scientists must try to observe as much as possible about the things and events they study.

Some observations describe something using only words. These observations are called qualitative observations. If you were making qualitative observations of a dog, you might use words such as furry, brown, short-haired, or short-eared.

Other observations describe how much of something there is. These are quantitative observations and use numbers as well as words in the description. Tools or equipment are used to measure the characteristic being described. Quantitative observations of a dog might include a mass of 45 kg, a height of 76 cm, ear length of 11 cm, and an age of 412 days.

Forming a Hypothesis

Suppose you wanted to earn a perfect score on a spelling test. You think of several ways to accomplish a perfect score. You base these possibilities on past observations. All of the following are hypotheses you might consider to explain how you could score 100% on your test:

If the test is easy, then I will get a good grade.
If I am intelligent, then I will get a good grade.
If I study hard, then I will get a good grade.

Scientists use hypotheses that they can test to explain the observations they have made. Perhaps a scientist has observed that fish activity increases in the summer and decreases in the winter. A scientist may form a hypothesis that says: If fish are exposed to warmer water, their activity will increase.

Designing an Experiment to Test a Hypothesis

Once you have stated a hypothesis, you probably want to find out whether or not it explains an event or an observation. This requires a test. To be valid, a hypothesis must be testable by experimentation. Let's figure out how you would conduct an experiment to test the hypothesis about the effects of water temperature on fish.

First, obtain several identical clear glass containers, and fill them with the same amount of water. Let the containers set. On the day of your experiment, you fill a container with an amount of aquarium water equal to that in the test containers. After measuring and recording the water temperature, you heat and cool the other containers, adjusting the water temperatures in the test containers so that two have higher temperatures and two have lower temperatures than the aquarium water temperature.

You place a guppy in each container. You count the number of horizontal and vertical movements each guppy makes during five minutes and record your data in a table. Your data table might look like this:

Number of Guppy Movements

Container	Temperature (°C)	Number of movements
aquarium water	38	56
A	40	61
B	42	70
C	36	46
D	34	42

From the data you recorded, you will draw a conclusion and make a statement about your results. If your conclusion supports your hypothesis, then you can say that your hypothesis is reliable. If it did not support your hypothesis, then you would have to make new observations and state a new hypothesis, one that you could also test. Do the data above support the hypothesis that warmer water increases fish activity?

Separating and Controlling Variables

When scientists perform experiments, they must be careful to manipulate or change only one condition and keep all other conditions in the experiment the

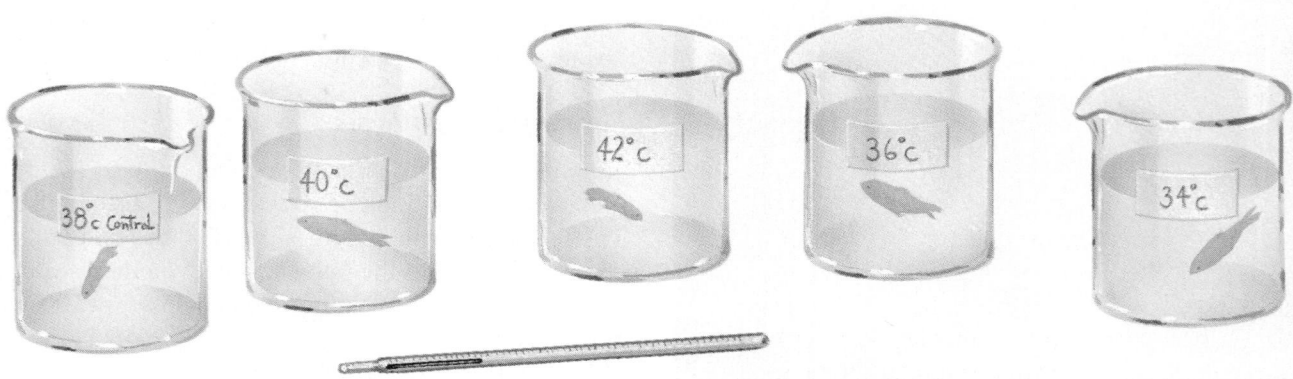

same. The condition that is manipulated is called the independent variable. The conditions that are kept the same during an experiment are called constants. The dependent variable is any change that results from manipulating the independent variable.

Scientists can only know that the independent variable caused the change in the dependent variable if they keep all other factors the same in an experiment. Scientists also use controls to be certain that the observed changes were a result of manipulation of the independent variable. A control is a sample that is treated exactly like the experimental group except that the independent variable is not applied to the control. After the experiment, the change in the dependent variable of the control sample is compared with any change in the experimental group. This allows scientists to see the effect of the independent variable.

What are the independent and dependent variables in the guppy experiment? Because you are changing the temperature of the water, the independent variable is the water temperature. Since the dependent variable is any change that results from the independent variable, the dependent variable is the number of movements the guppy makes during five minutes.

What factors are constants in the experiment? The constants are using the same size and shape containers, filling them with equal amounts of water, and counting the number of movements during the same amount of time. What was the purpose of counting the number of movements of a guppy in an identical container filled with aquarium water? The container of aquarium water is the control. The number of movements of the guppy in the aquarium water will be used to compare the movements of the guppies in water of different temperatures.

Interpreting Data

After doing a controlled experiment, you must analyze and interpret the collected data, form a conclusion, and compare the conclusion to your hypothesis. Analyze and interpret the data in the table. What conclusion did you form? The data indicate that the higher the temperature, the greater the number of movements of the guppy. How does the conclusion compare with your hypothesis? Was your hypothesis supported by the experiment or not?

TABLE OF CONTENTS

Appendices

APPENDIX A

Classification of Organisms

Kingdom Monera

Archaebacteria (Ancient Bacteria)

Phylum Aphragmabacteria (thermoacidophiles)
Shape irregular or bloblike due to lack of cell wall; found in hot springs of Yellowstone National Park at temperatures of 60°C and having a pH of 1 to 2, and in humans and domesticated animals where they cause certain types of pneumonia. **Examples:** *Mycoplasma pneumoniae, Thermoplasma acidophilum*

Phylum Halobacteria (halophiles)   Rod- or cocci-shaped; move by flagella; tolerate and live only in areas of high salt (sodium chloride) concentrations, such as in brine or salt flats. **Examples:** *Halobacterium halobium, Halobacterium salinarium*

Phylum Methanocreatrices (methanogens)   Rod, spiral, or coccus in shape; may move via flagella or are immobile; found worldwide in sewage, marine and freshwater sediments, swamps, and even cows' stomachs; responsible for producing marsh gas or methane. **Examples:** *Methanobacillus omelianski*

Eubacteria (True Bacteria)

Phylum Actinobacteria   Composed of rods, long complex filaments, or short multicellular filaments; filaments resemble hyphae of fungi; some cause diseases such as tuberculosis, leprosy, or skin lesions; others produce the antibiotic streptomycin; others found living in plant root nodules remove or fix nitrogen from the air. **Examples:** *Mycobacterium tuberculosis (tuberculosis), Mycobacterium leprae* (leprosy), *Streptomyces griseus*

Phylum Omnibacteria   A mixed group of bacteria having rod or comma-like shapes; can live with or without oxygen; many cause plant, human, and other animal diseases such as food poisoning, gonorrhea, meningitis; *Escherichia coli,* the most studied bacterium in the world, belongs to this phylum. **Examples:** *Salmonella typhi, Neisseria gonorrhoeae*

Phylum Spirochaetae (Spirochaetes)   Look like coiled springs, as suggested by the beginning of the phylum name (spiro); found in sewage, marine, and fresh water as well as in tooth plaque and the intestines of different animals; help to digest wood in the gut of termites; cause diseases in humans. **Examples:** *Treponema pallidum* (syphilis), *Leptospira icterohaemorrhagiae* (infectious jaundice)

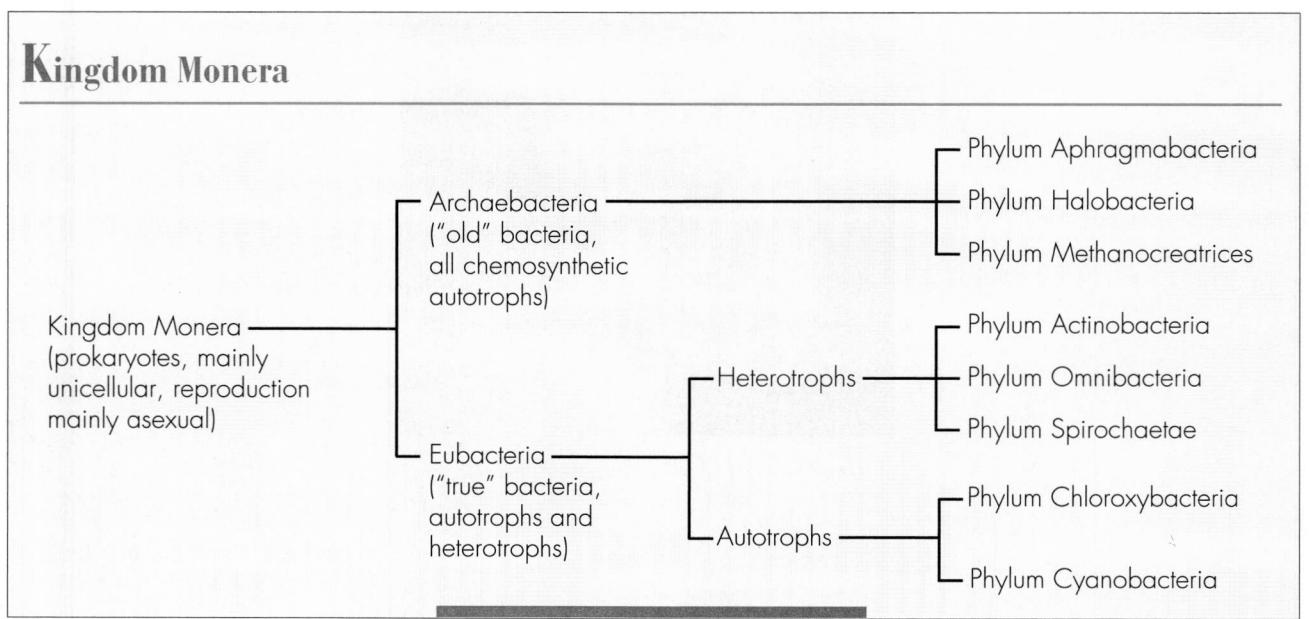

Phylum Chloroxybacteria (Grass-green Bacteria)

All spherical in shape and contain chlorophyll, thus making them autotrophic; found living on or in the body of marine organisms called tunicates, also known as sea squirts; structure strongly resembles chloroplasts, the photosynthetic organelles found in plants. **Examples:** *Prochloron sp., Prochlorothrix sp.*

Phylum Cyanobacteria (Blue-green Bacteria)

Autotrophic due to the presence of chlorophyll; have a bluish tinge due to blue-colored pigments; some cyanobacteria found as fossils dating as far back as 3 billion years ago (stromatolites); some found floating on or just below the surfaces of lakes and ponds, some fix nitrogen from the air. **Examples:** *Nostoc parmeloides, Anabaena subcylindrica*

Kingdom Protista

Animal-like Protists

Phylum Rhizopoda (Amoebas) Found in soil, fresh and marine waters; unicellular; move about by flowing cytoplasmic extensions called pseudopodia; many are parasitic. **Examples:** *Amoeba proteus, Entamoeba histolytica* (amoebic dysentery)

Phylum Ciliophora (Ciliates) Hairs called cilia that are used for locomotion; may reproduce asexually or sexually; two types of nuclei—a small micronucleus used in sexual reproduction and a larger macronucleus used for growth and asexual reproduction. **Examples:** *Paramecium caudatum, Stentor coeruleus*

Kingdom Protista

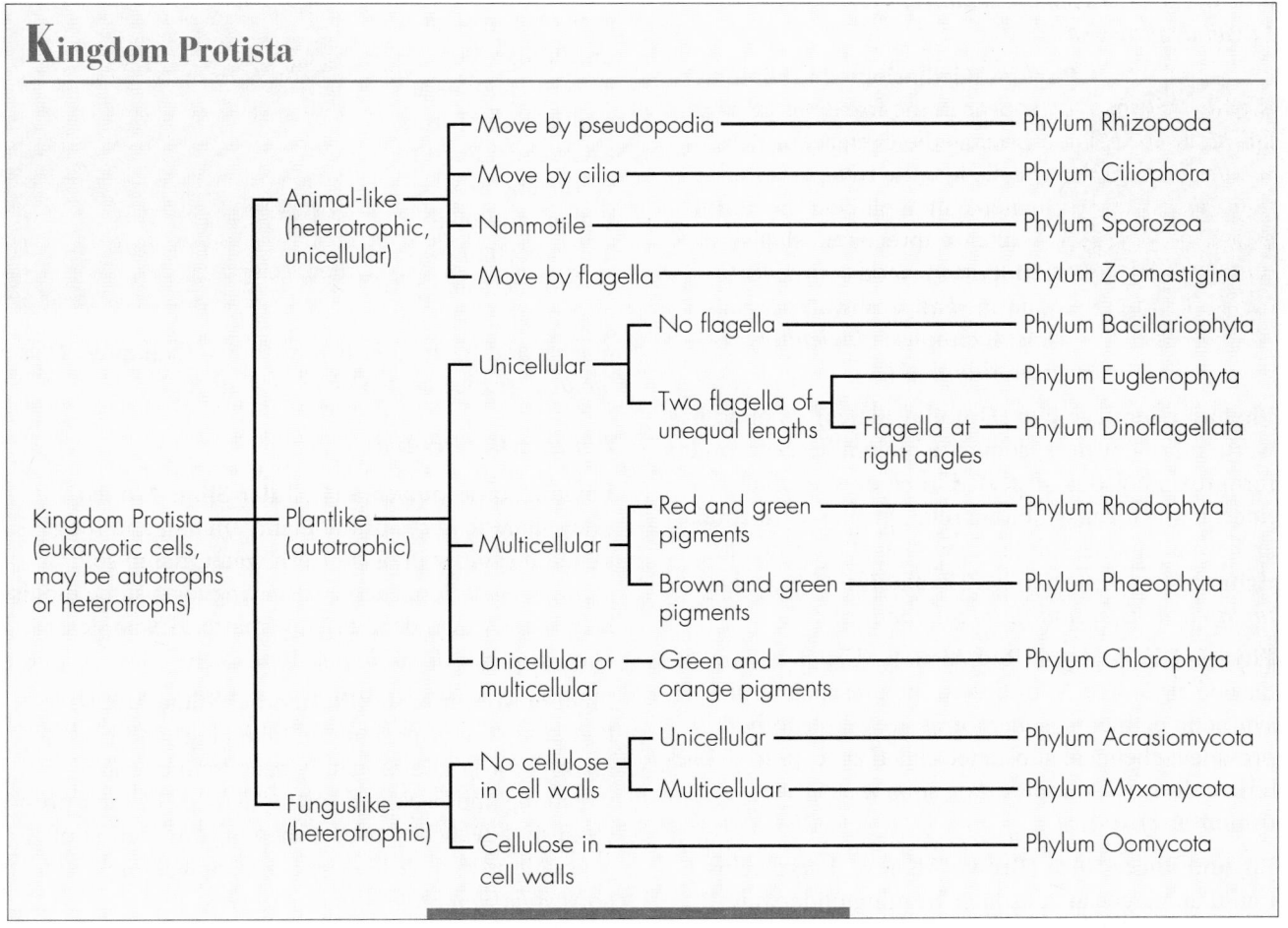

Phylum Sporozoa (Sporozoans) Protists with no means for locomotion; form spores, as their name suggests, that aid in their dispersal from host to host; many are bloodstream parasites. **Examples:** *Plasmodium malariae* (malaria), *Monocystis agilis*

Phylum Zoomastigina (Flagellates) One or many whiplike hairs called flagella that are used for locomotion; include human parasites; others are found in the gut of wood-eating termites and aid the digestion of cellulose. **Examples:** *Trichomonas muris*, *Trypanosoma gambiense* (sleeping sickness)

Plantlike Protists

Phylum Euglenophyta (Euglenoids) Lack cellulose cell walls; easily change shape; most live in fresh or stagnant waters; photosynthetic, but can become heterotrophic organisms. **Examples:** *Euglena gracilis*, *Euglena spirogyra*

Phylum Bacillariophyta (Diatoms) Appear in the fossil record as far back as the Cretaceous period; cells like the two parts that make up a pillbox; beautiful silica-impregnated shells; important food source for marine and freshwater organisms. **Examples:** *Navicula lyra*, *Frustulia rhomboides*

Phylum Dinoflagellata (Dinoflagellates) Found in warm marine waters; some are bioluminescent; others form toxic poisons dispersed in blooms, such as red tides; many form symbiotic relationships with corals and sea anemones; all have two flagella located at right angles to one another. **Examples:** *Gonyaulax tamarensis*, *Gymnodinium microadriaticum*

Phylum Rhodophyta (Red Algae) Usually complex marine algae; red color due to the presence of photosynthetic pigments; some forms are edible; others provide a chemical substance called agar, used in bacteria cultures. **Examples:** *Polysiphonia harveyi*, *Chondrus crispus*

Phylum Phaeophyta (Brown Algae) Large kelps found along ocean coastlines forming underwater

forests of seaweed beds; some are used as food or fertilizers; algin, a chemical formed by these algae, is used as a gel in soft ice creams. **Examples:** *Macrocystis pyrifera* (kelp), *Fucus vesiculosus*

Phylum Chlorophyta (Green Algae) Many green algae are unicellular, but colonies and large multicellular forms exist; considered to be the ancestors of higher land plants; are important oxygen-formers and provide food for many marine and freshwater heterotrophs. **Examples:** *Ulva lactuca* (sea lettuce), *Volvox aureus*

Funguslike Protists

Phylum Acrasiomycota (Cellular Slime Molds) They share features of the plant, animal, and fungi kingdoms; grow in or on fresh water, damp soil, or rotting vegetation, such as decaying logs; slime molds help break down dead organic matter. **Examples:** *Dictyostelium discoideum*, *Acrasia sp.*

Phylum Myxomycota (Plasmodial Slime Molds) Found as slimy wet scum on fallen logs and bark; a favorite for laboratory study because they show streaming and back-and-forth pulsating of their protoplasm; advanced protists showing alternation of generations in their life cycle. **Examples:** *Physarum polycephalum*, *Stemonitis splendens*

Phylum Oomycota (Water Molds, Mildews, Rusts)
Includes water mold that caused the infamous 19th-century blight that destroyed potato crops in Ireland and Germany; another parasitic form called "ich" is a common, often fatal pest of aquarium fish, seen as a white fuzz on their fins; many cause diseases of garden and crop plants. **Examples:** *Saprolegnia parasitica, Phytophthora infestans* (potato blight)

Kingdom Fungi

Phylum Zygomycota (Sporangium Fungi)   These fungi grow by feeding on any decaying vegetation or animal; they may reproduce by forming spores inside a round sporangium that is borne aloft by a long, thin hypha. **Examples:** *Rhizopus stolonifer* (black bread mold), *Mucor hiemalis*

Phylum Ascomycota (Cup Fungi and Yeasts)
Includes yeasts, morels, and truffles; causes athlete's foot and ringworm of the scalp; may reproduce by forming spores that are enclosed inside an ascus; important as decomposers; yeasts are essential for making wine and beer, and as an aid in baking.
Examples: *Saccharomyces cerevisiae* (bread yeast), *Neurospora sitophila*

Phylum Basidiomycota (Club Fungi)   Includes smuts, rusts, and puffballs; reproduce with a microscopic clublike part, called a basidium, that contains spores; fatal, hallucinogenic, and edible mushrooms all belong to this phylum. **Examples:** *Amanita phalloides* (death cap), *Agaricus brunnescens* (edible mushroom)

Phylum Deuteromycota (Imperfect Fungi)   Only asexual reproduction has been observed; the antibiotic penicillin is produced by a fungus in this phylum. One member of this phylum causes moniliasis, a vaginal infection of humans; other members cause root rot of many cultivated plants. **Examples:** *Penicillium chrysogenum* (penicillin antibiotic), *Candida albicans* (moniliasis)

Phylum Mycophycota (Lichens)   This phylum is made up of organisms formed through a symbiotic union of a fungus and a unicellular chlorophyte algae or cyanobacterium; are the first organisms that grow on volcanic areas or bare rock; ecologically important because they dissolve rocks, thus providing conditions suitable for the growth of rooted plants; are an important food for tundra reindeer. **Examples:** *Cladonia cristatella, Peltigera rufescens*

Kingdom Fungi

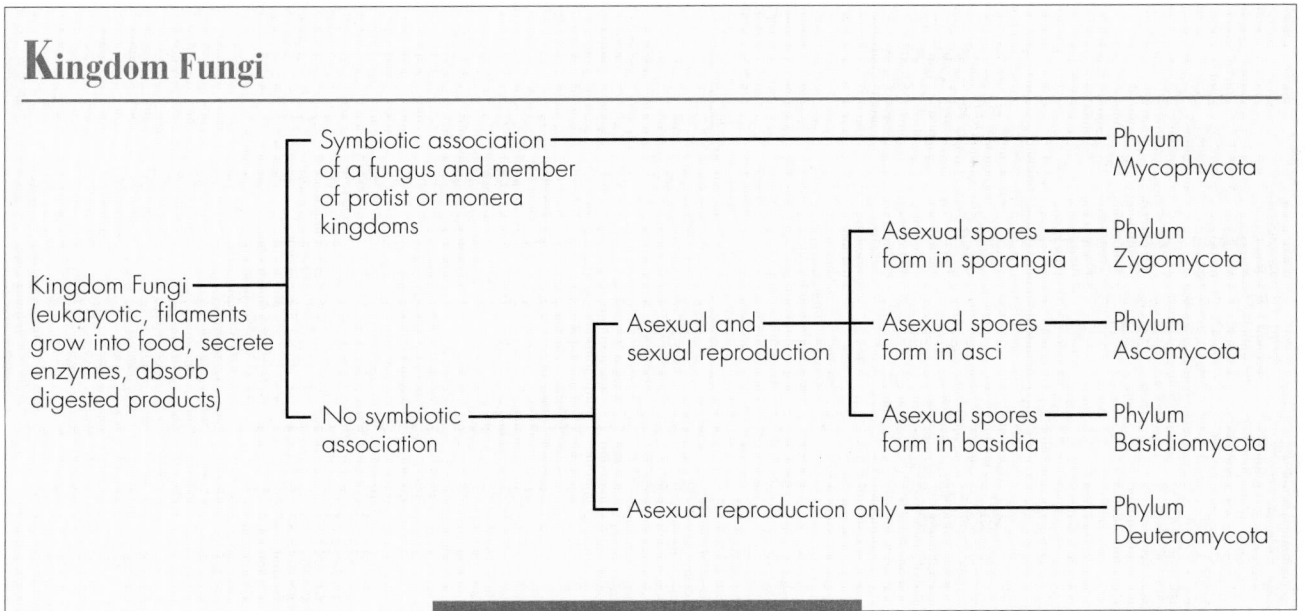

APPENDIX A

Kingdom Plantae

Spore Plants

Division Bryophyta (Mosses and Liverworts)   This division includes the mosses and liverworts, which grow in moist habitats; have no vascular tissue; the most conspicuous stage in the life cycle is a leaflike, green gametophyte generation; spores are formed in capsules; sporophyte dependent on gametophyte for water and nutrients.

> **Class Mucopsida (Mosses)**   More than 14 000 species; Gametophyte generations are creeping along ground or upright; multicellular rhizoids; ecologically important as soil contributors, increasing water retention in soil; peat moss can be burned as a fuel. **Examples:** *Polytrichum juniperinum, Sphagnum squarrosum*

> **Class Hepaticopsida (Liverworts)**   So-called because of the liverlike, flat shape of the gametophyte generation; about 9000 species of liverworts; most found in the tropics; obtain water from long, single-celled rhizoids. **Examples:** *Marchantia polymorpha, Pellia epiphylla*

Division Psilophyta (Whisk Ferns)   The conspicuous generation is a green sporophyte; sporophyte is a thin, leafless stem (about 30 cm); stem has small, leaflike scales along its length; no roots; found in tropical and subtropical regions of the world. **Examples:** *Psilotum nudum, Rhynia gwynnevaughanii* (*Rhynia* is an extinct plant.)

Division Lycophyta (Club Mosses)   The conspicuous generation is the sporophyte; most species extinct; fossil club mosses are giant treelike plants that lived more than 280 million years ago and helped form present-day coal beds; present-day species are very small and have often been used as holiday decorations—the main reason for their threatened condition. **Examples:** *Lycopodium obscurum, Selaginella lepidophylla*

Kingdom Plantae

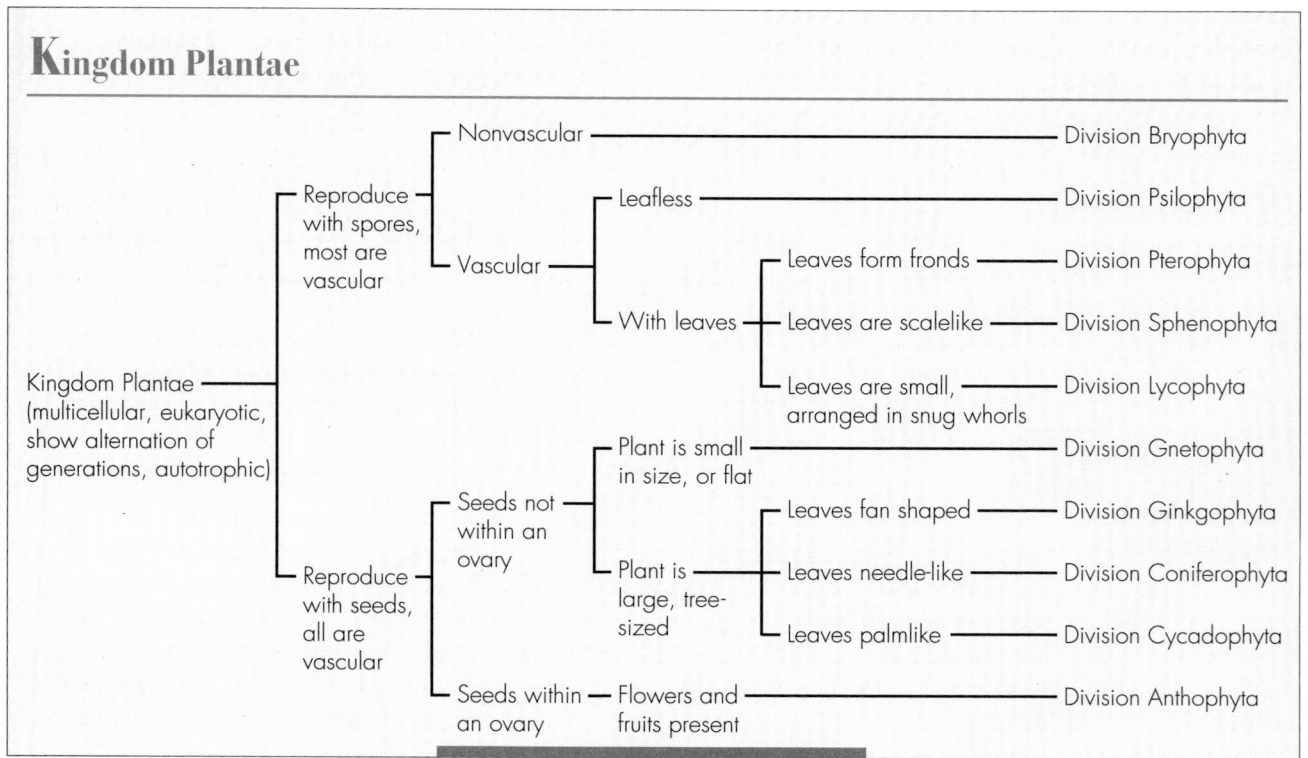

Division Sphenophyta (Horsetails) Stems jointed and have a rough, ribbed feel; stem looks like a pipe with small, scalelike leaves arranged in whorls at each node; commonly called "scouring rushes"; only one living genus with about five species; all present-day horsetails are short, whereas their fossil relatives were treelike; contributed to the formation of coal beds. **Examples:** *Equisetum arvense, Calamites carinatus* (*Calamites* is an extinct plant group.)

Division Pterophyta (Ferns) Most complex of all spore-forming plants; usually have lacy leaves called fronds; size of fronds varies from 1 cm to 500 cm long; two-thirds of the 12 000 present-day species are found in the tropics; many are cultivated for use as houseplants. **Examples:** *Polypodium virginianum, Osmunda cinnamonea*

Seed Plants

Division Ginkgophyta (Ginkgos) Only one living species of this division, which explains why it is called a living fossil; native to China; leaves deciduous and fan-shaped; trees are either male or female; very hardy and resistant to pollution and insects; popular choice for street plantings in many cities. **Examples:** *Ginkgo biloba* (maidenhair tree), *Ginkgoites digitata*. (*Ginkgoites* is an extinct plant group.)

Division Cycadophyta (Cycads) Some are small plants that look like pineapple fruits; others look like palm trees; stems are typically unbranched and covered with the woody bases of shed leaves; bear male and female reproductive organs in cones; grow only in tropical or subtropical climates; about 100 species. **Examples:** *Zamia floridana, Cycas revoluta*

Division Coniferophyta (Conifers) Reproduce by forming cones; commonly known as evergreen or conifer trees; includes the largest living plant, the giant Sequoia of California; leaves are usually needle-shaped and not deciduous; a major source of lumber, paper, turpentine, pitch, tar, amber, and resin; about 400 species. **Examples:** *Pinus virginiana* (Virginia pine), *Sequoiadendron gigantea* (giant redwood)

Division Gnetophyta Most are either desert-dwellers or found on mountainsides in Asia, Africa, and Central and South America; some are scrubby, cone-bearing, and pinelike; others are flat plants consisting of

two long, very large, thick leaves that lie on the ground; only 70 species are known; a drug called ephedrine comes from one of these plants and is used to treat asthma, emphysema, and hay fever. **Examples:** *Welwitschia mirabilis, Gnetum gnenom*

Division Anthophyta (Flowering Plants) Reproduction occurs in flowers; seeds are protected within an ovary in the flower; the most diverse and largest group of modern plants with over 230 000 species and 300 families; gametophyte generation is microscopic; sporophyte generation may reach tree size; plants grown commercially for food and decorations belong to this division.

Class Dicotyledones (Dicots) Flowering plants with two (di) cotyledons, or seed leaves, in their seeds; flower parts are in fours or fives.

Family Magnoliaceae (Magnolias) Found mainly in tropic or subtropic regions but a few species are found in temperate zones; about 100 species, most of which are magnolias; 2 species of tulip trees. **Examples:** *Magnolia virginiana, Liriodendron tulipifera*

Family Fagaceae (Beeches) Familiar members of this group are the oaks and beeches; used primarily for wood in building furniture, floors, and interior moldings; used almost exclusively by early shipbuilders; about 350 species. **Examples:** *Fagus grandifolia, Quercus alba*

Family Cactaceae (Cacti) Strictly desert dwellers, native to the new world; familiar succulents and spiny cacti that range in height from very small to treelike; flowers with numerous petals, many stamens often joined to the petals; fruit a many-seeded berry; about 2000 species. **Examples:** *Opuntia fragilis, Carnegiea gigantea*

Family Malvaceae (Mallows) Most important member commercially is the cotton plant; hollyhocks and hibiscus are common cultivated garden plants. Five separate petals; stalks of many stamens all fused and surrounding a long style; fruit a capsule; about

1500 species. **Examples:** *Gossypium hirsutum* (cotton), *Hibiscus tiliaceus*

Family Brassicaceae (Mustard) Commonly called the mustard group; most familiar as foods that are eaten throughout the world; includes cabbage, turnips, radish, Brussels sprouts, cauliflower, and broccoli; about 2000 species. **Examples:** *Brassica oleracea, Raphanus sativus*

Family Rosaceae (Rose) Worldwide in distribution; includes common garden rose, raspberries, strawberries; about 1200 species, 300 of which are found in the U.S. **Examples:** *Rosa alba, Rubus idaeus*

Family Fabaceae (Pea) Second largest family of plants; includes large trees, shrubs, vines, and perennial and annual herbs; includes clover, alfalfa, peas, beans, soybean, and peanuts; over 12 000 species. **Examples:** *Medicago sativa, Arachis hypogaea*

APPENDIX A

Family Aceraceae (Maple) Includes maples and boxelders; most abundant in the eastern half of U.S.; noted for their use in fine furniture, musical instruments, and as a source for sugar maple; about 100 species. **Examples:** *Acer saccharum, Acer negundo*

Family Lamiaceae (Mint) Best known for their fragrance and oils that are used as flavorings and in some medicines; includes plants such as peppermint, catnip, thyme, and sage; about 3000 species. **Examples:** *Mentha piperita, Thymus vulgaris*

Family Asteraceae (Daisies) The largest family; found growing worldwide and in almost every ecological condition; includes sunflower, goldenrod, lettuce, dandelion, ragweed, and chrysanthemum; more than 15 000 species. **Examples:** *Helianthus annuus, Lactuca sativa*

Class Monocotyledones (Monocots) Flowering plants with one (mono) cotyledon, or seed leaf, in their seeds; flower parts are in threes or multiples of three.

Family Poaceae (Grasses) Includes important herbaceous plants such as corn, rice, oats, wheat, sugarcane, and Kentucky bluegrass; flowers often inconspicuous; enclosed in two scalelike bracts; fruit a caryopsis; about 10 000 species. **Examples:** *Triticum aestivum* (wheat), *Zea mays* (corn)

Family Palmae (Palms) Tropical and semitropical trees and shrubs; includes date palms, coconut palms, and palmettos; about 1200 species. **Examples:** *Phoenix dactylifera, Cocos nucifera*

Family Liliaceae (Lilies) Many ornamental perennial plants such as lilies, tulips, and hyacinths that grow from bulbs or corms; also includes food plants such as onion and asparagus; leaves linear; flowers with three sepals, three petals, and six stamens; fruit a capsule or berry; about 6500 species. **Examples:** *Lilium philadelphicum, Asparagus officinalis*

Family Orchidaceae (Orchids) Largest family of monocotyledons; flower complexity surpasses everything in the plant kingdom; best known examples are orchids and ladyslippers; 15 000 species. **Examples:** *Cypripedium hirsutum, Spiranthes cernua*

APPENDIX A

Kingdom Animalia

Invertebrates

Phylum Porifera (Sponges)   Body tissues organized into pores and canals; found in either fresh or marine water; lack organs and symmetry; shapes may be like cups, fans, crusts, or tubes; sponge body is riddled with pores (porifera) and contain skeletal needles called spicules; larval stage free-swimming; adults are sessile; most are both male and female; feed by circulating water through pores and canals; about 10 000 species. **Examples:** *Spongilla lacustris, Tethya auvantitium* (sea orange)

Phylum Cnidaria (Corals, Jellyfish, Hydras)   First animal group with organ level of organization; saclike bodies include mouth surrounded by tentacles that sting and paralyze prey; have radial symmetry; almost all marine; certain species show alternation of generations between a polyp and medusa stage; others may spend lives as either medusas or polyps; about 10 000 species.

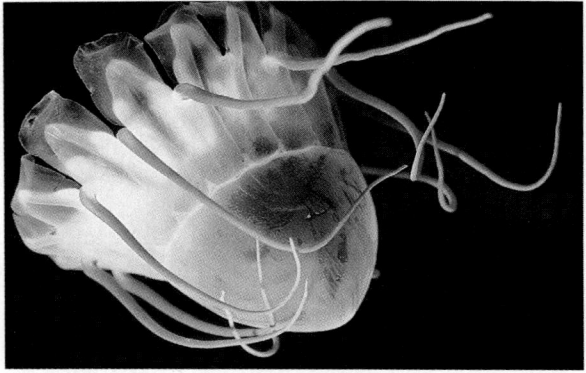

Class Hydrozoa (Hydroids)   About 3100 species known, including hydroid types and Portuguese man-of-war; tentacles give off a poison; paralyzed prey carried by tentacles to the mouth. **Examples:** *Hydra littoralis, Physalia physalis* (Portuguese man-of-war)

Class Scyphozoa (Jellyfish)   Marine species; often called medusae because this shape is dominant; 95-98 percent of body weight consists of water; about 200 species exist. **Examples:** *Aurelia aurita, Cyanea arctica*

Kingdom Animalia

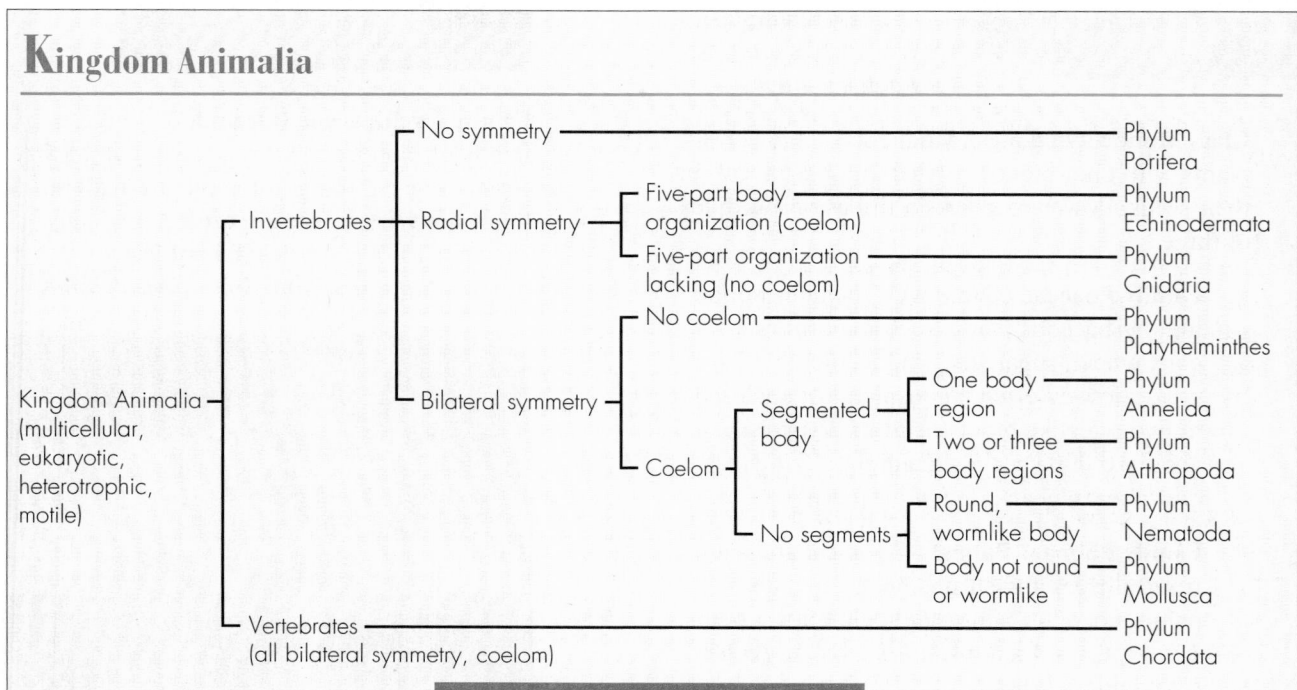

Kingdom Animalia (multicellular, eukaryotic, heterotrophic, motile)

- Invertebrates
  - No symmetry ———— Phylum Porifera
  - Radial symmetry
    - Five-part body organization (coelom) ———— Phylum Echinodermata
    - Five-part organization lacking (no coelom) ———— Phylum Cnidaria
  - Bilateral symmetry
    - No coelom ———— Phylum Platyhelminthes
    - Coelom
      - Segmented body
        - One body region ———— Phylum Annelida
        - Two or three body regions ———— Phylum Arthropoda
      - No segments
        - Round, wormlike body ———— Phylum Nematoda
        - Body not round or wormlike ———— Phylum Mollusca
- Vertebrates (all bilateral symmetry, coelom) ———— Phylum Chordata

Class Anthozoa (Sea Anemones, Corals)
Includes sea pens, sea fans, coral, and sea anemones; about 6200 species; polyp stage is dominant; coral reefs are home to thousands of different marine vertebrate and invertebrate species. **Examples:** *Corallium rubrum, Adamsia palliata* (sea anemone always found living with a species of hermit crab)

Phylum Platyhelminthes (Flatworms)   Includes all flat (platy), ribbon-shaped worms; about 15 000 species; all bilaterally symmetrical; have soft body and mouths but no anuses; organs forming simple systems are present; can regenerate and reproduce sexually; most species both male and female; some forms free-living; others parasitic.

Class Turbellaria (Free-living Flatworms)
Flatworms that are free-living; found on land in damp soil, and in fresh or marine waters; body shows a head region with eyespots; familiar example is *Planaria,* often used in regeneration experiments. **Examples:** *Dugesia tigrina, Bipalium kewense* (tropical land planarian)

Class Trematoda (Flukes)   All are parasitic; most live inside vertebrates, including humans; commonly called flukes; have a head region with suckers that attach to a host; can be found in host's liver, lungs, intestines, and blood; some cause diseases in hosts. **Examples:** *Fasciola hepatica* (liver fluke of cattle), *Clonorchis sinensis* (Oriental liver fluke)

Class Cestoda (Tapeworms)   Includes tapeworms; bodies very long, flat, and formed of many short sections; some longer than 7 meters; all parasitic; found in intestines of dogs, cats, fish, cattle, pigs, and humans; lack digestive system and mouth, but have suckers for attaching to host. **Examples:** *Taenia solium, Diphyllobothrium latum* (broad tapeworm)

Phylum Nematoda (Roundworms)   Made up of worms that are bilaterally symmetrical; have round bodies; have complete digestive systems with mouths and separate anuses; about 80 000 species; both free-living and parasitic species exist; pork worm (responsible for trichinosis) and hookworm are parasitic species of phylum; majority nonparasitic soil dwellers. **Examples:** *Trichinella spiralis* (roundworm that causes trichinosis), *Necator americanus*

Phylum Mollusca (Mollusks)   Includes soft-bodied animals such as clams, snails, octopuses, and squid; about 110 000 species; have bilateral symmetry; have either external shell, internal shell, or no shell; all have a mantle; locomotion achieved through muscular foot; many species used as human food; at least 110 000 species.

Class Gastropoda (Snails and Slugs) Includes snails, slugs, whelks, and conches; most members have a spiral shell and a head; called "univalves" because of one shell; land slugs and nudibranchs have no shell. **Examples:** *Helix aspersa* (European edible snail), *Achatina achatina* (African land snail)

Class Bivalvia (Bivalves) All have rigid shell of two (bi) parts; includes clams, oysters, and mussels; lack head; have wedge-shaped muscular foot for locomotion or protection; most species marine, but a few live in fresh water. **Examples:** *Arca zebra* (zebra mussel), *Ostvea edulis* (European flat oyster)

Class Cephalopoda (Octopuses, Squid) Most highly developed mollusks with large eyes, arms, and tentacles surrounding the mouth; shell may be external, internal, or lacking; includes squid, octopuses, and chambered nautilus; can move forward or backward with great speed by expelling water through a siphon. **Examples:** *Nautilus macromphalus* (nautilus), *Octopus vulgaria* (common odepus)

Phylum Annelida (Annelids) Segmented worms; have bilateral symmetry, closed circulatory system, and complete digestive system; almost 9000 species live on land or in marine or freshwater bodies; earthworm is an annelid; about 12 000 species.

Class Polychaeta (Polychaetes) Mainly marine; have distinct head with sensory appendages and eyes; each body segment has many setae (bristles) extending from it; most are either male or female; includes clam worms, tube worms, and sandworms. **Examples:** *Nereis virens, Pectinaria gouldii*

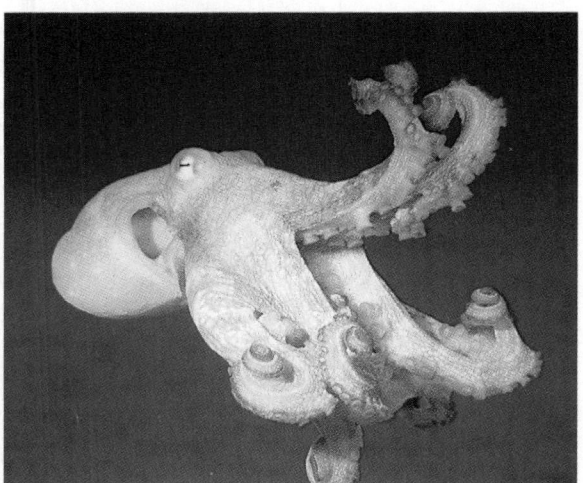

Class Oligochaeta (Earthworms) Found mainly in soil or fresh water; no distinct head region; setae are short and few per segment; worms are both male and female; includes the common earthworm. **Examples:** *Lumbricus terrestris* (earthworm), *Allolobophora caliginosa*

Class Hirudinea (Leeches) Both male and female; usually predators or parasitic; have a large sucker at the front end, enabling them to attach to their hosts; includes leeches; feed off the blood of hosts; one meal can last worms several months. **Examples:** *Hirudo medicinalis* (European medicinal leech), *Macrobdella decora*

Phylum Arthropoda (Arthropods)   Largest phylum in Animal Kingdom; bilaterally symmetrical; have appendages that are jointed; have tough exoskeletons of chitin; sense organs are well developed; open circulatory systems; includes insects, crayfish, crabs, millipedes, centipedes, and spiders; group is important as pollinators of flowers, carriers of disease, and as food; about 1 million species known at present.

Class Arachnida (Spiders, Mites, Scorpions)
Two body regions; have four pairs of walking legs, a pair of chelicerae (fangs), and a pair of pedipalps used to squeeze and chew food; carnivorous; no antennae; respire through book lungs or tracheae; includes spiders, scorpions, ticks, and mites; about 57 000 species. **Examples:** *Latrodectus mactans* (black widow spider), *Mastigo proctus giganteus* (American whip scorpion), *Dermacentor variabilis* (wood tick)

Class Merostomata (Horseshoe Crabs)   Two body regions; have five pairs of walking legs; have chelicerae like Arachnida; marine organisms; appendages all found on bottom of abdomen; have book gills for respiration; have compound eyes; front body region covered by a large horseshoe-shaped carapace, thus common name horseshoe crabs; only four species exist. **Examples:** *Limulus polyphemus* (horseshoe crab), *Eurypterus fischeri* (Eurypterus is extinct)

Class Crustacea (Lobsters, Crayfish, Crabs)
Two body regions; appendages two-branched and found on each segment; five pairs of walking legs and three pairs of mandibles (jaws) for chewing; two pairs of antennae; respire through gills; mainly aquatic; many have compound eyes; examples are crabs, lobster, shrimp, crayfish, barnacles, and land sowbugs or pillbugs; about 35 000 species. **Examples:** *Macrocheira kaempferi* (Japanese giant spider crab), *Homarus americanus* (American lobster)

Class Chilopoda (Centipedes)   Two body regions; have head with pair of long antennae and a pair of mandibles; two pair of maxillae (which hold their captured food); carnivorous; head followed by 15 to 177 body segments; first body segment has pair of poisonous claws; all other segments have one pair of jointed legs; tracheae used for respiration; class of centipedes; do not have 100 legs (as their common name suggests). **Examples:** *Lithobius forficatus, Scutigerella immaculata*

Class Diplopoda (Millipedes)   Three body regions; head followed by 20 to 200 segments; one pair of antennae; pair of mandibles and one pair of maxillae; two pairs of jointed legs on each segment; tracheae used for respiration; mostly herbivorous; class of millipedes; do not have 1000 legs (as the common name suggests). **Examples:** *Julus terrestris* (garden millipede), *Oxidus gracilis* (greenhouse millipede)

APPENDIX A

Class Insecta (Insects) Three body regions (head, thorax, and abdomen); one pair of antennae on head; have complex mouth parts with one pair of maxillae and mandibles; three pairs of walking legs located on thorax; only group of invertebrates capable of flight; have one or two

pairs of wings on thorax (when wings are present); tracheae used for respiration; have compound eyes; includes bees, flies, grasshoppers, lice, butterflies, moths, and beetles; single largest class of organisms with 750 000 known species; most insects are 2 to 40 mm in length, but some can reach lengths of 250 mm or have wingspans of 280 mm; most live on land; many have complete metamorphosis. **Examples:** *Bombus americanorum* (bumblebee), *Vanessa virginiensis* (American painted lady)

Phylum Echinodermata (Echinoderms) Marine invertebrates; includes sea stars, brittle stars, sea urchins, sea cucumbers, and sand dollars; usually have radial symmetry in the adult stage; skin covered with spines; only phylum having a water-vascular system that uses tube feet for locomotion; about 6000 species, all marine; many can regenerate lost parts.

Class Crinoidea (Sea Lilies, Feather Stars) Filter-feeders with mouth and anus in a disc on upper surface; includes sea lilies and feather stars; have a flowerlike appearance; marine, living below low tide line to depths of over 3000 meters; most have a long stalk with five to 200 arms around disc that forks into narrow, feather

like appendages; 2100 species of class are extinct; about 600 living species. **Examples:** *Ptilocrinus pinnatus, Antedon spinifera* (sea lily)

Class Asteroidea (Sea Stars) Star-shaped sea stars, commonly called starfish, make up this class; have five to 50 arms surrounding a central disc; mouth and anus on lower surface; have endoskeleton formed from flexible calcareous plates; have ability to bend and turn with ease; move by means of tube feet; about 1500 species. **Examples:** *Asterias forbesi, Acanthaster planci* (crown of thorns starfish)

Class Ophiuroidea (Brittle Stars) Also star-shaped; includes basket stars and brittle stars; arms very long, slender, jointed, and fragile; found in shallow to deep ocean waters; usually hide beneath stones or seaweed or bury themselves in sand; most active at night; about 2000 species. **Examples:** *Ophiura sarsi, Amphipholis squamata*

Class Echinoidea (Sea Urchins and Sand Dollars)
No distinct arms; have rigid external covering; includes sea urchins, sand dollars, and sea biscuits; bodies rounded and covered with many long, flexible spines (*Echinoidea* means "like a hedgehog"); most live on rocks and mud of seashore or buried in sand; move by tube feet or jointed spines; about 950 species. **Examples:** *Arbacia punctulata* (common sea urchin), *Heterocentrotus mammillatus* (slate-pencil sea urchin)

Class Holothuroidea (Sea Cucumbers)   Soft, slug-like echinoderms that lie on their sides; includes sea cucumbers; almost leathery body has an elongated cucumber shape; no arms or spines present; some species have small tentacles surrounding mouth; sluggish animals that burrow into ocean sand; when disturbed, sea cucumbers shoot out long, sticky tubules from anuses that entangle and often kill enemies; others completely throw up digestive and respiratory tracts (these systems regenerate); about 1500 species. **Examples:** *Thyone briaereus, Cucumaria frondosa* (sea cucumber)

Vertebrates

Phylum Chordata (Chordates)   Best known of all animal phyla, numbering some 45 000 species; all chordates have bilateral symmetry; four chordate characteristics that appear at some stage of development include (1) a single, dorsal nerve cord, (2) a cartilaginous, dorsal rod called a notochord, (3) gill slits, and (4) a tail; includes tunicates, lancelets, fish, amphibians, reptiles, birds, and mammals; includes three subphyla.

Subphylum Urochordata (Tunicates)   Free-swimming larvae have notochord and nerve cord, structures absent in sessile adults; includes tunicates (sea squirts); bodies covered by saclike coverings called "tunics" (thus, tunicates); look like soft potatoes; marine organisms that obtain food using cilia; gill slits also remain in adults; about 1250 species. **Examples:** *Polycarpa pomaria, Ecteinascidia turbinata*

Subphylum Cephalochordata (Lancelets)   Marine, fishlike chordates with permanent notochord; no internal skeleton; includes small animals called lancelets or amphioxus; have bladelike shape; gill slits and nerve cord remain in adult; filter feeders that draw in food with cilia; about 23 species. **Examples:** *Branchiostoma virginiae, Branchiostoma californiense*

Subphylum Vertebrata (Vertebrates)   Found on land and in fresh and marine waters; notochord replaced by cartilage or bone, forming segmented vertebral column (the backbone); have distinct head and brain inside a skull; gill slits may remain or be modified into other structures during development; hollow dorsal nerve cord remains, protected by backbone; *vertebrata* refers to the vertebrae (segments of backbone) surrounding dorsal nerve cord; about 43 700 species.

Class Agnatha (Lampreys and Hagfish) Aquatic organisms with smooth skin; includes eel-like animals called lampreys and hagfish; all lack jaws; have slender bodies; are parasitic or scavengers; have no scales; skeleton made of cartilage; notochord present throughout life. **Examples:** *Petromyzon marinus* (sea lamprey), *Polistotrema stouti*

Class Chondrichthyes (Sharks, Skates, Rays) Cartilagenous fish that are mostly marine; includes sharks, skates, and rays; skin covered with small, toothlike scales; have paired fins; skeleton remains cartilagenous even in adults; notochord remains in adults; dorsal nerve cord protected by individual vertebrae; all are predators; do not have air bladders. **Examples:** *Squalus acanthias, Raja undulata* (skate)

Class Osteichthyes (Bony Fish) Bony fish (except sturgeons) that have skeletons made from bone (ostei); abundant in marine and fresh waters; includes sea horses, goldfish, eels, catfish, trout, gar, salmon; skin covered with scales; have paired fins; notochord usually has disappeared; gills covered by flap (operculum); have air bladders to regulate their densities in water; about 18 000 species. **Examples:** *Perca flavescens* (yellow perch), *Hippocampas erectus* (sea horse)

Subclass Crossopterygii (Lobe-finned Fish) Called lobe-finned fish because of the lobelike nature of their fins; only one living species—the rest are extinct; a species thought to be extinct was caught off the coast of Africa in 1938. **Examples:** *Macropoma sp.* (extinct), *Latimeria chalumnae*

Subclass Dipneusti (Lungfish) Fish with a modified air bladder that enables them to breathe air; found living in Australia, Africa, and South America; six species. **Examples:** *Neoceratodus forsteri, Proptoperus annectens*

Subclass Actinopterygii (Ray-finned Fish) So called because the fins themselves are supported on fin rays, which appear like thin bones but really are extensions of the skin; the largest group of bony fish; over 20 000 species. **Examples:** *Acipenser sturio, Ameiurus melas*

Class Amphibia (Amphibians) Includes salamanders, toads, and frogs; were first vertebrates to live on land; most species spend part of life cycle in fresh water and part on land; larvae use gills and adults use lungs for respiration; skin is smooth with no scales; have two pairs of limbs (except for a few species; have three-chambered hearts; skeleton is bone; no trace of notochord found in adults; lay eggs; are cold-blooded (ectothermic); about 2800 species. **Examples:** *Rana pipiens, Agalychnis spurrelli* (Costa Rican flying tree frog)

Order Gymnophiona (Caecilians)   Slender, wormlike bodies with no limbs; order name means "naked of a snake"; burrow in moist ground; live mainly in the tropics; about 160 known species. **Examples:** *Ichthyophis glutinosus, Gymnopis sp.*

Order Urodela (Salamanders, Newts) Includes mud puppies, salamanders and newts; distinct head, trunk and tail regions; limbs about equal in size; about 300 species. **Examples:** *Triturus viridescens, Necturus punctatus*

Order Salientia (Frogs, Toads)   Order name means "leaping"; front legs are usually short, while the rear legs are larger and aid in leaping; no tail; about 2000 species. **Examples:** *Hyla arenicolor, Bufo boreas*

Class Reptilia (Reptiles)   Includes lizards, alligators, turtles, snakes, and extinct dinosaurs; skin is dry with scales that protect skin from drying out; most are land dwellers, though a few are aquatic; have imperfect, four-chambered hearts; use lungs for respiration; lay amniotic eggs; are ectothermic; legs absent in snakes and a few lizards; many are poisonous; about 6000 species. **Examples:** *Anolis carolinensis* (green anole), *Chelonia mydas* (green turtle)

Order Testudines (Turtles)   Turtles, tortoises, and terrapins; bodies enclosed in a rigid shell of plates; lack teeth; marine, fresh water, or terrestrial; about 330 species. **Examples:** *Chelydra serpentina, Trionyx ferox*

Order Squamata (Snakes, Lizards)   Horny skin with scales; lizards typically have four limbs, as seen in newts, skinks, iguanas, and chameleons; snakes do not have legs but show vestiges (remains) of hind

leg bones; about 2700 species of snakes, and 3000 species of lizards. **Examples:** *Chamaeleo chamaeleon, Crotalus viridis*

Order Crocodilia (Crocodiles, Alligators) Front legs with five toes, rear legs with four; crocodiles have a narrower snout than alligators and are considered more dangerous; about 25 species. **Examples:** *Crocodylus americanus, Alligator mississippiensis*

Class Aves (Birds)   Includes all birds; first group of Chordates to be warm-blooded (endothermic); have complete four-chambered hearts; respire with lungs; bodies covered with feathers; forelimbs modified into wings; most are capable of flight; mainly land dwellers; fertilization is internal; amniotic eggs are deposited externally for incubation by a parent; have complete double circulation; ostrich is largest bird (over 200 cm tall with mass of 140 kg) while hummingbirds are smallest (6 cm long with mass of 35 g); about 9000 species. **Examples:** *Casuarius casuarius* (Australian cassowary), *Mellisuga helenae* (bee hummingbird)

Order Anseriformes (Ducks, Geese, Swans) Broad bill covered with a soft skin; legs and tails tend to be short with webbing between toes; more than 200 species. **Examples:** *Olor columbianus, Chen hyperboreus*

APPENDIX A

Order Falconiformes (Hawks, Eagles)
Vultures, kites, hawks, condors, falcons, and eagles; all are predators with keen vision and sharp, curved claws for capturing food; all are daytime hunters and fliers; more than 250 species. **Examples:** *Falco peregrinus, Haliaeetus leucocephalus*

Order Galliformes (Ground Birds) Pheasants, grouse, partridges, and quails; worldwide in distribution; often desirable game birds; chicken-like vegetarians with strong, short beaks and feet well adapted for running and scratching; about 275 species. **Examples:** *Perdix perdix, Bonasa umbellus*

Order Charadriiformes (Shore Birds) Shore birds or waders; includes gulls, terns, puffins, auks, plovers, sandpipers, woodcocks, and snipe; usually live in colonies and are strong fliers; found worldwide; over 300 species. **Examples:** *Oxyechus vociferus, Philohela minor*

Order Passeriformes (Perching Birds, Including Songbirds) Largest of all bird groups; contains over 60 percent of all bird species; larks, swallows, crows, jays, nuthatches, creepers, wrens, robins, bluebirds, vireos, orioles, meadowlarks, and sparrows; perching birds with three toes in front and one in back; more than 5000 species. **Examples:** *Spizella passerina, Turdus migratorius*

Class Mammalia (Mammals) Endothermic organisms; have hair at some stage of development; provide milk to young through mammary glands; have muscular diaphragm; have four-chambered heart; respire with lungs; forelimbs sometimes modified into flippers in aquatic animals or wings (in bats); give birth to live young (except monotremes); about 4500 species in 22 orders.

Order Monotremata (Monotremes) Egg-laying mammals; includes duck-billed platypus and species of echidna; the spiny anteaters; native to Australia, New Guinea, and Tasmania only; platypuses mostly aquatic, having duck-like bill, flattened tail, and webbed feet; echidnas mostly found on land; eggs similar to eggs of reptiles; have some other reptilian features; are toothless as adults. **Examples:** *Ornithorhynchus anatinus* (duck-billed platypus), *Zaglossus bruijnii* (long-nosed echidna)

Order Marsupialia (Marsupials) Pouch for rearing their young; only one species found in the U.S.—the opossum; all other species found in Australia, where they comprise the dominant mammal order; order includes kangaroos, opossum, koala, wombats, and bandicoots; about 250 species. **Examples:** *Didelphis marsupialus, Phascolarctos cinereus*

Order Insectivora (Insect Eaters) Small with a long, tapered snout; feed on insects; includes the shrews, moles, and hedgehogs; typically five clawed toes for burrowing; about 400 species. **Examples:** *Scapanus latimanus, Suncus etruscus*

Order Chiroptera (Bats) Includes bats, the only mammals capable of sustained flight; most are nocturnal; most feed on insects; a few eat fruit, pollen, and nectar; true vampire bats feed on blood of large birds and mammals such as livestock; use echoes to orient themselves (echolocation); smallest bat (Philippine bamboo bat) only 1.5 g and has wingspan of 15 cm; largest bat (flying fox) 1 kg with wingspan of 1.5 m; about 900 species. **Examples:** *Desmodus rotundus* (vampire bat), *Noctilio leporinus* (fishing bat)

Order Carnivora (Carnivores) Includes flesh eaters (carnivores) such as bears, dogs, walruses, seals, skunks, otters, wolves, raccoons, and cats; canine teeth usually large and designed to tear flesh; some eat plants primarily (bears and raccoons); includes domestic cats and dogs

kept as pets; most live on land; lower jaws can only move vertically, but are very strong; have large brains; have simple stomachs; about 274 species. **Examples:** *Ursus arctos* (grizzly bear), *Castor canadensis* (beaver)

Order Rodentia (Rodents) Squirrels, beavers, rats, mice, porcupines, and woodchucks; gnawing mammals with long, chisel-like incisors; typical rodent pets include guinea pigs, hamsters, and gerbils; about 1700 species. **Examples:** *Cavia porcellus, Lepus californicus*

Order Cetacea (Whales, Dolphins) Aquatic mammals; toothed whales, baleen whales, and dolphins; three distinct groups; streamlined bodies; tails (flukes) and dorsal fins lacking bones; breathe through blowholes on top of head; use echolocation; sperm whale reaches lengths of 20 meters and has a mass equal to that of 10 elephants; baleen whales reach lengths of 30 meters; blue whale is largest animal on Earth; about 80 species. **Examples:** *Delphinus delphis* (common dolphin), *Orcinus orca* (killer whale)

Order Primates (Primates) Includes lemurs, baboons, gorillas, monkeys, and humans; large brain with complex cortex; binocular vision; varied teeth; ability to hold body upright; prehensile (grasping) hands with opposable thumbs; most primates live in trees; finger- and toenails; about 197 species. **Examples:** *Gorilla gorilla* (gorilla), *Homo sapiens* (human)

APPENDIX B

Plant Life Cycles

In their life cycles, plants exhibit alternation of a sporophyte generation with a gametophyte generation. Ferns, mosses, gymnosperms, and angiosperms all go through alternation of generations. In Chapter 18, you studied the life cycle of a representative angiosperm. Life cycles of ferns, mosses, and gymnosperms are presented here in Appendix B. As you study these life cycles, note the similarities and dif-

ferences. Similarities include alternation of diploid (2n) and haploid (n) generations; presence of sporophyte and gametophyte at some point during the life cycle; production of spore in some form during the life cycle; production of gametes during the life cycle. The most notable difference is the relative size and duration of the gametophyte generation as compared to the sporophyte generation.

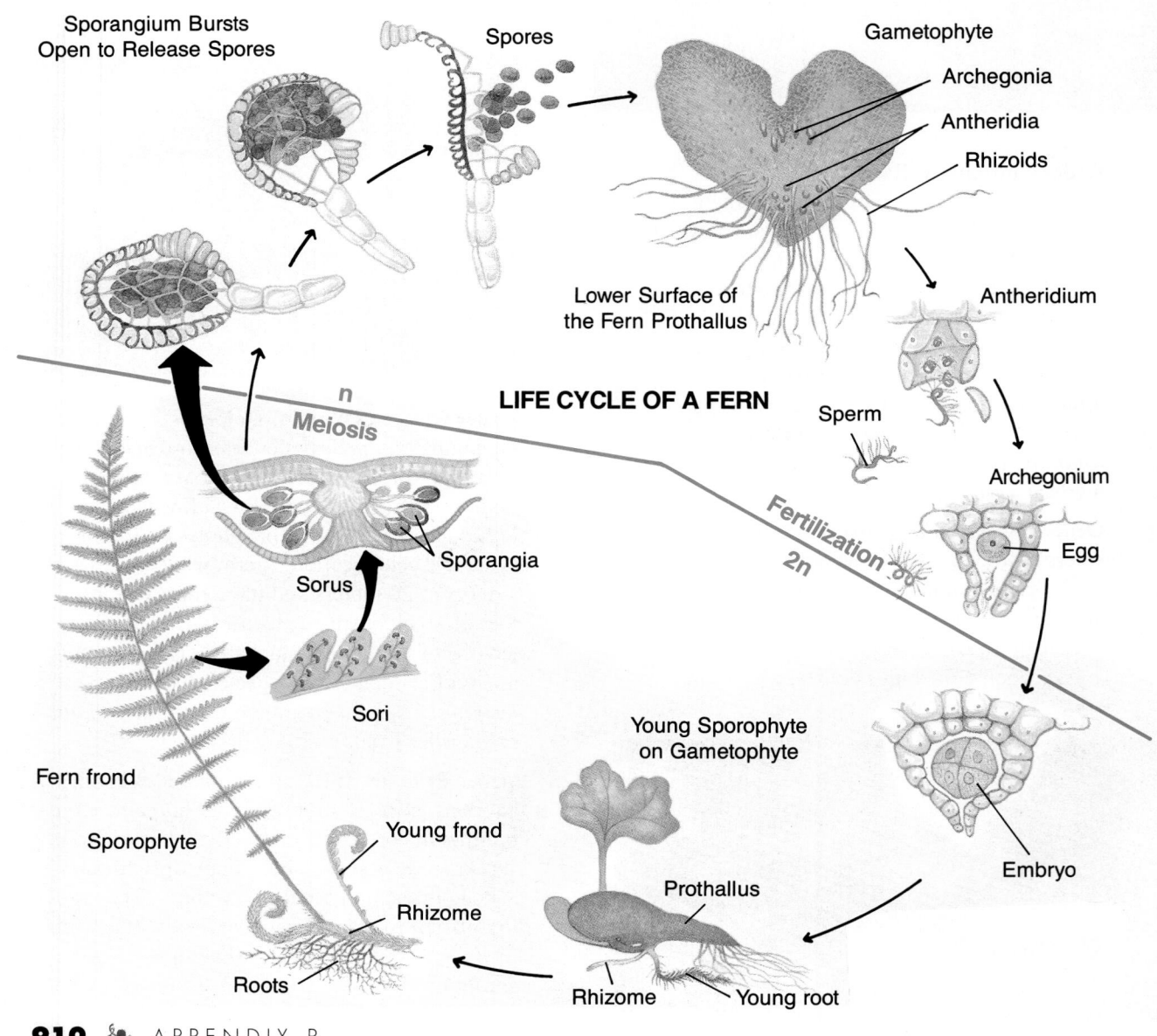

Sporangium Bursts Open to Release Spores

Spores

Gametophyte

Archegonia

Antheridia

Rhizoids

Lower Surface of the Fern Prothallus

Antheridium

Sperm

Archegonium

Egg

n
Meiosis

LIFE CYCLE OF A FERN

Fertilization

2n

Sporangia

Sorus

Sori

Young Sporophyte on Gametophyte

Embryo

Fern frond

Sporophyte

Young frond

Prothallus

Rhizome

Roots

Rhizome

Young root

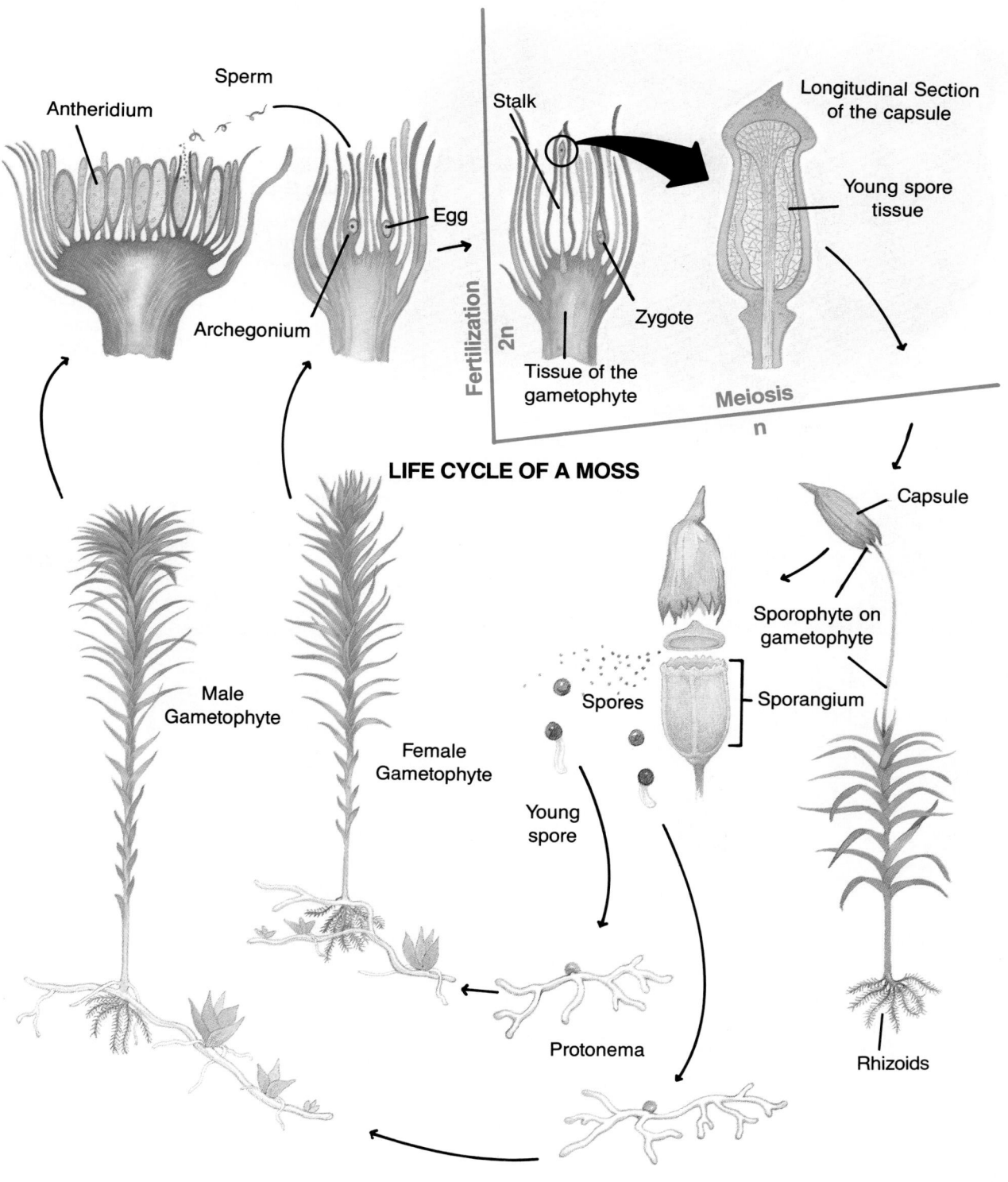

Sperm

Antheridium

Stalk

Longitudinal Section of the capsule

Egg

Young spore tissue

Archegonium

Fertilization

2n

Zygote

Tissue of the gametophyte

Meiosis

n

LIFE CYCLE OF A MOSS

Capsule

Male Gametophyte

Female Gametophyte

Sporophyte on gametophyte

Sporangium

Spores

Young spore

Protonema

Rhizoids

LIFE CYCLE OF A GYMNOSPERM

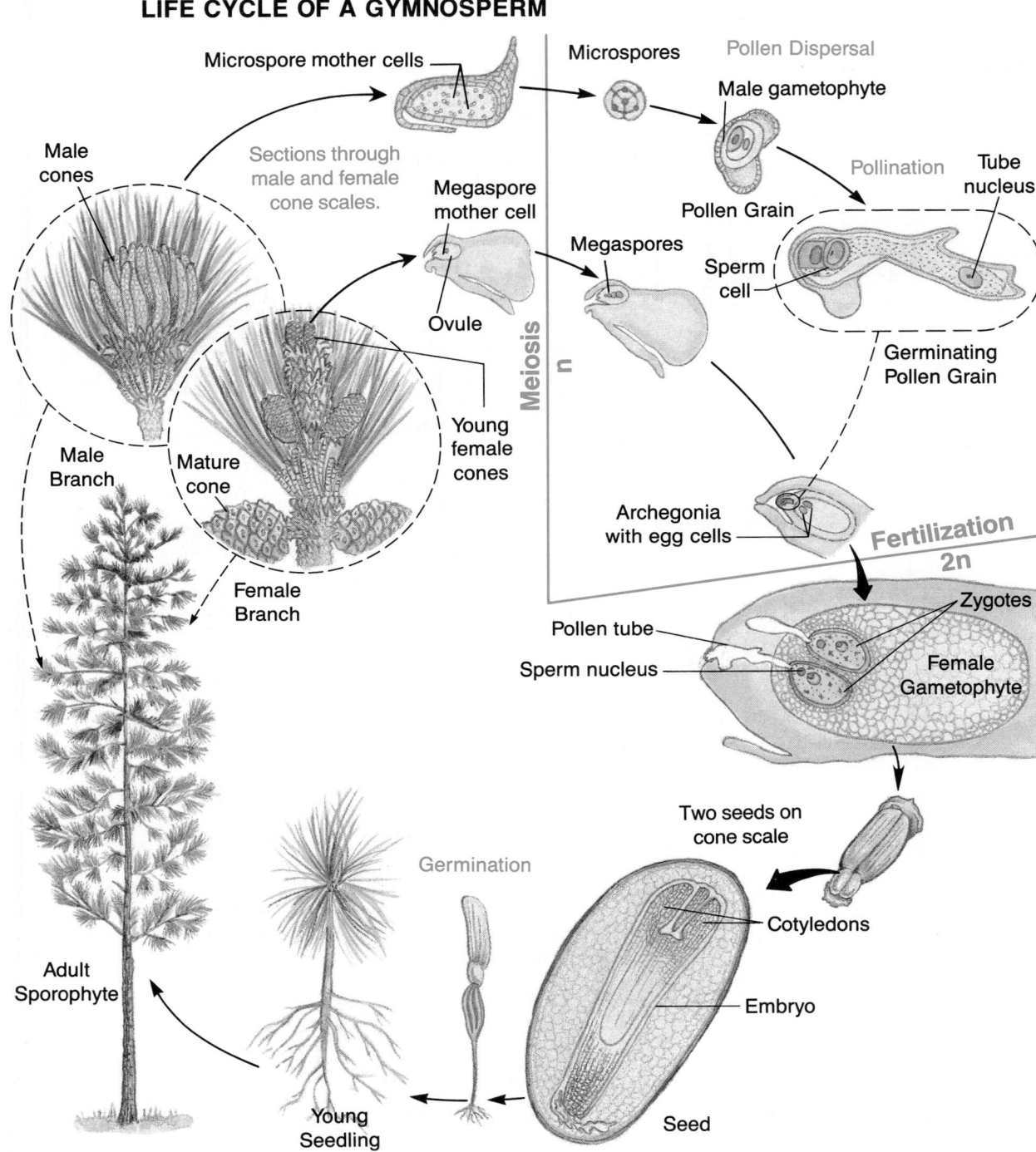

Microspore mother cells

Microspores

Pollen Dispersal

Male gametophyte

Sections through male and female cone scales.

Male cones

Megaspore mother cell

Pollen Grain

Pollination

Tube nucleus

Ovule

Megaspores

Sperm cell

Meiosis

n

Young female cones

Germinating Pollen Grain

Male Branch

Mature cone

Archegonia with egg cells

Fertilization

2n

Female Branch

Zygotes

Pollen tube

Sperm nucleus

Female Gametophyte

Two seeds on cone scale

Germination

Cotyledons

Adult Sporophyte

Embryo

Young Seedling

Seed

APPENDIX C

Cell Chemistry

Cellular Respiration

During aerobic respiration, the energy in the bonds of glucose (or other energy-rich compounds) is transferred to the bonds of ATP. How does this transfer of energy occur? Remember that the process involves many separate chemical reactions. Those reactions can be grouped into four major stages: glycolysis, pyruvic acid oxidation, citric acid cycle, electron transport chain. As you study them, focus on the main point: that energy is transferred from bonds of glucose to bonds of ATP.

Glycolysis

The first stage of aerobic respiration occurs in the cytoplasm. Because the glucose molecule is split apart, this stage is known as **glycolysis.** As a result of many enzyme-controlled reactions, one molecule of glucose, a six-carbon (C_6) compound is changed to two molecules of pyruvic acid, a C_3 compound. Two molecules of ATP are used early in glycolysis, because some of the first steps are endergonic. However, as the reactions occur, chemical bonds are broken and energy is released. In one important energy-releasing step, four hydrogen atoms are removed and join two each with a coenzyme, NAD, forming 2NADH + $2H^+$. Part of the energy released is used to make the NADH molecules and H^+ ions, whose importance

will be discussed later. Most of the other energy released is used to form two molecules of ATP from 2ADP + $2P_i$. The last step, which forms pyruvic acid, releases more energy, which is used to form two more ATP. Thus, there is a net "profit" of 2ATP. Some of the energy released is not trapped and escapes as heat. Note that glycolysis is not an aerobic process even though it is the first step in aerobic respiration.

Pyruvic Acid Oxidation

The next stage of aerobic respiration occurs in mitochondria. You have learned that a mitochondrion has an outer membrane and an inner, folded membrane. The folds of the inner membrane are called cristae. The space between the outer and inner membranes is known as the outer compartment. The rest of a mitochondrion, bordered all around by the inner membrane, is called the inner compartment. The two pyruvic acid molecules formed during glycolysis move into the inner compartment, where, in several steps, each is converted to acetic acid, a C_2 molecule. The carbon atoms and two oxygen atoms removed from each pyruvic acid molecule form CO_2, which you know is released during respiration. Each acetic acid molecule combines with a molecule of coenzyme A (CoA) forming acetyl-CoA. As this step occurs, two more hydrogen atoms are removed from each acetic acid, forming two more NADH + $2H^+$. Because removal of hydrogen with its electron is a type of oxidation reaction, this stage is known as pyruvic acid oxidation. Let's

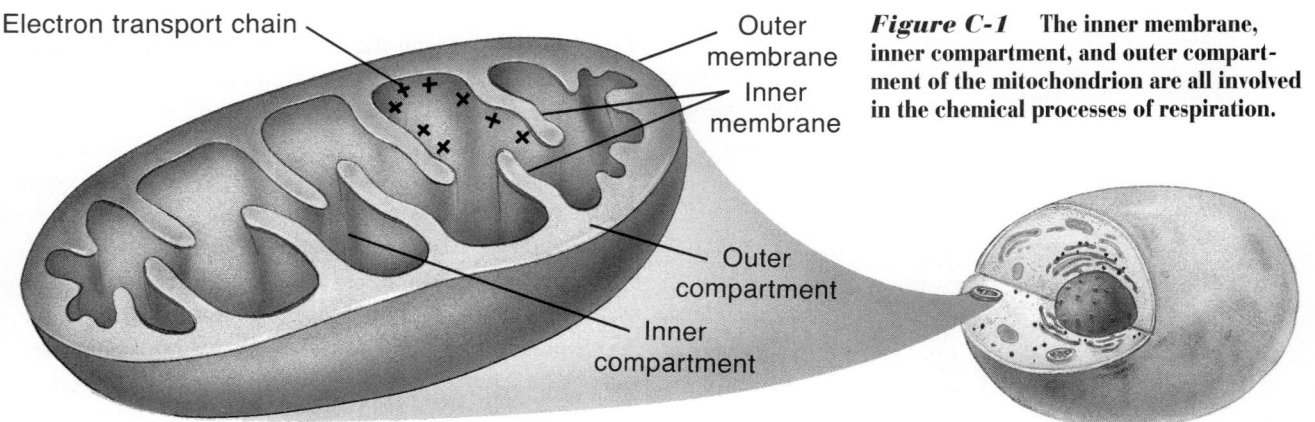

Electron transport chain
Outer membrane
Inner membrane
Outer compartment
Inner compartment

Figure C-1 The inner membrane, inner compartment, and outer compartment of the mitochondrion are all involved in the chemical processes of respiration.

stop for a moment and take account. So far, a glucose molecule has been converted to two acetyl-CoA molecules and two molecules of CO_2. Also produced were 2ATP (net) and four units of NADPH + H⁺.

Citric Acid Cycle

In the inner compartment, each C_2 molecule of acetyl-CoA combines with a C_4 compound, oxaloacetic acid, yielding a molecule of citric acid, a C_6 compound. The next series of reactions is known as the citric acid cycle. It is also called the Krebs cycle in honor of Hans Krebs, the biochemist who worked out many of its details. At the beginning of the cycle, each citric acid is broken down to a C_5 molecule and then finally to the original C_4 molecule, giving off $2CO_2$ in the process. Since two acetyl-CoA molecules enter the cycle for each original molecule of glucose, a total of $4CO_2$ molecules are given off. These four CO_2 plus the two CO_2 from the previous stage make a total of $6CO_2$, which accounts for all of the carbon atoms in the original glucose. For each acetyl-CoA that enters the citric acid cycle, $3H_2O$ are used and 8 hydrogens are removed. Because 2 acetyl-CoA enter for each glucose molecule being broken down, $6H_2O$ are used and 16 hydrogens are removed in the Krebs cycle. Twelve hydrogens combine with 6NAD to form 6NADH + 6H⁺, and 4 hydrogens

combine with two molecules of another coenzyme, FAD, forming $2FADH_2$. Much energy is released during the Krebs cycle. Most has been used to make molecules of NADH and $FADH_2$. A small amount is used to produce 2ATP from $2ADP + 2P_i$. The entire process so far has produced $6CO_2$, 4ATP, 10NADPH + 10H⁺, and $2FADH_2$. Six water molecules have been used.

Four ATP do not seem like a very impressive energy payoff for this complex process. Keep in mind, though, that a great deal of the energy released from the breakdown of glucose has been used to make NADPH + H⁺ and $FADH_2$. As you will see, it is in the final stage of aerobic respiration that the major energy payoff occurs.

Electron Transport Chain

Molecules of NADPH and $FADH_2$ produced during respiration contain electrons at high energy levels. However, the energy of these electrons cannot be released all at once because the amount would be too large for normal biological processes. Instead, the energy of the electrons is released in a stepwise fashion as the electrons are passed down through a series of electron acceptors. Each successive acceptor takes in electrons at a lower energy level. The energy released by the electrons as they pass through these acceptors is used to form several molecules of ATP from ADP + P_i. This series of electron acceptor molecules is located in the cristae and is known as the **electron transport chain** or **respiratory chain.** For each glucose molecule broken down, a total of 12 pairs of electrons, 10 pairs from NADH and 2 pairs from $FADH_2$, travel along the chain. Some of the electrons do not enter the chain at the first electron acceptor, but at a later one. The energy released can convert a maximum of $32ADP + 32P_i$ to 32ATP. Each pair of electrons, at the end of the chain, combines an oxygen atom, which combines with two hydrogen ions to form water. A total of 12 water molecules are thus formed. Because six water molecules were used in the citric acid cycle, there is a net production of six water molecules in aerobic respiration of one glucose molecule. You can now understand the need for oxygen for aerobic respiration. It is necessary as the final electron acceptor in the electron transport chain.

Figure C-2 Fats and proteins can be converted to simpler compounds that may enter the respiration pathway at several points and release energy for ATP formation.

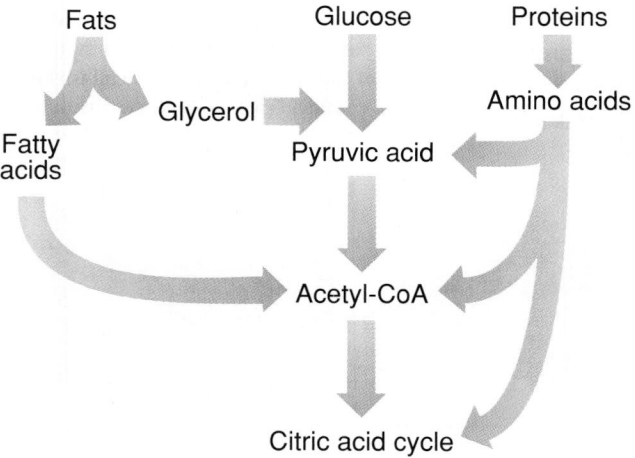

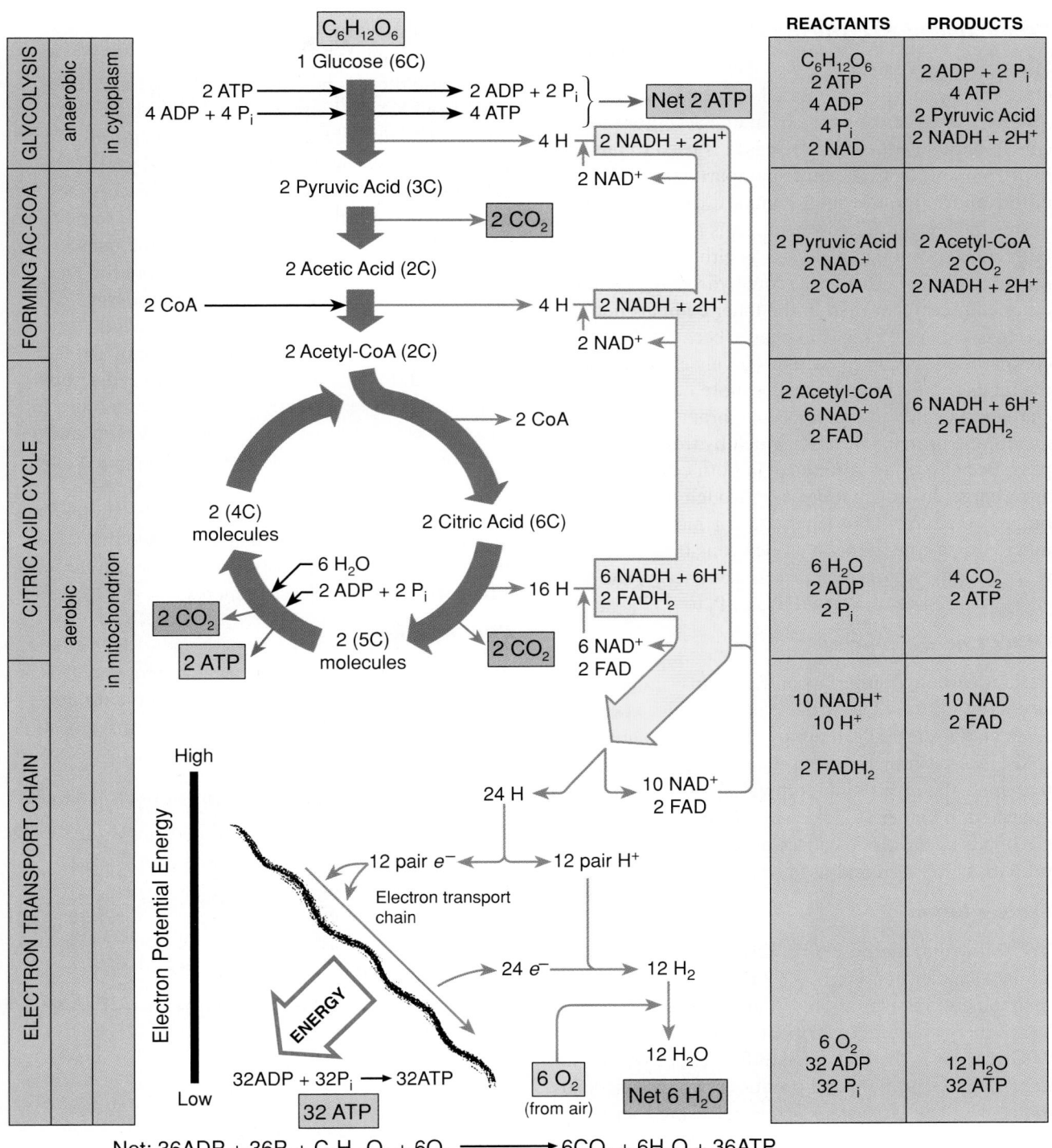

	REACTANTS	PRODUCTS
	$C_6H_{12}O_6$ 2 ATP 4 ADP 4 P_i 2 NAD	2 ADP + 2 P_i 4 ATP 2 Pyruvic Acid 2 NADH + 2H$^+$
	2 Pyruvic Acid 2 NAD$^+$ 2 CoA	2 Acetyl-CoA 2 CO_2 2 NADH + 2H$^+$
	2 Acetyl-CoA 6 NAD$^+$ 2 FAD	6 NADH + 6H$^+$ 2 FADH$_2$
	6 H_2O 2 ADP 2 P_i	4 CO_2 2 ATP
	10 NADH$^+$ 10 H$^+$ 2 FADH$_2$	10 NAD 2 FAD
	6 O_2 32 ADP 32 P_i	12 H_2O 32 ATP

GLYCOLYSIS — anaerobic — in cytoplasm

$C_6H_{12}O_6$
1 Glucose (6C)

2 ATP → 2 ADP + 2 P_i
4 ADP + 4 P_i → 4 ATP } → Net 2 ATP

→ 4 H ⊢ 2 NADH + 2H$^+$
2 NAD$^+$

2 Pyruvic Acid (3C)

FORMING AC-COA — aerobic — in mitochondrion

→ 2 CO_2

2 Acetic Acid (2C)

2 CoA → → 4 H ⊢ 2 NADH + 2H$^+$
2 NAD$^+$

2 Acetyl-CoA (2C)

CITRIC ACID CYCLE

→ 2 CoA

2 (4C) molecules

2 Citric Acid (6C)

6 H_2O
2 ADP + 2 P_i → 16 H ⊢ 6 NADH + 6H$^+$
2 FADH$_2$

2 CO_2

2 ATP

2 (5C) molecules → 2 CO_2

6 NAD$^+$
2 FAD

ELECTRON TRANSPORT CHAIN

Electron Potential Energy — High / Low

24 H ← ⊢ 10 NAD$^+$
2 FAD

12 pair e^- ← → 12 pair H$^+$

Electron transport chain

ENERGY

24 e^- → 12 H$_2$

12 H_2O

32 ADP + 32 P_i → 32 ATP

32 ATP

6 O_2 (from air)

Net 6 H$_2$O

Net: 36 ADP + 36 P_i + $C_6H_{12}O_6$ + 6O_2 ⟶ 6CO_2 + 6H_2O + 36 ATP

Figure C-3 In aerobic respiration, energy in the bonds of glucose is released and used to produce ATP. The process consists of four interrelated stages.

Production of ATP During Respiration

How is the movement of pairs of electrons along the electron transport chain related to production of ATP? As the electrons pass from one acceptor molecule to another, the energy released is used to pump hydrogen ions from the inner compartment of the mitochondrion to the outer compartment. The result of this movement of hydrogen ions is a difference in both concentration and charge on either side of the inner membrane. There are more hydrogen ions in the outer compartment than in the inner compartment. There is also a difference in charge because the outer compartment becomes more positively charged and the inner compartment becomes more negatively charged. Located in the inner membrane are complex membrane proteins through which hydrogen ions can pass. Because of the difference in both concentration and charge, hydrogen ions pass through the membrane protein from the outer compartment to the inner compartment, much as electrons flow in a battery. The passage of the ions releases energy, and that energy is used to convert ADP and P_i to ATP.

Other Energy Sources

Besides carbohydrates, fats and proteins can be used as energy sources. Fats are broken down into fatty acids and glycerol. Proteins are converted to amino acids. These simpler molecules are changed to others that enter the respiration pathway at various points, indicated in Figure C-2. As with glucose, energy in the bonds of these substances produced from fats and proteins is released and used to produce ATP.

Fermentation

You know that the process of fermentation results in a much smaller number of ATP molecules produced than in aerobic respiration. What is the reason for the lesser energy harvest of fermentation?

Glycolysis is the first stage of fermentation, as it is in aerobic respiration, and produces two pyruvic acid molecules. The same steps occur, releasing energy used to produce a total of 2NADH + 2H+ and a net of 2ATP. When oxygen is available, the NADH, as you have just learned, will pass electrons to the electron transport chain, and those electrons are eventually accepted by oxygen. However, when oxygen is not available, fermentation occurs.

Because there is no oxygen to accept them, electrons from NADH cannot be passed to the electron transport chain. Instead, they are donated to the molecules of pyruvic acid. In plant cells and yeasts, the addition of electrons to pyruvic acid molecules results in production of ethyl alcohol and CO_2. In some bacteria and in your skeletal muscle cells, the reactions produce two molecules of lactic acid from the two molecules of pyruvic acid. Much of the energy originally present in the bonds of glucose remains "locked" in molecules such as ethyl alcohol and lactic acid. The difference in energy yield between aerobic respiration and fermentation is like the difference in energy released when a ball bounces down an entire flight of stairs and when it bounces down just a few. More energy is released if the ball is allowed to bounce all the way down (as in aerobic respiration), than if the ball stops after bouncing down just a few steps (fermentation). Aerobic respiration is a far more efficient process of releasing energy.

Photosynthesis

Photosynthesis is a process by which light energy is trapped as chemical energy and used in the production of simple sugars. In many ways, photosynthesis is

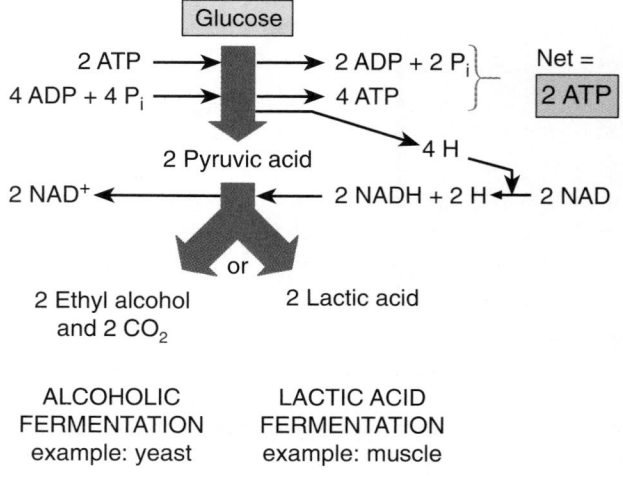

Figure C-4   Anaerobic respiration may result in the production of ethyl alcohol or lactic acid.

Thylakoid membrane, site of photosystems I and II

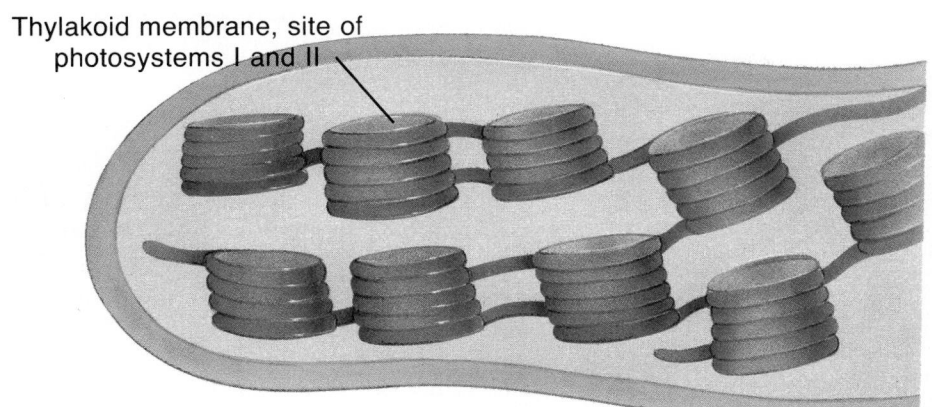

Figure C-5 Thylakoid membranes within the chloroplast are the site of light reactions.

a process opposite to aerobic respiration. Respiration starts with simple sugars and produces CO_2, H_2O, and energy from them. Photosynthesis starts with CO_2, H_2O, and energy and produces simple sugars. Whereas aerobic respiration is exergonic, photosynthesis is endergonic. Like aerobic respiration, photosynthesis is a process involving many steps, which can be divided into two related series of events, the light reactions and the Calvin cycle.

The Light Reactions

How does light interact with chlorophyll, and how is light energy transformed to chemical energy? The reactions in which these events occur are the light reactions. In eukaryotes, they occur in membranes, called *thylakoids,* within chloroplasts. Recall that these membranes may be separate in the stroma or stacked together to form grana.

Embedded in the thylakoid membranes are groups of chlorophyll and carotenoid molecules. These groups are of two types, called **photosystem I** and **photosystem II.** Chlorophyll and carotenoid pigments absorb specific wavelengths of light energy and pass that energy along to a particular molecule of chlorophyll known as the *reaction center.* The energy moves certain electrons of the reaction center molecule to a higher energy level. Such electrons, having gained energy, are said to be excited electrons.

In photosystem I, excited electrons are passed along a chain of acceptor molecules similar to those in the electron transport chain of a mitochondrion. In

some organisms, the following events occur. The electrons return finally to the chlorophyll molecule in their original energy level or ground state. As the energy-rich electrons travel along the acceptor molecules, their energy is released and used to form ATP from ADP and P_i. Because the electrons return to the chlorophyll molecule of which they were originally a part, this passage of electrons is known as the *cyclic pathway.* The cyclic pathway is the only form of photosynthesis in some bacteria (though they have no chloroplasts), and it does not use water or release oxygen. The cyclic pathway can be studied by following the red arrows in Figure C-6.

In other organisms, such as algae and plants, excited electrons most often follow a *noncyclic pathway* in which water is a reactant and oxygen is produced. Follow the blue arrows in Figure C-6 to study this pathway.

Both photosystems I and II play a role in the noncyclic pathway. Light strikes photosystem II, exciting electrons of the reaction center molecule. These electrons are transferred along a different path of acceptor molecules. As the electrons move from one acceptor to the next, they release energy that is used to make ATP. In the noncyclic pathway, electrons from photosystem II do not return to their original chlorophyll molecule. Instead they are transferred to the reaction center chlorophyll molecule of photosystem I to replace electrons that have left there after being excited by light. This process leaves the reaction center molecule of photosystem II with electron vacancies. These vacancies must be filled if photosynthesis is to continue. From where do the needed electrons come? The answer reveals the importance of water in photosynthesis. To provide electrons, water molecules are split into two H^+ ions, an oxygen atom, as well

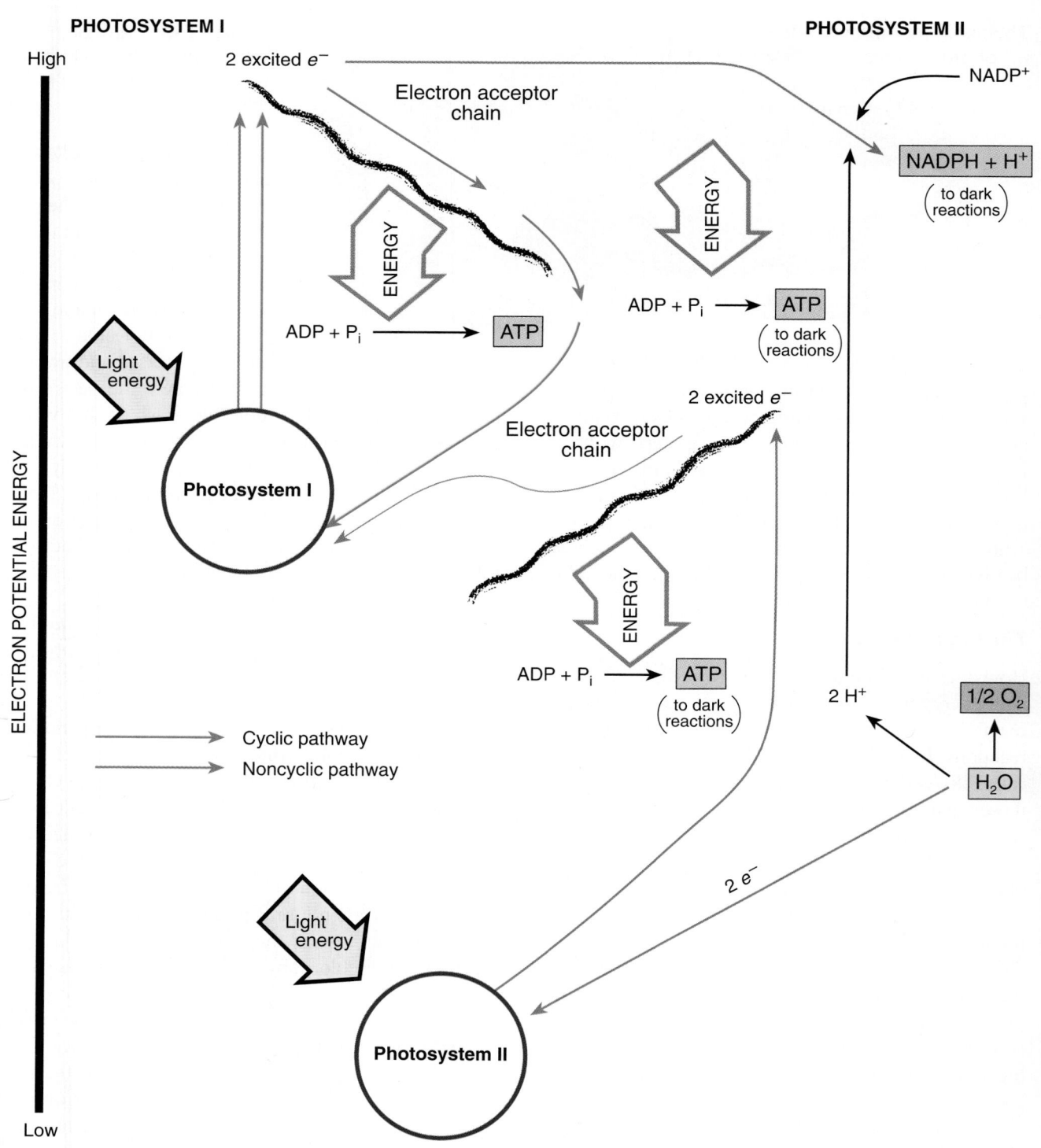

Figure C-6   In the light reactions of photosynthesis, light energy excites electrons. In the returning of these electrons to their ground state, energy is given off as ATP.

as electrons. The oxygen atoms form O_2 molecules, which are released during photosynthesis. Each water molecule provides two electrons, which are carried to the waiting chlorophyll molecule in photosystem II. Meanwhile, in photosystem I, light excites electrons, which enter another chain of acceptor molecules. Pairs of these high-energy electrons plus pairs of H^+ ions from the water are picked up by the coenzyme NADP, forming NADPH + H^+. The NADPH + H^+ and the ATP formed during the passage of electrons in the noncyclic pathway will be used in the Calvin cycle.

Production of ATP During Photosynthesis

Note that ATP production in the light reactions involves movement of excited electrons along electron acceptor molecules. These same events occur in aerobic respiration as electrons travel the electron transport chain. Synthesis of ATP occurs in the light reactions in much the same way it does during the last stage of aerobic respiration. In the light reactions, the energy released by excited electrons is used to pump hydrogen ions from the stroma to the inside of the thylakoid. This pumping sets up the same kind of concentration and charge differences that occur across the inner membrane of a mitochondrion. As the hydrogen ions pass back across the thylakoid membrane into the stroma, they release energy, which is used to convert ADP and P_i to ATP.

The Calvin Cycle

In the noncyclic pathway of the light reactions, light energy is transformed to chemical energy in molecules of ATP and NADPH + H^+. These two kinds of molecules are essential for building sugars. Manufacture of sugars from carbon dioxide, an endergonic process, occurs within the stroma of a chloroplast.

In a series of reactions known as the **Calvin cycle,** each CO_2 molecule combines with a C_5 molecule, ribulose bisphosphate, which is usually abbreviated RuBP, to form an unstable C_6 molecule that quickly breaks down to form two C_3 molecules. These molecules enter other reactions in which ATP and NADPH + H^+ from the light reactions provide the necessary energy and hydrogen atoms. The products of these

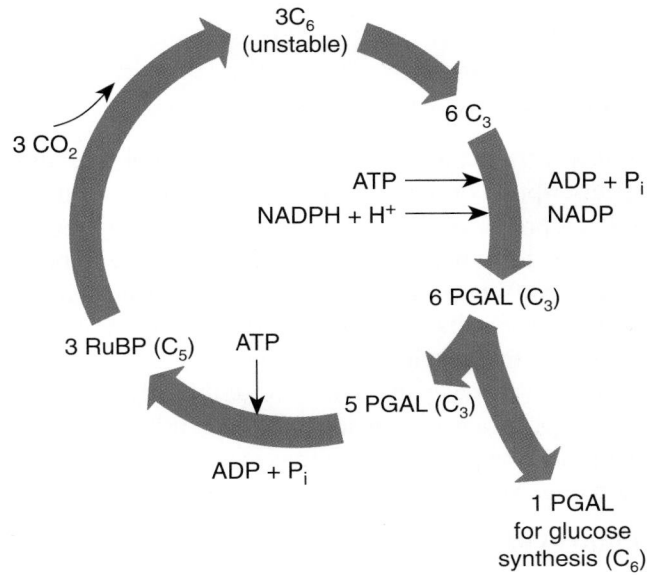

Figure C-7   In the Calvin cycle, CO_2 is converted to PGAL. Two "spins" of the cycle are needed to produce two PGALs for glucose synthesis.

reactions are two molecules of phosphoglyceraldehyde (PGAL), which are simple C_3 sugars that can be considered the end products of photosynthesis. Two PGAL molecules may be combined to make glucose. The other PGAL molecules, using energy from ATP, are converted back to RuBP, thus continuing the cycle. Note that Figure C-7 shows three CO_2 molecules entering the cycle at a time. In all, six CO_2 molecules must enter for each molecule of glucose produced.

Although glucose is a molecule that may be made from PGAL, it is not the only one. Of the glucose that is made, most is quickly changed to other forms. Some is used to make sucrose, the form in which sugars are transported away from the site of photosynthesis. Glucose is also used to make starch, which can be stored as an energy reserve, or to manufacture cellulose necessary for cell wall production.

Much of the PGAL produced in the Calvin cycle is not used to make glucose at all. It can be used directly as an energy source in aerobic respiration. It is also used in making many other organic molecules including lipids, amino acids, and nucleic acids.

Expression of Genotype

Control of Gene Expression

A gene carries the code for synthesis of a polypeptide. Transcription of the gene and then translation of the mRNA produced by transcription are necessary for the code to be deciphered and the polypeptide to be produced. When a polypeptide has been made and a phenotype results, the gene is said to be expressed. Processes that result in expression of a gene do not occur continuously in cells. Instead, the polypeptide coded for by a gene is made only when it is needed. Thus, there must be factors that regulate the expression of genes.

Control of Gene Expression in Prokaryotes

Bacteria called *E. coli* thrive on a culture medium containing glucose, which is their normal energy source. Lactose, a disaccharide, is not normally used by these bacteria as an energy source. Therefore, *E. coli* usually do not produce the enzyme lactase, which breaks down lactose. However, if *E. coli* are grown on a culture medium containing lactose instead of glucose, they will begin to produce lactase. Lactase enables the bacteria to convert lactose to monosaccharides that can be used as energy sources. Thus, the expression of the gene that codes for lactase depends on the presence or absence of lactose. This means of controlling gene expression, in which a substance causes (induces) a gene to be "switched on," is called an **inducible system.** The substance that causes the activation of the genes is known as an inducer. In this case, lactose is the **inducer.** If lactose is not present, the gene for lactase is inactivated or "switched off."

An inducible system consists of several adjacent regions of DNA along a bacterial chromosome. The DNA region that codes for lactase is known as a **structural gene.** Associated with it are two other structural genes that produce other enzymes also

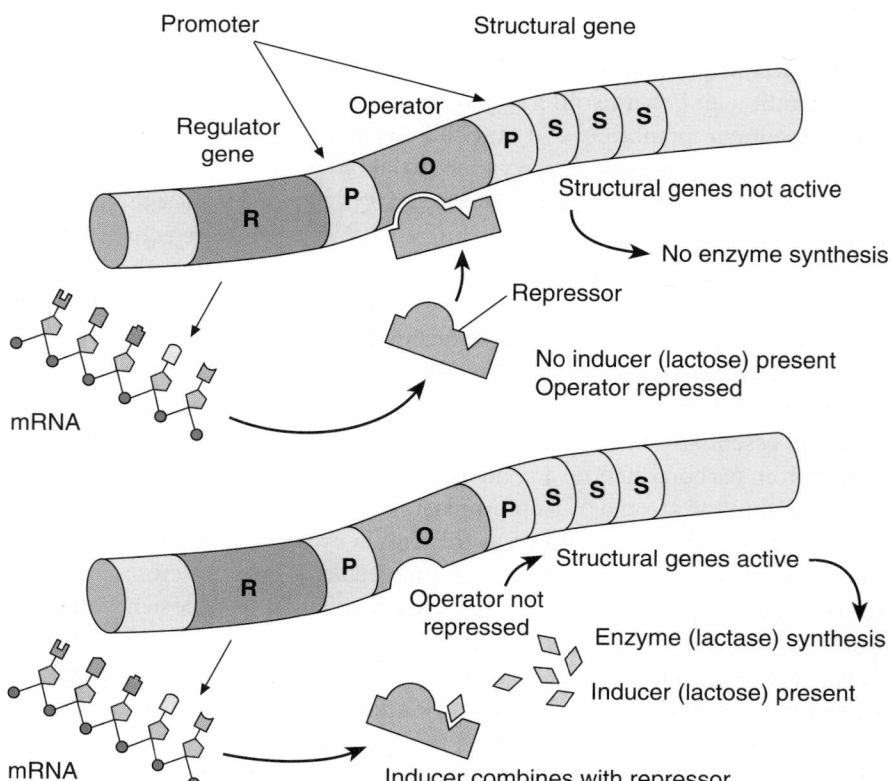

Figure D-1 When no inducer is present (top), a repressor molecule binds to the operator and prevents the synthesis of the enzyme. When an inducer is present (bottom), the inducer binds with the repressor molecule. Thus, the operator is no longer blocked, the structural genes are functional, and enzyme synthesis occurs.

needed for the cells to utilize lactose. All three structural genes are either activated or inactivated as a unit. Other DNA regions play a role in the regulation of the structural genes. A **regulator gene** codes for proteins called **repressors.** Next to the regulator gene is the promoter region, the area of DNA to which the enzyme RNA polymerase may bind, beginning the transcription process. The promoter consists of two separated DNA regions. Between them lies the **operator** region. The operator and its related structural genes are known as an **operon.** Locate these areas of DNA in Figure D-1.

In the absence of lactose, the regulator gene is transcribed and translated, producing repressor molecules. These repressors bind to the operator region. As a result of this binding of repressors and operator, RNA polymerase is blocked from binding to the promoter. Thus, transcription of the structural genes cannot occur, and lactase and the other enzymes are not produced. The structural genes are not expressed. They are "switched off."

When lactose, the inducer, is present, it enters the cells and combines with the repressor molecules. When that happens, the operator is no longer blocked, RNA polymerase can bind to the promoter, and transcription can occur. The mRNA produced from the structural genes is translated, and lactase and the other related enzymes are synthesized. The structural genes have been expressed.

There are several other means by which gene expression is controlled in bacteria. All of them confer an advantage. They insure that enzymes are produced only when they are necessary. Thus, the cell conserves both energy and materials by not producing unnecessary enzymes.

Control of Gene Expression in Eukaryotes

Gene control in eukaryotes is much more complex than in prokaryotes. Unlike bacterial DNA, you know that the DNA of eukaryotes is associated with proteins on the chromosomes. The DNA is wound like thread around spools of these proteins, called histones, and is highly coiled. Before transcription can begin, the DNA must somehow be uncoiled and unwound from the histones. Another type of chromosomal protein, the nonhistones, may function in gene regulation by causing DNA to unwind from the histone spools. This unwinding would be a necessary step leading to transcription.

Although more complicated, control of gene expression by regulating transcription is known to occur in eukaryotic cells. As in bacteria, such control depends upon chemical signals. These chemical signals may take several forms. Steroid hormones, for example, enter a cell, move into its nucleus, and activate a gene. Methylation, the addition of groups of atoms called methyl groups ($—CH_3$), is known to deactivate certain genes. Most important, though, are proteins that regulate transcription. However, these proteins do not function in the simple manner of the repressors (also proteins) of bacteria. In eukaryotes, two different proteins must be zippered together in order to regulate some genes. To further complicate matters, some DNA control regions, called enhancers and silencers, may lie hundreds or thousands of base pairs away from the genes whose activity they help regulate. Much research lies ahead before the picture of control of transcription in eukaryotes becomes clear.

Regulating transcription is only one means of controlling gene expression in eukaryotes. For example, some mRNA molecules can be processed in different ways, depending upon which nucleotides are removed and how the rest are spliced together. Each way of processing results in a different, final mRNA, and, thus, a different polypeptide and a different phenotype.

Expression of a gene may also depend upon changes in a polypeptide after translation has occurred. Consider these examples. Many proteins do not become functional until moved to a certain area within or outside the cell. Some hormones and enzymes cannot function until certain of their amino acids have been removed. The shape of proteins, and thus their functions, is influenced by physical factors such as temperature and pH. Thus, a protein exposed to a pH or temperature that changes its shape cannot function, and the gene coding for that protein will not be expressed.

Origin of Science Words

This list of Greek and Latin roots will help you interpret the meaning of biological terms. The column headed *Root* gives many of the actual Greek (GK) or Latin (L) root words used in science. If more than one word is given, the first is the full word in Greek or Latin. The letter groups that follow are forms in which the root word is most often found combined in science words. In the second column is the meaning of the root as it is used in science. The third column shows a typical science word containing the root from the first column. Most of these words can be found in your textbook.

Root	Meaning	Example	Root	Meaning	Example
a, an (GK)	not, without	anaerobic	bursa (L)	purse, bag	bursa
abilis (L)	able to	biodegradable			
ad (L)	to, attached to	appendix	caedere, cide (L)	kill	insecticide
aequus (L)	equal	equilibrium	capillus (L)	hair	capillary
aeros (GK)	air	anaerobic	carn (L)	flesh	carnivore
agon (GK)	assembly	glucagon	carno (L)	flesh	carnivore
aktis (GK)	ray	actin	cella, cellula (L)	small room	protocells
allas (GK)	sausage	allantois	cervix (L)	neck	cervix
allelon (GK)	of each other	allele	cetus (L)	whale	cetacean
allucinari (L)	to dream	hallucinate	chaite, chaet (GK)	bristle	oligochaeta
alveolus (L)	small pit	alveolus	cheir (GK)	hand	chiropteran
amnos (GK)	lamb	amnion	chele (GK)	claw	chelicerae
amoibe (GK)	change	amoebocyte	chloros (GK)	pale green	chlorophyll
amphi (GK)	both, about, around	amphibian	chondros (GK)	cartilage	Chondrichthyes
amylum (L)	starch	amylase	chondros (GK)	grain	mitochondrion
ana (L)	away, onward	anaphase	chorda (L)	cord	urochordata
andro (GK)	male	androgens	chorion (GK)	skin	chorion
anggeion, angio (GK)	vessel, container	angiosperm	chroma, chrom (GK)	colored	chromosome
anthos (GK)	flower	anthophyte	chronos (GK)	time	chronometer
anti (GK)	against, away, opposite	antibody	circa (L)	about	circadian
			cirrus (L)	curl	cirri
aqua (L)	water	aquatic	codex (L)	tablet for writing	codon
archaios, archeo (GK)	ancient, primitive	archaebacteria	corpus (L)	body	corpus luteum
			cum, col, com, con (L)	with, together	convergent
arthron (GK)	joint, jointed	arthropod	cuticula (L)	thin skin	cuticle
artios (GK)	even	artiodactyl			
askos (GK)	bag	ascospore	daktylos (GK)	finger	perissodactyl
aster (GK)	star	Asteroidea	de (L)	away, from	decompose
autos (GK)	self	autoimmune	decidere (L)	to fall down	deciduous
			degradare (L)	to reduce in rank	biodegradable
bakterion (GK)	small rod	bacterium	dendron (GK)	tree	dendrite
bi, bis (L)	two, twice	bipedal	dens (L)	tooth	edentate
binarius (L)	pair	binary fission	derma (GK)	skin	epidermis
bios (GK)	life	biology	deterere (L)	loose material	detritus
blastos (GK)	bud	blastula	dia, di (GK)	through, apart	diastolic
bryon (GK)	moss	bryophyte	dies (L)	day	circadian

APPENDIX E

Root	Meaning	Example	Root	Meaning	Example
diploos (GK)	twofold, double	diploid	gyne (GK)	female, woman	gynoecium
dis, di (GK)	twice, two	disaccharide			
dis, di (L)	apart, away	disruptive	haima, emia (GK)	blood	hemoglobin
dormire (L)	to sleep	dormancy	halo (GK)	salt	halophile
drom, drome (GK)	running, racing	dromedary	haploos (GK)	simple	haploid
ducere (L)	to lead	oviduct	haurire (L)	to drink	haustorium
			helix (L)	spiral	helix
echinos (GK)	spine	echinoderm	hemi (GK)	half	hemisphere
eidos, oid (GK)	form, appearance	rhizoid	herba (L)	grass	herbivore
ella (GK)	small	organelle	hermaphroditos (GK)	combining both sexes	hermaphrodite
endon, en, endo (GK)	within	endosperm	heteros (GK)	other	heterotrophic
engchyma (GK)	infusion	parenchyma	hierarches (GK)	rank	hierarchy
enteron (GK)	intestine, gut	enterocolitis	hippos (GK)	horse	hippopotamus
entomon (GK)	insect	entomology	histos (GK)	tissue	histology
epi (GK)	upon, above	epidermis	holos (GK)	whole	Holothuroidea
equus (L)	horse	Equisetum	homo (L)	man	hominid
erythros (GK)	red	erythrocyte	homos (GK)	same, alike	homologous
eu (GK)	well, true, good	eukaryote	hormaein (GK)	to excite	hormone
evolutus (L)	rolled out	evolution	hydor, hydro (GK)	water	hydrolysis
ex, e (L)	out	extinction	hyper (GK)	over, above	hyperventilation
exo (GK)	out, outside	exoskeleton	hyphe (GK)	web	hypha
extra (L)	outside, beyond	extracellular	hypo (GK)	under, below	hypotonic
ferre (L)	to bear	porifera	ichthys (GK)	fish	Osteichthyes
fibrilla (L)	small fiber	myofibril	instinctus (L)	impulse	instinct
fissus (L)	a split	binary fission	insula (L)	island	insulin
flagellum (L)	whip	flagellum	inter (L)	between	internode
follis (L)	bag	follicle	intra (L)	within, inside	intracellular
fossilis (L)	dug up	microfossils	isos (GK)	equal	isotonic
fungus (L)	mushroom	fungus	itis (GK)	inflammation, disease	arthritis
gamo, gam (GK)	marriage	gamete	jugare (L)	join together	conjugate
gaster (GK)	stomach	gastropoda			
ge, geo (GK)	earth	geology	kardia, cardia (GK)	heart	cardiac
gemmula (L)	little bud	gemmule	karyon (GK)	nut	prokaryote
genesis (L)	origin, birth	parthenogenesis	kata, cata (GK)	break down	catabolism
genos, gen, geny (GK)	race	genotype	kephale, ceph (GK)	head	cephalopoda
gestare (L)	to bear	progesterone	keras (GK)	horn	chelicerae
glene (GK)	eyeball	euglenoid	kinein (GK)	to move	kinetic
globus (L)	sphere	hemoglobin	koilos, coel (GK)	hollow cavity, belly	coelom
glotta (GK)	tongue	epiglottis	kokkus (GK)	berry	streptococcus
glykys, glu (GK)	sweet	glycolysis	kolla (GK)	glue	colloid
gnathos (GK)	jaw	Agnatha	kotyl, cotyl (GK)	cup	cotylosaur
gonos, gon (GK)	reproductive, sexual	gonorrhea	kreas (GK)	flesh	pancreas
gradus (L)	a step	gradualism	krinoeides (GK)	lilylike	Crinoidea
graphos (GK)	written	chromatograph	kyanos, cyano (GK)	blue	cyanobacterium
gravis (L)	heavy	gravitropism	kystis, cyst (GK)	bladder, sac	cystitis
gymnos (GK)	naked, bare	gymnosperm	kytos, cyt (GK)	hollow, cell	lymphocyte

APPENDIX E

Root	Meaning	Example	Root	Meaning	Example
lagos (GK)	hare	lagomorph	organon (GK)	tool, implement	organelle
leukos (GK)	white	leukocyte	ornis (GK)	bird	ornithology
libra (L)	balance	equilibrium	orthos (GK)	straight	orthodontist
logos, logy (GK)	study, word	biology	osculum (L)	small mouth	osculum
luminescere (L)	to grow light	bioluminescence	osteon (GK)	bone	osteocyte
luteus (L)	orange-yellow	corpus luteum	ostrakon (GK)	shell	ostracoderm
lyein, lysis (GK)	to split, loosen	lysosome	oura, ura (GK)	tail	anura
lympha (L)	water	lymphocyte	ous, oto (GK)	ear	otology
			ovum (L)	egg	oviduct
makros (GK)	large	macrophage			
marsupium (L)	pouch	marsupial	palaios, paleo (GK)	ancient	paleontology
meare (L)	to glide	permeable	pan (GK)	all	pancreas
megas (GK)	large	megaspore	para (GK)	beside	parenchyma
melas (GK)	black, dark	melanin	parthenos (GK)	virgin	parthenogenesis
meristos (GK)	divided	meristem	pathos (GK)	disease, suffering	pathogenic
meros (GK)	part	polymer	pausere (L)	to rest	decompose
mesos (GK)	middle	mesophyll	pendere (L)	to hang	appendix
meta (GK)	after, following	metaphase	per (L)	through	permeable
metabole (GK)	change	metabolism	peri (GK)	around	peristalsis
meter (GK)	a measurement	diameter	periodos (GK)	a cycle	photoperiodism
mikros, micro (GK)	small	microscope	pes, pedis (L)	foot	bipedal
mimos (GK)	a mime	mimicry	phagein (GK)	to eat	phagocyte
mitos (GK)	thread	mitochondrion	phainein (GK)	to show	phenotype
molluscus (L)	soft	mollusk	phaios (GK)	dusky	phaeophyta
monos (GK)	single	monotreme	phase (GK)	stage, appearance	metaphase
morphe (GK)	form	lagomorph	pherein, phor (GK)	to carry	pheromone
mors, mort (L)	death	mortality	phloios (GK)	inner bark	phloem
mucus (L)	mucus, slime	mucosa	phos, photos (GK)	light	phototropism
multus (L)	many	multicellular	phyllon (GK)	leaf	chlorophyll
mutare (L)	to change	mutation	phylon (GK)	related group	phylogeny
mykes, myc (GK)	fungus	mycorrhiza	phyton (GK)	plant	epiphyte
mys (GK)	muscle	myosin	pinax (GK)	tablet	pinacocytes
			pinein (GK)	to drink	pinocytosis
nema (GK)	thread	nematology	pinna (L)	feather	pinniped
nemato (GK)	thread, threadlike	nematode	plasma (GK)	mold, form	plasmodium
neos (GK)	new	Neolithic	plastos (GK)	formed object	chloroplast
nephros (GK)	kidney	nephron	platys (GK)	flat	platyhelminthes
neuro (GK)	nerve	neurology	plax (GK)	plate	placoderm
nodus (L)	knot, knob	internode	pleuron (GK)	side	dipleurula
nomos, nomy (GK)	ordered knowledge	taxonomy	plicare (L)	to fold	replication
noton (GK)	back	notochord	polys, poly (GK)	many	polymer
			poros (GK)	channel	porifera
oikos, eco (GK)	household	ecosystem	post (L)	after	posterior
oisein, eso (GK)	to carry	esophagus	pous, pod (GK)	foot	gastropoda
oligos (GK)	few, little	oligochaeta	prae, pre (L)	before	Precambrian
omnis (L)	all	omnivore	primus (L)	first	primary
ophis (GK)	serpent	Ophiuroidea	pro (GK and L)	before, for	prokaryote
ophthalmos (GK)	referring to the eye	ophthalmologist	proboskis (GK)	trunk	proboscidean

APPENDIX E

Root	Meaning	Example	Root	Meaning	Example
producere (L)	to bring forth	reproduction	taxis, taxo (GK)	to arrange	taxonomy
protos (GK)	first	protocells	telos (GK)	end	telophase
pseudes (GK)	false	pseudopod	terra (L)	land, Earth	terrestrial
pteron (GK)	wing	chiropteran	thele (GK)	cover a surface	epithelium
punctus (L)	a point	punctuated	therme (GK)	heat	endotherm
pupa (L)	doll	pupa	thrix, trich (GK)	hair	trichocyst
radius (L)	ray	radial	tome (GK)	cutting	anatomy
re (L)	again	reproduction	trachia (GK)	windpipe	tracheid
reflectere (L)	to turn back	reflex	trans (L)	across	transpiration
rhiza (GK)	root	mycorrhiza	trematodes (GK)	having holes	monotreme
rhodon (GK)	rose	rhodophyte	trope (GK)	turn	gravitropism
rota (L)	wheel	rotifer	trophe (GK)	nourishment	heterotrophic
rumpere (L)	to break	disruptive	turbo (L)	whirl	turbellaria
			tympanon (GK)	drum	tympanum
saeta (L)	bristle	Equisetum	typos (GK)	model	genotype
sapros (GK)	rotten	saprobe			
sarx (GK)	flesh	sarcomere	uni (L)	one	unicellular
sauros (GK)	lizard	cotylosaur	uterus (L)	womb	uterus
scire (L)	to know	science			
scribere, script (L)	to write	transcription	vacca (L)	cow	vaccine
sedere, ses (L)	to sit	sessile	vagina (L)	sheath	vagina
semi (L)	half	semicircle	valvae (L)	folding doors	bivalvia
skopein, scop (GK)	to look	microscope	vasculum (L)	small vessel	vascular
soma (GK)	body	lysosome	venter (L)	belly	ventricle
sperma (GK)	seed	angiosperm	ventus (L)	a wind	hyperventila-tion
spirare (L)	to breathe	spiracle			
sporos (GK)	seed	microspore	vergere (L)	to slant, incline	convergent
staphylo (GK)	bunch of grapes	staphylococcus	villus (L)	shaggy hair	villus
stasis (GK)	standing, staying	homeostasis	virus (L)	poisonous liquid	virus
stellein, stol (GK)	to draw in	peristalsis	vorare (L)	to devour	carnivore
sternon (GK)	chest	sternum			
stinguere (L)	to quench	extinction	xeros (GK)	dry	xerophyte
stolo (L)	shoot	stolon	xylon (GK)	wood	xylem
stoma (GK)	mouth	stoma			
streptos (GK)	twisted chain	streptococcus	zoon, zo (GK)	animal	zoology
syn (GK)	together	systolic	zygotos (GK)	joined together	zygote
synapsis (GK)	union	synapse			
systema (GK)	composite whole	ecosystem			

APPENDIX E **925**

APPENDIX F

Periodic Table

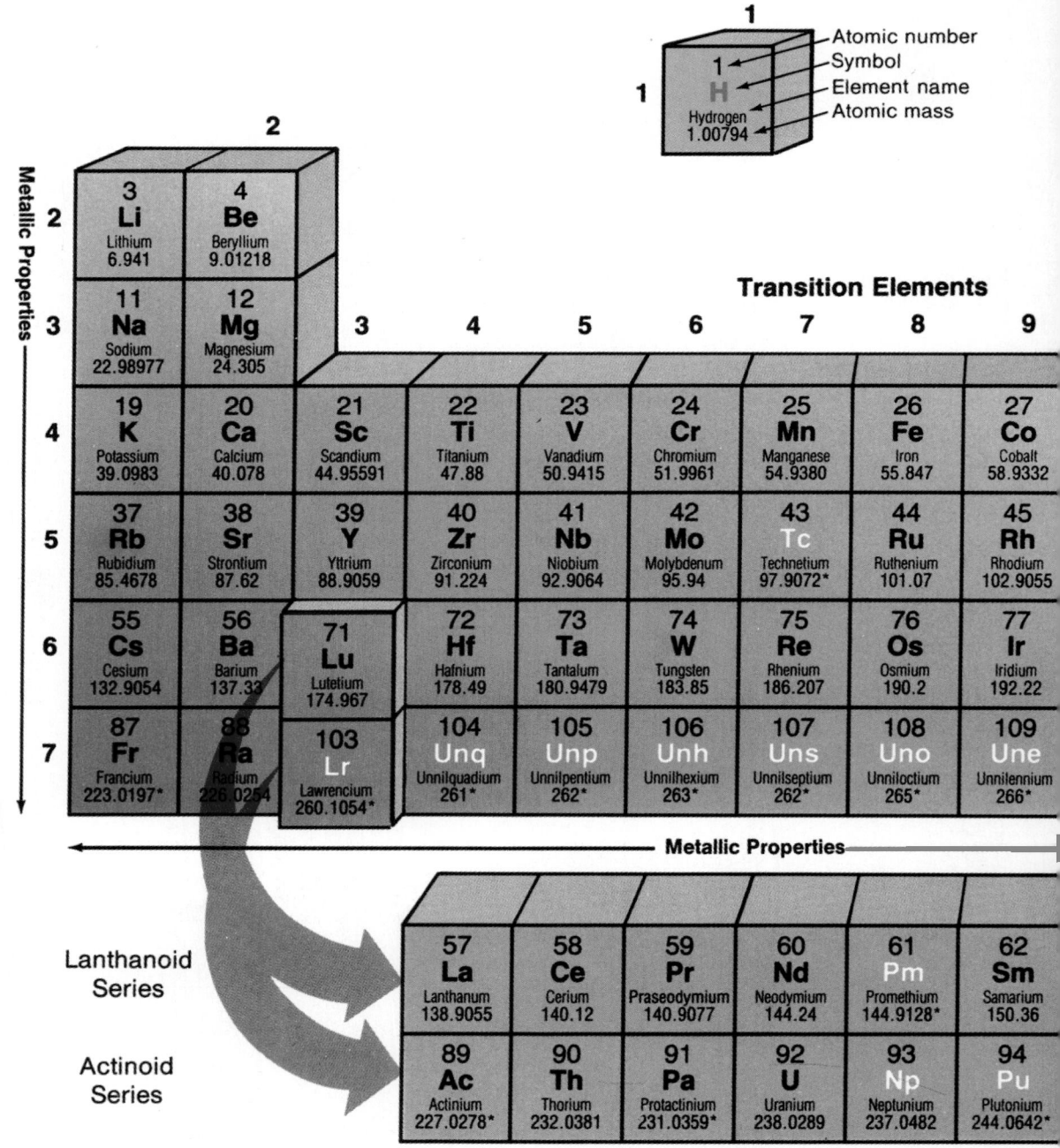

Atomic number
Symbol
Element name
Atomic mass

1
1
H
Hydrogen
1.00794

Metallic Properties

Transition Elements

	1	2	3	4	5	6	7	8	9
2	3 Li Lithium 6.941	4 Be Beryllium 9.01218							
3	11 Na Sodium 22.98977	12 Mg Magnesium 24.305							
4	19 K Potassium 39.0983	20 Ca Calcium 40.078	21 Sc Scandium 44.95591	22 Ti Titanium 47.88	23 V Vanadium 50.9415	24 Cr Chromium 51.9961	25 Mn Manganese 54.9380	26 Fe Iron 55.847	27 Co Cobalt 58.9332
5	37 Rb Rubidium 85.4678	38 Sr Strontium 87.62	39 Y Yttrium 88.9059	40 Zr Zirconium 91.224	41 Nb Niobium 92.9064	42 Mo Molybdenum 95.94	43 Tc Technetium 97.9072*	44 Ru Ruthenium 101.07	45 Rh Rhodium 102.9055
6	55 Cs Cesium 132.9054	56 Ba Barium 137.33	71 Lu Lutetium 174.967	72 Hf Hafnium 178.49	73 Ta Tantalum 180.9479	74 W Tungsten 183.85	75 Re Rhenium 186.207	76 Os Osmium 190.2	77 Ir Iridium 192.22
7	87 Fr Francium 223.0197*	88 Ra Radium 226.0254	103 Lr Lawrencium 260.1054*	104 Unq Unnilquadium 261*	105 Unp Unnilpentium 262*	106 Unh Unnilhexium 263*	107 Uns Unnilseptium 262*	108 Uno Unniloctium 265*	109 Une Unnilennium 266*

◀—— Metallic Properties ——

Lanthanoid Series

57 La Lanthanum 138.9055	58 Ce Cerium 140.12	59 Pr Praseodymium 140.9077	60 Nd Neodymium 144.24	61 Pm Promethium 144.9128*	62 Sm Samarium 150.36

Actinoid Series

89 Ac Actinium 227.0278*	90 Th Thorium 232.0381	91 Pa Protactinium 231.0359*	92 U Uranium 238.0289	93 Np Neptunium 237.0482	94 Pu Plutonium 244.0642*

*Mass of isotope with longest half-life, that
is, the most stable isotope of the element

APPENDIX F

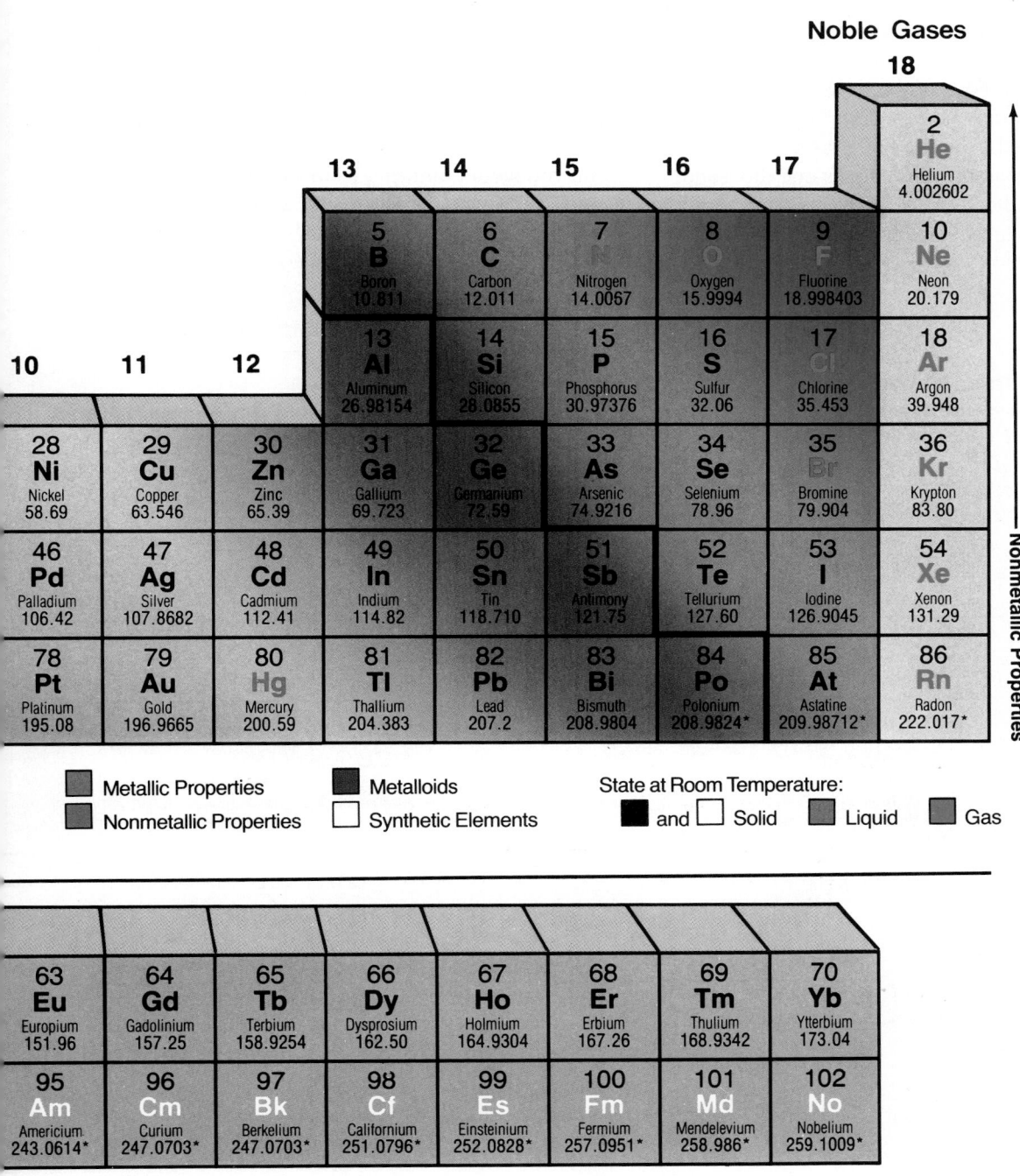

Noble Gases

					18
13	14	15	16	17	2 **He** Helium 4.002602
5 **B** Boron 10.811	6 **C** Carbon 12.011	7 **N** Nitrogen 14.0067	8 **O** Oxygen 15.9994	9 **F** Fluorine 18.998403	10 **Ne** Neon 20.179

10	11	12	13 **Al** Aluminum 26.98154	14 **Si** Silicon 28.0855	15 **P** Phosphorus 30.97376	16 **S** Sulfur 32.06	17 **Cl** Chlorine 35.453	18 **Ar** Argon 39.948
28 **Ni** Nickel 58.69	29 **Cu** Copper 63.546	30 **Zn** Zinc 65.39	31 **Ga** Gallium 69.723	32 **Ge** Germanium 72.59	33 **As** Arsenic 74.9216	34 **Se** Selenium 78.96	35 **Br** Bromine 79.904	36 **Kr** Krypton 83.80
46 **Pd** Palladium 106.42	47 **Ag** Silver 107.8682	48 **Cd** Cadmium 112.41	49 **In** Indium 114.82	50 **Sn** Tin 118.710	51 **Sb** Antimony 121.75	52 **Te** Tellurium 127.60	53 **I** Iodine 126.9045	54 **Xe** Xenon 131.29
78 **Pt** Platinum 195.08	79 **Au** Gold 196.9665	80 **Hg** Mercury 200.59	81 **Tl** Thallium 204.383	82 **Pb** Lead 207.2	83 **Bi** Bismuth 208.9804	84 **Po** Polonium 208.9824*	85 **At** Astatine 209.98712*	86 **Rn** Radon 222.017*

Nonmetallic Properties

- Metallic Properties
- Nonmetallic Properties
- Metalloids
- Synthetic Elements

State at Room Temperature: ■ and ☐ Solid ■ Liquid ■ Gas

63 **Eu** Europium 151.96	64 **Gd** Gadolinium 157.25	65 **Tb** Terbium 158.9254	66 **Dy** Dysprosium 162.50	67 **Ho** Holmium 164.9304	68 **Er** Erbium 167.26	69 **Tm** Thulium 168.9342	70 **Yb** Ytterbium 173.04
95 **Am** Americium 243.0614*	96 **Cm** Curium 247.0703*	97 **Bk** Berkelium 247.0703*	98 **Cf** Californium 251.0796*	99 **Es** Einsteinium 252.0828*	100 **Fm** Fermium 257.0951*	101 **Md** Mendelevium 258.986*	102 **No** Nobelium 259.1009*

APPENDIX G

SI Measurement

The International System (SI) of Measurement is accepted as the standard for measurement throughout most of the world. Four of the base units in SI are the meter, liter, kilogram, and second. The size of a unit can be determined from the prefix used with the base unit name. For example: *kilo* means one thousand; *milli* means one thousandth; *micro* means one millionth; and *centi* means one hundredth. The tables below give the standard symbols for these SI units and some of their equivalents.

Larger and smaller units of measurement in SI are obtained by multiplying or dividing the base unit by some multiple of ten. Multiply to change from larger units to smaller units. Divide to change from smaller units to larger units. For example, to change 1 km to m, you would multiply 1 km by 1000 to obtain 1000 m. To change 10 g to kg, you would divide 10 g by 1000 to obtain 0.01 kg.

Table G-1 Common SI Units

Measurement	Unit	Symbol	Equivalents
Length	1 millimeter	mm	1000 micrometers (μm)
	1 centimeter	cm	10 millimeters (mm)
	1 meter	m	100 centimeters (cm)
	1 kilometer	km	1000 meters (m)
Volume	1 milliliter	mL	1 cubic centimeter (cm^3 or cc)
	1 liter	L	1000 milliliters (mL)
Mass	1 gram	g	1000 milligrams (mg)
	1 kilogram	kg	1000 grams (g)
	1 tonne	t	1000 kilograms (kg) = 1 metric ton
Time	1 second	s	
Area	1 square meter	m^2	10 000 square centimeters (cm^2)
	1 square kilometer	km^2	1 000 000 square meters (m^2)
	1 hectare	ha	10 000 square meters (m^2)
Temperature	1 Kelvin	K	1 degree Celsius (°C)

APPENDIX H

Care and Use of a Microscope

Care of a Microscope

1. Always carry the microscope holding the arm with one hand and supporting the base with the other hand.
2. Don't touch the lenses with your finger.
3. Never lower the coarse adjustment knob when looking through the eyepiece lens.
4. Always focus first with the low-power objective.
5. Don't use the coarse adjustment knob when the high-power objective is in place.
6. Store the microscope covered.

Using a Microscope

1. Place the microscope on a flat surface that is clear of objects. The arm should be toward you.
2. Look through the eyepiece. Adjust the diaphragm so that light comes through the opening in the stage.
3. Place a slide on the stage so that the specimen is in the field of view. Hold it firmly in place by using the stage clips.
4. Always focus first with the coarse adjustment and the low-power objective lens. Once the object is in focus on low power, turn the nosepiece until the high-power objective is in place. Use ONLY the fine adjustment to focus with this lens.

Making a Wet Mount Slide

1. Carefully place the item you want to look at in the center of a clean glass slide. Make sure the sample is thin enough for light to pass through.
2. Use a dropper to place one or two drops of water on the sample.
3. Hold a clean coverslip by the edges and place it at one edge of the drop of water. Slowly lower the coverslip onto the drop of water until it lies flat.
4. If you have too much water or a lot of air bubbles, touch the edge of a paper towel to the edge of the coverslip to draw off extra water and force air out.

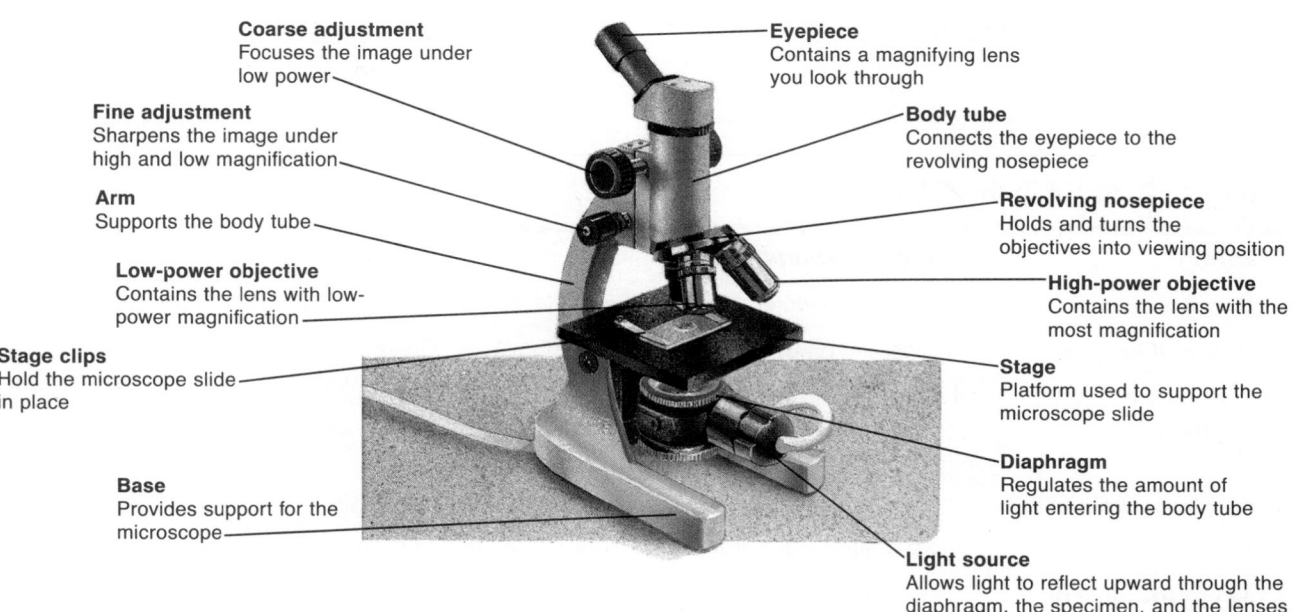

Coarse adjustment
Focuses the image under low power

Fine adjustment
Sharpens the image under high and low magnification

Arm
Supports the body tube

Low-power objective
Contains the lens with low-power magnification

Stage clips
Hold the microscope slide in place

Base
Provides support for the microscope

Eyepiece
Contains a magnifying lens you look through

Body tube
Connects the eyepiece to the revolving nosepiece

Revolving nosepiece
Holds and turns the objectives into viewing position

High-power objective
Contains the lens with the most magnification

Stage
Platform used to support the microscope slide

Diaphragm
Regulates the amount of light entering the body tube

Light source
Allows light to reflect upward through the diaphragm, the specimen, and the lenses

APPENDIX I

Safety in the Laboratory

The biology laboratory is a safe place to work if you are aware of important safety rules and if you are careful. You must be responsible for your own safety and for the safety of others. The safety rules given here will protect you and others from harm in the lab. While carrying out procedures in any of the **Biolabs,** notice the safety symbols and caution statements. The safety symbols are explained in the chart on the next page.

1. Always obtain your teacher's permission to begin a lab.
2. Study the procedure. If you have questions, ask your teacher. Be sure you understand all safety symbols shown.
3. Use the safety equipment provided for you. Goggles and a safety apron should be worn when any lab calls for using chemicals.
4. When you are heating a test tube, always slant it so the mouth points away from you and others.
5. Never eat or drink in the lab. Never inhale chemicals. Do not taste any substance or draw any material into your mouth.
6. If you spill any chemical, wash it off immediately with water. Report the spill immediately to your teacher.

7. Know the location and proper use of the fire extinguisher, safety shower, fire blanket, first aid kit, and fire alarm.
8. Keep all materials away from open flames. Tie back long hair.
9. If a fire should break out in the classroom, or if your clothing should catch fire, smother it with the fire blanket or a coat, or get under a safety shower. **NEVER RUN.**
10. Report any accident or injury, no matter how small, to your teacher.

Follow these procedures as you clean up your work area.

1. Turn off the water and gas. Disconnect electrical devices.
2. Return materials to their places.
3. Dispose of chemicals and other materials as directed by your teacher. Place broken glass and solid substances in the proper containers. Never discard materials in the sink.
4. Clean your work area.
5. Wash your hands thoroughly after working in the laboratory.

Table I.1 First Aid in the Laboratory

Injury	Safe Response
Burns	Apply cold water. Call your teacher immediately.
Cuts and bruises	Stop any bleeding by applying direct pressure. Cover cuts with a clean dressing. Apply cold compresses to bruises. Call your teacher immediately.
Fainting	Leave the person lying down. Loosen any tight clothing and keep crowds away. Call your teacher immediately.
Foreign matter in eye	Flush with plenty of water. Use eyewash bottle or fountain.
Poisoning	Note the suspected poisoning agent and call your teacher immediately.
Any spills on skin	Flush with large amounts of water or use safety shower. Call your teacher immediately.

APPENDIX I

Safety Symbols

DISPOSAL ALERT
This symbol appears when care must be taken to dispose of materials properly.

ANIMAL SAFETY
This symbol appears whenever live animals are studied and the safety of the animals and the students must be ensured.

BIOLOGICAL HAZARD
This symbol appears when there is danger involving bacteria, fungi, or protists.

RADIOACTIVE SAFETY
This symbol appears when radioactive materials are used.

OPEN FLAME ALERT
This symbol appears when use of an open flame could cause a fire or an explosion.

CLOTHING PROTECTION SAFETY
This symbol appears when substances used could stain or burn clothing.

THERMAL SAFETY
This symbol appears as a reminder to use caution when handling hot objects.

FIRE SAFETY
This symbol appears when care should be taken around open flames.

SHARP OBJECT SAFETY
This symbol appears when a danger of cuts or punctures caused by the use of sharp objects exists.

EXPLOSION SAFETY
This symbol appears when the misuse of chemicals could cause an explosion.

FUME SAFETY
This symbol appears when chemicals or chemical reactions could cause dangerous fumes

EYE SAFETY
This symbol appears when a danger to the eyes exists. Safety goggles should be worn when this symbol appears.

ELECTRICAL SAFETY
This symbol appears when care should be taken when using electrical equipment.

POISON SAFETY
This symbol appears when poisonous substances are used.

PLANT SAFETY
This symbol appears when poisonous plants or plants with thorns are handled.

CHEMICAL SAFETY
This symbol appears when chemicals used can cause burns or are poisonous if absorbed through the skin.

GLOSSARY

A

abiotic factor: physical environmental factor, such as water, temperature, soil, and light, that influences the composition and growth of an ecosystem. (Ch. 28, p. 795)

abscission: dropping of leaves or mature fruit from a plant that results from a lowered level of auxin. (Ch. 24, p. 678)

absorption spectrum: characteristic pattern of wavelengths, or colors, of light absorbed by a particular pigment. (Ch. 6, p. 154)

acid: solution with a higher concentration of hydrogen ions than hydroxide ions; solutions with a pH below 7 are acids. (Ch. 3, p. 68)

acid precipitation: precipitation with a pH under 5.6; can damage leaves and plant tissues, kill plankton, and contribute to human diseases such as asthma and emphysema. (Ch. 30, p. 854)

action potential: temporary change in a neuron's polarity, which creates an electrical current that moves through the dendrites, into the cell body, and along the axon. (Ch. 25, p. 704)

activation energy: minimum amount of energy needed to initiate a chemical reaction. (Ch. 3, p. 80)

active immunity: disease resistance as a result of memory cells recognizing antigens to which they have been previously exposed. (Ch. 23, p. 657)

active transport: energy-requiring process in which particles move from an area of lower concentration to an area of higher concentration. (Ch. 4, p. 105)

adaptation: physical trait of a living organism that helps it survive in its environment; for example, deer are adapted to browse on leaves. (Ch. 1, p. 20)

adaptive radiation: divergent evolution and adaptation of a species over time in response to a new environment. (Ch. 13, p. 348)

adenine: one of four nitrogen bases of DNA; the others are guanine, thymine, and cytosine. (Ch. 9, p. 228)

aerobic respiration: cellular respiration that uses oxygen, sequentially releasing energy and storing it in ATP. (Ch. 6, p. 147)

alcoholic fermentation: anaerobic process whereby enzymes break down glucose into ethanol and CO_2 and transfer energy to ATP. (Ch. 6, p. 151)

algal bloom: explosive population increase in algae that occurs when large amounts of phosphates and/or nitrates enter a body of water in the presence of warm temperatures; can cause stagnation. (Ch. 28, p. 804)

allantois: membrane sac that collects wastes from an embryo. (Ch. 19, p. 528)

allele: either of two alternative forms of a gene for a particular trait such as flower color. (Ch. 8, p. 203)

alternation of generations: reproductive cycle of all plants involving the gametophyte (haploid) and sporophyte (diploid) generations. (Ch. 16, p. 429)

alveoli: within the lungs, numerous, small, moist air sacs where diffusion of oxygen and carbon dioxide into and out of the lungs occurs. (Ch. 22, p. 620)

amino acid: simple compound that is the building block of proteins. (Ch. 3, p. 75)

amniocentesis: withdrawal of amniotic fluid to evaluate the fetal cells for a genetic disorder (for example, Down syndrome) prior to birth. (Ch. 11, p. 291)

amnion: fluid-filled sac developed from the innermost membrane surrounding the embryo; amniotic fluid provides moisture and protection. (Ch. 19, p. 528)

anaerobic: processes that occur in the absence of oxygen (for example, lactic acid fermentation and alcoholic fermentation). (Ch. 6, p. 149)

anaphase: third stage of mitosis during which the two sister chromatids separate and are pulled to opposite poles and only one set of single-stranded chromosomes is seen at each end of the cell. (Ch. 7, p. 176)

angiosperm: flowering plant (for example, wild and cultivated flowers, oats, strawberries, broccoli) whose seeds are contained in a fruit. (Ch. 16, p. 439)

anterior: front end of animals with bilateral symmetry (for example, mollusks); generally contains a head. (Ch. 17, p. 454)

anther: the pollen-bearing part of the stamen in flowering plants. (Ch. 16, p. 441)

antibody: protein contained in plasma that can unite with a specific antigen. (Ch. 21, p. 604)

anticodon: in a tRNA molecule, the set of three bases specific to the kind of amino acid carried by the tRNA; complementary to the mRNA codon. (Ch. 10, p. 254)

antigen: foreign substance that stimulates an immune response in the blood. (Ch. 21, p. 604)

antiserum: solution containing blood serum and antibodies extracted from an animal and injected into another animal in order to fight disease. (Ch. 23, p. 660)

anus: body opening through which undigested materials are eliminated. (Ch. 20, p. 551)

aorta: in arthropods, the only blood vessel; empties blood into the organism's body cavity; in vertebrates, the largest blood vessel in the body; directs blood into arteries and to all body parts. (Ch. 21, p. 590)

apical dominance: hormone-mediated inhibition of lateral bud development; can be used by gardeners to produce bushier plants. (Ch. 24, p. 677)

appendage: in arthropods, pairs of jointed structures, such as a leg, that extend from the body and enable complex movement. (Ch. 17, p. 459)

artery: thick-walled vessel composed of three layers of cells; carries blood away from the heart. (Ch. 21, p. 591)

ascus: saclike structure that produces sexual spores by meiosis; found in the sac fungi (for example, yeasts). (Ch. 15, p. 413)

GLOSSARY

asexual reproduction: involves only one parent—produces a new organism genetically identical to the parent. (Ch. 15, p. 396)

ATP: adenosine triphosphate; energy-carrying molecule, which, when hydrolyzed to ADP and inorganic phosphate, releases free energy. (Ch. 6, p. 145)

atrioventricular (A-V) node: small bundle of tissue in the vertebrate heart through which impulses are passed from the S-A node to the ventricles. (Ch. 21, p. 596)

atrium: thin-walled chamber in the heart that receives the body's incoming blood and passes it to the ventricle. (Ch. 21, p. 591)

autocidal control: release of sterile males into an area to help control a pest population. (Ch. 30, p. 850)

autoimmune disease: disease, such as rheumatoid arthritis, caused by the abnormal response of the immune system against the individual's own body. (Ch. 23, p. 654)

autonomic system: in humans, two groups of nerves—the sympathetic and parasympathetic—that control such reflexes as heartbeat and peristalsis. (Ch. 25, p. 718)

autosomes: chromosomes other than sex chromosomes. (Ch. 8, p. 213)

autotroph: organism that makes its own food through chemosynthesis or photosynthesis; blue-green bacteria are autotrophs. (Ch. 12, p. 327)

auxin: plant hormone with growth-regulating effects, including the growth of stems toward light. (Ch. 24, p. 674)

axon: portion of a neuron that conducts impulses away from the cell body. (Ch. 25, p. 701)

B

B cell: lymphocyte that matures in the bone marrow; can form two types of cells: plasma cells and memory cells. (Ch. 23, p. 646)

bacteriophage: a virus that attacks bacteria. (Ch. 9, p. 228)

base: solution with a higher concentration of hydroxide ions than hydrogen ions; solutions with a pH above 7 are bases. (Ch. 3, p. 68)

basidium: in club fungi, the clublike structure in which sexual spores are formed. (Ch. 15, p. 411)

behavioral adaptation: results from the response of an organism to its external environment (for example, bird migration). (Ch. 13, p. 340)

benthos: aquatic organisms that live attached to the ocean floor or crawl over it; some benthic animals are starfish, barnacles, and clams. (Ch. 29, p. 833)

bilateral symmetry: body plan of organisms, such as flatworms, in which there is only one plane that will divide the organism's body into nearly identical (mirror symmetry) parts. (Ch. 17, p. 454)

bile: digestive fluid produced by the liver, stored in the gallbladder, and released into the small intestine, where it breaks up fat globules. (Ch. 20, p. 563)

binary fission: method by which prokaryotes reproduce whereby their DNA is replicated and distributed equally to daughter cells. (Ch. 7, p. 181)

binomial nomenclature: two-part system of biological classification for naming organisms in which the first part of the name is the genus and the second part is the specific name (species name). (Ch. 14, p. 368)

biological control: preservation or introduction of predators to limit a pest population. (Ch. 30, p. 849)

biological magnification: increasing concentration of non-biodegradable chemicals (for example, the pesticide DDT) from one step in the food chain to the next. (Ch. 30, p. 848)

biology: branch of science that deals with living organisms and processes vital to life. (Ch. 1, p. 11)

biomass: total dried weight of all organic matter at different trophic levels. (Ch. 28, p. 791)

biome: one of the distinctive geographic regions of the world characterized by dominant plant and animal life forms and maintained by similar climates. (Ch. 29, p. 823)

biosphere: thin layer of Earth's surface, composed of the world's biomes and ocean and freshwater ecosystems, that is occupied by living things; highest level of biological organization. (Ch. 29, p. 836)

biotic factors: interspecies and intraspecies relationships of organisms living within a community. (Ch. 28, p. 786)

biotic potential: highest reproductive capacity of a population to increase indefinitely under ideal conditions; most organisms never reach their biotic potential. (Ch. 27, p. 761)

birthrate: number of organisms born or produced in a given time span. (Ch. 27, p. 763)

blastocoel: hollow cavity in the interior of a blastula. (Ch. 19, p. 513)

blastula: hollow ball of cells formed by cleavage during the early development of an embryo. (Ch. 19, p. 513)

Bowman's capsule: cup-shaped structure of the nephron that narrows into a tubule; involved in filtration. (Ch. 22, p. 627)

bronchus: one of two large tubes branching and rebranching from the trachea into smaller bronchial tubes and ending as alveoli. (Ch. 22, p. 620)

budding: in animals such as *Hydra,* asexual reproduction in which a new organism is produced by means of an outgrowth that breaks off from the parent. (Ch. 18, p. 485)

C

Calvin cycle: set of reactions in photosynthesis in which carbon dioxide is converted into sugars; takes place in the stroma. (Ch. 6, p. 158)

capillary: small, thin-walled blood vessel where exchange of carbon dioxide and oxygen occurs between blood and body cells. (Ch. 21, p. 590)

carbohydrate: organic compound containing carbon, hydrogen, and oxygen; carbohydrates such as sugars and starches provide sources of energy for life processes. (Ch. 3, p. 72)

carcinogen: agent such as cigarette smoke or X rays that is capable of causing cancer. (Ch. 10, p. 267)

cardiac muscle: in vertebrates, involuntary muscle found only in the heart. (Ch. 26, p. 744)

carnivore: meat eater; consumer, such as a snake or owl, that feeds on other consumers. (Ch. 28, p. 788)

carotenoid: pigment molecules—usually yellow, orange, and red—that interact with chlorophylls to absorb light energy needed in photosynthesis. (Ch. 6, p. 154)

carrying capacity: greatest number of individuals in a specific population that a specific environment is capable of supporting under particular conditions; can vary, for example, with time of year or access to food. (Ch. 27, p. 763)

cartilage: in a vertebrate fetus, flexible tissue that makes up most of the skeleton. (Ch. 26, p. 734) also found in the notochord of chordates, and as the primary means of internal support in cartilagenous fish. (Ch. 17, pp. 465, 467)

cell cycle: cycle of cell growth and division involving interphase and mitosis. (Ch. 7, p. 173)

cell plate: in the dividing cells of most plants, gives rise to the plasma membranes of the two new plant cells. (Ch. 7, p. 176)

cell theory: the theory that (1) all living things are composed of one or more cells or cell fragments, (2) the cell is the organism's basic unit of structure and function, and (3) all cells arise from other cells. (Ch. 4, p. 91)

cell wall: structure that lies outside the plasma membrane and provides support and protection. (Ch. 4, p. 109)

cellular respiration: process used by producers and consumers to convert the energy in glucose and other sugars into a usable form of energy. (Ch. 1, p. 15)

central nervous system: in bilaterally symmetrical animals, the brain and one or more nerve cords. (Ch. 25, p. 712)

centriole: found in the cells of animals and some algae and fungi; two pairs of centrioles play an important role in cell reproduction. (Ch. 5, p. 133)

centromere: region of attachment for the two chromatids of a chromosome (sister chromatids) during prophase; also attaches the chromosome to the microtubules of the spindle. (Ch. 7, p. 175)

cerebellum: motor area of the brain; coordinates muscular activity. (Ch. 25, p. 714)

cerebrum: largest portion of the human brain; controls speech, reasoning, emotions, and personality. (Ch. 25, p. 713)

chemical formula: a group of symbols showing the number and kind of each atom in a compound; for example, CO_2 is the chemical formula for carbon dioxide. (Ch. 3, p. 62)

chemosynthesis: process by which energy released from the breakdown of inorganic materials is used to synthesize organic compounds. (Ch. 12, p. 327)

chlorofluorocarbon (CFC): chemical used as a coolant in appliances and in spray-can propellants that causes air pollution and thinning of the ozone layer. (Ch. 30, p. 854)

chlorophyll: green substance in producers that traps light energy from the sun, which is then used to combine carbon dioxide and water into sugars in the process of photosynthesis. (Ch. 1, p. 14)

chloroplast: double-walled organelle found in plants and some algae; functions in photosynthesis to trap light energy. (Ch. 5, p. 131)

chorion: thin membrane lining an eggshell that functions with the allantois in gas exchange. (Ch. 19, p. 529)

chorionic villus biopsy: removal of cells from the chorion to determine whether disease-causing genes are present in the fetus. (Ch. 11, p. 293)

chromatid: each strand of the double-stranded chromosome resulting from replication during prophase. (Ch. 7, p. 175)

chromatin: dense mass of material within the nucleus that is composed of individual chromosomes. (Ch. 5, p. 130)

chromosome: structure that carries the genes; composed of proteins and DNA. (Ch. 5, p. 130)

cilia: flexible projections extending outward from a cell that enable locomotion by whiplike motion; *Paramecium* move by the motion of cilia. (Ch. 5, p. 133)

class: taxonomic grouping of related (similar) orders of living organisms. (Ch. 14, p. 376)

cleavage: process of rapid cell division that splits the zygote into many smaller cells to form a multicellular blastula. (Ch. 19, p. 513)

climax community: in ecological succession, the final community, which is complex, stable, and tends to remain basically the same unless disturbed by nature or humans. (Ch. 29, p. 817)

clones: genetically identical copies derived from a single cell. (Ch. 10, p. 269)

codominance: occurs when one allele is not dominant; both alleles are expressed equally. (Ch. 8, p. 210)

codon: basic unit of the genetic code formed by sets of three adjacent bases representing an amino acid. (Ch. 9, p. 239)

coelom: fluid-filled body cavity formed within the mesoderm in which internal organs are suspended and protected. (Ch. 17, p. 457)

coenzyme: nonprotein, reusable helper molecule that helps enzymes activate chemical reactions. (Ch. 3, p. 83)

colon: organ that reclaims water by absorption during digestion, resulting in the remaining material being formed into feces, which is eliminated through the anus. (Ch. 20, p. 569)

commensalism: relationship in which only one organism benefits and the other organism (the host) is unaffected; both plants and animals can exhibit commensalism. (Ch. 28, p. 793)

community: assortment of life forms (for example, bacteria, fish, and algae) living together in a particular place and interacting with and depending on one another in various ways. (Ch. 1, p. 13)

comparative anatomy: study comparing structures of different living organisms (for example, forelimbs of a bird, horse, and human); provides evidence of evolution. (Ch. 12, p. 309)

comparative biochemistry: study of organisms on a biochemical level; compares amino acid sequences, providing evidence of evolution. (Ch. 12, p. 312)

comparative embryology: study of developing organisms showing relationships not obvious in the organism's adult stage. (Ch. 12, p. 312)

compound: substance with two or more atoms of different elements combined chemically; sucrose and table salt are compounds. (Ch. 3, p. 60)

cone cell: type of photoreceptor cell in the retina; distinguishes colors. (Ch. 25, p. 723)

conifer: gymnosperm whose seeds develop in cones; pine trees, junipers, and giant redwoods are conifers. (Ch. 16, p. 436)

conjugation: sexual process in some simple organisms such as fungi in which genetic material is transferred from one cell to another by cell-to-cell contact. (Ch. 18, p. 488)

consumer: organism in a community that is unable to produce its own food. (Ch. 1, p. 15)

continuous variation: gradations in traits (for example, varied lengths of ears of corn) as a result of the presence of many different phenotypes. (Ch. 8, p. 215)

control group: in a controlled experiment, this is the group in which all the variables remain constant, including the independent variable. (Ch. 2, p. 37)

controlled experiment: one in which all variables are kept constant except the one being tested, so the results observed are due to changes in the variable being manipulated. (Ch. 2, p. 36)

convergent evolution: independent development of similar traits between unrelated species as a result of adaptations to similar environments. (Ch. 13, p. 348)

cornea: front surface of the eye; admits light, which then passes through the aqueous humor to the pupil. (Ch. 25, p. 722)

corpus luteum: yellowish body developed from the ruptured follicle after ovulation. (Ch. 24, p. 691)

corpus luteum stage: stage of the menstrual cycle during which the level of progesterone is high and FSH and LH are inhibited by the pituitary. (Ch. 24, p. 691)

cotyledon: leaflike food-storage organ of the embryo of a flowering plant. (Ch. 19, p. 519)

covalent bond: occurs when atoms combine by sharing electrons. (Ch. 3, p. 59)

crop: stretchable food-storage organ found in animals such as earthworms. (Ch. 20, p. 552)

crop rotation: alternating crop growth in a field in order to replace nutrients such as nitrogen and to control pests. (Ch. 30, p. 865)

cultural control: agricultural alternative to pesticides involving growing methods to discourage pests and breeding pest-resistant plants. (Ch. 30, p. 849)

cuticle: waxy layer in plants such as ferns and conifers that helps protect against water loss. (Ch. 16, p. 425)

cytoplasm: substance contained by living cells that is composed of a complex mixture of enzymes, sugars, and amino acids; these substances are constantly involved in chemical reactions. (Ch. 5, p. 123)

cytosine: a nitrogen-containing base of DNA; the other three are adenine, guanine, and thymine. (Ch. 9, p. 228)

cytotoxic T cell: specialized lymphocyte that can kill infected body cells directly by releasing proteins and indirectly by attracting macrophages. (Ch. 23, p. 652)

data: collection of bits of information about the natural world; for example, measuring the height of plants in sunny and shady fields provides data. (Ch. 2, p. 33)

death rate: number of organisms that die during a given time span. (Ch. 27, p. 763)

decomposer: organism that derives nourishment by feeding on dead organisms and waste products of living organisms. (Ch. 1, p. 16)

deletion: type of gene mutation in which a nucleotide is left out; results spontaneously from errors in copying. (Ch. 10, p. 261)

dendrite: portion of a neuron that transmits impulses toward the cell body. (Ch. 25, p. 701)

dependent variable: in an experiment, any change that results from the manipulation of the independent variable. (Ch. 2, p. 37)

desert: biome characterized by less than 25 cm of rainfall yearly, rapid evaporation, poor soil, and little plant or animal life; desert organisms show many adaptations such as large, shallow root systems and estivation. (Ch. 29, p. 831)

development: series of changes an organism undergoes in reaching its final, adult form; one of four functions that distinguishes organisms as living creatures. (Ch. 1, p. 8)

diaphragm: wide, flat band of muscle stretched across the base of the rib cage that flattens during inhalation and pushes upward during exhalation. (Ch. 22, p. 622)

differentiation: developmental process by which generalized cells become specialized in order to form such structures as the brain and spinal cord. (Ch. 19, p. 517)

diffusion: random movement of ions and other particles from an area of higher concentration to an area of lower concentration; through air, water and other liquids, and across membranes. (Ch. 4, p. 98)

digestion: chemical, enzymatic process that breaks down large, complex, organic molecules into simpler, smaller ones capable of passing through plasma membranes. (Ch. 20, p. 545)

diploid: cells having two of each chromosome ($2n$). (Ch. 7, p. 183)

divergent evolution: the evolution of a single species into two or more species with different characteristics. (Ch. 13, p. 345)

dominant: form of a trait that dominates another trait and appears in the F_1 generation. (Ch. 8, p. 200)

dorsal: top or back of animals with bilateral symmetry (for example, segmented worms). (Ch. 17, p. 454)

dynamic equilibrium: in diffusion, the point reached when the numbers of particles entering and leaving a region are equal—movement is continuous but produces no overall change. (Ch. 4, p. 98)

E

ecological succession: gradual process by which communities change over time. (Ch. 29, p. 814)

ecosystem: relationship of a community of organisms interacting with its environment. (Ch. 28, p. 786)

ectoderm: outermost tissue layer of an organism. (Ch. 17, p. 454)

ectotherm: animal whose body temperature can change with the environment; ectotherms such as turtles develop behaviors like sunning to regulate their body temperature. (Ch. 17, p. 471)

electron cloud: space that electrons occupy while moving around the nucleus. (Ch. 3, p. 57)

element: substance composed of only one type of atom; examples include gold, lead, and oxygen. (Ch. 3, p. 58)

embryo: a living organism's early stage of development. (Ch. 12, p. 312)

embryonic induction: occurs when one part of an embryo induces (for example, by hormones) the development of another part. (Ch. 19, p. 524)

endergonic: chemical reaction requiring free energy in addition to enzymes and activation energy. (Ch. 6, p. 144)

endocrine gland: in animals, a ductless gland such as the pituitary that secretes hormones directly into the bloodstream. (Ch. 24, p. 680)

endocytosis: process by which the plasma membrane of a cell surrounds material, engulfs it, and takes in the material from its environment. (Ch. 4, p. 106)

endoderm: innermost tissue layer of an organism. (Ch. 17, p. 454)

endoplasmic reticulum: network of interconnected structures found in all eukaryotes; functions include intracellular transport of proteins and the breakdown of harmful substances. (Ch. 5, p. 126)

endoskeleton: in echinoderms (for example, starfish), calcium carbonate spines that project through the skin; provides support. (Ch. 17, p. 463) any internal framework or support system. (Ch. 26, p. 732)

endosperm: in plants, the $3n$ tissue that is the product of double fertilization and provides the embryo with food. (Ch. 18, p. 495)

endospore: highly resistant, dormant structure formed from a bacterial cell during unfavorable environmental conditions. (Ch. 15, p. 396)

endotherm: animal that maintains a constant body temperature through internal metabolic activity, regardless of external conditions; birds are endotherms. (Ch. 17, p. 471)

endotoxin: poisonous chemical released by dead bacteria when their cell walls rupture, producing disease such as tuberculosis. (Ch. 23, p. 642)

energy: the ability to do work. (Ch. 1, p. 14)

energy level: any of several different regions in which electrons travel about the nucleus. (Ch. 3, p. 57)

enzyme: reusable protein that lowers the required activation energy and allows reactions to happen at the normal temperatures of cells. (Ch. 3, p. 81)

epicotyl: in plants, a structure that forms a pair of small leaves that will open early in development. (Ch. 19, p. 519)

erosion: process by which soil is worn away by rain and wind. (Ch. 30, p. 865)

esophagus: muscular tube found in animals that pushes food by means of muscular contractions through the digestive system. (Ch. 20, p. 552)

estivation: period of summer dormancy that allows organisms to survive heat or drought. (Ch. 28, p. 803)

estrus: hormone-regulated receptive mating periods in many female mammals. (Ch. 18, p. 500)

eukaryote: cell containing a nuclear membrane and a membrane-bound nucleus; the vast majority of living organisms (for example, plants, animals, protists, and fungi) are eukaryotic. (Ch. 5, p. 125)

evolution: the change in living organisms over time. (Ch. 1, p. 20)

excretion: in animals and protozoans, the removal of toxic nitrogenous wastes from the body that are produced by metabolism. (Ch. 22, p. 623)

exergonic: chemical reaction that releases free energy. (Ch. 6, p. 145)

exocytosis: reverse process of endocytosis in which cell products or wastes are enclosed in vesicles and released to the surroundings. (Ch. 4, p. 108)

exons: sets of nucleotides that code for amino acids; during mRNA processing, exons are spliced together. (Ch. 10, p. 253)

exoskeleton: hard, external body covering that provides support for tissues and organs and protects the organism from predators; found in arthropods. (Ch. 17, p. 459)

exotoxin: poisonous chemical produced by living bacteria that causes disease (for example, tetanus). (Ch. 23, p. 641)

experiment: scientific test of a hypothesis used to determine whether or not the hypothesis (or explanation) is correct. (Ch. 2, p. 35)

experimental group: in a controlled experiment, the group in which the independent variable is manipulated. (Ch. 2, p. 37)

extensor: muscle (for example, the triceps) that causes straightening at a joint by contracting. (Ch. 26, p. 740)

extracellular digestion: in multicellular heterotrophs, complete or partial digestion of large pieces of food outside the cells. (Ch. 20, p. 547)

F

facilitated diffusion: use of transport proteins to help the passage of particles across the plasma membrane; a form of passive transport. (Ch. 4, p. 102)

family: taxonomic group of related (similar) genera of living organisms. (Ch. 14, p. 376)

fertilization: process of fusion of two haploid sex cells (egg and sperm) to form a zygote. (Ch. 7, p. 183)

fetoscopy: diagnostic technique used to directly view the developing fetus and surrounding tissue by means of an endoscopic tube. (Ch. 11, p. 295)

fiber: long, multinucleated type of cell surrounded by connective tissues, which together make up striated muscle. (Ch. 26, p. 743)

fibril: threadlike structure composed of filaments of actin and myosin; fibers are made up of fibrils. (Ch. 26, p. 745)

first-order consumer: consumer that feeds directly on a producer (for example, a cow grazing on grass); also called a herbivore and occupies the second trophic level. (Ch. 28, p. 788)

flagella: long, whiplike strands extending from a cell that enable cell locomotion. (Ch. 5, p. 133)

flexor: muscle (for example, the biceps) that causes bending at a joint by contracting. (Ch. 26, p. 740)

fluid mosaic model: a current model of cell membrane structure in which the membrane is a phospholipid bilayer with proteins embedded in it. (Ch. 4, p. 95)

follicles: the groups of cells within the ovary where eggs develop. (Ch. 18, p. 500)

follicle stage: during menstruation, the stage lasting about ten days when a new egg begins to mature in the follicle. (Ch. 24, p. 690)

food chain: series of steps from producers to consumers to decomposers by which food and energy are transferred through the environment. (Ch. 1, p. 15)

food web: all feeding relationships of organisms in an ecosystem. (Ch. 28, p. 788)

fossil: evidence (for example, bones or casts) of organisms that lived long ago. (Ch. 12, p. 306)

fragmentation: asexual reproduction in which whole, new adults are formed from fragments of the original organism. (Ch. 18, p. 486)

free energy: energy available to do biological work; for example, free energy is needed to synthesize proteins from single amino acids. (Ch. 6, p. 144)

fruit: in angiosperms (for example, corn, strawberries, onions), the vessel that contains the seeds; an enlarged ovary. (Ch. 16, p. 439)

G

gallbladder: small, saclike digestive organ that stores bile from the liver and secretes it into the small intestine. (Ch. 20, p. 563)

gamete: a sex cell—a sperm in the male parent and an egg in the female parent. (Ch. 7, p. 183)

gametophyte generation: gamete-producing haploid (*n*) stage of all plants. (Ch. 16, p. 429)

gastrovascular cavity: fluid-filled digestive cavity with only one body opening. (Ch. 20, p. 551)

gastrula: cup-shaped form of the developing embryo; contains ectoderm, endoderm, and mesoderm. (Ch. 19, p. 515)

gene: factor containing information for every trait; genes can be dominant or recessive. (Ch. 8, p. 201)

gene pool: total collection of genes, including the new alleles that occur and are recombined, within a specific population. (Ch. 12, p. 322)

gene therapy: treatment of genetic disorders through the introduction of normal genes into body cells. (Ch. 11, p. 296)

genetic counseling: gives parents information concerning the chances of having a child with a genetic disorder based on family history, patterns of heredity, and new diagnostic techniques. (Ch. 11, p. 290)

genetic drift: random changes in the gene pool of a population; a driving force in speciation. (Ch. 13, p. 345)

genetic recombination: the formation of new gene combinations during meiosis and sexual reproduction. (Ch. 7, p. 189)

genetics: science of inheritance of traits. (Ch. 8, p. 198)

genotype: combination of alleles for a certain trait. (Ch. 8, p. 203)

genus: taxonomic grouping of related species. (Ch. 14, p. 368)

geographic isolation: geological change that isolates segments of a population. (Ch. 13, p. 344)

geothermal energy: energy of ocean floor communities that is released as heat energy from Earth's interior; the hot water reacts with minerals in the molten rock to produce hydrogen sulfide, which is used by chemosynthetic bacteria to make carbohydrates. (Ch. 29, p. 835)

germination: development of a seed into a new plant; occurs when environmental conditions are favorable. (Ch. 19, p. 519)

gibberellin: plant hormone that promotes growth and development (for example, stimulates the germination rate of seeds). (Ch. 24, p. 679)

gill: in aquatic animals (for example, fish), the respiratory organ in which gas exchange takes place. (Ch. 22, p. 616)

GLOSSARY

gizzard: muscular, thick-walled chamber found in animals such as earthworms; with the help of coarse particles ingested with food, grinds food into smaller pieces. (Ch. 20, p. 552)

glomerulus: cluster of capillaries in Bowman's capsule; involved in filtration. (Ch. 22, p. 627)

Golgi body: flattened, saclike organelle that functions as a processing, packaging, and delivery system in eukaryotes. (Ch. 5, p. 127)

gradualism: evolutionary change in a species that progresses at a slow and steady rate. (Ch. 13, p. 350)

grana: in chloroplasts, stacked internal membranes (thylakoids) at which the light reactions of photosynthesis take place. (Ch. 6, p. 156)

grassland: biome in which grasses are the major species; also known as steppe, prairie, or savanna, and can include such animals as antelopes, bison, and prairie dogs. (Ch. 29, p. 829)

greenhouse effect: results when carbon dioxide traps heat near Earth's surface; could be producing global warming, which may eventually have a serious effect on climate. (Ch. 30, p. 859)

growth: an increase in the amount of living material in an organism; a function carried out by all organisms. (Ch. 1, p. 8)

guanine: a nitrogen-containing base of DNA; the other three are adenine, thymine, and cytosine. (Ch. 9, p. 228)

guard cells: specialized cells in plants that surround each stoma, regulating its size and preventing the plant from drying out. (Ch. 21, p. 586)

gymnosperms: "naked seeds"; plants in which seeds develop on the woody scales of reproductive structures that form cones (for example, pine and spruce trees). (Ch. 16, p. 436)

H

haploid: cells having only one chromosome (*n*) from each homologous pair. (Ch. 7, p. 183)

Haversian canal: tiny channel in compact bone that carries blood vessels and nerves. (Ch. 26, p. 735)

helper T cell: specialized lymphocyte that binds with B cells and can release chemicals to stimulate B cells to form plasma cells. (Ch. 23, p. 650)

herbivore: plant-eating animal, such as a cow, that occupies the second trophic level in an ecosystem. (Ch. 28, p. 788)

heredity: transmission of traits from parents to offspring. (Ch. 8, p. 198)

hermaphrodite: an organism with both male and female reproductive organs; earthworms are hermaphrodites. (Ch. 18, p. 497)

heterotroph: organism requiring complex organic molecules for energy, and that must feed on matter produced by other organisms. (Ch. 12, p. 327)

heterozygous: organism in which each cell contains two different alleles for a certain trait. (Ch. 8, p. 203)

homeostasis: steady state of the internal operation of a living organism regardless of external changes. (Ch. 1, p. 9)

homologous: similar in structure, in function, or both; the more homologous structures shared by different kinds of organisms, the closer their relationship. (Ch. 12, p. 310)

homologues: in diploid cells, the two members of each pair of chromosomes that carry DNA with information for corresponding traits. (Ch. 7, p. 183)

homozygous: organism in which each cell contains two of the same alleles for a particular trait. (Ch. 8, p. 203)

hormone: chemical secreted by one part of an organism that affects another part of the organism. (Ch. 18, p. 500)

humus: decayed remains of organisms that contain organic material, enrich soil, and increase the capacity of the soil to hold air and water. (Ch. 28, p. 798)

hyphae: multinucleate mass of cytoplasm forming the basic structure of fungi. (Ch. 15, p. 410)

hypocotyl: stalklike part of embryo that will eventually become the upper part of the primary root and the lower part of the stem. (Ch. 19, p. 519)

hypothalamus: brain structure that secretes releasing factors that cause the anterior pituitary to release its own hormones. (Ch. 24, p. 686)

hypothesis: statement formed by analyzing observations and sorting through possible causes of the effect noted. (Ch. 2, p. 34)

I

incomplete dominance: pattern of inheritance in which neither allele is dominant and three totally different phenotypes can occur. (Ch. 8, p. 209)

independent variable: the only variable to be manipulated (or changed) in a controlled experiment. (Ch. 2, p. 37)

insertion: type of mutation in which an extra nucleotide is added during replication; produces very different amino acid sequences from those coded by normal genes. (Ch. 10, p. 261)

integrated pest management: use of biological, autocidal, and cultural controls in addition to chemical pesticides to control pest populations. (Ch. 30, p. 850)

interferon: protein made by a virus-infected host cell that prevents viral reproduction in healthy cells. (Ch. 23, p. 660)

International System of Measurement, (SI): a universal language of measurements and its symbols used by scientists; a decimal system based on tens, multiples of tens, and fractions of tens. (Ch. 2, p. 46)

interneuron: neuron that carries impulses from sensory neurons to motor neurons. (Ch. 25, p. 701)

interphase: cell division stage in which each chromosome and the DNA it contains replicates, producing a second set of chromosomes. (Ch. 7, p. 173)

GLOSSARY

interspecific competition: interaction that takes place when populations of different species compete for the same resources (for example, food); density-dependent limiting factor. (Ch. 27, p. 771)

intracellular digestion: efficient means of digestion that takes place inside a cell; used by heterotrophs such as amoebas to break down microorganisms and starch molecules. (Ch. 20, p. 547)

intraspecific competition: density-dependent competition between members of the same species for the same resources (for example, water). (Ch. 27, p. 771)

introns: sets of nucleotides that do not code for amino acids; during mRNA processing, introns are cut out. (Ch. 10, p. 253).

invertebrates: animals without backbones—examples are sponges and cnidarians. (Ch. 17, p. 450)

ion: a charged atom (or group of atoms) resulting from an atom's loss or gain of one or more negative charges (electrons). (Ch. 3, p. 61)

ionic bond: force of attraction between ions; for example, because sodium and chloride ions are oppositely charged, they attract one another. (Ch. 3, p. 61)

isomers: organic molecules with the same chemical formula but different structural formula. (Ch. 3, p. 71)

J

joint: in arthropods, the flexible connector located between segments of appendages and between segments of exoskeleton; in vertebrates, the points where bones connect with one another; makes body movement possible. (Ch. 26, p. 731)

K

karyotype: appearance of chromosomes in an organism regarding number, size, and shape; used to detect certain chromosomal abnormalities. (Ch. 11, p. 288)

kidney: main excretory organ that filters out nitrogenous wastes and helps maintain osmotic balance. (Ch. 22, p. 627)

kinetic energy: the energy of motion. (Ch. 3, p. 79)

kingdom: broadest taxon in the classification of living organisms; divided into more specific taxa: phylum, class, order, family, genus, and species. (Ch. 14, p. 376)

L

lactic acid fermentation: anaerobic process whereby enzymes break down glucose into lactic acid and transfer energy to ATP; sauerkraut is produced by lactic acid fermentation. (Ch. 6, p. 150)

large intestine: digestive organ that collects undigested waste, absorbs water from the waste, and compacts the waste. (Ch. 20, p. 569)

larva: immature stage that usually does not resemble the adult form but will eventually change into that form; tadpoles are larvae. (Ch. 19, p. 525)

law: a statement about events that always occur in nature; for example, the law of gravitation. (Ch. 2, p. 42)

lens: eye structure that focuses light on the retina. (Ch. 25, p. 723)

ligament: strong band of tissue composed of collagen and elastic fibers that connects bones. (Ch. 26, p. 736)

light reaction: in photosynthesis, the series of changes that occurs only in the presence of light, converting light energy into chemical energy. (Ch. 6, p. 156)

limiting factor: unfavorable factor such as temperature, disease, or predation that prevents organisms from achieving their biotic potential. (Ch. 27, p. 761)

lipid: organic compound composed of carbon, hydrogen, and oxygen, with the number of hydrogen atoms per molecule being much greater than the number of oxygen atoms; fats, oils, and waxes are lipids. (Ch. 3, p. 73)

littoral zone: in the ocean, the region of the neritic zone that is closest to the shore; organisms adapted to changing environmental conditions abound including algae, clams, and crabs. (Ch. 29, p. 833)

lung: in terrestrial vertebrates, organ used for gas exchange. (Ch. 22, p. 619)

lymph node: small, spongy capsule scattered through the lymphatic system in which destruction of bacteria occurs. (Ch. 21, p. 599)

lymphocyte: white blood cell involved in the immune response; may be either a B cell or T cell. (Ch. 23, p. 646)

lysogenic cycle: occurs when a cell containing a provirus reproduces—the host cell lyses and new viruses are released to become part of the chromosomes of other cells. (Ch. 15, p. 392)

lysosome: organelle containing digestive enzymes that break down food and digest worn out cell parts. (Ch. 5, p. 132)

lytic cycle: reproductive process in which a virus takes over all metabolic activities of a cell, replicates itself many times, and destroys the host cell. (Ch. 15, p. 391)

M

macrophage: white blood cell that engulfs and digests pathogens; plays an important role in inflammatory and immune responses. (Ch. 23, p. 645)

Malpighian tubules: in a grasshopper, stringlike excretory system that absorbs uric acid and water from blood. (Ch. 22, p. 627)

medulla oblongata: center portion of the brain; controls heartbeat rate and many other internal responses. (Ch. 25, p. 715)

meiosis: cellular process that occurs only in certain cells of reproductive organs by which haploid nuclei are formed from diploid nuclei; ensures genetic continuity between generations and promotes variation. (Ch. 7, p. 183)

memory cell: produced during the first exposure to an antigen; if a pathogen reappears, memory cells recognize its antigens and the immune system responds quickly. (Ch. 23, p. 657)

menstrual cycle: in human females, the series of monthly hormone-regulated changes that results in the maturation of an egg and preparation of the uterus for possible pregnancy. (Ch. 18, p. 502)

menstruation: phase of the menstrual cycle involving the loss of the egg and uterine tissue through the vagina. (Ch. 18, p. 502)

meristems: regions of cell division that in many plants can develop new structures throughout the plant's lifetime. (Ch. 19, p. 519)

mesoderm: in flatworms, the tissue layer between the ectoderm and endoderm containing several primitive organ systems; in animals with coeloms, the tissue layer in which the coelom forms and from which some organs develop. (Ch. 17, p. 455)

messenger RNA (mRNA): carries the genetic information representing a certain order of amino acids from the nucleus to the cytoplasm where translation takes place; RNA is transcribed from DNA. (Ch. 10, p. 252)

metabolism: sum of all chemical reactions that occur in cells; hydrolysis is one such chemical reaction. (Ch. 5, p. 123)

metamorphosis: various patterns of development that involve a series of changes in body structure—for example, some insects have three stages of change: egg, nymph, and adult. (Ch. 19, p. 525)

metaphase: second stage of mitosis during which the chromosomes gather at the equator of the spindle and become aligned. (Ch. 7, p. 176)

microenvironment: during each stage of succession, the slightly different environment created as organisms grow and affect their surroundings, which allows for the growth of other kinds of organisms better adapted to the changed conditions. (Ch. 29, p. 816)

microfilament: structure of the cytoskeleton in eukaryotes; composed of actin and myosin, it provides structural support and assists cell movement (for example, muscle movement). (Ch. 5, p. 133)

microtubule: structure of the cytoskeleton in eukaryotes that helps certain organelles move within the cell. (Ch. 5, p. 133)

middle lamella: in plants, the secondary wall, which lies inside the primary wall; composed of pectin; it strengthens the cell wall and provides support. (Ch. 4, p. 110)

mitochondria: organelles, bound by a double membrane, found in all eukaryotes; mitochondria break down organic molecules to release energy for cell reactions. (Ch. 5, p. 128)

mitosis: process by which chomosomes are duplicated and distributed to daughter nuclei; each daughter nucleus carries the same set of genetic information as the parent nucleus. (Ch. 7, p. 173)

modifier genes: genes interacting with each other to control particular traits, such as eye color. (Ch. 8, p. 217)

molecule: combination of two or more atoms joined by a covalent bond; water is a molecule. (Ch. 3, p. 59)

monoclonal antibody: specific antibody produced in the laboratory from the fusion of cancer cells and lymphocytes. (Ch. 23, p. 655)

morphogenesis: development of the species-specific form or structures of an organism. (Ch. 19, p. 514)

motor neuron: specialized cell that carries impulses from the brain or spinal cord to an effector (typically a muscle or gland). (Ch. 25, p. 701)

multiple alleles: set of three or more different alleles that control a trait—for example, the three alleles in humans that determine blood type are multiple alleles. (Ch. 8, p. 210)

multiple genes: pattern of inheritance (continuous variation) produced by many genes affecting a particular trait (for example, skin color in humans). (Ch. 8, p. 216)

mutagen: external chemical or physical agent such as radiation that causes mutations; some mutagens play a role in cancer development. (Ch. 10, p. 262)

mutation: a change (mistake) in the replication of genetic material; mutations occur in all living organisms as a result of changes to individual genes or the gene arrangement on chromosomes. (Ch. 10, p. 261)

mutualism: mutually beneficial relationship between different kinds of organisms. (Ch. 15, p. 405)

mycelium: mat of filaments composed of a mass of hyphae that form the body of fungi. (Ch. 15, p. 410)

N

natural selection: process resulting in the survival of the best-adapted individuals in a population. (Ch. 12, p. 320)

negative feedback: control mechanism whereby various body processes are kept at levels suitable to the body's changing needs; often involves the release of hormones to speed up or slow down other chemical reactions in the body. (Ch. 24, p. 682)

nekton: aquatic organisms that swim freely through the ocean; whales, fish, shrimp, and squid are nektonic animals. (Ch. 29, p. 833)

nephridium: tubular excretory organ of an earthworm, which acts as a filter, removing wastes and returning usable materials to the circulation system. (Ch. 22, p. 626)

nephron: functional unit of the kidney; the human kidney has about 1 million nephrons. (Ch. 22, p. 627)

neritic zone: shallow, lighted zone of the ocean, above the continental shelf, with diverse and plentiful animal life. (Ch. 29, p. 833)

nerve: bundle of neuron axons with accompanying connective tissue. (Ch. 25, p. 701)

neural tube: hollow tube formed by the folding and fusing of the neural plate; develops into the brain and spinal cord in the early vertebrate embryo. (Ch. 19, p. 517)

neuron: specialized nerve cell, consisting of a cell body, dendrites, and an axon, that transmits impulses from one part of the body to another. (Ch. 25, p. 700)

neurotransmitter: chemical agent that conducts nerve impulses across the synapse into specific sites on nearby dendrites. (Ch. 25, p. 707)

niche: in an ecosystem, the unique place occupied by an organism in relationship to all the biotic and abiotic factors. (Ch. 28, p. 787)

nonbiodegradable: chemical or other substance that living organisms cannot break down or degrade into harmless compounds; DDT, a pesticide, is nonbiodegradable. (Ch. 30, p. 848)

nonrenewable resource: resource such as fossil fuels that cannot be replaced. (Ch. 30, p. 864)

nucleic acid: biological compound (for example, DNA and RNA) that forms a code to determine a living thing's appearance and behavior. (Ch. 3, p. 76)

nucleoli: prominent bodies found in the nucleus of eukaryotes; the site of RNA synthesis. (Ch. 5, p. 130)

nucleotide: molecule composed of a phosphate group, a sugar molecule, and one of four nitrogen-containing bases (adenine, guanine, thymine, or cytosine). (Ch. 9, p. 229)

nucleus: control area of eukaryotic cells; contains chromatin and a nucleolus. (Ch. 5, p. 130)

nutrients: complex organic molecules and simpler substances needed by living organisms to carry out such life processes as respiration, protein synthesis, and reproduction. (Ch. 20, p. 544)

nymph: immature developmental stage of some insects (for example, grasshoppers) that closely resembles the adult stage. (Ch. 19, p. 528)

O

oceanic zone: deepest region of the ocean in which there is no light, no photosynthesis; populated by some nektonic animals, mostly fish. (Ch. 29, p. 833)

omnivore: organism that eats both plants and animals; humans are omnivores. (Ch. 28, p. 788)

oncogene: gene that has mutated into a cancer-causing gene; descendant cells will inherit the oncogene, which will disrupt the organism's cell cycle, causing cancer. (Ch. 10, p. 267)

order: taxonomic grouping of related (similar) families of living organisms. (Ch. 14, p. 376)

organ: a group of different tissues that work together to carry out a particular function; the stomach is an example. (Ch. 5, p. 122)

organism: anything capable of carrying on the life processes of reproduction, growth and development, homeostasis, and organization; more than 2 million organisms have been identified. (Ch. 1, p. 12)

organization: orderly, interrelated system of vital life processes or functions performed by a living organism; for example, different kinds of cells group together to make up structures with complex organization. (Ch. 1, p. 9)

osmosis: diffusion of water into and out of cells through a selectively permeable membrane from a region of higher concentration to a region of lower concentration. (Ch. 4, p. 100)

ossification: mineral-requiring process by which bone replaces cartilage as the skeleton develops. (Ch. 26, p. 734)

osteocyte: type of bone cell found in most formed bone tissue; the mature form of an osteoblast. (Ch. 26, p. 735)

ovary: structure in a flower's pistil in which the female gametes are produced. (Ch. 16, p. 441) in sexually reproducing animals, the structure that produces eggs. (Ch. 18, p. 500)

oviduct: tube that transports eggs to the uterus. (Ch. 18, p. 501)

ovulation: release of the immature egg from a follicle. (Ch. 18, p. 501)

ovule: part of a plant ovary that will develop into a seed if fertilization occurs. (Ch. 18, p. 493)

oxygen debt: amount of oxygen needed to metabolize lactic acid after exercise. (Ch. 26, p. 748)

ozone layer: found at the top of the stratosphere, the layer of ozone gas that absorbs almost all the UV radiation that reaches Earth from the sun. (Ch. 12, p. 329)

P

pancreas: digestive organ behind and partly below the stomach that secretes enzyme-containing pancreatic juice to break down polypeptides, convert nucleic acids to nucleotides, etc. (Ch. 20, p. 564)

parasite: organism living on or in another organism (the host) in order to obtain nourishment; parasites always harm the host. (Ch. 15, p. 379)

parasitism: density-dependent (often limiting) factor for population size in which harmful organisms live on or in a host organism, from which they draw nourishment. (Ch. 27, p. 769)

passive immunity: temporary immunity acquired by transfer of antibodies to the recipient. (Ch. 23, p. 660)

passive transport: movement of particles across membranes from an area of greater concentration to an area of lesser concentration as a result of a net random motion; diffusion and osmosis are examples. (Ch. 4, p. 100)

pathogen: disease-causing microorganism (for example, bacteria such as *Treponema pallidum,* which causes syphilis). (Ch. 23, p. 640)

peripheral nervous system: the sensory and motor neurons connected to the CNS in bilaterally symmetrical animals. (Ch. 25, p. 712)

peristalsis: alternating waves of muscular contraction and relaxation that push food through the esophagus and into the stomach. (Ch. 20, p. 561)

permafrost: permanently frozen soil layer of the tundra. (Ch. 29, p. 824)

pesticide: chemical agent used to kill pests such as termites, moths, and aphids; pesticides often destroy not only the target pests but also their natural predators. (Ch. 30, p. 845)

petals: a flower's inner row or rows of leaflike structures; often highly perfumed, brightly colored structures that dispense nectar. (Ch. 16, p. 440)

phagocytosis: process by which the cell takes in solid chunks of material; a form of endocytosis. (Ch. 4, p. 107)

pharynx: in animals such as earthworms, the muscular tube that moves food from the mouth to the esophagus by means of muscular contractions. (Ch. 20, p. 552)

phenotype: physical appearance of an organism as determined by a particular genotype. (Ch. 8, p. 203)

phloem: one of two types of vascular tissue in plants; conducts some of the sugars produced during photosynthesis. (Ch. 21, p. 580)

photosynthesis: process by which carbon dioxide and water are combined in the presence of sunlight and water to produce simple sugars and give off oxygen as a by-product. (Ch. 1, p. 14)

phylogeny: evolutionary history of a living organism based on such factors as fossil evidence and biochemical comparisons of enzymes. (Ch. 14, p. 374)

phylum: taxonomic grouping of related (similar) classes of living organisms. (Ch. 14, p. 376)

physiological adaptation: adaptation with a chemical basis that is associated with an organism's function; the spider's web is an example. (Ch. 13, p. 340)

pinocytosis: process by which a cell takes liquid droplets into itself; a form of endocytosis. (Ch. 4, p. 107)

pioneer stage: first stage of primary succession that begins with hardy organisms that can grow and reproduce under adverse conditions. (Ch. 29, p. 815)

pistil: a flower's female reproductive organ; produces female gametes. (Ch. 16, p. 441)

pituitary gland: three-lobed endocrine gland that works with the hypothalamus to release activity-regulating hormones such as vasopressin and oxytocin. (Ch. 24, p. 685)

placenta: tissue formed for exchange of gases, nutrients, and wastes between the mother and embryo. (Ch. 19, p. 530)

plankton: aquatic organisms, such as copepods and jellyfish, that are carried by ocean currents rather than swimming freely. (Ch. 29, p. 832)

plasma cell: antibody-producing cell resulting from division of a B cell. (Ch. 23, p. 649)

plasma membrane: outer boundary of a cell that regulates which particles enter and leave the cell. (Ch. 4, p. 92)

plasmid: structure in the cytoplasm of a bacterium with which foreign DNA is combined in genetic engineering techniques. (Ch. 10, p. 268)

plasmodium: slimy multinucleate mass with an amoeba-like feeding stage that is produced by a slime mold as part of its life cycle; engulfs bacteria and organic matter by phagocytosis. (Ch. 15, p. 407)

platelet: cell fragment involved in blood clotting. (Ch. 21, p. 603)

point mutation: type of gene mutation affecting only one amino acid in the polypeptide. (Ch. 10, p. 262)

pollen: protective structure enclosing the sperm of all seed plants; the pollen grain is often distributed by wind or air currents. (Ch. 16, p. 437)

pollination: sexual reproduction in plants in which pollen is transferred from anther to stigma either by cross-pollination or self-pollination. (Ch. 18, p. 494)

pollution: contamination of the air, water, or soil by such industrial or agricultural by-products as pesticides, automobile exhausts, and factory emissions. (Ch. 30, p. 844)

polypeptide: many amino acids bonded together form a polypeptide. (Ch. 3, p. 75)

polysaccharide: the most complex carbohydrate—such as cellulose, glycogen, or starch—which is formed by long chains of monosaccharides. (Ch. 3, p. 73)

population: all organisms within a breeding group. (Ch. 12, p. 313)

population density: number of one kind of organism in a given area. (Ch. 27, p. 764)

population growth: increase in population size over time; under ideal conditions (without being held in check by limiting factors), growth would increase indefinitely. (Ch. 27, p. 760)

population growth curve: used to graph changes in the size of a population. (Ch. 27, p. 762)

population growth rate curve: graph to show how fast a population grows. (Ch. 27, p. 763)

population size: number of organisms belonging to a population group; increases as organisms reproduce under favorable conditions. (Ch. 27, p. 760)

posterior: back end of animals with bilateral symmetry (for example, roundworms); generally contains a tail. (Ch. 17, p. 454)

potential energy: stored energy or energy of position; an example is the potential chemical energy of unlit wood in a fireplace. (Ch. 3, p. 79)

predation: feeding of one organism on another; the size of most populations is at least partially controlled by predation. (Ch. 27, p. 767)

primary succession: succession that begins in areas in which there is no life, such as barren rock. (Ch. 29, p. 815)

producer: food-producing member of a community; examples include plants and algae. (Ch. 1, p. 14)

GLOSSARY

prokaryote: cell that lacks a membrane-bound nucleus and membrane-bound organelles; bacteria are prokaryotes. (Ch. 5, p. 125)

prophase: first stage of mitosis in which the nucleoli disintegrate and the chromatin shortens and thickens. (Ch. 7, p. 175)

proteins: organic molecules used to build living material, carry out chemical reactions, fight disease, or transport particles out of cells. (Ch. 3, p. 75)

provirus: occurs when viral nucleic acid becomes part of the host cell's chromosomes; may cause phenotypic changes in the host but does not destroy the host. (Ch. 15, p. 392)

pseudopodia: "false feet," or extensions of the plasma membrane of amoebas that are used for locomotion and to engulf food. (Ch. 15, p. 404)

punctuated equilibrium: rapid evolutionary change in a species as the result of a sudden change in the environment. (Ch. 13, p. 350)

pupa: developmental stage of most insects (for example, butterflies) during which time the tissues are reorganized; the adult form emerges from the pupa. (Ch. 19, p. 527)

pyramid of biomass: diagram that shows how much living matter can be supported by an ecosystem. (Ch. 28, p. 791)

pyramid of energy: diagram of the transfer of energy in a food chain that shows the energy levels within an ecosystem. (Ch. 28, p. 789)

pyramid of numbers: diagram of energy loss in a food chain that shows how many individual organisms an ecosystem can support. (Ch. 28, p. 790)

R

radial symmetry: body plan in which all body parts of an organism are arranged equally around a center point; cnidarians exhibit radial symmetry. (Ch. 17, p. 454)

radicle: first part of the embryo to emerge from the seed; becomes the primary root of the plant. (Ch. 19, p. 519)

receptor: a protein embedded in the membrane of T cells that can recognize self proteins on B cells. (Ch. 23, p. 650)

recessive: the form of a trait that is hidden in the F_1 generation. (Ch. 8, p. 200)

recombinant DNA: DNA of an altered bacterium through insertion of a foreign gene; recombinant DNA techniques can engineer *E. coli* to produce insulin. (Ch. 10, p. 268)

red blood cell: most numerous blood cell; carries hemoglobin, which functions in oxygen transport. (Ch. 21, p. 602)

reflex: involuntary or automatic response to a stimulus—for example, blinking. (Ch. 25, p. 715)

reflex arc: in the nervous system, the path a nerve impulse takes during a reflex action. (Ch. 25, p. 718)

regeneration: replacement or regrowth of tissue in plants and animals. (Ch. 18, p. 486)

renewable resource: resource such as food that can be replaced. (Ch. 30, p. 864)

reproduction: production of offspring; living things reproduce, but nonliving things do not. (Ch. 1, p. 8)

reproductive isolation: results from the inability to exchange genes among species. (Ch. 13, p. 345)

resolving power: ability of a microscope to distinguish two objects as separate things. (Ch. 2, p. 44)

respiration: in complex animals, the intake of oxygen and the discharge of carbon dioxide. (Ch. 22, p. 615)

resting potential: unequal distribution of ions inside and outside the cell membrane (difference in electrical charge) that polarizes a neuron. (Ch. 25, p. 704)

retina: innermost layer of the eye; contains rod cells and cone cells. (Ch. 25, p. 723)

retrovirus: RNA virus that performs transcription from RNA to DNA; the virus that causes AIDS is a retrovirus. (Ch. 15, p. 394)

rhizoid: rootlike organ of zygote fungi; rhizoid enzymes break down food molecules into simpler molecules. (Ch. 15, p. 411)

rhizome: in vascular plants, a thick, underground stem that functions as a food storage organ. (Ch. 16, p. 434)

ribosomal RNA (rRNA): the RNA present in ribosomes; rRNA is transcribed from DNA. (Ch. 10, p. 252)

ribosomes: organelle composed of RNA and protein; site where proteins are made from amino acids. (Ch. 5, p. 126)

rod cell: type of photoreceptor cell in the retina; sensitive to very dim light but does not distinguish colors. (Ch. 25, p. 723)

root hairs: tiny, fine outgrowths of the epidermal cells of plants; provide a large surface area for absorbing water and dissolved ions. (Ch. 21, p. 581)

S

salivary gland: gland of the digestive system that secretes saliva into the mouth through ducts. (Ch. 20, p. 560)

saprophyte: organism that feeds on decaying organic matter; saprophytes such as some heterotrophic bacteria help recycle nutrients and prevent decayed matter from enveloping Earth. (Ch. 15, p. 397)

sarcomere: in a skeletal muscle, the structural subunit from one Z line to the next. (Ch. 26, p. 745)

scanning electron microscope (SEM): produces a three-dimensional image by sweeping a beam of electrons across an object and then bouncing off; used to study cell interaction and cell surfaces. (Ch. 2, p. 45)

science: looking at and learning about the natural world and producing a body of knowledge about nature; a process based on observing, interpreting observations, explaining observations, and testing the explanation. (Ch. 2, p. 33)

scrotum: sac of skin outside the male body wall that contains the testes, thereby providing the lower body temperature needed by sperm. (Ch. 18, p. 501)

second messenger hypothesis: model used to describe a cellular response wherein the first messenger never enters the target cell and the second messenger causes the response. (Ch. 24, p. 688)

second-order consumer: carnivorous consumer such as a snake that feeds directly on first-order consumers such as mice. (Ch. 28, p. 788)

secondary succession: new succession beginning when the dominant plants of a plant community are removed and the environment is left untended; can be caused by natural destruction such as fire or activities of humans such as logging. (Ch. 29, p. 817)

seed: embryonic plant consisting of the embryo, a food supply, and a protective coat; seed-bearing plants are the most numerous and diverse plants today. (Ch. 16, p. 435)

selective breeding: directing evolution by selecting only individuals with the most desirable traits and breeding them to pass these traits to the offspring. (Ch. 12, p. 313)

selectively permeable membrane: a membrane that is able to expel or admit particles used by the cell. (Ch. 4, p. 92)

semen: sperm and sperm-carrying fluids. (Ch. 18, p. 501)

sensory neuron: specialized cell that transmits impulses from sensory receptors to the brain or spinal cord for interpretation and processing. (Ch. 25, p. 700)

sensory-somatic system: part of the human peripheral nervous system that controls both conscious and involuntary responses to stimuli. (Ch. 25, p. 715)

sepal: a flower's outer row of leaves whose function is to enclose the bud and protect it. (Ch. 16, p. 440)

sex chromosomes: X and Y chromosomes that are different between the sexes and that are involved in sex determination. (Ch. 8, p. 213)

sex-linked trait: an inherited trait, such as eye color in fruitflies, determined by a gene on the sex chromosome. (Ch. 8, p. 214)

sexual reproduction: reproduction involving production and fusion of egg and sperm to form a diploid zygote, which will develop into a mature multicellular adult. (Ch. 7, p. 183)

sieve tube: series of sieve cells that form a pipeline and conduct sugars in plants. (Ch. 21, p. 580)

sinoatrial (S-A) node: small bundle of tissue (the pacemaker) in the right atrium of the vertebrate heart in which the heartbeat originates. (Ch. 21, p. 596)

sliding filament hypothesis: a model suggesting that muscle filaments slide over each other, shortening the fibrils during contraction of muscles. (Ch. 26, p. 746)

small intestine: longest part of the digestive tract; completes the chemical digestion of food and most of the absorption of nutrients. (Ch. 20, p. 563)

smooth muscle: nonstriated, involuntary muscle of vertebrates that lines the hollow internal organs of the body. (Ch. 26, p. 744)

social hierarchy: system of authority relationships ("pecking order") that is based on dominance and spread throughout a society. (Ch. 27, p. 773)

soil depletion: occurs when soil nutrients are not replaced in soil cultivated for crops. (Ch. 30, p. 865)

solution: a mixture of different substances in which every part of the mixture is the same. (Ch. 3, p. 65)

speciation: process by which new species are formed. (Ch. 13, p. 344)

species: group of organisms that can interbreed and produce fertile offspring; can include a broad range of variations. (Ch. 13, p. 342)

spindle: in newly dividing cells, the football-shaped structure formed from microtubules; functions to move the chromosomes during mitosis. (Ch. 7, p. 175)

spontaneous generation: disproved hypothesis that living organisms can arise from nonliving things. (Ch. 7, p. 170)

sporangium: reproductive structure of plants that produces spores. (Ch. 16, p. 432)

spore: reproductive cell capable of developing into a new haploid organism without first fusing with another cell. (Ch. 7, p. 184)

sporophyte generation: spore-producing diploid ($2n$) stage of all plants. (Ch. 16, p. 429)

stamen: a flower's male reproductive organ; produces male gametes and pollen. (Ch. 16, p. 441)

stigma: in a flower, the rough or sticky structure found at the tip of the style. (Ch. 16, p. 441)

stimulus: any change in the environment (for example, light) that influences an organism's activity. (Ch. 24, p. 672)

stoma: tiny opening in the epidermis of plants through which gases are exchanged. (Ch. 21, p. 583)

stomach: large, hollow organ in the digestive tract that churns food, secretes acid and enzymes to begin breaking up protein, and stores food. (Ch. 20, p. 561)

striated muscle: voluntary muscles with a striped appearance that are attached to bone and are important in vertebrate locomotion. (Ch. 26, p. 742)

stroma: fluid part of the chloroplast containing molecules used to synthesize sugars. (Ch. 6, p. 158)

structural adaptation: internal or external anatomic adaptation by an organism that promotes its fitness in its environment; the long, narrow tongue of woodpeckers, which is helpful in obtaining food, is an example. (Ch. 13, p. 339)

style: the neck of a flower's ovary. (Ch. 16, p. 441)

suppressor T cell: specialized lymphocyte that can release chemicals that reduce the response of macrophages and B cells. (Ch. 23, p. 651)

symbiosis: living together of two dissimilar organisms for their mutual benefit. (Ch. 5, p. 136)

synapse: small, fluid-filled space between neurons across which impulses are chemically transmitted. (Ch. 25, p. 707)

GLOSSARY

system: group of organs that depend on other organs to complete a biological process such as digestion, respiration, or reproduction. (Ch. 5, p. 122)

T

T cell: lymphocyte that matures in the thymus gland; can defend against body cells in which viruses are reproducing. (Ch. 23, p. 646)

taiga: biome south of the tundra characterized by coniferous forests, frequent fog, and acidic soil; typical animals include moose, caribou, and mink. (Ch. 29, p. 825)

taxa: particular categories in the biological classification system created for all living organisms and used to help organize information about the natural world. (Ch. 14, p. 366)

taxonomy: science of classification of living organisms that creates a hierarchy based on their similarities and differences. (Ch. 14, p. 366)

technology: application of knowledge gained through scientific study to real problems. (Ch. 2, p. 33)

telophase: fourth state of mitosis during which two cells begin to form from one; the nucleoli reappear and a nuclear membrane encloses each set of chromosomes. (Ch. 7, p. 176)

temperate forest: biome south of the taiga characterized by definite seasons (a growing season at least six months long), layers of vegetation, and, commonly, deciduous trees; typical animals include white-tailed deer, opossums, and squirrels. (Ch. 29, p. 826)

tendon: in vertebrates, the tough connective tissue that attaches most muscles to bone. (Ch. 26, p. 739)

territoriality: possession and defense of specific territories; reduces conflict within a species. (Ch. 27, p. 775)

testes: sperm-producing organs of animals. (Ch. 18, p. 497)

thallus: a liverwort's flat, liverlike body, which is only a few cells thick; produces sugars (which move by diffusion) during photosynthesis. (Ch. 16, p. 427)

theory: a major hypothesis that has withstood the test of time; although always subject to revision, theories—which are based on observations—are the closest to a complete explanation that science can offer. (Ch. 2, p. 42)

theory of natural selection: explanation for how evolution occurs, stressing the importance of variations among organisms and that variations can be inherited. (Ch. 12, p. 320)

third-order consumer: carnivorous consumer such as an owl that feeds directly on second-order consumers such as snakes. (Ch. 28, p. 788)

thymine: one of four nitrogen bases of DNA; the others are adenine, guanine, and cytosine. (Ch. 9, p. 228)

thymus gland: organ in which lymphocytes mature into T cells. (Ch. 23, p. 646)

tissue: in multicellular organisms, a specialized group of cells with the same basic structure that perform the same function. (Ch. 5, p. 121)

trachea: in some arthropods, branching and rebranching tube ending in tiny air sacs (tracheoles); in terrestrial vertebrates, the tube that conducts air to the lungs. (Ch. 22, p. 618)

tracheoles: tiny, moisture-containing air sacs of some arthropods that provide a large surface area for gas exchange. (Ch. 22, p. 618)

transcription: process whereby RNA is made from DNA; in prokaryotes, transcription is relatively simple but in eukaryotes is more complex. (Ch. 10, p. 252)

transfer RNA (tRNA): brings amino acids to the ribosomes, which are the site of translation in protein synthesis; acts as an interpreter translating nucleic acid language into protein language. (Ch. 10, p. 254)

translation: process by which genetic information present in a molecule of mRNA directs the synthesis of a polypeptide; depends on the correct joining of mRNA codons and tRNA anticodons. (Ch. 10, p. 253)

transmission electron microscope (TEM): used to view parts of very small, nonliving structures such as thin sections of cells embedded in plastic; forms an image by passing a beam of electrons through the object. (Ch. 2, p. 45)

transpiration: in plants, the loss of water through the pores in stomata. (Ch. 21, p. 586)

transport proteins: membrane proteins that play a role in the passage of particles—such as certain ions, sugars, and amino acids—through the plasma membrane. (Ch. 4, p. 102)

trisomy: having three of one kind of chromosome resulting in inherited defects such as Down syndrome (trisomy 21). (Ch. 11, p. 288)

trophic level: in an ecosystem, an organism's position in a food chain—for example, herbivores make up the second trophic level and carnivores make up higher trophic levels. (Ch. 28, p. 788)

tropical rain forest: biome characterized by heavy rainfall and constantly warm temperatures; has the most abundant plant growth and greatest species diversity of any biome and is populated, for example, by monkeys, parrots, beetles, and snakes. (Ch. 29, p. 827)

tropism: positive or negative response to a specific stimulus (e.g., positive phototropism causes plants to grow toward light). (Ch. 24, p. 673)

tundra: biome characterized by harsh winters, permafrost, short summers, little precipitation; typical life forms include mice, deer, wolves, true mosses, and stunted shrubs. (Ch. 29, p. 824)

U

ultrasonography: diagnostic technique using high-frequency sound waves and a screen image (echogram) to determine, for example, the position and anatomy of a fetus. (Ch. 11, p. 294)

GLOSSARY

umbilical cord: contains large blood vessels that transport blood between the fetus and placenta; connects the fetus to the placenta. (Ch. 19, p. 531)

uracil: the nucleic acid RNA contains uracil as one of its four bases; uracil in RNA takes the place of thymine in DNA. (Ch. 10, p. 250)

ureter: tube carrying urine from each kidney to the urinary bladder. (Ch. 22, p. 629)

urethra: structure through which urine is passed when the bladder fills. (Ch. 22, p. 629)

urinary bladder: muscular urine-storage sac; together with the kidneys and ureters makes up the human urinary system. (Ch. 22, p. 629)

urine: waste filtered by the kidney from the blood; composed of urea, salts, and a small amount of water. (Ch. 22, p. 629)

uterus: thick-walled, muscular organ of female mammals in which embryonic development takes place. (Ch. 18, p. 501)

V

vaccine: preparation created from weakened or killed microorganisms, viruses, or toxins that causes an immune response and prevents specific diseases. (Ch. 23, p. 658)

vacuole: membrane-bound, fluid-filled structure functioning in storage, digestion, and maintenance of osmotic balance. (Ch. 5, p. 132)

vagina: muscular tube that serves as the birth canal, leading to the outside of a female mammal's body, and also allows the passage of sperm to the uterus. (Ch. 18, p. 501)

vascular tissue: tissue that transports food, water, and other nutrients from one part of a plant to another. (Ch. 16, p. 426)

vasopressin: hormone that controls the amount of water excreted in urine—a reduced level results in a more dilute urine being produced. (Ch. 22, p. 630)

vegetative reproduction: asexual reproduction in plants that produces new plants from leaves, roots, or stems. (Ch. 18, p. 485)

vein: thick-walled vessel that carries blood back to the heart; contains one-way valves and is composed of three layers of cells. (Ch. 21, p. 591)

ventral: belly or bottom side of animals with bilateral symmetry (for example, segmented worms). (Ch. 17, p. 454)

ventricle: muscular chamber in the heart that receives blood from the atrium and pumps it to the body. (Ch. 21, p. 591)

vertebrae: bony or cartilaginous segments making up the backbone in vertebrates that provide protection and allow flexibility. (Ch. 17, p. 466)

vertebrates: animals with a backbone; subdivided into two major groups—the fish and the tetrapods. (Ch. 17, p. 450)

vesicle: small substance-containing sac formed from the plasma membrane in the process of endocytosis. (Ch. 4, p. 107)

vessel: tubelike element of the xylem formed from dead cells linked end to end; conducts water. (Ch. 21, p. 580)

vestigial: structure once useful for an organism's lifestyle but now has no function. (Ch. 12, p. 311)

villi: fingerlike projection on the intestinal lining that increases the intestine's absorptive area. (Ch. 20, p. 566)

visible spectrum: range of colors that make up white light. (Ch. 6, p. 152)

W

water-vascular system: series of water-filled canals and muscular tube feet that allows starfish to control movement. (Ch. 26, p. 739)

wavelength: a measure of radiant energy (for example, light, radio waves) calculated by the distance from one wave crest to the next. (Ch. 6, p. 152)

white blood cell: one of five types of colorless blood cells that are formed within lymph nodes and bone marrow and guard against foreign invaders such as viruses. (Ch. 21, p. 603)

X

xylem: one of two types of vascular tissue in plants; conducts water and dissolved minerals from the roots to the leaves. (Ch. 21, p. 580)

Z

zero population growth: condition where the birthrate equals the death rate and the population growth rate equals zero. (Ch. 27, p. 778)

zygote: diploid ($2n$) cell that results from fertilization of an egg by sperm. (Ch. 7, p. 183)

INDEX

R

S

336, ©Renee Lynn, Davis/Lynn Photography
337, (tr) ©Bruce Iverson Photomicrography, (cr) Animals Animals/Zig Leszczynski, (bl) Thomas Kitchin/First Light Toronto, (br) Dwight R. Kuhn;
338, ©Bruce Iverson Photomicrography;
340, (l) Lynn M. Stone, (r) Thomas Kitchin/First Light Toronto;
341,344, ©Tom & Pat Leeson;
345, ©Daniel J. Cox;
346, Leonard Wolfe/Photo Researchers, Inc.;
348, (l) Animals Animals/David C. Fritts, (r) Stephen J. Krasemann/DRK Photo;
349, Robert and Linda Mitchell;
350, Alvin E. Staffan;
351, (l) Richard Surall/Photo-Op, (r) Luis Castabeda, The Image Bank;
353, (bl) Frans Lanting/Minden Pictures, (bm) ©Gerry Ellis, (br) ©Renee Lynn, Davis/Lynn Photography;
354, THE FAR SIDE copyright 1987 FARWORKS, INC. Reprinted with permission of UNIVERSAL PRESS SYNDICATE. All rights reserved.;
355, (l) Michael Holford, (r) Visuals Unlimited/©John D. Cunningham;
356, ©Art Wolf;
357, (t) Kevin O'Farrell, (b) Stan Ries/Leo de Wys;
358, Boltin Picture Library;
359, (tl) Charles Gupton/Tony Stone Images, (tm) Ed & Chris Kunler, (tr) Brain Brake/Photo Researchers, Inc.;
360, Robert and Linda Mitchell;
361, Lynn M. Stone;
363, Frans Lanting/Minden Pictures;
364-365, Joseph Nettis/Photo Researchers, Inc.;
365, Rod Joslin;
366, Visuals Unlimited/©D. Foster, Woods Hole Observatory;
367, (l) Stephen J. Krasemann/DRK Photo, (m) James H. Robinson/Photo Researchers, Inc., (r) Runk/Schoenberger from Grant Heilman;
368, National Audubon Society;
369, (l) Jany Sauvanet/Photo Researchers, Inc., (r) ©Daniel J. Cox;
371, Stephen J. Krasemann/DRK Photo;
372, MAK-1 Photodesign;
373, (b) UPI/Bettmann, (t) ©Gerry Ellis;
374, (tl,r) ©Daniel J. Cox, (c) ©Gerry Ellis, (bl) ©Art Wolfe;
379, ©S. Nielsen/DRK Photo;
382, (tl) ©Clinton Ferris/Stock Imagery, (tr) Elaine Shay, (b) Tracy I. Borland;
383, Gaillarde/Gamma Liaison Network;
384, Visuals Unlimited/©J. R. Shute;
385, Kjell B. Sandved/Photo Researchers, Inc.;
388, Matt Meadows;
389, Biophoto Associates/Photo Researchers, Inc.;
395, Visuals Unlimited/©Fred E. Hossler;
396, (tl) Dr. Tony Brian/Science Photo Library/Photo Researchers, Inc., (bl,r) CNRI/Science Photo Library/Photo Researchers, Inc.;
397, Dr. Tony Brian/Science Photo Library/Photo Researchers, Inc.;
398, (l) Paul W. Johnson/Biological Photo Service, (r) Dr. Jeremy Burgess/Science Photo Library/Photo Researchers, Inc.;
399, Frans Lanting/Minden Pictures;
400, (l) John Colwell from Grant Heilman, (r) ©Gerry Souter/Tony Stone Images;
401, Visuals Unlimited/©Tom E. Adams;
402, (t) Biophoto Associates/Science Source/Photo Researchers, Inc., (bl) Dr. E. R. Degginger/Color-Pic, Inc., (cl) Dwight R. Kuhn, (cr) Alex Rakosy/Custom Medical Stock Photo, (br) Paul W. Johnson/Biological Photo Service;
403, (l) Visuals Unlimited/©Ron Dengler, (r) ©Bob Evans/Peter Arnold, Inc.;
404, (t) Larry Tackett/Tom Stack & Associates, (b) Visuals Unlimited/©Michael Abbey;
405, Eric V. Graves/Photo Researchers, Inc.;
406, Dr. David Patterson/Science Photo Library/Photo Researchers, Inc.;
408, (l) Dr. E. R. Degginger/Color-Pic, Inc., (r) Dwight R. Kuhn;
411, Barry L. Runk from Grant Heilman;
412, Matt Meadows;
413, (tl) William D. Popejoy, (tr) Visuals Unlimited/©David M. Phillips, (b) Robert Miller Gallery, New York;
415, Dwight R. Kuhn;
416, ©1989 Karen Jettmar/AllStock;

417, courtesy of Pfizer, Inc.;
419, ©Ed Reschke/Peter Arnold, Inc.;
420, Terry Donnelly/Tony Stone Images;
421, David Cavagnaro/Peter Arnold, Inc.;
422, (l) Science Photo Library/Custom Medical Stock Photo, (r) Dr. Jeremy Burgess/Science Photo Library/Photo Researchers, Inc.;
423, (l) Dr. E. R. Degginger/Color-Pic, Inc., (r) Carolina Biological Supply/Phototake NYC;
424, Dr. Jeremy Burgess/Science Photo Library/Photo Researchers, Inc.;
426, (l) Steve Lissau, (r) Dr. E. R. Degginger/Color-Pic, Inc.;
427, (t) Gwen Fidler, (m) John A. Lynch, PHOTO/NATS Inc., (b) Runk/Schoenberger from Grant Heilman;
430, (t) Visuals Unlimited/©John D. Cunningham, (b) Runk/Schoenberger from Grant Heilman;
433, (l) Alvin E. Staffan/Photo Researchers, Inc., (r) Barry L. Runk from Grant Heilman;
434, (l) Virginia Twinam-Smith, PHOTO/NATS Inc., (r) Harvey Lloyd/The Stock Market;
435, (t) Visuals Unlimited/©John D. Cunningham, (bl) Alan & Sandy Carey, (br) Biophoto Associates/Photo Researchers, Inc.;
436, Brian Parker/Tom Stack & Associates;
437, Christine Douglas/Photo Researchers, Inc.;
439, (t) Elaine Shay, (b) Earth Scenes/Wendy Neefus;
440, Patricia Gonzalez;
441, (l) Runk/Schoenberger from Grant Heilman, (m) Philip Beaurline, PHOTO/NATS Inc., (r) Alan Pitcairn from Grant Heilman;
442, John Gerlach/Tom Stack & Associates;
443, Earth Scenes/Zig Leszczynski;
444, The Granger Collection;
445, Animals Animals/Patti Murray;
446, Tom Branch/Photo Researchers, Inc.;
447, Alvin E. Staffan;
448-449, Norbert Wu;
449, Fred Bavendam/Peter Arnold, Inc.;
450, (l) Sobel/Klonsky, The Image Bank, (r) file photo;
452, Dwight R. Kuhn;
453, Norbert Wu;
456, Eric V. Grave/Photo Researchers, Inc.;
457, Visuals Unlimited/©John D. Cunningham;
458, (tl) Runk/Schoenberger from Grant Heilman, (bl) Dave B. Fleetham/Tom Stack & Associates, (r) Han Pfletschinger/Peter Arnold, Inc.;
459, (l) Animals Animals/Michael Fogden, (b) Animals Animals/Stephen Dalton, Oxford Scientific Films;
460, Visuals Unlimited/©Daniel W. Gotshall;
461, James R. Fisher/DRK Photo;
463, (l) Michael Collier, (r) Visuals Unlimited/©Dave B. Fleetham;
464, Biological Photo Service;
465, (t) Jeffrey Rotman/Peter Arnold, Inc., (b) Animals Animals/G.I. Bernard, Oxford Scientific Films;
466, Carleton Ray/Photo Researchers, Inc.;
467, (t) Animals Animals/Breck P. Kent, (b) Animals Animals/W. Gregory Brown;
468, Larry Lipsky/Tom Stack & Associates;
469, (t) Animals Animals/Zig Leszczynski, (bl) Animals Animals/Fred Whitehead, (br) Michael Fogden/DRK Photo;
471, (l) Dr. E. R. Degginger/Color-Pic, Inc., (r) Rod Plank/Photo Researchers, Inc.;
472, Warren & Kenny Garst/Tom Stack & Associates;
473, (tl) Stephen J. Krasemann/DRK Photo, (tr) Johnny Johnson, (bl) David Stoecklein/The Stock Market;
475, Andrew J. Martinez/Photo Researchers, Inc.;
476,477, Matt Meadows;
480, Ken W. Davis/Tom Stack & Associates;
480-481, Science Photo Library/Photo Researchers, Inc.;
482, Stephen P. Parker/Photo Researchers, Inc.;
483, Visuals Unlimited/©Bruce S. Cushing;
484, Kwang Shing Kim/Peter Arnold, Inc.;
485, (l) Animals Animals/Oxford Scientific Films, (r) Visuals Unlimited/©Robert E. Lyons;
486, (l) Animals Animals/Herb Segars, (r) Farrell Grehan/Photo Researchers, Inc.;
487, Treat Davidson/Photo Researchers, Inc.;
488, Biophoto Associates/Photo Researchers, Inc.;
489, Custom Medical Stock Photo;
490, Barry L. Runk from Grant Heilman;
493, Visuals Unlimited/©David M. Phillips;
495, (t) Dwight R. Kuhn, (b) Elaine Shay;
497, Stephen J. Krasemann/DRK Photo;
498, David M. Dennis;
499, Charles Palek/Tom Stack & Associates;

502, THE FAR SIDE copyright 1992 FARWORKS, INC. Reprinted with permission of UNIVERSAL PRESS SYNDICATE. All rights reserved.;
503, Lawrence Migdale;
504, (t) Francis Leroy, Biocosmos/Science Photo Library/Photo Researchers, Inc., (b) Boltin Picture Library;
506, Rawlins/Custom Medical Stock Photo;
507,508, Dwight R. Kuhn;
509, Visuals Unlimited/©David M. Phillips;
510, Animals Animals/Frits Prenzel;
511, Jean-Paul Ferrero/Auscape International;
512, Dr. E. R. Degginger/Color-Pic, Inc.;
513, Visuals Unlimited/©Cabisco;
517, Biophoto Associates/Photo Researchers, Inc.;
524, Barry L. Runk from Grant Heilman;
525, (l) Animals Animals/Alastair Shay, (r) Animals Animals/Tayrona Cedro;
529, Animals Animals/Zig Leszczynski;
530, Animals Animals/Gary Guiffen;
532, Farrell Grehan/Photo Researchers, Inc.;
533, "Storyteller", by Helen Cordero, Cochiti Pueblo, New Mexico. Slipped and painted earthenware, 1964. From the Girard Foundation in the Museum of International Folk Art, a unit of the Museum of New Mexico, Santa Fe. Photo by Michel Monteaux;
534, Jon Feingersh/The Stock Market;
535, Karen McCunnall/Leo de Wys;
536, (tl,bl) ©Lennart Nilsson, The Incredible Machine/The National Geographic Society, (tc,tr) ©Lennart Nilsson, A Child is Born, (br) ©SuperStock, Inc.;
538, MAK-1 Photodesign;
539, Michael Heron;
541, Charles E. Zirkle;
542-543, Bob Talbot/Allstock;
543, Howard Hall;
546, Visuals Unlimited/M. Abbey;
547, Animals Animals/Oxford Scientific Films/Peter Parks;
549, Earth Scenes/Oxford Scientific Films;
550, Dr. E. R. Degginger/Color-Pic, Inc.;
552, Matt Meadows;
554, Mike Peres/Custom Medical Stock Photo;
555, Jeffrey Rotman/Peter Arnold, Inc.;
556, Gary G. Gibson/Photo Researchers, Inc.;
559, Matt Meadows;
562, Dr. E. R. Degginger/Color-Pic, Inc.;
566, (t) Visuals Unlimited/©R. Kessel-C. Shih, (b) Visuals Unlimited/©Don Fawcett;
569, State Historical Society of Wisconsin;
571-573, Matt Meadows;
576, Earth Scenes/Fred Whitehead;
577, Visuals Unlimited/©R.G. Kessel-C.Y. Shih;
579, (l) Animals Animals/Lewis Trusty, (r) Peter Arnold, Inc.;
581,582, Dwight R. Kuhn;
583, (tr) Visuals Unlimited/©John D. Cunningham, (bl) Walter Hodge/Peter Arnold, Inc.;
584, Galeria Arvil, Mexico City, photo courtesy The Bronx Museum;
586, Brian Yarvin/Peter Arnold, Inc.;
587, Dr. Jeremy Burgess/Science Photo Library/Photo Researchers, Inc.;
591, (br) Visuals Unlimited/©Fred E. Hossler, (bc) Visuals Unlimited/©David M. Phillips, (bl) Visuals Unlimited/©Robert Caughey;
594, Duomo/Dan Helms;
596, Matt Meadows;
599, Jon Riley/Medichrome;
603, David M. Phillips/Photo Researchers, Inc.;
604, Don Fawcett/Science Source/Photo Researchers, Inc.;
606, ©Lennart Nilsson, Behold Man/Little Brown and Company;
607, CNRI/Science Photo Library/Photo Researchers, Inc.;
608, Visuals Unlimited;
610-611, Stephen Frink/Allstock;
611, Fred McConnaughey/Photo Researchers, Inc.;
612, (l) Helmut Grutscher/Peter Arnold, (r) ©Bruce Iverson Photomicrography;
613, (l) Earth Scenes/Patti Murray, (r) Runk/Schoenberger from Grant Heilman;
614, (l) Visuals Unlimited/©John Gerlach, (c) Jeff Lepore/Photo Researchers, Inc., (r) ©Bruce Iverson Photomicrography;
615, Fred McConnaughey/Photo Researchers, Inc.;

623, David Frazier;

624, Dr. E. R. Degginger/Color-Pic, Inc.;

628, CNRI/Science Photo Library/Photo Researchers, Inc.;

631, (t) Animals Animals/Holt Studios International, (b) Bill Curtsinger;

632, Alford W. Cooper/Photo Researchers, Inc.;

633, Doug Martin;

634, Hank Morgan/Science Source/Photo Researchers, Inc.;

635, David Brownell, The Image Bank;

637, Johnny Johnson;

638-639, ©Don Fawcett & E. Shelton/Science Source/Photo Researchers, Inc.;

639, Michael Kevin Daly/The Stock Market;

640, (bl) Visuals Unlimited/©David M. Phillips, (bc) London School of Hygiene & Tropical Medicine/Science Photo Library/Photo Researchers, Inc., (br) CNRI/Science Photo Library/Photo Researchers, Inc.;

641, (t) Matt Meadows, (b) Science VU/Visuals Unlimited;

642, (t) Science VU/Visuals Unlimited, (bc) ©Dr. Dennis Kunkel/Phototake NYC, (br) Visuals Unlimited/©Hans Gelderblom;

643, (l) Tim Davis/AllStock, (c) James N. Westwater, (r) Alvin E. Staffan;

644, Prof. Motta, Correr & Nottola/University "La Sapienza", Rome/Science Photo Library/Photo Researchers, Inc.;

646, Robert Becker/Custom Medical Stock Photo;

647,653, Visuals Unlimited/©David M. Phillips;

654, (t) Dr. C. Lowe, Oregon Health Sciences University/AllStock, (b) Biomedical Communications/Photo Researchers, Inc.;

656, (l) Dr. A. Liepins/Science Photo Library/Photo Researchers, Inc., (r) ©Cecil Fox/Science Source/Photo Researchers, Inc.;

658, David M. Stone, PHOTO/NATS Inc.;

659, UPI/Bettmann;

660, ©1991 Lawrence Migdale;

661, Prof. Luc Montagnier, Institute Pasteur/CNRI/Science Photo Library/Photo Researchers, Inc.;

664, Steve Liss/Time Magazine;

665, ©Hank Morgan/Science Source/Photo Researchers, Inc.;

667, file photo;

668, Richard Himelsen/Medichrome;

668-669, Fred Bavendam/Peter Arnold;

670, Tomomi Saito/Dunq/Photo Researchers, Inc.;

671, Doug Martin;

672,675, Runk/Schoenberger from Grant Heilman;

676, Matt Meadows;

677, (l) Visuals Unlimited/©Link, (r) Aaron Haupt;

678, Runk/Schoenberger from Grant Heilman;

679, BLT Production;

680, (tc,bc) M.H. Sharp/Photo Researchers, Inc., (l) J.H. Robinson/Photo Researchers, Inc., (r) J.L. Lepore/Photo Researchers, Inc.; 682, Matt Meadows;

693, Animals Animals/Ed Degginger;

694, Aaron Haupt;

695, David Woods/Tony Stone Images;

698, Superstock;

699, Jim Zapp/Photo Researchers, Inc.;

700, CNRI/Science Photo Library/Photo Researchers, Inc.;

702, ©Lennart Nilsson, The Incredible Machine/The National Geographic Society;

705, Matt Meadows;

706, Jim Pickerell/FPG International;

707, Don Fawcett/Science Source/Photo Researchers, Inc.;

709, Robert Mullenix;

716, Matt Meadows;

717, Dr. E. R. Degginger/Color-Pic, Inc.;

721, (c) ©Lennart Nilsson, The Incredible Machine/The National Geographic Society, (l,r) ©Lennart Nilsson, Behold Man/Little, Brown and Company;

723, ©Lennart Nilsson, The Incredible Machine/The National Geographic Society;

724, Bill Ross/AllStock;

725, John Welzenbach/The Stock Market;

728-729, Bob McKeever/Tom Stack & Associates;

729, Renee Lynn/Photo Researchers, Inc.;

731, (t) Dr. E. R. Degginger/Color-Pic, Inc., (b) Visuals Unlimited/©William C. Jorgensen;

733, Dr. E. R. Degginger/Color-Pic, Inc.;

734, (l) Biophoto Associates/Science Source/Photo Researchers, Inc., (r) Steinhart Aquarium/Photo Researchers, Inc.;

735, Animals Animals/Breck P. Kent;

737, Visuals Unlimited/©Don W. Fawcett;

738, Dwight R. Kuhn;

742, Matt Meadows;

743, Dr. E. R. Degginger/Color-Pic, Inc.;

744, (l) Visuals Unlimited/©J. Venable & D. Fawcett, (m) Dr. E. R. Degginger/Color-Pic, Inc., (r) Michael Abbey/Photo Researchers, Inc.;

748, Robert Mullenix;

749, Doug Martin;

750, Visuals Unlimited/©David M. Phillips;

751, Richard Hineisen/Medichrome;

752, Ted Rice;

753, Robert Mullenix;

755, Dave B. Fleetham/Tom Stack & Associates;

756-757, Animals Animals/Zig Leszczynski;

756, David Ball/AllStock;

758, Visuals Unlimited/David S. Addison;

759, M.E. Warren/Photo Researchers, Inc.;

760, (l) John Lynch, PHOTO/NATS Inc., (r) Lee Saucier, PHOTO/NATS Inc.;

762, Rich Brommer;

764, Visuals Unlimited/©Joe McDonald;

765, Don & Esther Phillips/Tom Stack & Associates;

768, Tom & Pat Leeson/Photo Researchers, Inc.;

769, Cath Ellis, Dept. of Zoology, University of Hull/Science Photo Library/Photo Researchers, Inc.;

770, Dr. E. R. Degginger/Color-Pic, Inc.;

772, Animals Animals/Patti Murray;

773, Tom Ulrich/AllStock;

774, Alan & Sandy Carey;

775, Ben Simmons/The Stock Market;

777, (l) Nick Gunderson/AllStock, (r) R. Hamilton Smith/AllStock;

779, David Weintraub/Photo Researchers, Inc.;

780, Elaine Shay;

781, Ray Pfortner/Peter Arnold, Inc.;

784-785, Jane Grushow from Grant Heilman Photography;

785, Paul W. Johnson/Biological Photo Service;

786, Dwight R. Kuhn;

786-787, Grant Heilman/Grant Heilman Photography;

787, (tr) Charles W. Melton, (br) Labat Jacana/Photo Researchers, Inc.;

793, (t) Viviane Holbrooke/The Stock Market, (b) Visuals Unlimited/©C.P. Vance;

796, Grant Heilman/Grant Heilman Photography;

797, Carmen Lomas Garza;

798, Robert Mullenix;

800, Earth Scenes/E. R. Degginger;

801, (tl) Animals Animals/G.L. Kooyman, (tr) Animals Animals/E. R. Degginger, (m) Bill Larkin, PHOTO/NATS Inc., (bl) Jack S. Grove/Tom Stack & Associates;

802, Dr. E. R. Degginger/Color-Pic, Inc.;

803, Animals Animals/Michael Fogden;

804, Animals Animals/C.C. Lockwood;

806, Animals Animals/Fran Allan;

807, (t) Sovfoto, (b) Brian & Cherry Alexander;

808, Shattil-Rozinski/Tom Stack & Associates;

809, Alan Pitcairn from Grant Heilman;

812, Earth Scenes/Thomas Long;

813, ©Krafft-Explorer/Photo Researchers, Inc.;

814, Chip & Jill Isenhart/Tom Stack & Associates;

815, (l) Aaron Haupt, (r) Dr. E. R. Degginger/Color-Pic, Inc.;

816, St. Meyers/Okapia/Photo Researchers, Inc.;

818, Steve Allen/Peter Arnold, Inc.;

824, (t) Stephen J. Krasemann/Photo Researchers, Inc., (c) ©1988 Tom Bean/AllStock, (b) Dr. E. R. Degginger/Color-Pic, Inc.;

825, (l) Animals Animals/Dan Suzio, (tr) Visuals Unlimited/©Joe & Carol McDonald, (br) Animals Animals/Michael Leach, Oxford Scientific Films;

827, (t) ©1984 Gregory G. Dimijian, M.D./Photo Researchers, Inc., (c) ©James Martin/AllStock, (b) Visuals Unlimited/©1992 David Matherly;

829, (t) Dr. E. R. Degginger/Color-Pic, Inc.,(b) ©Tim Davis/AllStock;

830, ©1987 Tom Bean/AllStock;

831, Animals Animals/Breck P. Kent;

832, (l) Joey Jacques, (r) Dr. E. R. Degginger/Color-Pic, Inc.;

833, ©Jett Britnell/DRK Photo;

834, ©Norbert Wu/Peter Arnold, Inc.;

835, (l) Science VU - WHOI/©Visuals Unlimited, (r) Visuals Unlimited/D. Foster, Wood's Hole Oceanographic Institution;

836, NASA;

837, William J. Weber;

838, Van Vucher/Photo Researchers, Inc.;

839, ©Jeff Henry/Peter Arnold, Inc.;

842-843, Robert Perron;

843, Richard Laird/FPG International;

844, Will & Deni McIntyre/Photo Researchers, Inc.;

845, (t) Runk/Schoenberger from Grant Heilman, (b) Dr. E. R. Degginger/Color-Pic, Inc.;

847, (l) John Paul Kay/Peter Arnold, Inc., (r) UPI/Bettmann;

849, (t) Animals Animals/Rudie H. Kuiter, (b) Louis Quitt/Photo Researchers, Inc.;

850, Animals Animals/Doug Wechsler;

851, (l) Jacques Jangoux/Peter Arnold, Inc., (r) Joe Towers/The Stock Market;

852, Tom McHugh/Photo Researchers, Inc.;

854, Aaron Haupt;

855, Barth Falkenberg/Stock Boston;

856, Robert Mullenix;

859, Vince Streano/The Stock Market;

860, Robin Pranhe/The Stock Market;

861, Martha Cooper/Peter Arnold, Inc.;

862, Mark Reinstein/FPG International;

863, Gary Milburn/Tom Stack & Associates;

864, Inga Spence/Tom Stack & Associates;

865, Larry LeFever from Grant Heilman;

866, H.P Merten/The Stock Market;

867, (t) C. Allan Morgan/Peter Arnold, Inc., (b) D. Holden Bailey/Tom Stack & Associates;

870, Robert Mullenix;

871, Greg Vaughn/Tom Stack & Associates;

875, Matt Meadows;

876, George Disario/The Stock Market;

879, Rod Joslin;

882, (l) Robert Mullenix, (tr) Dr. Jeremy Burgess/Science Photo Library/Photo Researchers, Inc., (br) James N. Westwater;

883, (tl) John Gerlach/Tony Stone Images, (tr) Roger K. Burnard, (br) Matt Meadows;

884, Robert Mullenix;

885, Matt Meadows;

887, (l) courtesy Ohio Disaster Services Agency, (r) Animals Animals/Margot Conte;

889, (l) Robert Mullenix, (r) Doug Martin;

891, Sharon M. Kurgis;

894, (l) Manfred P. Kage/Peter Arnold, Inc., (tr) Visuals Unlimited/©Daniel Gotshall, (br) Dr. E. R. Degginger/Color-Pic, Inc.;

897, (l) Alvin E. Staffan, (tr) Lefever/Grushow from Grant Heilman, (br) Earth Scenes/Michael Fogden;

898, (l) Jim Rorabaugh, (tr) Visuals Unlimited/©W. Ormerod, (br) Dr. E. R. Degginger/Color-Pic, Inc.;

899, (l) Morgan Photos, (tr) Earth Scenes/Richard Smiell, (br) Earth Scenes/E. R. Degginger;

900, William Curtsinger/Photo Researchers, Inc.;

901, (t) Geri Murphy, (b) Dave B. Fleetham/Tom Stack & Associates;

902, (tl) Dr. E. R. Degginger/Color-pic, Inc., (bl) Geri Murphy, (r) Joey Jacques;

903, (l) David M. Dennis, (tr) Animals Animals/T.A.L. Cooke, (br) David M. Dennis/Tom Stack & Associates;

904, (l) Animals Animals/Patti Murray, (tr) Denise Tackett/Tom Stack & Associates, (br) Dave B. Fleetham/Tom Stack & Associates;

905, (l) Dr. E. R. Degginger/Color-Pic, Inc., (r) Fred Bavendam/Peter Arnold, Inc.;

906, (l) Animals Animals/W. Gregory Brown, (tr) Brian Parker/Tom Stack & Associates, (br) Joey Jacques;

907, (t) Alan & Sandy Carey, (b) Alvin E. Staffan;

908, (tl) Animals Animals/Stan Osolinski, Oxford Scientific Films, (bl) Johnny Johnson, (br) Lynn M. Stone;

909, (tl) Stephen Dalton/Photo Researchers, Inc., (bl) Visuals Unlimited/©Leonard Lee Rue III, (r) Visuals Unlimited/© Barbara Gerlach.